The Geology of North America
Volume K-3

North America and Adjacent Oceans During the Last Deglaciation

Edited by

W. F. Ruddiman
Lamont-Doherty Geological Observatory
and
Department of Geological Sciences
of
Columbia University
Palisades, New York 10964

H. E. Wright, Jr.
Limnological Research Center
University of Minnesota
Minneapolis, Minnesota 55455

1987

Acknowledgment

Publication of this volume, one of the synthesis volumes of *The Decade of North American Geology Project* series, has been made possible by members and friends of the Geological Society of America, corporations, and government agencies through contributions to the Decade of North American Geology fund of the Geological Society of America Foundation.

Following is a list of individuals, corporations, and government agencies giving and/or pledging more than $50,000 in support of the DNAG Project:

ARCO Exploration Company
Chevron Corporation
Conoco, Inc.
Diamond Shamrock Exploration Corporation
Exxon Production Research Company
Getty Oil Company
Gulf Oil Exploration and Production Company
Paul V. Hoovler
Kennecott Minerals Company
Kerr McGee Corporation
Marathon Oil Company
McMoRan Oil and Gas Company
Mobil Oil Corporation
Pennzoil Exploration and Production Company
Phillips Petroleum Company
Shell Oil Company
Caswell Silver
Sohio Petroleum Corporation
Standard Oil Company of Indiana
Sun Exploration and Production Company
Superior Oil Company
Tenneco Oil Company
Texaco, Inc.
Union Oil Company of California
Union Pacific Corporation and its operating companies:
Union Pacific Resources Company
Union Pacific Railroad Company
Upland Industries Corporation
U.S. Department of Energy

Published by the Geological Society of America, Inc.
3300 Penrose Place, P.O. Box 9140, Boulder, Colorado 80301

Printed in U.S.A.

Front Cover: Klutlan Glacier, St. Elias Mountains, Yukon Territory in the foreground—a glacier with a history of multiple surges during the last few hundred years. Note the debris-covered stagnant ice in the middleground. This glacier and the small tributary glaciers in the background occupy valleys carved by much larger Late Wisconsin glaciers. Photo by H. E. Wright, Jr.

Library of Congress Cataloging-in-Publication Data

North America and adjacent oceans during the last deglaciation.

(Geology of North America ; K-3)
Includes bibliographies and index.
1. Glacial epoch—North America. I. Ruddiman, W. F. (William F.), 1943– . II. Wright, H. E. (Herbert Edgar), 1917– . III. Series.
QE71.G48 1987 vol. K-3 557.3 s 87-27652
[QE697] [551.7'92'097]
ISBN 0-8137-5203-5

Contents

THE NON-GLACIAL PHYSICAL RECORD ON THE CONTINENT

THE BIOLOGICAL RECORD ON THE CONTINENT

ANALYSIS AND SUMMARY

PLATES

(in pocket inside back cover)

Preface

The Geology of North America series has been prepared to mark the Centennial of The Geological Society of America. It represents the cooperative efforts of more than 1,000 individuals from academia, state and federal agencies of many countries, and industry to prepare syntheses that are as current and authoritative as possible about the geology of the North American continent and adjacent oceanic regions.

This series is part of the Decade of North American Geology (DNAG) Project which also includes eight wall maps at a scale of 1:5,000,000 that summarize the geology, tectonics, magnetic and gravity anomaly patterns, regional stress fields, thermal aspects, seismicity, and neotectonics of North America and its surroundings. Together, the synthesis volumes and maps are the first coordinated effort to integrate all available knowledge about the geology and geophysics of a crustal plate on a regional scale.

The products of the DNAG Project present the state of knowledge of the geology and geophysics of North America in the 1980s, and they point the way toward work to be done in the decades ahead.

In addition to the contributions from organizations and individuals acknowledged at the front of this book, major support has been provided to the editors of this volume by Lamont-Doherty Geological Observatory and the Department of Geology and Geophysics of the University of Minnesota.

A. R. Palmer
General Editor for the volumes
published by The Geological
Society of America

J. O. Wheeler
General Editor for the volumes
published by the Geological
Survey of Canada.

The Geology of North America
Vol. K-3, North America and adjacent oceans during the last deglaciation
The Geological Society of America, 1987

Chapter 1

Introduction

W. F. Ruddiman
Lamont-Doherty Geological Observatory and Department of Geological Sciences of Columbia University, Palisades, New York 10964
H. E. Wright, Jr.
Limnological Research Center, University of Minnesota, Minneapolis, Minnesota 55455

Despite the geological youthfulness of the Quaternary period, most of the surficial geologic record in high-latitude North America has been repeatedly erased by the erosive power of ice sheets. In the north, the record prior to the last 18,000 years is spatially scattered and discontinuous; across the entire continent, the older record is poorly constrained in time.

In contrast, North American sediments deposited since the last glacial maximum remain largely intact spatially and are relatively well dated by ^{14}C and other techniques. Thus, the Quaternary record of North America poleward of 40°N provides the basis for a detailed study of the last deglaciation, and particularly of the disintegration and retreat of the last ice sheet and subsequent revegetation and geomorphologic evolution of North America.

The theme of this volume is the timing, cause, and mechanism of the wastage of North American ice during the last deglaciation, as well as the accompanying environmental changes in the nonglaciated and deglaciated areas. Each chapter summarizes relevant geologic, geochemical, or biotic evidence from the interval 18 to 6 ka in order to present a coherent picture of the environmental conditions in and around North America during this period.

The volume particularly examines the mechanisms by which a mass of ice equivalent to some 100 m of global sea level was returned to the ocean within about 8,000 years. The ultimate cause of deglaciation is now widely acknowledged to be the orbital forcing proposed by Milankovitch (1941), involving the three cycles of earth/sun geometry that control the distribution of solar radiation on earth: the tilt (obliquity) of the earth's axis, the eccentricity of the earth's orbit around the sun, and the precession of the equinoxes. However, between that initial external forcing and the wide array of deglacial climatic responses observed on earth lie several fundamental climatic questions. How did the relatively modest impetus provided by insolation forcing produce such rapid and large-scale destruction of ice? What processes accelerated the effects of the initial insolation forcing? The detailed summary of the deglacial record of North America and adjacent oceans presented in this volume provides a basis for comparing and evaluating those processes, as well as their environmental consequences.

To provide a framework for understanding the last deglaciation, we first review the long-term history of the Northern Hemisphere ice age and North American ice sheet.

LONG-TERM HISTORY OF THE NORTH AMERICAN ICE SHEETS

Until the application of oxygen-isotope analysis to deep-sea cores by Emiliani (1955), the Pleistocene of North America was subdivided into four major glacial phases—Nebraskan, Kansas, Illinoian, and Wisconsin. When the technique of radiocarbon dating was introduced, many researchers concentrated on applying it to delineate the waning phases of the Late Wisconsin glaciation. Others worked to clarify the earlier glacial history of North America by tracing buried tills and interglacial weathering zones from region to region, but they were stymied by the lack of reliable dating tools for times earlier than radiocarbon limits. The land-mammal chronostratigraphies developed for the Pleistocene by vertebrate paleontologists were too crude to allow correlation with specific glacial cycles. In addition, pollen-stratigraphic sequences of interglacial deposits were too scarce to permit the kind of correlations attempted in Europe (Wright, 1977).

Thirty years ago, the marine oxygen-isotopic record showed that many cold episodes had occurred prior to the last glaciation (Emiliani, 1955), and dating later became possible through application of paleomagnetic chronostratigraphy (Shackleton and Opdyke, 1973). The discovery that the Pearlette ash—once considered a single stratigraphic unit and the cornerstone for correlation of early glacial episodes in the Kansas area—yielded fission-track ages ranging from 0.6 to 2.0 Ma at different locations (Wayne and others, 1987) provided further evidence that the North American record of Pleistocene climatic fluctuations was seriously incomplete or misinterpreted, presumably because of glacial erosion by successive ice advances or subaerial erosion during interglacial episodes. Recent research has attempted to subdivide the older glacial and periglacial sediments with the aid

Ruddiman, W. F., and Wright, H. E., Jr., 1987, Introduction, *in* Ruddiman, W. F., and Wright, H. E., Jr., eds., North America and adjacent oceans during the last deglaciation: Boulder, Colorado, Geological Society of America, The Geology of North America, v. K-3.

of paleomagnetic stratigraphy and tephrachronology to obtain a realistic correlation with ocean-core records, but the incompleteness of the continental records is still recognized (Wayne and others, 1987).

North American stratigraphic records in the western mountains have other limitations. Large deep lakes like Lake Bonneville, which might be expected to fill slowly with sediments and thus endure throughout the Pleistocene, all occur in tectonic basins that are subject to changing morphometry, and where the semi-arid to arid climate has caused periodic drying out and loss of record.

While continental geologists repeatedly encountered frustratingly incomplete long-range records, marine stratigraphers were able to exploit the demonstrably continuous sedimentary records in many areas of the world's ocean and develop a chronology of environmental change for the entire Quaternary that could be quantitatively correlated with the Milankovitch orbital variations.

Oceanic records contain three independent monitors of the amount of ice on land. Oxygen-isotope ratios ($\delta^{18}O$) from cores in open-ocean areas provide a first-order measure of changes in the total amount of global ice volume, and North American ice is the largest single component of those changes (Mix, Chapter 6). Continental detritus ice rafted into the high and middle latitudes of the North Atlantic is also a first-order measure of the size of ice sheets on surrounding continents, with North American ice dominating this signal because of geographic proximity and ocean circulation patterns (Ruddiman, Chapter 7). Estimates of sea-surface temperature in the high-latitude North Atlantic are a third indicator of ice volume; both geologic evidence and climate modeling experiments isolate North American ice as the key control on this response as well (Ruddiman, Chapter 7).

The $\delta^{18}O$ record by no means entirely reflects changes in North American ice volume; it also reflects volumetric changes in other ice sheets, as well as variations in local temperature signals (Mix, Chapter 6). Still, the very strong resemblance of the local $CaCO_3$ and SST responses to the $\delta^{18}O$ signal in high-latitude North Atlantic records is strong confirmation that all three signals are good first-order proxies for changes in North American ice volume during the Pleistocene and late Pliocene (Ruddiman, Chapter 7).

The initiation of moderate-sized ice sheets in the Northern Hemisphere occurred between 2.55 and 2.4 Ma (Backman, 1979; Shackleton and others, 1984; Zimmerman and others, 1985). This finding is based primarily on percent $CaCO_3$ records from Deep-Sea Drilling Project Site 552 in the North Atlantic; the uniformly high values of the early Pliocene suddenly yielded to periodically much lower values as a result of influxes of ice-rafted continental debris (Fig. 1). In addition, the abrupt increase in $\delta^{18}O$ values at 2.5 Ma (Fig. 1) partially reflects the growth of ice sheets on land. Subsequent coring by the Deep-Sea Drilling Project and Ocean Drilling Project across the subpolar North Atlantic and Labrador Sea and in the Norwegian Sea has confirmed 2.5 Ma as the age of sudden onset of ice rafting (Ruddiman and Kidd, 1986); Eldholm and others, 1987; Arthur and others, 1987).

Recent evidence shows that the early ice sheets fluctuated mainly at a period of 41,000 years (Ruddiman and others, 1986). Variations in percent $CaCO_3$, estimated sea-surface temperature, and benthic foraminiferal $\delta^{18}O$ were all dominated by the 41,000-year rhythm of orbital obliquity (tilt) during nearly the entire Matuyama magnetic chron from 2.5 to 0.735 Ma (Fig. 2). These three independent lines of evidence also specifically point to changes in North American ice sheets at the 41,000-year rhythm during this late-Pliocene and early-Pleistocene interval.

After 0.9 Ma, changes in $\delta^{18}O$, percent $CaCO_3$, and estimated sea-surface temperature increased in amplitude by a factor of about two, suggesting that ice sheets in the Northern Hemisphere during the Brunhes chron grew to maximum volumes roughly twice as large as those attained during most of the Matuyama. The first larger-scale $\delta^{18}O$ maximum occurred in isotopic stage 22 at 0.85–0.8 Ma (Shackleton and Opdyke, 1973), an age correlative with the Nebraskan glaciation in the paleomagnetically reversed interval between 0.91 and 0.735 Ma (Boellstorff, 1978; Rogers and others, 1985). After 0.9 Ma, all three paleoclimatic indicators begin to show increasing variance at or near the periods of orbital eccentricity (100,000 years) and precession (23,000 and 19,000 years), with the 100,000-year rhythm gradually attaining complete dominance in global $\delta^{18}O$ records (Imbrie and others, 1984; Fig. 3) and in North Atlantic $CaCO_3$ and SST signals (Ruddiman and McIntyre, 1984).

In summary, the first 1.6 to 1.7 m.y. of the Northern Hemisphere ice age was a sequence of some 40 full climatic cycles at the 41,000-year rhythm of orbital obliquity, whereas the last 0.9 to 0.7 m.y. was a time comparably dominated by the 100,000-year period of orbital eccentricity. In addition, numerous separate advances at periods of 41,000 and 23,000 years have been superimposed on the basic 100,000-year cycle during the Brunhes chron. By comparison with this overview from the marine record, it is clear that the classical continental scheme of four glaciations is far from complete.

This long-term perspective also raises four fundamental questions: (1) Why did the Northern Hemisphere ice age begin? (2) Why did the Matuyama ice-sheet variations occur at a rhythm of 41,000 years? (3) Why did the mid-Pleistocene shift in rhythmic response occur? (4) What is the origin of the 100,000-year rhythm of the late Pleistocene?

The first three questions are beyond the scope of this volume. The origin of the 100,000-year signal is, however, relevant to the theme of this volume, as outlined next.

ORIGIN OF ICE-SHEET RHYTHMS

Variations in insolation on earth caused by changes in the earth's orbit (Fig. 4) are the ultimate explanation for the long-term ice-sheet responses at 41,000, 23,000/19,000, and 100,000 years (Fig. 3). This was first demonstrated rigorously by evidence

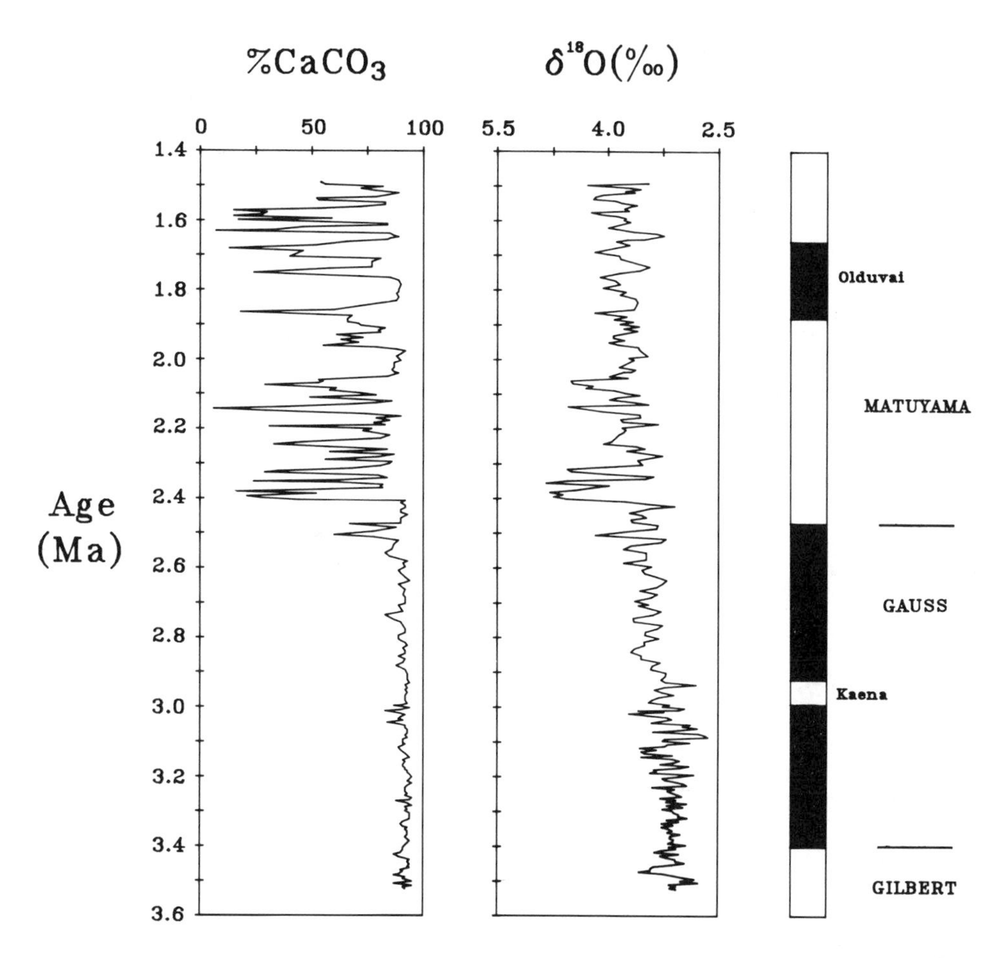

Figure 1. Late-Pliocene and early-Pleistocene records of percent $CaCO_3$ and benthic foraminiferal $\delta^{18}O$ from Site 552 in the North Atlantic at 56°03′N, 23°14′W (after Shackleton and others, 1984; Zimmerman and others, 1985). Abrupt decrease in $CaCO_3$ near 2.55 to 2.4 Ma marks the onset of ice rafting into the North Atlantic brought about by appearance of Northern Hemisphere ice sheets of moderate size.

showing the dominance of these astronomical rhythms in $\delta^{18}O$ signals and other paleoclimatic indicators (Hays and others, 1976). Imbrie and others (1984) recently refined the methodological basis for these calculations and extended the analysis to cover the entire Brunhes chron.

The form of these orbital changes over long time scales is known from astronomical calculations (Fig. 5; Berger, 1984). Significant uncertainty in the ages of the precessional and eccentricity terms only occurs prior to 1 Ma. Calculations of orbital obliquity variations are accurate to within 5,000 years even at the beginning of glaciation at 2.5 Ma.

As a consequence of these orbital changes, the insolation received at the top of the atmosphere varies with latitude, mainly on a seasonal basis. The 41,000-year tilt period is strongest at high latitudes, whereas the 23,000-year and 19,000-year precessional components are strongest at middle and low latitudes (Fig. 6). Eccentricity variations at periods near 100,000 years (94,000 and 125,000 years) and at 413,000 years do not appear as significant components of the insolation received (Fig. 6); instead, they modulate the amplitude of the precessional signal (Fig. 5).

The 41,000-year and 23,000-year ice-sheet rhythms

Both the origin and the phasing of the 41,000-year and 23,000-year rhythms in the $\delta^{18}O$ (ice-volume) records have a reasonably well-established physical basis (Imbrie and others, 1984). The origin of these rhythms is insolation forcing; the lag of

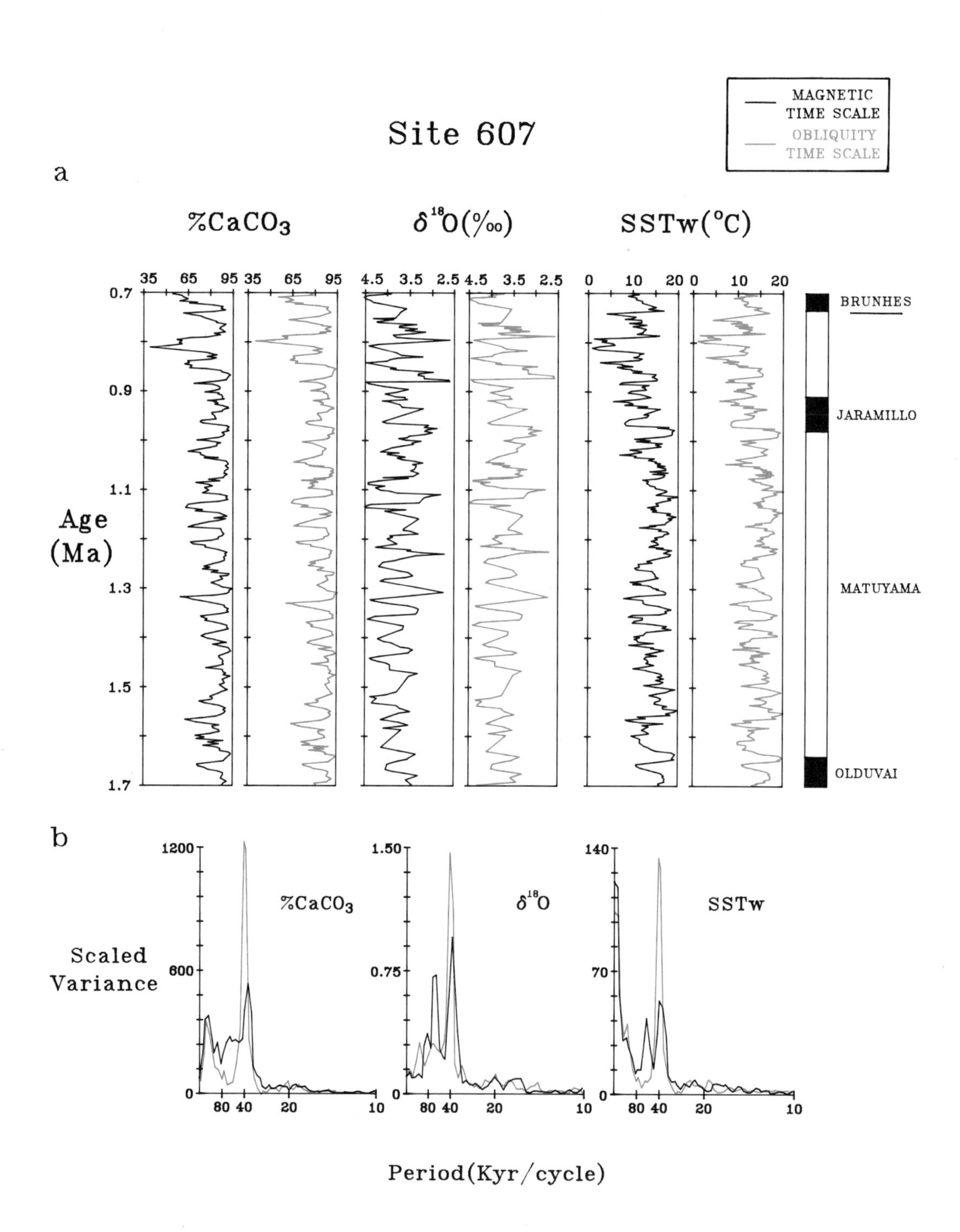

Figure 2. (a) Early Pleistocene records of percent $CaCO_3$, benthic foraminiferal $\delta^{18}O$, and estimated winter sea-surface temperature from Site 607 in the North Atlantic at 41°00′N, 32°58′W (after Ruddiman and others, 1986a). Records reveal 25 cycles during the 1-m.y. interval shown, indicating a 41,000-year rhythm. (b) Spectral analysis of records in Fig. 2a, with 41,000-year period dominant both in magnetic time scale and slightly adjusted obliquity time scale.

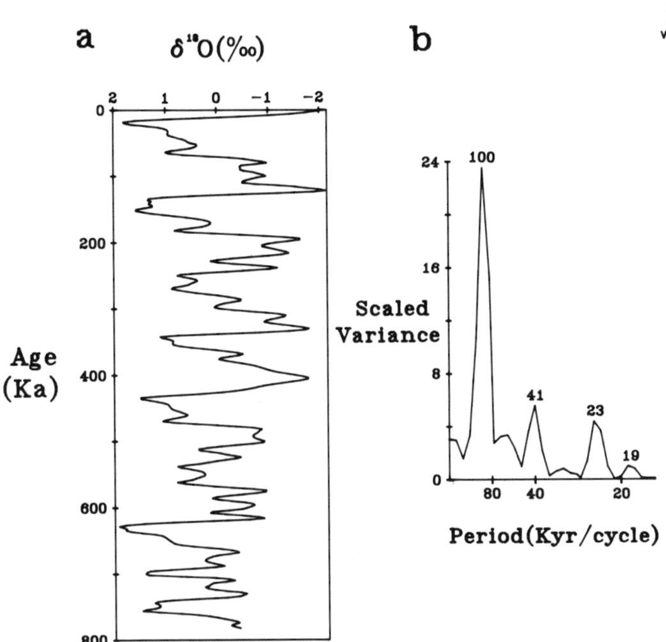

Figure 3. (a) Composite global $\delta^{18}O$ record from Imbrie and others (1984). Units are standard deviations from the mean. (b) Spectral analysis of $\delta^{18}O$ record in Fig. 3a (after Imbrie, 1985).

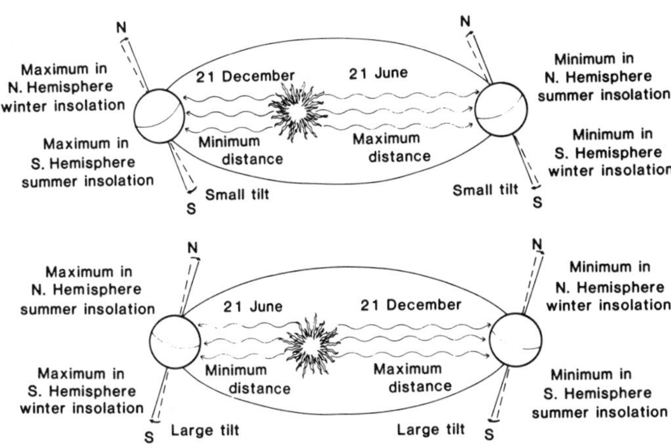

Figure 4. Schematic representation of the earth's orbital elements (eccentricity, tilt, and precession).

these signals behind the insolation forcing reflects the slow physical response of the ice sheets.

The 41,000-year rhythm that dominated ice-volume variations during the Matuyama chron and persisted through the Brunhes chron is explained by insolation forcing at high latitudes (Fig. 6). Summer is the time of maximum ablation of ice and snow and is widely regarded as the critical insolation season. Because summer insolation varies with a 41,000-year rhythm at high latitudes (Fig. 6), high-latitude ice responds at the same rhythm.

The late-Pleistocene phasing of ice growth and decay relative to insolation changes at the 41,000-year rhythm is also consistent with the concept of summer insolation forcing (Hays and others, 1976; Imbrie and Imbrie, 1980). The inherently slow physical response of ice sheets (Weertman, 1964) imposes a lag of 8,000 to 10,000 years behind summer insolation forcing at the 41,000-year period (Imbrie and others, 1984).

The onset of stronger 23,000-year (and 19,000-year) ice-sheet variations after 0.9 Ma also has a reasonably clear physical basis. Periodic excursions of the larger 100,000-year ice sheets into the middle latitudes of southern Canada and the United States brought them under the influence of stronger insolation forcing at these precessional periods (Fig. 6). Thus, larger ice-volume variations at 23,000 years are a natural corollary of the presence of larger ice-sheets in middle latitudes.

As with obliquity, the slow physical response of the ice sheets to precessional forcing imposes a lag of the $\delta^{18}O$ response behind summer insolation forcing. Because the precessional period is shorter, the lag of the $\delta^{18}O$ (ice-volume) response behind the insolation forcing is proportionately less, approximately 5,000 to 6,000 years (Imbrie and others, 1984).

The 100,000-year ice-sheet rhythm

The great enigma among the rhythms of ice-sheet response is the 100,000-year signal that has dominated the late Pleistocene (Fig. 3). Although eccentricity modulates the precessional cycle, the amount of direct insolation available at the eccentricity rhythms is so small as to be negligible (Hays and others, 1976; Berger, 1984). Why, then, is the response at this period the major rhythm of the late Pleistocene?

By analogy with musical theory, it has been suggested that this rhythm is in part a "beat frequency" produced by interaction of the 23,000-year and 19,000-year precessional signals (Wigley, 1976), but this explanation has not been widely accepted. A more likely possibility, suggested by Hays and others (1976), is that nonlinear responses in the climate system amplify the effects of direct insolation forcing in such a way as to give the observed ice-sheet response; one nonlinearity they proposed was that ice sheets decay faster than they grow.

In a study treating the ice-volume response in strictly numerical terms (that is, not attempting to model the actual physics of ice-sheets), Imbrie and Imbrie (1980) found that the summer insolation signal at 65°N could produce an encouraging approximation of the observed $\delta^{18}O$ signal if two assumptions are made: (1) ice sheets have a slow (17,000-year) average time constant of

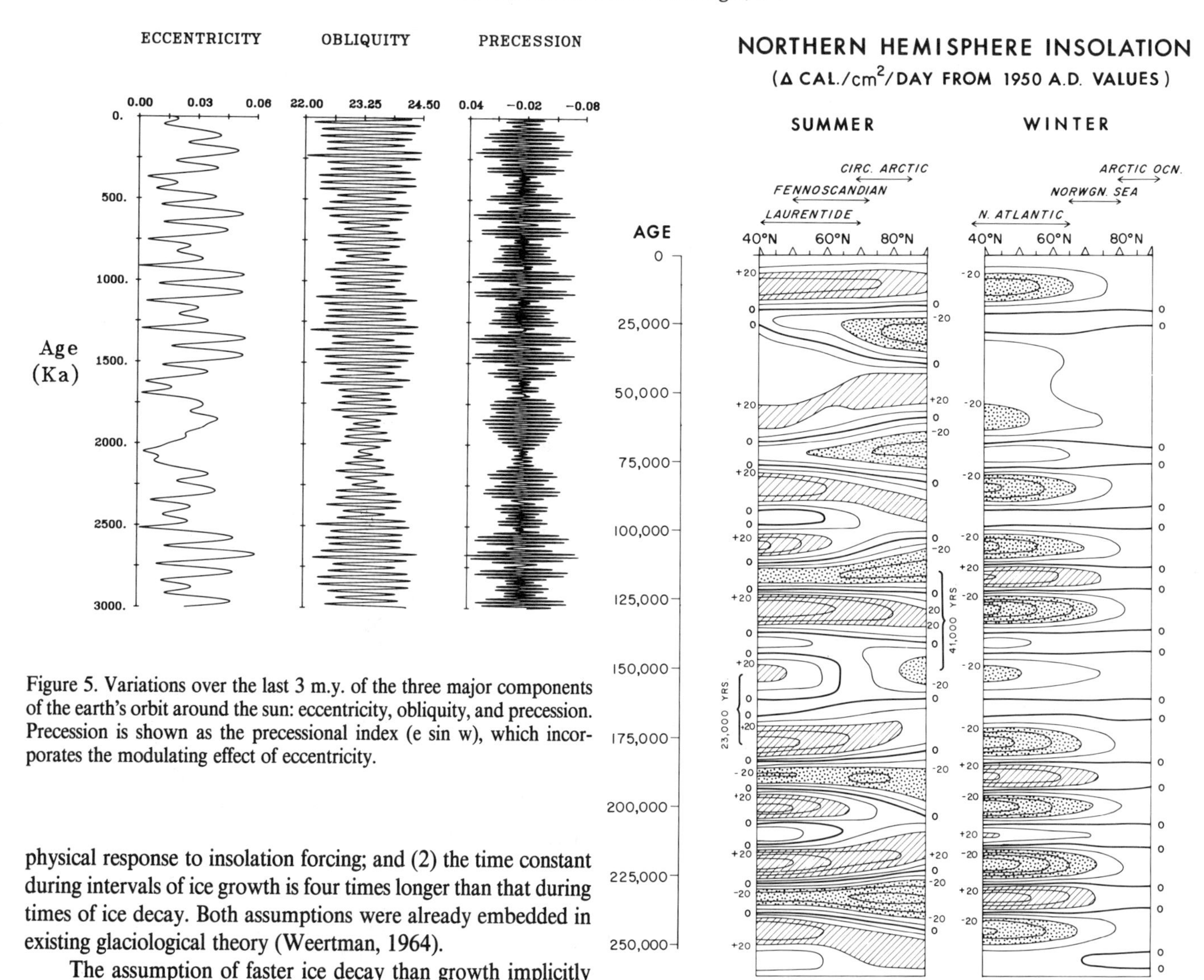

Figure 5. Variations over the last 3 m.y. of the three major components of the earth's orbit around the sun: eccentricity, obliquity, and precession. Precession is shown as the precessional index (e sin w), which incorporates the modulating effect of eccentricity.

physical response to insolation forcing; and (2) the time constant during intervals of ice growth is four times longer than that during times of ice decay. Both assumptions were already embedded in existing glaciological theory (Weertman, 1964).

The assumption of faster ice decay than growth implicitly incorporates a nonlinear response. Imbrie and Imbrie (1980) avoided specifying the kind of nonlinearity in the climate system, noting only that glaciological data show ablation to be a more powerful and rapid process than accumulation. With this nonlinearity included, their model succeeded in producing some response at the 100,000-year period, but not as much as that observed in $\delta^{18}O$ records. The same failure characterized other time-dependent models in which more realistic degrees of ice-sheet physics were included (Birchfield and others, 1981; Pollard, 1982). This suggested that additional sources of nonlinear behavior were needed from within the climate system to explain the strong 100,000-year rhythm.

The basic shape of late-Pleistocene $\delta^{18}O$ records is sawtoothed (Broecker and van Donk, 1970), with long intervals of relatively slow ice growth followed by short intervals of rapid ice decay (Fig. 3). Thus the most salient features of the late-Pleistocene $\delta^{18}O$ signal are the sudden deglaciations occurring roughly every 100,000 years. At these times, the nonlinear processes responsible for rapid ice decay must have been particularly active.

Figure 6. Changes in the rhythm of received insolation over the last 250,000 years with latitude in the Northern Hemisphere. Insolation values shown as departures from modern values for the winter and summer caloric half-year (after Berger, 1984). Tilt variations (41,000 years) are stronger at high latitudes, and precession (23,000 to 19,000 years) at middle and low latitudes. Summer and winter departures are opposite in sign.

Over the long term, all of the major deglaciations coincide, as predicted by theory, with summer insolation maxima (Fig. 7). However, some of the insolation maxima do not appear to be strong enough to explain the speed and amplitude of the corresponding deglaciations. The most recent deglaciation is a particularly good example of such a mismatch. Around 11,000 to 9,000 years ago, tilt and precession combined to create a moderate-sized summer insolation maximum at middle and high latitudes of the Northern Hemisphere (Fig. 8), and this maximum

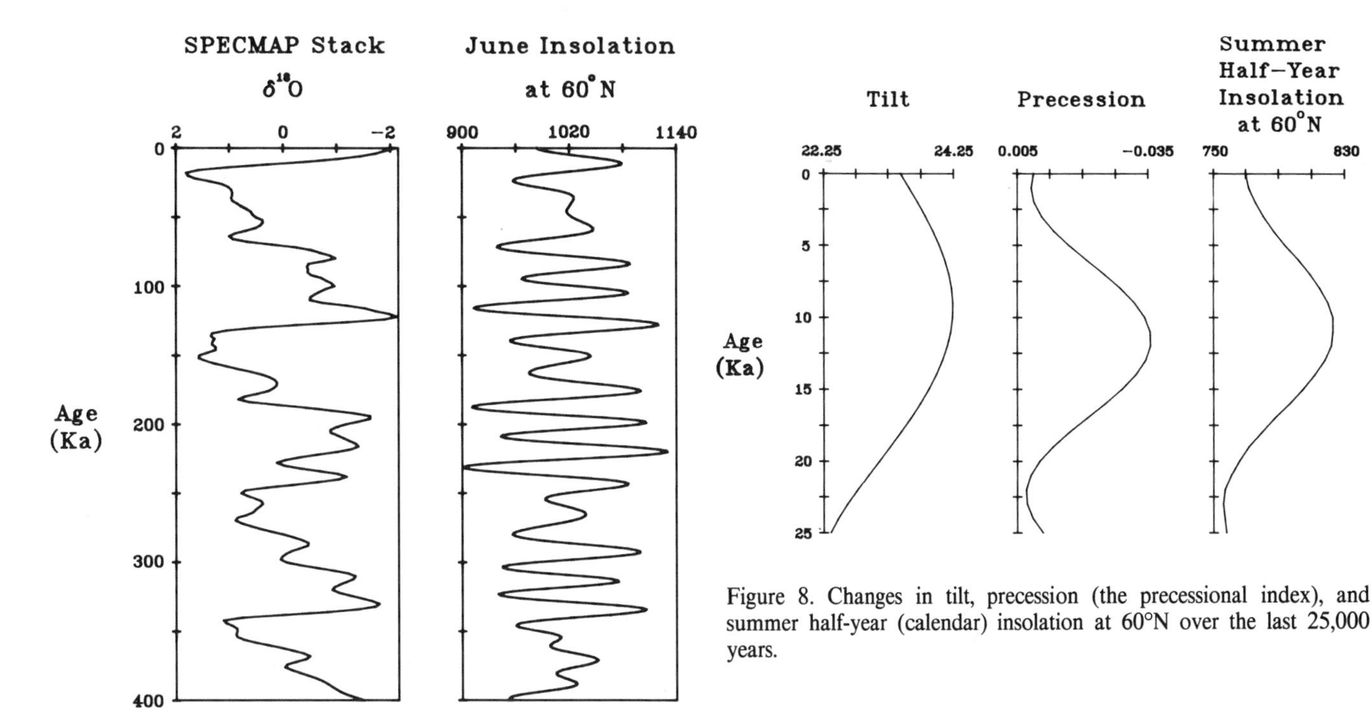

Figure 7. (Left) changes in $\delta^{18}O$ over the last 0.40 m.y. (after Imbrie and others, 1984). (Right) changes in June insolation at 60°N over the last 0.4 m.y.

Figure 8. Changes in tilt, precession (the precessional index), and summer half-year (calendar) insolation at 60°N over the last 25,000 years.

provided the first impetus toward deglaciation. However, the resulting total destruction of the ice sheets appears disproportionate to the relatively modest insolation forcing. This mismatch suggests that additional processes acting within the climate system accelerated the deglaciation.

This brings us back to the two main themes of this volume. What internal feedback processes within the climate system were the key sources of the accelerated ice-volume disintegration during the last deglaciation? On the continent beyond the limits of the ice sheets, were the patterns and rates of climatic change during the deglaciation a secondary effect of the ice sheet's existence, a manifestation of independent climatic change related in some other way to orbital forcing, or both?

INTERNAL FEEDBACK PROCESSES

Several processes acting within the climate system have been proposed as candidates for the nonlinearity required to accelerate deglaciations. We briefly review them here to set the stage for the chapters in this volume.

Bedrock rebound

The weight of ice sheets depresses the earth's surface and causes compensating lateral flow in the mantle. Removal of ice reverses that process. Oerlemans (1980) and Birchfield and others (1981) first used the concept that there is a delay in bedrock rebound following ice unloading as a basis for modeling experiments to explain accelerated ice-sheet disintegration (Peltier, Chapter 8).

The concept involves a positive feedback: as insolation begins to melt the ice, the ice surface is lowered in elevation, bringing it into the warmer layers of the lower atmosphere. This increases ablation, thus lowering the ice surface even more, and the process accelerates. If the bedrock behavior were completely elastic, it would immediately compensate for the removal of the ice load by lifting the ice surface back toward its initial elevation. This rebound would compensate roughly 30 percent of the ice melting, reflecting the difference in density between ice and rock. As a result, ablation would slow at the higher elevations. But if viscous bedrock responses delay the rebound, the net lowering of the ice surface can proceed more rapidly and ablation will be stronger.

Models of this process using the $\delta^{18}O$ curves as a proxy ice-volume "target" show that a time constant of about 3,500 to 5,000 years for bedrock rebound is required for an optimal match (Pollard, 1983). This conflicts with evidence for more rapid rebound time constants (1,000–1,500 years) based on the deglaciation of Baffin Island, Fennoscandia, and northern North America (Brotchie and Sylvester, 1969; Andrews, 1970; Walcott, 1973).

Peltier (1982) concluded from an array of geophysical evidence that the degree of viscosity of the bedrock response to ice loading and unloading depends on the scale of deformation, with very large ice sheets eliciting more viscous responses from deeper in the mantle. More recently, however, Weertman and Birchfield (1984) and Birchfield (1987) questioned these results and concluded that the integrated bedrock response to ice-sheet unloading still remains too short to provide the delayed rebound necessary for positive feedback to ice-sheet melting. In short, bedrock rebound is a candidate explanation of accelerated deglaciations, but published models have not fully solved the enigma of the 100,000-year power. This problem is examined by Peltier in Chapter 8.

Iceberg calving

A second proposed source of nonlinear feedback to ice sheets is through a link between sea level and ice ablation to the ocean via calving. Denton and Hughes (1983) noted that large portions of the Northern Hemisphere ice sheets were grounded below sea level in and around areas that presently are shallow seas (e.g., Hudson Bay, the Baltic Sea; Andrews, Chapter 2). They suggested that these parts of the ice sheets were vulnerable to rapid loss of mass to the ocean via iceberg calving (Hughes, Chapter 9).

The proposed feedback loop works as follows: rising summer insolation initiates melting along the southern margins of the Laurentide and Fennoscandian ice sheets, and global sea level begins to rise. Rising sea level increases calving of icebergs to the ocean either by breaking up ice shelves that had previously exerted back-pressure on ice flow from the interior or by lifting the shelves off bedrock pinning points. This enables ice streams to remove mass rapidly from the interior of the ice sheets. Accelerated flow in ice streams and iceberg calving into the ocean then lead to further increases in sea level and still more ablation. In addition, meltwater lubricates ice streams flowing to the ocean and also accelerates ablation.

As envisaged by Denton and Hughes (1983), this process also depends on ice-sheet size. Larger ice sheets depress more of the earth's surface below sea level and thus bring larger volumes of ice within potential reach of the sea-level/calving feedback link.

Andrews (1973) previously suggested a terrestrial version of this process, in which icebergs calve into proglacial lakes (Andrews, Chapter 2; Teller, Chapter 3). In this case, some positive feedback is provided by lake size and depth, both of which are linked to the degree of delayed bedrock rebound in regions adjacent to the margins of the melting ice sheet or to areas previously occupied by ice.

Modeling either kind of calving feedback process is a complex problem. Published attempts have used simplified *ad hoc* parameterizations of the calving mechanism (Pollard, 1984). Numerous uncertainties remain in constructing accurate models, including past ice elevation, the degree of equilibration of bedrock to the ice load, and the configurations of pinning points in the ice streams. Regardless, calving is a powerful ablative process that is a second obvious candidate to explain accelerated rates of deglaciation.

Carbon dioxide (CO_2)

More recently, feedback from CO_2 variations in the earth's atmosphere has been proposed as a factor in ice-age climate change (Broecker and Peng, 1984). CO_2 levels 70 ppm lower than those of the pre-industrial Holocene have been measured in ice cores from both polar regions (Berner and others, 1979; Neftel and others, 1982). These lower values imply cooler glacial air temperatures due to a reduced CO_2 greenhouse effect (Paterson, Chapter 5).

The deglacial increase in CO_2, not directly dated as yet, may thus have increased the greenhouse effect by trapping heat within the atmosphere and warming the earth. This in turn may have helped to melt the ice sheets. Indirect dating of at least one component of the deglacial CO_2 increase, by $\delta^{13}C$ differences between benthic and planktonic foraminifera (Shackleton and others, 1983), indicates that the CO_2 increase may have led the global ice-volume signal by several thousand years (Shackleton and Pisias, 1985). On the other hand, the largest CO_2 changes in Greenland ice cores appear to be largely in phase with global ice volume (Neftel and others, 1982; Paterson, Chapter 5).

CO_2 is a likely contributor to the global deglacial warming, but still unclear are the timing of the CO_2 increase relative to ice-volume changes, the magnitude of the CO_2-induced global warming, and the specific effects on the Northern Hemisphere (and particularly the North American) ice sheets.

Moisture feedback

Moisture feedback theories have long been part of the literature on ice ages (Stokes, 1955; Ewing and Donn, 1966). The most recent form of this concept involves self-starvation (Adam, 1975; Ruddiman and McIntyre, 1981; Ruddiman, Chapter 7).

One form of the moisture-feedback concept focuses on oceanic moisture sources. The deglacial side of this proposed feedback works as follows: summer insolation initiates melting of ice sheets, which increases the flux of icebergs and meltwater to the North Atlantic ocean. This flux chills the sea surface, and the low-density meltwater increases the salinity stratification, thus enhancing sea-ice formation in winter. These effects, combined with minimal levels of winter insolation, in turn suppress winter evaporation from the ocean and cut off the moisture flux to the ice sheets, which thus disintegrate faster because of moisture starvation.

This theory is also difficult to test with models. General circulation models do not reproduce modern precipitation patterns as well as they do the fields of temperature, sea-level pressure, and winds. The more simplified time-dependent ice-sheet models use parameterizations that relegate accumulation to a secondary role, primarily because it varies within fairly small limits relative to the more powerful ablative processes needed to accelerate deglaciation. One test of the oceanic moisture-feedback concept with a simple ice-sheet model (Pollard, 1984) showed that the oceanic moisture flux back to the ice sheets can only be a critical factor if it is very sharply curtailed above some threshold value of meltwater influx. Although such a nonlinear response seems unlikely, the North Atlantic polar-front oscillations are very abrupt in spatial terms (Ruddiman and McIntyre, 1984) and could imply threshold effects. This makes it difficult to entirely rule out reduced moisture flux from the oceans as a positive feedback factor during deglaciation, although it is probably a second-order one.

The most powerful factor in reducing moisture flux to ice sheets at the glacial maximum is probably ice elevation, rather than suppression of moisture flux from the oceans; this is a result of the lower moisture content of air at higher elevations of the middle and high latitudes, as observed today. Because this factor would lose its impact as a positive feedback as ice elevations began to lower during deglaciation, it is not clear how it could act as a positive feedback. In general, both kinds of moisture feedback appear to be at most secondary processes during deglaciations.

In summary, any of the factors summarized above, or combinations of them, could provide the necessary positive feedback to accelerate the deglacial ice-sheet destruction initiated by summer insolation. This volume summarizes the data most relevant to this topic and evaluates these data in light of these possible explanations.

THE CONTINENTAL RECORD

The chronology of deglacial environmental changes on the continent can be closely controlled by radiocarbon dating—more closely than much of the oceanic record. The abundant and easily accessible geologic features of the last deglaciation, ranging from moraines and lake deposits to cave fossils and cultural remains, provide a rich and varied record that generally can be dated with relatively little uncertainty. The paleoclimatic interpretation of this record is complicated by such local factors as topography, lithology, tectonics, and microclimate, which may modify the physical and chemical processes on the landscape and affect the flora and fauna in individualistic ways, resulting in vegetation types and animal communities for which modern analogues do not exist or are at best imperfect.

Today, vegetation across the continent east of the Rocky Mountains has a broad latitudinal zonation controlled primarily by the seasonal frequencies of northern and southern air masses, modified on the western plains by the influence of dry Pacific air that has lost moisture in the western mountains. With the emplacement of the vast ice sheet across the continent, the climatic zones were not only displaced but were altered in character. Cold Arctic air masses, which in modern winters occasionally sweep down across the plains and bring freezing temperatures to Texas and Florida, were then trapped in the Arctic Basin, with outbreaks restricted to the North Atlantic and the Siberian plains where no such barriers existed. When Arctic air did flow over the lower parts of the ice sheet and descended to the mid-continent plains of North America, it would have been adiabatically warmed by 20°C or more.

The result of these changes is that the extreme cold that limits the ranges of certain plants and animals in the mid-continent today may not have prevailed during the glacial maximum and the early phases of deglaciation. At the same time, in the summer the proximity of the ice sheet had a cooling effect on the periglacial climate, and the steepened temperature gradient from north to south resulted in winds so strong that sand dunes and loess were formed over broad areas.

Thus the seasonality of climate in the mid-continent during the glacial maximum was different principally because of the presence of the ice sheet. The precipitation patterns were also affected. Anticyclonic circulation around the ice sheet brought dry easterly winds to the Pacific Northwest, whereas the southward displacement of the jet stream and its associated storm track brought more moisture to the American Southwest.

Superimposed on all of these periglacial climatic effects—or actually underlying them through the intermediation of the ice sheets—were the changes in solar radiation driven by the Milankovitch mechanism. The tilt and precessional cycles combined to produce maximum summer radiation at 11 to 10 ka and not only brought about accelerated recession of the ice sheets but increased the temperatures in western North America upwind of the ice sheets and also enhanced the monsoonal circulation in the Southwest, bringing greater summer rainfall from marine sources to the south.

The principal picture that one can derive from considering these factors is that the deglacial scene was dynamic on the time scale of a few thousand years. As summer radiation increased, the ice sheet ceased to advance, except locally. Wastage then set in and accelerated as the summer radiation increased further, and the various feedback mechanisms affecting the ice sheet, as recounted above, resulted in instabilities of the ice margin (and in periglacial climates affected by these instabilities).

These rapid environmental changes, culminating between 13 and 9 ka, had profound effects on the vegetation and the fauna and thereby on cultural history—for major human immigration to the mid-continent occurred just at this time, as the Laurentide ice sheet retreated from the Rocky Mountain front. The catastrophic extinction of the Pleistocene megafauna at this time is perhaps the most dramatic manifestation of these events, for it was unequaled at any time in the past. Were the environmental changes so rapid that the megafauna could not adjust, and if so, why was this deglaciation different from earlier ones? Or was the immigration of human hunters the critical factor in extinction?

These complexities in environmental history are well established for the last deglaciation because so much of the geologic record for this phase is intact, suitable for radiocarbon dating, and thus amenable to intensive interdisciplinary study. One key objective of this volume is to describe and interpret the geologic record and thus the climatic history of the continent, with particular attention to the relative timing and rates of change of the major components of this dynamic and complex system.

OVERVIEW OF THE VOLUME

This volume is organized in five parts. Part I (Chapters 2 through 4) summarizes evidence of the deglacial history of the North American ice on the basis of glacial geology and glaciology. Chapter 2 traces the history of the Laurentide ice sheet, starting with the problem of glacial growth and the evidence for a mid-Wisconsin interval of major deglaciation. It emphasizes the evidence for the relative stability of the northeastern sector of the ice sheet during the early phases of deglaciation, compared to the large fluctuations of the southern margin after 16 ka.

Chapter 3 examines the sequence of proglacial lakes that developed to the south and west of the Laurentide ice sheet, once retreat had opened up the basins produced by glacial erosion or isostatic depression. The sequential drainage is traced for the large proglacial lakes in Saskatchewan and Manitoba into Lake Agassiz and thence either down the Mississippi River or eastward through the Laurentian Great Lakes to the Gulf of St. Lawrence. Chapter 3 also considers the role of proglacial lakes in accelerating wastage by the calving mechanism, and the importance of fine-grained lake sediments in providing a substrate that permitted accumulation of subglacial water to facilitate rapid advances of unstable ice lobes.

Chapter 4 summarizes the responses of the southern margin of the Cordilleran ice sheet to climatic change, with special attention to the nonsynchronous behavior of nearby alpine glaciers in the Cascade Range.

Part II (Chapters 5 through 9) summarizes an array of paleoclimatic data relevant to the history of North American ice during the last deglaciation. Chapter 5 reviews ice-core data that reveal climatic conditions at the former ice-sheet surfaces. Stable-isotope ($\delta^{18}O$) data show evidence of step-like warming of air temperatures at 10 ka and roughly 13 ka. Simultaneous increases in CO_2 values suggest that this factor played a role in ice ablation.

Chapter 6 reviews all the factors that can complicate direct interpretation of ice volume from oceanic $\delta^{18}O$ records. It then summarizes $\delta^{18}O$ data showing that the last (isotopic) deglaciation occurred between 14 and 6 ka, with fastest changes occurring in steps centered at 13 and 10 ka. This contrasts with the relative smooth integrated areal retreat of Northern Hemisphere and North American ice sheets.

Chapter 7 reviews climatic indicators (planktonic foraminifera and other planktonic fossils, and ice-rafted sand) from ocean surfaces that lie adjacent to North America and that respond to changes in ice-sheet size. As with the $\delta^{18}O$ data, the first-order picture is one of oceanic warming and productivity increases broadly synchronous with the ice-sheet retreat from 14 to 6 ka. In the North Atlantic, rapid episodes of warming occurred at or near 13 and 10 ka, with a major cooling at about 10.5 ka. Ice rafting slowed markedly after 13 ka.

Chapter 8 reviews various models of the bedrock response to unloading of the ice, based on an array of independent data that constrain the viscosity of the mantle. The results are compared to sea-level evidence in several deglaciated areas.

Chapter 9 summarizes glaciological and geomorphological constraints used in several models of the Laurentide ice sheet during the last deglaciation. The marine nature of the Laurentide ice sheet in the northeastern sector is contrasted with the terrestrial margins in the southwest.

Part III (Chapters 10 through 12) reviews evidence of environmental change from the physical and chemical record on the continents beyond the areas of glaciation. Chapter 10 deals with river systems. After reviewing the mechanisms by which river morphology and sedimentation are affected by local factors and by regional climatic and vegetational changes, it concentrates on factors influencing the major components of the Mississippi River system during phases of deglaciation. Inputs of glacial outwash in the headwaters of the Missouri and Ohio rivers, as well as the main stream, resulted in an episode of braided-stream deposition traceable to the Mississippi Lowland. This caused backwater lakes to form, even on major nonglacial tributaries like the Arkansas River, and on small tributaries in the Middle West. As the ice sheet withdrew, the drainage of proglacial lakes resulted in the deep dissection of outwash valley fills, and, as the drainage of the proglacial lakes shifted to the east out of the Mississippi River system, the braided pattern shifted to a meandering pattern. The base-level effects of sea-level fluctuations are discounted as major factors in the Mississippi River profile because of the great length and shallow slope of the delta.

Chapter 11 treats the pattern of lake-level changes in the basins of the semi-arid West. It includes an evaluation of the components of the hydrologic budget that are affected by climatic change and presents a chronology of lake-level fluctuations in Lake Lahontan, Lake Bonneville, and the Mono Lake area. The importance of increases in winter precipitation is emphasized as the critical factor in lake expansion during the time of maximum glaciation.

Chapter 12 treats the history of western lakes in the context of biological changes recorded especially in the stratigraphy of fossil ostracodes, which are sensitive in many cases to water chemistry and water temperature. The palaeolimnological interpretations are backed up by consideration of the geographic and limnological distributions of diagnostic species as well as life-history observations.

Part IV (Chapters 13 through 18) brings together the paleoecological records from different parts of the country and for different biological groups. Chapter 13 deals with the vegetational changes in central and eastern North America during the course of deglaciation and subsequently, by the presentation of colored maps showing the shifts in percentages of major pollen types at 2,000-year intervals from 18 to 6 ka. It also presents a compilation to show how the rates of vegetational change throughout the region accelerated, particularly during the interval from 13 to 9 ka and during the last 1,000 years.

Chapter 14 treats the vegetational history of the western mountain region and Alaska, showing how low temperatures and precipitation affected the Pacific Northwest during the glacial maximum. Subsequent to ice recession, warm conditions in that region dominated the early Holocene, whereas in the Rocky Mountains the time of maximum warmth was substantially later in some areas. In Alaska the chronology and pattern of tree invasion of tundra is traced in three different provinces—the southeastern panhandle, the interior, and the Brooks Range. The chapter includes a tabulated summary of the results of paleoclimatic modeling for the entire region.

Chapter 15 concerns the vegetation history of the Southwestern deserts, including separate discussions of the Chihuahuan and Mohave deserts, the Colorado Plateau, and the Great Basin. The discussion is based almost entirely on the occurrence of plant macrofossils in fossilized packrat middens, which are radiocarbon-dated to provide the basis for diagrams showing the relative abundance of key plant types over time. The climatic interpretations of the sequences consider the importance of the seasonal distribution of precipitation as well as the levels of temperature.

Chapter 16 considers the paleoecological record provided by fossil beetles, many of which are highly diagnostic of temperature and vegetational conditions, according to data on their modern geographic and ecologic distributions. The sequence of beetle faunas in eastern North America is traced in 1,000-year intervals during the phases of deglaciation.

Chapter 17 concerns the vertebrate faunas. Emphasis is placed on the co-occurrence of many species that are now widely separated in their geographic ranges, implying that climatic conditions during the glacial maximum and the deglacial period have no modern analogues.

In Chapter 18, several phases are summarized in the cultural Paleoindian sequence characterized by bone tools, blades, fluted points, or lanceolate points. Compilations of dated sites are presented to show the rapid rate of cultural change between 12 and 10 ka, presumably in response to the rapid changes in seasonal climate, vegetation, and animal resources.

Part V (Chapters 19 through 22) provides paleoclimatic overviews of the deglaciation period. Chapter 19 analyzes results from experiments with a general circulation model under past boundary conditions of ice-sheet area and height, albedo, sea level, sea ice, sea-surface temperatures, and solar-radiation distributions. Maps of simulated atmospheric pressure, temperature, precipitation, surface winds, and winds aloft are presented for January and July for 18 ka, 9 ka, and today, and implications for ice-sheet mass balance during deglaciation are discussed.

Chapter 20 compares deglacial climatic parameters estimated from general circulation models with pollen data from south of the ice sheet in central and eastern North America to evaluate the reliability of the two approaches in reconstructing climate. The striking shift in inferred vegetational patterns between 12 and 10 ka is a manifestation of the waning influence of the retreating ice sheet.

Chapters 21 and 22 summarize and integrate results from the rest of the volume, citing important areas of consensus and remaining points of disagreement or uncertainty requiring future work.

REFERENCES CITED

Adam, D. P., 1975, Ice Ages and the thermal equilibrium of the earth: Quaternary Research, v. 5, p. 161–171.

Andrews, J. T., 1970, A geomorphological study of postglacial uplift with particular reference to Arctic Canada: Institute of British Geographers Special Publication 2, 156 p.

——, 1973, The Wisconsin Laurentide ice sheet; Dispersal centers, problems of rates of retreat, and climatic implications: Arctic and Alpine Research, v. 5, p. 185–199.

Arthur, M. A., and Srivastava, S., ed., 1987, Initial reports of the Ocean Drilling Project: Washington, D.C., U.S. Government Printing Office, v. 105 (in press).

Backman, J., 1979, Pliocene biostratigraphy of DSDP Sites 111 and 116 from the North Atlantic Ocean and the age of the Northern Hemisphere glaciation: Stockholm Contributions to Geology, v. 32, p. 115–137.

Berger, A., 1984, Accuracy and frequency stability of the earth's orbital elements during the Quaternary, *in* Berger, A. L., and others, eds., Milankovitch and climate: Boston, Massachusetts, D. Reidel, p. 3–39.

Berner, W., Stauffer, B., and Oeschger, H., 1979, Past atmospheric composition and climate; Gas parameters measured on ice cores: Nature, v. 275, p. 53–55.

Birchfield, G. E., 1987, Ice-sheet dynamics and the Pleistocene ice ages, *in* NATO Proceedings on Irreversible Phenomena and Dynamical Systems Analysis in Geosciences (in press).

Birchfield, G. E., Weertman, J., and Lunde, A., 1981, A paleoclimate model of Northern Hemisphere ice sheets: Quaternary Research, v. 15, p. 126–142.

Boellstorff, J., 1978, North American Pleistocene stages reconsidered in light of probable Pliocene-Pleistocene glaciation: Science, v. 202, p. 305–307.

Brotchie, J. F., and Sylvester, R., 1969, On crustal flexure: Journal of Geophysical Research, v. 74, p. 5240–5252.

Broecker, W. S., and Peng. T.-H., 1984, The Climate-chemistry connection, *in* Hansen, J. E., and Takahashi, T., eds., Climate processes and climate sensitivity: American Geophysical Union Geophysical Monograph 29, Maurice Ewing Series, v. 5, p. 327–336.

Broecker, W. S., and van Donk, J., 1970, Insolation changes, ice volumes, and the O^{18} record in deep-sea cores: Reviews of Geophysics and Space Physics, v. 8, p. 169–197.

Denton, G. H., and Hughes, T., 1983, Milankovitch theory of ice age; Hypothesis of ice-sheet linkage between regional insolation and global climates: Quaternary Research, v. 20, p. 125–144.

Eldholm, O., and Thiede, J., eds., 1987, Initial reports of the Ocean Drilling Project: Washington, D.C., U.S. Government Printing Office, v. 104 (in press).

Emiliani, C., 1955, Pleistocene temperatures: Journal of Geology, v. 63, p. 538–578.

Ewing, W. M., and Donn, W. L., 1966, A theory of ice ages: Science, v. 23, p. 1061–1066.

Hays, J. D., Imbrie, J., and Shackleton, N. J., 1976, Variations in the earth's orbit; Pacemaker of the ice ages. Science, v. 194, p. 1121–1132.

Imbrie, J., 1985, A theoretical framework for the Pleistocene ice ages: Journal Geological Society of London, v. 142, p. 417–432.

Imbrie, J., and Imbrie, J. Z., 1980, Modeling the climatic response to orbital variations: Science, v. 207, p. 943–953.

Imbrie, J., and 8 others, 1984, The orbital theory of Pleistocene climate; Support from a revised chronology of the marine $\delta^{18}O$ record, *in* Berger, A. L., and others, eds., Milankovitch and climate: Boston, Massachusetts, D. Reidel, p. 269–305.

Milankovitch, M. M., 1941, Canon of insolation and the ice-age problem: Koniglich Serbische Akademie, Beograd. (English translation by the Israel Program for Scientific Translations; published by the U.S. Department of Commerce and the National Science Foundation, Washington, D.C.).

Neftel, A., Oeschger, H., Schwander, J., Stauffer, B., and Zumbrunn, R., 1982, Ice core sample measurements give atmospheric CO_2 content during the past

40,000 yr: Nature, v. 295, p. 220–223.

Oerlemans, J., 1980, Model experiments on the 100,000-year glacial cycle: Nature, v. 287, p. 430–432.

Peltier, W. R., 1982, Dynamics of the ice-age earth: Advances in Geophysics, v. 24, p. 2–146.

Pollard, D., 1982, A simple ice-sheet model yields realistic 100 kyr glacial cycles: Nature, v. 296, p. 334–338.

—— , 1983, A coupled climate-ice sheet model applied to the Quaternary ice ages: Journal of Geophysical Research, v. 88, p. 7705–7718.

—— , 1984, Some ice-age aspects of a calving ice-sheet model, *in* Berger, A. L., and others, eds., Milankovitch and climate: Boston, Massachusetts, D. Reidel, p. 541–564.

Rogers, K. L., and 7 others, 1985, Middle Pleistocene (Late Irvingtonian: Nebraskan) climatic changes in south-central Colorado: National Geographic Research, v. 1, p. 535–563.

Ruddiman, W. F., and McIntyre, A., 1981, Oceanic mechanisms for the amplification of the 23,000-year ice-volume cycle: Science, v. 212, p. 617–627.

—— , 1984, Ice-age thermal response and climatic role of the surface North Atlantic Ocean, 40° to 63°N: Geological Society of America Bulletin, v. 95, p. 381–396.

Ruddiman, W. F., and Kidd, R., eds., 1986, Initial reports of the Deep-Sea Drilling Project: Washington, D.C., U.S. Government Printing Office, v. 94, 1261 p.

Ruddiman, W. F., Raymo, M., and McIntyre, A., 1986, Matuyama 41,000-year cycles; North Atlantic Ocean and Northern Hemisphere ice sheets: Earth and Planetary Science Letters, v. 80, p. 117–129.

Shackleton, N. J., and Opdyke, N. D., 1973, Oxygen isotope and paleomagnetic stratigraphy of equatorial Pacific core V28–238; oxygen isotope temperatures and ice volumes on a 10^5 and 10^6 year scale: Quaternary Research, v. 3, p. 39–55.

Shackleton, N. J., and Pisias, N. D., 1985, Atmospheric CO_2, orbital forcing, and climate, *in* Sundquist, E., and Broecker, W. S., eds., The carbon cycle and atmospheric CO_2—natural variations Archean to present: Geophysical Monograph 32, p. 303–317.

Shackleton, N. J., Hall, M. A., Line, J., and Shuxi, C., 1983, Carbon isotope data in core V19–30 confirm reduced carbon dioxide concentration in the ice age atmosphere: Nature, v. 306, p. 319–322.

Shackleton, N. J., and others, 1984, Oxygen isotope calibration of the onset of ice-rafting and history of glaciation in the North Atlantic region: Nature, v. 307, p. 216–219.

Stokes, W. L., 1955, Another look at the ice ages: Science, v. 155, p. 815–821.

Walcott, R. I., 1973, Structure of the earth from glacio-isostatic rebound: Annual Reviews of Earth and Planetary Science, v. 1, p. 15–37.

Wayne, W. J., Aber, J. D., Bluemle, J., and Martin, J. E., 1987, Quaternary extra-glacial geology of the northern Great Plains, *in* Morrison, R. B., ed., Quaternary nonglacial geology; Conterminous U.S.: Boulder, Colorado, Geological Society of America, The Geology of North America, v. K-2 (in press).

Weertman, J., 1964, Rate of growth or shrinkage of non-equilibrium ice sheets: Journal of Glaciology, v. 5, p. 145–158.

Weertman, J., and Birchfield, G. E., 1984, Ice-sheet modeling: Annals of Glaciology, v. 5, p. 180–184.

Wigley, T.M.L., 1976, Spectral analysis and the astronomical theory of climatic change: Nature, v. 264, p. 629–631.

Wright, H. E., Jr., 1977, Quaternary vegetation history; Some comparisons between Europe and America: Annual Reviews of Earth and Planetary Science, v. 5, p. 123–158.

Zimmerman, H., and others, 1985, History of Plio-Pleistocene climate in the Northeast Atlantic, Deep Sea Drilling Project Hole 552A, *in* Roberts, D., Schnitker, D., and others, Initial reports of the Deep Sea Drilling Project: Washington, D.C., U.S. Government Printing Office, v. 81, p. 861–875.

MANUSCRIPT ACCEPTED BY THE SOCIETY MARCH 20, 1987

Printed in U.S.A.

The Geology of North America
Vol. K-3, North America and adjacent oceans during the last deglaciation
The Geological Society of America, 1987

Chapter 2

The Late Wisconsin Glaciation and deglaciation of the Laurentide Ice Sheet

John T. Andrews
INSTAAR and Department of Geological Sciences, University of Colorado, Boulder, Colorado 80309

INTRODUCTION

During the Late Wisconsin Glaciation, the Laurentide Ice Sheet covered an area approximately equivalent to that of the present Antarctic Ice Sheet, about 12.6×10^6 km^2. The ice sheet extended between 75° and ca. 45° N and between 64° and 120°W. It existed across major climatic gradients and was an important modifier of climate (e.g., Kutzbach and Wright, 1985; Manabee and Broccoli, 1985). In this chapter the deglacial history of this massive Northern Hemisphere ice mass is examined via four topics that constrain its area/volume relation and that indicate boundaries against which glaciologic, climatologic, and oceanographic reconstructions of the late-glacial world must be tested and verified. These topics are (1) the evidence of glacial extent during the middle Wisconsin interstade, (2) the data on the timing of the last glacial maximum and the extent of this glaciation, (3) the field evidence for ice-marginal profiles and hence the thickness of the ice sheet, and (4) the chronology, rates, and mechanisms of deglaciation.

This chapter is concerned with the glacial-geological evidence for ice-sheet extent and thickness. Additional evidence on the complex problem of reconstructing the Laurentide Ice Sheet is contained in the chapters by Hughes and by Peltier, who examine glaciological and glacial-isostatic aspects of the problem. Chapters by Ruddiman and by Mix also concern deglacial history; specifically, the influence and interaction between the ice sheets and different oceanographic parameters. The chronology and mechanisms of deglaciation for the southern and southwestern margins are also dealt with by Teller (this volume).

This chapter draws heavily on recent syntheses for the whole or parts of the Laurentide Ice Sheet, including reports of the I.G.C.P. Project #24 (Fulton, 1984), Dyke and others, (1987, and other chapters therein), Michelson and others (1983), and chapters in Teller and Clayton (1983).

VIEWS OF THE LAURENTIDE ICE SHEET

Prest (1969) and Bryson and others (1969) produced isochrone maps on the deglacial history of the North American ice sheets. These maps were based on radiocarbon-dated samples that delimited ice margins. Although differing in detail, both papers represent important contributions to our knowledge of the chronology of ice retreat. In retrospect, these two papers changed our way of viewing former ice sheets in that they stressed the importance of ice-sheet dynamics on a continental scale, as opposed to responses on a regional scale of 1000 km^2. The isochrone maps were used by several authors to comment on ice-sheet thickness, volume of water stored in the ice sheets, and the style of deglaciation (e.g., Bloom, 1971; Paterson, 1972; Andrews, 1973). In the above studies, an underlying assumption was that the Laurentide Ice Sheet consisted of a single dome, its axis oriented northwest-southeast across Hudson Bay. Shilts (1982, 1985), among others, has commented on the history of geological thought that led to that premise.

The advent of radiocarbon dating and glacial-geological reconnaissance in northern Canada in the 1960s produced the first evidence that deglaciation consisted of at least two episodes of catastrophic collapse; one dated at ca. 8 ka consisted of the rapid deglaciation of Hudson Strait and Hudson Bay (Falconer and others, 1965; Andrews and Falconer, 1969; Prest, 1970), and the other, dated at 6.7 ka, resulted in the deglaciation of Foxe Basin (Fig. 1) (Blake, 1966; Andrews, 1970). In both cases, the development of marine calving bays was called upon as a mechanism compatible with rapid deglaciation.

From the 1970s to the present, two contrasting views of the Laurentide Ice Sheet have developed. In the first, the extensive surficial-geological mapping throughout Canada and the northern United States resulted in the perception that deglaciation from a single-ridged ice sheet was not capable of explaining extensive dispersal trains around Hudson Bay, Foxe Basin, and the Arctic mainland coast (Shilts and others, 1979; Andrews and Miller, 1979; Dyke, 1983, 1984; Tippett, 1985), nor could it explain multiple ice sources for lobes west of the Great Lakes. The views of the ice sheet stemming from these field studies are not entirely compatible (cf. Prest, 1984; Shilts, 1980. Dyke and others, 1982), nor have the dynamics been modeled to any great extent.

Andrews, J. T., 1987, The Late Wisconsin Glaciation and deglaciation of the Laurentide Ice Sheet, *in* Ruddiman, W. F., and Wright, H. E., Jr., North America and adjacent oceans during the last deglaciation: Boulder, Colorado, Geological Society of America, The Geology of North America, v. K-3.

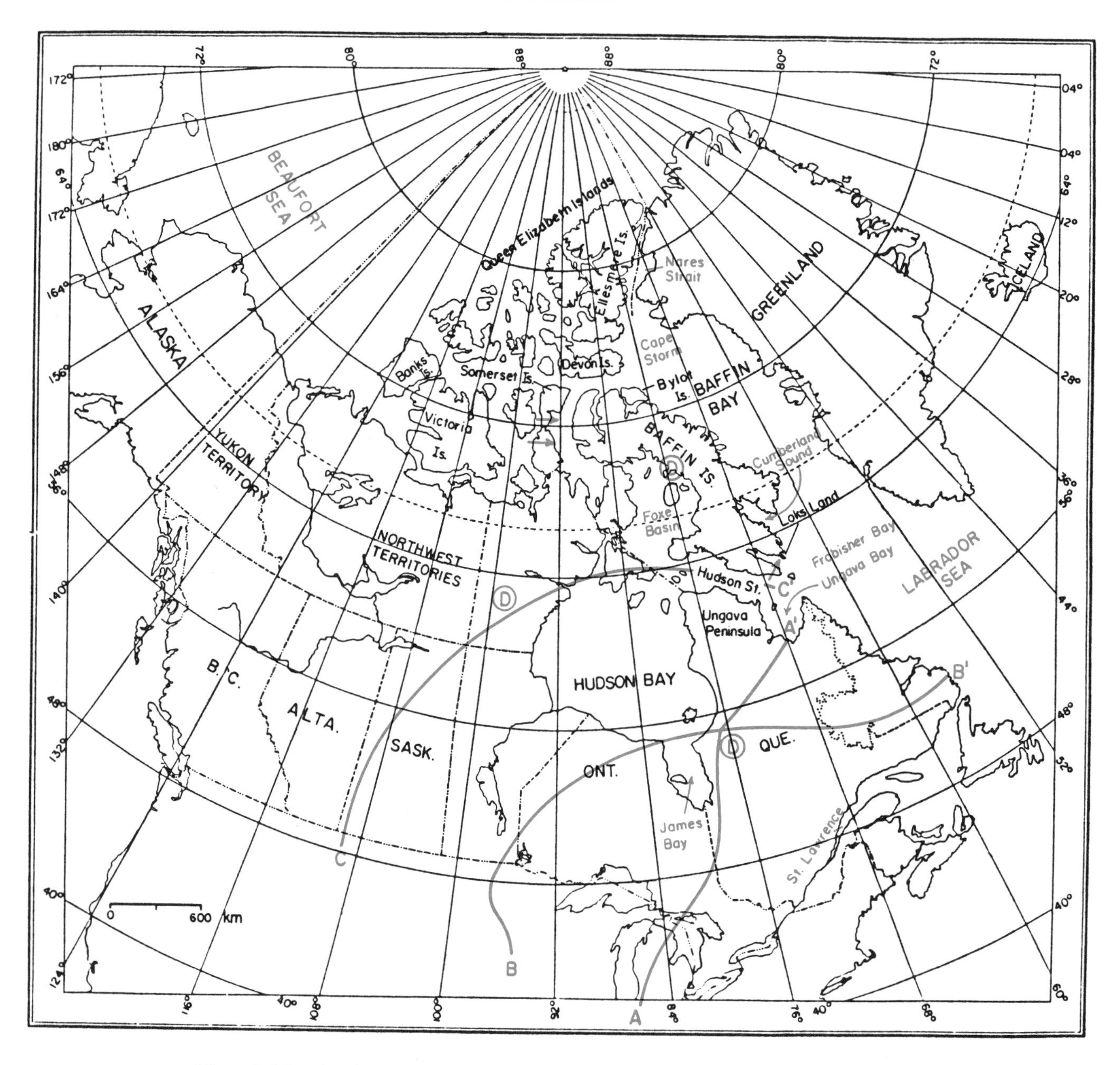

Figure 1. Map showing major geographic features and location of cross sections shown in Figure 5. D = Dome of ice sheet (Boulton and others, 1985).

In a variant of this view, Prest (1984) presented the Laurentide Ice Sheet as a complex of ice masses (Fig. 2A), the ice sheet dominated by flow from interior, rather than marine, ice domes. In this reconstruction, primary weight is given to trains of erratic stones. Boulton and others (1985) and Boulton (1984) modeled the Laurentide Ice Sheet and the time required to transport some of the distinctive rock types. The results indicate that erratic trains of the extent mapped in northern and southern Canada probably do not represent late-glacial transient flow. In commenting on the disagreement between one of their glaciological reconstructions and the mapped flow trajectories, Boulton and others (1985, p. 470) stated: ". . . In North America two very large-scale patterns of erratic dispersal (Shilts, 1980, 1982) cut dramatically across the predicted dispersal pathways *in such a way as to deny the validity of our model*" (italics added), thus indicating the importance of field data in model verification.

The second important approach to understanding the dynamics of the Laurentide Ice Sheet stemmed from the increased understanding of the current Antarctic Ice Sheet (Denton and Hughes, 1981a; Stuiver and others, 1981; Hughes and others, 1985a; Drewry and others, 1985; McIntyre, 1985) and the use of this system as an analogue for Northern Hemisphere glaciation (Hughes and others, 1985a; Hughes, 1985). This approach formalized the Laurentide Ice Sheet into marine and terrestrial sys-

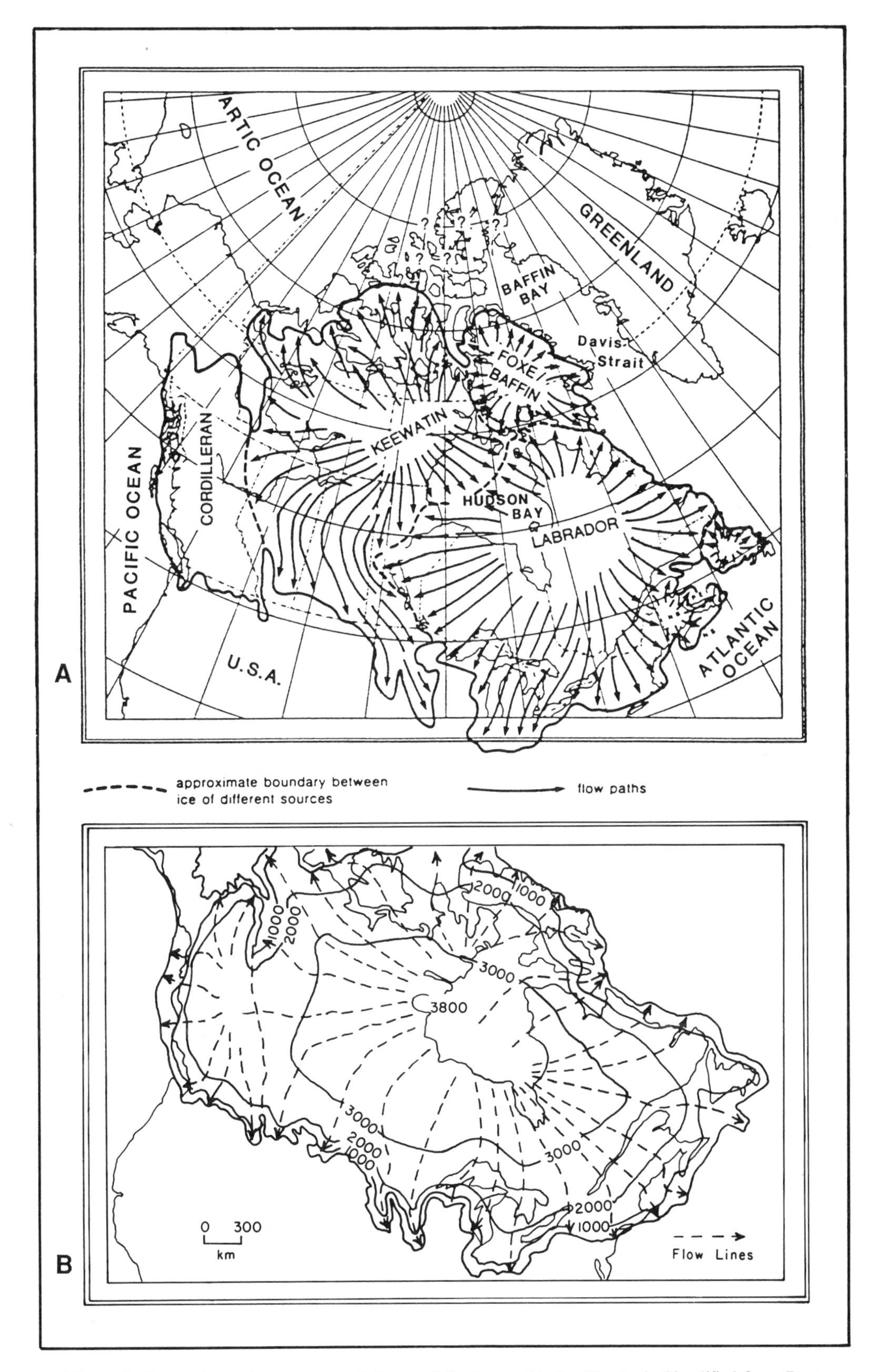

Figure 2. Comparison of two suggested forms of the Laurentide Ice Sheet. A: Simplified from Prest (1984); B: From Denton and Hughes (1981a).

tems, the former being inherently unstable unless its ice streams were buttressed by ice shelves. In one such reconstruction, the ice divide extends across Hudson Bay (Fig. 2B), and major ice streams drain the ice sheet through the Gulf of St. Lawrence, Hudson Strait, and the major channels of the Arctic Archipelago. In the Antarctic analogue, the deglacial response of the marine-based part of the Laurentide Ice Sheet is largely driven by sea-level changes, which cause retreat of the grounding line and the collapse of the marine portion of the ice sheet (Denton and others, 1981b, 1986).

Illustrations such as Figure 2A and 2B, however, are static representations of a dynamic system. There is considerable evidence that during glacial buildup and retreat, ice centers shifted in response to regional glaciological and climatological changes (Michelson and others, 1983; Teller and Clayton, 1983; Dyke and others, 1987; Klassen, 1985a).

The two views discussed above have radically different interpretations of the origin of the megageomorphology of the Canadian Shield and the bed of the ice sheet. Hughes and others (1985b) argued (see also Hughes, this volume) that ice-sheet reconstructions should be based on first-order glacial-geological landforms. In their view, the basins and channels of the shield and peripheral regions result from large-scale glacial erosion. The opposing view can be represented by a sentence from a structural geologist, who stated (Kerr, 1970, p. 570): "Overall shapes of lands and seas in northeastern Canada are the result of tectonic events that were part of continental drift."

Of concern here are two questions: (1) were the troughs created by glacial erosion; and (2) were they occupied by ice streams during the late Wisconsin? The answer to the first is probably not; the answer to the second is maybe. Denton and Hughes (1981a) implicitly ascribed the profiles of the major sounds in eastern Canada to glacial erosion (see also Groswald, 1984). However, Cumberland Sound (Fig. 1) is a 1200-m-deep graben floored with Cretaceous shales, and the other large sounds, such as Lancaster Sound, Frobisher Bay, and Hudson Strait, are fault-bounded and floored with Paleozoic limestones (Grant and Manchester, 1970; MacLean and others, 1986a, 1986b; Andrews and others, 1985a, 1985b).

Ice-sheet reconstructions based on ice discharging through the major channels perforce show the ice converging into such features. However, field mapping around Somerset Island and Boothia Peninsula (Dyke, 1983, 1984) and outer Hudson Strait (Osterman and others, 1985; Stravers, 1986) indicates that regional ice flow cut across the topographic grain (Fig. 1). In outer Hudson Strait, the north-northeast trend of striations and bedforms continues for at least 100 km from Resolution Island westward and does not appear to represent the expected divergence of ice flow near an ice-stream terminus. In the first-order view of glacial geology (Hughes and others, 1985b) such cross-cutting relationships are difficult to explain, especially for areas that should be strongly influenced by downdraw and convergence of ice flow.

One of the major differences between the two end-member reconstructions is the question of ice-sheet symmetry. Hughes (1985) thought that symmetry is called for on glaciological grounds (see also Dyke and others, 1982). Dyke and others (1987) have taken an intermediate stand between the alternatives shown in Figure 2. Although they favor a multidomed ice sheet, they differ from Prest (1984) in that they place a major north-south divide extending from the Arctic coast southward into Keewatin (McClintock Ice Divide) and propose a major divide over Hudson Bay. With time, the McClintock Ice Divide migrates to the position of the Keewatin Ice Divide. The presence of ice centered over western Hudson Bay also appears to be necessary to explain the late-glacial Cochrane readvances immediately prior to 8 ka. Thus Dyke and others (1987) propose a four-dome model compared to the three major divides of Prest (1984). As Denton (1986, personal commun.) has noted, the question of ice-sheet symmetry is tied in with questions of the chronology of the erratic trains as well as the process of interpreting till-provenance data. However, it is also tied in with the potential significance of deformable ice-sheet beds (Boulton and others, 1985; Fisher and others, 1985), and only Boulton (1984) and Boulton and others (1985) have made an effort to model the process of basal glacial transportation rates.

Implications

The two end-member views of the ice sheet have significantly different volume/time characteristics. Denton and Hughes (1981a, Table 6-3) computed the volume of their maximum and minimum ice-sheet models and arrived at global sea-level equivalents of 85 and 76 m. The maximum model (Fig. 2B) still constitutes an upper limit to the Laurentide's volume, but the volume of the ice sheet shown in Figure 2A would be less than that calculated for the minimum model above (e.g., Fisher and others, 1985).

Against this background of active debate, this chapter examines the Late Wisconsin history of the Laurentide Ice Sheet and deals with questions of ice-sheet thickness, symmetry, chronology, and rate of deglaciation.

THE MIDDLE WISCONSIN: WHERE WAS THE ICE?

Why is an understanding of the middle Wisconsin important for estimates of ice-sheet volume and history? The Late Wisconsin Glaciation cannot be correctly understood unless we have an understanding of ice extent during the previous interval. Weertman (1964) showed that the buildup of an ice sheet is slower than its wastage (even without calving); therefore, the thickness and extent of the Late Wisconsin ice sheet(s) is controlled by their history. Estimates of ice-sheet volume by Peltier and Andrews (1976), Paterson (1972), and Denton and Hughes (1981a, b) were based on ice sheets that were in glacial-isostatic equilibrium, thus under the tacit assumption that there was no significant middle Wisconsin deglaciation. In some scenarios (e.g., Andrews and Barry, 1978; Prest, 1970), the Laurentide Ice

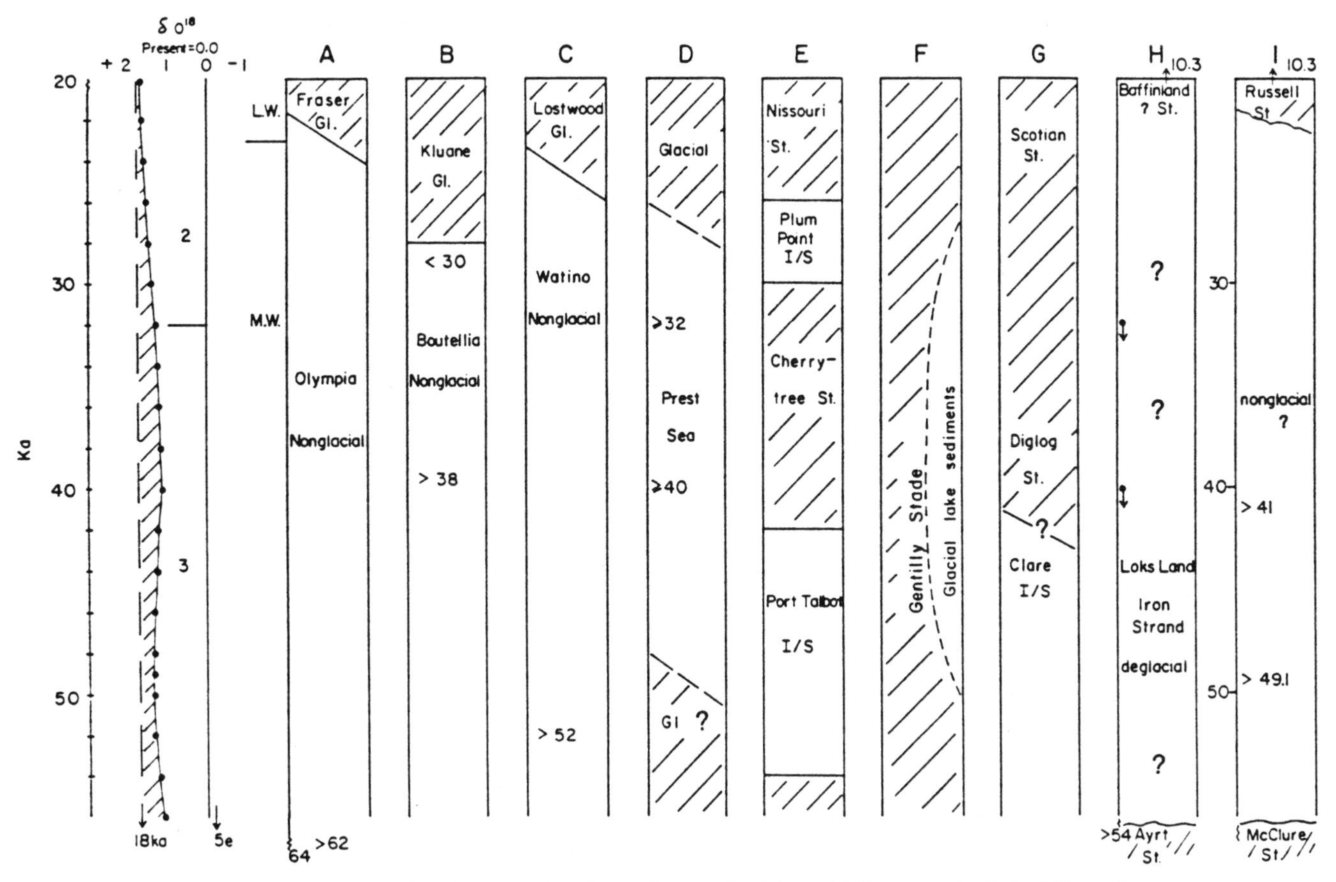

Figure 3. Stratigraphic columns (taken from chapters in Fulton, 1984; see text) of the Wisconsin Glaciation in Canada showing the evidence for conditions during the middle Wisconsin. (Sections are located in Figure 4.) The δ^{18}O ocean record is also shown.

Sheet was explicitly viewed as existing as a major ice sheet throughout the Wisconsin.

Marine isotope stage 3 (Fig. 3) consists of a low plateau surrounded by the heavier δ^{18}O events of stages 4 and 2. In sea-level terms, the oxygen ^{18}O/^{16}O ratios suggest the presence of substantial land-based ice, although less than in either stage 2 or 4. However, the summer insolation curve for 65°N, approximately the latitude of Hudson Strait, is marked by an insolation maximum ca. 50 ka that is 83% of the Holocene maximum. If the orbital variations noted by Milankovitch (1941) are the major forcing function controlling ice volume, then some response to the middle Wisconsin summer insolation maxima is expected. Glaciological models that incorporate solar radiation predict substantial deglaciation between 40 and 50 ka (Budd and Smith, 1981).

In 1975, Dreimanis and Raukas reviewed our knowledge of the middle Wisconsin and concluded that there is substantial stratigraphic and radiometric evidence to indicate that certain ice sheets were either much reduced or absent. For example, Fulton and Westgate (1975) reported a sequence of nonglacial sediments and interbedded ashes in the British Columbian trench that date from >50 to 25 ka, and Alley and others (1986, p. 1156) noted that "... by 34,000 years ago (temperatures) had ameliorated and were similar or slightly cooler than present." Throughout the Yukon Territory, the Boutellier nonglacial interval records an extensive cold interstadial dated between ca. 50 and 25 ka (Denton, 1974); similar events are recorded in Alaska (Hopkins, 1982).

One of the persistent problems of interpreting terrestrial stratigraphy for the interval >20 ka is the accuracy of ^{14}C dates, especially dates on marine shells (e.g., Szabo and others, 1981). However, if such dates are combined with other relative or radiometric dating methods, the "absolute" age can be better delimited. Figures 3 and 4 show stratigraphic columns, location of these sites, and a proposal for the outline of a middle Wisconsin ice sheet. Recent and forthcoming surveys of the Canadian glacial-geological literature (e.g., Fulton, 1984; Geology of North America volumes, in prep.) also testify that the middle Wisconsin was of interstadial rank for significant portions of the ice sheet (Alley and others, 1986). In the Prairie Provinces (Fenton, 1984; Fulton and others, 1944), the Watino nonglacial interval spans all of stage 3 (Fig. 3). The climate is considered to have been "... temperate boreal, similar in many respects to that of present-day

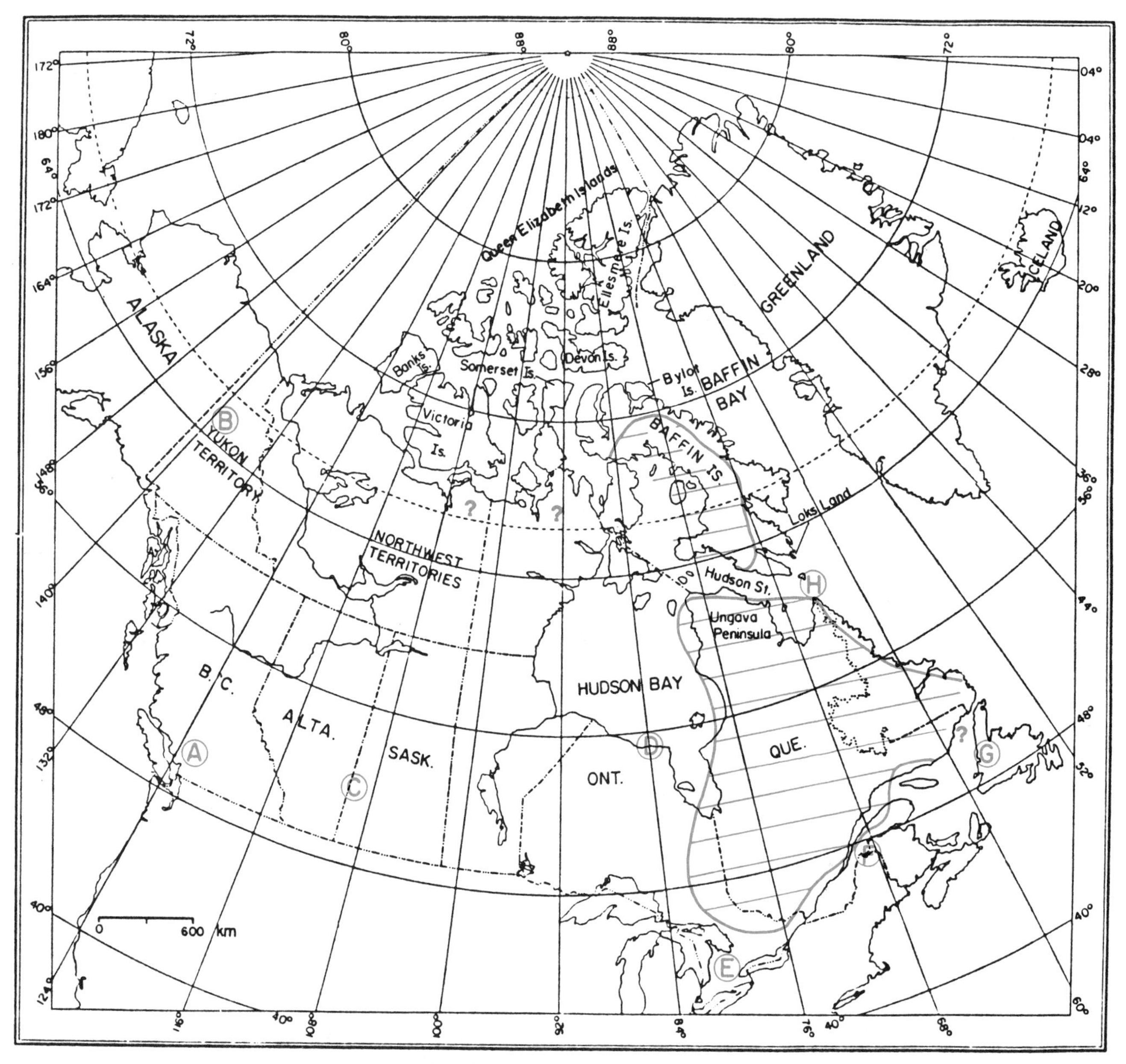

Figure 4. Suggested extent of middle Wisconsin ice over northern North America based on the evidence compiled in Figure 3.

central Alberta . . ." (Fenton, 1984, p. 63). However, the stratigraphy in southern Ontario, southern Quebec, and the Atlantic provinces (Karrow, 1984; Lasalle, 1984; Grant and King, 1984) indicates that the middle Wisconsin was largely glacial, with restricted nonglacial intervals (Vernal and others, 1986) (Fig. 3). Recent work on Anticosti Island in the outer Gulf of St. Lawrence and in areas near Cape Breton provides evidence for nonglacial conditions during the middle Wisconsin (Vernal and others, 1983; Bigras and others, 1985; St. Pierre and others, 1985). Bigras and others (1985) dated more than 40 samples on Anticosti Island. They stated that the sea-level history between 35 and 12 ka was derived from shells contained in glacial-marine sediments, and they recognized no Late Wisconsin overriding of the sections.

In northern Canada, a middle Wisconsin deglacial event has been recognized along the western side of Baffin Bay and the Labrador Sea from southern Ellesmere Island to northernmost Labrador (Ives, 1977; Blake, 1980; Clark, 1984; Miller, 1985; Klassen, 1985b). This event is associated with finite ^{14}C dates between 28 and 40 ka and amino acid ratios only slightly higher than Holocene levels but substantially lower than those of the Kogaul aminozone. The Kogaul aminozone is associated with

^{14}C dates older than 54 ka and U-series dates of ca. 70 ka (Szabo and others, 1981; Miller, 1985).

In northwestern Canada, Vincent (1984) has not been able to identify deposits that are clearly interbedded between deposits of the Russell (late Wisconsin) and McClure stades. On the basis of shell dates on in situ molluscs, Vincent (1984, p. 95) dated the McClure Stade as older than 37 ka (GSC-3698). However, in order to explain the existence of peats on the Beaufort shelf at depths of 150 m associated with ^{14}C dates of 27.4 ka, Hill and others (1985) proposed that ice existed in northwest Canada for at least part of the middle Wisconsin. England (1985) proposed that a seaway was formed by glacial-isostatic depression between 8.2 and >33 ka, in the area between northwest Greenland and northeast Ellesmere Island.

Research into the Quaternary stratigraphy of the Hudson Bay Lowlands (Shilts, 1982; Andrews and others, 1983; Dredge and Cowan, 1987; Wyatt and Thorleifson, 1986; Thorleifson and Wyatt, 1986) has resulted in the discovery of marine sediments stratigraphically above the Missinaibi Formation, which is inferred to be of last interglacial age (Prest, 1970). These findings indicate that the lowlands were deglaciated at least twice after deposition of the Missinaibi Formation (Andrews and others, 1983). A marine unit on the Attibi River consists of a prolific in situ shell bed. Amino acid ratios (total) on these shells gave a tight grouping of 0.09 compared to 0.033 for the Holocene Tyrrell Sea. Shells from the last interglacial(?) Bell Sea had ratios of 0.2. ^{14}C dates on the Attibi River shells gave finite ages of 36 ka and 40 ka. The underlying Fawn River Gravels (Shilts, 1982, 1985; Andrews and others, 1983) not only had a freshwater clam assemblage, but thermoluminesence dates averaged 74 ka (Forman and others, 1987).

Conclusions

Published data (Figs. 3 and 4) indicate that the middle Wisconsin glaciation was less extensive than the Late Wisconsin glacial maximum. Limiting dates in the west indicate that glaciation commenced about 25–27 ka and thus allow less than 10,000 yr for ice buildup prior to 18 ka. The stratigraphy of glacial/nonglacial indicators suggests that ice-free conditions existed across the western and southwestern margins of the former ice sheet, while glacial conditions prevailed over Labrador, much of the eastern Canadian arctic, and probably northern Keewatin.

A much reduced middle Wisconsin ice sheet implies that the ice sheet probably did not attain complete isostatic equilibrium during the Late Wisconsin interval of ice-sheet growth, and that the ice sheet may not have been in dynamic equilibrium at the 18-ka isotopic minimum.

The outline of a middle Wisconsin ice sheet (Fig. 4) conforms in many respects to the ice extent in North America at ca. 8 ka (Bryson and others, 1969; Prest, 1969), except for the presence of more ice over southern Quebec. The extent of ice is also similar to the outline of the Laurentide ice sheet after 10,000 yr of simulated growth (Andrews and Mahaffy, 1976).

THICKNESS ESTIMATES AND GEOMETRY OF THE ICE SHEET

Ice-Sheet Profiles

Isochrone maps (Prest, 1969; Bryson and others, 1969) provide a quantitative measure of the changes in the area of the ice sheet during deglaciation. However, the relationships between area and ice thickness are not as easily deduced from field evidence. Bloom (1971) and Paterson (1972) considered the relation between changes in area and integrated thickness (volume); however, Denton and Hughes (1981b) and Ruddiman and McIntyre (1981) suggested that there may be strong nonlinearities in the relation, especially if the ice sheet underwent periods of collapse (see Ruddiman, Chapter 7, this volume). This section examines glacial-geological evidence for the thickness and geometry of the ice sheet against which model results can be tested.

Hughes (this volume) discusses in detail the various approaches to glaciological reconstructions based on the analogues of the Greenland and Antarctic ice sheets which assume basal shear stresses close to 1 bar (100 kPa) (Paterson, 1972; Sugden, 1978; Hughes, 1985). Glaciological models founded on this premise produce an ice sheet similar in form to that in Figure 2B (Budd and Smith, 1981; Denton and Hughes, 1981a; Paterson, 1972; Boulton and others, 1985).

Current ice sheets, however, may not be complete analogues for Northern Hemisphere Pleistocene ice sheets (Matthews, 1974; Boulton and others, 1985; Fisher and others, 1985; Beget, 1986, 1987). Profiles for the former southern margin of the Laurentide Ice Sheet give low basal shear stresses; this means that the ice was thin and that the thickness increases up a flowline at a very slow rate (e.g., Fig. 5). Matthews (1974) and Nielson and Thorleifson (1985) indicated that shear stresses as low as 4 kPa existed at the base of the ice sheet over parts of the Canadian prairies. The ages of these profiles are between the late-glacial maximum and ca. 12 ka. Similar low values may explain the lobate form of the ice sheet draining through the Great Lake basins (Johnson and others, 1986). The time/distance diagrams for Illinois suggest extremely rapid glacial advances (cf. Wright, 1973), which might also indicate low basal shear stress. Beget (1986) concluded that the till underlying the Lake Michigan lobe had a yield strength of 9 kPa, and he suggested that the lobe may have been very thin. Beget (1987) also presented evidence that the northwest margin of the Laurentide Ice Sheet had low basal shear stresses of ca. 4 kPa.

These low yield strengths from various profiles around the southern, southwestern, and northwestern margins of the Laurentide Ice Sheet are incompatible with a simple ice sheet model having a uniform basal shear stress. The apparent extent of the phenomenon indicates that current volume estimates of the ice sheet are probably too high.

Profiles of the eastern and northeastern margin of the Laurentide Ice Sheet have been reconstructed from landforms in Maine (Shreve, 1985), Labrador (Clark, 1984), and several

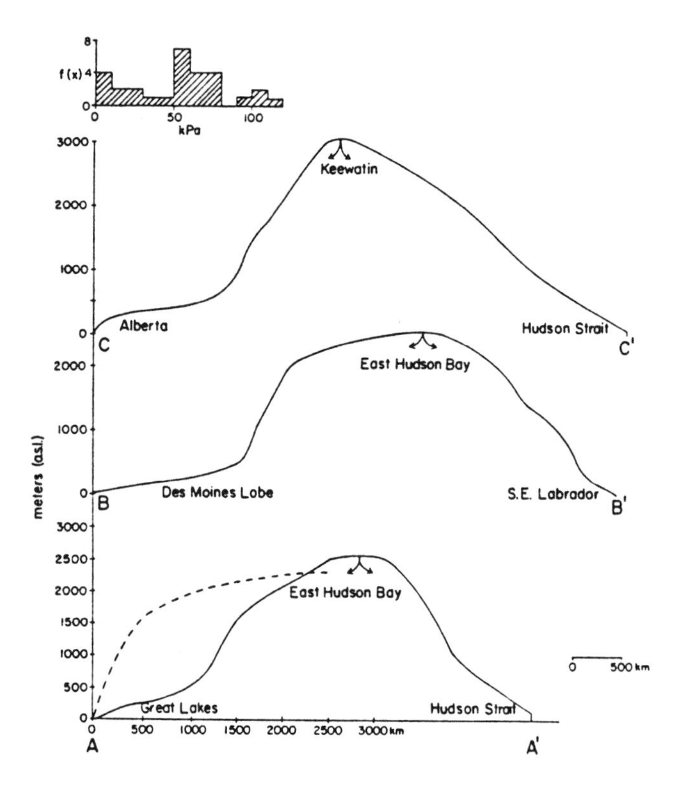

Figure 5. Series of cross profiles (from Boulton and others, 1985) across Laurentide Ice Sheet modelled with a deformable bed (see Fig. 1 for location). Also shown is a histogram of reported basal shear stresses as calculated from glacial landforms (see text for discussion).

major fiord glaciers along eastern Baffin Island (e.g., Smith, 1966; Buckley, 1969; Boyer, 1972; Mears, 1972). In an ingenious analysis, Shreve (1985, p. 34) calculated that the basal shear stress on an ice profile in Maine dated ca. 12.7 ka and, extending southeast-northwest, ranged from 20 kPa near the terminus to 70 kPa at 140 km behind the terminus. In northernmost Labrador, Clark (1984) and Clark and Josenhans (1986) mapped lateral moraines associated with the Saglek Glaciation (late Wisconsin) and estimated that the basal shear stress at several outlets ranged between 20 and 120 kPa or generally about 50 kPa. Boyer (1972), Mears (1972), and Pheasant (1971) calculated shear stresses at the bases of former outlet glaciers on eastern Baffin Island as between 170 and 8 kPa or generally between 50 and 70 kPa.

If extensive areas of thin ice existed along the northwest, southwest, and south margins of the ice sheet, ice divides must have been displaced to the north and east, and the ice sheet would not have radial symmetry. A theoretical basis for an asymmetric ice sheet has been advanced by Boulton and Jones (1979), Boulton and others (1985, Fig. 23), and Fisher and others (1985). It is based on the notion that the bed of the glacier deforms at a low yield strength. If the basal yield stress is forced to vary regionally and is high on the Precambrian shield but low on the flanking sedimentary series, a strongly asymmetric ice sheet can develop (see Hughes, this volume, for comments). The importance of a deformable bed on ice streams has recently been demonstrated by the finding of supersaturated tills beneath Ice Stream B in west Antarctica (Rooney and others, 1986).

Figure 5 shows cross sections along selected flowlines of the Laurentide Ice Sheet (see Fig. 1 for locations). The configuration of this ice sheet is close to that predicted on the basis of erratic trains around Hudson Bay (Shilts and others, 1979; Shilts, 1982, 1985), but it does not account for the west-east transport of erratics *across* major channels near Somerset Island and Boothia Peninsula (Dyke, 1983, 1984).

I suggest there are several reasons to propose that the Late Wisconsin Laurentide Ice Sheet was relatively thin, asymmetric, and multidomed. The deformable-bed ice-sheet reconstructions (Boulton and others, 1985; Fisher and others, 1985) show significant ice-sheet asymmetry and thus agree in broad terms with reconstructions based on erratic trains (e.g., Shilts, 1982). It is certain that yield stresses were as low as 4 to 9 kPa in some areas and between 50 and 100 kPa in others (see inset, Fig. 5).

It is still not certain, however, that these data totally rule out the *brief* existence of a massive, single-domed ice sheet. It is not at all obvious how such a condition could be verified by field data, especially if the central dome only lasted for 1,000 to 4,000 yr (Denton, 1986, personal commun.).

Data on Shoreline Deformation

Warped lacustrine and marine shorelines constitute an important body of data on late Quaternary ice-sheet dynamics (see Peltier, this volume) and on mantle rheology. If the ice sheet rested on a flat bed, the amount of isostatic depression would be a good guide to the position of the former ice divide. In a complex topographic setting, maximum ice thickness need not coincide with the location of ice divides, and this may explain the discrepancy between free-air gravity data and glacial-geological ice-sheet reconstructions (Peltier and Andrews, 1983). However, it is important to note that isobases on postglacial rebound show *two* centers of recovery, northwest and southeast of Hudson Bay.

Slopes of late-glacial shorelines (Fig. 6) are useful indicators of ice-sheet geometries. In the North American data, there is good coverage for the southern and eastern margins, but coverage is poor for the western margin. Upglacier projections of the strandlines do not converge on Hudson Bay but rather define a series of intersections (cf. Andrews and Barnett, 1972) that coincide with ice divides over eastern and central Labrador and Foxe Basin (Fig. 6).

Conclusion

Field observations of the gradients of former lobes and outlet glaciers of the Laurentide Ice Sheet indicate that the ice

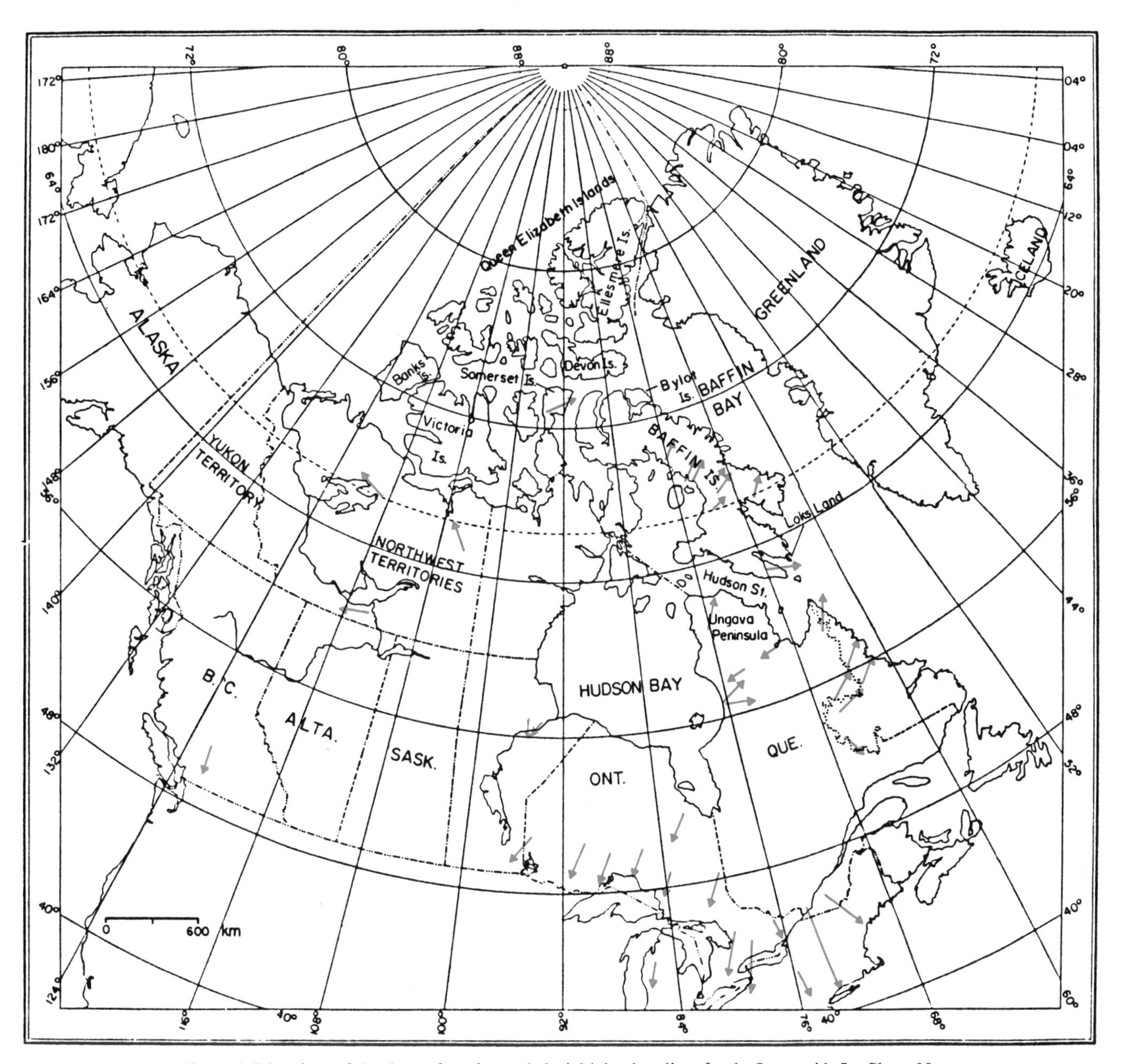

Figure 6. Directions of the slope of marine and glacial-lake shorelines for the Laurentide Ice Sheet. Note that because of the map projection these lines cannot be simply projected upslope to delimit centers of glacial unloading (after Andrews and Barnett, 1972).

sheet may have been thinner in the south, southwest, and northwest than would be expected on the basis of existing large ice sheets. The resulting asymmetry in the location of ice divides agrees reasonably well with field mapping of erratic trains, but not in detail. The slope projections on lake and marine strandlines from around the ice sheet intersect over Labrador and Foxe Basin; i.e., within the terrestrial portion of the ice sheet. In addition, isobases define rebound centers on either side of Hudson Bay, and a saddle between (Peltier and Andrews, 1983).

EXTENT OF GLACIATION AND THE DATE OF THE GLACIAL MAXIMUM

The 18 ka age for the last glacial maximum is derived in part from the peak in marine isotopic stage 2 and from dates around the southern margins of the Laurentide and Scandinavian ice sheets (cf. Denton and Hughes, 1981a). The radiocarbon-dated evidence from around the margins of the Laurentide Ice Sheet (e.g., Prest, 1969, 1970, 1984; Bryson and others, 1969;

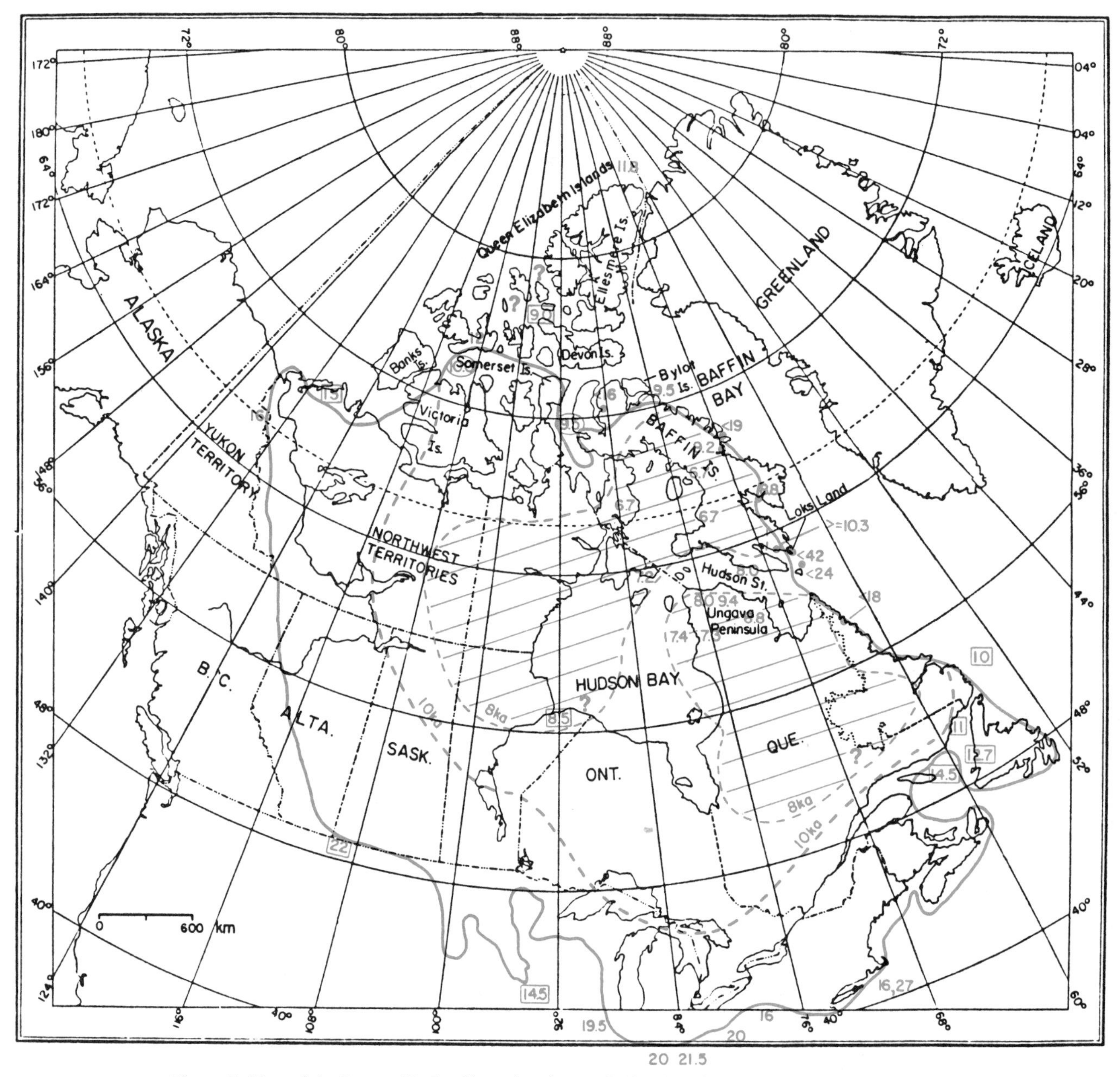

Figure 7. Map of the Laurentide Ice Sheet showing available dates (in ka) on the commencement of deglaciation (from Prest, 1984), with additions.

Mayewski and others, 1981; Michelson and others, 1983; Fulton, 1984) shows that the chronology of deglaciation differs between the southern and northern margins. Along the southern and western margins of the ice sheet, dates associated with the maximum Late Wisconsin glacial advance or initial retreat are between 24 and 14 ka. Conversely, studies around the northern and eastern margins of the ice sheet indicate that glacial retreat from the maximum occurred much later, between 12 and 8 ka (Fig. 7). The asymmetry in marginal response is evident in the isochrone patterns of deglaciation (Prest, 1969; Bryson and others, 1969), which show that the ice sheet retreated toward the north and east; even today, the Barnes Ice Cap on Baffin Island is a relict of the Laurentide Ice Sheet (Hooke, 1976).

It is evident from the isochrone maps that the ice sheet did not react in a uniform manner to the complex climatic/oceanographic/lacustrine forces that drive deglaciation. Denton and Hughes (1981b) proposed that disintegration of the marine-based sectors of the ice sheets was initiated by a global rise in sea level associated with the melting of terrestrial portions of the ice sheet. The initial cause of the melting might be the Milankovitch radiation perturbations discussed by Ruddiman and Wright (this volume) and Ruddiman (this volume). Disintegration of the

Laurentide Ice Sheet and the rearrangements of ice divides and saddles may have been caused by downdrawing of the marine-based center of the ice sheet (cf. Denton and Hughes, 1981a, b). A partial alternative hypothesis (Andrews, 1973) is that a major deglacial process was the calving into the extensive proglacial lakes that lay against the ice sheet in the south, southwest, and west (Pollard, 1983; see Teller, this volume).

The marine-based or marine-influenced sectors of the Laurentide Ice Sheet extended from New England northward to the Canadian Maritime Provinces, Labrador, and Baffin Island, and thence westward to Banks Island and the McKenzie Valley. The portion of the perimeter influenced by proglacial lakes extended from the Connecticut/Champlain valleys of New England west to the Great Lakes, and thence west and north into the basin of glacial Lake Agassiz and numerous glacial lakes in Alberta and the Canadian territories. This area coincides with the thin, fast-flowing ice lobes and deformable beds discussed earlier in this chapter.

Virtually all of the evidence on deglaciation from the east, northeast, and north sectors of the ice sheet is based on ^{14}C dates on marine shells and is associated with studies of raised beaches and the mapping of marine limits. The south, southwest, and west margins have been dated primarily by buried wood in a stratigraphic context, or by basal dates on lake sediments. Given this broad dichotomy between marine or glacial-lake influences and between the different materials used for controlling the glacial chronology, what can be said about differences in the rates of ice retreat and thinning?

There are two aspects to the problem of ice extent at 18 ka. The first is the ability to distinguish Late Wisconsin glacial deposits from those of earlier glacial events; the second is the actual dates on these glacial boundaries. The Late Wisconsin Glaciation limit is reasonably well defined in the northern United States and western Canada (Michelson and others, 1983; Fulton, 1984), although the limits in southwest Alberta and Montana are debated. Along the fiords and channels of northern and eastern Canada, soil development and the weathering of surface boulders on moraines have been used to distinguish Late Wisconsin deposits from earlier glacial episodes (Bockheim, 1979; Birkeland, 19718; Evans and Cameron, 1979; Dyke, 1977; Locke, 1985, 1986; Evans and Rogerson, 1986; Clark, 1984). Typically, Late Wisconsin drift shows ca. 10 cm of soil development compared with 50 cm for the next mappable glacial limit. Studies of soil development on moraines have largely replaced research on weathering zones (cf. Ives, 1978) as a method of defining former glacial limits. The studies noted above in the eastern Arctic define a Late Wisconsin glacial limit that favors the minimum ice-extent model (Mayewski and others, 1981; Prest, 1984). Problem areas still exist, and the probability that some weathering zones reflect former basal thermal conditions of the ice sheet (Denton and Hughes, 1981a) is certainly correct in some areas (Andrews and others, 1985b) (see Brooks, 1985, and Locke, 1985 for reviews of weathering-zone data and interpretations).

Although the Canadian syntheses on the late-glacial history of the Laurentide Ice Sheet largely subscribe to the minimum ice-sheet model (Fulton, 1987), research on the continental shelf (Scott and others, 1984; King and Fader, 1986; Fillon and Harmes, 1982; Josenhans and others, 1986) indicates that fringing ice shelves occupied the shelf from Nova Scotia to northernmost Labrador. Andrews (1985) and Osterman (1982) suggested that ice shelves existed on the Baffin Island continental shelf, and Dyke and Dredge (1987) propose that ice shelves extended across the Canadian arctic channels. Thus, although several workers now propose that fringing ice shelves existed along the eastern margin of the ice sheet, the concept of a massive buttressing ice shelf over Baffin Bay and the northern Labrador Sea is not supported in the literature (cf. Aksu and Mudie, 1985; Fillon and Aksu, 1985).

The problem of ice extent over the Queen Elizabeth Islands is still unresolved and will remain so for some time (Blake, 1970; England, 1976, 1985; Paterson, 1977; Retelle, 1986; Funder, 1987; Kelly and Bennike, 1985). The argument can be reduced to two questions. What processes other than glacial isostasy can explain the moderately large postglacial rebound? Why is direct evidence for glaciation lacking over much of the region? Hodgson (1985, p. 365) expressed the problem: "Prior to ca. 9000 BP, there is no firm evidence of whether it was a transgressing sea, a high stable sea . . ., or glacial ice that occupied the outer fiords and Eureka Sound. . . .What I did find were scattered concentrations of shells that provided old dates lying 15 m above the highest proven Holocene sediments. . . .The enclosing beach sediments, in open terrain, either were preserved by a nonerosive basal glacier regime, remained unglaciated (implying a full glacial or transgressive sea), or are a mix of Holocene and old shells."

The problems are further typified by data from Loughead Island, a small island (77°30′N, 105°W) in the archipelago. Weathered glacial deposits suggest some antiquity for glaciation, but this must be balanced by a late-glacial uplift of 90 m in 10,500 yr (Hodgson, 1981). There is also biological evidence for a refugium that supported a variety of life forms, e.g., mammals, spiders, beetles, and mosses (the review by Danks, 1981, p. 347), within the Queen Elizabeth Archipelago.

Although cold-based ice might have covered the archipelago and left no trace of its existence, the mapping of glacial limits and the ^{14}C chronology along the Ellesmere/Greenland corridor (England, 1985; Retelle, 1986) delimits part of the margin of the ice sheet and indicates that some ice-free areas existed. Funder (1987) argues that a 500-m-thick trunk glacier filled Nares Strait during the Late Wisconsin Glaciation, but Kelly and Bennike (1985, p. 114), working on the Greenland side of Nares Strait, stated, "The general featureless nature of the low lying ice sheet deposits and high degree of clast weathering, suggest a significant age for them, perhaps preceding the last interglacial or earlier." A modeling effort is called for to try and resolve the important question of ice extent in the high Canadian arctic and northern Greenland (e.g., Quinlan and Beaumont, 1982).

The specific outline of the Laurentide Ice Sheet at 18 ka cannot be drawn for about one-half its perimeter. A glacial maximum of ca. 18 ka has not been demonstrated in the eastern and northern margins of the ice sheet, although glacial-isostatic modeling (Andrews, 1975; Quinlan, 1985) suggests that the northeast sector was well developed by ca. 18 ka (Fig. 7). It is possible that the maximum was reached close to 18 ka, but the important question is why was evidence for deglaciation delayed until between 12 and 8 ka? Possible explanations are: (1) workers have failed to find and date the 18-ka glacial maximum; (2) the ice sheet reached its maximum at or before 18 ka but retreat was delayed until between 12 and 8 ka; (3) the ice was behind the 12–8 ka margins at 18 ka but then advanced so that the late-glacial maximum is recorded by these late-glacial limits; (4) sea levels between 18 and 13 ka were *below* present (even taking into account glacial-isostatic loading) and hence material for dating is not available for land-based geological surveys; and (5) the entry of datable materials into the proximal glacial marine environment was delayed by the presence of fringing ice shelves within the Arctic channels and along the coast of Baffin Island (e.g., Hughes and others, 1977).

Proposal #1 above is unlikely. There are hundreds of ^{14}C dates on shells, peat, and other materials from arctic Canada (e.g., see GSC date lists); the vast majority of these fall into two age classes, namely younger than 12 ka and older than 20 ka. No one can distinguish $<$12-and $>$20-ka shells by appearance, and bias is impossible in sample collections. Grosswald (1984) suggested that cold-based glaciers could advance over deposits and leave no stratigraphic trace. In his model, this explains the widespread existence throughout the region of raised glacial-marine deltas with ^{14}C dates on included faunas of older than 30 ka (e.g., Retelle, 1986; Loken, 1966; Miller and others, 1977), which are cut by younger Holocene sea levels (Andrews, 1975). Although cold-based glaciers most certainly preserved older deposits in areas where the ice was thin and slow-moving, this is more difficult to envisage in coastal areas flanking fast-flowing outlet glaciers.

Recent research has started to find some sites older than 12 and younger than 20 ka. One important site is at Arctic Bay, northern Baffin Island (Fig. 7), where Short and Andrews (in prep.) report ^{14}C ages between 14 and 16.5 ka on a 5-m-thick bank of peat. This is the first site in northeast Canada with reliable (peat) dates older than 12 ka and younger than 20 ka. The area is beyond the proposed limit of late Wisconsin ice (Dyke, 1983, 1984). Other ^{14}C dates in the 12 to 24 ka age range are being obtained on shells and foraminifera in piston cores within the fiords and on the shelf of eastern Baffin Island (Praeg and others, 1985; Andrews and others, 1985b). These dates are from basins beyond the belt of late-glacial moraines that extend some 1200 km along eastern Baffin Island (Ives and Andrews, 1963; Falconer and others, 1965).

Proposals 2 and 3 above have been discussed for some time (Andrews, 1975; Quinlan, 1985). Under present isostatic modeling, proposal 2 is more probable, although a slight advance to the late-glacial limits cannot be ruled out.

Proposal 4 carries with it the corollary that proposal 3 occurred. If the northern ice sheet was reduced in extent at 18 ka, then relative sea levels may have been *below* present sea level. However, with a late-glacial increase in ice thickness, glacial-isostatic depression would cause relative sea level to rise toward the altitude of the late-glacial marine limit. There is no firm evidence against proposal 4 except that glacial-isostatic reconstructions for the eastern Canadian Arctic argue for little change in ice-sheet extent between 8 and 18 ka (Quinlan, 1985). However, Clark (1980, 1985), in an inverse glacial-isostatic reconstruction of Late Wisconsin thickness changes, showed significant thinning and thickening of 3 of the 8 sectors of the Laurentide Ice Sheet and smaller oscillations for an additional 3 sectors. In the case of the Keewatin disc, maximum ice thickness of ca. 1.8 km is attained at 11 ka; thereafter the ice thins rapidly; this is in agreement with field data (Dyke and Dredge, 1987).

The dramatic increase in the number of proximal glacial-marine deposits between 12 and 10 ka in northwest Canada (Vincent, 1984; Dyke and Dredge, 1987), around 9.5 ka along the Arctic mainland coast (Dyke, 1983, 1984), and between 8 and 10 ka along Baffin Island, may reflect breakup of fringing ice shelves, similar in origin to the ice shelves 40–100 m thick off the coast of northern Ellesmere Island (Crary, 1960; Hughes and others, 1977; Jeffries, 1986). Even at present, the fast ice off eastern Baffin Island may survive individual summers (Andrews, 1985), and the northern Canadian channels are nearly always jammed with ice. If ice shelves existed in the seaways and on the shelf, this would restrict habitat availability for marine molluscs; the absence of open water would curtail the development of landforms and sediments associated with seasonally open water; hence, no landforms or datable marine sediments would be produced.

Conclusions

The late deglaciation of the northern and eastern margins of the Laurentide Ice Sheet indicates that the ice margin in and around the various seaways was relatively stable for about half the duration of the last deglacial cycle. The onset of retreat and the sudden increase in datable marine organisms most probably reflects the late Pleistocene/early Holocene breakup of fast ice and ice shelves (Hughes and others, 1977) and suggests that the stability of this margin was associated with the presence of ice shelves.

DEGLACIATION CHRONOLOGY FOR THE NORTHERN AND SOUTHERN MARGINS OF THE ICE SHEET

This section evaluates retreat rates around the ice sheet and comments on the style and chronology of deglaciation in the marine versus terrestrial parts of the ice sheet. The discussion is focused on the transects shown in Figure 8. Profiles of parts of these are illustrated in Figure 5.

Hudson Strait–Hudson Bay transect, A-A′ (Figs. 8 and 9)

An asymmetry in the chronology and rates of deglaciation around the perimeter of the Laurentide Ice Sheet (Figs. 7 and 9)

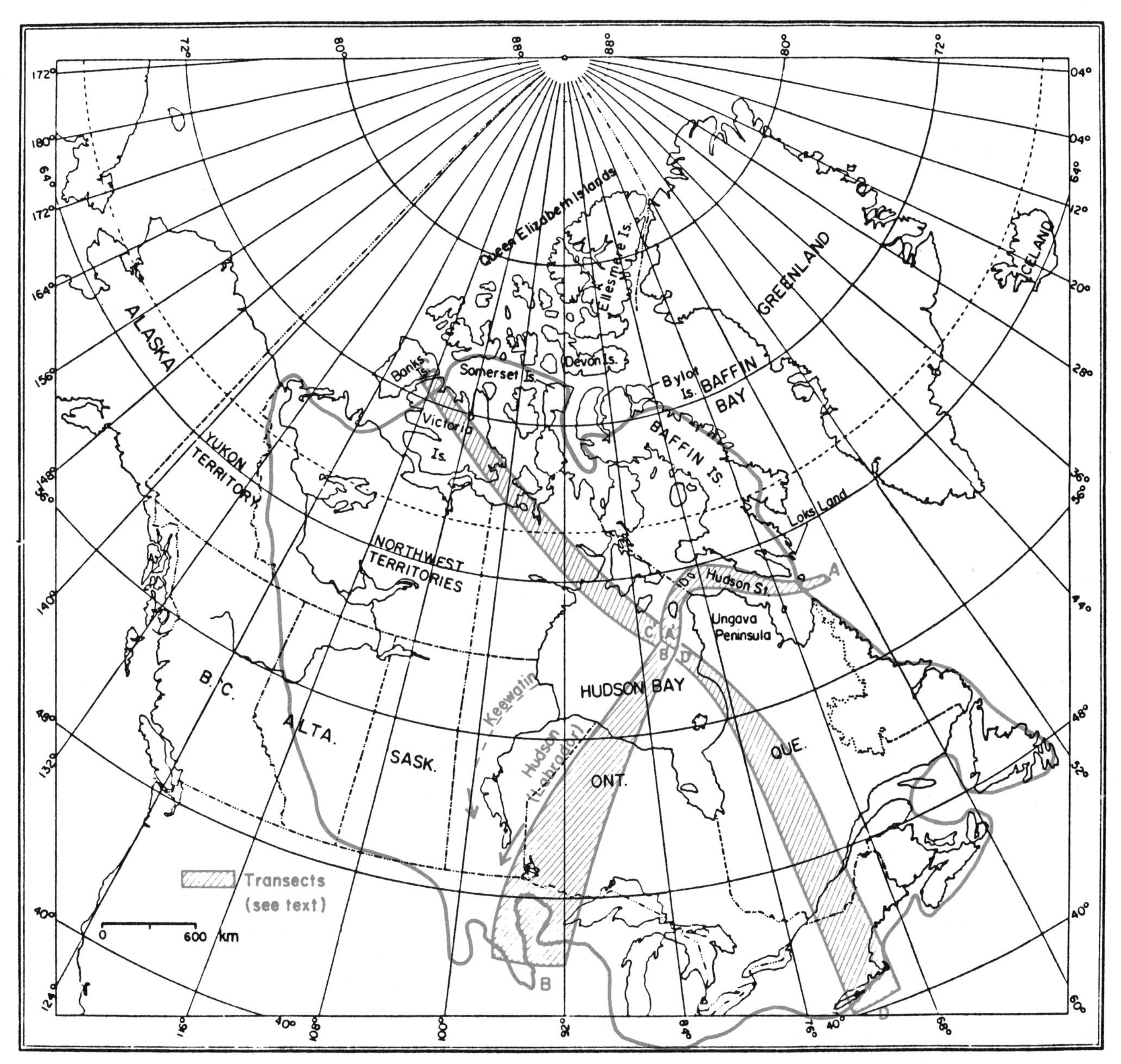

Figure 8. The Laurentide Ice Sheet showing the two major transects across the ice sheet discussed in the text (see Figs. 9 and 12).

has been alluded to. What is the response of the ice sheet in the vicinity of the major ice stream, Hudson Strait, which figures prominently in comments about deglaciation of the Laurentide Ice Sheet? Hughes and others (1985b, p. 143) stated, "Observationally, a great volume of ice was downdrawn through the Hudson Strait ice stream between 18,000 and 16,000 yr B.P. (Mayewski and others, 1981; Denton and Hughes, 1981b; Ruddiman and McIntyre, 1981)." This is a hypothesis that requires testing. Because of the importance of Hudson Strait in the deglaciation of the ice sheet, I review the increasing body of data from land and marine surveys.

Hudson Strait is a complex fault-bounded graben floored with Paleozoic carbonates (Sanford and others, 1979; MacLean and others, 1986b). The western end of the strait is blocked by a series of large islands; water depths are between 200 and 400 m (Fig. 10). A sill at 400-m depth extends from Resolution Island to Button Island. The Resolution Basin extends northeast off Resolution Island and is enclosed with the 500-m depth contour. A small basin, 900 m deep, is on the landward side of the sill.

The acoustic stratigraphy of the southeast Baffin Island Shelf (MacLean, 1985; Praeg and others, 1986), the northern Labrador Shelf (Josenhans and others, 1986; Fillon and Harmes, 1982),

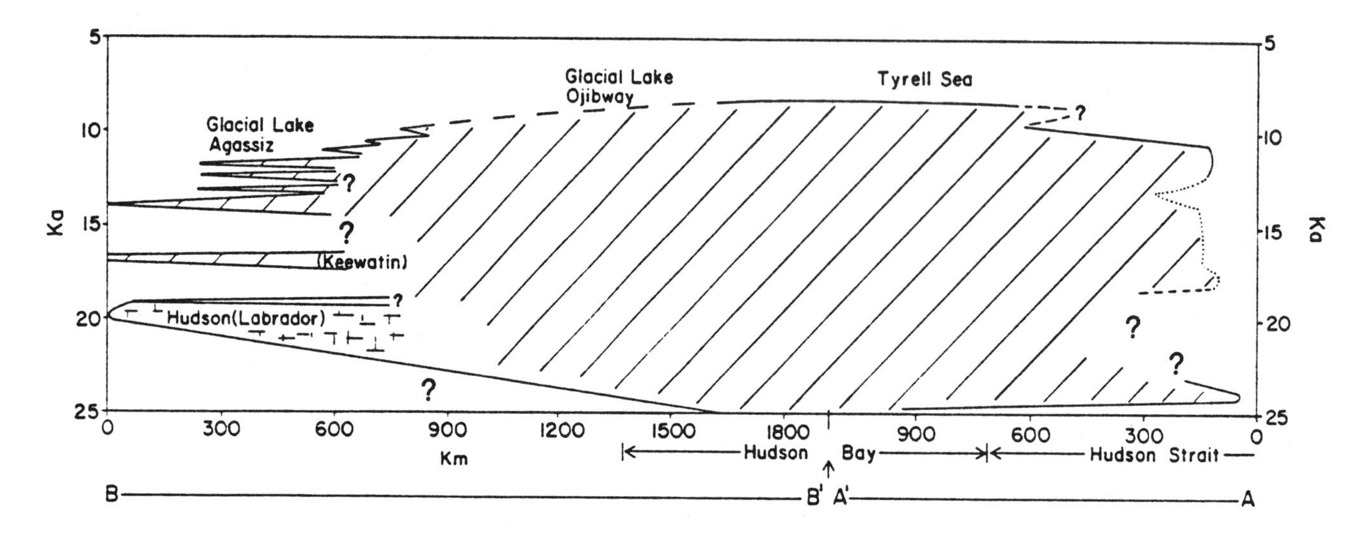

Figure 9. Time-distance diagram of the ice sheet along the transect A-A′ and B-B′ (see Fig. 8).

and Hudson Strait (MacLean and others, 1986b) consists of diamictons, acoustically laminated sediments, and acoustically transparent sediments. Major oscillations of ice margins can be inferred from the acoustic records.

Piston cores (Fig. 10) have been taken from the mouth of the strait and from the Resolution Basin (Osterman and others, 1985; Praeg and others, 1986). High-resolution acoustic stratigraphy shows a succession of diamictons (Baffin Drift) interbedded with acoustically layered sediments. At the site of HU82-57, a date of 24.8 ka was obtained by accelerator mass spectrometry (AMS) on foraminifera *above* the uppermost diamicton (till?) within the Davis Silt unit (Praeg and others, 1986). (Note: ^{14}C dates on shells and foraminifera are quoted with a reservoir correction factor of 450 yr). Closer inshore at HU77-156 (Fig. 10), a date of 27.3 ka on the acid-insoluble organic matter is considered too old because of contamination by old carbon (Fillon and others, 1981; Andrews and others, 1985b). An age of ca. 18 ka is proposed for the 2-m level in this core (Osterman and others, 1985) on the basis of calibrated shell and organic dates (Andrews and others, 1985b) and on amino acid racemization ratios on foraminifera. This assumption is partly validated by an AMS date of >27 ka on foraminifera at 3.0 m. AMS dates on small shells toward the base of cores HU82-68 and HU77-159 give ages ranging between 12 and 8 ka. The net accumulation of sediment at both HU77-156 and HU75-58 is low, especially considering their projected location close to the perimeter of a major ice lobe in Hudson Strait. An estimate of the accumulation of ice-rafted detrital (IRD) sediment between 25 and 13 ka is 43 $kg/m^2/ka$, and from the present to 13 ka is 11 $kg/m^2/ka$ (Fillon, 1985). For the same two time periods, estimates from HU77-156 are 40 and 14 $kg/m^2/ka$.

The dates discussed above constrain the biostratigraphy of benthic foraminiferal assemblages from a suite of cores (Praeg and others, 1986; Osterman, 1982, 1984) (Fig. 11). The assemblages consist of low-diversity zones that contain *Elphidium excavatum* (Terquem) forma *clavata* (Cushman); there is also evidence for more open ocean conditions at the base and top of the core (Fig. 11). Stable O and C isotopic ratios on planktonic foraminifera from core HU75-58 (Fillon, 1985) (Fig. 10) show a typical open-ocean $\delta^{18}O$ record (Fillon and Williams, 1984). The data from HU77-156, located northeast of Resolution Island (Fig. 10), however, can be divided into two response modes: an early stage similar to HU75-57, and a later phase between ca. 18 and 10 ka, dominated by meltwater dilution (Andrews and others, 1987). These events are illustrated in Figure 11 by the difference between the two cores. Excursions in light isotopic ratios coincide in two cases with *E. excavatum* f. *clavata* zones and in the third case occur at a foraminiferal assemblage boundary (Osterman, 1982; Osterman and others, 1985), suggesting that these variations are of regional importance. The uppermost light isotope phase is correlated with an event recorded in marine molluscs from outer Frobisher Bay and is dated ca. 10.3 ka (Andrews and others, 1987).

How does the offshore marine lithostratigraphy and biostratigraphy correspond with onshore glacial mapping and chronology (cf. Clark and Josenhans, 1986)? In particular, is the assumption valid that Hudson Strait was the site of an ice stream? Glacial-geologic studies from Ungava (Bouchard and Marcotte, 1986) and from northeastern Hudson Strait (Blake, 1966; Osterman and others, 1985; Stravers, 1986) indicate that the region consisted of a complex of ice masses during the late-glacial interval. Evidence for a strong regional flow to the north-northeast across the tip of southeast Baffin Island (Osterman and others, 1985; Stravers, 1986; Miller, unpublished data) does not fit comfortably into present conceptual models of ice-stream flow in Hudson Strait.

The onset of glacial retreat in Frobisher Bay, north of Hudson Strait, is well dated at 10.3 ka (Miller, 1980). Beyond the

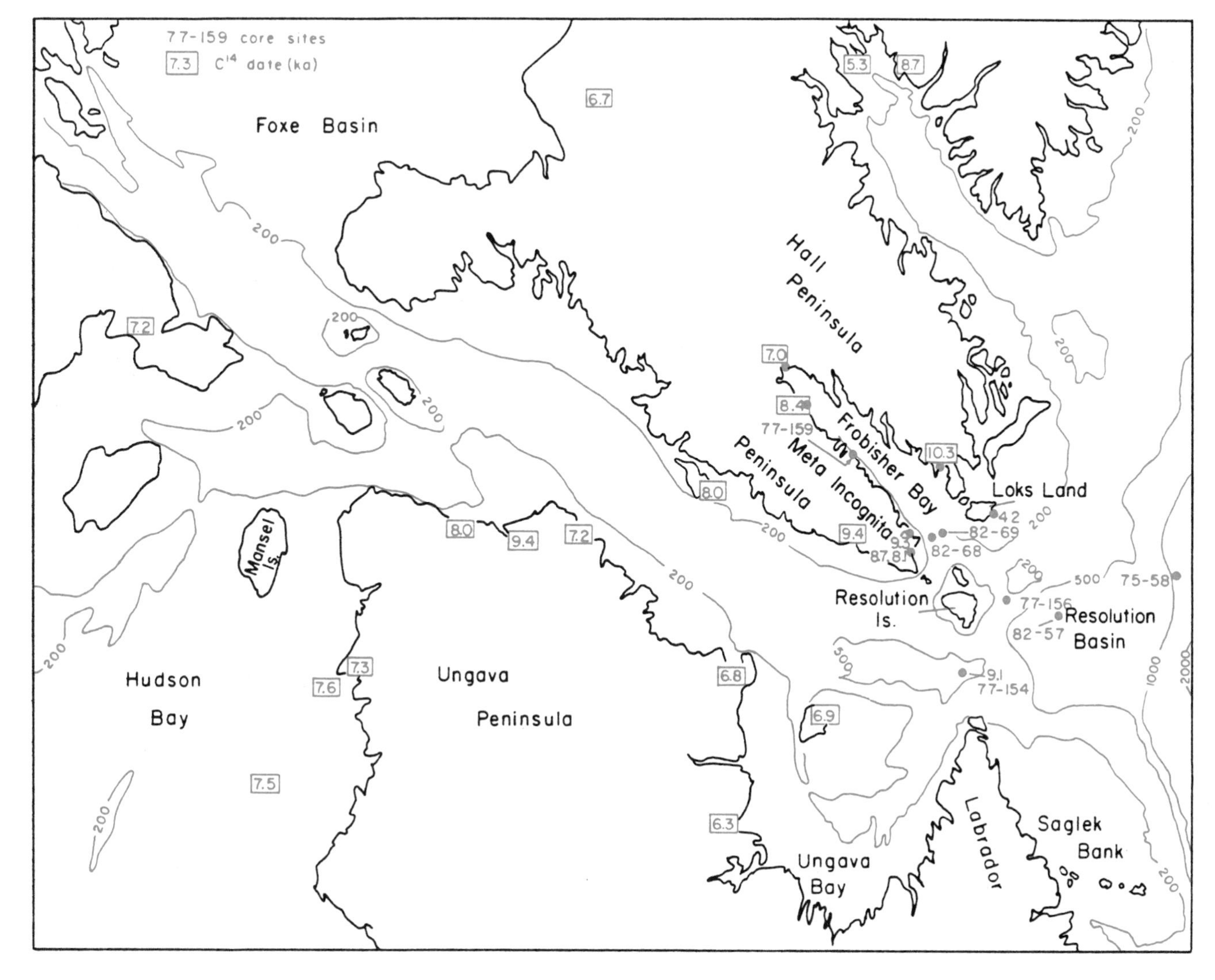

Figure 10. Bathymetry, core locations, and critical ^{14}C dates in and around Hudson Strait.

limits of that glacial margin, soils are thicker and ^{14}C dates on molluscs in situ in raised marine sediments are 40 ka or older. Between 10.3 and 8.6 ka, rapid deglaciation of two-thirds of Frobisher Bay occurred, as shown by numerous dates on shells in ice-proximal glacial-marine sediments (Blake, 1966; Miller, 1980). Recent ^{14}C dates of 9.4 and 9.1 ka on shells in situ on the coast of northeast Ungava Peninsula (Gray and Lauriol, 1985) indicate that substantial deglaciation of southern Hudson Strait occurred by ca. 9.4 ka. This episode of deglaciation is seen in the $\delta^{18}O$ records as a pronounced light isotope event (Fig. 11) and coincides with the lift-off of shelf ice along the northern Labrador coast (Josenhans and others, 1986).

After this initial phase of fast deglaciation, when net retreat rates were 500 m/yr in Frobisher Bay and Hudson Strait (Miller, 1980), a glacial readvance/stillstand occurred ca. 8.2 to 8.6 ka (Fig. 11). It was correlative with a major episode of moraine-building throughout northeastern and northern Canada (Falconer and others, 1965). In Hudson Strait, ice may have surged across the outer strait from Ungava Bay and impinged on the extreme southeast tip of Baffin Island (Stravers, 1986), introducing detrital carbonate onto the northern Labrador Shelf (Josenhans and others, 1986). Moraines overlying glacial marine sediments within Hudson Strait (MacLean and others, 1986b) indicate that ice readvanced from Ungava and Baffin Island to deposit glacial sediment parallel to the coastline during the late glacial. Final deglaciation of Hudson Strait and Hudson Bay shortly thereafter was in part caused by calving in Hudson Strait but may have been strongly influenced by the northward development of glacial Lake Ojibway-Barlow along the southern margin of the ice sheet (Vincent and Hardy, 1979; Hillaire-Marcel, 1979).

The extent of glaciation and the chronology of deglaciation reported in this chapter differs from that proposed by Hughes and

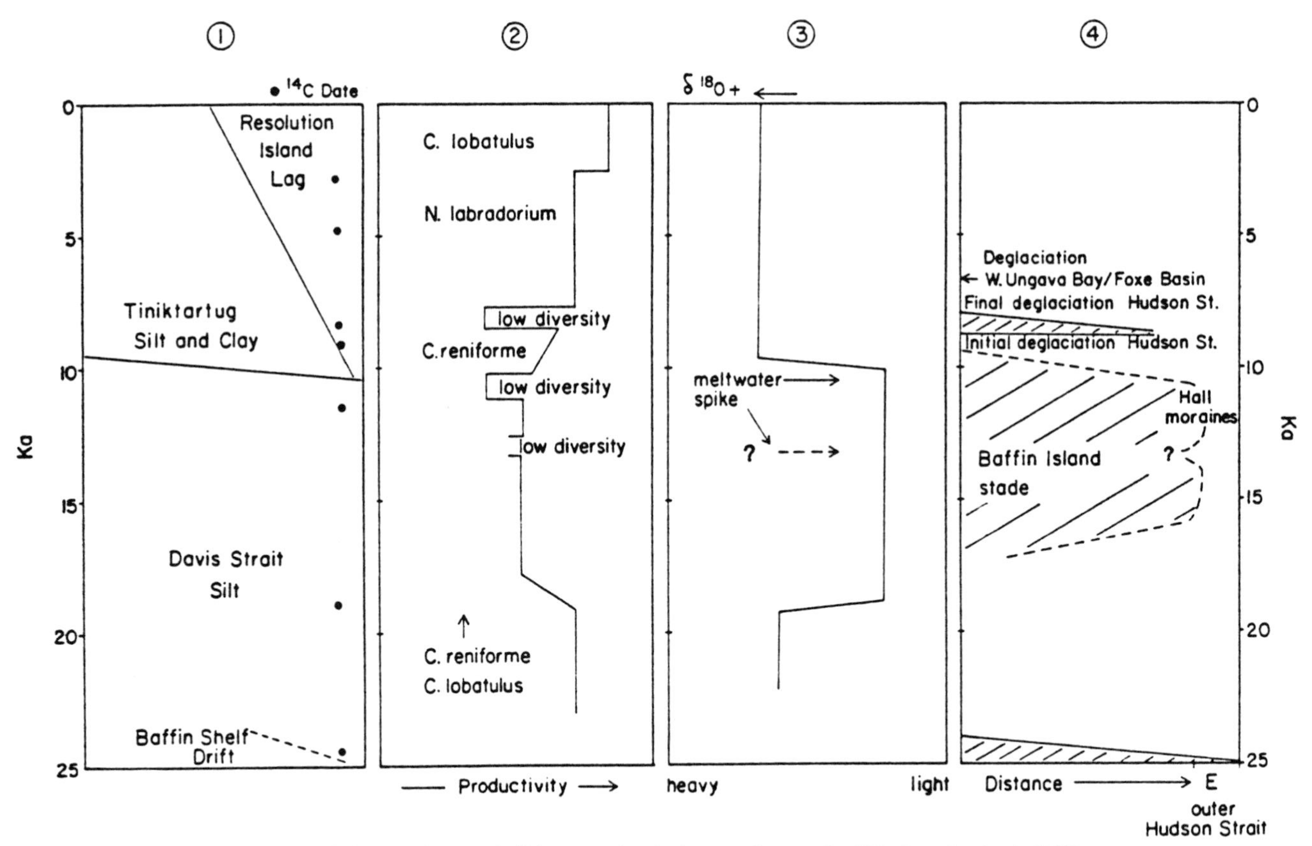

Figure 11. Glacial chronology and offshore marine geology at the mouth of Hudson Strait. 1: Offshore marine stratigraphy and available ^{14}C dates (from Praeg and others, 1986; Osterman and others, 1985). 2: Benthic foraminiferal zonation of cores in outermost north Hudson Strait (Osterman, 1982; Osterman and others, 1985; Praeg and others, 1986). 3: Difference in $\delta^{18}O$ between HU75-58 (Fillon, 1985) and HU77-156 (Andrews and others, 1987). 4: Glacial events in Hudson Strait (from Osterman and others, 1985; Gray and Lauriol, 1985; Praeg and others, 1986).

others (1985b). The combined offshore and onshore mapping indicates that the ice stream/ice shelf at 18 ka is too extensive in the Hughes and others (1985b, Fig. 2) scenario and, rather than being the site of early deglaciation by 12 ka, Ungava Bay was the location of residual ice that did not start to retreat from the coast until 7.5 ka or earlier (Prest, 1984; Hillaire-Marcel, 1979).

Prairies/Hudson Bay transect B-B′ (Figs. 8 & 9)

I once suggested (Andrews, 1973) that the rapid retreat of the southern margin of the ice sheet represented a problem in energy balance and that calving into extensive proglacial lakes might be a mechanism for rapid deglaciation. This process may have been important, although the evidence for a thinner southern margin (see above) would significantly reduce the proposed energy imbalance (cf. Beget, 1986a; Teller, this volume).

The time-distance diagram for transect B-B′ (Fig. 9) indicates an important element in the ice sheet's regime. The glacial history of the Dakotas/Minnesota/Iowa region involved major shifts in the source and direction of flow (see Fenton, 1984, p. 64). The first flow to reach the prairies came from the northeast and is ascribed to Labradorean (Hudson) ice (Clayton and Moran, 1982; Fenton and others, 1983); this was followed by south and southwest flow from ice over Keewatin. These data support the general concept of an ice sheet with several ice divides. However, the chronology of glacial events is confused because of arguments on the validity of ^{14}C dates from fine-grained organic lake sediments (cf. Clayton and Moran, 1982) and on the stratigraphic interpretation of wood in tills. Figure 9 follows Fenton and others (1983) and Mickelson and others (1983) in advocating the younger (wood-based) chronology. The moraines of the oldest advance, with wood dates of ca. 20 ka (Hudson/Labrador ice), were cross-cut by the Bemis Moraine (Keewatin ice) at ca. 14 ka. Glacial Lake Agassiz (see Teller, this volume) was initiated in the Red River Valley ca. 11.7 ka and continued to exist until the final emptying of the lake into Hudson Bay ca. 7.5–8 ka (Fig. 10; Dredge, 1983; Klassen, 1983).

Rates of glacial retreat and advance can be computed from

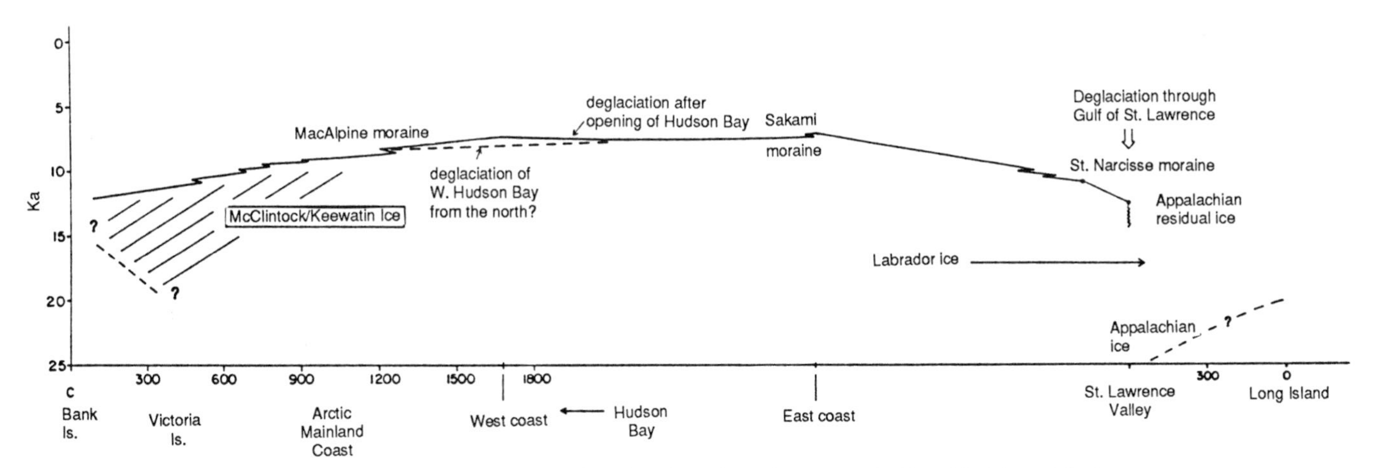

Figure 12. Time-distance diagram of the ice sheet along the transect C-C′ and D-D′ (see Fig. 8).

published time-distance diagrams (Fenton and others, 1983) and from ^{14}C-dated margins. These were extremely rapid and include advances and retreats of 500 km or so in a few hundred years. Calculated rates of 1700 m/yr (Fig. 9) indicate highly dynamic ice lobes south of the Canadian boundary. Recession rates in Manitoba were slower, and Klassen (1983) suggested an average of 300 m/yr during the last 2000 yr of the ice sheet's history.

During deglaciation, the glacial lakes along the southern margin of the ice sheet had a complex history associated with the opening of new outlets, the effect of glacial-isostatic adjustments, and glacial advances (Teller, this volume; Teller and Clayton, 1983). Nevertheless, starting about 14 ka around the southern Great Lakes and 11.7 ka for glacial Lake Agassiz, the southern margin of the ice sheet was largely terminating in water bodies of considerable dimensions and depths (see Fig. 21 in Teller, this volume).

By about 10 ka, the ice sheet was largely within the boundaries of the Precambrian shield (Boulton and others, 1985; Fisher and others, 1985). It readvanced ca. 9.9 ka along a broad front from the prairies (Cree Lake moraine), to Ontario (Hartman and Dog Lake moraines), into Lake Superior and northern Michigan, and along large sections of the north shore of the St. Lawrence. Other moraines subsequently formed as the ice retreated northward. Over the next 2000 yr, the ice margin retreated ca. 1000 km, averaging 500 m/yr. Glacial Lake Agassiz eventually emptied into glacial Lake Ojibway, which expanded northward across the James Bay Lowland and well into eastern Hudson Bay (Vincent and Hardy, 1979; Hillaire-Marcel and Vincent, 1980) to the vicinity of Great Whale River (56°N). The final drainage of the southern proglacial lakes was through Hudson Bay and thence Hudson Strait. An estimated 100,000 km^3 of lake water was discharged at that time (Andrews, 1984).

Northern Canada/Keewatin transect (C-C′, Figs. 8 and 12)

The next profile extends from the northwestern margin of the ice sheet and continues as D-D′ southeastward across Hudson Bay and the St. Lawrence Valley. In neither the northwest nor southeast sectors has unequivocal evidence of *extensive* middle Wisconsin deglaciation been noted (see Fig. 4 and earlier discussions), although the Plum Point interstadial testifies to glacier recession during this interval. Current thinking indicates that early Wisconsin ice persisted over Labrador, Foxe Basin, and northern Keewatin and expanded during the late Wisconsin (i.e., Fulton, 1984). Data for the northwest transect are summarized by Dyke and Dredge (1987) and Vincent (1984, 1987). The time/distance diagram (Fig. 12) is in places oblique to the patterns of ice flow.

The maximum extent of Late Wisconsin ice is not known with certainty. Vincent (1987) concludes that most of Banks Island and northwestern Victoria Island were ice-free. Hodgson and Vincent (1984) presented evidence for a late surge of ice into Viscount Melville Sound and suggest that deglaciation was delayed until between 9.8 and 10.4 ka. New shell dates (Dyke, 1987) obtained upglacier from the Viscount Melville Sound ice shelf require deglaciation by ca. 11 ka and a more extensive Late Wisconsin ice cover.

Regardless of these important areas of debate, the oldest shell date from this region is 12.6 ka. East of McClintock Channel, from Somerset Island to northern Baffin Island, is an impressive unity in shell dates—9.5–9.3 ka—associated with deposition into the late-glacial sea (Dyke, 1983, 1984; Klassen, 1985b). The extreme northwestern margin of the ice sheet extended into the eastern foothills of the cordillera and reached its maximum extent between 25 and 16 ka (or younger) (Vincent, 1987). Thus, dates derived from terrestrial and marine margins of the northwest sector of the ice sheet differ significantly, the response of the marine portion lagging by 4 to 12 ka.

Dyke and Dredge (1987) remark that their reconstruction of ice margins during deglaciation differs little from Prest's (1969) isochrone map, although they propose that the Keewatin ice sheet had disappeared by 8 ka rather than 7 ka. Several moraines mark stillstands or readvances between 11 and 8 ka. The formation of these moraine segments may be associated with climatic (mass-balance) changes but, alternatively, they may reflect the response of the ice sheet to fast calving at times of higher sea level, punctuated by intervals when the ice was grounded or pinned on islands or along the mainland coast (Andrews, 1973; Hillaire-Marcel and others, 1981; Dyke and Dredge, 1987). One of the longest moraine systems in the world is that which Falconer and others (1965) outlined from eastern Baffin Island northward and westward along the Arctic mainland coast. This interval of moraine formation dates from ca. 8.4 ka and falls within the Cockburn substage.

Rates of glacial retreat along the northwest transect averaged 120 m/yr. The development of calving bays was an important element in deglaciation of the McClintock/Keewatin ice sheet, but this was in part mitigated by vigorous late-glacial flow, especially from the ice over Foxe Basin (Dyke and Dredge, 1987).

Southeast Transect (D-D', Figs. 8, 12)

The extent of ice during the middle Wisconsin is poorly known in New England. In southern Quebec, the St. Pierre interval includes early Wisconsin and "... much of the Middle Wisconsinan" (Ochietti, 1987); in a more conservative scenario, the Trois Rivieres Stade began ca. 75 ka and involved retreat and glacial-lake formation during the middle Wisconsin.

Glaciation along the southeast transect involved the advance and retreat of ice on the Labrador plateaus and the development of glaciers in the Appalachians between the St. Lawrence Valley and southern New England. At the glacial maximum, the Appalachian glaciers were overwhelmed by ice from the northwest. Deglaciation of the southeast transect was complicated by the drawdown of ice into the St. Lawrence Valley (Thomas, 1977) and the reestablishment of Appalachian ice caps (Borns, 1973; Gadd, 1986). Thus we need to consider the evidence from both New England and the Canadian Maritimes.

Michelson and others (1983, p. 28) concluded that the ice in southern New England "... reached its southernmost position at various times between 21 and 15 ka." King and Fader (1986) evaluate a wealth of data on offshore marine data from the Canadian Maritime continental margin. In their interpretation, a thick ice shelf, supported by glacier flow, receded from a maximum position between 45 and 32 ka. Recession along re-entrants may explain the data from Anticosti Island (Bigras and others, 1985) and the evidence for warmer intervals in offshore marine cores (Alam and others, 1984). The grounded portions of the ice shelf retreated to coastal areas by 30 ka except in the eastern Gulf of Maine. The Late Wisconsin Scotian Stade is coeval with the classical Late Wisconsin glacial deposits of southern New England.

In southern New England, isostatic depression adjacent to the ice margin was minimal, and the ice and sea were probably not in contact. No glacial-isostatic modeling has been carried out along the southeast margin as compared to the northeast and east margin (Andrews, 1975; Quinlan and Beaumont, 1982; Quinlan, 1985). On the basis of the agreement between model results and field data in the Canadian Maritimes and eastern Arctic, the limited glacial-isostatic rebound in southern New England is most reasonably explained by invoking a short-lived glacial event, as compared to eastern Arctic data, where a long-lived, stable margin resulted in considerable isostatic depression at the margin of the ice sheet (Quinlan, 1985). However, along the Maine coast, marine limits are commonly associated with large deltas and moraine complexes (Borns and others, 1980). The oldest date on the deglaciation of the Maine coast is on shells at 13.4 ka (with 450 yr reservoir correction) (Andrews, unpublished data).

A complete record of the last glacial hemicycle is contained in core HU78-023-20 off eastern Newfoundland (Scott and others, 1984) in a water depth of 286 m. The core penetrated till. Detailed micropaleontological studies on foraminifera, pollen, and dinoflagellates, along with four ^{14}C dates, suggest the till was emplaced 25 ka or more recently. This was succeeded by an interval when the Labrador Current was present close inshore, but this gave way to glacial-marine and possibly ice-shelf conditions (Scott and others, 1984, p. 211). Around 10 ka, the inner Labrador Current was reestablished along the eastern Newfoundland shelf.

In the St. Lawrence Valley, there is considerable discussion of the importance of a calving bay in the process of deglaciation (Thomas, 1977). If the Anticosti Island data are accepted, then the marine margin across the Gulf of St. Lawrence was relatively stable for at least 16,000 yr, from 30 to 14 ka. Dates of deglaciation along the St. Lawrence Valley (Ochietti, 1987) range from 13.4 to 12 ka. Karrow (1981) and Gadd (1980) disagreed about the importance of a calving bay on deglaciation; this argument is largely centered on the interpretation of a controversial 12.8 ka date for deglaciation of the Ottawa Valley (Richard, 1975). If dates of 13.4 and 12 ka are used, the rate of retreat along the axis of the valley was ca. 500 m/yr.

Large end-moraine complexes were constructed along parts of the Maine coast and the north shore of the St. Lawrence Valley (Dubois and Dionne, 1985). Although these moraines have been interpreted in a climatic context to represent cooler conditions (cf. LaSalle and Elson, 1975), their close association with the marine limit suggests that they may be "re-equilibrium moraines" (cf. Hillaire-Marcel and others, 1981); i.e., moraines formed by a change in glacier regime as the ice moves from a calving to noncalving mode. The St. Narcisse Moraine has been traced for a considerable distance along the north shore and is dated at 10.8 ±0.2 ka and older than 10.3 ka (Ochietti, 1987).

Retreat from the north shore was relatively slow. Between the north shore and the final residual ice in central Nouveau Quebec (Dubois and Dionne, 1985; King, 1985), the retreat rate averaged 160 m/yr. By ca. 10 ka, deglaciation was influenced by the development of glacial lakes to the northwest and by the deglaciation of Hudson Strait to the north (i.e., Fig. 11; Vincent and Hardy, 1979). However, there was a significant interval of moraine formation along the north shore of the St. Lawrence ca.

9.9 ka (Dubois and Dionne, 1985; King, 1985), and again at ca. 7.8 ka along the southeast coast of Hudson Bay/James Bay, when the Sakami moraine was formed (Hillaire-Marcel and others, 1981). The Cochrane advances, or surges (Skinner, 1973; Prest, 1970; Vincent and Hardy, 1979), represent late-glacial events in southern Hudson Bay/James Bay that may have been associated with surges of ice into glacial Lake Ojibway-Barlow. These surges consisted of 50–75 km readvances from residual ice over Hudson Bay (Dredge and Cowan, 1987).

The process of deglaciation within Hudson Bay needs further study. Initial work (Andrews and Falconer, 1969; Hardy, 1976) suggests that a single narrow calving bay split the ice in two. Dyke and others (1982) suggested that this occurred along the contact of Labrador and Hudson ice. Dredge and Cowan (1987) indicate that another calving bay may have developed farther north and led to relatively early deglaciation (8.5–8.2 ka?) of western Hudson Bay. Although most workers consider Hudson Bay to have been deglaciated via Hudson Strait, Dyke and Dredge (1987) suggest that marine water might have penetrated the Bay from the Gulf of Boothia to the north.

CONCLUSIONS

Over the past two decades there has been tremendous progress in our understanding of the late Quaternary dynamics of the Laurentide Ice Sheet. This can be attributed to (1) the availability of radiometric dating, mainly ^{14}C; (2) a better understanding of the physical basis of glaciology; (3) clearer models of glacial sedimentation; and (4) the geographic expansion of the data base, especially for the northern and eastern margins of this great ice sheet. However, the various methods that can be applied to ice-sheet reconstructions (cf. Andrews, 1982) have not yet produced a wholly self-consistent model.

Because the Laurentide Ice Sheet contained more than 50% of the interglacial/glacial change in the global water budget, it is evident that changes in volume of this ice sheet exerted an influence well beyond its margins. This is particularly true of the oceanic $\delta^{18}O$ balance, which is strongly affected by both ice volume and oceanic temperature.

The review of the Late Wisconsin Glaciation earlier in this chapter stressed the difference in the timing of deglaciation between the southern and northern margins. This, together with the evidence for middle Wisconsin ice-free conditions over certain regions, indicates that a substantial part of the south and southwest sectors of the ice sheet advanced rapidly during the Late Wisconsin. In contrast, much of the north and east margins probably existed throughout the middle Wisconsin interstadial and experienced only a limited advance during the Late Wisconsin. These two different regimes can be differentiated on the basis of the associated glacial-isostatic response, and they can explain the difference in strandline warping and glacial-isostatic depression (see discussion in Beget, 1986) between southern New England and the Canadian prairies, when compared with the Canadian north and northeast margins.

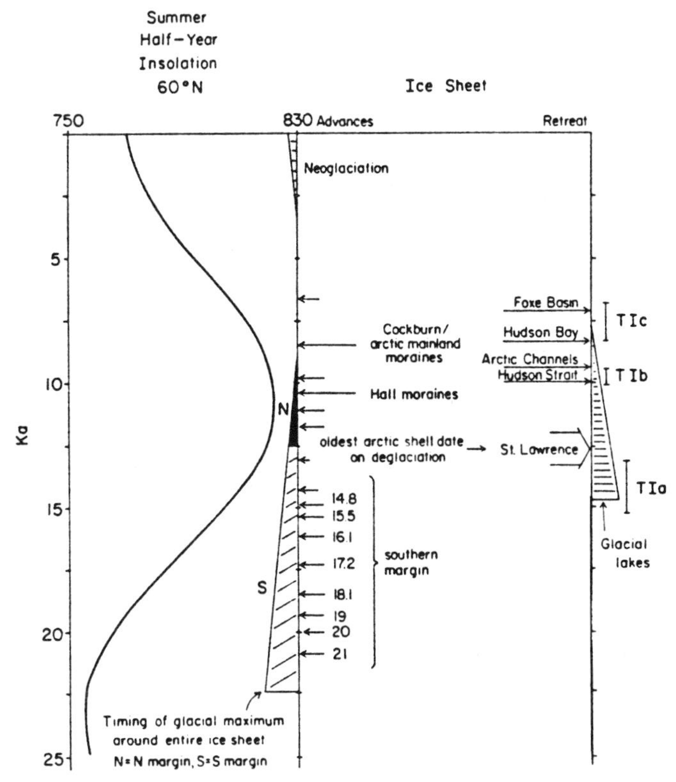

Figure 13. Comparison of the event stratigraphy, glacial advances and retreats, for the Laurentide Ice Sheet with the summer half-year insolation curve (from Ruddiman and Wright, this volume) for lat 60°N—note this is close to the latitude of Hudson Strait. TIa, TIb, TIc are the steps in Termination I (see Mix and Ruddiman, 1985).

The causes of deglaciation can be divided into possible global influences and regional controls. At the global level, the coincidence between the summer insolation at high northern latitudes and the glacial/interglacial climatic changes is reasonably well established (see Ruddiman and Wright, this volume). Melting, particularly of the southern margins, would transfer water back into the ocean and thus cause the global rise in sea level. Because the marine portions of ice sheet are sensitive to changes in relative sea level at their margins (Thomas, 1977; Denton and others, 1986), the proposed rise in sea level might then trigger rapid retreat of the grounding line and collapse of the ice sheet. However, the coupling between melting and regional *relative* sea level changes is complicated by the glacial-isostatic response (see Peltier, this volume), which operates on both a longer time scale and at a faster rate than eustatic sea-level changes (Andrews, 1987; Lingle and Clark, 1985). Around the north and east margins of the ice sheet, the history of relative sea level since deglaciation (i.e., between 12 and 6 ka) has been dominated by rapid regional emergence that would act toward stabilizing grounding-line retreat (Lingle and Clark, 1985).

Figure 13 presents many of the major events around the

Laurentide Ice Sheet in the context of Milankovitch (1941) summer insolation. A striking feature is the long-continued history of glacial readvances well past the insolation maximum of 11 ka. Michelson and others (1983) proposed that the southern margin of the Laurentide Ice Sheet readvanced (surged? cf. Clayton and others, 1985) at 21, 20, 19, 18.1, 17.2, 16.1, 15.5, and 14.8 ka, or approximately once every 1,000 yr. This response continued as deglaciation progressed, as noted in the earlier commentary. Of particular note are the continental-scale readvances that are dated close to 10 ka and between 8 and 8.5 ka. These occur at a time of rising temperatures on the continent of North America; they may represent either increased snow accumulation or reequilibrium adjustments of regional sectors of the ice sheet from marine to terrestrial margins (Andrews, 1973; Hillaire-Marcel and others, 1981).

The importance of glacial lakes and meltwater drainage along the southern margin has been discussed by Andrews (1973), Pollard (1983), and Teller (this volume), and the tracing of the meltwater signal offers a powerful means of directly linking the record of the ice sheet with the oceans (see Ruddiman, this volume). As noted by Teller (this volume, his Fig. 21), the initial drainage was largely to the Mississippi River and Gulf of Mexico, but between 11 and 10 ka the bulk of the discharge was directed toward the Atlantic Ocean (considerable discharge toward the Arctic Ocean would have also occurred via the Mackenzie River).

Of considerable importance in all discussions is the volume of the Laurentide Ice Sheet and the way in which that volume was transferred to the oceans (e.g., Paterson, 1972; Bloom, 1971; Denton and Hughes, 1981a; Ruddiman and McIntyre, 1981). Mix and Ruddiman (1985) proposed a three-step global deglaciation with terminations at 14–12 (Ia), 10–9 (Ib), and 8–6 ka (Ic). These represented 25–33%, 11–17%, and ca. 22%, respectively, of the glacial/interglacial isotopic signal. However, Duplessy and others (1986), on the basis of an exceptionally well dated deep-sea record, dated termination Ia as 15.8–13.8 ka. This involved a $\delta^{18}O$ decrease of 1.05‰ in planktonic foraminifera. Terminations 1b and 1c correlate well with intervals of rapid change in area (and volume?) of the Laurentide Ice Sheet). However, the 14–12 ka (15.8–13.8 ka?) step is more difficult to explain. It takes place prior to any evidence for deglaciation along approximately one-third of the ice-sheet margin (Fig. 7), although the southern and western margins were in fluctuating retreat/advance during this time. Isochrones on ice margins in the southern sector indicate a retreat of only 100–200 km between 20 and 15 ka, and between 14 and 12 ka it varied from 50 to 200 km or more (20–40 km/yr on average) (Mickelson and others, 1983, Figs. 1–9).

If the deforming-bed model of the ice sheet is correct (Boulton and others, 1985; Fisher and others, 1985), the volume of ice that was lost between 18 and 11 ka can only have been considerable if the repeated rapid advances and retreats were fed from the bulk of the ice sheet located over the Precambrian shield. Early work suggested only a single light isotope peak in the Gulf of Mexico during deglaciation (Kennett and Shackelton, 1975), but the data of Leventer and others (1982) indicate a more complex record which, although poorly dated, indicates two events ca. 16.5–15 and 14.5–12.2 ka. At 18 ka, the deformable-bed model of Boulton and others (1985) had an ocean-level volume equivalent of 50 m in the Precambrian sector of the ice sheet (from Paterson, 1972, Table 5, for 11.8–10.8 ka) and less than 2 m in the flanking apron of ice (errors on these estimates may be as high as ±15%). This estimate of ca. 52 m of global sea level contrasts with Paterson's (1972) estimate of 72 m of global sea-level change. Fisher and others (1985) computed the volume of the soft-bed ice-sheet reconstruction as 18×106 km^3 (=50 m sea-level change) as compared with 25.9×106 km^3 for the maximum hard-bed model. This figure compares favorably with my estimate, but whether a $\delta^{18}O$ mass-balance approach can resolve the differences in volume estimates is not known.

In ice-sheet units (Andrews, 1982; 1 unit = sea-level equivalent of the present Greenland Ice Sheet), termination Ia represents the melting of five Greenland ice sheets, but as we have seen, the period in question (Figs. 9 and 12) is not notable for massive changes in the area of the ice sheet (cf. Prest, 1969; Bryson and others, 1969; Paterson, 1972). It is thus appropriate to consider the notion that this event reflects a substantial volume change caused by the discharge of ice through Hudson Strait, through the St. Lawrence River, or southward as meltwater into the Gulf of Mexico. The discharge of ice into the St. Lawrence drainage is not considered a possibility for the termination Ia event. The timing is approximately correct, but I suggest that the volume of ice available was limited because readvances of the Labrador/Hudson ice into the eastern Great Lakes occurred at 12–13 ka, and thus restricts the area that could be affected by downdraw. The evidence from Hudson Strait (Fig. 11) shows one low-diversity *E. clavatum* f. *clavata* zone prior to 10.3 ka; this coincided with a single-point light $\delta^{18}O$ minimum. As yet no inshore $\delta^{18}O$ planktonic records have been presented from the area of the Gulf of St. Lawrence/Canadian Maritimes, and the link between projected discharge and the $\delta^{18}O$ record remains to be established. Thus, the glacial-stratigraphic evidence around the Laurentide Ice Sheet suggests that termination Ia may not be a primary Laurentide signal. Other ice sheets (Barents, Scandinavian, Antarctic) must therefore be important contributors to the termination Ia event; however, I would attribute terminations Ib and Ic largely to Laurentide/Scandinavian and Laurentide contributions, respectively.

In my view, the great need at the present time is for research on the deposits of the continental margins so that land/deep-sea correlations can be more firmly based on records from this intermediate environment.

REFERENCES CITED

Aksu, A. E., and Mudie, P. J., 1985, Late Quaternary stratigraphy and paleoceanography of northwest Labrador Sea: Marine Micropaleontology, v. 9, p. 537–557.

Alam, M., Piper, D.J.W., and Cooke, H.B.S., 1983, Late Quaternary biostratigraphy, isotope stratigraphy, paleoclimatology, and sedimentation on the Grand Banks continental margin: Boreas, v. 12, p. 253–261.

Alley, N. F., Valentine, K.W.G., and Fulton, R. J., 1986, Paleoclimatic implications of middle Wisconsin pollen and paleosol from the Purcell Trench, south central British Columbia: Canadian Journal of Earth Sciences, v. 23, p. 1156–1168.

Andrews, J. T., 1970, Differential crustal recovery and glacial chronology (6,700-0 BP), west Baffin Island, N.W.T., Canada: Arctic and Alpine Research, v. 2, p. 115–134.

—— , 1973, The Wisconsin Laurentide Ice Sheet; dispersal centers, problems of rates of retreat, and climatic implications: Arctic and Alpine Research, v. 5, p. 185–199.

—— , 1975, Support for a stable late Wisconsin ice margin (14,000 to ca. 9,000 BP); a test based on glacial rebound: Geology, v. 4, p. 617–620.

—— , 1982, On the reconstruction of Pleistocene ice sheets; a review: Quaternary Science Reviews, v. 1, p. 1–30.

—— , 1984, The Laurentide Ice Sheet; evidence from the eastern Canadian Arctic on its geometry, dynamics and history: England, University of Reading, Geographical Papers No. 86, Norma Wilkinson Memorial Lecture 1983, 61 p.

—— , 1985, Reconstruction of environmental conditions in the eastern Canadian arctic during the last 11,000 years: National Museums of Canada, Syllogeus, no. 55, p. 423–451.

—— , 1987, Glaciation and sea level, *in* Devoy, R. J., ed., Sea surface studies; a global review: London, Chapman (in press).

Andrews, J. T., and Barnett, D. M., 1972, Analysis of strandline tilt direction in relation to ice centres and postglacial crustal deformation: Geografiska Annaler, v. 54A, p. 1–11.

Andrews, J. T., and Barry, R. B., 1978, Glacial inception and disintegration during the last glaciation: Annual Review of Earth and Planetary Sciences, v. 6, p. 205–228.

Andrews, J. T., and Falconer, G., 1969, Late glacial and postglacial history and emergence of the Ottawa Islands, Hudson Bay, N.W.T.; evidence on the deglaciation of Hudson Bay: Canadian Journal of Earth Sciences, v. 6, p. 1263–1276.

Andrews, J. T., and Mahaffy, M. A., 1976, Growth rates of the Laurentide Ice Sheet and sea level lowering (with emphasis on the 115,000 BP sea level low): Quaternary Research, v. 6, p. 167–183.

Andrews, J. T., and Miller, G. H., 1979, Glacial erosion and ice sheet divides, northeastern Laurentide Ice Sheet, on the basis of the distribution of limestone erratics: Geology, v. 7, p. 592–596.

Andrews, J. T., Shilts, W. W., and Miller, G. H., 1983, Multiple deglaciation of the Hudson Bay Lowlands since deposition of the Missinaibi (Last Interglacial?) formation: Quaternary Research, v. 19, p. 18–37.

Andrews, J. T., Stravers, J. A., and Miller, G. H., 1985a, Patterns of glacial erosion and deposition around Cumberland Sound, Frobisher Bay, and Hudson Strait, and the location of ice streams in the eastern Canadian Arctic, *in* Waldenburg, M., ed., Models in geomorphology: London, Allen and Unwin, p. 93–117.

Andrews, J. T., and four others, 1985b, Sedimentation rates in Baffin Island fiord cores from comparative radiocarbon dates: Canadian Journal of Earth Sciences, v. 22, p. 1827–1834.

Andrews, J. T., and five others, 1987, Deglaciation and meltwater events in Hudson Strait and the eastern Canadian arctic: Geomarine Letters (in press).

Beget, J., 1986, Influence of till rheology on Pleistocene glacier flow in the southern Great Lakes area, U.S.A.: Journal of Glaciology, v. 32, p. 235–241.

—— , 1987, Low profile of the Laurentide Ice Sheet: Arctic and Alpine Research, v. 20 (in press).

Bigras, P.J.G., Dubois, J.-M., and Gwyn, Q.H.J., 1985, Relative sea level during the last 35,000 years, northern Gulf of St. Lawrence, Quebec: Geological Association of Canada Abstracts, v. 10, p. A5.

Birkeland, P. W., 1978, Soil development as an indication of relative age of Quaternary deposits, Baffin Island, N.W.T., Canada: Arctic and Alpine Research, v. 10, p. 733–747.

Blake, W., Jr., 1966, End moraines and deglaciation chronology in northern Canada, with special reference to southern Baffin Island: Geological Survey of Canada Paper 66-21, p. 1–31.

—— , 1970, Studies of glacial history in Arctic Canada; I. Pumice, radiocarbon dates, and differential postglacial uplift in the eastern Queen Elizabeth Islands: Canadian Journal of Earth Sciences, v. 7, p. 634–664.

—— , 1980, Mid-Wisconsin interstadial deposits beneath Holocene beaches, Cape Storm, Ellesmere Island, arctic Canada: Orono, Maine, American Quaternary Association, 6th Biennial Meeting, Abstracts, p. 26–27.

Bloom, A. L., 1971, Glacial-eustatic and isostatic controls of sea level since the last glaciation, *in* Turekian, K. K., ed., The late Cenozoic glacial ages: New Haven, Connecticut, Yale University Press, p. 355–380.

Bockheim, J. G., 1979, Properities and relative age of soils of southwestern Cumberland Peninsula, Baffin Island, N.W.T., Canada: Arctic and Alpine Research, v. 11, p. 289–306.

Borns, H. W., Jr., 1973, Late Wisconsin fluctuations of the Laurentide Ice Sheet in southern and eastern New England: Geological Society of America Memoir 136, p. 37–45.

Borns, H. W., Jr., Huges, T. J., and Kellogg, T. B., 1980, Glacio-marine geology of the eastern coastal zone: 1980 Field Trip Guide, Orono, Maine, University of Maine, American Quaternary Association, 18 p.

Bouchard, M. A., and Marcotte, C., 1986, Regional glacial dispersal patterns in Ungava, Nouveau-Quebec: Geological Survey of Canada Paper 86-18, p. 275–318.

Boulton, G. S., 1984, Development of a theoretical model of sediment dispersal by ice sheets, prospective in areas of glaciated terrain: London, England, Institution of Mining and Metallurgy, p. 213–223.

Boulton, G. S., and Jones, A. S., 1979, Stability of temperate ice caps and ice sheets resting on beds of deformable sediment: Journal of Glaciology, v. 24, p. 29–43.

Boulton, G. S., and three others, 1985, Glacial geology and glaciology of the last mid-latitude ice sheets: Geological Society of London Journal, v. 142, p. 447–474.

Boyer, S. J., 1972, Pre-Wisconsin, Wisconsin, and Neoglacial ice limits in Maktak Fiord, Baffin Island; a statistical analysis [M.S. thesis]: Boulder, University of Colorado, 117 p.

Brooks, I. A., 1985, VIII. Weathering, *in* Rutter, N. W., ed., Dating methods of Pleistocene deposits and their problems: Geoscience Canada, Reprint Series 2, p. 61–72.

Bryson, R. A., Wendland, W. M., Ives, J. D., and Andrews, J. T., 1969, Radiocarbon isochrones on the disintegration of the Laurentide Ice Sheet: Arctic and Alpine Research, v. 1, p. 1–14.

Buckley, J. T., 1969, Gradients of past and present outlet glaciers: Geological Survey of Canada Paper 69-29, 13 p.

Budd, W. F., and Smith, I. N., 1981, The growth and retreat of ice sheets in response to orbital radiation changes, *in* Sea level, ice, and climatic change: International Association of Hydrological Sciences, v. 131, p. 369–409.

Clark, J. A., 1980, The reconstruction of the Laurentide Ice Sheet of North America from sea-level data; method and preliminary results: Journal of Geophysical Research, v. 85, p. 4307–4323.

—— , 1985, Forward and inverse models in sea-level studies, *in* Woldenberg, M. J., ed., Models in geomorphology: London, Allen and Unwin, p. 119–138.

Clark, P. U., 1984, Glacial geology of the Kangalaksiovik-Abloviak region, northern Labrador [Ph.D. thesis]: Boulder, University of Colorado, 240 p.

Clark, P. U., and Josenhans, H. W., 1986, Late Quaternary land-sea correlations, northern Labrador and Labrador shelf: Geological Survey of Canada Paper 86-1B, p. 171–178.

Clayton, L., and Moran, S. R., 1982, Chronology of late Wisconsin glaciation in middle North America: Quaternary Science Reviews, v. 1, p. 55–82.

Clayton, L., Teller, J. T., and Attig, J. W., 1985, Surging of the southwestern part of the Laurentide Ice Sheet: Boreas, v. 14, p. 235–241.

Crary, A. P., 1960, Arctic ice island and ice shelf studies; part II: Arctic, v. 13, p. 32–50.

Danks, H. V., 1981, Arctic arthropods: Ottawa, Entomological Society of Canada, Tyrell Press Ltd., 605 p.

Denton, G. H., 1974, Quaternary glaciations of the White River Valley, Alaska, with a regional synthesis for the northern St. Elias Mountains, Alaska and Yukon Territory: Geological Society of American Bulletin, v. 85, p. 871–892.

Denton, G. H., and Hughes, T. J., editors, 1981a, The last great ice sheets: New York, John Wiley and Sons, 484 p.

—— , 1981b, The Arctic Ice Sheet: an outrageous hypothesis, *in* Denton, G. H., and Hughes, T. J., eds., The last great ice sheets: New York, John Wiley and Sons, p. 437–467.

Denton, G. H., Hughes, T. J., and Karlen, W., 1986, Global ice-sheet system interlocked by sea level: Quaternary Research, v. 26, p. 3–26.

Dredge, L. A., 1983, Character and development of northern Lake Agassiz and its relation to Keewatin and Hudsonian ice regimes, *in* Teller, J. T., and Clayton, L., eds., Glacial Lake Agassiz: Toronto, University of Toronto Press, Geological Association of Canada Special Paper no. 26, p. 117–132.

Dredge, L. A., and Cowan, W. R., 1987, Quaternary geology of the southwestern Canadian Shield, *in* Fulton, R. J., Heginbottom, J. A., and Funder, S., eds., Quaternary geology of Canada and Greenland: Geological Survey of Canada, Geology of Canada No. 1 [also Geological Society of America, The Geology of North America, v. K-1] (in press).

Dreimanis, A., and Raukas, A., 1975, Did middle Wisconsin, middle Weischelian, and their equivalents represent an interglacial or an interglacial complex in the Northern Hemisphere?, *in* Quaternary studies: Royal Society of New Zealand Bulletin, v. 13, p. 109–120.

Drewry, D. J., McIntrye, N. F., and Cooper, P., 1985, The Antarctic Ice Sheet; a surface model for satellite altimeter studies, *in* Woldenberg, M. J., ed., Models in geomorphology: London, Allen and Unwin, p. 1–24.

Dubois, J-M.M., and Dionne, J-C, 1985, The Quebec North Shore Moraine System: A major feature of late Wisconsin deglaciation, *in* Borns, H. W., Jr., et al., eds., Late Pleistocene history of northeastern New England and adjacent Quebec: Geological Society of America Special Paper 197, p. 125–133.

Duplessy, J-C. Arnold, and four others, 1986, Direct dating of the oxygen-isotope record of the last deglaciation by ^{14}C accelerator mass spectrometry: Nature, v. 320, 32 p.

Dyke, A. S., 1977, Quaternary geomorphology, glacial chronology, and climatic and sea-level history of southwestern Cumberland Peninsula, Baffin Island, Northwest Territories, Canada [Ph.D. thesis]: Boulder, University of Colorado, 185 p.

—— , 1983, Quaternary geology of Somerset Island, District of Franklin: Geological Survey of Canada Memoir 404, 32 p.

—— , 1984, Quaternary geology of Boothia Peninsula and northern District of Keewatin, central arctic Canada: Geological Survey of Canada Memoir 407, 26 p.

—— , 1987, A reinterpretation of glacial and marine limits around the northwestern Laurentide Ice Sheet: Canadian Journal of Earth Sciences (in press).

Dyke, A. S., and Dredge, L. A., 1987, Quaternary geology of the northwestern Canadian Shield, *in* Fulton, R. J., Heginbottom, J. A., and Funder, S., eds., Quaternary geology of Canada and Greenland: Geological Survey of Canada, Geology of Canada, No. 1 [also Geological Society of America, The Geology of North America, v. K-1] (in press).

Dyke, A. S., Dredge, L. A., and Vincent, J. S., 1982, Configuration of the Laurentide Ice Sheet during the late Wisconsin maximum: Geographie Physique et Quaternaire, v. 36, p. 5–14.

Dyke, A. S., and four others, 1987, Quaternary geology of the Canadian Shield, *in* Fulton, R. J., Heginbottom, J. A., and Funder, S., eds., 1987, Quaternary geology of Canada and Greenland: Geological Survey of Canada, Geology of Canada, No. 1 [also Geological Society of America, The Geology of North America, v. K-1] (in press).

England, J. H., 1976, Late Quaternary glaciation of the eastern Queen Elizabeth Islands, N.W.T., Canada; alternative models: Quaternary Research, v. 6, p. 185–203.

—— , 1985, The late Quaternary history of Hall Land, northwest Greenland: Canadian Journal of Earth Sciences, v. 22, p. 1394–1408.

Evans, D.J.A., and Rogerson, R. J., 1986, Glacial geomorphology and chronology in the Selamuit Range-Nachvak Fiord area, Torngat Mountains, Labrador: Canadian Journal of Earth Sciences, v. 23, p. 66–76.

Evans, L. J., and Cameron, B. H., 1979, A chronosequence of static cryosols developed on granite gneiss, Baffin Island, N.W.T.: Canadian Journal of Soil Sciences, v. 59, p. 203–211.

Falconer, G., Ives, J. D., Loken, O. H., and Andrews, J. T., 1965, Major end moraines in eastern and central arctic Canada: Geographical Bulletin, v. 7, p. 137–153.

Fenton, M. M., 1984, Quaternary stratigraphy of the Canadian prairies: Geological Survey of Canada Paper 84-10, p. 57–68.

Fenton, M. M., Moran, S. R., Teler, J. T., and Clayton, L., 1983, Quaternary stratigraphy and history in the southern part of the Lake Agassiz Basin, *in* Teller, J. T., and Clayton, L., eds., Glacial Lake Agassiz: Toronto, University of Toronto Press, Geological Association of Canada Special Paper no. 26, p. 49–74.

Fillon, R. H., 1985, Northwest Labrador Sea stratigraphy, sand input and paleoceanography during the last 160,000 years, *in* Andrews, J. T., ed., Late Quaternary environments; eastern Canadian arctic, Baffin Bay, and western Greenland: London, Allen and Unwin, p. 210–247.

Fillon, R. J., and Aksu, A. E., 1985, Evidence for subpolar influence in the Labrador Sea and Baffin Bay during marine isotopic stage 2, *in* Andrews, J. T., ed., Quaternary environments; eastern Canadian arctic, Baffin Bay, and western Greenland: London, Allen and Unwin, p. 248–262.

Fillon, R. H., and Harmes, R. A., 1982, Northern Labrador shelf glacial chronology and depositional environments: Canadian Journal of Earth Sciences, v. 19, p. 162–192.

Fillon, R. H., and Williams, D. F., 1984, Dynamics of meltwater discharge from Northern Hemisphere ice sheets during the last deglaciation: Nature, v. 310, p. 674–677.

Fillon, R. H., and others, 1981, Labrador shelf; shelf and total organic matter ^{14}C date discrepancies: Geological Survey of Canada Paper 81-1B, p. 105–111.

Fisher, D. A., Reeh, N., and Langley, K., 1985, Objective reconstructions of the late Wisconsin Laurentide Ice Sheet and the significance of deformable beds: Geographie Physique et Quaternaire, v. 39, p. 229–238.

Forman, S. L., Wintle, A. G., Thorleifson, H. L., and Wyatt, P. H., 1987, Thermoluminescence properties and preliminary dates for Quaternary raised-marine sediments, Hudson Bay Lowland, Canada: Canadian Journal of Earth Sciences (in press).

Fulton, R. J., editor, 1984, Quaternary stratigraphy of Canada; a Canadian contribution to IGCP Project 24: Geological Survey of Canada Paper 84-10, 210 p.

Fulton, R. J., Heginbottom, J. A., and Funder, S., eds., 1987, Quaternary geology of Canada and Greenland: Geological Survey of Canada, Geology of Canada, No. 1 [also Geological Society of America, The Geology of North America, v. K-1], in press.

Fulton, R. J., and Westgate, J. A., 1975, Tephrostratigraphy of Olympia Interglacial sediments in south-central British Columbia, Canada: Canadian Journal of Earth Sciences, v. 12, p. 489–502.

Fulton, R. J., Fenton, M. M., and Rutter, N. W., 1984, Summary of Quaternary stratigraphy and history, western Canada, *in* Fulton, R. J., ed., Quaternary stratigraphy of Canada; a Canadian Contribution to IGCP Project 24: Geological Survey of Canada Paper 84-10, p. 69–86.

Funder, S., 1987, Quaternary geology of Greenland, *in* Fulton, R. J., Heginbottom, J. A., and Funder, S., eds., 1987, Quaternary geology of Canada and

Greenland: Geological Survey of Canada, Geology of Canada, No. 1 [also Geological Society of America, The Geology of North America, v. K-1] (in press).

Gadd, N. R., 1980, Late-glacial regional ice-flow patterns in eastern Ontario: Canadian Journal of Earth Sciences, v. 17, p. 1439–1453.

—— , 1986, The basin, the ice, the Chaplain Sea: Geological Association of Canada Abstracts with Program, no. 11, p. 71.

Grant, A. C., and Manchester, K. S., 1970, Geophysical investigation in the Ungava Bay-Hudson Strait region of northern Canada: Canadian Journal of Earth Sciences, v. 7, p. 1062–1076.

Grant, D. R., and King, L. H., 1984, A stratigraphic framework for the Quaternary history of the Atlantic provinces, Canada, *in* Fulton, R. J., ed., Quaternary stratigraphy of Canada; a Canadian contribution to IGCP Project 24: Geological Survey of Canada Paper 84-10, p. 173–192.

Gray, J. T., and Lauriol, B., 1985, Dynamics of the late Wisconsin Ice Sheet in the Urgava Peninsula, interpreted from geomorphological evidence: Arctic and Alpine Research, v. 17, p. 289–310.

Grosswald, M. G., 1984, Glaciation of the continental shelf, part 1: Polar Geography and Geology, v. 8, p. 196–258.

Hardy, L., 1976, Contribution a l'etude geomorphologique de la portion quebecoise des basses terres de la baie de James [Ph.D. thesis]: Montreal, McGill University, 264 p.

Hill, P. R., Mudie, P. J., Moran, K., and Blasco, S. M., 1985, A sea-level curve for the Canadian Beaufort Shelf: Canadian Journal of Earth Sciences, v. 22, p. 1383–1393.

Hillaire-Marcel, C., 1979, Les Mers post-glaciers due Quebec; quelques aspects [Ph.D. thesis]: Paris, France, L'Universite Pierre et Marie Cure, 293 p.

Hillaire-Marcel, C., and Vincent, J-S, 1980, Holocene stratigraphy and sea level changes in southeastern Hudson Bay, Canada: Trois-Rivieres, Paleo-Quebec II, Musee d'Archeologie, Universite du Quebec a Trois-Rivieres, 165 p.

Hillaire-Marcel, C., Occhietti, S., and Vincent, J. S., 1981, Sakami moraine, Quebec; a 500-km-long moraine without climatic control: Geology, v. 9, p. 210–214.

Hodgson, D. A., 1981, Surficial geology, Lougheed Island, northwest Arctic Archipelago: Geological Survey of Canada Paper 81-1C, p. 27–34.

—— , 1985, The last glaciation of west-central Ellesmere Island, Arctic Archipelago, Canada: Canadian Journal of Earth Sciences, v. 22, p. 347–368.

Hodgson, D. A., and Vincent, J. S., 1984, A 10,000 year B.P. extensive ice shelf over Viscount of Melville Sound, arctic Canada: Quaternary Research, v. 22, p. 18–30.

Hooke, R. de B., 1976, Pleistocene ice at the base of the Barnes ice cap, Baffin Island, N.W.T., Canada: Journal of Glaciology, v. 17, p. 49–59.

Hopkins, D. M., editor, 1982, Paleoecology of Beringia: New York, Academic Press, 489 p.

Hughes, T. J., 1985, The great Cenozoic ice sheet: Palaeogeography, Palaeoclimatology, Palaeoecology, v. 50, p. 9–43.

Hughes, T. J., Denton, G. H., and Grosswald, M. G., 1977, Was there a late-Wurm Arctic ice sheet?: Nature, v. 266, p. 596–602.

Hughes, T. J., Denton, G. H., and Fastook, J. L., 1985a, The Antarctic ice sheet; an analogy for Northern Hemisphere paleo ice sheets, *in* Woldenberg, M. J., ed., Models in geomorphology: London, Allen and Unwin, p. 25–72.

Hughes, T. J., and four others, 1985b, Models of glacial reconstruction and deglaciation applied to eastern maritime Canada and New England, *in* Barns, H. W., Jr., et al., eds., Late Pleistocene history of northeastern New England and adjacent Quebec: Geological Society of America Special Paper 197, p. 139–150.

Ives, J. D., 1977, Were parts of the north coast of Labrador ice-free at the Wisconsin glacial maximum?: Geographie Physique et Quaternaire, v. 31, p. 401–403.

—— , 1978, The maximum extent of the Laurentide Ice Sheet along the east coast of North America during the last glaciation: Arctic, v. 32, p. 24–53.

Ives, J. D., and Andrews, J. T., 1963, Studies in the physical geography of north-central Baffin Island, N.W.T.: Geographical Bulletin, v. 19, p. 5–48.

Jeffries, M. O., 1986, Ice Island calving and ice shelf changes, Milne ice shelf and Ayles ice shelf, Ellesmere Island, N.W.T.: Arctic, v. 39, p. 15–19.

Johnson, W. H., Moore, D. W., and McKay, D. E., III, 1986, Provenance of late Wisconsin (Woodfordian) till and origin of the Decatur sublobe, east-central Illinois: Geological Society of America Bulletin, v. 97, p. 1098–1105.

Josenhans, H. W., Zevenhuizen, J., and Klassen, R. A., 1986, The Quaternary geology of the Labrador shelf: Canadian Journal of Earth Sciences, v. 23, p. 1190–1214.

Karrow, P. F., 1981, Late-glacial regional ice-flow patterns in eastern Ontario; discussion: Canadian Journal of Earth Sciences, v. 18, p. 1386–1390.

—— , 1984, Quaternary stratigraphy and history, Great Lakes-St. Lawrence region: Geological Survey of Canada, Paper 84-10, p. 137–154.

Kelley, M., and Bennike, O., 1985, Quaternary geology of parts of central and western North Greenland; a preliminary account: Grönlands Geologiske Undersölgelse Report, v. 126, p. 111–116.

Kerr, J. W., 1970, Today's topography and tectonics in northeastern Canada: Canadian Journal of Earth Sciences, v. 7, p. 570.

Kennett, J. P., and Shackleton, N. J., 1975, Laurentide Ice Sheet meltwater recorded in Gulf of Mexico deep-sea cores: Science, v. 188, p. 147–150.

King, G. A., 1985, A standard method for evaluating radiocarbon dates of local deglaciation: Appolication to the deglaciation history of southern Labrador and adjacent Quebec: Geographie Physique et Quaternaire, v. 39, p. 163–182.

King, L. H., and Fader, G.B.J., 1986, Wisconsinan glaciation of the Atlantic Continental Shelf of southeast Canada: Geological Survey of Canada Bulletin 363, 72 p.

Klassen, R. W., 1983, Lake Agassiz and the late glacial history of northern Manitoba, *in* Teller, J. T., and Clayton, L., eds., Glacial Lake Agassiz: Toronto, University of Toronto Press, Geological Association of Canada Special Paper no. 26, p. 97–116.

—— , 1985a, Dispersal centers of the Laurentide Ice Sheet in Labrador and eastern Quebec: Geological Association of Canada Abstracts, v. 10, p. 31.

—— , 1985b, An outline of glacial history of Bylot Island, District of Franklin, N.W.T., *in* Andrews, J. T., ed., Late Quaternary environments; eastern Canadian arctic, Baffin Bay, and western Greenland: London, Allen and Unwin, p. 428–460.

Kutzbach, J. E., and Wright, H. E., Jr., 1985, Simulation of the climate of 18,000 years BP; results from the North American/North Atlantic/European sector and comparison with the geologic record of North America: Quaternary Science Reviews, v. 4, p. 147–188.

LaSalle, P., 1984, Quaternary stratigraphy of Quebec; a review, *in* Fulton, R. J., ed., Quaternary stratigraphy of Canada; a Canadian contribution to IGCP Project 24: Geological Survey of Canada Paper 84-10, p. 155–172.

LaSalle, P., and Elson, J. A., 1975, Emplacement of the St. Narcisse Moraine as a climatic event in eastern Canada: Quaternary Research, v. 5, p. 621–625.

Leventer, A., Williams, D. F., and Kennett, J. P., 1982, Dynamics of the Laurentide Ice Sheet during the last deglaciation; evidence from the Gulf of Mexico: Earth and Planetary Science Letters, v. 59, p. 11–17.

Lingle, C. S., and Clark, J. A., 1985, A numerical model of interaction between a marine ice sheet and the solid earth; application to a west Antarctic ice stream: Journal of Geophysical Research, v. 90, p. 1100–1114.

Locke, W. W. II, 1985, Weathering and soil development on Baffin Island, *in* Andrews J. T., ed., Quaternary environments; eastern Canadian arctic, Baffin Bay and western Greenland: London, Allen and Unwin, p. 331–353.

—— , 1986, Fine particle translocation in soils developed on glacial deposits, southern Baffin Island, N.W.T., Canada: Arctic and Alpine Research, v. 18, p. 33–43.

Loken, O. H., 1966, Baffin Island refugia older than 54,000 years: Science, v. 153, p. 1378–1380.

MacLean, B., 1985, Geology of the Baffin Island Shelf, *in* Andrews, J. T., ed., Quaternary environments; eastern Canadian arctic, Baffic Bay, and western Greenland: London, Allen and Unwin, p. 154–177.

MacLean, B., Williams, G. L., Jennings, A. E., and Blakeney, C., 1986a, Cumberland Sound, N.W.T.; investigations of bedrock and surficial geology: Geological Society of Canada Paper 86-1B, p. 605–616.

MacLean, B., and five others, 1986b, A reconnaissance of the bedrock and surficial geology of Hudson Strait; preliminary results: Geological Survey of Canada Paper 86-1B (in press).
Manabee, S., and Broccoli, A. J., 1985, A comparison of climate model sensitivity with data from the last glacial maximum: Journal of the Atmospheric Sciences, v. 42, p. 2643–2651.
Matthews, W. H., 1974, Surface profiles of the Laurentide Ice Sheet in its marginal areas: Journal of Glaciology, v. 13, p. 37–43.
Mayewski, P. A., Denton, G. H., and Hughes, T. J., 1981, Late Wisconsin ice sheets of North America, *in* Denton, G. H., and Hughes, T. J., eds., The last great ice sheets: New York, John Wiley and Sons, p. 67–178.
McIntrye, M. F., 1985, The dynamics of ice sheet outlets: Journal of Glaciology, v. 31, p. 99–109.
Mears, A. I., 1972, Glacial geology and crustal properties in the Nedlukseak Fiord region, eastern Baffin Island, Canada [M.S. thesis]: Boulder, University of Colorado, 60 p.
Michelson, D. M., Clayton, L., Fullerton, D. S., and Borns, H. W., Jr., 1983, The late Wisconsin glacial record in the United States, *in* Wright, H. E., Jr., ed., Late Quaternary environments of the United States, Volume 1, The late Pleistocene: Minneapolis, University of Minnesota Press, p. 3–37.
Milankovitch, M. M., 1941, Canon of insolation and the Ice-Age problem: Beograd, Koniglich Serbische Akademie [English translation by the Israel Program for Scientific translations]: Washington, D.C., U.S. Department of Commerce and National Science Foundation, 484 p.
Miller, G. H., 1980, Late Foxe glaciation of southern Baffin Island, N.W.T., Canada: Geological Society of America Bulletin, Part I, v. 91, p. 399–405.
—— , 1985, Aminostratigraphy of Baffin Island shell-bearing deposits, *in* Andrews, J. T., ed., Quaternary environments; eastern Canadian arctic, Baffin Bay, and western Greenland: London, Allen and Unwin, p. 394–427.
Miller, G. H., Andrews, J. T., and Short, S. K., 1977, The last interglacial-glacial cycle, Clyde foreland, Baffin Island, N.W.T.; stratigraphy, biostratigraphy, and chronology: Canadian Journal of Earth Sciences, v. 14, p. 2824–2857.
Mix, A. C., and Ruddiman, W. F., 1985, Structure and timing of the last deglaciation; oxygen-isotope evidence: Quaternary Science Reviews, v. 4, p. 59–108.
Nielson, E., and Thorliefson, L. H., 1985, Glaciological reconstruction of a late Wisconsin sublobe of the Laurentide Ice Sheet, southern Manitoba, Canada: Geological Association of Canada Abstracts, v. 10, p. A43.
Ochietti, S., 1987, St. Lawrence Valley and adjacent Appalachian subregion, *in* Fulton, R. J., Heginbottom, J. A., and Funder, S., eds., 1987, Quaternary geology of Canada and Greenland: Geological Survey of Canada, Geology of Canada, No. 1 [also Geological Society of America, The Geology of North America, v. K-1] (in press).
Osterman, L. E., 1982, Late Quaternary history of southern Baffin Island, Canada; a study of foraminifera and sediments from Frobisher Bay [Ph.D. thesis]: Boulder, University of Colorado, 380 p.
—— , 1984, Benthic foraminiferal zonation of a glacial/interglacial transition from Frobisher Bay, Baffin Island, N.W.T., Canada: Benthos, 2nd International Symposium, Benthic foraminifera, 1983, p. 471–476.
Osterman, L. E., Miller, G. H., and Stravers, J. A., 1985, Middle and late Foxe glacial events in southern Baffin Island, *in* Andrews, J. T., ed., Quaternary environments; eastern Canadian arctic, Baffin Bay, and western Greenland: London, Allen and Unwin, p. 520–454.
Paterson, W.S.B., 1972, Laurentide Ice Sheet; estimated volumes during the late Wisconsin: Review of Geophysics and Space Physics, v. 10, p. 885–917.
—— , 1977, Extent of the late-Wisconsin glaciation in northwest Greenland and northern Ellesmere Island; a review of the glaciological and geological evidence: Quaternary Research, v. 8, p. 180–190.
Peltier, W. R., and Andrews, J. T., 1976, Glacial-isostatic adjustment-I; the forward problem: Royal Astronomical Society Geophysical Journal, v. 46, p. 605–646.
—— , 1983, Glacial geology and glacial isostasy, Hudson Bay, Canada, *in* Smith, D. E., ed., Shorelines and isostasy: New York, Academic Press, p. 285–319.
Pheasant, D. R., 1971, The glacial chronology and glacio-isostasy of the Narpaing-Quajon fiord area, Cumberland Peninsula, Baffin Island [Ph.D. thesis]: Boulder, University of Colorado, 232 p.
Pollard, D., 1983, Ice-age simulations with a calving ice-sheet model: Quaternary Research, v. 20, p. 30–48.
Praeg, D. B., MacLean, B., Hardy, I. A., and Mudie, P. J., 1986, Quaternary geology of the southeast Baffin Island continental shelf, N.W.T.: Geological Survey of Canada Paper 85-14, 38 p.
Prest, V. K., 1969, Retreat of Wisconsin and Recent ice in North America: Geological Survey of Canada Map 1257A, scale 1:5,000,000.
—— , 1970, Quaternary geology in Canada, *in* Douglas, R. J., ed., Geology and economic minerals in Canada, 5th edition: Ottawa, Department of Energy, Mines and Resources, p. 676–764.
—— , 1984, The Late Wisconsin glacier complex, *in* Fulton, R. J., ed., Quaternary stratigraphy of Canada; a Canadian contribution to IGCP Project 24: Geological Survey of Canada Paper 84-10, p. 21–38, map 1584A (in pocket).
Quinlan, G., 1985, A numerical model of postglacial relative sea level change near Baffin Island, *in* Andrews, J. T., ed., Quaternary environments; eastern Canadian arctic, Baffin Bay, and western Greenland: London, Allen and Unwin, p. 560–584.
Quinlan, G., and Beaumont, C., 1982, The deglaciation of Atlantic Canada as reconstructed from the postglacial relative sea-level record: Canadian Journal of Earth Sciences, v. 19, p. 2232–2248.
Retelle, M. J., 1986, Glacial geology and Quaternary marine stratigraphy of the Robeson Channel area, northeastern Ellesmere Island: Canadian Journal of Earth Sciences, v. 23, p. 1001–1012.
Richard, S. H., 1975, Surficial geology mapping; Ottawa Valley Lowlands (Parts of 31G, B, and F): Geological Survey of Canada Paper 75-1B, p. 113–117.
Rooney, S. T., Blankenship, D. D., and Alley, R. B., 1986, Structure and continuity of ba till layer beneath Ice Stream B, west Antarctica: Chapman Conference on Fast Glacier Flow: American Geophysical Union Abstracts with Program, p. 11.
Ruddiman, W. F., and McIntyre, A., 1981, The mode and mechanism of the last deglaciation; oceanic evidence: Quaternary Research, v. 16, p. 125–134.
Sanford, B. V., Grant, A. C., Wade, J. A., and Barss, M. A., 1979, Geology of eastern Canada and adjacent areas: Geological Survey of Canada Map 1401A, 4 sheets, scale 1:2,000,000.
Scott, D. B., Mudie, P. J., Vilks, G., and Younger, C., 1984, Latest Pleistocene–Holocene paleoceanographic trends on the continental margin of eastern Canada; forminiferal, dinoflagellate and pollen evidence: Marine Micropaleontology, v. 9, p. 181–218.
Shilts, W. W., 1982, Flow patterns in the central North American ice sheet: Nature, v. 286, p. 213–218.
—— , 1985, Geological models for the configuration, history and style of disintegration of the Laurentide Ice Sheet, *in* Waldenberg, M. J., ed., Models in geomorphology: London, Allen and Unwin, p. 73–92.
Shilts, W. W., Cunningham, C. M., and Kaszycki, C. A., 1979, Keewatin Ice Sheet—Re-evaluation of the traditional concept of the Laurentide Ice Sheet: Geology, v. 7, p. 537–541.
Shreve, R. L., 1985, Late-Wisconsin ice-surface profile calculated from esker paths and types, Katahdin esker system, Maine: Quaternary Research, v. 23, p. 27–37.
Skinner, R. G., 1973, Quaternary stratigraphy of the Moose River Basin, Ontario: Geological Survey of Canada Bulletin, v. 225, 77 p.
Smith, J. E., 1966, Sam Ford Fiord; a study in deglaciation [M.S. thesis]: Montreal, McGill University, 93 p.
St. Pierre, L., Gwyn, P.H.J., Dubois, J-M., 1985, Dynamique des ecoulements glacieres Wisonsiniens, ile d'Anticost, Golfe du Saint-Laurent: Geological Association of Canada Abstracts, v. 10, p. A60.
Stravers, J. A., 1986, Glacial geology of the outer Meta Incognita Peninsula, southern Baffin Island, Arctic Canada [Ph.D. thesis]: Boulder, University of Colorado, 231 p.
Stuiver, M., Denton, G. H., Hughes, T. J., and Fastook, J. L., 1981, History of the marine ice sheet in west Antarctica during the last deglaciation; a working hypothesis, *in* Denton, G. H., and Hughes, T. J., eds., The last great ice

sheets: New York, John Wiley and Sons, p. 319–392.

Sugden, D. E., 1978, Glacial erosion by the Laurentide Ice Sheet: Journal of Glaciology, v. 20, p. 367–392.

Szabo, B. J., Miller, G. H., Andrews, J. T., and Stuiver, M., 1981, Comparison of uranium-series, radiocarbon, and amino acid data from marine molluscs, Baffin Island, Arctic Canada: Geology, v. 9, p. 451–457.

Teller, J. T., and Clayton, L., editors, 1983, Glacial Lake Agassiz: Geological Association of Canada Special Paper 26, 451 p.

Thomas, R. H., 1977, Calving bay dynamics and ice sheet retreat up the St. Lawrence valley system: Geographie Physique et Quaternaire, v. 31, p. 347–356.

Thorleifson, L. H., and Wyatt, P. H., 1986, Sedimentology of Quaternary glacial deposits in the central Hudson Bay lowland, northern Ontario: Geological Association of Canada Abstracts with Program, v. 11, p. 136.

Tippett, C. R., 1985, Glacial dispersal train of Paleozoic erratics, central Baffin Island, N.W.T., Canada: Canadian Journal of Earth Sciences, v. 22, p. 1818–1826.

Vernal, A. de., Richard, P.J.H., and Ochietti, S., 1983, Palynologie et paleoenvironments du Wisconsinien de la region baie Saint-Laurent, ile du Cap-Breton: Geographie Physique et Quaternaire, v. 37, p. 307–322.

Vernal, A. de., Causse, C., Hillaire-Marcel, C., Mott, R. J., and Ochietti, S., 1986, Palynostratigraphy and Th/U ages of upper Pleistocene interglacial and interstadial deposits on Cape Breton Island, eastern Canada: Geology, v. 14, p. 554–557.

Vincent, J. S., 1984, Quaternary stratigraphy of the western Canadian Arctic Archipelago, *in* Fulton, R. J., ed., Quaternary stratigraphy of Canada; a Canadian contribution to IGCP Project 24: Geological Survey of Canada Paper 84-10, p. 87–100.

——, 1987, Quaternary geology of the northern interior plains: Ottawa, Ontario Geological Survey of Canada, Geology of North America, Canadian Quaternary (in press).

Vincent, J-S, and Hardy, L., 1979, The evolution of Glacial Lake Barlow and Ojibway, Quebec and Ontario: Geological Survey of Canada Bulletin 136, 18 p.

Weertman, J., 1964, Rate of growth or shrinkage of non-equilibrium ice sheets: Journal of Glaciology, v. 5, p. 145–158.

Wright, H. E., Jr., 1973, Tunnel valleys, glacial surges, and sub-glacial hydrology of the Superior lobe, Minnesota, *in* Black, R. F., Goldthwait, R. P., and Willman, H. B., eds., The Wisconsinan Stage: Geological Society of America Memoir 136, p. 251–276.

Wyatt, P. H., and Thorliefson, L. H., 1986, Provenance and geochronology of Quaternary glacial deposits in the central Hudson Bay lowlands: Geological Association of Canada Abstracts with Program, v. 11, p. 147.

MANUSCRIPT ACCEPTED BY THE SOCIETY FEBRUARY 2, 1987

ACKNOWLEDGMENTS

I thank colleagues at the Geological Survey of Canada for providing advance copies of chapters from Fulton (1987) and my colleagues at INSTAAR for discussions and ideas over the course of several years. In particular I would like to thank Drs. Denton, Dyke, Miller, Osterman, Teller, Fillon, MacLean, Meier, Lingle, and Stravers for reading and commenting on this chapter. Dr. A. S. Dyke and Dr. G. H. Denton in particular contributed thoughtful reviews which are much appreciated. The editors of this volume are thanked for their considerable efforts to improve this manuscript. Dr. S. Funder discussed the problem of ice extent in northwest Greenland and Ellesmere Island.

Preparation of this chapter was supported by National Science Foundation Grants EAR-84-09915 and DPP-83-06581.

Printed in U.S.A.

The Geology of North America
Vol. K-3, North America and adjacent oceans during the last deglaciation
The Geological Society of America, 1987

Chapter 3

Proglacial lakes and the southern margin of the Laurentide Ice Sheet

James T. Teller
Department of Geological Sciences, University of Manitoba, Winnipeg, Manitoba R3T 2N2, Canada

INTRODUCTION

During the advances and retreats of the Laurentide Ice Sheet in North America, drainage routes were disrupted and dammed, and new basins were formed by erosion and deposition. Some basins were completely enclosed by rock or sediment, whereas others formed closed depressions only as long as glacial ice served as one of their margins. During the early history of nearly all of these basins, ice bounded the lakes and, in many cases, helped control their level. Across most of central North America the Laurentide Ice Sheet advanced upslope for hundreds of kilometers from the Hudson Bay Lowland before crossing the continental divide into the Mississippi and Atlantic drainage basins. As a result, water was impounded over a large region of the Hudson Bay watershed, being confined partly by the glacial margin and partly by the elevation of land to the south.

Strictly speaking, only lakes that lay in direct contact with ice or its marginal deposits are considered proglacial lakes. However, most workers would include lakes where meltwater was a significant part of their hydrological budget, even though they may have been some distance beyond the glacial margin.

In this chapter I describe the main factors controlling the formation of proglacial lakes and explain how these factors, and the processes within the lakes, influenced their sedimentation and history. I also speculate on how these lakes may have influenced ice flow and deglaciation. A large part of the chapter is devoted to a discussion of the histories of major proglacial lakes in North America associated with the southern Laurentide Ice Sheet and to their interrelationship with each other and with the retreating glacial margin. The timing of meltwater flow into the Gulf of Mexico and Atlantic Ocean during deglaciation is examined and compared to the isotopic record of sediments in these basins.

CONTROLS ON THE DISTRIBUTION AND MORPHOLOGY OF PROGLACIAL LAKES

Hundreds of thousands of lakes today dot the glaciated landscape of North America. Some of these lie within depressions in glacial drift. Most of the large lakes, however, are located in glacially scoured bedrock basins, which are controlled by the lithology and structure of the bedrock. For example, the fringe of large basins around the Precambrian Shield in western Canada, including those of Great Bear Lake, Great Slave Lake, and Lake Winnipeg, owe their origin to differential glacial scouring of the Proterozoic and Paleozoic sedimentary rocks that lie adjacent to the tough core of Shield rocks. Similarly, the softer Proterozoic rocks of the Superior tectonic basin, set entirely within the Shield, have been extensively eroded to form the Lake Superior Basin. Differing resistances to erosion within the Paleozoic rocks—probably first exploited by rivers—were responsible for the location of most of the Great Lakes basins (Spencer, 1890; Shepard, 1937; Hough, 1958). Glaciers not only scoured and deepened these preglacial depressions but had their flow partly controlled by them.

Of fundamental importance to the development of Quaternary proglacial lakes in North America was the presence of a large structural and topographic basin in central Canada, part of which now contains Hudson Bay. This basin still retains a thick sequence of lower Paleozoic rocks, and it probably has been topographically low at least since Mesozoic time (Norris and Sanford, 1969; Cumming, 1969). Thus, whenever ice has accumulated across the high latitudes of central Canada, drainage toward Hudson Bay has been impeded. Many of the proglacial lakes that developed during glaciation were directly related to the centripetal drainage toward Hudson Bay from all but the marginal fringe of the southern Laurentide Ice Sheet.

The boundaries and volume of proglacial lakes were controlled by the location and height of the ice margin and by the configuration and elevation of the rock or sediment surface. These components normally varied through the life of a proglacial lake. In some situations, climate and the ratio of water inflow to outflow in the basin (i.e., hydrological budget) also played a role.

As Laurentide ice retreated and exposed the subglacial

Teller, J. T., 1987, Proglacial lakes and the southern margin of the Laurentide Ice Sheet, *in* Ruddiman, W. F., and Wright, H. E., Jr., North America and adjacent oceans during the last deglaciation: Boulder, Colorado, Geological Society of America, The Geology of North America, v. K-3.

landscape, lower routes for overflow frequently were uncovered, resulting in a decline in lake level or even complete drainage. The floors of overflow outlets, especially where located in soft sediment, commonly were deepened by overflow, resulting in a reduction in volume and areal extent of the lake through time. In some areas, glacial and glacially related sedimentation in the lake also influenced its size, shape, and volume. Periodic readvances during deglaciation also redammed overflow outlets of many lakes, causing them to rise and expand in area.

Glacial isostatic depression and rebound played a major role in proglacial lake development. Although the rate and exact nature of response of the bedrock to loading and unloading by glacial ice has not been well established, its importance is unquestioned (e.g., Andrews, 1970, 1974; Walcott, 1970; Hillaire-Marcel and Fairbridge, 1978; Smith and Dawson, 1983). In many bedrock models, for example those of Peltier (1982), isostatic adjustment depends primarily on the history of glaciation and the mantle viscosity profile. In order to understand the extent to which the data constrain either one of these imperfectly known components in the model, it is necessary to understand simultaneously the contribution of the other (Peltier and Andrews, 1983). Because more rebound occurred toward the former centers of glaciation, where the ice was thickest, and because the crust began to rebound last in those areas where thick ice remained the longest, there is a complex time relationship of isostatic rebound at any one place. In general the resulting crustal rebound caused ice-marginal areas to begin rising first but to experience less total rebound than locations more central in the glaciated region. The net result over the postglacial history of a large lake was that the side of the lake nearest to the center of maximum isostatic rebound (typically the northern side) was differentially raised above the opposite side. The appropriateness of this simple relationship is difficult to appraise because of significant uncertainties in the rheology of the crust and mantle. For example, the viscoelastic models of Peltier (1982) and Birchfield and Grumbine (1985) predict a trough immediately ahead of an advancing ice margin as deep as 200–300 m and a crustal forebulge beyond the trough that may have been as much as 70 m high. Interpretations are also complicated by assumptions about the viscosity of the mantle and by the variability of the ice mass in space and time, including a late-glacial shift in the centers of outflow (cf. Andrews, 1982), an increase in lobation, and rapid fluctuations of hundreds of kilometers in the ice margin. Furthermore, the isostatic response of the crust to unloading of large masses of water, such as the abrupt discharges from Lake Agassiz, North America's largest late-glacial lake, also complicates the calculations of rebound rates. Crittenden (1963), for example, estimated that up to 64 m of isostatic adjustment resulted from the loss of water from pluvial Lake Bonneville.

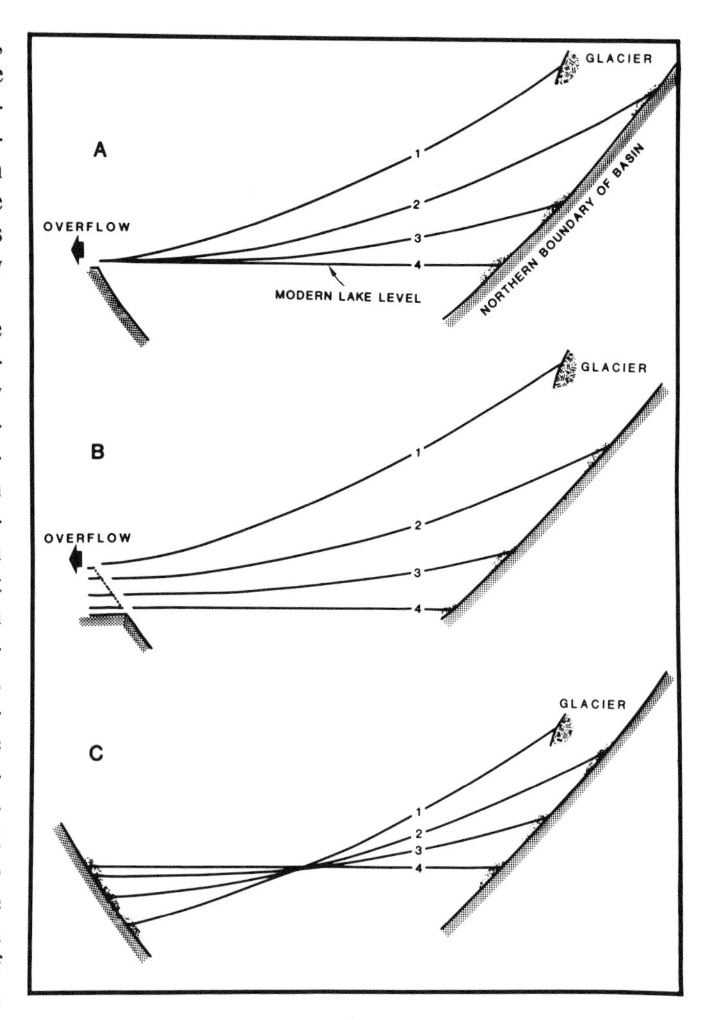

Figure 1. Strandline histories in rebounding proglacial lake basins that have three different types of overflow outlets. Numbers on the strandlines indicate relative age. A. Outlet in southern end that is not eroding. As differential rebound progresses, former water planes rise more in the north and merge at the outlet, where a constant level is maintained. Northern shorelines only develop after ice retreats from basin. B. Outlet in southern end that is eroding. Similar to A except depth of lake declines, resulting in a decrease in lake volume relative to A. C. Outlet not in southern end, but somewhere along eastern, western, or even northern margin of the basin. As differential rebound elevates the northern side of the basin and water shallows there, water deepens in the southern part, resulting in a shift in strandlines to lower levels in the north and to higher levels in the south. These water planes cross in the central part of the basin at the elevation of the overflow outlet and, as in A and B, they have curvatures related to the amount of differential crustal rebound. In this example, the outlet is not eroding.

In some basins, such as that of Lake Agassiz, the level of the lake was controlled by an outlet in the southern end, opposite from the retreating ice margin, and the depth of water at any one place in the lake remained the same (if controlled by a stable outlet level) or declined through time (if the outlet was eroded) (Figs. 1A and 1B). In other basins, such as some of the Great Lakes, there was a progressive shift of the lake mass toward the south as isostatic tilting progressed, and former shorelines along that side were gradually submerged (Fig. 1C). In all cases the resulting beaches and cut strandlines, which represent former contemporary water levels, today rise toward the former glacial center. The older shorelines are more deformed and have steeper gradients than the younger ones because the rate of differential crustal rebound decreased through time.

As can be seen in Figure 2, a vast region of glaciated North America is overlain by lake sediments. Throughout much of late glacial time the proglacial lake system that deposited these sediments was interconnected. The age of these lacustrine materials, however, is diachronous, with the youngest sequences in the north, where ice retreat and crustal rebound occurred last, and the oldest sequences along the southern periphery. These lakes not only left a sedimentary record that has provided some of the most important information on late-Quaternary paleoenvironmental conditions, but they also helped to control the dynamics of ice flow and glacier disintegration, climate, and recolonization of the new postglacial landscape by plants and animals.

GENERAL CHARACTERISTICS OF PROGLACIAL LAKE SEDIMENT

Most proglacial lakes along the southern Laurentide ice margin probably were covered by ice for many months each year. Some of the larger lakes, because of their great fetch and volume, may only have been frozen solidly during a few late-winter months. Present-day Hudson Bay, for example, which is similar in size and volume to glacial Lake Agassiz at its maximum, today only freezes solidly around its periphery, albeit for about eight months each year; most of the surface, however, is covered by ice pack during the winter. Lake Superior, the largest and northernmost of the Great Lakes, has seldom frozen completely in historic times.

The presence of ice on a proglacial lake can influence beach and shoreline development, offshore sedimentation, organisms, and, both directly and indirectly, runoff into the basin. The heat budget, water stratification, oxygen content, turbidity, light penetration, wave power and duration, currents, debris rafting, and postdepositional sediment deformation (by meltout, ice push, or ice drag) are all influenced by the presence, distribution, thickness, and duration of ice on a lake. As discussed later, the duration of the highly reflective ice cover on large proglacial lakes may have influenced the regional late-glacial climate.

Glaciolacustrine sediments have been discussed by many, and Jopling and McDonald (1975) and Smith and Ashley (1985) provide good summaries of both the sediments and the processes responsible for their deposition. In some ways, proglacial lake deposits are similar to those in nonglacial lake basins and, of course, many proglacial lakes evolved into nonglacial ones. In general, however, there are some distinctive characteristics of proglacial lakes: (1) the presence of ice-rafted debris, "flow tills", iceberg furrows, and beach "abnormalities"; (2) rhythmic sedimentation with a high rate of accumulation; (3) rapid variations in lake size, depth, and borders of confinement; (4) presence of isostatically deformed shorelines; and (5) scarcity of biological components.

Although ice-rafted detritus is common in proglacial lake sediments, it must be remembered that winter ice on lakes in nonglacial environments may also raft nearshore sediment into the deeper part of the basin during spring breakup. Ovenshine (1970), Edwards (1978), Anderson and others (1980), Thomas and Connell (1985), and others have described the nature of detritus in the proglacial environment and how it affects the laminae of lacustrine muds (see Fig. 3). Taylor and McCann (1983), Dionne and others (1974), Dionne (1979), Mollard (1983), and Nichols (1961) summarize the various features produced by ice on the floor and shore of proglacial lakes.

The introduction of ice-rafted detritus into proglacial lacustrine muds may be so great that even distinguishing it from glacially deposited sediment may not be easy. The boundary between what constitutes true glacially deposited sediment (till) and that which was sedimented through water continues to be debated. In outlining the criteria for distinguishing till from other sediments, Goldthwait (1970) observes that there are "conservatives," who require that glacial ice must be the last medium to handle the sediment if it is to be called till, and "liberals," who accept a varying amount of reworking of ice-transported debris by water and mass movement. The problems of "waterlain tills" and other diamictons of proglacial lakes have been discussed by Dreimanis (1979), Eyles and Eyles (1983), Evenson and others (1977), and others.

Not only may lacustrine sediment in a proglacial environment look like till, but so also may till appear like lacustrine mud. For example, the Huot and Falconer Formations of North Dakota and Minnesota, which lie near the southern end of the glacial Lake Agassiz Basin, are indistinctly laminated and typically contain less than 10% sand and coarser grains. However, they can be directly correlated northward to good loamy tills, and are interpreted as till, having been deposited by ice after it had flowed across more than 500 km of lacustrine muds (Harris and others, 1974; Fenton and others, 1983).

The high influx of sediment to proglacial lakes was a function of the availability of detritus from the melting glacier and the unstable watershed, which was initially unvegetated and under attack by both meltwater and wave erosion. This was compounded by the frequent changes in lake depth and boundaries and by large seasonal variations in the hydrologic budget. For example, concentrations of suspended sediment in modern proglacial lakes on Baffin Island typically are 5 to 50 mg/1 but range up to 700 mg/1 (Gilbert and others, 1985). In contrast, streams entering proglacial Lake Peyto in Alberta have sediment concentrations that almost always exceed 200 mg/1, with a maximum of more than ten times that amount (Smith and others, 1982). Sediment concentrations in rivers draining from glacial ice bordering Malaspina Lake, Alaska, typically exceed 1000 mg/1 and range up to 4700 mg/1, while the lake itself contains 100–700 mg/1 of suspended sediment (Gustavson, 1975).

Sediment is introduced into proglacial lakes mainly in meltwater, although subaqueous slumping and wave action may contribute through resuspension. If the sediment suspension is less dense than the lake waters, then it will spread out across the lake

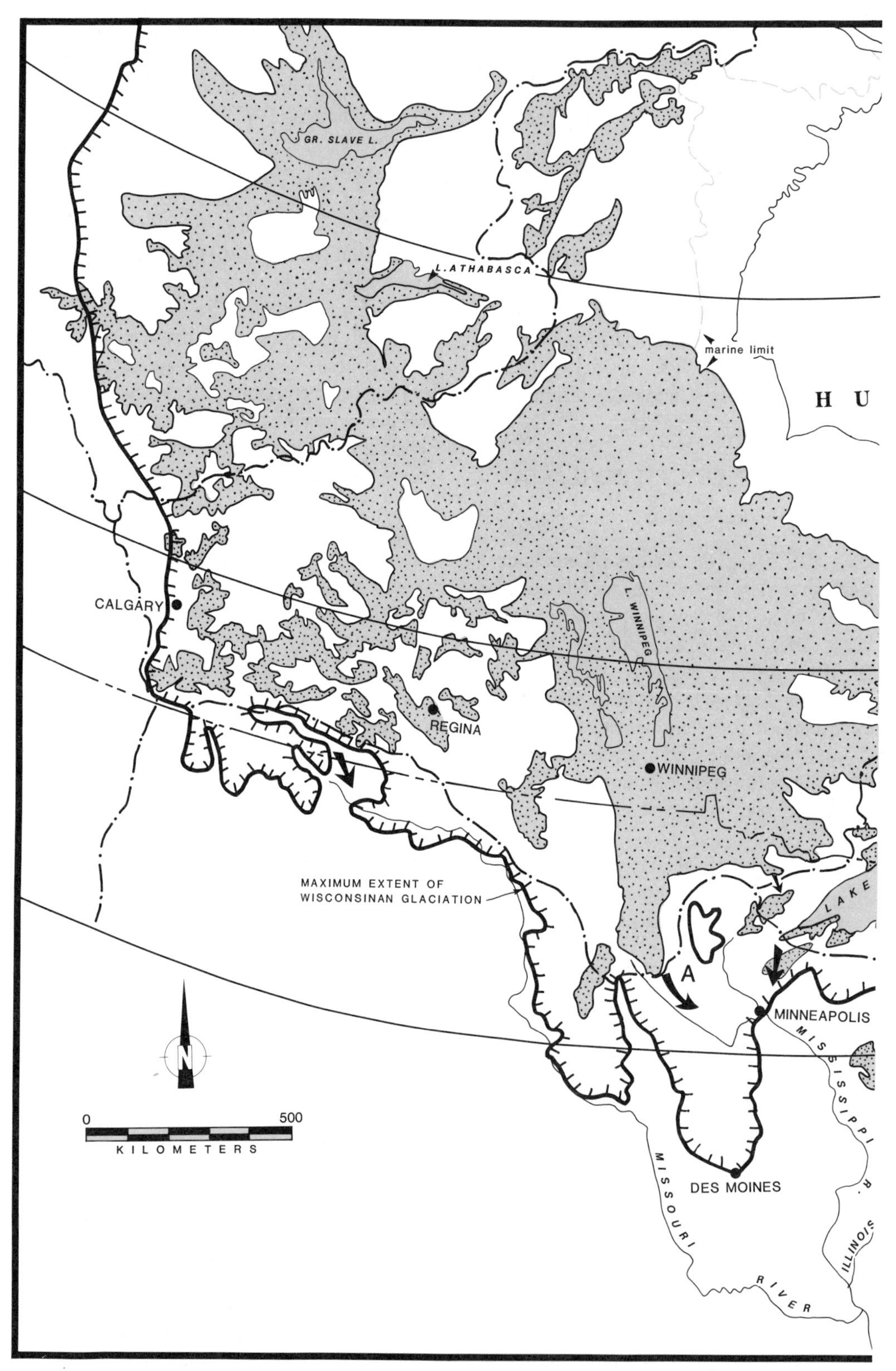

Figure 2 (this and facing page). Area covered by proglacial lacustrine sediment (stippled) deposited during the last retreat of the Laurentide Ice Sheet (after Flint and others, 1959; Prest and others, 1967; Teller and others, 1983). Maximum extent of Wisconsinan ice (hachured line) after Prest (1984). Major avenues of overflow into the Mississippi River and Atlantic Ocean basins shown by arrows. Letters identify names of the major overflow channels used in this paper, as follows: A = Minnesota River

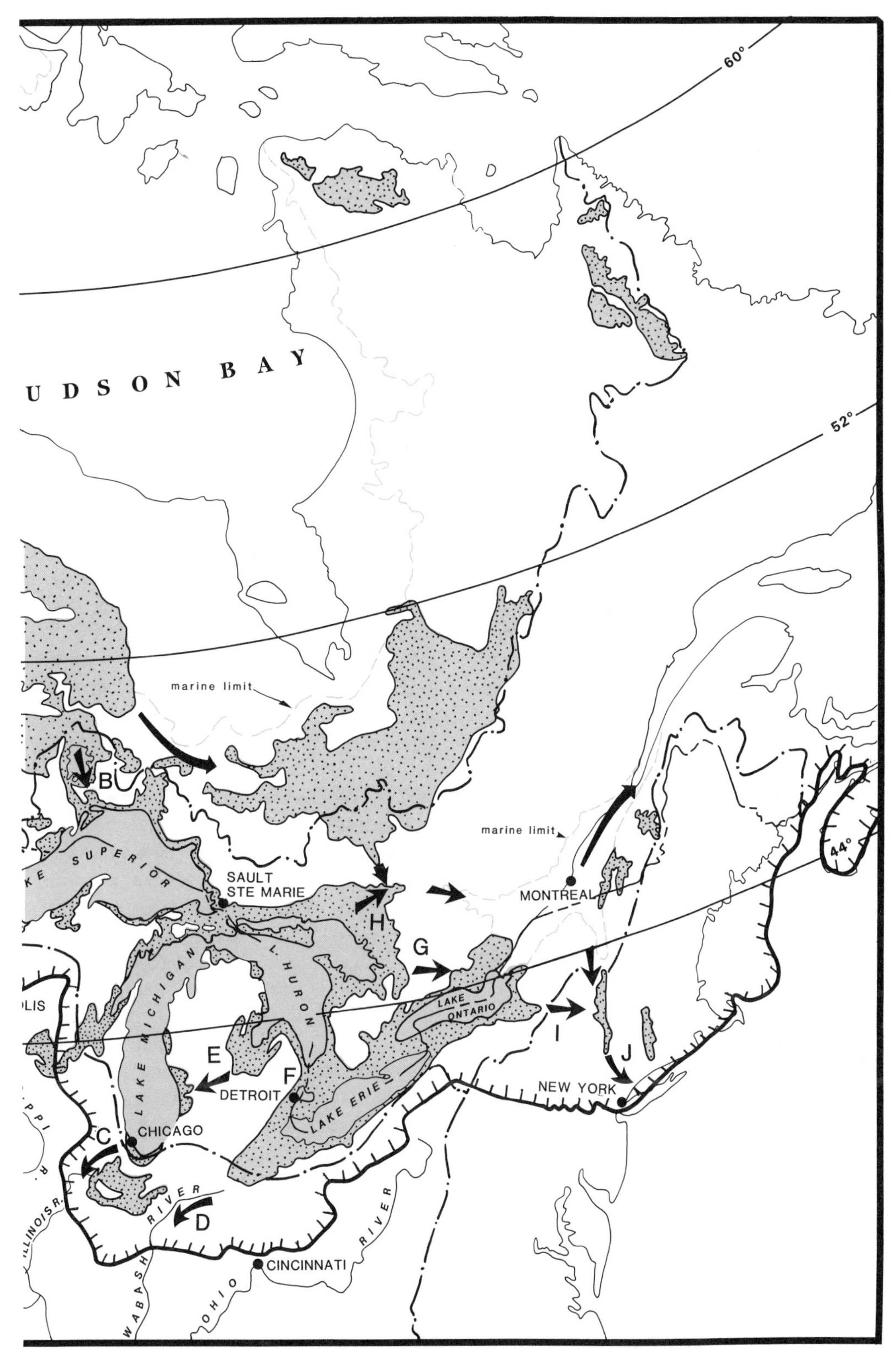

Valley, B = Eastern Agassiz outlets, C = Chicago outlet, D = Wabash River Valley, E = Grand River Valley, F = Port Huron outlet, G = Kirkfield (Fenelon Falls) outlet, H = North Bay outlet, I = Mohawk Valley, J = Hudson Valley. The divides between major watersheds are shown by dash–dot line. Lacustrine sediment is not shown in the St. Lawrence Valley and Hudson Bay Lowland, where it is overlain by marine sediment (outlined by dashed line).

Figure 3. Poorly laminated proglacial sediment in Lake Agassiz that contains a high content of ice-rafted detritus.

surface as an overflow. If the lake is density stratified, the influx of sediment-laden water may sink through the less-dense epilimnion and travel along the top of the hypolimnion as an interflow (Fig. 4). Although these currents gradually slow and mix with the lake waters, allowing much of their entrained sediment load to settle out, wave-generated turbulence may keep the fine particles in suspension for long periods. In some lakes, these fines may remain in the epilimnion until they eventually are carried out of the basin (Sturm and Matter, 1978; Pharo and Carmack, 1979; Smith and others, 1982). Even under the low-shear-stress conditions, such as below a cover of winter ice, clay-sized particles do not easily settle from suspension; the processes responsible for depositing clays in fresh waters during a single season are not fully understood. Mathews (1956) and Sturm and Matter (1978) suggest that seasonal overturn of the water column might bring the clays closer to the lake floor, thus reducing the long settling time. Ferrante and Parker (1977), Smith and Syvitski (1982), and Smith and others (1982) suggest that settling may occur as clay is aggregated into fecal pellets by zooplankton.

When the concentration of incoming sediment exceeds the density of the entire column of lake water, it will remain in contact with the solid sediment surface and move as an underflow plume or density current across the lake floor, depositing sediment as it decays (Fig. 4). These underflows may continue uninterrupted for long periods and are regarded as being responsible for depositing the bulk of sediment in proglacial lakes (e.g., Lajtai, 1967; Banerjee, 1973; Ashley, 1975; Gilbert, 1975; Shaw, 1977; Fenton and others, 1983; Smith and Ashley, 1985). Current-bedded sands and silts commonly develop (Fig. 5) and include climbing ripples that indicate high concentrations of sediment in the flow. These currents are also capable of eroding sediment on the lake floor.

Normally there is a trend from coarse to fine particles away from the point of influx, but the topography of the lake floor tends to control the overall depositional pattern. Fans of sediment may accumulate away from the mouths of rivers. Although these deposits can accumulate as Gilbert-type deltas, with topset, bottomset, and foreset beds, most rivers deposit their load as subaqueous splays with concave-up surfaces and bedding that dips gently toward the center of the basin. In the central part of large lake basins, far from the influx of sediment, accumulation may remain relatively constant over long periods of time, and the resultant sedimentary sequence may be poorly laminated and fine grained.

As with ice-rafted detritus, which increases in concentration in lake muds during the summer period of iceberg calving and winter-ice breakup, sediment transported by density currents tends to be episodically deposited. In some lakes, sediment from these flows is deposited nearly continuously throughout the summer. Although this summer component may be fairly uniform in grain size, detailed sediment studies commonly reveal frequent and substantial variations that reflect fluctuations in the current (Figs. 5 and 6) (e.g., Quigley, 1980; Peach and Perrie, 1975; Lajtai, 1967; Shaw and others, 1978). Studies of modern underflow currents show that velocities fluctuate widely within short time spans (Smith and Ashley, 1985).

Closely related to the periodicity of density flows into a proglacial lake is the problem of identifying varved sediments (Fig. 7). As defined by De Geer (1912), a varve is a couplet comprised of alternatively coarse and fine sediment that was deposited in a quiet body of water during one year. Although used by De Geer in the glacial context, this term has come to represent any annual couplet of lake sediment, whether glacial or nonglacial. Unfortunately, the term varve also has been extended by some to include any couplet in a sedimentary sequence that is rhythmically repeated, whether or not it represents an annual event. The danger in this usage is apparent and significant. Multiple seasonal couplets (see Fig. 6)—deposited, for example, as a result of varying wave or river conditions that produce fluctuating density current underflows—may be misinterpreted and used to count "years" in a sequence. Unless the annual nature of the couplet can be established radiometrically or by firm sedimentological analysis (e.g., by identifying seasonal biota), the term "varve" should be replaced by a term such as "rhythmite."

In spite of these complications, seasonal laminations (varves) are common in proglacial lake sediment, and they have been effectively used in establishing a glacial chronology (e.g., Antevs, 1922, 1951; Ashley, 1975). O'Sullivan (1983) reviews the nature of glacial and nonglacial freshwater varves and discusses their distribution, composition, modes of deposition, and relationship to vegetation. Smith and Ashley (1985) discuss both the processes that lead to rhythmites and to true varves as well as the distinguishing characteristics between the two. Deep lakes are particularly amenable to the formation and preservation of varves. Shallow lakes, or those that have a large fetch in relation to their depth, tend not to have annual couplets preserved because of physical mixing. And, as noted previously, the variable influx

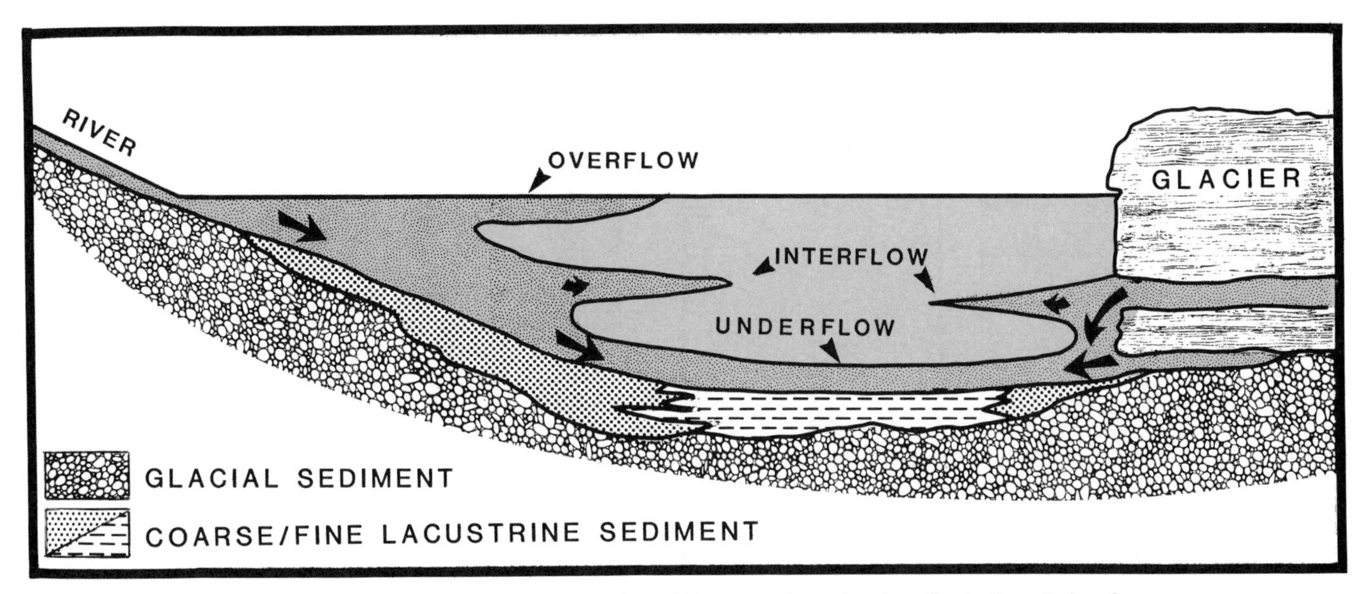

Figure 4. Sediment may enter a proglacial lake and flow across its surface (overflow), along the interface between water strata of two differing densities (interflow), or across the floor of the lake (underflow) (after Gustavson, 1975). Sediment will become progressively finer toward the center of the lake.

of sediment-laden water to a lake may obscure a true seasonal variation in sediment deposition. Thus, sediments in many parts of the Great Lakes, Lake Agassiz, and other large proglacial basins in Canada contain a sequence of poorly or variably laminated clays, silts, and sands, which has little direct relationship to annual events.

The combination of cold, turbid conditions and a high rate of sedimentation in proglacial lakes accounts for the paucity of biological remains in the early record of postglacial lake basins. Even the influx of pollen from outside the basin may be highly diluted by the large influx of clastic sediment, so that pollen concentrations in the sedimentary record, especially in the earliest stages of the lake's history, may be too low to permit standard palynological analyses. Although plant and animal remains are not abundant in most proglacial lakes, organic materials in the more recent, postglacial record in these basins commonly provide the best means for reconstructing the paleoclimatic and paleoenvironmental conditions of the basin and region. In addition to pollen, studies of diatoms, plant macrofossils, Cladocera, chironomids, beetles, ostracods, molluscs, and vertebrates have been useful in interpreting the history of proglacial lake basins. Birks and Birks (1980) provide a good general summary of the paleoecology of these organisms, including a topical bibliography.

Trace fossils such as burrows, tracks, and trails, are occasionally found in proglacial lake sediments (e.g., Gibbard and Stuart, 1974; Gibbard and Dreimanis, 1978). In rhythmically laminated sequences, they are mainly confined to the coarser-grained (silt) laminae and the top of the underlying clay laminae, suggesting that their presence can be used to distinguish annual from nonannual rhythmites (Smith and Ashley, 1985). Ekdale

Figure 5. Three rhythmites of clay (dark) and silty clay (light) in lower half overlain by a graded sand turbidite in the Lake Superior Basin. Note the many fine laminae within the light-colored part of the couplet. Scale in centimeters.

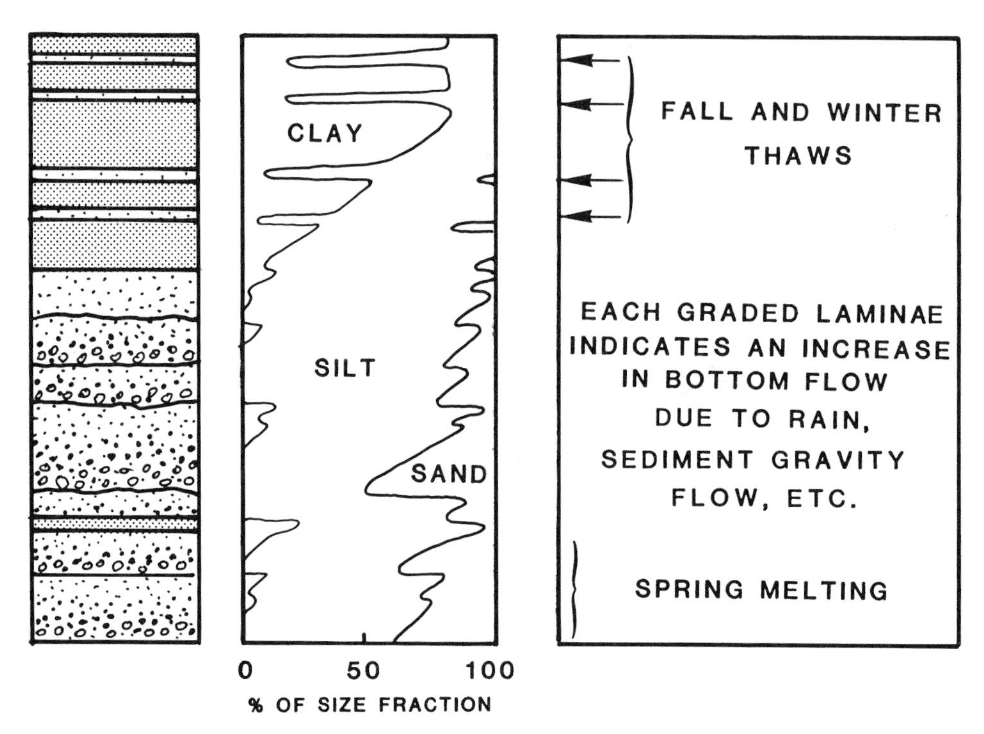

Figure 6. Schematic representation of grain-size variation in a proglacial bed deposited during one year (after Quigley, 1980).

and others (1984) and Chamberlain (1975) discuss the origin and distribution of trace-making organisms in lacustrine environments.

THE LAST DEGLACIATION AND ITS PROGLACIAL LAKES

Introduction

In this section I describe the history of major proglacial lakes along the southern Laurentide Ice Sheet from the northern Prairies of Canada across central and eastern North America to New England. The literature is vast, but it has been summarized in a number of recent publications, which are cited. The history of these lakes and their interrelationship to the ice margin are discussed in three time periods, which seem to reflect three different phases in proglacial lake development: 18 to 15 ka, 15 to 11 ka, and 11 to 7.5 ka. In most areas, little is known about the first period because glacial readvances after 15 ka destroyed much of the record.

The Early Period, 18 to 15 ka

Proglacial lakes developed, and were progressively overridden, as Laurentide ice advanced up the regional slope from central Canada. Once ice reached the Atlantic and Mississippi River drainage basins, all water previously impounded along its margin was completely displaced, and, except for local drainageways ponded by or encircled by ice, meltwater freely drained away from the glacial margin (Fig. 2). The maximum extent of Late Wisconsinan glaciation was reached along most of the southern margin of the Laurentide Ice Sheet by 22 to 17 ka (e.g., Mickelson and others, 1983; Prest, 1984; Clayton and Moran, 1982). In places, such as coastal regions of eastern Canada and the north central states of the U.S., a readvance after 15 ka pushed beyond the previous maximum.

As the Laurentide Ice Sheet retreated northward from its maximum into the Hudson Bay and Great Lakes–St. Lawrence watersheds, a fringe of proglacial lakes developed again. Prest (1970), Clayton and Moran (1982), Mickelson and others (1983), Mayewski and others (1981), and Fulton and others (1987) summarize our knowledge of the fluctuations along the Laurentide ice margin from the Rockies to the Atlantic coast during the early period of deglaciation. In general, ice wasted a short distance back across the divide into the Great Lakes depressions for a few hundred years at about 18, 16, and 15 ka, allowing water to be ponded in the Erie, southern Huron, Michigan, and, probably, the western Ontario basins (Mickelson and others, 1983). It is also possible that ice retreated north into the Agassiz Basin before 15 ka. This suggestion is based on the dramatic change in till lithology in that basin related to a shift in ice-flow direction from the Labradorean center, which lay to the northeast, to the Keewatin center, which lay to the north (Clayton and Moran, 1982; Fenton and others, 1983). The Laurentide margin

Figure 7. The typical coarse-grained (light) and fine-grained (dark) couplets that represent summer and winter variation (respectively) in sediment influx and deposition.

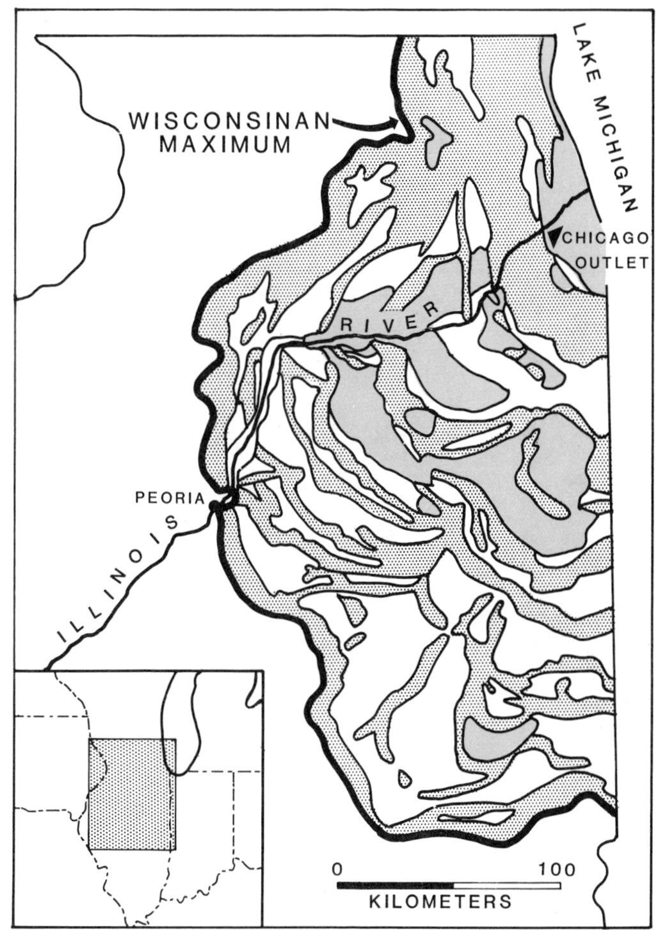

Figure 8. Proglacial lake sediment that was deposited between end moraines (stippled) formed by the retreating Lake Michigan Lobe in northeastern Illinois (Willman and Frye, 1970, Plate 2). Overflow from the Michigan Basin later established the Chicago outlet to the Illinois River Valley.

may also have receded far enough north of its maximum near the Missouri River in North Dakota and Montana to have allowed ponding within the Hudson Bay watershed of western Canada at this time. Subsequent readvances back into the Mississippi drainage basin overrode all proglacial lake deposits related to this phase of deglaciation, and the record of these lakes is poorly known.

Throughout this period, most meltwater and overflow from proglacial lakes west of the Appalachians flowed south to the Mississippi River and to the Gulf of Mexico, while only the eastern regions drained into the Atlantic Ocean.

The Great Lakes and Eastern Regions, 15 to 11 ka

Ice readvanced all along the Laurentide margin between 15 and 14 ka. In the Great Lakes this advance is referred to as the Port Bruce Stade (Dreimanis and Karrow, 1972), and its lobate margin pushed south through the Michigan, Huron, Erie, and Ontario basins, depositing fine-grained till as it overrode proglacial lake deposits associated with the previous Erie Interstade (e.g., see 1:1,000 000 maps of Goebel and others, 1983; Lineback and others, 1983; and Farrand and others, 1984). Mickelson and others (1983) conclude that most clayey till around the southern part of the region covered by Laurentide ice was probably derived from lake sediment.

Retreat from this new terminal position fluctuated, giving rise to lakes along its margin that were partly or entirely overridden within a few hundred years. Sediment was deposited in a number of local basins between the edge of the retreating ice and newly deposited moraines (Fig. 8), as well as in the larger Great Lakes basins. Increasing lobation of the ice margin, probably controlled by regional topography as the ice thinned, helped generate more ice-marginal lakes.

The development of large lakes, both in this region and to the west, probably influenced the dynamics of flow along the margin of the ice sheet. The rapid glacial readvances (surges) that interrupted the life of many proglacial lakes (e.g., Clayton and others, 1985; Fulton and others, 1984; Wright, 1973) probably were promoted by the presence of these lakes and their relatively impermeable sediment. Although the overall trend after 14 ka was a retreating ice margin and a northward expansion of proglacial lakes, this was punctuated by rapid glacial readvances, most of which fell short of the preceding one. Thus, ice-marginal lakes were caught in the battle between warming climate and the destabilizing effect of proglacial lakes on the Laurentide Ice Sheet.

The first major lake to form, as ice of the Port Bruce Stade retreated, was Lake Maumee in the western part of the Lake Erie

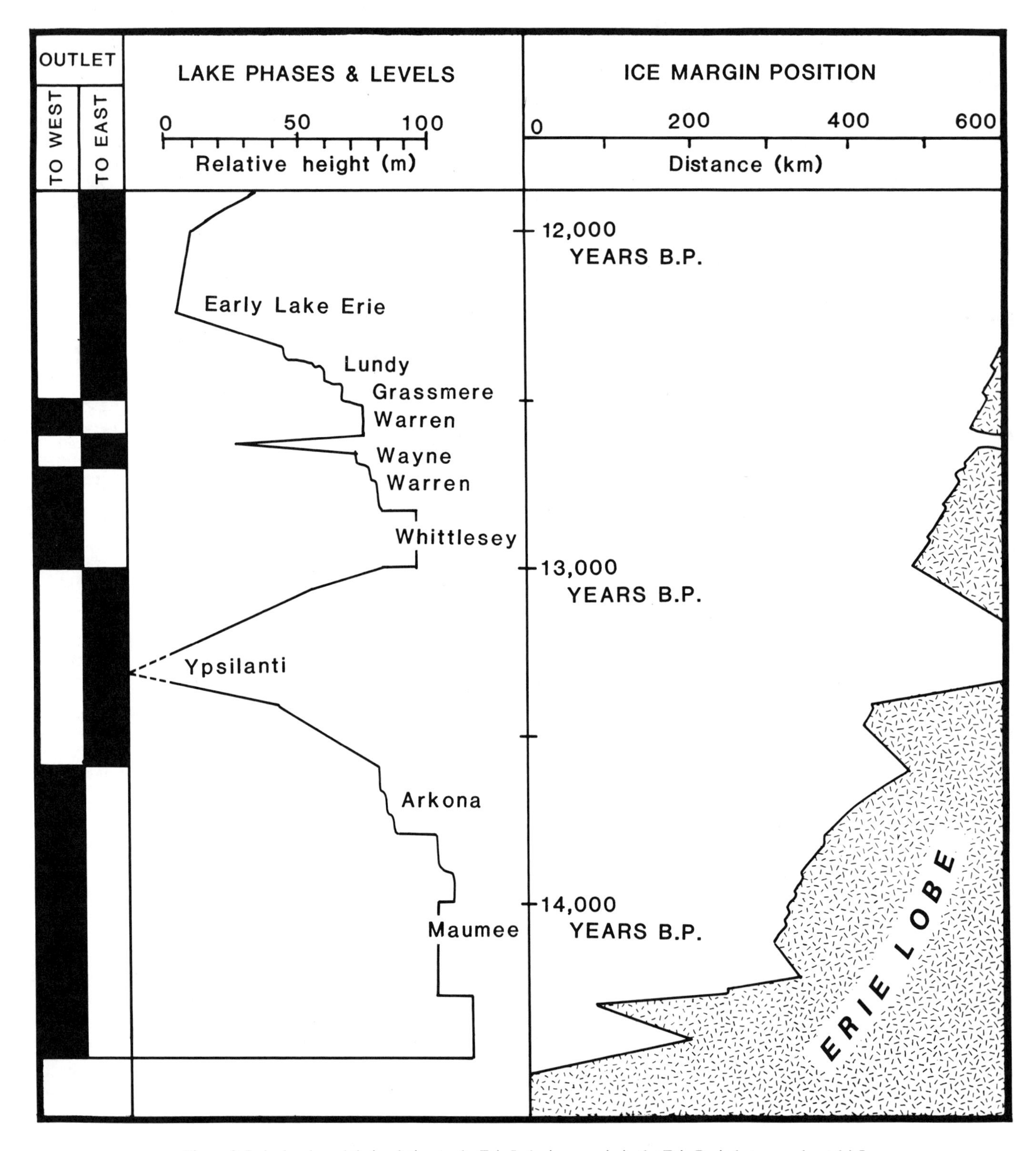

Figure 9. Lake levels and their relation to the Erie Lobe ice margin in the Erie Basin between about 14.5 and 12 ka. Eastward overflow was to the Ontario Basin. Westward overflow was either to the Michigan Basin or through the Wabash River Valley (Calkin and Feenstra, 1985).

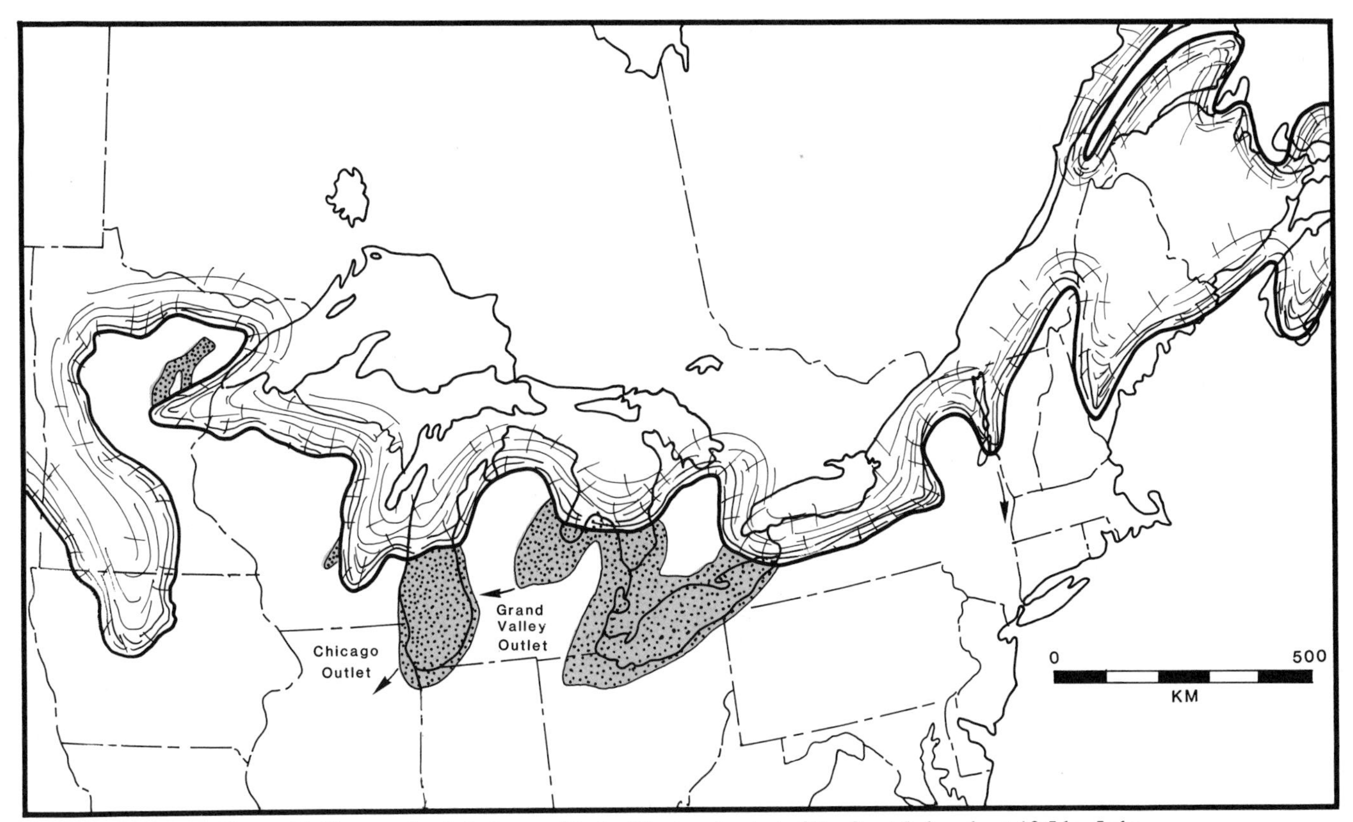

Figure 10. The margin of the Laurentide Ice Sheet and extent of the Great Lakes about 13.5 ka. Lake Arkona in the Erie and Huron basins spilled west into Lake Chicago. All overflow west of New York State was south through the Illinois River Valley via the Chicago outlet to the Mississippi River at this time. The eastern area drained downslope to the Atlantic Ocean (after Eschman and Karrow, 1985; Hansel and others, 1985; Clayton and Moran, 1982; Mayewski and others, 1981, Map 2-2).

Basin. A series of fluctuations in lake level occurred over the next several hundred years, and overflow to the Mississippi River alternated between south, via the Wabash River of Indiana, and west, into the Lake Michigan Basin, which overflowed through the Chicago outlet (Fig. 2) (Karrow, 1984). By about 14 ka, water in the Erie Basin had expanded northward into the central Huron Basin, and Lake Arkona developed as westward overflow to the Michigan Basin through the Grand Valley of central Michigan was established. Calkin and Feenstra (1985) summarize the history of proglacial waters in the Erie Basin (Fig. 9), and Figure 10 shows the outline of water in the Great Lakes at about 13.5 ka.

Concurrent with the development of Lakes Maumee and Arkona (Fig. 9) was proglacial Lake Chicago (Glenwood Phase I) in the Michigan Basin. The northward expansion of this lake, and perhaps the cutting of its overflow channel (the Chicago outlet), may have been expedited by the influx of overflow from proglacial Lake Arkona (Fig. 10).

Shortly after 13.5 ka, ice in the Great Lakes had retreated far enough north to allow overflow into an independent lake in the Ontario Basin, which discharged eastward through the Mohawk and Hudson Valleys to the Atlantic Ocean (Calkin and Feenstra, 1985). Lake levels fell substantially at this time, and ponded water was mainly confined to the deepest part of the Erie Basin (Lake Ypsilanti) during this, the Mackinaw Interstadial period (Fig. 9). To the west in the Michigan Basin, isostatic rebound forced the abandonment of the Chicago outlet, and overflow began to spill out the northern end into the Huron Basin, first through the Indian River Lowland and then through the Straits of Mackinac (Farrand and others, 1969; Hansel and others, 1985), with a resulting decline in the level of the lake. Hansel and others (1985) suggest that ice may have retreated far enough north at this time to allow overflow from the Superior Basin, although only small ice-marginal lakes are known from the southern side of that basin until later (*cf.* Leverett, 1929, p. 55–57; Farrand and Drexler, 1985; Attig and others, 1985).

About 13 ka, ice in the Great Lakes region began to readvance, pushing its highly lobate margin through the Ontario and Huron basins (Calkin and Feenstra, 1985) and into the central Michigan Basin as far south as about Milwaukee (Hansel and others, 1985), initiating the Port Huron Stadial. Water levels rapidly rose in the Great Lakes as overflow east through the

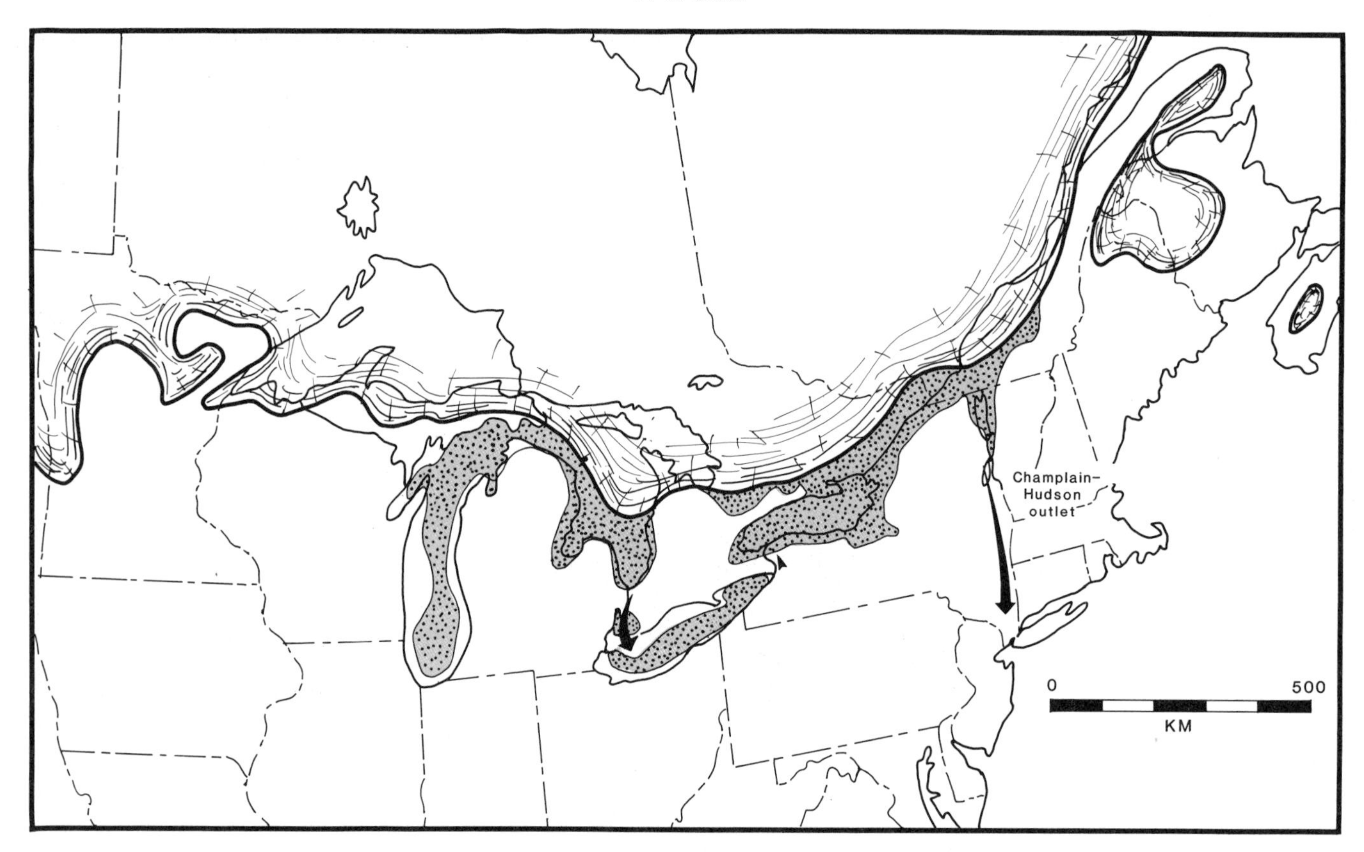

Figure 11. The margin of the Laurentide Ice Sheet and extent of the Great Lakes about 12 ka. Overflow had now shifted eastward, through proglacial waters in the St. Lawrence Valley and into the Champlain–Hudson Valley system on its way to the Atlantic Ocean (after Eschman and Karrow, 1985; Hansel and others, 1985; Prest, 1970; Clayton and Moran, 1982; Mayewski and others, 1981, Map 2-2).

Mohawk and Hudson valleys of New York came to an end and the Chicago outlet was reactivated. Glenwood Phase II of glacial Lake Chicago was initiated. Ice in the Huron Basin pushed far enough south to isolate water along its western side in Saginaw Bay (Lake Saginaw), while completely displacing water in the rest of the basin (Calkin and Feenstra, 1985). Lake Whittlesey formed in the Erie Basin, reaching levels comparable to those during the preceding Maumee and Arkona stages (Fig. 9), and overflowed through Lake Saginaw into the Michigan Basin (Leverett and Taylor, 1915; Eschman and Karrow, 1985).

During retreat from the Port Huron maximum, the Great Lakes again expanded northward into the Michigan and Huron basins. The interplay of differential crustal rebound, deepening of outlets, and the development of new outlets in newly deglaciated areas led to a complex sequence of lake levels, with overflow in the Erie–Huron Basin spilling variably west into the Michigan Basin and east through the Mohawk valley of New York State to the Atlantic Ocean (Fig. 9) (e.g., Calkin and Feenstra, 1985). The relationship of these lakes to each other, to glacial sediments, and to avenues of overflow is shown in a correlation chart by Fullerton (1980) and is discussed in Karrow and Calkin (1985).

Between 13 and 12 ka, water levels in the Huron Basin near the glacial margin remained high, although they did decline (Eschman and Karrow, 1985). Water in the Michigan Basin may have fluctuated substantially during this period as ice repeatedly surged and retreated across the outlet into the Huron Basin (Mickelson and others, 1983; Wright, 1971). Crustal rebound and outlet deepening combined to produce falling lake levels in the Michigan Basin (the Two Creeks low-water stage) and Erie Basin (Early Lake Erie stage) (Fig. 9), and their areal extent constricted substantially from earlier times (Fig. 11). "Early Lake Erie may have initially consisted of an eastward-descending and deepening series of up to three sub-lakes," which were connected by channels cut across the moraine ridges that separated these sub-basins (Calkin and Feenstra, 1985, p. 163). Continued isostatic uplift may have diverted outflow from the Huron Basin away from the Erie Basin and through the more northerly Kirkfield outlet into the Ontario Basin, thus leaving Early Lake Erie in a cul-de-sac for a short time after 11.5 ka (Karrow and others, 1975; Calkin and Feenstra, 1985).

East of the Great Lakes the Laurentide ice margin retreated into central New York state by about 13.5 ka, although it still extended east to the Maine coast at this time (Fig. 10). Narrow proglacial lakes developed in the Finger Lakes region of New York. After the Port Huron readvance to the Valley Heads moraine about 13 ka, these lakes again formed as ice retreated (Calkin and Miller, 1977; Fullerton, 1980). By 12.7 ka, the active margin of the Laurentide Ice Sheet had retreated into the St. Lawrence Lowland, leaving residual ice caps to the south over the northern Appalachian uplands (Prest and Grant, 1969; Borns, 1973, 1985; Gadd, 1976).

Although most drainage was away from the ice in seaward-sloping valleys across Pennsylvania and New England during this time, proglacial lakes did develop against the ice front in many north–south valleys, because their preglacial southward gradient had been reversed by isostatic depression; newly deposited glacial sediment also helped control the level of water in these valleys. In addition to the Finger Lakes and a number of small debris- and ice-dammed valleys of New England (Caldwell and others, 1985; Lowell, 1985), glacial Lake Hitchcock formed in the Connecticut River valley (Antevs, 1922; Ashley, 1975), Lake Albany in the Hudson River valley, and glacial Lake Champlain in the Champlain valley (Connally and Sirkin, 1973).

Lake Albany was the first lake to form in the Hudson valley, and successive lakes developed at lower levels as the glacial margin retreated northward (Fig. 12). Because of differential crustal rebound, waters in the southern (Hudson) part of the valley were no longer confluent with those in the northern (Champlain) part of the valley by the time the ice had retreated to the New York–Quebec border about 12.6 ka (Fig. 12) (Fullerton, 1980).

Overflow from all basins east of the Great Lakes continued to be south across the Appalachians into the Atlantic until water in the St. Lawrence Lowland could drain east around the glacial margin to the North Atlantic Ocean and establish the modern divide (Fig. 2).

In the Ontario Basin, Lake Iroquois received waters from the Great Lakes to the west, overflowing through the Rome, New York, outlet into the Mohawk Valley and, in turn, to the Hudson Valley. By 12.3 ka the Laurentide margin had retreated enough to allow water in the Ontario Basin to spill through a lower outlet into the Champlain Valley, which was still overflowing south into the Hudson Valley (Fig. 11). Glacial Lake Frontenac was established in the Ontario Basin at this time (Muller and Prest, 1985) and was succeeded by a brief period when fresh water in that basin was confluent with fresh water in the St. Lawrence–Champlain Valley (Lake Belleville) (Fig. 11) (Muller and Prest, 1985; Prest, 1970, p. 717, 726; Fullerton, 1980; Clark and Karrow, 1984).

Shortly after 12 ka, the glacial margin had retreated far enough to allow Atlantic Ocean waters to enter the St. Lawrence Valley and establish the Champlain Sea westward past Montreal and Ottawa and south into the Champlain Valley (Muller and Prest, 1985; Fullerton, 1980; Prest, 1970). The level of early Lake Ontario declined substantially following its separation from waters in the St. Lawrence Lowland, reaching its lowest levels by about 11.4 ka before isostatic rebound raised its outlet into the Champlain Sea (Sly and Prior, 1984; Anderson and Lewis, 1985). Overflow from the Huron Basin entered the Ontario Basin through the Kirkfield outlet during part of the early Lake Ontario low-water phase between 12 and 11 ka (Sly and Prior, 1984). In the Erie Basin the lowest level was reached just before 12 ka for Early Lake Erie (Calkin and Feenstra, 1985). The subsequent history of water levels in both of these basins is related almost entirely to differential crustal rebound.

In contrast, Lakes Huron and Michigan, which continued to be under the direct influence of the Laurentide ice margin, rose during another glacial readvance (the Greatlakean Stade) from their 12 ka low stage (Early Lake Algonquin) to the main Lake Algonquin stage before 11.5 ka (Eschman and Karrow, 1985; Fullerton, 1980). By 11 ka—after the Straits of Mackinac had been deglaciated and differential uplift had reestablished drainage east through the Kirkfield (Fenelan Falls) outlet into the Ontario Basin—lake levels in both the Michigan and Huron basins, which had become confluent, began to fall rapidly to new lows (Hansel and others, 1985; Eschman and Karrow, 1985).

The Western Regions and the Initiation of Lake Agassiz, 15 to 11 ka

Ice readvancing from the Keewatin center pushed south out of the Hudson Bay drainage basin into the Mississippi River watershed to a new maximum position across South Dakota and Iowa about 14 ka. To the west, the ice nearly reached its previous maximum along the Missouri River of North Dakota and Montana (Fig. 2) (Clayton and Moran, 1982).

In the southwestern Canadian Prairies, many proglacial lakes formed when this glacial margin began retreating downslope. The sequential development of lakes in western Alberta and Saskatchewan is shown in Figures 13 and 14. The size and areal extent of these lakes changed rapidly. Some were interconnected during catastrophic flooding from glacial lakes upslope in the watershed. Kehew (1982), Kehew and Clayton (1983), and Kehew and Lord (1986) describe the domino effect of flood bursts from glacial Lake Regina on glacial Lakes Souris and Hind (see Fig. 15) and on the construction of an underflow fan at its terminus in Lake Agassiz.

Other proglacial lakes developed in the irregular interlobate area of northern Minnesota, between the Superior Lobe of the Labradorean ice center and the Red River–Des Moines Lobe of the Keewatin center, which advanced from the northeast and northwest, respectively (Figs. 10 and 11). One of these lakes, glacial Lake Grantsburg, developed north of Minneapolis at about 12–13 ka when a sublobe of the Des Moines Lobe pushed east across the Mississippi River (Cooper, 1935; Wright and others, 1973), nearly enclosing the headwaters of this river in northeastern Minnesota (Fig. 16). Other lakes were impounded in this interlobate region at later times, as the St. Louis sublobe disrupted the drainage (Wright, 1972; Hobbs, 1983). In the Superior Basin, a series of glacial surges pushed southwestward,

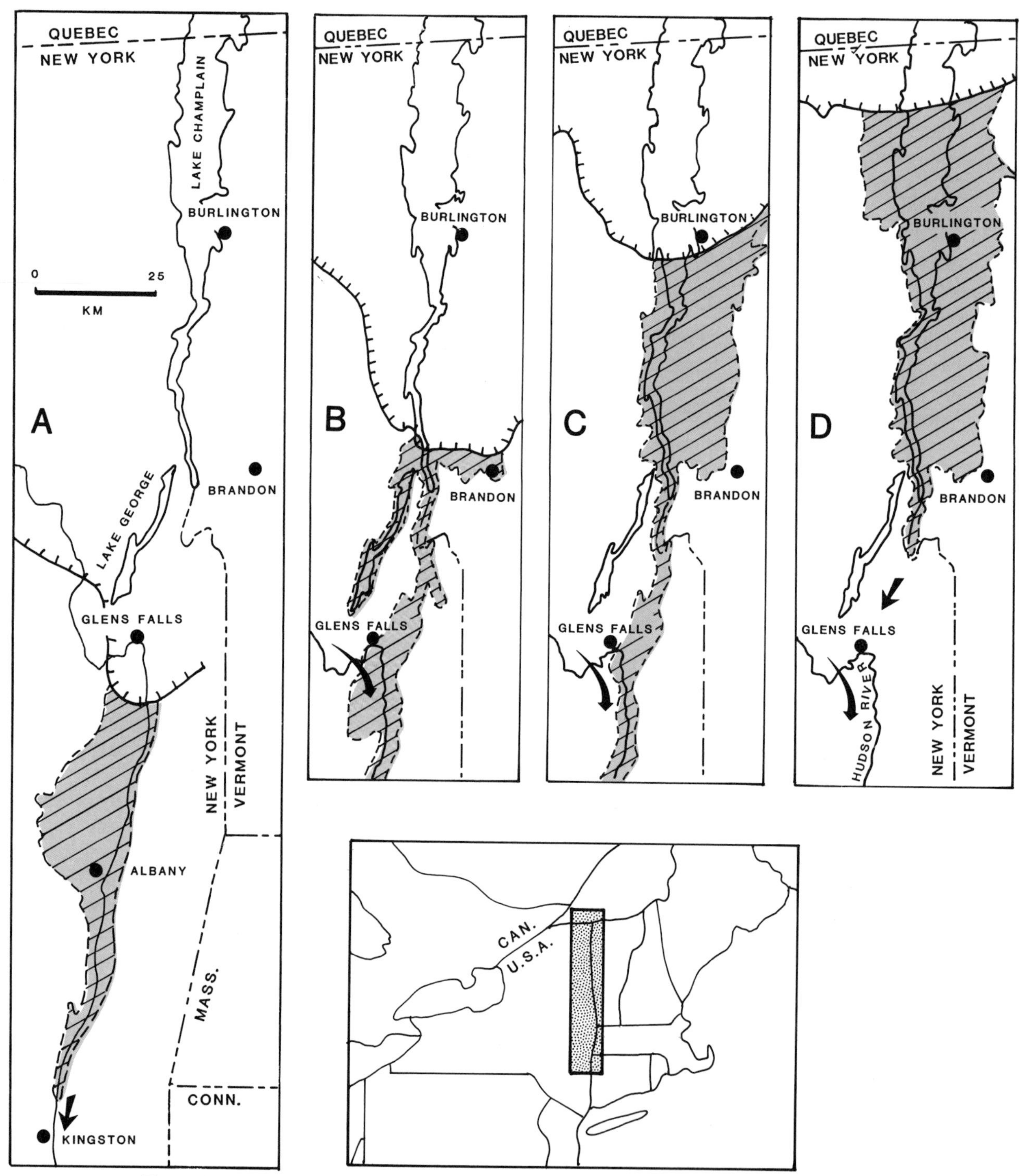

Figure 12. Proglacial lake development in the Champlain and Hudson River valleys between about 13 and 12.6 ka (Connally and Sirkin, 1973). A. Impoundment of glacial Lake Albany in the Hudson Valley during the time when the southward gradient of the valley was reversed in the Glens Falls area by isostatic depression. B. Expansion of a successor of proglacial Lake Albany (Lake Quaker) into the Champlain Valley of the St. Lawrence watershed. C. Further expansion into the Champlain Valley (Lake Coveville). Crustal rebound and erosion continued to reduce the closure on the Hudson Valley south of Glens Falls. D. Continued ice retreat and crustal rebound eventually restricted proglacial waters to the Champlain Valley, forming Lake Fort Ann. Overflow from this lake was south into the Hudson River Valley, where it joined with overflow from the Great Lakes. By about 12.2 ka the ice margin had retreated enough so that Great Lakes' overflow was directed into the north end of glacial Lake Fort Ann, which spilled south to the Hudson Valley.

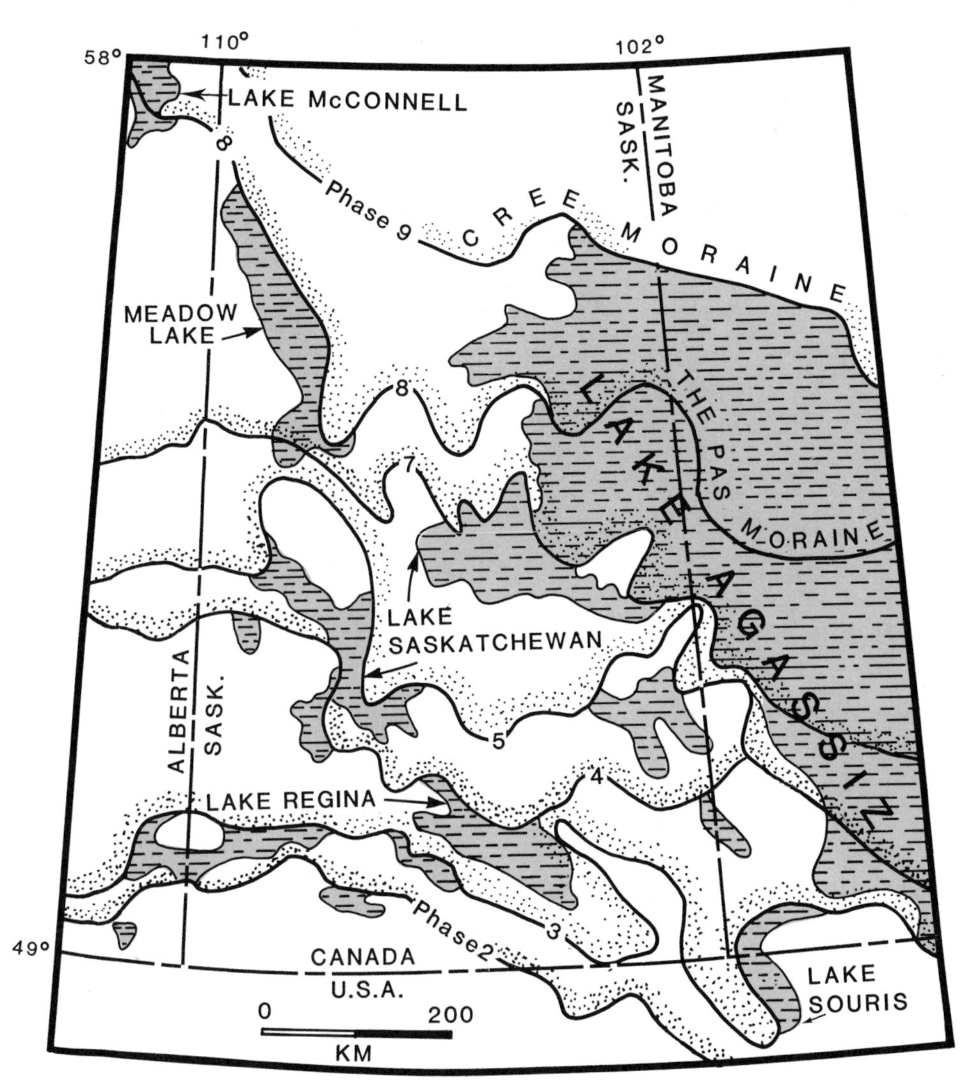

Figure 13. Sequence of proglacial lakes in Saskatchewan and adjacent areas (Quigley, 1980, after Christiansen, 1979). The Phase 2 glacial boundary represents the Late Wisconsin position about 13 ka. By about 11 ka the ice had retreated to the Phase 8 position.

displacing the proglacial lakes that may have encouraged the surging (Wright, 1973).

Across much of North Dakota and Minnesota, the glacial margin still lay south of the Hudson Bay–Mississippi divide at 12 ka (see Fig. 11). An earlier deglacial chronology and associated proglacial lake development advocated by Klassen (1972; 1983a; 1984) and Christiansen (1979) has been rejected because radiocarbon dates on wood indicate that ice was still advancing into South Dakota and Iowa as late as 12.3 ka (Teller and others, 1980; Clayton and Moran, 1982; Fenton and others, 1983).

After retreating several hundred kilometers north of its 14 ka maximum in Iowa, a series of at least eight rapid readvances (surges) and retreats of the Red River–Des Moines Lobe occurred (Fenton and others, 1983; Clayton and others, 1985). During this period each successive glacial advance fell short of its predecessor, and each intervening glacial retreat appears to have receded slightly farther north. Several times before 12 ka the glacial margin wasted north of the divide between Hudson Bay and Mississippi watersheds at the southern end of the Red River Lowland (Fenton and others, 1983). At times, lakes developed in the southern (upslope) end of the lowland for a few hundred years before being overridden by a new surge of ice. Finally, about 11.7 ka, ice wasted north across the divide into the Red River Valley for the last time, establishing Lake Agassiz.

Initially, Lake Agassiz may have overflowed east through glacial Lake Koochiching into glacial Lake Aitkin-Upham, which was located between the Superior and Red River–Des Moines Lobes (Fig. 17). Lake Aitkin-Upham overflowed into the Mississippi watershed, and subsequent drainage across its floor became

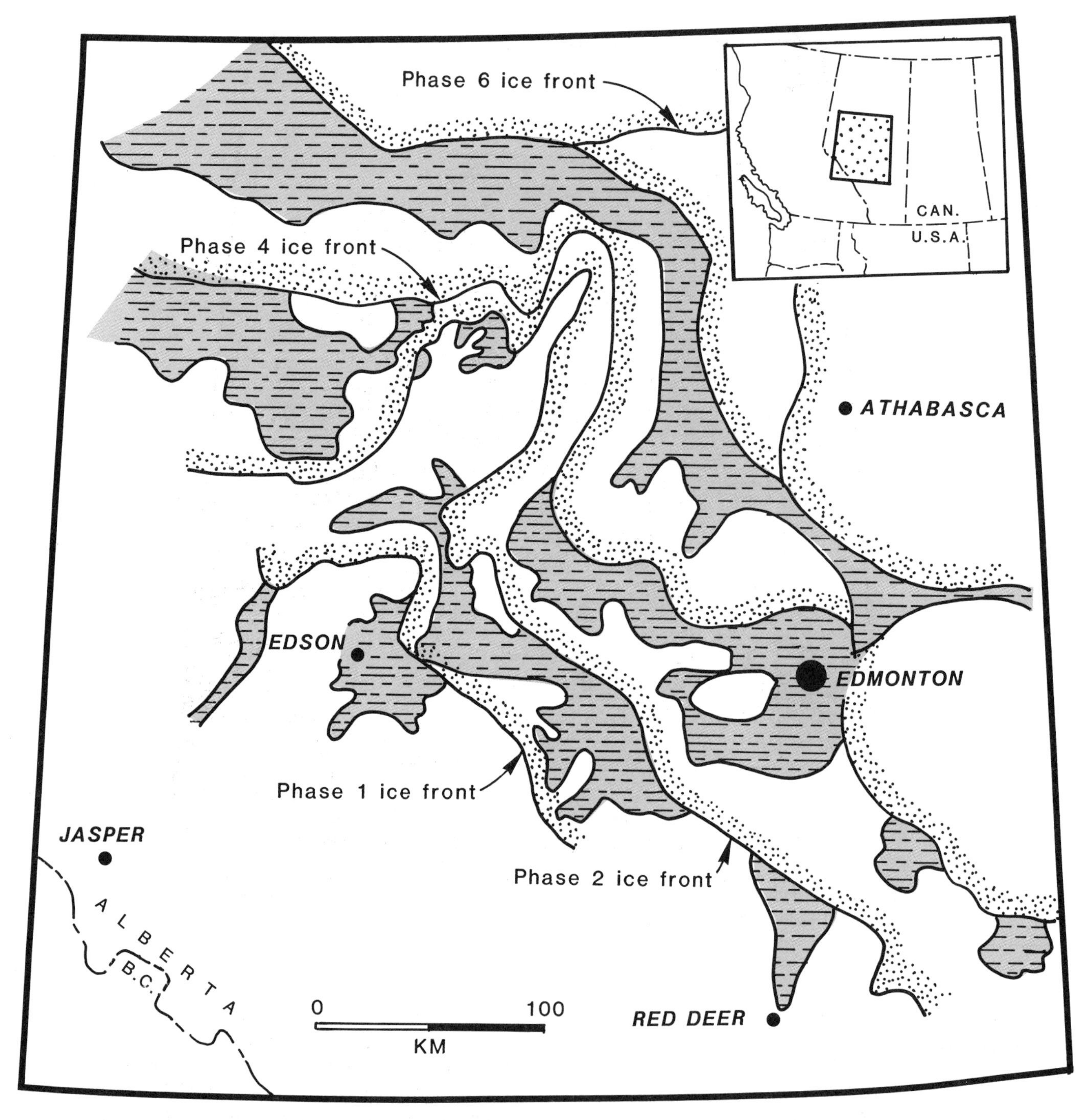

Figure 14. Sequence of proglacial lakes in western Alberta between about 14 and 12 ka (Quigley, 1980, after St. Onge, 1972). Phase 4 in this figure was correlated with Phase 4 in Figure 13 by St. Onge (1980).

integrated into the modern course of the Mississippi River (Hobbs, 1983; Fenton and others, 1983).

Within a few hundred years, overflow from Lake Agassiz was established from the southern end of the Red River Valley through the Minnesota River spillway and into the Mississippi River near Minneapolis (Fig. 15). The margin of the Red River Lobe rapidly wasted northward, with Lake Agassiz following. Thick silty clay containing ice-rafted clasts was deposited in the axis of the valley. Coarser underflow fan sediment was deposited near the mouths of rivers that drained the huge Agassiz wa-

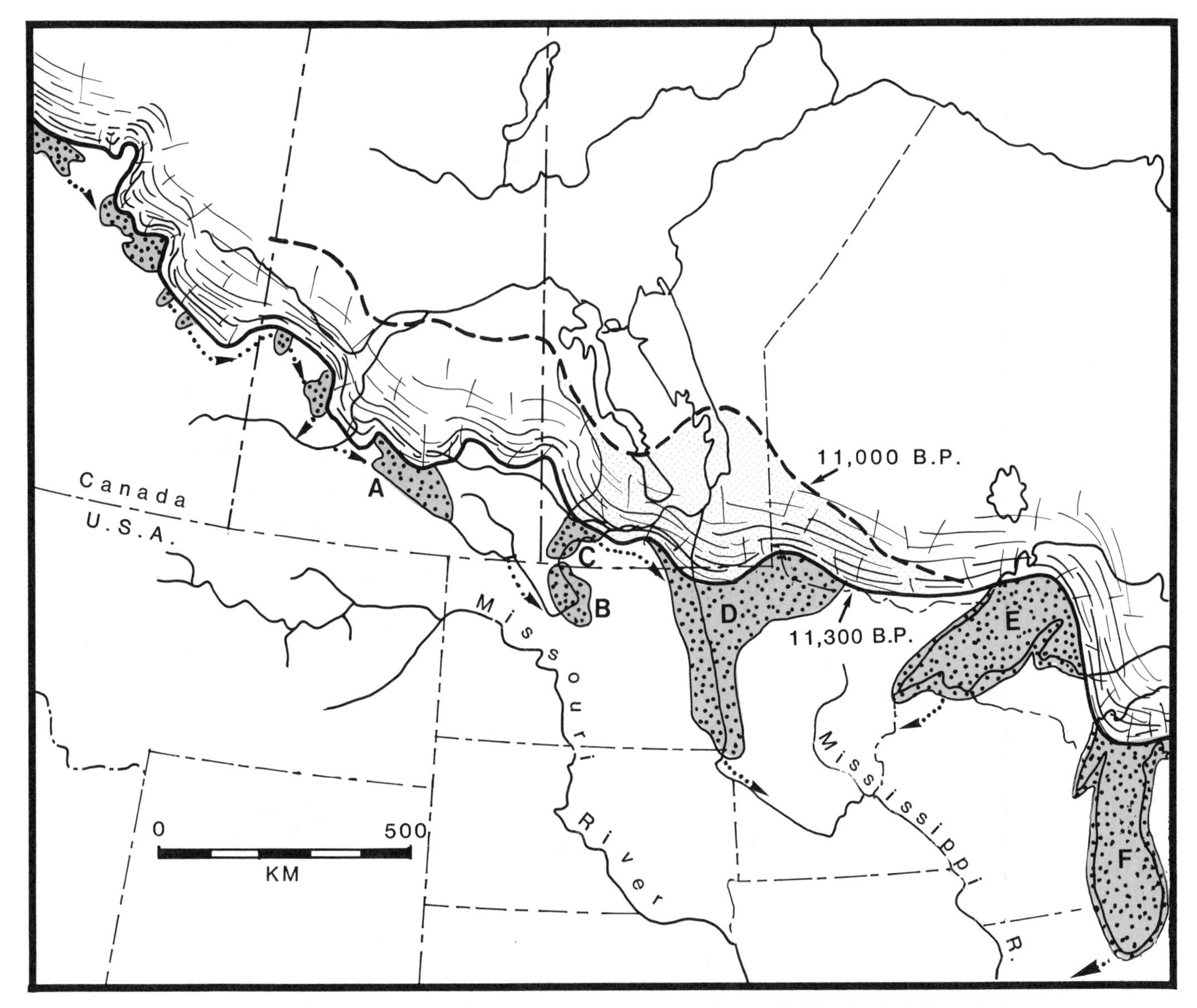

Figure 15. Proglacial lakes along the western Laurentide ice margin about 11.3 ka; the glacial boundary and additional area of Lake Agassiz at 11 ka are shown by heavy dashed line and fine blue stipple, respectively. All lakes in this region overflowed south into the Mississippi River at this time A = Lake Regina, B = Lake Souris, C = Lake Hind, D = Lake Agassiz (Lockhart Phase), E = Lake Superior (Duluth Phase), F = Lake Michigan (Calumet Phase) (after Clayton and Moran, 1982; Christiansen, 1979; St. Onge, 1980; Kehew and Clayton, 1983; Teller and others, 1983).

tershed, which stretched northwest for more than 1200 km to the Rockies. A series of beaches formed as lake levels fluctuated in response to isostatic rebound, catastrophic flood bursts from glacial lakes to the west, erosion of the southern outlet, and new surges of the Red River Lobe. By 11 ka Lake Agassiz had expanded more than 700 km north into central Manitoba, and ice along the eastern margin in Ontario had retreated nearly far enough to allow overflow from Lake Agassiz into the western Lake Superior Basin (Fig. 15) (Teller and others, 1983).

History and Interrelationship of Proglacial Lakes of Eastern and Western North America, 11 to 7.5 ka

Overflow from proglacial Lake Agassiz into the Superior Basin began shortly after 11 ka, initiating a link between west and east and integrating the drainage and lake systems from the Rockies to the Atlantic Ocean—a distance of more than 3000 km (Fig. 18). When this link developed, about 10.8 ka, water in the Superior Basin must have been below the ice-marginal Duluth

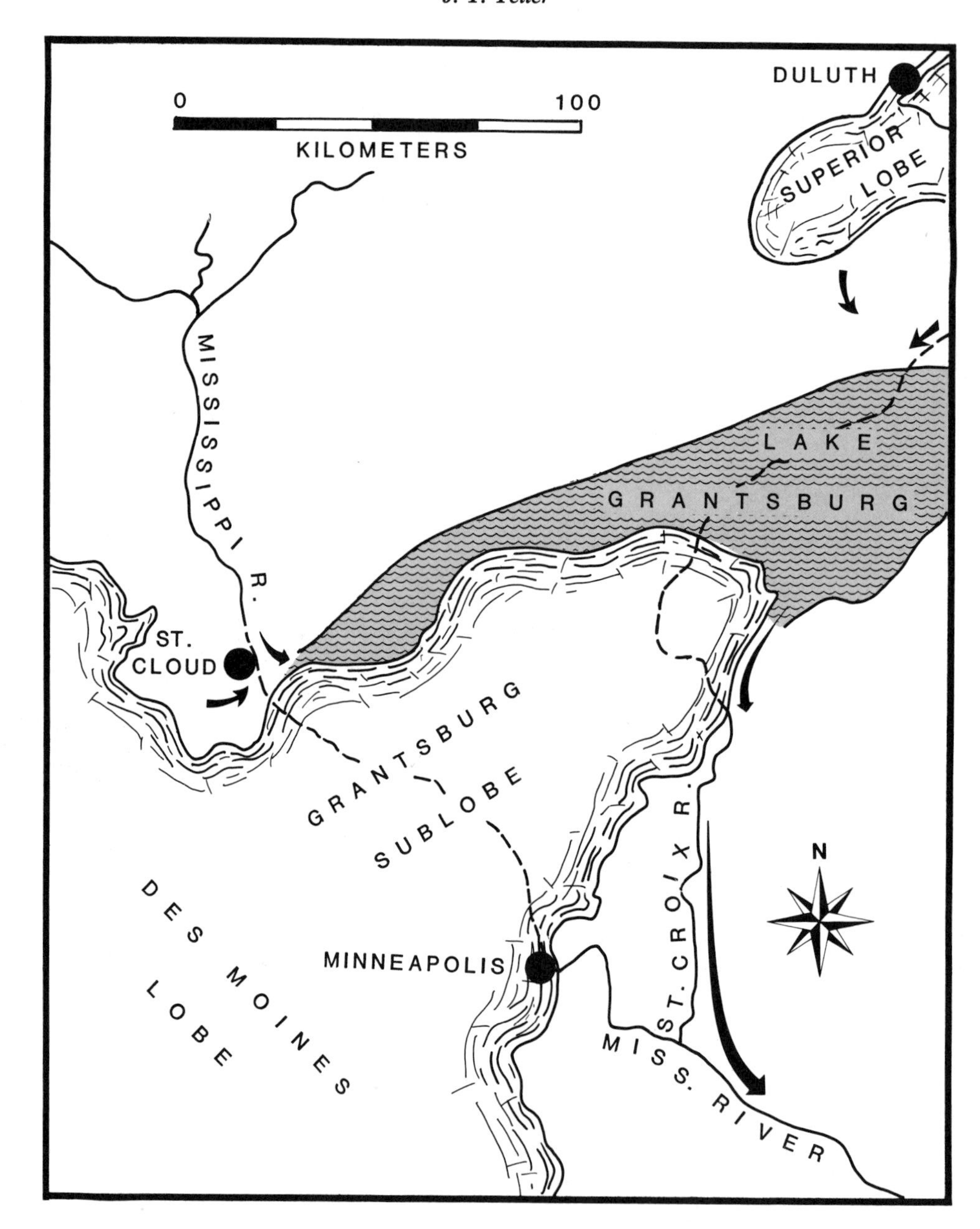

Figure 16. Glacial Lake Grantsburg formed when a sublobe of the Des Moines Lobe dammed the headwaters of the Mississippi River 12–13 ka (modified after Cooper, 1935; Goebel and others, 1983). The absence of red clay in the western Lake Grantsburg sediments suggests that the Superior Lobe did not advance out of the Superior Basin at this time (Wright, 1972).

level (Clayton, 1983) and was confluent with the so-called post-Lake Algonquin stage in the Michigan and Huron basins (e.g., Farrand and Drexler, 1985; Eschman and Karrow, 1985). During the next several hundred years, the water level in these Great Lakes basins was controlled by the Kirkfield (Fenelon Falls) outlet into the Ontario Basin (Eschman and Karrow, 1985) and possibly through the Chicago outlet from Lake Michigan. Hansel and others (1985), however, have presented evidence that the Chicago outlet was not used at this time. "The history of Algonquin and post-Algonquin lake phases is extremely complicated" (Fullerton, 1980, p. 17), as can be seen from the recent summaries on the Michigan and Huron basins in Karrow and Calkin (1985). C. E. Larsen (personal communication, 1987) summarizes and evaluates data on the Algonquin Lake stages using an updated model of glacio-isostatic recovery to reinterpret the complex history of the Great Lakes during this time.

In spite of the complexities and uncertainties of this period in the history of the Great Lakes, the monograph by Leverett and

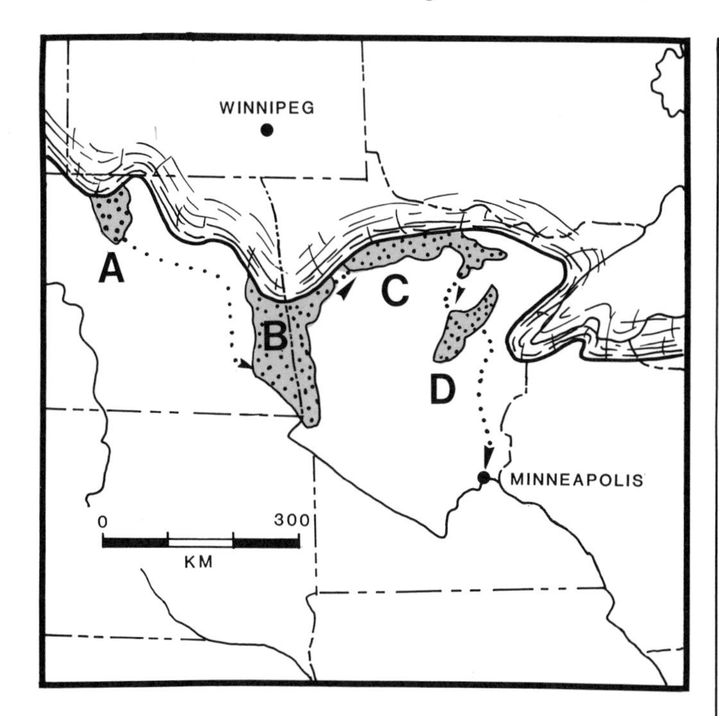

Figure 17. Initial phase of Lake Agassiz (B) about 11.6 ka showing overflow east through Lake Kootchiching (C) and Lake Aitkin-Upham (D) to the Mississippi River near Minneapolis. Lake Souris (A) received runoff from much of the newly deglaciated region to the west and overflowed into the southern end of Lake Agassiz at this time (after Fenton and others, 1983; Clayton, 1983). Alternatively, water may have initially overflowed south from Lake Agassiz through the Minnesota River Valley to Minneapolis as it did during much of the early history of the lake (see Fig. 15).

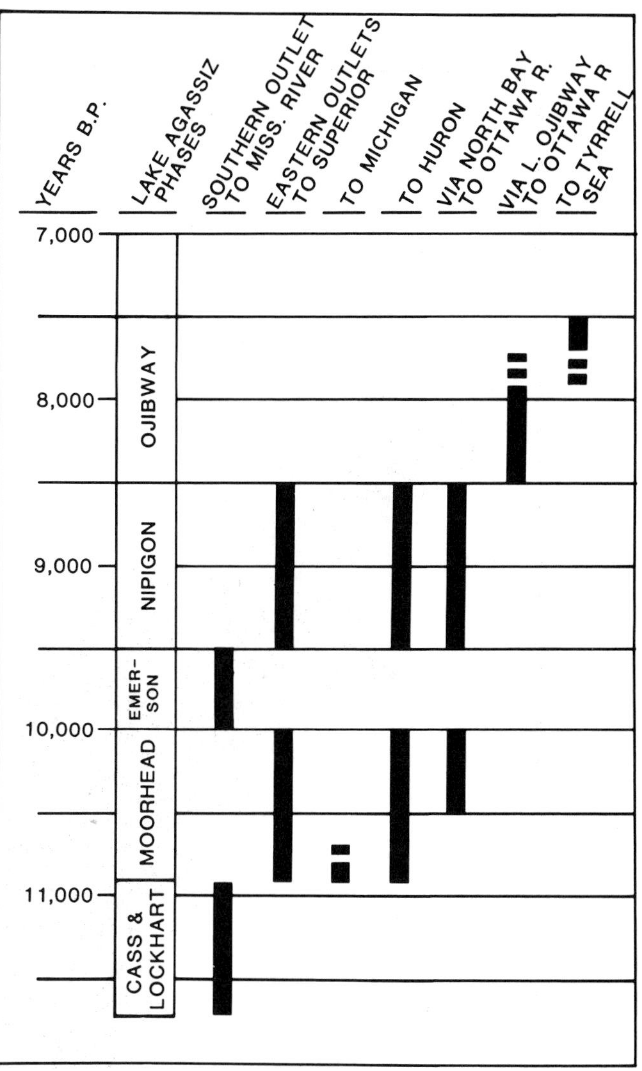

Figure 18. Routing of overflow from glacial Lake Agassiz (after Teller, 1985).

Taylor (1915) is still a useful and comprehensive reference on the Algonquin and post-Algonquin history of the Great Lakes; Larsen (personal communication, 1987) and Fullerton (1980, p. 17) cite a number of other useful references. Most authors agree that, under the combined influences of ice retreat and differential crustal rebound on outlet development, lake levels fell between 11 and 10 ka in the Superior, Huron, and Michigan basins, as well as in the Lake Agassiz Basin. However, as Teller and Thorleifson (1983) show, periodic catastrophic bursts of overflow from Lake Agassiz into the Superior Basin were superposed on a baseline overflow that is estimated by Teller (1987) to have been about 33,000 m^3s^{-1} at this time, and only slightly less between 9.5 and 8.5 ka. Farrand and Drexler (1985) estimate that a large Agassiz flood could have rapidly and briefly raised the level of water in the Superior, Huron, and Michigan basins by 20 m during the Algonquin stage, although field evidence for this is lacking. Figure 19 shows the substantially reduced size of Lake Agassiz during the Moorhead low-water phase that was initiated when the lower eastern outlets were opened between about 10.8 and 10 ka.

A major readvance of ice across the Superior Basin about 10 ka (the Marquette advance) isolated small lakes in the western part of that basin at elevations hundreds of meters above post–Lake Algonquin levels and dammed the eastern Lake Agassiz outlets (Fig. 19). Overflow from Lake Duluth through the Brule and Portage outlets at the western end of the Superior Basin was south into the Mississippi system (Clayton, 1983), whereas that from Lake Kam along the northwestern side rose and spilled westward into Lake Agassiz (Fig. 19) (Zoltai, 1961, 1963; Teller and Thorleifson, 1983).

Lake Agassiz expanded at this time from its Moorhead Phase low to its maximum size, 350,000 km^2, becoming the largest Pleistocene lake ever in North America (Fig. 19). Water in the Agassiz Basin rose at least to the Campbell level, as the southern outlet through the Minnesota River Valley was reactivated. Because 24 red varves are typically found in the eastern region of the Lake Agassiz Basin, overflow of Lake Kam, from

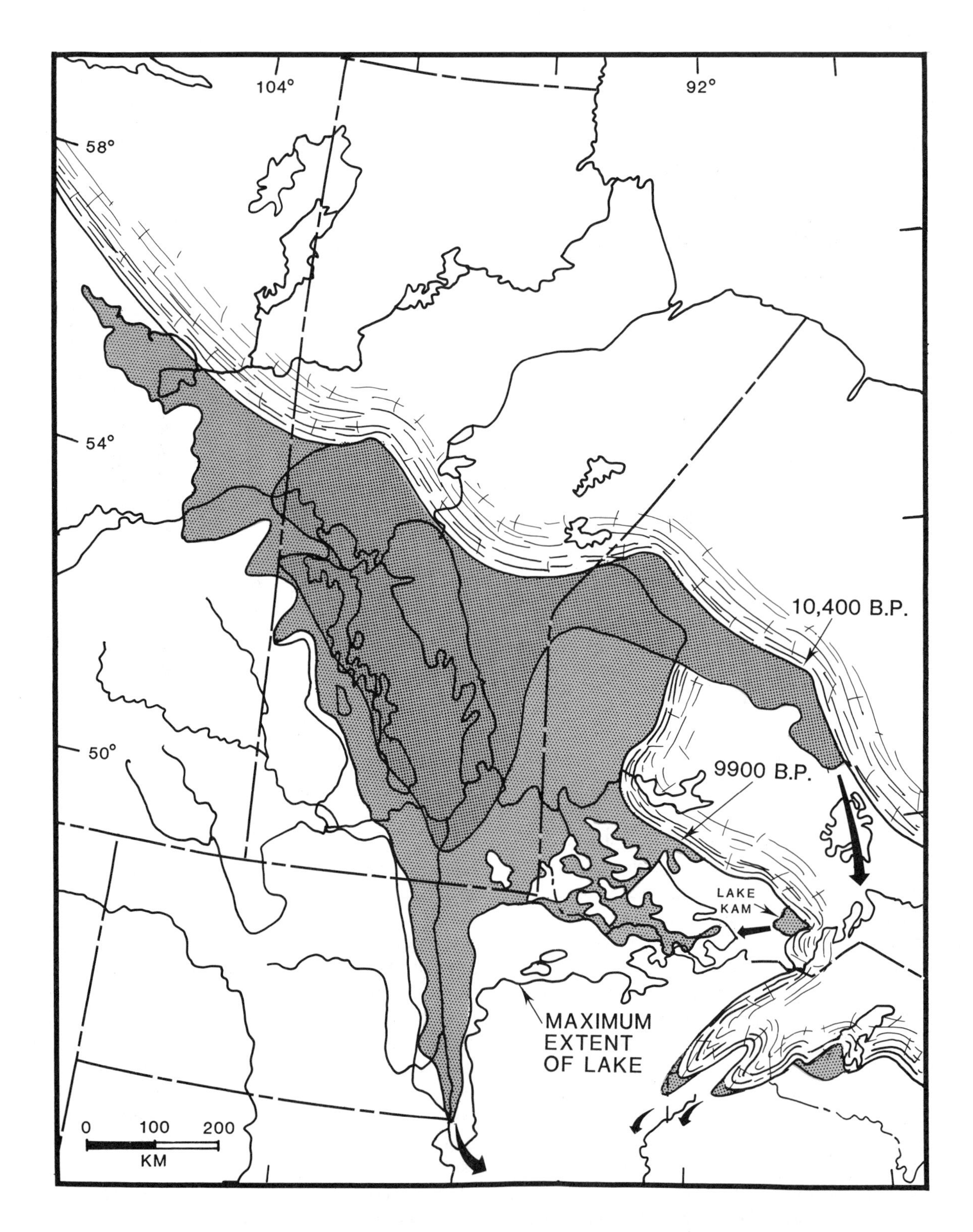

Figure 19. Lake Agassiz during the low-water Moorhead Phase (heavy stippled area), about 10.4 ka when overflow was east into the Superior Basin, and during the Emerson Phase, 9.9 ka, after the Marquette ice advance filled the Superior Basin and blocked the eastern outlets of Lake Agassiz. Note the small proglacial lakes along the 9.9 ka margin in the Superior Basin. Arrows show the main overflow routes at these two times (after Teller, 1985; Teller and others, 1983; Attig and others, 1985).

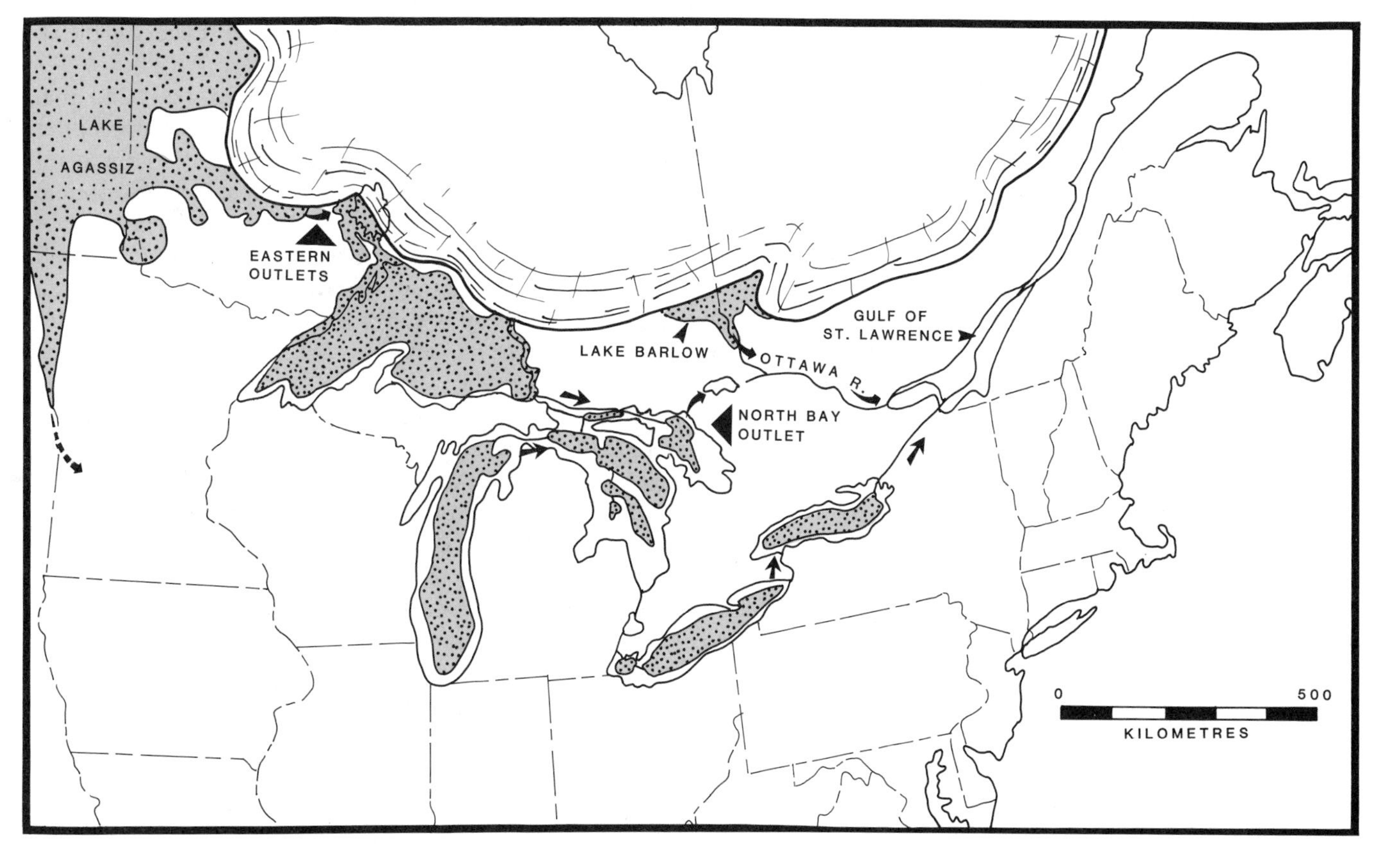

Figure 20. The Great Lakes about 9.5 ka. Ice of the Marquette advance had just retreated from the Superior Basin, reopening the eastern outlets of Lake Agassiz and allowing the 2,000,000 km^2 western drainage basin to be integrated with the Great Lakes system. Water levels in the Michigan and Huron basins were near their lowest levels ever and overflowed through the North Bay outlet to the St. Lawrence Lowland via the Ottawa River Valley. Lakes Erie and Ontario overflowed northeastward through the St. Lawrence River (after Eschman and Karrow, 1985; Vincent and Hardy, 1979; Prest, 1970; Elson, 1967).

the red-clay Superior Basin into the gray-clay Agassiz Basin, is known to have lasted at least 24 years. Levels in Agassiz remained high for at least another century after the influx of red clay ceased, and 100 gray varves were deposited over the red ones (Teller and Thorleifson, 1983; Antevs, 1951).

At about the same time as the Marquette readvance in the Superior Basin, glacial retreat along the northern side of the Huron Basin opened the lower North Bay outlet, which allowed overflow through the Ottawa River Valley to the St. Lawrence Lowland (see Fig. 20). After this the glacial margin itself no longer played a role in determining the level of the Great Lakes east of the Superior Basin.

Lake levels in the Michigan and Huron basins fell substantially after the North Bay outlet was opened, initiating the Chippewa and Stanley low phases, respectively (Hough, 1955, 1958, 1962; Hansel and others, 1985; Eschman and Karrow, 1985). Estimates are that water in the Michigan and Huron basins fell to levels that were more than 60 m, and perhaps more than 100 m, below the present lake surface (Hough, 1955, 1962; Hansel and others, 1985). Waters from the Michigan and nonglaciated eastern Superior basins overflowed into Lake Stanley in the Huron Basin, which in turn drained north of Manitoulin Island into Lake Hough, an isolated lake in the Georgian Bay Basin. In turn, Lake Hough overflowed northeastward through the North Bay outlet (Lewis and Anderson, 1985; Harrison, 1972).

After retreating north of the eastern Great Lakes basins, a reentrant developed in the glacial margin near the Ontario–Quebec border, as the Laurentide Ice Sheet progressively separated into the Labrador and Hudson ice domes (Hardy, 1977). Initially, glacial Lake Barlow formed in this re-rentrant, draining south through the Ottawa River valley to the St. Lawrence Lowland (Fig. 20). This lake grew northward and westward, eventually becoming glacial Lake Ojibway (Vincent and Hardy, 1979), which received overflow from Lake Agassiz after about 8.5 ka.

By 9.5 ka overflow into the Superior Basin through the eastern outlets of Lake Agassiz was resumed (Fig. 20) (Zoltai, 1965; Mahnic and Teller, 1985). This influx of water may have expedited the wastage of ice from the Superior Basin, which only

500 years earlier had been completely filled by the Superior Lobe (Fig. 19). For the next 1000 years, the entire proglacial drainage system of western North America, from the Rockies to the Huron Basin, overflowed through the North Bay outlet to the Gulf of St. Lawrence and to the North Atlantic Ocean (Figs. 18 and 20). Baseline overflow from Lake Agassiz between 9.5 and 8.5 ka is estimated on the basis of the rate of ice margin retreat in western Canada, plus precipitation, to have been about 27,000 m^3s^{-1} or 850 km^3 yr^{-1} (Teller, 1987).

Superposed on this was a series of catastrophic lake bursts that occurred as ice of the Marquette advance, which had dramatically raised water levels in the Agassiz Basin for several hundred years, repeatedly failed along the continental divide, allowing enormous volumes of water to flood through the eastern outlets of Lake Agassiz into the Superior Basin. As ice wasted north along the continental divide, which sloped northward as it does today, progressively lower gaps were uncovered, and each successive catastrophic flood rapidly lowered Lake Agassiz to a new, temporary equilibrium level. As described by Teller and Thorleifson (1983), a series of beaches, linked to 17 flood channels containing boulders 1 to 5 m in diameter, has allowed estimates of the rate and volume of discharge of these floods. At times, more than 3000 km^3 of water passed from Lake Agassiz into the Great Lakes in only about a year, and discharge rates commonly exceeded 100,000 m^3 s^{-1} (Teller and Thorleifson, 1983). These floods delivered gray clay to the Superior Basin, replacing the typical red clays deposited prior to this influx (Farrand and others, 1969; Farrand and Drexler, 1985; Teller, 1985).

Farrand and Drexler (1985) conclude that the St. Marys outlet that linked the Superior and Huron basins was constricted by drift across Whitefish Bay until sometime after the Marquette advance, holding water in the Superior Basin at the Minong level (Fig. 20). Depending on the cross-sectional area of this outlet, some flood bursts from Lake Agassiz may have raised the water level in the Superior Basin by as much as 50 m in only a few months (Teller, 1985; Farrand and Drexler, 1985). These floods were responsible for erosion of the drift barrier at Whitefish Bay and the resulting decline in water in the Superior Basin from the Minong to the Houghton level in post-Marquette time (Farrand and Drexler, 1985). The Houghton low level came into existence well after the lowest Chippewa and Stanley levels in the Michigan and Huron basins, which, in fact, had already begun to rise by this time because of crustal rebound of the North Bay outlet (Farrand and Drexler, 1985).

Interpretation of the sedimentary record and the beach and outlet chronology of the Huron and Michigan basins indicates that water levels in these basins remained low for several thousand years after the North Bay outlet began functioning (Eschman and Karrow, 1985; Hansel and others, 1985). During this period, between 9.5 and 8.5 ka, overflow from Lake Agassiz and its 2,000,000-km^2 watershed was into the Superior Basin. Teller (1985) concluded that Agassiz–Superior overflow must have bypassed the rest of the Great Lakes at this time, spilling into North Channel (north of Manitoulin Island), rather than into the main Huron Basin, on its route east to the North Bay outlet and the St. Lawrence Valley (Fig. 20).

By about 8.5 ka, Lake Agassiz had fallen nearly 200 m in the southern regions. As ice continued to retreat downslope, a lower avenue of overflow developed along the ice margin north of Lake Superior, and the eastern outlets from Lake Agassiz into the Superior Basin were abandoned. This ended the influence of Lake Agassiz and the western watershed on the Great Lakes, with all further overflow passing along the glacial margin through glacial Lake Ojibway in northern Ontario to the Ottawa River valley and into the St. Lawrence valley (Teller and Thorleifson, 1983).

Progressive differential uplift of the North Bay region elevated the outlet there, submerging broad areas that had been dry, vegetated land during the Stanley and Chippewa low stages in the Huron and Michigan basins (Lewis, 1969). By about 8 ka, water in the Huron and Michigan basins had become confluent, and by about 5.5 ka, water in the Superior Basin had risen to the same level, forming the Nipissing Great Lakes. For the next thousand years, overflow was concurrently through the Chicago outlet into the Mississippi, through the Port Huron outlet into Lake Erie, and through the North Bay outlet into the St. Lawrence Valley (e.g., Lewis, 1969; Eschman and Karrow, 1985; Hansel and others, 1985; Farrand and Drexler, 1985). By about 4.5 ka, the North Bay outlet had rebounded to elevations that were too high to carry overflow, and within another few hundred years the Port Huron outlet had cut deeply enough to force the abandonment of the higher Chicago outlet (Hansel and others, 1985; Lewis, 1969; Eschman and Karrow, 1985).

Large proglacial lakes continued to lie along the southern Laurentide glacial margin from the Northwest Territories of Canada, across northern Alberta, Saskatchewan, Manitoba, and Ontario to Quebec, until the Hudson Bay Lowland was deglaciated enough to allow outflow through the Tyrrell Sea to the North Atlantic Ocean.

In the far western region, isostatic depression reversed the drainage in part of the headwaters of the Mackenzie River system, ponding water all along the Laurentide margin. A chain of interconnected lakes, named Glacial Lake McConnell, stretched for more than 1100 km, from north of the Arctic Circle to the Peace River of Alberta, and overflowed northwestward through the Mackenzie River to the Arctic Ocean (Craig, 1965; St. Onge and Dredge, 1985). Today, Great Bear Lake, Great Slave Lake, and Lake Athabasca, whose basins were produced by differential glacial erosion of the softer Phanerozoic rocks at the edge of the Precambrian Shield, occupy the largest remaining closed basins of Glacial Lake McConnell.

In central Canada, Lake Agassiz, which after about 8.5 ka overflowed across northern Ontario into glacial Lake Ojibway, continued to occupy the lowlands southwest of the ice margin in Hudson Bay, into which nearly 2,000,000 km^2 of western North America drained. Vincent and Hardy (1979) describe the complex history of Lakes Barlow and Ojibway, which developed adjacent to the retreating ice margin north of the Great Lakes after about 10 ka, and which overflowed south through the Ottawa River valley to the St. Lawrence. Antevs (1925) measured more than 2000 varves in the northern areas of this lake system. All of these proglacial lakes responded to the retreating ice margin and isostatic rebound by shifting northward (downslope) through time (Prest, 1970; Vincent and Hardy, 1979; Klassen, 1983b; Dredge, 1983), although this retreat was periodically interrupted by glacial surges into the lakes (Hardy, 1977; Dyke and others, 1982; Dredge, 1983). The Cochrane readvances in the James Bay Lowland at 8.2–8.0 ka are the best known of the surges (e.g., Prest, 1970).

Thinning and stagnation may have allowed the initial drainage of Lake Agassiz into the Tyrrell Sea of the Hudson Bay Lowland through subglacial channels (Klassen, 1983b). Until the drainage of this lake into the Tyrrell Sea at 8–7.5 ka, water depths along the ice margin commonly exceeded 100 m (Klassen, 1983b; Vincent and Hardy, 1979).

As a result of crustal depression, marine sediments overlie the freshwater proglacial lake sediments of Lake Agassiz and Ojibway throughout the Hudson Bay Lowland (Fig. 2). Isostatic rebound over the past 7500 years has now raised the oldest of these marine deposits more than 150 m above sealevel in places (e.g., Hillaire-Marcel and Fairbridge, 1978; Klassen, 1983b) and has displaced the modern marine shoreline 100–400 km to the north.

INTERRELATIONSHIP BETWEEN PROGLACIAL LAKES AND THE LAURENTIDE ICE SHEET AND THE POSSIBLE INFLUENCE OF MELTWATER ON THE OCEANS

Introduction

The major controls on most proglacial lakes in North America were the location and configuration of the margin of the Laurentide Ice Sheet and the effect of this ice on isostasy and outlet location. Meltwater erosion and deposition played an important role in modifying the size, depth, and distribution of the proglacial waters.

Climate was the driving force behind the accumulation and demise of the Laurentide ice. In turn, the presence of large areas of ice on the continent influenced the earth's crust, oceans, and atmosphere, and a complicated process of interaction developed between them and the ice sheets (see Ruddiman and Wright, this volume), which in turn helped control and nourish the marginal lakes. Many have discussed the various interactive components in this global feedback process (e.g., Saltzman, 1985; Wright, 1984; Ruddiman and McIntyre, 1981), but quantification and the actual timing of feedback between the variables has so far eluded resolution.

The response time of an ice sheet to climatic change is large, usually being measured in thousands of years. An increasing amount of chronostratigraphic evidence is accumulating, however, to indicate that most ice-marginal fluctuations during deglaciation were nonclimatic in origin. Andrews (1982) notes that with some numerical models of the late-glacial history of the Laurentide Ice Sheet the ice cannot be forced to maintain an equilibrium profile, and the southern margin keeps surging. The vast area of proglacial lakes that developed along the ice margin, both during the initial glacial invasion and upon retreat in the late stages, probably influenced the dynamics of glacier flow. These lakes may even have contributed to the growth of the Laurentide Ice Sheet by providing local sources of atmospheric moisture, and to its final demise by accelerating marginal wastage and surging.

Retreat of the Laurentide Ice Sheet

Glacial retreat seems to have begun only after Laurentide ice had thinned substantially. By 13 ka, more than 4000 years after the "glacial maximum" was reached in North America, the ice margin still lay less than 300 km from its maximum position. Bloom (1971) shows that the areal extent of the Laurentide Ice Sheet had been reduced by only 20% at 13 ka, whereas calculations of ice volume—based on ice thickness decreasing at a linear or exponentially declining rate in relation to its radius—yielded a >50% decrease by this time. Similarly, studies of North Atlantic Ocean cores by Ruddiman and McIntyre (1981) prompted them to conclude that ". . . the bulk of volumetric deglaciation in the Northern Hemisphere occurred considerably earlier than the main areal retreat of ice-sheet limits" (p. 145), with about 50% of the ice having melted by 13 ka. This reduction in volume/area represented a progressive flattening in the profile of the Laurentide Ice Sheet after it had reached its maximum position. Mathews (1974), for example, postulated that the lateral slope of the western Laurentide ice during late-glacial time was much gentler than that of modern glaciers, which flow by normal plastic deformation, averaging less than 2 m/km. Because thinning was accompanied by a warming of the ice, Whillans (1978) speculated that the ice would have responded by increasing its flow toward the margin; this would have delayed or slowed the retreat of the margin in the early stages of deglaciation. With a reduction in total ice volume to area (thinning), less insolation (heat) would have been needed to produce marginal retreat than would have been the case during the early stages of deglaciation when the ice was thicker and colder. This helps to overcome the problem that Andrews (1973) and Hare (1976) noted on the shortfall of calories needed to produce the known rate of ice-margin retreat later in deglaciation.

Although fluctuations in the outer glacial margin did occur in the first few thousand years after it had reached its maximum, it remained close to or south of the divide between the Mississippi and Great Lakes–Hudson Bay watersheds near its maximum position until after 13 ka. Few large proglacial lakes developed until the margin had retreated north of this divide, long after the climate had begun to warm. As the ice retreated downslope into the Hudson Bay and Great Lakes basins, the area covered by these lakes expanded northward (Fig. 21). By the time the glacial barrier to drainage across the high latitudes of Canada had melted

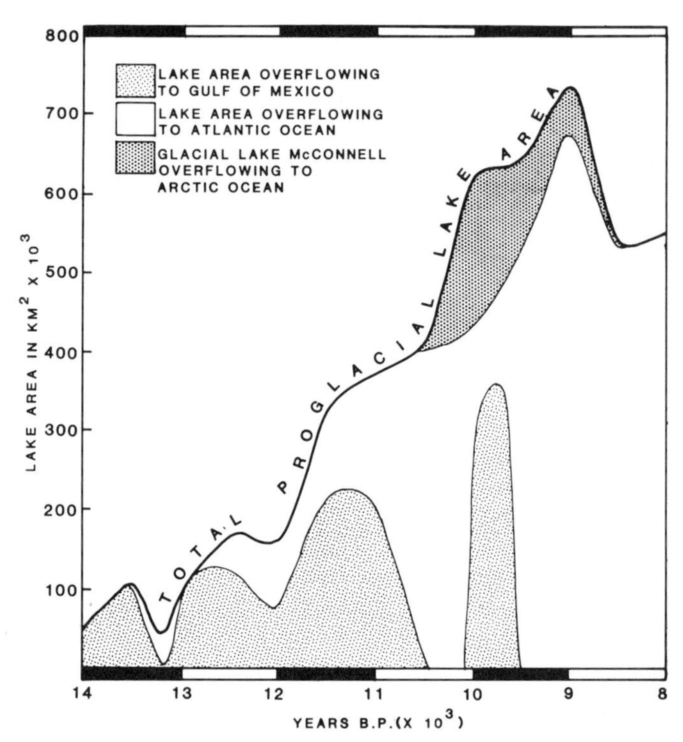

Figure 21. Change of proglacial lake area along the southern Laurentide Ice Sheet between 14 and 8 ka. Prior to this the ice margin lay mainly south of the Hudson Bay and Great Lakes basins with most meltwater flowing directly to the oceans. After 8 ka the Laurentide ice no longer served as a dam to waters draining north into Hudson Bay. The Great Lakes are treated as proglacial lakes throughout this period even though all were not receiving meltwater after 11.5 ka. Compiled from many references cited in this paper.

away, about 8 ka, most of the region north of the Mississippi River watershed had been part of one or more proglacial lake basins, and a vast region was blanketed by lacustrine sediment (Fig. 2).

After about 13 ka the southern and western Laurentide ice was thin and retreating downslope in contact with standing water, and this influenced its flow. The margin of the ice became highly lobate, and flow was more strongly controlled by the subglacial topography. This was particularly evident in the Great Lakes basins and along the southwestern side of the Laurentide Ice Sheet, where even small bedrock elevations were able to divert ice flow.

Marginal areas of the ice began to surge periodically or to stream rapidly into proglacial depressions, displacing many of the lakes. In some areas overflow outlets were dammed, causing lakes to expand and rise to previous levels, while in other areas completely new lakes were formed. This surging also increased the ratio of surface area to volume of the Laurentide Ice Sheet and helped speed its retreat by melting. The presence of proglacial lake waters in contact with the ice margin may have promoted the decoupling of the glacier from its bed, initiating surging, although warming of the ice and the associated increase in basal meltwaters may have been the major controls. As noted by Clayton and others (1985), the presence of relatively impermeable bedrock in the zone beneath the glacial margin at this time, as well as the newly deposited clayey lacustrine sediment in fringing proglacial basins, probably contributed to both the initiation of surging and to its extension into lacustrine depressions. Boulton and Jones (1979) also suggest that rapid flow developed where the ice readvanced across fine-grained sediment of proglacial lake beds. For example, a rise in the level of Lake Agassiz after its eastern outlets were dammed may have produced instability, initiating surging of the glacial margin bounding the lake in central Manitoba; a subsequent decline in lake level when the eastern outlets were re-opened may have increased the stability of the marginal zone of the Laurentide Ice Sheet.

Because ice of these glacier streams and surges was thin and occupied lake basins where water promoted calving and melting, retreat also occurred quickly. Clayton and others (1985) estimate that a total of 3000 km of advances (surges) and retreats occurred in the southern Lake Agassiz Basin between 12.3 and 10.8 ka. Surges into the Lake Michigan Basin are estimated to have advanced and retreated as much as 1000 km in the 1200 years after 13 ka (Mickelson and others, 1981). Surging in the Lake Superior Basin has also been postulated by Wright (1973). As ice retreated farther north into the Hudson Bay Lowland of northern Manitoba, Ontario, and Quebec, surging continued (e.g., Hardy, 1977; Dredge, 1983; Dyke and others, 1982). A nonclimatic cause for fluctuations in the retreating Laurentide ice margin seems likely, and this, in turn, helps explain the nonsynchroneity between the readvances of many lobes during deglaciation (*cf.* Wright, 1973).

The downdraw concept of glacier surging into the oceans around the Laurentide and Antarctic Ice Sheets advocated by Denton and Hughes (1981) and others may be applicable to the Laurentide ice where its margin stood in the deeper proglacial waters of Lakes Agassiz and Ojibway and in the Great Lakes. Denton and Hughes (1981, p. 462) state that surging ". . . from the saddle between the Quebec and Keewatin domes may have fed the Superior, Michigan, and Huron Lobes, which rapidly melted to create the initial stages of the Great Lakes."

The Laurentide Ice Sheet appears to have retreated at a fairly constant rate from its position south of the Great Lakes and Agassiz Basin at 14 ka, even though there were numerous surges into proglacial basins during deglaciation. Because the glacial surge lobes were thin and of short duration, there may have been no significant isostatic response to them. Therefore, glacial rebound may also have progressed northward at a constant rate during retreat of the thicker Laurentide ice mass, from which the surges emanated, rather than to have coincided with the actual position of the fluctuating ice margin. Because surging expedited thinning of the main (thick) Laurentide ice mass, marginal retreat after surging into a proglacial lake may have been rapid both in the surge lobes and in the peripheral zone of the main ice mass, which had been drawn down to supply ice to the surge. In turn, this may have led to deep proglacial waters along the ice margin that accelerated calving and melting of the main Laurentide ice mass (Pollard, 1983; see Peltier, this volume; Ruddiman and Wright, this volume), and may have led to initiation of the next surge.

The thin late-glacial ice margin also set the stage for huge outbursts of overflow into the Great Lakes from the largest lake in North America, Lake Agassiz. After a readvance (surge?) had closed the link between these two watersheds at 10 to 9.5 ka (and also possibly prior to 11 ka), causing the level of Lake Agassiz to rise nearly a hundred meters over an area of 350,000 km^2, the glacial dam failed just west of Lake Superior (Teller and Thorleifson, 1983). The resulting catastrophic bursts, which were 5 to 10 times the volume of present-day Lake Erie, not only accelerated the retreat of the glacial margin in the Superior Basin but changed its area, volume, length of ice-free season, circulation, and water properties (e.g., turbidity, temperature, and stratification). In only a few hundred years, from shortly before 10 ka to shortly after 9.5 ka, Lake Agassiz more than doubled in size to about 350,000 km^2 and then constricted by a similar amount as its eastern outlets reopened. In turn, the climate of the Agassiz region, as well as that of the Superior Basin, must have been affected by this linkage. In water bodies the size of these lakes, a major change in size, length of the open-water season, or temperature of the lake surface may have significantly influenced evaporation and albedo, both of which helped control temperature and the nature and amount of precipitation in the region. The flow dynamics of glaciers that were marginal to these lakes may have been affected. Furthermore, the rapid loading and unloading of the earth's crust, both by surging ice and by the periodic confinement and release of large masses of proglacial waters, may have influenced isostatic rebound, which would have influenced the rate of ice retreat and melting (see Ruddiman and Wright, this volume).

The influence of proglacial lakes on the southern Laurentide Ice Sheet may have been related to the area of water that actually bordered the glacial margin. The increasing proportion of lake water to ice probably helped speed glacial retreat by melting, calving, and perhaps by encouraging surging. The area of ice-marginal lake waters is plotted in Figure 22, and is less than, but proportional to, the total area of proglacial lakes shown in Figure 21. Another even more significant indicator of the impact of lakes on deglaciation may be the volumes of water in the ice marginal lakes. As can be seen in Figure 22, the rising volume after 14 ka closely paralleled that of ice-marginal lake area, and was inversely proportional to the shrinking area of the Laurentide Ice Sheet. The decline in area and volume after 9.5 ka was directly related to the fall in level of Lake Agassiz as it began to overflow into the Great Lakes. Shortly after 8 ka, Laurentide ice ceased to be a barrier to northward drainage into Hudson Bay, and nearly all ice-marginal lakes drained, including Lakes Agassiz and Ojibway.

In Figure 23, three ratios are plotted—ice-marginal lake area to glacier area, glacier area to ice-marginal lake area, and

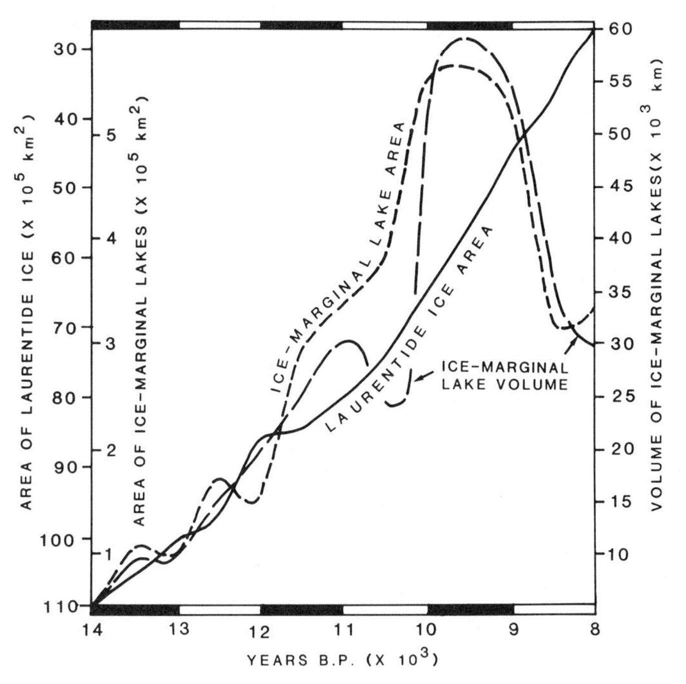

Figure 22. Curves showing the change through late glacial time in (1) area of the Laurentide Ice Sheet (Bloom, 1971), (2) area of ice-marginal lakes, and (3) volume of ice-marginal lakes. Volume calculated by multiplying the measured area of individual ice-marginal lakes at 500-year intervals by their estimated mean depths at those times. Depths used average about 100 m, but vary from 20 to 250 m. Data mainly from Karrow and Calkin (1985), Teller and Clayton (1983), Prest (1970), Craig (1965), and maps compiled in this chapter.

glacial area to ice-marginal lake volume. The percentage of lake area along the southern Laurentide ice margin to the total Laurentide ice area rose from 1% at 14 ka to 12% at 9 ka. Although the overwhelming majority of proglacial lakes lay along the southern edge of the ice and are included in this calculation, the influence of marine waters on the eastern and northern ice margin must also have been significant in deglaciation. During this same period, both the ratios of glacier area to lake area and glacier area to lake volume declined by a factor of more than 20. In general, the lake area and volume differ consistently by an order of magnitude, reflecting the fact that the average depth of ice-marginal lakes was 100 m (0.1 km).

Figure 24 identifies the possible interaction of some of the variables that may have led to climatic change in the Lake Agassiz region. No attempt is made to quantify these parameters or even to weight them. Although the overall trend of deglaciation is the result of global climate change, the complexities of glacial retreat and its associated proglacial lake system are partly attributable to regional phenomena that influenced the dynamics of ice flow and climate. Not only did proglacial lakes play a role in controlling these phenomena, but their sediments provide the best

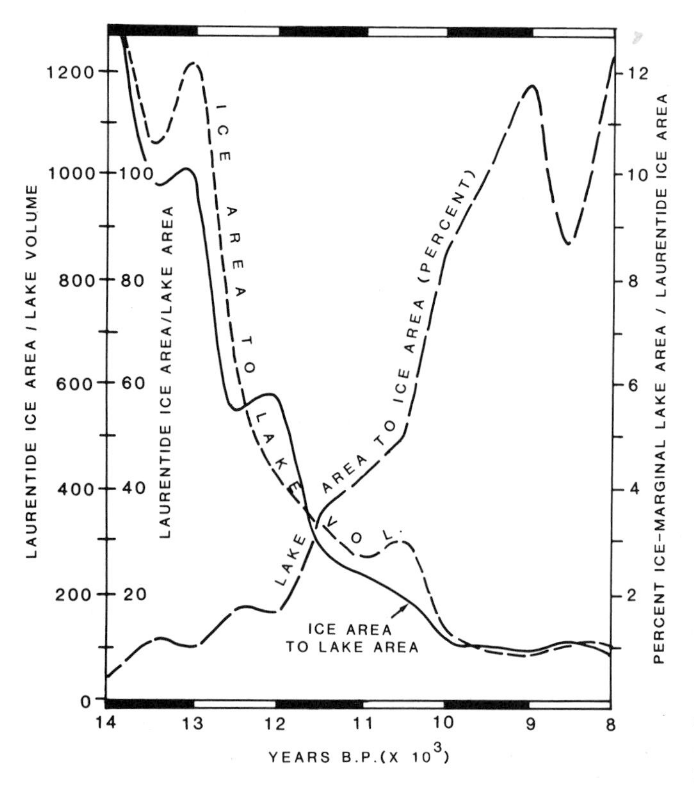

Figure 23. Relationship of expanding ice-marginal lake area and volume to the shrinking Laurentide Ice Sheet between 14 and 8 ka. Ice area from Bloom (1971). Lake area and volume from Figure 22 calculations.

record of both regional and continent-wide changes in the newly deglaciated environment.

Meltwater and the Ocean Record

Figure 21 shows the changing areal extent of proglacial lakes through time and indicates the relative proportion contributing to the Gulf of Mexico and the North Atlantic Ocean. Overflow to the oceans occurred mainly through four outlet systems (Fig. 2): (a) Chicago and Wabash valleys to the Mississippi Valley and Gulf of Mexico; (b) Minnesota River valley to the Mississippi valley and Gulf of Mexico; (c) Mohawk, Champlain, and Hudson valleys to the Atlantic Ocean; and (d) Ottawa River valley through the St. Lawrence lowland to the North Atlantic Ocean. Meltwater from the eastern Laurentide Ice Sheet entered the North Atlantic along a broad line from New England to Labrador prior to 12 ka.

As can be seen in Figure 21, there was a nearly steady growth in the size of the region covered by proglacial lake waters between 14 and 9 ka; north-flowing rivers resumed their outflow into Hudson Bay shortly after 8 ka. Although a rapidly fluctuating glacial margin during this time periodically displaced hundreds of cubic kilometers of water, forcing additional outflow, in some cases there was a corresponding increase in lake area and volume within the basin because the proglacial water body deepened and expanded southward in response to the damming of its lower northern outlets. In other instances, the glacial advance completely filled the lake basin, displacing all water and forcing added volumes of overflow to the oceans.

The relative influx of isotopically light water to the Atlantic Ocean and to the Gulf of Mexico from the Laurentide Ice Sheet varied through the deglacial period, and there have been attempts to estimate the volume of meltwater responsible for the isotopic anomalies in the ocean sediment record in those areas (Fillon and Williams, 1984). Until about 13.5 ka, most meltwater and precipitation from central North America discharged south through the Mississippi River into the Gulf of Mexico (Fig. 21), with much of the water from the eastern Laurentide margin entering the North Atlantic directly from the ice margin itself. After the ice margin retreated north of the St. Lawrence valley about 12 ka, most overflow from the eastern areas, plus that from the Great Lakes region, discharged through the Gulf of St. Lawrence to the North Atlantic Ocean. The Gulf of Mexico continued to receive water from the large Lake Agassiz Basin until that watershed was integrated into the St. Lawrence system at 10.8 ka. After that, all meltwater and precipitation from the Hudson Bay Basin overflowed eastward into the North Atlantic—either through the St. Lawrence (until 8 ka) or northward into Hudson Bay (after 8 ka)—except for a brief return of discharge to the Gulf of Mexico from the Agassiz Basin at 10–9.5 ka.

The "meltwater spike," of negative oxygen isotopic values in Gulf of Mexico sediments estimated at between 16.5 and 11 ka (Leventer and others, 1982; Kennett and Shackleton, 1975; Emiliani and others, 1978), approximately coincides with the period of maximum meltwater discharge into the Mississippi River watershed from the Laurentide Ice Sheet. The fluctuating nature of the isotopic record there probably reflects the variability of melting, lake displacement by surging, and the interplay of ice retreat, isostasy, and shifting watershed divides. Based on changes in planktonic foraminifera preserved in the anoxic sediments of the Orca Basin of the northern Gulf of Mexico, the salinity of surface waters there also decreased between 16.5 and 12 ka (Kennett and others, 1985). The 2000-year-long decline in influx of isotopically light meltwater into the Gulf after the peak at 13 ka (Leventer and others, 1982) may reflect the redirection of meltwater eastward into the North Atlantic via the St. Lawrence Valley.

Many problems exist in identifying and dating "meltwater spikes" in the ocean record (e.g., Jones and Ruddiman, 1982; Berger, 1985; Mix and Ruddiman, 1985), and numerous chronologies of meltwater influx and ocean warming have been proposed. Mix (this volume) and Ruddiman (this volume) discuss the ocean record during deglaciation and address many of these concerns, which are important in interpreting the varying discharge of meltwater to the oceans by the Laurentide Ice Sheet. Some have interpreted the isotopic, foraminiferal, and carbonate record of deep-sea cores from the Atlantic Ocean as indicating that there was a rapid rise in meltwater influx centered on 11 ka

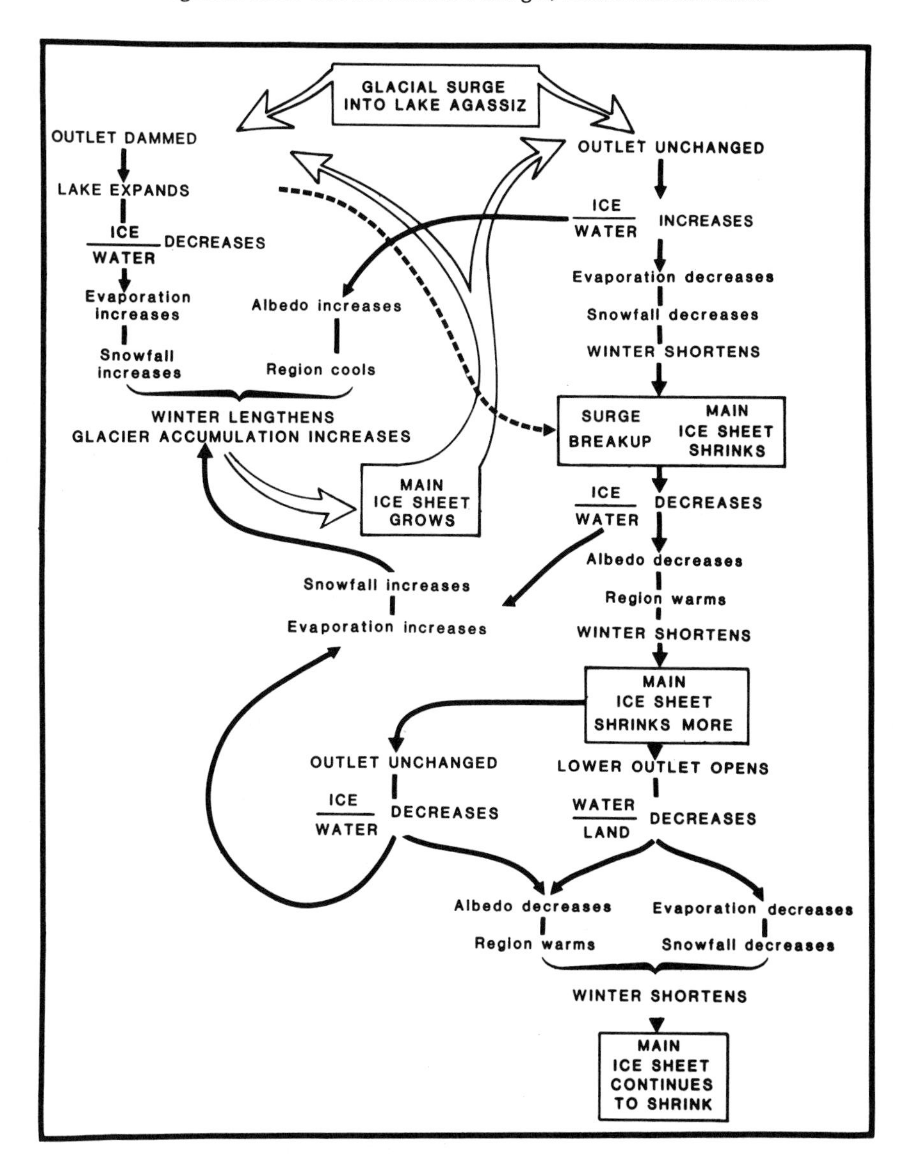

Figure 24. Possible interaction and feedback due to variations in lake, land, and glacier size (resulting from surging into Lake Agassiz during deglaciation) that may lead to climatic change. Although the long-term net result was a progressive shift downward along the right-hand side, there may have been temporary interruptions through other feedback loops.

(e.g., Broecker and others, 1960; Sancetta and others, 1973). McIntyre and others (1972) placed this rise between 13 and 11 ka. More recent studies indicate that the major influx of meltwater occurred farther back in time at 16–13 ka (Ruddiman and McIntyre, 1981). Duplessy and others (1981) suggest that deglaciation and meltwater influx were most rapid during two stages, 16–13 ka and 10–8 ka. A recent compilation of $\delta^{18}O$ data by Mix and Ruddiman (1985) constrains major isotopic changes between 14 and 6 ka, and indicates that the most rapid rates occurred at 14–12, 10–9, and 8–6 ka.

The continental record indicates that meltwater discharge from the southern Laurentide ice margin into the North Atlantic Ocean increased about 12 ka, after the Great Lakes began to overflow through the St. Lawrence valley. When Lake Agassiz overflow was added to the Great Lakes about 10.8 ka, this discharge more than doubled. All runoff from the southern Laurentide margin continued to flow through the St. Lawrence valley until 8 ka, except for a brief period at 10–9.5 ka when waters from the Lake Agassiz watershed were diverted to the Gulf of Mexico. If meltwater from the southern Laurentide region had a significant affect on sediments of the Atlantic Ocean, perhaps the initial rise in isotopically light water recorded there in part corresponds to the initial phase of glacier thinning, with subsequent rises related to the opening of the St. Lawrence valley and to the influx of water from the Lake Agassiz basin. Of course the northern Laurentide, Greenland, and Scandinavian ice sheets strongly affected the sedimentary record in the North Atlantic, and deglaciation along the southern Laurentide margin was only partly responsible for the isotopic, biological, and mineralogical signals in the ocean record.

REFERENCES CITED

Anderson, J. B., Domak, E. W., and Kurtz, D. D., 1980, Observations of sediment-laden icebergs in Antarctic waters; Implications to glacial erosion and transportation: Journal of Glaciology, v. 25, p. 387–397.

Anderson, T. W., and Lewis, C.F.M., 1985, Postglacial water-level history of the Lake Ontario basin, *in* Karrow, P. F., and Calkin, P. E., eds., Quaternary evolution of the Great Lakes: Geological Association of Canada Special Paper 30, p. 231–253.

Andrews, J. T., 1970, A geomorphological study of post-glacial uplift, with particular reference to Arctic Canada: Institute of British Geographers Special Publication no. 2, 156 p.

—— , 1973, The Wisconsin Laurentide Ice Sheet; Dispersal centers, problems of rates of retreat, and climatic implications: Arctic and Alpine Research, v. 5, p. 185–199.

—— , ed., 1974, Glacial isostasy: Stroudesburg, Pennsylvania, Dowden, Hutchinson, and Ross, 491 p.

—— , 1982, On the reconstruction of Pleistocene ice sheets; A review: Quaternary Science Reviews, v. 1, p. 1–30.

Antevs, E., 1922, The Recession of the last ice sheet in New England: American Geographical Society Research Series no. 11, 120 p.

—— , 1925, Retreat of the last ice-sheet in eastern Canada: Geological Survey of Canada Memoir 146, 142 p.

—— , 1951, Glacial clay in Steep Rock Lake, Ontario, Canada: Geological Society of America Bulletin, v. 62, p. 1223–1262.

Ashley, G. M., 1975, Rhythmic sedimentation in glacial Lake Hitchcock, Massachusetts–Connecticut, *in* Jopling, A. V., and McDonald, B. C., eds., Glaciofluvial and glaciolacustrine sedimentation: Society of Economic Paleontologists and Mineralogists Special Publication no. 23, p. 304–320.

Attig, J. W., Clayton, L., and Mickelson, D. M., 1985, Correlation of late Wisconsin glacial phases in the western Great Lakes: Geological Society of America Bulletin, v. 96, p. 1585–1593.

Banerjee, I., 1973, Sedimentology of Pleistocene glacial varves in Ontario, Canada: Geological Survey of Canada Bulletin, v. 226, 44 p.

Berger, W. H., 1985, On the time-scale of deglaciation; Atlantic deep-sea sediments and Gulf of Mexico: Palaeogeography, Palaeoclimatology, Palaeoecology, v. 50, p. 167–184.

Birchfield, G. E., and Grumbine, R. W., 1985, "Slow" physics of large continental ice sheets and underlying bedrock and its relation to the Pleistocene ice ages: Journal of Geophysical Research, v. 90, p. 11294–11302.

Birks, H.J.B., and Birks, H. H., 1980, Quaternary palaeoecology: London, Edward Arnold Publications, 289 p.

Bloom, A. L., 1971, Glacial-eustatic and isostatic controls of sea level since the last glaciation, *in* Turekian, K. K., ed., The late Cenozoic glacial ages: New Haven, Connecticut, Yale University Press, p. 355–379.

Borns, H. W., 1973, Late Wisconsin fluctuations of the Laurentide ice sheet in southern and eastern New England, *in* Black, R. F., Goldthwait, R. P., and Willman, H. B., eds., The Wisconsinan stage: Geological Society of America Memoir 136, p. 37–45.

—— , 1985, Changing models of deglaciation in northern New England and adjacent Canada, *in* Borns, H. W., LaSalle, P., and Thompson, W. B., eds., Late Pleistocene history of northeastern New England and adjacent Quebec: Geological Society of America Special Paper 197, p. 135–138.

Boulton, G. S., and Jones, A. S., 1979, Stability of temperate ice caps and ice sheets resting on beds of deformable sediment: Journal of Glaciology, v. 24, p. 29–43.

Broecker, W. S., Ewing, M., and Heezen, B. C., 1960, Evidence for an abrupt change in climate close to 11,000 years ago: American Journal of Science, v. 258, p. 429–440.

Caldwell, D. W., Hanson, L. S., and Thompson, W. B., 1985, Styles of deglaciation in central Maine, *in* Borns, H. W., LaSalle, P., Thompson, W. B., eds., Late Pleistocene history of northeastern New England and adjacent Quebec, Geological Society of America Special Paper 197, p. 45–58.

Calkin, P. E., and Feenstra, B. H., 1985, Evolution of the Erie Basin, Great Lakes, *in* Karrow, P. F., and Calkin, P. E., eds., Quaternary evolution of the Great Lakes: Geological Association of Canada Special Paper 30, p. 149–170.

Calkin, P. E., and Miller, K. E., 1977, Late Quaternary environment and man in western New York: Annals of the New York Academy of Sciences, v. 288, p. 297–315.

Chamberlain, C. K., 1975, Recent lebensspuren in nonmarine aquatic environments, *in* Frey, R. W., ed., The study of trace fossils: New York, Springer--Verlag, p. 431–458.

Christiansen, E. A., 1979, The Wisconsinan deglaciation of southern Saskatchewan and adjacent areas: Canadian Journal of Earth Sciences, v. 16, p. 913–938.

Clark, P., and Karrow, P. F., 1984, Late Pleistocene water bodies in the St. Lawrence Lowland, New York, and regional correlations: Geological Society of America Bulletin, v. 95, p. 805–813.

Clayton, L., 1983, Chronology of Lake Agassiz drainage to Lake Superior, *in* Teller, J. T., and Clayton, L., eds., Glacial Lake Agassiz: Geological Association of Canada Special Paper 26, p. 291–307.

Clayton, L., and Moran, S. R., 1982, Chronology of Late Wisconsinan glaciation in middle North America: Quaternary Science Reviews, v. 1, p. 55–82.

Clayton, L., Teller, J. T., and Attig, J. W., 1985, Surging of the southwestern part of the Laurentide Ice Sheet: Boreas, v. 14, p. 235–241.

Connally, G. G., and Sirkin, L. A., 1973, Wisconsinan history of the Hudson–Champlain Lobe, *in* Black, R. F., Goldthwait, R. P., and Willman, H. B., eds., The Wisconsinan Stage: Geological Society of America Memoir 136, p. 47–69.

Cooper, W. S., 1935, The history of the upper Mississippi River in late Wisconsin and postglacial time: Minnesota Geological Survey Bulletin 26, 116 p.

Craig, B. G., 1965, Glacial Lake McConnell, and the surficial geology of parts of Slave River and Redstone River Map-Areas, District of MacKenzie: Geological Survey of Canada Bulletin 122, 33 p.

Crittenden, M. D., 1963, New data on the isostatic deformation of Pleistocene Lake Bonneville: U.S. Geological Survey Professional Paper 454–F, p. 1–31.

Cumming, L. M., 1969, Rivers of the Hudson Bay Lowlands, *in* Hood, P. J., ed., Earth science symposium on Hudson Bay: Geological Survey of Canada Paper 68–53, p. 144–168.

DeGeer, G., 1912, A geochronology of the last 12,000 years; 11th International Geological Congress, Stockholm: Compte Rendu, v. 1, p. 241–258.

Denton, G. H., and Hughes, T. J., 1981, The arctic ice sheet; An outrageous hypothesis, *in* Denton, G. H., and Hughes, T. J., eds., The last great ice sheets: New York, J. Wiley and Sons, p. 437–467.

Dionne, J.-C., 1979, Ice action in the lacustrine environment; A review with particular reference to subarctic Quebec, Canada: Earth Science Review, v. 15, p. 185–212.

Dionne, J.-C., Drapeau, G., and Reeves, G., 1974, Symposium on the geological action of drift ice: Institut National de la Recherche Scientifique, Universite Quebec, 51 p.

Dredge, L. A., 1983, Character and development of northern Lake Agassiz and its relation to Keewatin and Hudsonian ice regimes, *in* Teller, J. T., and Clayton, L., eds., Glacial Lake Agassiz: Geological Association of Canada Special Paper 26, p. 117–131.

Dreimanis, A., 1979, The problem of waterlain till, *in* Schluchter, C., ed., Moraines and varves: Rotterdam, A. A. Balkema, p. 167–177.

Dreimanis, A., and Karrow, P. F., 1972, Glacial history of the Great Lakes–St. Lawrence region, the classification of the Wisconsin(an) Stage, and its correlatives: 24th International Geological Congress, Montreal, Quaternary Geology, Section 12, p. 5–15.

Duplessy, J. C., Delibrais, G., Turon, J. L., Pujol, C., and Duprat, J., 1981, Deglacial warming of the northeastern Atlantic Ocean; Correlation with the paleoclimatic evolution of the European continent: Palaeogeography, Palaeoclimatology, and Palaeoecology, v. 35, p. 121–144.

Dyke, A. S., Dredge, L. A., and Vincent, J-S., 1982, Configuration and dynamics of the Laurentide Ice Sheet during the Late Wisconsin maximum: Geographie Physique et Quaternaire, v. 36, p. 5–14.

Edwards, M. B., 1978, Glacial environments, *in* Reading, H. G., ed., Sedimentary environments and facies: Oxford, Blackwell Scientific Publications, p. 416–438.

Ekdale, A. A., Bromley, R. G., and Pemberton, S. G., 1984, Ichnology; The uses of trace fossils in sedimentology and stratigraphy: Society Economic Paleontologists and Mineralogists, Short Course no. 15, 317 p.

Elson, J. A., 1967, Geology of glacial Lake Agassiz, *in* Mayer–Oakes, W., ed., Life, land and water: Winnipeg, University Manitoba Press, p. 36–95.

Emiliani, C., Rooth, C., and Stipp, J. J., 1978, The late Wisconsin flood into the Gulf of Mexico: Earth and Planetary Science Letters, v. 41, p. 159–162.

Eschman, D. F., and Karrow, P. F., 1985, Huron basin glacial lakes; A review, *in* Karrow, P. F., and Calkin, P. E., eds., Quaternary evolution of the Great Lakes: Geological Association of Canada Special Paper 30, p. 79–93.

Evenson, E., Dreimanis, A., and Newsome, J., 1977, Subaquatic flow tills; A new interpretation for the genesis of some laminated till deposits: Boreas, v. 6, p. 115–133.

Eyles, C. H., and Eyles, N., 1983, Sedimentation in a large lake; A reinterpretation of the late Pleistocene stratigraphy at Scarborough Bluffs, Ontario, Canada: Geology, v. 11, p. 146–152.

Farrand, W. R., and Drexler, C. W., 1985, Late Wisconsinan and Holocene history of the Lake Superior basin, *in* Karrow, P. F., and Calkin, P. E., eds., Quaternary evolution of the Great Lakes: Geological Association of Canada Special Paper 30, p. 17–32.

Farrand, W. R., Zahner, R., and Benninghoff, W. S., 1969, Cary–Port Huron Interstade; Evidence from a buried bryophyte bed, Cheboygan County, Michigan: Geological Society of America Special Paper 123, p. 249–262.

Farrand, W. R., Mickelson, D. M., Cowan, W. R., and Goebel, J. E., 1984, Quaternary geologic map of the Lake Superior 4° × 6° Quadrangle, United States and Canada: U.S. Geological Survey Map I–1420(NL–16), scale 1:1,000,000.

Fenton, M. M., Moran, S. R., Teller, J. T., and Clayton, L., 1983, Quaternary stratigraphy and history in the southern part of the Lake Agassiz basin, *in* Teller, J. T., and Clayton, L., eds., Glacial Lake Agassiz: Geological Association of Canada Special Paper 26, p. 49–74.

Ferrante, J. G., and Parker, J. I., 1977, Transport of diatom frustules by copepod fecal pellets to the sediments of Lake Michigan: Limnology and Oceanography, v. 22, p. 92–98.

Fillon, R. H., and Williams, D. F., 1984, Dynamics of meltwater discharge from Northern Hemisphere ice sheets during the last deglaciation: Nature, v. 310, no. 5979, p. 674–677.

Flint, R. F., Colton, R. B., Goldthwait, R. P., and Willman, H. B., 1959, Glacial map of the United States east of the Rocky Mountains: Geological Society of America, scale 1:1,750,000.

Fullerton, D. S., 1980, Preliminary correlation of post-Erie Interstadial events (16,000–10,000 radiocarbon years before present), central and eastern Great Lakes region, and Hudson, Champlain, and St. Lawrence Lowlands, United States and Canada: U.S. Geological Survey Professional Paper 1089, 52 p.

Fulton, R. J., Karrow, P. F., LaSalle, P., and Grant, D. R., 1984, Summary of Quaternary stratigraphy and history, eastern Canada, *in* Fulton, R. J., ed., Quaternary stratigraphy of Canada; A Canadian contribution to IGCP Project 24: Geological Survey Canada Paper 84–10, p. 194–210.

Fulton, R. J., Heginbottom, J. A., and Funder, S., eds., 1987, Quaternary geology of Canada and Greenland: Geological Survey of Canada, Geology of Canada No. 1 (Geological Society of America, Geology of North America, v. K-1) (in press).

Gadd, N. R., 1976, Quaternary stratigraphy in southern Quebec, *in* Mahaney, W. C., ed., Quaternary stratigraphy of North America: Stroudsburg, Pennsylvania, Dowden, Hutchinson, and Ross, p. 37–63.

Gibbard, P. L., and Dreimanis, A., 1978, Trace fossils from late Pleistocene glacial lake sediments in southwestern Ontario, Canada: Canadian Journal of Earth Sciences, v. 15, p. 1967–1976.

Gibbard, P. L., and Stuart, A. J., 1974, Trace fossils from proglacial lake sediments: Boreas, v. 3, p. 69–74.

Gilbert, R., 1975, Sedimentation in Lillooet Lake, British Columbia: Canadian Journal of Earth Sciences, v. 12, p. 1697–1711.

Gilbert, R., Syvitski, J., and Taylor, R., 1985, Reconnaissance study of proglacial Stewart Lakes, Baffin Island, District of Franklin: Geological Survey of Canada Paper 85–1A, Current Research, Part A, p. 505–510.

Goebel, J. E., Mickelson, D. W., Farrand, W. R., Clayton, L., Knox, J. C., Cahow, A., Hobbs, H. C., and Walton, M. S., 1983, Quaternary geologic map of the Minneapolis 4° × 6° Quadrangle, United States: U.S. Geological Survey Map I-1420(NL–15), scale 1:1,000,000.

Goldthwait, R. P., 1970, Introduction to till, today, *in* Goldthwait, R. P., ed., Till; A symposium: Columbus, Ohio State University Press, p. 3–26.

Gustavson, T. C., 1975, Sedimentation and physical limnology in proglacial Malaspina Lake, southeastern Alaska, *in* Jopling, A. V., and McDonald, B. C., eds., Glaciofluvial and glaciolacustrine sedimentation: Society of Economic Paleontologists and Mineralogists Special Publication 23, p. 249–263.

Hansel, A. K., Mickelson, D. M., Schneider, A. F., and Larsen, C. E., 1985, Late Wisconinan and Holocene history of the Lake Michigan basin, *in* Karrow, P. F., and Calkin, P. E., eds., Quaternary evolution of the Great Lakes: Geological Association of Canada Special Paper 30, p. 39–53.

Hardy, L., 1977, La deglaciation et les episodes lacustre et marin sur le versant québéccois des basses terres do la baie de James: Geographie Physique et Quaternaire, v. 31, p. 357–372.

Hare, F. K., 1976, Late Pleistocene and Holocene climates; Some persistent problems: Quaternary Research, v. 6, p. 507–517.

Harris, K. L., Moran, S. R., and Clayton, L., 1974, Late Quaternary stratigraphic nomenclature, Red River Valley, North Dakota and Minnesota: North Dakota Geological Survey Miscellaneous Series 52, 47 p.

Harrison, J. E., 1972, Quaternary geology of the North Bay–Mattawa region: Geological Survey of Canada Paper 71–26, 37 p.

Hillaire-Marcel, C., and Fairbridge, R. W., 1978, Isostasy and eustasy of Hudson Bay: Geology, v. 6, p. 117–122.

Hobbs, H. C., 1983, Drainage relationships of Glacial Lakes Aitkin and Upham and early Lake Agassiz in northeastern Minnesota, *in* Teller, J. T., and Clayton, L., eds., Glacial Lake Agassiz: Geological Association of Canada Special Paper 26, p. 245–259.

Hough, J. L., 1955, Lake Chippewa, a low stage of Lake Michigan indicated by bottom sediments: Geological Society of America Bulletin, v. 66, p. 957–968.

——, 1958, Geology of the Great Lakes: Urbana, University Illinois Press, 313 p.

——, 1962, Lake Stanley, a low stage of Lake Huron indicated by bottom sediments: Geological Society of America Bulletin, v. 73, p. 613–619.

Jones, G. A., and Ruddiman, W. F., 1982, Assessing the global meltwater spike: Quaternary Research, v. 17, p. 148–172.

Jopling, A. V., and McDonald, B. C., eds., 1975, Glaciofluvial and glaciolacustrine sedimentation: Society of Economic Paleontologists and Mineralogists Special Publication 23, 320 p.

Karrow, P. F., 1984, Quaternary stratigraphy and history, Great Lakes–St. Lawrence region, *in* Fulton, R. J., ed., Quaternary stratigraphy of Canada; A Canadian contribution to IGCP Project 24: Geological Survey of Canada Paper 84–10, p. 138–153.

Karrow, P. F., and Calkin, P. E., eds., 1985, Quaternary evolution of the Great Lakes: Geological Association of Canada Special Paper 30, 258 p.

Karrow, P. F., Anderson, T. W., Clarke, A. H., Delorme, L. D., and Sreenivasa,

M. R., 1975, Stratigraphy, paleontology, and age of Lake Algonquin sediments in southwestern Ontario, Canada: Quaternary Research, v. 5, p. 49–87.

Kehew, A. E., 1982, Catastrophic flood hypothesis for the origin of the Souris spillway, Saskatchewan and North Dakota: Geological Society of America Bulletin, v. 93, p. 1051–1058.

Kehew, A. E., and Clayton, L., 1983, Late Wisconsinan floods and development of the Souris–Pembina spillway system in Saskatchewan, North Dakota, and Manitoba, *in* Teller, J. T., and Clayton, L., eds., Glacial Lake Agassiz: Geological Association of Canada Special Paper 26, p. 187–209.

Kehew, A. E., and Lord, M. L., 1986, Origin and large-scale erosional features of glacial lake spillways in the northern Great Plains: Geological Society of America Bulletin, v. 97, p. 162–177.

Kennett, J. P., and Shackleton, N. J., 1975, Laurentide Ice Sheet meltwater recorded in Gulf of Mexico deep-sea cores: Science, v. 188, p. 147–150.

Kennett, J. P., Elmstrom, K., and Penrose, N., 1985, The last deglaciation in Orca Basin, Gulf of Mexico; High resolution planktonic foraminiferal changes: Palaeoecology, Palaeoclimatology, and Palaeoecology, v. 50, p. 189–216.

Klassen, R. W., 1972, Wisconsin events and the Assiniboine and Qu'Appelle valleys of Manitoba and Saskatchewan: Canadian Journal of Earth Sciences, v. 9, p. 544–560.

—— , 1983a, Assiniboine delta and the Assiniboine–Qu'Appelle valley system; Implications concerning the history of Lake Agassiz in southwestern Manitoba, *in* Teller, J. T., and Clayton, L., eds., Glacial Lake Agassiz: Geological Association of Canada Special Paper 26, p. 211–229.

—— 1983b, Lake Agassiz and the late glacial history of northern Manitoba, *in* Teller, J. T., and Clayton, L., eds., Glacial Lake Agassiz: Geological Association of Canada Special Paper 26, p. 97–115.

—— , 1984, Dating methods applicable to late glacial deposits of the Lake Agassiz basin, Manitoba, *in* Mahaney, W. C., ed., Quaternary dating methods: Amsterdam, Elsevier Science Publication, p. 375–388.

LaFleur, R. G., 1968, Glacial Lake Albany, *in* Fairbridge, R. W., ed., Encyclopedia of Geomorphology: New York, Reinhold Book Corp., p. 455–456.

Lajtai, E. Z., 1967, The origin of some varves in Toronto, Canada: Canadian Journal of Earth Sciences, v. 4, p. 633–639.

Leventer, A., Williams, D. F., and Kennett, J. P., 1982, Dynamics of the Laurentide ice sheet during the last deglaciation; Evidence from the Gulf of Mexico: Earth and Planetary Science Letters, v. 59, p. 11–17.

Leverett, F., 1929, Moraines and shorelines of the Lake Superior basin: U.S. Geological Survey Professional Paper 154–A, 72 p.

Leverett, F., and Taylor, F. B., 1915, The Pleistocene of Indiana and Michigan, and the history of the Great Lakes: U.S. Geological Survey Monograph 53, 529 p.

Lewis, C.F.M., 1969, Late Quaternary history of lake levels in the Huron and Erie basins: 12th Great Lakes Research Conf., Ann Arbor, Proceedings International Association of Great Lakes Research, p. 250–270.

Lewis, C.F.M., and Anderson, T. W., 1985, Postglacial lake levels in the Huron basin; Comparative uplift histories of basins and sills in a rebounding glacial marginal depression, *in* Karrow, P. F., and Calkin, P. E., eds., Quaternary evolution of the Great Lakes: Geological Association of Canada Special Paper 30, p. 147–148.

Lineback, J. A., Bleuer, N. K., Mickelson, D. M., Farrand, W. R., and Goldthwait, R. P., 1983, Quaternary geologic map of the Chicago 4° × 6° Quadrangle, United States: U.S. Geological Survey Map I-1420(NK–16), scale 1:1,000,000.

Lowell, T. V., 1985, Late Wisconsin ice-flow reversal and deglaciation, northwestern Maine, *in* Borns, H. W., LaSalle, P., Thompson, W. B., eds., Late Pleistocene history of northeastern New England and adjacent Quebec: Geological Society of America Special Paper 197, p. 71–83.

Mahnic, P., and Teller, J. T., 1985, History of sedimentation in the northwestern Lake Superior basin as related to Lake Agassiz overflow: CANQUA Symposium on the paleoenvironmental reconstruction of the Late Wisconsin deglaciation and the Holocene: Program with abstracts, p. 44.

Mathews, W. H., 1956, Physical limnology and sedimentation in a glacial lake: Geological Society of America Bulletin, v. 67, p. 537–552.

—— , 1974, Surface profiles of the Laurentide Ice Sheet in its marginal areas: Journal of Glaciology, v. 13, p. 37–43.

Mayewski, P. A., Denton, G. A., and Hughes, T. J., 1981, Late Wisconsin Ice Sheets of North America, *in* Denton, G. H., and Hughes, T. J., eds., The last great ice sheets: New York, John Wiley and Sons, p. 67–178.

McIntyre, A., Ruddiman, W. F., and Jantzen, R., 1972, Southward penetrations of the North Atlantic Polar Front; Faunal and floral evidence of large-scale surface water mass movements over the last 225,000 years: Deep-Sea Research, v. 19, p. 61–77.

Mickelson, D. M., Acomb, L. J., and Bentley, C. R., 1981, Possible mechanisms for the rapid advance and retreat of the Lake Michigan Lobe between 13,000 and 11,000 years B.P.: Annals of Glaciology, v. 2, p. 185–186.

Mickelson, D. M., Clayton, L., Fullerton, D. S., and Borns, H. W., 1983, The Late Wisconsin glacial record of the Laurentide Ice Sheet in the United States, *in* Wright, H. E., ed., Late Quaternary environments of the United States, v. 1: Minneapolis, University of Minnesota Press, p. 3–37.

Mix, A. C., and Ruddiman, W. F., 1985, Structure and timing of the last deglaciation; Oxygen isotope evidence: Quaternary Science Reviews, v. 4, p. 59–108.

Mollard, J. D., 1983, The origin of reticulate and orbicular patterns on the floor of the Lake Agassiz basin, *in* Teller, J. T., and Clayton, L., eds., Glacial Lake Agassiz: Geological Association of Canada Special Paper 26, p. 355–374.

Muller, E. H., and Prest, V. R., 1985, Glacial lakes in the Ontario basin, *in* Karrow, P. F., and Calkin, P. E., eds., Quaternary evolution of the Great Lakes: Geological Association of Canada Special Paper 30, p. 213–229.

Nichols, R. L., 1961, Characteristics of beaches formed in polar climates: American Journal of Science, v. 259, p. 694–708.

Norris, A. W., and Sanford, B. V., 1969, Paleozoic and Mesozoic geology of the Hudson Bay Lowlands, *in* Hood, P. J., ed., Earth science symposium on Hudson Bay: Geological Survey of Canada Paper 68–53, p. 169–205.

O'Sullivan, P. E., 1983, Annually-laminated lake sediments and the study of Quaternary environmental changes—a review: Quaternary Science Reviews, v. 1, p. 245–313.

Ovenshine, A. T., 1970, Observations on iceberg rafting in Glacier Bay, Alaska, and the identification of ice-rafted deposits: Geological Society of America Bulletin, v. 81, p. 891–894.

Peach, P., and Perrie, L., 1975, Grain-size distribution within glacial varves: Geology, v. 3, p. 43–46.

Peltier, W. R., 1982, Dynamics of the ice age earth: Advances in Geophysics, v. 24, p. 1–146.

Peltier, W. R., and Andrews, J. T., 1983, Glacial geology and glacial isostasy of the Hudson Bay region, *in* Smith, D. E, and Dawson, A. G., eds., Shorelines and isostasy: Institute of British Geographers Special Publication 16, p. 285–319.

Pharo, C. H., and Carmack, E. C., 1979, Sedimentation processes in a short residence-time intermontane lake, Kamloops Lake, British Columbia: Sedimentology, v. 26, p. 523–541.

Pollard, D., 1983, Ice-age simulations with a calving ice-sheet model: Quaternary Research, v. 20, p. 30–48.

Prest, V. K., 1970, Quaternary geology, *in* Douglas, R.J.W., ed., Geology and economic minerals of Canada: Geological Survey of Canada Economic Geology Report no. 1, p. 675–764.

—— , 1984, The Late Wisconsinan glacier complex, *in* Fulton, R. J., ed., Quaternary stratigraphy of Canada; A Canadian contribution to IGCP Project 24: Geological Survey of Canada Paper 84–10, p. 22–36.

Prest, V. K., and Grant, D. R., 1969, Retreat of the last ice sheet from the Maritime Provinces Gulf of St. Lawrence Region: Geological Survey of Canada Paper 69–33, 15 p.

Prest, V. R., Grant, D. R., and Rampton, V. N., 1967, Glacial map of Canada: Geological Survey of Canada Map 1253A, scale 1:5,000,000.

Quigley, R. M., 1980, Geology, mineralogy, and geochemistry of Canadian soft soils; A geotechnical perspective: Canadian Geotechnical Journal, v. 17, p. 261–285.

Ruddiman, W. F., and McIntyre, A., 1981, The North Atlantic Ocean during the last deglaciation: Palaeogeography, Palaeoclimatology, and Palaeoecology, v. 35, p. 145–214.

St. Onge, D. A., 1972, Sequence of glacial lakes in north-central Alberta: Geological Survey of Canada Bulletin 213, 16 p.

—— , 1980, The Wisconsin deglaciation of southern Saskatchewan and adjacent areas; Discussion: Canadian Journal of Earth Sciences, v. 17, p. 287–288.

St. Onge, D. A., and Dredge, L. A., 1985, Northeast extension of Glacial Lake McConnell in the Dease River basin, District of MacKenzie: Geological Survey of Canada Paper 85–1A, Current Research, Part A, p. 181–186.

Saltzman, B., 1985, Paleoclimatic modeling, *in* Hecht, A. D., ed., Paleoclimate analysis and modelling: New York, John Wiley and Sons, p. 340–396.

Sancetta, C., Imbrie, J., and Kipp, N. G., 1973, Climatic record of the past 130,000 years in North Atlantic deep-sea core V23-82; Correlation with the terrestrial record: Quaternary Research, v. 3, p. 110–116.

Shaw, J., 1977, Sedimentation in an alpine lake during deglaciation, Okanagan Valley, British Columbia, Canada: Geografiska Annaler 59A, p. 221–240.

Shaw, J., Gilbert, R., and Archer, J., 1978, Proglacial lacustrine sedimentation during winter: Arctic and Alpine Research, v. 10, p. 689–699.

Shepard, F. P., 1937, Origin of the Great Lakes basins: Journal of Geology, v. 45, p. 76–88.

Sly, P. G., and Prior, J. W., 1984, Late glacial and postglacial geology in the Lake Ontario basin: Canadian Journal of Earth Sciences, v. 21, p. 802–821.

Smith, D. E., and Dawson, A. G., eds., 1983, Shorelines and isostasy: Institute of British Geographers Special Publication 16, 398 p.

Smith, N. D., and Ashley, G., 1985, Proglacial lacustrine environment, *in* Ashley, G., Shaw, J., and Smith, N., eds., Glacial sedimentary environments: SEPM Short Course No. 16, Society of Economic Mineralogists and Paleontologists, p. 135–215.

Smith, N. D., and Syvitski, J. P., 1982, Sedimentation in a glacier-fed lake; The role of pelletization on deposition of fine-grained suspensates: Journal of Sedimentary Petrology, v. 52, p. 503–513.

Smith, N. D., Venol, M. A., and Kennedy, S. K., 1982, Comparison of sedimentation regimes in four glacier-fed lakes of western Alberta, *in* Davidson-Arnott, R., Nickling, W., and Fahey, B. D., eds., Research in glacial, glacio-fluvial, and glacio-lacustrine systems: Norwich, Geo Books, p. 203–238.

Spencer, J. W., 1890, Origin of the basins of the Great Lakes of America: Quarterly Journal of the Geological Society London, v. 46, p. 523–533.

Sturm, M., Matter, A., 1978, Turbidites and varves in Lake Brienz (Switzerland); Deposition of clastic detritus by density currents, *in* Matter, A., and Tucker, M. E., eds., Modern and ancient lake sediments: International Association of Sedimentologists Special Publication 2, p. 147–168.

Taylor, R. B., and McCann, S. B., 1983, Coastal depositional landforms in northern Canada, *in* Smith, D. E., and Dawson, A. G., eds., Shorelines and isostasy: Institute of British Geographers Special Publication 16, p. 53–75.

Teller, J. T., 1985, Glacial Lake Agassiz and its influence on the Great Lakes, *in* Karrow, P. F., and Calkin, P. E., eds., Quaternary evolution of the Great Lakes: Geological Association of Canada Special Paper 30, p. 1–16.

—— , 1987, Lake Agassiz and its contribution to flow through the Ottawa–St. Lawrence system, *in* Gadd, N., ed., The Champlain Sea: Geological Association of Canada Special Paper (in press).

Teller, J. T., and Clayton, L., eds., 1983, Glacial Lake Agassiz: Geological Association of Canada Special Paper 26, 451 p.

Teller, J. T., and Thorleifson, L. H., 1983, The Lake Agassiz–Lake Superior connection, *in* Teller, J. T., and Clayton, L., eds., Glacial Lake Agassiz: Geological Association of Canada Special Paper 26, p. 261–290.

Teller, J. T., Moran, S. R., and Clayton, L., 1980, The Wisconsinan deglaciation of southern Saskatchewan and adjacent areas; Discussion: Canadian Journal of Earth Sciences, v. 17, p. 539–541.

Teller, J. T., Thorleifson, L. H., Dredge, L. A., Hobbs, H. C., and Schreiner, B. T., 1983, Maximum extent and major features of Lake Agassiz, *in* Teller, J. T., and Clayton, L., eds., Glacial Lake Agassiz: Geological Association of Canada Special Paper 26, p. 43–45.

Thomas, G., and Connell, R., 1985, Iceberg drop, dump, and grounding structures from Pleistocene glacio-lacustrine sediments, Scotland: Journal of Sedimentary Petrology, v. 55, p. 243–249.

Vincent, J-S., and Hardy, L., 1979, The evolution of glacial Lakes Barlow and Ojibway, Quebec and Ontario: Geological Survey of Canada Bulletin 316, 18 p.

Walcott, R. I., 1970, Isostatic response to loading of the crust in Canada: Canadian Journal of Earth Sciences, v. 7, p. 716–726.

Whillans, I. M., 1978, Inland ice sheet thinning due to Holocene warmth: Science, v. 201, p. 1014–1016.

Willman, H. B., and Frye, J. C., 1970, Pleistocene stratigraphy of Illinois: Illinois Geological Survey Bulletin 94, 204 p.

Wright, H. E., 1971, Retreat of the Laurentide ice sheet from 14,000 to 9,000 years ago: Quaternary Research, v. 1, p. 316–330.

—— , 1972, Quaternary history of Minnesota, *in* Sims, P. K., and Morey, G. B., eds., Geology of Minnesota: Minnesota Geological Survey, p. 515–548.

—— , 1973, Tunnel valleys, glacial surges, and subglacial hydrology of the Superior Lobe, Minnesota: Geological Society of America Memoir 136, p. 251–276.

—— , 1984, Sensitivity and response time of natural systems to climatic change in the late Quaternary: Quaternary Science Reviews, v. 3, p. 91–131.

Wright, H. E., Matsch, C. L., Cushing, E. J., 1973, Superior and Des Moines Lobes, *in* Black, R. F., Goldthwait, R. P., and Willman, H. B., eds., The Wisconsinan stage: Geological Society of America Memoir 136, p. 153–185.

Zoltai, S. C., 1961, Glacial history of part of northwestern Ontario: Proceedings, Geological Association Canada, v. 13, p. 61–83.

—— , 1963, Glacial features of the Canadian Lakehead area: Canadian Geographer, v. 7, p. 101–115.

—— , 1965, Glacial features of the Quetico–Nipigon area, Ontario: Canadian Journal of Earth Sciences, v. 2, p. 247–269.

MANUSCRIPT ACCEPTED BY THE SOCIETY FEBRUARY 23, 1987

ACKNOWLEDGMENTS

My thanks to J. Andrews, G. Ashley, W. Ruddiman, E. Birchfield, and H. E. Wright, Jr. for reviewing this paper and offering suggestions for improvement. Thanks also to C. Peters and R. Pryhitko.

Printed in U.S.A.

The Geology of North America
Vol. K-3, North America and adjacent oceans during the last deglaciation
The Geological Society of America, 1987

Chapter 4

Timing and processes of deglaciation along the southern margin of the Cordilleran ice sheet

Derek B. Booth
U.S. Geological Survey, Quaternary Research Center, University of Washington AK-60, Seattle, Washington 98195

INTRODUCTION

Approach

The Cordilleran ice sheet covered the northwest part of the North American continent during the last glaciation (Fig. 1). It developed from an initial core of coalescing mountain glaciers on Vancouver Island and the British Columbia mainland, spreading outward over a period of 10,000 to 15,000 yr. South of the ice-sheet limit, isolated alpine glaciers fluctuated in size, leaving a similar but not identical record of glacial advance and retreat. From the behavior of these glaciers, three questions have been posed and, in part, addressed by previous studies (including Crandell, 1965; Porter, 1976; Hicock and others, 1982; Clague, 1981; Waitt and Thorson, 1983). (1) Why did the Cordilleran ice sheet attain its maximum 5,000 yr later than its smaller counterparts to the south? (2) What was the extent of the alpine glaciers during the ice-sheet maximum, and why were most apparently well back from their maximum position? (3) What physical factors determined the rate and character of final ice-sheet retreat?

This chapter approaches these questions by applying current knowledge of glacial mechanics, both theoretical and empirical, to various aspects of the inferred or reconstructed Cordilleran glaciers. The record of advance and retreat should reflect changes in the external environment (regional climate, sea-level changes), the glaciers' physical responses to those changes, and changes that in turn result from ice growth and decay (isostasy, local climate). Existing data on the late-glacial advance-retreat chronology and climate constrain this analysis and provide an independent check on the conclusions suggested by this approach.

Although the primary focus here is on the mechanics of deglaciation, this chapter considers the record of Cordilleran ice advance as well for several reasons. Retreat of most of the other North American ice sheets (Prest, 1969) coincided with the Cordilleran ice's achievement of maximal advance, about 15 ka (Clague and others, 1980). Retreat of Cordilleran glaciers was neither monotonic nor entirely synchronous, in that some glaciers had retreated from their maximum positions 5,000 yr before others had finished advancing. Finally, the full advance-retreat record characterizes the physical behavior of the Cordilleran glaciers more completely, because the same physical principles determine glacier behavior regardless of the direction of motion of the ice terminus.

The most detailed and best studied record of ice advance and retreat is found along the southern margin of the Cordilleran ice sheet, particularly in the Puget and Fraser Lowlands. Because the chronology, physical setting, and environmental changes are particularly well constrained in this region during late-glacial time, the geologic record here offers an excellent opportunity to evaluate the responses of glaciers of various sizes to late Pleistocene environmental changes and interactions among them.

The following analysis of those responses is obviously indebted to nearly a century of geologic investigations of this region, together with recent advances in climatology, palynology, and glaciology. I have also benefited greatly from prior efforts to quantify the physical behavior of Pleistocene glaciers in western North America (Pierce, 1979; Thorson, 1981).

Regional Setting

The southern part of the Cordilleran ice sheet occupied a distinctive physiographic region (Fig. 2). In British Columbia a broad topographic basin extends southeastward as the Georgia Depression and the Fraser Lowland, bounded by Vancouver Island and the mainland coast mountains. The basin continues south into Washington State at the Puget Lowland, bordered by the Olympic Mountains on the west and the Cascade Range on the east. Low hills at about 46°45′ define the southern limit of glacial advance in this basin, but the lowland province itself extends south for at least several hundred kilometers more. The basin has persisted for millions of years, as shown by the great thickness of its Cenozoic fill (e.g., Weaver, 1937). East of the Cascade Range the ice sheet extended from the mountainous highlands in Canada onto the northern edge of the Columbia Basin, primarily along major south-trending valleys.

Booth, D. B., 1987, Timing and processes of deglaciation along the southern margin of the Cordilleran ice sheet, *in* Ruddiman, W. F., and Wright, H. E., Jr., eds., North America and adjacent oceans during the last deglaciation: Boulder, Colorado, Geological Society of America, The Geology of North America, v. K-3.

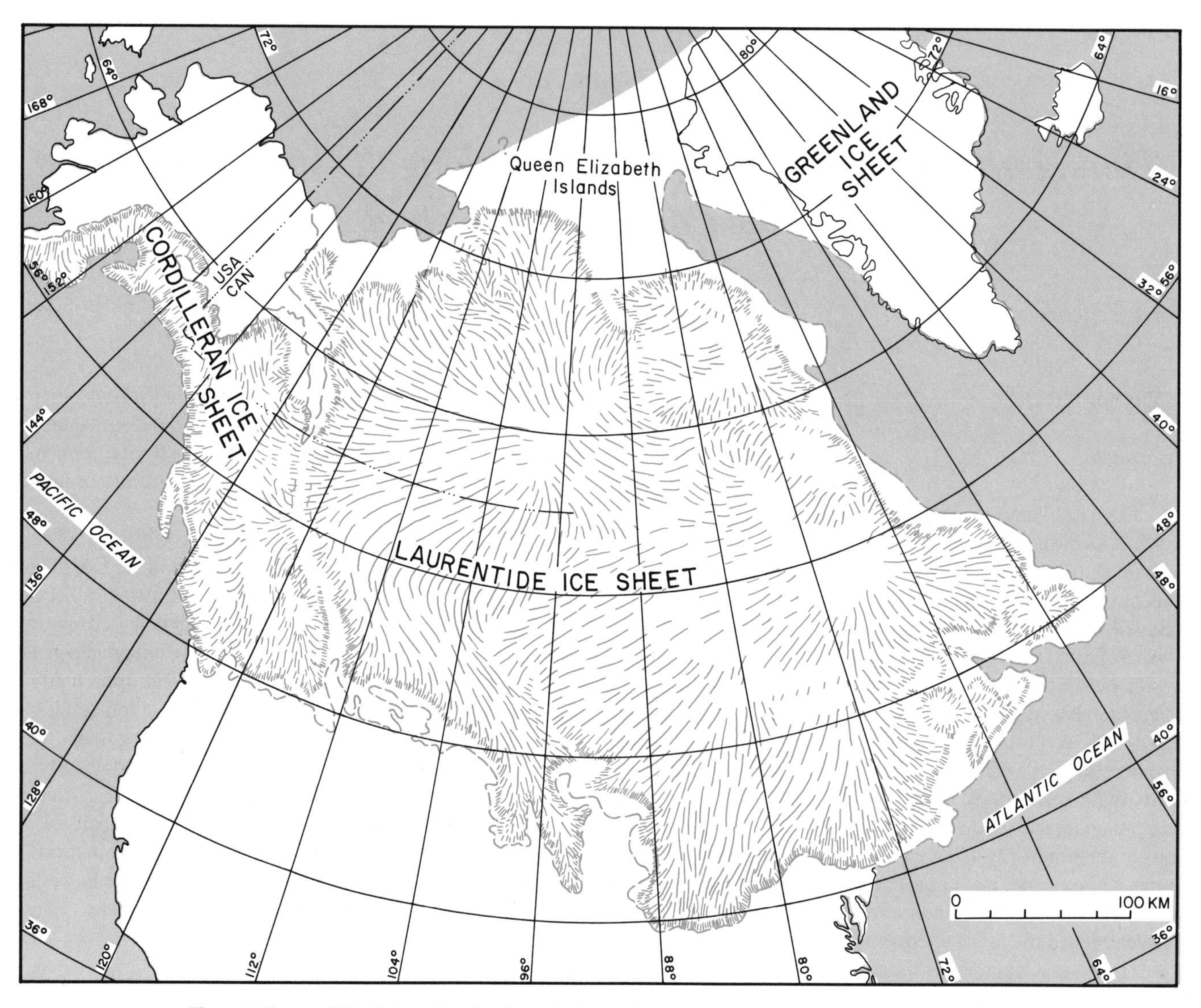

Figure 1. Extent of North American ice sheets during the last glaciation. Dashed lines indicate maximum proposed reconstructions along the northern, eastern, and southern boundaries.

In the west at latitude 48°15′, the Strait of Juan de Fuca separates Vancouver Island from the Olympic Mountains and opens westward to the continental shelf. Bottom depths in the strait, typically 100 to 200 m below modern sea level, are comparable to those throughout the Georgia Depression and in the deeper parts of the Puget Lowland that are occupied by Puget Sound. They lie more than 2,000 m below the summits of the adjacent coastal mountains. In contrast, the continental shelf is a low-relief surface over 50 km wide, gently sloping seaward to a depth of about 200 m before dropping precipitously down the continental slope.

Growth of the Cordilleran ice sheet has occurred several times during the Pleistocene, leaving a discontinuous record of glacial advances throughout the Puget and Fraser Lowlands (Crandell and others, 1958; Armstrong and others, 1965; Easterbrook and others, 1967). By analogy with the well-documented pattern of the last glaciation (the Fraser glaciation of Armstrong and others, 1965), coalescing mountain ice caps advanced into the Georgia Depression and split around the Olympic Mountains into the Juan de Fuca and Puget lobes (Fig. 2). East of the Cascade Range, ice from the eastern slopes of the Canadian Coast Ranges probably met west-flowing ice from the Rocky Mountains of Canada, extending south as a series of lobes and sublobes into eastern Washington, Idaho, and Montana. The Cascade and Olympic mountains south of the ice-sheet limit were locally high enough to support isolated mountain glaciers as well.

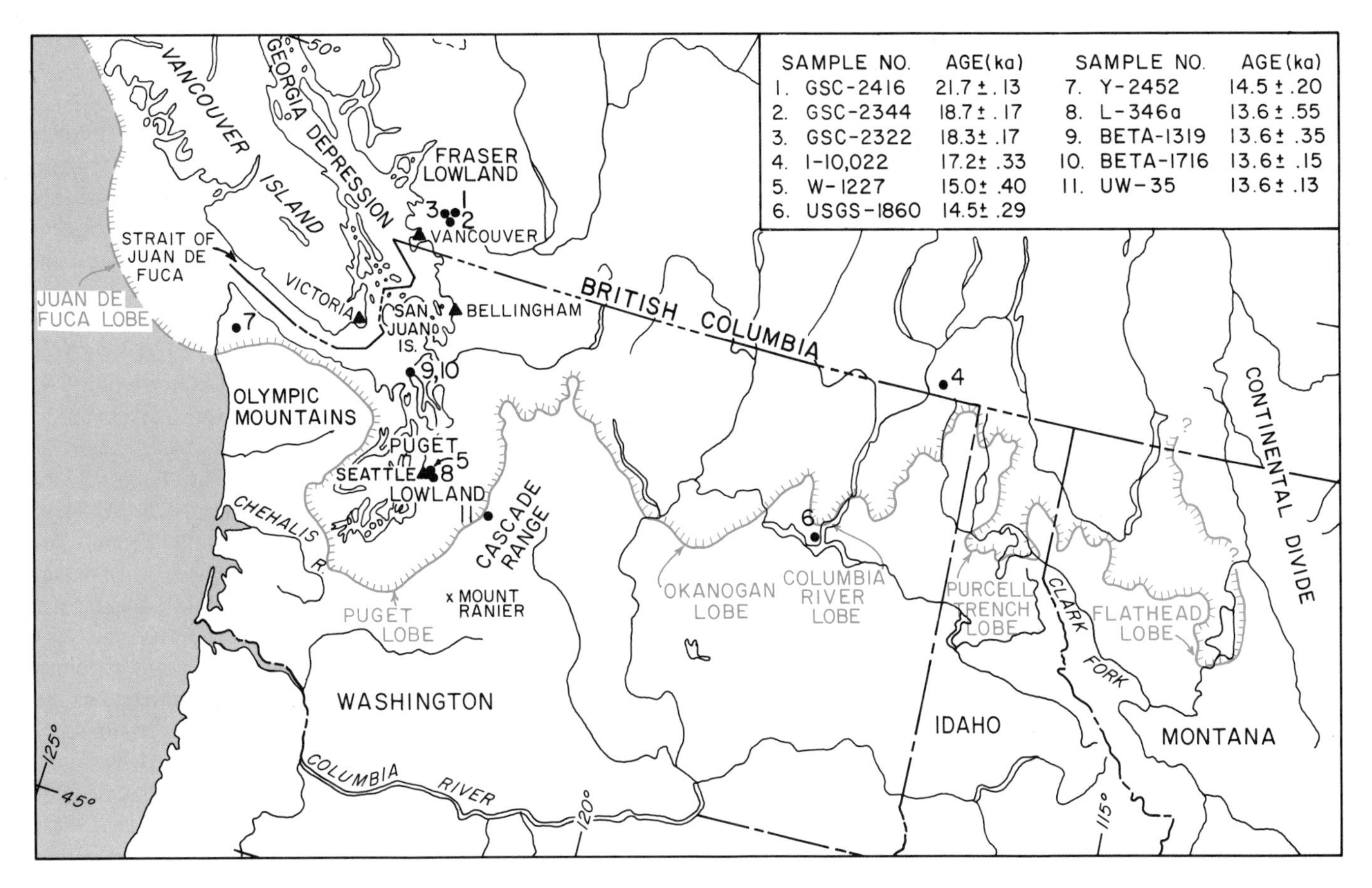

Figure 2. Index map of the southern part of the Cordilleran ice sheet. Location of critical radiocarbon dates discussed in text are indicated by sample number and age in thousands of years (ka).

GEOLOGIC AND CLIMATOLOGIC DATA

Chronology of the Fraser glaciation

Introduction. Although the history of glaciations and interglaciations for the Cordilleran ice sheet is poorly constrained for most of the Pleistocene, the record of the latest glaciation is well documented and extensively dated. Armstrong and others (1965) named and subdivided the Fraser glaciation into three stades and one named interstade (named glaciations, interglaciations, stades, and interstades are used informally in this report, although their original use had presumed them to be formal geologic-climate units). Although the relative ranking of these intervals continues to be deliberated, their overall stratigraphy provides a useful and widely accepted framework for this period. Other recent summaries of the late Pleistocene glaciation are found in Armstrong (1981), Clague (1981), Waitt and Thorson (1983), and Pessl and others (1987).

Pre-Fraser interval. Preceding the Fraser glaciation, nonglacial conditions prevailed in the lowlands of the Pacific Northwest. Named the Olympia interglaciation by Armstrong and others (1965), deposits associated with this period are as old as 40.5 ±1.7 ka and perhaps 58.8 +2.9/−2.1 ka (^{14}C sample numbers GSC-2167 and QL-195; Clague, 1981, p. 5). The end of the Olympia interval is marked stratigraphically by glacial deposits derived from either the expanding Cordilleran ice sheet or from local mountain glaciers in the Cascade Range and Olympic Mountains. Because of the great distance between these glacial-sediment sources, local dates indicating the end of the Olympia nonglacial interval span nearly 10,000 yr.

Growth of interior ice cap. In response to climatic deterioration at the close of the Olympic nonglacial interval, mountain ice caps on Vancouver Island and the British Columbia mainland grew and coalesced. Proglacial outwash, particularly along the Georgia Depression, records the slow southward expansion of this ice. Minimum dates for the outwash include 28.8 ±0.74 ka (GSC-95) in the northern Georgia Depression, 18.3 ±0.17 and 18.7 ±0.17 ka (GSC-2322 and GSC-2344) near Vancouver, and 15.1 ±0.4 ka (W-1227) near Seattle (Dyck and Fyles, 1963, p. 49–50; Armstrong and Clague, 1977, Fig. 1; Mullineaux and others, 1965, p. 7).

Early Fraser maximum stage. During the buildup of ice in the British Columbia mountains, a more short-lived expansion of alpine ice occurred in the mountains of Washington and southern British Columbia. During this interval, named the Evans Creek stade (of the Fraser glaciation) by Armstrong and others (1965), valley glaciers on Mount Rainier and elsewhere in the Cascade Range and Olympic Mountains expanded to their late Pleistocene maximum positions (Crandell, 1963; Crandell and Miller, 1974; Carson, 1970; Porter, 1976). No absolute ages directly date this advance in Washington, but it is correlated by Porter and others (1983) with a glacial advance into the eastern Fraser Lowland from the adjacent coast mountains in British Columbia. Although this correlation implies a regional episode of climatic cooling or increased precipitation, nowhere north of Vancouver or east of the Cascade Range have any conclusive signs been found of other correlative glacial episodes that similarly punctuate the slow growth of the main Cordilleran ice sheet (Clague, 1981). The advance of alpine ice into the Fraser Lowland, depositing the Coquitlam Drift, culminated about 20 ka, according to limiting dates in fluvial sediment of 21.7 ± 0.13 ka (GSC-2416) and 18.7 ± 0.17 ka (GSC-2344; Hicock and Armstrong, 1981).

Lowland nonglacial interval. Following the Evans Creek and Coquitlam advances, alpine glaciers retreated to undetermined positions in their respective mountain valleys. Cary and Carlston (1937) and Mackin (1941) first inferred that these lower alpine valleys were ice-free at some time during the (subsequent) lowland occupation by the Cordilleran ice sheet. Direct evidence, however, for the retreat of Evans Creek ice prior to the *first* arrival of the southward-expanding ice sheet has been reported only in the Fraser Lowland (dated fluvial sediments overlying Coquitlam Drift; Hicock and Armstrong, 1981, sample GSC-2344) and just east of the central Puget Lowland (undated fluvial sediments between alpine and ice-sheet tills; Booth, 1987a).

Ice-sheet advance to maximum. The Vashon stade of the Fraser glaciation of Armstrong and others (1965) encompasses the growth of the Cordilleran ice sheet to its late Pleistocene maximum position (Figs. 3 and 4). In the Fraser and northern Puget Lowlands, this ice occupied the terrain probably covered previously by glaciers of the Evans Creek/Coquitlam advance after 18.3 ka (GSC-2322; Armstrong and Clague, 1977, Fig. 2; see also Clague and others, 1980, Table 1). In this region the ice sheet probably merged with the Vashon-age remnants of these alpine glaciers. At the lateral margins of the central and southern Puget Lowland, tongues of the ice sheet extended up into alpine valleys over areas previously occupied by Evans Creek glaciers (Knoll, 1967; Williams, 1971; Booth, 1986a, 1987a), but at no time did ice from these two sources meet. A readvance of alpine glaciers, of smaller magnitude than during Evans Creek time, has been tentatively correlated with the Vashon advance (Porter, 1976).

Precise dating of the Vashon maximum suffers from a paucity of dates near the terminus. Postglacial sediments on the northwestern Olympic Peninsula, about 50 km behind the maximum ice limit of the Juan de Fuca lobe, specify a minimum retreat date of 14.46 ±0.20 ka (Y-2452; Heusser, 1973). In the Seattle area, 100 km north of the Puget-lobe limit, the Vashon maximum is bracketed between dates of 15.0 ±0.4 ka (W-1227) and 13.65 ±0.55 ka (L-346a) (Mullineaux and others, 1965, p. 7; Rigg and Gould, 1957). East of Seattle at the edge of the lowland, wood in lake sediments dated at 13.57 ±0.13 ka (UW-35; Porter and Carson, 1971) provides the only other possible constrain on the Puget-lobe maximum. The lake, however, was probably debris-dammed long after ice had retreated from the adjacent lowland areas (Booth, 1987b); active Vashon-age ice had probably retreated from the Seattle area before 13.6 ka.

Limits on the Fraser maximum are broader east of the Cascade and Coast ranges. They include a date of 17.24 ±0.33 ka (I-10,022) in proglacial (advance) outwash just north of 49°N, 100 km north of the ice limit, followed by advance to maximum and retreat of at least 80 km in the following 6,000 yr, based on the distribution of Glacier Peak tephra-layer G at 11.18 ±0.15 ka (WSU-2668) (Clague and others, 1980; Porter, 1978, p. 38–39; Mehringer and others, 1984). Lake Missoula flood deposits, interbedded with varves of glacial Lake Columbia that contain detrital wood dated at 14.49 ±0.29 ka (USGS-1860), suggest that the Purcell Trench lobe of the Cordilleran ice sheet lay across the Clark Fork for 2,000 to 3,000 years and reached its terminus about 15 ka or later (Atwater, 1986). These interpretations are consistent with the presence of Mount St. Helens tephra-set S between Lake Missoula flood deposits (Waitt, 1980), which indicates blockage of the Clark Fork by the Purcell Trench lobe until at least 13.08 ±0.30 ka (W-3404; Mullineaux and others, 1978).

Thus the Fraser maximum on both sides of the Cascade Range was at least broadly synchronous (Waitt and Thorson, 1983, p. 67). The Fraser maximum coincides with the maximum advance of parts of the Laurentide ice sheet in central North America but lags 2,000 to 6,000 yr behind the culmination of most of the rest of the Laurentide ice sheet (Prest, 1969; Mickelson and others, 1983).

Ice-sheet retreat and the Everson interval. Rates and timing of the initial retreat of the Puget and Juan de Fuca lobes are relatively well constrained by radiocarbon dates. The Juan de Fuca lobe retreated 50 km by 14.46 ka (Y-2452; see above). The next-oldest sediments are found in the north-central Puget Lowland (48°30′N latitude), 150 km north of the Puget-lobe limit and 250 km east of the Juan de Fuca limit. Dates of 13.60 ±0.15 ka (BETA-1716) and 13.65 ±0.35 ka (BETA-1319) (D. P. Dethier and others, written communication, 1986) from shells in glacial-marine sediment specify the maximum southern limit of grounded ice at this time.

During the preceding 1,000 years, the rate and pattern of ice retreat are known only indirectly from undated landforms and deposits. North of about 47°30′N latitude, the western part of the Puget lobe left little erosional or depositional record of its retreat, suggesting a rapid and chaotic decay (Thorson, 1980). South of this line and near the eastern Puget-lobe margin, abundant marginal channels and sequential recessional deposits imply a systematic and probably slower marginal recession (Newcomb, 1952; Crandell, 1963; Curran, 1965; Anderson, 1965; Knoll, 1967; Thorson, 1980; Booth, 1987a, 1987b). Conversely, irregu-

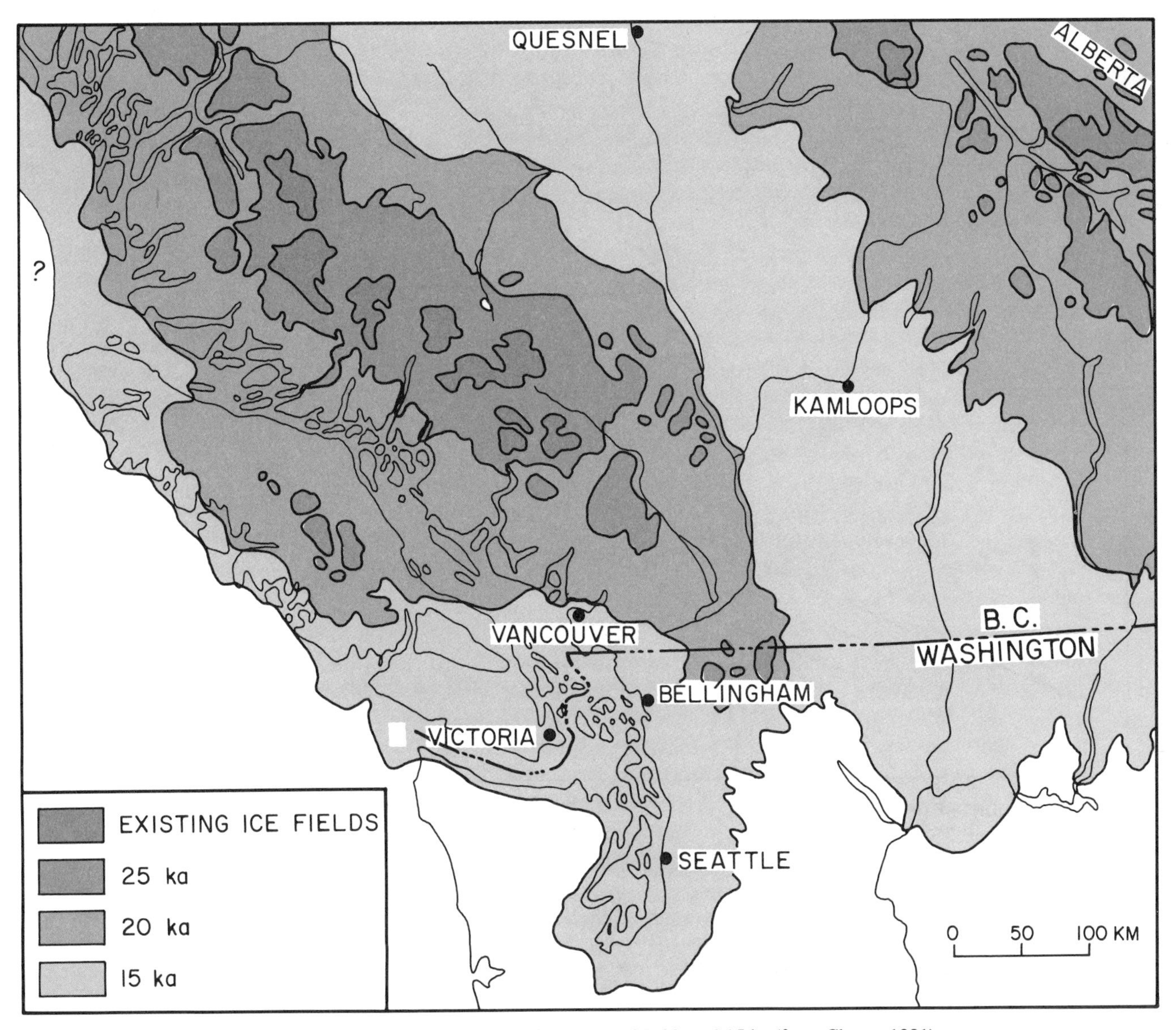

Figure 3. Extent of the Cordilleran ice sheet at 25, 20, and 15 ka (from Clague, 1981).

lar constructional landforms in the eastern Strait of Juan de Fuca (Anderson, 1968; Chrzastowski, 1980) suggest rapid retreat of the Juan de Fuca lobe, constrained by the radiocarbon dates to be at least 200 m/yr.

In the central and northern Puget Lowland, incursion of marine water via the Strait of Juan de Fuca resulted in deposition of abundant glacial-marine drift (Easterbrook, 1963; Thorson, 1980; Pessl and others, 1987; Dethier and others, written communication, 1986). Dated marine and glacial-marine sediments in this region are as old as 13.65 ±0.35 ka (BETA-1319) in the north-central Puget Lowland and 13.5 ±0.2 ka (GSC-3124) just south of the International Boundary. Deposition of these sediments continued until at least 11.30 ±0.07 ka (USGS-124) in the central Puget Lowland (Armstrong, 1981; Dethier and others, written communication, 1986).

This suite of dates marks the end of an interval that probably lasted less than 1,000 years, during which the ice lobe retreated 100 km primarily by calving, with local backwasting and stagnation (Pessl and others, 1987). Net retreat was therefore at least 100 m/yr and may have been several times more rapid. Marine sedimentation, with and without glacial input via submarine flows and ice-rafted debris (Domack, 1983), continued for at least another 1,000 yr in areas subsequently lifted above sea level by isostatic rebound.

East of the Cascade Range, ice-sheet retreat probably occurred more slowly (Fig. 4). The limiting dates on the lobes in the

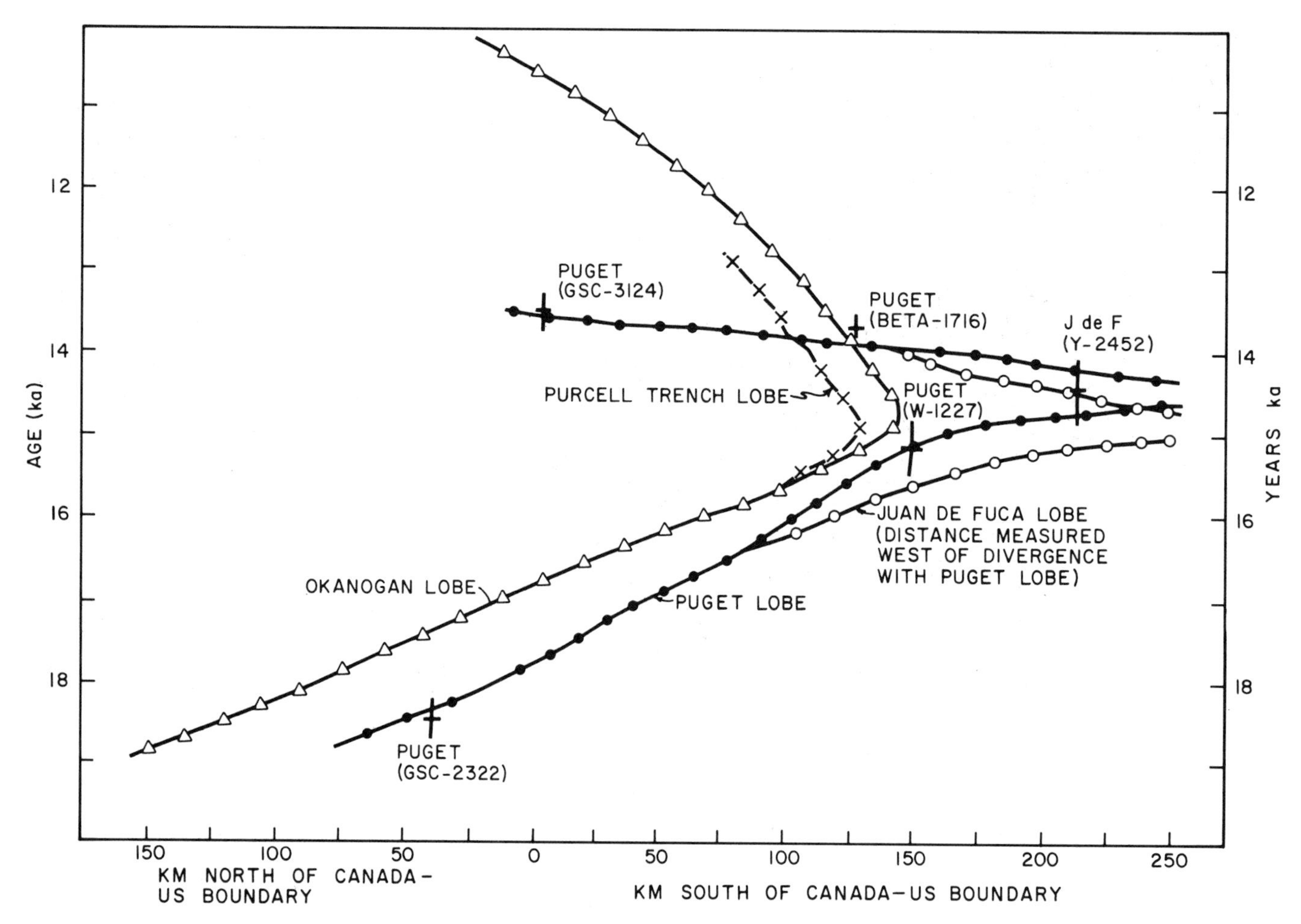

Figure 4. Advance and retreat of the Juan de Fuca, Puget, Okanogan, and Purcell Trench lobes of the Cordilleran ice sheet. Distances are measured relative to the International Boundary except for the Juan de Fuca lobe, which is measured from its western divergence from the Puget lobe. Crosses locate critical dates and reported standard errors on nonglacial deposits in the Puget and Fraser Lowlands. The curve for the Puget lobe differs from Waitt and Thorson (1983, Fig. 3-2) in assuming a relatively uniform advance rate during the period 19 to 15 ka and in more recent limiting dates on the retreat at 13.6 ka (BETA-1716; Dethier and others, written communication, 1986). Curves for the two eastern lobes (Okanogan and Purcell Trench) are from Atwater (1986).

northwest interior indicate 180 km of advance and retreat during the interval 17.24 to 11.18 ka (see above; Mullineaux and others, 1978; Clague, 1981; Mehringer and others, 1984). They permit an interpretation of slower ice retreat in the east (as little as 30 m per year) but do not constrain a hypothesized lag in eastern ice-sheet retreat (Waitt and Thorson, 1983; e.g., see Atwater, 1986, Fig. 28).

Sumas interval and final deglaciation of the lowlands. Glacial sediment in the eastern Fraser Lowland near the International Boundary was initially interpreted as recording a widespread readvance of Cordilleran ice during the Sumas stade of Armstrong and others (1965). Dates on its maximum position range from 11.70 ±0.15 ka (L-3313) to 11.40 ±0.17 ka (GSC-1695), with final deglaciation of the lowland complete by 11.1 ka (Armstrong, 1981). This later work, however, has questioned the climatic significance of the Sumas interval, suggesting instead that the readvance marked only the glacier's response to isostatic emergence of the terminus by reduction in the rate of calving.

Elsewhere in the Pacific Northwest, evidence for a glacial readvance ca. 11.5 ka is ambiguous. East of the Cascade Range the oldest nonglacial date north of 49°N latitude is only 11.00 ±0.18 ka (GSC-909; Lowden and Blake, 1970, p. 71). Additional evidence suggests that retreat across the International Boundary in most areas followed deposition of Glacier Peak tephra-layer G at 11.18 ±0.15 ka (WSU-2668; Mehringer and others, 1984; Clague, 1981, p. 17). Recessional stillstands or readvances of the eastern Cordilleran lobes are recorded by glacial deposits (Waitt and Thorson, 1983, p. 67) but remain undated and uncorrelated.

Alpine glaciers in the Cascade Range show widespread evidence of a late-glacial readvance, the Rat Creek advance of Page (1939; summarized in Porter and others, 1983, p. 86). Originally inferred to predate 11.05 ±0.05 ka (UW-321) and to postdate the 12-ka Glacier Peak tephra-layer M, this advance was correlated provisionally with the Sumas interval by Porter (1978). Additional data on tephra distribution and age, however, cast doubt on both the upper age constraint and its applicability here; they instead suggest only a pre–11.2 ka age that may have no correlation to the regional climate of Sumas time (Porter and others, 1983, p. 87; Mehringer and others, 1984).

Fraser-age lowland climate

Plant-fossil record. The climate of the Pacific Northwest during the Fraser glaciation has been reflected not only in the growth and wastage of glaciers but also by the changing influx of pollen and plant macrofossils into lakes and wetlands of the region. Such data provide an independent picture of temperature and precipitation during this period. The value of this information to glacial geology and reconstruction lies in the spatially discrete and easily dated nature of the samples, which, when correlated among multiple sites over a region, provide a fairly complete picture of the near-glacier climate. The response of local plant communities to climate should also be more rapid than that of a continent-scale ice sheet (Wright, 1984).

Yet as a companion to a study of a glacier's response to climate, the pollen record also has two significant limitations. First, it may be influenced by the physical proximity of the ice, implying a set of climatic conditions that exist only locally. Second, the conditions of temperature and precipitation over the main body of the ice sheet may correlate only poorly with those recorded at or beyond its margins.

Despite these potential problems, a reasonably consistent picture of glacial-age climate has emerged from palynologic study throughout the region that provides a climatic framework to complement the previously described chronology of ice advance and retreat. Barnosky's (1984) synthesis of data from sites in south-central Washington identifies several climatically distinct periods in the Puget Lowland during the Fraser glaciation (see also Barnosky and others, this volume): (1) 26 to 19 ka: cold and dry parkland-tundra environment; (2) 19 to 17 ka: cold and drier than before; (3) 17 to 15 ka: possible warming, indicated by fossil-plant and insect data at some sites (4) 15 to 12.5 ka: cool and increasingly wet maritime conditions; and (5) post-12.5 ka: widespread warming; possibly beginning as much as 1,000 years earlier east of the Cascades (Barnosky, 1985).

Most of these climatic conclusions are in broad accord with direct records of glacial growth and wastage. Both the Evans Creek and Vashon maxima coincide with cold ("glacial") conditions, and the final wastage of Cordilleran ice roughly coincides with the late-glacial warming. Some of the details of these two chronologies, however, fit less readily. The main western lobe of the Cordilleran ice sheet must have advanced south into the Fraser Lowland during either the period of enhanced aridity before 17 ka or immediately following it, during a time of possibly increased warmth in the southern Puget Lowland (the "unnamed interstade" of Barnosky, 1981, 1984). To explain this apparent contradiction, Hicock and others (1982) suggest an accompanying increase in precipitation. Retreat of the Juan de Fuca and Puget lobes at ca. 15 ka appears to have occurred during a time of increasingly cool and wet climate, conditions that intuitively should *favor* ice-sheet stability and growth. The initial and most rapid retreat of ice west of the Cascade Range preceded general warming by several thousand years, an asynchrony that is difficult to explain without postulating a lag in the vegetational response of 2,000 yr or more (see also Barnosky and others, this volume).

East of the Cascades, pollen-recorded climatic fluctuations are quite sparse. Barnosky's reconstruction for Fraser time, based on a single site in the Columbia Basin, only specifies a long interval of periglacial steppe or tundra from 23.5 to 10 ka. Glacier fluctuations also are not as well constrained as those west of the Cascades, rendering any present effort to make a detailed comparison of eastern and western chronologies nearly meaningless.

Global climate models. Numerical simulations of late Quaternary climate provide a supplementary view of some of the external controls on ice-sheet growth and decay. All available simulations share the disadvantage of low spatial resolution; the entire region of the Cordilleran ice sheet is typically represented by only a few grid points. Nevertheless, these efforts provide an independent view of probable large-scale climatic conditions during glacial time and help identify certain determinants of ice-sheet behavior.

Model simulations of the ice-age climate, typically at ca. 18 ka, generally agree with one another in their characterization of conditions in the Pacific Northwest region (e.g., Gates, 1976; Manabe and Hahn, 1977; Kutzbach, this volume). Relative to present conditions here, sea-surface temperatures estimated from ocean-core data were 2° to 4°C colder (CLIMAP, 1976), land temperatures were less than 5° colder beyond the ice-sheet limit and up to 20° colder toward its interior (Manabe and Hahn, 1977, Fig. 20; Kutzbach and Guetter, 1986, Figs. 5 and 10), and both precipitation and runoff differed by less than 1 mm/day (Manabe and Hahn, 1977, Figs. 8b and 9; Kutzbach and Guetter, 1986, Figs. 9 and 14). The primary deglacial change in the region was an increase in temperature (Kutzbach, this volume); neither precipitation nor prevailing winds show dramatic change during the period 18 to 12 ka.

THE PHYSICAL BEHAVIOR OF THE CORDILLERAN GLACIERS

The ice build-up phase

The alpine-glacial maximum. Alpine glaciers from the mountains flanking the Puget and Fraser Lowlands advanced to their maximum position during the Fraser glaciation, probably about 20 ka. Although the main Cordilleran ice sheet was still several hundred kilometers and 5,000 years away from its maximum stand, this widespread alpine-glacial maximum must record a period of regional climate change that favored ice growth. Their subsequent, pre-Vashon retreat in turn required no more than a minor warming or drying.

The most useful parameter to quantify the effect of climate on a glacier is the glacier's equilibrium-line altitude (ELA). This value defines the altitude at which the yearly gain of snow equals its yearly loss by ablation. Above the ELA lies the accumulation zone, where a year's supply of snow is not completely melted; below lies the ablation zone, where a net yearly loss of mass occurs. The rate at which mass is added or subtracted from the glacier is a function of position on the glacier and is proportional to glacier-surface altitude relative to the ELA (the "balance gradient"; cf. Schytt, 1967).

The ELA of a glacier can be estimated by several methods. A regional relationship of height to mass balance has been compiled from data on seven modern Pacific Northwest maritime glaciers (Fig. 5; Meier and others, 1971; Porter and others, 1983). If a glacier lies in a similar climatic regime and its boundaries and ice-surface contours can be reconstructed at equilibrium conditions, this curve can be used directly to find an ELA that brings the glacier into balance (see below). A simpler approximation is to assume a ratio of the accumulation area to the total glacier area (the accumulation-area ratio, or AAR) of 0.60 ± 0.05, observed on modern maritime glaciers (Porter, 1975). Finally, the regional ELA can also be estimated by the glaciation threshold method (e.g., Porter, 1977).

These techniques have been applied to various regional and local areas in the Cascade Range and Olympic Mountains (Porter, 1977; Williams, 1971; Waitt, 1977; Booth, 1986b, 1987b). Median ELA values in northern and central Washington lie near 1,000 m for Evans Creek time, about 900 m lower than present. A pronounced east-west decrease in ELA across the Cascade Range, inferred to reflect greater precipitation on the Pacific side of the mountains, complicates this picture. At latitude 47°30′N, for example, the glaciation threshold descends westward from over 1,500 m at the Cascade crest to 750 m at the western rangefront, a gradient of about 10 m/km (Porter, 1977, Fig. 7).

Glaciers that occupied the west-draining alpine valleys probably integrated the effect of traversing a range of local ELA values. Inspection of Porter's (1964, 1977) data suggests that such composite values probably lay in the range of 1,100 ± 100 m for glaciers that extended from the crest of the range to their base in the Puget Lowland. Those glaciers that headed in subsidiary ridges west of the main Cascade crest, such as documented by Williams (1971) and Booth (1987a, 1987b), experienced wetter conditions and had ELA values about 100 to 200 m lower. Conversely, those glaciers that did not extend as far west as the Cascade rangefront (e.g., the reconstructed Skagit Valley glacier in Waitt, 1977, Fig. 2) had median ELAs one to several hundred meters higher.

The rate and magnitude of alpine-glacier retreat is poorly documented. Porter (1976) suggests that moraines in the southern North Cascade Range lying 20 km upvalley of the Fraser ice-sheet maximum may indicate the minimum amount of retreat that occurred during the millennia intervening before the ice-sheet maximum, equivalent to the removal of a layer of ice 300 to 400 m thick (Porter, 1976, Fig. 6). In the time interval available, this would have necessitated an annual deficit of 1 m of ice or less, corresponding to an approximate 100-m rise in the ELA (Fig. 3; also Østrem, 1975). An ELA rise of this magnitude represents only about 10 percent of the maximum Fraser-age depression. Even with an additional rise in the ELA of about 100 m to account for these reduced glaciers (Porter, 1976, Fig. 6), the climatic warming or drying represented by post–Evans Creek glacial retreat represents but a small fraction of the maximum Fraser-age climatic change. The potential effect of isostatic depression by the ice lobe in the adjacent Puget Lowland does not substantially affect this conclusion (see below).

Ice advance along the Georgia Depression. *Introduction.* While most alpine glaciers in the Pacific Northwest probably reached their late-glacial maxima about 20 ka, the Cordilleran ice sheet was still 5,000 years from reaching its farthest southern limit. Plausible explanations for this asynchrony include differences in glacier response times or in terminus position relative to sea level. Although the ice sheet typically is characterized as advancing steadily down the Georgia Depression during this 5,000-year interval, no evidence supports this assumption (Clague, 1981, p. 10–11). Such data, however, might be recorded only along the axis of the Georgia Depression in sediments that either are now under hundreds of meters of water or were completely eroded away after their deposition.

Response times. To any perturbation in mass balance, such as that caused by a change in temperature or precipitation, a glacier will respond by a net increase or decrease in its mass that ultimately results in movement of its terminus. Climatic change and the resulting glacier adjustments, however, are not related simply. The flowing ice averages the effects of short-term climatic fluctuations (Kuhn, 1981), and the response at the terminus lags behind the long-term climatic trend. Because the geologic record typically documents only the position of the terminus through time, distortions and delays characterize this geologic reflection of paleoclimate.

Nye (1960, 1963, 1965a, 1965b) has attempted to quantify most comprehensively the relationship between climatic change and glacial fluctuation. He considers a two-dimensional glacier resting on a sloping planar bed. The discharge of ice at any cross section depends on the position downglacier, ice thickness, and ice-surface slope. By considering a small perturbation of the mass balance, he shows that the disturbance should move along the glacier's length as a kinetic wave—a "wave" of constant discharge per unit of cross-sectional area that moves through the glacier.

Nye's theory predicts not only changes in thickness and discharge but also response times needed to accomplish these

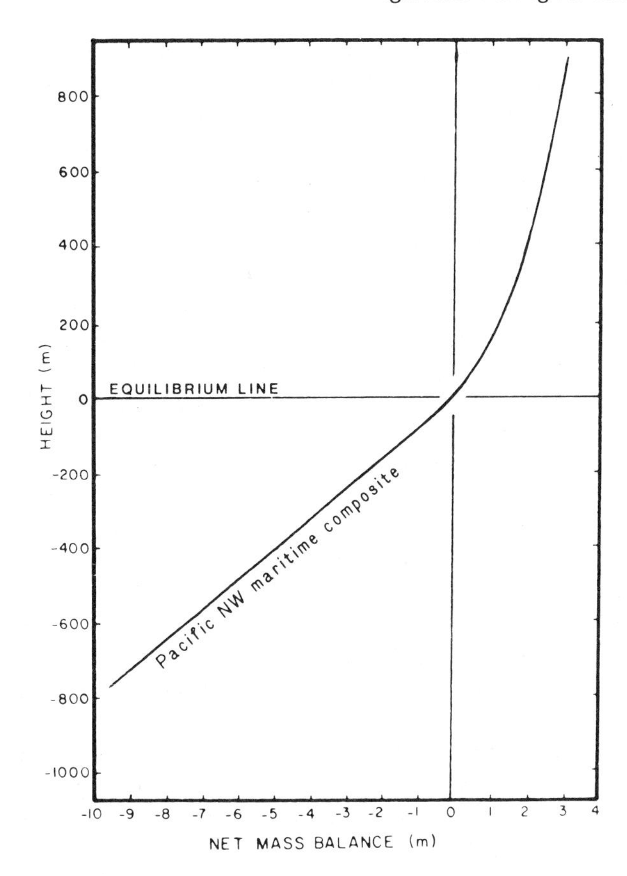

Figure 5. Height–mass balance relationship compiled from seven modern maritime glaciers in the Pacific Northwest and Alaska. Height is referenced to the equilibrium-line altitude; net mass balance is in cubic meters of equivalent water volume per square meter of ice surface per year. Original data are from Pacific Northwest glaciers (Meier and others, 1971), summarized in Porter and others (1983).

changes by making several simplifying assumptions. The response time is equated with how long such a wave takes to traverse the length of the glacier. This velocity in turn depends on the assumed velocity of the ice at the terminus and on the diffusion of the kinematic wave as it propagates downglacier. Later calculations for actual glaciers, however, suggest that the predicted response time may be significantly too long (Meier and Tangborn, 1965; Nye, 1965a; Johannesson, 1986).

Despite uncertainties in this analysis, the theory predicts that response times should increase linearly with glacier length and inversely with ice velocity. These results are intuitively reasonable and provide a quantitative basis to compare disparate glacial advance-retreat histories.

An alternative calculation of response time is presented by Oerlemans and van der Veen (1984), whose calculations for a simplified ice sheet are based on "typical" values of the ice thickness, ice velocity, and mass balance. Weertman's (1957) expression for sliding velocity is used to express the velocity as a function of ice thickness and length. Changes in ice thickness are then evaluated by incremental changes in the mass balance, defining a "relaxation time" in which the glacier responds. Although the magnitude of the relaxation time depends on poorly quantified sliding parameters, the form of its dependence on physical parameters of the glacier is fairly simple to obtain. Because most ice-sheet profiles can be approximated by a parabola (e.g., Mathews, 1974; Thorson, 1980; Paterson, 1981, p. 157), ice thickness is generally proportional to the square root of the distance from the terminus. Substituting this relationship in the expression for relaxation time predicts that relaxation time is proportional simply to the length of the glacier. This result resembles Nye's, in that the response time depends linearly on length. An implicit dependence on velocity also exists via the sliding parameter.

A third method of evaluating response times exists (Johannesson, 1986; Johannesson, Raymond, and Waddington, written communication, 1986). A particular flow law is not assumed; rather, mass is simply assumed to be conserved over the glacier as a whole. A uniform perturbation over the glacier is imposed on the equilibrium steady-state mass balance. In response, the glacier grows (or shrinks) to a new length but is assumed to change only a little in height. The change in volume to achieve the new equilibrium shape divided by the rate of gain (or loss) of ice volume defines a minimum response time, which can be expressed simply as the maximum ice-sheet height divided by the ablation rate at the terminus. This is not a "dynamic" response time, because no formulation of ice velocity is included. Yet the calculated time should both rank glaciers by relative rate of response and specify the minimum time needed to achieve final equilibrium. The expression resembles the preceding two by a dependence of response time on glacier length, here proportional to the square root of glacier length for parabolic ice-surface profiles, and an implicit dependence on the ice velocity, which is driven by the rate of ablation for an ice sheet at or near equilibrium.

Opportunities to verify any of these methods are sparse. In one example, this last method predicts a response time of 20 yr using four years of data from 100-m-thick South Cascade Glacier (Meier and Tangborn, 1965). From the same data, Nye (1962) calculated a response time by yet another technique of correlating measured accumulation rates with observed changes in glacier thickness (a method possible only where continuous measurements are available). He determined a response time of 10 yr, which suggests that model predictions are correct at least to an order of magnitude.

In summary, all three approaches relate the response time to the glacier's size and its rate of mass transfer. The first two explicitly include dynamic considerations, although the glaciologic theory underpinning such formulation is still only approximate. The last technique estimates minimum response times and also permits simple comparison of such estimates among different glaciers.

During early Fraser time, typical ice-sheet thicknesses probably approached or exceeded 1,000 m (see below), and ablation rates at the terminus, by analogy with modern glaciers, ranged between 5 and 10 m/yr. The response-time formulation of Johannesson (1986) therefore yields minimum values of 100 to 200 yr. Alpine glaciers, by comparison, were as much as an order of magnitude less in thickness (e.g., Porter, 1976) but experienced nearly equivalent ablation rates at their termini during their probable advance into major trunk valleys and perhaps beyond the rangefront down to lowland altitudes. Their predicted response times were thus in the range of several tens of years.

Geologic evidence supports these predictions. Thorson's (1980) reconstructed profile of the Puget lobe closely approximates the "ideal" parabolic form of a glacier with uniform basal shear stress throughout. He argues by analogy that the Puget lobe was at or close to equilibrium at maximum. Because the southernmost 100 km of advance and retreat occurred in less than 1,500 years, the inferred equilibrium ice profile at maximum requires a substantially shorter response time.

The Evans Creek/Coquitlam advance of alpine glaciers at ca. 20 ka preceded the maximum stand of the main Cordilleran ice sheet by up to 5,000 yr, at least several times longer than the ice sheet's predicted response time. Thinner, shorter alpine glaciers fluctuated more rapidly in response to changing climate, and thus their 20-ka maximum probably marks the most extreme Fraser-age ELA lowering. These simplified calculations demonstrate, however, that the much later maximum advance of the Cordilleran ice sheet, culminating after nearly 10,000 yr of ice buildup in central and coastal British Columbia and flow into the southern lowlands (Clague, 1981), cannot simply be attributed to a sluggish response to the *same* climatic episode.

Environments of glacier termini. Differences in the physical environments of the alpine glaciers and the Cordilleran ice sheet may also have contributed to their asynchronous development. During Evans Creek/Coquitlam time, eustatic sea level was close to its late Pleistocene minimum, 120 m below present (see below), and the alpine glaciers emerging onto the Puget and Fraser Lowlands all terminated on land.

In contrast, many parts of the Georgia Depression are now 200 to 300 m below sea level, and thus any ice sheet attempting to advance directly into this tidewater environment would have experienced greatly increased ablation rates from calving. This part of the ice sheet probably could not have maintained a rate of advance equivalent to its nontidewater counterparts in the Cascade Range and Olympic Mountains. Advance rates would have been controlled less by net mass balance, which could have been driven quickly to zero or negative values by calving, than by the rate at which proglacial outwash infilled the water-filled depressions, thereby allowing the now-grounded ice front to advance. Stratigraphic evidence is consistent with this inference; Clague (1976, 1977) argues that the advance outwash of the ice sheet in British Columbia (Quadra Sand) probably filled the Georgia Depression an indeterminant distance in front of the advancing margin. Mullineaux and others (1965) arrived at the identical conclusion for the correlative deposit (the Esperance Sand Member of the Vashon Drift) in the Puget Lowland.

Advance of the eastern Cordilleran ice sheet. Advance of the lobes east of the Cascades was generally synchronous with that of the Puget and Juan de Fuca lobes to the west (Waitt and Thorson, 1983) and not with the Evans Creek/Coquitlam alpine-glacial advance. Because none of the eastern lobes reached tidewater, this delay with respect to the alpine glaciers must reflect differential response of the ice sheet as a whole to climate and not simply a difference in terminal environments. Either (1) the supply of moisture was limited so that significant growth was possible only by smaller alpine glaciers at 20 ka (Hicock and others, 1982), (2) the climatic change was not uniform over the region, being more pronounced in the south where alpine-glacial fluctuations were recorded, or (3) the change that drove the alpine-glacial advance at ca. 20 ka was sufficiently brief (i.e., centuries) that only these shorter, faster-reacting glaciers had time to respond fully.

Evidence for and against these options is limited and circumstantial. Porter (1977) argues for a change in Evans Creek precipitation of less than 30 percent relative to present, weakening support for the first alternative; global climate models similarly suggest little relative change in precipitation or adjacent sea-surface temperatures throughout this period. Existing data and climate-model reconstructions are too coarse in scale to support (or reject) the second alternative. The third option, although unsupported by the detail of present vegetation or climate models, would in fact be best recorded only by a system that could respond rapidly to such change. The record of alpine-ice advance and ELA depression during this time, augmented by analysis of likely glacier-response times, thus may offer the best available evidence for a substantial, but necessarily brief, period of regional climatic cooling at 20 ka.

The Vashon stade—Ice advance to maximum

Introduction. Whereas the Cordilleran ice sheet reached its maximum extent during the Vashon stade (ca. 15 ka), alpine glaciers experienced a readvance (Porter, 1976) that was both generally smaller than their own earlier Evans Creek/Coquitlam advance and dwarfed by the rate and magnitude of the advance of the main Cordilleran ice sheet. The relative sizes of Evans Creek/Coquitlam alpine glaciers, Vashon-age alpine glaciers, and the ice sheet at maximum stand reflect most closely the ELAs for these various glaciers. The ELA during Evans Creek/Coquitlam time is deduced from direct evidence of glaciated cirques (see above; Porter, 1977). The ELA during Vashon time requires more circuitous analysis of the ice-sheet equilibrium itself because, of all the glaciers in the Cordillera, only the ice sheet's position can be well established during this time.

Reconstruction of the southwestern Cordilleran ice sheet. *Method.* Not only the regional ELA but also many of the mechanics of ice-sheet behavior can be derived from the physical dimensions and mass balance of the ice sheet, including its rates of advance and retreat, magnitude of isostatic loading, and response time to climatic change. A reconstruction of these param-

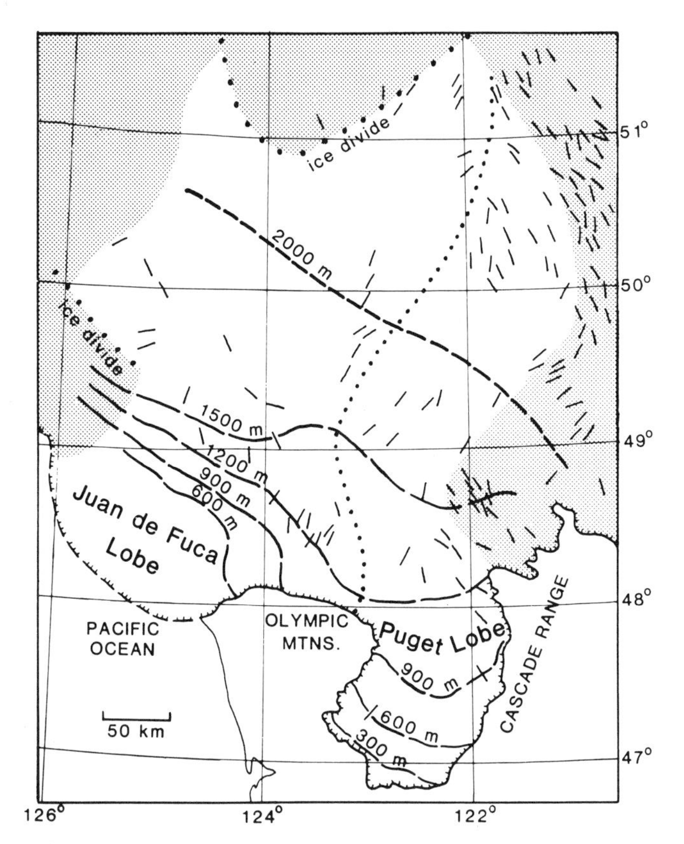

Figure 6. Reconstruction of the Puget and Juan de Fuca lobes of the Cordilleran ice sheet at maximum stage. Short lines show orientation of representative striations and other glacial lineations; hachured line shows maximum extent of ice. Heavy lines are contours of ice-surface altitude above modern sea level, uncorrected for glacial-age sea level or isostatic depression. Unshaded ice-covered area includes all ice of the Puget and Juan de Fuca lobes. Ice of these two lobes is inferred to have been separated along the dotted line. Sources of data include Wilson and others (1958), Prest and others (1968), Thorson (1980), Heller (1980), Dethier and others (1981), Fred Pessl, Jr. (personal communication, 1983), and Booth (1987a, 1987b, and unpublished data).

eters is possible only in a few special instances where sufficient data are available. The southwest part of the Cordilleran ice sheet is one such example.

Because the terminal configuration of the Puget lobe is well determined and a plausible mass-balance relationship is available, the ELA and basal sliding velocity of this lobe can be calculated (Booth, 1986b). For a glacier in equilibrium, net accumulation above any transect perpendicular to flow must be transferred by ice discharge through that transect to replenish the net ablation downglacier from it. This mass-balance method requires a reconstruction of the physical boundaries of the ice mass, topographic contours of its surface, and a relationship between specific net balance and altitude on the ice sheet. An equilibrium-line altitude (ELA) can then be found that brings the reconstructed glacier as a whole into balance. The rate of ice flux through any cross section is then easily calculated.

Reconstruction of the ice sheet. Ice limits for the southern boundary of the ice sheet on land have been rather accurately determined from extensive geologic mapping (Fig. 6; see Thorson, 1980, and Waitt and Thorson, 1983, for recent summaries). The Juan de Fuca lobe terminus was probably at tidewater and its locale subsequently submerged (Clague, 1981); it has been inferred (Alley and Chatwin, 1979) to coincide with the edge of the continental shelf southwest of Vancouver Island. Flow-direction indicators (striations, elongated topography) define a consistent pattern of ice flow and permit discrimination between ice in the Puget–Juan de Fuca portion of the Cordilleran ice sheet and that which lay to the northwest and east. Ice-surface contours shown on Figure 6 largely follow Thorson (1980) for the Puget lobe and Wilson and others (1958) over Canada. This present compilation attempts to reconcile these and other reconstructions (e.g., Waitt and Thorson, 1983), following several basic principles: the ice-surface contours should lie perpendicular to the direction of flow, flow lines should not converge or diverge without commensurate changes in ice thickness or net balance, and the local shear stress (proportional to the product of ice thickness and slope) should vary only gradually along the glacier's length.

Mass-balance calculations. The local mass balance of modern Pacific Northwest maritime glaciers is plotted in Figure 5 as a function of height above or below the equilibrium line (Meier and others, 1971; Porter and others, 1983). These data allow quantitative estimates of ice discharge through transverse cross sections of any glacier having an equivalent height–mass balance relationship.

If the ELA is known independently, the balance of any glacier can be calculated by defining areas on the ice surface with a representative altitude for each, converting each area into a rate of accumulation or ablation by using the height–mass balance curve and the given ELA, and integrating over the glacier surface to determine the predicted state of equilibrium or nonequilibrium for the chosen ELA value. The areas of each region shown in Figure 6 that are bounded by contour lines, ice limits, and lobe-dividing flow lines are assigned representative mid-point altitudes. In Table 1 the net balance is calculated for three alternative vlaues of the ELA.

Conversely, if equilibrium is assumed, the ELA can be determined by choosing an arbitrary initial ELA value and iterating by the same procedure. Equilibrium for the Puget–Juan de Fuca portion of the Cordilleran ice sheet, assumed to approximate conditions during ice maximum, is attained with an ELA of about 1,225 m. The balance contribution from calving along the tidewater terminus of the Juan de Fuca lobe is ignored in this analysis; based on measured calving rates from modern glaciers (Brown and others, 1982), this process could have consumed at most the excess yearly mass of a 25-m drop in the ELA.

Sliding velocity. On any glacier the ice accumulated above the equilibrium line must be transported into the ablation area. The flux will be greatest at the equilibrium line and decrease

TABLE 1. NET BALANCE VALUES IN 10^{10} m^3/a OF EQUIVALENT WATER VOLUME*

Altitude interval (m)	ELA = 1250 m J de F	ELA = 1250 m Puget	ELA = 1200 m J de F	ELA = 1200 m Puget	ELA = 1150 m J de F	ELA = 1150 m Puget
0-300	} -12.9	-1.2	} -12.3	-1.1	} -11.6	-1.0
300-600		-2.4		-2.2		-2.1
600-900	-2.0	-2.4	-1.8	-2.1	-1.6	-1.9
900-1200	-1.1	-1.5	-0.8	-1.1	-0.5	-0.8
1200-1500	0.7	0.8	0.8	0.9	0.9	1.1
1500-2000	5.4	2.0	5.6	2.1	5.9	2.1
>2000	9.5	4.3	9.8	4.4	10.1	4.5
NET BALANCE	-0.4	-0.4	+1.3	+0.9	+3.2	+1.9

*Positive values indicate net accumulation; negative values are net ablation. Calculations are based on areas between contours (Fig. 6) and specific net mass balance relative to ELA shown in Figure 5. Both lobes are brought into balance by an ELA between 1250 and 1200 m. An ELA value of 1225 m is therefore assumed for all subsequent calculations. Note these values differ from those used in the discussions of the ice volume by a density factor of approximately 10%.

through successive downglacier cross sections, because increasing volumes of glacier mass are lost through surface ablation. The inferred pattern of ice flow in the terminal part of the Juan de Fuca lobe is sufficiently complex to introduce large lateral variability in this flux. The simpler flow pattern of the Puget lobe, however, is well suited to flow calculations based on accumulation and ablation volumes. With an ELA of 1,225 m, 7.9×10^{10} m^3 of ice must have crossed the equilibrium line each year for the Puget lobe to remain in a balanced state (Table 2). The cross-sectional area at the equilibrium line is 1.2×10^8 m^2, giving an average ice speed of 660 m/yr. As only about 10 m/yr can be accounted for by internal deformation of the ice (based on the reconstructed ice thickness and surface slope; see Paterson, 1981, p. 87), the sliding velocity is approximately 650 m/yr, or 98 percent of the total velocity. Average rates of this order are estimated downglacier as well; for example, where the ice-surface altitude is 900 m (about the latitude of Seattle), the average ice speed is still over 500 m/yr.

Each element in this reconstruction is sufficiently constrained that the final conclusion, that is, basal sliding rates of several hundred meters per year, is scarcely affected by the uncertain precision of this analysis (see Booth, 1986b, for a more detailed discussion). The mass-balance relationship is least verifiable, because of the uncertainties involved in scaling up an empirical relationship derived from modern valley glaciers to a part of a subcontinental ice sheet. The most probable error is that high-altitude accumulation rates are overestimated, suggested by observational and theoretical evidence of an "elevation desert" high on an ice sheet (e.g., Budd and Smith, 1981; however, their data from a polar area may not be equally applicable to a maritime regime). To assess these uncertainties, the consequences of an imposed yearly accumulation limit of 2 m/yr can be evaluated. In response, the ELA of the reconstructed Puget lobe must drop by 100 m to remain in equilibrium. Yet this lowered ELA falls outside the range of geologically plausible values based on observed convergence and divergence of flow-line indicators (Booth, 1986b); this contradiction suggests a lower limit for the accumulation rates far upglacier. Using a variety of ELAs constrained by such geologic data, alternate ice velocities can be calculated and are shown in Table 3. Sliding velocity at the equilibrium line was consequently at least 500 m/yr and exceeded the internal deformation rate by nearly two orders of magnitude.

Implications for ice-sheet behavior. Predicted high mass flux, expressed as high sliding rates, is consistent with the rapid advance and brief occupancy of the Puget lobe in western Washington during the Vashon stade. Limiting radiocarbon dates in the Seattle area (Rigg and Gould, 1957; Mullineaux and others, 1965) require that the ice sheet advanced and then retreated more than 80 km during a period of 1500 ± 600 ^{14}C yr. If the advance rate is assumed to have been half that of the retreat (Weertman, 1964), the ice margin advanced at least 80 to 200 m/yr. Although these rates are high relative to many modern glaciers, they represent only a fraction of the equilibrium mass-transfer rate of the ice sheet. The Vashon stade is thus a modest perturbation in a much larger and longer depression of the ELA. For example, a rise of the ELA that recovers only 3 percent of the maximum Fraser-age ELA depression (900 m; Porter, 1977), from 1,225 m to 1,250 m, would generate a 5-km^3/yr deficit of the ice-maximum Puget lobe (Table 1), and, consequently, about 50 m/yr of retreat (using a typical cross-sectional area of $10^8 m^2$; see Table 2). The sensitivity of the lobe to small climatic changes demonstrates the plausibility of rapid ice-front movements without equally precipitous climatic changes.

Isostatic depression. As the Puget lobe advanced to its maximum position, it may have affected neighboring alpine gla-

TABLE 2. ICE DISCHARGE AND VELOCITY THROUGH TRANSVERSE SECTIONS OF THE PUGET LOBE*

Contour Interval (m)	Ice Width (10^5 m)	Average Thickness (10^3 m)	Cross-sectional Area (10^8 m^2)	Ice Discharge (10^{10} m^3/yr of ice)	Ice Velocity (m/yr)
2000	1.1	1.0	1.1	4.8	430
1500	1.1	1.1	1.2	7.0	580
1200	1.0	1.2	1.2	7.9	660
900	1.3	0.9	1.2	6.4	540
600	1.1	0.7	0.8	3.8	470
300	1.0	0.4	0.4	1.2	310
TERMINUS	1.0	0	0	0	0

*ELA = 1225 m. Sections are taken along contour lines shown on Figure 6. Ice velocity through these sections will be overwhelmingly by basal sliding (see text).

ciers by isostatically depressing the Cascade and Olympic mountains (Thorson, 1981). Any glacier on this depressed terrain would sink relative to its predepression ELA (Porter and others, 1983, p. 87). Depending on the areal extent of the depression, particularly the involvement of features affecting orographic snowfall, the ELA may have remained fixed relative to the glacier (i.e., no change in mass balance) or it may have remained fixed relative to a regional, undepressed datum (i.e., a decrease in net mass balance, reflected by an apparent rise in ELA on the glacier's surface). The actual effect would probably lie between these extremes. Thorson (1979, Fig. 31) estimates the rebound following deglaciation of the Cascade Range in areas not covered by the Puget lobe; it ranges from 0 to 80 m. He also argues (see below) for a maximum depression during full-glacial conditions that was at most twice this value. If this isostatic depression caused an effective rise in alpine-glacier ELA in the Cascade Range, the rise was negligible in the southern part of the Cascades to at most 160 m in the central Northern Cascades. ELA rise can be translated into changes in glacial length for an assumed accumulation-area ratio (0.6; Porter, 1975) and measured average valley slope. For example, in the South Fork Snoqualmie valley, a maximum ELA rise of 80 m and an average valley slope of 0.01 yields:

$$(80 \text{ m})/(0.01) = 8{,}000 \text{ m lateral (upvalley) ELA shift; thus}$$
$$(8{,}000 \text{ m})/(0.6) = \sim 10 \text{ km upvalley retreat of terminus.}$$

These values for ELA rise and terminus retreat account for only a fraction of the change in the likely position of the South Fork Snoqualmie glacier terminus between Evans Creek and Vashon times (Porter, 1976, Fig. 6), in spite of representing the *maximum* estimate of relative ELA rise due to isostatic depression for this area. Just north, in the Middle Fork Snoqualmie River valley, Williams' (1971) reconstruction of the Evans Creek and Vashon-age alpine glaciers is even less well explained by isostatic depression by the ice sheet. On the east slope of the Cascade Range, this effect would have been further diminished because of the increased distance from any major ice sheet, yet an equivalent

TABLE 3. ICE VELOCITIES AT THE EQUILIBRIUM LINE OF THE PUGET LOBE FOR VARIOUS ESTIMATES OF THE ELA

ELA	Average Ice Velocity (m/yr)	Ice Velocity Due To Internal Deformation (m/yr)	% of Total Flow Due to Basal Slip
1125	500	11	98
1200	610	11	98
1225	660	11	98
1500	900	13	99

substantial rise in inferred post–Evans Creek ELAs remains (Porter, 1976). Thus the influence of ice-sheet depression on the mass balance of contemporaneous alpine glaciers was probably small and primarily limited to the west slope of the north-central Cascade Range.

Fraser-age ELA variation. Barring major impact of ice-sheet depression on the ELAs for alpine glaciers, the reduced extent of these mountain glaciers during the Vashon maximum can only reflect a climatically controled rise in the ELA of a few hundred meters or less relative to Evans Creek/Coquitlam time. This Vashon-age alpine-glacier ELA does in fact compare quite well with the regional value of 1,200 to 1,250 m calculated for the Puget lobe at maximum stage. On the basis of likely response times and the reconstructed ice-sheet profile, both alpine glaciers and the ice sheet approached or attained equilibrium form as the Vashon stade came to its climax. Their respective limits thus accurately reflected a regional, long-term climatic depression of the ELA relative to postglacial conditions. Over the entire Fraser glaciation, the magnitude of this depression was apparently exceeded only once in the Pacific Northwest, during the earlier and short-lived Evans Creek/Coquitlam interval.

The final retreat of the Cordilleran ice sheet

Isostatic and eustatic changes. The process and rate of the retreat of the southwestern Cordilleran ice sheet was strongly

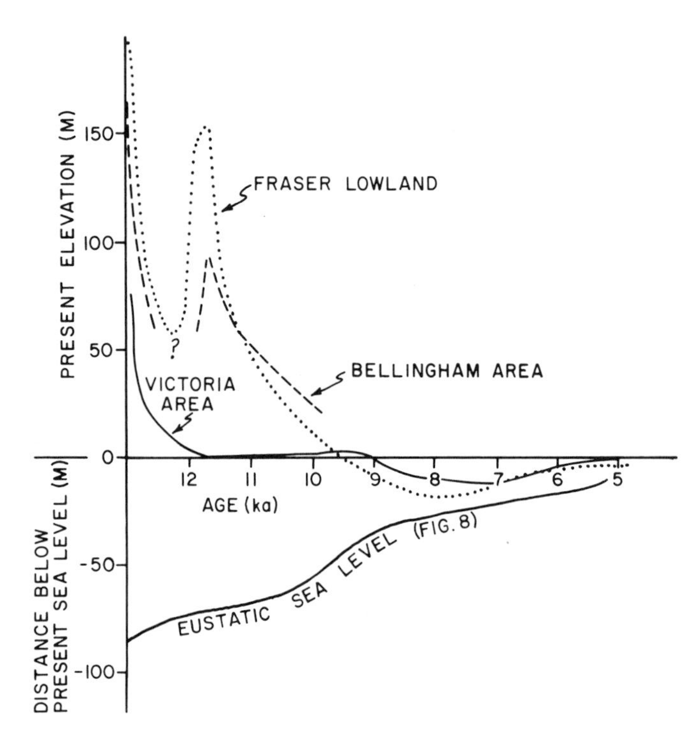

Figure 7. Proposed regional uplift curves for the Victoria, Fraser, and northern Puget Lowland areas (from Mathews and others, 1970, Fig. 4). The eustatic sea level curve in the lower section of the diagram is part of the data for the entire glacial period shown in Figure 8. The vertical distance between eustatic and shoreline curves at any given age indicates the amount of isostatic rebound that has subsequent occurred.

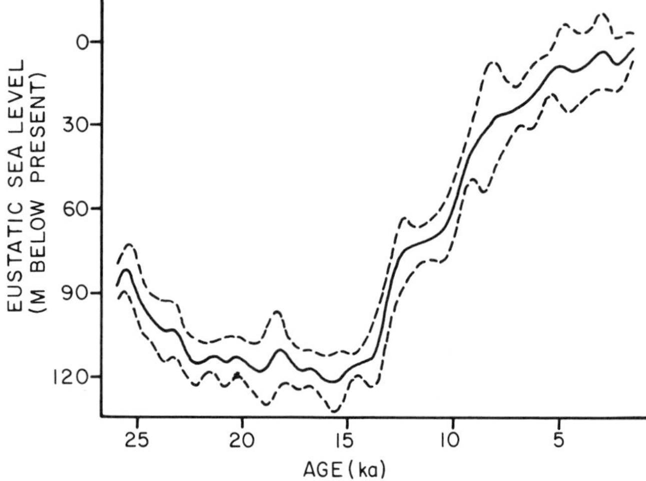

Figure 8. Proposed changes in sea level during the last glaciation, based on a maximum lowering of 120 m (modified from Mix, this volume). Dashed lines show error bounds on the best-fit curve.

influenced by concurrent changes in relative sea level. Dated marine deposits in the Puget and Fraser lowlands have yielded several regional uplift curves for the late-glacial and postglacial interval (Fig. 7). These curves reflect the combined influence of both eustatic rise (Fig. 8) and isostatic adjustment, plus an indeterminant component due to tectonic movement.

Shoreline data in the central and southern Puget Lowland are consistent with simple, monotonic uplift following deglaciation (Thorson, 1979, 1981). The rate of this uplift is unknown, and no known deposits lying between the maximum marine limit and present sea level suggest temporary cessation of relative uplift. Total rebound increases from south to north with a gradient of about 9 m/km and correlates extremely well with the reconstructed mass distribution of the Puget lobe at its maximum stage (Thorson, 1979, Fig. 28). South of the observed Everson-age marine limit, which is exposed above modern sea level only north of 47°40′N latitude (just north of Seattle; Thorson, 1979, Fig. 22), deltas associated with temporary lowland lakes that were dammed by the retreating ice extend this uplift gradient and allow its southward extension to a line of zero deformation near the ice-maximum limit (the data are insufficiently precise to identify any crustal bulge beyond the ice margin).

The uplift at Victoria also appears to have been monotonic, fitting into the pattern of the regional gradient with a maximum value of 75 m on shorelines cut immediately after deglaciation. Virtually all of the rebound here occurred in the period 13 to 11.5 ka (Mathews and others, 1970), thus averaging about 5 cm/yr. Because the rebound rate is generally presumed to be proportional to the magnitude of the isostatically uncompensated mass, initial rates were probably several times faster.

Data from the northern Puget and Fraser Lowlands suggest at least one interruption of such uplift. Citing dated shorelines and stratigraphic sequences near Bellingham, Washington, Easterbrook (1963) proposed a midrecessional resubmergence of 150 to 200 m prior to final emergence. Mathews and others (1970) proposed an equivalent pattern for the Fraser Lowland, with the maximum marine limit now at altitude 175 m and a 100-m transgression and regression at 12 ka. Although Thorson (1979) doubted whether the radiocarbon and stratigraphic evidence demanded such a complex history, subsequent scrutiny (Armstrong, 1981, p. 24–25; Clague, 1981, 1983) reaffirmed that at least one marine transgression, dated at 11.50 ±0.13 ka (BETA-1324) near Bellingham and indicating at least 35 m of resubmergence (Dethier and others, written communication, 1986), interrupted isostatic emergence of the Fraser and northern Puget Lowlands.

The pattern of maximum emergence in these northern lowlands continues the pattern inferred for the south. Emergence began as the ice thinned; the now-highest glacial-marine drift and shore deposits were formed immediately after deglaciation, at ca. 13.5 ka in the north. Sea-level altitudes increase to the north on a gradient of 1.25 to 1.50 m/km (Dethier and others, written communication, 1986) to a maximum reported value of 200 m just north of Vancouver (Clague, 1983, Fig. 3). Uplift was largely complete by 9 ka (Mathews and others, 1970), giving average rates of 4 cm/yr in the Fraser Lowland. This value, however, was temporarily exceeded in the northern lowlands because (1) initial rates probably decayed with time as the magnitude of the uncompensated mass decreased, and (2) the inferred transgression and regression required the net uplift to be accomplished in a shorter time. Initial rates also decrease from north to south; 30 cm/yr calculated on eastern Vancouver Island contrasts with 10 cm/yr in the northern Puget Lowland and with only 1 to 2 cm/yr in the central Puget Lowland (Dethier and others, written communication, 1986).

Relative sea-level changes were even more complex along the northern British Columbia coast. Here the mainland was uplifted following deglaciation, but the outer islands, beyond the limit of the main ice sheet, subsided during deglaciation until 10 to 9.5 ka (Clague, 1983, p. 335). The rate of subsidence was about 1.25 cm/yr during the period 12 to 8.5 ka and contrasted with a mainland emergence rate several times higher (Clague, 1983, Fig. 13).

Eustatic sea level. Estimates of maximum global ice volume during the late Pleistocene range from 75 m of sea-level lowering (Clark and others, 1978) through 110 to 115 m (Paterson, 1972), 120 m (Curray, 1965), and 130 m (Flint, 1971). Recent studies tend to favor eustatic sea level near −120 m, attained about 18 ka (Chappell, 1981; Duplessy and others, 1981).

Summaries in Ruddiman and Duplessy (1985) and Mix (this volume) suggest that global deglaciation did not occur uniformly. Although there is no consensus for the exact timing of eustatic sea-level change or of fluctuations in the rate of eustatic rise, a growing number of well-dated ^{18}O curves, analyzed in light of the problems and ambiguities raised by earlier studies, are in remarkably good agreement (Berger and others, 1985; Mix and Ruddiman, 1985). They indicate a global decrease in ice volume beginning about 14 ka, with a possibly more rapid 1,000 to 2,00-year interval of sea-level rise centered at about 13 ka (Fig. 8).

Isostatic rebound. *Isostatic response time.* From the available data in the Fraser Lowland, Thorson (1979, p. 130) calculated values of 800 to 1,600 years as the time necessary for the crust to compensate for one-half of the imposed ice load. Thus the crustal response was probably grossly in phase with the several-thousand-year interval of advance and retreat of the Cordilleran ice sheet. However, isostatic depression probably did not fully compensate for the imposed loads during the glacier's relatively brief maximum stand (Thorson, 1979, p. 139).

Victoria area. In the Victoria area, the shoreline data depict simple monotonic uplift during the period 13 to 11 ka. Global $\delta^{18}O$ data (Fig. 8) indicate that eustatic sea level may have risen up to 20 to 30 m during the interval 14 to 13 ka. The magnitude of the rebound following deglaciation equals the sum of the present marine limit (e.g., 75 m at Victoria) plus the magnitude of eustatic sea-level lowering at the time of deglaciation. This value probably ranged from about 90 to 110 m. Total rebound in the Victoria area, therefore, was probably 160 m or more, with an average uplift rate of 10 cm/yr (Mathews and others, 1970), several times greater than the predicted rate of sea-level rise.

Northern lowlands. In the northern Puget Lowland and in the Fraser Lowland, deglaciation has left a more complex record. Maximum marine-limit altitudes of 200 m imply total rebound in the range 270 to 310 m. Average rebound rates were several times greater than those of eustatic sea-level rise. Thus the only plausible mechanism for the inferred late-Everson transgression and regression is a temporary change in not only the rate but also the direction of crustal movement. Dethier and others (written communication, 1986) propose a migrating forebulge (Clark and others, 1978) to explain these data. Predictions of the magnitude of such a bulge, however, suggest that at maximum it would only have been about one-tenth that of the initial deformation (Walcott, 1972) and should further decay as it migrates with the retreating ice margin. Its effects, therefore, should have been even more pronounced farther south. In combination, these criteria do not fit well the requirements near Bellingham for at least 35 m of relative subsidence, although changes inferred from $\delta^{18}O$ data of Mix and Ruddiman (1985) for this period may account for nearly half of this change. The 100-m resubmergence postulated for the Fraser Lowland by Mathews and others (1970) appears to be well beyond the expected range of either isostatic or eustatic adjustment. If these data are correct, otherwise unrecognized tectonic movements must have occurred during this time. In contrast, the submergence noted by Clague (1983) along the northern British Columbia coast was sufficiently slow that eustatic rise alone, with or without the added effects of a colapsing forebulge, is sufficient to explain the data there.

In summary, deglaciation of the Puget and Fraser Lowlands occurred during a time of eustatic sea-level rise, at a rate of up to a few centimeters per year, at about 14 to 13 ka. Over the following several thousand years, isostatic rebound elevated the lowland areas at average rates of several centimeters per year, typically several times faster than concurrent rates of sea-level rise. Total isostatic rebound increased nearly linearly from zero at or near the southern ice margin to perhaps 300 m north of Vancouver, 250 km farther north. Resubmergence of at least 35 m probably interrupted the isostatic uplift of the northern Puget Lowland and the Fraser Lowland at about 11.5 ka. According to present geophysical theories, migration of a collapsing forebulge cannot fully account for such a drastic reversal in the postglacial isostatic rebound and suggests that tectonism may have been active as well.

Retreat in the southern and near-marginal areas of the Puget Lowland. As the margins of the Puget lobe first retreated from their Vashon-age maximum positions, a sequence of recessional meltwater deposits was laid down by streams issuing from the ice and in lakes impounded by the ice (Bretz, 1913). The pattern and preservation of these deposits owes much to the physiography of the southern Puget Lowlands. A drainage divide, nearly coincident with the Vashon-age ice limit, separates southward and (ultimately) westward flow into the Chehalis River and then into the Pacific Ocean from predominantly northward flow into Puget Sound. During the deglaciation, thick ice over the central and northern sound blocked northward drainage, impounding lakes whose waters spilled southward over the drainage divide into the Chehalis River. Because of the north-south grain of the topography in the lowlands, itself largely created by glacial erosion, several such postglacial lakes occupied adjacent valleys simultaneously. These lakes were interconnected by temporary spillways, with the lowest, Glacial Lake Russell, draining south into the Chehalis River for as long as the ice blocked drainage routes to the north (Fig. 9). Such lakes include Hood, Puyallup, Sammamish, Cedar, and Snoqualmie (Bretz, 1913; Thorson, 1980). Deposits associated with the progressive opening of channels by retreating ice can be identified and correlated with specific spillways from the southern edge of the ice-occupied lowland (Crandell, 1963; Lea, 1984) at least as far north as 48°10′N latitude in the eastern Puget Lowland (Booth, 1987a), 150 km north of the southern ice limit. Equivalent deposits in the east- and west-draining alpine river valleys from the Olympic Mountains and Cascade Range also are present this far north along the periphery of the Lowland (Thorson, 1980; Booth, 1986a, 1987b).

In total, the pattern of these deposits requires that the early retreating Puget lobe largely maintained an active, nonstagnant ice front. Discrete spillways separated by less than a few kilometers in the Snoqualmie and Skykomish river valleys (Booth, 1987b) demonstrate that stagnation zones were probably of negligible extent during this time, an inference supported by the general paucity of extensive upland dead-ice topography (Crandell, 1963; Thorson, 1980) in all but the most southerly plains of the Puget Lowland.

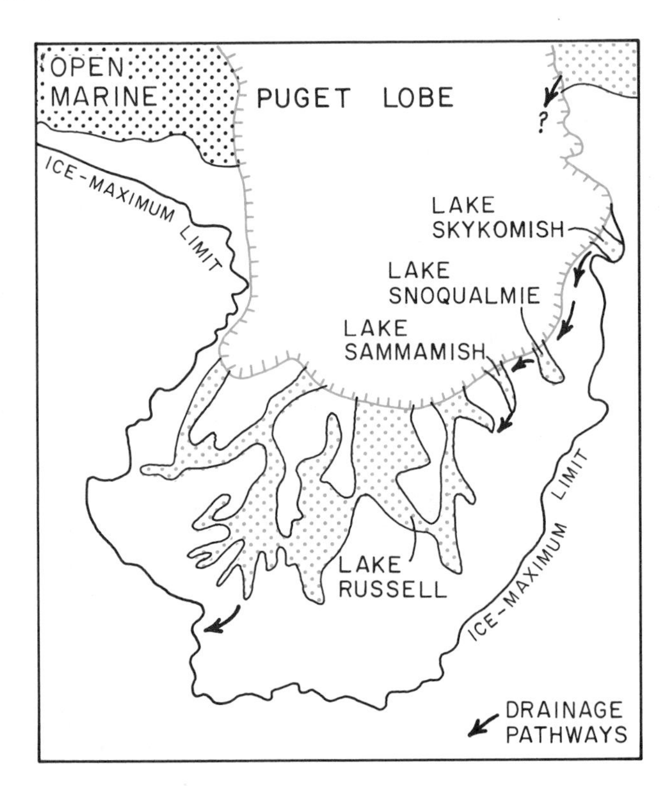

Figure 9. Glacial lakes during deglaciation of the Puget Lowland (after Thorson, 1980, Fig. 9C).

Retreat in the central and northern Puget Lowland. In contrast to the grounded, systematically retreating ice margin in the southern Puget Lowland, the Puget and Juan de Fuca lobes farther north calved extensively into tidewater (Armstrong and Brown, 1954; Easterbrook, 1963; Anderson, 1968). Even during the advance of the western Cordilleran ice lobes (17 to 15 ka), local and global conditions already were determining the character of their brief maximum stand and rapid subsequent retreat. Eustatic sea level may have already begun rising from its minimum glacial level (Ruddiman and Duplessy, 1985). Simultaneously, the earth's crust beneath and adjacent to the Puget Lowland and Strait of Juan de Fuca was sinking in response to the increased ice load. These factors, in combination, created a progressively less favorable environment for this part of the ice sheet independent of all climatic factors, by increasing water depths along the extensive seaward ice margin.

The magnitude and significance of these effects can be evaluated more precisely. Brown and others (1982) summarized calving-rate data from twelve modern Alaskan glaciers and derived the simple relationship:

$$v_c = 27.1\ h_w,$$

where v_c is the calving rate in m/yr and h_w is the water depth in meters. Eustatic sea level is estimated to have risen at most a few cm/yr around 15 to 14 ka; by analogy to later rates of isostatic rebound, the rate of crustal depression was probably several centimeters per year near the margin of the ice lobes. The net result would have been a deepening of tidewater conditions of up to 10 cm/yr. Deeper water would translate into an *acceleration* of the calving rate by 2 to 3 m/yr per year, assuming only that a minor fluctuation of the terminus occurred to remove the ice from any protective shoal that may have been deposited. Over the interval of several hundred to a thousand years that spanned the maximum stand of the ice (ca. 15 ka), the Juan de Fuca lobe would thus have increased its calving rate by perhaps 1,000 m/yr. An average terminus thickness of 150 m is suggested by the median value of observed Alaskan tidewater glaciers (Brown and others, 1982); the ice front was 150 km wide (Fig. 2). The product of these factors yields an increase in mass-loss rate during this interval, due solely to probable water-depth changes at the terminus, of approximately 10^{10} m^3/yr. This additional loss represents a significant fraction of the yearly mass budget of the Juan de Fuca lobe (Table 1) and is alone of sufficient magnitude to remove most of the approximately 10^{13} m^3 of this lobe occupying the Strait of Juan de Fuca and adjacent continental slope in less than 1,000 yr. The rate of retreat represented by these values is also the required order of magnitude for the grounding line of the Puget lobe to have retreated to its radiocarbon-dated (tidewater) positions during the Everson interval. This retreat rate is several times faster than retreat rates observed on modern nontidewater glaciers throughout the world (International Association of Hydrological Sciences, 1977). It is also over 10 times faster than the inferred retreat rate of eastern parts of the Cordilleran ice sheet (Fig. 4).

Rapid retreat of the western ice lobes was therefore a likely consequence of environmental changes at their termini, with or without concurrent, equivalent climatic amelioration. The speed of the Juan de Fuca lobe's retreat, in particular, is suggested geologically by the absence of recognized terminal or recessional features on the continental shelf or along the Strait of Juan de Fuca. This initial eastward retreat influenced the adjacent Puget lobe as well, because the opening of the Strait to ocean water led to accelerated decay of the western margin of the retreating Puget lobe (Thorson, 1980).

Climatic factors and the eastern Cordilleran ice sheet. Despite convincing evidence and consistent reconstruction of rapid ice-sheet retreat along the extensive tidewater margins of the western ice lobes, the inferred histories of both marine and nonmarine parts of the Cordilleran ice sheet are remarkably synchronous. The terminal zone of the Puget lobe was not exposed to tidewater; its early retreat history, probably initiated within a few hundred years of that of the Juan de Fuca lobe, does not reflect any marine influence. Within the broad constraints of available radiocarbon dates, the retreat of all lobes, marine and nonmarine alike, was approximately synchronous (Waitt and Thorson, 1983). Because terminal environments varied so widely among different parts of the ice sheet, this synchrony must reflect a regional climatic change at about 14.5 ± 0.5 ka. The *rates* at which individual lobes retreated, however, were closely tied to terminal conditions and correlated quite well with the presence or absence of a calving margin. Because the pollen record does not record an equivalent climatic change for 2,000 yr or more, either the lag in vegetation response was substantial or the ice sheet may have responded initially to a drier accumulation area rather than a warmer terminus.

SUMMARY

Three fundamental questions arise from the 20,000-yr history of the latest advance and retreat of the Cordilleran ice sheet. First, why was the alpine-glacial maximum at ca. 20 ka not matched by an equivalent maximum stand of the main ice sheet? Second, why had the alpine glaciers withdrawn from their farthest limits during the ice-sheet maximum 5,000 yr later? Finally, what environmental factors determined the timing and rate of final deglaciation?

Variation in the regional ELA, based on glacier reconstruction, and the physical behavior and environment of the ice-sheet terminus address these questions. During Evans Creek/Coquitlam time (20 ka), ELAs of individual alpine glaciers on the west slopes of the Cascade Range averaged close to 1,000 m. In contrast, the ELA of the southwestern part of the Cordilleran ice sheet at its maximum was probably near 1,200 m. Alpine glaciers responded much more rapidly to changes in climate than did the ice sheet, with characteristic times of about 10 to 100 yr. In contrast, the ice sheet required at least several hundred years to respond fully. The tidewater terminus of the western part of the ice sheet contrasted with land-based alpine glaciers and the largely land-based eastern ice-sheet lobes. Nevertheless, ice-sheet advance and retreat of all lobes coincided to within a millennium, although the rates of retreat were in some cases dramatically different.

The alpine-glacier maximum at ca. 20 ka reflects the lowest depression of the regional ELA during the last glaciation. It roughly coincided with a time of maximum glacial advance elsewhere in North America. The more limited extent of the western ice-sheet terminus at this time may express the slowing effect of a marine trough down which the glacier advanced. Yet the equivalent, limited extent of the eastern lobes virtually requires that this episode of lower ELA was simply too short for lobes of the ice sheet to respond fully. The predicted duration is therefore hundreds rather than thousands of years, consistent (though not required) by 2,000-yr limiting dates on the Coquitlam advance.

The ice-sheet maximum at 15 ka reflected a subsequent, long-term ELA depression over the region. The near-parabolic profile of the Puget lobe suggests that these conditions persisted for 1,000 yr or more; both the ice sheet and alpine glaciers attained equilibrium forms. Yet the reconstructed ELA value for the Puget lobe is over 100 m higher than during Evans Creek/Coquitlam time. Thus the relatively reduced extent of alpine glaciers simply reflects the shrinkage of glaciers under somewhat less favorable long-term climatic conditions.

The list of plausible determinants on ice-sheet retreat is severely limited by the relative timing of the individual lobes' wastage. In spite of a variety of terminal environments, including marine tidewater, lacustrine, and land-based on both sides of the Cascade Range, initial retreat began everywhere within 1,000 yr and progressed rapidly without substantial pause. Once initiated, rates of retreat correlated well with the terminal environments of individual lobes; the tidewater western ice sheet wasted back several times more rapidly than its eastern counterpart. However, only a regional climatic change adequately accounts for the fundamental relationship among the disparate ice-sheet lobes. The delayed expression of warming climate in the pollen record, up to 2,000 yr later, may indicate a lag in vegetation response but also suggests that the ice sheet may have responded initially not to any significant warming at its terminus but instead to reduced accumulation in its headward reaches.

REFERENCES CITED

Alley, N. F., and Chatwin, S. C., 1979, Late Pleistocene history and geomorphology, southwestern Vancouver Island, British Colombia: Canadian Journal of Earth Sciences, v. 16, p. 1645–1657.

Anderson, C. A., 1965, Surficial geology of the Fall City area, Washington [M.S. thesis]: Seattle, University of Washington, Department of Geological Sciences, 70 p.

Anderson, F. E., 1968, Seaward terminus of the Vashon continental glacier in the Strait of Juan de Fuca: Marine Geology, v. 6, p. 419–438.

——, 1981, Post-Vashon Wisconsin glaciation, Fraser Lowland, British Columbia: Geological Survey of Canada Bulletin 322, 34 p.

Armstrong, J. E., and Brown, W. L., 1954, Late Wisconsin marine drift and associated sediments of the lower Fraser valley, British Columbia, Canada: Geological Society of America Bulletin, v. 65, p. 349–364.

Armstrong, J. E., and Clague, J. J., 1977, Two major Wisconsin lithostratigraphic units in southwest British Columbia: Canadian Journal of Earth Sciences, v. 14, p. 1471–1480.

Armstrong, J. E., Crandell, D. R., Easterbrook, D. J., and Noble, J. B., 1965, Late Pleistocene stratigraphy and chronology in southwestern British Columbia and northwestern Washington: Geological Society of America Bulletin, v. 76, p. 321–330.

Atwater, B. F., 1986, Pleistocene glacial-lake deposits of the Sandpoil River valley, northeastern Washington, and their implications for glacial Lake Missoula, the Cordilleran ice sheet, and the Grand Coulee: U.S. Geological Survey Bulletin 1661, 39 p.

Barnosky, C. W., 1981, A record of late Quaternary vegetation from Davis Lake, southern Puget Lowland, Washington: Quaternary Research, v. 16, p. 221–239.

——, 1984, Late-Pleistocene and early-Holocene environmental history of southwestern Washington State: Canadian Journal of Earth Sciences, v. 21, p. 619–629.

——, 1985, Late quaternary vegetation in the southwestern Columbia Basin, Washington: Quaternary Research, v. 23, p. 109–122.

Berger, W. H., Killingley, J. S., Metzler, C. V., and Vincent, E., 1985, Two-step deglaciation; ^{14}C-dated high-resolution $\delta^{18}O$ records from the tropical Atlantic Ocean: Quaternary Research, v. 23, p. 258–271.

Booth, D. B., 1986a, The formation of ice-marginal embankments into ice-dammed lakes in the eastern Puget Lowland, Washington, U.S.A., during the late Pleistocene: Boreas, v. 15, p. 247–263.

——, 1986b, Mass balance and sliding velocity of the Puget lobe of the Cordilleran ice sheet during the last glaciation: Quaternary Research, v. 29, p. 269–280.

——, 1987a, Surficial geology of the Granite Falls 15-minute Quadrangle, Snohomish County, Washington: U.S. Geological Survey Miscellaneous Investigations Map I-1852, scale 1:50,000.

——, 1987b, Surficial geology of the Skykomish and Snoqualmie Rivers area, Snohomish and King counties, Washington: U.S. Geological Survey Miscellaneous Investigations Map I-1745, scale 1:50,000.

Bretz, J. H., 1913, Glaciation of the Puget Sound region: Washington Geological Survey Bulletin 8, 244 p.

Brown, C. S., Meier, M. F., and Post, A., 1982, Calving speed of Alaska tidewater glaciers, with applications to Columbia Glacier: U.S. Geological Survey Professional Paper 1258-C, 13 p.

Budd, W. F., and Smith, I. N., 1981, The growth and retreat of ice sheets in response to orbital radiation changes: International Association of Hydrologic Sciences Publication 131, p. 368–409.

Carson, R. J., III, 1970, Quaternary geology of the south-central Olympic Peninsula, Washington [Ph.D. thesis]: Seattle, University of Washington, Department of Geological Sciences, 67 p.

Cary, A. S., and Carlston, C. W., 1937, Notes on the Vashon stage glaciation of the South Fork of the Skykomish River valley, Washington: Northwest Science, v. 11, p. 61–62.

Chappell, J., 1981, Relative and average sea level changes, and endo-, epi-, and exogenic processes on the Earth: International Association of Hydrologic Sciences Publication 131, p. 411–430.

Chrzastowski, M. J., 1980, Submarine features and bottom configuration in the Port Townsend Quadrangle, Puget Sound region, Washington: U.S. Geological Survey Open-File Map 80-14, scale 1:100,000.

Clague, J. J., 1976, Quadra Sand and its relation to the late Wisconsin glaciation of southwest British Columbia: Canadian Journal of Earth Sciences, v. 13, p. 803–815.

——, 1977, Quadra Sand; A study of the late Pleistocene geology and geomorphic history of coastal southwest British Columbia: Geological Survey of Canada Paper 77-17, 24 p.

——, 1981, Late Quaternary geology and geochronology of British Columbia; Part 2, Summary and discussion of radiocarbon-dated Quaternary history: Geological Survey of Canada Paper 80-35, 41 p.

——, 1983, Glacio-isostatic effects of the Cordilleran ice sheet, British Columbia, Canada, *in* Smith, D. E., and Dawson, A. G., eds., Shorelines and isostasy: London, Institute of British Geographics Special Publication 16, p. 321–343.

Clague, J. J., Armstrong, J. E., and Mathews, W. H., 1980, Advance of the late Wisconsin Cordilleran ice sheet in southern British Columbia since 22,000 yr B.P.: Quaternary Research, v. 13, p. 322–326.

Clark, J. A., Farrell, W. E., and Peltier, W. R., 1978, Global changes in postglacial sea level; A numerical calculation: Quaternary Research, v. 9, p. 265–287.

CLIMAP, 1976, The surface of the ice-age earth: Science, v. 191, p. 1131–1137.

Crandell, D. R., 1963, Surficial geology and geomorphology of the Lake Tapps Quadrangle, Washington: U.S. Geological Survey Professional Paper 388A, 84 p.

——, 1965, The glacial history of western Washington and Oregon, *in* Wright, H. E., Jr., and Frey, D. G., eds., The Quaternary of the United States: Princeton, Princeton University Press, p. 341–353.

Crandell, D. R., and Miller, R. D., 1974, Quaternary stratigraphy and extent of glaciation in the Mount Rainier region, Washington: U.S. Geological Survey Professional Paper 847, 59 p.

Crandell, D. R., Mullineaux, D. R., and Waldron, H. H., 1958, Pleistocene sequence in the southeastern part of the Puget Sound Lowland, Washington: American Journal of Science, v. 256, p. 384–397.

Curran, T. A., 1965, Surficial geology of the Issaquah area, Washington [M.S. thesis]: Seattle, University of Washington, Department of Geological Sciences, 57 p.

Curray, J. R., 1965, Late Quaternary history, continental shelves of the United States, *in* Wright, H. E., Jr., and Frey, D. G., eds., The Quaternary of the United States: Princeton, Princeton University Press, p. 723–735.

Dethier, D. P., Saffioles, S. A., and Pevear, D. R., 1981, Composition of tills from the Clear Lake Quadrangle, Skagit and Snohomish counties, Washington: U.S. Geological Survey Open-File Report 81-517, 55 p.

Domack, E., 1983, Facies of late Pleistocene glacial-marine sediments on Whidbey Island, Washington; An isostatic glacial-marine sequence, *in* Molnia, B. F., ed., Glacial-marine sedimentation: New York, Plenum, p. 535–570.

Duplessy, J. C., Delibrias, G., Turon, J. L., Pujol, C., and Duprat, J., 1981, Deglacial warming of the northeastern Atlantic Ocean; Correlation with the paleoclimatic evolution of the European continent: Palaeogeography, Palaeoclimatology, Palaeoecology, v. 35, p. 121–144.

Dyck, W., and Fyles, J. G., 1963, Geological Survey of Canada radiocarbon dates I and II: Geological Survey of Canada Paper 63-21, p. 15–31.

Easterbrook, D. J., 1963, Late Pleistocene glacial events and relative sea level changes in the northern Puget Lowland, Washington: Geological Society of America Bulletin, v. 74, p. 1465–1484.

Easterbrook, D. J., Crandell, D. R., and Leopold, 1967, Pre-Olympia Pleistocene stratigraphy and chronology in the central Puget Lowland, Washington: Geological Society of America Bulletin, v. 78, p. 13–20.

Flint, R. F., 1971, Glacial and Quaternary geology: New York, Wiley, 892 p.

Gates, W. L., 1976, The numerical simulation of ice-age climate with a global general circulation model: Journal of Atmospheric Sciences, v. 33, p. 1844–1873.

Heller, P. L., 1980, Multiple ice-flow directions during the Fraser Glaciation in the lower Skagit Valley drainage, North Cascade Range, Washington: Arctic and Alpine Research, v. 12, p. 299–308.

Heusser, C. J., 1973, Environmental sequence following the Fraser advance of the Juan de Fuca lobe, Washington: Quaternary Research, v. 3, p. 284–306.

Hicock, S. R., and Armstrong, J. E., 1981, Coquitlam Drift; A pre-Vashon glacial formation in the Fraser Lowland, British Columbia: Canadian Journal of Earth Sciences, v. 18, p. 1443–1451.

Hicock, S. R., Hebda, R. J., and Armstrong, J. E., 1982, Lag of the Fraser glacial maximum in the Pacific Northwest; Pollen and macrofossil evidence from western Fraser Lowland, British Columbia: Canadian Journal of Earth Sciences, v. 19, p. 2288–2296.

International Association of Hydrological Sciences, 1977, Fluctuations of glaciers, v. 3: International Commission on Snow and Ice, Paris, IAHS-UNESCO, 269 p.

Johannesson, T., 1986, The response time of glaciers in Iceland to change in climate: Annals of Glaciology, v. 8, p. 100–101.

Knoll, K. M., 1967, Surficial geology of the Tolt River area, Washington [M.S. thesis]: Seattle, University of Washington, Department of Geological Sciences, 91 p.

Kuhn, M., 1981, Climate and glaciers, *in* Allison, I., ed., Sea level, ice, and climatic change: International Association of Hydrologic Sciences Publication 131, p. 3–20.

Kutzbach, J. E., and Guetter, P. J., 1986, The influence of changing orbital parameters and surface boundary conditions on climate simulations for the past 18,000 years: Journal of Atmospheric Sciences, v. 43, no. 16, p. 1726–1759.

Lea, P. D., 1984, Pleistocene glaciation at the southern margin of the Puget lobe, western Washington [M.S. thesis]: Seattle, University of Washington, Department of Geological Sciences, 96 p.

Lowden, J. A., and Blake, W., Jr., 1970, Geological Survey of Canada radiocarbon dates IX: Geological Survey of Canada Paper 70-2B, 41 p.

Mackin, J. H., 1941, Glacial geology of the Snoqualmie-Cedar area, Washington: Journal of Geology, v. 49, p. 449–481.

Manabe, S., and Hahn, D. F., 1977, Simulation of the tropical climate of an ice age: Journal of Geophysical Research, v. 82, p. 3889–3911.

Mathews, W. H., 1974, Surface profile of the Laurentide ice sheet in its marginal areas: Journal of Glaciology, v. 13, p. 37–43.

Mathews, W. H., Fyles, J. G., and Nasmith, H. W., 1970, Postglacial crustal movements in southwestern British Columbia and adjacent Washington State: Canadian Journal of Earth Sciences, v. 7, p. 690–702.

Mehringer, P. J., Jr., Sheppard, J. C., and Foit, F. F., Jr., 1984, The age of Glacier Peak tephra in west-central Montana: Quaternary Research, v. 21, p. 36–41.

Meier, M. F., and Tangborn, W. V., 1965, Net budget and flow of South Cascade Glacier, Washington: Journal of Glaciology, v. 5, p. 547–566.

Meier, M. F., Tangborn, W. V., Mayo, L. R., and Post, A., 1971, Combined ice and water balances of Gulkana and Wolverine glaciers, Alaska, and South Cascade Glacier, Washington, 1965 and 1966 hydrologic years: U.S. Geological Survey Professional Paper 715A, 23 p.

Mickelson, D. M., Clayton, L., Fullerton, D. S., and Borns, H. W., Jr., 1983, The late Wisconsin glacial record of the Laurentide ice sheet in the United States, *in* Wright, H. E., Jr., and Porter, S. C., eds., The Quaternary of the United States: Minneapolis, University of Minnesota Press, v. 1, p. 3–37.

Mix, A. C., and Ruddiman, W. F., 1985, Structure and timing of the last deglaciation; Oxygen-isotope evidence: Quaternary Science Reviews, v. 4, p. 59–108.

Mullineaux, D. R., Waldron, H. H., and Rubin, M., 1965, Stratigraphy and chronology of late interglacial and early Vashon glacial time in the Seattle area, Washington: U.S. Geological Survey Bulletin 1194-0, 10 p.

Mullineaux, D. R., Wilcox, R. E., Ebaugh, W. F., Fryxell, R., and Rubin, M., 1978, Age of the last major scabland flood of the Columbia Plateau in eastern Washington: Quaternary Research, v. 10, p. 171–180.

Newcomb, R. C., 1952, Groundwater resources of Snohomish County, Washington: U.S. Geological Survey Water Supply Paper 1135, 133 p.

Nye, J. F., 1960, The response of glaciers and ice-sheets to seasonal and climatic changes: Proceedings of the Royal Society of London, Series A, v. 256, p. 559–584.

——, 1962, Implications of mass balance studies (summary): Journal of Glaciology, v. 4, p. 264–265.

——, 1963, On the theory of advance and retreat of glaciers: Geophysical Journal of the Royal Astronomical Society, v. 7, p. 431–456.

——, 1965a, The frequency response of glaciers: Journal of Glaciology, v. 5, p. 567–587.

——, 1965b, A numerical method of inferring the budget history of a glacier from its advance and retreat: Journal of Glaciology, v. 5, p. 589–607.

Oerlemans, J., and van der Veen, C. J., 1984, Ice sheets and climate: Dordrecht, Holland, D. Reidel Publishing Company, 217 p.

Østrem, G., 1975, ERTS data in glaciology; An effort to monitor glacier mass balance from satellite imagery: Journal of Glaciology, v. 15, p. 403–415.

Page, B. M., 1939, Multiple glaciation in the Leavenworth area, Washington: Journal of Geology, v. 47, p. 785–815.

Paterson, W.S.B., 1972, Laurentide ice sheet; Estimated volumes during late Wisconsin: Review of Geophysics and Space Physics, v. 10, p. 885–917.

——, 1981, The physics of glaciers: Oxford, Pergamon Press, 380 p.

Pessl, F., Jr., Dethier, D. P., Booth, D. B., and Minard, J. P., 1987, Surficial geologic map of the Port Townsend 30-minute by 60-minute Quadrangle, northern Puget Sound region, Washington: U.S. Geological Survey Miscellaneous Investigations Map I-1198, scale 1:100,000 (in press).

Pierce, K. L., 1979, History and dynamics of glaciation in the northern Yellowstone National Park area: U.S. Geological Survey Professional Paper 729F, 90 p.

Porter, S. C., 1964, Composite Pleistocene snow line of Olympic Mountains and Cascade Range, Washington: Geological Society of America Bulletin, v. 75, p. 477–482.

——, 1975, Equilibrium-line altitudes of late Quaternary glaciers in the southern Alps, New Zealand: Quaternary Research, v. 5, p. 27–47.

——, 1976, Pleistocene glaciation in the southern part of the North Cascade Range, Washington: Geological Society of America Bulletin, v. 87, p. 61–75.

——, 1977, Present and past glaciation threshold in the Cascade Range, Washington U.S.A.; Topographic and climatic controls, and paleoclimatic implications: Journal of Glaciology, v. 18, p. 101–116.

——, 1978, Glacier Peak tephra in the North Cascade Range, Washington; Stratigraphy, distribution, and relationship to late-glacial events: Quaternary Research, v. 10, p. 30–41.

Porter, S. C., and Carson, R. J., III, 1971, Problems of interpreting radiocarbon dates from dead-ice terrain, with an example from the Puget Lowland of Washington: Quaternary Research, v. 1, p. 410–414.

Porter, S. C., Pierce, K. L., and Hamilton, T. D., 1983, Late Wisconsin mountain glaciation in the western United States, *in* Wright, H. E., Jr., and Porter, S. C., eds., Late Quaternary environments of the United States: Minneapolis, University of Minnesota Press, p. 71–111.

Prest, V. K., 1969, Retreat of recent and Wisconsin ice in North America: Geological Survey of Canada Map 1257A, scale 1:5,000,000.

Prest, V. K., Grant, D. R., and Rampton, V. N., 1968, Glacial map of Canada: Geological Survey of Canada Map 1253A, scale 1:3,801,600.

Rigg, G. B., and Gould, H. R., 1957, Age of Glacier Peak eruption and chronology of postglacial peat deposits in Washington and surrounding areas: American Journal of Science, v. 255, p. 341–363.

Ruddiman, W. F., and Duplessy, J. C., 1985, Conference on the last deglaciation; Timing and mechanism: Quaternary Research, v. 23, p. 1–17.

Schytt, V., 1967, A study of "ablation gradient": Geografiska Annaler, v. 49A, p. 327–332.

Thorson, R. M., 1979, Isostatic effects of the last glaciation in the Puget Lowland, Washington [Ph.D. thesis]: Seattle, University of Washington, Department of Geological Sciences, 154 p.

Thorson, R. M., 1980, Ice sheet glaciation of the Puget Lowland, Washington, during the Vashon Stade: Quaternary Research, v. 13, p. 303–321.

Thorson, R. M., 1981, Isostatic effects of the last glaciation in the Puget Lowland, Washington: U.S. Geological Survey Open-File Report 81-370, 100 p.

Waitt, R. B., Jr., 1977, Evolution of glaciated topography of upper Skagit drainage basin, Washington: Arctic and Alpine Research, v. 9, p. 183–192.

Waitt, R. B., Jr., 1980, About forty last-glacial Lake Missoula jokulhlaups through southern Washington: Journal of Geology, v. 88, p. 653–679.

Waitt, R. B., Jr., and Thorson, R. M., 1983, The Cordilleran ice sheet in Washington, Idaho, and Montana: *in* Porter, S. C., and Wright, H. E., Jr. (eds.), Late-Quaternary environments of the United States: Minneapolis, University of Minnesota Press, v. 1, p. 53–70.

Walcott, R. I., 1972, Late Quaternary vertical movements in eastern North America: quantitative evidence of glacio-isostatic rebound: Review of Geophysics and Space Physics, v. 10, p. 849–884.

Weaver, C. E., 1937, Tertiary stratigraphy of western Washington and northwestern Oregon: Seattle, University of Washington, Publications in Geology, v. 4, 266 p.

Weertman, J., 1957, On the sliding of glaciers: Journal of Glaciology, v. 3, p. 33–38.

Weertman, J., 1964, Rate of growth or shrinkage of non-equilibrium ice sheets: Journal of Glaciology, v. 5, p. 145–158.

Williams, V. S., 1971, Glacial geology of the drainage basin of the Middle Fork of the Snoqualmie River [M.S. thesis]: Seattle, University of Washington, Department of Geological Sciences, 45 p.

Wilson, J. T., Falconer, G., Mathews, W. H., and Prest, V. K., 1958, Glacial map of Canada: Geological Association of Canada, scale 1:3,801,600.

Wright, H. E., Jr., 1984, Sensitivity and response time of natural systems to climatic change in the late Quaternary: Quaternary Science Reviews, v. 3, p. 91–131.

Manuscript Accepted by the Society April 10, 1987

Printed in U.S.A.

The Geology of North America
Vol. K-3, North America and adjacent oceans during the last deglaciation
The Geological Society of America, 1987

Chapter 5

Ice core and other glaciological data

W.S.B. Paterson
Paterson Geophysics Inc., Box 303, Heriot Bay, British Columbia V0P 1H0, Canada
C. U. Hammer
Department of Glaciology, Geophysical Institute, University of Copenhagen, Haraldsgade 6, DK 2200 Copenhagen, Denmark

Ice cores contain all of the essential data on factors forcing climatic changes . . . and on parameters describing the new climatic state. Oeschger (1985)

INTRODUCTION

A core taken from a polar ice sheet provides a record of snowfall and, by analysis of its isotopic composition and chemistry, of past climate and environmental conditions. Ice cores also contain atmospheric fallout such as wind-blown dust, volcanic deposits, sea salts, extraterrestrial particles, pollen, and trace elements resulting from natural causes, pollution, or nuclear bomb tests. Such cores, extending in one instance back 160,000 yr, have been obtained from Antarctica, Greenland, and arctic Canada. If the core is from an area where the surface never melts, or at least where any meltwater refreezes within the snowpack, the record is continuous, except for a bottom layer that may be disturbed by folding, faulting, or displacement along shear planes as the ice moves over uneven bedrock. The lowest 5 to 10 percent of the ice column may be disturbed if the bedrock is mountainous. In addition, melting and refreezing at the bed may have destroyed the oldest part of some records. The record becomes progressively less detailed with depth because ice flow thins the annual layers. Nevertheless, ice cores provide a much more detailed record of the last glaciation and its end than ocean cores provide. Moreover, unlike ocean cores, ice cores are not disturbed by burrowing fauna. In this chapter we discuss what ice cores, principally those from Greenland, tell us about the timing and mechanism of the deglaciation.

DATA PRESENTATION

Core sites

Table 1 gives details of all sites from which continuous cores extending back into the last glaciation have been obtained. Figure 1 shows the locations of core sites and other places in the northern hemisphere where samples of old ice have been obtained from near the ice margin. Although the cores from the thin ice caps in arctic Canada extend as far back in time as the Greenland cores, all the ice older than 10 ka is concentrated in the lowest 10 m. The resolution at the Holocene/Wisconsin transition in these cores is therefore poor. Moreover, there are gaps in the Wisconsin record, as comparison of cores from adjacent boreholes has shown (Paterson and others, 1977). The lowest parts of the Dye 3 and Camp Century records are also discontinuous; however, all the disturbed ice is much older than the time of the last glacial maximum. Because the Dye 3 core was in better condition and has been sampled in greater detail for more parameters than the Camp Century core, most of our discussion of the deglaciation is based on it.

Measurement of climatic and other parameters

Temperature. The most important climatic parameter in an ice core is the oxygen-isotope ratio ($^{18}O/^{16}O$) or similarly the ratio of deuterium to hydrogen (D/H). These ratios depend on the condensation temperature of the air when the snow fell. The measure used is

$$\delta = 1000\,(R - Ro)\,/Ro$$

where R and Ro are the ratios of the concentrations of the heavy to the light isotope in the sample and in "standard mean ocean water." It is measured in parts per thousand, denoted ‰. The relation of δ to temperature results from the slightly reduced vapor pressure of the heavy isotope. Thus, during natural cycles of evaporation and condensation, molecules of $H_2{}^{18}O$ evaporate less readily and condense more readily than molecules of $H_2{}^{16}O$. As an air mass containing oceanic water vapor moves toward the polar regions, it is cooled and loses water as precipitation. It therefore becomes progressively more depleted in the heavy isotope, that is, $\delta^{18}O$ becomes more negative. Dansgaard and others (1973) discussed the details of this process and concluded that the

Paterson, W.S.B., and Hammer, C. U., 1987, Ice core and other glaciological data, *in* Ruddiman, W. F., and Wright, H. E., Jr., eds., North America and adjacent oceans during the last deglaciation: Boulder, Colorado, Geological Society of America, The Geology of North America, v. K-3.

TABLE 1. CORE SITE DATA

Site	Region	Position	Core length (m)	Base of of ice reached?	~Age at bottom of core (ka)
Camp Century	N.W. Greenland	77.2°N 61.1°W	1,390	yes	130
Dye 3	S.E. Greenland	65.8°N 43.8°W	2,037	yes	100
Devon ice cap	Devon Island	75.3°N 82.5°W	299	yes	130
			299	yes	130
Agassiz ice cap	Ellesmere Island	80.8°N 72.9°W	139	yes	120
			338	yes	120
Dome C	E. Antarctica	74.7°S 124.2°E	906	no	30
Vostok	E. Antarctica	78.5°S 106.8°E	2,083	no	160
Law Dome	E. Antarctica	66.2°S 111.0°E	350	no	?
			300	yes	?
			344	yes	?
Byrd	W. Antarctica	80.0°S 120.0°W	2,164	yes	65
J9	Ross ice shelf	82.4°S 168.6°W	416	yes	?

most important factor is the amount of cooling since the last substantial uptake of water vapor from the ocean. Because sea-surface temperatures are much more stable than air temperatures at high latitudes, the $\delta^{18}O$ value of polar snow depends strongly on the temperature at the site and time of deposition. Values of $\delta^{18}O$ in polar ice cores show both seasonal oscillations and long-term trends.

The amount by which an air mass is cooled as it moves over an ice sheet depends primarily on the gain in elevation and secondarily on the distance from the source of water vapor. Values of $\delta^{18}O$ are highly correlated with mean annual air temperature at the surface, which in turn depends largely on elevation in both Greenland and Antarctica. Moreover, the relation between $\delta^{18}O$ and temperature appears to be linear, although the slope of the line varies from one region to another (Dansgaard and others, 1973, p. 12). The relation breaks down in the Canadian arctic islands, except on the slopes facing Baffin Bay (Koerner, 1979), and at elevations below 1,000 m in Antarctica (Dansgaard and others, 1973). In these cases, the value of $\delta^{18}O$ appears to depend mainly on distance from the source of water vapor.

Fisher and Alt (1985) argued that $\delta^{18}O$ is not a simple function of temperature, and so conditions along the entire path of the air mass must always be taken into account. Their conclusions, however, are based on a model of the transport of atmospheric water vapor that is zonally averaged and has only one vertical layer. We believe that, although this model may be realistic for conditions near sea level in certain areas, it does not represent conditions at the higher elevations on the Greenland and Antarctic ice sheets. The $\delta^{18}O$ values of precipitation on these ice sheets seem to be determined largely by the temperature distribution on them.

Factors other than climate can also change the value of $\delta^{18}O$: (1) Unless the core is from an ice divide, the ice at depth originated as snow farther inland and at higher elevations than the core site. Thus, even in a stable climate, $\delta^{18}O$ would become more negative with increasing depth; (2) changes in ice thickness with time; (3) changes in ice-flow pattern; (4) changes in the source area of precipitation resulting from changes in atmospheric circulation or in the extent of sea ice; and (5) changes in the seasonal distribution of precipitation, which will change the mean annual value of $\delta^{18}O$.

A $\delta^{18}O$ record can be corrected for the first factor by using a numerical model of present ice flow to calculate particle paths upstream from the core site. Such a model may be satisfactory for the Holocene period. It should not be used to correct the ice-age part of a record because precipitation rate, ice thickness, and ice temperature probably differed greatly from present values at that time. Lack of information about past conditions usually prevents any attempt to correct a record for the other four factors.

If both oxygen and hydrogen isotopes are measured, additional climatic information can be obtained by calculating the deuterium excess

$$d = \delta D - 8\,\delta^{18}O.$$

The factor 8 is the slope of the regression line of δD on $\delta^{18}O$. Nonequilibrium processes in the water cycle cause changes in d. For example, Jouzel and others (1982) interpreted an increase in d from 5 to 8 ‰ at the Wisconsin/Holocene transition in the core from Dome C, Antarctica, as a sign of increased relative humidity during the ice age. Because deuterium data from Greenland have not yet been published, we do not discuss this topic further.

Figure 2 shows two oxygen-isotope profiles from Greenland. The Holocene/Wisconsin boundary is prominent, and the high (less negative) values near the base of the Camp Century record almost certainly correspond to the preceding interglacial.

Table 2 shows measured changes in $\delta^{18}O$ at the end of the glaciation or, more specifically, the difference between the value for the coldest part of the Late Wisconsin and the mean for the first few thousand years of the Holocene (see Fig. 1 and Table 1 for locations). These values could be changed by about 1‰ according to the way in which the maximum and minimum values are chosen. No corrections have been made for any of the

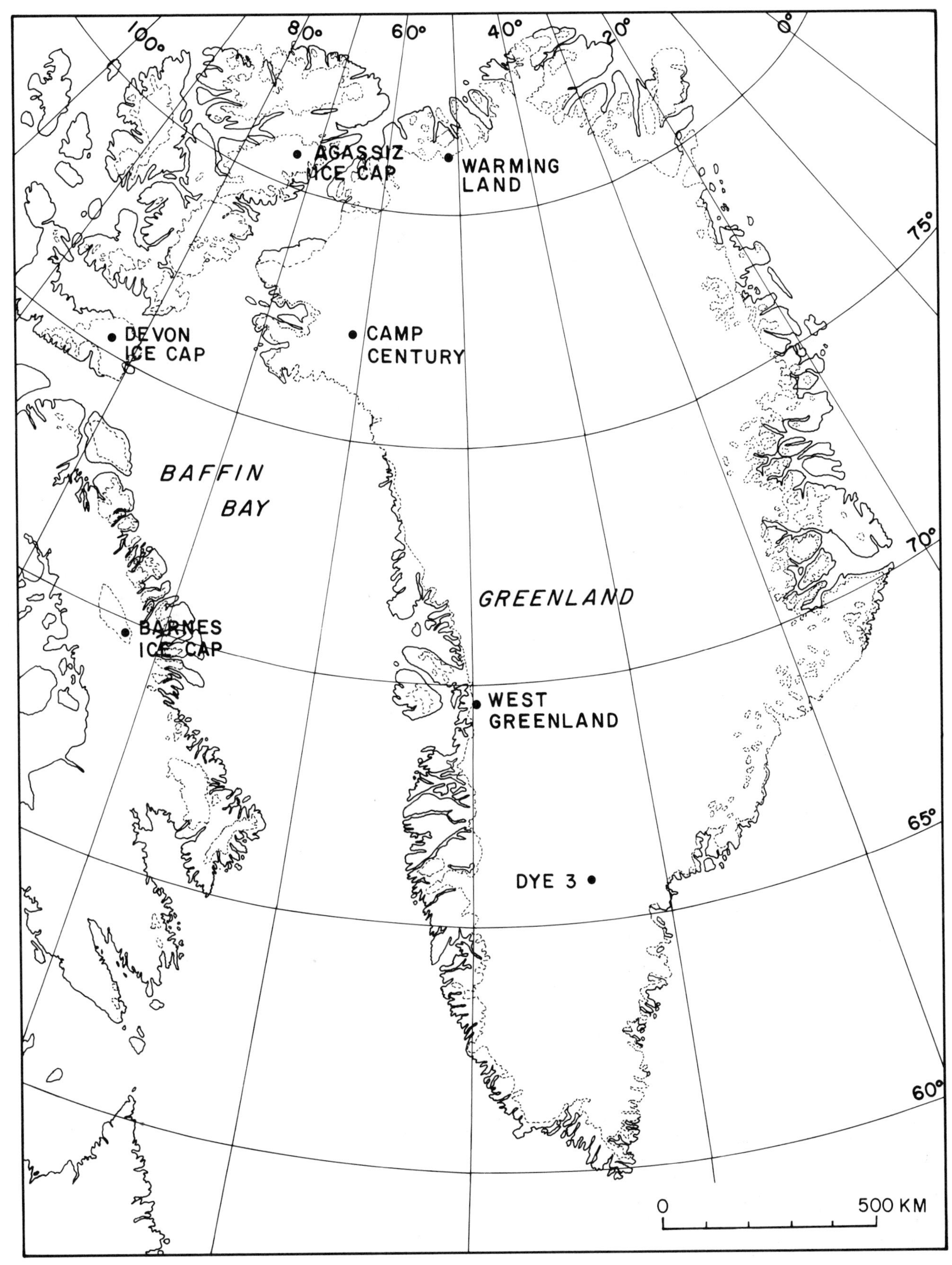

Figure 1. Locations of northern hemisphere sites.

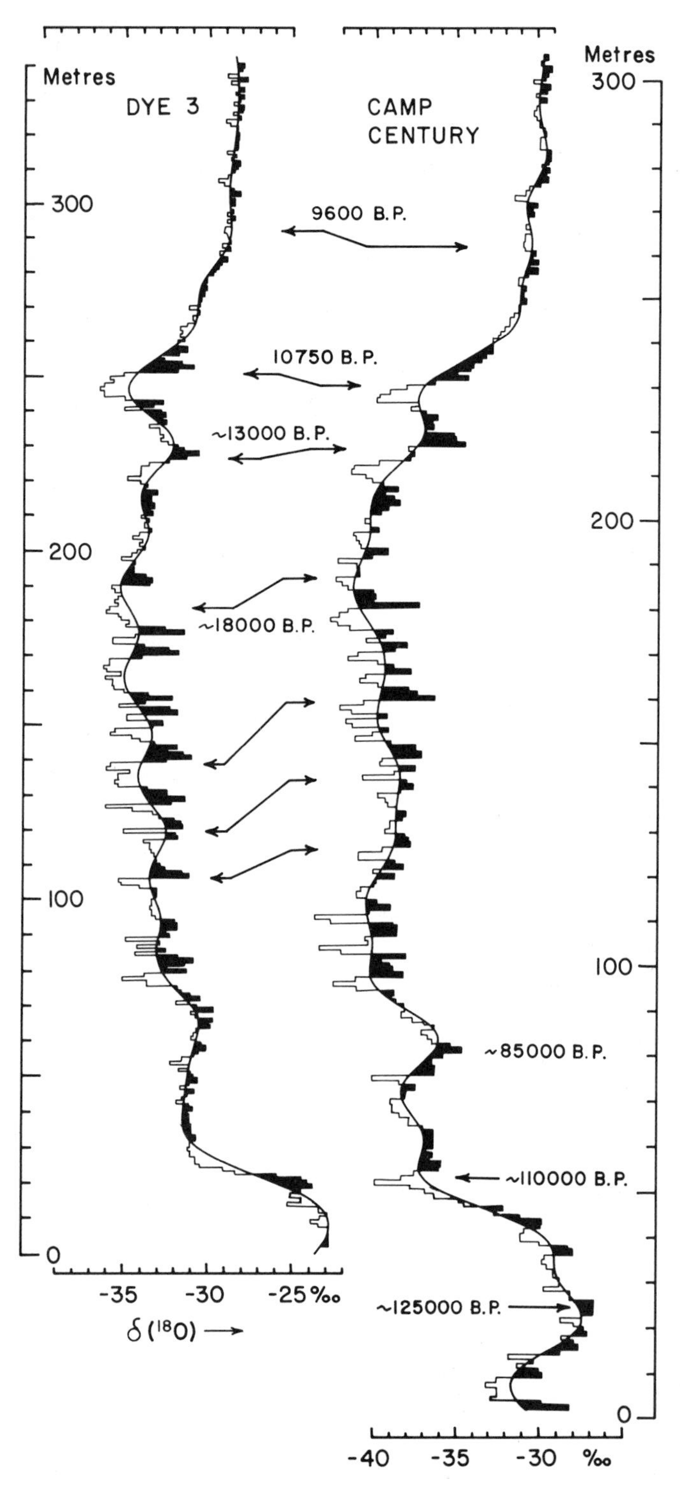

Figure 2. Oxygen-isotope profiles for Dye 3 and Camp Century, Greenland. Measurements in silty basal ice have been omitted. The Dye 3 record is probably discontinuous below 87 m. Arrows relate what are believed to be identical climatic events. Ages younger than 18 ka are discussed in the text. Older ages are rough estimates based on comparison with an oceanic climatic record. Adapted from Dansgaard and others (1982). Reproduced here with the permission of the authors and the American Association for the Advancement of Science. Copyright (c) 1982 by A.A.A.S.

five factors listed above or for other factors, such as the change in isotopic composition of the oceans at the end of the glaciation and the rise in sea level and corresponding decrease in surface elevation. These last two factors, which tend to cancel each other, should be the same at all sites.

These values, particularly those in Antarctica, follow a consistent pattern, with a smaller change in central East Antarctica than at lower elevations near the coast. Interpretation of the Ross ice shelf value is difficult because the source area of the ice-age ice is uncertain. Much of the change of 15 ‰ at the Barnes ice cap in Baffin Island must result from a change in surface elevation. This region was covered by the Laurentide ice sheet, so the Wisconsin ice there must have been deposited at elevations much higher than any on the present ice cap. The other northern-hemisphere values fall into two groups: stations south of latitude 76°N with values of 7 to 8 ‰, and stations north of that latitude with higher values. The geographic consistency of the pattern of values in Table 2 increases our confidence in using $\delta^{18}O$ as an indicator of past temperature.

It is tempting to use the present decrease of $\delta^{18}O$ with mean annual air temperature (about 0.62 ‰ per degree on the Greenland ice sheet and about 0.75 ‰ per degree in the interior of Antarctica) to convert the differences in Table 2 to temperature changes. Although this may be justified in certain cases, in general it is not, because the nonclimatic factors listed above may account for some of the change in $\delta^{18}O$. Nevertheless, the value of 13 K derived for Dye 3 compares well with the 12 K obtained from an analysis of the present temperature distribution in the borehole (Dahl-Jensen and Johnsen, 1986).

Precipitation rate. The fact that annual layers can be distinguished in many cores suggests a way of measuring past precipitation rates. However, to convert measured annual-layer thickness to precipitation rate, one must know how much each layer has been thinned since it was deposited at the surface. This requires detailed modeling of ice flow based on present precipitation, ice thickness, and temperature along the upstream flowline. However, present values may not be typical of the past few thousand years, let alone of the ice age, and any model involves simplifying assumptions, some of which may be invalid. Although this method might give reliable estimates for places where the flow pattern is particularly simple and stable, existing studies have not done so (Paterson and Waddington, 1984; Reeh and others, 1985). Moreover, very few measurements of annual layers in ice-age ice have been made.

Measurement of the concentration of ^{10}Be in an ice core is a promising alternative method for finding past variations in precipitation rate. ^{10}Be is produced in the atmosphere by cosmic radiation and has a half life of 1.5 m.y. It becomes attached to atmospheric aerosols and is removed by precipitation within 1 to 2 years of its formation. Concentrations in ice samples as small as 1 kg can be measured by accelerator mass spectrometry. There are at least three possible reasons for temporal variations of ^{10}Be concentration in ice cores (Raisbeck and others, 1981): (1) Variations in the incoming cosmic ray flux. Because this flux is influ-

TABLE 2. CHANGE IN $\delta^{18}O$ AT END OF GLACIATION

Site	Change in δ (o/oo)	Reference
Warming Land	11-13	Reeh, 1987
Camp Century	11	Dansgaard and others, 1971
West Greenland	7-8	Reeh, 1987
Dye 3	8	Dansgaard and others, 1982
Agassiz ice cap	10	Fisher and others, 1983
Devon ice cap	8	Paterson and others, 1977
Barnes ice cap	15	Hooke and Clausen, 1982
Dome C	5.5	Lorius and others, 1979
Vostok	5	Lorius and others, 1985
Law Dome	7	Morgan and McCray, 1985
Byrd	7	Johnsen and others, 1972
Ross ice shelf (J9)	10	Grootes and Stuiver, 1986

enced by the magnetic properties of the solar wind plasma, it varies with solar activity. (2) Variations in atmospheric circulation patterns. Because most of the ^{10}Be is produced in the stratosphere, changes in atmospheric circulation could change the fraction of the total production of ^{10}Be that goes into the troposphere over the polar regions. (3) Variations in precipitation rate. If the fallout rate of ^{10}Be is constant, the lower the precipitation rate, the higher the concentration of ^{10}Be in the snow will be.

Precipitation rate appears to be the most important variable (Oeschger and others, 1984; Yiou and others, 1985). Solar variations undoubtedly have some effect: ^{10}Be concentrations show the 11-year sunspot cycle (Beer and others, 1985), and the concentration increased by 50 percent during the Maunder minimum of solar activity (Raisbeck and others, 1981). However, the fact that the concentration of ^{10}Be in Wisconsin ice is two to three times that in Holocene ice cannot be explained by variations in solar activity. Moreover, the variations in concentration are highly correlated with the oxygen-isotope variations during the glaciation; the highest concentrations occur during the coldest periods, when minimum precipitation is expected (Fig. 3). (Precipitation rate is highly correlated with surface temperature in central Antarctica at present.) Again, changes in sulfate concentration, attributed primarily to changes in precipitation rate, show the same trend (Finkel and Langway, 1985). Measuring ^{10}Be concentrations therefore seems to provide valid estimates of long-period trends in precipitation rate.

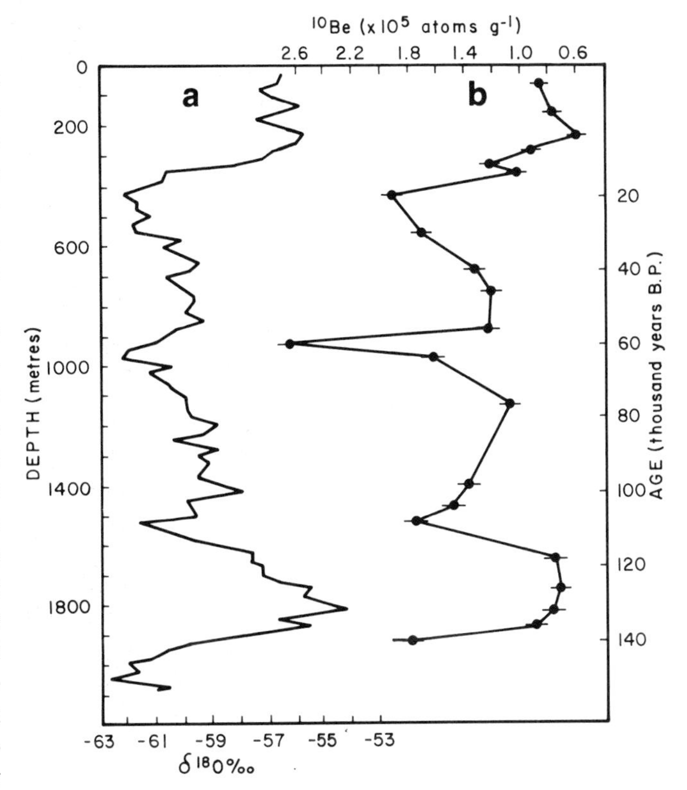

Figure 3. a. Concentration of ^{10}Be and b. oxygen-isotope ratio as functions of depth and age at Vostok, Antarctica. From Yiou and others (1985). Reprinted by permission of the authors and the publishers of *Nature.* Copyright (c) 1985 Macmillan Journals Limited.

Storminess. Ice cores contain microparticles, commonly referred to as "dust." About 95 to 99 percent of the particles have radii between 0.1 and 2 μm. Most of this dust has been carried from distant ice-free land as a tropospheric aerosol. In Greenland more than 90 percent of it reaches the surface in precipitation, rather than as dry fallout. The dry fallout fraction is higher in the low-precipitation areas of East Antarctica. The particles are insoluble in water, except in the Wisconsin ice in Greenland, where the dust is partly soluble due to the presence of alkaline material. The dust concentration in Wisconsin ice is 3 to 70 times that in Holocene ice. Moreover, the temporal variations are much greater in the Wisconsin (Fig. 4). Reduced precipitation in arctic regions would account for a factor of 2 to 3. In addition, the source areas of dust in North and South America and in Australia are believed to have been more extensive during the arid ice-age climate than they are now. Moreover, continental shelves, exposed by the fall in sea level, would provide an additional source, as suggested by the observation that the ice-age dust in Greenland

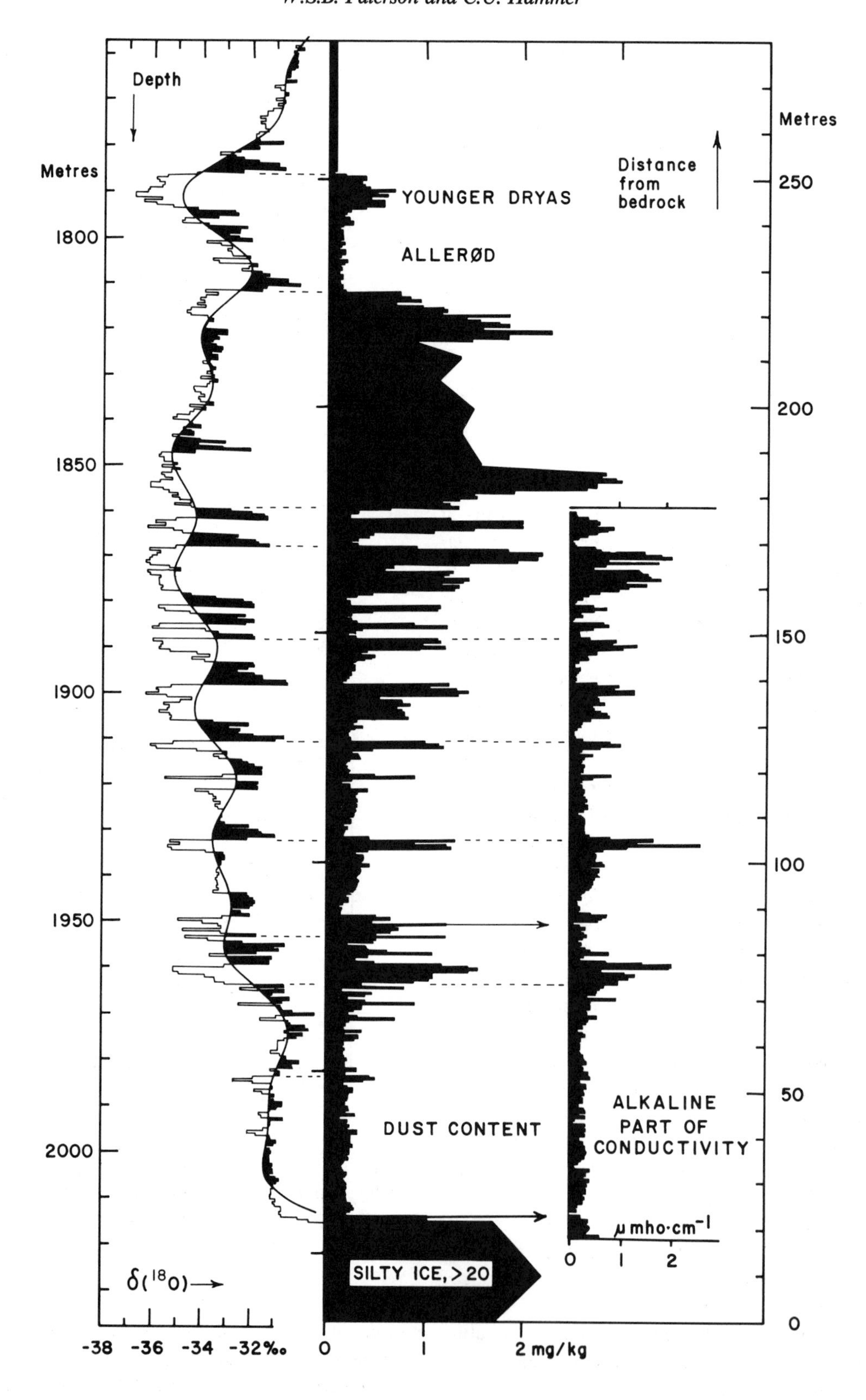

Figure 4. Oxygen-isotope ratio, dust concentration, and alkalinity versus depth in the ice-age part of the Dye 3 core. Only selected segments were sampled for dust between 1,821 and 1,852 m. See Figure 2 for estimated ages. Reproduced from Hammer and others (1985) by permission of the authors and the American Geophysical Union.

is rich in calcium and alkaline material (Cragin and others, 1977; Hammer and others, 1985). However, these factors are considered to be inadequate to account for the size of the observed increase in dust; hence it is also interpreted as evidence for more vigorous atmospheric circulation during the ice age than at present (Petit and others, 1981; Dansgaard and others, 1984).

Atmospheric carbon dioxide. Air bubbles in glacier ice provide samples of the atmosphere at the time they were formed. The snow near the ice-sheet surface is permeable. However, as each layer is buried by subsequent snowfalls, compaction and recrystallization increase its density. Eventually the interconnecting passageways between the ice grains become sealed off, and the remaining air is present only as bubbles. Techniques have recently been developed whereby the concentration of CO_2 in the bubbles can be measured in as little as one gram of ice (Stauffer and others, 1985). The air sample, which dates from the time when the bubble is formed, is younger than the surrounding ice. Moreover, there is a spread in the age of the bubbles, because all are not formed at exactly the same depth. The age difference and spread depend mainly on firn temperature and snow-accumulation rate (Schwander and Stauffer, 1984). At present, the age difference in Greenland ranges from 100 to 400 yr, with spreads of 20 to 75 yr. At Vostok, east Antarctica, it is 3,000 to 4,000 yr with a spread of 600 yr (Paterson, 1981, p. 15; Schwander and Stauffer, 1984). These values were probably appreciably greater during the ice age. Figure 5 shows the CO_2 record from Byrd Station.

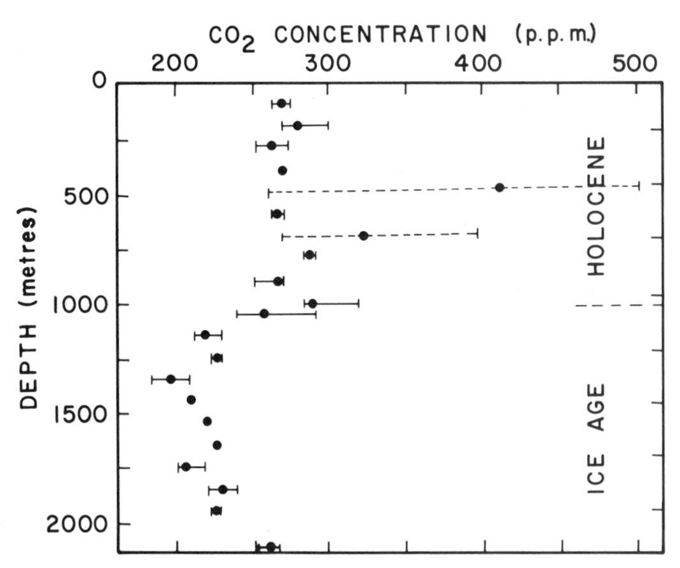

Figure 5. Concentration of carbon dioxide versus depth at Byrd Station, West Antarctica. Lowest, median, and highest values for each depth are shown. The samples at 500 and 700 m were contaminated by drilling fluid. The Holocene/ice age transition is at about 1,030 m, and at 1,360 m the age is about 20 ka. From Neftel and others (1982). Reprinted by permission of the authors and the publishers of *Nature.* Copyright (c) 1982 Macmillan Journals Limited.

There are two possible sources of error in these measurements. Because CO_2 dissolves readily in water, if there is any melting at the ice-sheet surface the dissolved CO_2 increases the concentration in the ice when the meltwater refreezes. Concentrations measured in Holocene ice from Dye 3 in Greenland show seasonal variations, with maxima in the summer melt layers (Stauffer and others, 1985). At present some melting and refreezing occurs at all the northern-hemisphere core sites; melting at any of them during the glaciation, although unlikely, cannot be ruled out. In Antarctica the snow never melts, even at present, except near the coast. It is therefore a better place for CO_2 measurements than the existing Greenland sites.

Another potential source of inaccuracy arises from interactions between CO_2 and impurities in the ice. At Dye 3, hydrostatic pressure causes the bubbles to diffuse into the ice lattice and form a clathrate hydrate at a depth of about 1,300 m, which is nearly 500 m above the Holocene/Wisconsin transition (Shoji and Langway, 1982). Bubbles start to reappear after the core has been brought to the surface, and they tend to form at places where there are microparticles in the ice. A chemical reaction at the surface of the bubble between the CO_2 and water vapor in the bubble and the carbonate in the dust might reduce the concentration of CO_2 in the air in the bubble.

Changes in ice-surface elevation. Data on past variations in surface elevation of an ice sheet would be of great value both as a record of ice extent and also to correct for one of the major uncertainties in interpreting oxygen-isotope records. Raynaud and Lorius (1973) proposed a method, based on the air bubbles in the ice, for measuring elevation changes. When the pores close off to form air bubbles, the amount of air trapped per unit pore *volume* should depend on the atmospheric pressure and thus on the elevation of the site at that time. In practice, only the volume of air per unit *mass* of an ice sample can be measured, because the pore volume in unit mass at the time of close-off is unknown. However, the volume of air per unit mass (usually called "total gas content") in near-surface ice is in fact highly correlated with present site elevation (Raynaud and Lebel, 1979). Use of such measurements in old ice to estimate past surface elevations, on the other hand, rests on the assumption that the present rate of change of gas content with surface elevation is typical of the past, including the ice age. This is doubtful. Total gas content depends on pore volume, which in turn depends on the grain size of the ice crystals; this was reduced during the glaciations as a result of reduced temperature and increased atmospheric dust concentration. Again, gas content can vary seasonally by amounts that correspond to elevation differences of several hundred meters (Dansgaard and others, 1973; Paterson, 1981, p. 343; Dansgaard and others, 1985).

We do not use gas content data in this review because we believe that, although the method is potentially useful, the conclusions from most studies so far published are unreliable. In particular, we disregard the conclusion that the ice-age ice in the Camp Century core originated about 800 m higher than the present elevation at the station (Raynaud and Lebel, 1979) because it is based on only seven samples, none of which was thicker than one-quarter of an annual layer. We do not, however, rule out the possibility of some ice-thickness change there.

Dating of ice cores

General. Methods of finding the age-depth relationship for an ice core are, in decreasing order of precision: (1) counting annual layers, distinguished by seasonal variations in one or more properties; (2) defining horizons of known age, such as acid layers from identifiable volcanic eruptions; (3) radioactive dating (^{14}C from CO_2 in air bubbles, ^{32}Si, $^{36}Cl/^{10}Be$); (4) matching features of the $\delta^{18}O$ record with another dated climatic record, such as those contained in ocean or lake sediments; and (5) making calculations based on an ice-flow model. The best time scales for ice cores have precisions of about 2 percent back to 10,000 yr B.P. Precision declines beyond that, mainly because it is difficult to distinguish annual layers in ice-age ice.

Paterson (1981, p. 328–336) described the different dating methods and their limitations. Here we discuss the dating of the Dye 3 and Camp Century cores with particular attention to the precision for the period of the deglaciation. There are no flow-induced discontinuities in the deglacial section of either of these cores.

Greenland. The Dye 3 core has been dated back to the Holocene/Wisconsin boundary by counting annual layers, which were distinguished by seasonal variations in $\delta^{18}O$, acidity, and dust. Use of more than one criterion increases the precision. Seasonal variations in acidity and dust were detected back to the Holocene/Wisconsin boundary, but those in $\delta^{18}O$ died out by about 8,000 yr B.P. Deconvolution techniques were not needed. The Holocene/Wisconsin boundary is the limit of acidity measurements in Greenland because the Wisconsin ice is alkaline. Seasonal variations in dust content should be detectable in older ice, but so far only a few spot measurements have been made.

These methods give a date of 10,750 yr B.P. $\pm$ 150 yr for the Holocene/Wisconsin boundary, defined here as the Younger Dryas/Preboreal transition (Hammer and others, 1986). This boundary shows clearly in most core properties; Figure 6 shows the profiles of $\delta^{18}O$ and dust content. This date agrees with that of 10,750 +50/−150 yr B.P. for the Younger Dryas/Preboreal transition based on the newly revised Swedish varve chronology (Stromberg, 1985). The points in the $\delta^{18}O$ record that correspond to dates of 8,500, 9,000, and 9,500 yr B.P. as determined by counting annual layers, are also shown. Note that all these dates are in true, not radiocarbon, years. Carbon-14 dating of the Dye 3 core by accelerator mass spectrometry has so far been confined to Holocene ice, and the results have uncertainties of 170 to 850 yr (Andrée and others, 1986).

The dating can be extended, though less reliably, by comparing the ice-core values of $\delta^{18}O$ with another climatic record. Variations of $\delta^{18}O$ of precipitation are also recorded in lake sediment. Figure 6 includes such a record from Gerzensee in Switzerland. Similar records have been obtained from two other European lakes (Siegenthaler and others, 1984). The resemblance between the two profiles is so close that they must be recording the same climatic events. The two major rapid warmings in the lake record (the beginning of the Bølling interval and the end of the Younger Dryas) have been dated at 13,000 and 10,000 yr B.P. (radiocarbon years) by comparison with dated pollen profiles (Eicher and Siegenthaler, 1976). We have therefore transferred the older date to the ice-core record. The 750-yr difference between ice-core and pollen dates for the end of the Younger Dryas is the difference between real and radiocarbon years at that time (Hammer and others, 1986). We do not distinguish between the two in the older record.

We tentatively extend the time scale to the Late Wisconsin maximum, which we believe occurs at 1,854 m, the lowest point in Figure 6B. This is a $\delta^{18}O$ minimum, and the highest concentrations of ^{10}Be (Beer and others, 1985) and dust (Fig. 4) also occur at this depth. The annual-layer thickness in the cores immediately above the Holocene/Wisconsin transition is 30 mm (Hammer and others, 1986, Fig. 2). Measurements of ^{10}Be and sulfate concentrations (Beer and others, 1985; Herron and Langway, 1985) both suggest that the mean precipitation rate in the Late Wisconsin was about one-third of the Holocene value. The mean annual-layer thickness between a depth of 1,812 m, where the age is estimated at 13 ka (Fig. 6), and 1,854 m is therefore 10 mm. This gives an estimated age of 17.2 ka for the Late Wisconsin maximum (1,854 m). We round this to 18 ka to take into account the progressive thinning of the annual layers as their depth increases.

We emphasize that our time scale beyond the Holocene/Wisconsin boundary depends on the correct identification of the Bølling interval and the glacial maximum, combined with the date of 13,000 yr B.P. in the Gerzensee profile. Revision of that date would change our time scale.

Measurements of annual-layer thickness are also important for calculating the rapidity of climatic changes. At Dye 3 the measured annual-layer thickness changes from 30 mm just above the Holocene/Wisconsin boundary to less than 20 mm just below (Hammer and others, 1986). The change in $\delta^{18}O$ from −36 to −30.5 ‰ at this boundary is complete in four 0.5 m samples, i.e., within 2 m of core. This temperature change therefore took place in about 100 years.

The poor condition of much of the Camp Century core prevented continuous counting of annual layers. A combination of spot measurements of annual-layer thickness and a semiempirical flow model was used. The end of the Younger Dryas was dated at 10,100 yr B.P. $\pm$ 300 yr (Hammer and others, 1978). The apparent discrepancy with the date for the same event at Dye 3 is not believed to be real; it is ascribed to inaccuracies in the Camp Century time scale (Hammer and others, 1986). Dating of the Camp Century core beyond this boundary is uncertain.

Greenland data

In this section we present the data, draw attention to some important features, and discuss some difficulties in interpretation.

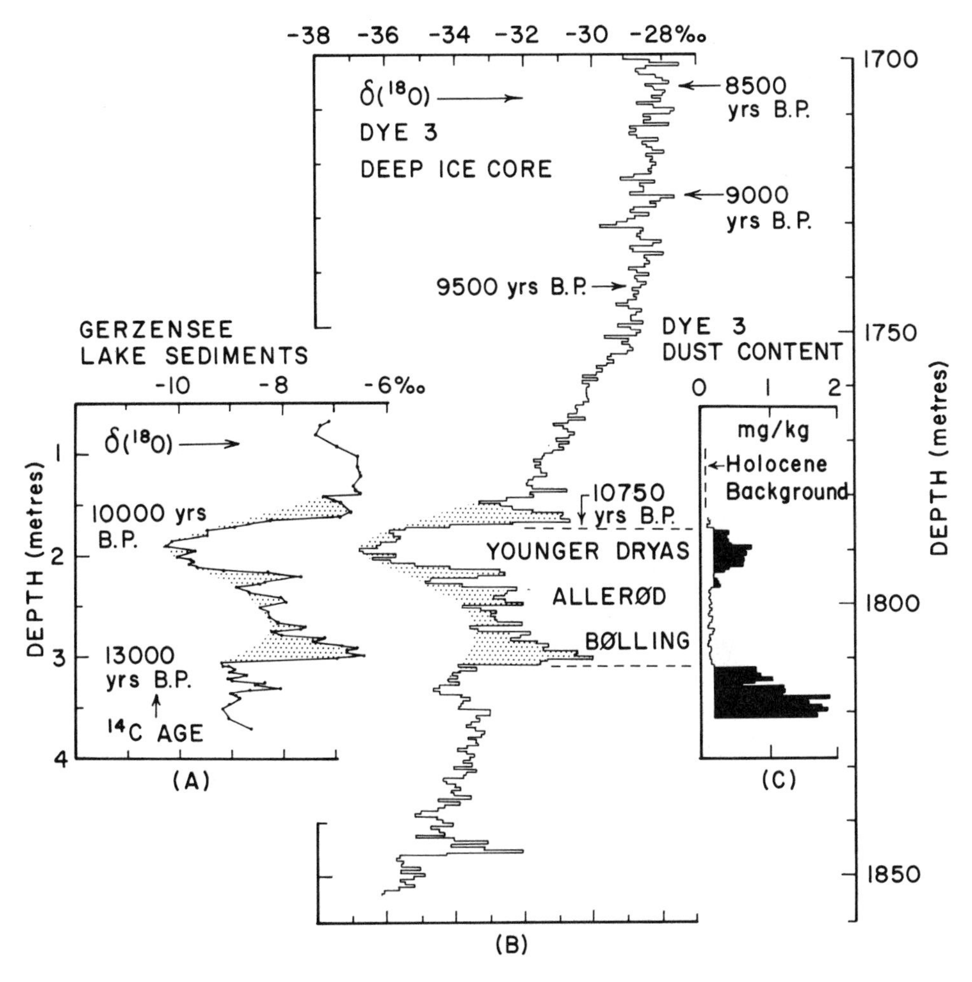

Figure 6. Depth variation of: A, $\delta^{18}O$ in lake sediments, Gerzensee, Switzerland (Siegenthaler and others, 1984); B, $\delta^{18}O$ at Dye 3, Greenland; and C, dust concentration at Dye 3. Note that ages in A are in ^{14}C years and in B are in real years.

Relating these to other types of data, and to possible mechanisms of deglaciation, is postponed until the discussion section.

Figure 7 shows the Dye 3 record of $\delta^{18}O$ (temperature), ^{10}Be concentration (inversely proportional to precipitation rate), and atmospheric CO_2 concentration from the late Wisconsin maximum (~ 18,000 yr B.P.) to 8,500 yr B.P. Note that the $\delta^{18}O$ record is based on continuous sampling, whereas only spot measurements of ^{10}Be and CO_2 were made. References are Dansgaard and others (1985), Beer and others (1985), and Stauffer and others (1985).

One of the most striking features of the $\delta^{18}O$ record (Fig. 7a) is its complexity. Its major features are: (1) An approximately linear warming trend occurred from 18,000 to about 14,000 yr B.P. with a change in $\delta^{18}O$ of about 35 percent of the total Wisconsin/Holocene transition. (2) Superimposed on this trend was a warmer interval, which began about 1,000 yr after the glacial maximum and lasted for a few hundred years. (3) The warming trend was ended about 14,000 yr B.P. by a cooler interval of roughly 1,000 yr. (4) A rapid warming at about 13,000 yr B.P., by the end of which $\delta^{18}O$ was close to its Holocene value. (5) The Bølling-Allerød period, lasting about 2,000 yr, which was generally warm, although there was an overall cooling trend with cold "spikes" superimposed. (6) The Younger Dryas period which lasted less than 1,000 yr. Values of $\delta^{18}O$ were comparable with those at the glacial maximum. (7) Rapid warming occurred at 10,750 yr B.P., the end of the Wisconsin glaciation. As noted previously, the change of $\delta^{18}O$ from –36 ‰ to –30.5 ‰ takes place in 2 m of core, which corresponds to a time interval of about 100 yr. (8) After a short rapid cooling, temperature increased slowly for another 1,000 to 1,500 yr until typical Holocene values were reached.

Continuous measurements of direct-current conductivity along the core, which have a much higher resolution than the oxygen-isotope measurements, also show a rapid change at the Holocene/Wisconsin boundary and confirm that it is completed in 2 m of core (Neftel and others, 1985). Moreover, Siegenthaler

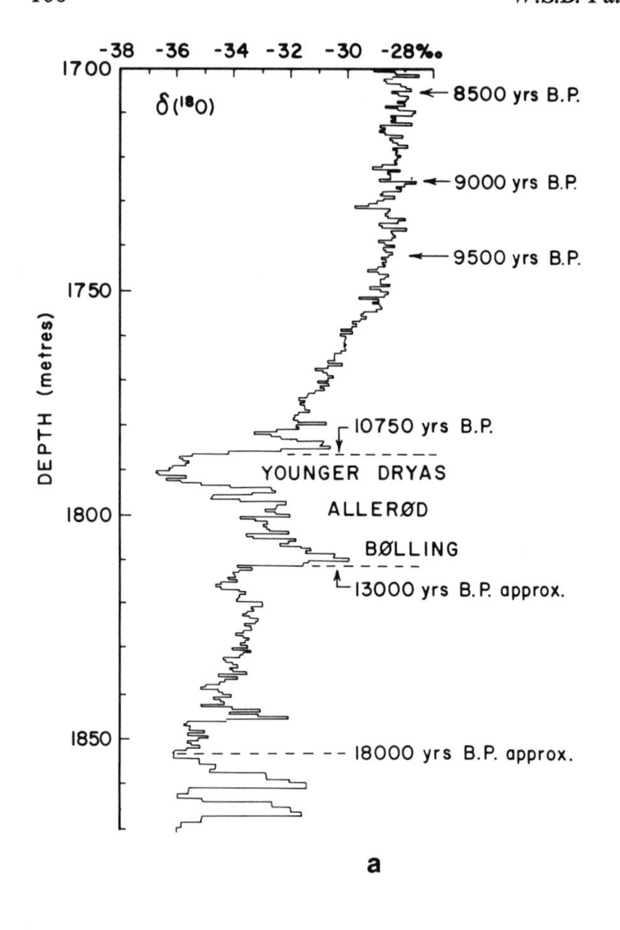

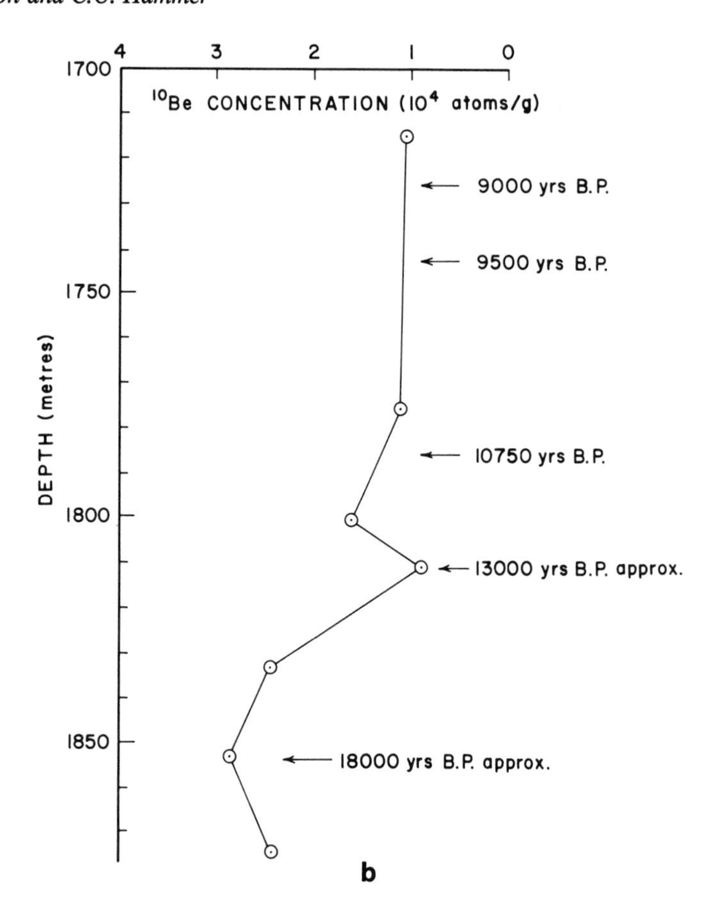

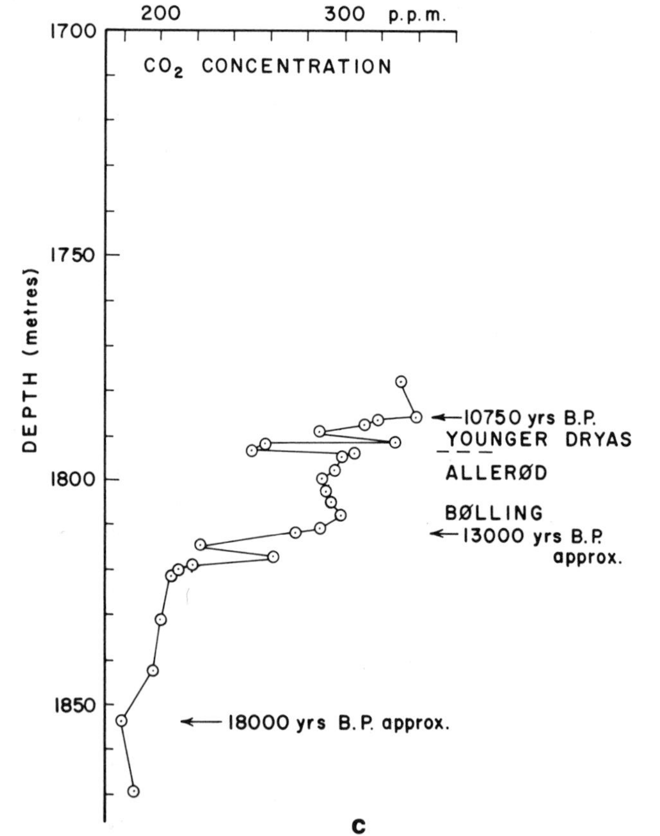

Figure 7. Dye 3, Greenland. Depth and age variation of a, $\delta^{18}O$ (temperature); b, ^{10}Be concentration (precipitation rate); and c (next page), atmospheric CO_2 concentration. The time scale is explained in the text. Data sources are given in the text.

and others (1984, p. 150) concluded that this climatic change, as recorded in sediments in Gerzensee and other lakes in central Europe, occurred "within a century or even less." The warming at about 13,000 yr B.P. appears to have been equally rapid (Fig. 6), and there are many other similar features throughout the core (Fig. 2). The fact that these changes are recorded in lake sediments as well as ice cores rules out the possibility that they result from some glaciological phenomenon such as a surge; they must represent rapid climatic changes. Moreover, these changes must have affected areas the size of continents; they cannot represent minor changes in storm tracks that might have produced an atypical record at one particular site.

The Camp Century record of the deglaciation (Fig. 2) shows the same general features as observed at Dye 3, but the relative magnitudes of the cold and warm "spikes" differ. At Camp Century, the pre-Bølling cold period was colder than the Younger

Dryas, whereas the reverse is true at Dye 3. The value of $\delta^{18}O$ at the peak of the Bølling interval is lower than Holocene values at Camp Century, and the rapid warming at the end of the Younger Dryas lasted about 400 rather than the 100 yr at Dye 3 (Hammer and others, 1978, Fig. 11). Some of these differences probably represent differences between the climates of the two sites; others may result from the more detailed sampling of the Dye 3 core.

We now discuss the $\delta^{18}O$ record in the light of the complications described in the section above on measurement of temperature. At least part of the apparent deglacial warming might result from ice thinning or from the fact that the older the ice, the higher the elevation at which it originated. We first discuss the change from –36 to –33 ‰ between about 18,000 and 14,000 yr B.P. at Dye 3 (Fig. 7a). Numerical modeling of ice flow suggests that ice of this age originated on or near the crest of the ice sheet, where the surface slope is about 2×10^{-3} and the ice velocity is not more than 1 m/yr (Reeh and others, 1985). This corresponds to a decrease in elevation of 8 m by downslope movement in 4,000 yr. Because the average change of $\delta^{18}O$ with elevation in Greenland is about 0.62 ‰ per 100 m, this movement would have a negligible effect on the value of $\delta^{18}O$. Another possibility is that the increase of 3 ‰ in the period 18,000 to 14,000 yr B.P. results entirely from ice thinning; the amount required is about 450 m. We believe that this amount of thinning near the crest of the ice sheet in the initial stages of deglaciation is unlikely.

The other major gradual trend in δ is an increase from –32 to –29 ‰ between about 10,500 and 9,500 yr B.P. Again, and for a similar reason, we believe that downslope movement cannot account for this. Alternatively, 450 m of thinning would again be required. These two gradual trends account for nearly all of the total change of 8 ‰ between the Late Wisconsin maximum and the Holocene. The greater part of this must certainly represent climatic warming. Ice thinning may have contributed; however, recent work has challenged the conventional view that the Greenland ice sheet was thicker during the glaciation than it is now (Reeh, 1985). The amount of thinning, if any, could be determined by measuring total gas content of the Dye 3 core.

Strong support for a climatic interpretation of the long-term changes in $\delta^{18}O$ comes from analysis of the present temperature distribution in the borehole at Dye 3 (Dahl-Jensen and Johnsen, 1986). Changes in surface temperature are propagated into the ice by conduction and by the downward flow of ice, the rate of which depends on the precipitation rate. Seasonal changes are attenuated rapidly, but long-period changes, such as the oscillation between glacials and interglacials, can be propagated to depths of several thousand meters, although their amplitude is progressively reduced. The present temperature at the ice-rock interface at Dye 3 is 5 K colder than expected from the present surface temperature; this represents surface cooling during the Wisconsin. Numerical modeling of the temperature distribution in the borehole suggested that the mean ice-age surface temperature was 12 K ± 2 K colder than at present and that the mean precipitation rate was 50±25 percent of the present one. This temperature difference is close to the value of 13 K obtained by converting the 8 ‰ change in $\delta^{18}O$ to temperature by the commonly accepted factor of 0.62 ‰ per degree.

We therefore believe that the long-term trends in the $\delta^{18}O$ record of the deglaciation at Dye 3 represent temperature changes; ice-thickness changes and the effects of downslope movement are minor. In particular, at the Late Wisconsin maximum, the surface temperature at the crest of the Greenland ice sheet at latitude 65°N, where the Wisconsin ice in the Dye 3 core originates, was about 13 K colder than at present. This can be compared with the maximum summer sea-surface temperature anomaly of 18 K in the North Atlantic (CLIMAP, 1976). Another important conclusion is that the increase in $\delta^{18}O$ between the glacial maximum and about 14,000 yr B.P. represents warming early in the deglaciation.

The interpretation of the large "steps" in $\delta^{18}O$, such as those at about 13,000 yr B.P. and at the beginning and end of the Younger Dryas, is less certain. These coincide with movements of the polar front in the North Atlantic, southeast of Greenland, over 10° to 15° of latitude, with corresponding changes in the extent of winter sea ice (Ruddiman, this volume). These major changes would certainly affect temperatures on the Greenland ice sheet. Thus, part of each $\delta^{18}O$ step might result from changes n sea-surface temperature. A reduction in the distance from moisture sources might also have changed the value of $\delta^{18}O$. However, W. Dansgaard (personal communication, 1987) believes, on the basis of measurements of the deuterium excess in Greenland ice cores, that most of the present-day precipitation on the ice sheet originates in the Atlantic Ocean between latitudes 35° and 40°N. This region lies south of the position of the polar front even at the glacial maximum (Ruddiman and McIntyre, 1981). Thus, movements of the polar front may have had little effect on the position of the moisture source.

Measurements of ^{10}Be concentration, which we interpret as a measure of precipitation rate, are shown in Figure 7b. For comparison, the mean ^{10}Be concentration for the period 1900 to 1976 A.D. is 0.93×10^4 atoms/gram, and the mean of six spot measurements covering the period 3,500 to 10,000 yr B.P. is almost the same. Figure 7b shows that precipitation rate was lowest at the glacial maximum, about one-third of the Holocene value if ^{10}Be concentration depends only on precipitation rate. Precipitation rate reached the Holocene value by the start of the Bølling interval. There was a reduction during the Allerød period. There are no measurements for the Younger Dryas. However, a sudden change in the concentration of sulphate ions at the Holocene/ Wisconsin boundary has been interpreted as evidence of a 60 percent reduction in precipitation rate during the Younger Dryas (Herron and Langway, 1985). These trends are consistent with the mean value deduced from the temperature distribution in the Dye 3 borehole, discussed above.

Part of the great increase in concentration of windblown dust during the Wisconsin (a factor of 3 to 70 times the mean Holocene value) is attributed to more vigorous atmospheric circulation, particularly during the coldest periods. Moreover, the

TABLE 3. DATES (ka B.P.) OF TERMINATIONS IN ICE-VOLUME RECORD

Termination 1a	1b	1c	Reference
16-13	10-7		Duplessy and others, 1981
16-13.5	10-8.5		Sarnthein and others, 1982
13	10		Berger and others, 1985
14-12	10-9	8-6	Mix and Ruddiman, 1985

changes in dust content, though not as rapid as those in $\delta^{18}O$, take place within a few hundred years. They suggest that major rapid changes in atmospheric circulation occurred repeatedly during the Wisconsin.

Figure 7c shows the record of atmospheric CO_2. For comparison, the concentration before the industrial era, as measured in ice cores, was about 270 p.p.m. by volume (Neftel and others, 1982). The value in 1958, when systematic measurements began, was 313 p.p.m., and this rose to 344 p.p.m. by 1985. Figure 7c shows that the concentration was less than 200 p.p.m. at the glacial maximum. The trend thereafter is an increase, interrupted by three periods of reduced concentration, each lasting less than 1,000 yr. It appears that the concentration can change by as much as 70 p.p.m. in about 100 yr, in the Younger Dryas, for example. Several of the concentrations exceed the preindustrial value; this is attributed to the effect of summer melting described in the above section on measurement of atmospheric CO_2. The mean Holocene value at Byrd Station, where there is no melting, is about 270 p.p.m. (Fig. 5).

Comparison of Figures 7a and 7c shows that changes in CO_2 concentration during the deglaciation are accompanied by corresponding changes in $\delta^{18}O$. This is also the case during the glaciation (Oeschger, 1985, Fig. 7). Published data, however, do not show any consistent phase difference between changes in CO_2 and $\delta^{18}O$. To establish such a relation would require more detailed sampling of CO_2 and improved modeling of the process of bubble closure in the firn. Moreover, there are still technical problems with the CO_2 measurements. The apparent large rapid changes need to be checked with data from Antarctica. The ice-age ice there, unlike the Greenland ice, is acid, not alkaline, so there is no possibility of a reduction in CO_2 concentration in a bubble as a result of a chemical reaction with carbonate at its surface. Because changes in atmospheric CO_2 concentration must be world-wide, if the large rapid changes are not observed in Antarctica, the Greenland results must be an artifact. We believe that further discussion of these changes, and of possible mechanisms for them, is premature until they have been confirmed by measurements on Antarctic ice.

DISCUSSION

Comparison of ice-core and oceanic records

In this section we compare the ice-core record of climate during the deglaciation with the oceanic record of climate and ice volume.

Ruddiman (this volume) describes the sea-surface temperature changes in the North Atlantic (north of latitude 45°N) during the deglaciation:

18,000 to 13,000 yr B.P.: A cold surface water mass of low salinity extended at least as far south as 45°N. Winter sea ice probably extended to near 50°N.

13,000 yr B.P.: Warmer and more saline water from the subtropics flowed into the eastern and central North Atlantic, so that the boundary of the polar water retreated to the north coast of Iceland.

11,000 yr B.P.: The polar front readvanced to within 350 to 600 km of its glacial maximum position and stayed there for at most 1,000 yr.

10,000 yr B.P.: The polar front retreated rapidly to the southeastern Labrador Sea.

These temperature changes correspond closely to the major rapid warmings and coolings in the temperature record from Dye 3, Greenland (Fig. 7a), particularly when the discrepancy of 700 yr between true and radiocarbon ages at 10,000 yr B.P. (see section on dating of ice cores in Greenland) and the problems with ^{14}C dating of ocean sediments (redistribution of sediments and contamination with old carbon from the continents) are taken into account.

The oceanic oxygen-isotope record is of world ice volume, not North American. However, North America contributed an estimated 60 percent of the excess ice volume at the late Wisconsin maximum, and the Antarctic ice sheet another 20 percent (Hughes and others, 1981, Table 6.3). Moreover, because the Antarctic ice sheet extends to the edge of the continent, its extent is controlled by world sea level, which in turn depends on the amount of ice in the northern hemisphere. Thus, variations in world ice volume should reflect, to a first approximation, changes in North American ice.

Table 3 lists the estimated radiocarbon dates of the "steps", also referred to as terminations, in the oceanic oxygen-isotope record. These are the times when world ice volume was decreasing most rapidly. The Greenland ice-core record shows that the two major rapid warmings occurred at 13,000 and 10,000 yr B.P. (radiocarbon years). These dates lie within all the limiting dates for terminations 1a and 1b. Because increases in temperature would increase ice-ablation rates, the ice-core data support the idea of "steps" in the deglaciation. The ice-core data give no evidence of termination 1c tentatively determined by Mix and Ruddiman (1985); temperatures had reached Holocene values by

9,000 yr B.P. However, this termination, if real, may have resulted from the disintegration of the ice sheet in Hudson Bay and Hudson Strait between 9,000 and 8,200 yr B.P. (Prest, 1969).

Figure 8 shows the estimated rate of decrease of volume of the Laurentide ice sheet during the deglaciation. The volumes are those that Paterson (1972, Fig. 7, curve A) calculated from the ice-sheet areas (Prest, 1969) using a steady-state ice-flow model. In such a model, the surface profile of the ice sheet, and thus the relation between its area and volume, does not change with time. In reality, the profile of an ice sheet does change while it is adjusting to changes in precipitation and ablation (Paterson, 1981, Chapter 12). Use of a more refined model seems pointless, however, in view of the imprecision of the data on ice-sheet areas. The ice margins on Prest's map are "speculative" and based on extrapolation over long distances between dated moraines. Moreover, some sectors of the Laurentide ice sheet did not attain their maxima until about 14,000 yr B.P.; the position of these margins before that is unknown.

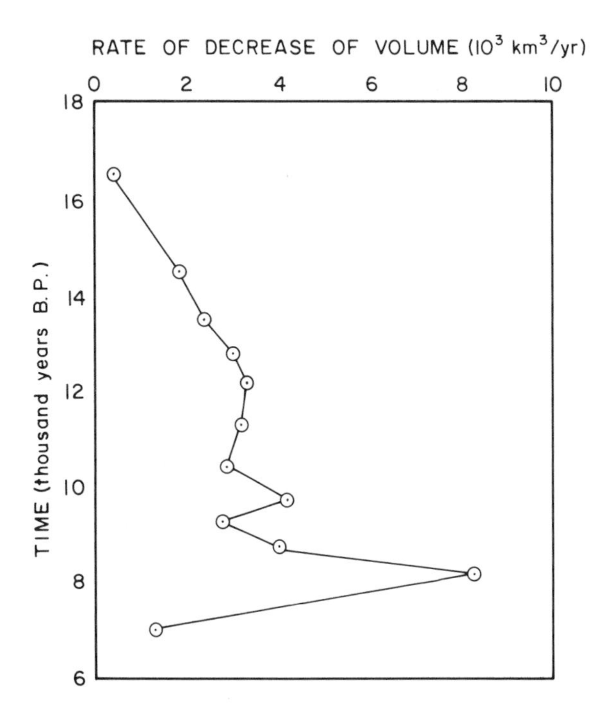

Figure 8. Rate of reduction of volume of Laurentide ice sheet, derived from volume estimates of Paterson (1972).

The rate of decrease of ice volume has three maxima, at about 12,200, 9,800, and 8,300 yr B.P.. These dates are within Mix and Ruddiman's (1985) limiting dates for the three terminations in the world ice-volume record. However, 12,200 yr B.P. is younger than other estimates of the date of the first termination (see Table 3). On the other hand, this date relates well to the rapid warming at 13,000 yr B.P. recorded in the ice cores, when the response time of a large ice sheet to climatic change is taken into account.

We conclude that, within the accuracy of the present data, the oceanic, ice-core, and terrestrial records of the deglaciation show the same major features at about the same times. The discrepancies are that (1) most oceanic records do not show termination 1c, and (2) the date of termination 1a in the Laurentide ice-volume record is younger than in some oceanic records. More precise dating is needed to assess whether these discrepancies are significant.

Feedback mechanisms

Introduction. Ruddiman and Wright (this volume) describe the Milankovitch hypothesis and summarize the evidence for it. They also explain the apparent need for feedback mechanisms within the climate system to account for the rapid deglaciations that occur every 100,000 yr.

Proposed feedback mechanisms are: (1) iceberg calving accelerated by rising sea level, (2) increase in concentration of atmospheric carbon dioxide, (3) moisture starvation of the decaying ice sheets, (4) ice-albedo feedback, and (5) delayed rebound of the earth's mantle as the ice melts.

We now discuss what ice cores, and some other glaciological data, tell us about these mechanisms.

Iceberg calving. This feedback mechanism works in the following way. Melting and retreat of the southern margins of the Laurentide and Fennoscandian ice sheets early in the deglaciation raise sea level. The rate of iceberg calving from grounded glacier tongues, which is proportional to the water depth (Brown and others, 1983), will increase. As the icebergs melt, they will raise sea level further and accelerate the process. Melting of ice from floating tongues will change sea level only by a small amount proportional to the density difference between ice and water. However, a rise in sea level may cause grounded portions of such tongues to float free and break up.

Glaciers that extend to tidewater in fiords often maintain stable positions for long periods at a narrow or shallow part of the fiord or at its mouth. A slight retreat from such a position, brought about by a slight increase in melting or calving, usually results in a catastrophic breakup as the terminus retreats rapidly to the next anchor point. This happens even if the water behind the original anchor point is not deep enough to float the ice (Mercer, 1961). During retreat, the surface slope of the lower part of the glacier increases, with a consequent increase in velocity. The increased ice flux is usually inadequate to compensate for the increase in calving rate, however, so that retreat continues and a wave of thinning (a "kinematic wave") travels up the glacier. The glacier eventually attains a new equilibrium profile with its terminus at the new anchor point.

This process makes the relation between glacier response and climatic forcing nonlinear: if the terminus is in a critical state, a small increase in melting or calving can trigger a substantial retreat. The process might also change the relation between ice-sheet area and volume, because a tidewater glacier does not have a steady-state profile during rapid retreat or for some time afterward. However, for this effect to be observable in the present imprecise data on area and volume of the Late Wisconsin ice sheets, it would have to operate over an appreciable part of the Laurentide ice sheet at one time. This is unlikely, because much of the ice margin was on land, even at the glacial maximum.

The propagation of a wave of thinning up the glacier during retreat could be described as "downdraw." However, there are important differences between this process and "downdraw" as used in the current literature (e.g., Denton and Hughes, 1983).

Hughes and others (1985) use the present Antarctic ice sheet as an analogue for the Late Wisconsin ice sheet in North America. Most of the ice from central Antarctica flows to the coast in fast-moving ice streams and outlet glaciers. Most of these drain into the floating ice shelves that surround much of the continent. The shelves exert a back pressure on the ice streams and restrict their flow. Much of this back pressure arises where the ice shelves are grounded on shoals. Rising sea level, or thinning as a result of increased surface or basal melting, would reduce the grounded area. The resulting decrease in back pressure would cause the ice streams to speed up, and this would reduce the thickness of the central part of the ice sheet.

The essential feature of the downdraw hypothesis, as we understand it, is that ice streams are inherently unstable; flow through ice streams thins the central regions of an ice sheet not only when its margin is retreating but also when the margin is stationary or even advancing. Moreover, according to the hypothesis, downdraw is irreversible, so that the whole ice sheet will eventually disappear. Increases in ice-stream velocity, as the back pressure of the ice shelves is reduced, appear to be regarded as a secondary effect. We emphasize that the inherent instability of ice streams has not been established: they may merely maintain the ice sheet in a steady state.

Denton and Hughes (1981, 1983) and Hughes and others (1985) assert that downdraw was the prime cause of the decay of the North American ice sheet. They believe that the ice sheet, at maximum, was fringed by extensive ice shelves in the Labrador Sea, Baffin Bay, and the Arctic Ocean, and that it was drained by major ice streams in Hudson Strait and the Gulf of St. Lawrence. There is little geological evidence for or against the existence of these ice shelves, as Denton and others (1986) now admit.

The strongest case for downdraw came from Ruddiman and McIntyre's (1981) result, based on North Atlantic core data, that the volume of the North American ice sheet had been halved by 13,000 yr B.P., although the area was still about 80 percent of maximum. This would of course imply extensive thinning with relatively little reduction in area. The composite $\delta^{18}O$ curve in Mix (this volume) would support a reduction of global ice volume of between 30 and 40 percent by 12,500 yr B.P. (Ruddiman, this volume). This reduces the implied discrepancy with the 25 percent predicted by the steady-state ice-flow model, which assumes a constant relationship between area and volume, that is, no downdraw (Paterson, 1972, Table 5). Allowing for effects other than North American ice volumes in the $\delta^{18}O$ curves, there may be no discrepancy between the two estimates.

In addition, Hughes and others (1985) state that downdraw can explain the "steps" in the deglaciation. Our demonstration that the steps coincide, within the precision of the dating, with strong climatic warmings further weakens the case for downdraw.

Ruddiman and McIntyre (1981) observed a "barren zone" in the North Atlantic cores, characterized by low carbonate production. This zone was originally believed to span the period 16,000 to 13,000 yr B.P. The low productivity was attributed to a large influx of glacial meltwater and icebergs. Ruddiman and Duplessy (1985) argued that icebergs predominated because there would be little atmospheric heat for melting early in the deglaciation. They interpreted the large iceberg flux as evidence for downdraw. However, Ruddiman (this volume) noted that the starting date for this barren zone is not well defined, and could be as young as 14,000 yr B.P. This weakens the argument considerably, because the ice cores show that there had been appreciable warming by then.

The North American ice sheet, at its maximum, reached the sea along its northern and eastern margins. At this time iceberg calving, rather than melting, was probably the major ablation mechanism. Calving would continue until the ice retreated from the coast. Moreover, rising sea level would tend to increase the calving rate. In addition, there would be episodes of increased iceberg production when the ice retreated rapidly, as described at the beginning of this section, in places such as the Gulf of St. Lawrence, Hudson Strait and Hudson Bay, and the channels among the Canadian Arctic Islands. To this extent, we agree with Denton and Hughes. On the other hand, we are not convinced that ice streams are inherently unstable or that there is evidence for major thinning of the Laurentide ice sheet by "downdraw" early in the deglaciation. Indeed, Denton and others (1986, p. 22) now state that the data of Mix and Ruddiman (1985) "suggest strongly that volumetric deglaciation began at 14,000 ^{14}C yr B.P., concurrent with extensive areal recession."

Changes in atmospheric carbon dioxide concentration. The ice-core data (Figs. 5 and 7c) show that the atmospheric CO_2 concentration increased from a minimum value of 180 to 200 p.p.m. at the glacial maximum to a typical Holocene value of about 270 p.p.m. at the Wisconsin/Holocene transition. The difference in concentration of ^{13}C betwen mean surface water and deep ocean water, which is an indirect indicator of atmospheric CO_2 concentration, shows a similar trend (Shackleton and others, 1983). The Greenland ice-core data show, in addition, a reduction in CO_2 concentration during the Younger Dryas interval.

General circulation models of the earth's atmosphere predict that doubling the preindustrial CO_2 concentration would produce a mean global warming of between 1.5 and 4.5 K and two or three times that amount in the polar regions. Extrapolation of this result to lower concentrations suggests that the increase in CO_2 during deglaciation could have increased polar temperatures by between 1 and 4 K. The greenhouse effect of increasing CO_2 concentrations therefore appears to be an important feedback factor.

The mechanism of these large variations in atmospheric CO_2 concentration is obscure. The prime cause must lie in the

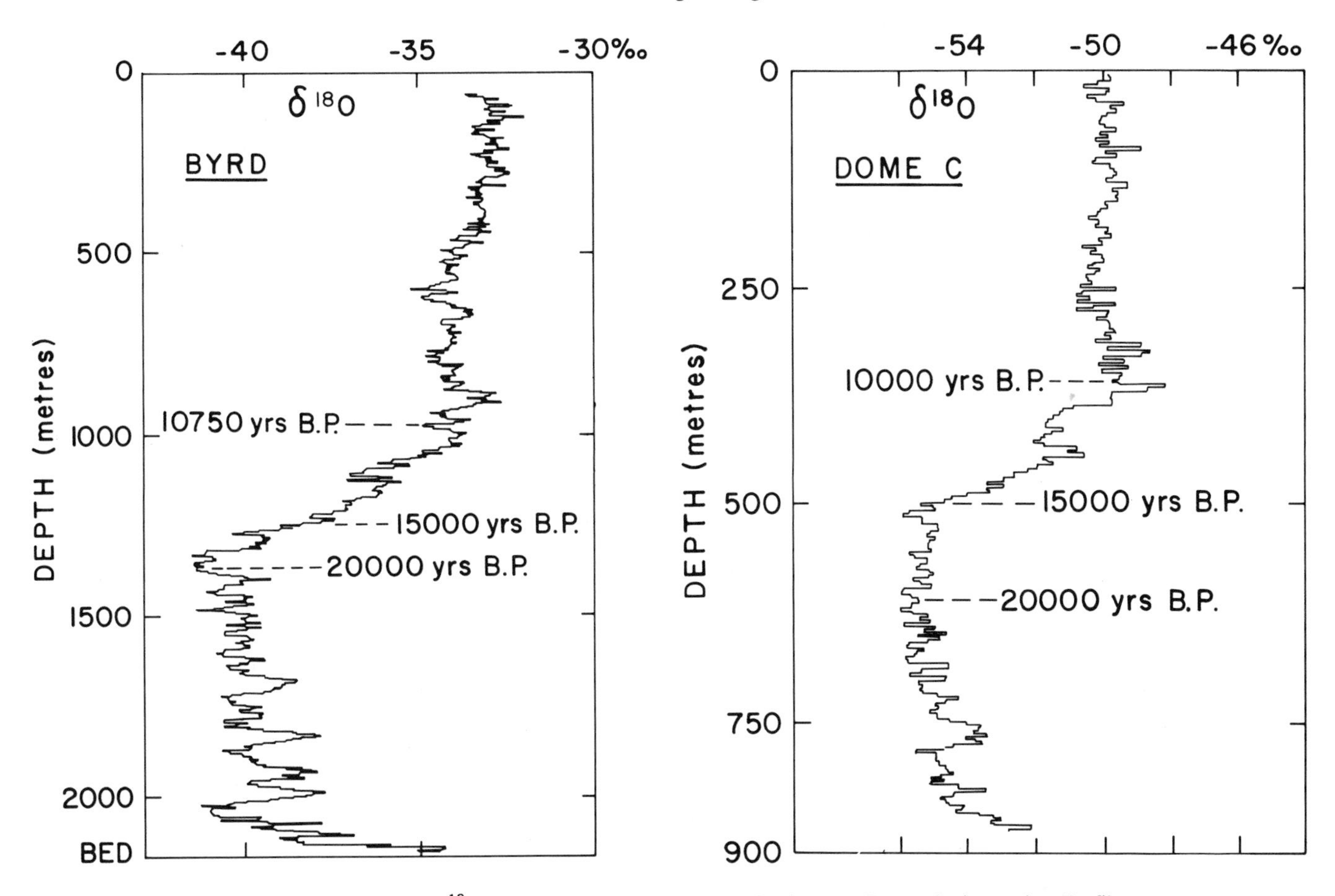

Figure 9. Variation of $\delta^{18}O$ with depth and age at Byrd Station and Dome C, Antarctica. Profiles adapted from Robin (1983) by permission of the author and the Cambridge University Press. Byrd time scale from Hammer and others (1987).

oceans; at present these store about 55 times as much carbon as the atmosphere (Clark and others, 1982). The atmospheric CO_2 reservoir therefore comes into equilibrium with the oceanic reservoir; the time lag is about 1,000 yr. Any mechanism that extracts additional carbon from the ocean surface will reduce the partial pressure of CO_2 over it. Several such mechanisms have been proposed (Berger, 1982; Broecker, 1982). Discussion of these is outside the scope of this chapter.

Moisture starvation, ice-albedo feedback, and mantle rebound. Ruddiman and McIntyre (1981) suggested that winter sea ice, which apparently covered the North Atlantic as far south as 50°N until 13,000 yr B.P., would cut off the supply of winter moisture to the ice sheets. They believed that this "moisture starvation" speeded up deglaciation, particularly between 14,000 and 12,500 yr B.P. (Ruddiman, this volume). The ice-core record of precipitation rate (Fig. 7b) shows three low values (about one-third of the Holocene mean) around the glacial maximum. There are no data between these points and a typical Holocene value early in the Bølling interval at 13,000 yr B.P. We believe that, in contrast to the linear trend shown in Figure 7b, it is more likely that the precipitation rate remained low until the rapid warming associated with the retreat of the polar front at about 13,000 yr B.P. (Fig. 7a). Ruddiman (this volume) no longer believes that moisture starvation was a major factor in deglaciation. We think that it may have been important, not so much in the period 14,000 to 12,500 yr B.P. but in the earlier stages of deglaciation. A combination of low precipitation and the slow warming indicated by the $\delta^{18}O$ data (Fig. 7a) may have been enough to cause the slow retreat of the ice edge between 17,000 and 14,000 yr B.P. (Prest, 1969). The slow warming may have resulted from an increase in atmospheric CO_2 (Figs. 5 and 7c).

Kutzbach (this volume) discusses ice-albedo feedback, and Peltier (this volume) deals with mantle rebound.

The rapid changes in the Greenland isotopic record

The ice-core oxygen-isotope records (Figs. 2, 3, and 9) show the same general features as the oceanic record: the Holocene, the Wisconsin glaciation, at least part of the Sangamon interglacial, and, in the Vostok record, the last part of the preceding glaciation. However, as Broecker and others (1985) have pointed out, the Greenland cores show no obvious sign of the 23 and 41 ka cycles.

In the oceanic record, these cycles appear as three peaks, each larger than the preceding one, so that the maximum ice volume was attained immediately before the deglaciation. The Vostok record (Fig. 3) has cold peaks, each about the same size, at roughly 20,000, 60,000, and 100,000 yr B.P. These correspond to the Milankovitch tilt cycle, which predominates at high latitudes. In both Greenland cores, in contrast, $\delta^{18}O$ switches between two values (about –36 and –32 ‰ at Dye 3) that are separated by about half the difference between mean Wisconsin and mean Holocene values (Fig. 2). These oscillations also appear in other core properties, such as concentration of ^{10}Be (Beer and others, 1984), dust (Dansgaard and others, 1984), sulfate, nitrate, and chlorine (Finkel and Langway, 1985). The switch from a "cold" to a "warm" value of $\delta^{18}O$ takes place in about 100 yr, whereas the switch from warm to cold usually takes longer (Dansgaard and other 1984). The Younger Dryas, and the cold period immediately before the Bølling interval, appear to be typical in this respect. Dansgaard and others (1984) argue convincingly that these oscillations must be climatic and not due to folding in the ice, repeated surges, or other glaciological peculiarities. They also show that the rapidity of the changes is real and not the result of discontinuities in the time scale.

The period of these oscillations has been estimated at about 2.5 ka, based on the number of cycles during the Wisconsin in the Camp Century core. However, dating uncertainties make it more realistic to assume that the average period lies between about 1,200 and 2,600 yr. The lower figure is based on a few measurements of annual-layer thickness in Wisconsin ice from Dye 3 and Camp Century (C.U. Hammer, unpublished data). Indeed, the oscillations may not have a regular period at all. For example, the cold peaks at the end of the glaciation occur at about 18,000, 14,000 and 11,000 yr B.P. However, these changes may be of different character from those during the glaciation.

Antarctic cores (Fig. 9) also show oscillations in $\delta^{18}O$ during the glaciation. However, their amplitudes are smaller and the changes appear to be less abrupt than in Greenland. This suggests that the oscillations result from a North Atlantic phenomenon, perhaps related to movements of the polar front and/or changes in the rate of deep-water production. However, it is pointless to discuss phase relations between the oscillations in Greenland and Antarctica as long as the dating of Wisconsin ice is so uncertain.

Large ice sheets can respond to climatic oscillations with periods of 2 to 2.5 ka. The southern lobes of the Laurentide ice sheet advanced and retreated several times during the first few thousand years of the deglaciation. Numerical modeling suggests that, although the response time of the central part of the Antarctic ice sheet to a rapid rise in sea level is about 8 ka, most of the reduction in volume, which takes place at the edge, is completed within 100 yr (Alley and Whillans, 1984). We therefore believe that the rapid oscillations could have produced significant fluctuations in world ice volume. These do not appear in most oceanic oxygen-isotope records, because bottom fauna, by stirring up the sediments, effectively filter out oscillations at such relatively high frequencies.

Existing data are inadequate for assessing the importance of the rapid oscillations relative to the 23 and 41 ka cycles observed in oceanic records. This will require not only older cores, but also greatly improved methods of dating ice older than 10 ka. However, the large rapid changes in $\delta^{18}O$, dust concentration, and soluble impurities certainly exist. They cannot bear any relation to orbital forcing.

Ice-core data and the Milankovitch hypothesis

In order to assess the Milankovitch theory of glaciations, it is necessary to compare its predictions to proxy records of Quaternary climate. Deep-sea sedimentary records have been used to test the hypothesis (Hays and others, 1976), while well-dated deep ice cores present a future potential for scrutinizing it.

Greenland ice cores have been accurately dated back to the Wisconsin/Holocene boundary (Hammer and others, 1986), and recently it has been shown that the transition from the late-glacial maximum to the Holocene occurred simultaneously in the two hemispheres (Hammer and others, 1987). This conclusion is based on a comparison between ice-core datings and variations in $\delta^{18}O$, annual layer thicknesses, and acidity in the deep ice cores from Dye 3, Camp Century, and Byrd. Further, the rapid and substantial climatic oscillations observed in Greenland ice cores during glacial times and the retreats of the Laurentide and Scandinavian ice sheets during the transition add to the evidence that the northern-hemisphere environmental and climatic changes may have influenced southern-hemisphere climate. The rapid climatic transition from Younger Dryas to Holocene, as inferred from Greenland ice cores, also indicates such a coupling, because this North Atlantic phenomenon apparently caused drastic changes in the acidity of the Antarctic air as well as a rapid change of the annual-layer thicknesses in the Byrd core.

The change in $\delta^{18}O$ at Byrd occurs before the change in acidity and annual layer thickness at 10,750 yr B.P. (Fig. 9). Moreover, it is less rapid than in Greenland. The total change in $\delta^{18}O$ over the Wisconsin/Holocene transition is, however, the same at Byrd and Dye 3. A priori, we would not expect such a climatic scenario. In this context, the Milankovitch hypothesis is of little help: rather, these interesting facts indicate a lack in our understanding of the climatic coupling between the two hemispheres.

At present, the main support for the Milankovitch hypothesis comes from the high coherence between orbital and oceanic $\delta^{18}O$ (world ice volume) signals at each of the main orbital periods (19, 23, 41, and 100 ka) over the last 800 ka (Imbrie and others, 1984). However, these authors assumed that (1) variations in orbital precession and tilt cause changes in world climate, (2) the rate of climatic response at any instant is proportional to the magnitude of the orbital forcing, and (3) the time constant of the system is 17 ka. They then adjusted the time scale of the oceanic record so as to lock the variations in $\delta^{18}O$ to those in orbital forcing. The only check of this time scale prior to the late-Wisconsin maximum consisted of radiometric dates at 127,000 and 730,000 yr B.P. We would like to see if the coherence remains high when an independently-dated climatic record is used.

We believe that ice-core dating, when developed to its full potential, will provide an excellent way of testing the Milankovitch hypothesis. This potential has not yet been realized because of technical difficulties and the peculiarities of existing core sites. For example, the well-dated Byrd core covers only about 60 ka (Fig. 9), while indirect datings, such as those for Greenland (Dansgaard and others, 1982) and Vostok, Antarctica (Lorius and others, 1985), leave ample scope for speculation. We suggest that the first step toward a further test is to obtain a deep core from central Greenland, where ice older than 120 ka, which can be dated by seasonal stratigraphy with a precision of ±1 ka, may be expected if the core is continuous and in good physical condition.

SUMMARY OF CONCLUSIONS

Cores from polar ice sheets provide the most detailed records of climatic parameters such as temperature and precipitation rate, and of forcing factors such as concentration of atmospheric CO_2. The core from Dye 3, Greenland, has been dated with a precision of ±1.5 percent back to the Holocene/Wisconsin boundary at 10,750 yr B.P. by counting annual layers. Dating beyond that, based on comparisons with other climatic records, is much less reliable.

The most marked features of the Greenland temperature record of the deglaciation are major rapid warmings (increases of several degrees in a period of the order of 100 yr) at about 13,000 and 10,700 yr B.P., and a major rapid cooling, the start of the Younger Dryas interval, at about 11,500 yr B.P. These features are superimposed on an overall warming trend that began about 1,000 yr after the glacial maximum. Typical Holocene temperatures were reached by about 9,000 yr B.P.

The direction and timing of these temperature changes agree with the inferred changes in sea-surface temperature and movements of the polar front in the North Atlantic north of latitude 45°N. Moreover, the two major "steps" in the deglaciation (times of rapid decrease in world ice volume) coincide, within the precision of the dating, with the two major warmings. These correlations suggest a direct link, via the atmosphere, between temperatures over the ice sheet and in the high-latitude North Atlantic Ocean. The third "step," tentatively identified by Mix and Ruddiman (1985), may result from the final outflow of ice from Hudson Strait and Hudson Bay.

The rate of decrease of volume of the Laurentide ice sheet, estimated from its area, has three maxima. The first two correspond to the major rapid warmings at 13,000 and 10,700 yr B.P. in the ice-core records, when the delay in the response of a large ice sheet is taken into account. The dates of all three maxima are within the limiting dates of the three terminations of Mix and Ruddiman (1985).

Precipitation rate during the deglaciation was highly correlated with temperature. The minimum rate, which occurred at the glacial maximum, was roughly one-third the Holocene value. "Moisture starvation" may have been a significant factor in the deglaciation, particularly in the initial stages.

The concentration of atmospheric CO_2 at the glacial maximum was about 190 p.p.m. compared with the mid-nineteenth century value of about 270 p.p.m. The concentration during the deglaciation is highly correlated with temperature (Figs. 5 and 9). Atmospheric general circulation models suggest that CO_2 changes of the size observed would be an important feedback mechanism. The CO_2 changes must result from changes in the ocean, but the detailed mechanism is uncertain.

Iceberg calving accelerated by rising sea level was probably an important feedback factor. The warming early in the deglaciation, recorded in the ice cores, seems adequate to account for the initial thinning of the ice sheet. The 25 percent reduction in volume of the Laurentide ice sheet by 12,500 yr B.P., as estimated from the reduction in area, is somewhat less than the 30 to 40 percent shift in $\delta^{18}O$ values by 12,500 yr B.P. (Mix, this volume), but factors other than North American ice volume affect the $\delta^{18}O$ signature. We therefore see no need to invoke the Denton-Hughes downdraw hypothesis, whereby inherently unstable ice streams thin the central part of an ice sheet when its terminus is stationary or even advancing.

Large rapid changes in $\delta^{18}O$ and in concentrations of dust and soluble impurities occur throughout the ice-age ice in both Greenland cores. The period may be irregular; its average value appears to be between 1.2 and 2.6 ka. Antarctic cores also show oscillations in $\delta^{18}O$. (The other parameters have not been measured in sufficient detail.) However, they have smaller amplitudes and are less abrupt than in Greenland. This suggests that the oscillations result from a North Atlantic phenomenon.

Techniques for distinguishing annual layers in ice cores are now sufficiently well developed to permit dating of a core from a favorable location with a precision of 1,000 yr over the past 120 ka or more. This would provide a definitive test of the Milankovitch hypothesis.

REFERENCES CITED

Alley, R. B., and Whillans, I. M., 1984, Response of the East Antarctica Ice Sheet to sea-level rise: Journal of Geophysical Research, v. 89. p. 6487–6493.

Andrée, M., and 10 others, 1986, Dating polar ice by ^{14}C accelerator mass spectrometry: Radiocarbon, v. 28, p. 417–423.

Beer, J., and 6 others, 1984, Temporal variations in the ^{10}Be concentration levels found in the Dye 3 ice core, Greenland: Annals of Glaciology, v. 5, p. 16–17.

Beer, J., and 9 others, 1985, ^{10}Be variations in polar ice cores. American Geophysical Union Geophysical Monograph 33, p. 66–70.

Berger, W. H., 1982, Deglacial CO_2 build-up; constraints on the coral reef model: Palaeogeography, Palaeoclimatology, Palaeoecology, v. 40, p. 235–253.

Berger, W. H., Killingly, J. S., Metzler, C. V., and Vincent, E., 1985, Two-step deglaciation; ^{14}C high-resolution $\delta^{18}O$ records from the tropical Atlantic Ocean: Quaternary Research, v. 23, p. 258–271.

Broecker, W. S., 1982, Glacial to interglacial changes in ocean chemistry: Progress in Oceanography, v. 11, p. 151–197.

Broecker, W. S., Peteet, D. M., and Rind, D., 1985, Does the ocean-atmosphere system have more than one stable mode of operation?: Nature, v. 315, p. 21–25.

Brown, C. S., Sikonia, W. G., Post, A., Rasmussen, L. A., and Meier, M. F., 1983, Two calving laws for grounded iceberg-calving glaciers: Annals of Glaciology, v. 4, p. 295.

Clark, W. C., and 7 others, 1982, The carbon dioxide question; Perspectives for 1982, *in* Clark, W. C., ed., Carbon dioxide review 1982: New York, Oxford University Press, p. 3–44.

CLIMAP Project Members, 1976, The surface of the ice-age earth: Science, v. 191, p. 1131–1137.

Cragin, J. H., Herron, M. M., Langway, C. C., and Klouda, G., 1977, Interhemispheric comparison of changes in the composition of atmospheric precipitation during the late Cenozoic era, *in* Dunbar, M. J., ed., Polar oceans: Calgary, Alberta, Canada, Arctic Institute of North America, p. 617–631.

Dahl-Jensen, D., and Johnsen, S. J., 1986, Palaeotemperatures still exist in the Greenland ice sheet: Nature, v. 320, p. 250–252.

Dansgaard, W., Johnsen, S. J., Clausen, H. B., and Langway, C. C., Jr., 1971, Climatic record revealed by Camp Century ice core, *in* Turekian, K. K., ed., The late Cenozoic glacial ages: New Haven, Connecticut, Yale University Press, p. 37–56.

Dansgaard, W., Johnsen, S. J., Clausen, H. B., and Gundestrup, N., 1973, Stable isotope glaciology: Meddelelser om Grønland, v. 197, no. 2, p. 1–53.

Dansgaard, W., and 6 others, 1982, A new Greenland deep ice core: Science, v. 218, p. 1273–1277.

Dansgaard, W., and 6 others, 1984, North Atlantic climatic oscillations revealed by deep Greenland ice cores: American Geophysical Union Geophysical Monograph 29, p. 288–298.

Dansgaard, W., Clausen, H. B., Gundestrup, N., Johnsen, S. J., and Rygner, C., 1985, Dating and climatic interpretation of two deep Greenland ice cores: American Geophysical Union Geophysical Monograph 33, p. 71–76.

Denton, G. H., and Hughes, T. J., 1981, The Arctic ice sheet; An outrageous hypothesis, *in* Denton, G. H., and Hughes, T. J., eds., The last great ice sheets: New York, Wiley, p. 437–467.

—— , 1983, Milankovitch theory of ice ages; Hypothesis of ice-sheet linkage between regional insolation and global climate: Quaternary Research, v. 20, p. 125–144.

Denton, G. H., Hughes, T. J., and Karlen, W., 1986, Global ice-sheet system interlocked by sea level: Quaternary Research, v. 26, p. 3–26.

Duplessy, J.-C., Delibrias, G., Turon, J. L., Pujol, C., and Duprat, J., 1981, Deglacial warming of the northeastern Atlantic Ocean; Correlation with the paleoclimatic evolution of the European continent: Palaeogeography, Palaeoclimatology, Palaeoecology, v. 35, p. 121–144.

Eicher, U., and Siegenthaler, U., 1976, Palynological and oxygen-isotope investigations on late-glacial sediment cores from Swiss lakes: Boreas, v. 5, p. 109–117.

Finkel, R. C., and Langway, C. C., Jr., 1985, Global and local influences on the chemical composition of snowfall at Dye 3, Greenland; The record between 10 ka B.P. and 40 ka B.P.: Earth and Planetary Science Letters, v. 73, p. 196–206.

Fisher, D. A., and Alt, B. T., 1985, A global oxygen isotope model; Semi-empirical, zonally-averaged: Annals of Glaciology, v. 7, p. 117–124.

Fisher, D. A., and 5 others, 1983, Effect of wind scouring on climatic records from ice-core oxygen-isotope profiles: Nature, v. 301, p. 205–209.

Grootes, P. M., and Stuiver, M., 1986, Ross Ice Shelf oxygen isotopes and west Antarctic climate history: Quaternary Research, v. 26, p. 49–67.

Hammer, C. U., and 5 others, 1978, Dating of Greenland ice cores by flow models, isotopes, volcanic debris, and continental dust: Journal of Glaciology, v. 20, p. 3–26.

Hammer, C. U., and 5 others, 1985, Continuous impurity analysis along the Dye 3 deep core: American Geophysical Union Geophysical Monograph 33, p. 90–94.

Hammer, C. U., Clausen, H. B., and Tauber, H., 1986, Ice core dating of the Pleistocene/Holocene boundary applied to a calibration of the ^{14}C time scale: Radiocarbon, v. 28, p. 284–291.

Hammer, C. U., Clausen, H. B., and Langway, C. C., 1987, Acidity along the Byrd ice core; 70,000 years of seasonal changes and southern hemisphere volcanism: Nature (in press).

Hays, J. D., Imbrie, J., and Shackleton, N. J., 1976, Variations in the earth's orbit; Pacemaker of the ice ages: Science, v. 194, p. 1121–1132.

Herron, M. M., and Langway, C. C., Jr., 1985, Chloride, nitrate, and sulfate in the Dye 3 and Camp Century, Greenland, ice cores: American Geophysical Union Geophysical Monograph 33, p. 77–84.

Hooke, R. LeB., and Clausen, H. B., 1982, Wisconsin and Holocene $\delta^{18}O$ variations, Barnes ice cap, Canada: Geological Society of America Bulletin, v. 93, p. 784–789.

Hughes, T. J., and 5 others, 1981, The last great ice sheets; A global view, *in* Denton, G. H., and Hughes, T. J., eds., The last great ice sheets: New York, Wiley, p. 263–317.

Hughes, T. J., Denton, G. H., and Fastook, J. L., 1985, The Antarctic ice sheet; An analog for northern hemisphere paleo-ice sheets?, *in* Woldenberg, M. J., ed., Models in geomorphology: Boston, Allen and Unwin, p. 25–72.

Imbrie, J., and 8 others, 1984, The orbital theory of Pleistocene climate: support from a revised chronology of the marine $\delta^{18}O$ record, *in* Berger, A., Imbrie, J., Hays, J., Kukla, G., and Saltzman, B., eds., Milankovitch and Climate, Part 1: Dordrecht, Reidel Publishing Company, p. 269–305.

Johnsen, S. J., Dansgaard, W., Clausen, H. B., and Langway, C. C., Jr., 1972, Oxygen isotope profiles through the Antarctic and Greenland ice sheets: Nature, v. 235, p. 429–434.

Jouzel, J., Merlivat, L., and Lorius, C., 1982, Deuterium excess in an East Antarctic ice core suggests higher relative humidity at the oceanic surface during the last glacial maximum: Nature, v. 299, p. 688–691.

Koerner, R. M., 1979, Accumulation, ablation, and oxygen isotope variations on the Queen Elizabeth Islands ice caps, Canada: Journal of Glaciology, v. 22, p. 25–41.

Lorius, C., Merlivat, L., Jouzel, J., and Pourchet, M., 1979, A 30,000-yr isotope climatic record from Antarctic ice: Nature, v. 280, p. 644–648.

Lorius, C., and 6 others, 1985, A 150,000-year climatic record from Antarctic ice: Nature, v. 316, p. 591–596.

Mercer, J. H., 1961, The response of fjord glaciers to changes in the firn limit: Journal of Glaciology, v. 3, p. 850–858.

Mix, A. C., and Ruddiman, W. F., 1985, Structure and timing of the last deglaciation; Oxygen-isotope evidence: Quaternary Science Reviews, v. 4, p. 59–108.

Morgan, V. I., and McCray, A. P., 1985, Enhanced shear zones in ice flow; Implications for ice cap modelling and core dating: Australian National Antarctic Research Expeditions Research Notes, v. 28, p. 4–9.

Neftel, A. Oeschger, H., Schwander, J., Stauffer, B., and Zumbrunn, R., 1982, Ice core sample measurements give atmospheric CO_2 content during the past 40,000 yr: Nature, v. 295, p. 220–223.

Neftel, A., Andrée, M., Schwander, J., Stauffer, B., and Hammer, C. U., 1985, Measurements of a kind of DC-conductivity on cores from Dye 3: American Geophysical Union Geophysical Monograph 33, p. 32–38.

Oeschger, H., 1985, The contribution of ice core studies to the understanding of environmental processes: American Geophysical Union Geophysical Monograph 33, p. 9–17.

Oeschger, H., and 5 others, 1984, Late glacial climate history from ice cores: American Geophysical Union Geophysical Monograph 29, p. 299–306.

Paterson, W.S.B., 1972, Laurentide Ice Sheet; Estimated volumes during late Wisconsin: Reviews of Geophysics and Space Physics, v. 10, p. 885–917.

—— , 1981, The physics of glaciers (2nd edition): Oxford and New York, Pergamon, 380 p.

Paterson, W.S.B., and Waddington, E. D., 1984, Past precipitation rates derived from ice core measurements; Methods and data analysis: Reviews of Geophysics and Space Physics, v. 22, p. 123–130.

Paterson, W.S.B., and 7 others, 1977, An oxygen-isotope climatic record from the Devon Island ice cap, arctic Canada: Nature, v. 266, p. 508–511.

Petit, J. -R., Briat, M., and Royer, A., 1981, Ice age aerosol content from East Antarctic ice core samples and past wind strength: Nature, v. 293, p. 391–394.

Prest, V. K., 1969, Retreat of Wisconsin and Recent ice in North America: Ottawa, Geological Survey of Canada Map 1257A, scale 1:5,000,000.

Raisbeck, G. M., and 6 others, 1981, Cosmogenic ^{10}Be concentrations in Antarctic ice during the past 30,000 years: Nature, v. 292, p. 825–826.

Raynaud, D., and Lebel, B., 1979, Total gas content and surface elevation of polar ice sheets: Nature, v. 281, p. 289–291.

Raynaud, D., and Lorius, C., 1973, Climatic implications of total gas content in ice at Camp Century: Nature, v. 243, p. 283–284.

Reeh, N., 1985, Was the Greenland ice sheet thinner in the late Wisconsinian than now?: Nature, v. 317, p. 797–799.

——, 1987, The Greenland ice-sheet margin; A mine of ice for paleo-environmental studies: Palaeogeography, Palaeoclimatology, Palaeoecology (in press).

Reeh, N., Johnsen, S. J., and Dahl-Jensen, D., 1985, Dating the Dye 3 deep ice core by flow model calculations: American Geophysical Union Geophysical Monograph 33, p. 57–65.

Reeh, N., Thomsen, H. H., and Clausen, H. B., 1987, The Greenland ice-sheet margin; A mine of ice for paleo-environmental studies: Palaeogeography, Palaeoclimatology, Palaeoecology, v. 58, p. 229–234.

Robin, G. de Q., 1983, Profile data, inland Antarctica, *in* Robin, G. de Q., ed., The climatic record in polar ice sheets: Cambridge, England, Cambridge University Press, p. 112–118.

Ruddiman, W. F., and Duplessy, J. -C., 1985, Conference on the last deglaciation; Timing and mechanism: Quaternary Research, v. 23, p. 1–17.

Ruddiman, W. F., and McIntyre, A., 1981, The North Atlantic Ocean during the last deglaciation: Palaeogeography, Palaeoclimatology, Palaeoecology, v. 35, p. 145–214.

Sarnthein, M., Erlenkeuser, H., and Zahn, R., 1982, Termination 1: The response of continental climate in the subtropics as recorded in deep-sea sediments: Bordeaux, Bulletin Gèologique Bassin d'Aquitane, v. 31, p. 393–407.

Schwander, J., and Stauffer, B., 1984, Age difference between polar ice and the air trapped in its bubbles: Nature, v. 311, p. 45–47.

Shackleton, N. J., Hall, M. A., Line, J., and Shuxi, C., 1983, Carbon isotope data in core V19-30 confirm reduced carbon dioxide concentration in the ice-age atmosphere: Nature, v. 306, p. 319–322.

Shoji, H., and Langway, C. C., Jr., 1982, Air hydrate inclusions in fresh ice core: Nature, v. 298, p. 548–550.

Siegenthaler, U., Eicher, U., Oeschger, H., and Dansgaard, W., 1984, Lake sediments as continental $\delta^{18}O$ records from the glacial/post-glacial transition: Annals of Glaciology, v. 5, p. 149–152.

Stauffer, B., Neftel, A., Oeschger, H., and Schwander, J., 1985, CO_2 concentration in air extracted from Greenland ice samples: American Geophysical Union Geophysical Monograph 33, p. 85–89.

Strømberg, B., 1985, Revision of the late-glacial Swedish varve chronology: Boreas, v. 14, p. 101–105.

Yiou, F., Raisbeck, G. M., Bourles, D., Lorius, C., and Barkov, N. I., 1985, ^{10}Be in ice at Vostok, Antarctica, during the last climatic cycle: Nature, v. 316, p. 616–617.

MANUSCRIPT ACCEPTED BY THE SOCIETY APRIL 15, 1987

ACKNOWLEDGMENTS

The final version of this chapter was written while the first author was enjoying six months in the Department of Glaciology at the Geophysical Institute, University of Copenhagen. He would like to thank Professor W. Dansgaard for making possible this most stimulating visit. Comments by Sigfus Johnsen, Alan Mix, and Bill Ruddiman on earlier versions resulted in substantial improvements.

Printed in U.S.A.

The Geology of North America
Vol. K-3, North America and adjacent oceans during the last deglaciation
The Geological Society of America, 1987

Chapter 6

The oxygen-isotope record of glaciation

A. C. Mix
College of Oceanography, Oregon State University, Corvallis, Oregon 97331

INTRODUCTION

This chapter reviews the use of oxygen-isotope ratios (i.e., $^{18}O/^{16}O$) from fossil foraminifera in paleoclimate studies.[1] It focuses on the constraints given by oxygen-isotope measurements for the timing, structure, and mechanisms of glaciation. The literature on stable isotopes in foraminifera has expanded considerably since the review of early work by Duplessy (1978). The emphasis here is on new developments and what they say about oxygen-isotope measurements as an ice-volume proxy. The last deglaciation is compared with long-term patterns of Quaternary glaciation. The evidence is examined for the structure of the last deglaciation recorded by marine $\delta^{18}O$, its timing as determined by radiocarbon dating, and the constraints this gives for interpreting mechanisms of glacial/interglacial climate change.

EARLY WORK

Early work on foraminiferal $\delta^{18}O$ by Emiliani (1955) centered on its use as a paleothermometer, although he recognized that $\delta^{18}O$ depends on the isotopic composition of the ambient water as well as the temperature. Water $\delta^{18}O$ relates both to local water-mass effects and to the amount of ^{18}O-depleted water stored on land as ice. Of the measured glacial/interglacial $\delta^{18}O$-range of about 1.8‰ in Caribbean planktonic foraminifera, Emiliani attributed about 0.4‰ to the ice-volume effect. This was equivalent to 100 m of sea-level change, assuming glacier ice had a mean $\delta^{18}O$ of −15‰ (relative to Standard Mean Ocean Water, SMOW), leaving the bulk of the signal, about 1.4‰, to be explained by glacial/interglacial temperature changes, amounting to about 6°C in the Caribbean.

As it became possible to measure $\delta^{18}O$ in small samples of calcite, oxygen-isotopes in benthic foraminifera became an important piece of the ice-volume puzzle. Early data of Emiliani (1955) showed that the glacial/interglacial difference in $\delta^{18}O$ from benthic foraminifera was at least 1.2‰. Emphasizing that benthics have essentially the same pattern of down-core change as planktonics, Shackleton (1967) argued that the ice-volume effect dominated the record, and that the temperature effect was relatively minor (Fig. 1). Although the scatter is large, Shackleton (1967) felt that the benthic and planktonic foraminifera had the same variations, offset by a constant. He inferred that deep-water temperatures could not have varied more than about 2 to 3°C, and that the ice-volume effect changed oceanic $\delta^{18}O$ by 1.4 to 1.6‰.

The data from benthic foraminifera are widely quoted as being constrained by the freezing point of seawater, which at the ocean surface is −1.9°C. The temperature of modern Caribbean deep waters, however, is about 4°C, so up to 6°C of temperature change ($\simeq$1.4‰ in $\delta^{18}O$) would be possible in the Caribbean benthic foraminiferal data (Shackleton, 1967). The constraint envisaged by Shackleton was that identical temperature changes in surface and deep waters are unlikely, as even small temperature changes in the deep ocean would, because of its large volume, imply a huge redistribution of heat. Emiliani (1970, 1971) continued to argue for smaller ice-volume effects of about 0.5‰, with most of the oxygen-isotope signal related to temperature. More recent work, to be examined below, has centered on resolving this argument by learning more about foraminiferal ecology and the recording of $\delta^{18}O$ as a proxy for both ice volume and temperature.

DISEQUILIBRIUM EFFECTS

Benthic foraminifera. In order to use foraminiferal $\delta^{18}O$ as a proxy for ice volume, a first requirement is that foraminifera are reliable recorders of $\delta^{18}O$. That is, they must secrete calcite either in equilibrium with the temperature and isotopic composition of the water, or they must have a known or predictable offset. The existence of such nonequilibrium "vital effects" offset-

[1]By convention, $^{18}O/^{16}O$ ratios are reported as

$$\delta^{18}O = 1000 \times \frac{(^{18}O/^{16}O)_{sample} - (^{18}O/^{16}O)_{REF}}{(^{18}O/^{16}O)_{REF}}$$

where, for calcite, REF is PDB, a belemnite from the Pee Dee Formation, South Carolina. In practice, as this standard no longer exists, most laboratories calibrate with the National Bureau of Standards references NBS-19 or NBS-20.

Mix, A. C., 1987, The oxygen-isotope record of glaciation, *in* Ruddiman, W. F., and Wright, H. E., Jr., eds., North America and adjacent oceans during the last deglaciation: Boulder, Colorado, Geological Society of America, The Geology of North America, v. K-3.

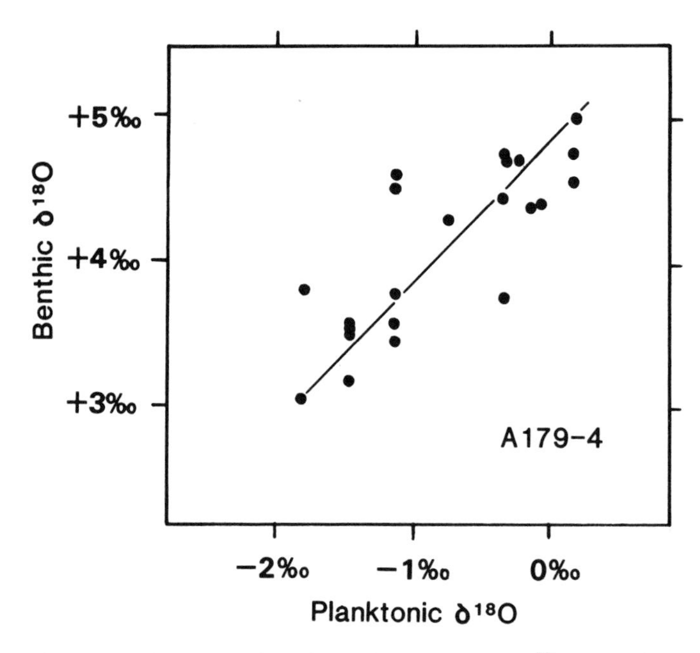

Figure 1. The first comparison of benthic and planktonic $\delta^{18}O$ time series, plotted as $\delta^{18}O$ time series, plotted as $\delta^{18}O$ of *G. sacculifer* (vertical axis) versus $\delta^{18}O$ of mixed benthic species (horizontal axis). (From Shackleton, 1967.)

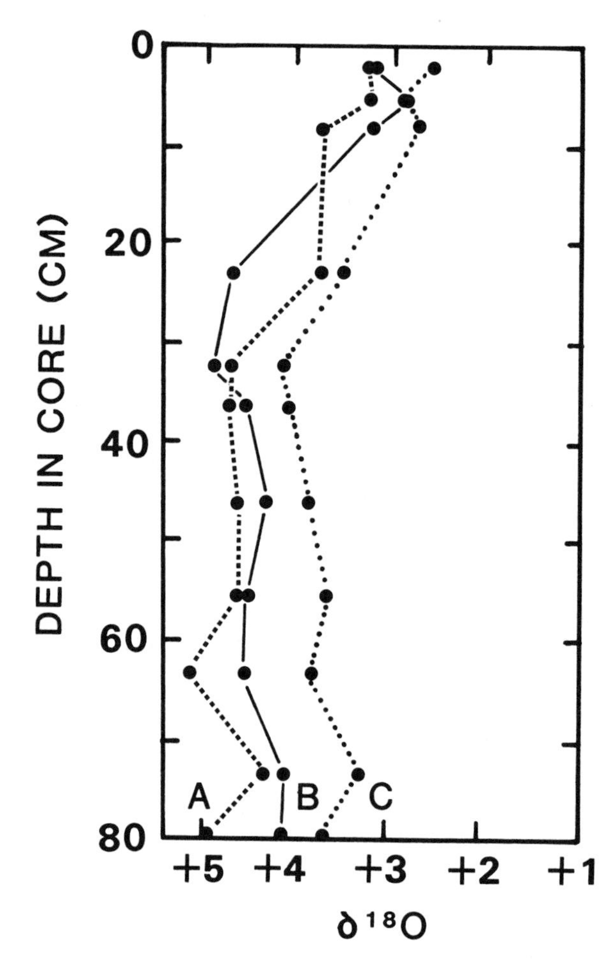

Figure 2. Disequilibrium vital effects, showing different $\delta^{18}O$ values for different species of benthic foraminifera. (From Duplessy and others, 1970.)

ting $\delta^{18}O$ values for different genera of benthic foraminifera was discovered by Duplessy and others (1970) (Fig. 2). Similar offsets in $\delta^{13}C$ among species may be caused by benthic foraminiferal microhabitats in the sediment, coupled to pore-water $\delta^{13}C$ gradients (Belanger and others, 1981; McCorkle and others, 1985; Zahn and others, 1986). The $\delta^{18}O$ offsets cannot be attributed to this effect, however, as pore-water gradients do not display temperature or $\delta^{18}O$ gradients similar to those of $\delta^{13}C$.

Although oxygen-isotope offsets between species are generally assumed to reflect metabolic processes, the exact mechanism for their occurrence is unknown. It is also not certain that the offsets are constant. Although metabolic effects would likely depend on environment, some available data tend to support the constancy of species offsets for $\delta^{18}O$, within the uncertainty of measurement and the effects of bioturbation (Graham and others, 1981). On the other hand, Vincent and others (1981) argue that $\delta^{18}O$ offsets among benthic species are not constant but are related to productivity of surface waters. Studies of living benthic foraminifera (Corliss, 1985) may shed some light on this topic.

Planktonic foraminifera. Planktonic foraminifera may also exhibit disequilibrium effects. Different species of planktonic foraminifera from the same deep-sea sediment samples yield different $\delta^{18}O$ values. It was first suggested that this was a temperature effect, with the species more enriched in ^{18}O living in cold subthermocline waters, and those depleted in ^{18}O living in warm surface waters (Emiliani, 1954; Lidz and others, 1968).

Shackleton and others (1973) suggested that some species in surface-water plankton tows, such as *Globigerinoides ruber, Globigerinoides sacculifer,* and *Neogloboquadrina dutertrei,* are more depleted in ^{18}O (by up to 0.6‰ for *G. ruber*) than even the warmest water would allow and thus must have disequilibrium effects, perhaps caused by photosynthetic symbionts that live within the protoplasm of some of these species (*G. ruber* and *G. sacculifer*). This conclusion was supported by Vergnaud-Grazzini (1976). Shackleton and Vincent (1978), however, revised the standard calibration used by Shackleton and others (1973) and suggested that although *G. ruber* may have a small disequilibrium effect, the other species measured are in oxygen-isotope equilibrium. Like Emiliani (1954), they stressed depth habitats of foraminifera, inferring that *Globigerina rubescens, G. ruber, G. sacculifer, Globigerinoides conglobatus,* and *Globigerinita glutinata* live near the surface, that *Pulleniatina obliquiloculata, N. dutertrei, Orbulina universa, Globigerinella siphonifera* (= *Globigerinella aequitaleralis*) live in subsurface (oxygen-minimum?) waters, and that *Globorotalia* spp. live below the thermocline in relatively cold waters. They noted, however, that this was really more an assumption than fact, because deviations of $\delta^{18}O$ in the positive direction from values expected for surface waters would give the appearance of living in deeper (and therefore colder) waters.

More recent studies of plankton tow samples (Fairbanks and others, 1982; Williams and others, 1981) give similar results, with apparent surface-dwelling foraminifera (such as *G. ruber* and *G. sacculifer*) slightly depleted in ^{18}O relative to equilibrium with surface waters. In contrast, studies of sediment-trap samples (Deusser and others, 1981; Curry and others, 1983) and core-top sediments (Williams and Healy-Williams, 1980; Durazzi, 1981; Curry and Matthews, 1981a,b) generally indicate near-equilibrium with surface waters for the same species. Erez and Luz (1983) detected a pattern in which the examples of apparent oxygen-isotope disequilibrium are from living foraminifera (collected in towed nets) that contain large quantities of protoplasm. They caution that removal of organic matter in these samples prior to isotopic analysis may cause isotopic exchange between the skeletal calcite and ^{18}O-depleted CO_2 produced during combustion of the protoplasm. This would result in measured calcite $\delta^{18}O$ values that are lower than expected equilibrium levels, as observed. They conclude that planktonic foraminifera probably secrete calcite in equilibrium with their surroundings.

An alternative hypothesis to explain the apparent mismatch between the tow samples and the sea-floor samples is that foraminifera continue to calcify as they sink to deep waters (Vergnaud-Grazzini, 1976). Duplessy and others (1981a) suggest that *G. sacculifer* adds 20 to 25 percent of its calcite between 300 and 800 m depth during gametogenesis. Under this interpretation, the apparent equilibrium of core-top foraminifera with surface waters is fortuitous, caused by the integration of initial ontogenetic calcite more depleted in ^{18}O than surface waters (due to disequilibrium) and late gametogenic calcite more enriched in ^{18}O than surface waters (due to equilibrium with colder subsurface waters). This is consistent with the findings of Berger and others (1978) that there is ^{18}O enrichment in larger specimens versus smaller specimens of many species. Subsurface gametogenic calcification would be particularly troubling for the interpretation of down-core oxygen-isotope data, because $\delta^{18}O$ would depend on variations in the surface temperature, magnitude of disequilibrium, vertical temperature gradient, amount of secondary calcification, depth of secondary calcification, and relative dissolution effects on gametogenetic crusts versus porous early calcite. All of these effects would likely change through time, yielding down-core planktonic foraminiferal isotope records that are difficult to interpret.

Fairbanks and others (1982) attempted to quantify depth preferences of foraminiferal species and determine isotopically their depths of calcification using depth-stratified plankton-tow samples (Fig. 3). As with other plankton-tow studies, these authors found apparent disequilibrium effects, with ^{18}O depletion for *G. ruber* and *G. sacculifer* (Fig. 3A, B). They assumed equilibrium calcification for all other species. As with the Shackleton and Vincent (1978) study, both equilibrium and disequilibrium remain unproven because the exact depth of calcification cannot be determined independently, and because of the uncertainties noted by Erez and Luz (1983) regarding combustion of protoplasm-full specimens. Fairbanks and others (1982) demonstrated variations in $\delta^{18}O$ of foraminifera across the thermocline (especially for *G. sacculifer* and *N. dutertrei*) due to calcification at various depths. Disagreeing with Duplessy and others (1981a), however, they stated that all secondary calcification occurs within the euphotic zone (upper 100 m), and that if gametogenic calcification is important at all, it would alter foraminiferal $\delta^{18}O$ only in areas of very shallow thermoclines (i.e., tropical upwelling areas).

In summary, disequilibrium effects are well established for benthic foraminifera. Most available evidence suggests that these effects are approximately constant through space and time. It is possible, however, that slightly variable disequilibrium effects in benthic foraminifera must ultimately be considered. For planktonic foraminifera, the existence of disequilibrium effects remains controversial. Although apparent disequilibrium effects have been detected in plankton-tow studies, some investigators believe that they are an artifact of sample-preparation techniques. A further uncertainty is the quantitative importance of secondary "gametogenic" calcification of planktonic foraminifera at depth. If this process is important, foraminiferal calcite may yield an integrated picture of the upper water column, which may prove difficult to relate to conditions at the ocean surface.

TEMPERATURE EFFECTS

Paleotemperature equations. All work on oxygen isotopes as climate indicators builds on the work of Urey (1947), who established the thermodynamic properties of light isotopes and predicted the utility of $\delta^{18}O$ measurements as a paleothermometer. Equations relating temperature to $\delta^{18}O$ (Table 1) were calibrated empirically by McCrea (1950) from calcite precipitated inorganically, and by Epstein and others (1953) from mollusks grown at known temperature. Slight improvements in the mollusk equation were made by Craig (1965) and Horibe and Oba (1972) and in the inorganic equation by O'Neil and others (1969). A similar equation developed for the planktonic foraminiferan *Globigerinoides sacculifer* grown in culture (Erez and Luz, 1983) adds confidence that oxygen-isotope equilibrium is attained by some planktonic foraminifera, but is at variance with some field data suggesting disequilibrium. All of the equations are similar enough in slope so that relative temperature changes predicted from down-core data are not significantly different for the various equations. Thus, any of the standard temperature equations yields acceptable results for planktonic foraminifera within the range of 15 to 25°C.

Benthic foraminifera, however, calcify in the deep sea, where water temperatures less than 5°C are common. The absolute differences in the equations at cold temperatures are large enough that significant errors may occur when the freezing point

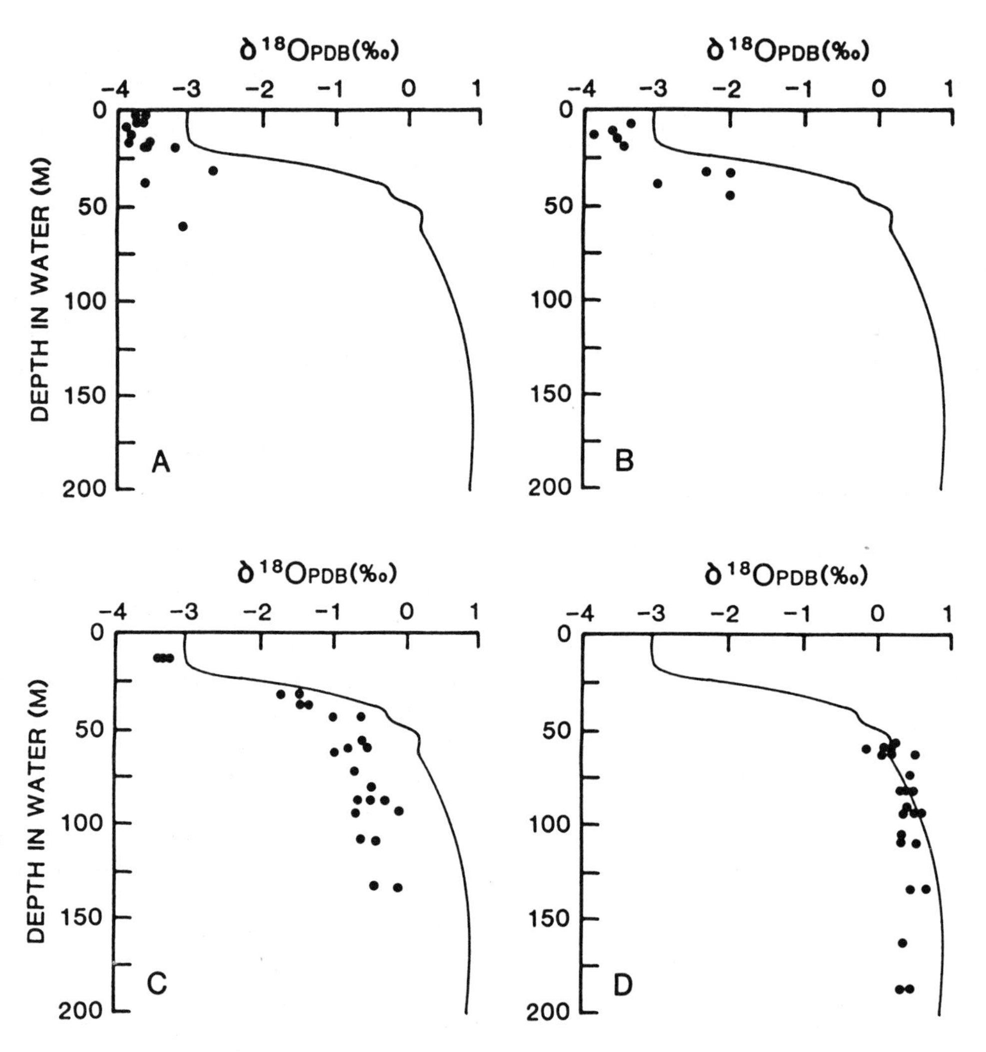

Figure 3. Depth-stratified plankton tows from the Panama Basin demonstrate variable depth habitats of foraminifera and continued calcification as foraminifera sink across the thermocline. (From Fairbanks and others, 1982.) Solid lines are the expected equilibrium $\delta^{18}O$ values at various depths. Symbols in each plot are the measured $\delta^{18}O$ values of plankton collected from a fixed depth. (a) *G. ruber.*, (b) *G. sacculifer,* (c) *N. dutertrei,* and (d) *G. theyeri.*

TABLE 1. EMPIRICAL EQUATIONS FOR CALCITE-WATER $\delta^{18}O$ EQUILIBRIUM*

Reference	a	b	c	Comments
McCrea (1950)	16.0	-5.1	0.09	Inogranic
Epstein and others (1953)	16.5	-4.3	0.14	Mollusk
Craig (1965)	16.9	-4.2	0.13	Mollusk
O'Neil and others (1969)	16.9	-4.68	0.10	Inorganic
Horibe and Oba (1972)	17.04	-4.34	0.16	Mollusk
Shackleton (1974)	16.9	-4.0	0.00	Foram <16°C
Erez and Luz (1983)	16.998	-4.52	0.028	Pl. Foram

*All equations are of the form $T = a + b(\delta c - \delta w) + c(\delta c - \delta w)^2$, where T is temperature in °C, δc is the $\delta^{18}O$ of calcite, and δw is the $\delta^{18}O$ of water, measured on the same scale.

of sea water is used as a constraint for interpretation of down-core records. Under these conditions the various equations (Table 1) can yield temperatures differing by 3°C (Craig, 1965 versus Erez and Luz, 1983) to 5°C (McCrea, 1950 vs. Craig, 1965). Shackleton (1974) evaluated paleotemperature equations in this range, using core-top benthic foraminifera. Because deep-sea sediments are bioturbated, the core-tops used for this analysis contained foraminifera dating from the present to several thousand years old. Shackleton assumed relatively stable deep-sea temperatures and no contamination with glacial-age foraminifera for his samples. Using these assumptions, he suggested that the linear equation $T = 16.9 - 4.0\ (\delta_c - \delta_w)$ is the best equation available at present for temperatures below 16°C. For more rigorous consideration of temperature effects in benthic foraminifera, further work on the calibration of isotopic temperature equations at low temperatures is necessary.

Planktonic $\delta^{18}O$ and sea-surface temperatures. Plankton-tow studies clearly demonstrate that tropical planktonic foraminifera calcify at a range of temperatures both with depth (Fairbanks and others, 1982; Fig. 3 above) and with season (Deusser and others, 1981; Williams and others, 1981; Ganssen and Sarnthein, 1983) (Fig. 4a). Foraminifera also have large seasonal variations in abundance, so the isotopic composition of the population as a whole may not reflect the mean temperatures of the surface waters. For example, Williams and others (1981) show that *G. ruber* in the subtropical Atlantic is most abundant in the summer (Fig. 4b). Thus the mean isotopic signature of the population of this species reaching the sea floor in this area is weighted toward summer conditions. This weighting can change through time as the climate changes. If temperatures warmed by a few degrees, *G. ruber* may suppress its summer reproduction and enhance its winter reproduction (assuming that foraminiferal ecology is related to temperature); thus the mean population would record little or no isotopic change.

This seasonal weighting of signals recorded by foraminiferal populations is assessed here quantitatively in a model that considers the degree to which foraminiferal species fluxes are related to the range of environments available at a site. If the environmental preferences of a species are known, we can calculate the recorded environmental parameters sensed by the average population of a foraminiferal species. The purpose of this test is to illustrate the difference between how individual foraminifera and populations of foraminifera record their environment.

The model run here for temperature (Fig. 5) is equally applicable to other environmental parameters. At any site, the distribution of water temperatures W(T) has a mean (W_m) and some variability (W_σ), which could reflect either temporal variability or depth stratification. For modeling purposes, a Gaussian distribution of temperatures is assumed. W_σ is the standard deviation. Similarly, a foraminiferal species has a preferred mean temperature (F_m) and a standard deviation (F_σ), which defines the Gaussian distribution F(T). The flux distribution of the species as a function of temperature is given by $SF(T) = W(T) \times F(T)$. The total flux $F_T = \int SF(T)\ dT$. The recorded temperature T_r (i.e., the temperature sensed by the mean population of a species) is the weighted mean of the distribution SF(T), or $T_r = 1/F_T \times (\int SF(T) \times T\ dT)$. For a given species, if the distribution of preferred temperatures, F(T), is offset from the actual distribution of temperatures, W(T), then T_r will lie between the mean water temperature W_m and mean preferred by the species, F_m.

Using this model, Figure 6a demonstrates variations in the recorded temperature T_r (which would control isotopic compositions) versus true mean temperature for three hypothetical species. All are assumed to prefer the same mean temperature of 15°C, but each selects a different range around the mean. The integrated fluxes of these species as a function of mean temperature are shown in Figure 6b. For this model, the variation of water temperature around the mean is arbitrarily assumed to be constant (W_σ = 4°C).

Species A is very selective to temperature, with F_σ = 1°C. It demonstrates little change in recorded temperature (and thus would have little $\delta^{18}O$ change) with changing mean water temperatures (Fig. 6a), because it chooses to calcify only at its narrowly defined temperatures, presumably by adjusting its depth and/or season of growth (Fig. 6b). Species B, which is less selective to temperature (F_σ = 2°C), senses more change in its recorded temperature than species A (Fig. 6a). Species C, which is not very selective about temperature (F_σ = 10°C), has a broader distribution of fluxes versus temperature (Fig. 6b) and recorded temperatures closer to the annual mean temperatures (Fig. 6a). This occurs because species C does not restrict its calcification to warmer episodes in the cold regions (and colder episodes in the warmer regions) as much as the other species. Note, however, that none of the species records the full amplitude of mean annual temperature variations. Exact recording of mean annual temperatures by T_r (or $\delta^{18}O$ at the equilibrium slope) would occur only for species with no temperature preferences. In this model, the slope of the line relating recorded temperature to mean annual temperature is a function of the range of temperatures in which a species calcifies.

The example above held W_σ constant. That is, it assumed a constant range of water temperatures around the variable mean. In the real world, near-surface temperature is less variable in the warmer low latitudes than at the colder and highly seasonal high latitudes. To illustrate this effect, as would occur in core-top surveys of foraminifera, the model input is modified to have W_σ range from 0°C at a mean temperature of 30°C, to 10°C at a mean temperature of 0°C. Although this oversimplifies the true patterns of annual and interannual variability of near-surface temperatures, it illustrates the potential effects that spatial patterns of temperature variability around a mean value have on the recorded temperature (i.e., that sensed by the population). Again, it is this recorded temperature that controls the isotopic composition that would be measured in a sample.

The results of this more realistic model, which as in the earlier model assumes an optimal temperature of 15°C for all species, are given in Figure 6c (recorded temperatures) and 6d (species fluxes). Because the range of near-surface temperatures is

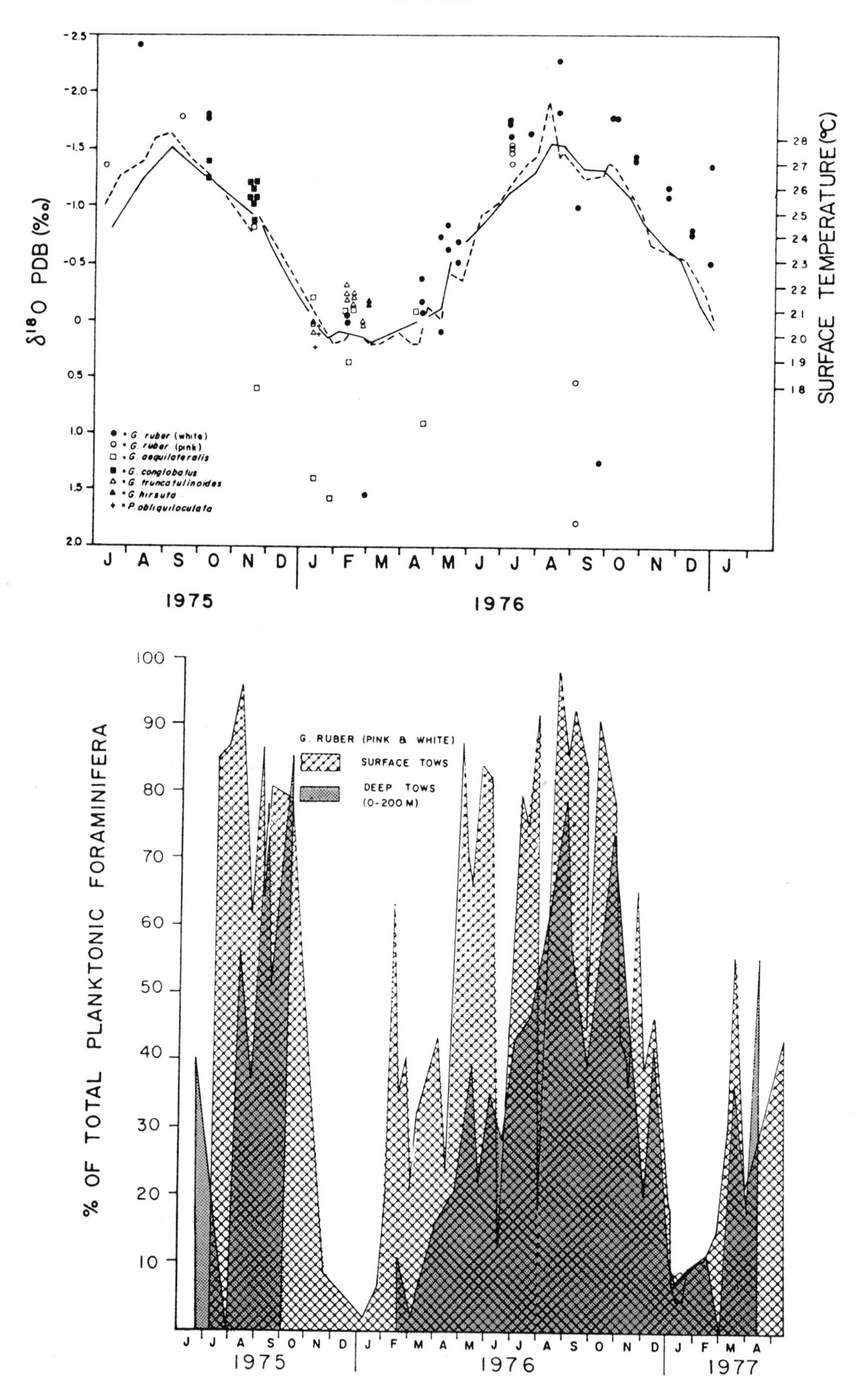

-2.5
-2.0
-1.5
-1.0
-0.5
0
0.5
1.0
1.5
2.0
δ18O PDB (‰)
28
27
26
25
24
23
22
21
20
19
18
SURFACE TEMPERATURE (°C)
● = G. ruber (white)
○ = G. ruber (pink)
□ = G. aequilateralis
■ = G. conglobatus
△ = G. truncatulinoides
▲ = G. hirsuta
+ = P. obliquiloculata
J A S O N D J F M A M J J A S O N D J
1975
1976
100
90
80
70
60
50
40
30
20
10
% OF TOTAL PLANKTONIC FORAMINIFERA
G. RUBER (PINK & WHITE)
SURFACE TOWS
DEEP TOWS (0-200 M)
J J A S O N D J F M A M J J A S O N D J F M A
1975
1976
1977

Figure 4. (a) Seasonal variations of $\delta^{18}O$ of living planktonic foraminifera collected off Bermuda. Solid line is surface-water temperature measured at the time of plankton collection. Dashed line is the predicted isotopic composition of calcite deposited in isotopic equilibrium with surface waters (from Williams and others, 1981). (b) Seasonal variations in relative abundance (percent of total fauna) of living *G. ruber* collected off Bermuda at two different water-depth ranges. (From Williams and others, 1981.)

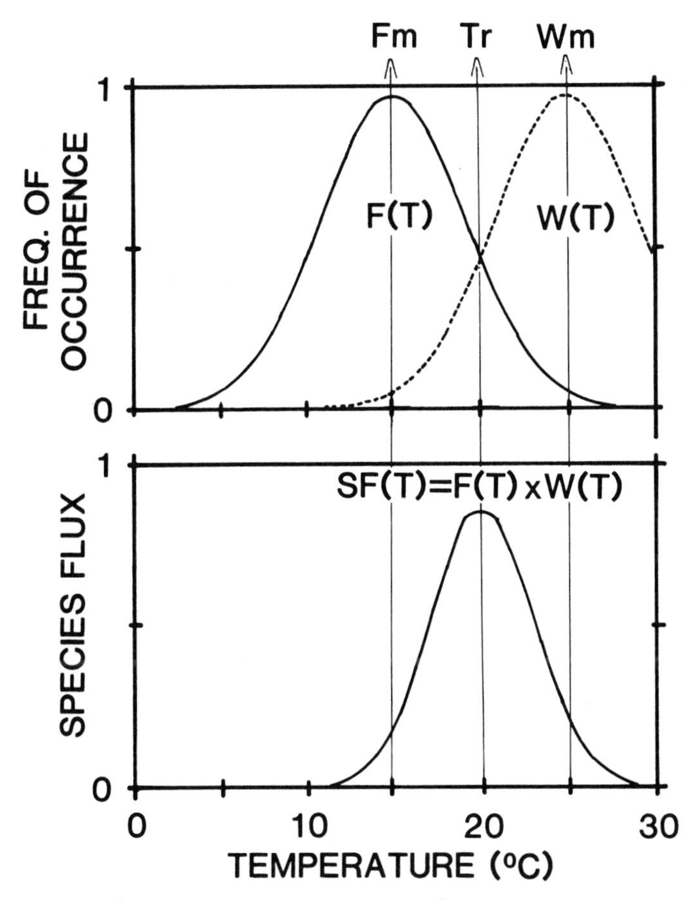

Figure 5. A model showing how the temperature preferences of a species F(T) and the range of temperatures available at a site W(T) determines its recorded temperature, i.e., the mean temperature recorded by the population. The net flux of the species SF(T) is F(T) x W(T). The recorded temperature T_r lies between the mean preferred temperature (F_m) and the true mean water temperature (W_m).

large at low mean temperatures, all of the species' optimum temperatures are available with this range. Thus, effective temperatures are insensitive to mean temperature changes in colder regions (Fig. 6c), and species fluxes are higher than in the constant W_σ cases above (compare Figs. 6b, d). At high mean temperatures, where the range of available temperatures is low, the optimum temperature of 15°C for all three species never occurs. The species still exist but are on the warm "tail" of their preferred temperature distribution, so they are relatively sensitive to changing mean temperatures (Fig. 6c) and have relatively low flux (Fig. 6d). The net result is that the lines relating recorded temperature (or $\delta^{18}O$) to mean temperature curve have low slopes where there is large variability of water temperatures (such as in the high latitudes) and high slopes where there is little variability of water temperatures (such as in the tropics).

Thus, if the productivity of a foraminiferal species is sensitive to temperature, the $\delta^{18}O$ composition of its mean population should not record mean annual temperatures at the expected thermodynamic slope of about 0.22‰ per °C, even if individual specimens are in isotopic equilibrium with ambient waters. This occurs because a population sensitive to temperature would adjust its season (and/or depth) of calcification to choose water temperatures within its preferred range. The relationship between mean annual temperatures and the recorded temperature, which controls the mean $\delta^{18}O$ of the species, is not simple or linear but depends on the degree to which species flux is sensitive to temperature, as well as on the mean and variance of available temperatures at a site. This means that attempts to correct downcore $\delta^{18}O$ curves for temperature change by subtracting transfer-function temperature estimates using the paleotemperature equation (Imbrie and others, 1973; Mix and Ruddiman, 1985) are not appropriate, at least for any species whose abundance is sensitive to temperature.

Core-top calibration. Core-top samples assumed to reflect modern conditions have been used to assess the degree to which foraminiferal $\delta^{18}O$ reflects spatial patterns of temperature (Shackleton and Vincent, 1978; Williams and Healy-Williams, 1980; Curry and Matthews, 1981a; Durazzi, 1981). Data from four species in these studies and that of Mix and Ruddiman (1985) are shown in Figure 7, plotted versus predicted $\delta^{18}O$ for calcite in equilibrium with annual average surface waters at each site. For this calculation, the equation of O'Neil and others (1969) was used. The oxygen-isotope composition of water is estimated from atlas salinity data (Levitus, 1982), using local relationships between salinity and $\delta^{18}O$ of surface waters (Craig and Gordon, 1965).

The isotopic composition of some species reflects modern near-surface temperature/salinity variations better than others. The $\delta^{18}O$ values of *O. universa* (Fig. 7a) best reflect the modern patterns of mean annual sea-surface conditions. The fit of measured $\delta^{18}O$ values to those predicted for mean annual conditions is good over a range of more than 4‰ (~9°C). In light of the model results discussed above, this means that *O. universa* is not very selective to temperature.

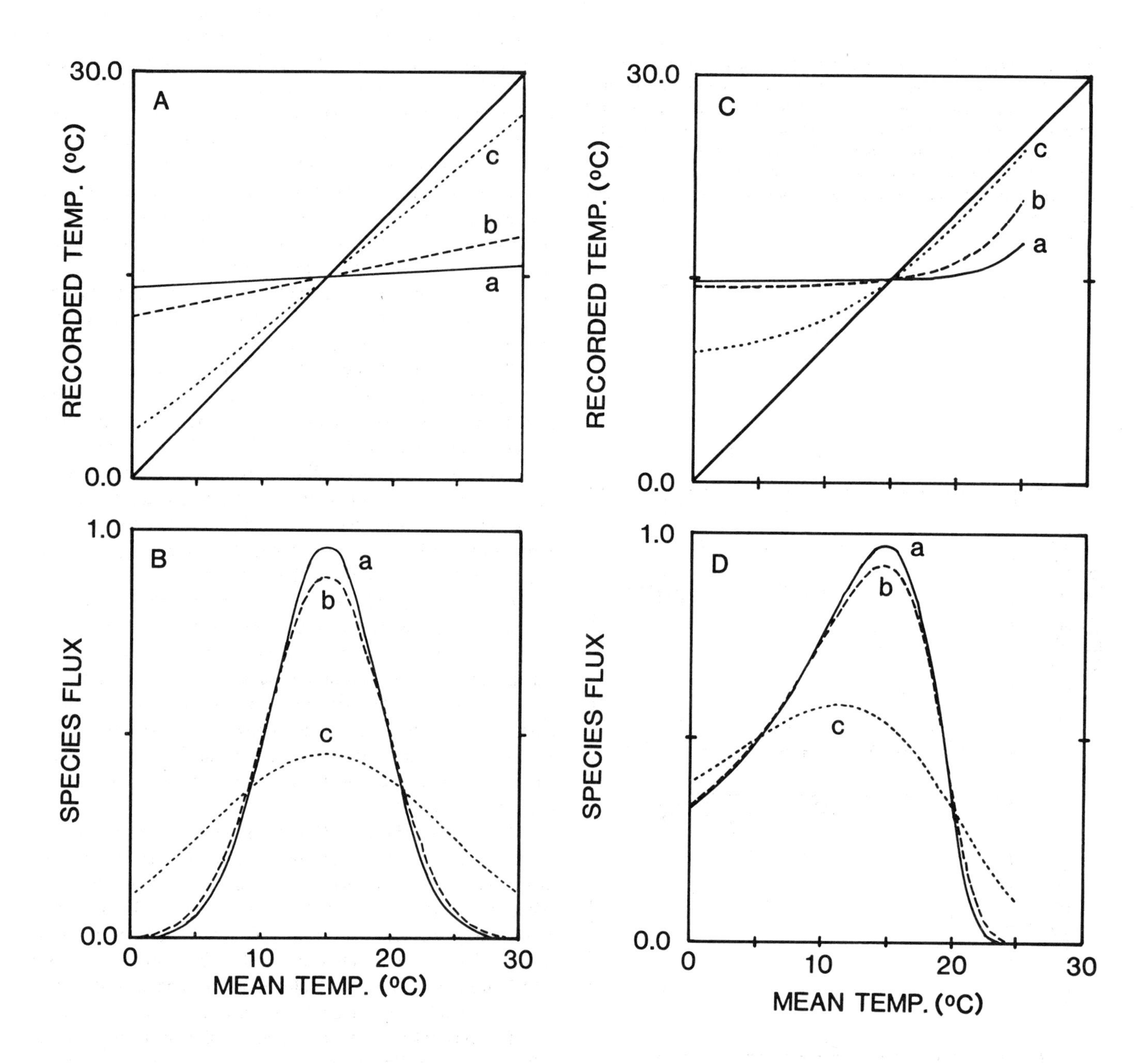

Figure 6. Output from species flux model (Fig. 5) demonstrating the effect of population dynamics on the recording of temperature by a foraminiferal species. (a) Effective temperature (which controls $\delta^{18}O$ composition) is different from mean temperature, because species tends to choose a particular temperature. (b) Species flux distributions for three hypothetical species, each choosing a mean temperature of 15°C but having differing widths of distribution. Species A is very selective to temperature ($F_\sigma = 1$), Species C is not selective to temperature ($F_\sigma = 10$), Species B is intermediate ($F_\sigma = 2$). (c) Same as 6a, but with lower variance of climates in the tropics than at the poles. This demonstrates that response of T_r to W_m also depends on W_σ. (d) Same as 6a. Low variability of oceanic temperatures in the tropics skews the flux distributions toward regions with high variability of climate.

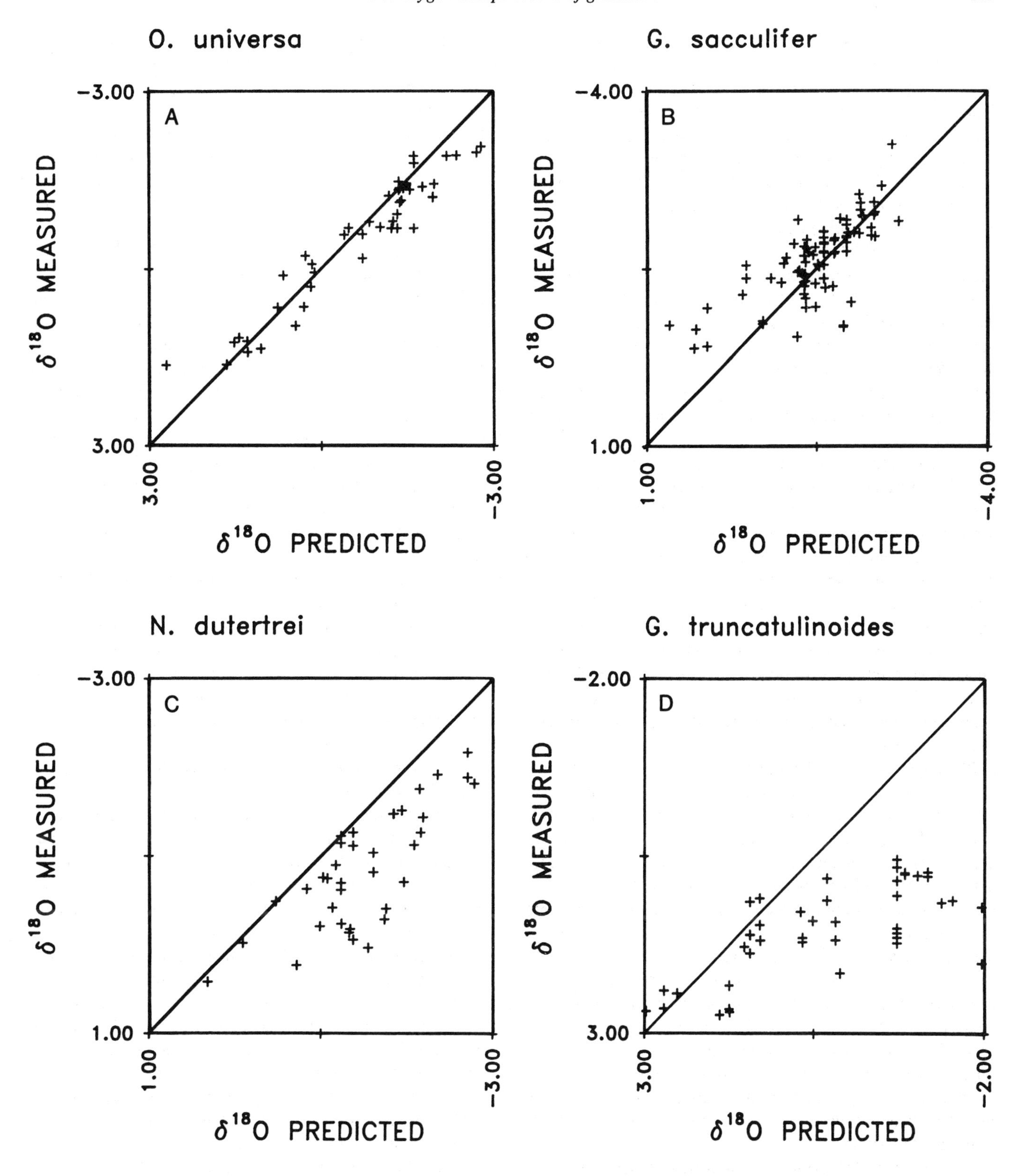

Figure 7. Core-top data on different species of planktonic foraminifera, compared to expected $\delta^{18}O$ at annual mean temperatures and salinities at the sea surface. Data from Shackleton and Vincent (1978), Williams and Healy-Williams (1980); Curry and Matthews (1981a); Durazzi (1981).

The $\delta^{18}O$ values from *G. sacculifer* (Fig. 7b) correspond well with those predicted for mean annual conditions, but over a smaller range (~2‰) than that of *O. universa.* At higher predicted $\delta^{18}O$ values (i.e., colder conditions) measured values are relatively depleted in ^{18}O, suggesting that *G. sacculifer* is preferentially growing in the warmer seasons at these sites. This is consistent with the seasonal sediment-trap results of Thunnell and Reynolds (1984), who found that the flux of *G. sacculifer* was highest during the warmest seasons in the Panama Basin.

Measured $\delta^{18}O$ values of *N. dutertrei* (Fig. 7c) are near equilibrium with mean annual conditions at the cold (high $\delta^{18}O$) end, but more enriched in ^{18}O at the warm (low $\delta^{18}O$) end of the measured range. This may mean that the flux of *N. dutertrei* is more sensitive to temperature than that of *O. universa* or *G. sacculifer.* Although there is some scatter to the data, the pattern suggests curvature to the line relating measured and predicted $\delta^{18}O$ values, as suggested for temperature in the model discussed earlier (Fig. 6c).

G. truncatulinoides demonstrates the least change in measured $\delta^{18}O$ values relative to values predicted for mean annual conditions (Fig. 7d). This means either that *G. truncatulinoides* has a relatively narrow range of preferred temperatures (if it responds to temperature) and/or that it lives in sub-thermocline waters (for reasons unknown) so that its isotopic composition reflects the relatively constant temperatures at depth. The $\delta^{18}O$ data alone do not constrain a choice between these hypotheses.

If *G. truncatulinoides* does selectively calcify in a narrow range of temperatures, down-core isotope records from this species would be ideal ice-volume indicators (they would give temperature-free signals). Arguing against the constant temperature hypothesis for *G. truncatulinoides,* Curry and Matthews (1981a) speculate that this species responds to water density, calcifying at the 26.8 σ-t surface. If this is true, the down-core $\delta^{18}O$ record of *G. truncatulinoides* would underestimate ice-volume effects, because the increase in salinity related to increased glacial ice-volume would cause it to seek warmer temperatures in order to remain at the 26.8 σ-t surface. If either of these hypotheses is correct, the glacial/interglacial $\delta^{18}O$ record from *G. truncatulinoides* should provide a lower limit for the ice-volume effect on $\delta^{18}O$ of sea water.

For all of the species shown in Figure 7, the scatter of $\delta^{18}O$ is so large that these relationships cannot be used to assess in detail the reality of the recorded temperature model discussed above. This scatter may come in part from the use of core tops of uncertain age. The combination of flux data (now being collected from sediment traps) and $\delta^{18}O$ data, along with comparison to model predictions such as those made above, will eventually help to narrow the options.

Downcore $\delta^{18}O$ data. Ice volume and temperature. It is not yet clear whether significant temperature effects are recorded in down-core $\delta^{18}O$ records from planktonic foraminifera. Opinions range from that of Emiliani (1955, 1970, 1971), who believed that the bulk of the tropical $\delta^{18}O$ signal was temperature, to that of Matthews and Poore (1981), who believed that surface-dwelling tropical planktonic foraminifera experienced essentially no temperature changes and thus are optimal ice-volume indicators.

Attempts to remove the effects of temperature from down-core isotope records have been made by Imbrie and others (1973) and Mix and Ruddiman (1985). Both used the transfer-function approach of Imbrie and Kipp (1971) applied to foraminiferal species abundances to generate temperature estimates independently of the $\delta^{18}O$ measurements. Imbrie and others (1973) made this attempt on Caribbean core V12-122, in which a 2.2‰ glacial/interglacial change in $\delta^{18}O$ was measured (on *G. ruber*), and a 2.2°C temperature change was estimated from the transfer function. Removal of this temperature effect left 1.8‰ as the estimated change in water $\delta^{18}O$. Imbrie and others (1973), however, felt that this was too large an isotopic change to be accounted for by ice volume alone. Thus they hypothesized a 0.4‰ effect due to an increase in the evaporation/precipitation ratio in the region, leaving 1.4‰ as an ice-volume effect.

Mix and Ruddiman (1985) analyzed $\delta^{18}O$ in planktonic foraminifera, and Mix and others (1986) estimated annual mean sea-surface temperatures via the transfer-function approach. After noting large spatial variability of the nonisotopic estimates of temperature change, Mix and Ruddiman (1985) hypothesized that removing mean annual temperature variations from the down-core isotope records should make the corrected isotope records in the various cores more similar to each other. By removing the temperature estimates from the $\delta^{18}O$ records in many cores with different patterns of temperature change, a best estimate of the ice-volume contribution to the down-core $\delta^{18}O$ records should result. Surprisingly, Mix and Ruddiman (1985) demonstrated that the temperature-corrected $\delta^{18}O$ records of *G. sacculifer* were much more dissimilar than the original uncorrected $\delta^{18}O$ records (Fig. 8).

There are four possible explanations for this dissimilarity. First, the transfer-function temperature estimates may be wrong. Second, the faunal temperature estimates may be correct, but the foraminiferal species analyzed may have a narrow range of preferred temperatures and thus change their depth and/or season of growth to maintain a constant temperature (the recorded temperature model, above). Third, the estimated temperature variations may be nearly correct but may apply to thermocline rather than sea-surface temperatures. In this case, the surface-dwelling foraminifera (such as *G. sacculifer* analyzed by Mix and Ruddiman, 1985) would not necessarily sense any temperature change. Fourth, both the temperature estimates and the isotope data may be correct, but the effects of local salinity changes (related to flow of ^{18}O-depleted fresh water from the continents) may fortuitously offset the isotopic corrections for temperature.

The solution to this problem is not completely known at present. The core-top data (Fig. 7b) suggest, however, that the second hypothesis of a narrow range of preferred temperatures is not correct for *G. sacculifer.* It is clear from the model discussed earlier (Figs. 5 and 6) that removing temperature effects from down-core $\delta^{18}O$ records is not straightforward. Much more

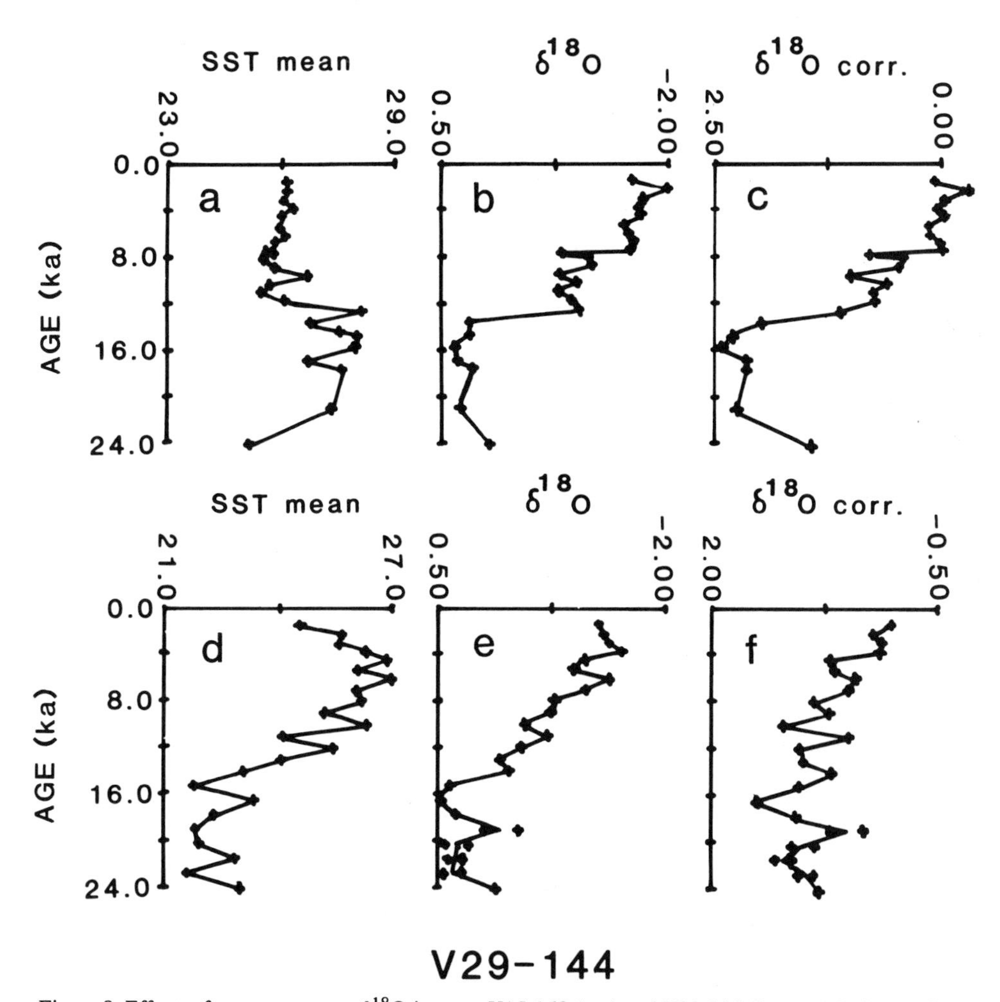

Figure 8. Effects of temperature on $\delta^{18}O$ in cores V15-168 (top) and V29-144 (bottom). Left: transfer-function temperature estimates of sea-surface temperature. Center: $\delta^{18}O$ data measured on *G. sacculifer* (with "sac", 355–415μm). Right: temperature-corrected $\delta^{18}O$ data. Removing temperature estimates from two similar $\delta^{18}O$ curves yields dissimilar corrected $\delta^{18}O$ curves. This suggests that attempts to remove temperature from planktonic $\delta^{18}O$ signals using transfer-function temperature estimates do not improve estimates of changing oceanic $\delta^{18}O$. From Mix and Ruddiman (1985).

work must be done to test and refine nonisotopic methods for estimating temperature changes, and the recording of $\delta^{18}O$ by a foraminiferal population must be better understood before temperature and ice volume effects can be isolated.

It was noted earlier that $\delta^{18}O$ of *G. sacculifer* seems to reflect modern variations of sea-surface conditions, whereas that of *G. truncatulinoides* does not. One test of whether or not significant temperature effects mask the ice-volume effect in a down-core record is to analyze both species from the same samples. By analogy with the core-top data, temperature changes should be recorded differently by the two species. In Figure 9, $\delta^{18}O$ data from *G. sacculifer* (with "sac," 415 to 500 μm) and *G. truncatulinoides* (right coiling, 415 to 500 μm) are compared. Although the signals are not identical, there are no systematic differences in the glacial/interglacial $\delta^{18}O$ records. This supports the generally held view that the ice-volume effect dominates down-core records, and that temperature effects are relatively minor.

MORE COMPLICATIONS

If a $\delta^{18}O$ record of glaciation is to be obtained, one must be certain that local salinity-driven variations in watermass $\delta^{18}O$ do not bias down-core signals. Berger and others (1977) and Berger (1978) infer that a global transient low-$\delta^{18}O$ meltwater spike

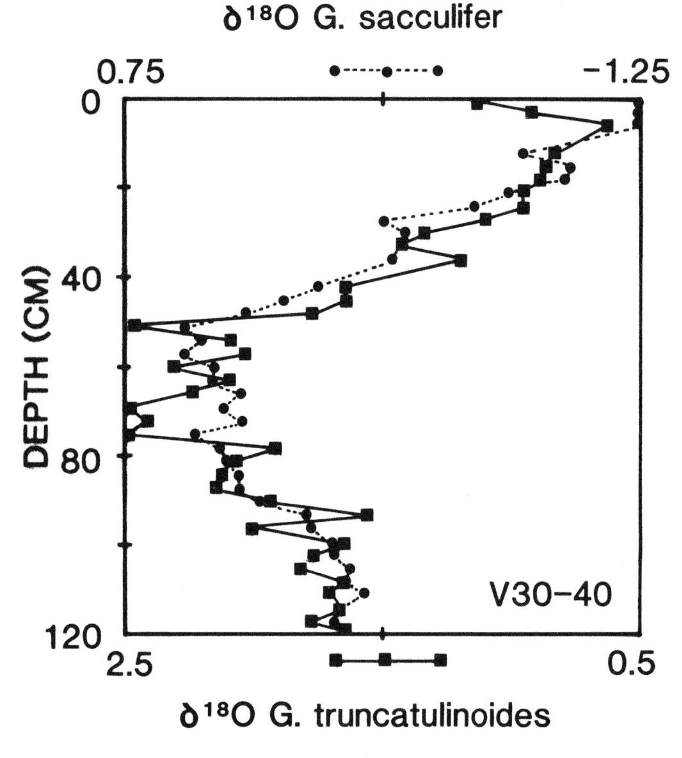

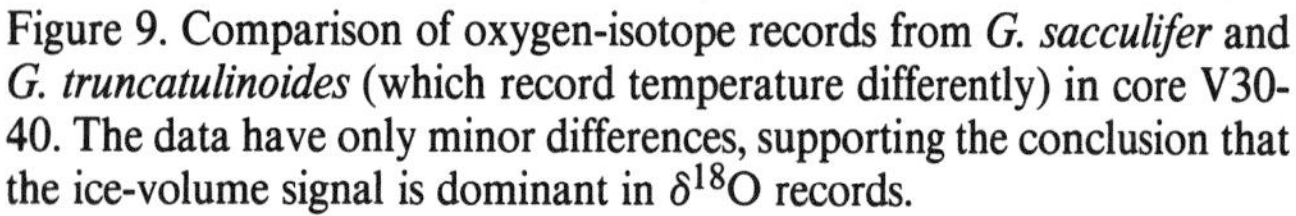

Figure 9. Comparison of oxygen-isotope records from *G. sacculifer* and *G. truncatulinoides* (which record temperature differently) in core V30-40. The data have only minor differences, supporting the conclusion that the ice-volume signal is dominant in $\delta^{18}O$ records.

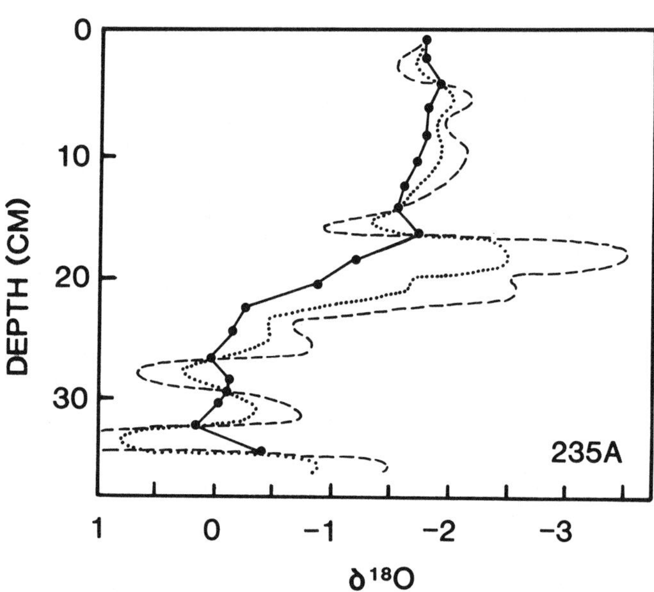

Figure 10. Attempts to remove the effect of bioturbation from the record (Berger, 1978). Deconvolution processing suggests the presence of a meltwater spike on deglaciation; other studies do not confirm this large a meltwater effect.

covered the surface oceans during rapid deglaciation (Fig. 10). If this spike occurred, no down-core planktonic $\delta^{18}O$ data would yield a reliable ice-volume history. Although these events are seldom seen in $\delta^{18}O$ data, Berger and others (1977) suggest that this was due to stirring of sediments by deep-sea fauna, or bioturbation, which would smooth out brief events. They detect the deglacial spike by mathematically unmixing (deconvolving) measured $\delta^{18}O$ data. Jones and Ruddiman (1982), however, suggest that the large global meltwater spike is an artifact of improper deconvolution processing. This conclusion is supported by $\delta^{18}O$ analysis of cores with higher sedimentation rates, which do not show a large global meltwater event (Duplessy and others, 1981b; Mix and Ruddiman, 1985). Although small meltwater transients are likely, it remains unclear how large this effect is on a global basis.

Evidence of glacial meltwater effects does exist locally in regions close to the ice sheets, such as the Gulf of Mexico (Kennett and Shackleton, 1975; Emiliani and others, 1975; Leventer and others, 1982) and possibly the North Atlantic (Keigwin and others, 1984). Evidence for local river-water effects on marine isotope records unrelated to glacial meltwater has also been found (Pastouret and others, 1978; Showers and Margolis, 1985). These local effects must be avoided when an isotopic record of ice-volume is constructed.

Carbonate dissolution could also contribute to variations in stable-isotope records. Berger and Killingley (1977) suggest that relatively large dissolution effects may increase the $\delta^{18}O$ values of dissolution-sensitive species such as *G. sacculifer* by as much as 0.3‰. Their reasoning is that dissolution selectively removes the smaller, thin-walled specimens that calcify in shallower (warmer) water. Erez (1979) uses this idea to explain why down-core planktonic $\delta^{18}O$ records from the Pacific generally have lower amplitudes than those from the Atlantic (Fig. 11). He suggests that interglacial dissolution in the Pacific reduced the amplitude of isotope records, and glacial dissolution in the Atlantic increased it. Analyzing a similar set of isotope data, Broecker (1986) infers relative warmth or low salinity of glacial Pacific surface waters, and slight cooling of glacial Atlantic surface waters, rather than dissolution effects. Both of these hypotheses are speculative. The data of Williams and others (1981) argue against the effects of dissolution, for direct dissolution of foraminifera in the laboratory yields no change in $\delta^{18}O$ values. On the other hand, Bonneau and others (1980) demonstrate that real-world samples of planktonic foraminifera do reflect enrichment of ^{18}O with dissolution at greater water depths.

Even if the complicating effects of temperature, bioturbation, and perhaps dissolution can be legitimately ignored or removed from the $\delta^{18}O$ record, a further complication limits the

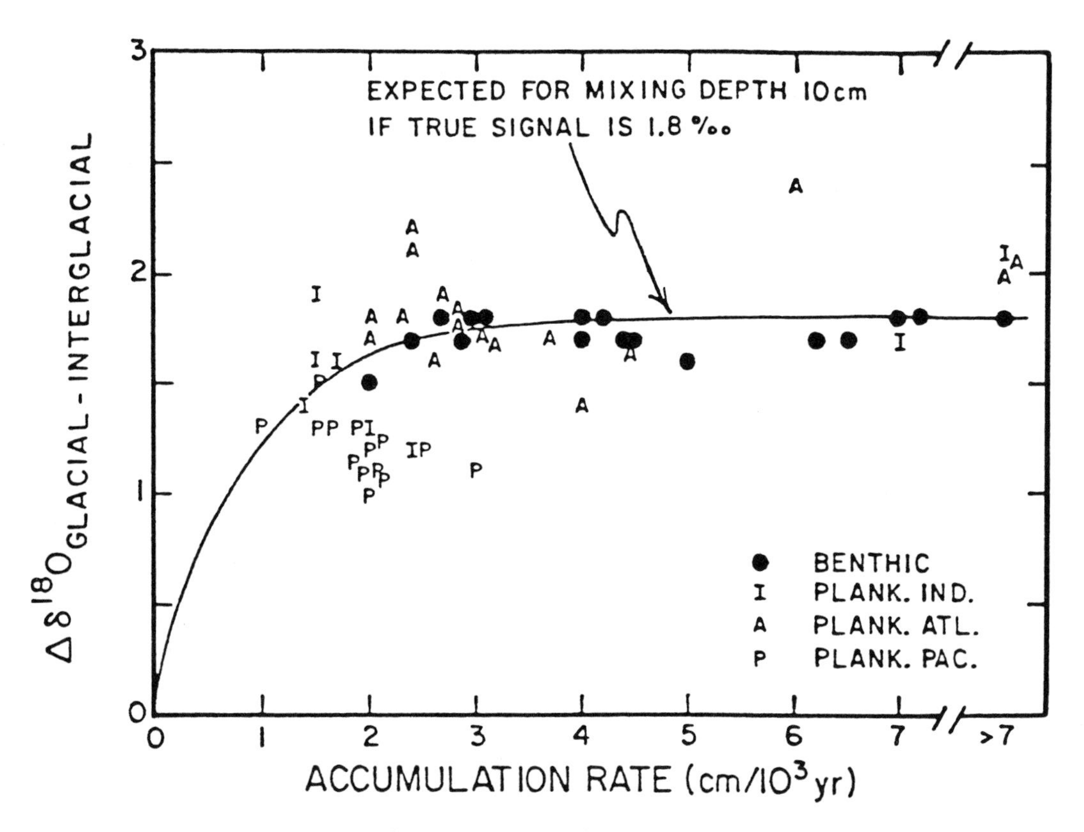

Figure 11. Even after consideration of smoothing of the record at different sedimentation rates, the Pacific and Atlantic $\delta^{18}O$ records have different amplitudes. The relatively low-amplitude Pacific records, compared to high-amplitude Atlantic records, suggest to Erez (1979) dissolution effects and to Broecker (1986) temperature effects and moisture transport.

reconstruction of ice-volume history from $\delta^{18}O$ time series. The isotopic effect of ice volume on the ocean depends not just on the volume of glacier ice, but also on its isotopic composition. This ice $\delta^{18}O$ probably changes through time, resulting in nonlinear recording of ice volume by $\delta^{18}O$.

With a relatively simple numerical model, Mix and Ruddiman (1984) infer that the changing isotopic composition of glacier ice would induce a systematic lag of oceanic $\delta^{18}O$ change 1,000 to 3,000 yr behind ice-volume changes at major glacial/interglacial transitions (Fig. 12). With a more complicated ice sheet and vapor-transport model, Covey and Schneider (1985) support the likelihood of this effect but suggest even larger lags of the isotopic response of the oceans to ice-volume change. Mix and Ruddiman (1985) note that very rapid (step-like) events would be misrepresented not in timing but in amplitude in the marine record by this mechanism. Not enough is known at present about the history or dynamics of nonequilibrium ice sheets to remove these effects from marine $\delta^{18}O$ curves. Some hope comes from the inclusion of isotope tracers in global climate models (Jousaume and others, 1984) that will be used in the future for paleoclimatic studies.

ICE VOLUMES

Despite the complications discussed above, there is a remarkable similarity among most down-core benthic and planktonic $\delta^{18}O$ records from many areas. The only possible explanations of this fact are that the down-core variations reflect mostly ice-volume changes, or other signals that contaminate the ice-volume signal in $\delta^{18}O$ are fortuitously similar in phase and amplitude in most oceanic records. Shackleton's (1967) belief in the improbability of the second possibility is the major reason that $\delta^{18}O$ records from foraminifera have been used as proxies for ice volume. Because deep-sea sediments offer a relatively continuous record of climate, this proxy record has been invaluable for study of Quaternary paleoclimatology. Many use the oxygen-isotope signal to define the ice ages. The exact value for the ice-volume effect on marine $\delta^{18}O$, however, remains poorly constrained. In this section, the $\delta^{18}O$ evidence is compared to predictions of ice-sheet and sea-level models to test the reliability of $\delta^{18}O$ as an ice-volume proxy.

Comparison to ice-sheet models. A first check of the importance of $\delta^{18}O$ measurements as an ice-volume proxy is to

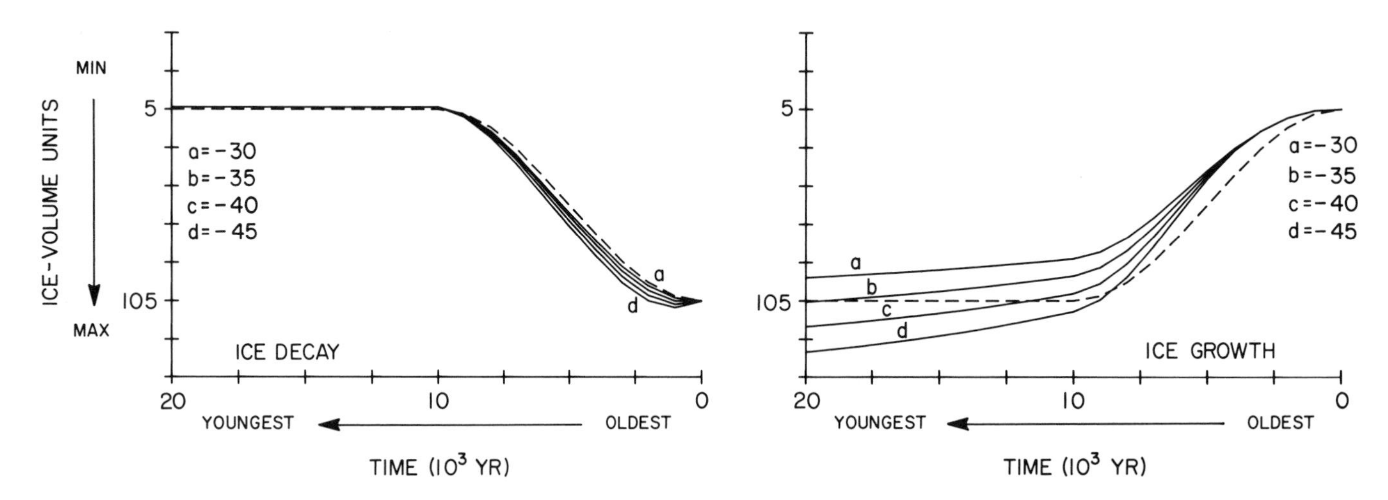

Figure 12. The effect of changing isotopic composition of glacier ice may induce a lag of oceanic $\delta^{18}O$ of about 1,000 to 3,000 yr behind true ice-volume change (Mix and Ruddiman, 1984). This model tests the sensitivity of the lag to the glacial-maximum isotopic composition of the mean snow on the global ice sheets.

compare measured glacial/interglacial ranges of marine $\delta^{18}O$ records of about 1.7‰ with glaciological estimates of ice volume and its isotopic effects. One of the early attempts (Dansgaard and Tauber, 1969) calculated the isotope effect, using glaciological estimates of glacial maximum ice (47 × 10^6 km^3 more than at present) and assuming a mean ice $\delta^{18}O$ composition of -30‰ (as an upper limit). It also assumed that the 29 × 10^6 km^3 of existing ice in Greenland and Antarctica was 5‰ more depleted in ^{18}O at the glacial maximum than it is at present. This model yields a total change in oceanic $\delta^{18}O$ from glacial-to-interglacial time of 1.2‰ (as a lower limit) and implies at least a 2°C glacial cooling in typical marine $\delta^{18}O$ records.

This calculation is updated here with the more recent estimates of ice volume and isotopic composition. The effects of each ice sheet on oceanic $\delta^{18}O$ are calculated in Table 2 for both the "minimum" and "maximum" reconstructions of Hughes and others (1981) in order to address the range of possible solutions. Not all Quaternary geologists accept these minimum and maximum models as limiting cases. For example, in the minimum reconstruction, the Laurentide ice-sheet has a volume of 30.5 × 10^6 km^3. Paterson (1972) estimates the volume of the Laurentide at 26.5 × 10^6 km^3. Paterson (personal communication, 1985) states that, if anything, he would now reconstruct the Laurentide with somewhat less volume than he did in 1972. Despite this uncertainty, the Hughes and others (1981) minimum and maximum reconstructions are used here because they are the most recent published models that are global in scale.

For the isotopic composition of ice, values consistent with ice-core data from North America (Hooke and Clausen, 1982), Greenland (Dansgaard and others, 1982), west Antarctica (Epstein and others, 1970), and east Antarctica (Barkov and others, 1977; Lorius and others, 1985), which are more depleted in ^{18}O than present ice, are used. Disagreeing with this strategy, Emiliani (1970) and Yapp and Epstein (1977) suggest that snow falling on the margins of glacial ice sheets was less depleted in ^{18}O than at present. These data, however, are not relevant to the calculation of ice $\delta^{18}O$. Snow falling on the margins of an ice-sheet would be in the ablation zone, and thus would not be incorporated in a significant way into the ice sheet. The estimates of ice $\delta^{18}O$ made here are appropriate for the accumulation zones of large ice sheets. For some of the ice sheets, the isotopic composition is little more than a guess. In some cases (e.g., southern Barents, Kara, and Putorana ice sheets), the existence of the ice sheet is in doubt, let alone the isotopic composition. In any case, a range of estimates is considered by carrying through error bars in the calculation. Undoubtedly, these estimates will be subject to further revision as more is learned about ice-sheet volumes and their isotopic composition.

The sums of the isotopic effects of individual ice sheets (Table 2) indicate that the maximum reconstruction of Hughes and others (1981) would change the mean oceanic $\delta^{18}O$ by 2.0 ± .4‰, whereas the minimum reconstruction would change it by 1.5 ± .2‰. For both reconstructions the major contributors to the isotopic signal are the Laurentide ice sheet (45% of maximum model, 55% of minimum model) and excess Antarctic ice (16% of maximum model, 23% of minimum model). Within the uncertainty, either the minimum or maximum models (or something in between) could produce the observed (global average) glacial/interglacial ranges of down-core foraminiferal $\delta^{18}O$ (about 1.7‰) without other contributing effects.

TABLE 2. ISOTOPIC EFFECTS OF DENTON AND HUGHES (1981) ICE-SHEET RECONSTRUCTIONS

Ice Sheet	$\delta^{18}O$ Ice (‰ SMOW)	Maximum Model Volume (10^6 km^3)	Maximum Model $\delta^{18}O$ Ocean (‰ SMOW)	Minimum Model Volume (10^6 km^3)	Minimum Model $\delta^{18}O$ Ocean (‰ SMOW)
Laurentide*	-35 ± 5	34.2	.91 ± .13	30.5	.81 ± .11
Cordilleran	-30 ± 5	1.8	.04 ± .01	0.3	.01 ± .00
Innuitian	-45 ± 5	1.0	.03 ± .01	0.0	.00 ± .00
Greenland	-35 ± 3	2.6	.07 ± .01	0.3	.01 ± .00
Iceland	-25 ± 5	0.2	.00 ± .00	0.1	.00 ± .00
Britain	-25 ± 5	0.8	.02 ± .01	0.8	.02 ± .01
Scandinavia	-30 ± 5	7.3	.16 ± .03	7.1	.16 ± .02
Barents/Kara	-40 ± 10	6.3	.19 ± .04	0.9	.03 ± .00
Putorana	-45 ± 10	0.6	.02 ± .00	0.0	.00 ± .00
E. Antarctic	-60 ± 3	3.3	.15 ± .01	3.3	.15 ± .01
W. Antarctic	-40 ± 3	6.5	.19 ± .02	6.5	.19 ± .02
Miscellaneous	-25 ± 5	0.7	.01 ± .00	1.8	.03 ± .01
Ice shelf	-40 ± 10	4 ± 2**	.12 ± .09	0.0	.00 ± .00
Present Ice	Δ = -5 ± 2	32.0	.12 ± .05	32.0	.07 ± .05
Total			2.03 ± .41		1.48 ± .24

*Paterson (1972) believes that the Laurentide was smaller than the Hughes and others (1981) minimum model. His volume estimate, 25 x 10^6 km^3, would imply an isotopic effect of 0.66 ‰, bringing the total isotopic effect of the minimum model to 1.33 ± .24 ‰.

**This corresponds to an ice shelf 700 ± 300 m thick over the entire Arctic Ocean. Hughes (personal communication, 1986) believes that if the maximum model is correct, an ice shelf in inevitable, as it must be present to buttress the marine ice sheets surrounding the Arctic. He states that a minimum reasonable thickness for an ice shelf to maintain itself over this area would be about 300 to 400 m. Hughes gives an upper limit of 1,000 m thickness for an Arctic ice shelf.

This does not prove that temperature (and other) effects are absent from foraminiferal $\delta^{18}O$ curves. It means that other data must be sought to constrain these effects. Through comparison of downcore $\delta^{18}O$ data with the sea-level record of the Huon Peninsula, New Guinea, Shackleton and Duplessy (1986) infer ice-volume effects on $\delta^{18}O$ of ~1.2‰, implying relatively large cooling of the deep-sea at the last glacial maximum. If glacial oceanic temperatures were significantly cooler than at present, and if these cooler temperatures were reflected in foraminiferal calcite, the minimum model of Hughes and others (1981) is probably closer to correct than the maximum model for the peak of the last glaciation.

Comparison to sea levels. Another constraint for interpreting foraminiferal $\delta^{18}O$ records as ice volume comes from direct indicators of sea level. According to Hughes and others (1981), the maximum and minimum ice-sheet models imply 117 and 91 m, respectively, of sea-level fall, assuming complete isostatic compensation of the sea floor (163 and 127 m respectively without compensation).

The first line of evidence for paleo–sea levels is radiocarbon-dated shells and peats on stable continental margins. On the Gulf of Mexico continental margin, Curray (1965) estimates 125 m of sea-level fall at about 19 ka. Milliman and Emery (1968) estimate 130 m of sea-level fall relative to modern on the Atlantic American margin about 15 ka. Dillon and Oldale (1978) correct this estimate for the effects of continental shelf subsidence and suggest that only about 90 m of sea-level fall occurred. If this study sampled the true glacial maximum sea-level, it supports the Hughes and others (1981) minimum model or something even smaller. Contrary to this conclusion, Veeh and Veevers (1970) found apparent sea-level terraces off Australia at up to 175 m depth. They sampled shallow-water coral and beachrock deposits from these terraces and obtained reproducible radiocarbon dates of about 13.7 ka, and ^{230}Th dates of about 17 ka. They conclude that these terraces were caused by sea levels 175 m below present at the last glacial maximum. If these data are correct, this large sea-level change would be consistent with ice volumes equal to or greater than the Hughes and others (1981) maximum ice-sheet model.

On tectonically active margins, coral terraces have been used to estimate sea levels. Uncertainties arise regarding the assumption of constant uplift rates (Fairbanks and Matthews, 1978; Bloom and others, 1974), but relatively consistent estimates are obtained for sea-level high stands during isotopic substage 5a (~82 ka) and 5c (~103 ka) from a number of transects with different uplift rates on New Guinea (Bloom and others, 1974) and on various islands (Dodge and others, 1983).

The coral terraces of the Huon Peninsula of New Guinea have been studied extensively by Chappell (1983), who inferred both high and low stands of sea level from radiometrically dated terraces. This record is compared to the most detailed $\delta^{18}O$ records available from the benthic foraminifera *Uvigerina* spp. in

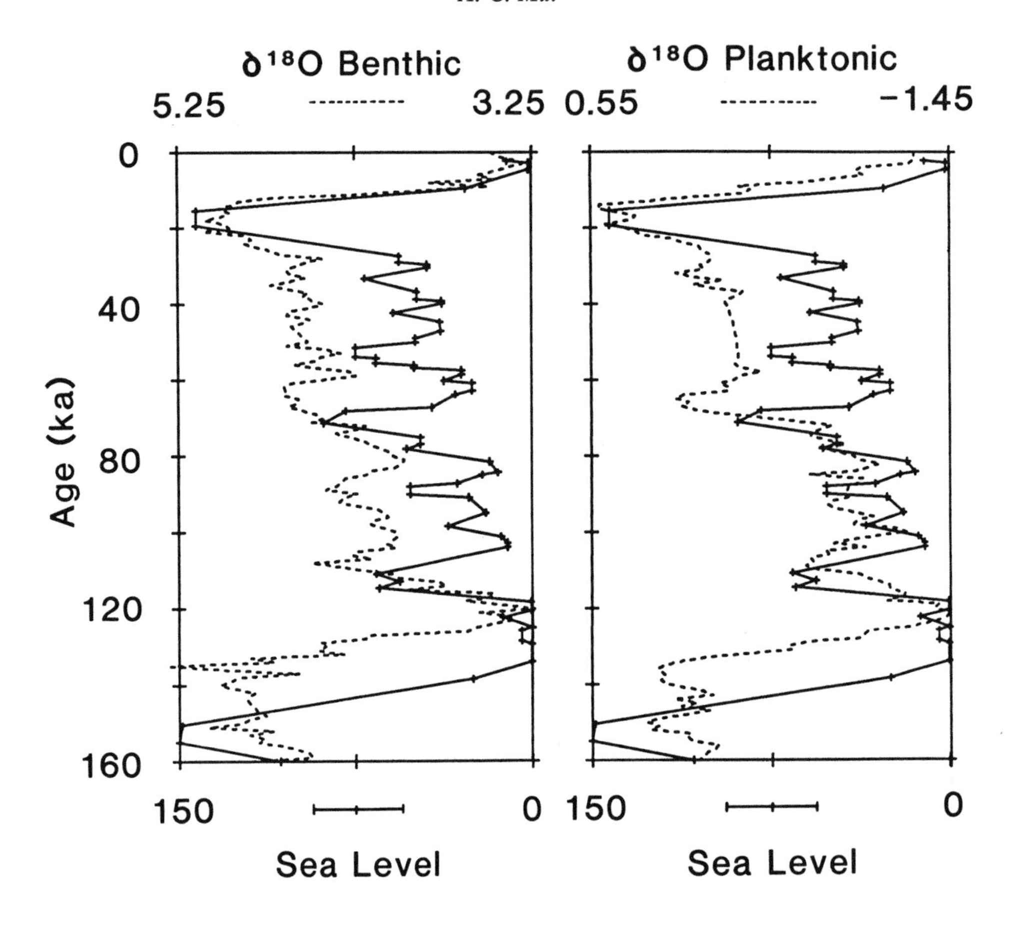

Figure 13. Comparison of $\delta^{18}O$ curves to sea levels estimated from coral terraces on New Guinea (Chappell, 1983). (a) Comparison of sea-level record to benthic $\delta^{18}O$ data on *Uvigerina* spp. in V19-30 (Shackleton and Pisias, 1985). (b) Comparison of sea-level record to planktonic $\delta^{18}O$ data on *G. sacculifer* in V30-40.

eastern Pacific core V19-30 (Shackleton and Pisias, 1985) (Fig. 13a) and from the planktonic foraminifera *G. sacculifer* (with "sac," 415 to 500 μm) in tropical Atlantic core V30-40 (Fig. 13b). The similarity of the sea-level and $\delta^{18}O$ records is striking. There is a mismatch in timing of some of the events, such as the stage 6/5 transition, inferred to occur about 150 ka in the terrace data and about 125 ka in the isotopic data. Further, amplitude mismatches occur between the terrace data and the benthic $\delta^{18}O$ data in stages 3 and 4 and substages 5a–5d, and between the terrace data and the planktonic $\delta^{18}O$ data in stages 3 and 4.

Chappell and Shackleton (1986) attribute the amplitude offsets between the terrace data and the benthic $\delta^{18}O$ data to a combination of time scale errors in the sea-level record (especially at the stage 6/5 transition), and a 1.5°C cooling of the deep sea following $\delta^{18}O$ stage 5e about 115 ka and an equivalent warming during the last deglaciation about 11 ka. This implies a similar cooling in tropical Atlantic surface waters during stages 2–4. The inferred $\delta^{18}O$ ranges due to ice volume of 1.2‰ (Shackleton and Duplessy, 1986) fall at the low end of values produced by the minimum ice-sheet construction and are insufficient to account for the ~150 m sea-level changes inferred by Chappell (1983) or the ~175 m sea-level variations inferred by Veeh and Veevers (1970). They would be able to account for reduced sea-level variation of ~130 m, suggested by Chappell and Shackleton (1986).

Summary: upper and lower limits to ice-volume effects. The constraints given by $\delta^{18}O$, ice-sheet reconstructions, and sea-level data do not yield a unique reconstruction of ice-volume history. If temperature effects are not significant in the isotope records, $\delta^{18}O$ is consistent with either minimum or maximum models of ice volume at the last glacial maximum. The presence of temperature effects, even if small, would favor the minimum models. Different analysts of sea-level data reconstruct glacial-maximum sea-level lowering of 90 to 175 m. The higher values favor the maximum models. Amplitude mismatches between parts of the sea level and $\delta^{18}O$ chronologies imply the presence of temperature effects on $\delta^{18}O$, but it is still not possible to separate reliably the effects of temperature and ice volume on the isotope records.

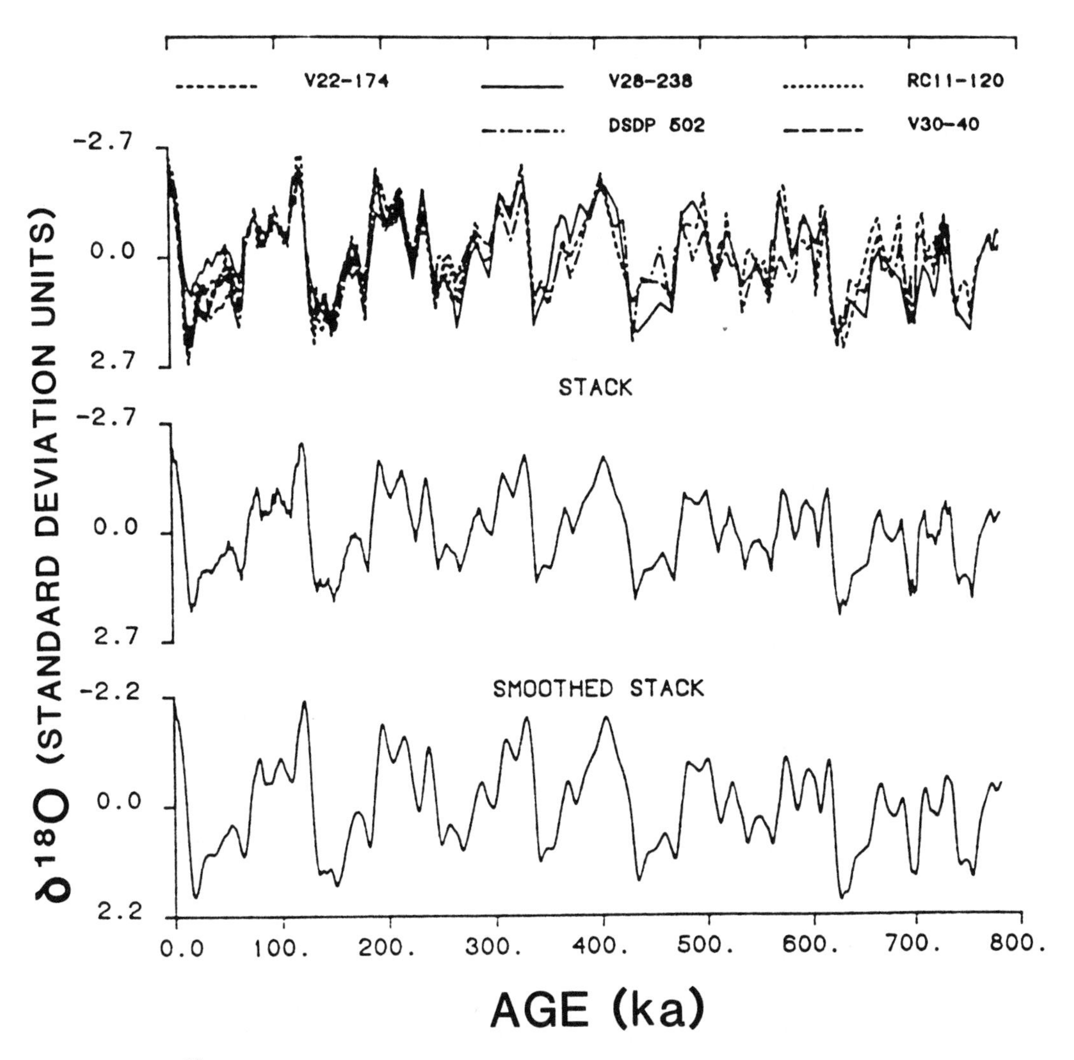

Figure 14. $\delta^{18}O$ record of the last 700,000 yr, generated by stacking multiple records and tuning the time scale to orbital periods (Imbrie and others, 1984). Values are normalized to zero mean and unit variance. Top: original data. Middle: data interpolated at 1 ka intervals. Bottom: data after stacking and smoothing with a 9-pt. Gaussian filter.

$\delta^{18}O$ CHRONOLOGIES

The last 730,000 yr. Much effort has gone into developing an absolute time scale for $\delta^{18}O$ variations in the Quaternary, as the cyclic patterns of ice-volume change have significant implications for mechanisms of climate change. The $\delta^{18}O$ record has been one of the key indicators demonstrating the importance of orbital variations in forcing earth's climate (the Milankovitch hypothesis). Even the earliest records produced by Emiliani (1955) demonstrate that the ice ages were cyclic. A significant revolution in thinking, due primarily to the oxygen-isotope record, was the realization that instead of just four ice ages (e.g., Wisconsinan, Illinoian, Kansas, and Nebraskan, or equivalents) more than 20 ice ages occurred within the Pleistocene, and more in the late Neogene. Indeed, the concept of discrete ice ages has become blurred, as it is clear that ice ages have occurred cyclically with various periods and amplitudes, such that a large variety of glacial states has existed.

Early efforts to determine a time scale for marine $\delta^{18}O$ records used radiometrically dated paleomagnetic boundaries and assumptions of constant sedimentation rates (Emiliani and Shackleton, 1974; Shackleton and Opdyke, 1973) along with correlations to coral terraces dated with uranium-series isotopes (Broecker and Van Donk, 1970). Hays and others (1976) noted the existence of orbital periods in the $\delta^{18}O$ data, and they made the first efforts to refine the time scale by tuning to orbital variations.

Through many iterations, this tuning effort, combined with stacking of many $\delta^{18}O$ records, has yielded the current SPECMAP time scale (Imbrie and others, 1984; Fig. 14). This time

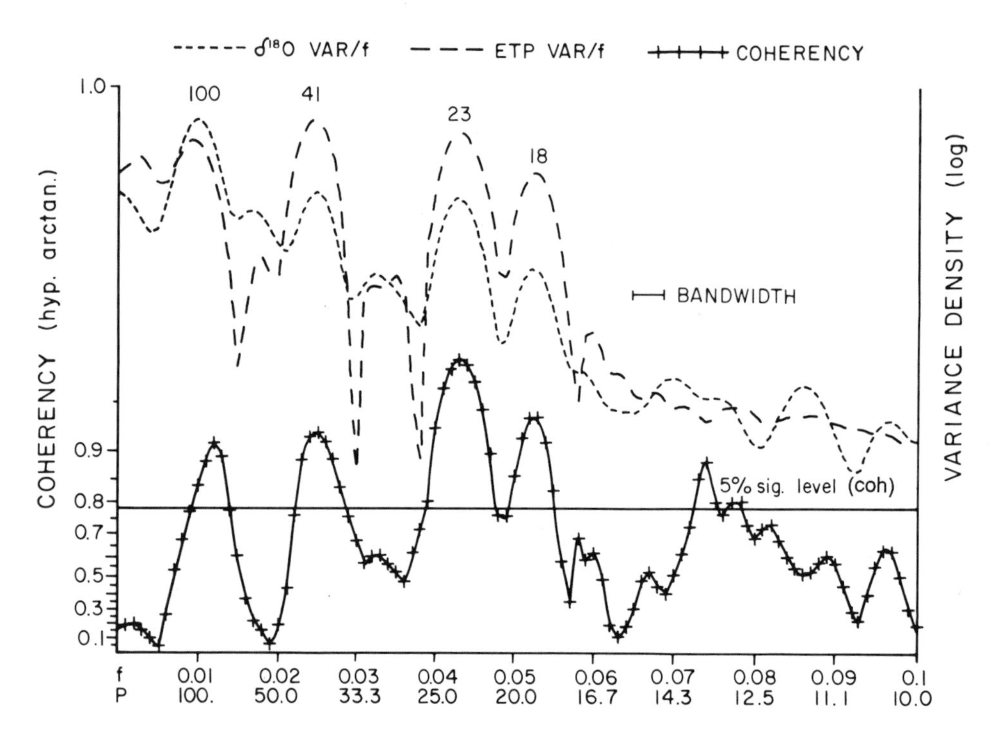

Figure 15. Coherency and variance spectra calculated from $\delta^{18}O$ curves over the past 780,000 yr. Two signals have been processed: (1) ETP, a signal formed by normalizing and adding variations in eccentricity, obliquity, and precession, and (2) $\delta^{18}O$, the unsmoothed unstacked isotope record plotted on the SPECMAP time scale. Top: variance spectra for the two signals, on arbitrary log-log scale. Bottom: coherency spectrum plotted on a hyperbolic arctangent scale and provided with a 5 percent significance level. Frequencies are in cycles per thousand years. From Imbrie and others (1984).

scale allowed orbital periods present in the climate system (Hays and others, 1976) to be used in statistical tests to demonstrate that isotopic variations are phase-locked and strongly coherent with orbital variations, not only at the main periods of precession (19 ka, 23 ka) and obliquity (41 ka), but also in the untuned eccentricity band (100 ka) (Fig. 15). This statistical evidence of a close relationship between the time-varying amplitudes of the orbital forcing and the time-varying amplitudes of the isotopic response in both tuned and untuned bands yields confidence that the relationship is not an artifact of peak matching between the records. The implication is that orbital variations are the main external cause of the succession of late Pleistocene ice ages.

This analysis, of course, remains subject to the reservation that the time scale was generated by tuning to the orbital forcing. That is, because the record was forced to have a consistent relationship to the orbitals, fine-scale deviations from this pattern could have been tuned out of the record. The best hopes for finding these deviations, which could point to climate mechanisms not related directly to orbital forcing, are within the period that can be dated radiometrically with adequate precision. As radiocarbon is currently the best tool for independent dating of the marine record, significant effort has gone into dating the record of the last 20,000 yr, to assess the structure and timing of the last deglaciation as recorded by $\delta^{18}O$. Of course, any leads or lags found over this short transition integrate processes operating separately at the different Milankovitch frequencies. That is, the phases at ~100-, 41-, and 23-ka periods cannot be assessed from a 20,000-yr record. Nonetheless, consistency of the ice-sheet record with a model of Milankovitch forcing can be considered.

The last deglaciation. Two basic approaches have been used to study the structure and timing of the last deglaciation (isotopic termination 1). Each has its strengths and weaknesses. The first is to analyze a few cores with high sedimentation rates in great detail to determine isotopic fine structure and to avoid the problems of studying sections smoothed by bioturbation. The weakness in this approach is that individual cores may contain local signals irrelevant to the global signal and may yield questionable dates due to sediment reworking and contamination. The second approach, stacking and averaging multiple records from normal pelagic sediments, has the advantage that local noise is diminished, while global signal is reinforced. The problem with this technique is that normal pelagic sedimentation rates (1–4 cm/ka) fall in the range that may be sensitive to bioturbational smoothing, and inevitable slight miscorrelations would tend to smooth fine structure to the point that it may be undetectable.

Duplessy and others (1981) emphasized the possibility of

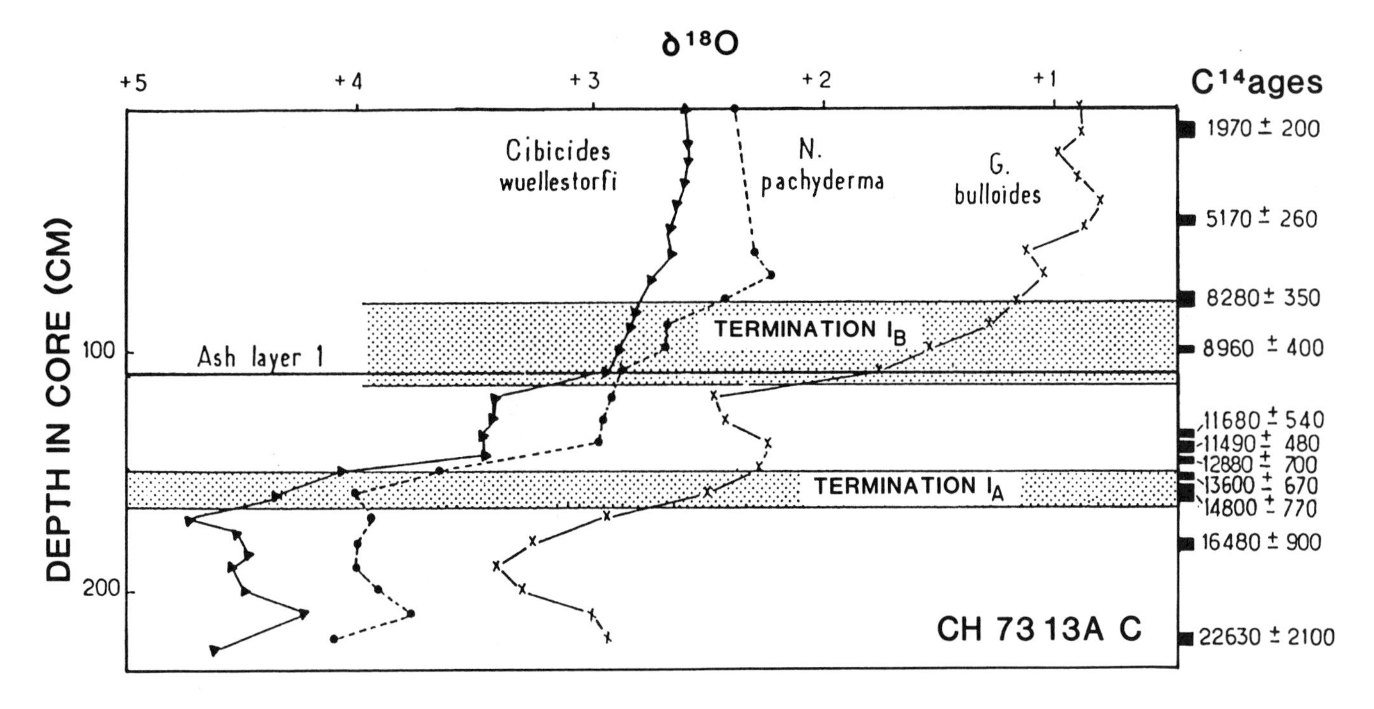

Figure 16. Radiocarbon-dated view of the last deglaciation from high-latitude North Atlantic core CH73139. From Duplessy and others (1981).

fine structure in the isotopic record of deglaciation with implications for climate mechanisms (Fig. 16). They found two steps to the isotopic termination in a high-resolution study of core CH73139 (sedimentation rate ~10 cm/ka) and adjacent cores from the North Atlantic. The first (termination 1-a) dated between 16 and 13 ka, and the second (termination 1-b) dated between 10 and 8 ka. The possibility of rapid deglaciation prior to 14 ka raised considerable interest because it implied that the highest rates of deglaciation were not directly linked to highest northern hemisphere summer insolation, which peaked about 11,000 yr ago. That is, no direct linkage existed between the rate of change in ice volume and the assumed forcing, as required by simple climate models (Imbrie and Imbrie, 1980). If the chronology of Duplessy and others (1981b) was correct, it required additional mechanisms for deglaciation, either as feedback internal to the climate system or as external forcing unrelated to orbital forcing. If insolation acted as the sole trigger for deglaciation, the chronology of Duplessy and others (1981b) implied enormous sensitivity of Earth's climatic system to relatively small perturbations, at least when ice sheets were large.

Duplessy and others (1986) later added radiocarbon dates analyzed by accelerator mass spectrometry to the chronology. This new technique (Andree and others, 1984) allowed the species of foraminifera being analyzed for $\delta^{18}O$ to be dated directly, thus minimizing contamination effects. The dates support the early age for deglaciation, with the youngest full-glacial $\delta^{18}O$ value dating at 15.8 ka. Bard and others (1987), however, consider the effects of bioturbation on this record, and demonstrate that full-glacial conditions persisted until at least 15.0 ka. The earliest deglaciation could not be discerned, as it seemed to occur in a zone barren of foraminifera in CH 73139, as in most other high-latitude North Atlantic cores (Ruddiman and McIntyre, 1981).

Other high-resolution chronologies followed that of Duplessy and others (1981b), and new features were discovered. Sarnthein and others (1982) first proposed three steps to the deglaciation. Their chronology based on three $\delta^{18}O$ records from cores with high sedimentation rates on the northwest African continental margin, suggests deglaciation even earlier than that of Duplessy and others (1981b), with Termination 1-a between 16 and 13.5 ka, a second melting event between 12.4 and 11.1 ka, and Termination 1-b between 10 and 8.5 ka. Keigwin and others (1984) also suggest three steps in the termination, but they were unable to date the events reliably because of sediment reworking in their high-sedimentation-rate core from the Bermuda Rise.

Berger (1982) stacked planktonic $\delta^{18}O$ data from eight Pacific cores from the Ontong Java Plateau. The average sedimentation rate of these cores is about 2 cm/ka, so they are rather sensitive to the effects of bioturbation. The isotopic termination begins about 15 ka, and $\delta^{18}O$ values change rapidly between 12 and 10 ka to a peak at 9 ka. Considering the problems of bioturbation and radiocarbon-dating, however, Berger (1982) suggests that the true ^{14}C age of the isotopic termination may lie between 10 and 9 ka.

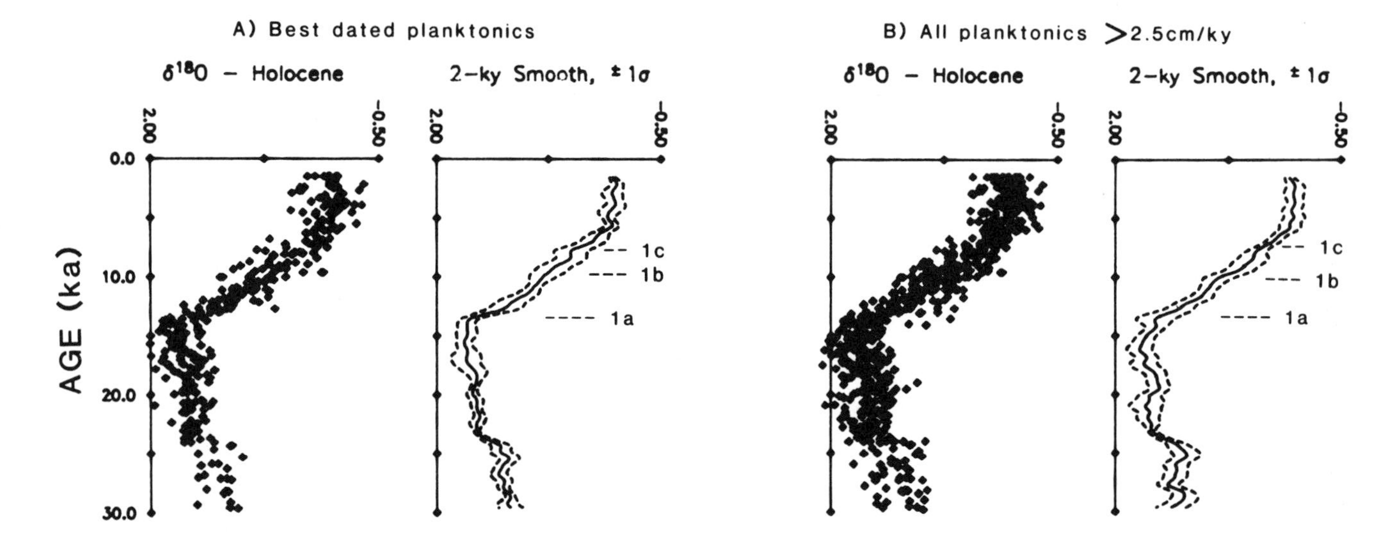

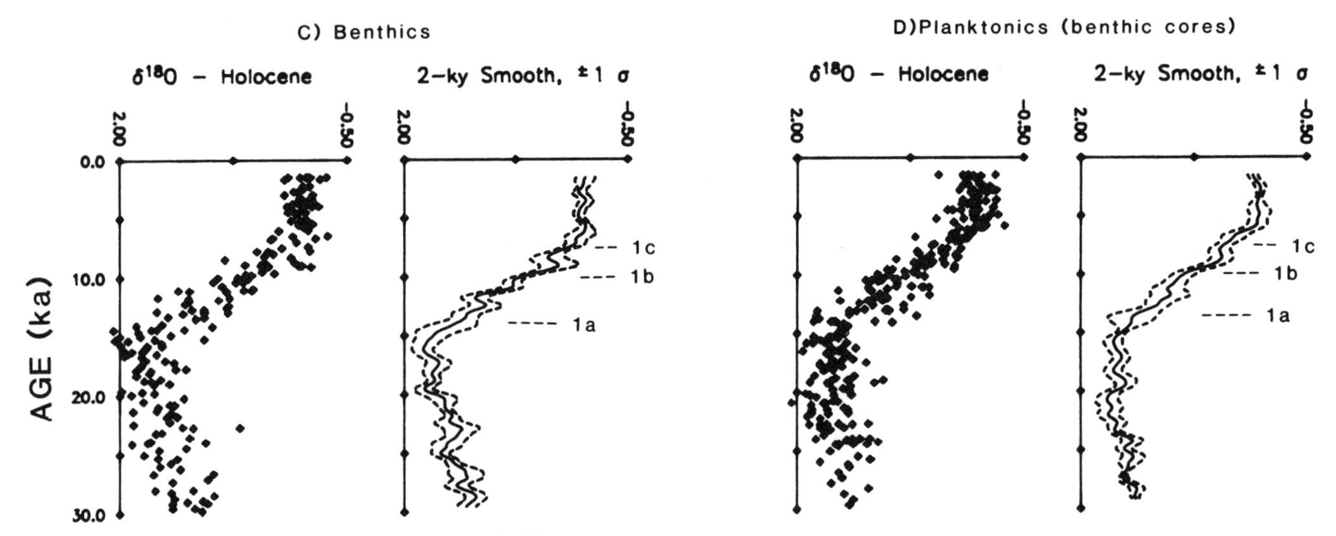

Figure 17. Stacked and smoothed $\delta^{18}O$ data (from Mix and Ruddiman, 1985). On each plot, crosses are stacked data points. Solid lines is composite record, smoothed with a 2,000-yr Gaussian filter. Dashed lines show $\pm$ 1σ error envelope. (a) Best-dated planktonic data. (b) All planktonic data. (c) Benthic data. (d) Planktonic data from same cores as benthics.

In a stack of $\delta^{18}O$ data from three cores with low sedimentation rates (~1.5 cm/ka) from the tropical Atlantic, Berger and others (1985) infer two steps to the deglaciation. After considering potential influences of contamination on their radiocarbon dates, they conclude that the first deglacial step occurred about 13 ka and the second at 10 ka, followed by an overshoot at 9 ka, and rebound to higher $\delta^{18}O$ values near 8 ka.

The most comprehensive stacked isotope chronology for the last deglaciation is that of Mix and Ruddiman (1985), with 77 ^{14}C dates and roughly 1,000 $\delta^{18}O$ analyses. They examined $\delta^{18}O$ of planktonic and benthic foraminifera from 24 tropical Atlantic cores and made four separate composite records (average sedimentation rates of 4–5 cm/ka) to test the sensitivity of the resulting chronology to the amount of data analyzed (Fig. 17). Two steps to the deglaciation were confirmed, dating at 14–12 ka (Termination 1-a), 10–9 ka (Termination 1-b), and a possible

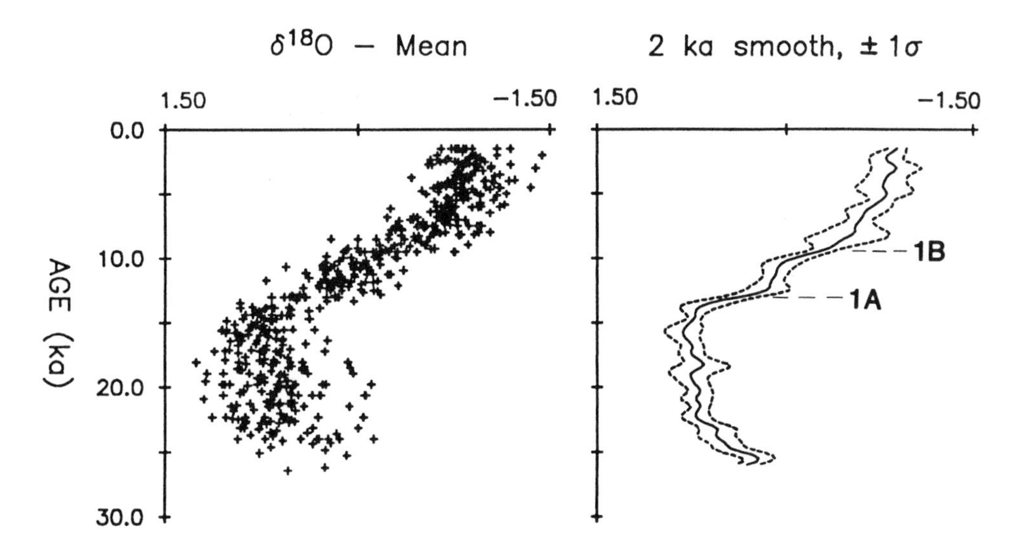

Figure 18. A revised stack of deglacial $\delta^{18}O$ data, modified from that of Mix and Ruddiman (1985) by correlating Terminations 1-a and 1-b in individual cores and fixing the timing of mid-points of these steps at 13 ka and 9.5 ka, respectively. In no cases were ^{14}C dates violated by more than 1,000 yr.

third step was detected at 8–6 ka (tentatively termed Termination 1-c). The error envelopes calculated for the composite records, however, indicate that the steps were at the limits of detection in a composite record. The smoothing induced by the stacking process means that composite records underrepresent the severity of steps. The best representation of isotopic fine structure must come from individual time series. The best estimate of dates for major features, however, will probably come from an objectively dated composite record.

The Mix and Ruddiman (1985) isotopic chronology is updated here, using the same $\delta^{18}O$ data as in their stack of "all planktonic data," but modified by adjusting dates slightly to correlate the apparent steps of Termination 1-a or 1-b in the raw data (Fig. 18). Although this is less objective than the original procedure, it allows for the possibility of errors in radiocarbon dates from individual cores while tuning the chronology to the average timing of steps in the original composite. To do this, ages were fixed at 1.5 ka for the core tops, 9.5 ka for Termination 1-b, 13 ka for Termination 1-a, and 24 ka for the $\delta^{18}O$ stage 2/3 boundary. No attempt was made to correlate Termination 1-c, as it was only tentatively identified in the original stack.

Another difference between this composite and that of Mix and Ruddiman (1985) is that all records are superimposed after removing the 0–20 ka mean $\delta^{18}O$ value, rather than that of the core tops (0–5 ka). This reduces slightly the statistical errors in the deglacial sections, as some of the scatter in the Mix and Ruddiman (1985) composite was due to small amplitude differences among records. A further improvement made here is that the Gaussian smoothing was done on the actual data points rather than on evenly spaced interpolated values. Error envelopes were calculated as in Mix and Ruddiman (1985)[2].

In this revised stack (Fig. 18), Termination 1-a and 1-b stand out clearly. The composite does not preclude the existence of Termination 1-c, at about 6 ka, but this step does not stand out above the error envelope. These data cannot be used to support the existence of a third step of deglaciation.

This isotopic chronology for the last deglaciation is compared with time sequences of volumetric deglaciation calculated by Paterson (1972) for the Laurentide ice sheet (Fig. 19). Paterson (1972) used radiocarbon-dated areal ice retreat and made assumptions about ice thickness to reconstruct volumetric deglaciation of the continents. On average, the volumetric estimates lead the isotopic termination by 1,000–2,000 yr and lack the steps present in the $\delta^{18}O$ transition.

This apparent mismatch may reflect the fact that the model (Paterson, 1972) is not global but is based just on North American ice, which accounts for only about half of the total isotopic signal

[2]The equation for calculating the error envelopes in Mix and Ruddiman (1985) contains a typographical error. The correct equation should be:

$$\sigma_i = \sqrt{\frac{\sum_{j=i-N}^{i+N} w_j (\hat{Y}_j - Y_j)^2}{1 - 1/n}}$$

where σ_i is the standard deviation at each time step i, Y_j is the interpolated data, $\hat{Y}_j$ is the smoothed estimate, w_j is the Gaussian filter weights, and $n = 2N + 1$ is the number of filter elements.

(Table 2). If correct, this would mean that deglaciation of other areas (especially Antarctica) must lag behind the deglaciation of North America and Europe. If anything, the evidence from the southern hemisphere points to early rather than late deglaciation there (Hays, 1978; Stuiver and others, 1981; Lorius and others, 1985), so this does not seem to be the source of the isotopic/volumetric mismatch. If temperature contributions to the $\delta^{18}O$ record are responsible for the lack of fit, average tropical Atlantic temperatures on deglaciation must have been 1–2°C colder than at present or at the glacial maximum and varied cyclically to produce steps in the isotope record. While this is not impossible, it is not consistent with faunal transfer-function estimates of mean annual temperatures for the region (Mix and others, 1986). Nonlinear recording of ice volume by $\delta^{18}O$, due to early melting of ice less depleted in ^{18}O, followed by later melting of ice more depleted in ^{18}O (Mix and Ruddiman, 1984), could account for the lag of the $\delta^{18}O$ transition behind those of the area/volume models, but could not alone generate isotopic steps on the deglaciation.

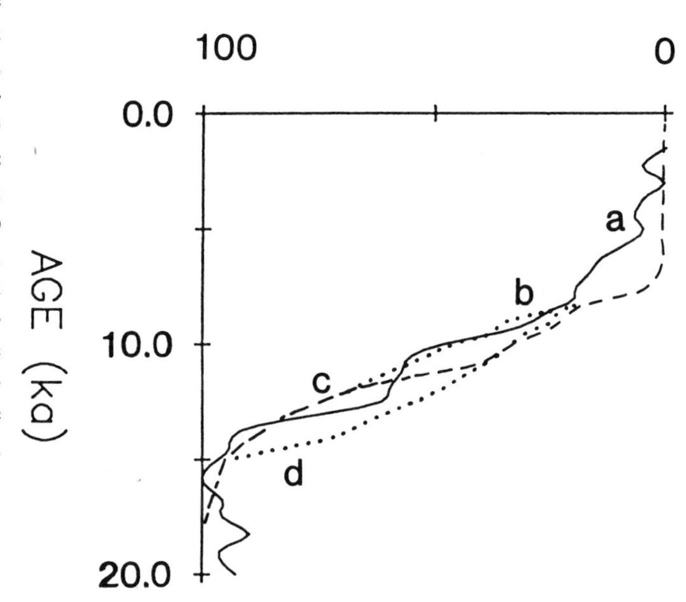

Figure 19. Comparison of the revised $\delta^{18}O$ stack of this chapter (solid lines) with ice-volume estimates made using ice-area data (from Paterson, 1972). Curve a: $\delta^{18}O$ composite. Curve b: equilibrium-profile ice sheet. Curve c: stagnant ice sheets after 11.8 ka. Curve d: stagnant ice sheets after 15 ka.

IMPLICATIONS FOR MECHANISMS OF GLACIATION

The finding of coherent relationships between the $\delta^{18}O$ record and orbital variations over the last 730,000 yr has provided convincing evidence for the astronomical theory of climatic change (Milankovitch, 1941). Imbrie and others (1984) calculate that at least 60 percent of the variance of the $\delta^{18}O$ record is related to orbital forcing. The exact mechanisms translating orbital variations into glaciation, however, are hotly debated.

The search for mechanisms has led to better-defined details of the last deglaciation, where time control can be provided by radiocarbon dating. The chronologies for this time period discussed above are basically consistent with the traditional view that deglaciation (centered near 11 ka) was driven by the seasonal distribution of insolation. Complications, however, are that the most rapid rates of deglaciation predate the maximum insolation forcing, contrary to predictions of simple climate models (Imbrie and Imbrie, 1980). Thus other mechanisms must exist, either as feedbacks internal to the climate system or as external forcing unrelated to orbital variations.

One mechanism that may account for both relatively early deglaciation and steps in the isotopic transition is rapid thinning of ice sheets via unstable downdraw (Denton and Hughes, 1981), perhaps due to calving of ice to the ocean (Hoppe, 1948; Blake, 1966; Andrews, 1973), or grounding-line retreat caused by sea-level rise relative to marine-ice pinning points (Weertman, 1974; Thomas, 1979). Other mechanisms related to isostatic adjustment of the earth to the ice sheets (Peltier and Hyde, 1984) or variations in atmospheric CO_2 content (Pisias and Shackleton, 1984) may also provide feedback mechanisms. Each of these mechanisms, when put into climate models, will make some predictions about the structure and timing of glaciation. The $\delta^{18}O$ record will provide tests of some of these models. As we refine our abilities to detect global ice-volume changes from $\delta^{18}O$ and other means, we will also narrow our choice of mechanisms controlling large-scale climate change.

SUMMARY AND CONCLUSIONS

The development of an oxygen-isotope chronology in deep-sea sediments provides a continuous time series of variations in continental ice sheets that could never be obtained from the fragmentary continental record, and thus it has led to a revolution in thinking about the Quaternary ice ages. The patterns of isotopic change over the last 730,000 yr strongly support the astronomical theory of climate change. A closer look at the patterns, however, suggests that various feedback mechanisms must contribute to climatic changes.

Despite its value in defining the ice ages, the isotope record is not without its deficiencies. Questions remain about the extraction of local effects from global effects in the oxygen-isotopic record. For example, estimates of the contribution of temperature to down-core $\delta^{18}O$ records range from 0 to 30 percent of the signal. We can not at present exclude the possibility that isotopic steps on deglaciation reflect temperature overprints on the ice volume record, rather than pauses in deglaciation. If the isotopic steps are recorded globally, the ice volume interpretation is more probable. Although it is clear that individual foraminifera do record temperature in the isotopic content of their calcite shells, extracting this effect from populations of foraminifera is difficult, as it is tied to understanding ecological responses of the foraminiferal populations. Refining our knowledge of foraminiferal ecology and our ability to make nonisotopic estimates of marine temperature changes will be critical to understanding the ice-volume signal.

The data from $\delta^{18}O$, sea level, and glaciology do not yet converge on a single view of glaciation. With reasonable estimates of the isotopic composition of glacier ice, the $\delta^{18}O$ record is consistent with either the minimum or maximum reconstructions of Hughes and others (1981). If temperature effects on the $\delta^{18}O$ record are significant, the minimum model is favored. Sea-level changes between the glacial maximum and the present are estimated by various workers to be between 90 m (consistent with a minimum model with isostatic compensation) and 175 m (consistent with a maximum model without isostatic compensation) and thus do not constrain the detailed interpretation of either glaciological models or the isotopic record.

REFERENCES

Andree, M., and 12 others, 1984, ^{14}C measurements on foraminifera of deep-sea core V28–238 and their preliminary interpretation: Nuclear Instruments and Methods in Physics Research B5, p. 340–345.

Andrews, J. T., 1973, The Wisconsin Laurentide ice sheet; Dispersal centers, problems of rates of retreat, and climatic implications: Arctic and Alpine Research, v. 5, p. 185–199.

Bard, E., Arnold, A., Duplessy, J.-C., Duprat, J., and Moyes, J., 1987, Reconstruction of the last deglaciation; Deconvolved records of $\delta^{18}O$ profiles, micropaleontological variations and accelerator mass spectrometric ^{14}C dating: Climate Dynamics (in press).

Barkov, N. I., Korotkevich, E. S., Gordienko, F. G., and Kotlyakov, V. M., 1977, The isotope analysis of ice cores from Vostok station (Antarctica) to the depth of 950 m, *in* Isotopes and impurities in snow and ice: International Association of Hydrological Sciences Publication 118, 382–387.

Belanger, P. E., Curry, W. B., and Matthews, R. K., 1981, Core-top evaluation of benthic foraminiferal isotopic ratios for paleooceanographic interpretations. Palaeogeography, Palaeoclimatology, Palaeoecology, v. 33, p. 205–220.

Berger, W. H., 1978, Oxygen-18 stratigraphy in deep-sea sediments; Additional evidence for deglacial melt water effect: Deep-Sea Research, v. 25, p. 473–480.

——, 1982, On the definition of the Pleistocene–Holocene boundary in deep-sea sediments: Sveriges Geologiska Undersokning, v. 76c, p. 270–280.

Berger, W. H., and Killingley, J. S., 1977, Glacial Holocene transitions in deep-sea carbonates; Selective dissolution and the stable isotope signal: Science, v. 197, p. 563–566.

Berger, W. H., Johnson, R. F., and Killingley, J. S., 1977, "Unmixing" of the deep-sea record and the deglacial meltwater spike: Nature, v. 269, p. 661–663.

Berger, W. H., Killingley, J. S., and Vincent, E., 1978, Stable isotopes in deep-sea carbonates; Box core ERDC–92, West Equatorial Pacific: Oceanologica Acta, v. 1, p. 203–216.

Berger, W. H., Killingley, J. S., Metzler, C. V., and Vincent, E., 1985, Two step deglaciation; ^{14}C dated high-resolution $\delta^{18}O$ records from the tropical Atlantic Ocean: Quaternary Research, v. 23, p. 258–271.

Blake, W., 1966, End moraines and deglaciation chronology in northern Canada, with special reference to southern Baffin Island: Geological Survey of Canada Paper 66-29, p. 1–31.

Bloom, A. L., Broecker, W. S., Chappell, J., Matthew, R. K., and Mesolella, K. J., 1974, Quaternary sea level fluctuations on a tectonic coast; New $^{230}Th/^{234}U$ dates from the Huon Peninsula, New Guinea: Quaternary Research, v. 4, p. 185–205.

Bonneau, M.-C., Vergnaud-Grazzini, C., and Berger, W. H., 1980, Stable isotope fractionation and differential dissolution in recent planktonic foraminifera for Pacific Box Cores: Oceanologica Acta, v. 3, p. 377–382.

Broecker, W. S., 1986, Oxygen isotope constraints on surface ocean temperatures: Quaternary Research, v. 26, p. 121–135.

Broecker, W. S., and Van Donk, J., 1970, Insolation changes, ice volumes, and the 0-18 record in deep-sea cores: Reviews in Geophysics and Space Physics, v. 8, p. 169–198.

Chappell, J., 1983, A revised sea-level record for the last 300,000 years from Papua New Guinea: Search, v. 14, p. 99–101.

Chappell, J., and Shackleton, N. J., 1986, Oxygen isotopes and sea level: Nature, v. 324, p. 137–138.

Corliss, B. H., 1985, Micro-habitats of benthic foraminifera within deep-sea sediments: Nature, v. 314, p. 435–438.

Covey, C., and Schneider, S. H., 1985, Models for reconstructing temperature and ice volume from oxygen isotope data, *in* Berger, A., Imbrie, J., Hays, J., Kukla, G., and Saltzman, B., eds., Milankovitch and climate, Part 2: Reidel, Dordrecht, p. 699–706.

Craig, H., 1965, The measurement of oxygen isotope paleotemperatures, *in* Tongiorgi, E., ed., Stable isotopes in oceanic studies and paleotemperatures; Third SPOLETO Conference on Nuclear Geology: Pisa, Consiglio Nazionale delle Richerche, Laboratorio di Geologia Nucleare, p. 161–182.

Craig, H., and Gordon, L. I., 1965, Deuterium and oxygen-18 variations in the ocean and marine atmosphere, *in* Tongiorgi, E., ed., Stable isotopes in oceanic studies and paleotemperatures; Third SPOLETO Conference on Nuclear Geology: Pisa, Consig. Nazion. Richerche, Laborat. Geol. Nucl., p. 9–130.

Curray, J. R., 1965, Late Quaternary history, continental shelves of the United States, *in* Wright, H. E., and Frey, D. C., eds., The Quaternary of the United States: Princeton, New Jersey, Princeton University Press, p. 723–735.

Curry, W. B., and Matthews, R. K., 1981a, Paleo-oceanographic utility of oxygen isotopic measurements on planktonic foraminifera; Indian Ocean core-top evidence: Paleogeography, Palaeoclimatology, Palaeoecology, v. 33, p. 173–191.

——, 1981b, Equilibrium ^{18}O fractionation in small size fraction planktonic foraminifera; Evidence from recent Indian Ocean sediments: Marine Micropaleontology, v. 6, p. 327–337.

Curry, W. B., Thunell, R. C., and Honjo, S., 1983, Seasonal changes in the isotopic composition of planktonic foraminifera collected in Panama Basin sediment traps: Earth and Planetary Science Letters, v. 64, p. 33–43.

Dansgaard, W., and Tauber, H., 1969, Glacier oxygen-18 content and Pleistocene ocean temperatures: Science, v. 166, p. 499–502.

Dansgaard, W., and 6 others, 1982, A new Greenland deep ice core: Science, v. 218, p. 1273–1275.

Denton, G. H., and Hughes, T. J., 1981, The last great ice sheets: New York, Wiley Interscience, 484 p.

Deusser, W. G., Ross, E. H., Hemleben, C., and Spindler, M., 1981, Seasonal changes in species composition, numbers, mass, size, and isotopic composition of planktonic foraminifera settling into the deep Sargasso Sea: Palaeogeography, Palaeoclimatology, Palaeoecology, v. 33, p. 103–127.

Dillon, W. R., and Oldale, R. N., 1978, Late Quaternary sea-level curve; Reinterpretation based on glaciotectonic influence: Geology, v. 6, p. 56–60.

Dodge, R. E., Fairbanks, R. G., Binninger, L. K., and Maurrasse, F., 1983, Pleistocene sea levels from raised coral reefs of Haiti: Science, v. 219, p. 1423–1425.

Duplessy, J.-C., 1978, Isotope studies, *in* Gribben, J., ed., Climatic change: Cambridge, Massachusetts, Cambridge University Press, p. 46–67.

Duplessy, J.-C., Lalou, C., and Vinot, A. C., 1970, Differential isotopic fractionation in benthic foraminifera and paleotemperatures reassessed: Science, v. 168, p. 250–251.

Duplessy, J.-C., Blanc, P.-L., and Be, A.W.H., 1981a, Oxygen-18 enrichment of planktonic foraminifera due to gametogenic calcification below the euphotic zone: Science, v. 213, p. 1247–1250.

Duplessy, J.-C., Delibrias, G., Turon, J. L., Pujol, C., and Duprat, J., 1981b, Deglacial warming of the northeastern Atlantic Ocean; Correlation with the paleoclimate evolution of the European continent: Palaeogeography, Palaeoclimatology, Palaeoecology, v. 35, p. 121–144.

Duplessy, J.-C., Arnold, M., Maurice, P., Bard, E., Duprat, J., and Moyes, J., 1986, Direct dating of the oxygen-isotope record of the last deglaciation by ^{14}C accelerator mass spectrometry: Nature, v. 320, p. 350–352.

Durazzi, J. T., 1981, Stable-isotope studies of planktonic foraminifera in North Atlantic core tops: Palaeogeography, Palaeoclimatology, Palaeoecology, v. 33, p. 157–172.

Emiliani, C., 1954, Depth habitats of some species of pelagic foraminifera as indicated by oxygen isotope ratios: American Journal of Science, v. 252, p. 149–158.

——, 1955, Pleistocene temperatures: Journal of Geology, v. 63, p. 538–578.

——, 1970, Pleistocene paleotemperatures: Science, v. 168, p. 822–825.

——, 1971, The amplitude of Pleistocene climatic cycles at low latitudes and the isotopic composition of glacial ice, *in* Turekian, K. K., ed., Late Cenozoic glacial ages: New Haven, Connecticut, Yale University Press, p. 183–197.

Emiliani, C., and Shackleton, N. J., 1974, The Brunhes Epoch; Isotopic paleotemperatures and geochronology: Science, v. 183, p. 511–514.

Emiliani, C., and 7 others, 1975, Paleoclimatological analysis of late Quaternary cores from the northeastern Gulf of Mexico: Science, v. 189, p. 1083–1087.

Epstein, S., Buchsbaum, R., Lowenstam, H. A., and Urey, H. C., 1953, Revised carbonate water isotopic temperature scale: Geological Society of America Bulletin, v. 64, p. 1315–1326.

Epstein, S., Sharp, R. P., and Gow, A. J., 1970, Antarctic ice sheet; Stable isotopic analyses of Byrd Station cores and interhemispheric climatic implications: Science, v. 168, p. 1570–1572.

Erez, J., 1979, Modification of the oxygen-isotope record in deep-sea cores by Pleistocene dissolution cycles: Nature, v. 281, p. 535–538.

Erez, J., and Luz, B., 1983, Experimental paleotemperature equation for planktonic foraminifera: Geochimica et Cosmochimica Acta, v. 47, p. 1025–1031.

Fairbanks, R. G., and Matthews, R. K., 1978, The marine oxygen isotope record in Pleistocene coral, Barbados, West Indies: Quaternary Research, v. 10, p. 181–196.

Fairbanks, R. G., Sverdlove, M., Free, R., Wiebe, P. H., and Be, A.W.H., 1982, Vertical distribution and isotopic composition of living planktonic foraminifera from the Panama Basin: Nature, v. 298, p. 841–844.

Ganssen, G., and Sarnthein, M., 1983, Stable-isotope composition of foraminifers; The surface and bottom water record of coastal upwelling, *in* Suess, E., and Thiede, J., eds., Coastal upwelling; Its sediment records; Part A, Responses of the sedimentary regime to present coastal upwelling: New York, Plenum Press, p. 99–121.

Graham, D. W., Corliss, B. H., Bender, M. L., and Keigwin, L. E., Jr., 1981, Carbon and oxygen isotopic disequilibria of recent deep-sea benthic foraminifera: Marine Micropaleontology, v. 6, p. 483–497.

Hays, J. D., 1978, A review of the Late Quaternary climatic history of Antarctic Seas, *in* Van Zinderen Bakker, E. M., ed., Antarctic glacial history and world paleoenvironments: Rotterdam, Balkeena, p. 57–71.

Hays, J. D., Imbrie, J., and Shackleton, N. J., 1976, Variations in the Earth's orbit; Pacemaker of the ice ages: Science, v. 194, p. 1121–1132.

Hooke, R. LeB., and Clausen, H., 1982, Wisconsin and Holocene $\delta^{18}O$ variations, Barnes Ice Cap, Canada: Geological Society of America Bulletin, v. 93, p. 784–789.

Hoppe, G., 1948, Isrecessionen fran Norbottens Kustland i belysning av de glaciala formelementon: Geographica Skriften, Uppsala University Geografiska Inst., v. 20, p. 1–112.

Horibe, Y., and Oba, T., 1972, Temperature scales of aragonite-water and calcite-water systems: Fossiles, v. 23/24, p. 69–74.

Hughes, T. J., Denton, G. H., Andersen, B. G., Schilling, D. H., Fastook, J. L., and Lingle, C. S., 1981, The last great ice sheets; A global view, *in* Denton, G. H., and Hughes, T. J., eds., The last great ice sheets: New York, Wiley Interscience, p. 263–317.

Imbrie, J., and Imbrie, J. Z., 1980, Modeling the climatic response to orbital variations: Science, v. 207, p. 943–953.

Imbrie, J., and Kipp, N. J., 1971, A new micropaleontological method for quantitative paleoclimatology; Application to a Lake Pleistocene Caribbean core, *in* Turekian, K. K., ed., Late Cenozoic glacial ages: New Haven, Connecticut, Yale University Press, p. 71–147.

Imbrie, J., Van Donk, J., and Kipp, N. G., 1973, Paleoclimatic investigation of a late Pleistocene Caribbean deep-sea core; Comparison of isotopic and faunal methods: Quaternary Research, v. 3, p. 10–38.

Imbrie, J., and 8 others, 1984, The orbital theory of Pleistocene climate; Support from a revised chronology of the marine $\delta^{18}O$ record, *in* Berger, A., Imbrie, J., Hays, J., Kukla, G., and Saltzman, B., eds., Milankovitch and climate, Part 1: Dordrecht, Reidel, p. 269–305.

Jones, G. A., and Ruddiman, W. F., 1982, Assessing the global meltwater spike: Quaternary Research, v. 17, p. 148–172.

Joussaume, S., Sadourny, R., and Jouzel, J., 1984, A general circulation model of water isotope cycles in the atmosphere: Nature, v. 311, p. 24–29.

Keigwin, L. D., Corliss, B. H., Druffel, E. M., and Laine, E. P., 1984, High resolution isotopic study of the latest deglaciation based on Bermuda Rise cores: Quaternary Research, v. 22, p. 383–386.

Kennett, J. P., and Shackleton, N. J., 1975, Laurentide ice sheet meltwater recorded in Gulf of Mexico deep-sea cores: Science, v. 188, p. 147–150.

Leventer, A., Williams, D. F., and Kennett, J. P., 1982, Dynamics of the Laurentide ice sheet during the last deglaciation; Evidence from the Gulf of Mexico: Earth and Planetary Science Letters, v. 59, p. 11–17.

Levitus, S., 1982, Climatological atlas of the world ocean: National Oceanic and Atmospheric Administration Professional Paper 13, p. 1–173.

Lidz, B., Kehm, A., and Miller, H., 1968, Depth habitats of pelagic foraminifera during the Pleistocene: Nature, v. 217, p. 245–247.

Lorius, C., and 6 others, 1985, A 150,000-year climatic record from Antarctic ice: Nature, v. 316, p. 591–596.

Matthews, R. K., and Poore, R. Z., 1981, Tertiary $\delta^{18}O$ record and glacio-eustatic sea-level fluctuations: Geology, v. 8, p. 501–504.

McCorkle, D. C., Emerson, S. R., and Quay, P., 1985, Stable carbon isotopes in marine porewaters: Earth and Planetary Science Letters, v. 74, p. 13–26.

McCrea, J. M., 1950, On the isotopic chemistry of carbonates and a paleotemperature scale: Journal of Chemical Physics, v. 18, p. 849–857.

Milankovitch, M., 1941, Kanon der Erdbestrahlung und seinde Andwendung auf das Eiszeitproblem: Belgrade, Royal Serbian Academy Special Publication 133.

Milliman, J. D., and Emery, K. O., 1968, Sea levels during the last 35,000 years: Science, v. 162, p. 1121–1123.

Mix, A. C., and Ruddiman, W. F., 1984, Oxygen-isotope analyses and Pleistocene ice volumes: Quaternary Research, v. 21, p. 1–20.

——, 1985, Structure and timing of the last deglaciation; Oxygen isotope evidence: Quaternary Science Reviews, v. 4, p. 59–108.

Mix, A. C., Ruddiman, W. F., and McIntyre, A., 1986, Late Quaternary paleoceanography of the tropical Atlantic 1; Spatial variability of annual mean sea-surface temperatures, 0–20,000 years B.P.: Paleoceanography, v. 1, p. 43–66.

O'Neil, J. R., Clayton, R. N., and Mayeda, T. K., 1969, Oxygen isotope fractionation in divalent metal carbonates: Journal of Chemical Physics, v. 51, p. 5547–5558.

Pastouret, L., Chamley, H., Delibrias, G., Duplessy, J.-C., and Thiede, J., 1978, Late Quaternary climatic changes in western tropical Africa deduced from deep-sea sedimentation off the Niger delta: Oceanologica Acta, v. 1, p. 217–232.

Paterson, W.S.B., 1972, Laurentide ice sheet; Estimated volumes during the late Wisconsin: Reviews in Geophysics and Space Physics, v. 10, p. 885–917.

Peltier, W. R., and Hyde, W., 1984, A model of the ice age cycle, *in* Berger and others, eds., Milankovitch and climate, Part 2: Dordrecht, Reidel, p. 565–580.

Pisias, N. G., and Shackleton, N. J., 1984, Modelling the global climate response to orbital forcing and atmospheric carbon dioxide changes: Nature, v. 310, p. 757–759.

Ruddiman, W. F., and McIntyre, A., 1981, The North Atlantic during the last deglaciation: Palaeogeography, Palaeoclimatology, Palaeoecology, v. 35, p. 145–214.

Sarnthein, M., Erlenkeuser, H., and Zahn, R., 1982, Termination I; The response of continental climate in the subtropics as recorded in deep-sea sediments: Bulletin de l'Institut de Geologie du Bassin d'Aquitaine, v. 31, p. 393–407.

Shackleton, N. J., 1967, Oxygen isotope analyses and Pleistocene temperatures, reassessed: Nature, v. 215, p. 15–17.

—— , 1974, Attainment of isotopic equilibrium between ocean water and the benthonic foraminiferal genus Uvierina; Isotopic changes in the ocean during the last glacial: Colloques Int. Centr. Nat. Rech. Sci., v. 219, p. 203–219.

Shackleton, N. J., and Duplessy, J.-C., 1986, Temperature changes in ocean deep waters during the late Pleistocene [abs.]: Woods Hole, Massachusetts, Second International Conference on Paleoceanography, p. 63.

Shackleton, N. J., and Opdyke, N. D., 1973, Oxygen isotope and paleomagnetic stratigraphy of equatorial Pacific core V28-238; Oxygen isotope temperatures and ice volumes on a 10^5 and 10^6 year scale: Quaternary Research, v. 3, p. 39–55.

Shackleton, N. J., and Pisias, N. G., 1985, Atmospheric carbon dioxide, orbital forcing, and climate: American Geophysical Union Geophysics Monograph 32, p. 303–318.

Shackleton, N. J., and Vincent, E., 1978, Oxygen and carbon isotope studies in recent foraminifera from the southwest Indian Ocean: Marine Micropaleontology, v. 3, p. 1–13.

Shackleton, N. J., Wiseman, J.D.H., and Buckley, H. A., 1973, Non-equilibrium isotopic fractionation between seawater and planktonic foraminiferal tests: Nature, v. 242, p. 177–179.

Showers, W. J., and Margolis, S. V., 1985, Evidence for a tropical freshwater spike during the last glacial/interglacial transition in the Venezuela Basin; $\delta^{18}O$ and $\delta^{13}C$ of calcareous plankton, Marine Geology, v. 68, p. 145–165.

Stuiver, M., Denton, G. H., Hughes, T. J., and Fastook, J. L., 1981, History of the marine ice sheet in west Antarctica during the last glaciation; A working hypothesis, in Denton, G. H., and Hughes, T. J., eds., The last great ice sheets: New York, Wiley, p. 319–436.

Thomas, R. H., 1979, The dynamics of marine ice sheets: Journal of Glaciology, v. 24, p. 167–177.

Thunnell, R. C., and Reynolds, L. A., 1984, Sedimentation of planktonic foraminifera; Seasonal changes in species flux in the Panama Basin: Micropalentology, v. 30, p. 243–262.

Urey, H. C., 1947, The thermodynamic properties of isotopic substances: Journal Chem. Soc., p. 169–182.

Veeh, H. H., and Veevers, J. J., 1970, Sea level at –175 m off the Great Barrier Reef 12,600 to 17,000 years ago: Nature, v. 226, p. 536–537.

Vergnaud-Grazzini, C., 1976, Non-equilibrium isotopic compositions of shells of planktonic foraminifera in the Mediterranean Sea: Palaeogeography, Palaeoclimatology, Palaeoecology, v. 320, p. 263–276.

Vincent, E., Killingley, J. S., and Berger, W. H., 1981, Stable isotope composition of benthic foraminifera for the equatorial Pacific: Nature, v. 289, p. 639–643.

Weertman, J., 1974, Stability of the junction of an ice sheet and an ice shelf: Journal of Glaciology, v. 13, p. 3–11.

Williams, D. F., and Healy-Williams, N., 1980, Oxygen isotope-hydrogeographic relationships among recent planktonic foraminifera from the Indian Ocean: Nature, v. 283, p. 848–852.

Williams, D. F., Be, A.W.H., and Fairbanks, R. G., 1981, Seasonal stable isotopic variations in living planktonic foraminifera from Bermuda plankton tows: Palaeogeography, Palaeoclimatology, Palaeoecology, v. 33, p. 71–102.

Yapp, C. J., and Epstein, S., 1977, Climatic implications of D/H ratios of meteoric water over North America (9,500–22,000 B.P.) as inferred from ancient wood cellulose C-H hydrogen: Earth and Planetary Science Letters, v. 34, p. 333–350.

Zahn, R., Winn, K., and Sarnthein, M., 1986, Benthic foraminiferal $\delta^{13}C$ and accumulation rates of organic carbon; Uvigerina peregrina group and Cibicidoides wuellerstorfi: Paleoceanography, v. 1, p. 27–42.

Manuscript Accepted by the Society March 20, 1987

ACKNOWLEDGMENTS

I am grateful for discussions with many colleagues about stable isotopes and paleoclimatology. Constructive reviews by N. J. Shackleton, W. H. Berger, and N. G. Pisias and editing by W. F. Ruddiman and H. E. Wright improved the manuscript. C. Peterson assisted with the figures. Support from National Science Foundation Grant ATM83-19371 (Project SPECMAP) is gratefully acknowledged.

Printed in U.S.A.

The Geology of North America
Vol. K-3, North America and adjacent oceans during the last deglaciation
The Geological Society of America, 1987

Chapter 7

Northern oceans

W. F. Ruddiman
Lamont-Doherty Geological Observatory of Columbia University, Palisades, New York 10964

INTRODUCTION

This chapter reviews our knowledge of changes in northern hemisphere surface oceans during the last deglaciation. To varying degrees, these oceans were all affected by the presence of the ice sheets, and deep-sea cores thus provide evidence pertinent to changes in ice-sheet size during the deglaciation. In addition, several theories that call on the oceans to provide feedback to ice sheets during deglaciation have been proposed (see Introduction, this volume) and are evaluated here.

Emphasis is directed to the North Atlantic Ocean, which, by virtue of its location, is the ocean likely to be most sensitive to the presence of ice on North America. The North Atlantic also has an extensive coverage of cores with relatively good time control and with well preserved, climatically sensitive microplanktonic fossils.

Other oceanic regions (the North Pacific, marginal seas around the periphery of the Atlantic, and lower-latitude oceans) are reviewed in lesser detail, due to their greater geographic (and climatologic) remoteness from North American ice and, in some cases, to the larger uncertainties in their deglacial chronologies.

The focus of this chapter is on cores obtained beyond the edge of the continental shelf. Shallow-water records collected from the continental shelf are complicated by local changes in the margins of the ice sheets and by vertical movements resulting from local ice-loading histories. Some of this material is discussed in Chapters 2 and 8 (this volume).

BACKGROUND: ORBITAL FORCING OF NORTHERN OCEANS

Because of the dominance of earth–sun orbital rhythms in Pleistocene climatic signals, it is important to see how the last deglaciation fits into longer-term trends. Orbital responses provide information about linkages between different parts of the climate system that cannot be derived from analysis of shorter intervals.

North Atlantic Ocean

Orbital-scale responses in the subpolar North Atlantic during the late Pleistocene are summarized in Ruddiman and McIntyre (1981a, 1984). Signals of estimated sea-surface temperature (SST), generated by methods described in Imbrie and Kipp (1971), contain all the major orbital rhythms: eccentricity (approximately 100 ka), obliquity (41 ka), and precession (23 and 19 ka). The absolute and relative strengths of these SST rhythms at orbital periods show significant variation with latitude (Figs. 1 and 2).

Two kinds of evidence indicate that the North Atlantic SST responses at the 41,000-year and 100,000-year periods represent orbital signals initially registered in the ice sheets and then directly transferred to the ocean via the atmosphere (Fig. 3a, b). Geologic data show that both the 41,000-year and 100,000-year components of SST are in phase with the same components in $\delta^{18}O$ signals, or lag only slightly behind them (Ruddiman and McIntyre, 1984). Because $\delta^{18}O$ is in large part a proxy for ice volume (see Chapter 6, this volume), the North Atlantic ocean responds very nearly in phase with the ice sheets at both of these periods, indicating direct transferral of the signal from ice to ocean via the atmosphere (Fig. 3a).

This ice–ocean link also makes sense geographically. Because the 41,000-year insolation signal is strongest at high latitudes (Berger, 1978; see Introduction, this volume), the higher-latitude portions of northern hemisphere ice sheets are the components most likely to have responded at this period. Similarly, the 41,000-year SST signal is strongest at higher latitudes in the North Atlantic. This suggests direct transferral of this high-latitude insolation rhythm from high-latitude ice to high-latitude ocean (Ruddiman and McIntyre, 1984). This signal may also be recorded by *Cycladophora davisiana,* a radiolarian species indicative of salinity-stratified polar water (Morley and Hays, 1979, 1983).

The 100,000-year SST rhythm dominates all latitudes of the North Atlantic (Fig. 2), in accordance with the overwhelming dominance of the same rhythm in the $\delta^{18}O$ signal and in northern hemisphere ice volume (Ruddiman and McIntyre, 1984). There is, however, no direct insolation forcing of the ice sheets at the 100,000-year period. The main insolation change at the 100,000-year rhythm is a modulation of the amplitude of the 23,000-year precessional signal. The apparent origin of the

Ruddiman, W. F., 1987, Northern oceans, *in* Ruddiman, W. F., and Wright, H. E., Jr., North America and adjacent oceans during the last deglaciation: Boulder, Colorado, Geological Society of America, The Geology of North America, v. K-3.

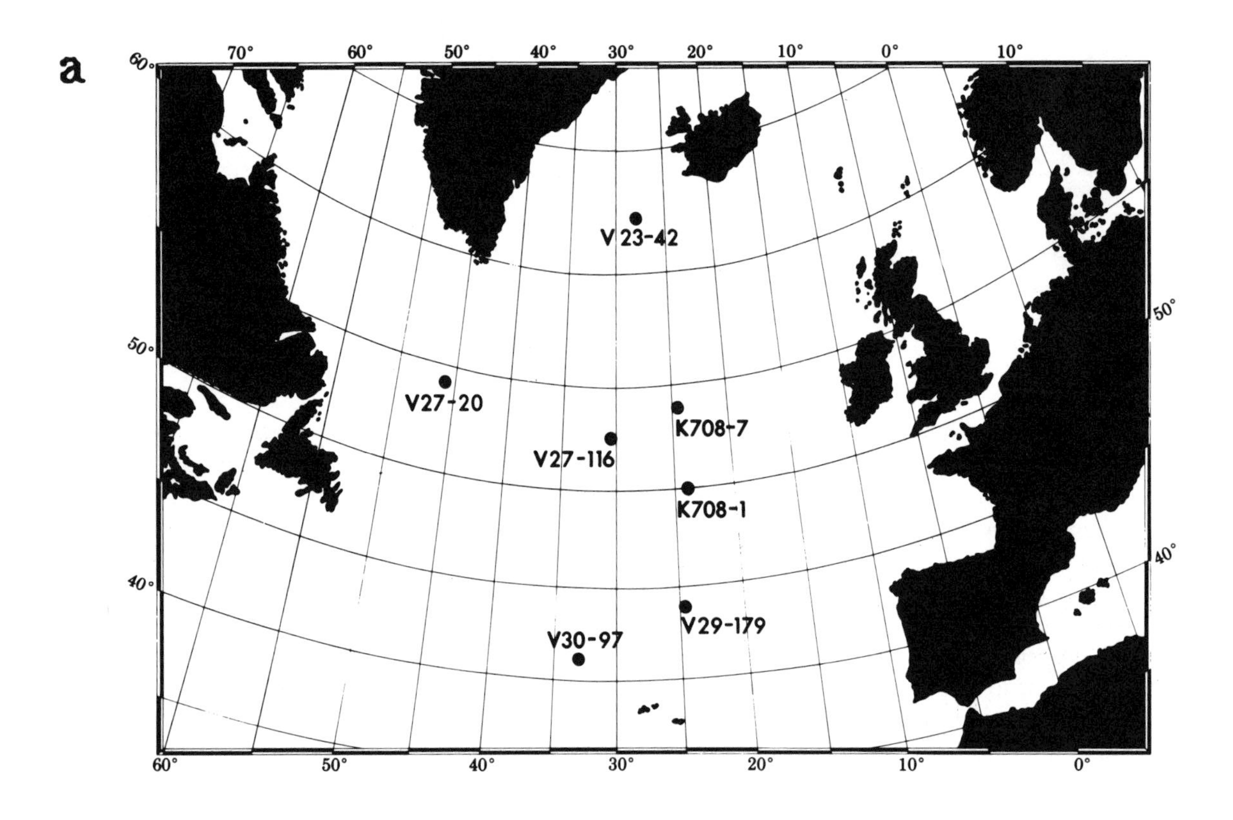

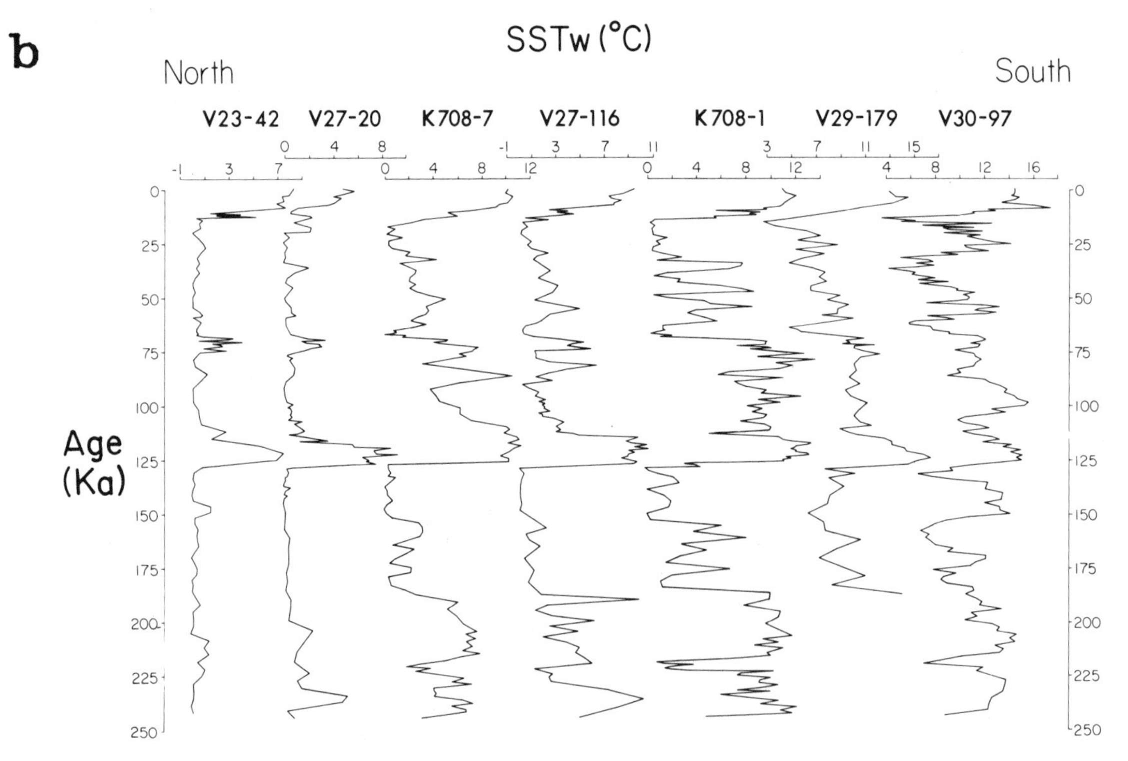

Figure 1. Changes in estimated winter sea-surface temperature (SST_W) in North Atlantic cores: (a) locations of cores; (b) north–south transect of SST trends during the last 250,000 years (after Ruddiman and McIntyre, 1984).

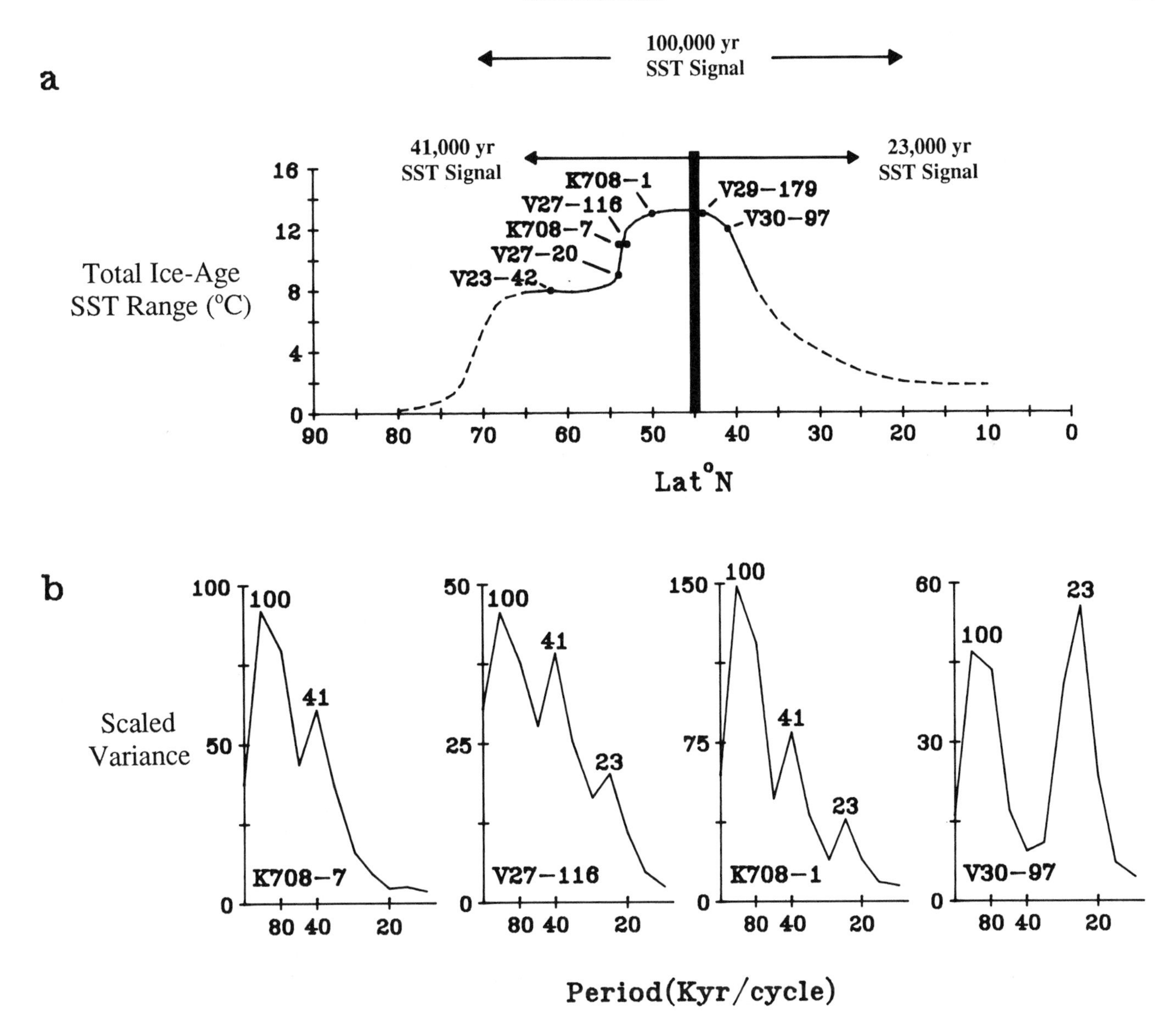

Figure 2. Concentrations of North Atlantic SST responses at orbital rhythms: (a) summary of latitudinal changes in SST response at orbital rhythms (after Ruddiman and McIntyre, 1984); (b) spectral analysis of four SST records from Figure 1.

100,000-year rhythm within the ice sheets (see Introduction, this volume) suggests that the North Atlantic SST response is controlled by the ice sheets. Because the two signals are in phase, the linkage must again be via the atmosphere (Fig. 3b).

General-circulation modeling experiments are the second line of evidence in support of the ice–atmosphere–ocean link. Manabe and Broccoli (1985) inserted large glacial-maximum ice sheets into a model world with boundary conditions otherwise identical to that today and ran a sensitivity test designed to isolate the unique impact of ice sheets on global ocean temperatures. The resulting effect on the North Atlantic north of 45°N was a large cooling and sea-ice advance that were closely comparable to the 18-ka North Atlantic reconstruction of CLIMAP (1981; Fig. 4). This experiment suggests that much of the very large glacial-maximum SST cooling of the North Atlantic north of 45–50°N was due to the thermal impact of the ice sheets on the atmosphere and subsequent transferral of the signal to the ocean. Because the general circulation model lacked ocean dynamics, the comparison is only a first approximation.

The specific way in which the ice controlled the North Atlantic ocean is suggested both by the Manabe and Broccoli (1985) experiment and by the recent reconstruction of atmospheric circulation in the 18-ka glacial world by Kutzbach and Guetter (1986). In both experiments, the large North American

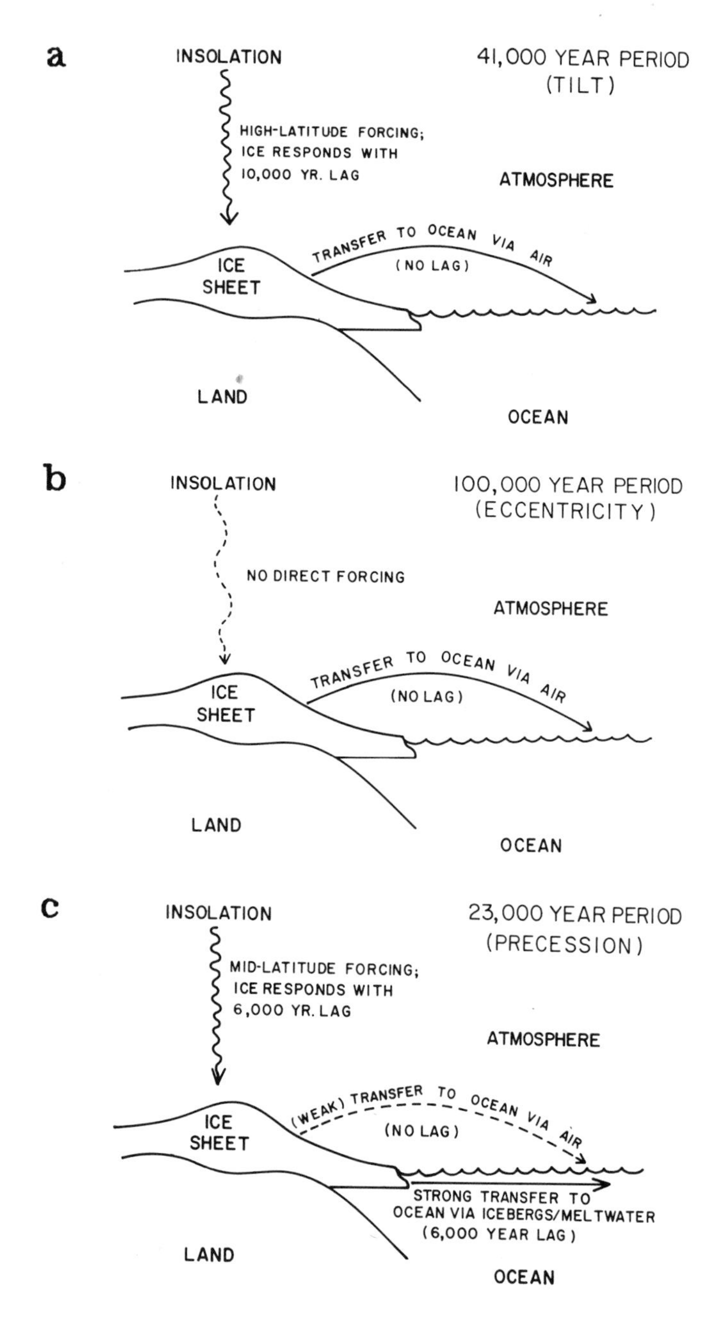

Figure 3. Schematic representation of insolation forcing of ice sheets, and subsequent ice-sheet forcing of the North Atlantic Ocean, for the three orbital components: (a) obliquity; (b) eccentricity; (c) precession. Transfer of signal from ice to ocean via the atmosphere occurs with no lag; transfer via the meltwater–iceberg flux introduces large lag in SST response.

ice sheet split the jet stream in two parts, one passing to the north and one to the south of the ice sheet. At the surface, very strong westerly winds blew along the northern flank of the ice sheet, turned to the southeast between the Laurentide and Greenland ice sheets, and blew out into the Labrador Sea and subpolar North Atlantic (Fig. 5). Cold winds also blew southward along the east coast of Greenland. These northerly winds extracted heat from the subpolar ocean, chilling its surface and freezing it at higher latitudes.

In summary, both geologic data and modeling results indicate that North American ice sheets chilled and seasonally froze the North Atlantic Ocean north of 45–50°N at the 41,000-year and 100,000-year rhythms and cooled the ocean southward into middle latitudes at the 100,000-year rhythm. These two signals were transferred from the ice sheets to the ocean via the atmosphere with little or no lag (Fig. 3a, b).

The explanation for the 23,000-year signal in mid-latitude North Atlantic SST is not quite as clear. As reviewed in the Introduction to this volume, precessional forcing has been concentrated at the 23,000-year period during the last 250,000 years of the Pleistocene (Berger, 1978). And there are variations in $\delta^{18}O$ at the 23,000-year period that, at least in part, record ice-volume changes (Hays and others, 1976; Imbrie and Imbrie, 1980; Imbrie and others, 1984). Geographically, the strength of precessional insolation forcing below 60–65°N suggests that mid-latitude ice, especially in North America, would respond at this rhythm (Ruddiman and McIntyre, 1981a). Thus insolation forcing of the ice sheets at this rhythm appears to be reasonably well demonstrated.

However, the 23,000-year SST signal lags far behind the 23,000-year $\delta^{18}O$ (ice-volume) signal. Ruddiman and McIntyre (1981a) initially calculated a lag of 5500 ±1500 years. A recent recalculation of this lag using newer time scales and methodology (J. Imbrie, personal communication, 1986) indicates a comparably large lag (5900 ±1500 years). A delay this long of the oceanic response behind the ice sheets does not appear to be consistent with direct transferral of signals from ice to ocean via the atmosphere.

Ruddiman and McIntyre (1981a) suggested two possible sources for this large lag. The first is meltwater/iceberg influxes from the mid-latitude margins of the ice sheets (Fig. 3c). In theory, this explanation could create a 5,750-year lag of SST behind ice volume, because the ocean could respond in phase with the negative first derivative of ice volume, that is, by lagging one fourth of the 23,000-year wavelength behind ice volume. Thus, for example, the coldest ocean might occur not at the time of largest ice sheets, but 5,750 years later, during the time of fastest net ice wastage (and conversely).

The second possible source suggested for the 23,000-year SST signal is advection from lower latitudes of the Atlantic. Ruddiman and McIntyre (1981a) noted that this 23,000-year SST signal need not have been driven by northern hemisphere ice sheets, but could have been an independent signal.

Several new lines of evidence favor, but do not prove, the

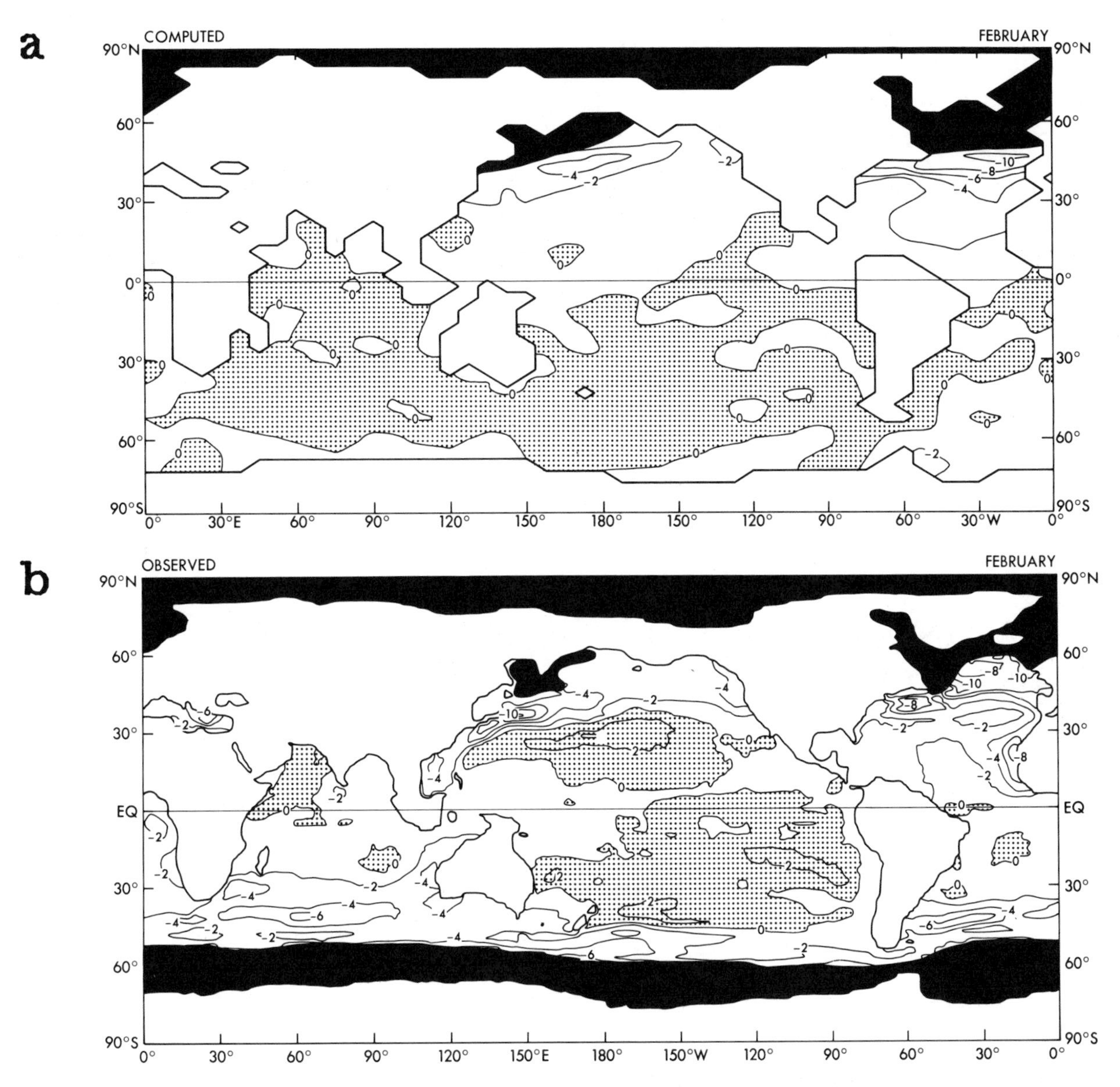

Figure 4. Results of general circulation model experiment designed to isolate cooling caused by glacial ice sheets (from Manabe and Broccoli, 1985): (a) computed decrease in February mean sea-surface temperature (°C) due to glacial-maximum ice sheets; (b) estimated decrease in February sea-surface temperatures at the last glacial maximum calculated by CLIMAP (1981). Note the close match to Figure 4a in the high-latitude oceans of the northern hemisphere. Black areas are sea ice, stippled areas are regions where glacial SST increased.

meltwater–iceberg explanation. First, a previous inconsistency in that explanation has been resolved. Because the 23,000-year precessional signal is not very strong in $\delta^{18}O$ records (Imbrie and others, 1984), Ruddiman and McIntyre (1981a) noted that it was difficult to understand how the large North Atlantic SST signal at this rhythm could be driven by the much smaller ice-sheet changes at the same tempo. However, two new lines of evidence reduce or eliminate this problem.

First, the rhythms of iceberg/meltwater flux to the oceans are not proportional to those of ice volume (A. Mix, personal communication, 1985). Instead, the fluxes of meltwater and icebergs at shorter periods must be enhanced relative to those at longer periods by an amount inversely proportional to their wavelengths. For example, the loss of a unit volume of ice at the 23,000-year period will generate a meltwater flux to the ocean more than 4 times stronger than the loss of the same amount of ice at the 100,000-year period. This occurs because the same volume of ice must be disintegrated four times faster within the

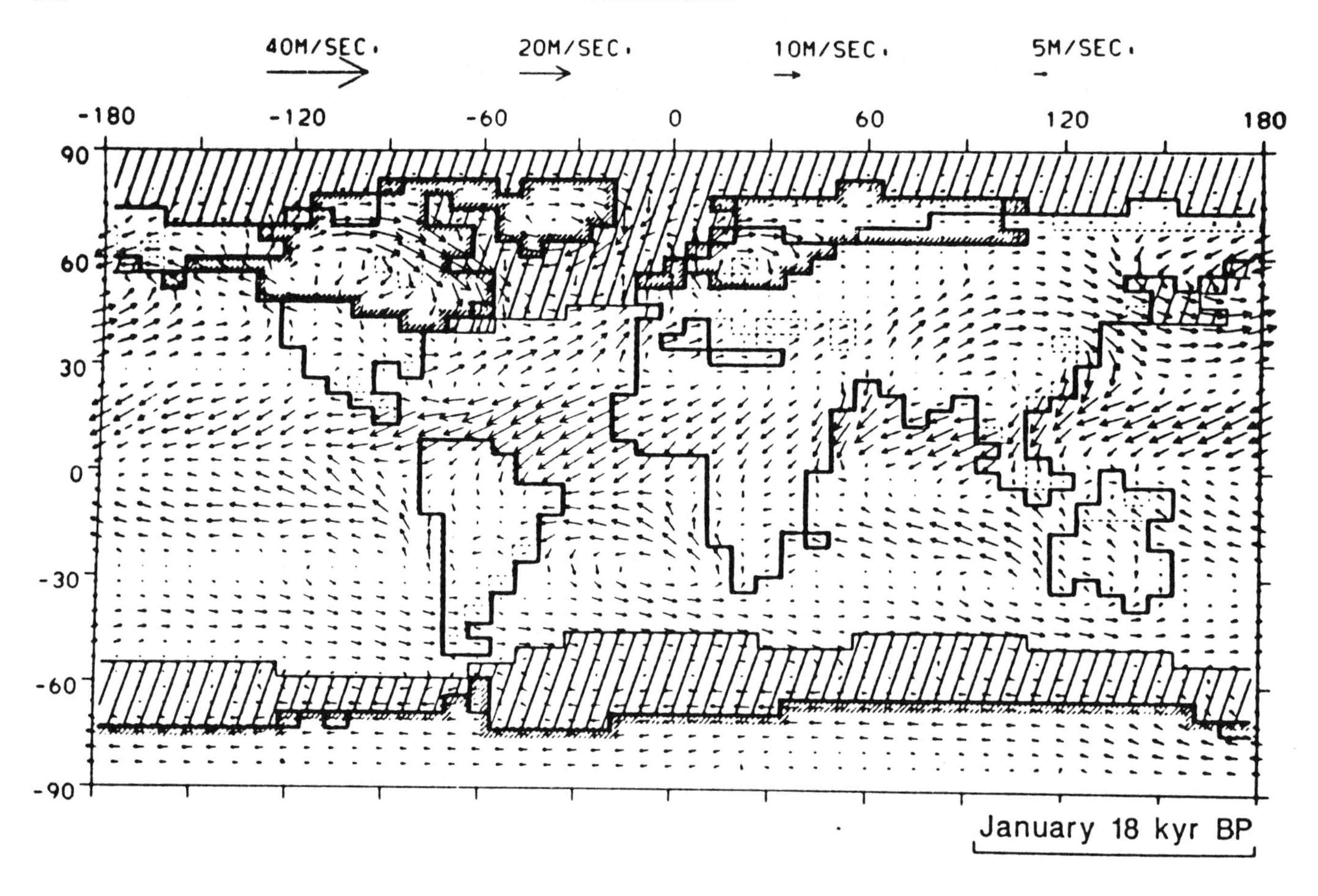

Figure 5. Wind strength and direction at the earth's surface in winter at the last glacial maximum, 18 ka (from Kutzbach and Guetter, 1986). Strong winds blow from northern flank of Laurentide ice sheet into the North Atlantic, cool the ocean, and freeze the sea surface in winter.

shorter 23,000-year cycle than within the longer 100,000-year cycle. As a result, although the spectra of ice volume for the last 780,000 years is dominated by the 100,000-year rhythm (see Fig. 3 in Introduction, this volume), the 23,000-year period is the strongest signal in the negative first derivative of $\delta^{18}O$, used here as an index of the net rate of meltwater discharge to the ocean (Fig. 6).

Second, because mid-latitude ice sheets were isotopically less negative than high-latitude ice, mid-latitude ice volume is probably underrepresented in the global $\delta^{18}O$ signal by as much as a factor of two (Mix and Ruddiman, 1984). Consequently, the 23,000-year fluctuations of the mid-latitude ice margins are probably also underrepresented. Thus the 23,000-year peak in the spectrum of Figure 6 is probably even larger relative to the other orbital rhythms. These two factors resolve the seeming inability of the relatively weak 23,000-year $\delta^{18}O$ signal to drive the strong 23,000-year North Atlantic SST signal.

New evidence from long-term records also points to a specific link between the 23,000-year SST signal and large northern hemisphere ice sheets. There was little or no 23,000-year power in North Atlantic SST variations during the Matuyama (Ruddiman and others, 1986), even through precessional insolation forcing was very strong. The Matuyama was a time of smaller high-latitude ice sheets oscillating mainly at 41,000 years, with little or no 23,000-year signal (Ruddiman and others, 1986). The initial appearance of 23,000-year power in North Atlantic SST spectra occurred at 0.9–0.8 Ma, coincident with the first isotopic and continental indications of larger ice-volume changes (Shackleton and Opdyke, 1973; Rogers and others, 1985). The coincidence of these changes suggests that the 23,000-year North Atlantic SST signal is causally linked to fluctuations of large mid-latitude ice sheets under 23,000-year insolation forcing.

In summary, northern hemisphere ice sheets large enough to reach middle latitudes (beyond about 55°N) appear to control the North Atlantic SST response at 23,000 years during the late Pleistocene. The nearly 6000-year lag of SST behind $\delta^{18}O$ is

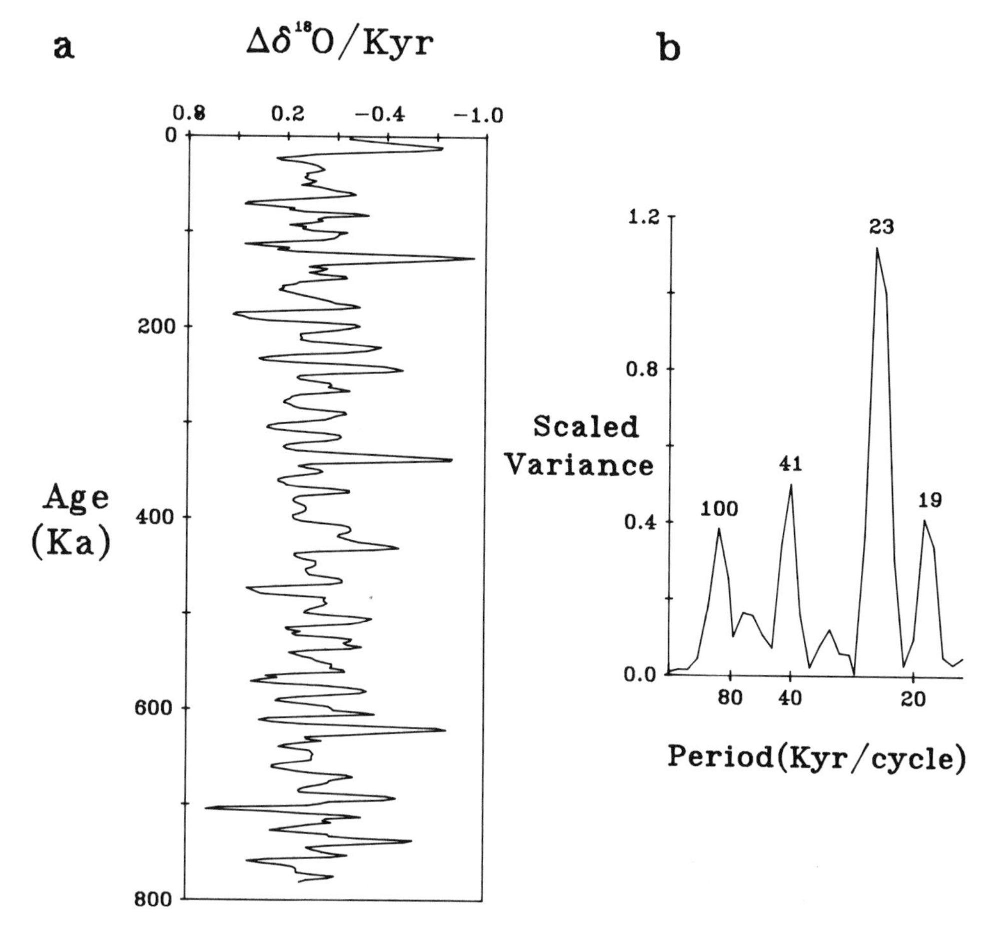

Figure 6. Evaluation of relative strengths of orbital rhythms in the meltwater runoff signal. (a) First derivative of the composite global $\delta^{18}O$ curve assembled by SPECMAP (Imbrie and others, 1984). This curve should approximate the rate of meltwater return to the oceans. Unit is rate of change in $\delta^{18}O$ per 1,000 years; (b) spectral analysis of the first derivative of $\delta^{18}O$ shown in Figure 6a. The 23,000-year rhythm is dominant in this index of meltwater flux.

consistent with the explanation that this control occurs via the metlwater–iceberg flux (Fig. 3c), but this explanation needs independent testing by modeling experiments.

North Pacific Ocean

Late Pleistocene climatic records from the northwest Pacific Ocean have also been investigated for responses at orbital rhythms (Pisias and Leinen, 1984; Morley and others, 1987). Because $CaCO_3$ in northwest Pacific sediments is generally too scarce to yield $\delta^{18}O$ records, time scales in this region are established by correlating percentage variations of the radiolarian *C. davisiana* with similar variations in Antarctic and North Atlantic sediments. Phase relationships versus ice volume ($\delta^{18}O$) thus cannot be directly calculated from records within cores and must be calculated indirectly via timescale comparisons.

Records of estimated sea-surface temperature and other climatic signals from the northwest Pacific Ocean contain significant variance at the periods of orbital variations. The phase lag of the dominant 41,000-year component of the *C. davisiana* record in the northwest Pacific with respect to obliquity is similar to that determined from analyses of cores with isotopic records from other high-latitude oceans (Morley and others, 1987). These results and others (Hays and others, 1976) indicate that the 41,000-year response of *C. davisiana* to variations in tilt is nearly synchronous in the high-latitude regions of both hemispheres and has a phasing nearly identical to that of the 41,000-year component of $\delta^{18}O$. Several other radiolarian assemblage factors also contain a 41,000-year component and are also in phase with the 41,000-year component of $\delta^{18}O$. This phasing suggests significant control of the high-latitude North Pacific response by northern hemisphere ice sheets via the atmosphere at the period of orbital obliquity (Fig. 3a).

Several climatic signals from the northwest Pacific Ocean

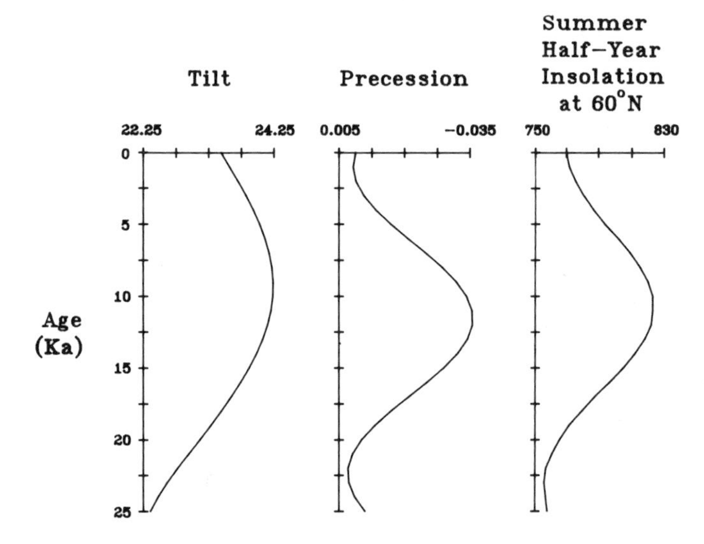

Figure 7. Variations of tilt (left), the precessional index e sin w (middle), and summer half-year insolation at 60°N over the last 25,000 years. Tilt and the precessional index have different wavelengths but come into phase around 10,000 years ago, producing the large summer insolation peak at 60°N.

also contain a prominent 100,000-year component that is in phase with the 100,000-year component of ice volume (Pisias and Leinen, 1984; Morley and others, 1987). As in the North Atlantic, the presence of this rhythm, and its in-phase relationship with $\delta^{18}O$, suggests a significant impact of the ice sheets on the North Pacific Ocean via the atmosphere at the period of orbital eccentricity (Fig. 3b).

The GCM ice-sheet experiment of Manabe and Broccoli (1985) supports the idea of a significant ice-sheet impact on the North Pacific (Fig. 4). The model predicts a 2–4°C February cooling of the ocean surface due to the presence of the ice sheets, and over most of the North Pacific this prediction roughly matches the February sea-surface cooling estimated by CLIMAP (1981) for the last glacial maximum.

The cooling of the North Pacific attributable to the ice sheets is, however, only 30 to 50 percent as large as that at comparable latitudes of the North Atlantic (Fig. 4b). In addition, areas such as the far northwest Pacific appear to have cooled considerably more than predicted by ice sheets alone. Nevertheless, the model results confirm that a significant portion of the long-term response of the North Pacific Ocean to orbital variations can be understood in terms of direct ice–ocean transfer via the atmosphere. The specific form of that atmospheric link is not known, but it presumably involves changes in strength and position of the Siberian High and Aleutian low (Sancetta and Sylvestri, 1986) as part of the Rossby-wave adjustments to the presence of the ice sheets.

LAST DEGLACIAL WARMING OF NORTHERN OCEANS

What do these long-term relationships specifically predict about the responses of the northern hemisphere oceans to orbital forcing during the last deglaciation? We could attempt to use the phase relationships between insolation forcing (Fig. 7) and SST response over the last 25,000 years to predict the response of the North Atlantic to insolation forcing during the last deglaciation. However, the above discussion shows that insolation does not directly drive the northern ocean SST responses. Instead, the ice sheets mediate between the insolation forcing and the SST response.

Therefore, to predict more exactly the pattern of deglacial warming in the northern oceans, we would have to incorporate knowledge of the ice sheets, particularly the volume of ice lying immediately up-wind over North America. That, of course, is precisely the critical unknown that this volume seeks to resolve. We thus invert this approach and turn to North Atlantic sediments for evidence of the ice-sheet behavior and signal transfer to the oceans. In the discussion that follows, all ages are expressed in ^{14}C years before present and are thus offset several hundred years later than calender ages (see Chapter 5, this volume).

High-Latitude North Atlantic, Norwegian Sea, and Labrador Sea

The basic pattern of polar-front retreat and surface-ocean warming across the high-latitude North Atlantic is mapped in Figure 8. From prior to 18 ka until 13 ka, a cold surface-water mass of low salinity extended southward into the subpolar North Atlantic to at least 45°N. This is shown by distinctive polar-water indicators in assemblages of foraminifera (Ruddiman and McIntyre, 1981b; Duplessy and others, 1981; Kellogg, 1984a) and radiolaria (Morley and Hays, 1979), by the absence of coccoliths (McIntyre and others, 1972), and by the abundant ice-rafted debris deposited far southward into the mid-latitude North Atlantic (Ruddiman, 1977).

The seasonal limits of sea ice at this time are not well known. The foraminiferal evidence permits, but does not prove, a winter sea-ice cover southward to near 50°N. The radiolarian fauna indicates surface waters with strong salinity stratification (Morley and Hays, 1983); such waters are likely to freeze in winter if air temperatures are very cold. The modeling results of Manabe and Broccoli (1985) predict very cold winds blowing from the ice sheets and the formation of extensive winter sea ice, even to latitudes south of the limits set by CLIMAP (1981).

There may, however, have been substantial year-to-year variability in the extent of winter sea ice in the North Atlantic, with large ice-free areas in some winters. Summer sea-ice limits during the glacial maximum were probably much farther north, as inferred by CLIMAP (1981) and suggested by modeling results from Manabe and Broccoli (1985). Cores from the Norwegian Sea also indicate cold, low-productivity waters and probably

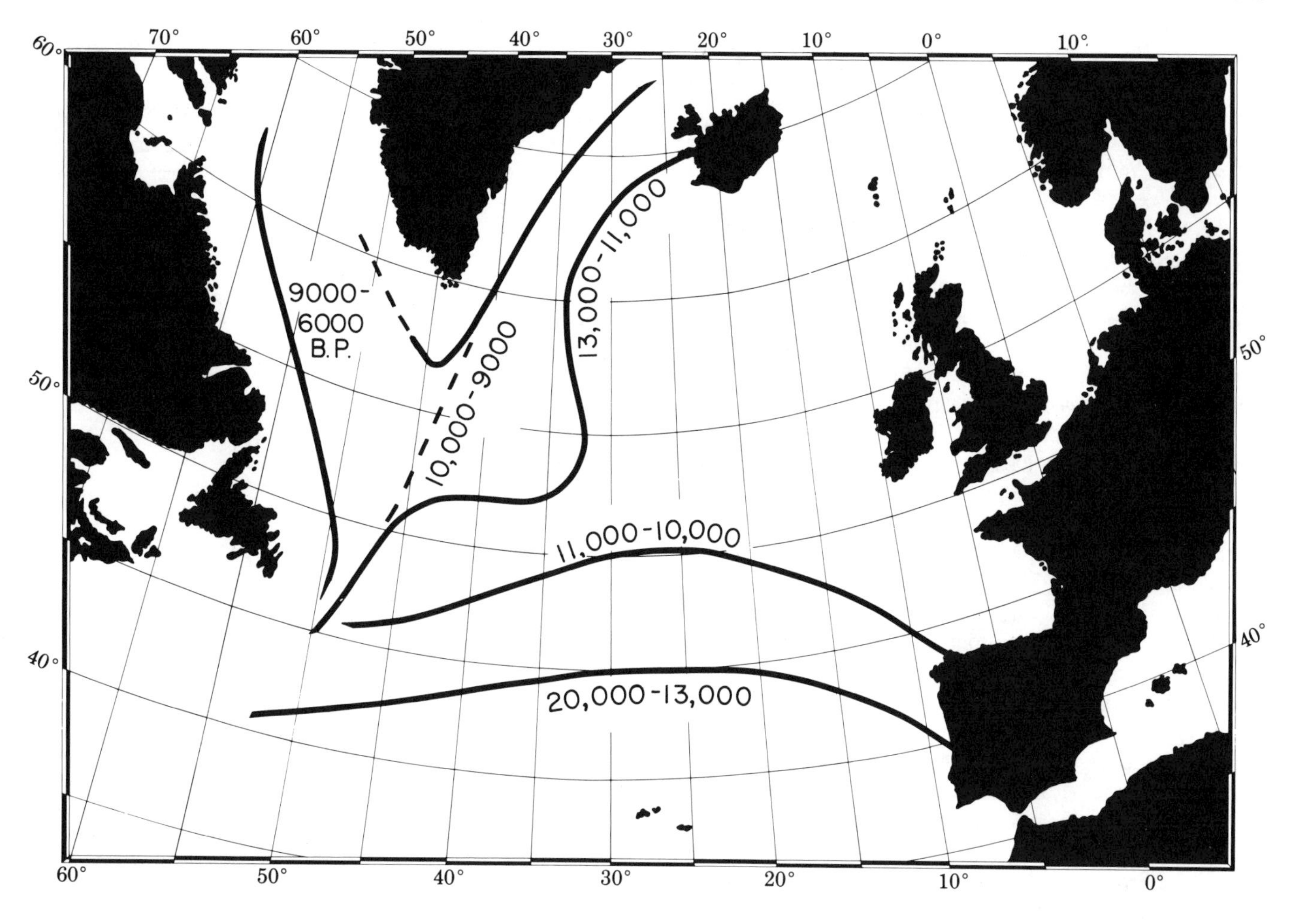

Figure 8. Map summary of deglacial polar front retreat (from Ruddiman and McIntyre, 1981b).

year-round ice cover during the glacial maximum (Kellogg and others, 1978; Jansen and others, 1983; Kellogg, 1984a).

There is less agreement about conditions in the Labrador Sea and Baffin Bay. Increased abundances of subpolar species of small size (63 to 150 μm) have been interpreted as indicating intervals of less frigid and relatively ice-free conditions in the Labrador Sea during portions of $\delta^{18}O$ stage 2 (Aksu and Piper, 1979; Fillon and Duplessy, 1980; Aksu, 1983, 1985; Fillon and Aksu, 1985). This evidence has been challenged as an artifact of complications involving fluctuating dissolution intensity on the sea floor and lack of standardized taxonomic procedures (Kellogg, 1984b, 1986). In this latter view, the evidence does not rule out year-round sea-ice cover in the Labrador Sea during $\delta^{18}O$ stage 2.

Two other factors suggest that the Labrador sea must have been very cold during stage 2 and ice covered at least in winter. The modeling results of Manabe and Broccoli (1985) and Kutzbach and Guetter (1986), which show strong cold winter winds blowing off the northern Laurentide ice sheet down Baffin Bay and into the Labrador Sea, are incompatible with any major surface-ocean warming or decrease of sea-ice cover for the winter season in the Labrador Sea during the stage 2 $\delta^{18}O$ maximum. In the Manabe and Broccoli (1985) results, Baffin Bay and the Labrador Sea are filled with sea ice in winter. At most, there may have been a narrow band of frigid ice-free waters along coasts where strong winds drove sea ice away from the coast. The very cold winter temperatures prevailing in the North Atlantic Ocean far to the south and east (Ruddiman and McIntyre, 1981b) also argue against significant amelioration of winter temperatures in the Labrador Sea during the peak of $\delta^{18}O$ stage 2.

It is more difficult to rule out ice-free conditions and small temperature increases during summer, when the northwesterly winds off the ice sheet were weaker. The modeling results of Manabe and Broccoli (1985) indicate summer sea-ice cover in the northern Labrador Sea and Baffin Bay, but they suggest that sea ice retreated in summer from its more extensive winter limits in the subpolar North Atlantic and southern Labrador Sea. It may be that the increased percentages of immature subpolar species during portions of $\delta^{18}O$ stage 2 represent summers in which sea ice melted farther back than normal, allowing a thin surface ocean layer to warm for a short interval. Such conditions were probably more likely during the early part of $\delta^{18}O$ stage 2 (Fillon and Aksu, 1985), when the mid-latitude North Atlantic was moderately warm (Ruddiman and McIntyre, 1981a). Later in stage 2, the North Atlantic was very cold and the Labrador Sea evidence also suggests severe cold (Fillon and Aksu, 1985). In any case, the above results are all consistent with a cold Labrador Sea with winter sea-ice cover throughout $\delta^{18}O$ stage 2.

Around 13 ka, the polar front abruptly swung back to the northwest (Fig. 8), and warmer, more saline water from the subtropical gyre flowed into the eastern and central parts of the North Atlantic (Ruddiman and others, 1977; Ruddiman and McIntyre, 1981b). The Denmark Straits area did not warm at this time, and the polar front remained to the south (Kellogg, 1984a). Some evidence suggests that the extent of sea ice in the southeastern Norwegian Sea was reduced significantly after 13 ka (Jansen and others, 1983) and possibly even earlier (Grousset and Duplessy, 1983).

Near 11 ka, the polar front readvanced to the south, and between 11 and 10 ka it reached a position only a few degrees north of the glacial-maximum position (Fig. 8). Very cold surface waters again filled the subpolar gyre, but productivity remained at intermediate levels, suggesting no extreme suppression of productivity by sea ice (Ruddiman and McIntyre, 1981b). Duplessy and others (1981) found a major cooling at this time as far southeast as the Bay of Biscay. To the north, the southeastern Norwegian Sea apparently remained largely ice-free during this interval (Jansen and others, 1983).

This was abruptly followed about 10 ka by a second major polar-front retreat to a position in the southeastern Labrador Sea. This shift sent temperate saline North Atlantic drift waters into the subpolar North Atlantic (except for the Labrador Sea) and into the east–central Norwegian Sea (Kellogg and others, 1978; Jansen and others, 1983; Kellogg, 1984a). The southwestern Norwegian Sea warmed slightly later, in the early Holocene (Jansen and others, 1983). The surface waters of the Labrador Sea and Baffin Bay finally reached interglacial warmth and productivity levels by 9 to 7 ka (Andrews, 1972; Miller and others, 1977; Aksu and Piper, 1979; Fillon and Duplessy, 1980).

This detailed pattern of response of the high-latitude North Atlantic has been compiled despite major complications from mixing of sediments on the sea floor by burrowing animals. Figure 9 summarizes the general effect of mixing on records from the eastern and central high-latitude North Atlantic as compiled by Ruddiman and McIntyre (1981b). The foraminiferal polar-water indicator *Neogloboquadrina pachyderma* (left-coiling) was abundant near the glacial maximum, whereas the temperate-water (subpolar/transitional) species were abundant only after 13 ka. For a few thousand years prior to 13 ka, planktonic foraminifera were very scarce or even absent, and most of the specimens found in those levels of the core today were mixed in from above and below. This indicates an interval of exceptionally low surface-water productivity and/or sea-ice cover just prior to 13 ka. A second short pulse of the polar species occurred between 11 and 10 ka, but warm-water forms then predominated throughout the Holocene (Fig. 9).

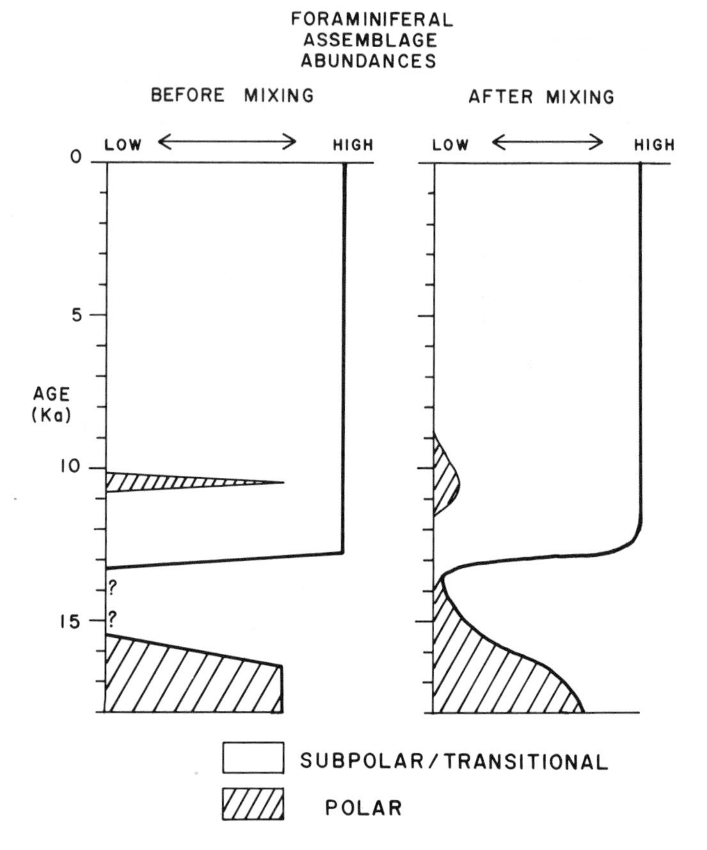

Figure 9. Schematic cartoon of first-order deglacial trends in typical core from the eastern and central subpolar North Atlantic north of 45°N. Left: reconstruction of the flux of cold (polar) water and warm (subpolar) water foraminifera to the sea floor, with little or no flux during interval immediately preceding 13 ka. Reappearance of cold-water assemblage at 11 to 10 ka marks the sea-surface cooling during the Younger Dryas. Right: the record preserved in typical North Atlantic core after mixing of sediments at the sea floor. Cores with high deposition rate preserve records closer to that shown at left (Ruddiman and others, 1977; Duplessy and others, 1981). No attempt is made to show smaller-scale faunal changes during the Holocene.

The bioturbation modeling efforts, as well as records from cores with high deposition rates, suggest that the major polar-front movements mapped in Figure 8 occurred in at most 1,000 to 2,000 years, and could even have been instantaneous (Ruddiman and others, 1977; Ruddiman and McIntyre, 1981b; Duplessy and others, 1981). If these changes are linked to isotopic, chemical, and physical transitions recorded in Greenland ice cores and European lakes (see Chapter 5, this volume), they probably occurred in less than 100 years. In summary, the warming at 13 ka, the cooling around 11 ka, and the warming at 10 ka

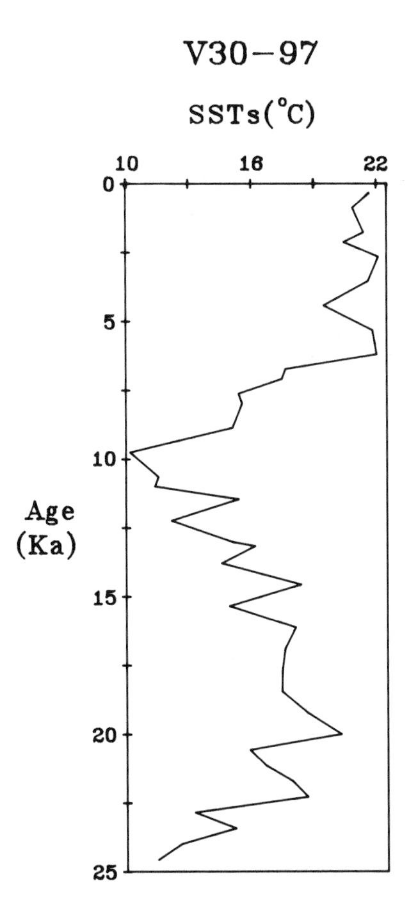

Figure 10. Estimated deglacial summer sea-surface temperature in core V30-97 from northern subtropical latitudes south of the polar front (41° 00N; 32° 50W).

are the most distinctive changes, both for their speed and geographic scope.

Recently, Mangerud and others (1984) suggested a revised age for a zone of dispersed silicic ash (zone "1" of Ruddiman and Glover, 1972) dated at 9.8 ka by Ruddiman and McIntyre (1981b) and Duplessy and others (1981). The new estimate, based on lake records from Norway, indicates an age of 10.6 ka for airfall deposition of the ash. This revised age does not, however, force a revision of the deglacial polar-front retreat pattern (Fig. 8), because the ash zone was not used in the final compilation of that story. Ruddiman and McIntyre (1981b) pinned the basic chronology of polar-front movements of the distinctive cold event marked by the polar-front readvance and equated to the European "Younger Dryas." This relationship still appears valid.

In any case, the basic deglacial polar-front pattern defined by Ruddiman and McIntyre (1981b) is supported by other paleoceanographic studies in the North Atlantic (Duplessy and others, 1981), by interpretations of pollen sequences and other climatic indicators in the maritime parts of Europe (Watts, 198), and by the synchronous Younger Dryas ice-sheet readvances in Norway (Mangerud, 1980). It also matches records of air-temperature changes from ice cores (see Chapter 5, this volume). Recent particle-accelerator dating of planktonic foraminifera confirms the basic chronology shown in Figures 8 and 9 (W. Broecker, personal communication, 1987).

Mid-Latitude North Atlantic

Polar water masses did not reach south of 45°N; there, the North Atlantic responded entirely differently during the last deglaciation. The basic pattern, shown in Figure 10, was a cooling that began around 20 ka, deepened through the glacial maximum, and reached peak intensity around 13 to 9 ka. This was followed by a moderately rapid warming to modern temperature levels by the early Holocene (7 ka). This pattern differs from that in the north in that the coldest temperatures occur not at the glacial maximum or just after it, but on the deglaciation. This is compatible with the long-term phase relationship in which coldest SST values occur during fastest deglaciations (Fig. 3c).

Farther to the south, near the east coast of the United States, the pattern is different still. Balsam (1981) dated the major warming of the western subtropical Atlantic at 14 to 6 ka, with the fastest warming between 14 and 10 ka. Problems with ^{14}C dating and dissolution caution against any more detailed interpretations in this region.

Gulf of Mexico

The most detailed study of the deglacial paleoceanography of the Gulf of Mexico (Brunner, 1982) again found the sea-surface warming broadly correlative with ice disintegration. In this instance, the very wide sampling intervals and the lack of ^{14}C dates prohibit more detailed comments about the deglacial response.

Another deglacial response registered in Gulf of Mexico sediments is the lightening of $\delta^{18}O$ values by deglacial meltwater. There is a range of estimates for the age of maximum runoff defined in this way by $\delta^{18}O$ trends. Some estimates place the maximum runoff rates relatively early in the deglaciation: 16.5–12.5 ka in Kennett and Shackleton (1975) and Leventer and others (1982), and 15–11 ka in Emiliani and others (1975).

At the other extreme, Berger (1985a) proposed a chronology in which the low-$\delta^{18}O$ spike in the Gulf of Mexico occurred at 9.5 ka. This estimate, however, post-dates the interval of maximum glacial drainage to the Gulf of Mexico defined from continental work (see Chapter 3, this volume).

Recent particle accelerator dating (W. Broecker, personal communication, 1987) suggests that the large meltwater influxes to the Gulf of Mexico began around 14 ka (later than previously published), peaked sharply around 12.5 ka (in accordance with earlier estimates), and subsided sharply after that time. This new dating confirms the earlier chronologies in placing the Gulf of Mexico meltwater influx on the first rapid change in the global $\delta^{18}O$ curve (termination "1a" of Duplessy and others, 1981; dated at 14–12 ka by Mix and Ruddiman, 1985).

North Pacific and Bering Sea

Evidence of deglacial circulation changes in the Pacific Ocean and Bering Sea is less detailed than that in the Atlantic. As noted earlier, dissolution of the $CaCO_3$ fraction in most areas of the Pacific eliminates much of the $CaCO_3$ fraction needed for isotopic stratigraphy and ^{14}C dating and also reduces sedimentation rates and hence stratigraphic resolution. This problem is particularly acute in the eastern North Pacific adjacent to North America. Studies of the northern Pacific Ocean and Bering Sea rely primarily on siliceous diatoms and radiolarians and on dating of the acid-insoluble organic carbon fraction or on stratigraphies based on the radiolarian *C. davisiana.*

In general, available deglacial records in the central North Pacific Ocean and Bering Sea indicate that surface waters were cold and highly stratified at the glacial maximum, with seasonality increased (Sancetta, 1979; Morley and Hays, 1983; Heusser and Morley, 1985). At high latitudes, the sea-ice cover was prolonged through much of the year, with only a brief cool summer season of relatively intense productivity.

In both the North Pacific and Bering Sea, productivity and temperature both increased somewhere within the interval between 18 and 7 ka, with the productivity increase appearing to slightly precede the warming (Sancetta, 1979, 1983; Sancetta personal communication, 1986). However, the few available ^{14}C dates do not closely constrain the timing of these changes. Other detailed deglacial records still more remote from the Laurentide ice sheet indicate deglacial warmings between 15 and 12 ka in the far northwestern Pacific (Heusser and Morley, 1985) and between 18 and 9.5 ka in the Sea of Japan (Morley and others, 1987).

With the presently limited geographic coverage of the North Pacific and Bering Sea, as well as the limited dating resolution, it is difficult to draw any conclusions beyond the fact that the sea-surface warmings and productivity increases in those regions are, to first order, correlative with the disintegration of the ice sheets.

LAST DEGLACIAL ICE-SHEET FORCING OF NORTHERN OCEANS

The northern ocean responses during the last deglaciation are all linked to ice-sheet disintegration: the various sea-surface warmings, sea-ice retreats, and productivity increases are broadly synchronous with the retreat of the ice sheets between 16 and 6 ka. But what is the actual nature of the linkage between ice and ocean? There are two alternatives, both already discussed in connection with the long-term variations in the North Atlantic (Fig. 3).

Ice-Sheet Forcing of the Ocean via the Atmosphere

The first linkage is via the atmosphere. The cooling of the ice-age atmosphere by the ice sheets led to heat extraction from the thin, warm surface-ocean layer at high latitudes (Manabe and Broccoli, 1985). This ice-to-atmosphere-to-ocean transfer occurred with no significant lag. In far northern oceans, the increased extraction of heat during glaciations led to the persistence of sea-ice cover through most or all of the year.

Atmospheric transfer appears to be the major explanation for the first-order synchroneity of most of the oceanic responses with those of the ice sheets. The extent of sea-ice cover and the amount of sea-surface cooling predicted by the Manabe and Broccoli (1985) model results (Fig. 4a) match fairly closely those estimated by CLIMAP (1981) in each of the regions discussed (Fig. 4b). Other factors (e.g., reduced CO_2) no doubt contributed to the intensity of the glacial cooling across the globe, but direct transfer of the ice-sheet impact to the ocean via the atmosphere appears to predominate at high northern latitudes. Even the very late deglacial warming of the Labrador Sea is broadly consistent with this explanation: the ocean immediately adjacent to the last remnants of North American ice should have been the last to warm.

This raises a crucial point about the North Atlantic polar-front response during the deglaciation. Because the long-term behavior of the high-latitude North Atlantic is driven by the ice sheets with no lag (Fig. 3a, b), it is worth examining whether the same cause-and-effect relationship could also be true for the more abrupt movements in the last deglacial polar-front response shown in Figure 8.

Such an explanation appears plausible for the two major ocean warmings and polar front retreats at 13 and 10 ka. These large oceanic changes are synchronous with the fastest steps in the best-dated $\delta^{18}O$ (ice-volume) proxies (Mix and Ruddiman, 1985; see Chapter 6, this volume). The $\delta^{18}O$ data suggest that global ice volume underwent large and rapid losses at these times, and there is reason to expect that North American ice followed the same trend. As noted by Mix (see Chapter 6, this volume), North American ice volume represents an estimated 60 percent of the glacial excess in global ice volume relative to today, and it probably influences via direct sea-level linkages the estimated 20 percent of the total global variation contributed by the marine-based west Antarctic ice sheet (Hollin, 1962; Denton and Hughes, 1983). Thus it is difficult to avoid the conclusion that Laurentide ice participated significantly in the abrupt change of $\delta^{18}O$ values 33 to 40 percent of the way toward interglacial levels at about 13 ka. Isotopic evidence from the Gulf of Mexico cited earlier also suggests an early phase of rapid melting around 13 ka.

Individual $\delta^{18}O$ curves show a second rapid change toward interglacial values at and after 10 ka (termination "1b" of Duplessy and others, 1981), with the amplitude of change roughly comparable to that of the earlier step. Again, Laurentide ice should have been a significant component of this change.

Areal retreat of North American ice was not, however, unusually rapid during either of these two intervals (Prest, 1969; see Chapter 2, this volume). The seeming contradiction between the abrupt reductions of ice volume and the limited areal shrinkage of ice sheets could be reconciled if there were large and rapid decreases in Laurentide ice height at 13 and 10 ka (Mix and Ruddiman, 1985). Large decreases in ice-sheet height over North America would diminish the strength of cold winds sent out to the North Atlantic and thus allow the ocean surface to warm. Thus the retreats of the North Atlantic polar front at 13 and 10 ka are consistent with the inferrence of large ice-elevation decreases over North America during intervals of modest shrinkage of ice area.

An important and perplexing question remains: what caused the Younger Dryas ocean cooling and polar-front readvance around 11 to 10.5 ka? Calling on ice-sheet height as the explanation would apparently require that the Laurentide ice sheet regained much of its glacial-maximum height in order to once again send cold strong winds from its northern flanks out across the North Atlantic.

There is no evidence in the global $\delta^{18}O$ signal, however, for any significant increase in ice volume during the Younger Dryas. Many $\delta^{18}O$ signals show a leveling out of isotopic values from 13 or 12.5 ka to about 10 ka (see Chapter 6, this volume). Although existing ice sheets readvanced in parts of southern and western Norway and ice reappeared in the Scottish highlands, these are very small changes relative to the full scale of glacial-interglacial variations. The North American record generally indicates at most a stabilization of ice-sheet areal limits at this time (Prest, 1969; see Chapter 2, this volume). In short, there is neither isotopic nor glacial geological evidence favoring the growth of a large volume of ice to explain the Younger Dryas polar-front readvance in the North Atlantic.

One possible explanation for the Younger Dryas polar-front response in the North Atlantic involves changes in elevation of Laurentide ice without any change in ice volume. Because of the delay in mantle rebound (see Introduction, this volume; see Chapter 8, this volume), the surface of the Laurentide ice sheet could have been raised to a higher elevation during the apparent pause in deglaciation between the end of isotopic step "1a" (14–12.5 ka) and the start of isotopic step "1b" (10 ka).

The maximum impact of this effect can be evaluated with the following assumptions: (1) a glacial-maximum Laurentide ice sheet 3900 m thick and 3000 m high at the center, using the assumption of 1:4 isostatic compensation of Denton and Hughes (1983); and (2) a rapid 33 percent loss of Laurentide ice volume between 14 and 12.5 ka, proportional to the fractional change of the composite global $\delta^{18}O$ curve indicated in Chapter 6 (this volume). This would result in a 1300-m loss of thickness at the center of the Laurentide ice sheet. If uncompensated by rebound, this would also mean a 1300-m loss in elevation of the ice surface.

Approximately 25 percent (or 325 m) of the 1300-m loss in ice elevation resulting from this initial decrease in ice volume would be compensated over time by bedrock rebound, depending on assumptions about the visco-elastic bedrock behavior. Assuming that 23 percent of the response is elastic (Birchfield and others, 1981), roughly 75 m of the lost elevation would be restored immediately upon ice unloading (by 12.5 ka). The remaining 77 percent (250 m) of uplift would occur more slowly. If the half-response time of the viscous component were as short as 1760 years (Peltier, 1982), another 150 to 175 m of rebound could occur during the apparent pause in deglaciation between 12.5 ka and 10.5 ka (the time of the Younger Dryas).

Combining all these assumptions, the ice surface would drop from 3,000 m late in the glacial maximum (14 ka) to 1,830 m following the early phase of rapid deglaciation (12.5 ka), and then rise to nearly 2,000 m during the pause in deglaciation ending at 10.5 ka. The 2,000-m ice elevation in the Younger Dryas would represent 66 percent of the glacial-maximum height.

The obvious question is whether such a modest increase in ice elevation can explain the large polar-front response. The calculated 150- to 175-m increase in ice elevation due to rebound from 12.5 to 10.5 ka is only 5 percent of the glacial-interglacial range, whereas the polar-front migration during that period (Fig. 8) covers 60–70 percent of the glacial-interglacial areal range. In addition, several of the assumptions made above maximize the possible amount of rebound between 12.5 and 10.5 ka. There would thus have to be a very strongly nonlinear relationship between ice elevation and the thermal impact on the North Atlantic. This might occur if there were some threshold of ice elevation around 2,000 m at which the Laurentide ice sheet impacts the jet-stream flow sufficiently to force a large-scale rearrangement of lower atmospheric circulation. GCM sensitivity tests are needed to address this question.

There are other possible explanations for the Younger Dryas polar-front response. It is possible that the central or eastern parts of the Laurentide ice sheet did increase in thickness somewhat during the 12.5–10.5 ka interval, but only if this accumulation was balanced by disintegration on the southern and western margins or in other ice sheets. Clark (cited in Ruddiman and Duplessy, 1985) found that the postglacial sea-level rise around Hudson Bay could be modeled by a major thinning of the ice sheet prior to 13 ka, a significant thickening between 13 and 11 ka, and a subsequent thinning after 10 ka. Most other areas of North America were modeled best by single-step deglaciations. This interesting result has not been confirmed by other models (see for example Chapter 2, this volume).

High-frequency variations in northern-hemisphere climate, driven by conjectured changes in solar emissivity at the 2500-year cycle (Denton and Karlén, 1973) are a possible contributor to the Younger Dryas oscillation. These variations might also contribute to the warmings associated with the prominent deglacial steps at 13 and 10 ka.

Other explanations for the Younger Dryas oscillation previously suggested are: (1) break-up of ice shelves in the Arctic Ocean and massive outflow of tabular bergs to the North Atlantic (Mercer, 1969); and (2) reduced atmospheric CO_2 levels due to slower oceanic mixing during deglaciation (Berger, 1985b; Broecker and others, 1985). The latter explanation appears to be contradicted by recent work by Andree and others (1986) showing no change in deglacial ocean circulation but slower overturn during the last glacial maximum. In any case, the Younger Dryas still remains a mystery.

Whatever the cause of the Younger Dryas, the succeeding rapid volumetric and areal deglaciation at and after 10 ka appears to have quickly dropped the elevation of the Laurentide ice sheet to levels at which it could no longer effect the east–central North Atlantic Ocean via the atmosphere. Some effect of the remaining ice probably continued to be felt in the westernmost North Atlantic and Labrador Sea.

In summary, the concept of long-term control of the high-latitude North Atlantic by the size of the Laurentide ice sheet can be applied with partial success to the pattern of North Atlantic polar-front retreat during the last deglaciation. Major retreats of the polar front at 13 and 10 ka can be interpreted as evidence of significant decreases in size of Laurentide ice, in agreement with times of fastest decrease in global $\delta^{18}O$ (see Chapter 6, this volume). In the absence of large-scale areal retreat of Laurentide ice at these times, either significant thinning of the Laurentide ice sheet occurred at 13 and 10 ka (Mix and Ruddiman, 1985), or other factors overwhelmed the basic ice-sheet control of the North Atlantic polar front.

Ice-Sheet Forcing of the Ocean Via Meltwater–Iceberg Fluxes

A second possible kind of linkage between the ice sheets and the North Atlantic is via the meltwater–iceberg flux to the mid-latitude ocean. We have already shown that at long time scales the 23,000-year SST response in the mid-latitude North Atlantic has a phasing consistent with this kind of ice–ocean linkage (Fig. 3c).

What is the evidence from the most recent deglaciation? The most recent SST minimum in core V30-97 from the northern subtropical North Atlantic occurs at approximately 13 to 9 ka (Fig. 10). Consistent with the long-term phase relationships, this interval coincides with the fastest rates of deglaciation, and the SST minimum lags 5,000 to 6,000 years behind the $\delta^{18}O$ maximum at 19 to 14 ka (see Chapter 6, this volume). Because of sediment mixing, however, it is not possible to ascertain whether there are separate SST minima at 13 and 10 ka coincident with the intervals in which Laurentide ice may have been melting most rapidly (Fig. 10).

A second potential index of the meltwater–iceberg impact, the rate of deposition of ice-rafted detritus, was high in sediments across most of the eastern and central North Atlantic prior to 13 ka and then fell to very low levels (Ruddiman, 1977; Ruddiman and McIntyre, 1981b). Large influxes of ice-rafted sand at and prior to 13 ka are consistent with evidence of extensive calving in the St. Lawrence embayment at that time (Borns, 1973).

Cessation of ice rafting over much of the North Atlantic at 13 ka does not mean that calving was no longer an important mode of deglaciation for the Laurentide ice sheet. Active calving may have continued later in the deglaciation, but with the locus of deposition of the ice-rafted debris shifted into the Labrador Sea. Over longer intervals of time, deposition of ice-rafted debris in the Labrador Sea appears to be partially out of phase with deposition in the central North Atlantic (Fillon and others, 1981; Fillon, 1985). Unfortunately, there is inadequate chronologic control and core coverage to determine fully the late-deglacial influx of ice-rafted detritus to the Labrador Sea. Ice-rafting evidence thus argues for abundant iceberg influxes early in the deglaciation, but is inconclusive about the later stages.

A third possible indicator of the meltwater–iceberg influx is the effect of glacial meltwater in suppressing the vertical overturn of ocean surface waters that leads to renewal of nutrients from below. In an examination of the last deglaciation in North Atlantic cores north of 50°N, Ruddiman and McIntyre (1981b) found intervals barren (or nearly barren) of foraminifera dated at, and prior to, 13 ka (Fig. 9). They ascribed these barren zones to suppression of foraminiferal productivity by a very large influx of meltwater and icebergs from 16 to 13 ka. They further estimated that this influx could represent a loss of as much as half of the glacial-maximum ice volume from North America. The timing and amplitude of this estimate roughly matched the change of $\delta^{18}O$ values nearly half way from glacial to interglacial values at 15 to 13 ka found in core Ch73-139 by Duplessy and others (1981).

This interpretation appears to require substantial modification for two reasons. First, the chronology of Duplessy and others (1981) does not appear to represent the global signal. Second, the assumed link between meltwater–iceberg influx and degree of productivity suppression may have been too simplistic.

Extensive $\delta^{18}O$ evidence compiled by Mix and Ruddiman (see Chapter 6, this volume) favors an age for the first major deglacial $\delta^{18}O$ change of 14–12.5 ka, later by 500 to 1000 years than that of Duplessy and others (1981). This shifts the apparent time of fastest ice-sheet disintegration to a period somewhat later than that originally estimated for the barren zone, requiring a reassessment of its cause.

One possibility is that the original explanation for the barren zone is correct, but the earlier estimates of its age were wrong. Thus the barren zone may still be synchronous with the rapid $\delta^{18}O$ rise, but may date at 14–12.5 ka not at 16–13 ka. Radiocarbon ages of bulk samples in and near the barren zone in North Atlantic cores are probably contaminated by ice-rafted carbonate detritus; even the foraminifera found in this interval were mainly deposited higher or lower in the sediment column and then mixed in by bioturbation (Fig. 9). Because of these problems, the age of initiation of the barren zone is poorly constrained and could have begun late enough (15 ka or even 14 ka) to be in phase with the first rise of the composite $\delta^{18}O$ curve (see Chapter 6, this volume). The end of the barren zone at about 13 ka and the end of the first rapid rise of $\delta^{18}O$ at 12.5 ka are effectively synchronous within the dating uncertainty of each signal.

The second possible explanation of the barren zone is that the published age is still correct, but that the explanation was wrong. Ruddiman and McIntyre (1981b) tacitly assumed that a unit influx of meltwater would lead to a unit suppression of foraminiferal productivity, but this need not be so. One possibility is that lower foraminiferal productivity also reflects the regime of increased salinity stratification in the high-latitude North Atlantic polar gyre during glacial times. Such a water mass, continuing in existence late into the glacial maximum, may have registered the first modest increases of deglacial meltwater influx very strongly and led to an overestimation of the amount of iceberg–meltwater influx prior to 13 ka (Ruddiman and McIntyre, 1981b).

Either of these interpretations requires abandonment of the more extreme form of a provocative conclusion reached by Ruddiman and McIntyre (1981b): there is no compelling evidence that 50 percent of the North American ice sheet had disappeared by 13 ka. And, by extension, there is no reason to call on the drastic decrease of Laurentide ice elevation between 16 and 13 ka that this amount of volumetric change would require.

The lines of evidence summarized above (mid-latitude SST, ice rafting, foraminiferal productivity suppression) all lead to the interpretation that the interval of strongest metlwater–iceberg influxes to the North Atlantic began shortly before 13 ka and continued until 9 ka. To a first approximation, this interval coincides with, or slightly precedes, the time of fastest areal retreat of the ice sheet. But it is also consistent with the scenario proposed in the previous section in which episodes of abrupt thinning of the Laurentide ice sheet allowed retreat of the polar front around 13 and 10 ka. Thus, in effect, thinning of the Laurentide ice sheet early in the deglaciation (14–12.5 ka) is still suggested, but not so early and not to so large a degree as suggested by Ruddiman and McIntyre (1981b).

IMPLICATIONS FOR ICE-SHEET FEEDBACK

There are three ways in which the ocean could provide significant feedback to northern hemisphere ice sheets during the last deglaciation (see Introduction, this volume): (1) through sea-level change and calving, (2) via the moisture flux, and (3) by effects on CO_2. We review here evidence bearing on the importance of the first two factors. Chapter 5 (this volume) addresses the impact of CO_2.

Sea Level and Calving

Denton and Hughes (1983) proposed that the ocean can accelerate disintegration of ice sheets through a link in which water from the melting ice returns to the ocean and raises sea level, which attacks vulnerable portions of the ice sheets either by removing fringing ice shelves or by lifting the shelves off bedrock pinning points. They proposed that this causes additional loss of ice from broad interior regions of the ice sheet because of accelerated flow in ice streams that deliver ice toward the calving margins. It is not yet possible to constrain the importance of this effect to the last deglaciation in ice models; many key parameters that would have to be specified in such a model are poorly understood in the available glacial geologic record.

The glacial record on land does indicate that a calving bay developed around and just prior to 13 ka in the Gulf of St. Lawrence, isolating New England ice from the main Laurentide ice to the north (Borns, 1973). This argues for thinning of ice by calving on at least the southeastern sector of the Laurentide ice sheet. The synchroneity of this episode with the first rapid $\delta^{18}O$ rise, with the barren zone and ice-rafting influx in the high-latitude North Atlantic, and with the SST minimum in the mid-latitude North Atlantic are all consistent with the conclusion that calving in this region provided significant feedback to ice-sheet decay early in the last deglaciation.

How important this calving episode was in the larger scheme of the deglaciation is not yet known. One critical unresolved question is whether the calving on the southeastern margin of the ice sheet affected a broad interior portion of the ice sheet via the ice-stream downdraw mechanism.

Hudson Strait must be the real key to the importance of the calving process during deglaciation, because it is potentially the largest marine outlet for the Laurentide ice sheet and because it leads into the very center of the ice sheet. Because no sediment from the early part of the deglacial interval (14–12.5 ka) has been dated in Hudson Strait (see Chapter 2, this volume), the rates of calving through Hudson Strait during the period of initial $\delta^{18}O$ rise unfortunately are unknown.

Final clearance of ice from most of Hudson Bay occurred rapidly around 8 to 7.5 ka (Prest, 1969; Andrews and Falconer, 1969), but the calving history prior to that date is largely unconstrained. It is likely that final penetration of an open marine embayment into the center of the former North American ice sheet was preceded by a longer interval of active calving, but the duration of that interval is not known. An embayment appears to have developed part of the way into Hudson Strait by 9.5 ka (see Chapter 2, this volume). This raises the possibility that the rapid rise of $\delta^{18}O$ values that began around 10 ka and continued until 8 or 7 ka primarily reflects iceberg calving to the ocean through Hudson Strait, combined with calving into proglacial lakes that drained to the ocean (see Chapters 2 and 3, this volume).

In summary, calving is implicated as a potentially key feedback process during the two intervals of most rapid deglaciation suggested by $\delta^{18}O$ trends. Calving is particularly pertinent to the evidence cited above for ice-sheet thinning on the first step of the deglaciation (14–12.5 ka). There is little evidence of strong caloric attack on the southern margin of the Laurentide ice sheet at that time (see Chapter 20, this volume); the oceans provide an alternative source of heat (Ruddiman and McIntyre, 1981b).

During the second deglacial step suggested by $\delta^{18}O$ (10–7 ka), calving appears to be less uniquely demanded as an explanation; other factors (CO_2, calving into proglacial lakes, delayed mantle rebound) may have played important or even dominant roles in the second period of accelerated ice-volume loss. However, it is entirely possible that calving through Hudson Strait was a deglaciation mechanism of first-order importance at this phase of the deglaciation as well.

Moisture

Ruddiman and McIntyre (1981a) proposed that the flux of meltwater and icebergs into the ocean from rapidly disintegrating ice sheets helped accelerate deglaciation by choking off the moisture flux necessary for continued snow accumulation. This hypothesis was a variant on several earlier theories featuring moisture fluxes from other regions or at different tempos (Stokes, 1955; Ewing and Donn, 1966; Adam, 1973; Johnson and McClure, 1976). This hypothesis requires chilling of the surface ocean by melting icebergs and, in its extreme form, freezing of the ocean surface because of strong stratification and reduced vertical mixing brought about by the low-salinity meltwater layer.

It now seems likely that moisture feedback is at most a minor factor during deglaciations. GCM modeling results (see Chapter 19, this volume) do show much diminished precipitation over the Laurentide ice sheet at 18 ka and 15 ka, but this appears to be due mainly to the ice-elevation effect (decreased moisture content of colder air at higher elevations) and only secondarily to the effects of the North Atlantic in directing storm tracks away from the eastern part of the ice sheet. Because the Atlantic was coldest when the Laurentide ice sheet was highest, the quantitative impact of oceanic moisture starvation early in the deglaciation was probably secondary. This is particularly likely in view of the excessive ablation rates required to explain the speed of deglaciation.

In addition, the revised chronology discussed previously no longer equates the time of fastest initial ice-sheet shrinkage with a period when the high-latitude North Atlantic was cold and seasonally ice covered. The revised age of the initial $\delta^{18}O$ increase (14–12.5 ka) is very close to the age of the initial warming of the high-latitude North Atlantic (13 ±0.5 ka). Within the dating uncertainties, a case could even be made that the ice sheets melted fastest when the ocean warmed, opposite of the requirements of the moisture-feedback hypothesis.

It is still possible that the moisture flux from the Atlantic and Labrador Sea was important to the mass balance of small secondary ice centers at higher Arctic latitudes. Deglacial warming of the North Atlantic may have accelerated the moisture flux to these regions, allowing small ice advances prior to final ice disintegration (Boulton, 1979). However, these represent small volumes of ice relative to the central focus of this volume, the Laurentide ice sheet.

CONCLUSIONS

1. During the last deglaciation, all high-latitude oceans of the northern hemisphere showed trends toward increased warmth and salinity of the surface waters, decreased areal and seasonal extent of sea ice, and increased productivity. Because of dating uncertainties and lack of detailed resolution, these changes can only be bracketed as roughly synchronous with the disintegration of northern hemisphere ice sheets.

2. General Circulation Model sensitivity tests show that the direct thermal impact of the ice sheets on the atmosphere explains much of the basic glacial cooling (and freezing) of high-latitude surface waters. Thus the first-order synchroneity of the surface-ocean changes with the demise of the ice sheets probably reflects the removal of the ice-sheet cooling.

3. In the North Atlantic Ocean, the specific mechanism of surface-ocean cooling is via cold winds blowing from the northern flanks of the North American ice sheet. Long-term data from ocean cores show that such a mechanism can explain both the rhythms and the phasing of polar-front movements in the North Atlantic at the 41,000-year and 100,000-year orbital periods.

4. High-resolution studies of the North Atlantic during the last deglaciation show abrupt surface-ocean warmings (and polar-front retreats) at 13 and 10 ka. These correlate with steps "1a" and "1b" in the composite global $\delta^{18}O$ record. Lack of rapid areal ice retreat at these times could be reconciled if these are episodes of relatively abrupt thinning of North American ice, resulting in less vigorous chilling of the North Atlantic and large-scale retreats of the polar front.

5. The Younger Dryas polar-front readvance at 10.5 ka remains a mystery. There is no evidence for growth of the Laurentide ice sheet, although available data do not rule this out. It may have been in part caused by rebound of the central area of the North American ice sheet to elevations sufficient to again impact the atmospheric circulation, but only if the relationship between ice elevation and impact on the atmospheric circulation is highly nonlinear. High-frequency climatic oscillations or other factors may also have contributed to or caused the Younger Dryas cooling.

6. Evidence of meltwater–iceberg fluxes to the North Atlantic suggests that calving was an important feedback factor during at least the early phase of ice shrinkage at 13 ka. Little evidence exists to constrain its importance during the later phase of shrinkage at 10 ka.

7. Moisture feedback during the deglaciation appears to be a secondary feedback factor.

REFERENCES CITED

Adam, D. P., 1973, Ice ages and the thermal equilibrium of the earth: U.S. Geological Survey Journal of Research, v. 1, p. 587–596.

Aksu, A. E., 1983, Holocene and Pleistocene dissolution cycles in deep-sea cores of Baffin Bay and Davis Strait; Paleoceanographic implications: Marine Geology, v. 53, p. 331–340.

——, 1985, Climatic and oceanographic changes over the past 400,000 years; Evidence from deep-sea cores on Baffin Bay and Davis Strait, *in* Andrews, J. T., ed., Late Quaternary environments; Eastern Canadian Arctic, Baffin Bay and West Greenland: London, Allen and Unwin, p. 181–209.

Aksu, A. E., and Piper, D.J.W., 1979, Baffin Bay in the past 100,000 yr: Geology, v. 7, p. 245–248.

Andree, M., Oeschger, H., Broecker, W. S., Beavan, N., Klas, M., Mix, A. C., Bonani, G., Hoffman, H. J., Suter, M., Woelfli, W., and Peng, T.-H., 1986, Limits on the ventilation rate for the deep ocean over the last 12,000 years: Climate Dynamics, v. 1, p. 53–62.

Andrews, J. T., 1972, Recent and fossil growth rates of marine bivalves, Canadian Arctic, and Late Quaternary Arctic marine environments: Palaeogeography, Palaeoclimatology, Palaeoecology, v. 11, p. 157–176.

Andrews, J. T., and Falconer, G., 1969, Late glacial and postglacial history and emergence of the Ottawa Islands, Hudson Bay, N. W. T.; Evidence on the deglaciation of Hudson Bay: Canadian Journal Earth Science, v. 6, p. 1263–1276.

Balsam, W., 1981, Late Quaternary sedimentation in the western North Atlantic; Stratigraphy and paleoceanography: Palaeogeography, Palaeoclimatology, Palaeoecology, v. 35, p. 215–240.

Berger, A. L., 1978, Long-term variations of caloric insolation resulting from the earth's orbital elements: Quaternary Research, v. 9, p. 139–167.

Berger, W. H., 1985a, On the time scale of deglaciation; Atlantic deep-sea sediments and Gulf of Mexico: Palaeogeography, Palaeoclimatology, Palaeoecology, v. 50, p. 167–184.

—— , 1985b, CO_2 increase and climate prediction; Clues from deep-sea carbonates: Episodes, v. 8, p. 163–168.

Birchfield, G. E., Weertman, J., and Lunde, A. T., 1981, A paleoclimate model of northern hemisphere ice sheets: Quaternary Research, v. 15, p. 126–142.

Borns, H. W., Jr., 1973, Late Wisconsin fluctuations of the Laurentide ice sheet in southern and eastern New England, *in* Black, R. F., Goldthwait, R. P., and Wilman, H. B., eds., The Wisconsin Stage: Geological Society of America Memoir 136, p. 37–46.

Boulton, G. S., 1979, A model of Weichselian glacier variation in the North Atlantic region: Boreas, v. 8, p. 373–395.

Broecker, W. S., Peteet, D. M., and Rind, D., 1985, Does the ocean–atmosphere system have more than one stable mode of operation?: Nature, v. 315, p. 21–26.

Brunner, C. A., 1982, Paleoceanography of surface waters of the Gulf of Mexico during the Late Quaternary: Quaternary Research, v. 17, p. 105–119.

CLIMAP Project Members, 1981, Seasonal reconstructions of the Earth's surface at the last glacial maximum: Geological Society of America Map series MC–36.

Denton, G. H., and Hughes, T., 1983, Milankovitch theory of Ice Age; Hypothesis of ice-sheet linkage between regional insolation and global climates: Quaternary Research, v. 20, p. 125–144.

Denton, G. H., and Karlén, W., 1973, Holocene climatic variations; Their pattern and possible cause: Quaternary Research, v. 3, p. 155–205.

Duplessy, J.-C., Delibrias, G., Turon, J. L., Pujol, C., and Duprat, J., 1981, Deglacial warming of the northeastern Atlantic Ocean; Correlation with the paleoclimatic evolution of the European continent: Palaeogeography, Palaeoclimatology, Palaeoecology, v. 35, p. 121–144.

Emiliani, C., Gartner, S., Lidz, B., Eldridge, K., Elvey, D. K., Huang, T. C., Stipp, J. I., and Swanson, M. F., 1975, Paleoclimatological analysis of Late Quaternary cores from the northeastern Gulf of Mexico: Science, v. 189, p. 1083–1088.

Ewing, W. M., and Donn, W. L., 1966, A theory of ice ages: Science, v. 23, p. 1061–1066.

Fillon, R. H., 1985, Northwest Labrador Sea stratigraphy, sand input, and paleoceanography during the last 160,000 years, *in* Andrews, J. T., ed., Late Quaternary environments; Eastern Canadian Arctic, Baffin Bay, and West Greenland: London, Allen and Unwin, p. 210–247.

Fillon, R. H., and Aksu, A. E., 1985, Evidence for a subpolar influence in the Labrador Sea and Baffin Bay during marine stage 2, *in* Andrews, J. T., ed., Late Quaternary environments; Eastern Canadian Arctic, Baffin Bay, and West Greenland: London, Allen and Unwin, p. 248–262.

Fillon R. H., and Duplessy, J.-C., 1980, Labrador Sea bio–, tephro–, oxygen isotopic stratigraphy and late Quaternary paleoceanographic trends: Canadian Journal Earth Sciences, v. 17, p. 831–854.

Fillon, R. H., Miller, G. H., and Andrews, J. T., 1981, Terrigenous sand in Labrador Sea hemipelagic sediments and paleoglacial events on Baffin Island over the last 100,000 years: Boreas, v. 10, p. 107–124.

Grousset, F., and Duplessy, J. C., 1983, Early deglaciation of the Greenland Sea during the last glacial to interglacial transition: Marine Geology, v. 52, p. M11–M17.

Hays, J. D., Imbrie, J., and Shackleton, N. J., 1976, Variations in the earth's orbit; Pacemaker of the Ice Ages: Science, v. 194, p. 1121–1132.

Heusser, L. E., and Morley, J. J., 1985, Pollen and radiolarian records from deep-sea core RC14–103; Climatic reconstructions of northeast Japan and northwest Pacific for the last 90,000 years: Quaternary Research, v. 24, p. 60–72.

Hollin, J. T., 1962, On the glacial history of Antarctica: Journal of Glaciology, v. 4, p. 173–195.

Imbrie, J., and Imbrie, J. Z., 1980, Modeling the climatic response to orbital variations: Science, v. 207, p. 943–953.

Imbrie, J., and Kipp, N. D., 1971, Paleoclimatic investigation of a late Pleistocene Caribbean deep-sea core, *in* Turekian, K. K., ed., The late Cenozoic glacial ages: New Haven, Connecticut, Yale University Press, p. 71–181.

Imbrie, J., Hays, J. D., Martinson, D. G., McIntyre, A., Mix, A. C., Morley, J. J., Pisias, N. G., Prell, W. L., and Shackleton, N. J., 1984, The orbital theory of Pleistocene climate; Support from a revised chronology of the marine $\delta^{18}O$ record, *in* Berger, A. L., and others, eds., Milankovitch and climate: Boston, Massachusetts, D. Reidel, p. 269–305.

Jansen, E., Sejrup, H. P., Fjaeran, T., Hald, M., Holtedahl, H., and Skarbø, O., 1983 Late Weichselian paleoceanography of the southeastern Norwegian Sea: Norsk Geologisk Tidsskrift, v. 2–3, p. 117–146.

Johnson, R. G., and McClure, B. T., 1976, A model for northern hemisphere continental ice sheet variation: Quaternary Research, v. 6, p. 325–353.

Kellogg, T. B., 1984a, Late-glacial–Holocene high-frequence climatic changes in deep-sea cores from the Denmark Strait, *in* Mörner, N.-A., and Karlén, W., eds., Climatic changes on a yearly to millenial basis: Boston, Massachusetts, D. Reidel Publishing, p. 123–133.

—— , 1984b, Paleoclimatic significance of subpolar foraminifera in high-latitude marine sediments: Canadian Journal of Earth Sciences, v. 21, p. 189–193.

—— , 1986, Late Quaternary paleoclimatology and paleo-oceanography of the Labrador Sea and Baffin Bay; An alternative viewpoint: Boreas, v. 15, p. 331–343.

Kellogg, T. B., Duplessy, J. C., and Shackleton, N. J., 1978, Planktonic foraminiferal and oxygen isotope stratigraphy and paleoclimatology of Norwegian Sea cores: Boreas, v. 7, p. 61–73.

Kennett, J. P., and Shackleton, N. J., 1975, Laurentide ice sheet meltwater recorded in Gulf of Mexico deep-sea cores: Science, v. 188, p. 147–150.

Kutzbach, J. E., and Guetter, P. J., 1986, The influence of changing orbital parameters and surface boundary conditions on climate simulations for the past 18,000 years: Journal of Atmospheric Sciences, v. 43, p. 1726–1759.

Leventer, A., Williams, D. W., and Kennett, J. P., 1982, Dynamics of the Laurentide ice sheet, during the last deglaciation; Evidence from the Gulf of Mexico: Earth and Planetary Science Letters, v. 59, p. 11–17.

Manabe, S., and Broccoli, A. J., 1985, The influence of continental ice sheets on the climate of an icea age: Journal of Geophysical Research, v. 90, p. 2167–2190.

Mangerud, J., 1980, Ice-front variations of different parts of the Scandinavian ice sheet, 13,000–10,000 years B.P., *in* Lowe, J. J., Gray, J. M., and Robinson, J. E., eds., Studies of the Lateglacial of Northwest Europe: Oxford, Pergamon Press, p. 23–30.

Mangerud, J., Lie, S. E., Furnes, H., Kristiansen, I. L., and Lomo, L., 1984, A Younger Dryas ash bed in western Norway, and its possible correlations with tephra in cores from the Norwegian Sea and the North Atlantic: Quaternary Research, v. 21, p. 85–104.

McIntyre, A., Ruddiman, W. F., and Jantzen, R., 1972, Southward penetrations of the North Atlantic polar front; Faunal and floral evidence of large-scale surface water mass movements over the last 225,000 years: Deep-Sea Research, v. 19, p. 61–77.

Mercer, J. H., 1969, The Allerød oscillation; A European climatic anomaly?: Arctic Alpine Research, v. 1, p. 227–234.

Miller, G. H., Andrews, J. T., and Short, S. K., 1977, The last glacial–interglacial cycle, Clyde Foreland, Baffin Island, N. W. T.; Stratigraphy, biostratigraphy, and chronology: Canadian Journal of Earth Sciences, v. 14, p. 2824–2827.

Mix, A. C., and Ruddiman, W. F., 1984, Oxygen-isotope analyses and Pleistocene ice volume: Quaternary Research, v. 20, p. 1–20.

——, 1985, Structure and timing of the last deglaciation: Quaternary Science Reviews, v. 4, p. 59–108.

Morley, J. J., and Hays, J. D., 1979, *Cycladophora davisiana;* A stratigraphic tool for Pleistocene North Atlantic and interhemispheric correlations: Earth and Planetary Science Letters, v. 44, p. 383–389.

——, 1983, Oceanographic conditions associated with high abundances of the radiolarian *C. davisiana:* Earth and Planetary Science Letters, v. 66, p. 63–72.

Morley, J. J., Pisias, N. G., and Leinen, M., 1987, Late Pleistocene time series of atmospheric and oceanic variables recorded in sediments from the subarctic Pacific: Paleoceanography, v. 2, p. 49–62.

Peltier, W. R., 1982, On the dynamics of the ice-age earth: Advances in Geophysics, v. 24, p. 1–146.

Pisias, N. G., and Leinen, M., 1984, Milankovitch forcing of the oceanic system; Evidence from the northwest Pacific, *in* Berger, A., and others, eds., Milankovitch and climate: Boston, Massachusetts, D. Reidel Publishing Co., p. 307–347.

Prest, V. K., 1969, Retreat of Wisconsin and recent ice in North America: Geological Survey of Canada Map 1257A (with text).

Rogers, K. L., Repenning, C. A., Forester, R. M., Larson, E. E., Hall, S. A., Smith, G. R., Anderson, E., and Brown, T. J., 1985, Middle Pleistocene (Late Irvingtonian: Nebraskan) climatic changes in south-central Colorado: National Geographic Research, v. 1, p. 535–563.

Ruddiman, W. F., 1977, Late Quaternary deposition of ice-rafted sand in the subpolar North Atlantic (lat. 40° to 65°N) Geological Society of America Bulletin, v. 88, p. 1813–1827.

Ruddiman, W. F., and Duplessy, J.-C., 1985, Conference on the last deglaciation; Timing and mechanism: Quaternary Research, v. 23, p. 1–17.

Ruddiman, W. F., and Glover, L. K., 1972, Vertical mixing of ice-rafted volcanic ash in North Atlantic sediments: Geological Society of America Bulletin, v. 83, p. 2817–2836.

Ruddiman, W. F., and McIntyre, A., 1981a, Oceanic mechanisms for the amplification of the 23,000-year ice-volume cycle: Science, v. 212, p. 617–627.

——, 1981b, The North Atlantic Ocean during the last deglaciation: Palaeogeography, Palaeoclimatology, Palaeoecology, v. 35, p. 145–214.

——, 1984, Ice-age thermal response and climatic role of the surface North Atlantic Ocean, 40°N to 63°N: Geological Society of America Bulletin, v. 95, p. 381–396.

Ruddiman, W. F., Sancetta, C. D., and McIntyre, A., 1977, Glacial/interglacial response rate of subpolar North Atlantic waters to climatic change; The record left in deep-sea sediments: Philosophical Transactions of the Royal Society of London, B280, p. 119–142.

Ruddiman, W. F., Raymo, M., and McIntyre, A., 1986, Matuyama 41,000-year cycles; North Atlantic Ocean and northern hemisphere ice sheets: Earth and Planetary Science Letters, v. 80, p. 117–129.

Sancetta, C., 1979, Oceanography of the North Pacific during the last 18,000 years; Evidence from fossil diatoms: Marine Micropaleontology, v. 4, p. 103–123.

——, 1983, Fossil diatoms and the oceanography of the Bering Sea during the last glacial event, *in* Iijima, A., Hein, J. R., and Siever, R., eds., Siliceous deposits in the Pacific region: Amsterdam, Elsevier Publishing Co., p. 333–346.

Sancetta, C., and Silvestri, S., 1986, Plio–Pleistocene evolution of the North Pacific ocean-atmosphere system, interpreted from fossil diatoms: Paleoceanography, v. 1, p. 163–180.

Shackleton, N. J., and Opdyke, N. D., 1973, Oxygen isotope and paleomagnetic stratigraphy of equatorial Pacific core V28–238; Oxygen isotope temperatures and ice volumes on a 10^5 and 10^6 year scale: Quaternary Research, v. 3, p. 10–38.

Stokes, W. L., 1955, Another look at the ice age: Science, v. 122, p. 815–821.

Watts, W. A., 1980, Regional variation in the response of vegetation to late glacial climatic events in Europe, *in* Lowe, J. J., Gray, J. M., and Robinson, J. E., eds., Studies in the Lateglacial of Northwest Europe: Oxford, Pergamon Press, p. 1–22.

MANUSCRIPT ACCEPTED BY THE SOCIETY FEBRUARY 23, 1986

ACKNOWLEDGMENTS

I thank J. Andrews, T. Growley, D. Fillon, T. Kellogg, J. Morley, and C. Sancetta for reviews. This research was funded by National Science Foundation Grant OCE 85-21514. This is LDGO contribution number 4151.

The Geology of North America
Vol. K-3, North America and adjacent oceans during the last deglaciation
The Geological Society of America, 1987

Chapter 8

Glacial isostasy, mantle viscosity, and Pleistocene climatic change

W. R. Peltier
Department of Physics, University of Toronto, Toronto, Ontario M5S 1A7, Canada

INTRODUCTION

Throughout the Pleistocene epoch of earth history, terrestrial ice-sheet volume has been highly variable with time. Our knowledge of this variability has been markedly improved by the recognition that $\delta^{18}O$ variations in deep-sea sedimentary cores constitute an excellent indicator of changes in land ice volume (Shackleton, 1967). These data demonstrate that throughout the past 500,000 yr, terrestrial ice-volume variations have consisted of successive episodes of glaciation and deglaciation separated by interglacials spaced at rather precise 100,000-yr intervals (Broecker and Van Donk, 1970; Hays and others, 1976). Each slow oscillatory increase of ice volume during the glaciation phase has been followed by a relatively fast collapse, or termination, during the deglaciation phase. These observations are illustrated in Figure 1a, which shows the SPECMAP record of Imbrie and others (1984).

The considerable interest in such data stimulated by publication of the seminal paper of Hays, Imbrie, and Shackleton in 1976 is made clear by Figure 2a, which shows the power spectrum of the time series in Figure 1a. The $\delta^{18}O$ spectrum is dominated by five spectral lines, the most important one corresponding to the frequency of 1 cycle/100 kyr and generated by the 100 kyr spacing between successive interglacials (Fig. 1a). Equally important, however, is the additional spectral line at the period of 41 kyr and the closely spaced triplet of peaks corresponding to periods of 24.1, 23.4, and 19.1 kyr. The reason for the importance of the presence of these particular periods as dominant constituents of the proxy record of continental ice-volume variations has to do with the fact that they are precisely predicted by the conventional astronomical theory of ice ages, which has usually been attributed in the modern literature to Milankovitch (e.g., 1941), although elements of it had been earlier suggested by Croll (1875) and Köppen (e.g., Köppen and Wegener, 1924). Prior to the paper by Hays and others (1976), which first demonstrated that the power spectrum of continental ice-volume variations was dominated by these terms, the astronomical theory of ice ages was in considerable disrepute. Afterward, it was considered to be essentially established, although incomplete.

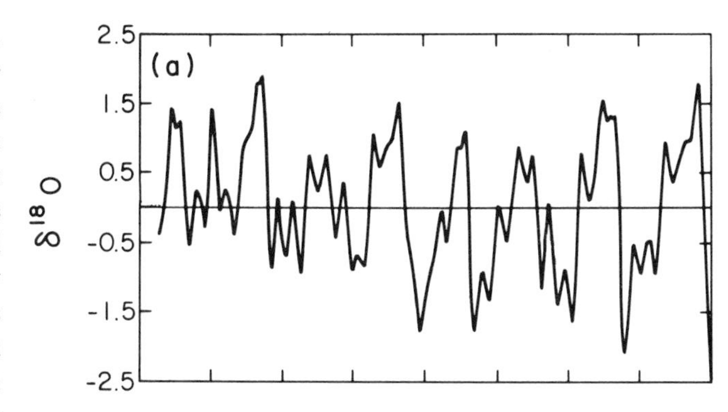

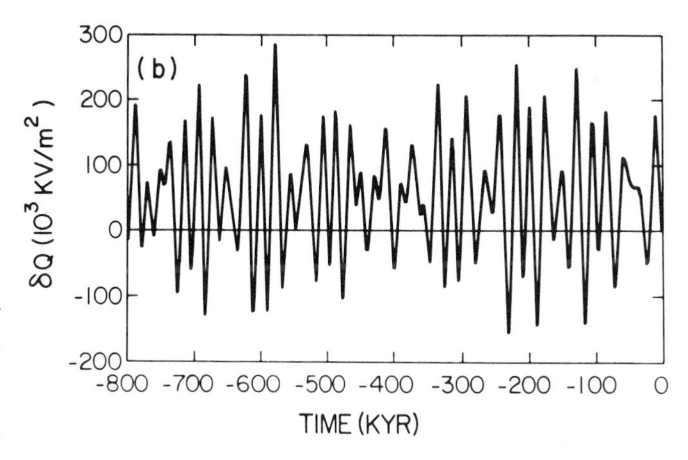

Figure 1. (a) The SPECMAP time series relates the ratio of ^{18}O concentration to that of ^{16}O in open-ocean sediments to the time (in kyr) before present. (b) The change in caloric summer seasonal insolation from present at 65°N latitude over the past 800 kyr. Computed from the Fourier series expansions of Berger (1978).

Peltier, W. R., 1987, Glacial isotasy, mantle viscosity, and Pleistocene climatic change, *in* Ruddiman, W. F., and Wright, H. E., Jr., eds., North America and adjacent oceans during the last deglaciation: Boulder, Colorado, Geological Society of America, The Geology of North America, v. K-3

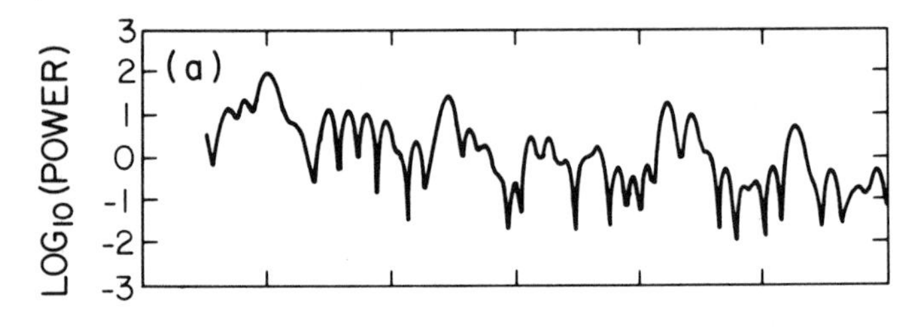

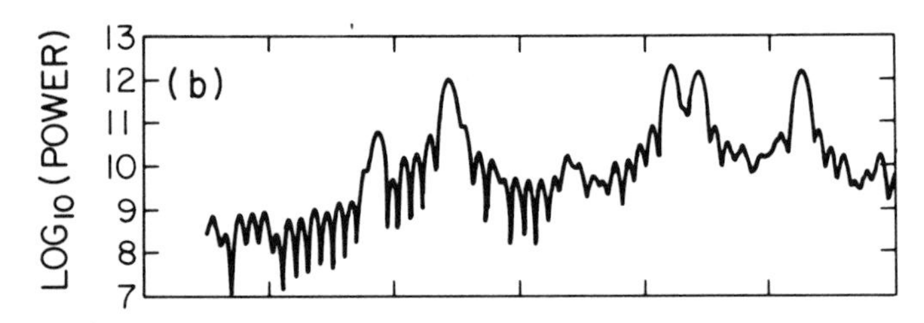

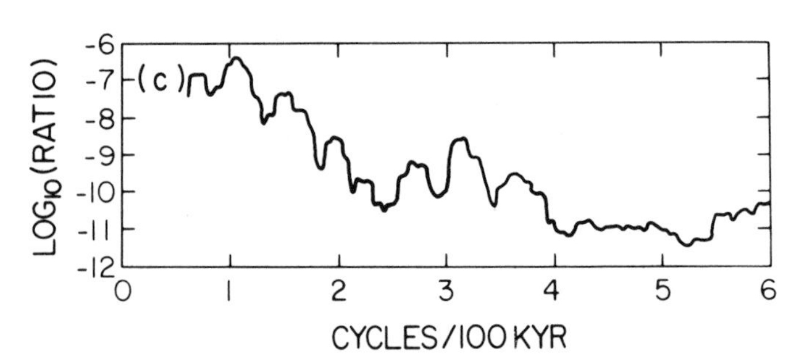

Figure 2. (a) The power spectrum of the SPECMAP time series expressed in cycles per 100 kyr. The four dominant peaks are near 100, 41, 23, and 19 kyr. (b) The power spectrum of the last 1000 kyr of the Milankovitch forcing for 65°N latitude. Note that there is negligible power at 100 kyr, whereas strong peaks exist at 41, 23, and 19 kyr. (c) The term-by-term ratio of the power spectrum of the SPECMAP time series to that of the insolation time series. The result has been smoothed with a running average of width 4 kyr.

That difficulties remained with the astronomical theory was immediately apparent on the basis of comparisons of the power spectrum of the climatic response to that of the astronomical forcing. According to Köppen, the cryosphere should respond principally to variations in summertime seasonal insolation, winters always being sufficiently cold to maintain an ice sheet if one already existed. Ice sheets are therefore expected to retreat when insolation is anomalously high and to advance when it is anomalously low. Figure 1b shows a typical summertime insolation anomaly time series (for 65°N latitude) based on the reconstruction of the earth's orbit over the same time period spanned by the SPECMAP record, according to a reconstruction of the analysis of Berger (1978). Figure 2b shows the power spectrum of this record. Four of the five lines prominent in the power spectrum of $\delta^{18}O$ variations are duplicated in the astronomical signal. The most important line however, corresponding to the period of 100 kyr, is entirely absent. Although ice volume fluctuates, with a dominant period of 100 kyr, there is no power at this period in the astronomical forcing. Figure 2c shows the ratio of the $\delta^{18}O$ spectrum to the insolation anomaly spectrum and thus the spectral transfer function of the paleoclimatic system. Clearly the input energy is dominantly of high frequency and the response of low frequency.

The origin of the four prominent spectral lines in the astronomical input may be seen most simply from the form of the summertime seasonal insolation anomaly ΔQ_s, which has the following mathematical form:

$$\Delta Q_s = \Delta R_s \Delta\epsilon + m\Delta\left(e \sin(\hat{\omega})\right) \tag{1}$$

where

$$m = \frac{2TS\cos(\phi)}{\pi^2 (1-e^2)^{1/2}},$$

in which T is the duration of the tropical year, S the solar constant, ϕ the latitude, and R_s the total insolation received at the top of the atmosphere over the caloric summer. The remaining parameters ϵ, e, and $\hat{\omega}$ are respectively the obliquity of the earth's orbit, the orbital eccentricity, and the longitude of the perehelion with respect to the vernal equinox (Fig. 3). The origin of the 41-kyr line in the astronomical input is due to changes of ϵ at this period (first term in Equation 1). The remaining three peaks in the input are associated with the eccentricity modulation of the precessional cycle (second term in Equation 1). Although e does vary with a dominantly 100-kyr period, this period does not appear in the power spectrum of ΔQ_S because e simply modulates the precessional amplitude, the period of which is near 22 kyr. The e modulation splits this precessional peak to produce the dominant triplet of closely spaced lines near the precessional period.

This is the source of the enigma posed by the conventional astronomical theory of ice ages: according to the conventional astronomical theory, the 100-kyr cycle should not exist at all, yet it is observed to dominate the record of late Pleistocene climate change.

My purpose in this chapter will be to address the enigma posed by these data and in so doing illustrate a model that has been employed to successfully explain the process whereby the climate system forced by astronomically induced insolation variations delivers a cryospheric volume oscillation represented by the SPECMAP record shown in Figure 1a. As I attempt to show, the process responsible for effecting the observed spectral transformation in the climate system appears to be the same process that governs glacial-isostatic adjustment, that is, the slow viscous flow of material within the interior of the planet that is forced by the unbalanced gravitational forces set up by large surface accumulations of ice. This explanation is a plausible one

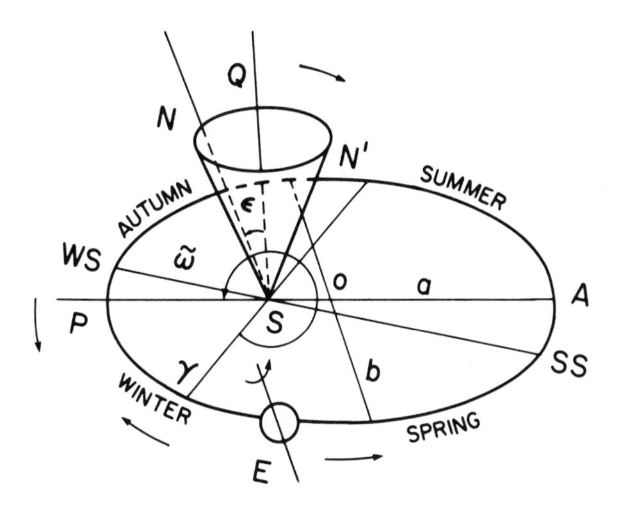

Figure 3. Elements of the earth's orbit. The orbit of the earth (E) around the sun (S) is represented by the ellipse (PEA), P being the perihelion, and A the aphelion. Its eccentricity $e = \sqrt{a^2 - b^2}/a$, with a the semi-major and b the semi-minor axes of the ellipse. WS and SS are respectively the current winter and summer solstices and γ is the vernal equinox. SQ is perpendicular to the ecliptic and the obliquity ϵ is the angle between SQ and the earth's axis of rotation SN. The angle $\omega = \pi + \Psi$ is the longitude of perihelion relative to the moving vernal equinox, γ. The motion of γ is the annual general precession in longitude Ψ, which describes the absolute motion of γ along the earth's orbit relative to the fixed stars. The longitude of perihelion π is measured from the reference vernal equinox and describes the motion of the perihelion relative to the fixed stars.

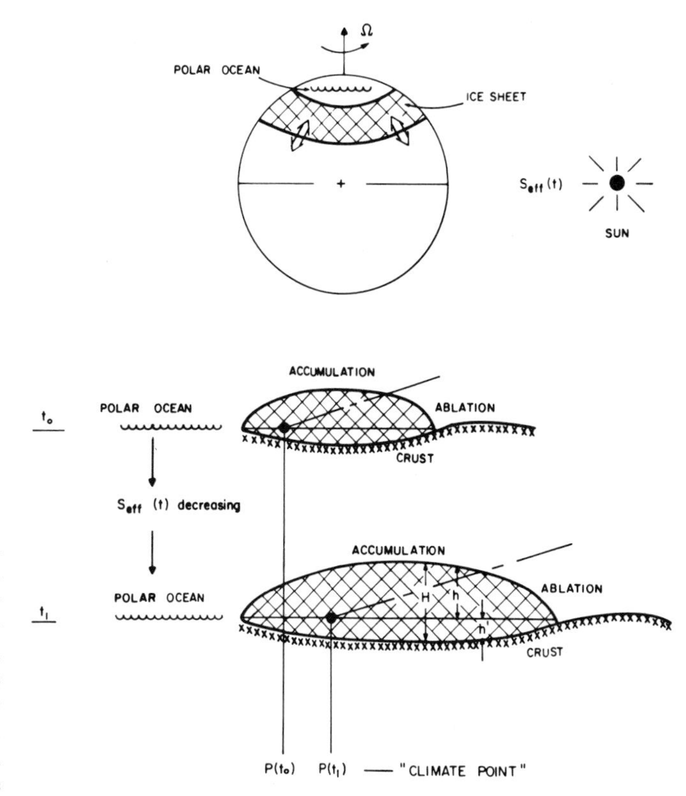

Figure 4. Schematic diagram of the paleoclimate model, which consists of an active ice sheet that expands and contracts in response to variations of effective solar insolation caused by changes in the parameters controlling the geometry of the earth's orbit around the sun (see Fig. 3). Variations of insolation are assumed to cause latitudinal shifts in the location of the climate point and thus to cause changes in the relative areas of accumulation and ablation zones separated by the model firn line. Because accumulation and ablation rates are strong functions of elevation, the process of glacial-isostatic adjustment strongly influences ice-volume variability.

only if the effective viscosity of the planetary interior is close to a rather well constrained value. After establishing what this value must be in the next section of this chapter by explicit analysis of a climate model that includes the isostatic adjustment process, I devote the remainder of the chapter to a discussion of the constraints that may be placed on the viscosity of the mantle through analysis of various signatures of the earth's response to the last termination (deglaciation event) of the current ice age. These signatures include: relative sea-level histories, the observed free-air gravity anomalies remaining over present-day centers of postglacial rebound, and two anomalies of earth rotation that are also explicable as memories of the planet of the 100-kyr glaciation--deglaciation cycle.

A MODEL OF CLIMATE CHANGE WITH GLACIAL ISOSTASY

The model of long-term climatic change is described schematically in Figure 4. It consists of two active and explicitly modeled components, representing respectively a continental ice sheet and the earth upon which the ice sheet rests. The model is zonally symmetric, and the single ring-shaped ice sheet surrounds a circular Arctic ocean whose "coast" is coincident with the northern boundary of the ice sheet. The accumulation and flow of the ice mass are coupled to the delayed sinking of the earth under the weight of the ice that is effected by the isostatic adjustment process.

The reason for the coupling between these two fundamentally distinct physical ingredients has to do with the nature of the accumulation process. In the model, the rates at which snow accumulates on the surface of the ice sheet in the accumulation zones or melts in the ablation zone (Fig. 4) are both strong functions of the elevation of the surface of the ice sheet with respect to sea level. In the accumulation zone the accumulation rate decreases with increasing elevation (the elevation desert effect). In the ablation zone, the ablation rate decreases with increasing elevation. These rates of change may be accurately estimated on the basis of the observed rate of decrease of the saturation mixing ratio of water in air as a function of increasing elevation (i.e., colder air at higher altitudes holds less water vapor than warmer air; thus the higher the elevation of an ice sheet, the colder the air above it, and the smaller the available precipitation).

Glacial-isostatic adjustment strongly influences the variation of ice volume by changing the surface elevation of the ice sheet relative to sea level and thus the accumulation and ablation rates, the balance of which determines the change of ice-sheet volume.

Astronomically induced variations of summertime seasonal insolation are introduced into the model through the north–south motion of the "climate point" which these variations are assumed to induce. This point is also marked on Figure 4. In the absence of significant accumulation of ice, the climate point denotes the southern boundary of the perennial snow field. There is an annually averaged accumulation of snow north of that point but no accumulation to the south. As the net accumulation of ice increases, however, and the sheet thickens so as to raise its upper surface into colder air, the boundary separating the accumulation and ablation zones migrates southward, essentially following the equatorward tilt of some standard atmospheric isothermal surface (the −10°C surface, say). This "climate surface" intersects the ice sheet at the model firn line, separating the northern accumulation zone from the southern ablation zone. This scheme is essentially identical to that introduced by Weertman (1976) in analyzing the factors controlling ice-sheet volumes at the glacial maximum at 18 ka.

Insolation variations are assumed to induce proportional north–south displacements of the climate point and therefore to cause the relative areas of the accumulation and ablation zones to change and thus to produce increases or decreases of ice volume. The amount of lateral movement of the climate point is tied to the present insolation gradient. The goal of experiments with this model is to show that insolation fluctuations, such as those associated with variations in the earth's orbit illustrated in Figure 1b, can indeed induce ice-sheet volume fluctuations such as those inferred from the SPECMAP $\delta^{18}O$ observations shown in Figure 1a. To the extent that the model is able to explain these data, it will provide an answer to the enigma posed by the dominance of the 100-kyr oscillation in a climate system that is not subject to any significant forcing at this period.

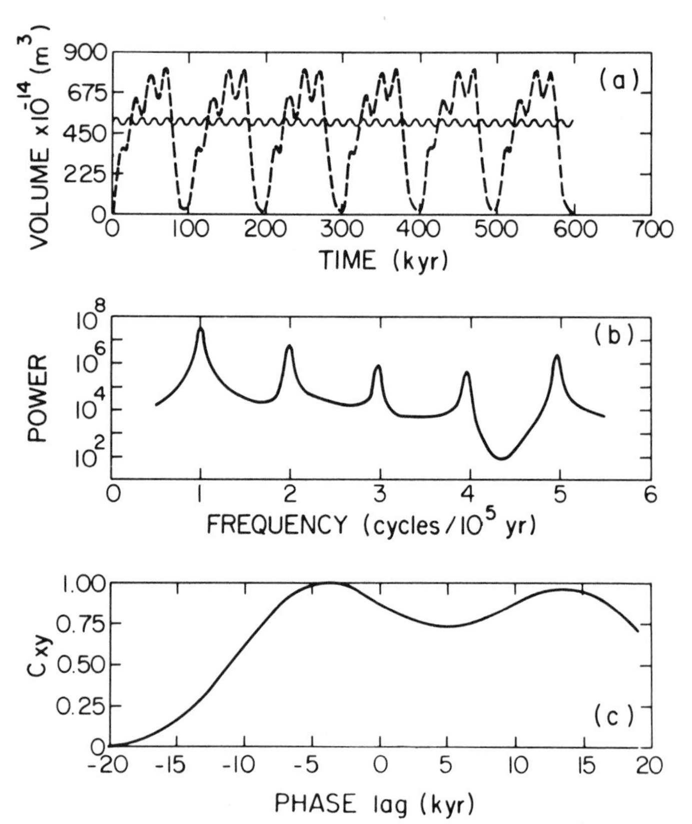

Figure 5. (a) Synthetic ice-volume history predicted by the paleoclimate model with the astronomical forcing approximated by a simple harmonic oscillation of period 20 kyr. Note that the response is dominated by a periodic signal with period 100 kyr. (b) The power spectrum of the time series shown in Figure 5a illustrates the dominance of the 100-kyr line. Other peaks are harmonics of this line, the fifth of which, of period 20 kyr, is enhanced because it coincides with the frequency of the forcing. (c) The cross-correlation vector of the synthetic ice volume time series with the 20-kyr forcing (shown as the solid line on Fig. 5a for phase reference). This demonstrates that the phase lag between the input and output at this period is 4–5 kyr in accord with the observation of Imbrie and others (1984) based on the SPECMAP data.

The mathematical structure of the model described above in physical terms was derived by Peltier (1982), and a first integration of its governing equations was presented by Peltier and Hyde (1984). For the purpose of this initial calculation, the astronomically induced insolation variations were approximated by a simple sinusoidal function with a period of 20 kyr intended to mimic the precessional component of the astronomical input. The result obtained in this first simulation of such ice-volume fluctuations is reproduced in Figure 5; parts a, b, and c illustrate the synthetic ice-volume time series itself, its power spectrum, and the cross-correlation vector between the 20-kyr periodic forcing function and the ice-volume response. The ice-volume time series produced by the model is dominated by a 100-kyr oscillation, even though it is forced entirely at a period of 20 kyr. Furthermore, the individual ice-volume pulses have a very similar shape to that revealed in the $\delta^{18}O$ data of Figure 1a, with a slow oscillatory glaciation phase followed by a fast collapse or termination. Because the individual pulses have identical shape, the power spectrum shown in Figure 1b is a simple line spectrum with power confined to the fundamental 100-kyr period and its harmonics. The power in the fifth harmonic is enhanced because this period coincides with that at which the model is directly forced. The 100-kyr cycle in the model is an example of what is called a subharmonic resonant relaxation oscillation (e.g., Peltier, 1986b), a fundamentally nonlinear oscillation with a period that is an integer multiple (in this case ×5) of the period of the forcing. Figure 5c demonstrates that the phase lag of the 20-kyr component of the response behind the 20-kyr component of the forcing

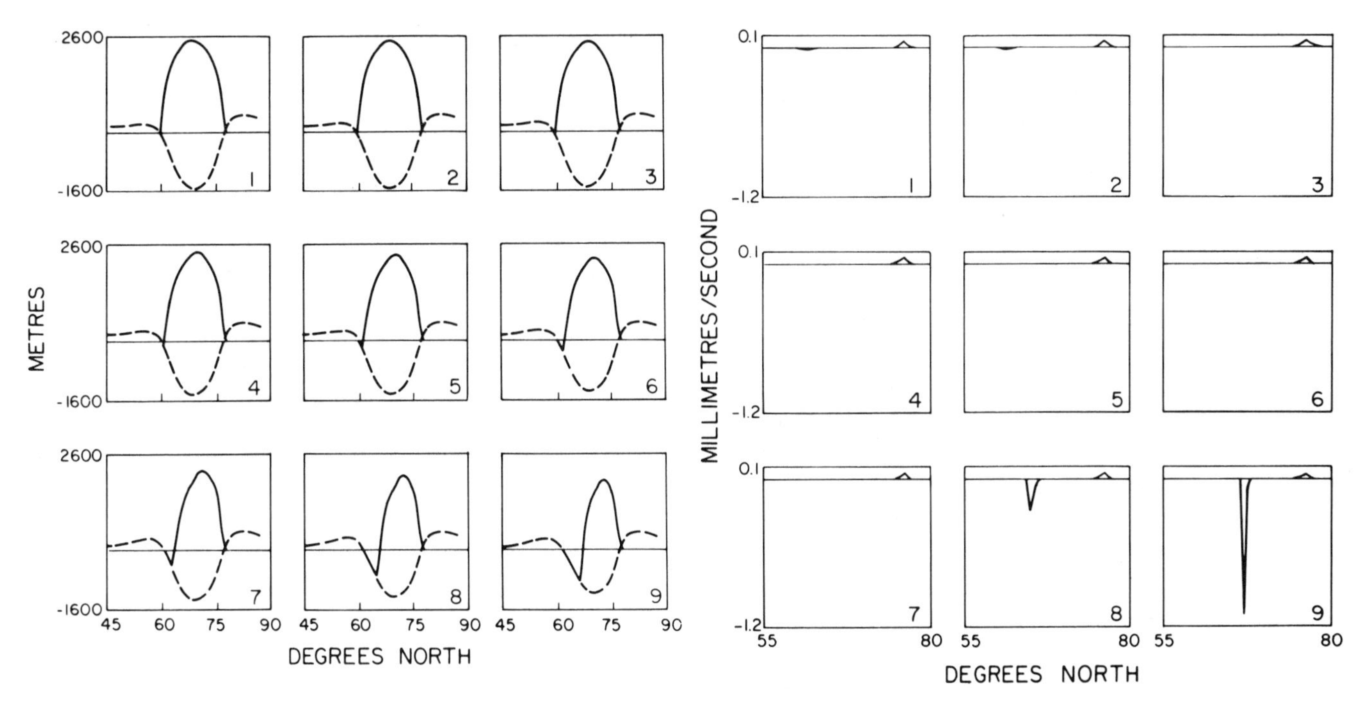

Figure 6. Cross sections of ice-sheet height and earth depression at the beginning of a typical termination in the synthetic ice-volume history of Figure 5. Figures 6-1 to 6-9 are at 1-kyr intervals from 71 kyr to 79 kyr.

Figure 7. Cross sections of north–south velocity at the beginning of a typical termination for the same sequence of times as the ice-height cross sections shown on Figure 6.

is 4–5 kyr, which is in accord with that inferred by Imbrie and others (1984) from the data shown in Figure 1.

Fundamental to the physical understanding of the 100-kyr cycle that this model predicts are the interactions within it that are responsible for causing the sharp terminations characterizing this cycle. These interactions are illustrated in Figures 6 and 7, which respectively show profiles of ice height (solid) and earth deflection (dashed) through one of the terminations of the ice-volume history prediction of Figure 5, and ice velocity versus latitude profiles for the same sequence of times. These diagnostic data show that the termination is produced when a sufficiently large ice sheet melts back into the depression of the crust induced by glacial-isostatic adjustment (sinking) of the earth beneath it. When this occurs, the increased rate of ablation due to the decrease in elevation of the ablation zone causes the southern slope of the ice sheet to steepen, which in turn (Fig. 7) causes the southward rate of ice flow to rise precipitously due to the highly nonlinear nature of the Glen flow law, which shows that ice flux depends upon the cube of the surface slope. Ice thereafter continues to flow from the cold northern accumulation zone to the warm southern ablation zone at a rate sufficient to melt the entire ice mass on the time scale of half a precessional cycle.

This detailed explanation of the process inducing ice-sheet collapse in the climate model with coupled ice dynamics and glacial isostasy was first provided in the paper by Hyde and Peltier (1985), who also presented a detailed discussion of the sensitivity of the predictions of the climate model to plausible variations of its parameters. Although similar models had previously been constructed (e.g., Oerlemans, 1980; Pollard, 1983, 1984; Birchfield and others, 1981, 1982), the mechanism of oscillation was never explained, and the models contained erroneous descriptions of the glacial-isostatic adjustment process that strongly affected their behavior, as pointed out by Peltier (1982). Completely dissimilar models of the 100-kyr oscillation have also been recently proposed (e.g., Ghil and Letreut, 1983; Saltzman and others, 1984).

That the 100-kyr oscillation persists when the new climate model is forced by more realistic astronomical forcing is demonstrated in Figure 8a, which shows the synthetic ice-volume history predicted by a model forced by the Milankovitch variations appropriate to 65°N latitude. The insolation anomaly time series used to drive the system was computed from the formulae in Berger (1978) and is shown for phase reference above the ice-volume time series. The response to this forcing is again dominated by an oscillation with a period very near 100 kyr, demonstrating that the basic physical mechanism responsible for the long-term ice-volume oscillation is not eliminated when the considerably more complicated forcing function of Equation 1 is applied, nor is the period of the oscillation substantially modified. This is clear from the power spectrum of the response (Fig. 8b) and the direct overlay of the SPECMAP time series with the synthetic (Fig. 8c).

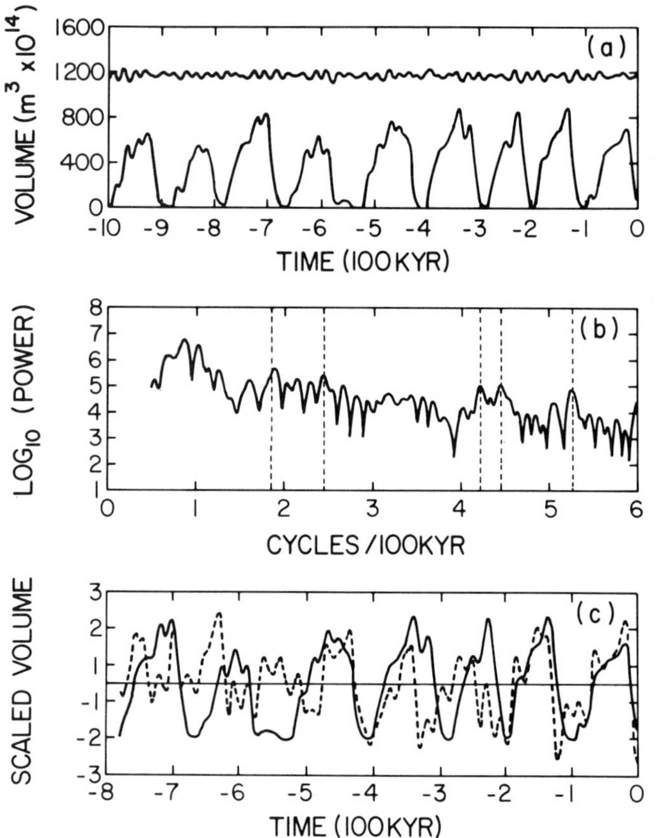

Figure 8. (a) Synthetic ice-volume history predicted by a Milankovitch experiment employing the summertime insolation anomaly appropriate to 65°N latitude to force the model. The forcing is shown above for phase reference. (b) The power spectrum of the time series in part (a). The dashed lines mark each of the forcing periods. (c) Overlay of the last 800 kyr of the synthetic (solid curve) with the SPECMAP curve (dashed) demonstrating the reasonable qualitative fit of the predictions of the model to the observational data.

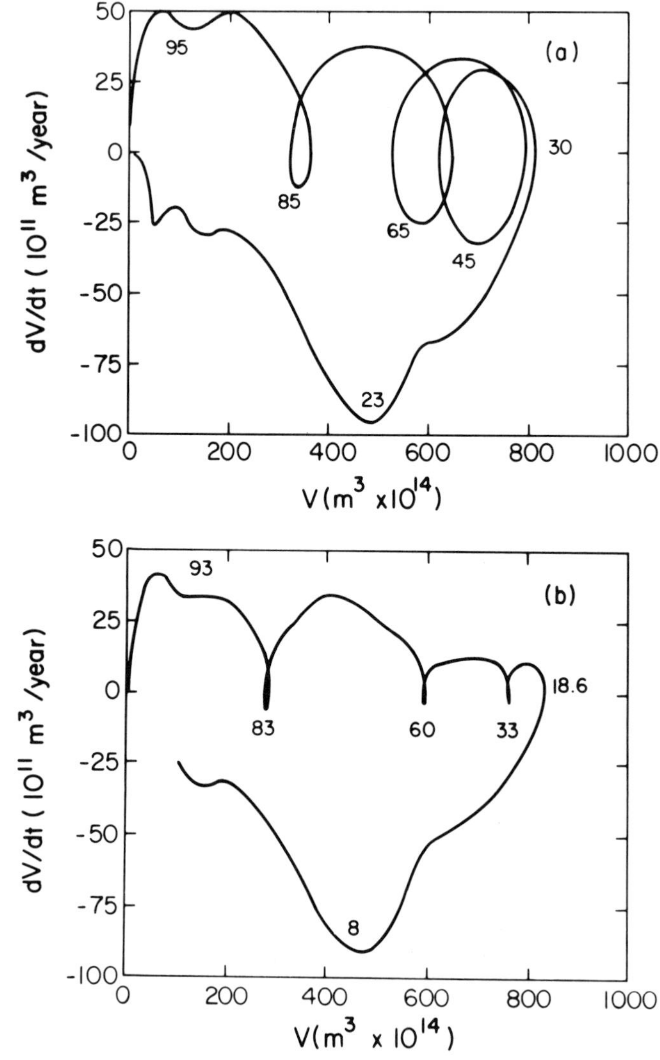

Figure 9. (a) Phase space plot of the time derivative of ice volume as a function of ice volume from a harmonically forced experiment similar to that shown in Figure 5. The numbers refer to time in kyr before the end of the cycle. Note the gradual increase of the strength of the decay mechanism from one precessional cycle to the next. (b) A similar plot for a realistically forced Milankovitch experiment using the 55°N insolation anomaly as input. On this figure the numbers are in kyr before present.

An important success of this model is the skill with which it is apparently able to predict the correct timing of the individual pulses of glacial ice and thus the previously inexplicable correlation of the timing of the glacial cycles with the 100-kyr variation of eccentricity (e.g., Imbrie and others, 1984). It has often been assumed that this correlation was only explicable if the climate system was directly subject to insolation forcing at eccentricity periods; according to the simplest formulation of the Milankovitch forcing, it is not (e.g., see Fig. 2b). The way in which the new climate model is able to explain the phase locking of the dominant ice-volume oscillation to the eccentricity cycle has to do with the fact first pointed out in Hyde and Peltier (1985) that both the amplitude and period of the predicted response are *inversely* proportional to the amplitude of the forcing. Since the variation of orbital eccentricity does modulate the amplitude of the precessional component of the forcing (Equation 1), terminations are most often induced in that phase of the eccentricity cycle that corresponds to maximum amplitude of the precessional component of the forcing (e.g., Fig. 8a). The reason for the existence and dominance of the 100-kyr cycle, however, has nothing to do with the 100-ka eccentricity cycle. This is established by the previously described results shown in Figure 7. The 100-kyr cycle dominates the response even when the forcing is a simple harmonic with a period of 20 kyr.

Figure 9 provides a further synopsis of the fundamental processes at work during a single 100-kyr cycle for the harmonically forced model (Fig. 9a) and for the realistically forced model

(Fig. 9b). Both display the time rate of change of ice volume dV/dt as a function of volume V with individual points on these phase space plots corresponding to different instants of time, labeled in kyr at diagnostic points on the trajectories. These data show that as the volume of the ice sheet increases it becomes more and more unstable until it is finally forced to completely disintegrate via the previously described collapse process. This tendency toward increasing degree of instability with increasing volume is more evident in Figure 9a, for which the amplitude of the forcing is constant, than in Figure 9b, in which it is somewhat concealed by the fact that the amplitude of the forcing is actually decreasing as the ice sheet undergoes the final collapse.

Figure 10a plots volume versus surface-area data for the synthetic ice sheets, with appropriate corrections for their two dimensionality as described in Hyde and Peltier (1987). The regression line is derived by Paterson (1981) from data from the Barnes ice cap (B), Roosevelt Island (R), Vantanajökull (V), the Wilkes ice dome (W), the Greenland ice sheet (G), and the Antarctic ice sheet (A). These data presumably refer to stable ice caps. The crosses on this figure (marked 1–6) correspond to synthetic data from six points through two 100-kyr cycles of a single realistically forced simulation (Fig. 10b). The model ice sheets are closest to the Paterson line when they are small but move farther away from this stability line as they grow and become more unstable, according to the phase space trajectories shown on Figure 9.

Of course this theoretical climate model has many adjustable parameters, even for the geometrically simplified version used for actual numerical experiments. In order to evaluate the proposed explanation of the 100-kyr cycle of the Pleistocene ice age, we must assess the extent to which values of the model parameters are known. In general these parameters comprise three groups that respectively control the glaciological, meteorological/hydrological, and geophysical components of the model. All of these elements are debatable to a certain degree. For example, the ice-sheet dynamics employed are appropriate for the motion of a one-dimensional ice sheet frozen to its bed. No effects are included for sliding at the base, which might be expected for temperate ice sheets like the Laurentide ice mass. Similarly the sea-level rates of ablation and accumulation, although selected on the basis of the observed rates for Greenland and Antarctica, are also imperfectly known meteorological/hydrological parameters. Even more debatable may be the assumed feedback between ice-sheet elevation and the rates of accumulation and ablation. The single parameter controlling this dependence is fixed in the model on the basis of the observed rate of decrease of the saturation mixing ratio of water in air as a function of increasing height in a standard atmosphere. Lindzen (1986) has recently questioned the plausibility of this feedback, although he has not given any reason for his skepticism. The sensitivity analyses of Hyde and Peltier (1985) certainly demonstrate that this posited feedback is crucial to the model's success in explaining the 100-kyr cycle. General Circulation Model experiments might be performed to test this basic assumption of the climate model, although great care would have to be taken to ensure the best possible treatment of water-vapor transport (e.g., use of positive definite advection schemes in grid-point space) and precipitation processes, both of which are poorly simulated in the current generation of GCMs, particularly in polar regions.

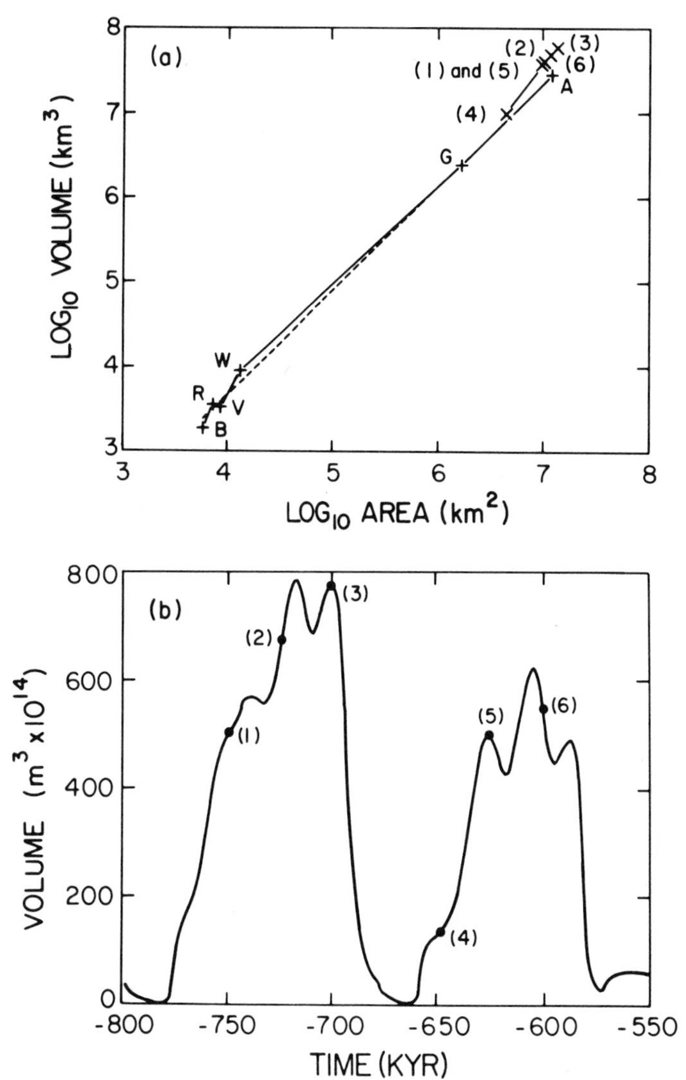

Figure 10. (a) The Paterson regression line (dashed) is compared with the observed volume/area relation of existing ice sheets and ice caps (the letters refer to the ice masses named in the text) and the ice sheets generated in a Milankovitch experiment. The numbers refer to the points in volume history shown in (b). (b) Location of the points plotted in (a) on the evolutionary histories from which they were taken.

Although both the glaciological and meteorological/hydrological elements of the model are indeed crucial to its success and debatable in detail, no 100-kyr cycle would be produced at all in the absence of the geophysical component of the model, no matter what the settings of the parameters that control the other

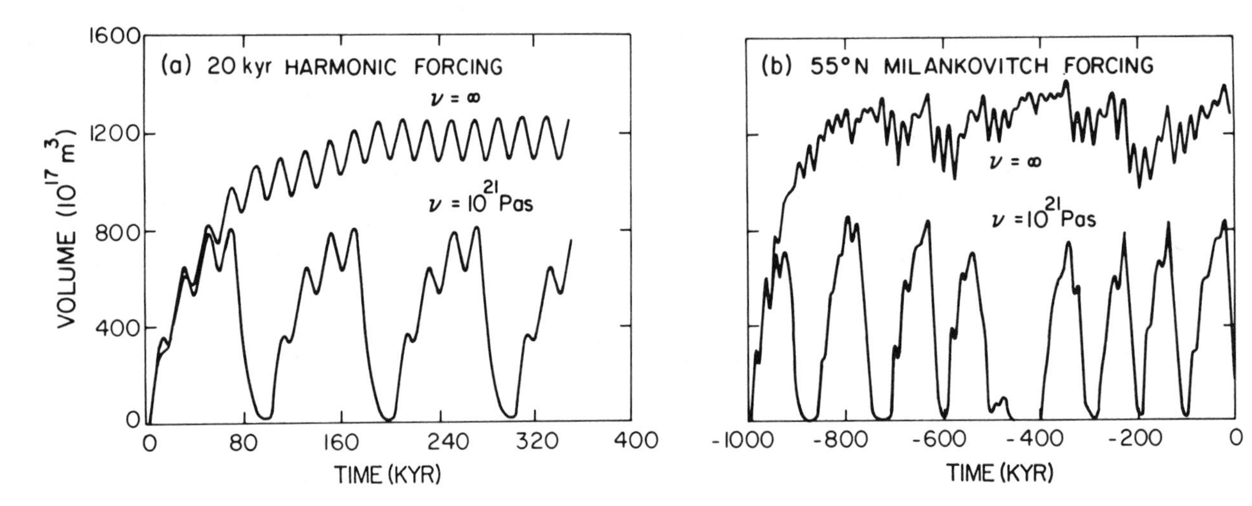

Figure 11. (a) The synthetic ice-volume history predicted by a harmonically forced model, in which the isostatic-adjustment process is governed by a realistic viscosity of 10^{21} Pa s, compared with that for the same model when isostatic adjustment is suppressed by taking the mantle viscosity to be infinite. (b) Same as (a) but for a realistic Milankovitch experiment. When isostatic adjustment is suppressed, the system never deglaciates.

processes. This is demonstrated explicitly in Figure 11, which shows ice-volume histories predicted by the model under the assumptions that (1) the earth's mantle has a viscosity fitting the requirements of the data of postglacial rebound and (2) mantle viscosity is infinite so that no glacial-isostatic adjustment occurs. Figure 11a is from the model forced by a harmonic insolation anomaly time series with period of 20 kyr, whereas Figure 11b shows similar data for the model forced by realistic Milankovitch input. Both experiments demonstrate that when the glacial-isostatic adjustment process is eliminated from the model ($\nu = \infty$) the system simply enters a glaciated state in which the ice volume fluctuates about some average in accord with the astronomical forcing. Large-scale deglaciation never occurs. Clearly, then, the process of glacial-isostatic adjustment is a required ingredient for the success of the model.

The strength of the model's sensitivity to the time scale governing isostatic readjustment is made clear by Figure 12. This shows synthetic ice-volume histories from harmonically forced experiments that differ from one another only by the time scale governing the isostatic adjustment process (and therefore by the effective viscosity of the planetary mantle, which controls the time scale). These model results demonstrate that variations of the effective viscosity of the mantle by even a factor of 2 from the value of 10^{21} Pa s, required to explain postglacial rebound data, would lead to unacceptably large shifts of the period of the dominant oscillation away from the observed period of 100 kyr. Decreasing the viscosity decreases the predicted period. Although one may trade off variations of viscosity with variations in other parameters of the model to a certain extent, it is clear that if the viscosity of the mantle were much different from 10^{21} Pas the model could not explain the 100-kyr cycle.

In order to establish that the physical/mathematical model described in this section provides the explanation of observed ice-volume fluctuations through the Pleistocene ice age, we are clearly obliged to establish that the earth possesses parameter values appropriate to make this not only plausible but inevitable. We have not yet reached the point at which such a claim would be warranted. We have not proven, for example, that the posited feedback between accumulation/ablation rates and ice-sheet height must have prevailed during the ice age. However, we are in a much better position to claim that the geophysical components of the climate model have the values required to make this explanation fully plausible. This derives from the marked progress achieved over the past decade in refining our knowledge of the viscosity of the earth's mantle through analysis of observations on glacial-isostatic adjustment.

In the remaining sections of this chapter I review the theory and observations upon which our present understanding of mantle viscosity is based. First I review the controversial history of attempts to infer the value of this quantity, and then I explore the constraints placed upon it by a variety of different geophysical observations. The last deglaciation event of the current ice age might easily be viewed as a natural large-scale experiment whose objective was to determine the constitutive relation of the material of the earth's mantle. In the words of Reginald Daly (1934), this was in fact "Nature's Great Experiment."

MANTLE VISCOSITY AND THE ICE AGE

It has been generally understood since the pioneering work of Haskell (1935, 1936, 1937) and Vening-Meinesz (1937) on the uplift of Fennoscandia that the observed response of the earth

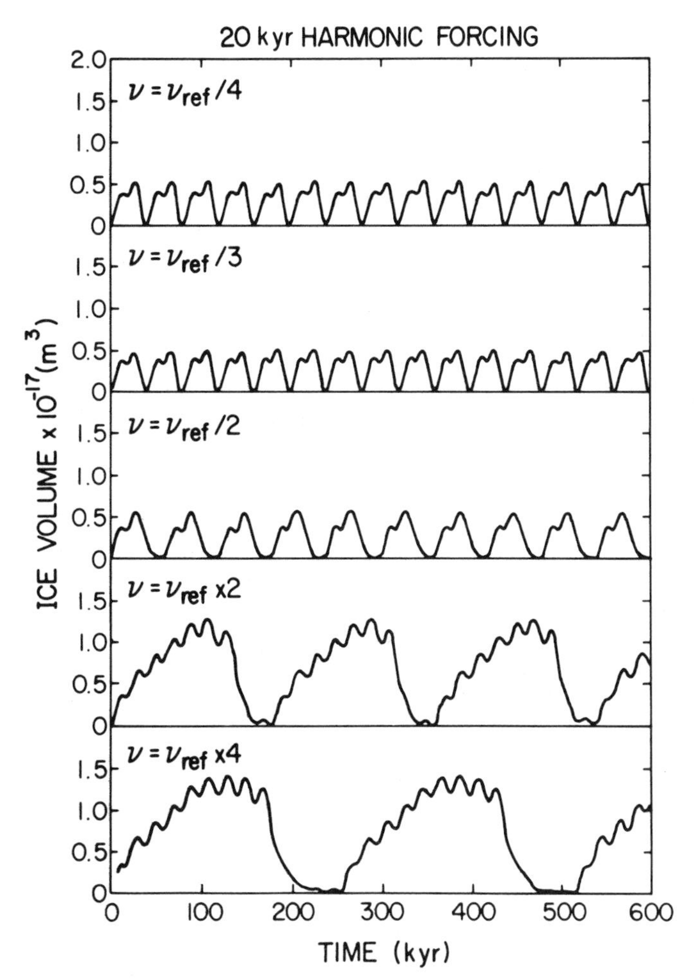

Figure 12. A sequence of ice-volume histories predicted by the model when subjected to 20-kyr simple harmonic forcing for different choices of the viscosity governing the isostatic-adjustment process. ν_{ref} is the viscosity required by the geophysical evidence discussed in later sections of this paper.

to deglaciation could be employed to infer the effective viscosity of the planetary mantle, but the subject has remained intensely controversial. It is increasingly recognized that postglacial rebound observations play an important role within the overall fabric of theoretical geodynamics, for the geological and geophysical communities have gradually accepted the convection hypothesis of continental drift over this period of time. Few informed geoscientists are not presently adherents of some version of this hypothesis, basic to which is the assumption that the earth's solid mantle behaves as a fluid when it is subjected to shear stresses of sufficient duration. In the fluid regime, the rate of flow strongly depends upon the viscosity, so that the convection hypothesis of continental drift may be construed as viable only if the effective viscosity of the mantle lies within rather well constrained bounds (Peltier, 1976; Sharpe and Peltier, 1978; Peltier, 1980b, 1985c). The number required by the convection hypothesis is in fact the same as that required by the previously discussed theory of the 100-kyr Pleistocene climate oscillation, namely about 10^{21} Pa s. This value was first inferred by Haskell from Fennoscandian uplift data on emerged strandlines dated by the varve chronology of Lidén (1938). Such differences as might exist between the requirements of glacial isostasy and convection are understandable in terms of a relatively modest transient component of the relaxation spectrum (Peltier 1985c, 1986a; Peltier and others, 1986).

Given that the most recent analyses of glacial-isostatic adjustment observations continue to deliver estimates of the viscosity of the mantle that are essentially identical to that of Haskell, one might legitimately ask why this problem has been controversial. In fact the debate has been driven by the apparent incompatibility of viscosity estimates based either on similar data from different geographic locations, or different data from the same location. The first such controversy to arise concerned a conflict between the work of Haskell and Vening-Meinesz on the one hand and that of Van Bemmelen and Berlage (1935) on the other. Using a Newtonian viscous half-space model with constant density, the former inferred a mantle viscosity near 10^{21} Pa s from the Fennoscandian emergence observations, whereas Van Bemmelen and Berlage assumed that the adjustment was confined to a thin channel of depth h. Assuming h = 100 km and fitting this model to the same Fennoscandia data analyzed by Haskell and Vening-Meinesz, they inferred a viscosity for the material in the channel of $\nu = 3 \times 10^{18}$ Pa s. Both models were considered equally plausible until the subject was revived in the early 1960s.

Prior to the advent of plate tectonics and the subsequent general acceptance of the convection hypothesis, Crittenden (1963) compiled data on the surface relaxation associated with the disappearance of Pleistocene Lake Bonneville. He discovered that, although the spatial extent of this region differed by an order of magnitude from that of the Fennoscandian uplift, the apparent relaxation times for the two regions differed only slightly—both being about 5,000 years. These results seemed to Crittenden to support the idea of thin channel flow, and he then followed Van Bemmelen and Berlage in advocating that the viscous component of the response to deglaciation was confined to the outermost few hundred kilometers of the earth.

This controversy over the Lake Bonneville and Fennoscandia data inspired the later work of McConnell (1968), who realized that the increasing confinement of the flow to the near-surface region as wavelength decreased could be accommodated by a model in which the viscosity increased smoothly as a function of depth. This represented the first attempt to use the relaxation-time information contained in the Fennoscandia data to directly constrain the viscosity (depth) profile, and many of his conclusions remain valid today. One additional feature that McConnell's model contained was a "lithosphere" at the planetary surface in which the viscosity was assumed infinite but whose thickness was a parameter of the model. He found that the relaxation spectrum extracted from the data required such a layer with a thickness constrained to be near 120 km. The presence of the thin surficial elastic lithosphere modifies Haskell's model in such a way that, for wavelengths shorter than a few times the lithospheric thickness, relaxation time is forced to decrease with increasing deformation wavenumber.

The results obtained by McConnell (1968) for the viscosity stratification in the sublithosphere region have not proven so immune to later analysis. McConnell found that the $\tau(k_H)$ data from Fennoscandia required an upper-mantle viscosity essentially equal to the value of 10^{21} Pa s previously inferred by Haskell. He was also aware that the $\tau(k_H)$ data were largely insensitive to changes of mantle viscosity below a depth of about 600 km, as later clarified by Parsons (1972). In order to constrain the viscosity at greater depth, McConnell was obliged to invoke other information. He assumed, following Munk and MacDonald (1960), that the earth's shape possessed an equatorial bulge in excess of that which would be in equilibrium with the present-day rate of rotation. This line of argument led McConnell to conclude that the earth's lower mantle had a viscosity in excess of 10^{23} Pa s, a view that was strongly reinforced by McKenzie (1966, 1967, 1968) on the basis of similar reasoning. This argument was completely undermined by Goldreich and Toomre (1969), who pointed out that the assumed nonhydrostatic component of the equatorial bulge did not in fact exist. By the early 1970s, there was therefore no reliable estimate of the viscosity of the earth's mantle beneath the transition zone.

Near the turn of the decade, data on ^{14}C–dated shorelines from Canada first became available in sufficient quantity and quality to promise a considerable improvement in the viscosity/depth resolution. This promise derives from the fact that the scale of the Laurentide ice sheet was several times in excess of the Fennoscandian. Although papers describing detailed observations of crustal rebound in North America began to appear in the early 1960s (Loken, 1962; Farrand, 1962; Washburn and Stuiver, 1962; Bloom, 1963), no detailed attempt at geophysical interpretation was made prior to that by Brotchie and Sylvester (1969) and Walcott (1970). Brotchie and Sylvester were among the first to consider the isostatic recovery problem with a spherical model, and they noted that the relative sea level (RSL) curves were characterized by shorter relaxation times than those in Fennoscandia (1,000 to 1,500 yr). Short relaxation times (2,000 yr) were also reported by Andrews (1970) for Baffin Island and other sites in the Canadian Arctic.

The validity of this general observation was also made clear by the extensive set of North America RSL data compiled by Walcott (1972). If one applies the uniform half-space model of Haskell for a load of Laurentide scale, one is again led to the conclusion that the effective viscosity of the mantle is near 10^{21} Pa s and that there is no substantial increase of viscosity from the upper to the lower mantle. This conclusion was reinforced by Cathles (1975), Peltier (1974), and Peltier and Andrews (1976), who employed different spherical viscoelastic models to show that ^{14}C–controlled RSL data from North America could not be fit by models that had any substantial increase in viscosity from the upper to the lower mantle. This was interpreted by Peltier (1976) as implying that the convective circulation in the mantle had a vertical scale equal to the mantle thickness, a notion still capable of inducing the liveliest of debates!

The result itself, however, also served to reignite rather than resolve the debate that first began with the conflicting interpretations of mantle viscosity by Haskell and by Van Bemmelen and Berlage. The reason was the incompatibility between the uniform viscosity required by the sea-level data and the very large free-air gravity anomaly observed over the Hudson Bay region which is well correlated with the topography of the Laurentide ice sheet. Walcott (1970) presented a free-air gravity-anomaly map for Canada, based on the data of Innes and others (1968), with a clear elliptical anomaly centered on Hudson Bay and trending NW with maximum amplitude near –35 mgal. As Walcott argued, it seems unreasonable to suppose, as O'Connell (1971) and Cathles (1975) subsequently did, that this anomaly was unrelated to the presently existing degree of isostatic disequilibrium over Canada. The strongest counterargument is that the zero anomaly is virtually coincident with the edge of the former ice sheet, a highly unlikely circumstance if the gravitational field in the region were strongly contaminated by other effects. Yet if one does ascribe this anomaly to isostatic disequilibrium, then one is led to an impasse in terms of conventional isostatic adjustment models. An anomaly of this size implies an amount of uplift remaining of $\Delta h = \Delta g / 2\pi G \rho$ (where Δg is the observed gravity anomaly, G the gravitational constant, and ρ the density of the material displaced to form the depression), which gives approximately 250 m. As Walcott correctly argued, such a large remaining uplift is incompatible with exponential relaxation governed by the 2,000-yr time constant inferred from the RSL data in this region.

Prior to the appearance of new analyses of the isostatic adjustment problem during the past decade, all authors who have concluded that the viscosity of the deep mantle did not differ significantly from that of the upper mantle have had to ignore the free-air gravity observations. Cathles (1975), for example, argued along with O'Connell (1971) that there is no substantial deglaciation-related free-air anomaly associated with either the Laurentian or Fennoscandian depressions. The correlation demonstrated by Walcott's analysis was considered by Cathles to be merely coincidental. At the same time O'Connell (1971) followed Dicke (1969) and attributed the nontidal component of the earth's acceleration of rotation to Pleistocene deglaciation. He excluded the possibility of a relaxation time near 100 kyr on the basis of the argument (using satellite observations) that there was *no* significant gravity anomaly remaining over Hudson Bay, in direct contradiction of Walcott (1970).

The apparent inability of Newtonian viscoelastic earth models to simultaneously satisfy both RSL and free-air gravity data has been used by several authors as a point of departure from which to launch arguments that the fault lay with the Newtonian constitutive relation itself. Jeffreys (1972), for example, maintained that the difficulty is due to the assumption that the material behavior was steady-state Newtonian viscous in the limit of long time. He argued that *any* such steady-state deformation of mantle matieral was impossible—a corrolary to which is the notion that thermal convection in the mantle cannot occur! Although somewhat less extreme in their conclusions, Post and Griggs (1973) argued that the observed relation between the uplift remaining and the rate of uplift demands that the mantle be considered non-Newtonian in its mode of steady-state deformation. Their conclusion, however, is based on the assumption that the present-day free-air anomaly associated with the Fennoscandian deglaciation is about –30 mgal. Best current estimates of the amplitude of this anomaly suggest values of about –15 mgal (e.g., Balling, 1980; Walcott, 1973, accepts –17 mgal), and this completely undermines the conclusion of Post and Griggs. Cathles (1975), arguing for a mantle of relatively uniform viscosity, is obliged to claim that the free-air gravity anomaly over Fennoscandia is only about –3.5 mgal (again in sharp discord with the observations).

A different modification of the rheology was proposed by Weertman (1978) to address the difficulty posed by the RSL-derived uniform viscosity model for his view, based on microphysical considerations of the solid-state creep process, that the stress-strain relation of the mantle was non-Newtonian. Because a-priori considerations suggested that a Newtonian analysis of the rebound of a non-Newtonian mantle should deliver a radial viscosity profile characterized by a sharp increase with depth, he suggested that the low values apparently required for the deep-mantle viscosity represented a transient component rather than the steady-state component of the mantle viscosity spectrum. This idea does not appear to address the problems posed by the failure of Newtonian models to simultaneously reconcile the free-air gravity and RSL data with the same values of the model parameters.

In the past decade the issue of mantle rheology has remained vital, as one would expect of a question that lies at the heart of modern geodynamics and its central concern for a detailed understanding of the nature of the deep-seated process of mantle convection and its surface manifestation as plate tectonics. In the following sections of this chapter, I review progress since the publication of Cathles (1975), particularly new models of the rebound process that effect a reconciliation between the requirements of RSL and free-air gravity data. I also discuss the way in which new observations of earth rotation and polar motion have been employed to corroborate inferences based on RSL and free-air gravity information.

MANTLE VISCOSITY FROM POSTGLACIAL CHANGES OF RELATIVE SEA LEVEL

One of the more significant advances in our understanding of Quaternary geological processes achieved over the past decade concerns the interactions between the aquasphere, cryosphere, and solid earth that are responsible for determining variations of relative sea level induced by the last deglaciation beginning about 18 ka. The account of this interaction is based on the theoretical analysis of the response of a spherical viscoelastic planet to surface loading (Peltier, 1974). This theory was employed by Peltier and Andrews (1976) to predict the RSL variations caused by a realistic global melting event involving several ice sheets under the assumed constraint that the glacial meltwater enters the global ocean uniformly.

Farrell and Clark (1976) showed how this assumption could be relaxed to account for the nonuniform filling of the ocean basins required to keep the ocean surface as an equipotential surface. This extended model was used to make RSL predictions for a realistic melting event by Peltier and others (1978) and Clark and others (1978). Probably the most significant result obtained was the prediction of raised beaches in the southern oceans beginning at roughly 6 ka and achieving maximum elevations above present-day sea level of about 3 m. Prior to this (e.g., Russell, 1961), there had been widespread debate as to whether these raised beaches were produced by a "eustatic" fall of sea level or were caused by local tectonic uplift. In order to obtain a good fit to the 6-ka beaches, however, Clark and others (1978) delayed the melting of the northern hemisphere ice sheets in their model by about 2 kyr, a shift that violated the ^{14}C–controlled disintegration isochrons of the Laurentian and Fennoscandian ice sheets. Only very recently has this flaw in the model been rectified (e.g., see Wu and Peltier, 1983, and Peltier, 1987, for discussions).

The use of this model to constrain radial viscoelastic earth structure remains a major topic of research. The model is illustrated here through recent work designed to infer both lithospheric thickness and the radial profile of sublithospheric viscosity on the basis of ^{14}C–controlled RSL data from North America. The model requires two inputs to make such predictions. The first is a detailed model of the deglaciation history from the maximum glaciation to the present. Peltier and Andrews (1976) provided a first model of the disintegration of northern hemisphere ice sheets called ICE-1. This was subsequently refined by Wu and Peltier (1983) in detailed analyses of the fits of the RSL predictions of the model with Northern Hemisphere observations. The refined model (called ICE-2) included an Antarctic component of the melting history based upon Clark and Lingle (1979), who had this event leading the melting of northern hemisphere ice. On the basis of detailed calculations with this model, Wu and Peltier (1983) suggested that the time of the emergence of the 6-ka beach in the southern oceans could best be fit by delaying the west Antarctic component of the disintegration history from that of Clark and Lingle (1979) by several thousand years rather than by delaying the melting of Northern Hemisphere ice, as proposed by Clark and others (1978).

This scenario, recently investigated in detail by Peltier (1987), proves to be in much closer accord with the basic idea underlying the climate model discussed previously, namely that the basic 100-kyr oscillation of terrestrial ice volume is fundamentally driven by Northern Hemisphere processes. In this scenar-

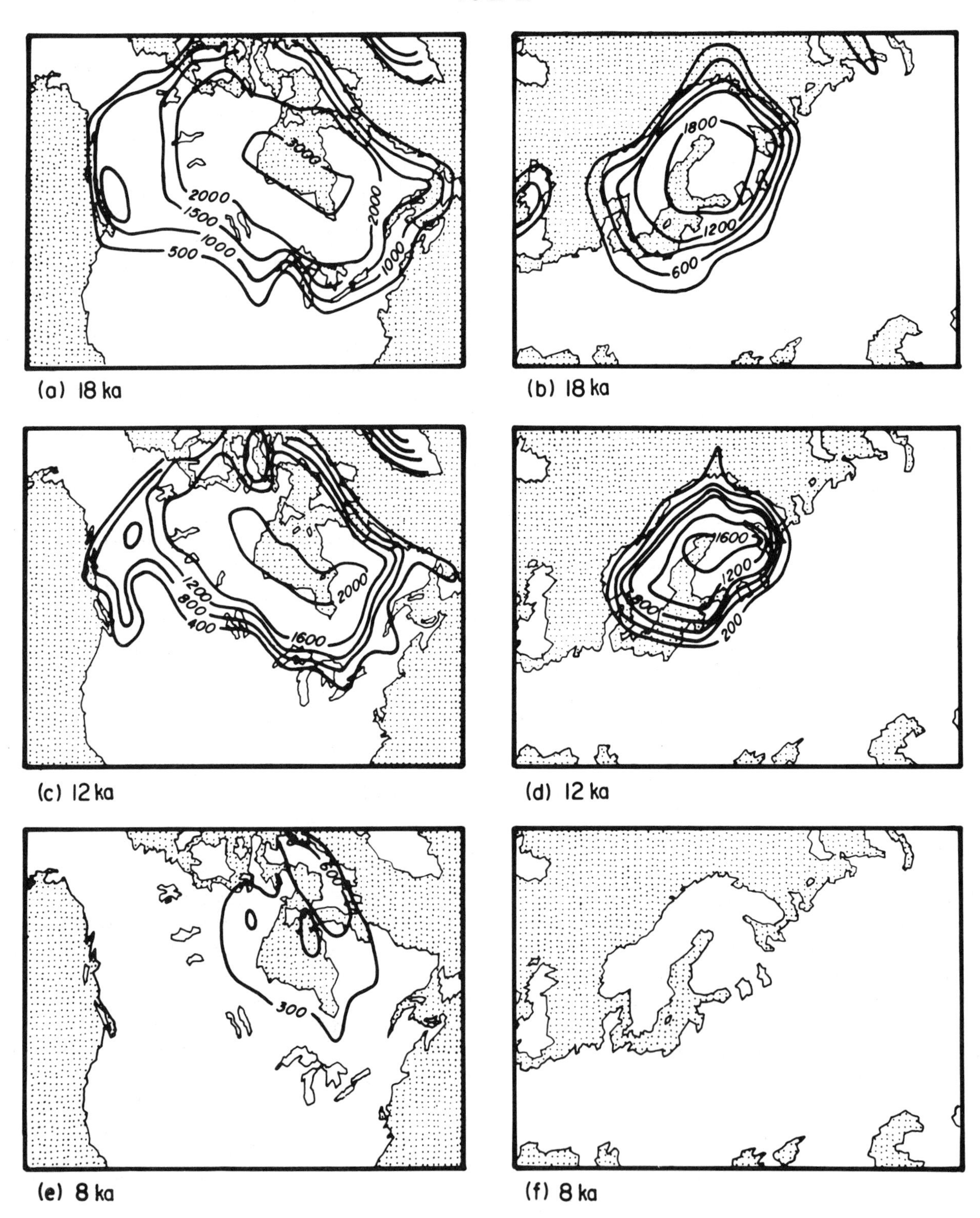

Figure 13. Three time slices through the ICE-2 melting chronology tabulated in Wu and Peltier (1983). Maps of ice-sheet topography are shown for both the Laurentian (a, c, e) and Fennoscandian (b, d, f) complexes.

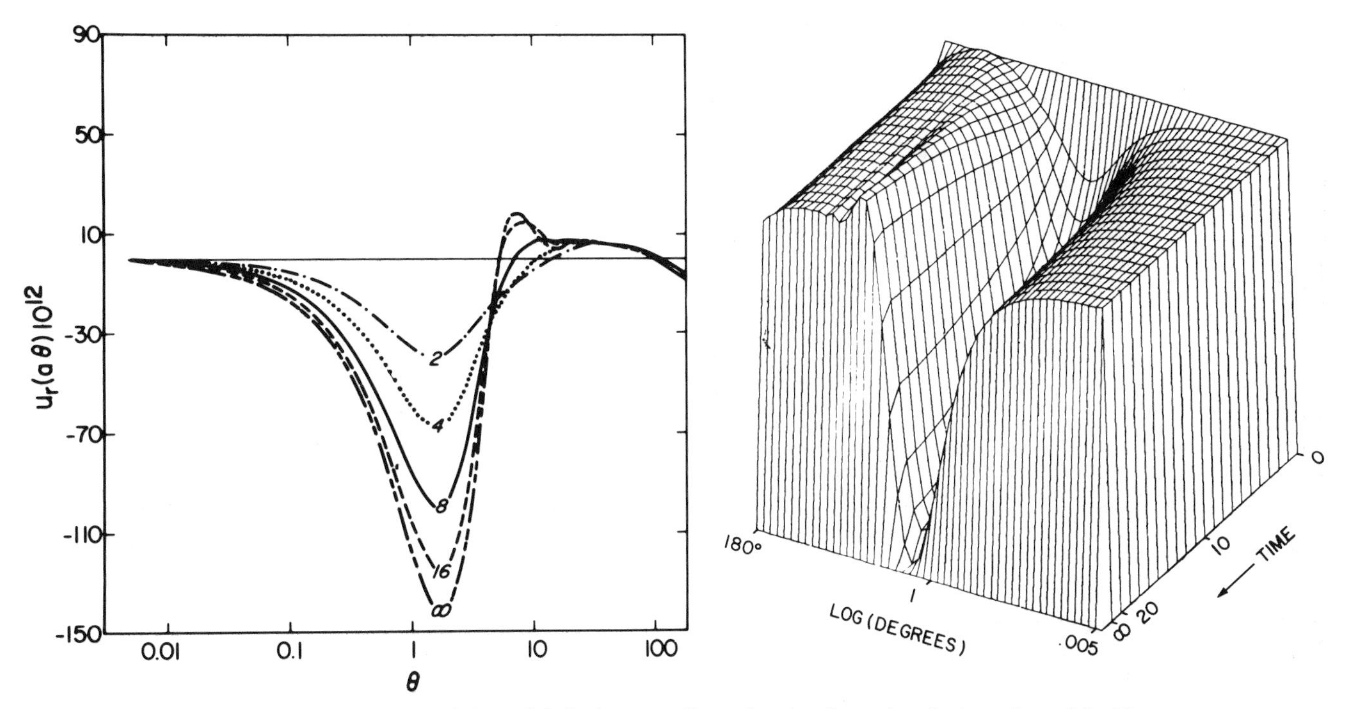

Figure 14. Viscous part of the radial displacement Green function for a viscoelastic earth model with 1066B elastic structure, a lithospheric thickness of 120.7 km, and upper-mantle and lower-mantle viscosity of 10^{21} Pa s.

io, the observed synchronization of Northern and Southern Hemisphere advances and retreats is explained by the hypothesis that the marine nature of the west Antarctic ice sheet allows it to respond rapidly and sympathetically to expansions and contractions of Northern Hemisphere continental glaciation through the sea-level variations associated with the latter. Rising sea levels associated with the Northern Hemisphere termination, itself induced by orbital insolation changes, provokes a disintegration of west Antarctic ice by floating the ice sheet's grounding line.

The ICE-2 model of Wu and Peltier (1983), with delayed Antarctic melting, is the best available model (according to RSL and other geophysical data) of the disintegration of continental ice sheets that occurred during the last termination. The Northern Hemisphere components of this model are illustrated in Figure 13, which shows three time slices through the thickness isopachs of the Laurentian-Innuitian-Cordilleran and Fennoscandian complexes beginning at the 18-ka maximum ice advance. The maximum thickness of the Laurentide ice sheet at 10 ka is close to 3,500 m, which is about 1,500 m thinner than the preferred models in the reconstructions of Denton and Hughes (1981). This is explained by the Denton and Hughes assumption that the ice sheet was in equilibrium at 18 ka, an assumption unlikely to be correct. In fact the climate model described previously predicts (see Fig. 11) that the ice volume will be lower than the equilibrium volume at glacial maximum by about 30 percent. The climate model therefore correctly predicts the degree of disequilibrium required to understand the difference between the model of ice sheet thickness derived by considerations of glacial isostasy (Peltier and others) and that deduced by steady-state ice mechanical considerations (Denton and Hughes).

The second input required by the model consists of the set of space- and time-dependent viscoelastic Green functions required to calculate signatures of the response of the earth to an imposed ice-sheet load. These Green functions are computed from the theory developed in Peltier (1974), which itself requires as input a model of the assumed spherically symmetric elastic and viscous components of earth structure. The elastic part of this structure is assumed to be fixed, so the viscosity, as the only free parameter, is determined by fitting the model to RSL and other observations. An example of a viscoelastic Green function for a realistic earth model is illustrated in Figure 14, which shows the viscous part of the variation of radial displacement that would be induced by the addition onto the earth's surface of a 1-kg point mass load at angular position $\theta = 0$. The time-dependent viscous displacement from zero at $t = 0$ is shown at 1-kyr intervals.

Given a model of surface glaciation and deglaciation, such as the ICE-2 history described previously, convolution of this history with the Green function constitutes a mathematical prediction of the variation of planetary shape (rebound) that would be induced by the variation of surface mass (ice). Given a model of the radial viscoelastic structure and a model of the deglaciation history, the theory delivers predictions of the RSL history at any point on the earth's surface that may be of interest and that can be compared with observations.

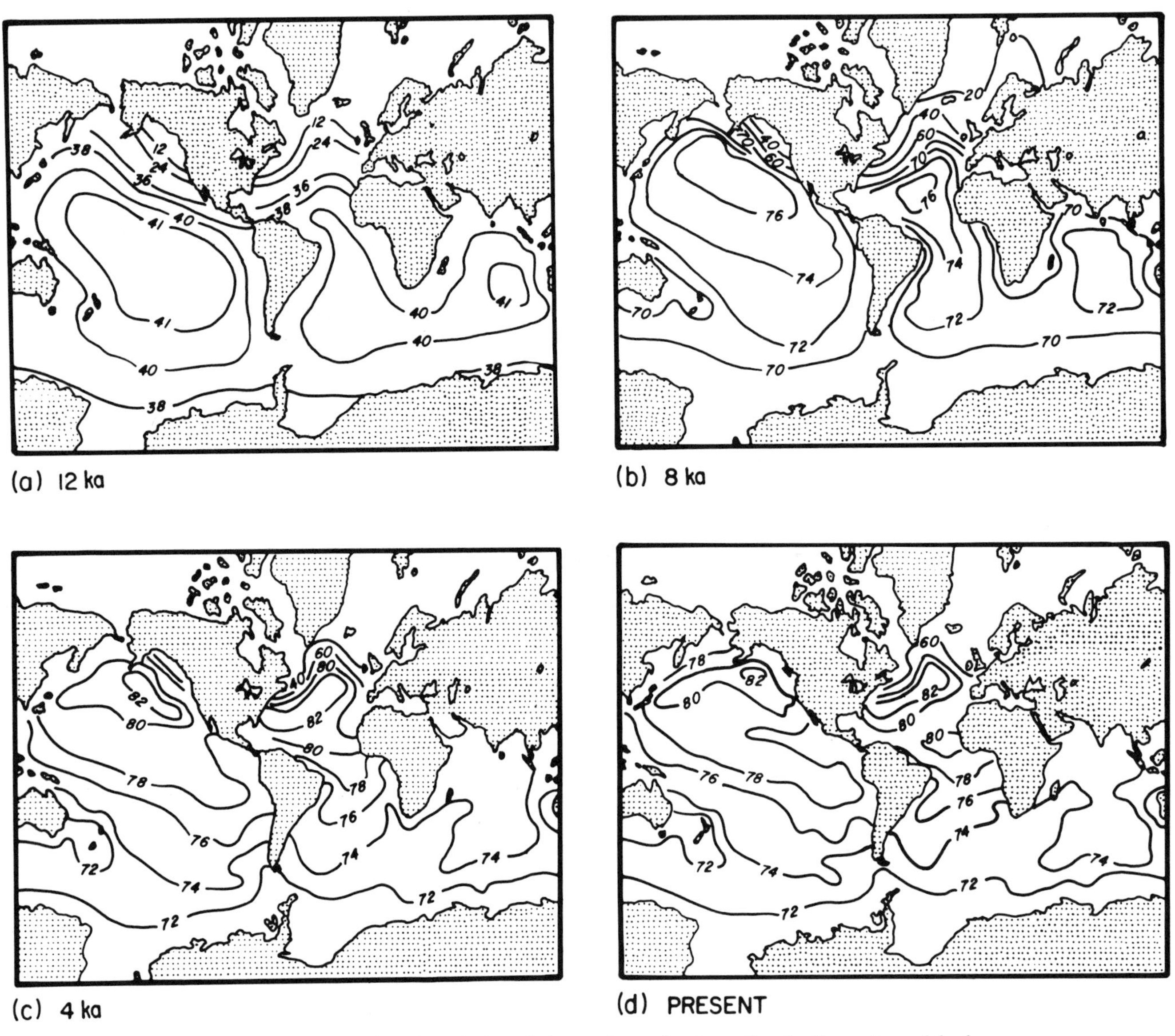

Figure 15. Four time slices through the solution to the sea-level equation for the earth model whose radial-displacement Green function was shown in Figure 14, based on the deglaciation history shown in Figure 13 and neglecting entirely the melting of the west Antarctic ice sheet.

An example of the output obtained by solution of the sea-level equation is shown in Figure 15 for a calculation performed with the ICE-2 disintegration chronology. The solution is illustrated by contours of RSL rise in meters at four times from 12 ka to the present. In order to make a prediction of the RSL history that should be observed at any specific location, we pierce the stack of such maps, which are delivered at 1-kyr intervals, at any particular point from which data are available. The current data base compiled for such analyses contains ^{14}C–controlled RSL information from 374 locations on the earth's surface. The detailed systematic and formal inversion of these data to deliver viscoelastic earth structure is a current research subject.

For present purposes it will suffice to illustrate the main results of such analyses already obtained from RSL histories for North American sites. All of these illustrations are from Peltier (1986a) and pertain to sites whose locations are in Figure 16. The elastic earth structure employed in all these model calculations is a simple multi-homogeneous layer approximation to the seismically realistic structure of 1066-B, which I have previously called CML64 (Peltier, 1986a). Figures 17 through 21 show analyses for 15 sites illustrating the sensitivity of the model predictions to variations of both lithospheric thickness (L) and lower mantle viscosity (NULM), with the upper mantle viscosity fixed to the 10^{21} Pa s value of Haskell. Inspection of these analyses demonstrates the following points: (1) data from sites within the ice margin (e.g., Hudson Bay) are insensitive to lithospheric thickness

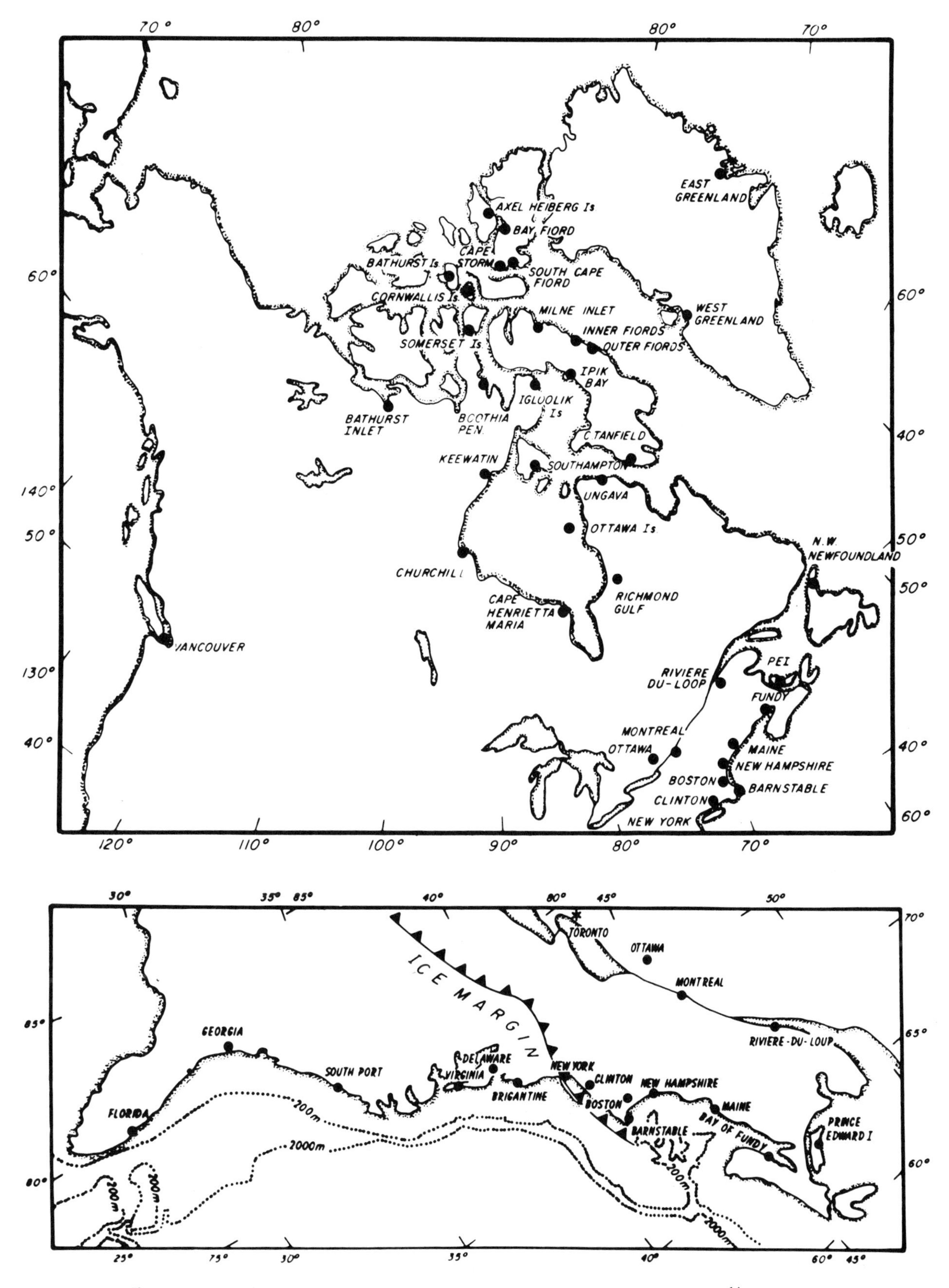

Figure 16. Location maps for sites on the North American continent at which ^{14}C-controlled relative-sea-level data are available.

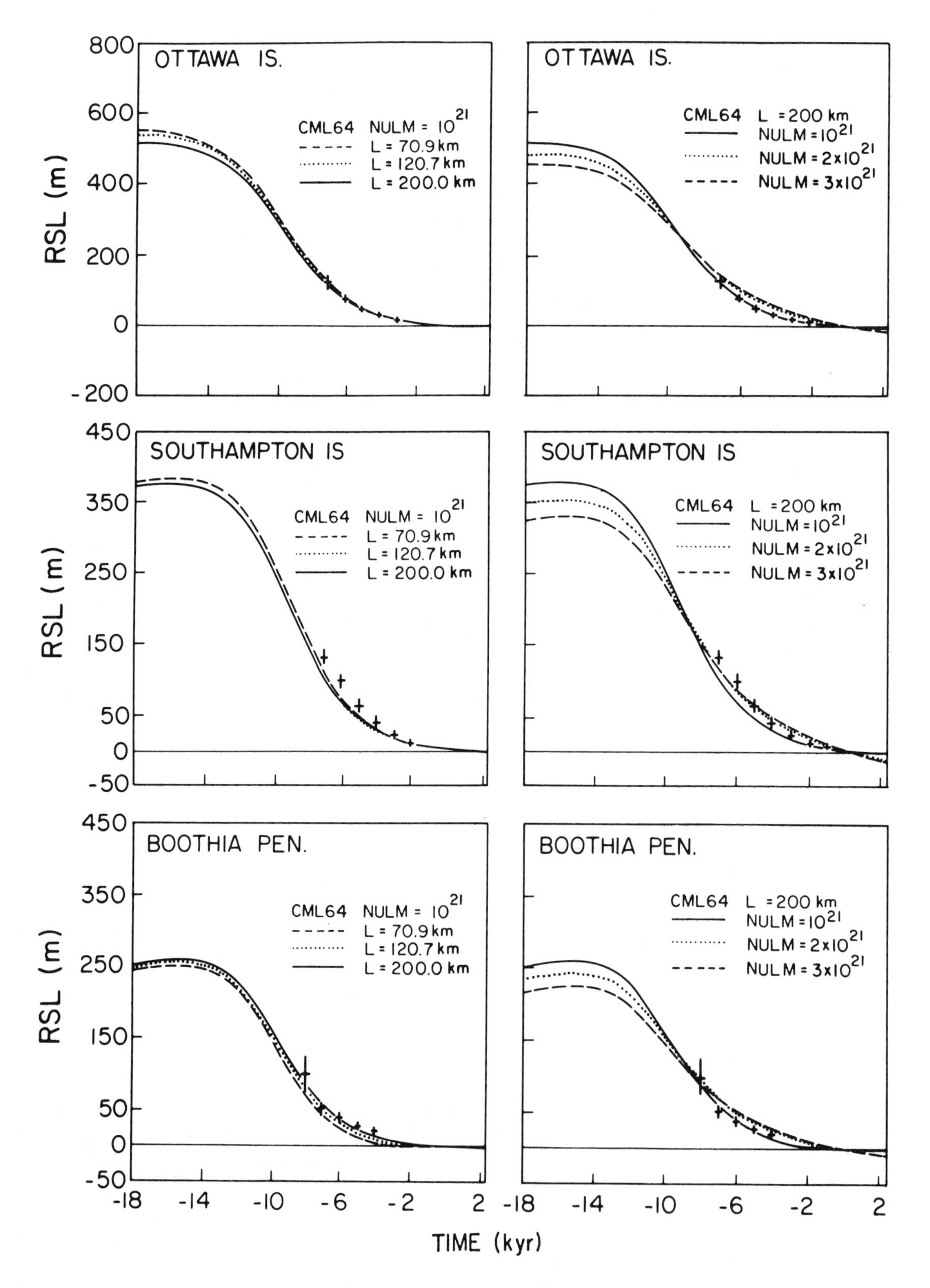

Figure 17. Comparisons between predicted and observed RSL histories at three of the North American sites on the map shown in Figure 16. The left column illustrates the influence of variations of lower-mantle viscosity, and the right column demonstrates the sensitivity of the model predictions to changes of lithospheric thickness.

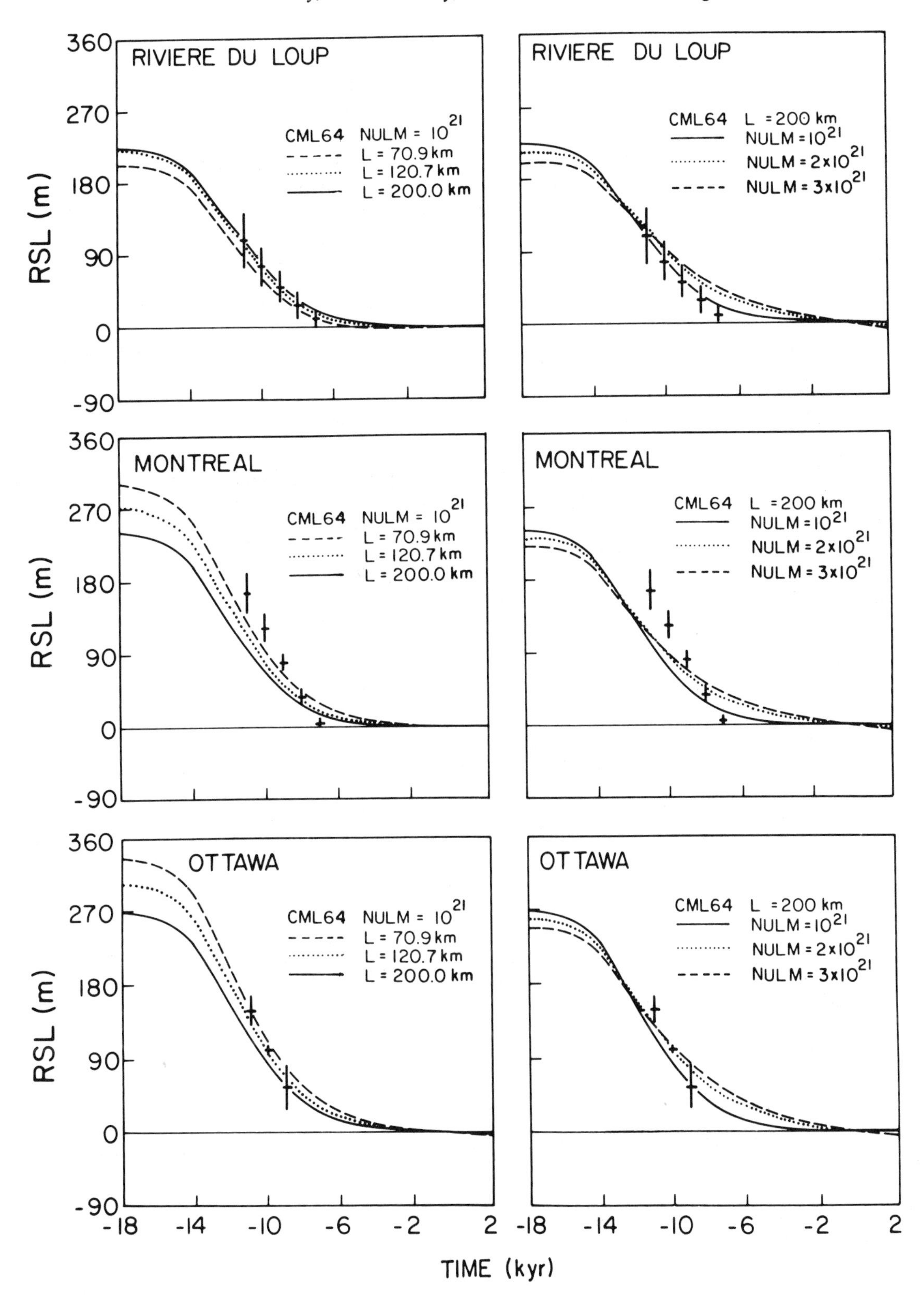

Figure 18. Same as Figure 17 but for three additional sites.

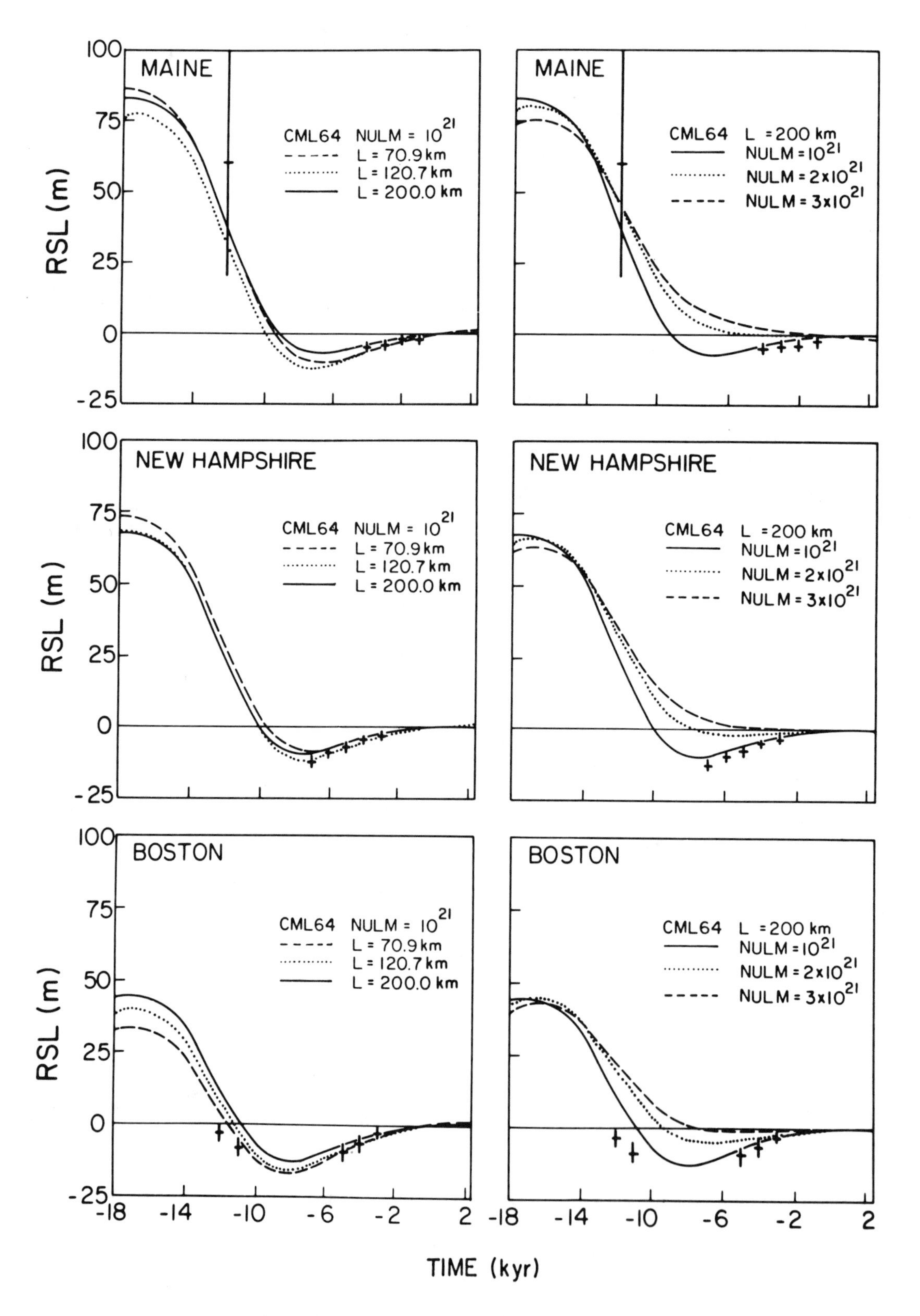

Figure 19. Same as Figure 17 but for three additional sites.

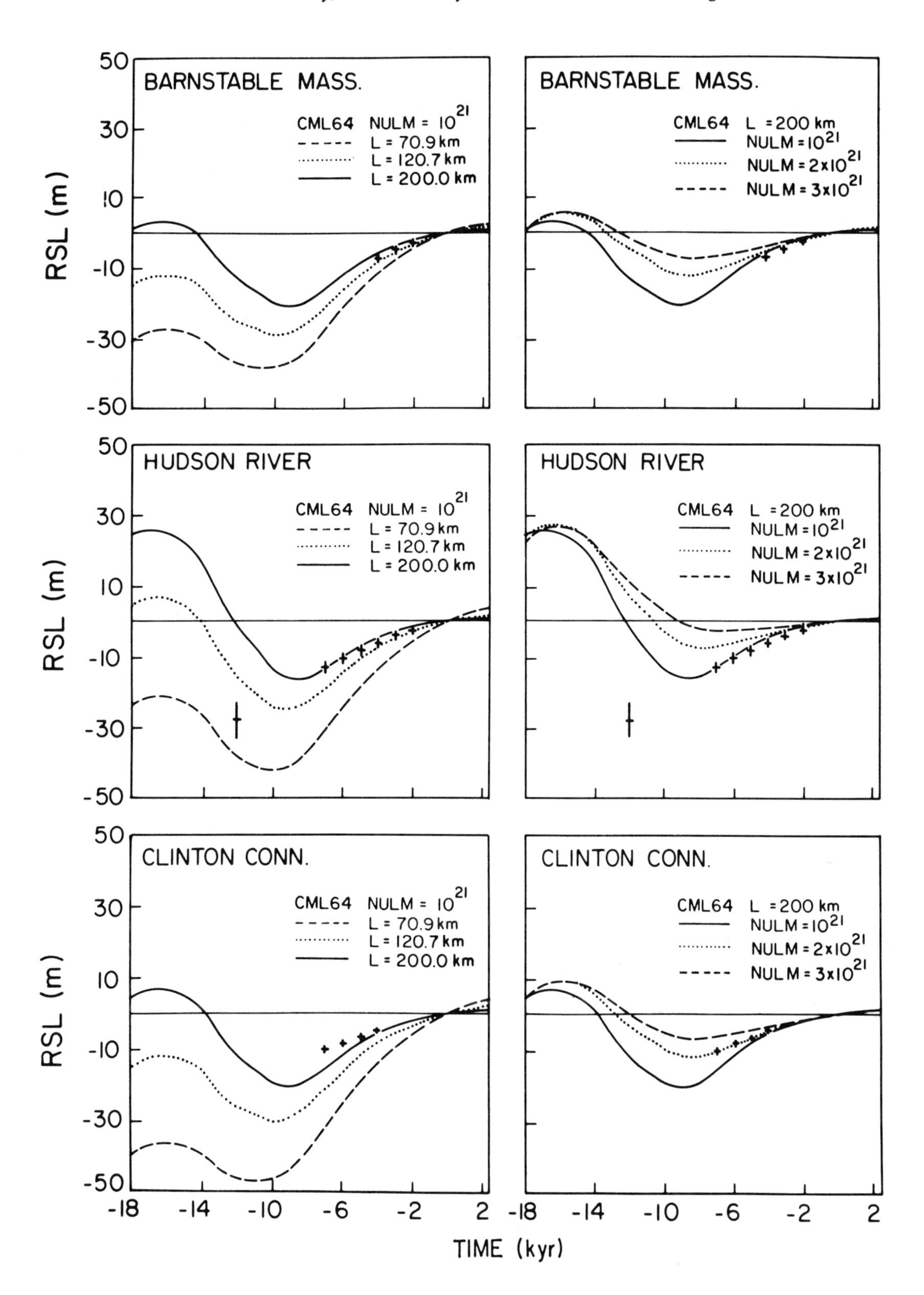

Figure 20. Same as Figure 17 but for three additional sites.

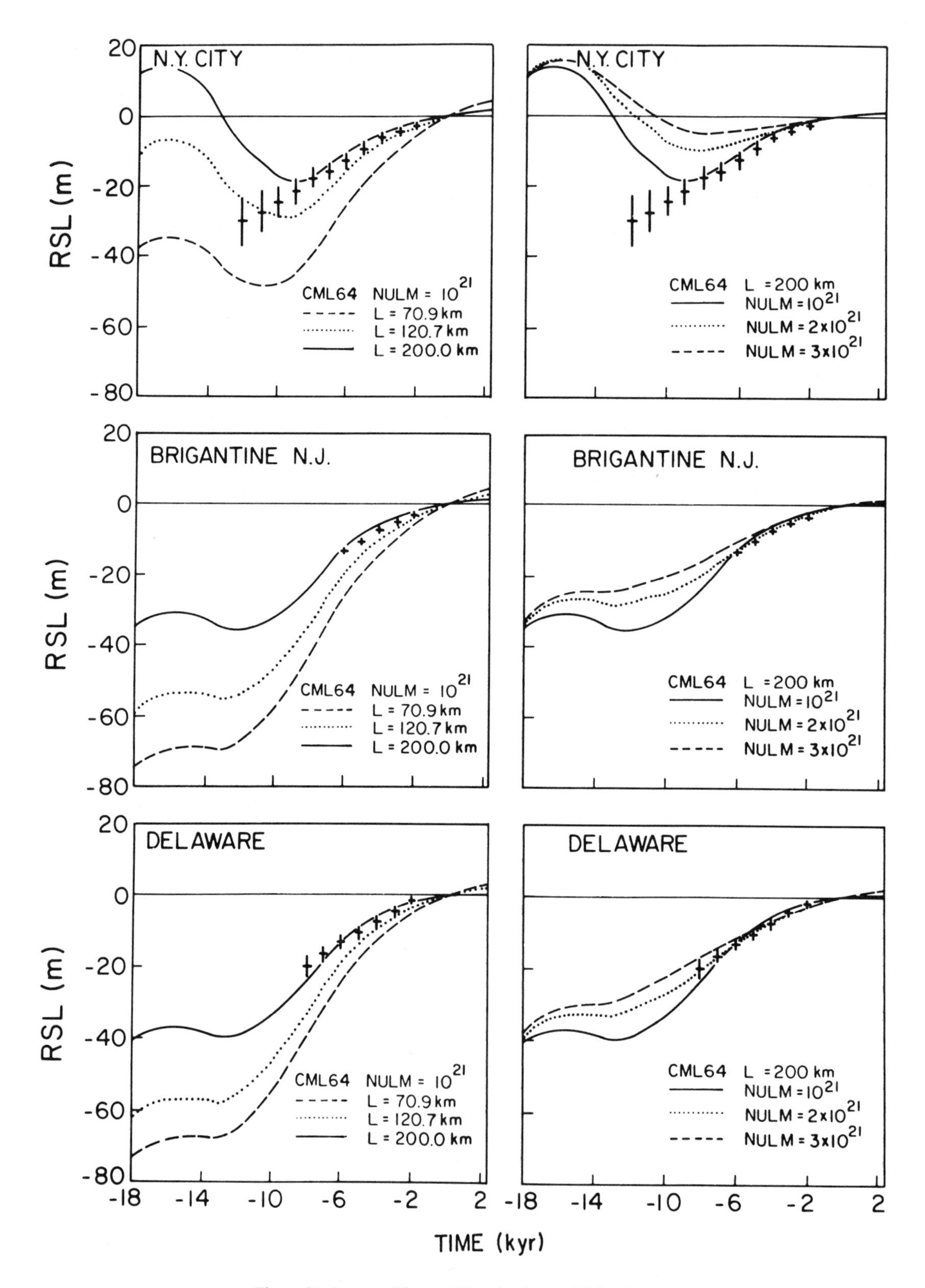

Figure 21. Same as Figure 17 but for three additional sites.

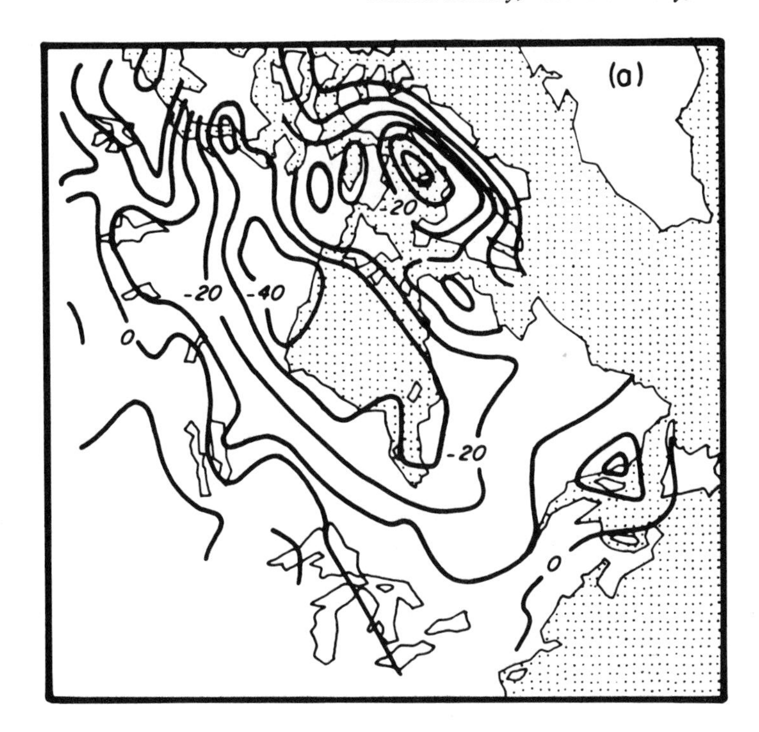

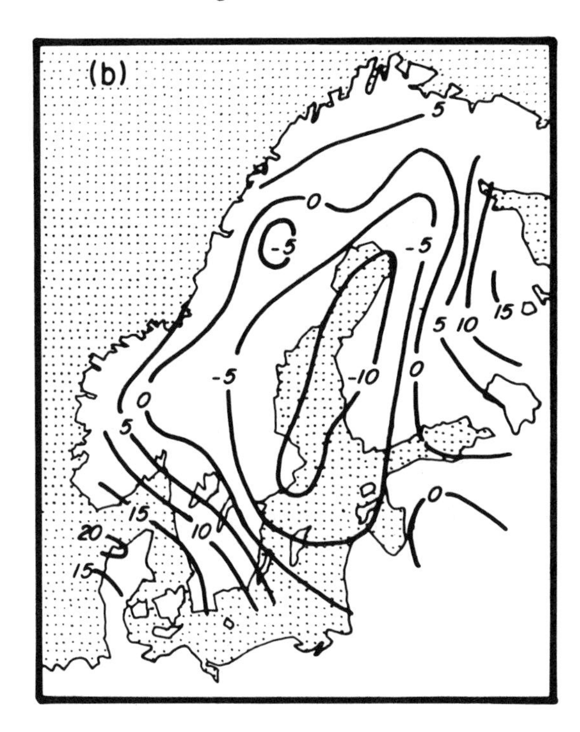

Figure 22. Free-air gravity (in mgals) for the Laurentide (a) and Fennoscandian (b) regions. Data sources are discussed in the text.

but quite sensitive to lower-mantle viscosity, and they serve to restrict the latter to a value less than about 3×10^{21} Pa s, with the upper mantle value fixed to 10^{21} Pa s; (2) data from the immediate vicinity of the ice margin like Maine, New Hampshire, and Boston, Massachusetts, are also very sensitive to lower-mantle viscosity and also constrain its value to a number less than about 3×10^{21} Pa s; (3) data from points farther south along the U.S. East Coast, on the other hand, are very sensitive to lithospheric thickness and may be invoked (Peltier, 1984) to constrain this parameter to a value near 200 km, if possible effects of transient rheology (Peltier, 1985b) and to errors in the icesheet melting history are neglected.

The main conclusions are therefore that the viscosity of the earth's mantle is rather uniform and near the Haskell value of 10^{21} Pa s (except for an increase by a factor of about two in the lower mantle) and that the thickness of the continental lithosphere (at least of the N.E. section of the N.A. lithosphere) is about 200 km, somewhat greater than previously assumed by some (e.g., Sclater and others, 1980, preferred 120 km based on heat-flow considerations). In the following sections of this chapter, I investigate the extent to which other geophysical observations are compatible with these values, which are based on RSL information.

MANTLE VISCOSITY FROM FREE-AIR GRAVITY-ANOMALY OBSERVATIONS

Figure 22 displays free-air gravity-anomaly maps for both Canada and Fennoscandia based on the previously cited analyses of Walcott (1970) and Balling (1980) respectively. In Canada the peak negative anomaly over the Hudson Bay region is somewhat less than −35 mgal. Over the Gulf of Bothnia, it is near −10 mgal, which Balling (1980) argues should be increased to about −15 mgal due to bias traceable to geoidal undulations of larger spatial scale. Accepting these anomalies as observations to be reconciled by the model of glacial isostasy used to predict RSL histories, we are obliged to generalize somewhat the basic ingredients of this theory. The reason has to do with a critical and fundamental difference between RSL observations and free-air gravity data.

This fundamental difference is as follows. RSL histories clearly constitute records of the separation of the surface of the solid earth from the surface of the ocean (the geoid) extending backward into the past and relative to a zero determined by present-day sea level. They do not constitute absolute measures of either the deformation of the solid earth or of the geoid but only histories of their relative local displacement from some time subsequent to the onset of deglaciation up to the present. Free-air gravity anomalies, on the other hand, provide an absolute measure of the presently existing degree of isostatic disequilibrium, since they are determined by the surface deformation that has yet to occur before isostatic equilibrium will be restored. It is therefore clear that RSL data "look into the past," whereas free-air gravity data "look into the future." This is a fundamental point, for it ensures that if each wavelength in the deformation spectrum is characterized by a number of distinct relaxation times, then RSL data are governed principally by the shortest decay times, whereas free-air gravity data are extremely sensitive to the longest

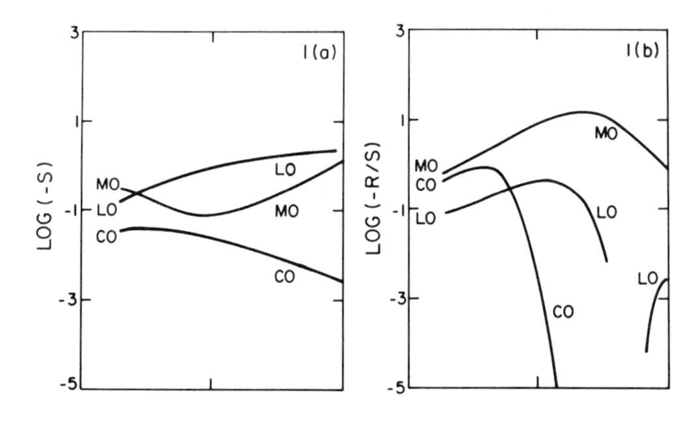

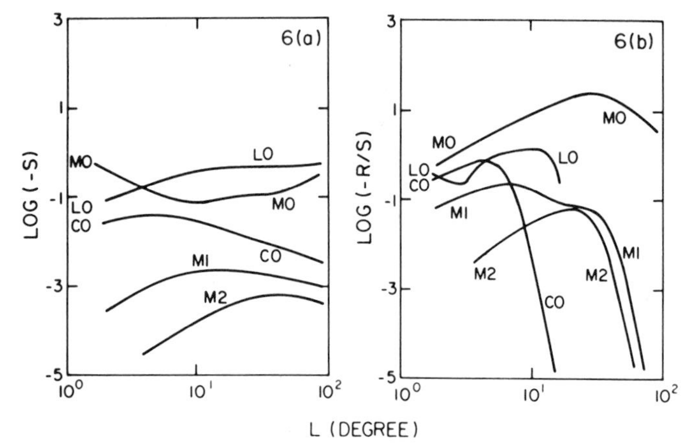

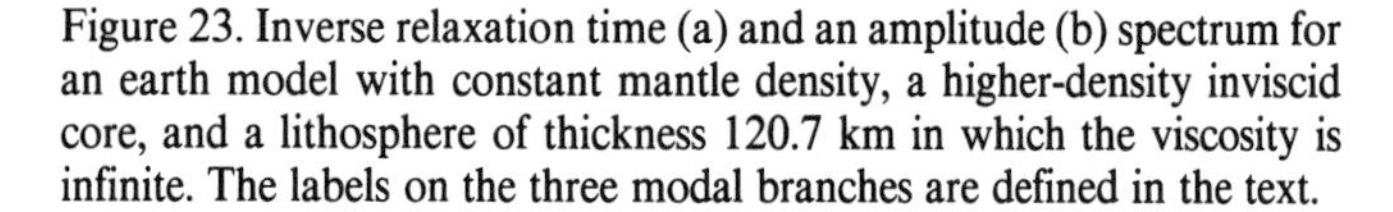
Figure 23. Inverse relaxation time (a) and an amplitude (b) spectrum for an earth model with constant mantle density, a higher-density inviscid core, and a lithosphere of thickness 120.7 km in which the viscosity is infinite. The labels on the three modal branches are defined in the text.

Figure 24. As in Figure 23 but for an Earth model whose elastic structure is that of the seismically realistic model PREM.

decay times if the loading history is of sufficient duration for significant relaxation to have been excited in the longest time-scale modes. In order to reconcile the previously cited dichotomous requirements of the RSL and free-air gravity data, we have only to understand how the relaxation spectrum of a realistic earth model might possess several distinct relaxation times for each deformation wavelength rather than the single unique time that governs models of the Haskell type even when mantle viscosity is a function of depth.

As should be clear on physical grounds, different modes of viscous gravitational relaxation are introduced into the spectrum of the viscoelastic model by the action of different horizons within the earth's interior that are capable of inducing a buoyant restoring force when they are deflected from their equilibrium depths by the gravitational perturbations associated with glacial loading and unloading. The core/mantle boundary is one internal interface that thereby resists any tendency to deform it, because the material on the two sides of the boundary is of different chemical composition and state (molten iron compared to solid iron–magnesium sililcate). However, this is not the only internal horizon capable of adding such complexity to the modal relaxation spectrum, as evidenced by Figure 23, which shows relaxation time and amplitude spectra for an earth model with a mantle of constant density and viscosity and an inviscid higher-density core. In addition the model includes a lithosphere with infinite viscosity and 120.7 km thickness.

Figure 23 shows (Peltier and others, 1986) that each wavenumber in the deformation spectrum (spherical harmonic degree labelled L) possesses three characteristic relaxation times: MO, CO, and LO, representing the fundamental mantle, core, and lithosphere modes respectively. In Figure 23a, inverse relaxation times (s) are nondimensionalized with a characteristic time scale of 10^3 years. The MO mode is dominant for all L, and the CO mode contributes significantly only for L< 8; deformation wavelengths shorter than this are unaffected by the presence of the core (Fig. 23b). The lithosphere mode LO never carries more than 10 percent of the variance for any value of L. The main effect of the lithosphere is to reverse the tendency for relaxation time to increase with increasing L beyond L = 25 for the MO mode.

These data are to be compared with those on Figure 24 for the seismically realistic model PREM of Dziewonski and Anderson (1981). The spectrum of the latter model differs by the appearance of two additional modes labelled M1 and M2, which exist (Peltier and others, 1986) because the phase boundaries at 420- and 650-km depth bracketing the mantle transition zone have been treated as non-adiabatic horizons for motions on the time scale of glacial rebound. Unlike the modes MO, CO, and LO, which have relaxation times that are at most 10^4 years, the M1 and M2 modes have relaxation times that are 10^6 years or greater. Given that the characteristic times for glacial loading is near 9×10^4 years, it is clear that the deformation carried by the combined effects of the three fundamental modes will be near isostatic equilibrium by the time deglaciation commences. It is equally clear, however, that this may not be the case for the M1 and M2 modes.

Figure 25 illustrates the important impact that the presence of such modes has on model predictions of the present-day peak free-air gravity anomalies expected over Laurentia and Fennoscandia. The observed anomalies from Figure 22 are shown as the hatched regions. As discussed in detail in Peltier and others (1986), these predictions were made on the basis of circular disk load approximations to the two ice sheets, with the loading history controlled by the $\delta^{18}O$ data shown in Figure 1. Five 100-kyr sawtooth cycles were assumed to constitute the entire prehistory of ice-sheet loading and unloading. In Figure 25, comparisons between theory and observation are shown for six mod-

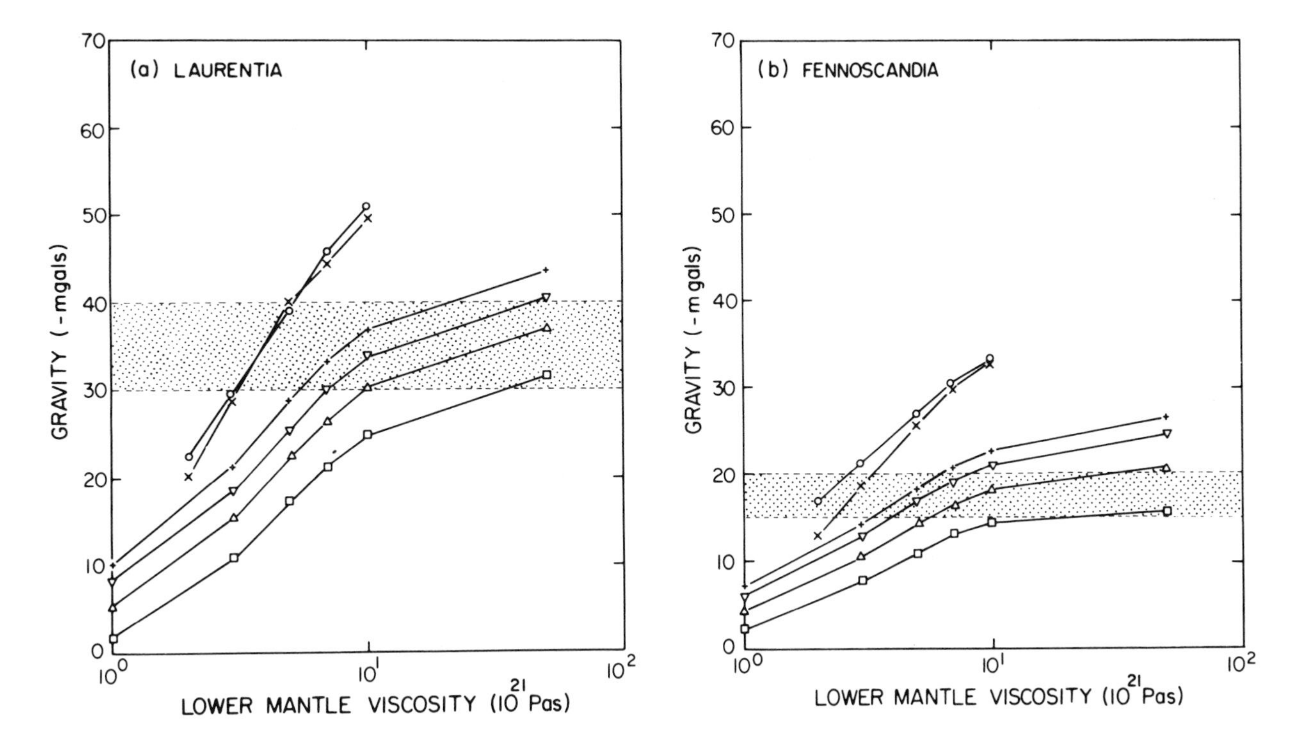

Figure 25. Comparisons of predicted and observed peak free-air gravity anomalies over Laurentia (a) and Fennoscandia (b). Predictions are shown for six different viscoelastic earth models, which differ from one another solely in terms of the amount of internal buoyancy in their mantles. The models each have lithospheric thicknesses of 120.7 km and upper-mantle viscosities of 10^{21} Pa s. The different symbols represent predictions for the following models: □, for a model with no internal mantle buoyancy; Δ, for a model with a 6.2 percent increase of density at 670-km depth; ∇ , for a model with 6.2 and 3.8 percent increases of density at 670- and 420-km depth respectively; +, for a model with a 12.4 percent increase of density at 670-km depth; x, for model 1066B; O, for PREM. See Peltier and others (1986) for details.

els, which differ from one another (as described in the figure caption) only by the amount of buoyancy assumed to act across the phase boundaries at 420- and 650-km depth. These models vary from one with zero internal mantle buoyancy (whose predictions are shown by the open squares) to the seismically realistic model PREM, which is assumed to behave in a completely non-adiabatic manner (predictions shown by the open circles).

The important conclusion following from this sequence of comparisons is that it is not possible to explain observed free-air gravity anomalies with the same almost isoviscous model of the mantle required to reconcile RSL data unless the mantle contains considerable internal buoyancy. However, if all of the density variation occurring across the 420- and 650-km seismic discontinuities is assumed to behave non-adiabatically on the time scale of glacial isostasy, then free-air gravity data are nicely fit by a model whose parameters are fixed by the RSL data.

Peltier (1985c) discusses the implications of this result to the interpretation of the 650-km discontinuity. Although it might appear that the explanation of this horizon as an equilibrium phase transformation would not be compatible with the requirement that it be able to induce a buoyant restoring force on deflection, this turns out to be not entirely accurate. It is in fact quite possible for the 650-km boundary to behave non-adiabatically on the time scale of glacial isostasy so long as the phase change is almost univariant. In this case, material will actually change phase only on the rather long time scale of thermal diffusion. In any event, the discovery that internal mantle buoyancy appears to be necessary to reconcile the otherwise conflicting requirements of isostatic adjustment observations effectively resolves what had become a major impediment to further progress in this field.

MANTLE VISCOSITY FROM EARTH-ROTATION OBSERVATIONS

Both of the above-described classes of geophysical data sample a rather wide range of spatial scales simultaneously. Earth-rotation observations are unique, for they are sensitive almost exclusively to the degree-two components of the deformation spectrum. They therefore offer the prospect of optimal depth resolution and of testing the validity of the mantle viscosity structure inferred on the basis of the glacial-isostatic adjustment data discussed above. There are two specific rotational observables that may be employed for this purpose: the so-called nontidal component of the acceleration of the earth's axial rotation rate, and the secular drift of the rotation pole relative to the surface geography upon which the dominant Chandler and Annual wobbles are superimposed. Only during the past five years has it been demonstrated that both of these processes are probably effects of the same large-scale deglaciation event responsible for inducing the previously discussed RSL variations and free-air gravity anomalies (Peltier, 1982, 1983, 1985c; Wu and Peltier, 1984). It is also only recently that these observations have been reconfirmed by modern space-based geodetic techniques. The original inference of the existence of a nontidal component of the earth's acceleration of rotation was based on the analysis of ancient eclipse data (e.g., Müller and Stephenson, 1975). Similarly, the first observation of a secular drift of the rotation pole with respect to the surface geography was made from the photo-zenith-tube data of the International Latitude Service (Vincente and Yumi, 1969, 19670). The former observation was recently confirmed by analysis of laser ranging data to the LAGEOS satellite (Yoder and others, 1983; Rubincam, 1984; Peltier, 1983). The latter is in the process of confirmation using modern VLBI observations (e.g., Carter and others, 1986).

A very detailed recent reanalysis of the constraints on mantle viscoelastic structure provided by the LAGEOS observation of the nontidal acceleration is found in Peltier (1985a), which extends the original analysis in Peltier (1983). This acceleration is not influenced substantially either by internal mantle buoyancy or by lithospheric thickness. Therefore it provides a useful constraint on the upper-mantle/lower-mantle viscosity contrast. Results of such analyses are summarized in Figure 26, which shows predictions of the present-day nontidal acceleration of rotation as a function of lower-mantle viscosity, with the upper-mantle value fixed at Haskell's value of 10^{21} Pa s. The earth model has 1066B elastic structure and a lithospheric thickness of 120.7 km. The LAGEOS observation is marked as the hatched region, and theoretical predictions are shown for three different loading models, which differ from one another only in the number of ice sheets assumed to force the rotational response. Curve L includes only the dominant Laurentide ice sheet, L + F incorporates the influence of Fennoscandia, while L + F + A adds the effect of west Antarctica and is the prediction to be compared to the LAGEOS result. In these calculations, each of the three ice sheets was approximated by a circular cap assumed to accrete and disintegrate on the 100-kyr time scale revealed by the δ^{18}O data and to have undergone five such oscillations prior to the last deglaciation event. Figure 26 shows that the observation may be fit by the theory for either one or the other of two widely spaced values for the viscosity of the lower mantle, one near 3×10^{21} Pa s and the other near 10^{23} Pa s. Clearly, the former is compatible with the requirements of the RSL data, whereas the latter is not and must

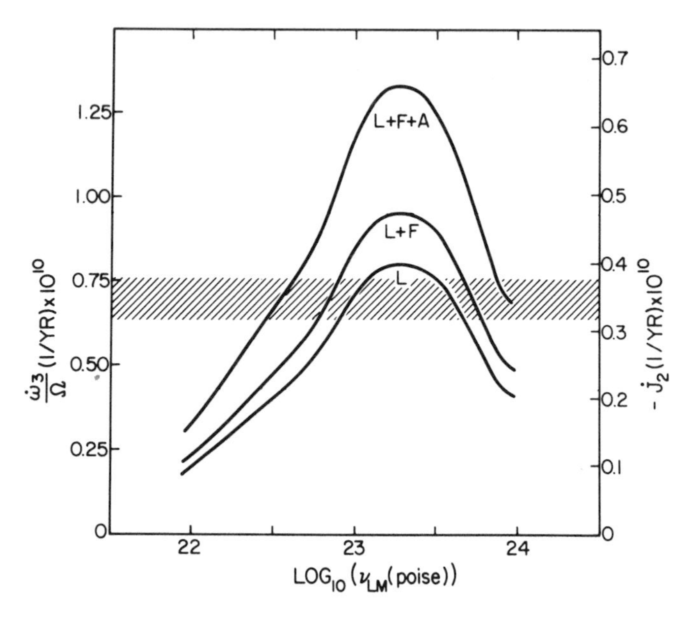

Figure 26. Predicted and observed nontidal acceleration of planetary rotation (or equivalently $\dot{J}_2$ for three different loading histories containing only Laurentide forcing (L), Laurentide plus Fennoscandian forcing (L+F), and including the forcing from the deglaciation of west Antarctica (L+F+A). The earth models have constant lithospheric thickness of 120.7 km and upper-mantle viscosity of 10^{21} Pa s. For each loading history the predictions are shown as a function of lower-mantle viscosity.

be rejected. The nontidal acceleration observation provides a very useful confirmation of the validity of the previously described analysis of the viscosity profile, but not an independent requirement for the unique ratio of upper-mantle to lower-mantle viscosity inferred from the RSL data.

The second of the two rotational observations mentioned above is less precise than the nontidal acceleration in constraining the ratio of the upper- and lower-mantle viscosities. As we await the results from VLBI observations of the secular drift of the rotation pole (a preliminary result was presented in Carter and others, 1986), the best-available constraint on the polar motion remains the ILS data, which are reproduced here in Figure 27. This shows the motion of the pole from A.D. 1900 in terms of x and y components in the coordinate system shown on the inset polar projection. The polar motion is dominated by an oscillatory signal with period near one year and amplitude modulated at a period of about seven years. The oscillatory signal is produced by the interference between the annual wobble and the Chandler wobble, whose periods are respectively 12 and 14 months. Of greater interest here is the slow secular drift of the pole on which these dominant fluctuations are superimposed. The rate of this drift is near 0.95 (± 0.15) degrees per m.y., while its direction is that shown by the arrow, which points toward eastern Canada on the inset polar projection. Also shown for reference purposes on

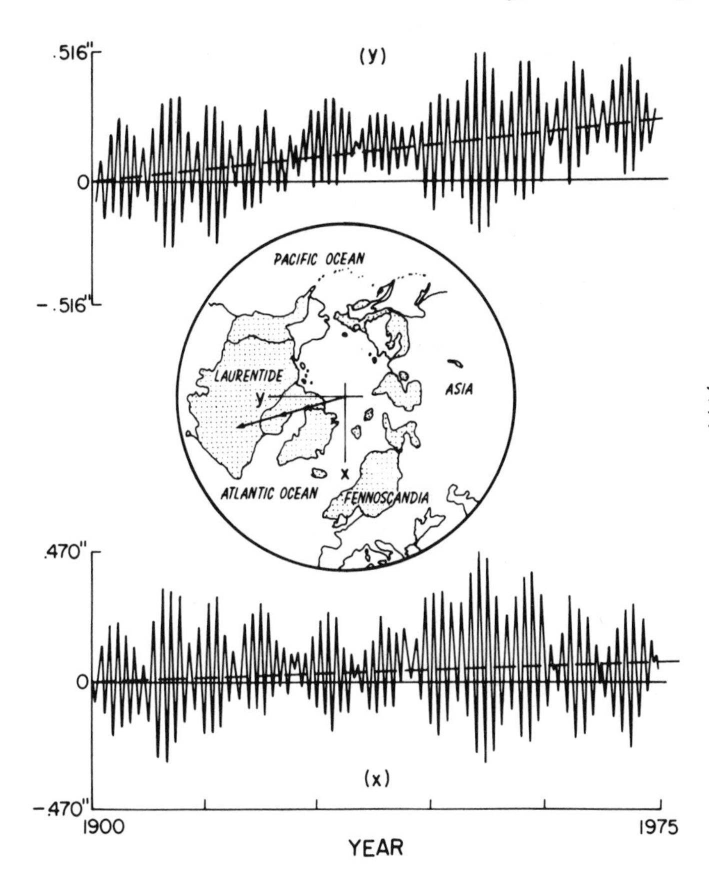

Figure 27. International Latitude Service (ILS) data showing the location of the rotation pole relative to the CIO (conventional international origin) as a function of time since A.D. 1900. The seven-year beat in the polar-wander signal is a consequence of the superposition of the 12-month annual wobble and the 14-month Chandler wobble. This oscillatory signal is superimposed on the slow secular drift denoted by the dashed lines, which corresponds to a polar-wander speed of 0.95 ± 0.15 degrees per 10^6 years in the direction shown on the inset polar projection.

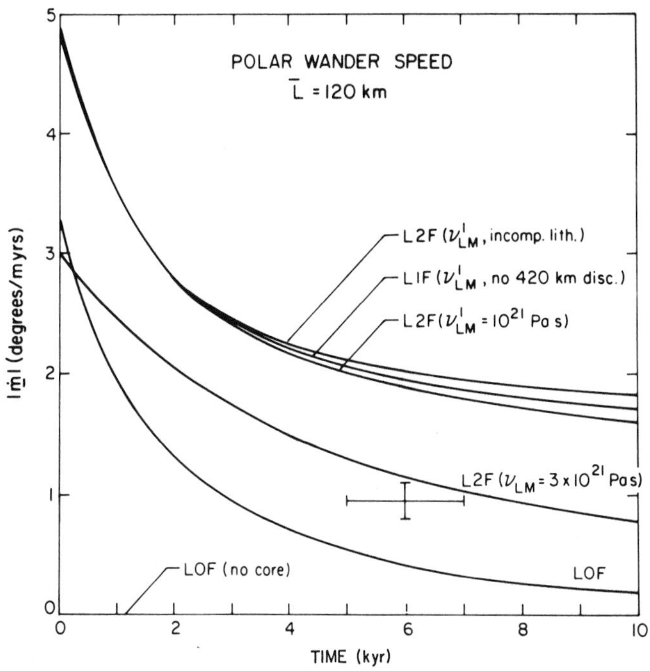

Figure 28. Predicted polar-wander velocity as a function of time for the series of simplified viscoelastic models described in the text. The zero of time is taken to coincide with the end of the final deglaciation phase of a glaciation history consisting of seven complete 100-kyr cycles. The polar-wander speed implied by the data in Figure 27 is shown as the cross at a time (t = 6 kyr) such as to provide a reasonable fit to the final deglaciation phase of the loading history described by the ICE-2 chronology of Wu and Peltier (1983).

this inset are the locations of the Laurentide and Fennoscandian ice masses.

Until recently the interpretation of this polar drift has been the source of some controversy in the geophysical literature. The first analysis by Munk and Revelle (1952) led to the suggestion that the drift was associated with the presently occurring melting of ice on Greenland and/or Antarctica. The problem with this conjecture is that such direct analyses of the mass balance of these ice sheets as are presently available (e.g., Meier, 1984) do not support the idea that any significant negative mass balance exists. Dickman (1977) hypothesized that the secular motion of the pole is associated with the process of continental drift. This explanation would appear to be excluded on the basis of the fact that on the long time scale of continental drift the system remains in isostatic equilibrium and therefore the mechanism is extremely inefficient at exciting polar wander.

Peltier (1982) and Wu and Peltier (1984) first demonstrated that the observed drift in the ILS path was completely explicable as a memory of the planet's response to the last deglaciation event of the current ice age. This analysis showed that the previous conclusion of Munk and Revelle (1952), to the effect that such polar drift required simultaneous redistribution of the surface mass load, was an artifact of their assumption that a homogeneous viscoelastic earth model could provide an adequate vehicle to address this problem. No current melting of Greenland or Antarctic ice was required to explain the observations. This is demonstrated by predictions of polar-wander speed shown in Figure 28 for a sequence of models. These are labelled LNF, in which N denotes the number of density discontinuities in the mantle, F denotes flat (the models are multiple homogeneous layer approximations to the seismically realistic model 1066B), and L represents lithosphere, so that each model in the set has a lithosphere of thickness $\bar{L}$ = 120 km. In each calculation, circular disk loads represent the Laurentide, Fennoscandian, and west Antarctic ice sheets, and the history of loading is assumed to consist of seven 100-kyr cycles of glaciation and deglaciation, each of which has a simple sawtooth form consisting of a 90-kyr glaciation phase and a 10-kyr termination.

The observed polar-wanter speed is shown as the cross centered on Time = 6 kyr. The speed prediction for model LOF, which has a constant mantle viscosity of 10^{21} Pa s and includes the density discontinuity across the core/mantle boundary, differs substantially from zero but is still too low to fit the data. This misfit is corrected in model L1F, which includes the density discontinuity across the seismic horizon at 670-km depth but still has a uniform mantle viscosity of 10^{21} Pa s. Addition of a second density discontinuity associated with the seismic horizon at 420-km depth (either with a compressible lithosphere, as is normally assumed, or with an incompressible lithosphere) does not significantly modify the speed prediction, which remains high for models with constant mantle viscosity. A modest increase in the viscosity of the lower mantle, however, completely eliminates the misfit (model L2F [$\nu_{LM} = 3 \times 10^{21}$ Pa s]), and this model is the same model previously shown to be required by observations of relative sea level, free-air gravity, and nontidal acceleration of rotation. That it also fits the observed direction of polar wander was demonstrated in Wu and Peltier (1984). Of paramount importance from the point of view of the ideas developed in this chapter, is the fact that this model of the internal viscoelastic structure of the earth's mantle was previously shown to fit the requirements of the explanation of the 10^5-yr cycle of ice-age ice-sheet volume fluctuations revealed through $\delta^{18}O$ data from deep-sea sedimentary cores.

CONCLUSIONS

In the preceding sections of this chapter, I reviewed and in places extended the ideas developed over the past decade of work on the problems of glacial-isostatic adjustment and Quaternary climate change. On the basis of these analyses, it would appear that the process of glacial isostatic adjustment plays an active role in the mechanism of long time scale climatic change. In fact, in the absence of this process, it does not appear to be possible to understand the most important manifestation of climatic change to have occurred during the past million or more years of earth history, that is to say the appearance and disappearance of the vast continental ice sheets, which, according to the $\delta^{18}O$ data from deep-sea cores, have come and gone at roughly 100-kyr intervals. A model of this ice age cycle has been developed which, when forced by realistic summertime seasonal insolation anomalies, delivers a rather accurate simulation of the ice-volume variations inferred from the SPECMAP time series of Imbrie and others (1984). This model predicts the 100-kyr period characteristic of this cycle only when the viscosity governing the isostatic adjustment process is fixed to the value required by the data of glacial isostasy. In the absence of the action of this physics in the climate system, the astronomical theory of climate change developed by Milankovitch and others would appear to be, at best, incomplete, for this theory otherwise does not predict the existence of the 100-kyr cycle that dominates the ice-volume record.

REFERENCES CITED

Andrews, J. T., 1970, A geomorphological study of postglacial uplift with particular reference to Arctic Canada: London and New York, Oxford University Press, 156 p.

Balling, N., 1980, The land uplift in Fennoscandia, gravity field anomalies and isostasy, *in* Mörner, N.-A., ed., Earth rheology, isostasy, and eustasy: New York, Wiley, p. 297–321.

Berger, A. L., 1978, Long term variations of daily insolation and Quaternary climatic changes: Journal of Atmospheric Science, v. 35, p. 2362–2367.

Birchfield, G. E., Weertman, J., and Lunde, A. T., 1981, A paleoclimatic model of Northern Hemisphere ice sheets: Quaternary Research, v. 15, p. 126–142.

——, 1982, A model study of the role of high-latitude topography in the climatic response to orbital insolation anomalies: Journal of Atmospheric Science, v. 39, p. 71–87.

Bloom, A. L., 1963, Late Pleistocene fluctuations of sea level and postglacial crustal rebound in coastal Maine: American Journal of Science, v. 261, p. 862–879.

Broecker, W. S., and Van Donk, J., 1970, Isolation changes, ice volumes, and the O^{18} record in deep sea cores: Reviews in Geophysics and Space Physics, v. 8, p. 169–198.

Brotchie, J. F., and Sylvester, R., 1969, On crustal flexure: Journal of Geophysical Research, v. 74, p. 5240–5252.

Carter, W. E., Robertson, D. S., Pyle, T. E., and Diamante, J., 1986, the application of geodetic radio interferometric surveying to the monitoring of sea level: Geophysical Journal of the Royal Astronomical Society, v. 87, p. 3–13.

Cathles, L. M., 1975, The viscosity of the Earth's mantle: Princeton, New Jersey, Princton University Press, 386 p.

Clark, J. A. and Lingle, C. S., 1979, Predicted relative sea level changes (18,000 years B.P. to present) caused by late glacial retreat of the Antarctic ice sheet: Quaternary Research, v. 11, p. 279–298.

Clark, J. A., Farrell, W. E., and Peltier, W. R., 1978, Global changes in postglacial sea level; A numerical calculation: Quaternary Research, v. 9, p. 265–287.

Crittenden, M. D., Jr., 1963, Effective viscosity of the earth derived from isostatic loading of Pleistocene Lake Bonneville: Journal of Geophysical Research, v. 68, p. 5517–5530.

Croll, J., 1875, Climate and time: New York, Appleton and Co., 388 p.

Daly, R. A., 1934, The changing world of the ice age: New Haven, Connecticut, Yale University Press, 271 p.

Denton, G. H., and Hughes, T. J., 1981, The last great ice sheets: New York, John Wiley and Sons, 484 p.

Dicke, R. H., 1969, Average acceleration of the earth's rotation and the viscosity of the deep mantle: Journal of Geophysical Research, v. 74, p. 5895–5902.

Dickman, S. R., 1977, Secular trend of the earth's rotation pole; Consideration of motion of the latitude observatories: Geophysical Journal of the Royal Astronomical Society, v. 57, p. 41–50.

Dziewonski, A. M., and Anderson, D. L., 1981, Preliminary reference earth model: Physics of the Earth and Planetary Interiors, v. 25, p. 297–356.

Farrand, W. R., 1962, Postglacial uplift in North America: American Journal of Science, v. 260, p. 181–199.

Farrell, W. E., and Clark, J. A., 1976, On post-glacial sea level: Geophysical Journal of the Royal Astronomical Society, v. 46, p. 647–667.

Ghil, M., and Le Treut, H., 1983, A climate model with cryodynamics and geodynamics: Journal of Geophysical Research, v. 36, p. 5262–5270.

Goldreich, P., and Toomre, A., 1969, Some remarks on polar wandering: Journal of Geophysical Research, v. 74, p. 2555–2567.

Haskell, N. A., 1935, The motion of a viscous fluid under a surface load: New York, 1. Physics, v. 6, p. 265–269.

——, 1936, The motion of a viscous fluid under a surface load: New York, 2. Physics, v. 7, 56–61.

——, 1937, The viscosity of the asthenosphere: American Journal of Science, v. 33, p. 22–28.

Hays, J. D., Imbrie, J., and Shackleton, N. J., 1976, Variations in the Earth's orbit; Pacemaker of the ice ages: Science, v. 194, p. 1121–1132.

Hyde, W. T., and Peltier, W. R., 1985, Sensitivity experiments with a model of the ice age cycle; The response to harmonic forcing: Journal of Atmospheric Science, v. 42, p. 2170–2188.

——, 1987, Sensitivity experiments with a model of the ice age cycle; The response to Milankovitch forcing: Journal of Atmospheric Science, v. 44, p. 1351–1374.

Imbrie, J., and 8 others, 1984, The orbital theory of Pleistocene climate; Support from a revised chronology of the marine $\delta^{18}O$ record, *in* Berger, A., Imbrie, J., Hays, J., Kukla, G., and Saltzman, B., eds., Milankovitch and climate: Dordrecht, Nederlands, D. Reidel, p. 269–305.

Innes, M.J.S., Goodacre, A. K., Weston, A., Argun, A., and Weber, J. R., 1968, Gravity and isostasy in the Hudson Bay region, *in* Beals, C. S., Science, history, and Hudson Bay: Ottawa, Ontario, Canada Department of Energy, Mines and Resources, v. 2, p. 703–728.

Jeffreys, H., 1972, Creep in the Earth and planets: Tectonophysics, v. 13, p. 569–581.

Köppen, W., and Wegener, A., 1924, Die klimate der geologischen vorzeit: 256 p.

Lidén, R., 1938, Den senkvartära strandförskjutningens förlopp och kronologi i Angermanland: Geologiska Föreningen i Stockholm, Förhandlingar, bd. 60, p. 397–404.

Lindzen, R. S., 1986, A simple model for 100K-year oscillations in glaciation: Journal of Atmospheric Science, v. 43, p. 986–996.

Löken, O. H., 1962, The late-glacial and postglacial emergence and deglaciation of northern-most Labrador: Geography Bulletin, v. 17, p. 23–56.

McConnell, R. K., 1968, Viscosity of the mantle from relaxation time spectra of isostatic adjustment: Journal of Geophysical Research, v. 73, p. 7089–7105.

McKenzie, D. P., 1966, The viscosity of the lower mantle: Journal of Geophysical Research, v. 71, p. 3995–4010.

——, 1967, The viscosity of the mantle: Geophysical Journal of the Royal Astronomical Society, v. 14, p. 297–305.

——, 1968, The geophysical importance of high temperature creep, *in* Phinney, R. A., ed., The history of the Earth's crust: Princeton, New Jersey, Princeton University Press, p. 28–44.

Meier, M. F., 1984, Contribution of small glaciers to global sea level: Science, v. 226, p. 1418–1421.

Milankovitch, M. M., 1941, Canon of isolation and the ice-age problem: Koniglich Serbische Akademie, Beograd. (English translation by the Israel Program for Scientific Translations; published by the U.S. Department of Commerce and the National Science Foundation, Washington, D.C.)

Müller, P. M., and Stephenson, F. R., 1975, The acceleration of the Earth and moon from early observations, *in* Rosenburg, G. D., and Runcorn, S. K., eds., Growth rhythms and history of the Earth's rotation: New York, Wiley, p. 459–534.

Munk, W. H., and MacDonald, G. F., 1960, The rotation of the Earth: London and New York, Cambridge University Press, 323 p.

Munk, W. H., and Revelle, R., 1952, On the geophysical interpretation of irregularities in the rotation of the Earth: Monthly Notices of the Royal Astronomical Society, Geophysics Supplement, v. 6, p. 331–347.

O'Connell, R. J., 1971, Pleistocene glaciation and the viscosity of the lower mantle: Geophysical Journal of the Royal Astronomical Society, v. 23, p. 299–327.

Oerlemans, J., 1980, Some model studies of the ice age problem [thesis]: Utrecht, Koninkliyk Nederlands Meteorologisch Instituut Publication 158.

Parsons, B. E., 1972, Changes in the Earth's shape [thesis]: Cambridge University.

Peltier, W. R., 1974, The impulse response of a Maxwell Earth: Reviews in Geophysics and Space Physics, v. 12, p. 649–669.

——, 1976, Glacial isostatic adjustment; II. The inverse problem: Geophysical Journal of the Royal Astronomical Society, v. 46, p. 669–706.

——, 1980a, Ice sheets, oceans, and the Earth's shape, *in* Mörner, N.-A., ed., Earth rheology, isostasy, and eustasy: New York, Wiley, p. 45–63.

——, 1980b, Mantle convection and viscosity, *in* Dziewonski, A. M., and Boschi, E., eds., Physics of the earth's interior: Amsterdam, North-Holland Publishers, p. 362–431.

——, 1982, Dynamics of the ice age Earth: Advances in Geophysics, v. 24, p. 1–146.

——, 1983, Constraint on deep mantle viscosity from LAGEOS acceleration data: Nature, v. 304, p. 434–436.

——, 1984, The thickness of the continental lithosphere: Journal of Geophysical Research, v. 89, p. 11303–11316.

——, 1985a, The LAGEOS constraint on deep mantle viscosity; Results from a new normal mode method for the inversion of viscoelastic relaxation spectra: Journal of Geophysical Research, v. 90, p. 9411–9421.

——, 1985b, New constraints on transient lower mantle rheology and internal mantle buoyancy from glacial rebound data: Nature, v. 318, p. 614–617.

——, 1985c, Mantle convection and viscoelasticity: Annual Review of Fluid Mechanics, v. 17, p. 561–608.

——, 1986a, Deglaciation induced vertical motion of the North American continent and transient lower mantle rheology: Journal of Geophysical Research, v. 91, p. 9099–9123.

——, 1986b, A relaxation oscillator model of the ice age cycle, *in* Nicholas, C., and Nicholas, G., eds., Irreversible phenomena and dynamical systems models in the geosciences: Dordrecht, Nederlands, D. Reidel (NATO ASI Series), p. 399–416.

——, 1987, Lithospheric thickness, Antarctic deglaciation history, and ocean basin discretization effects in a global model of postglacial sea level change, *in* Cloetigh, S., Wortel, R., and Vlaar, S., eds., Mathematical geophysics: Dordrecht, Netherlands, D. Reidel (in press).

Peltier, W. R., and Andrews, J. T., 1976, Glacial isostatic adjustment; I. The forward problem: Geophysical Journal of the Royal Astronomical Society, v. 46, p. 605–646.

Peltier, W. R., Farrell, W. E. and Clark, J. A., 1978, Glacial isostasy and relative sea level; a global finite element model: Tectonophysics, v. 50, p. 81–110.

Peltier, W. R., and Hyde, W. T., 1984, A model of the ice age cycle, *in* Berger, A., Imbrie, J., Hays, J., Kukla, G., and Saltzman, B., eds., Milankovitch and climate: Dordrecht, Nederlands, D. Reidel, p. 563–580.

Peltier, W. R., Drummond, R. A., and Tushingham, A. M., 1986, Postglacial rebound and transient lower mantle rheology: Geophysical Journal of the Royal Astronomical Society, v. 87, p. 79–116.

Pollard, D., 1983, Ice-age simulation with a calving ice sheet model: Quaternary Research, v. 20, p. 30–48.

——, 1984, Some ice-age aspects of a calving ice sheet model, *in* Berger, A., Imbrie, J., Hays, J., Kukla, G., and Saltzman, B., eds., Milankovitch and climate: Dordrecht, Netherlands, D. Reidel (NATO ASI Series), p. 591–564.

Post, R., and Griggs, D., 1973, The Earth's mantle; Evidence of non-Newtonian flow: Science, v. 181, p. 1242–1244.

Rubincam, D. P., 1984, Postglacial rebound observed by LAGEOS and the effective viscosity of the lower mantle: Journal of Geophysical Research, v. 89, p. 1077–1087.

Saltzman, B., Hansen, A. R., and Maasch, K. A., 1984, The late Quaternary glaciations as the response of a three component feedback system to Earth-orbital forcing: Journal of Atmospheric Science, v. 41, p. 3380–3384.

Sclater, J. G., Jaupart, C., and Galson, D., 1980, The heat flow through the oceanic and continental crust and the heat loss of the Earth: Reviews in Geophysics and Space Physics, v. 18, p. 269–311.

Sharpe, H. N., and Peltier, W. R., 1978, Parameterized mantle convection and the Earth's thermal history: Geophysical Research Letters, v. 5, p. 737–740.

Shackleton, N. J., 1967, Oxygen isotope analyses and Pleistocene temperatures re-addressed: Nature, v. 215, p. 15–17.

Van Bemmelen, R. W., and Berlage, H. P., 1935, Versuch einer mathematischen Behandlung Gestektonischer Bewegungen unter besonderer Beruchsichtigung der Undationstheories: Gerlands Beiträge zur Geophysik, v. 43, p. 19–55.

Vening-Meinesz, F. A., 1937, The determination of the Earth's plasticity from the postglacial uplift of Scandinavia; Isostatic adjustment: Koninklijke Nederlandse Akademie van Wetenschappen Proceedings, v. 40, p. 654–662.

Vincente, R. O., and Yumi, S., 1969, Co-ordinates at the pole (1899–1968), returned to the conventional international origin: Publications of the International Latitude Observatory, Mizusawa, v. 7, p. 41–50.

——, 1970, Revised values (1941–1961) of the co-ordinates of the pole referred to the CIO: Publications of the International Latitude Observatory, Mizusawa, v. 7, p. 109–112.

Walcott, R. I., 1970, Isostatic response to the loading at the crust in Canada: Canadian Journal of Earth Science, v. 7, p. 716–727.

——, 1972, Late Quaternary vertical movements in eastern North America: Reviews in Geophysics and Space Physics, v. 10, p. 849–884.

——, 1973, Structure of the Earth from glacio-isostatic rebound: Annual Reviews of Earth and Planetary Science, v. 1, p. 15–37.

Washburn, A. L., and Struiver, M., 1962, Radiocarbon-dated postglacial delevelling in northeast Greenland and its implications: Arctic, v. 15, p. 66–72.

Weertman, J., 1976, Milankovitch solar radiation variation and ice age sheet size: Nature, v. 261, p. 17–20.

——, 1978, Creep laws for the mantle of the Earth: Philosophical Transactions of the Royal Society of London, Series A., p. 110–125.

Wu, P., and Peltier, W. R., 1983, Glacial isostatic adjustment and the free air gravity anomaly as a constraint on deep mantle viscosity: Geophysical Journal of the Royal Astronomical Society, v. 74, p. 377–450.

——, 1984, Pleistocene deglaciation and the Earth's rotation; A new analysis: Geophysical Journal of the Royal Astronomical Society, v. 76, p. 753–791.

Yoder, C. F., and 5 others, 1983, Secular variation of Earth's gravitational harmonic J_2 coefficient from Lageos and nontidal acceleration of Earth rotation: Nature, v. 303, p. 757–762.

MANUSCRIPT ACCEPTED BY THE SOCIETY MARCH 20, 1987

Printed in U.S.A.

The Geology of North America
Vol. K-3, North America and adjacent oceans during the last deglaciation
The Geological Society of America, 1987

Chapter 9

Ice dynamics and deglaciation models when ice sheets collapsed

T. Hughes
Department of Geological Sciences and Institute for Quaternary Studies, University of Maine, Orono, Maine 04469

INTRODUCTION

The Quaternary Period is remarkable because of its dramatic environmental fluctuations, recorded most profoundly in the geological evidence for a succession of worldwide glaciations accompanying global climatic oscillations, especially in mid and high northern latitudes. North America, which spans these latitudes, has been a major stage for this continuing drama. Broecker and van Donk (1970), on the basis of oxygen-isotope records from nine core sites in the Atlantic and Caribbean basins, discovered a sawtooth pattern for Brunhes glaciation cycles (the last 730,000 yr): glaciation was prolonged and unsteady, the deglaciation was swift and sure. They used the word "terminations" to identify times of rapid deglaciation. The last termination took place from about 17 to 7 ka. It was a time when ice sheets collapsed.

But how did the ice sheets collapse, and from what size? These are the questions addressed in this chapter. Research aimed at understanding cycles of glaciation can be guided by two competing philosophical viewpoints. On the one hand, the succession of major glaciations and interglaciations may be the most extreme perturbations of a global climatic machine in stable equilibrium, and analogous to a ball in a hollow: if it is rolled and released, it rolls back to its starting position. For stable equilibrium, climate dynamics is understood best by focusing research on conditions during the extremes of climate, when the slope of a climate-versus-time curve is zero. On the other hand, the succession of major glaciations and interglaciations may be the manifestation of a global climatic machine in unstable equilibrium, and analogous to a ball on a hilltop: if it is rolled and released, it continues to roll forward. For unstable equilibrium, climate dynamics is understood by focusing research on conditions during the most rapid climatic change, when the slope of a climate-versus-time curve is maximum. These conditions exist when ice sheets collapse. By examining mechanisms during the last glacial and interglacial maxima, CLIMAP (Climate: Long-range Investigation, Mapping, and Prediction; a project of the International Decade of Ocean Exploration, 1970–1980) addressed the possibility of stable climatic equilibrium. This chapter examines deglaciation mechanisms that promote unstable climatic equilibrium.

Collapse of continental ice sheets was the main event of the last deglaciation. Collapse was total in North America and Eurasia, and partial in Antarctica and Greenland (Denton and Hughes, 1981). Mechanisms of collapse lowered the surface, and thereby decreased the area and volume. Understanding the collapse mechanisms is a key to understanding the dynamics of climatic change from conditions during the last glacial maximum at 18 ka to the world we inhabit today. Surface lowering is important because the change of ice elevation in space and time is the major orographic constraint on global atmospheric circulation during deglaciation. Areal decrease is important because it reduced continental albedo, thereby increasing the efficiency of continental-atmospheric heat exchange. Volume loss is important because deglaciation caused oceans to transgress onto land, thereby increasing the area of oceanic-atmospheric heat exchange. These consequences of collapse were most profound in North America, where ice sheets attained their most extensive advance and where retreat was most nearly complete. This review examines the collapse of North American ice sheets during the last deglaciation; it analyzes disagreements about the areal extent of the ice sheets, shows how these perceptions influence efforts to compute the elevation and volume distribution of the ice sheets, and investigates methods for using isostatic adjustments to distinguish between ice-sheet elevation and thickness in computing ice-volume changes.

Important concepts, processes, and features used in this chapter will now be defined. Some of them are illustrated in Figures 1 through 3.

Astronomical theory. The modern version of the Milankovitch (1941) theory states that cyclic variations in summer insolation at specific Northern Hemisphere latitudes trigger the advance and retreat of Pleistocene ice sheets spanning these latitudes. Hays and others (1976) employed spectral analysis to correlate ice-volume variations, obtained from oxygen-isotope stratigraphy in microfossils preserved as ocean-floor sediments, with mid- and high-latitude insolation variations arising from cyclic variations in the precession and tilt of Earth's rotation axis. Imbrie and others (1984) developed a time scale for Brunhes glaciation cycles in which, with appropriate phase lags, the strongest correlation between the ice-volume and insolation vari-

Hughes, T., 1987, Ice dynamics and deglaciation models when ice sheets collapsed, *in* Ruddiman, W. F., and Wright, H. E., Jr., eds., North America and adjacent oceans during the last deglaciation: Boulder, Colorado, Geological Society of America, The Geology of North America, v. K-3.

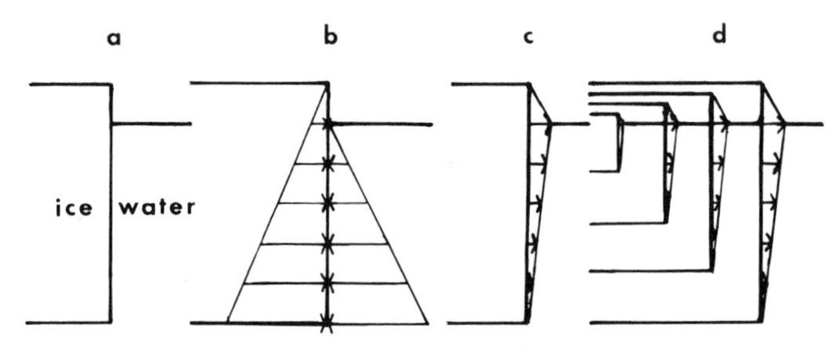

Figure 1. Pulling power in buoyant ice. A floating ice margin is a vertical ice cliff (a), against which horizontal hydrostatic forces that increase linearly with depth point seaward in ice and landward in water (b), and can be subtracted from top to bottom to give a net horizontal hydrostatic force that pulls the ice seaward (c) and increases with the square of buoyant ice thickness (d). Pulling power is this Force multiplied by the ice velocity.

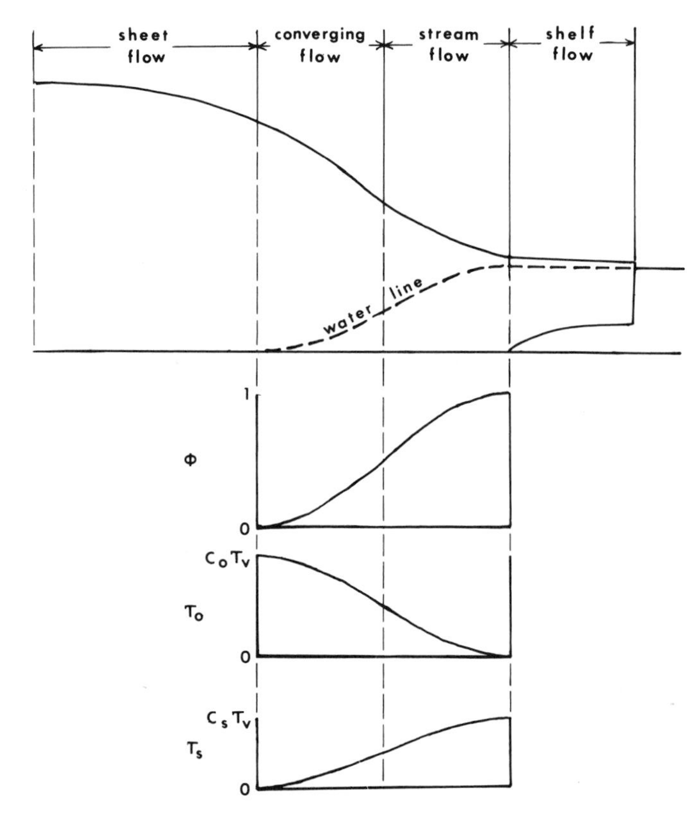

Figure 2. Possible conditions along a flowband of a marine ice sheet. Sheet flow near the ice divide has a convex surface that steepens as ice converges toward an ice stream, where a concave surface characterizes stream flow, beyond which an ice shelf floats and has a nearly flat surface. The water line marks the height to which water would rise in imaginary boreholes in this model, with ϕ being the fractional decrease in the pulling force of floating ice upstream from the grounding line, τ_0 and τ_s being the respective increase in basal shear stress and decrease in side shear stress upstream, τ_v being the viscoplastic yield stress of ice, and C_0 and C_s being constants to account for the fact that viscoplastic yielding is different at the basal ice-rock interface than at the side ice–ice interfaces. The point on the curves in Figure 5 where curvature is greatest is τ_v.

ations was obtained by adding normalized cycles of (1) axial precession having a weak amplitude with a period of 19,000 yr and a strong amplitude with a period of 23,000 yr, (2) axial tilt having a strong amplitude with a period of 41,000 yr, and (3) orbital eccentricity having very weak amplitudes with periods of 100,000 and 413,000 yr.

Termination. A rapid decrease in the oxygen-isotope ratio $^{18}O/^{16}O$ observed in the stratigraphy of ocean-floor sediments worldwide. It has occurred about every 100,000 yr for the past 700,000 yr and has been interpreted as a rapid reduction of global ice volume by about two-thirds in less than 10,000 yr, after a modulated increase in ice volume over an average of 90,000 yr (Broecker and van Donk, 1970).

Collapse. An extensive, irreversible loss in elevation of an ice sheet accompanying a termination. The areal extent of an ice sheet might increase temporarily during collapse, but ultimately it must decrease. This distinguishes collapse from shorter, recoverable deglaciation episodes observed as interstadials in the glacial-geological and oxygen-isotope records.

Downdraw. Surface lowering in an ice-stream drainage basin caused by the pulling power of the ice stream. Hence, downdraw is different from lowering by surface melting. It is most important in marine sectors of an ice sheet, because marine ice streams have the greatest pulling power. Denton and Hughes (1981) proposed downdraw as a major cause of ice-sheet collapse during the last deglaciation, because downdrawn ice-stream drainage basins must expand in area if pulling power lowers the surface faster than ice precipitation raises it.

Ice stream. A fast current of ice that develops toward the margin of an ice sheet, where thinning ice tends to follow depressions in the subglacial landscape. Terrestrial ice streams follow river valleys and produce terminal ice lobes. Marine ice streams follow inter-island channels and produce floating ice tongues. Stream-flow has a concave ice surface produced by the pulling power of the ice stream, in contrast to the convex surface of sheet-flow produced by basal traction.

Ice shelf. A nearly flat slab of floating ice thick enough to spread under its own weight and maintained by snow precipitation over its surface and by glacial ice moving into deep water. Antarctic ice shelves formed in marine embayments from the confluence of several ice streams are thought to be floating successors of marine sectors of the Antarctic Ice Sheet that collapsed during the last termination (Stuiver and others, 1981).

Disintegration. The deglaciation mechanism arising from calving along ice-sheet margins, where the transportation of calved ice to heat sources is a more efficient ablation mechanism than transportation of heat to melt the ice surface. Calving margins include the floating fronts of ice-stream tongues and ice shelves, as well as ice walls grounded on the shores of marine coasts and proglacial lakes. During the last deglaciation, the Ross Sea and Weddell Sea sectors of the Antarctic Ice Sheet retreated from a combination of interior collapse and peripheral disintegration, each reinforcing the other (Hughes, 1973).

Pulling power. The product of the pulling force and the ice

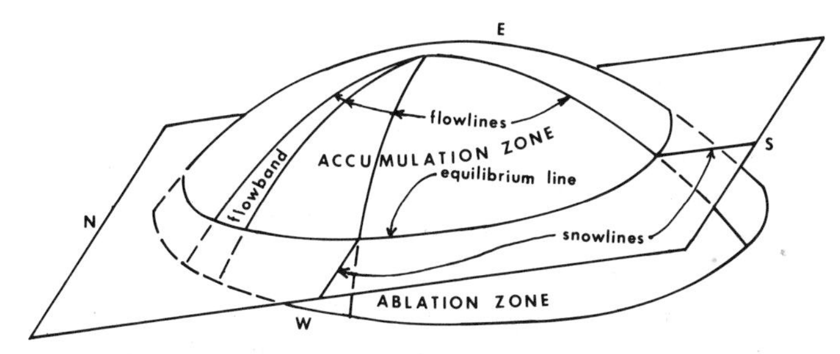

Figure 3. Mass-balance conditions in a North American ice sheet. A snow surface in the atmosphere separates mean annual precipitation as snowfall (above) from rainfall (below), and intersects the ice sheet along an equilibrium line that separates an annual net accumulation zone (above) from an annual net ablation zone (below). A snowline on the snow surface intersects a flowline on the ice sheet at a point on the equilibrium line. Each flowline is part of a flowband.

velocity. The pulling force is gravity acceleration acting horizontally on a mass of ice having a sloping surface and a buoyant bed. It increases as the square of the fraction of ice thickness supported fully by basal water pressure. Pulling power is greatest at the heads of ice streams, where surface slope is a maximum and where converging flow causes a rapid downslope increase in ice velocity (Hughes, 1987).

The Jakobshavns Effect. A hypothetical mechanism for enhancing downdraw when an ice-stream drainage basin has a large ablation zone, as observed in Jakobshavns Isbrae, located at 69°10′N, 50°10′W on the west-central margin of the Greenland Ice Sheet (Hughes, 1986). It postulates that summer meltwater pouring into surface crevasses, and subsequently reaching the bed, should increase basal water pressure and thereby allow the pulling power of the ice stream to reach further into the ice sheet.

Snowline. An imaginary line across which precipitation changes from snow to rain when averaged over several years. In computer models simulating Pleistocene glaciation cycles using the Weertman (1976) approach, connecting all intersections of snowlines with flowlines produces the equilibrium line of an ice sheet. In these models, snowlines dip toward the poles and rise or fall with Milankovitch insolation variations, causing migrations of the equilibrium line that give the ice sheet a positive or negative mass balance, so it must advance or retreat as ice volume changes.

Flowline. Any line on the surface of an ice sheet along which ice moves from an ice dome to the ice margin. Flowlines are usually directions of maximum surface slope, so they can be mapped by drawing lines perpendicular to surface contour lines. Some computer reconstructions of former ice sheets compute ice elevations and thicknesses along flowlines, entered as input or computed as output, and then plot surface contour lines by connecting equal elevations along flowlines.

Flowband. A bundle of flowlines that converges or diverges. For example, all flowlines that converge on an ice stream constitute a flowband that defines the drainage basin of the ice stream.

Surge. A periodic burst of speed observed in some mountain glaciers, during which a ten-fold to thousand-fold increase in ice velolcity transforms a smooth ice surface into a jumbled chaos of crevasses and seracs (Paterson, 1981). Several ice streams, notably Jakobshavns Isbrae in Greenland (Lingle and others, 1981) and Thwaites Glacier in Antarctica (Lindstrom and Tyler, 1984), now move at surge velocities, and the long Dibble and Dalton iceberg tongues grounded beyond inactive ice streams on the Wilkes Land coast of Antarctica may be evidence for spent surges. Surges last from months to years for mountain glaciers, but may last from decades to millennia in ice streams.

GLACIOLOGICAL THEORY

Areal components of an ice sheet

An ice sheet is a mantle of ice extensive enough to cover much of a continent, including areas above and below contemporary sea level, and thick enough to spread under its own weight. Under favorable conditions, an ice sheet can also spread as a floating ice shelf over offshore seas and embayments. Pinning points exist where the ice shelf is grounded locally on the continental shelf. The surface expression of a pinning point is an ice rise when ice must flow around it, and ice rumples when ice can scrape over it. Small islands also act as pinning points in an ice shelf.

Terrestrial parts of an ice sheet are grounded above sea level, or below sea level but with ice margins that are ice walls grounded less than 100 m below sea level. Changing sea level, then, has little or no effect on the ice margin. Advance and retreat of terrestrial ice is controlled primarily by lowering and raising the surface equilibrium line, which separates the area of net annual accumulation from the area of net annual ablation. Raising the equilibrium line, either by climatic warming or by isostatic sinking of the ice sheet in a stable climatic regime, increases the ablation area relative to the accumulation area, and the ice sheet retreats. Lowering the equilibrium line has the opposite effect. Ice walls grounded in water from 100 m to 500 m deep are transitional between terrestrial and marine margins, because changing sea level becomes important. Areas of the ice sheet that are grounded 500 m or more below sea level, both in the interior and along the margin, can be downdrawn into the sea along their floating margins by the pulling force of buoyant ice, which increases rapidly as sea level rises or as the grounding line retreats downslope. These areas are the marine part of the ice sheet, and pulling power tends to be concentrated in ice streams. Advance and retreat of marine ice is controlled primarily by lowering and raising sea level, which causes advance and retreat of the grounding line of ice floating beyond the marine ice margin. Rising sea level tends to lift the marine ice margin from its bed, causing the grounding line to retreat. Falling sea level has the opposite effect.

Areas of the ice sheet that are floating constitute its ice shelves. Their inner border is a grounding line and their outer border is a calving front. Ice shelves grow by retreat of their grounding lines and by advance of their calving fronts, and they shrink when opposite conditions prevail. The grounding line advances or retreats with falling or rising sea level and with a positive or negative mass balance of the ice sheet. The calving front advances when the ice discharge velocity exceeds the iceberg calving rate; otherwise, the calving front is carved back, and a calving bay develops. In summary, the areal extent of an ice sheet is controlled primarily by changes in the equilibrium-line elevation, in sea level, and in iceberg calving rates. Which of these predominates for a given part of the ice sheet depends on whether that part is terrestrial, marine, or afloat.

Over a given area, the thickness of an ice sheet in its terrestrial and marine components is greater if the ice is hard and if traction at the bed is great (Paterson, 1981). Softer ice with less basal traction can be more easily pulled down by gravity, so that a lesser thickness can spread over a greater area. Cold ice is harder than warm ice. Also, ice is hard when the optic axes (the c-axis of hexagonal symmetry) in individual ice crystals are randomly oriented. Cold ice with random crystal orientations prevails near the surface, so the upper part of an ice sheet generally is harder than the lower part, where geothermal heat, the great ice overburden, and basal traction can produce temperate ice with strongly oriented ice fabrics for which principal strain rates tend to align with principal stresses. This ice is soft with respect to the local stress field. But if the stress field changes (for example, if grounded ice becomes afloat, thereby eliminating basal shear stresses), recrystallization would be required to create an oriented ice fabric that is soft for the new stress field. Crystal size, texture, and purity also influence ice hardness and, as with ice fabric, these variables are not quantified adequately.

Basal traction is determined primarily by bed roughness, permeability, and softness, and by the hydrology of basal meltwater (Paterson, 1981). Bed roughness is most effective in retarding ice flow around bedrock obstructions on the scale of ice crystals (a few millimeters). Bed permeability regulates hydrostatic pressure of meltwater at the ice-bed interface, but how this affects the pulling power of an ice streams is unclear. One might argue that low permeability increases the pulling power of terrestrial ice streams by preventing basal meltwater from draining away, whereas high permeability increases the pulling power of marine ice streams by allowing sea water to seep in. On the other hand, Boulton and Jones (1979) argue that the lobate end moraines of the Laurentide Ice Sheet south of the Great Lakes resulted from rapid flow on a soft, permeable bed. High velocities during glacial surges occur when basal water pressure lifts ice from contact with the bed, either by drowning the most effective projections of a hard impermeable bed or by super-saturating a soft permeable bed (Kamb and others, 1985). More meltwater is needed to reduce basal traction if the bed is rough or permeable than if it is smooth or impermeable. A uniform meltwater distribution reduces traction more than a ponded or dendritic distribution. How the hydrology of basal meltwater interacts with bed roughness, permeability, and softness to reduce bed traction is not fully understood; it awaits a comprehensive theory of glacial sliding. This theory is crucial to understanding how the pulling power of ice streams can reach far into an ice sheet. In its pure form, pulling power can be observed in freely floating tongues of marine ice streams, as opposed to ice-stream tongues imbedded in confined and pinned ice shelves (Weertman, 1957; Hughes, 1987).

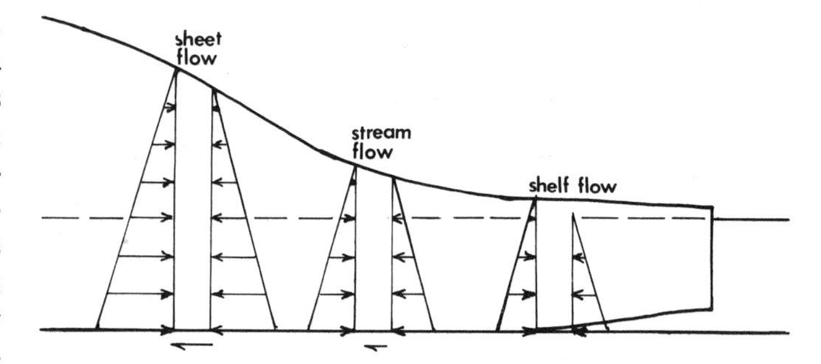

Figure 4. Relationship between the surface slope of an ice sheet and the horizontal component of the gravity force. Basal traction for sheet flow (arrow at base of columns) decreases downslope for stream flow and vanishes for shelf flow. See text for discussion.

Flow in an ice sheet is a response to two pulling forces, one arising from the surface slope of grounded ice and one arising from the buoyancy of floating ice. Both forces are illustrated in Figure 4, and they can be understood as the difference between horizontal hydrostatic forces acting on the upstream and downstream sides of imaginary columns along a flowband. Three columns are shown for sheet flow, stream flow, and shelf flow, and each one has a net horizontal hydrostatic force pulling it in the downstream direction. The pulling force caused by surface slope is zero at the ice divide, where surface slope is zero and sheet flow begins; is a maximum at the inflection line, where surface slope is a maximum and stream flow begins; and is a minimum at the grounding line, where surface slope is a minimum and shelf flow begins. The basal traction force must balance this pulling force, so it also is zero at the ice divide, rises to a maximum at the inflection line, and falls to a minimum of zero at the grounding line. A nonzero minimum-surface slope coexists with zero basal traction at the grounding line, because the pulling force due to ice buoyancy is a maximum and draws the ice down. The ice-buoyancy pulling force is also the reason why velocity peaks at the grounding line, even though surface slope approaches zero, as it does near the ice divide, where velocity is zero. This ice-buoyancy pulling force grows weaker up the ice stream, if decreasing buoyancy allows increasing basal traction and a corresponding increase in the surface slope, and in the pulling force arising from the surface slope. Pulling forces from ice buoyancy and surface slope combine to produce a maximum pulling force

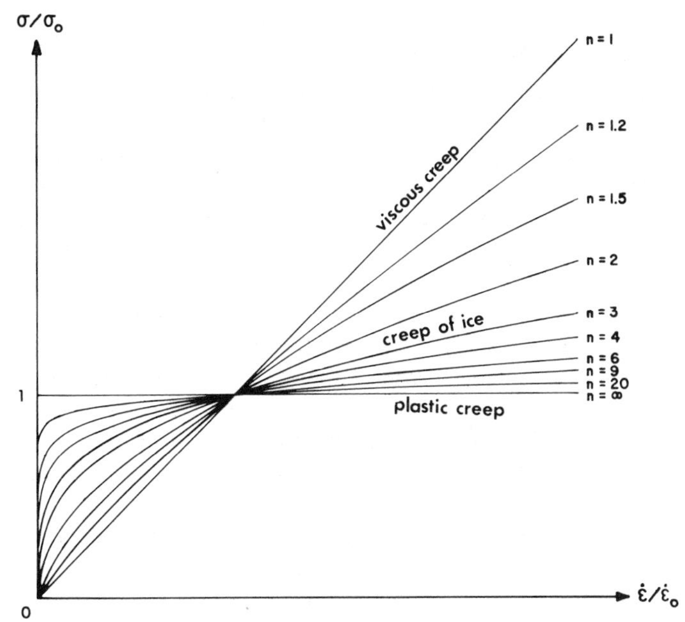

Figure 5. The viscoplastic creep spectrum. Curves are drawn relating effective stress σ to effective strain rate $\dot{\epsilon}$ through the creep equation $\dot{\epsilon} = \dot{\epsilon}_0 (\sigma/\sigma_0)^n$, in which strain rate $\dot{\epsilon}_0$ may vary with time, temperature, and crystal properties; σ_0 is the plastic yield stress, n is a viscoplastic exponent, and $(\dot{\epsilon}_0/\sigma_0{}^n)$ is a softness coefficient. Viscous creep occurs for n = 1, plastic creep occurs for n = ∞, and creep in ice occurs for n $\simeq$ 3.

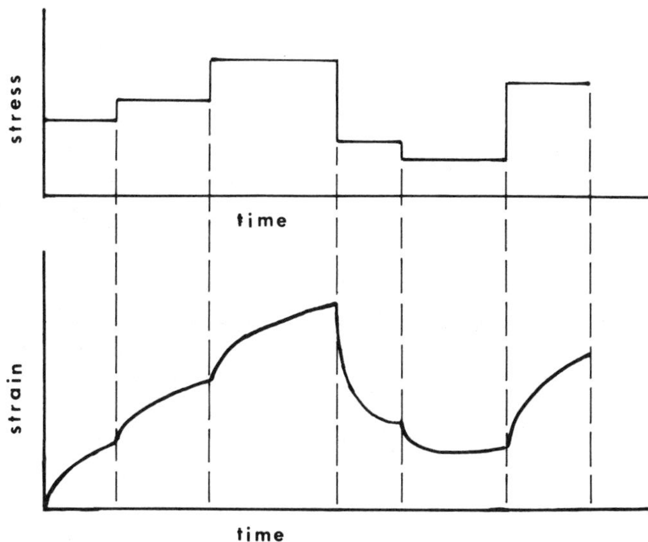

Figure 6. Transient creep perpetuated by large step changes in applied stress. If the magnitude of changes in basal shear stress over time compares with the average basal shear stress on a local scale (top), then transient creep is never replaced by steady-state creep (bottom).

at the head of the ice stream, so that ice is drawn into the ice stream, creating a zone of converging flow, as shown in Figure 2. Converging flow maximizes the surface slope, which maximizes basal traction and allows rapid erosion of the bed, so that ice-stream channels are characteristically foredeepened. Moreover, maximum erosion rates at the head of the ice stream extend the channel into the ice sheet. In this way the pulling power of an ice stream is able to reach far into the ice sheet (Hughes, 1987).

Pulling forces cause an ice sheet to deform in creep. The surface-slope pulling force exists with basal traction, so ice deforms by shearing over the bed. The ice-buoyancy pulling force exists without basal traction, so ice deforms by stretching along its length. Like deformation in other crystalline materials, deformation in ice exists within the viscoplastic creep spectrum depicted in Figure 5. The stress causing ice to deform in creep is normalized when the stress field, represented by effective shear stress σ, is divided by the effective yield stress σ_0 of purely plastic deformation. The strain rate of creep deformation is normalized when the strain-rate field, represented by effective strain rate $\dot{\epsilon}$, is divided by a strain rate $\dot{\epsilon}_0$ that includes the effect on $\dot{\epsilon}$ of variables other than stress (such as temperature, fabric, and purity variations in an ice sheet). Ratio $\dot{\epsilon}/\dot{\epsilon}_0$ varies exponentially with ratio σ/σ_0, such that $\dot{\epsilon} = \dot{\epsilon}_0 (\sigma/\sigma_0)^n$, where n = 3 is commonly employed in theoretical models of ice dynamics. Figure 5 shows that $\dot{\epsilon}$ increases linearly with σ when n = 1. This is the condition for Newtonian flow, and it represents the viscous end of the viscoplastic creep spectrum. The plastic end of the spectrum is reached when n = ∞. For plastic flow, $\dot{\epsilon} = 0$ when $\sigma < \sigma_0$, any value of $\dot{\epsilon}$ is possible when $\sigma = \sigma_0$, and $\sigma > \sigma_0$ is an impossible condition. The term $\dot{\epsilon}_0/\sigma_0{}^n$ is a softness coefficient, and n is a viscoplastic exponent.

Creep data from laboratory experiments on ice and from field measurements on glaciers point to a gradual change from n = 1 for $\sigma < 10$ kPa to n = 3 for 10 kPa $< \sigma <$ 100 kPa, and n > 3 for $\sigma >$ 100 kPa (Weertman, 1983). Regional stresses in ice sheets range from $\sigma <$ 10 kPa along ice divides to $\sigma >$ 100 kPa immediately upstream from the bedrock projections that are most effective in controlling basal sliding, but the usual range is 10 kPa $< \sigma <$ 100 kPa (Paterson, 1981). If changes in σ are comparable to the average value of σ over the time span when $\dot{\epsilon}$ is being measured, ice deforms in regimes of transient creep rather than in the steady-state creep regime, as illustrated in Figure 6. Under these conditions, n = 1 gives a better fit than n = 3 to creep data from ice sheets, provided that $\dot{\epsilon}_0$ is kept constant (Doake and Wolff, 1985; Weertman, 1985). As seen in Figure 6, however, $\dot{\epsilon}_0$ becomes strongly time-dependent when σ changes abruptly in a creep experiment. We therefore expect $\dot{\epsilon}_0$ to change with time during transient creep, so n = 3 with large $\dot{\epsilon}_0$ variations may be reporting creep in ice sheets more faithfully than does n = 1 with $\dot{\epsilon}_0$ being relatively constant (Weertman, 1985; Paterson, 1985). Also, σ varies over time, not abruptly as in Figure 6.

It would seem that using n = 1 is particularly appropriate in

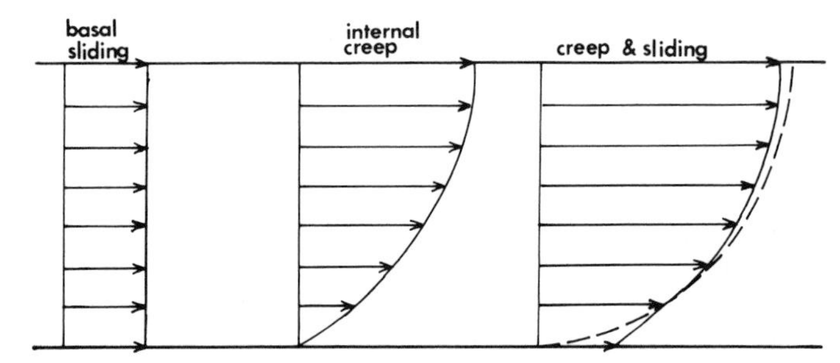

Figure 7. An illustration of how horizontal velocities in an ice sheet can be approximated by plastic flow. The actual vertical variation of horizontal velocity (arrows) compares well with the plastic flow variation for $n = \infty$ (dashed curve) when the velocities of basal sliding (left) and internal creep for $n = 3$ (center) are added (right). From Nye (1957).

models of ice-sheet dynamics for which rates of basal sliding are important, because transient creep should predominante in ice sliding over rough, impermeable bedrock (Kamb, 1970; Nye, 1970), and viscous creep may predominate for simple shear of a soft, permeable deforming bed (Boulton and Jones, 1979). Rough bed conditions guarantee a rapidly changing basal stress field and therefore the predominance of transient creep. On the other hand, glaciologists have obtained great insights into the dynamics of ice sheets by assuming that $n = \infty$, beginning with the classic calculation of stress and velocity fields by Nye (1951) and, most recently, with computations of surface flowline trajectories and elevations by Reeh (1982). Even the interaction between creep and sliding velocities can be represented reasonably well by perfectly plastic flow, as shown in Figure 7. These insights are possible because creep in ice sheets is approximated better by purely plastic flow than by purely viscous flow. Figure 8 compares flowline profiles of an ice sheet for viscous creep ($n = 1$), viscoplastic creep ($n = 3$), and plastic creep ($n = \infty$). The variation in surface elevation and slope of ice along the flowline is strikingly different in the three cases. Flowline profiles in Figure 8 are for sheet flow in terrestrial ice sheets. Instability of terrestrial ice sheets is largely controlled by the elevation at which snowlines intersect flowlines, and, as Figure 8 shows, this elevation depends strongly on the creep behavior of the ice sheet.

Stable ice sheet margins

The areal extent of an ice sheet is determined by environmental conditions that can stabilize the ice-sheet margins. These conditions are illustrated in Figure 9, which shows the transition from a terrestrial margin to a marine margin. The equilibrium line separates a broad interior zone, where ice accumulates by precipitation, from a peripheral zone, where ice ablates primarily by melting and calving. The snowline in the atmosphere above a given flowline on the ice sheet increases in elevation from the poles to the equator and intersects the flowline at a point on the equilibrium line (Weertman, 1961).

If the ice sheet advances in the direction of a rising snowline, such as toward the equator, the surface area of ice in the ablation zone will increase faster than the surface area in the accumulation zone. Eventually the annual volume of ice melted in the ablation zone will equal the annual ice volume deposited over the accumulation zone. Then advance of the ice sheet will stop, because the rate of forward ice motion is completely offset by the melting rate at the terminus. This is mass-balance equilibrium. Because it stabilizes the ice-sheet margins on land, it is the equilibrium condition for terrestrial components of an ice sheet. Even if the snowline slopes downward toward the ice margin, the margin is stable if no perturbations disturb mass-balance equilibrium. A stable terrestrial ice-sheet margin (Fig. 9a) can be observed today around much of the Greenland Ice Sheet.

If the snowline lowers in the direction of ice-sheet advance, such as toward one of the poles, then the ablation zone decreases in area while the area of the accumulation zone grows. Surface melting alone cannot stabilize the ice-sheet margin under these conditions, unless accumulation rates above the equilibrium line decrease markedly with respect to ablation rates below the equilibrium line as the ice sheet advances poleward. In the absence of this starvation effect, the ice sheet will continue to advance until a new ablation mechanism more powerful than melting is able to overcome the accumulation rate. The first opportunity for a new ablation mechanism appears when the ice sheet advances onto the beaches of polar seas and embayments. Wave action can undercut the ice margin on beaches, especially the periodic wave action between high and low tide. Ice will then calve from the overhanging edge above the undercut groove along the ice-sheet margin, and this introduces a new ablation mechanism, one which produces a grounded ice cliff called an ice wall. If this calving rate is combined with surface melting and melting by wave action and offsets the rate of ice accumulation, then a second stable ice-sheet margin can develop in the intertidal zone on a beach. Shoreline calving along an intertidal ice wall has been analyzed by Iken (1977). A stable intertidal ice-sheet margin (Fig. 9b) can be observed in Antarctica, particularly along the Antarctic Peninsula.

If ablation rates along the shoreline ice wall and inland on the ice surface do not offset accumulation rates above the equilibrium line, the ice sheet will advance into deeper water until other ablation mechanisms are initiated. The first of these mechanisms is calving of ice lying below the tidewater ice groove along the ice wall. Ice below the groove is always below sea level, even at low tide, so it is subjected to a buoyancy force that increases as the volume of ice below the groove increases, i.e., as the ice sheet advances into increasingly deeper water. This buoyancy force will eventually lift ice that is below the groove, causing calving below sea level. If the additional ice ablation caused by submarine calving can offset the excess accumulation rate driving ice-sheet advance, a stable margin is produced by the combined losses from surface melting and tidewater calving. Sikonia (1982)

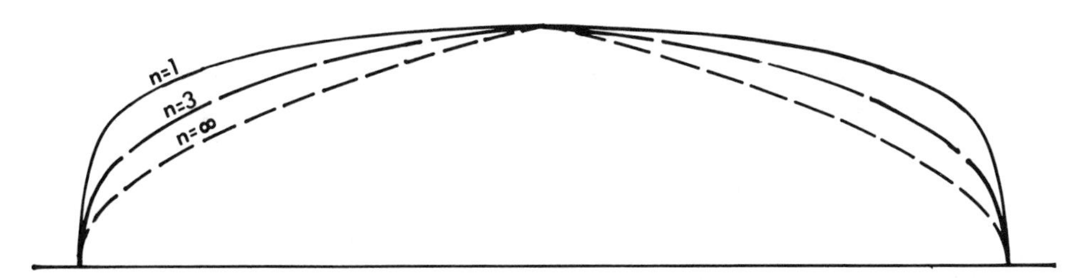

Figure 8. The effect of viscoplastic exponents on the surface profile of an idealized ice sheet. Profiles are shown for viscous creep with n = 1, plastic creep with n = ∞, and viscoplastic creep with n = 3, where n is the viscoplastic exponent in Figure 5.

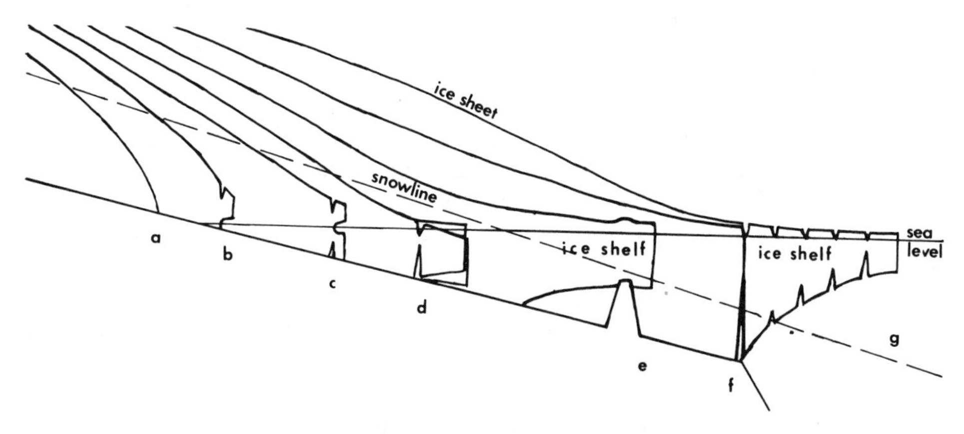

Figure 9. Stable margins of an ice sheet. As the ice sheet advances, stable positions exist at (a) terrestrial melting margin, (b) intertidal margin, (c) tidewater margin, (d) grounding line of a confined ice shelf, (e) calving front of a confined ice shelf, (f) grounding line of an unconfined ice shelf, and (g) calving front of an unconfined ice shelf.

obtained an empirical law for tidewater calving rates, but a theoretical model that includes the submarine calving mechanism is not available. Tidewater ice-sheet margins (Fig. 9c) are common around Greenland and east Antarctica. In North America, some Alaskan glaciers have tidewater margins.

When tidewater calving is unable to check the advance of an ice sheet, ice advances into deeper water and eventually becomes afloat. The floating ice cliff is not called an ice wall, and the tidewater ice groove disappears because the ice cliff rises and falls with the tide. Since the water angle between basal ice and the bed is very acute at the grounding line, the grounding line migrates back and forth with the rise and fall of the tide. Therefore, bending of the floating ice by tidal flexure occurs across a grounding zone defined by the tidal sweep of the grounding line. Surface and basal crevasses opened during the tensile half of bending cycles are lengthened by the pulling force acting on buoyant ice. If these crevasses meet near sea level before ice moves across the grounding zone, then the whole floating section at high tide eventually will be released as a tabular iceberg. Bottom crevasses can extend upward to sea level because they are filled with sea water, which exerts a positive hydrostatic pressure on crevasse walls. Surface crevasses can extend downward to sea level only if they are at least partly filled with surface meltwater. Tabular icebergs represent a huge volume loss, so grounding-zone calving is a powerful ablation mechanism. Weertman (1973, 1980) analyzed migration of water-filled surface and bottom crevasses. A stable grounding-zone ice-sheet margin is shown in Figure 9d. Calving by this mechanism may exist in some Greenland outlet glaciers that become afloat in coastal fjords, because both surface melting and crevassing are extensive.

An ice sheet can advance into deeper water if ice crosses the grounding zone faster than tidal-flexure crevasses can work through the ice thickness. New ablation mechanisms become possible in the floating ice shelf seaward of the grounding zone. Melting on its underside can be substantial, and calving can occur along transverse tensile crevasses that open in response to the horizontal pulling force. Basal melting can be caused by strong ice-shelf thickness gradients (Robin, 1979), tidal pumping near grounding zones (Hughes, 1982), and estuarine circulation beneath the ice shelf (MacAyeal, 1985). The ice shelf initially spreads across a polar continental shelf, where it is often pinned to the sea floor at islands or shoals and confined laterally within an embayment. These constraints provide braking forces that oppose the pulling force, so the ice shelf thins less rapidly seaward of its grounding zone than it otherwise would. A stable position of the ice-shelf calving front is often immediately seaward of a transverse line of pinning points on the continental shelf. As shown in Figure 1, the pulling force is a maximum at sea level along the calving front and, since sea level is much closer to the top than the bottom of the ice shelf, the pulling force produces a bending moment that can break off a slab of ice along a surface bending crack that develops one or two ice thicknesses behind the calving front. Reeh (1968) and Fastook and Schmidt (1982) analyzed this surface-bending mechanism of calving. Such a stable pinned ice-shelf margin (Fig. 9e) is common in Antarctica, where the positions of ice-shelf calving fronts are often just beyond lines of pinning points on the continental shelf (Swithinbank, 1955; Hughes, 1983a).

If the pulling force acting on floating ice at the grounding zone is largely nullified by braking forces at pinning points, ice will thicken slowly upstream from pinning points and may allow the grounding zone of the ice shelf to advance. The ultimate limit of grounding-zone advance is the edge of the continental shelf, where bed slope increases sharply and beyond which floating ice will be thinned rapidly by a pulling force unrestrained by pinning points. Here the water angle between basal ice and the bed is much less acute, so the tidal sweep of the grounding line narrows markedly. The grounding zone is then more properly a grounding line, and tidal bending becomes much more concentrated. Surface and especially basal tidal bending crevasses can migrate toward each other much more rapidly, and calving occurs when they meet. Lingle and others (1981) analyzed tidal bending along a grounding line where the floating tongue of Jakobshavns Isbrae in Greenland is grounded against the steep sidewalls and headwall of Jakobshavns Isfjord. A stable calving margin at the edge of the continental shelf (Fig. 9f) was probably common in Antarctica during the 18-ka glacial maximum, but the only modern examples of this calving mechanism are found where glaciers float in fjords.

Under unusual conditions an ice shelf may extend into deep ocean water beyond its grounding line along the edge of the continental shelf. The nearly land-locked Arctic Ocean may have been a site for ice shelves fed by marine components of Northern Hemisphere ice sheets during the last glaciation. Under these conditions an Arctic Ice Sheet might develop. The stable ice-shelf margin of this ice sheet would be the unglaciated polar continental shelves where the ice shelf ran aground, or it might be the open North Atlantic Ocean south of straits between Labrador, Greenland, Iceland, the Faeroes, and Scotland (Hughes and others, 1977; Denton and Hughes, 1981; Lindstrom and MacAyeal, 1986). The ice-shelf pulling force would be unrestrained south of the straits, where surface and basal melting would combine with longitudinal and lateral spreading to cause rapid ice thinning and open intersecting transverse and longitudinal crevasses that would section the thinned ice shelf into huge tabular icebergs. Holdsworth (1981) showed how storms could trigger the actual calving events. A stable oceanic ice-shelf margin is shown in Figure 9g. The closest present-day counterparts exist in Antarctica in freely floating tongues of ice streams and in regions where ice shelves extend well seaward of pinning points (Hughes, 1983a).

The areal extent of an ice sheet changes when its margins become unstable. Primary destabilizing factors are different for terrestrial, marine, and ice-shelf components. Atmospheric changes, notably in temperature and precipitation, are the main destabilizing factors at terrestrial margins, because the equilibrium line is so sensitive to these changes. Oceanic changes, notably in temperature and sea level, affect marine margins the most, because the grounding line is the sensitive boundary. Ice-shelf margins are destabilized by interactions with both atmosphere and oceans. Since ice shelves are nearly flat, the equilibrium line tends to lie either beyond the calving front or behind the grounding line, making the ice shelf either short or long (Oerlemans and van der Veen, 1984). Rising and falling sea levels compel the ice-shelf grounding line to retreat or advance. Increasing or decreasing the floating ice thickness can decrease or increase the iceberg calving rate.

Other factors that can destabilize an ice-sheet margin occur at the beds of terrestrial and marine components, and at pinning points of ice-shelf components. Most important among these factors are changes in the distribution of frozen and thawed regions at the bed and rates of isostatic uplift and sinking of bedrock beneath and beyond the ice sheet. Both these factors raise and lower the ice-sheet surface. The distribution of frozen and thawed regions affects ice-surface elevations in two ways for a given areal extent of an ice sheet. First, thawing a frozen bed lowers ice elevations because basal traction is reduced. Conversely, freezing the bed increases traction and raises the surface. Second, freezing and melting conditions at a thawed bed control basal erosion and deposition rates, which change surface elevations by changing bed topography, especially along grounded ice margins.

Deglaciation of terrestrial ice sheets

The most serious problem in treating instability of terrestrial ice sheets is relating the mass balance to the snowline. Weertman (1961) emphasized the importance of the snowline by considering two extremes, illustrated in Figure 10. In Figure 10a the snowline depends on latitude, so it is vertical. Retreat of the ice margin shrinks the ablation zone, so mass balance turns positive and the ice margin advances. This reversible response represents stable equilibrium. In Figure 10b, the snowline depends on altitude, so it is horizontal. Retreat of the ice margin shrinks the accumulation zone, so mass balance turns negative, and the ice margin retreats farther. This irreversible response represents unstable equilibrium. As seen in Figure 3, a horizontal snowline dependent only on elevation may exist in the east–west direction, and the most vertical snowline with the strongest dependence on

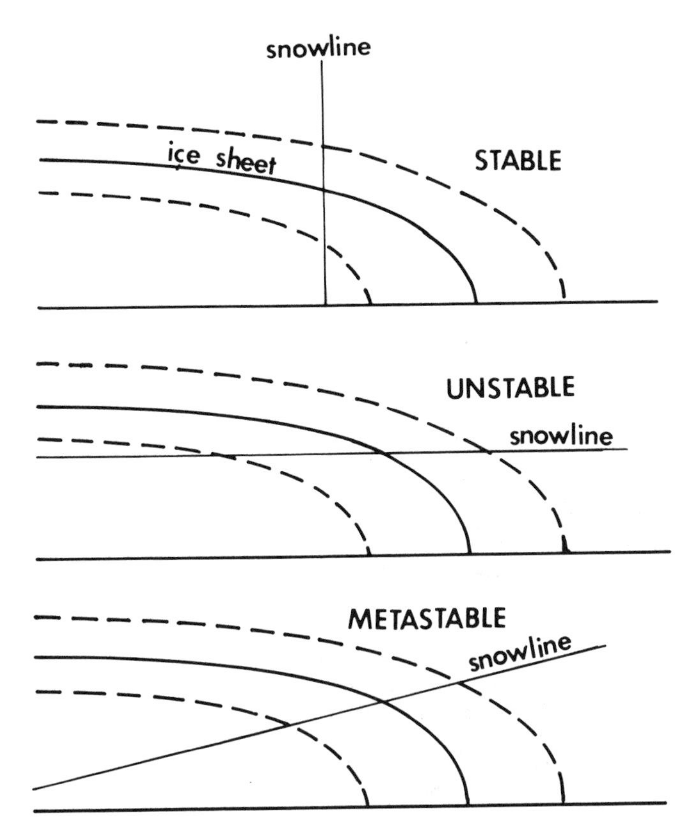

Figure 10. Snowline slope and stability of terrestrial ice-sheet margins. The snowline can vary with latitude (top), altitude (middle), or both latitude and altitude (bottom). A small advance or retreat (dashed profiles) of an ice sheet in initial mass-balance equilibrium (solid profile) increases or reduces the ablation zone when the snowline is vertical, so that advance or retreat is reversible (top); it increases or reduces the accumulation zone when the snowline is horizontal, so that advance or retreat is irreversible (middle); and it increases or reduces both the accumulation and ablation zone when the snowline is tilted, so the slope determines whether advance or retreat is reversible or irreversible (bottom). Equilibrium is stable for the vertical snowline, unstable for the horizontal snowline, and metastable for the tilting snowline. From Weertman (1961).

latitude exists in the north–south direction. However, most snowlines have slope between these extremes (Fig. 10c).

Weertman (1961) attempted to quantify the response of a terrestrial ice sheet to perturbations at its equatorward margin for a realistic snowline that decreases in altitude as latitude increases. The less the snowline slopes, the more irreversible will be an initial retreat of the ice-sheet margin. The response to initial retreat can vary along the margin, because the snowline slope decreases to zero as flow changes from moving north or south to east or west, east–west margins being the most unstable. Moreover, a snowline elevation that decreases with increasing latitude can have slopes that vary with longitude in response to geographical variables, such as continental topography and proximity to oceans.

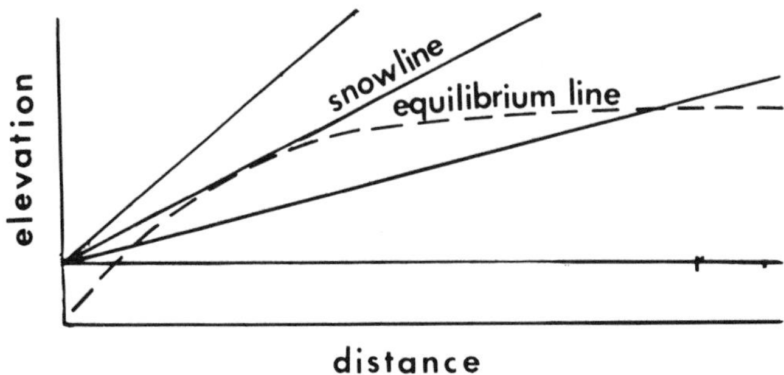

Figure 11. Stable and unstable responses of an ice sheet to changes in snowline slope. Variations in elevation of snowline (solid lines) and equilibrium line (dashed line) with distance from the center to the equilibrium line of an ice sheet in mass-balance equilibrium are plotted. Higher intersections are in stable equilibrium, so advance or retreat of the ice-sheet margin is reversible. Lower intersections are in unstable equilibrium, so advance or retreat of the ice-sheet margin is irreversible. Modified from Weertman (1961).

Figure 11 is a theoretical representation of snowlines having various slopes. The dashed curve shows various elevations of the surface equilibrium line. The elevations both of the snowlines and the equilibrium line are plotted to show the increase with distance from the center of an ice sheet in mass-balance equilibrium, so that the equilibrium-line elevation increases as the ice sheet becomes larger (Weertman, 1961). The horizontal snowline crosses the equilibrium-line curve at only one point. This is the condition of unstable equilibrium, depicted in Figure 10b, that exists for east–west snowlines. A slight retreat of the ice margin produces a negative mass balance, because the new point on the equilibrium-line curve is below the snowline, thereby keeping the mass balance negative and allowing more retreat. This situation is possible only toward the left side of Figure 11, meaning that it exists for ice caps or for small ice sheets. If the snowline has a small slope, it will cross the equilibrium-line curve at two points. The left-hand point is a state of unstable equilibrium qualitatively similar to Figure 10b. The right-hand point, however, is a state of stable equilibrium qualitatively similar to Figure 10a. Here a slight retreat of the ice margin produces a positive mass balance, because the new point on the equilibrium-line curve is above the snowline, thereby making the mass balance positive and allowing advance to the original ice margin. Since this situation is possible only toward the right side of Figure 11, it is most likely for large ice sheets. As the snowline slope increases, the two intersection points with the equilibrium-line curve move closer together and finally meet. At this point, an advance of the ice-sheet margin places the new point on the equilibrium-line curve below the snowline. This produces a negative mass balance, so the ice margin retreats to its original position. A retreat of the ice-sheet margin also places the new point on the equilibrium-line curve below the snowline. This produces a negative mass balance and

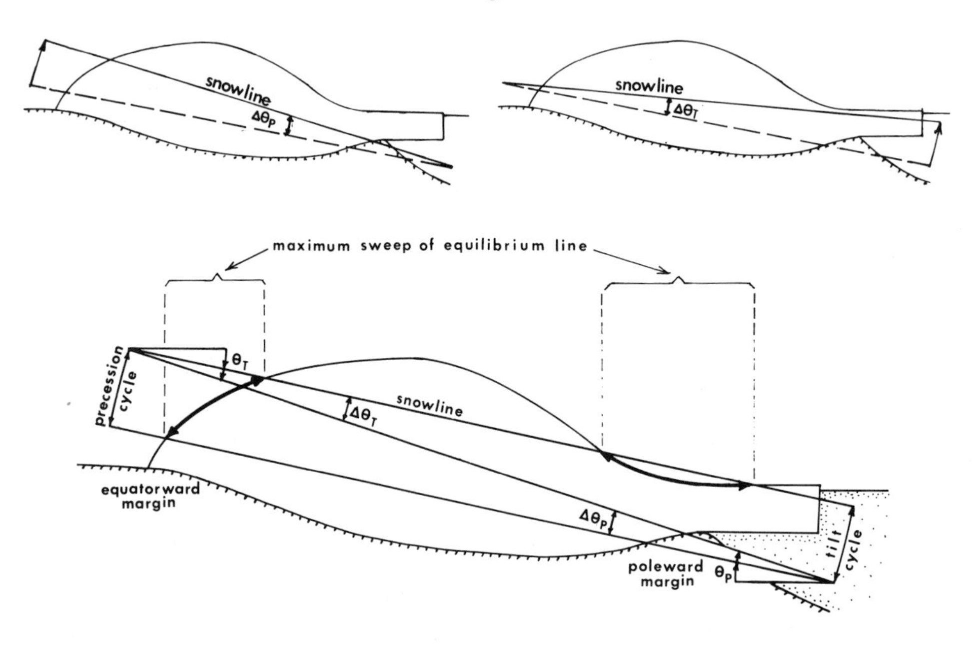

Figure 12. Migration of the equilibrium line on an ice sheet in response to orbital variations in insolation. Variations $\Delta\theta_P$ of an angle θ_P occur over the 23,000-yr cycle of axial precession and add to or subtract from variations $\Delta\theta_T$ of angle θ_T that occur over the 41,000-yr cycle of axial tilt, to determine the slope of the snowline. The insets show extreme cases for snowlines having maximum slope, with $\Delta\theta_P$ fully opened and $\Delta\theta_T$ fully closed (upper left), and minimum slope, with $\Delta\theta_P$ fully closed and $\Delta\theta_T$ fully opened (upper right). These are the closest approaches to stable and unstable equilibrium, respectively, in the ice sheet, as defined in Figures 10 and 11.

allows retreat to continue irreversibly. A still larger snowline slope places the snowline above the equilibrium-line curve everywhere, so there is no state of mass-balance equilibrium, either stable or unstable, and the ice sheet must shrink until it vanishes.

Figure 11 shows that a slight increase or decrease in snowline elevation can cause existing ice caps in the high Arctic to either shrink to nothing or grow to become a stable continental ice sheet, because the ice caps are small and therefore exist in a state of unstable mass-balance equilibrium. Weertman (1961) noted that a stable continental ice sheet can become unstable and retreat if its initially frozen bed becomes thawed and if isostatic sinking of its bed continues after the ice sheet is in stable mass-balance equilibrium. Both these effects lower the ice-sheet surface, which is equivalent to raising the snowline or increasing the ablation rate, either of which can initiate retreat.

Unstable advance or retreat of an ice sheet is particularly sensitive to snowline slope in the Weertman (1961) analysis of terrestrial ice sheets. He pointed out that the slope changes during periods of climatic change and that the smaller the slope the more unstable the ice sheet becomes. A completely unstable response exists for zero slope, as illustrated in Figure 10b. Of the astronomical variables controlling insolation, the 23,000-yr cycle of equinox precession changes insolation most in mid-latitudes, the 41,000-yr cycle of axial tilt changes insolation most in high latitudes, and the 100,000-yr cycle of orbital eccentricity modulates insolation changes caused by precession (Milankovitch, 1941; Imbrie and others, 1984). These and insolation changes related to the 2500-yr cycle of atmospheric CO_2 concentration may trigger terminations of major glaciations (Denton and Hughes, 1983; Denton and others, 1986).

The latitudinal dependence of the dominant variations in precession and tilt insolation implies that variations of snowline slope may control stability of North American ice sheets in the manner illustrated in Figure 12. The 23,000-yr cycle of precession variations changes the slope of a line that turns about a pivot near the northern ice margin. The southern end of this line is the pivot for the snowline, which turns about the southern pivot with the 41,000-yr cycle of tilt variations. The snowline, therefore, intersects the ice-sheet surface at two equilibrium lines, one in the south that migrates strongly with the precession cycle and weakly with the tilt cycle, and one in the north that migrates strongly with the tilt cycle and weakly with the precession cycle. The ice sheet is most stable when the slope of the snowline is a maximum and most unstable when this slope is a minimum.

This model implies that stability during the last glaciation was optimized when insolation was maximized by precession and minimized by tilt ($\Delta\theta_P$ is fully opened and $\Delta\theta_T$ is fully closed in Fig. 12). That is, when the ablation zone was widest in the south and narrowest or absent in the north, so the southern ice margin retreated reversibly and the northern ice margin advanced into the sea according to Figure 9. The model implies optimum instability when insolation was minimized by precession and maximized by tilt ($\Delta\theta_P$ is fully closed and $\Delta\theta_T$ is fully opened in Fig. 12). That is, when the ablation zone was narrowest in the south and widest in the north, so the southern ice margin could advance while the northern ice margin retreated, both irreversibly. The model predicts that both $\Delta\theta_P$ and $\Delta\theta_T$ are fully closed during a glacial maximum and fully opened during a glacial termination. Ice sheets would have metastable north and south margins at these times. East and west margins would always be unstable in this model.

All models of unstable conditions at terrestrial ice-sheet margins can be traced to the original analysis by Weertman (1961). It is the pioneering study and has the great advantage of introducing the essential concepts with a minimum of mathematics. These elementary concepts have been the basis for developing models of the interaction of terrestrial ice sheets with climate by Weertman (1964, 1966, 1976), Birchfield (1977), Birchfield and Weertman (1978, 1982), Birchfield and others (1981), Oerlemans (1981, 1982, 1983), Oerlemans and van der Veen (1984), Budd and Smith (1981), Källén and others (1979), Bindschadler and Gore (1982), and Pollard (1982, 1983), among others. Most of these models disregard the constraints of Figure 12 because their snowlines move up and down without rotating through $\Delta\theta_P$ and $\Delta\theta_T$. With rotation, the last glacial maximum was preceded by a stable, reversible migration of ice sheets from south to north, and the last glacial termination was preceded by an unstable, irreversible migration of ice sheets from north to south. Did these migrations occur? Even without the rotations, ice-sheet margins should have been more stable in the north and south than in the east and west, where snowlines should be horizontal. Were they? The record shows that some ice margins were unstable at their glacial maxima and some were stable during deglaciation (Mayewski and others, 1981). The Des Moines and James River ice lobes, for example, reached their maximum extent at 14 ka and were gone by 12 ka, so this southern sector of the Laurentide Ice Sheet margin reached its maximum extent in an unstable manner. Conversely, the northern margin of the Laurentide Ice Sheet on Baffin Island retreated little from 14 ka to 8 ka, a time when southern North American ice margins were in rapid retreat. Simple snowline models of terrestrial ice sheets cannot explain the complexity of the last North American deglaciation. We must also look to marine ice sheets and the pulling power of ice streams for a more satisfactory resolution of this problem.

Deglaciation of marine ice sheets

Instability at marine margins of an ice sheet is a consequence of total uncoupling of ice from the bed, to produce the pulling power of a floating ice shelf. Ice-bed uncoupling need not occur abruptly at the marine margin. Often it begins in marine ice streams, which are fast currents of ice that drain most of the marine ice sheet and which are the dynamic link between a marine ice sheet and its ice shelf. Progressive ice-bed uncoupling beneath ice streams gives stream flow a concave surface profile, which is a transition profile between the convex surface profile of sheet flow and the nearly flat surface profile of shelf flow, as shown in Figures 2 and 4. The pulling force allows the ice stream to literally pull ice out of the marine ice sheet. A downstream pulling force also exists at the terminal ice walls of terrestrial ice sheets, but it is weaker than at ice-shelf grounding lines because an upstream basal shear stress exists at ice walls and exerts a braking force. The pulling power of marine ice streams is much greater because they move so fast, and the reduced basal shear stress along the ice stream allows the pulling force to reach far into a marine ice sheet.

Weertman (1957) showed that the pulling force increases as the square of ice thickness at the grounding line. Pulling power at a grounding line that retreats downhill into the isostatically depressed subglacial basin beneath a marine ice sheet increases exponentially, and it is therefore the major destabilizing factor at the margin of a marine ice sheet. As Figure 1 shows, this powerful instability mechanism at marine margins of an ice sheet involves changing the thickness of the ice column with respect to the water column at the grounding line of the ice shelf. Figure 13 illustrates processes that accomplish this in ways that make the grounding line retreat. Reversing these processes would make the grounding line advance. For purposes of illustration, the grounding line in Figure 13 lies just beyond the sill separating the subglacial basin from the submarine continental slope. Initial retreat is up the sill, and it is therefore reversible and represents stable equilibrium.

Figure 13a shows that during an ice-stream surge the reduction of basal traction extends the ice stream inland, causing surface lowering of grounded ice and seaward extension of floating ice, in order to maintain conservation of ice volume. Surface lowering inland from the grounding line compels the grounding line to retreat upslope in order to satisfy the buoyancy requirement. Figure 13b shows that progressive isostatic sinking of the bed inland from the grounding line, combined with glacial erosion inland and glacial deposition seaward of the grounding line, lowers the crest of the sill to the level of the grounding line. This has the same effect as an uphill retreat of the grounding line. Figure 13c shows the effect of rising sea level on the grounding line. In order to maintain the buoyancy condition, the grounding line must retreat upslope as sea level rises. Figure 13d shows the effect of surface melting landward of the grounding line and both surface and basal melting seaward of the grounding line. Melting thins the ice and causes the grounding line to retreat upslope. Figure 13e shows what happens after the grounding line crosses the crest of the sill and finds itself on a downward slope inland. This is an unstable condition for the grounding line, and it retreats irreversibly until it finds a new uphill slope against which it can anchor itself.

Although instability mechanisms at marine margins of an ice sheet look straightforward in the simple representation in Figure 13, these mechanisms involve processes that are poorly understood and that complicate any simple model for the collapse and

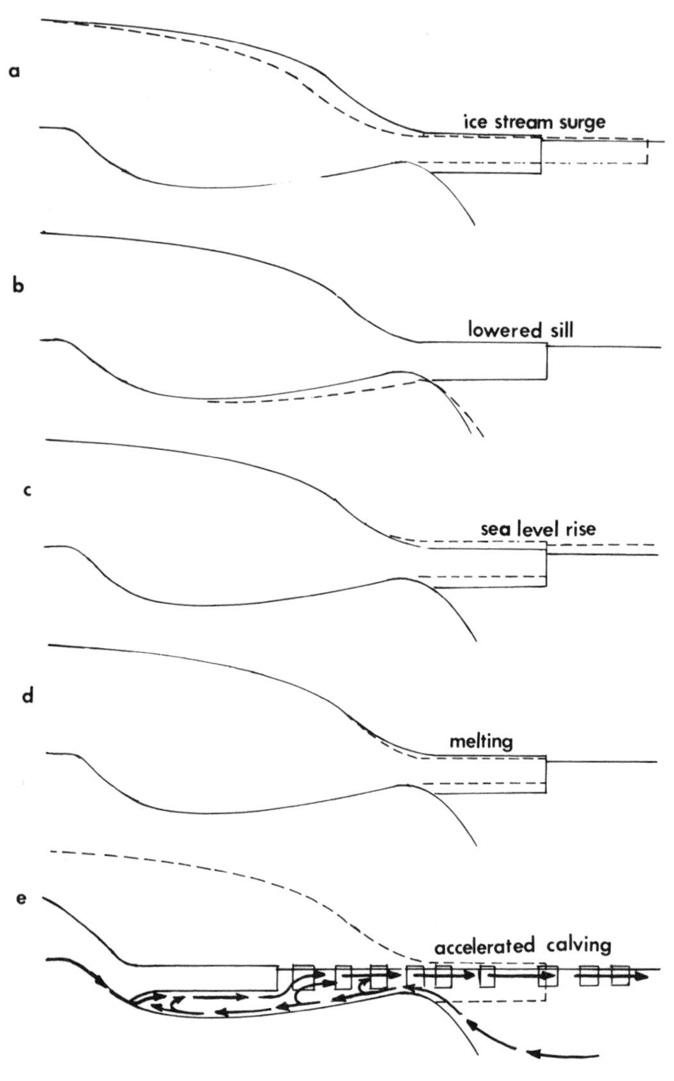

Figure 13. Mechanisms for destabilizing the marine margin of an ice sheet. Grounding-line retreat is facilitated by (a) an ice-stream surge that lowers the ice-sheet surface, (b) delayed isostatic sinking that lowers the bed, (c) rising sea level that raises the ice shelf, (d) surface and basal melting that thin the ice shelf, or (e) estuarine circulation and basal meltwater discharged at the grounding line that ferry icebergs out to sea as the grounding line retreats.

retreat of a marine ice sheet. It is worthwhile to examine briefly the more obvious complicating processes.

Figure 13a shows the effect at the grounding line when stream flow reaches farther into the heart of the marine ice sheet during a surge. The lengthening ice stream increases the portion of the flowline having a concave profile and thereby lowers the ice surface. Ice volume between the old and new surface profiles represents ice downdrawn into the ice stream and discharged onto the ice shelf. Enhanced downdraw therefore requires an increase in ice-stream velocity, which is accomplished by decreasing basal traction all along the ice stream as it lengthens. During a surge, the normally smooth ice surface is transformed into a jumbled chaos of crevasses and seracs as ice is stretched in the downstream direction by the pulling force. Stretching causes thinning at the grounding line, where basal ice lifts off the bed about nine times faster than surface ice lowers, owing to the fact that about 90 percent of floating ice lies below sea level. The grounding line must therefore retreat upslope toward the crest of the sill in Figure 13a. This retreat mechanism is bound up with the complex subglacial hydrology during a surge, as described by Kamb and others (1985).

Figure 13b shows the effect at the grounding line due to lagging isostatic depression of the bed and glacial erosion beneath the ice stream. Although details of isostatic adjustments associated with advancing and retreating ice sheets are complex and not fully understood, the broad features can be described. Isostatic depression occurs on a continental platform above sea level and beneath a growing ice sheet, whereas isostatic uplift occurs on the continental shelves below sea level and beyond the ice sheet. Depression, of course, is caused by the increasing ice burden, and uplift is caused by the decreasing water burden as sea level drops to supply precipitation over the ice sheet.

Exactly where the changeover point between depression and uplift lies in relation to the ice-sheet margin depends in part on whether creep in the Earth's mantle occurs near the viscous end or the plastic end of the viscoplastic creep spectrum (Hughes, 1981a). Most numerical models of these isostatic adjustments use a Newtonian viscous rheology for mantle creep (Cathles, 1975), with Walcott (1972) and Peltier and Andrews (1976) doing pioneering studies of isostatic adjustments in North America during the last deglaciation. These models do not consider isostatic depression beneath marine ice sheets that advance toward the edge of the continental shelves; in particular, they ignore isostatic adjustments beneath marine ice streams with their concave surface profiles. Lingle and Clark (1985), however, did model the isostatic response beneath a marine ice stream as the marine West Antarctic Ice Sheet collapsed over the Ross Sea after 17 ka. Clark (1980) concluded that glacioisostatic rebound data favored the marine Innuitian Ice Sheet that Blake (1970) postulated over the Queen Elizabeth Islands of Arctic Canada during the last glaciation. Existence of this ice sheet determines the changeover point from depression to uplift during deglaciation of Arctic Canada merely by its presence.

A major problem with isostatic depression beneath a marine ice sheet is to determine at what point along a flowline the ice load on the continental shelf becomes greater than the water load that was there before the marine ice sheet (Hughes and others, 1981, p. 295–311). Invariably this point is some distance in from the grounding line that defines the outer margin of the marine ice sheet. The question is, does isostatic depression begin inland from that point or does it continue seaward of it and, in either case, how far inland or seaward?

Grounding-line retreat initiated when glacial erosion lowers the sill in Figure 13b poses another question: At what point along an ice stream of a marine ice sheet is glacial erosion replaced by glacial deposition? Since fjords through coastal mountains and inter-island channels on continental shelves are both probable sites for ice streams, and both are typically foredeepened with a sill at their seaward end, the usual assumption is that glacial erosion occurs beneath the inner part of an ice stream and glacial deposition beneath the outer part. There are a few theoretical models based on this assumption (Crary, 1966; Sugden and John, 1982; McIntyre, 1985). Stream flow is a transition from sheet flow to shelf flow, in which basal traction progressively lowers to zero. A peak in basal traction can be expected near the inflection point in the surface flowline profile where the convex surface characteristic of sheet flow becomes the concave surface typical of stream flow (Hughes, 1981b). Conversely, a minimum in basal traction occurs at the grounding line, where the concave surface of stream flow becomes the nearly flat surface of shelf flow. Glacial erosion therefore reaches a maximum beneath the inflection line and a minimum beneath the grounding line. Eroded material entrained in the ice stream at the inflection line may be dumped immediately beyond the grounding line to produce an end moraine, perhaps akin to the Egga moraines on the outer Norwegian continental shelf. These moraines lie beyond prominent troughs that extend into coastal fjords and may have been occupied by ice streams (Andersen, 1981). This grounding-line deposition zone could contribute to the broad basal sills that mark the outer limit of a marine ice sheet. Deposition could be accomplished by several powerful mechanisms for basal melting at and immediately seaward of marine ice-sheet grounding lines (Robin, 1979; Hughes, 1982; MacAyeal, 1985).

There is a lag between advance of a marine ice sheet to the edge of a continental shelf and isostatic adjustments in the mantle beneath and beyond the marine grounding line. Moreover, glacial erosion is probably replaced by deposition in the vicinity of the grounding line. Just where isostatic uplift replaces isostatic sinking, and just where glacial deposition replaces glacial erosion, are unknown even for a steady-state ice sheet in mass-balance equilibrium. These crossover sites are mobile, and they are even more obscure in advancing and retreating ice sheets, which never attain steady-state equilibrium. We must therefore accept with extreme caution the marine ice-margin instability mechanism depicted in Figure 13b.

Figure 13c shows the most straightforward of the marine-margin instability mechanisms, namely the effect of rising sea level. Even here there are complications that are not understood fully. Clark (1976) showed how the gravitational attraction of an ice sheet can pull sea water up against marine margins of the ice sheet. This attraction can raise sea level several meters above mean global sea level at the grounding line, but attraction decreases as the marine ice sheet collapses (Lingle and Clark, 1979). Retreat of the grounding line during collapse, therefore, is not solely a response to rising sea level even when that is the only process under consideration. Other processes at the grounding line that may change as sea level rises are tidal effects, notably the tidal amplitude. This could be an important factor at the grounding line of marine ice streams in inter-island channels on the continental shelf, where amplification of tides might occur (Williams and Robinson, 1979). Although these amplitude changes occur with a short periodicity, they might nudge grounding-line oscillations past a critical threshold for catastrophic retreat.

Figure 13d relates retreat of the marine ice grounding line to ice melting. This is deceptively straightforward; but in reality retreat by melting is tied up with a complex set of positive feedback mechanisms observed today on Jakobshavns Isbrae in Greenland, and which I have called the Jakobshavns Effect (Hughes, 1983b, 1986). Jakobshavns Isbrae (69°10′N, 50°10′W) is one of 20 outlet glaciers that lie between 69°N and 72°N along the west-central coast of Greenland and drain about 22 percent by volume of the Greenland Ice Sheet (Bader, 1961; Carbonnell and Bauer, 1968). Terrestrial ice downdrawn through terrestrial ice streams into these outlet glaciers has bowed the north–south crest of the ice divide eastward. The Jakobshavns Effect has made Jakobshavns Isbrae the world's fastest known outlet glacier; its midsummer velocity of 23 m/d at the calving front, which floats in Jakobshavns Isfjord, is a typical velocity for glacial surges (Lingle and others, 1981; Paterson, 1981, p. 275–298). All the other west Greenland outlet glaciers between 69°N and 72°N experience the combination of positive feedback mechanisms that constitute the Jakobshavns Effect, although less dramatically than in Jakobshavns Isbrae itself.

Positive feedback mechanisms that combine to produce the Jakobshavns Effect are, in order, the pulling power of the ice stream after it becomes afloat, ubiquitous surface crevassing on the ice stream, rapid rates of summer melting over the crevassed surface, transport of surface meltwater through crevasses to warm the ice internally and lubricate the ice-bed interface beneath the ice stream, basal and lateral uncoupling as the ice stream becomes afloat in its coastal fjord, and rapid iceberg calving rates along its floating front. All these mechanisms feed back to maintain the rapid velocity of the ice stream.

These positive feedback processes have been observed and measured directly on Jakobshavns Isbrae (Lingle and others, 1981). Pulling power increases exponentially from the calving front to the grounding line of the portion floating in Jakobshavns Isfjord, owing to the steady increase in ice thickness over this 10-km distance and to the exponential increase in the pulling force with increasing thickness of freely floating ice (Weertman, 1957). The high longitudinal strain rates caused by the exponentially increasing pulling force transform the otherwise smooth surface of an ice stream into the jumbled chaos of crevasses and seracs observed on Jakobshavns Isbrae. Summer melting over this fractured surface produces much more meltwater than melting over a smooth surface, both because of the greater area exposed to solar radiation and because a greater percent of solar energy is absorbed from the multiple reflections between crevasse walls (Pfeffer, 1982; Pfeffer and Bretherton, 1987). Meltwater refreezing in crevasses releases latent heat that warms the ice, making it softer and therefore less able to resist the pulling force.

Meltwater reaching the bed lubricates the ice-rock interface, drowns bedrock projections, and increases basal water pressure. All of this tends to uncouple the ice stream from its bed, so the pulling force can reach far into the ice sheet.

Basal uncoupling is complete when Jakobshavns Isbrae becomes afloat in Jakobshavns Isfjord. Lateral uncoupling is also largely complete in Jakobshavns Isfjord, owing to water-filled bottom crevasses that open and migrate upward to meet surface crevasses in the lateral shear zone. Repeated tidal flexure along lateral grounding lines facilitates crevasse migration (Lingle and others, 1981). In Jakobshavns Isford, therefore, there is virtually no constraint on the pulling power of Jakobshavns Isbrae. Rapid Iceberg calving rates in Jakobshavns Isfjord keep pace with the forward velocity of Jakobshavns Isbrae at its calving front. Rapid calving is facilitated by transverse crevasses that open on the top and bottom surfaces and migrate toward one another, due to the imbalance of longitudinal hydrostatic forces at the calving front (Reeh, 1968) and the high longitudinal strain rates from the grounding line to the calving front (Weertman, 1980). Without rapid calving rates, the calving front would advance and the floating part of Jakobshavns Isbrae would have to conform to sidewall turns and branches along the full length of Jakobshavns Isfjord. Resistance to ice motion caused by conformity to sidewall geometry has been termed "form drag" by MacAyeal (1987). He shows that form drag can reduce substantially the pulling power of a freely floating ice stream.

Icebergs must also be transported through Jakobshavns Isbrae. In Figure 13e, ferrying icebergs to the open sea is accomplished by estuarine circulation of sea water in the fjord and by discharge of subglacial water at the grounding line.

Figure 14 shows a marine grounding line retreating from a bedrock sill at the edge of the continental shelf to a fjord headwall in coastal mountains. The marine ice sheet collapses during retreat of its grounding line, and what remains is the terrestrial ice sheet beyond the fjord headwall. This happened in the ice drainage system of Jakobshavns Isbrae over the past 17,000 yr (Andersen, 1981; Carbonnell and Bauer, 1968; Hughes, 1986). Grounding-line retreat is reversible so long as it occurs on an uphill slope, because reversing processes shown in Figure 13a through 13d will make the grounding line advance. Once the grounding line retreats over the crest of the bedrock sill and finds itself on the downhill slope, however, retreat is irreversible because ice at the grounding line now becomes thicker with further grounding-line retreat. Since the pulling power of a marine ice stream increases exponentially with increasing ice thickness at its grounding line, as illustrated in Figure 1, the ice stream pulls ice out of the ice sheet faster and causes marine ice to collapse. During collapse, the grounding line retreats rapidly to the floor of the marine subglacial basin and continues to retreat more slowly up the landward side of the basin. Retreat of the grouding line stops when it climbs to a landward elevation higher than the elevation of the seaward sill where catastrophic retreat began. If coastal mountains separate marine and terrestrial portions of an ice sheet, the grounding line will typically retreat to the bedrock headwall of fjords through the coastal mountains. This is the situation with Jakobshavns Isbrae, and similar situations may have been common along marine margins of North American ice sheets during the last deglaciation.

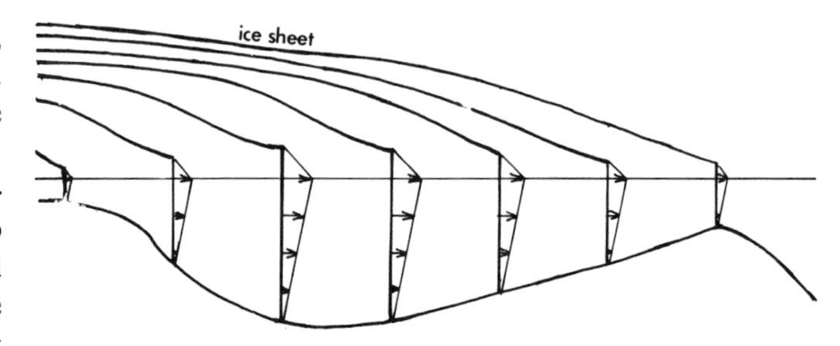

Figure 14. Collapse of an ice sheet caused by the pulling power of a retreating ice stream. Surface lowering (solid lines) is most rapid as the pulling force (arrows) increases during downslope retreat of the grounding line and calving front into a foredeepened submarine ice-stream trough.

In summary, the margin of a steady-state marine ice sheet is in stable equilibrium if the grounding line lies on a slope that is uphill inland, and in unstable equilibrium if it lies on a downhill slope in that direction. With stable equilibrium, the pulling force at the grounding line is reduced during retreat and the retreat rate decreases. With unstable equilibrium, however, the pulling force is increased during grounding-line retreat, and the retreat rate increases.

GEOMORPHOLOGICAL CONSTRAINTS

Weathering zones

Possible marine margiof North American ice sheets during the last deglaciation may have extended as far as the edge of the continental shelf, beyond which floating ice shelves might have formed. Marine ice grounded on the narrow Pacific continental shelf would have been an insignificant component of the Cordilleran Ice Sheet. Marine ice grounded on the broader Atlantic continental shelf, in Hudson Bay, in Foxe Basin, and in Arctic straits would have been a major component of the Laurentide Ice Sheet. Marine ice grounded in the inter-island channels of the Queen Elizabeth Islands, combined with isostatic depression of the islands themselves, could have made the Innuitian Ice Sheet postulated by Blake (1970) almost entirely marine. A key to assessing the extent of marine ice is to determine whether these ice-sheet margins extended into water deep enough to allow substantial pulling power to develop. Without strong pulling power at the ice margins, interior ice on a bed isostatically depressed far below sea level would behave dynamically as terrestrial ice, not marine ice. The marine character of Laurentide ice over Hudson Bay and Foxe Basin, for example, depends on whether Laurentide ice advanced into deep water along its northern and eastern margins.

A dispute over the seaward extent of northern and eastern North American ice during the last deglaciation is centered on the interpretation of weathering zones. Weathering zones are described in parts of Ellesmere Island (England, 1976; England and Bradley, 1978), Baffin Island (Pheasant and Andrews, 1972; Boyer and Pheasant, 1974), Labrador (Ives, 1957; Andrews, 1963), Newfoundland (Grant, 1977; Brookes, 1982), and Nova Scotia (Grant, 1971, 1976).

Three weathering zones usually can be distinguished. The lowest zone consists of fjords or valleys having the U cross-section ascribed to glacial erosion, plateaus landward of fjord headwalls, and end moraines seaward of valleys opening onto coastal plains. Glacially polished and striated bedrock and erratics are common, a thin till sheet mantles much of the surface, and weathering is minimal. The highest zone is found on headlands between fjords, on ridges between valleys, and on rounded summits of the interior plateau. At these sites, deeply weathered bedrock tors occasionally poke through a nearly ubiquitous mantle of mature felsenmeer, evidence of frost shattering and permafrost activity is common, and the few erratics have thick weathering rinds. The intermediate zone is a rolling landscape with scattered small lakes, patches of till, and numerous erratics. On exposed bedrock, weathering has obliterated glacial-erosion features smaller than deep grooves, but weathering on erratics can be quite variable. The intermediate weathering zone usually has a lower boundary between 300 m and 900 m and an upper boundary between 600 m and 1300 m above sea level. Its boundaries with the lower and upper zones can be quite distinct, even marked by a moraine, but more often they are diffuse.

Stratigraphic and geomorphic interpretations of weathering zones have been proposed. The stratigraphic interpretation allows little or no marine ice in northern and eastern North America, to the point of denying the existence of an Innuitian Ice Sheet and any significant marine character to the Laurentide Ice Sheet. The geomorphic interpretation allows substantial marine ice and guarantees largely marine behavior by both these ice sheets during the last deglaciation.

The stratigraphic interpretation of weathering zones originated with Boyer and Pheasant (1974) in Baffin Island, according to Ives (1978). Because the three weathering zones always have the same altitudinal sequence—with the upper zone being the most weathered, the lowest zone being the least weathered, and the intermediate zone having intermediate or mixed weathering characteristics—Andrews and Miller (1976) called them "horizontal stratigraphy" that could be correlated all along eastern North America to delineate the upper limits of the three most recent Pleistocene glaciations, with the highest weathering zone assigned to the oldest of these glaciations and the lowest assigned to the last glaciation.

Because reliable radiometric dating of weathering zones is almost nonexistent, there is no agreement as to their ages. The lowest zone is ascribed to either the Early or Late Wisconsin glaciation, the intermediate zone to either the Early Wisconsin or the last pre-Wisconsin glaciation, and the upper zone to the last pre-Wisconsin glaciation, an earlier glaciation, or no glaciation at all (Andrews and Miller, 1976; Grant, 1977; Mayewski and others, 1981, p. 106–117). If the low weathering zone was the upper limit of Late Wisconsin ice in northern and eastern North America, the ice margin would not have reached present-day sea level, let alone lower sea level at the last glacial maximum, except in a few fjords. North American ice sheets would have been terrestrial, according to the stratigraphic interpretation of weathering zones.

In the geomorphic interpretation, all three weathering zones were glaciated during the last glaciation and earlier glaciations as well; they identify different thermal conditions at the ice-rock interface, not different glaciations. Figure 15 illustrates this interpretation. The lowest zone reflects subglacial conditions at a completely thawed bed on which sheet flow over interior plateaus produced areal scouring, and stream flow toward the coast produced U-shaped valleys and fjords by selective linear erosion. These two distinctive glacial landscapes produced the freshly polished and striated surfaces with thin blankets of till and many unweathered erratics (Sugden and John, 1982).

The highest zone reflects subglacial conditions at a completely frozen bed covered by relatively thin and stagnant ice, so that the little frictional heat generated by basal shear was conducted rapidly to the surface. Pre-glacial landforms, notably tors and felsenmeer, are largely undisturbed. The few erratics (glacially transported rocks) are invariably of local origin, because local ice on summits of interior plateaus and on the headlands between coastal fjords had strongly divergent flow. Sugden and Watts (1977) postulated this condition for the high weathering zone on eastern Baffin Island.

The intermediate zone reflects a transition from frozen to thawed basal conditions, in which isolated thawed patches appeared at the boundary with the high zone. These patches become more numerous and larger and coalesce farther downslope, leaving isolated frozen patches that become less numerous, smaller, and separate still farther downslope, being absent altogether at the boundary with the low zone (Hughes, 1981b). Glacial quarrying occurred within the thawed patches and around the frozen patches, creating a rolling landscape in which postglacial lakes identify sites of intensive quarrying, erratics are locally quarried blocks, and till occurs where debris-charged refrozen ice originating at lake beds was subsequently melted. Occasional moraines near boundaries of the weathering zones were deposited during the last deglaciation at times of stillstands. If the high weathering zone was frozen and covered by ice during the last glacial maximum, then the northern and eastern margins of North American ice sheets could have extended far out into deep water on the continental shelves, so that pulling power was strong along these margins and could reach up ice streams into the heart of the ice sheets, giving them a strong marine character.

Glacial geology

Those who see major causes for the last deglaciation of

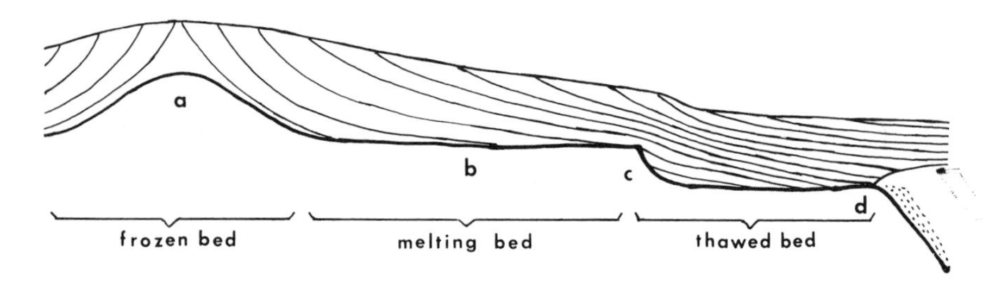

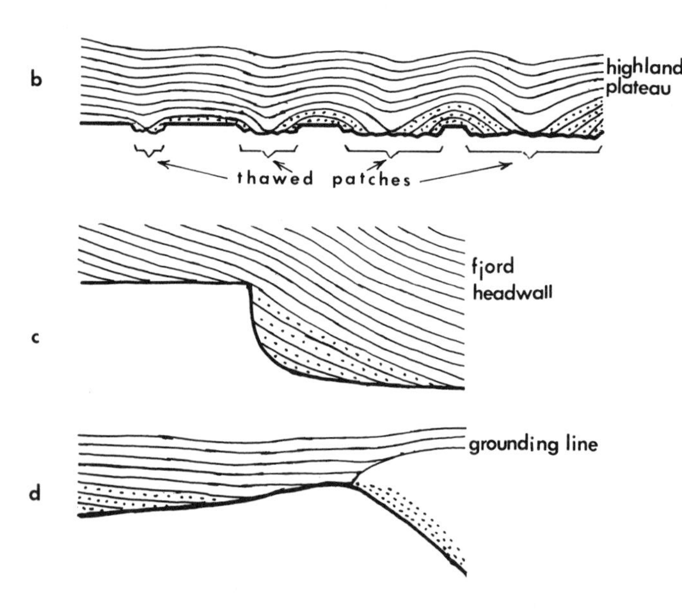

Figure 15. A geomorphic interpretation of weathering zones. Top: Generalized ice trajectories along a flowline for an ice sheet with flow diverging from a local ice dome over the frozen bed of a highland summit (left), spreading over the melting bed of a highland plateau (center), and converging over the thawed bed of a coastal fjord (right). Details of flow are shown below: (a) flowlines over and around tors on the highland summit; (b) flowlines entering and leaving thawed patches on the melting bed of the highland plateau, with dotted areas being debris-filled refrozen ice formed in the downstream part of thawed patches, which get bigger and more numerous in the downstream direction; (c) flowlines intersecting the melting bed upstream from the fjord headwall, originating at the freezing bed of the fjord headwall, or intersecting the melting bed downstream from the fjord headwall, with active quarrying at the fjord headwall to produce debris-filled refrozen ice shown as the dotted area; (d) flowlines intersecting the melting bed upstream from the grounding line and the melting underside of the floating ice shelf downstream from the grounding line as the ice stream leaves the fjord, with dotted areas showing refrozen ice (left) and a submarine terminal moraine (right).

North America in the unique dynamics of marine ice sheets tend to embrace the geomorphic interpretation of weathering zones. Those who prefer models of deglaciation based on raising the snowline relative to the surface of purely terrestrial ice sheets might be more comfortable with the stratigraphic interpretation. These disagreements force us to rethink our conceptions about how glacial geology should be interpreted, especially if the interpretation is to be used as geomorphic input to computer models that reconstruct ice sheets for the last glaciation.

An interpretation of glacial isostasy, glacial erosion, and glacial deposition was used to reconstruct a geomorphic model of former ice sheets at the last glacial maximum and during the last deglaciation for CLIMAP by making a distinction between first order and second order glacial geology (Denton and Hughes, 1981, chapters 2, 5, and 6). For North American glaciation, first-order and second-order glacial geology used in this interpretation are mapped in the National Atlas of Canada (Anonymous, 1974). First-order glacial geology is that which reveals steady-state conditions repeated at the base of an ice sheet during successive glaciations, so that an increasingly permanent regional imprint remains on the deglaciated landscape. This imprint is most easily recognized by remote sensing from a satellite in a high orbit around Earth. Second-order glacial geology is that which reveals transient conditions that were temporarily imprinted on the local landscape as ice retreated during the last deglaciation, but that will be either erased or reoriented by advancing ice during the next glaciation. Denton and Hughes (1981) saw first-order glacial geology in maps on pages 1–2, 7–8, 9–10, 11–12, 27–28, and 37–38 of the National Atlas of Canada (NAC). Figure 16 shows geomorphology that was used to identify first-order glacial geology on the basis of these maps. Significance was attached to the following features.

Topographic relief (NAC map 1–2). Hudson Bay is the heart of a low central region surrounded by raised beaches. This suggests that an ice sheet was centered over Hudson Bay for a long time, because isostatic rebound was substantial and is still incomplete. A broad upland separates the central lowland from a landscape of large elongated lakes, long inter-island channels, and deep coastal fjords that tend to radiate from the rebounding central lowland. This suggests prolonged streaming of ice flow to-

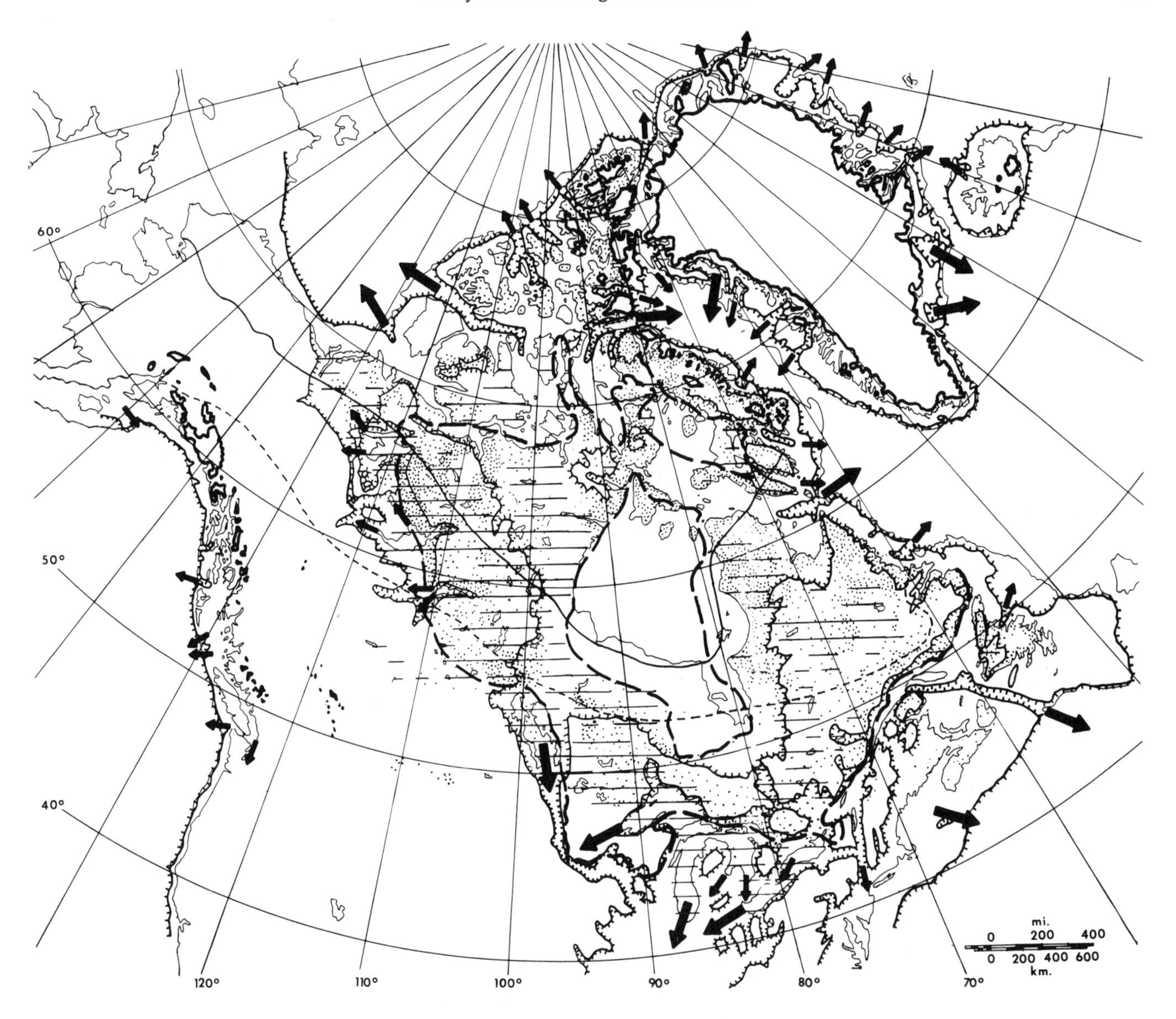

Figure 16. First-order glacial geology of North America. Deglaciated topography, shown by hachured lines along the 300-m (~1000-ft) contours below and above present sea level, reveals linear troughs where ice streams (arrows) may have radiated from a central subglacial basin (Hudson Bay). Deglaciated bedrock, shown by an exposed Precambrian shield (between heavy broken lines), by regions in which over half of the area is exposed bedrock (dotted areas), and by regions in which over 10 percent of the area is lakes (horizontal lines), reveals the extent, intensity, and nature of prolonged glacial erosion. Deglaciated permafrost, shown by the limit of continuous permafrost (north of the light solid line through Hudson Bay) and discontinuous permafrost (north of the light dashed line through James Bay), reveals regions where the bed may have been frozen and melting, respectively, during the glacial maximum. (After Anonymous, 1974.)

ward the margins of the ice sheet, with ice streams moving along the axes of these troughs.

Distribution and size of lakes (NAC maps 7–8 and 9–10). The distribution of lakes is quantified by mapping the proportion of the postglacial landscape occupied by freshwater lakes. This proportion and the size of lakes tends to increase with distance from Hudson Bay. Because bedrock lake basins result from glacial erosion of a thawed bed, this distribution suggests a transition from a frozen bed beneath the central ice divide over Hudson Bay to a thawed bed where the largest lakes lie along an arc at the outer edge of the crystalline Canadian Shield. These are Great Bear Lake, Great Slave Lake, Lake Athabasca, Lake Manitoba, Lake Winnipeg, Lake Superior, Lake Michigan, Lake Huron, Lake Erie, and Lake Ontario. Most of the smaller lakes lie in the broad uplands that arc around the Hudson Bay lowlands. This suggests that the transition from a frozen to a thawed bed consisted of isolated thawed patches that became larger and more numerous with distance from Hudson Bay, until they eventually merged to become basins for the large elongated lakes that were sites of stream flow near the ice-sheet margin. This is the geomorphologic interpretation of weathering zones depicted in Figure 15, but on a continental scale instead of a local scale. It implies a dome over Hudson Bay as the dominant center of spreading.

Distribution of permafrost and temperature (NAC map 11–12). A zone of nearly continuous permafrost surrounds Hudson Bay, and mean annual temperatures over Hudson Bay are below freezing, whereas discontinuous permafrost and slightly warmer temperatures tend to arc southward around the Hudson Bay lowlands. This suggests that sea-ice thickening in Hudson Bay at the beginning of a glaciation would ground on a bed that was already frozen, and that might remain frozen for most of the glaciation cycle, whereas ice spreading outward from Hudson Bay might advance over a bed having thawed patches that remained thawed, especially where lakes now exist and to the south. Permafrost does not exist beneath lakes, but it can floor marine embayments.

Geological provinces (NAC map 27–28). A crystalline Precambrian shield is exposed from the arc of large intracontinental lakes to a cap of Paleozoic bedrock over most of Hudson Bay and the southern Hudson Bay lowlands. This suggests that glacial erosion has stripped the cover of Paleozoic rocks between the central zone of nearly stagnant ice frozen to the bed beneath a Laurentide dome over Hudson Bay and the zone of ice streams sliding on a thawed bed near the ice-sheet margin.

Distribution of exposed bedrock and unconsolidated rocks (NAC map 37–38). Exposed bedrock is concentrated along the eastern and northwestern shores of Hudson Bay, along the arc of large intracontinental lakes, and in the fjordlands of Newfoundland, Labrador, and Baffin Island. Unconsolidated rocks are concentrated in the broad uplands between the Hudson Bay lowlands and the outer zone of intracontinental lakes, inter-island channels, and coastal fjords. This suggests the following model. An inner melting zone began in the Hudson Bay lowlands, where areal scouring by clean ice scraped off the unconsolidated rocks. Surrounding that was a freezing zone in the uplands and between ice streams, where unconsolidated rocks were frozen into a basal layer of refrozen ice, and where quarrying pitted the underlying bedrock to produce more unconsolidated rock and the lake-studded postglacial landscape. Unconsolidated rocks and quarried rock carried southward in the basal refrozen ice were melted out in ice ridges between ice streams and ice lobes beyond ice streams, to produce the broad zone of glacial deposition south of the Great Lakes and over the Great Plains. Basal refrozen ice moving northward was funneled into marine ice streams, and the entrained debris was probably deposited as ice-rafted sediments on the continental slopes and beyond.

There is an alternative geomorphic interpretation of the information summarized in Figure 16 from maps 1–2, 7–8, 9–10, 11–12, 28–29, and 37–38 of the National Atlas of Canada (Hughes, 1985). A high central ice dome over Hudson Bay, if sufficiently thick, could depress the pressure melting point enough to thaw the bed. Because ice was stagnant beneath the dome, however, no glacial erosion would occur, and the cap of Paleozoic bedrock would remain intact. Basal freezing would begin in the Hudson Bay lowlands, where unconsolidated rocks would be frozen into the basal layer of refrozen ice, thereby exposing bedrock. Basal melting in ice passing over the surrounding uplands would deposit unconsolidated rocks that were frozen into the ice. Basal melting would increase and intensify in the ice streams, but basal freezing would occur in the slow ice between ice streams, where unconsolidated rocks would be frozen into a second refrozen ice layer, leaving exposed bedrock. Unconsolidated rocks and other material transported by this second refrozen ice layer would be discharged in icebergs after it crossed the frozen bed along the northern marine margin and dovetailed into ice streams occupying inter-island channels, and it would be deposited as basal till sheets and end moraines as it crossed the melting bed along the southern terrestrial margin and dovetailed into ice streams ending in ice lobes. Freshwater lakes that were sites of intensive quarrying beneath the steady-state ice sheet, would be associated with freezing and melting beds where frozen and thawed patches coexisted.

The alternative interpretation of first-order glacial geology in the National Atlas of Canada applies if ice over Hudson Bay is thicker than in the first interpretation. However, thawing of the bed that is triggered by thickening ice reduces ice-bed traction, and this allows ice to spread outward faster, so the central ice dome must lower and perhaps even become a saddle on the ice divide. Latent heat is transferred upward to the surface faster as ice thins, so basal meltwater in Hudson Bay may refreeze, as shown by temperature-depth curves derived by Robin (1955). A newly frozen bed increases ice-bed traction and allows ice to thicken once again. We then have an inherent instability in the ice divide over Hudson Bay, producing a continuous flip-flop in freezing and thawing conditions at the bed, with both sets of conditions being compatible with the first-order glacial geology (Hughes, 1973, 1985). First-order glacial geology, therefore, need not provide a unique solution for the maximum thickness of a paleo ice sheet in geomorphic ice-sheet models.

Computer reconstructions of North American ice sheets during the last deglaciation should be based on what Denton and Hughes called second-order glacial geology. Boulton and others (1985) called such features drift lineations and noted, "Lineations dominantly reflect the latest movement pattern until, on final exposure beyond a retreating glacier margin, the majority assume the pattern of movement beneath the margin, although palimpsests of earlier patterns may survive. Thus, continent-wide patterns of glacier lineations do not reflect the ice-sheet-wide pattern of flow at any one stage, but a concentric succession of sub-marginal flow patterns." They distinguish two kinds of drift lineations: transverse and longitudinal, with respect to ice flow at the margin. Transverse lineations include "Rogen" moraines, push moraines, belts of stagnation terrain, dump moraines, and ice-contact fluviatile accumulations. Longitudinal lineations include longitudinal elements on "Rogen" moraines, striae, drumlins, flutes, and bedrock lineations.

Boulton and others (1985) combined detailed field maps of the smaller lineations, such as push moraines and striae, and Landsat imagery showing the larger lineations, such as long-term stillstand moraines and giant eskers, to produce a retreat map of Laurentide ice during the last deglaciation. Figure 17 reproduces the essential features of their map. These lineations are most reliable as ice-flow indicators for Laurentide ice, because the bed was relatively smooth and gently sloping; they are unreliable as flow indicators for Cordilleran ice, where ice moved down prominent valleys. Consequently, Figure 17 shows Cordilleran flow indicators on map 33–34 in The National Atlas of Canada. Stillstands of the Laurentide ice from 20 to 11 ka, at 10 ka, at 9.5 ka and from 8.0 to 7.9 ka left large-scale transverse lineations that have been dated. These lineations provide anchors for isochrons drawn normal to longitudinal lineations by Boulton and others (1985), as seen in Figure 17. They used a steady-state computer model developed by Boulton and others (1984) to reconstruct the Laurentide Ice Sheet at or near these stillstands, on the assumption that nearly steady-state flow was attained at these times. A single ice dome over Hudson Bay was the dominant center of spreading in all their reconstructed ice sheets.

Boulton and others (1985) assumed that ice-sheet retreat was a mirror image of ice-sheet advance and, on that basis, they noted that the dispersal of Laurentide erratics mapped by Shilts (1980, 1982) did not agree with the time-transgressive flow pattern in Figure 17. If the flow pattern was not time-transgressive, however, then the pattern of erratic dispersal could be ascribed to ice spreading from two major ice domes, one in Quebec and one in Keewatin, as shown by concentric transverse lineations and radiating longitudinal lineations in Figure 17. They concluded, "Thus the patterns [in Fig. 17] may not only give the pattern of retreat of the ice sheet but also reflect the long-term disposition of centres of mass." When they used their computer model to reconstruct a steady-state ice sheet with domes over these two centers, the resulting ice volume was far below what is indicated by oxygen-isotope stratigraphy in ocean-floor sediments. Moreover, contrary to their claim, this alternative ice sheet did not produce flowlines that corresponded with dispersal of erratics south of Hudson Bay, because its Quebec dome was too far south (compare Figs. 17 and 18).

Failure of the alternative model by Boulton and others (1985) to account for either the ice-volume record or the distribution pattern for erratics may be a consequence of using a steady-state model to reconstruct a transient Laurentide Ice Sheet during the last deglaciation. Their steady-state model always constructs a single major dome over Hudson Bay. That is reasonable at the glacial maximum, but not during the glacial termination. First-order glacial geology shown in Figure 16 supports a single Hudson Bay dome during the maximum, but second-order glacial geology shown in Figure 17 supports two domes, one in Quebec and one in Keewatin, during the termination. This implies that the Hudson Bay dome collapsed during deglaciation, leaving a saddle over Hudson Bay and separate domes over Quebec and Keewatin. Because their steady-state model is incapable of simulating this collapse, it led Boulton and others (1985) to conclude that the glacial geology was not time-transgressive. In reality, the model was incapable of reproducing time-transgressive glacial geology.

The main reason why ice volumes are so low in the Laurentide Ice Sheet preferred by Boulton and others (1985), and shown in Figure 18 at the last glacial maximum, is that their reconstruction is based on flow over a water-saturated deformable bed in most regions beyond the exposed Precambrian shield. As shown by Boulton and Jones (1979), this permits very low flowline profiles and produces an ice sheet somewhat like a British bowler, with a thin brim of ice over the soft sediments and a high crown of ice over hard shield rocks. Bed hardness in Figure 18 decreases more or less abruptly as ice passes from exposed shield bedrock to unconsolidated shield rocks to exposed sedimentary bedrock to unconsolidated sedimentary rocks. Boulton and Jones (1979) argue that low glacier profiles of this type are not common today because present-day glaciation exists in a core area from which most soft, permeable sediments have been glacially eroded, leaving hard, impermeable bedrock. Were modern glaciers to advance onto soft, permeable sediments beyond these core areas, they would develop the flowline profiles depicted in Figure 18.

Two problems exist with the Laurentide Ice Sheet in Figure 18. First, the ice surface is only 500 m above sea level up to 600 km from ice margins, making the ice brim little different from a floating ice shelf in its dynamics and in its response to snowline fluctuations. Ice-shelf dynamics typically develop stream flow around areas of local grounding, such as islands or shoals, yet no provision for stream flow around uplands along the ice margin is provided in Figure 18. Second, even minor changes of snowline elevation can cause the equilibrium line to sweep across the entire length of the nearly flat surface of an ice shelf, which implies only two stable equilibrium-line positions in Figure 18: at the front and at the rear of the ice brim. An equilibrium line at the front would put the ice brim in a surface accumulation zone, and rapid heat conduction through the thin ice to the surface would, in the Zotikov (1964) theory, virtually guarantee a frozen bed. Once frozen, the bed would transform from soft and permeable to hard and impermeable, so ice would thicken rapidly in from the margin. An equilibrium line at the rear would put the ice brim in a surface ablation zone, and rapid thinning by melting would, in the van der Veen theory, permit the ice brim to be only about 200 km wide (Oerlemans and van der Veen, 1984, p. 64–66). Therefore, the ice sheet in Figure 18 would seem to be highly unstable, and we can ask if it could exist at all, let alone exist over the long time span needed to distribute shield erratics in the pattern shown.

Fisher and others (1985) also assumed that soft deformable beds existed where the Laurentide Ice Sheet advanced over unconsolidated sediments, and they used a computer model based

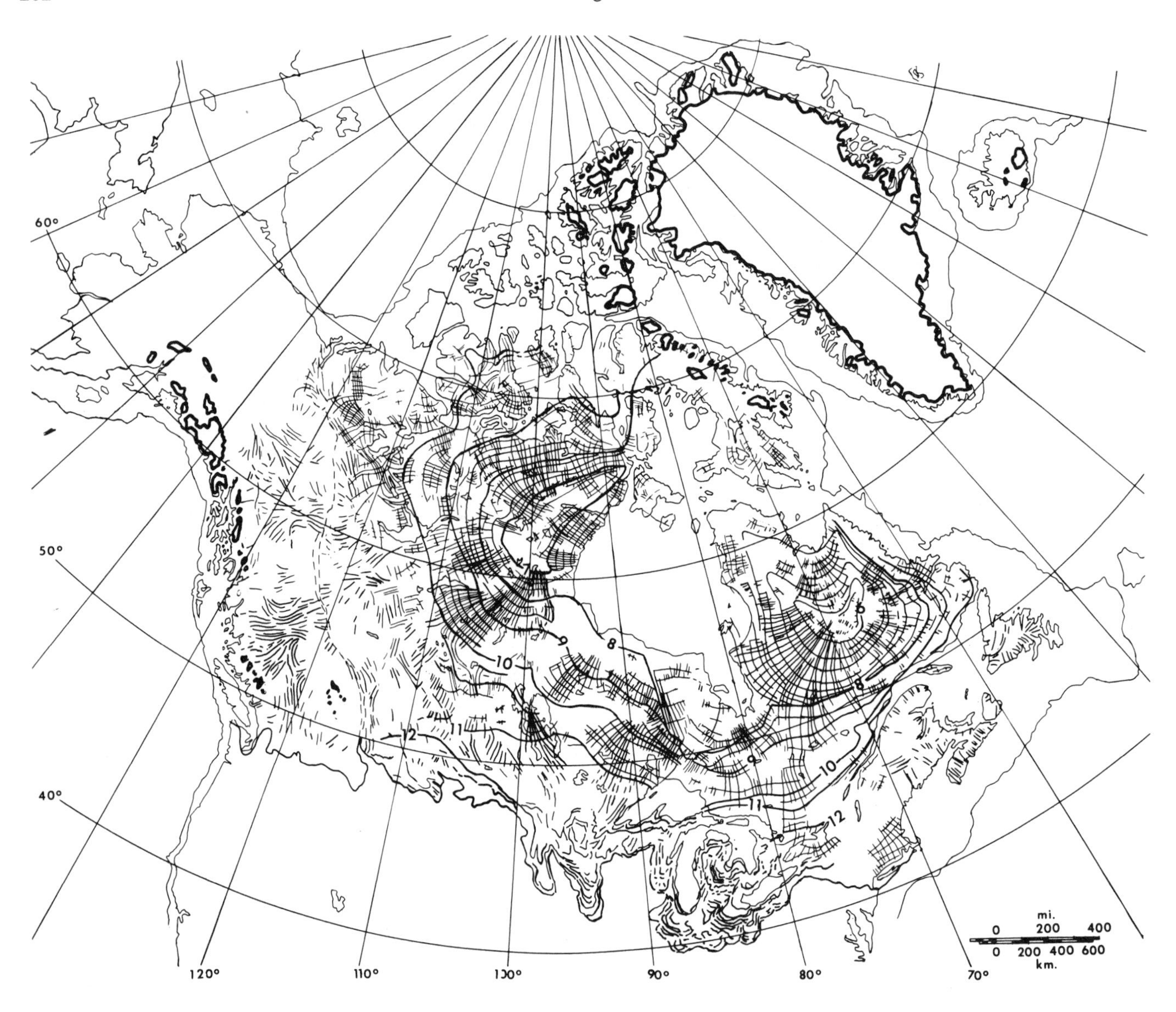

Figure 17. Second-order glacial geology of North America. Trends of two types of glacial lineations believed to be formed predominantly near or at the retreating ice margin are shown. Longitudinal lineations aligned predominantly with ice flow include eskers, drumlins, flutes, and striations. Transverse lineations aligned predominantly normal to ice flow include various kinds of recessional moraines, belts of stagnation terrain, and ice-contact fluviatile accumulations. Numbers are ages for ice-margin positions for prominent still-stands in millennia. After Anonymous (1974) and Boulton and others (1985).

on plasticity theory to reconstruct ice elevations at the glacial maximum. Two versions are shown in Figure 19. In version A, basal shear stress is constant in each of five zones, with 5 kPa ascribed to ablating ice covering a zone in the Great Plains and south of the Great Lakes, where soft deformable beds were assumed; 14 kPa in the Mackenzie River valley and Hudson Bay, where light precipitation fell on ice covering unconsolidated sedimentary rocks; 54 kPa in a zone around Hudson Bay, where moderate precipitation fell on ice covering the Precambrian shield; and 81 kPa in an eastern zone, where heavy precipitation fell on ice covering the Precambrian shield and the northern Appalachians. In version B, lowering basal shear stress to 14 kPa over Hudson Bay produced a central depression and a doughnut-shaped ice sheet, with a bite taken out in Hudson Strait. Fisher and others (1985) called their versions of the Laurentide Ice Sheet "objective" reconstructions, but their arbitrary selection of basal shear stresses made the reconstructions subjective.

The reconstructions by Fisher and others (1985) employed a

computer model by Reeh (1982) in which flowline trajectories are plotted as output using bed topography, basal shear stress, and ice margins as input. A feature of the model is that flowlines converge on re-entrant ice margins to produce ice streams, as seen in Figure 19A. This is realistic along marine margins where ice streams occupy straits or inter-island channels, but it is unrealistic along terrestrial margins where ice streams occupy valleys or elongated lake basins and end as ice lobes. Another problem with these reconstructions is that flowlines change directions abruptly when they angle across boundaries between zones having different basal shear stresses (Fig. 19A). This violates the requirement that flowline trajectories be normal to ice elevation contour lines. The problem cannot be solved by using more zones to minimize basal shear stress changes across zone boundaries, as that violates the basic requirement of plasticity theory that basal shear stress is constant.

Glacial isostasy

A major reason why Fisher and others (1985) produced a doughnut-shaped version of the Laurentide Ice Sheet at the last glacial maximum is because Andrews and others (1983) had postulated deglaciation of Hudson Bay sometime between 40 ka and 20 ka, on the basis of amino acid dates obtained from marine sequences in the Hudson Bay lowlands. They revived the idea of Tyrrell (1898) that the Laurentide Ice Sheet formed when Baffin, Patrician, Labrador, and Keewatin ice sheets merged over Hudson Bay from the north, south, east, and west to produce an interior depression. This idea was subsequently challenged in the theory of highland origin and windward growth by Flint (1943), which produced a Laurentide dome over Hudson Bay at the glacial maximum. Even if Tyrrell (1898) was correct, an ice dome would develop over Hudson Bay if time was sufficient and the bed was hard. A hard bed would be either rigid bedrock, frozen or thawed, or permeable sediments in the permafrost condition. In order to determine whether time was sufficient to produce a dome, Peltier and Andrews (1983) conducted modeling experiments based on glacial isostasy, an independent data set that provides geomorphological constraints to ice-sheet reconstructions at the last glacial maximum and during its termination.

Peltier and Andrews (1983) employed an improved version of a glacial isostasy model developed by Peltier (1976). In the so-called forward problem, Peltier and Andrews (1976) used as input to the Peltier (1976) model a mantle viscosity of 10^{22} Poise and a history of crustal unloading in North America during the last deglaciation. For crustal unloading, ice elevations of the Laurentide Ice Sheet were computed along flowlines drawn normal to retreating ice margins published by Bryson and others (1969). Following Paterson (1972), they assumed that ice flowed at the plastic end of the viscoplastic creep spectrum, so that flowline profiles were independent of changing rates of accumulation and ablation. Model output was a sea-level history that compared well with the record of relative sea-level variations dated along the east coast of North America. They called this the

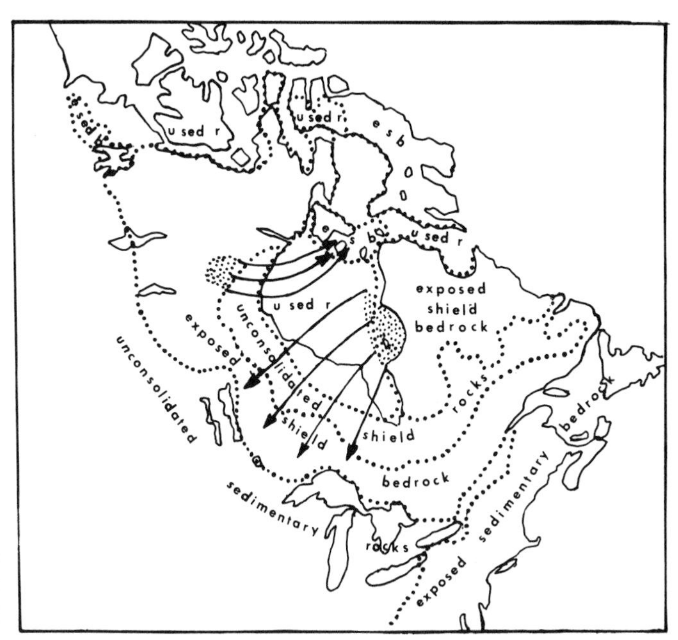

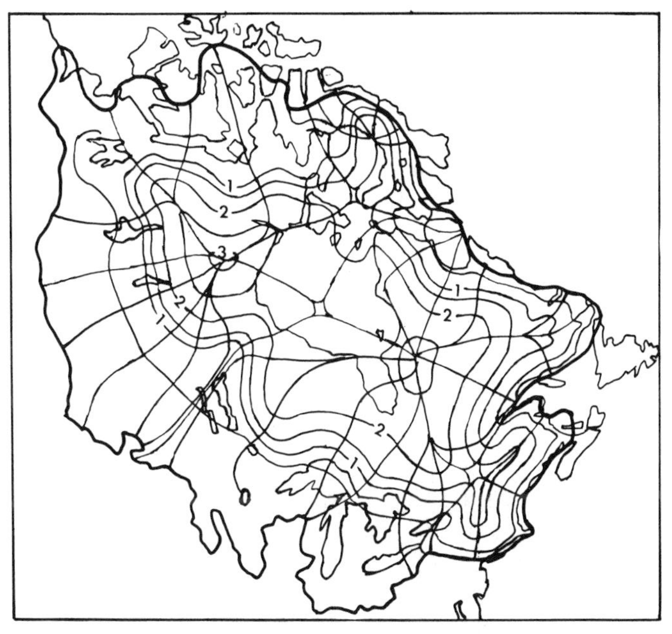

Figure 18. The relationship between a deformable bed and the Laurentide Ice Sheet according to Boulton and others (1985). Top: a thawed bed is least deformable for exposed shield bedrock (e s b), most deformable for unconsolidated sedimentary rocks (u sed r), and of intermediate deformability for unconsolidated shield rocks (u s r) and exposed sedimentary bedrock (e sed b). Arrows show the distribution of erratics from the dotted areas where the rock types outcrop. Bottom: ice elevations contoured at 0.5-km intervals and surface flowlines for the Laurentide Ice Sheet at the last glacial maximum.

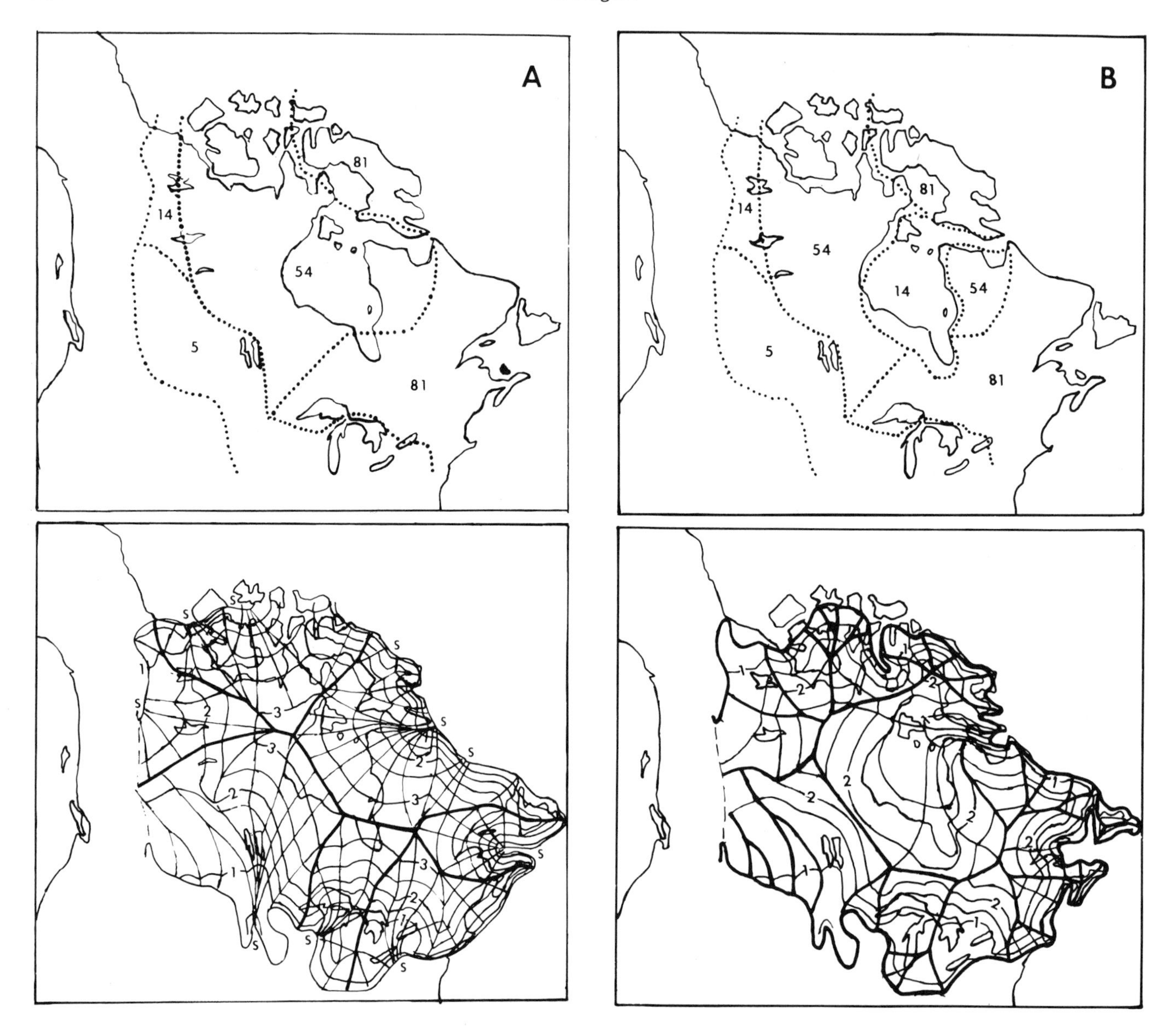

Figure 19. The relationship between a deformable bed and the Laurentide Ice Sheet according to Fisher and others (1985). Top: bed deformability is greatest where basal shear stress is lowest, with zones delineated for basal shear stresses of 5, 14, 54, and 81 kPa. Bottom: ice elevations contoured at 0.5-km intervals and surface flowlines for the Laurentide Ice Sheet at the last glacial maximum, with S identifying major ice streams at the ice margin and heavy lines identifying major ice divides. Paleozoic sedimentary rock beneath and south of Hudson Bay, and in Hudson Strait, is assumed to be rigid in version A and deformable in version B.

forward solution, and it encouraged Clark and others (1978) to compare dated records of relative sea-level variations worldwide with forward solutions based on instantaneous deglaciation of Northern Hemisphere terrestrial ice that was uniformly 1,000 m thick. Clark and Lingle (1979) obtained forward solutions that also included linear and exponential deglaciation histories for Antarctic marine ice.

None of the deglaciation inputs to these forward solutions included marine ice in the Northern Hemisphere. However, as dated records of relative sea-level variations worldwide became more complete, it became possible to enter these sea-level histories and mantle viscosities as input to the Peltier (1976) model, and to reconstruct ice sheets during deglaciation as model output. This was the inverse solution, and Clark (1980) applied it to North American deglaciation. His ice sheets consisted of simple discs whose thickness decreased during deglaciation, and he found that an ice disc covering the Queen Elizabeth Islands of arctic Canada was necessary. This was the marine Innuitian Ice Sheet that Blake (1970) had postulated on the basis of dated raised beaches throughout the islands. Because there was little or no glacial-geological evidence for this ice sheet (England, 1976), sheet flow must have occurred over a frozen bed on the islands, although stream flow on a thawed bed may have existed in the outer inter-island channels.

Peltier and Andrews (1983) called the Peltier and Andrews (1976) forward solution their Ice I model. They used its deglaciation history and various viscosity variations through the mantle to obtain a best fit with relative sea-level data worldwide as output in a forward solution. This fit and the viscosity data were entered as model input to refine the glacial history obtained as model output in an inverse solution. Forward and inverse iterations were repeated until the inverse solution converged on what they called the Ice II model of North American glacial history. Free-air gravity anomalies were also used to constrain the Ice II model. As in the Ice I model, a constant mantle viscosity of 10^{22} Poise was most satisfactory. Next, Peltier and Andrews (1983) relaxed the Ice I and Ice II assumption that Laurentide ice was in isostatic equilibrium at the 18-ka glacial maximum. They approximated the Shackleton and Opdyke (1973) oxygen-isotope record of ice volume changes over the last 500 ka by a crude sawtooth variation with time in which ice volume increased linearly for 100 ka during each glaciation, and then dropped to the interglacial volume in an instantaneous termination. Various degrees of isostatic equilibrium at the last glacial maximum and various viscosity-depth profiles in the mantle were linked to worldwide sea-level data to obtain an Ice III model in which the last termination was at 16 ka. However, the deglaciation history produced as model output was incompatible with the ice-margin isochrones of Bryson and others (1969) and with modern free-air gravity anomalies for Hudson Bay. Peltier and Andrews (1983) then produced an Ice IV model by forcing the Ice III model to give a best fit to a subset of relative sea-level data along Canadian coastlines. All problems that appeared in the Ice III model remained in the Ice IV model.

The Ice I and Ice II models provided the best fit with relative sea-level data, both close to and far from the Laurentide Ice Sheet, and were in best agreement with modern North American free-air gravity anomalies and ice-margin isochrones during deglaciation. Both models employed complete isostatic equilibrium at the glacial maximum, during which the Laurentide Ice Sheet had a single dome located over Hudson Bay. In contrast, the Ice III and IV models gave poorer fit to all these data, allowed various degrees of isostatic equilibrium, and a saddle over Hudson Bay separated Labrador and Keewatin domes at the glacial maximum, especially in the Ice IV model.

These results support Flint (1943) rather than Tyrrell (1898), and they question the amino acid chronology reported for Hudson Bay by Andrews and others (1983). Taken at face value, this chronology points to deglaciation of Hudson Bay about every 40,000 yr over the past 150 kyr, with the last deglaciation occurring sometime between 40 ka and 20 ka. A complete understanding of all variables involved in amino acid dating is not at hand, however, and the amino acid cycle may record the last four inter-glacial stages reported by Shackleton and Opdyke (1973) since 400 ka.

Results by Peltier and Andrews (1983) depend on key assumptions used in their glacio-isostasy model. This was the Peltier (1976) model, improved by covering a viscous mantle with an elastic lithosphere of variable thickness (Peltier, 1980 a, b). The first assumption is that mantle creep is viscous. Weertman and Weertman (1975) and Weertman (1978) summarize creep data for rocks, including mantle minerals, that show viscoplastic creep to be closer to the plastic than the viscous end of the viscoplastic creep spectrum (Fig. 5). However, Nye (1957) showed that if viscosity is taken as the ratio of a weak pair of deviator-stress and strain-rate components in a complex stress field containing much stronger pairs, then the ratio from the weak pair gives a Newtonian viscosity that appears to reflect purely viscous flow. Peltier (1976) may have used such a weak pair, since he used Green functions to compute radial strain rates along finite elements radiating from Earth's core. The strain rates were produced by changing crustal loads of ice and water at points where the finite elements intersected Earth's surface. The strain rates produced vertical velocities at each point that were linked to dated records of worldwide relative sea-level changes. Hence, it seems that Peltier (1976) used only the radial pair of deviator stresses and strain rates in a complex stress field that probably contained stronger pairs associated with mantle convection and needed to move crustal plates. If so, then the apparent Newtonian viscosity used by Peltier (1976) is justified for his purpose, even if mantle creep is non-Newtonian. However, Nye (1957) based his conclusion on the assumption that creep in ice is controlled by the second invariant of deviator stresses only. If the third invariant is also important, as may be the case in ice (Glen, 1958) and in the mantle, then we have no reliable creep law for complex states of stress.

The second assumption by Peltier (1976; 1980 a, b) is that viscosity increases by only two to ten times through the mantle. This seems to contradict the belief that temperature increases greatly with depth in the mantle and with the observation that creep rate increases exponentially with increasing temperature in ice and rocks (Weertman, 1970, 1978; Weertman and Weertman, 1975). Peltier (1981) countered this objection by developing a model for mantle-wide convection that would tend to homogenize mantle temperatures. More recently, Doak and Wolff (1985) showed that creep data from tilt of coreholes through the Greenland and Antarctic ice sheets would fit a creep law in which $n = 1$ and ice hardness was unchanged with depth, as well as the usual creep law in which $n = 3$ and ice hardness decreased exponentially with depth in response to increasing temperature. Weertman (1985) suggested that if the tilt data for glacial creep could fit both creep laws, then perhaps glacio-isostatic data for mantle creep could also fit, especially if both were transient creep superimposed on steady-state creep (Peltier,

1986). If so, the second assumption by Peltier (1980 a, b; 1986) is good, mainly because he made the first assumption.

The third assumption by Peltier (1980 a, b) is that only elastic deformation occurs in the lithosphere. This is an assumption of convenience, as elastic lithosphere deformation and viscous mantle deformation are both linear and can therefore be added to provide deformation along the radial finite elements. However, a case can be made that increasing temperature with depth through the lithosphere requires creep in the lower lithosphere, probably controlled by n = 3, so that overall lithosphere deformation is elastic-viscoplastic (Hughes, 1981a). Worldwide relative sea-level changes during and after deglaciation would be affected very little, but the crustal forebulge that migrates toward the center of a retreating ice sheet might be closer to the ice margin or even under it at the glacial maximum, depending on the value of n for lithosphere creep. This would account for the absence of pro-glacial lakes along the southern margin of North American ice sheets during the last glacial maximum. Peltier and Andrews (1983) were able to eliminate such lakes by increasing lithosphere thickness until excessive depression beyond the ice-sheet margin was suppressed. Along marine shorelines, a forebulge beyond a terrestrial ice margin may become a forebulge beneath a marine ice margin on the outer continental shelf. Peltier and Andrews (1983) did not consider marine ice sheets, however.

SIMULATING THE LAST DEGLACIATION

Identifying problems

Computer models designed to simulate the last deglaciation must cope with problems that have been identified in the preceding sections. Glaciological theory needs development. It must integrate dynamics in the sheet-flow, stream-flow, and shelf-flow areal components of an ice sheet. It needs to produce mechanisms that stabilize an ice-sheet margin as it advances from land into the sea. It needs to understand the relationship between the snowline and the equilibrium line more fully during retreat of terrestrial ice margins. It needs to explore the role of pulling power and the Jakobshavns Effect on ice streams during retreat of marine ice margins. Geomorphological constraints need to be examined more closely. They must resolve the different interpretations of weathering zones, so the maximum ice margin along northeastern North America can be identified. They must settle the question of whether different scales of glacial geology should be emphasized during a glacial maximum and during retreat, including the role of deformable beds during both phases. They must lead us to a more complete theory of glacial isostasy, especially concerning crustal displacements along ice-sheet margins during and following the glacial maximum. Other problems can be cited, but most of them exist within the problems identified here.

As an example of how problems are interrelated, consider the question of ice-sheet profiles over deformable beds. Boulton and Jones (1979) cited field evidence for low surface profiles at various sites along the southern Laurentide margin during the last glacial maximum and deglaciation, and they presented a theoretical model of deformable beds to account for low profiles. Boulton and others (1985) used this theory to reconstruct a Laurentide Ice Sheet in which a thin brim up to 600 km wide lay on a deforming bed, flowed much as an ice shelf, and drained ice from a high crown of ice over the undeformable Precambrian shield of Canada. This ice sheet is shown in Figure 18.

As was noted, the ice-shelf analogy is flawed because present-day Antarctic ice shelves have ice rises where they are grounded locally on a bed that does not deform because it is probably frozen. Stream flow develops in the ice shelf between ice rises only some tens of kilometers apart (Lang and MacAyeal, 1986; MacAyeal, 1987). Figure 17 shows lobate moraines along the southern Laurentide Ice Sheet margin that were termini of ice streams flowing between the uplands shown in Figure 16. These uplands were only slightly higher than surrounding lowlands, so they were probably over-ridden by ice at the glacial maximum. However, their beds could have been frozen, with faster ice on a thawed deforming bed streaming between them. This view places deformable beds beneath ice streams that discharge most of the ice but that are a relatively small fraction of the ice-sheet perimeter. The thin brim of ice around the Laurentide Ice Sheet in Figure 18 is not necessary if stream flow replaces shelf flow in this region, because the deformable bed extends up relatively narrow ice streams instead of along the while ice margin.

Fisher and others (1985) drew more attention to the problem with shelf flow in a thin brim of ice on a deformable bed, by applying plasticity theory to obtain the Laurentide flowlines shown in Figure 19. Plasticity exaggerates subtle differences in flow. For example, viscoplastic flow is faster down the flanks of an ice divide than along its crest, but plastic flow is entirely down the flanks, and ice is stagnant along the crests. Shelf flow over a deformable bed, shown where basal shear stress drops to 5 kPa in Figure 19, degenerates into local ice domes over lobate margins separated by ice streams in re-entrant margins. Basal shear stress could not possibly be uniform for such a complex flow regime. Moreover, the alternation of ice domes with ice streams along the margins can be sustained only if the equilibrium line hugs the ice margin around the relatively stagnant ice domes and extends far up the fast-moving ice streams. Looping of the equilibrium line to conform with such drastic changes in ice velocity is incompatible with the relatively smooth ice surface in the zone of low basal shear stress. Shelf flow over broad deformable beds along the southern ice margins is therefore incompatible with the dynamics of stream flow and sheet flow, and with atmospheric circulation that ties the snowline to the equilibrium line.

Two ways of dealing with the problems identified above have been developed for computer models used to simulate the last deglaciation of North America. One uses geomorphology as fundamental input that specifies the steady-state flow regime at the glacial maximum and the transient flow regime during deglaciation. Ice-sheet flowlines and margins are specified as input, along with surface accumulation or ablation rates, bed topography, and frozen, thawed, freezing, and melting conditions at the bed, as deduced from geomorphology in general, not just glacial geology. For example, permafrost and grabens are not formed by glacial erosion or deposition. Yet they are related to glaciation, because the limits of continuous and discontinuous permafrost shown in Figure 16 identify regions where ice sheets spread over beds that were initially frozen or partly frozen, and inter-island channels and straits in Figure 16 are mostly grabens where ice streams would develop along marine margins of an ice sheet. Since the full suite of landforms is used to specify input of basal conditions to these models, they are essentially geomorphic models.

The other way of dealing with these problems uses geomorphology only as a check on model output. Model input consists of bed topography and the surface mass-balance regime, both specified at points on a two-dimensional grid. Equations of ice dynamics and heat flow are then used to compute, as output, surface flowlines and basal frozen, thawed, freezing, and melting conditions (Radok and others, 1982). A diagnostic feature of output is the possibility of multiple solutions for the same input. Output from some models includes glacial surge cycles, during which prolonged periods of slow ice-sheet growth are punctuated by short bursts during which ice margins advance rapidly as interior ice lowers, all without substantial loss of ice volume (Budd and McInnes, 1979). Output from other models can be two states of stable climatic equilibrium, one with a stable continental ice sheet and one with no ice sheet at all, and in between these extremes exists a climate in unstable equilibrium for which an ice sheet is either growing irreversibly toward its stable continental dimensions or shrinking irreversibly until it vanishes (Oerlemans and van der Veen, 1984, p. 127–131). Multiple solutions for the same input of external boundary conditions result from mechanisms built into the models, so they are essentially mechanical models.

It is very difficult for mechanical models to generate the detailed flow regime that produces ice streams, local ice domes, and ice margins, because the cost of computer time dictates a grid that is too coarse. Geomorphology specifies these flow patterns quite accurately, but confusion arises when several flow directions overprint each other in an area, and when their time relationships cannot be sorted out. For these reasons, the full potential of both geomorphic and mechanical models has not been realized. With this precaution, the only two models that have produced time-dependent retreat of three-dimensional ice sheets during the last deglaciation of North America will now be presented. One is a geomorphic model and one is a mechanical model, so the strengths and weaknesses of both types can be compared. The geomorphic model is presented here for the first time. It includes both terrestrial and marine ice, with the pulling power of marine ice streams being the major deglaciation mechanism for marine ice sheets. It has its origins in the ice-sheet reconstruction and disintegration schemes developed for CLIMAP (Denton and Hughes, 1981, chapters 5, 6, and 7). The mechanical model was developed by Budd and Smith (1981). It deals with terrestrial ice only, and focuses on an irreversibly negative mass balance brought about by climatic warming and delayed isostatic sinking beneath ice sheets as the major deglaciation mechanism.

A geomorphic model of North American deglaciation

A three-dimensional character has been introduced into the CLIMAP model presented by Denton and Hughes (1981) by replacing flowlines with flowbands having variable widths, so the length, width, and height of each flowband are compatible from flowband to flowband around the ice sheet (Fastook, 1984). This approach has two advantages over simulating width by points on a uniform two-dimensional grid. First, flowbands could be selected to conform with topographic variations too localized to be simulated by the coarse grids that are used to economize computer time. This brought out a distinction between sheet flow and stream flow that is nearly impossible with grid points. Second, flowband widths could be varied to preserve intact individual ice drainage systems, especially for ice streams, in whch a narrow width for stream flow widens in the ice catchment area. This emphasizes the fact that about 90 percent of ice is discharged by ice streams, as observed today in the Greenland and Antarctic ice sheets.

Variable accumulation and ablation rates are specified along flowbands to allow for changing patterns of mass balance as an ice sheet advances and retreats. This permits both the equilibrium line and the line of maximum accumulation rates to migrate over time. For example, the "elevation desert effect" is simulated by adjusting decay constants in rates of accumulation and ablation that decrease exponentially from the margin to the center of an ice sheet, so that the maximum accumulation rate migrates from the center toward the margin of a growing ice sheet, and back toward the center for a shrinking ice sheet (Hughes, 1985).

Mass-balance changes over time are simulated by allowing the equilibrium line to migrate in conformity with Milankovitch insolation variations over time at each latitude. This can be done in two ways, either analytically by quantifying the snowline-slope changes depicted in Figure 12 for sinusoidal precession and tilt cycles, or numerically by locking the equilibrium-line migrations to actual insolation rates, as they vary with latitude and time. A retreat mechanism for southward-flowing terrestrial ice is thereby added to the retreat mechanism for northward-flowing marine ice for the last North American deglaciation (Fastook, 1985; Hughes and others, 1985).

Isostatic rebound rates that decrease exponentially with time during deglaciation are included in the model by using a time constant that is fixed for rebound of deglaciated areas beyond the retreating ice margin, but that increases from the margin to the center of the ice sheet as it retreats. This allows rebound at the center of the ice sheet to proceed at a slower rate than rebound beyond its margins, as was actually the case, because a heavy ice load retards rebound. The actual values of these time constants must be selected so that rebound is compatible with the dated record of raised beaches around North America. An exponen-

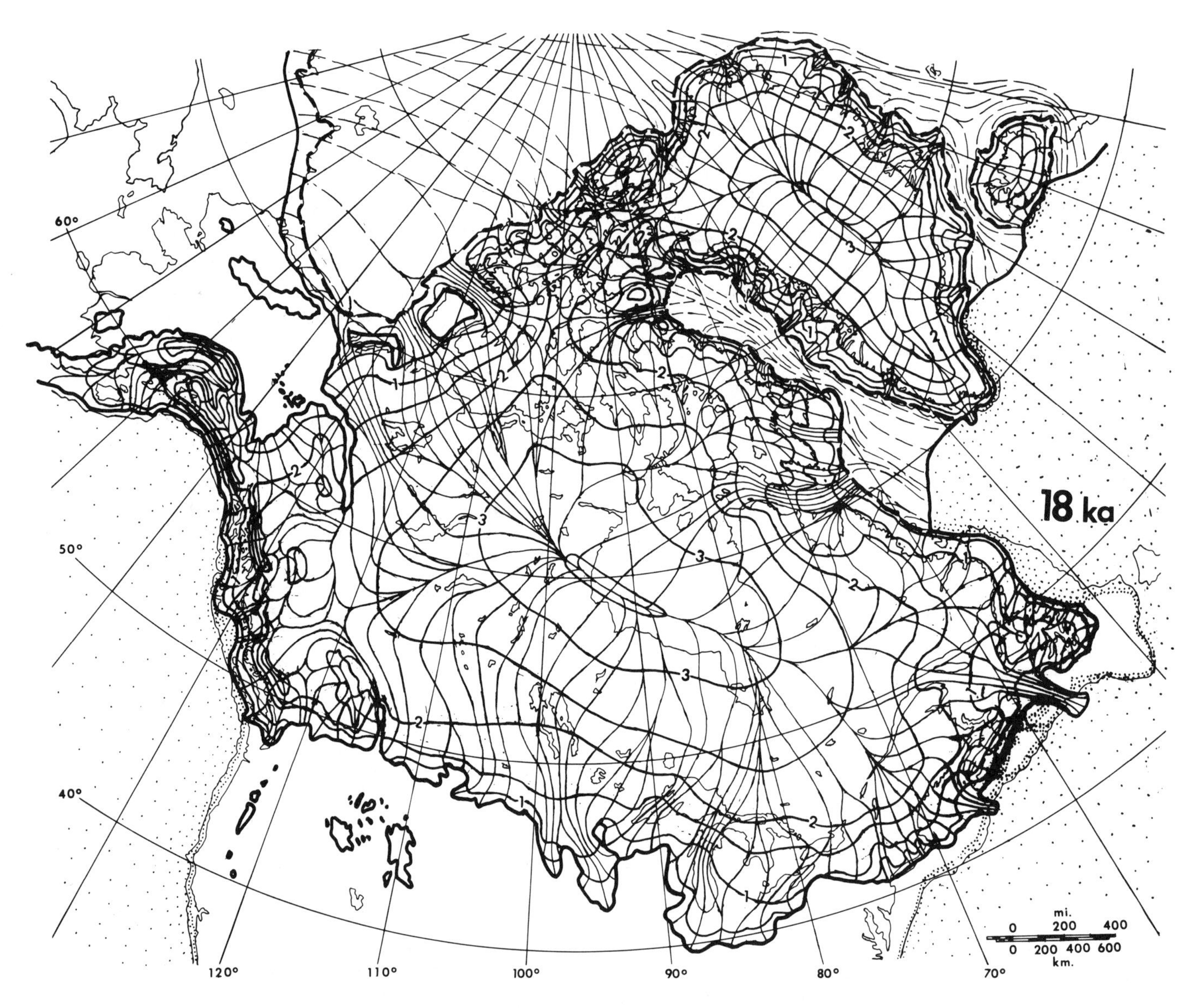

Figure 20. The last glacial maximum in North America simulated by a geomorphic model. Ice-elevation contour lines are shown at 0.5-km intervals at 18 ka. Surface flowlines drawn normal to surface contour lines are solid for grounded ice and dashed for floating ice. Ice-sheet margins are heavy solid lines and ice-shelf grounding lines are heavy broken lines. The edge of the continental shelf is taken as the 0.5-km bathymetric contour and is shown as a thin solid line beyond present-day coastlines. The extent of sea water is the dotted area. These designations also apply to Figures 21 through 25.

tially decreasing rate of isostatic sinking beneath an advancing ice sheet over time is also included in the model, so that sinking continues during the glacial maximum. This allows the ice sheet to lower with respect to the snowline, so that the equilibrium line migrates upslope and makes the mass balance negative. This contributes to termination of the last glaciation, just as in the mechanical model by Budd and Smith (1981).

In simulating the last deglaciation of North America with the geomorphic model, first order glacial geology shown in Figure 16 was used to reconstruct the 18-ka glacial maximum in Figure 20, and second-order glacial geology shown in Figure 17 was used to reconstruct deglaciation from 14 to 7 ka in Figures 21 through 24.

At 18 ka a frozen bed was specified inside the limit of continuous present-day permafrost, and a melting bed extended from there to the limit of discontinuous present-day permafrost. A freezing bed extended south of that line to the edge of the crystalline Precambrian shield, and a second melting bed ex-

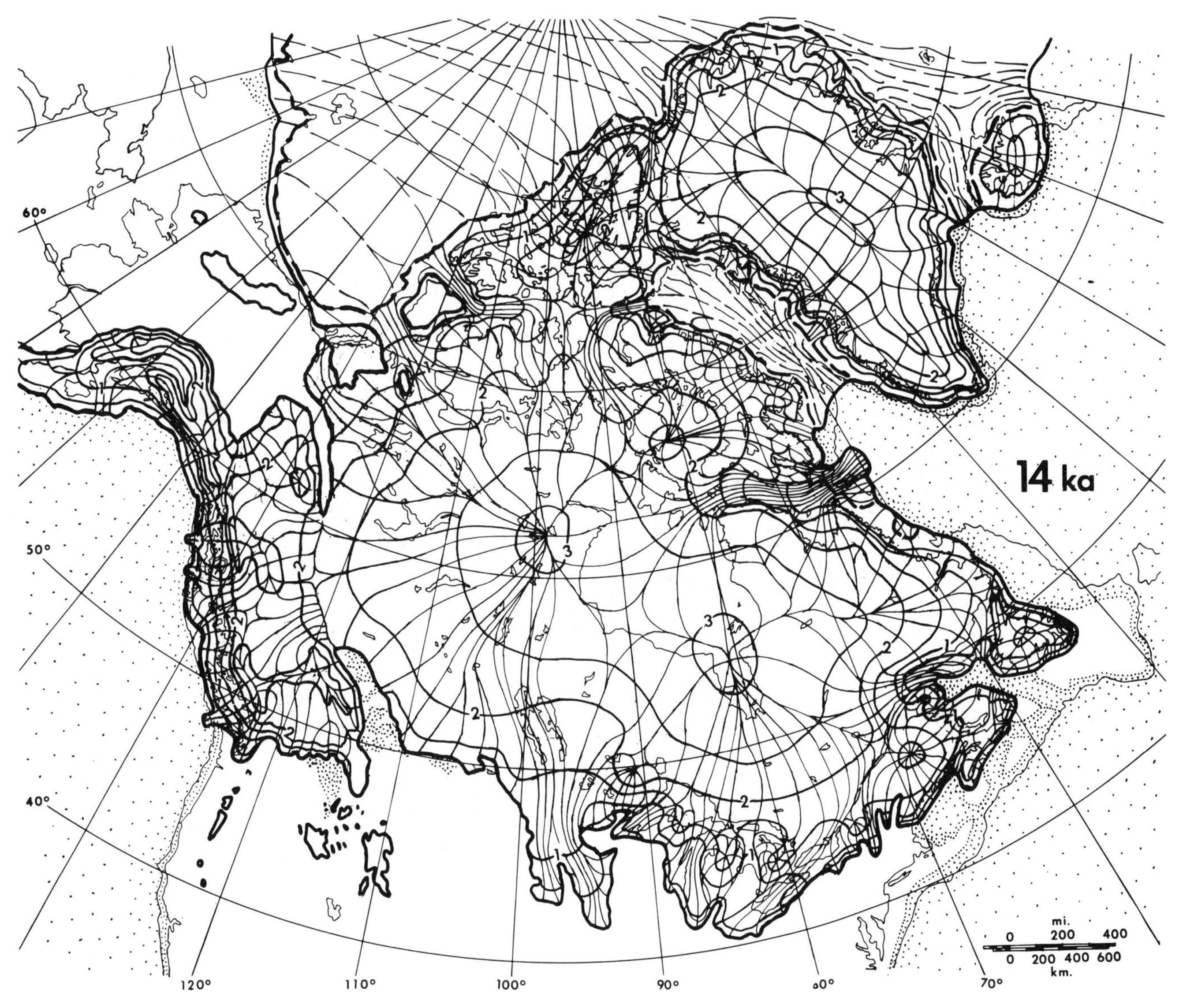

Figure 21. The last deglaciation of North America at 14 ka simulated by a geomorphic model. See text for discussion.

tended from there to the southern ice margin. Glacial erosion existed in thawed patches of the inner melting and freezing zones, creating a pitted landscape that became a land of lakes after deglaciation, and eroded material incorporated with refrozen ice in the freezing zone was deposited as a basal moraine and end moraines in the outer melting zone. Terrestrial ice streams occupied the Greak Lakes and produced the lobate moraines shown in Figure 17. Marine ice streams occupied inter-island channels in the Arctic and troughs across Atlantic and Pacific continental shelves. These ice streams are identified by arrows in Figure 16.

A single Laurentide dome was constructed over Hudson Bay, to conform with the glacio-isostatic modeling study by Peltier and Andrews (1983). However, a single dome is also preferable to a saddle between separate Keewatin and Labrador domes, because flow converges from saddles and diverges from domes. For comparable bed topography, flowbands are steeper if they widen upslope than if they narrow upslope, so flowbands to a saddle must be much shorter than flowbands to domes, if the saddle is to be lower than its flanking domes. However, a saddle over Hudson Bay would have longer flowbands than those for domes in Keewatin and Quebec; hence a dome had to exist over Hudson Bay. Several secondary domes were reconstructed along the margins of North American ice sheets, especially over highlands and between ice streams. A saddle over Kane Basin con-

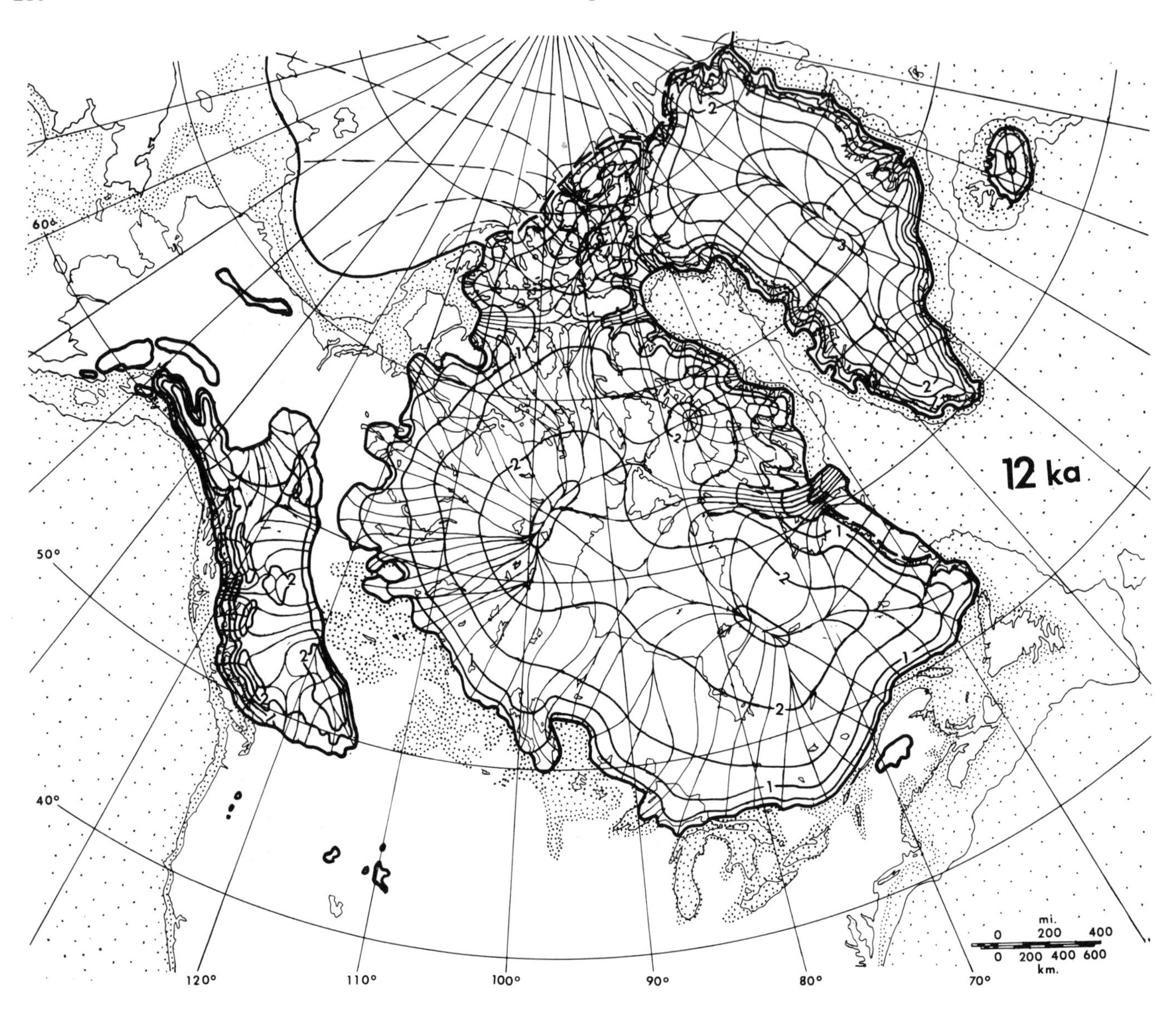

Figure 22. The last deglaciation of North America at 12 ka simulated by a geomorphic model. See text for discussion.

nects the Innuitian and Greenland ice sheets, as dictated by fresh striations showing ice flowing northward (Hudson, 1983) and southward (Blake, 1977) from the basin.

Isostatic sinking and thickening ice raised the temperature and lowered the melting point of basal ice in Hudson Bay, until the bed became thawed between 18 and 14 ka in the model. Removal of an ice shelf in the Labrador Sea, perhaps in response to a northward migration of the North Atlantic polar front, led to increased discharge by the ice stream in Hudson Strait. Basal sliding over the newly thawed bed and downdraw from enhanced ice-stream discharge lowered the dome over Hudson Bay, and it became a saddle between new Keewatin and Quebec domes by 14 ka. The portion of lowered ice that moved southward caused Laurentide ice to approach or exceed its maximum, after it had retreated slightly during the Erie interstade (Mayewski and others, 1981). In the east, calving bays carved away ice streams in the Gulf of Maine and the Laurentian Channel. Ice downdrawn by ice streams in the Laurentian Channel and Hudson Valley produced an ice dome over Maine. In the west, Cordilleran ice attained its maximum extent, but the saddle connecting Cordilleran and Laurentide ice may have lowered, so that north–northwest and south–southeast flow of ice on the Laurentide side of the saddle was replaced with flow to the northwest, west, and southwest. Figure 21 shows the model simulation.

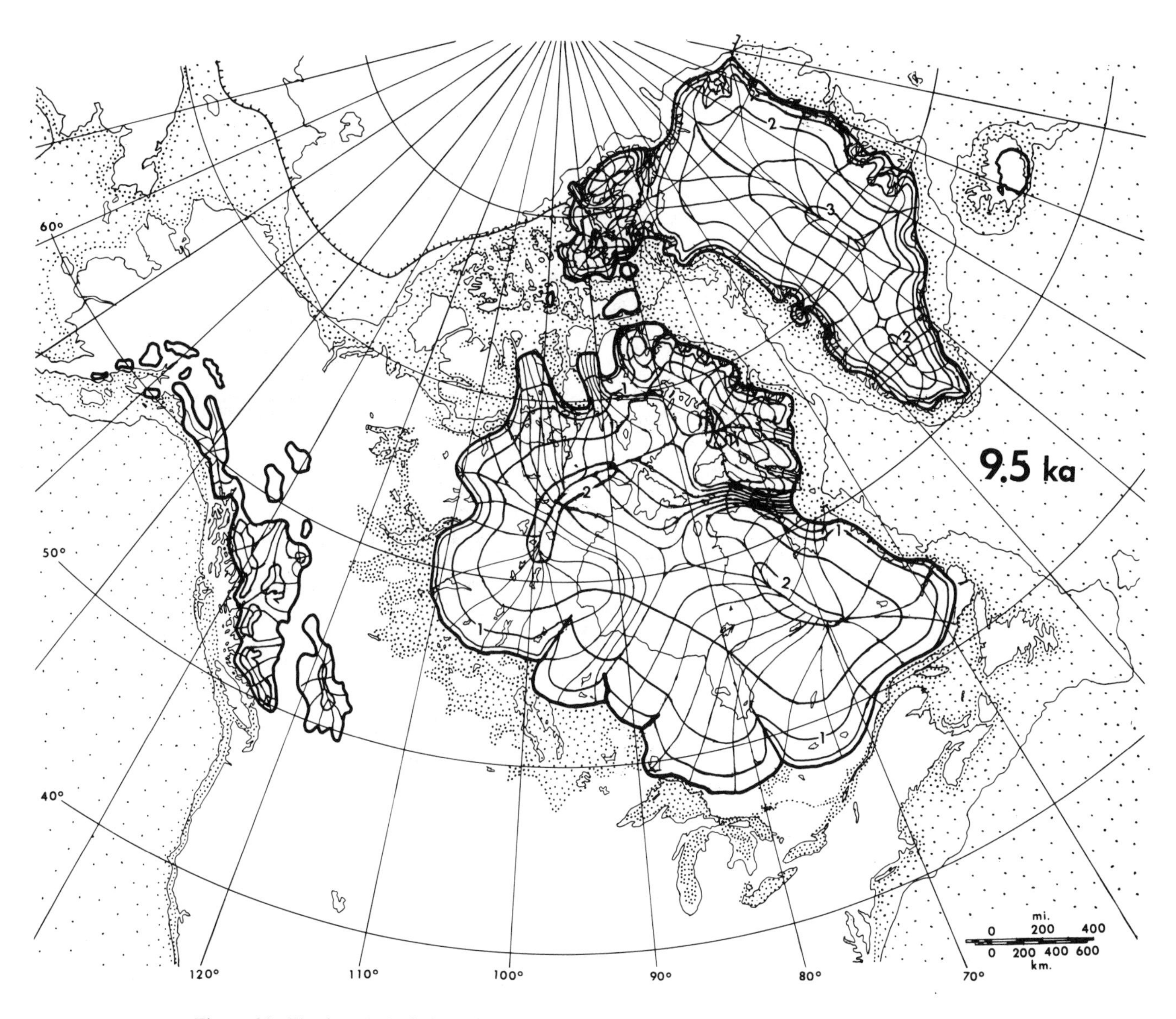

Figure 23. The last deglaciation of North America at 9.5 ka simulated by a geomorphic model. Perennial sea ice lies north of the hachured line in this figure and in Figures 24 and 25.

By 12 ka a corridor had opened between Laurentide and Cordilleran ice, and ice streams existed along the western Laurentide margin in Great Bear Lake, Great Slave Lake, and Lake Athabaska. Ice streams occupied Lakes Manitoba and Winnipeg, and Lake Superior along the southern Laurentide margin, but the margin had retreated north of the other Great Lakes. Pro-glacial lakes were almost continuous in the isostatically rebounding deglaciated bed beyond the western and southern Laurentide margins, and a degree of pulling power was exerted by ice streams discharging into these lakes. In the north, retreat of the polar front caused ice shelves in Baffin Bay and in the Norwegian and Greenland seas to disintegrate. Rising sea level introduced warm Pacific water across the Bering Strait into the Arctic Ocean, causing ice-shelf disintegration in the Beaufort Sea. Disintegration of these ice shelves and rising sea level increased discharge rates for ice streams in Amundsen Gulf, M'Clure Strait, Nares Strait, Lancaster Sound, Cumberland Sound, Frobisher Bay, and around all but northern Greenland. Enhanced discharge promoted interior downdraw and grounding-line retreat into subglacial basins. This increased the pulling power of ice streams and accelerated collapse of the Laurentide Ice Sheet. Figure 22 presents the reconstruction by the model.

Laurentide ice had retreated substantially by 9.5 ka. The southern margin had attained a temporary stillstand along the

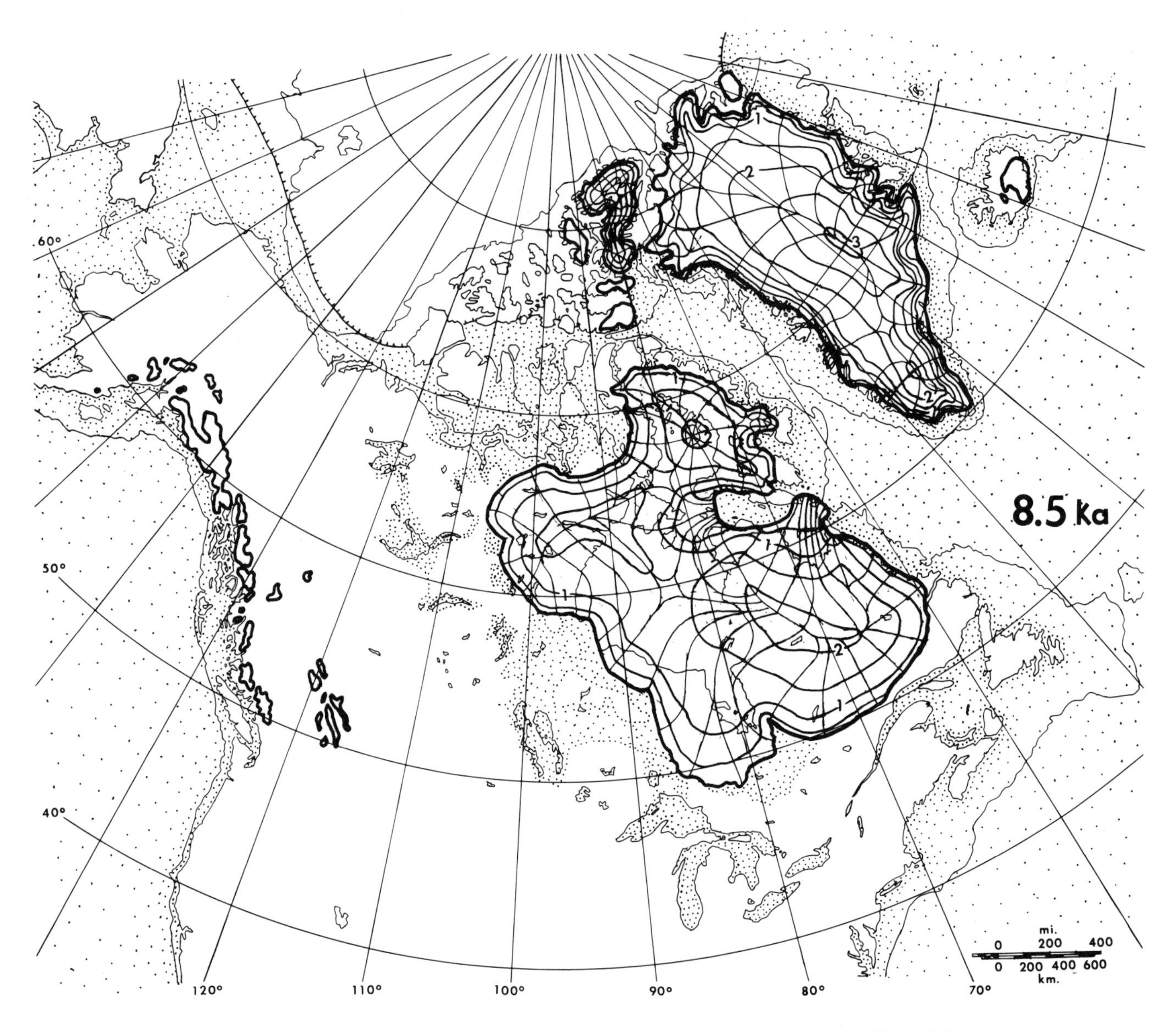

Figure 24. The last deglaciation of North America at 8 ka simulated by a geomorphic model.

Cree, Chapleau, and Monotou-Matamek moraines, which were separated by interlobate moraines, (Mayewski and others, 1981, Fig. 2.2). Ice streams occupied M'Clintock Channel and Prince Regent Inlet on the northern Laurentide margin. The Innuitian Ice Sheet was collapsing and the Cordilleran Ice Sheet had separated into a number of shrinking ice caps. Figure 23 displays a reconstruction having these features.

Innuitian and Cordilleran ice were merely enlarged versions of present-day ice fields and ice caps at 8.5 ka. The Laurentide Ice Sheet was still intact, but calving bays had carved away all marine ice streams. A glacial surge out of Ungava Bay may have crossed Hudson Strait and covered islands southeast of Meta Incognita Peninsula on Baffin Island at that time (Andrews and others, 1985). Nares Strait opened between Ellesmere Island and Greenland. An ice shelf in the Arctic Ocean may have thinned to become the perennial sea ice cover. Figure 24 shows this reconstruction.

Saddles on the Laurentide ice divide over Hudson Bay and Foxe Basin collapsed into ice shelves that were carved away by calving bays, so that ice domes over Keewatin, Quebec, and Baffin Island had become large independent ice caps by 7 ka. Figure 25 gives this modeling result.

An important result from using this geomorphic model to simulate the last deglaciation is that the distribution of erratics, as mapped by Shilts (1980, 1982) and shown in Figure 18, can be explained completely by the flow pattern from 12 to 8.5 ka

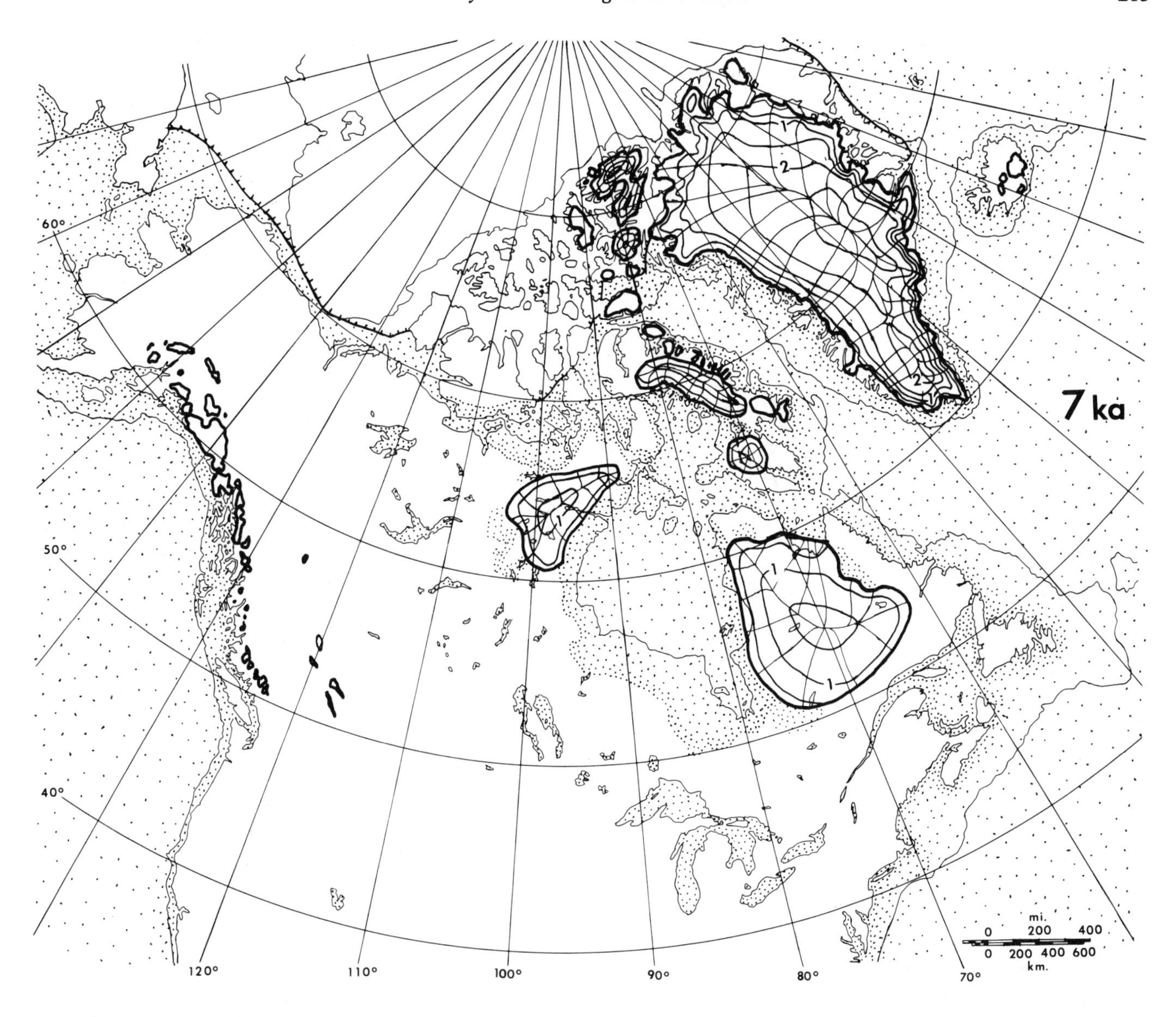

Figure 25. The last deglaciation of North America at 7 ka simulated by a geomorphic model.

during deglaciation, provided this pattern is repeated during each deglaciation. This 3,500-yr time span, and similar time spans for the three preceding late Quaternary deglaciations (Shackleton and Opdyke, 1973), is ample time to move erratics over the distances shown in Figure 18. In order to keep erratics inside this distribution area, their source areas would have been regions with a frozen bed at other times when different ice flow patterns existed during the last and preceding glaciations. Hence, the distribution of erratics in Figure 18 is used to specify frozen beds over their source areas at 18 ka (Fig. 20) and 14 ka (Figure 21). Erratics at the 18-ka ice margin on Banks Island had a source area in Bathurst Inlet (Jean-Serge Vincent, personal communication, 1987), so ice flowlines from there to Banks Island were input for producing Figure 18. These examples illustrate how geomorphic models should be used.

A mechanical model of North American deglaciation

Budd and Smith (1981) simulated the last glaciation cycle of North America using a three-dimensional time-dependent mechanical model in which input was specified at grid points 200 km apart. Time-dependent inputs were mass-balance data based on insolation variations (Milankovitch, 1941; Vernekar, 1972) and isostatic data obtained by fitting isostatic rebound curves in eastern North America by Walcott (1972) to an empirical equation that gave a nonlinear response time to changing ice loads.

This equation was based on their view that isostatic adjustments occurred primarily by viscoplastic creep in the asthenosphere, in contrast to the theoretical model for a purely viscoelastic mantle-wide response to changing ice loads employed by Peltier and Andrews (1976). Budd and Smith (1981) used their empirical isostasy equation to restore isostatic equilibrium to North America, which is still rebounding from the last glacial maximum, in order to obtain a landscape in complete isostatic equilibrium. Elevations on this completely rebounded landscape were then entered at each grid point to provide the landscape at 120 ka just prior to onset of the last (Wisconsin) glaciation. Foxe Basin and most of Hudson Bay were above sea level for the 120-ka landscape, so an ice sheet originating on Baffin Island could advance over North America as a purely terrestrial ice sheet.

Precipitation input to the Budd and Smith (1981) model was present-day precipitation over North America modified in ways predicted by atmospheric general circulation models for ice-age conditions. In addition, precipitation rates over the ice-sheet interior at elevations over 2 km were reduced by the "elevation-desert effect" observed over the Antarctic and Greenland ice sheets today. A precipitation parameter described both effects. Present-day ablation rates on glaciers at different latitudes and altitudes were used to construct empirical curves for ablation rates, which were then linked to an adiabatic lapse rate of 6.5°C/km for summer temperatures. The lapse rate was applied to mean summer orbital variations of insolation with latitude and time. In this way, ablation rates on the surface of North American ice sheets were made to vary directly with altitude, latitude, and time, through an insolation parameter. It was then varied to make advance and retreat of North American ice sheets in the model output fit best with marine oxygen-isotope records of ice-volume changes and the glacial-geological record of the last North American deglaciation. Budd and Smith (1981) introduced an albedo that decreased the elevation of mean summer isotherms on the ice-sheet surface in proportion to increases in the ice-sheet area, thereby linking changing surface ablation rates to changes in the glaciated area of North America. The insolation and albedo feedback parameters were assumed to have additive linear effects on the elevation of mean summer isotherms. By varying their effects independently, each over a range of reasonable values, North American ice sheets simulated by the model could display a range of advance and retreat histories.

Ice sheets could flow by both internal creep and basal sliding in the Budd and Smith (1981) model, but only creep was used in their simulations. However, the range of viscoplastic exponents used for creep velocities encompassed sliding velocities as well. The temperature-velocity feedback was parameterized by scanning through a range of softness coefficients in the rheological creep equation. This coefficient changes with ice temperature, and changing it allows the basal shear stress to change in a way that compensates for the feedback.

In summary, Budd and Smith (1981) employed four parameters to optimize the fit between North American ice sheets simulated by their model and the dated record of the last deglaciation; one to link present-day precipitation rates to former precipitation rates over the ice-sheet surface, one to link insolation variations with ice ablation rates, one to link the areal albedo over an ice sheet with ice ablation rates, and one to link temperature-velocity feedback with a rheological creep equation for ice. Basal shear stress was proportional to the product of ice thickness and surface slope. The surface slope at a grid point was the root mean square of surface slope components aligned in orthogonal directions on the grid, as formulated by Mahaffy (1976). The direction of surface slope was the direction of ice velocity, where surface slope depends on bed topography and ice-thickness gradients. Ice thickness at a grid point varied over time as the sum of the local accumulation or ablation rate and the ice-flux gradient, as prescribed by the continuity equation.

Budd and Smith (1981) produced maps of ice-surface elevation, ice thickness, and ice velocity at intervals of 4,000 yr as output from their model. Other output included basal shear stresses, bed depression, net mass balance, and ice flux at each grid point, as well as plots of ice area and volume with time for North American glaciation beginning 120,000 yr ago. Leads and lags between insolation, glaciation, and isostasy were monitored. They also produced a motion picture film that depicted advance and retreat of North American ice sheets in relation to color-coded ice-elevation contours. With only terrestrial ice permitted, the model predicted that Laurentide ice spread from the highlands of Baffin Island and Quebec–Labrador. The Laurentide Ice Sheet covered all of the weathering zones, thereby disagreeing with the stratigraphic interpretation that each zone represents a separate glaciation. The Cordilleran Ice Sheet was far too extensive because the 200-km spacing of grid points converted all of western North America into a uniformly high plateau that was ideal for nucleating ice sheets in the model. Ice retreat was favored by increased insolation and isostatic sinking and opposed by precipitation and albedo feedbacks that increase as ice sheets advance. By lowering the ice surface, and thereby increasing ablation rates, delayed isostatic sinking was found to be essential to total disappearance of the ice sheets.

Budd and Smith (1981) contend that a simple relationship between the areal and vertical extent of North American ice sheets does not exist, because insolation variations, glaciated area, and isostatic adjustments proceed at different rates, and each factor contributes to the surface elevation of ice sheets.

Figure 26 shows the last deglaciation of North America, as simulated by the mechanical model of Budd and Smith (1981).

Figure 26. North American deglaciation simulated by a mechanical model. Shown are (a) bedrock topography contoured at 0.5-km intervals at 120 ka, as approximated by grid points 200 km apart; and ice thickness contoured at 0.5-km intervals at (b) 18 ka, (c) 12 ka, (d) 10 ka, (e) 8 ka, and (f) 6 ka. These maps were provided by William F. Budd for this chapter. They were produced from the model of Budd and Smith (1981).

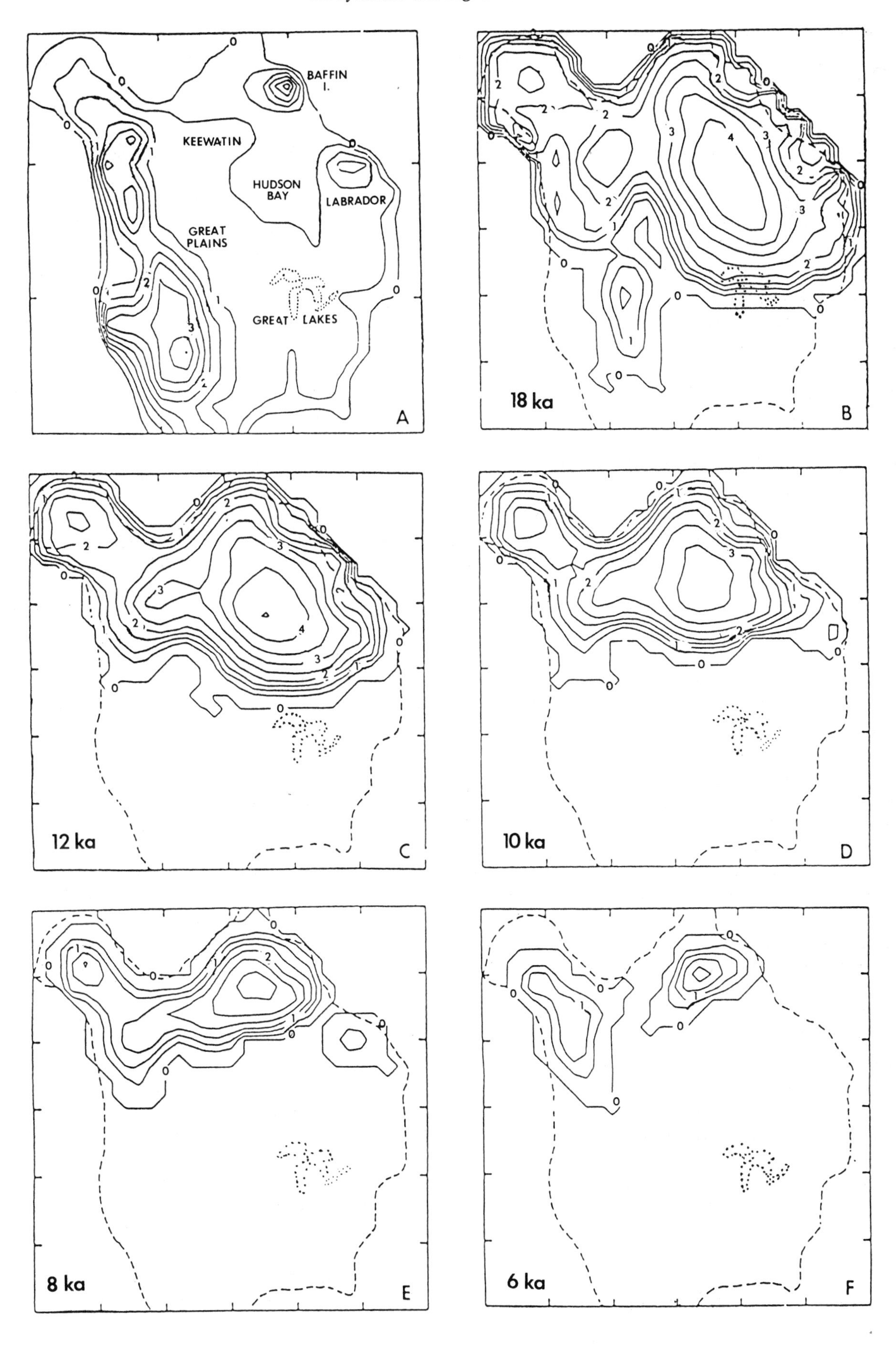
BAFFIN I.
KEEWATIN
HUDSON BAY
LABRADOR
GREAT PLAINS
GREAT LAKES
A
18 ka
B
12 ka
C
10 ka
D
8 ka
E
6 ka
F

Bedrock topography contoured at 500-m intervals is shown in part A, and ice thickness contoured at 500-m intervals are shown in parts B through F at 18, 12, 10, 8, and 6 ka, respectively. The inappropriateness of a 200-km grid spacing is apparent not only in converting western North America into a single plateau ranging from 1,000 to 4,000 m in elevation, but in converting Labrador into a plateau from 1,000 to 1,500 m high and by giving Baffin Island an average elevation of some 1,500 m. Actually, only a small part of Labrador is even 600 m high, and only a few scattered headlands on Baffin Island rise above 1500 m. We must wonder if the Budd and Smith (1981) model would nucleate a Labrador ice center on land only 600 m high, and if ice nucleated on Baffin Island headlands would not flow down the steep slopes into flanking fjords and be lost as icebergs, rather than spread southwestward down gentle slopes to become the major source of Laurentide ice. Cordilleran ice would certainly be much less extensive if the grid spacing allowed a better representation of bedrock topography. For example, an 18-ka ice dome 2,500 m thick was produced over the Yukon River valley in Alaska only because the grid spacing converted the valley into a plateau 1,000 m high. On the other hand, the model extends Laurentide ice northward over the Queen Elizabeth Islands, even though they have a fully rebounded elevation that is barely above the 120-ka sea level in the model.

Given imperfections linked to the 200-km grid spacing, the Budd and Smith (1981) model simulates the last North American deglaciation rather well, especially retreat of Laurentide ice. Glaciation at 18 ka is dominated by the Laurentide Ice Sheet, with a single dome over Hudson Bay, despite overly extensive Cordilleran ice. The 12-ka simulation is even better, because retreat has begun and the southern limit of both Laurentide and Cordilleran ice is close to the limit deduced from glacial geology. The high Laurentide dome over Hudson Bay persists, but a secondary Laurentide dome developed to the west can account for south–southeast and north–northwest flow lineations over the Canadian prairies during deglaciation, as shown in Figure 17. The high Laurentide dome has moved northward from Hudson Bay to Foxe Basin by 10 ka, and a Labrador ice dome has developed. The southern Laurentide ice margin is reasonable, but there is no corridor between Laurentide and Cordilleran ice, in violation of dated glacial geology. By 8 ka, the model produces a corridor between the Labrador ice dome and the main Laurentide ice dome over Foxe Basin, while retaining a strong connection between Laurentide and Cordilleran ice. The major agreement with glacial geology is deglaciation of Hudson Bay. The ice dome over central Alaska is gone by 6 ka and a corridor has finally developed between Cordilleran and Laurentide ice. However, Cordilleran ice is still much too extensive. Laurentide ice extends from Keewatin to Baffin Island, with a single dome over Foxe Basin. Glacial geology in Figure 17 does not support this distribution of ice.

CONCLUSIONS

Components of present-day ice sheets include terrestrial ice, marine ice, ice shelves, and ice streams. We can expect that North American ice sheets during the last glacial maximum and termination also had these components. The unanswered question is where were these components and which were most important in controlling deglaciation? An answer to this question lies in analyses of lake sediments on islands in arctic and subarctic Canada, similar to analyses of lake sediments in Scandinavia by Anundsen (1985) and in New England by Davis and Jacobson (1985), and analyses of continental-shelf sediments between these islands, similar to analyses of Ross Sea sediments on deglaciated Antarctic continental shelves by Kellogg and others (1979), Kellogg and Kellogg (1986), and Anderson and others (1980, 1984). A bonus of these analyses will be to provide the correct interpretation of weathering zones from Ellesmere Island to Nova Scotia. If the sediments point to extensive marine ice, the geomorphic interpretation is probably correct; otherwise, the stratigraphic interpretation makes more sense. Both interpretations may apply, but for different sectors.

The transition from terrestrial to marine ice occurs when grounded ice advances into deep water and acquires pulling power as it becomes more buoyant. Calving replaces melting as the dominant ablation mechanism during advance, and different calving mechanisms exist for ice terminating in water of different depths. An unanswered question is which mechanisms determined the seaward margins of North American ice sheets in which sectors during the last glacial maximum and termination? Answers to this question can be found by studying calving dynamics where each of these mechanisms can be observed today along margins of the Greenland and Antarctic ice sheets. Calving dynamics is a virtually unstudied process in glaciology. Yet it is crucial to developing a successful computer model of ice-sheet dynamics, because ice sheets will calve into marine embayments or pro-glacial lakes at some time during their existence.

Terrestrial ice-sheet margins, and even some marine margins, are acutely sensitive to migrations of the surface equilibrium line. The relationship between the equilibrium line and the slope of the snowline, the pattern of accumulation and ablation, the size of the ice sheet, and isostatic adjustments beneath the ice sheet is a highly complex unanswered question. A resolution of this question for one sector of an ice sheet may be ill-suited to other sectors, and for this reason models that examine the stability of two-dimensional ice sheets can give misleading results. One such result is the prediction that the same climatic boundary conditions can coexist with no North American ice sheets, an ice sheet covering most of the continent, and everything in between. Progress in resolving this question of stable versus unstable dynamic equilibrium will be made as we come to understand the present-day dynamics of the Greenland Ice Sheet, which has an equilibrium line that interacts with a range of snowline slopes, accumulation and ablation patterns, ice-sheet diameters, and isostatic-rebound rates (Radok and others, 1982).

Ice streams drain as much as 90 percent of Greenland and Antarctic ice. It is prudent to expect that the same was true of North American ice sheets. Ice-stream dynamics should therefore be a central component of any model designed to simulate the last deglaciation. At the heart of ice-stream dynamics is the pulling power that develops as thick ice becomes buoyant. How pulling power works and its role in the last deglaciation of North America is a major unanswered question. A number of ice streams are currently being investigated in Greenland and Antarctica, notably Jakobshavns Isbrae in Greenland, where pulling power may trigger a powerful deglaciation mechanism I called the Jakobshavns Effect (Hughes, 1986). In Antarctica, ice streams are being studied that have almost the full range of boundary conditions, particularly their interaction with ice shelves. No definitive conclusions have emerged from these studies, but they should provide the key to understanding the possibly continuing collapse of the marine ice sheet in West Antarctica that began 17,000 yr ago (Stuiver and others, 1981) and, by analogy, collapse of marine components of North American ice sheets during the last deglaciation.

The role of transient creep in the flow of ice sheets, and in perturbations of flow in Earth's mantle in response to advance and retreat of ice sheets on Earth's crust, is an unanswered question. In ice sheets, the basal sliding rates that depend on flow over and around bedrock bumps and across frozen and thawed patches may involve transient creep. Transient creep may also be important through the full ice thickness near ice divides, and at other sites where strain rates and their gradients are of comparable magnitude. In Earth's mantle, transient creep may control strain rates from glacio-isostatic adjustments if strain rates related to thermal convection are larger. The effect of transient creep may be to shift deformation to the viscous end of the viscoplastic creep spectrum. This can have a profound effect on computer models that simulate ice-sheet dynamics and global glacio-isostatic adjustments. A resolution of this question will come as the creep process in ice and rock becomes understood more fully.

Glacial geology holds a major key to understanding both the dynamics of ice sheets and the last deglaciation of North America. It is a direct record of processes at the base of an ice sheet, the most inaccessible region for glaciologists studying present-day ice sheets. It is also a geomorphic Rosetta Stone upon which the history of the last deglaciation is inscribed. However, that history has not been translated into a language we understand. This has led Radok to exclaim, "What is glacial geology? A hundred and one insignificant little moraines with a thousand and one interpretations!" (Uwe Radok, personal communication, 1979). How to resolve various interpretations of glacial geology is an unanswered question. It will be answered as glacial geology becomes more completed dated.

Answers to all of these questions will allow computer models to resolve the ultimate unanswered question: How did the last deglaciation take place? A fully three-dimensional time-dependent finite-element computer model will be needed to answer this question. Full three dimensionality will include computing temperature and stress fields throughout the ice sheet. This will specify steady-state thermal conditions at the bed that might sustain first-order glacial erosion and deposition processes that can make a permanent imprint on the subglacial landscape. Full time dependency will tell us if a first-order landscape can be overprinted by second-order glacial geology during the last deglaciation. Finite elements will allow us to employ a fine grid for rugged topography and stream flow, and a coarse grid for smooth topography and sheet flow, so that the full spectrum of topographic and dynamic conditions can be represented reliably. This kind of model is within our capability, but field data needed to calibrate it are lacking. Studies of glacial history in lake and continental-shelf sediments in the Canadian Arctic are crucial, along with studies of ice dynamics in Greenland and Antarctica.

Important insights into how the last deglaciation occurred are provided by existing computer models. Both the geomorphic and mechanical models reviewed here show that terrestrial margins began to retreat irreversibly as the continuing isostatic depression of the bed lowered the surface beneath snowlines, so that mass balance became irreversibly negative. The geomorphic model also showed that rising sea level from melting terrestrial margins triggered irreversible retreat of marine margins, which crossed a glaciological point of no return between 12 and 10 ka. This was a time of widespread extinctions among mammals and of cultural transformations by early man. A stable world in North America came to an end when ice sheets collapsed.

REFERENCES CITED

Andersen, B. G., 1981, Late Weichselian ice sheets in Eurasia and Greenland, *in* Denton, G. H., and Hughes, T. J., eds., The last great ice sheets: New York, Wiley-Interscience, p. 1–65.

Anderson, J. B., Kurtz, D. D., Domac, E. W., and Balshaw, K. M., 1980, Glacial and glacial marine sediments on the Antarctic continental shelf: Journal of Geology, v. 88, p. 399–414.

Anderson, J. B., Brake, C. F., and Myers, N. C., 1984, Sedimentation on the Ross Sea continental shelf, Antarctica: Marine Geology, v. 57, p. 295–333.

Andrews, J. T., 1963. End moraines and late-glacial chronology in the northern Nain-Okak section of the Labrador coast: Geografiska Annaler, v. 45, p. 645–665.

Andrews, J. T., and Miller, G. H., 1976. Quaternary glacial chronology of the eastern Canadian arctic; A review and a contribution on amino acid dating of Quaternary molluscs from the Clyde Cliffs, *in* Mahoney, W., ed., Quaternary stratigraphy of North America: Stroudsburg, Pennsylvania, Dowden, Hutchinson, and Ross, p. 1–32.

Andrews, J. T., Shilts, W. W., and Miller, G. H., 1983, Multiple deglaciation of the Hudson Bay lowlands since deposition of the Missinaibi (last interglacial?) Formation: Quaternary Research, v. 19, no. 1, p. 18–37.

Andrews, J. T., Stravers, J. A., and Miller, G. H., 1985, Patterns of glacial erosion and deposition around Cumberland Sound, Frobisher Bay, and Hudson Strait, and the location of ice streams in the eastern Canadian Arctic, *in* Woldenberg, M. J., ed., Models in geomorphology: Boston, Massachusetts, Allen and Unwin, p. 93–117.

Anonymous, 1974. The national atlas of Canada, 4th edition: Toronto, Macmillan.

Anundsen, K., 1985, Changes in shore-level and ice front positions in Late Weichselian and Holocene, southern Norway: Norsk Geografisk Tidskrift, v. 39, p. 205–225.

Bader, H., 1961, The Greenland ice sheet, *in* Sanger, F. J., ed., Cold regions science and engineering 1-B2: Hanover, New Hampshire, U.S. Army Cold Regions Research and Engineering Laboratory, p. 1–18.

Bindschadler, R. A., and Gore, R., 1982, A time-dependent ice sheet model; Preliminary results: Journal of Geophysical Research, v. 87, no. C12, p. 9675–9685.

Birchfield, G. E., 1977, A study of the stability of a model continental ice sheet subject to periodic variations in heat input: Journal of Geophysical Research, v. 82, no. 31, p. 4904–4913.

Birchfield, G. E., and Weertman, J., 1978, A note on the spectral response of a model continental ice sheet: Journal of Geophysical Research, v. 83, no. C8, p. 4123–4125.

——, 1982, A model study of the role of variable ice albedo in the climate response to orbital insolation anomalies: Icarus, v. 50, p. 462–472.

Birchfield, G. E., Weertman, J., and Lunde, A. T., 1981, A paleoclimate model of the northern hemisphere ice sheets: Quaternary Research, v. 15, p. 126–142.

Blake, W., Jr., 1970, Studies of glacial history in Arctic Canada; I. Pumice, radiocarbon dates, and differential postglacial uplift in eastern Queen Elizabeth Islands: Canadian Journal of Earth Sciences, v. 7, p. 634–664.

——, 1977, Glacial sculpture along the east-central coast of Ellesmere Island, Arctic Archipelago; Report of Activities, Part C: Geological Survey of Canada Paper 77-1C, p. 107–115.

Boulton, G. S., and Jones, A. S., 1979, Stability of temperate ice caps and ice sheets resting on beds of deformable sediment: Journal of Glaciology, v. 24, p. 29–43.

Boulton, G. S., Smith, G. D., and Morland, L. W., 1984, The reconstruction of former ice sheets and their mass balance characteristics using a non-linearly viscous flow model: Journal of Glaciology, v. 30, p. 140–152.

Boulton, G. S., Smith, G. D., Jones, A. S., and Newsome, J., 1985, Glacial geology and glaciology of the last mid-latitude ice sheets: Journal of the Geological Society of London, v. 124, p. 447–474.

Boyer, S. J., and Pheasant, D. R., 1974, Delineation of weathering zones in the fiord area of eastern Baffin Island, Canada: Geological Society of America Bulletin, v. 85, p. 805–810.

Broecker, W. S., and van Donk, J., 1970, Insolation changes, ice volumes, and the O^{18} record in deep sea cores: Reviews of Geophysics and Space Physics, v. 8, p. 169–198.

Brookes, I. A., 1982, Ice marks in Newfoundland; A history of ideas: Géographie physique et Quaternaire, v. 36, no. 1-2, p. 139–163.

Bryson, R. A., Wendland, W. M., Ives, J. D., and Andrews, J. T., 1969, Radiocarbon isochrones on the disintegration of the Laurentide ice sheet: Arctic and Alpine Research, v. 1, p. 1–14.

Budd, W. F., and McInnes, B. J., 1979, Periodic surging of the Antarctic ice sheet; An assessment by modelling: Bulletin des Sciences Hydrologiques, v. 24, nos. 1-3, p. 95–104.

Budd, W. F., and Smith, I. N., 1981, The growth and retreat of ice sheets in response to orbital radiation changes: International Association of Hydrological Sciences Publication 131, p. 369–409.

Carbonnell, M., and Bauer, A., 1968, Exploitation des couvertures photographiques aeriennes repetees due front des glaciers velant dans Disko Bugt en Umanak Fjord, Juin-Juillet 1964: Expedition Glaciologique Internatinale au Groenland, 1957–1960, Copenhagen, v. 2, no. 3, p. 1–78.

Cathles, L. M., III, 1975, The viscosity of the earth's mantle: Princeton, New Jersey, Princeton University Press, 386 p.

Clark, J. A., 1976, Greenland's rapid postglacial emergence; A result of ice-water gravitational attraction: Geology, v. 4, p. 310–312.

——, 1980, The reconstruction of the Laurentide ice sheet of North America from sea level data; Method and preliminary results: Journal of Geophysical Research, v. 85, no. B8, p. 4307–4323.

Clark, J. A., and Lingle, C. S., 1979, Predictive relative sea level changes (18,000 years BP to present) caused by late-glacial retreat of the Antarctic ice sheet: Quaternary Research, v. 11, p. 279–298.

Clark, J. A., Farrell, W. E., and Peltier, W. R., 1978, Global changes in postglacial sea levels; A numerical calculation: Quaternary Research, v. 9, p. 265–287.

Crary, A. P., 1966, Mechanism of fjord formation indicated by studies of an ice-covered inlet: Geological Society of America Bulletin, v. 77, p. 911–930.

Davis, R. B., and Jacobson, G. L., Jr., 1985, Late glacial and Early Holocene landscapes in northern New England and adjacent areas of Canada: Quaternary Research, v. 23, p. 341–368.

Denton, G. H., and Hughes, T., 1981, eds., The last great ice sheets: New York, Wiley-Interscience, 484 p.

Denton, G. H., and Hughes, T., 1983, Milankovitch theory of ice ages; Hypothesis of ice-sheet linkage between regional insolation and global climate: Quaternary Research, v. 20, p. 125–144.

Denton, G. H., Hughes, T. J., and Karlén, W., 1986, Global ice-sheet system interlocked by sea level: Quaternary Research, v. 26, p. 3–26.

Doake, C.S.M., and Wolff, E. W., 1985, Flow law for ice in polar ice sheets: Nature, v. 314, p. 255–257.

England, J., 1976, Late Quaternary glaciation of the eastern Queen Elizabeth Islands, N.W.T., Canada; Alternative models: Quaternary Research, v. 6, no. 2, p. 185–202.

England, J., and Bradley, R. S., 1978, Past glacial activity in the Canadian high arctic: Science, v. 200, p. 265–270.

Fastook, J. L., 1984, West Antarctica, the sea-level controlled marine instability; Past and Future, *in* Hansen, J. E., and Takahashi, T., eds., Climate processes and climate sensitivity: American Geophysical Union Geophysics Monograph 29, p. 275–287.

——, 1985, Ice shelves and ice streams; Three modeling experiments, *in* Glaciers, ice sheets, and sea level; Effect of a CO_2-induced climate change: Springfield, Virginia, National Technical Information Service, p. 297–300.

Fastook, J. L., and Schmidt, W. F., 1982, Finite-element analysis of calving from ice sheets: Annals of Glaciology, v. 3, p. 103–106.

Fisher, D. A., Reeh, N., and Langley, K., 1985, Objective reconstruction of the Late Wisconsinan Laurentide ice sheet and the significance of deformable beds: Géographie physique et Quaternaire, v. 39, no. 3, p. 229–238.

Flint, R. F., 1943, Growth of North American ice sheet during the Wisconsin age: Geological Society of American Bulletin, v. 54, p. 325–362.

Glen, J. W., 1958, The flow law of ice: International Association of Scientific Hydrology, no. 47, p. 171–183.

Grant, D. R., 1971, Glaciation of Cape Breton Island, Nova Scotia: Geological Survey of Canada Paper 71–1B, p. 100–102.

——, 1976, Reconnaissance of early and middle Wisconsinan deposits along the Yarmouth–Digby coast of Nova Scotia: Geological Survey of Canada Paper 76–1B, p. 363–369.

——, 1977, Altitudinal weathering zones and glacial limits in western Newfoundland with particular reference to Gros Morne National Park: Geological Survey of Canada Paper 77–1A, p. 455–463.

Hays, J. D., Imbrie, J., and Shackleton, N. J., 1976, Variations in the Earth's orbit; Pacemaker of the ice ages: Science, v. 194, no. 4270, p. 1121–1132.

Holdsworth, G., 1981, A mechanism for the formation of large icebergs: Journal of Geophysical Research, v. 86, no. C4, p. 3210–3222.

Hudson, R. D., 1983, Direction of glacial flow across Hans Island, Kennedy Channel, N.W.T., Canada (letter): Journal of Glaciology, v. 29, no. 102, p. 353–354.

Hughes, T., 1973, Glacial permafrost and Pleistocene ice ages, *in* Permafrost: The North American Contribution to the Second International Conference, Washington, D.C., Academy of Sciences, p. 213–223.

——, 1981a, Lithosphere deformation by continental ice sheets: Proceedings of the Royal Society of London, ser. A, v. 378, p. 507–527.

——, 1981b, Numerican reconstruction of paleo ice sheets, *in* Denton, G. H. and Hughes, T., eds., The last great ice sheets: New York, Wiley-Interscience, p. 221–261.

——, 1982, On the disintegration of ice shelves; The role of thinning: Annals of Glaciology, v. 3, p. 146–151.

——, 1983a, On the disintegration of ice shelves; The role of fracture: Journal of Glaciology, v. 101, p. 97–107.
——, 1983b, The stability of the west Antarctic ice sheet; What has happened and what will happen; CO2, v. 021: Springfield, Virginia, National Technical Information Service, part IV, p. 51–73.
——, 1985, The great Cenozoic ice sheet: Palaeogeography, Palaeoclimatology, Palaeoecology, v. 50, p. 9–43.
——, 1986, The Jakobshavns effect: Geophysical Research Letters, v. 13, p. 46–48.
——, 1987, On the pulling power of ice streams: Journal of Geophysical Research (in press).
Hughes, T., Denton, G. H., and Grosswald, M. G., 1977, Was there a late-Würm Arctic ice sheet?: Nature, v. 266, no. 5603, p. 596–602.
Hughes, T., Denton, G. H., Andersen, B. G., Schilling, D. H., Fastook, J. L., and Lingle, C. S., 1981, The last great ice sheets; A global view, *in* Denton, G. H., and Hughes, T., eds., The last great ice sheets: New York, Wiley-Interscience, p. 263–317.
Hughes, T., Denton, G. H., and Fastook, J. L., 1985, The Antarctic ice sheet; An analog for Northern Hemisphere paleo-ice sheets?, *in* Woldenberg, M. J., ed., Models in geomorphology: Boston, Massachusetts, Allen and Unwin, p. 25–72.
Iken, A., 1977, Movement of a large ice mass before breaking off: Journal of Glaciology, v. 19, no. 81, p. 595–605.
Imbrie, J., and 8 others, 1984, The orbital theory of Pleistocene climate; Support from a revised chronology of the marine $\delta^{18}O$ record, *in* Berger, A., Imbrie, J., Hays, J., Kukla, G., and Saltzman, B., Milankovitch and climate: Dordrecht, Reidel, Part I, p. 269–305.
Ives, J. D., 1957, Glaciation of the Torngat Mountains: Geographical Bulletin, no. 10, p. 67–87.
——, 1978, The maximum extent of the Laurentide ice sheet along the east coast of North America during the last glaciation: Arctic, v. 31, p. 24–53.
Källen, E., Crawford, C., and Ghil, M., 1979, Free oscillations in a climate model with ice-sheet dynamics: Journal of Atmospheric Sciences, v. 36, p. 2292–2303.
Kamb, W. B., 1970, Sliding motion of glaciers; Theory and observation: Reviews of Geophysics and Space Physics, v. 8, p. 673–728.
Kamb, W. B., and 7 others, 1985, Glacier surge mechanism; 1982–1983 surge of Variegated Glacier, Alaska: Science, v. 227, no. 4686, p. 469–479.
Kellogg, D. E., and Kellogg, T. B., 1986, Diatom biostratigraphy of sediment cores from beneath the Ross Ice Shelf: Micropaleontology, v. 32, p. 74–94.
Kellogg, T. B., Osterman, L. E., and Stuiver, M., 1979, Late Quaternary sedimentology and benthic foraminiferal paleoecology of the Ross Sea, Antarctica: Journal of Foraminiferal Research, v. 9, no. 4, p. 322–335.
Lange, M. A. and MacAyeal, D. R., 1986, Numerical models of the Filchner-Ronne Ice Shelf: an assessment of reinterpreted ice thickness distributions: Journal of Geophysical Research, v. 91, no. B 10, p. 10,457-10, 462.
Lindstrom, D. R., and MacAyeal, D. R., 1986, Paleoclimatic constraints on the maintenance of possible ice-shelf cover in the Norwegian and Greenland seas: Paleooceanography, v. 1, p. 313–337.
Lindstrom, D., and Tyler, D., 1984, Preliminary results of Pine Island and Thwaites Glaciers study: Antarctic Journal of the United States, v. 19, no. 5, p. 53–55.
Lingle, C. S., and Clark, J.A ., 1979, Antarctic ice sheet volume at 18,000 years BP and Holocene sea-level changes at the west Antarctic margin: Journal of Glaciology, v. 24, p. 213–230.
——, 1985, A numerical model of interactions between a marine ice sheet and the solid Earth; Application to a west Antarctic ice stream: Journal of Geophysical Research, v. 90, no. Cl, p. 1100–1114.
Lingle, C. S., Hughes, T., and Kollmeyer, R. C., 1981, Tidal flexure of Jakobshavns Glacier, west Greenland: Journal of Geophysical Research, v. 86, no. 5B, p. 3960–3968.
MacAyeal, D. R., 1985, Tidal-current rectification and tidal mixing fronts: Controls on the Ross Ice Shelf flow and mass balance [Ph.D. thesis]: Princeton, New Jersey, Princeton University, 287 p.
——, 1987, Ice-shelf back pressure; Form drag vs. dynamic drag, *in* van der Veen, C. J., and Oerlemans, J., eds., Ice dynamics of the west Antarctic ice sheet: Dordrecht, Reidel, p. 141–160.
Mahaffy, M. W., 1976, A three-dimensional numerical model of ice sheets; Tests on the Barnes ice cap, northwest territories: Journal of Geophysical Research, v. 81,, no. 6, p. 1059–1066.
Mayewski, P. A., Denton, G. H., and Hughes, T., 1981, Late Wisconsin ice sheets in North America, *in* Denton, G. H., and Hughes, T., eds., The last great ice sheets: New York, Wiley-Interscience, p. 67–178.
McIntyre, A., 1985, The dynamics of ice sheet outlets: Journal of Glaciology, v. 31, no. 108, p. 99–107.
Milankovitch, M. M., 1941, Canon of insolation and the ice-age problems: Belgrade, Koniglich Serbische Akademie Special Publication 133.
Nye, J. F., 1951, The flow of glaciers and ice sheets as a problem in plasticity: Proceedings of the Royal Society of London, ser. A, v. 207, p. 554–572.
——, 1957, The distribution of stress and velocity in glaciers and ice sheets: Proceedings of the Royal Society of London, ser. A, v. 239, p. 113–133.
——, 1970, Glacier sliding without cavitation in a linear viscous approximation: Proceedings of the Royal Society of London, ser. A, v. 315, no. 1522, p. 381–403.
Oerlemans, J., 1981, Modeling of Pleistocene European ice sheets; Some experiments with simple mass-balance parameterizations: Quaternary Research, v. 15, no. 1, p. 77–85.
——, 1982, Glacial cycles and ice-sheet modeling: Climate Change, v. 4, p. 353–374.
——, 1983, A numerical study on cyclic behaviour of polar ice sheets: Tellus, v. 35A, p. 81–87.
Oerlemans, J., and van der Veen, C. J., 1984, Ice sheets and climate: Dordrecht, The Netherlands, Reidel, 217 p.
Paterson, W.S.B., 1972, Laurentide ice sheet; Estimated volumes during late Wisconsin: Reviews of Geophysics and Space Physics, v. 10, p. 885–917.
——, 1981, The physics of glaciers (second edition): Oxford, Pergamon, 380 p.
——, 1985, Flow law for ice in polar ice sheets: Nature, v. 318, p. 82–83.
Peltier, W. R., 1976, Glacial isostatic adjustment: II, The inverse problem: Geophysical Journal of the Royal Astronomical Society, v. 46, p. 669–706.
——, 1980a, Models of glacial isostasy and relative sea level, *in* Bally, A. W., Bender, P. L., McGetchin, T. R., and Walcott, R. I., eds., Dynamics of plate interiors: American Geophysical Union Geodynamics Series, v. 1, p. 111–128.
——, 1980b, Ice sheets, oceans, and the earth's shape, *in* Mörner, N.-A., ed., Earth rheology, isostasy, and eustasy: Chichester, John Wiley and Sons, p. 45–64.
——, 1981, Surface plates and thermal plumes; Separate scales of the mantle convective circulation, *in* O'Connell, R. J., and Fyfe, W. F., eds., Evolution of the Earth: American Geophysical Union, Geodynamics Series, v. 5, p. 229–248.
——, 1986, Deglaciation induced vertical motion of the North American continent and transient lower mantle rheology: Journal of Geophysical Research, v. 87, p. 9099–9123.
Peltier, W. R., and Andrews, J. T., 1976, Glacial-isostatic adjustment; I, The forward problem: Geophysical Journal of the Royal Astronomical Society, v. 46, p. 605–646.
——, 1983, Glacial geology and glacial isostasy of the Hudson Bay region, *in* Shorelines and isostasy: Institute of British Geographers, p. 285–319.
Pfeffer, T., 1982, The effect of crevassing on the radiative absorptance of a glacier surface [abs.]: Annals of Glaciology, v. 3, p. 353.
Pfeffer, W. T., and Bretherton, C. S., 1987. The effect of crevasses on the solar heating of a glacier surface, *in* The physical basis of ice sheet modeling: International Association of Hydrological Sciences Publication no. 170, p. 1–14.
Pheasant, D. R., and Andrews, J. H., 1972, The Quaternary history of the northern Cumberland Peninsula, Baffin Island, N.W.T.; Pt. 8, Chronology of Narpaing and Quajon Fiords during the past 120,000 years: Montreal, 24th International Geological Congress, section 12, p. 81–88.

Pollard, D., 1982, A simple ice sheet model yields realistic 100 kyr glacial cycles: Nature, v. 296, p. 334–338.
——, 1983, Ice-age simulations with a calving ice-sheet mode: Quaternary Research, v. 20, no. 1, p. 30–48.
Radok, U., Barry, R. G., Jenssen, D., Keen, R. A., Kiladis, G. N., and McInnes, B., 1982, Climatic and physical characteristics of the Greenland ice sheet, Pts. I and II: Boulder, University of Colorado, Cooperative Institute for Research in Environmental Sciences, 193 p.
Reeh, N., 1968, On the calving of ice from floating glaciers and ice shelves: Journal of Glaciology, v. 7, p. 215–232.
——, 1982, A plasticity theory approach to the steady-state shape of a three-dimensional ice sheet: Journal of Glaciology, v. 28, no. 100, p. 431–455.
Robin, G. de Q., 1955, Ice movement and temperature distribution in glaciers and ice sheets: Journal of Glaciology, v. 2, p. 523–532.
——, 1979, Formation, flow, and disintegration of ice shelves: Journal of Glaciology, v. 24, no. 90, p. 259–271.
Shackleton, N. J., and Opdyke, N. D., 1973, Oxygen isotope and palaeomagnetic stratigraphy of equitorial core V28–238; Oxygen isotope temperatures and ice volumes on a 10^5 year and 10^6 year scale: Quaternary Research, v. 3, p. 39–55.
Shilts, W. W., 1980, Flow patterns in the central North American ice sheet: Nature, v. 286, no. 5770, p. 213–218.
——, 1982, Quaternary evolution of the Hudson/James Bay region: Naturaliste canadienne, v. 109, p. 309–332.
Sikonia, W. G., 1982, Finite-element glacier dynamics model applied to Columbia Glacier, Alaska: U.S. Geological Survey Professional Paper 1258–B, 74 p.
Stuiver, M., Denton, G. H., Hughes, T., and Fastook, J. L., 1981, History of the marine ice sheet in west Antarctica during the last glaciation; A working hypothesis, *in* Denton, G. H., and Hughes, T., eds., The last great ice sheets: New York, Wiley-Interscience, p. 319–436.
Sugden, D. E., and John, B., 1982, Glaciers and landscape; A geomorphological approach: London, Edward Arnold, 320 p.
Sugden, D. E., and Watts, S. H., 1977, Tors, felsenmeer, and glaciation in northern Cumberland Peninsula, Baffin Island: Canadian Journal of Earth Sciences, v. 14, p. 2817–2823.
Swithinbank, C., 1955, Ice shelves: The Geographical Journal, v. 121, pt. 1, p. 62–76.
Tyrrell, J. B., 1898, The glaciation of north-central Canada: Journal of Geology, v. 6, p. 147–160.
Vernekar, A. D., 1972, Long-period global variations of incoming solar radiation: American Meteorological Society Meteorological Monographs 12, no. 34.
Walcott, R. I., 1972, Late Quaternary vertical movements in eastern North America; Quantitative evidence of glacial-isostatic rebound: Reviews of Geophysics and Space Physics, v. 10, no. 4, p. 849–884.
Weertman, J., 1957, Deformation of floating ice shelves: Journal of Glaciology, v. 3, p. 38–42.
——, 1961, Stability of ice-age ice sheets: Journal of Geophysical Research, v. 66, no. 11, p. 3783–3792.
——, 1964, Rate of growth or shrinkage of nonequilibrium ice sheets: Journal of Glaciology, v. 5, p. 145–158.
——, 1966, Effect of a basal water layer on the dimensions of ice sheets: Journal of Glaciology, v. 6, p. 191–207.
——, 1970, The creep strength of the earth's mantle: Reviews of Geophysics and Space Physics, v. 8, no. 1, p. 145–168.
——, 1973, Can a water-filled crevasse reach the bottom surface of a glacier?: International Association of Scientific Hydrology Publication 95, p. 139–145.
——, 1976, Milankovitch solar radiation variations and ice age ice sheet sizes: Nature, v. 261, p. 17–20.
——, 1978, Creep laws for the mantle of the earth: Philosophical Transactions of the Royal Society of London, ser. A, v. 288, p. 9–26.
——, 1980, Bottom crevasses: Journal of Glaciology, v. 25, no. 91, p. 185–188.
——, 1983, Creep deformation of ice: Annual Reviews of Earth and Planetary Sciences, v. 11, p. 215–240.
——, 1985, Unsolved problems of creep: Nature, v. 314, p. 227.
Weertman, J., and Weertman, J. R., 1975, High temperature creep of rock and mantle viscosity, *in* Donath, F. A., Stehli, F. G., and Wetherill, G. W., eds., Annual reviews of earth and planetary sciences: Palo Alto, California, Annual Reviews, Inc., v. 3, p. 293–315.
Williams, R. T., and Robinson, E. S., 1979, Ocean tide and waves beneath the Ross Ice Shelf, Antarctica: Science, v. 203, p. 443–445.
Zotikov, I. A., 1964, Temperatures of Antarctic glaciers: Izd-vo Akad. Nauk SSSR, Moscow, p. 61–105.

Manuscript Accepted by the Society March 16, 1987

ACKNOWLEDGMENTS

Glacial-geological constraints used in the geomorphic ice-sheet model of deglaciation were assigned after consultations with George H. Denton and Jean-Serge Vincent. The variable flowband widths, accumulation and ablation rates, and time-dependency of this model were programmed by James L. Fastook. Flowlines and ice-elevation contours employed in deglaciation of the Cordilleran Ice Sheet using this model were determined by Mauri S. Pelto, who will publish the details independently. Information in Figure 26 was provided by William F. Budd. I am deeply grateful to all these colleagues. I also thank William F. Ruddiman and Herbert E. Wright, Jr., for multiple editorial and scientific reviews of this chapter, and John T. Andrews, Charles R. Bentley, Mauri S. Pelto, and Charles F. Raymond for solicited reviews of the chapter. Ideas about ice streams presented in the chapter are a result of field studies on Byrd Glacier, Antarctica, funded by NSF grants DPP-7722204 and DPP-7918681, and Jakobshavns Isbrae, Greenland, funded by NSF grants DPP-7707806 and DPP-8400886.

DEDICATION

A major theme of this chapter is the radically different understanding of the last glaciation that emerges when deglaciation mechanisms include the dynamics of marine components of continental ice sheets, in addition to terrestrial components. The late John H. Mercer was the father of the concept of marine ice sheets. He originated that term; he was the first to suggest that they are inherently unstable, and first to propose an unstable glacial history for the present-day marine ice sheet in West Antarctica. He was also among the first to postulate the former existence of extensive marine ice sheets in the Arctic, interconnected by floating ice shelves, and to recognize their prominent role in the glacial history of the Northern Hemisphere. He was a pioneer in linking glaciation and deglaciation mechanisms to the dynamics of climatic change and in correlating glaciation histories on a global scale. In appreciation of his contributions, I dedicate this chapter to the memory of John H. Mercer.

Printed in U.S.A.

The Geology of North America
Vol. K-3, North America and adjacent oceans during the last deglaciation
The Geological Society of America, 1987

Chapter 10

River responses

S. A. Schumm
Department of Earth Resources, Colorado State University, Fort Collins, Colorado 80523
G. R. Brakenridge
Department of Geological Sciences, Wright State University, Dayton, Ohio 45435

INTRODUCTION

Two main aspects of rivers relate to the interpretation of the last deglaciation. One involves changes of channel morphology, as affected by climatic and hydrologic controls. The other is the record preserved in the valley alluvium. Although intimately related, the two topics will be considered separately in this chapter because they involve different approaches to river history.

The fluvial record of the period 18 to 6 ka is complex and fragmentary, and therefore it does not provide an unequivocal basis for interpreting climate change and ice retreat. However, the current understanding of rivers is sufficiently detailed so that conclusions about past hydrologic conditions can be made if a paleochannel can be described quantitatively or if the sequence of valley fill deposits is complete. Herein lies the greatest problem, for much of the necessary information is commonly obscured or incomplete. Generally, paleochannels are destroyed by erosion or buried by deposition, and older fluvial deposits are removed by erosion, so that inferences regarding past hydrologic and climatic conditions can rarely be made. Even when this is possible the results are frequently inconsistent with conclusions reached elsewhere, partly because the nature and degree of climate change varied in North America. In addition, different rivers and even reaches of the same river respond differently to similar changes of discharge and sediment load. Hence, even if morphologic or sedimentologic changes can be completely documented, they do not necessarily provide a basis for a general interpretation of climatic or hydrologic change. Some of the problems of this type of interpretation are summarized below (for a discussion see Schumm, 1984, 1985).

1. Location: Different river reaches may be out of phase. That is, a reach of channel may be aggrading, but the deposition can be the result of major degradation upstream.

2. Convergence: Different causes or processes can produce similar effects. For example, incision and terrace formation can result from changes of runoff, sediment load, baselevel, or tectonic influences. A braided stream can result from high sediment load, aggradation, or flashy or seasonal runoff.

3. Divergence: Similar causes or processes can produce different effects. For example, a given climate change can either increase or decrease sediment yields, and an increase of stream power can either increase meandering or cause braiding, depending on channel sensitivity.

4. Sensitivity: Depending on the tendency to change, a change of runoff, sediment yield, or stream power can cause either minor adjustment or complete metamorphosis of a river.

These problems demonstrate why environmental interpretation is difficult, but they need not discourage interpretation. Rather, if the problems are recognized, then they can be solved. Uncertainties can largely be removed by careful study of river morphology and behavior.

FLUVIAL MORPHOLOGY

Rivers exhibit a great morphologic diversity because of variations of water discharge and sediment load as well as the presence of bedrock outcrops, human activities, and tectonic influences (Burnett and Schumm, 1983), and different types of channels respond differently to altered hydrologic conditions. Sensitive channels respond quickly and dramatically, whereas insensitive channels may have their adjustment delayed, and even then the adjustment may be relatively minor.

Five basic channel patterns exist (Fig. 1): (1) straight channels with downstream-migrating transverse or linguoidal bars; (2) sinuous-thalweg channels; (3) meandering channels that range from highly sinuous channels of equal width (pattern 3A) to lower-sinuosity channels that are wider at bends than crossings (pattern 3B); (4) transitional meandering-braided channels; and (5) braided channels. The relative stability of these channels and some of their other morphologic characteristics can be related to relative sediment size, sediment load, velocity of flow, and stream power (Fig. 1).

Modern rivers can be placed within these five general categories of stream patterns. However, within the meandering group

Schumm, S. A., and Brakenridge, G. R., 1987, River responses, *in* Ruddiman, W. F., and Wright, H. E., Jr., eds., North America and adjacent oceans during the last deglaciation: Boulder, Colorado, Geological Society of America, The Geology of North America, v. K-3.

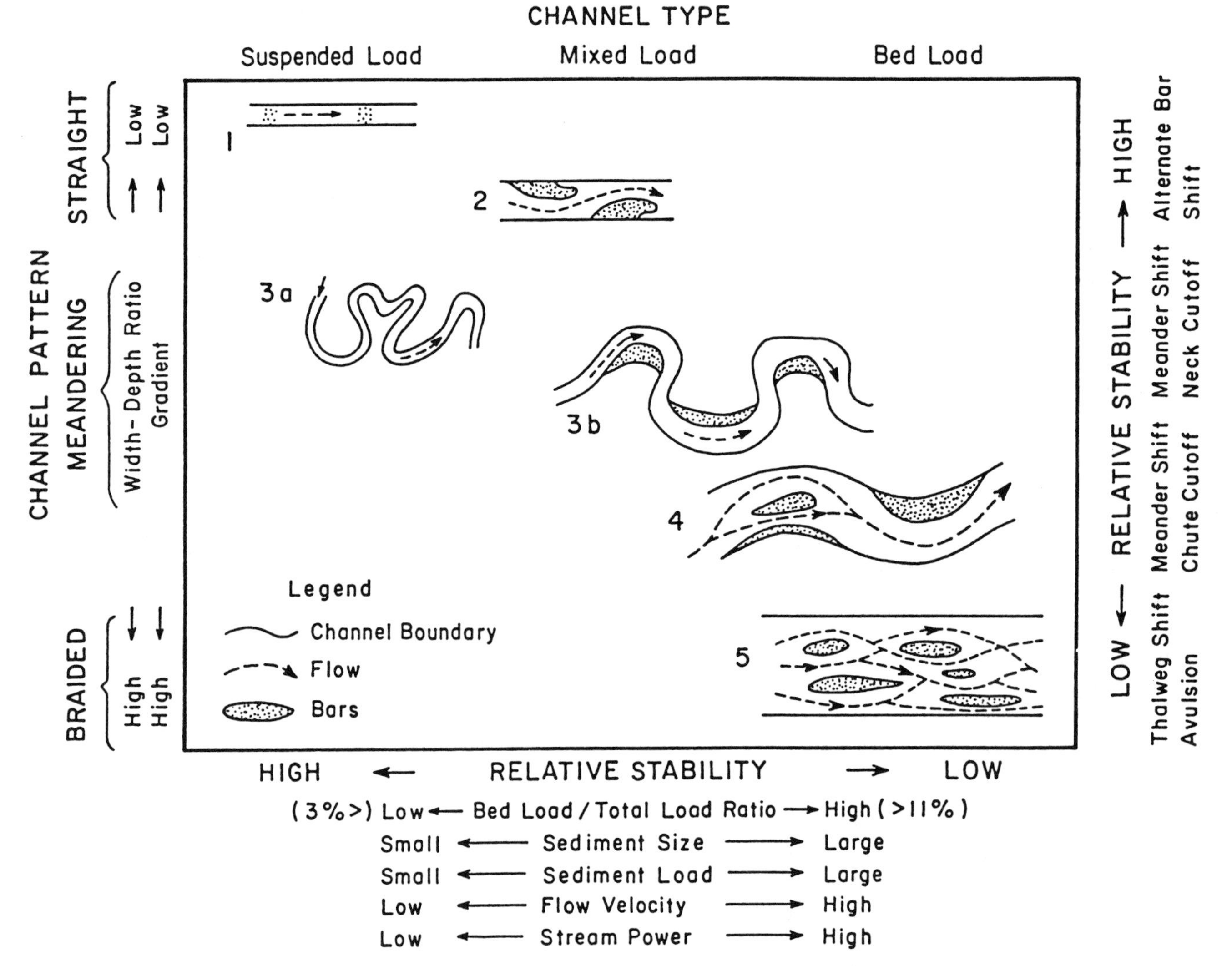

Figure 1. Channel classification based upon channel pattern and type of sediment load (from Schumm, 1981).

there is a considerable range of sinuosity (ratio of channel length to valley length) from relatively straight (1.2) to very sinuous (3.0). In addition, there are both bar-braided and island-braided channels. Islands are vegetated bars. Multiple-channel patterns are termed anastomosing or anabranching channels (Schumm, 1977, p. 155).

Alluvial channels can be further classified as suspended-load, mixed-load, and bed-load channels. Water discharge determines the dimensions of the channel (width, depth, meander dimensions), but the type of sediment load (Q_t) expressed as a ratio of bed load (sand and gravel) to total load (silt, clay, sand, and gravel) determines not only the shape (width-depth ratio) but to a large extent channel pattern. A suspended-load channel transports less than 3 percent bed load, and a bed-load channel is one transporting more than 11 percent bed load (Schumm, 1977; Fig. 1). The range of channels from straight through braided forms a continuum, but experimental work and field studies (Schumm and Khan, 1972; Leopold and Wolman, 1957) indicate that the changes in pattern occur at critical values of stream power, gradient, and sediment load (Fig. 2).

Variability along rivers

A single river can have great pattern variability. Most rivers increase in size downstream because a larger channel is required to convey the increasing discharge. In general, channel width and meander dimensions increase as the 0.5 power of discharge, and channel width increases as the 0.4 power of discharge. When river discharge decreases, downstream channel dimensions also decrease. For example, the Finke River in central Australia flows

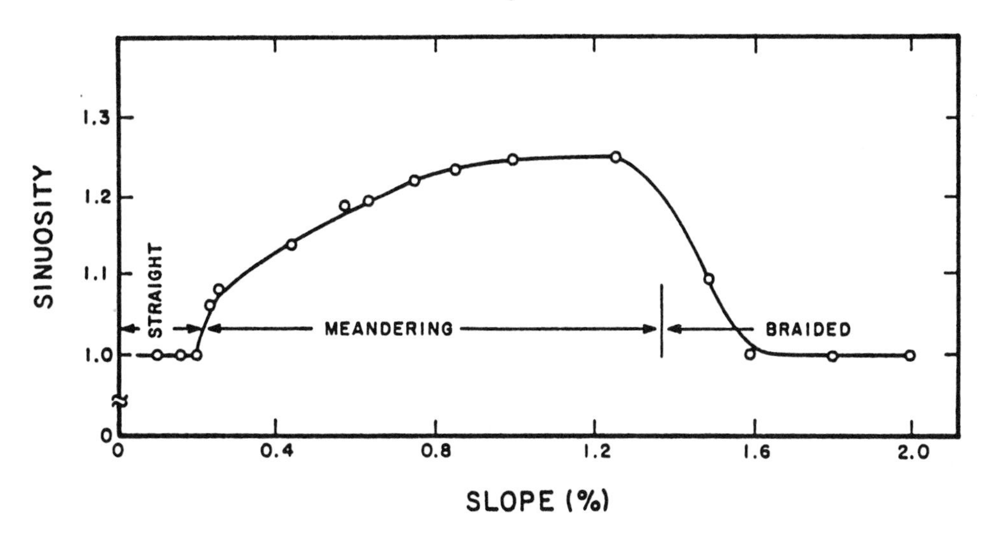

Figure 2. Relation between flume slope and sinuosity (ratio between channel length and valley length) during experiments at constant water discharge (from Schumm and Khan, 1972).

from its source in the McDonald Range as a wide channel, but the channel becomes smaller as the river flows into the Simpson Desert, and it eventually disappears as discharge decreases to zero. The Sacramento River also becomes smaller downstream over part of its course, as a result of water loss into adjacent flood basins. Meander cutoffs may convert a meandering reach of channel to a relatively straight reach that with time will regain its original sinuosity. Often the influence of tributary sediment contributions to the main channel change the channel from braided to meandering and vice versa. In addition, an increase of valley floor slope can result in increased sinuosity in order to maintain a constant channel gradient (Fig. 2). However, if the range is too great, the large increase of slope (stream power) may cause a change from meandering to braided (Fig. 2).

Variations of valley-floor slope and channel pattern can result from tectonic activity (Adams, 1980; Burnett and Schumm, 1983). In addition, a high-sediment-transporting tributary may build a valley fan, which will persist even after the tributary sediment load has decreased. When the main river crosses this fan, pattern changes may result (Fig. 2), as the river maintains a constant gradient.

Although the above discussion concentrates on channel patterns, the classification of Figure 1 indicates that channel dimensions, hydraulic characteristics, shape (width-depth ratio), and gradient all change as pattern changes. Therefore, as sediment loads or valley slopes change, there can be significant variations of channel morphology. The degree of change depends on channel sensitivity (for example, its position on a graph like Figure 2).

Those channels that lie near a pattern threshold may change their characteristics dramatically with only a slight change in the controlling variable. For example, some meandering rivers near a pattern threshold become braided with only a small addition of bed load, whereas less sensitive channels are not affected significantly (Fig. 2). The differences among rivers at different locations are the same as the differences that occur through time as climate, base level, and valley-floor slope are altered.

EFFECT OF CLIMATE ON HYDROLOGY

Water and sediment discharge are the primary independent variables that determine channel morphology (Fig. 1). The relations between runoff and precipitation are generally well established (Langbein, 1949), but seasonality of precipitation is also important. For example, Mediterranean wet-winter climates will produce more runoff than wet-summer climates because evapotranspiration losses are less. For the purposes of this discussion, however, it is sufficient to conclude that more precipitation produces more runoff.

The relation between sediment yield and precipitation is strongly influenced by vegetation and seasonality of precipitation (Walling and Webb, 1983). Following Fournier's (1949) earlier work, Langbein and Schumm (1958) developed a curve for sediment yield in the U.S., using only data from drainage basins with natural vegetation, so that land-use effects were eliminated (Fig. 3). The peak of sediment yield is found in the transition from poorly protected brush and bunchgrass of semiarid regions to the better grass cover and forest of more humid regions. Nevertheless, the locations of the peak of sediment yield varies with the data set used (Fournier, 1949, 1960; Douglas, 1967; Wilson, 1973; Ohmori, 1983). For the U.S. the maximum sediment yield exists between 200 and 500 mm of precipitation. Wilson's peak is very broad and extends from about 400 mm to 700 mm; however, above 500 mm it may reflect agricultural practices that greatly increase sediment yields. The various curves should not be used for quantitative estimates of sediment yield (Jansson, 1982), but they do provide an indication of the direction of change.

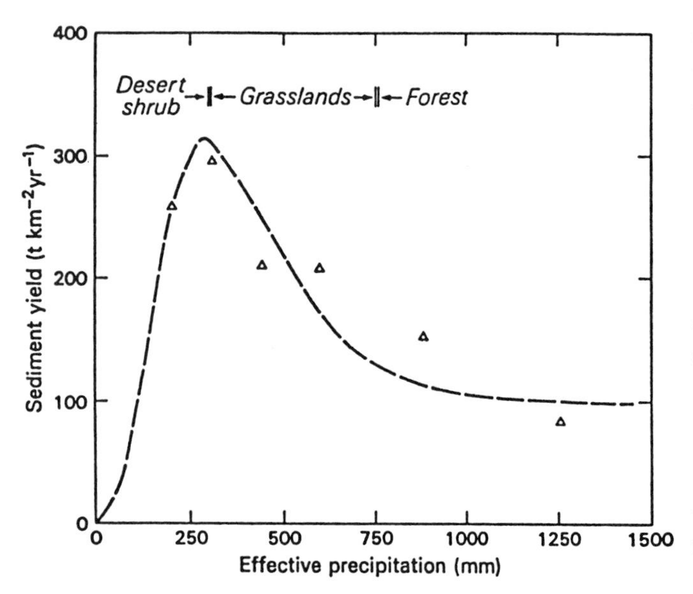

Figure 3. The relation between sediment yield and effective precipitation (adjusted for temperature variations; after Langbein and Schumm, 1958).

In North America the climate during deglaciation changed from glacial to temperate and from subhumid to semiarid. Knox (1983) stressed the importance of vegetation on sediment yield and divided the U.S. into six vegetation regions. In humid regions, as in High Elevation Western Woodlands and in Eastern Woodlands, the vegetation change had only a minor influence, whereas sediment yields in the other four vegetational regions may have been far more sensitive to climate change. In the Tall Grass Prairie of the Midwest the late-glacial spruce forest was replaced by grasses, with a major increase of sediment yield. In the Short Grass Prairie and Western Great Plains the vegetation type was not greatly altered, but, as in the shrubland of the Great Basin and the Southwest, any climate change could produce a significant change of sediment yield (Fig. 3).

Clearly the hydrologic changes in the period of concern would have varied greatly in magnitude and type within the different regions. For example, with climate change the hydrologic response in the eastern half of the U.S. would be dominated by runoff, and flood magnitudes might change, but sediment yields might be little affected except in glaciated terrain. Westward the effect of a similar climatic change on sediment yield would be more important, and in the semiarid grasslands and shrublands it would dominate. Of course an overriding condition is sediment production from active glaciers and under paraglacial and periglacial conditions. No matter what the vegetation conditions, meltwater and runoff highly charged with sediment from eroding glacial deposits will determine channel morphology.

In summary, a considerable range of river channels (Fig. 1) and climatic environments can be found. Therefore the types of channel change will depend on initial channel type (Fig. 1), channel sensitivity (Fig. 2), and initial climatic conditions and associated vegetation type and density (Fig. 3).

RIVER-CHANNEL CHANGE

Controls of morphology

With an increase of discharge (Q_W) the channel width (w), depth (d), and meander wavelength (L) increase and gradient (S) decreases:

$$Q_W \approx \frac{w,d,L}{S} \tag{1}$$

If at a constant discharge bed load (Q_s) increases, then width, meander wavelength, gradient, and width/depth ratio (F) increase, and depth and sinuosity (P) decrease:

$$Q_s \approx \frac{w,L,S,F}{d,P} \tag{2}$$

Equations 1 and 2 show that an increase of discharge (Q_W) alone changes the dimensions of the channel and its gradient, whereas a change of bed load (Q_s) at constant discharge changes not only channel dimensions but also channel slope and pattern (Schumm, 1969).

A change of Q_W is usually accompanied by a change of Q_s. A change of bed load may not greatly affect F and P if the Q_W change compensates for it. However, a change in type of load (Q_t, ratio of bed load to total load) will cause changes of channel morphology like the differences displayed by Figure 1.

Starkel (1983a) used these equations to suggest what may occur when the magnitude of the change of the two independent variables is greatly different. The rapid invasion of forests during the early Holocene of northern Europe was associated with a marked decrease of Q_s but a smaller decrease of Q_W ($Q_{W^-} < Q_{s^-}$). This imbalance leads to incision and/or meandering; if the situation is reversed ($Q_{W^-} > Q_{s^-}$) aggradation and perhaps braiding should result until a new balance is established. Therefore, during a period of transition the results may temporarily differ from those suggested by equations 1 and 2.

Although Q_W is an index of the volume of water passing through a channel, other hydrologic characteristics are important. For example, although Q_W can be the same for two areas, one may have high flood peaks or highly seasonal runoff in contrast to more uniform discharge in the other area. The effect on the channel can be significant (Gupta, 1983). The concentration of discharge as floods or as seasonal runoff can cause braiding, perhaps because the concentration of flow both increases total sediment transport (Q_s) and changes the type of load (Q_t).

Paleohydrology

Discharge can be estimated from channel dimensions (Dury, 1976; Wahl, 1984). The simplest approach is to use hydraulic geometry relations of the following type:

$$w = kQ^{0.5} \quad (3)$$
$$d = kQ^{0.4} \quad (4)$$
$$L = kQ^{0.5} \quad (5)$$

With a change of discharge the channel dimensions change accordingly, but only in unusual cases will discharge change without a change of sediment load and type, and this significantly affects the relations of equations 3, 4, and 5. For example, with an increase of Q_w, the width, depth, and meander dimensions should decrease, but the nature and magnitude of the change of sediment load (Q_t and Q_s) may reverse these relations (Schumm, 1977). Because of the difficulty of knowing the magnitude of Q_w and Q_s changes, attempts to estimate paleohydrology or channel change are open to large error (Ethridge and Schumm, 1978; Williams, 1984; Dury, 1985).

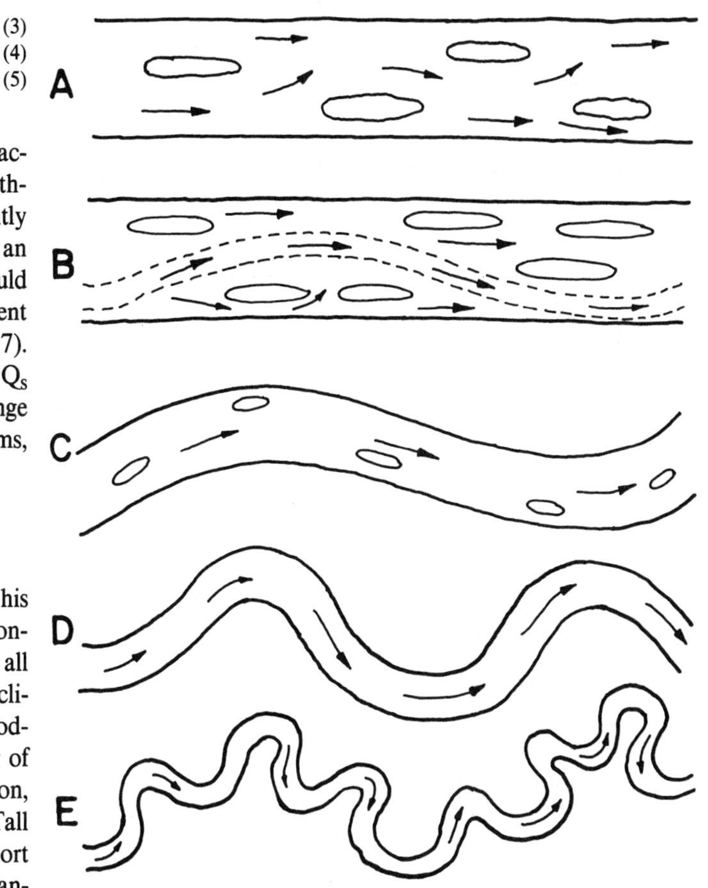

Figure 4. Sequence of channel changes as water discharge and sediment loads decrease. A, braided channel; B, transitional meandering-braided channel with well-defined thalweg; C, low-sinuosity channel; D, relatively narrow and deep moderately sinuous channel; E, multiphase meandering channel.

Channel response

Knox's review of the alluvial chronologies of four of his vegetation zones emphasizes the significance of the climatic conditions existing before a change occurs. Between 10 and 8 ka, all of the valleys studied were aggrading. With the subsequent climate change, the channels of the less sensitive Eastern Woodlands responded by lateral channel migration and reworking of flood-plain sediments. Some minor incision, slow aggradation, and lateral channel migration were also characteristic of the Tall Grass Prairie channels. The sensitive subhumid-semiarid Short Grass Prairie region and the Southwest Desert Shrubland channels show evidence of major deposition followed by major incision.

The major difference between the humid Eastern Woodlands and Midwest and the dry Great Plains and Southwest appears to be in the lateral shift of the eastern streams and the cut-and-fill episodes in the western valleys. Apparently the abundant runoff of the humid eastern area could transport much of the sediment out of the valleys, whereas in the West, sediment storage and then flushing were characteristic as alluvium was deposited and then remobilized at a later time.

An understanding of river variability and response to altered conditions makes it possible to speculate about river change during the period of continental deglaciation. Because of the high sediment loads and perhaps high flood peaks (high stream power), the oldest streams were wide, shallow, steep, braided bed-load channels (Fig. 4A). As sediment loads decreased, perhaps more rapidly than discharge, a meander-braided transition pattern developed with a well-defined single thalweg (Fig. 4B). The thalweg, in turn, became the channel as a new flood plain formed, and the channel narrowed with further reduction of sediment load (Fig. 4C). Finally, as bed load became a fraction of its former volume, a meandering mixed-load channel with large meanders formed (Fig. 4D).

With a further significant reduction of bed load and discharge, a highly sinuous channel with low width-depth ratio formed with a multiphase meandering pattern (Fig. 4E), as

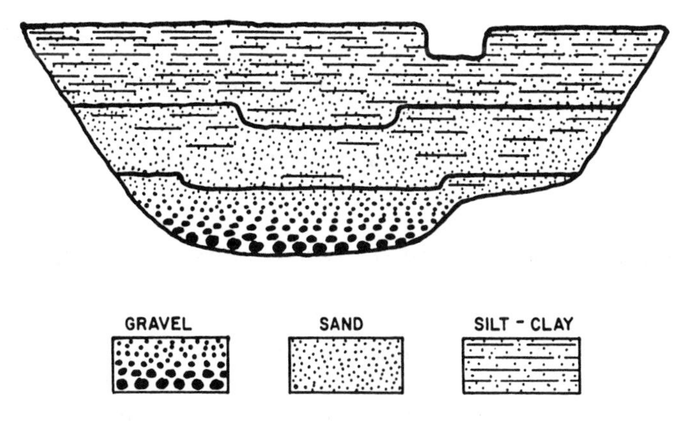

Figure 5. Cross section of a valley fill showing change of channel cross section as sediment load becomes progressively finer and as the straight bed-load channel at the base of the fill adjusts to become a sinuous suspended-load channel at the top of the fill (from Schumm, 1963).

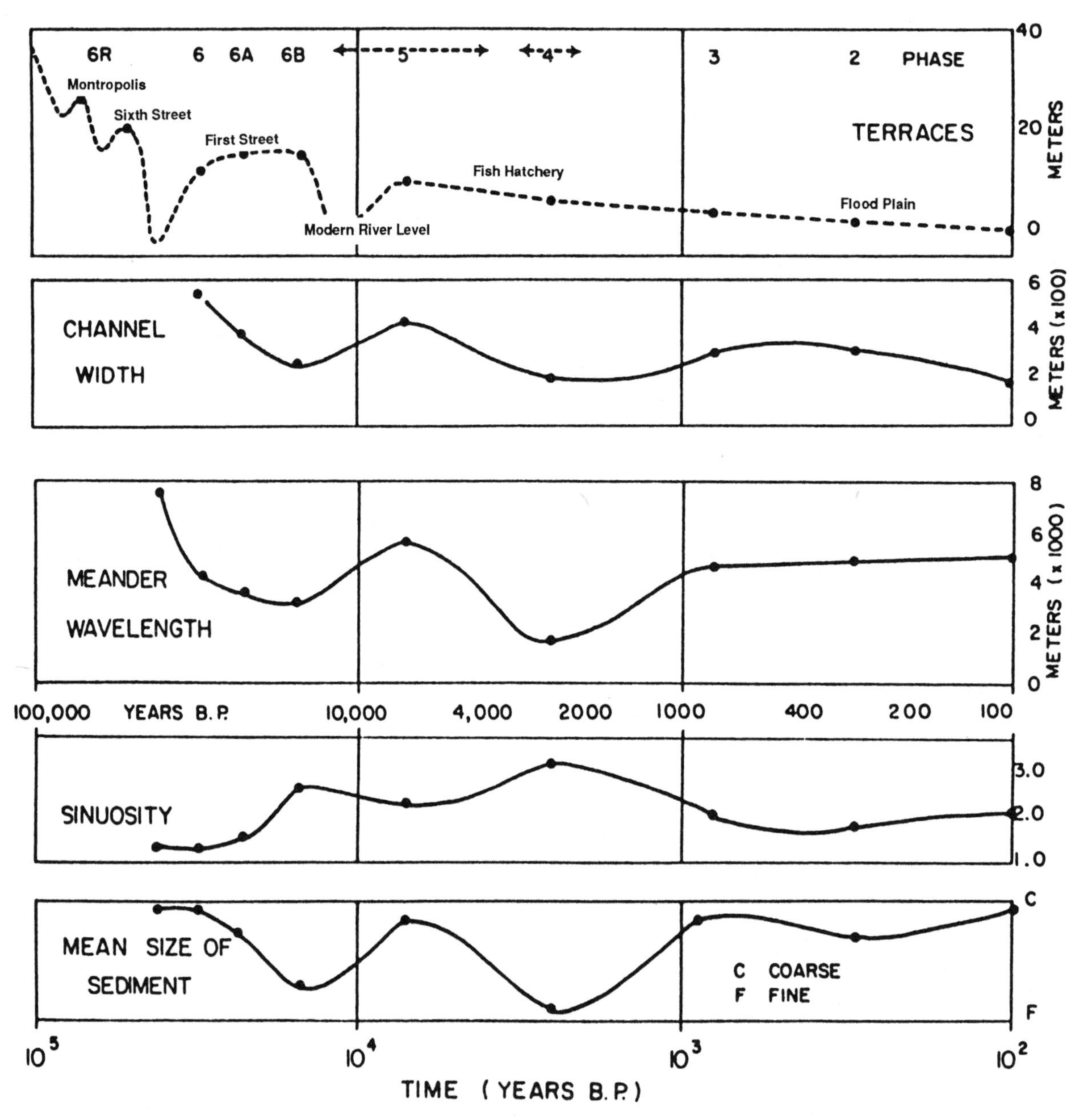

Figure 6. Changes of fluvial variables with time, Colorado River, Texas (from Baker and Penteado-Orellana, 1977).

smaller wavelength meanders were superimposed on the pattern of Figure 4D.

The channel changes displayed in Figure 4 could occur laterally on the surface of an alluvial deposit with channel shift or vertically in a valley fill, as the type of sediment load changed (Fig. 5). If the initial deposition was by a braided stream, both valley and channel gradients were the same, and both were relatively steep. However, as sediment load decreased and aggradation ceased, the river flowed on an alluvial surface that in many cases was too steep, and a sinuous course resulted. This decreased the gradient and increased the frictional resistance to flow within the channel (Dryden and others, 1956, p. 482).

Fisk's (1944) description of the post-Pleistocene changes of the Mississippi River valley between Cairo, Illinois, and the Gulf of Mexico provides a good example. The lowest alluvium in this valley is composed of coarse sand and gravel that may have been transported by a braided river during the last glaciation. In contrast, the modern river (prior to modification for navigation and flood control) meandered over the greater part of its course, and its load was predominantly silt, clay, and sand. Probably the same

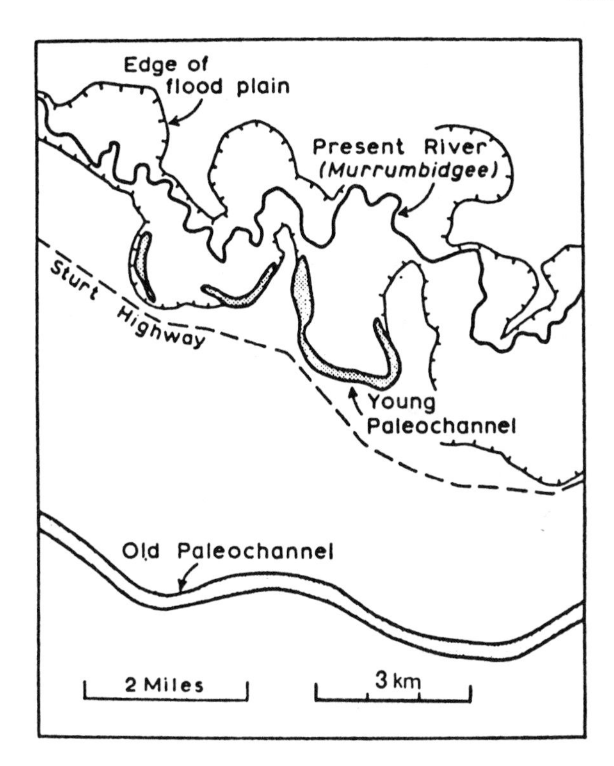

Figure 7. Riverine plains near Darlington Point, New South Wales, Australia. The sinuous Murrumbidgee River flows to the west (left) at the top of the figure (upper arrow). The irregular flood plain contains large meander scars and oxbow lakes (young paleochannel). The oldest paleochannel (lower arrow) crosses the lower part of the figure (from Schumm, 1968).

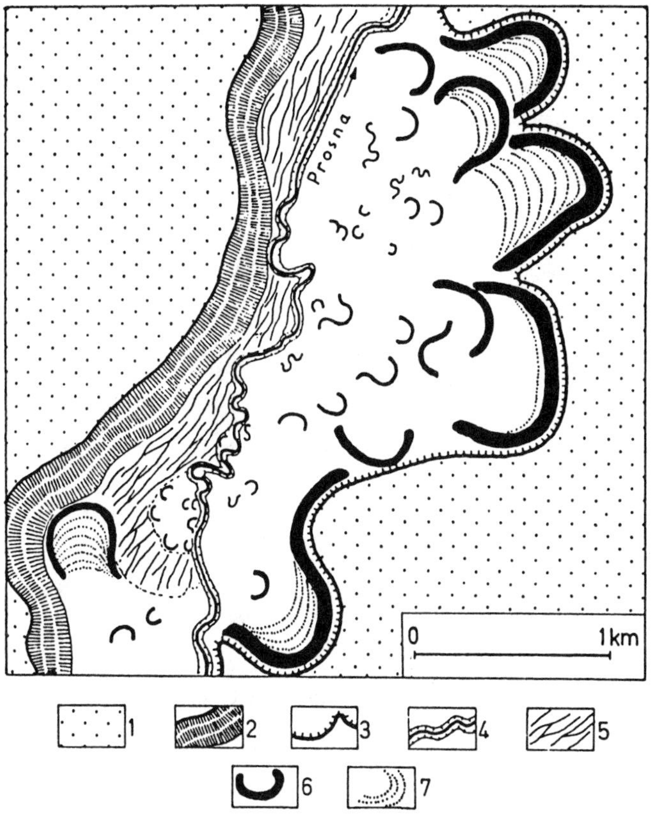

Figure 8. Paleochannels in the Prosna valley at Mirków, Poland. 1, high Pleistocene terrace with braided-channel pattern; 2, and 3, slope to valley floor; 4, modern Prosna River; 5, braided pattern on valley floor; 6, paleomeanders; 7, point bars (from Kozarski and Rotnicki, 1977).

sequence of events occurred in the valleys of the Great Plains, for the alluvium filling these valleys shows an upward decrease in sediment size. Shlemon (1972) describes similar changes of the American River in California as a result of both sea-level change and glaciation in the Sierra Nevada.

Modern rivers draining from glaciers (Maizels, 1983; Church and Gilbert, 1975) and paraglacial areas (Church and Ryder, 1972; Jackson and others, 1982) show how many late Pleistocene rivers appeared before climate change produced a different type of river. Descriptions of channel change as sediment loads change with the addition of suspended-sediment load (Schumm and Khan, 1972; Richards, 1979; Nadler and Schumm, 1981) and bed load (Smith and Smith, 1984) conform to the conclusions presented above. The careful study of the Colorado River valley near Austin, Texas, by Baker and Penteado-Orellana (1977) demonstrates what can be done when an exceptional record of channel change is preserved in an area transitional between moist and dry subhumid climate zones. Large paleomeander scars on the valley sides indicate late Pleistocene entrenchment during a period of high discharge (Fig. 6, phase 6R). This was followed by aggradation (Fig. 6, phase 6) during a period of dryness, when the channels transported coarse sediment and were relatively wide with low sinuosity. The later part of phase 6 (6A, 6B) shows an opposite trend, with a progressive change to high sinuosity and fine sediment loads. Coarse sediment was no longer contributed from eroding tributaries, and incision followed at the end of phase 6. Aggradation during phase 5 produced wide, low-sinuosity channels that were transporting coarse gravel. This was followed by a change to a narrow sinuous channel that transported clay, silt, and fine sand during stage 4. An apparent shift back to a dry climate produced a coarse gravel load and the wider, low-sinuosity channels of phases 3 2, and 1.

On the Gulf Coastal Plain, large paleomeanders of the late Pleistocene Deweyville terrace have been recognized (Gagliano and Thom, 1967; Saucier and Fleetwood, 1970; Alford and Holmes, 1985), which are in striking contrast to the much smaller modern river patterns.

Elsewhere in the world, similar patterns of channel change emerge with braided streams changing to large meandering streams and then to smaller modern channels. The Riverine Plain of New South Wales, Australia, provides an example from an unglaciated region where tectonics and sea-level changes were negligible. Climate change from relatively dry to wet to intermediate conditions is reflected in the changing patterns of the paleochannels (Fig. 7). The oldest paleochannel, which formed under a

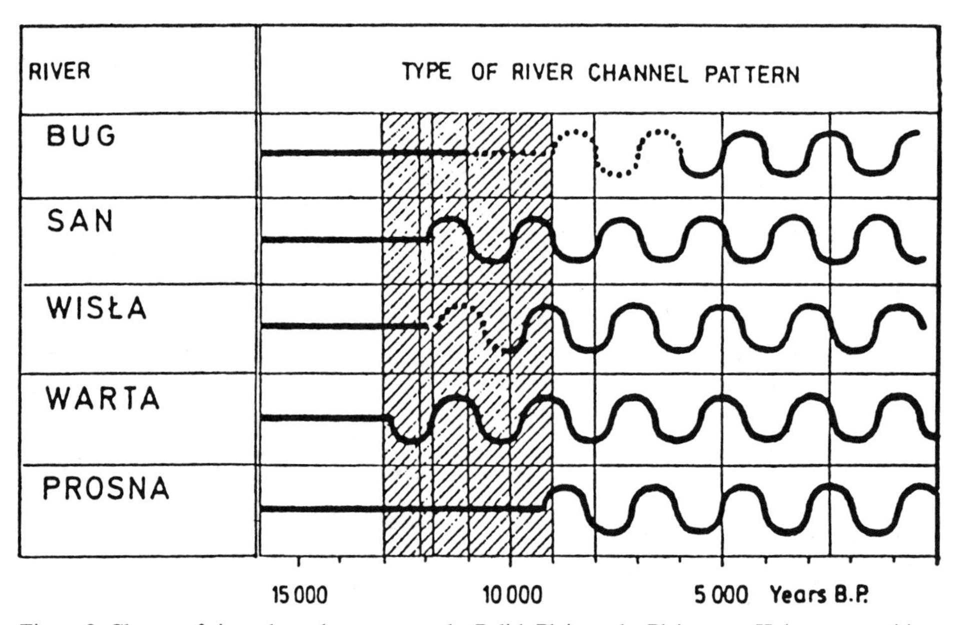

Figure 9. Change of river-channel patterns on the Polish Plain at the Pleistocene-Holocene transition. The straight line represents braiding, curved lines represent meandering, the dotted line represents the probable river condition. The cross-hatched zone represents the period of river change on the Polish Plain (from Kozarski and Rotnicki, 1977).

relatively dry climate, may have been of the pattern 4 type (Fig. 1) rather than braided, but it transported large amounts of sand, and it was a bed-load channel in contrast to the mixed-load younger paleochannel and the modern Murrumbidgee River, which formed under a more humid climate (Schumm, 1968).

The Polish Plain also provides an example of this type of change (Mycielska-Dowgiallo, 1977; Froehlich and others, 1977; Starkel, 1983b) in even more detail, as the Pleistocene braided rivers changed to meandering (Fig. 8). A transition from large meandering channels to the present river conditions can be documented by a decrease in size of the meanders that were preserved by cutoffs, although in some areas the modern rivers show a tendency to braid again as a result of agricultural activities, deforestation, and other factors.

The results of studies of six Polish rivers show that the change from braided to meandering took place between 13 and 9 ka and in some cases perhaps as late as 6 ka (Fig. 9). This lag indicates the great range of channel sensitivity and the influence of other variables in addition to climate and hydrology. This example indicates how variable river response to climate change can be, and how difficult it may be to relate river channel change to other deglacial events.

Alluvial fans can also show the same delayed response to changes of climate and sediment yield with fans in the same area of Idaho being deeply incised, shallowly incised, or not incised, although climatic and tectonic influences were the same (Weaver, 1982; Schumm and others, 1987, p. 343–350).

FLUVIAL SEDIMENTOLOGY

The preceding discussion concentrated on the morphology of measurable channels. This section deals with the sedimentary record of vanished paleochannels. If, as concluded above, deductions about climate change from changes of channel morphology is difficult, even more so is the interpretation of climate change from fragmentary sedimentary records over large areas. Nevertheless, a record of change is present, and this may be more important than the type of change. The following review of the fluvial sedimentary record of continental deglaciation is an attempt to compare fluvial chronologies from several sites within the Mississippi-Missouri-Ohio basin. Portions of this fluvial system are examined in order to describe how they changed during deglaciation. Specifically, parts of the river system were selected that are significantly affected by (1) climate alone, (2) meltwater discharge, and (3) base-level change (Fig. 10).

A simplified model of sedimentation in a meandering-river valley is useful in understanding the alluvial chronologies to be discussed (Fig. 11). The model illustrates how changes in lateral migration rate and the resulting fluctuation between active and stable sedimentary modes (Knox, 1976, 1985) can result in the emplacement of conformable "row terraces" separated by buried paleosols, even while, at other locations along the same river, erosional unconformities (inset terraces) are produced. The resulting stratigraphy could be the product of either system-external or system-internal causes. Also, the stratigraphy will include local

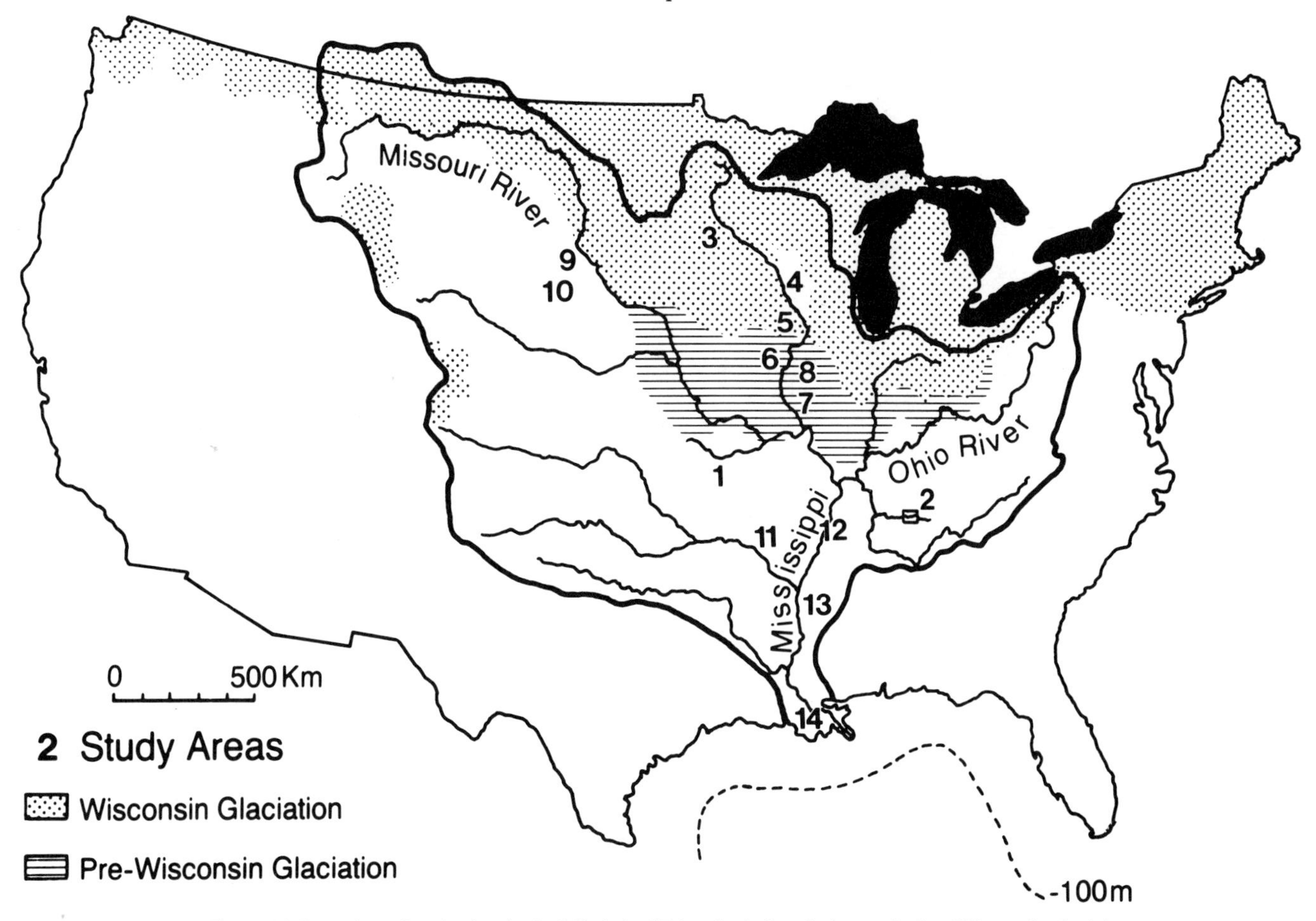

Figure 10. Location of study sites in the Mississippi River basin in relation to the late Wisconsin glacial margin (after Brakenridge, 1981).

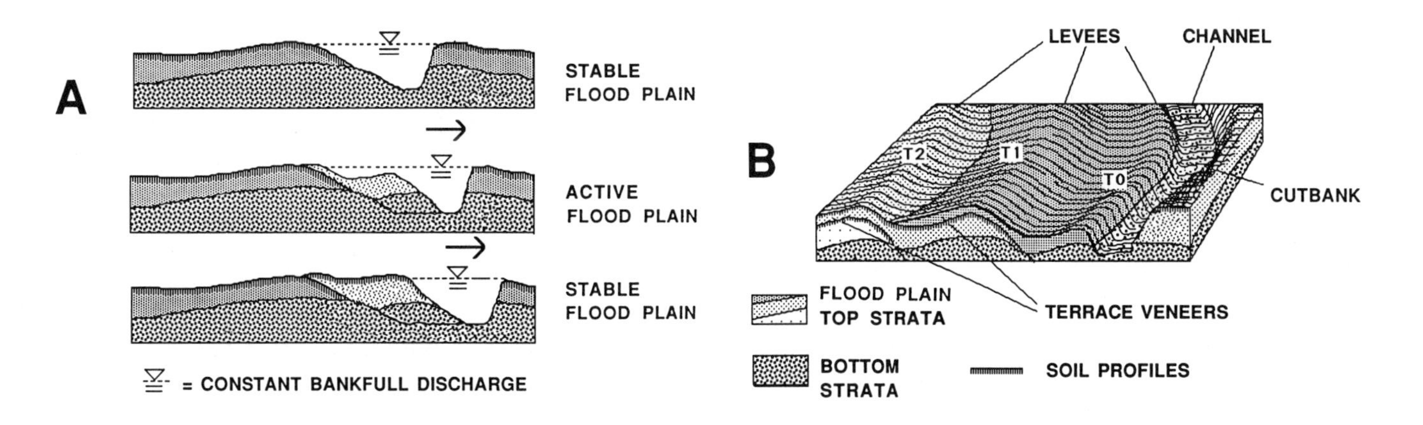

Figure 11. A process-response model of meandering river episodic flood-plain sedimentation and terrace formation. A, shows stable and active sedimentary modes in relation to bankfull discharge. The river is moving laterally to the right, resulting in a conformable sequence of deposits and the production of an erosional unconformity (the cut bank). Reversal of the sense of migration may preserve the unconformity by emplacing much younger sediment adjacent to it. B, shows the long-term sedimentary result of the alternation between these modes, which may be either the result of external perturbations or system-internal mechanisms.

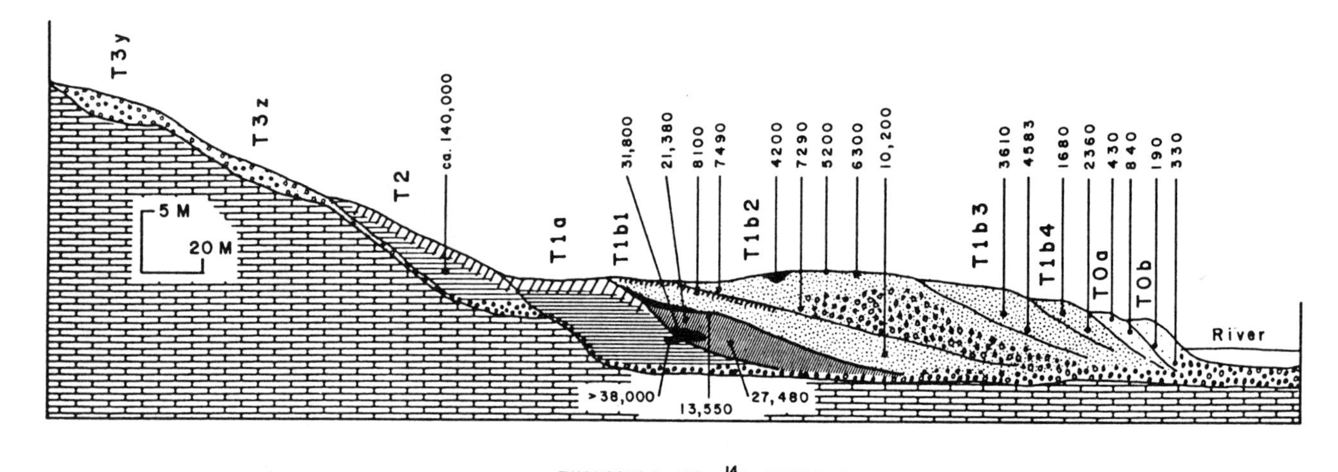

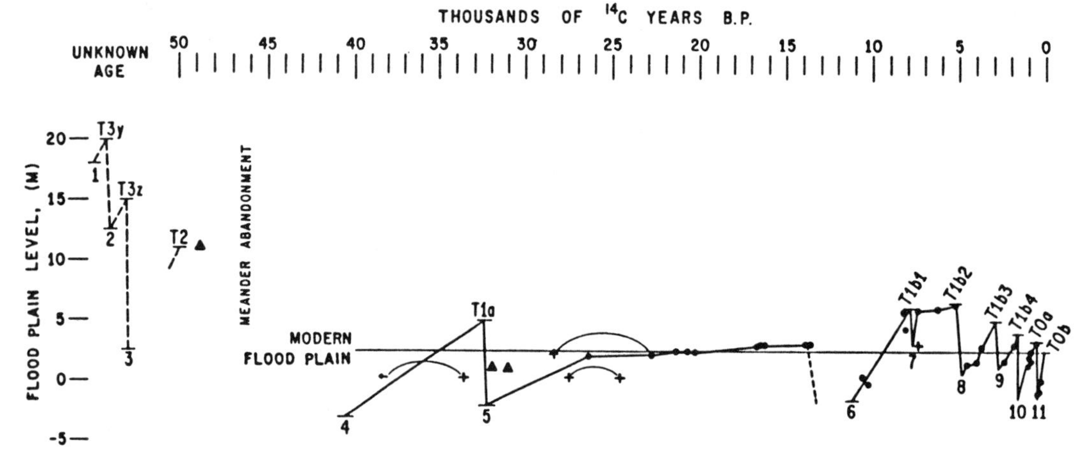

Figure 12. Top: composite cross-section for the Pomme de Terre River, southern Missouri, with radiocarbon dates. Bottom: time-level diagram (from Brakenridge, 1981 and 1983). Terraces and accumulation phases within each terrace are identified by letters and numbers as follows: capital letters and numbers indicate terraces (T0 is historic alluvium, T1, T2, etc.), lower case letters and numbers indicate accumulation phases within the terraces (a1, b2, etc.). See Brakenridge, 1984, Fig. 8, p. 19) for further explanation.

evidence of cut-and-fill cycles, whether or not valley-wide cut-and-fill cycles occurred. The model thus emphasizes the normal episodic pattern of sedimentation along the studied rivers, as well as the possibility of enhanced activity as a response to imposed environmental changes.

Effect of climatic change

Stratigraphic results have been analyzed along the Pomme de Terre River in the Ozark Highlands of southern Missouri (Fig. 10, Site 1) and along the middle Duck River valley in the Nashville Dome region of central Tennessee (Fig. 10, Site 2). Neither of these rivers was directly affected by glacial drainage, and the river channels are presently on or close to bedrock, at elevations of 100 m above the present Mississippi River channel. As a result, fluvial sedimentation in these upland areas was not controlled by changes in base level but instead by the direct and indirect effects of climatic changes in their watersheds.

Complex sequences of Holocene row terraces have been exposed in backhoe trenches and archaeological excavations along both rivers (Figs. 12 and 13). The possible relationship of these terraces to environmental changes depends on whether the documented alternations of active and stable sedimentary modes occurred only locally (at a site) or instead on a valley-wide basis (e.g. Brakenridge, 1987). In contrast, strong changes in the geometry and lithology of sedimentation occurred during the deglaciation period. These events appear to reflect environmental changes, as they cannot easily be explained by the model shown in Figure 11.

1. Along the lower Pomme de Terre River a low flood plain was present adjacent to the river from ca. 23 to 13 ka (Fig. 12). This stable flood-plain surface was underlain by about 5 m of lateral-accretion and overbank olive or gray silty clay, superposed on 2 to 3 m of channel-bed sand and gravel. The very fine grained alluvium, its reduced colors, the presence of carbonate concretion sheets more level than the (subdued) flood-plain surface, and the available radiocarbon dates (see also Haynes,

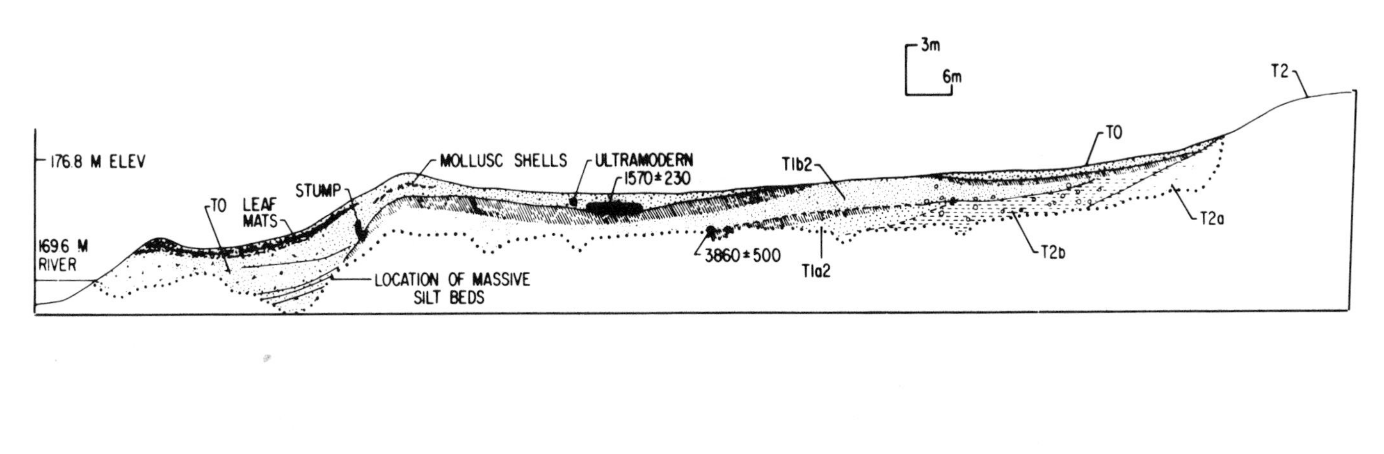

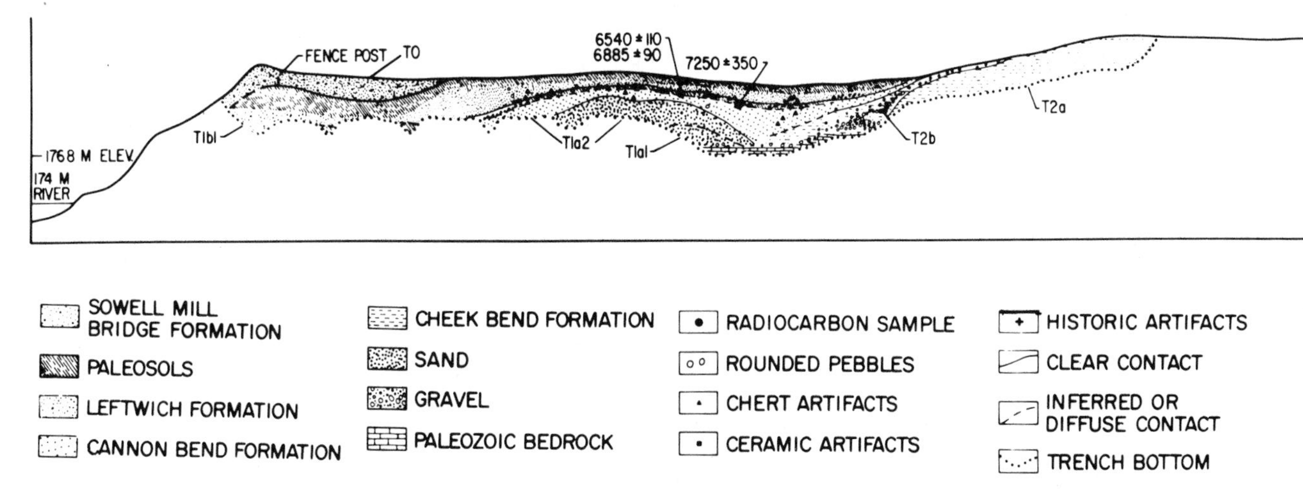

Figure 13. Two trench cross-sections across the Duck River Valley bottom, central Tennessee (from Brakenridge, 1984). For explanation of the T1b1, T2a, etc. nomenclature, see caption of Figure 12.

1985) all suggest very slow accumulation in a paludal environment with high water tables during the time of maximum glaciation (Brakenridge, 1983). The river appears to have been dominated by base flow instead of storm flow, and the ground may have been frozen during much of the year.

In contrast, by 10.5 ka deposition of brown silty bank and overbank facies and brown sandy and gravelly channel-bed facies was underway along the Pomme de Terre River. This alluvium was subsequently reworked at frequent intervals throughout the Holocene (Fig. 12). Relative flood-plain stability thus seems to have been contemporaneous with the glacial maximum, whereas fluvial erosion and sedimentation was much more vigorous during the Holocene (Brakenridge, 1981, 1983). Vertical accretion of flood-plain alluvium was slow during the Late Pleistocene (0.3 mm/yr) and fast in the early Holocene (2.0 mm/yr; Haynes, 1985). Early Holocene alluvium is commonly preserved within the valley, but remnants of the Late Pleistocene alluvium are rare; much of it has been removed (see map in Haynes, 1985). This most likely occurred between 14 and 11 ka.

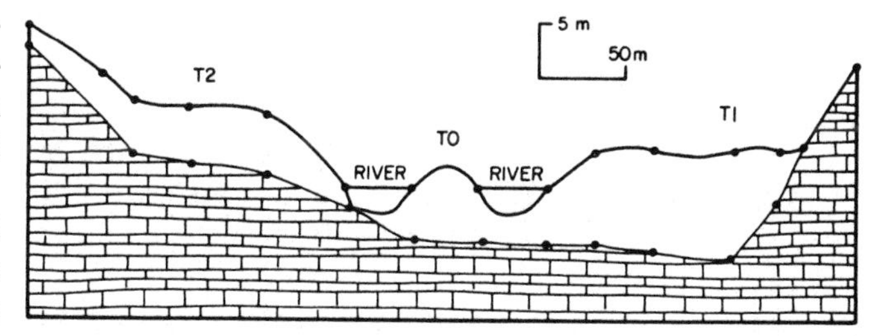

Figure 14. Buried bedrock topography below Holocene (T1 and T0) and late Pleistocene (T2) alluvium along the Duck River.

2. In central Tennessee along the Duck River, a Late Pleistocene (T2) alluvial terrace marks meandering-river sedimentation over a bedrock valley floor about 5 m higher than that of today (Fig. 14). Unlike the case of the Pomme de Terre, late Wisconsin terrace remnants were not identified along the Duck,

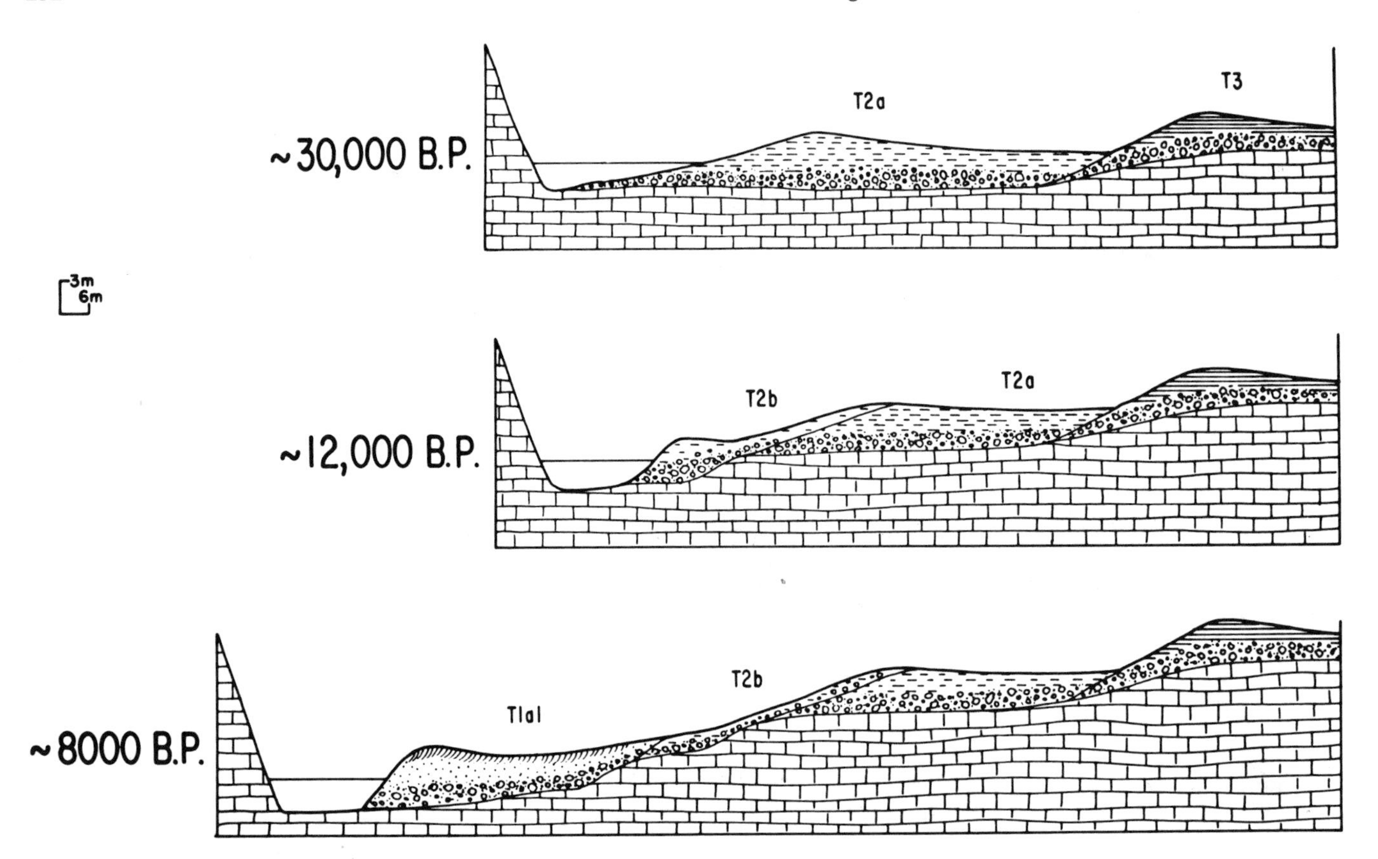

Figure 15. Sedimentary history of the middle reaches of the Duck River valley, central Tennessee during the time period of continental deglaciation. T2b is the terminal Pleistocene-age wedge-shaped unit (modified from Brakenridge, 1984). For explanation of the T2a, T3, etc. nomenclature, see caption of Figure 12.

despite numerous trenches dug during the exploratory stages of the work. Instead, at the T2 terrace scarp a wedge-shaped pebbly silty clay unit is commonly present, which is bounded at its base by a pebble gravel sheet that laterally truncates the T2 sediments. The wedge-shaped pebbly unit is itself overlain conformably by Holocene T1 sediments. The latter contain early Archaic (ca. 10 to 8 ka) artifacts. At several locations, artifacts correlated with the Paleoindian Period (ca. 12 to 10 ka) have been found within the wedge-shaped unit. The unit apparently represents the preserved outlying remnants of a latest Pleistocene flood-plain surface. A reasonable inference is that extensive fluvial erosion at the end of the Pleistocene removed much of the Late Wisconsin sedimentary record. Figure 15 illustrates the inferred sedimentary history as reconstructed from the preserved stratigraphy.

In summary, along both rivers it is easy to map Holocene sediments (<10 ka) inset against or burying sediments of late (but not terminal) Pleistocene age (~30 ka). Strong lithologic contrasts between such alluvia permit mapping of "late Pleistocene" and "Holocene" alluvial fills and surfaces. However, the actual chronology of events, as revealed by trenching and time-stratigraphic investigations, was more complex than such maps would indicate. Terminal Pleistocene units are difficult to locate in both valleys even by extensive trenching: a portion of the record appears to be missing. Fluvial erosion may have removed many terminal Pleistocene deposits along both valleys between 14 and 11 ka.

After this erosional phase, both rivers aggraded their valleys. In Missouri, the rapid deposition rates and brown silty nature of the early Holocene flood-plain sediments apparently reflect large inputs of reworked loess (Haynes, 1976, 1985; Brakenridge, 1983). Thus, the eolian transfer of fine-grained Late Wisconsin river sediment to the unglaciated carbonate uplands in southern Missouri may have set the stage for the subsequent early Holocene aggradation of this valley.

Effect of meltwater discharge

The upper portions of the Missouri, Mississippi, and Ohio rivers all received large amounts of meltwater and sediment during deglaciation. In the glaciated northern portions of their catchments the advance and retreat of continental ice sheets also resulted in periodic rearrangement of whole drainage networks. Aggrading trunk streams fed by glacial debris and meltwater impounded their tributary valleys, producing lakes. Intermittent draining of these lakes (or those formed by morainal dams) caused catastrophic floods that carved deep valleys. Younger valley networks were superimposed over earlier systems already filled by glacial or glacio-fluvial sediment. These palimpsest networks of buried valley systems form the complex framework for the events of the last deglaciation.

Mississippi River. A general model of deglacial events for the lower Mississippi River was described by Fisk (1944): entrenchment during the onset of continental glaciation caused by lower sea level was followed by braided river aggradation during deglaciation and then by metamorphosis to a meandering river sometime during the early Holocene (Baker, 1983). The braided-river phase caused damming of those major tributaries not fed by glacial meltwater, thereby inducing lake-clay deposition in the downstream reaches of these valleys (e.g., the Arkansas River; Saucier, 1974, 1978).

A major shift in the direction of meltwater drainage occurred between about 14 to 11 ka, when much meltwater began draining eastward through the Great Lakes to the Hudson River and the St. Lawrence Valley (Baker, 1983). Large amounts of meltwater flowed down the Mississippi valley prior to the diversion, and this influx significantly altered the Gulf of Mexico's surface-water oxygen-isotopic composition (Leventer and others, 1982). Braided-river aggradation may have ceased at approximately the time of diversion.

The actual chain of events during deglaciation were complex in detail. In particular, local chronologies of sedimentation and erosion are more easily related to the complexities of sediment routing and to deglacial flood events than to the perhaps too-general picture of sea-level rise and late Wisconsin aggradation painted above. We review several studies from north to south.

Flock (1983) indicates that the upper reach of the Mississippi during deglacial time was aggraded by outwash sediments now underlying the Savanna Terrace, which extends at decreasing heights from Minnesota at least as far downstream as southern Illinois. This braided-river aggradation was abruptly terminated about 12 ka by deep dissection (85 m near St. Paul, Minnesota; Fig. 10, Site 3; Wright, Chapter 22) caused by the discharge from the Glacial River Warren outlet of Lake Agassiz. Then, when the lake's outlet was diverted to the east, about 9.2 ka, deposition began, with an alluvial thickness of 50 m eventually being reached in the St. Paul area.

In southwestern Wisconsin (Fig. 10, Site 4) Knox (1985) concludes that 150 to 200 m of Mississippi valley entrenchment had occurred before Wisconsin time, that 21 to 12 ka witnessed major aggradation, and that a major degradational event occurred between 12 and 9.5 ka. Since then, he infers only a relative minor accumulation of alluvium.

In the upper Mississippi valley in eastern Iowa (Fig. 10, Site 5), Bettis and Hallberg (1986) conclude that aggradation dominated along the Mississippi when the late Wisconsin ice sheet occupied headwater areas (about 29 to 18 ka). This depositional phase was followed by a complex sequence of aggradation and degradation in response to outwash deposition and catastrophic lake-drainage events.

For the central Des Moines valley in Iowa (Fig. 10, Site 6), Bettis and Benn (1984) and Bettis and Hallberg (1986) describe episodic downcutting of some 48 to 68 m between 12.6 and 11 ka, followed by metamorphosis to a meandering stream as the input of glacial meltwater ended. No catastrophic lake drainage event is identified for this erosional interval. Between 10.5 and 4 ka, aggradation occurred, followed by an incision event at about 4 to 3.5 ka, and subsequent formation of row terraces.

Hajic (1986) describes how the Lake Michigan ice lobe diverted the Mississippi River out of the present Illinois River valley at about 20 ka (Fig. 10, Site 7). The Illinois valley then aggraded as a major proglacial system, and valley train deposits were being "extensively remodeled into several terrace levels" (Hajic, 1986, p. 624) by ca. 15 ka. But more rapid aggradation along the much larger Mississippi soon impounded the Illinois valley (at about 13.3 ka). Three episodes of Illinois River downcutting and aggradation between 12.2 and 9.8 ka, again in response to Mississippi River fluctuations, terminated the backwater lakes. Most of the Illinois River today exhibits a very low-gradient longitudinal profile and flows along its old lake plain; its broad shallow flood plains, yazoo streams, and well-developed levees are typical of the lower Mississippi valley, suggesting the river's aggradational response to a slowly aggrading Mississippi River (Hajic, 1986).

A tributary to the Illinois River, the Sangamon River (Fig. 10, Site 8), also drained the Late Wisconsin ice (Miller, 1973). Rapid aggradation of this valley by sand and silt in a meandering river was underway at ca. 30 to 20 ka. This deposition filled a valley ca. 25 m deep and dammed non-glacial tributaries (Units I and II, Fig. 16). The filling is attributed to contemporaneous rapid aggradation of the Illinois River, its local base level. Alluviation of the Sangamon valley slowed at about 25 to 20.5 ka, when the Mississippi River was diverted into its present valley. During the Late Wisconsin ice advance into this river's watershed, extensive scouring of these sediments occurred, even to bedrock in many areas (Miller, 1973). However, by about 20 ka enough outwash sediment was entrained at the glacier margin to cause extensive sand deposition over the scoured surfaces (the Henry Formation, Fig. 16). When this aggradation overtopped Unit I deposits, tributary valleys were impounded and lacustrine sediments accumulated (20 to 14 ka). Finally, after the glacier margin retreated out of the basin, the Sangamon River underwent a change to a meandering regime, the lakes were drained, and the silt facies of the early Holocene Cahokia Alluvium was deposited.

The clay mineralogy and overall grain-size characteristics of the Cahokia Alluvium along the Sangamon River are similar to those of the adjacent Peoria Loess of Late Wisconsin age

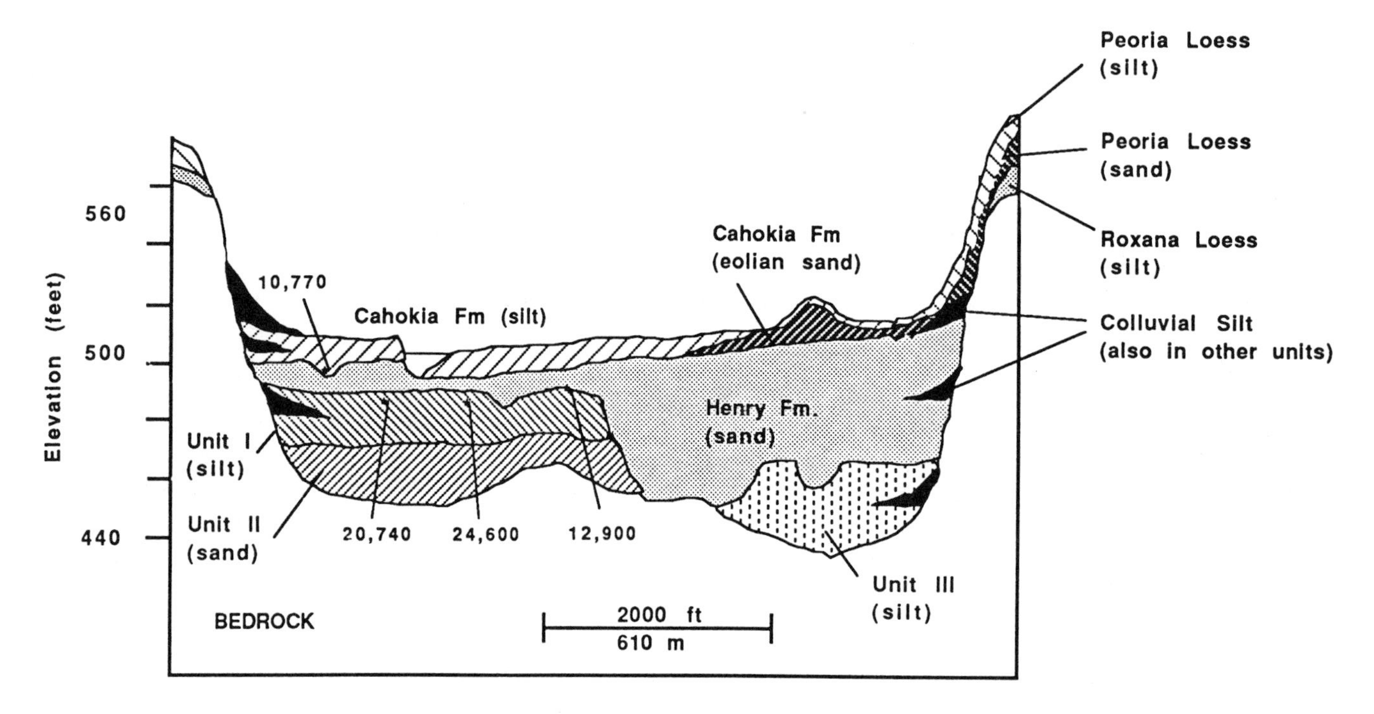

Figure 16. Diagrammatic cross section across the Sangamon River valley (from Miller, 1973).

(Miller, 1973). On this and other evidence, Miller (1973) concludes that the loess acted as the primary sediment source for the early Holocene alluvium.

In summary, it appears to be well established that Late Wisconsin glacial maximum and early deglacial time were the periods when a large amount of braided-river aggradation occurred along the upper Mississippi River and its major outwash tributaries. Such aggradation appears most clearly associated with general glacial retreat, as at least one study documents that during the late Wisconsin ice advance significant scouring, instead of deposition, occurred (Miller, 1973). Tributary rivers fed by little or no outwash were impounded as their base levels rose. The aggradational phase was followed during deglacial and early Holocene time by perhaps several intervals of valley-train entrenchment and formation of multiple outwash terraces.

Given the complex nature of the alluvial record, the clearest evidence for a large pulse of meltwater discharge down the Mississippi River at ca. 14 to 12 ka is still the Gulf of Mexico oxygen-isotope data (Leventer and others, 1982). This event clearly agrees with the intervals of terminal Pleistocene entrenchment reported at most of the sites. Large catastrophic lake-drainage events occurred during this period from glacial Lake Agassiz westward to Saskatchewan, creating distinctive long spillways with gross sedimentary features (Matsch, 1983; Kehew and Lord, 1986) Continued ice retreat eventually reduced the supply of sediment and meltwater, and extensive "remodeling" of valley landforms may have occurred immediately prior to the Holocene (Hajic, 1986).

Lastly, an early Holocene depositional interval of fine-grained sediment is inferred for several of the study sites. Such aggradation cannot be directly tied to ice-sheet dynamics. A vast amount of glaciation-related sediment, especially loess, was still available for sedimentary transport. As suggested by White (1982) for eastern South Dakota, but perhaps applicable over much of the upper Mississippi basin, loess deposits on slopes and interfluves may have been relatively stable or undergoing aggradation during the last part of the late Wisconsin glaciation. The early Holocene then witnessed the large-scale transport of mainly fine-grained sediment from the hillslopes to the valley floors. The Holocene meandering rivers that occupied the Mississippi and other valleys responded by forming a younger flood-plain deposit that is inset within the glacial-maximum flood-plain sediment or is superposed on it. The early Holocene deposits are commonly bounded at their base by a terminal Pleistocene erosional unconformity. The existence of these deposits as fill terraces suggests a complex response (entrenchment, then valley filling, and then entrenchment into the fill). The initial external perturbation to the system appears to have occurred at ca. 12 ka and is probably related to climatic change.

Missouri River. Because of large inputs of sediment from the Rocky Mountains, the upstream (and especially western) reaches of the Missouri valley and tributaries have been in an aggradational mode for the latter part of the Cenozoic. Such aggradation was intensified during glacial stages, and locally over-steepened reaches were incised (e.g., Moss, 1974). In contrast, reaches farther downstream, beginning in the Dakotas, oc-

Figure 17. The entrenched valley of the Cheyenne River near Pierre, South Dakota.

cupy a deep valley along the western margin of Pleistocene glacial topography. Steep-sided gullies and knife-edged ridges descend toward the floor of the trench. Several major tributaries to the Missouri River also flow within valleys trenched below the surrounding relatively level topography (e.g., the Moreau, Cheyenne, and White rivers in South Dakota; Fig. 17). It appears likely that these rivers responded to Missouri River base-level changes. The timing of incision of the "Missouri Trench" and the attainment of the modern channel level are thus an important regional question, but one that has not been adequately resolved.

Working along the Missouri River southeast of Pierre, South Dakota (Fig. 10, Site 9), Coogan and Irving (1959) describe three alluviual surfaces: one at about 30 m above river level (with 10 m of eolian cover), one at about 12 m (with a thinner eolian cover), and a third at about 3 m, which is the modern flood plain. Although the entrenchment is clearly of late Pleistocene age (Coogan and Irving, 1959), it is not known if late Wisconsin deposits occur buried below the lowest level or if only Holocene sediments are present. Coogan and Irving (1959) note that drilling at one dam site revealed the bedrock was 20 m deeper beneath the valley floor than beneath the Pleistocene terraces; thus, the entrenchment episode did not merely reexcavate a buried valley but deepened the valley to a level not previously attained. After this erosional interval, ca. 24 m of aggradation occurred at the present dam site.

Some radiocarbon control is provided by studies of South Dakota tributaries. Working close to the White River (Fig. 10, Site 10), White and Hannus (1985) describe three episodes of Pleistocene incision and a late Pleistocene (10.7 ka) channel level similar to that of today. As was the case at other locations in the Mississippi basin, the early Holocene was a time of net aggradation of fine-grained alluvium. Subsequent to 5 ka the late Holocene cut-and-fill cycles repeatedly eroded this alluvium down to about the level of the late Pleistocene channel. This study suggests the possibility that at least thin Late Wisconsin alluvium may be buried by Holocene alluvium at low elevations along the Missouri as well.

Effect of Base-Level Change

At its junction with the Ohio River, the lower Mississippi River enters the structural low known as the Mississippi Embayment. Net sedimentation by the Mississippi during the entire Cenozoic has constructed an extensive alluvial bottomland. The highest elevation reached by this accumulation lies less than 100 m above modern sea level, and the surface slopes gently to the south. The manner in which sedimentation along the lower Mississippi River was affected by the combined processes of glaciation in the northern basin and sea-level changes in the southern basin remains controversial.

Sedimentologic and oceanographic studies of the Gulf Coast, the adjacent continental shelf, and the Mississippi delta, beginning in the 1960s and continuing to the present, have demonstrated the limited effects of sea-level change on the river itself. Modern coastal facies models emphasize progradational shorelines and deltas in this region, instead of vertical aggradation, and the most significant effect of sea-level change is held to be changes in the positions of different coastal or deltaic facies. Because of the gradual slopes of the Gulf of Mexico continental shelf and the adjacent coastal plain, lowering of sea level less than 100 m does not significantly alter the longitudinal gradients of rivers emptying into it. Thus, although there is strong evidence that the lower Mississippi River is aggrading today, there is no reason to believe that this regime must be restricted to postglacial or deglacial time periods. Nevertheless, the following examples illustrate the degree to which workers studying the lower Mississippi have been influenced by the sea-level model of Fisk (1944) even while describing problems with it, especially the lack of documented vertical incision during the last low sea level (Saucier and Smith, 1986).

Delcourt and others (1980) dated organic sediments ranging from 22.3 to 17.2 ka in deposits within 2 to 3 m of the flood plain of Noconnah Creek, near Memphis, Tennessee (Fig. 10, Site 12). Noting that this result seems to imply an absence of net valley aggradation after that time, Grissinger and others (1982, p. 157) attempt to reconcile it with clear evidence for Holocene aggradation downstream. They write that Memphis "has been a fulcrum of changes in the flood-plain elevation of the Mississippi River over the past 12,000 years, the valley below Memphis aggrading and that north of Memphis degrading." Thus, Saucier (1974) inferred 23 to 24.5 m of post–12 ka (but early Holocene) Mississippi valley aggradation near the Gulf Coast, 6 to 8 m of aggradation near Vicksburg, Mississippi, and net degradation during the Holocene upstream from Memphis. The aggradation downstream was attributed by Grissinger and others (1982, p. 157) to the post–12 ka rise in sea level of ca. 30 m.

This seemingly reasonable conceptual model appears to lack abundant corroborating documentation. We located no clear evidence for Holocene net erosion in the lower Mississippi valley above Memphis. As noted above, the late-glacial sea-level rise may have resulted primarily in the inland movement of existing deltaic and coastal facies, rather than vertical fluvial aggradation upstream. It is more probable that other factors, such as slow changes in the lateral position of the Mississippi channel, rapid changes by avulsion, or changes in the sediment supply available to Nonconnah Creek, played important roles in determining how thickly the 22.3 to 17.2 ka sediments were locally covered.

In this respect, during the Farmdalian and early late Wisconsin substages, the Mississippi valley in eastern Arkansas contained prominent braided streams (Saucier, 1974, 1978). As was the case for upstream unglaciated tributary rivers, Mississippi valley aggradation dammed the lower end of the Ouachita River of Arkansas and Louisiana (Saucier and Fleetwood, 1970; Fig. 10, Site 11). The Ouachita River subsequently trenched the lacustrine plain and developed a set of terraces that are not well dated. At about 17 ka the Mississippi was diverted from its valley train location in eastern Arkansas; the zone of outwash deposition shifted eastward to its present position (Saucier, 1974). As a result, the Nonconnah Creek site could record the same aggradational surface that dammed tributaries such as the Ouachita much earlier.

In the Yazoo–Little Tallahatchie watershed east of the Mississippi Alluvian Plain (Fig. 10, Site 13), Grissinger and others (1982) and Grissinger and Murphey (1983) describe a massive early Holocene silt that is approximately 4 m thick. Meander-belt deposits (Harvey and others, 1981) were buried by the silt at ca. 10 ka. No dates are available from within the initial aggradation, but subsequent cut-and-fill cycles left wood ranging in age from 4.1 to 6.1 ka.

The massive silt is interpreted as a low-energy fluvial deposit (Grissinger and others, 1982) resulting from periodic near-impoundment by an also aggrading Mississippi River (with such aggradation being again related to sea-level rise). However, Harvey identified this deposit as lacustrine (Harvey and others, 1981; Schumm and others, 1984, p. 116). Reworked loess is part of this sediment, and brown or yellowish brown massive loesslike silts underlying early Holocene (10 to 8 ka) river terraces were reported at sites 1, 2, 8, and 10 discussed above, as well as at other localities not discussed in this report (e.g., streams in the Driftless Area of Wisconsin; Knox and others, 1981). Also, direct field relations (e.g., Fig. 16, Peoria Loess and Cahokia Alluvium) and mineralogical comparisons with known loess deposits (e.g., Miller, 1973; Brakenridge, 1983) support such alluvia as being largely of loessial origin. Therefore an eolian source for these silts is reasonable (Grissinger and others, 1982). Early Holocene aggradation may again have been related to a hillslope sediment supply greatly exceeding the capacity of these tributary streams to transport sediment. Subsequent cut-and-fill cycles could represent complex responses to this external perturbation.

Grissinger and others (1982) and Grissinger and Murphey (1983) also infer an important episode of net erosion immediately preceding the Holocene, and not long before 12 ka. They interpret this event as a combined result of low sea level and postglacial pluvial conditions; this erosion is coincident with enhanced meltwater discharge independently inferred for this time.

The most distal portion of the Mississippi basin is the delta itself. Yet the critical events of the deglacial time period are poorly known largely becuse late Wisconsin deltaic sediments are now submerged on the continental shelf and are also covered by enormous quantities of Holocene sediment.

Simplified bathymetric contour maps of the inferred late Pleistocene surface indicate that approximately 150 m of Holocene accumulation occurred in response to a sea-level rise of only about 100 m (Fisk and McFarlan, 1955). The implication is that at least 50 m of isostatic subsidence of the late Pleistocene surface has occurred in Holocene time due to sediment loading. This local subsidence, and other factors, complicate attempts to read sea-level change from Gulf Coast deltaic related facies.

Seismic profiles and core data (not reviewed here) hold the key to a better understanding of the evolution of this delta during deglaciation. However, it is possible to suggest that the following sequence of events must have occurred in the Mississippi Delta during the late Quaternary. (1) The Wisconsin glacial stage, with accompanying lower sea level, exposed a large deltaic accumulation as a positive topographic feature raised about 100 m over the surrounding, relatively flat portions of the emergent shelf. (2) Stream entrenchment, perhaps including the Mississippi, caused fluvial erosion of this topographic high during the Wisconsin. If significant quantities of sediment were removed, isostatic rebound enhanced the erosional nature of these streams. (3) Given the known Holocene "switching" behavior of the delta, it is probable that the main channel of the Mississippi was diverted, by avulsion or by capture, to a lower route to the sea, one that avoided the higher portions of the relict delta. (4) Also during the Wisconsin a new delta probably began to form near the present margin of the continental shelf. (5) The subsequent sea-level rise during deglaciation moved the locus of deltaic accumulation to the north (inland), probably to approximately the same location as older deltaic wedges. Any preserved earlier deltaic sediments and the bounding erosional topography were then covered by new accumulations.

This scenario emphasizes the possible internal complexity and long geologic history of the Mississippi Delta. It also suggests that the deltaic sedimentary records, especially of late-deglacial time, are in a difficult place to study and date—at the interface between the very thick Holocene deposits and those predating the last glaciation. They may also be nearly absent; rapid sea-level rise near the end of the late Wisconsin may have prohibited a long-lived locus of deltaic deposition until the early Holocene. A better strategy may be to search for proxy sedimentary records of the lower Mississippi River that represent the last deglaciation in adjacent marine environments of the modern continental slope and rise fringing this portion of the Gulf.

DISCUSSION

Any attempt to generalize about river responses to environmental change is difficult not only because of the great diversity of rivers (Fig. 1) and the different initial climatic conditions (Fig. 3), but also because of the varying "typology" of river systems (Starkel, 1979). Starkel identifies six major conditions and controls. These may be applied to North America as follows.

1. Polyzonal conditions: (a) rivers flowing from dry to wetter conditions (Great Plains rivers and parts of Missouri River); (b) rivers flowing from relatively wet to dry conditions (Rocky Mountain streams, Sacramento River).

2. Glacial conditions: (a) rivers with glaciers in the headwaters (Rocky Mountain streams; Mississippi, Ohio, and Missouri rivers); (b) rivers blocked by glaciers (Clark's Fork, Ohio and Missouri rivers); (c) waters diverted in and out of a river by glacial disruption/drainage (Columbia and Snake rivers).

3. Base-level change: (a) sea level (coastal rivers); (b) inland seas and pluvial lakes (Basin and Range valleys, Humboldt River, and aggraded tributary valleys of the Mississippi, Missouri, and Ohio rivers).

4. Glacioisostatic influences: (a) uplift at mouth, reducing gradients; (b) uplift in headwaters, steepening gradient.

5. Tectonics: changing valley slope (lower Mississippi River, Gulf Coast rivers).

6. Cultural disturbance: (no known North American examples for deglacial time).

As our review indicates, many of the above factors complicate the record and prevent correlation of terraces and alluvial deposits along the length of large rivers.

In addition, the problems discussed earlier create added difficulties: (1) different types of rivers (Fig. 1) will have different responses; (2) effects of a similar climate or hydrologic change (Figs. 3, 9) are different; (3) river sensitivity will greatly influence how a river adjusts to an external influence (Figs. 2, 9); and (4) once a change has been initiated the response will be complex, and this may greatly complicate the interpretation of the causes of change. All of the above problems complicate general statements about the effect of climate change on river morphology and alluvial stratigraphy and about the identification of climate change from channel changes. Nevertheless, with careful consideration of river typology and of the above four causes of variability, the type of river under consideration can be identified (Fig. 1) and its likely response estimated.

It cannot be expected that changes in fluvial system will be orderly and progressive. If a very complex system is perturbed by climate change, the response will be complex.

This drastic change caused by incision may institute a complex series of events that lead to complicated erosional and depositional histories, each component of which could be attributed to climatic change. However, such changes are inherent as a channel adjusts to changed base level and sediment loads. Such episodic behavior has been identified in the valleys of Douglas Creek, Colorado (Womack and Schumm, 1977); Little Sioux River, Iowa (Hoyer, 1980a, 1980b); the Bear and Yuba rivers, California (Wildman, 1981); on the Zeev Elem alluvial fan, Israel (Bowman, 1978); and possibly in Alaska (Ritter, 1982). The complexity of river response to climate change or other factors has also recently been described by Love (1979), McDowell (1983), Patton and Schumm (1981), and Boison and Patton (1985). This leads to problems of terrace and alluvial-fill correlation, for events in one valley may not be directly correlatable with those in another nearby valley (Waters, 1985). Nevertheless, changes in channel morphology or sedimentology can mark the times of significant hydrologic change related to glacial and climatic history. It is clear that both climatic change and complex response must be considered at each site.

Along the Pomme de Terre and Duck rivers, whose basins were not glaciated, strong lithologic contrasts permit surficial mapping of "late Pleistocene" and "Holocene" alluvial fills and surfaces (terraces). However, detailed radiocarbon dating, supported by subsurface trenching and cross-section mapping, demonstrate that common "late Pleistocene terraces" actually predate the last glacial maximum. Glacial-maximum sediments instead have been eroded away or, where preserved, are commonly covered by early Holocene alluvium. Along these rivers a degradational interval (between approximately 14 to 11 ka) removed these sediments, an early Holocene aggradation buried the remnants, and frequent episodes of Holocene fluvial activity subsequently occurred. the "style" of fluvial sedimentation during the glacial maximum was much different from that of the Holocene: aggradation rates were slower and water tables were higher. Also, the early Holocene alluvium appears to have been derived largely from loess. The sequences suggest the impact of environmental changes on hydrology and sediment yield: high-stage discharges may have been larger or more frequent at the close of the Pleistocene, and local hillslopes may have been stripped of previously stable loess deposits in early Holocene time, possibly in response to the same environmental change.

For the variety of Mississippi River sites affected by outwash, the alluvial chronologies are compatible with termination of braided-river aggradation (glacial outwash) about 12 ka. An interval of subsequent deep dissection was noted at several sites (60 m by St. Paul, Minnesota), followed by early Holocene aggradation (20 m at the same location). For this example, the degradation was apparently related to the draining of Lake Agassiz, but at other locations (e.g., central Des Moines valley in Iowa, with 48 to 68 m of erosion between 12.6 and 11 ka) it was not associated with catastrophic lake drainage. Such changes may be directly related to ablation of the terminus of the continental ice, to the direct effects of climatic change, or to aggradational/degradational trends underway along the regional base level, such as the Mississippi River.

Along rivers affected by base-level change, including the Mississippi itself, the sea-level model of Fisk (1944) does not appear well supported by recent evidence. According to Fisk, entrenchment of the lower Mississippi River during the onset of continental glaciation was caused by lower sea level, and subsequent aggradation by the braided river was the result of rising sea level coupled with abundant sediment supply during deglaciation. However, there is little evidence for the deep erosional phase (Saucier and Smith, 1986), nor does it appear that a regime characterized by net aggradation (due to the regional tectonics) has been restricted to postglacial or deglacial time periods. Because of the gradual slopes of both the lower Mississippi Valley and the Gulf of Mexico continental shelf, it is also not clear that an entrenched phase should have occurred along the lower Mississippi during the last glacial maximum. Recent detailed dating and stratigraphic studies by Grissinger and others (1982) in the adjacent Yazoo–Little Tallahatchie watershed do document a period of erosion not long before approximately 12 ka, followed by early Holocene aggradation by silt that was possibly derived from loess.

Grissinger and others (1982) interpret the degradational event to a combination of low sea level and "post-glacial pluvial conditions." We note that it agrees with (1) fluvial erosion dated along Pomme de Terre and Duck river valleys between 14 and 11 ka, (2) river entrenchment in outwash-dominated valleys at approximately 12 ka, and (3) the change in Gulf of Mexico oxygen-isotopic composition at between 15 and 12 ka (Kennett and Shackleton, 1975), suggesting an influx of meltwater. A possible inference is that the same climatic change that made possible deglaciation and the associated glacial-lake drainage also directly affected river hydrology, even in southern unglaciated watersheds.

In this respect it is important to note that a late-glacial "pluvial" during the time of the melt-water influx has been inferred by other studies, including those on the Colorado River in south Texas (Fig. 6). Detailed morphological and sedimentological studies of that river's late Quaternary history (Baker and Pentado-Orellana, 1977) delineated eight alluvial terraces. The terraces range in inferred age from Sangamon (Terrace 8) to Holocene (Terraces 2 to 5), and Terrace 5 is dated at ca. 11 to 7 ka by correlation with sequences along the San Gabriel River. Terrace 6 is believed to be of latest Pleistocene age, and it exhibits very large relict channel meanders of low sinuosity and relatively coarse sand and gravel. The younger terraces are characterized by high-sinuosity paleomeanders and common silt and clay facies. Baker and Penteado-Orellana (1977) infer that discharge was high during the deposition of Terrace 6 sediments when the climate shifted from a drier regime approximately contemporaneous with the time of glacial maximum. Discharge declined again near the end of the Pleistocene (the Terrace 6–Terrace 5 transition). They interpret the responsible climatic change as the result of rising Gulf of Mexico sea levels, overall enlargement of the Gulf, and a 1° to 2° C rise in the mean Gulf sea-surface temperature during Wisconsin deglaciation, all of which would favor a precipitation increase in regions affected by Gulf air masses. Also, at the westernmost extent reached by these Gulf air masses in southwestern U.S., Spaulding and Graumlich (1986) document a 12 to 8 ka pluvial climate from paleobotanical results. These workers suggest that the beginning of this interval marks the beginning of transport of Gulf-derived moisture to these desert areas, followed by progressive climatic warming. Thus several lines of evidence support the inference that rivers in the Mississippi valley exhibited strong responses not only to actual meltwater discharge but also to the direct effects of climatic change.

Finally, early Holocene net aggradation is inferred at a variety of study sites. It has long been known that the preserved and dated glacial-maximum loess deposits indicate late Wisconsin transfer of fine-grained sediment by eolian processes from braided rivers to regional hillslopes and uplands. The fluvial record also suggests, however, that loess deposition set the stage for subsequent reversed transport, by rainfall and runoff, of fine-grained sediment during the early Holocene. This resulted in net aggradation along streams. This change in one component of overall sediment routing was most likely the result of altered climatic/

vegetational conditions, which caused hillslope sediment supply to temporarily exceed the capacity of streams to transport sediment.

REFERENCES CITED

Adams, J., 1980, Active tilting of the United States Midcontinent; Geodetic and geomorphic evidence: Geology, v. 8, p. 442–446.

Alford, J. J., and Holmes, J. C., 1985, Meander scars as evidence of major climate change in southwest Louisiana: Association of American Geographers Annals, v. 75, p. 395–403.

Baker, V. R., 1983, Late-Pleistocene fluvial systems, *in* Porter, S. C., ed., Late Quaternary environments of the United States: Minneapolis, University of Minnesota Press, v. 1, p. 115–129.

Baker, V. R., and Penteado-Orellana, M. M., 1977, Adjustment to Quaternary climatic change by the Colorado River in central Texas: Journal of Geology, v. 85, p. 395–422.

Bettis, E. A., and Benn, D. W., 1984, An archaeological and geomorphological survey in the central Des Moines river valley, Iowa: Plains Anthropologist, v. 29, p. 211–227.

Bettis, E. A., and Hallberg, G. R., 1986, Evolution of valleys and valley fills in eastern Iowa; Insights into Mississippi valley history: Geological Society of America Abstracts with Programs, v. 18, p. 540.

Boison, P. J., and Patton, P. C., 1985, Sediment storage and terrace formation in Coyote Gulch basin, south-central Utah: Geology, v. 13, p. 31–34.

Bowman, D., 1978, Determination of intersection points within a telescopic alluvial fan complex: Earth Surface Processes and Landforms, v. 3, p. 265–276.

Brakenridge, G. R., 1981, Late Quaternary floodplain sedimentation along the Pomme de Terre River, southern Missouri: Quaternary Research, v. 15, p. 62–76.

—— , 1983, Late Quaternary sedimentation along the Pomme de Terre River, southern Missouri; Part II, Notes on sedimentology and pedogenesis: Geologishes Jahrbuch, series A, v. 71, p. 265–283.

—— , 1984, Alluvial stratigraphy and radiocarbon dating along the Duck River, Tennessee; Implications regarding flood-plain origin: Geological Society of America Bulletin, v. 95, p. 9–25.

—— , 1987, Floodplain stratigraphy and flood regime, *in* Baker, V. R., Kochel, R. C., and Patton, P. C., eds., Flood Geomorphology: New York, Wiley (in press).

Burnett, A., and Schumm, S. A., 1983, Alluvial river response to neotectonic deformation in Louisiana and Mississippi: Science, v. 222, p. 49–50, see p. 50.

Church, M., and Gilbert, R., 1975, Proglacial fluvial and lacustrine sediments: Society of Economic Paleontologists and Mineralogists Special Publication 23, p. 22–100.

Church, M., and Ryder, J. M., 1972, Paraglacial sedimentation; A consideration of fluvial processes conditioned by glaciation: Geological Society of America Bulletin, v. 83, p. 3059–3072.

Coogan, A. H., and Irving, W. N., 1959, Late Pleistocene and Recent Missouri River terraces in the Big Bend Reservoir, South Dakota: Iowa Academy of Science Proceedings, v. 66, p. 317–327.

Delcourt, P. A., Delcourt, H. R., Brister, R. C., and Lackey, L. E., 1980, Quaternary vegetation history of the Mississippi Embayment: Quaternary Research, v. 13, p. 111–132.

Douglas, I., 1967, Man, vegetation, and the sediment yield of rivers: Nature, v. 215, p. 925–928.

Dryden, H. L., Murnaghan, F. D., and Bateman, H., 1956, Hydrodynamics (reprint): New York, Dover Publications, 634 p.

Dury, G. H., 1976, Discharge prediction, present and former, from channel dimensions: Journal of Hydrology, v. 30, p. 219–245.

—— , 1985, Attainable standards of accuracy in the retrodiction of palaeodischarge from channel dimensions: Earth Surface Processes and Landforms, v. 10, p. 205–213.

Ethridge, F. G., and Schumm, S. A., 1978, Reconstructing paleochannel morphologic and flow characteristics; Methodology, limitations, and assessments, *in* Miall, A. D., ed., Fluvial sedimentology: Canadian Society of Petroleum Geologists Memoir 5, p. 703–721.

Fisk, H. N., 1944, Geological investigation of the alluvial valley of the lower Mississippi River: Vickburg, Mississippi River Commission, 78 p.

Fisk, H. N., and McFarlan, E. M., 1955, Late Quaternary deltaic deposits of the Mississippi River, *in* Poldevart, A., ed., Crust of the Earth: Geological Society of America Special Paper 62, p. 279–302.

Flock, M. A., 1983, The late Wisconsinan Savanna Terrace in tributaries to the upper Mississippi River: Quaternary Research, v. 20, p. 165.

Fournier, M. F., 1949, Les facteurs climatiques de l'erosion due sol: Association Geographic Francais Bulletin, v. 203, p. 97–103.

—— , 1960, Climate et erosion: Paris, Universite France Press, 201 p.

Froehlich, W., Kazowski, L., and Starkel, L., 1977, Studies of present-day and past river activity in the Polish Carpathians, *in* Gregory, K. G., ed., River channel changes: New York, Wiley Interscience, p. 411–428.

Gagliano, S. M., and Thom, B. G., 1967, Deweyville terrace, Gulf and Atlantic coasts: Louisiana State University Coastal Studies Bulletin, no. 1, p. 23–41.

Grissinger, E. H., and Murphey, J. B., 1983, Present channel stability and late Quaternary valley deposits in northern Mississippi: International Association of Sedimentologists Special Publication, v. 6, p. 241–250.

Grissinger, E. H., Murphey, J. B., and Little, W. C., 1982, Late-Quaternary valley-fill deposits in north-central Mississippi: Southeastern Geology, v. 23, p. 147–162.

Gupta, A., 1983, High magnitude floods and stream channel response, *in* Collinson, J. D., and Lewin, J., eds., Modern and ancient fluvial systems: Oxford, Blackwell, p. 219–227.

Hajic, E. R., 1986, Late-Quaternary alluvial history of the lower Illinois valley and adjacent middle Mississippi valley: Geological Society of America Abstracts With Programs, v. 18, no. 6, p. 624.

Harvey, M. D., Rentschler, R. E., and Schumm, S. A., 1981, Environments of deposition; Controls on channel erosion in northern Mississippi: Geological Society of America Abstracts with Programs, v. 13, p. 469.

Haynes, C. V., 1976, Late Quaternary geology of the lower Pomme de Terre River, Missouri, *in* Wood, W. R., and McMillan, R. B., eds., Prehistoric man and his paleoenvironments: New York, Academic Press, p. 47–61.

—— , 1985, Mastodon-bearing springs and late Quaternary geochronology of the lower Pomme de Terre valley, Missouri: Geological Society of America Special Paper 204, 35 p.

Hoyer, B.E., 1980a, The geology of the Cherokee Sewer Site, *in* Anderson, D. C., and Semken, H. A., Jr., eds., Holocene ecology and human adaptations in northwestern Iowa: New York, Academic Press, p. 21–66.

—— , 1980b, Geomorphic history of the Little Sioux River valley: Iowa Geological Survey, Fall field trip, September 1980, 94 p.

Jackson, L. E., MacDonald, G. M., and Wilson, M. C., 1982, Paraglacial origin for terraced river sediments in Bow valley, Alberta: Canadian Journal of Earth Sciences, v. 19, p. 2219–2231.

Jansson, M. B., 1982, Land erosion by water in different climates: Uppsala University, Department of Physical Geography, Naturegeografiska Institutionen Report 57, 151 p.

Kehew, A. E., and Lord, M. L., 1986, Origin and large-scale erosional features of glacial-lake spillways in the northern Great Plains: Geological Society of America Bulletin, v. 97, p. 162–177.

Kennett, J. P., and Shackleton, N. J., 1975, Laurentide ice sheet meltwater recorded in Gulf of Mexico deep-sea cores: Science, v. 188, p. 147–150.

Knox, J. C., 1976, Concept of the graded stream, *in* Flemal, R., and Melhorn, W., eds., Theories of landform development: Binghamton, State University of New York, Publications in Geomorphology, 306 p.

—— , 1983, Responses of river systems to Holocene climates, *in* Wright, H. E., Jr., ed., Late Quaternary environments of the United States, The Holocene: Minneapolis, University of Minnesota Press, p. 26–41.

—— , 1985, Responses of floods for Holocene climate change in the upper Mississippi valley: Quaternary Research, v. 23, p. 287–300.

Knox, J. C., McDowell, P. F., and Johnson, W. C., 1981, Holocene fluvial stratigraphy and climate change in the Driftless Area of Southwestern Wisconsin, *in* Mahaney, W. C., ed., Quaternary paleoclimate: Norwich, U.K., Geobooks, p. 107–127.

Kozarski, S., and Rotnicki, K., 1977, Valley floors and changes of river channel patterns in the north Polish Plain during the late Würm and Holocene: Quaestiones Geographicae, v. 4, p. 51–93.

Langbein, W. B., 1949, Annual runoff in the United States: U.S. Geological Survey Circular 52, 14 p.

Langbein, W. B., and Schumm, S. A., 1958, Yield of sediment in relation to mean annual precipitation: EOS American Geophysical Union Transactions, v. 39, p. 1076–1084.

Leopold, L. B., and Wolman, M. G., 1957, River channel patterns; Braided, meandering, and straight: U.S. Geological Survey Professional Paper 282-B, p. 39–84.

Leventer, A., Williams, D. F., and Kennett, J. P., 1982, Dynamics of the Laurentide ice sheet during the last deglaciation; Evidence from the last deglaciation: Earth and Planetary Sciences Letters, v. 59, p. 11–17.

Love, D. W., 1979, Quaternary fluvial geomorphic adjustments in Chaco Canyon, New Mexico, *in* Rhodes, D. D., and Williams, G. P., eds., Adjustments of the fluvial system: Dubuque, Kendall-Hunt, p. 277–280.

Maizels, J. K., 1983, Proglacial channel systems; Change and thresholds for change over long, intermediate, and short-time scales, *in* Collinson, J. D., and Lewin, J., eds., Modern and ancient fluvial systems: Oxford, Blackwell, p. 251–266.

Matsch, C., 1983, River Warren, The southern outlet to glacial Lake Agassiz: Geological Association of Canada Special Paper 26, p. 231–244.

McDowell, P. F., 1983, Evidence of stream response to Holocene climate change in a small Wisconsin watershed: Quaternary Research, v. 19, p. 100–116.

Miller, J. A., 1973, Quaternary history of the Sangamon River drainage system, central Illinois: Illinois State Museum Reports of Investigations, no. 27, 36 p.

Moss, J. H., 1974, The relation of river terrace formation to glaciation in the Shoshone Basin, western Wyoming, *in* Coates, D. R., ed., Glacial geomorphology: State University of New York at Binghamton Publications in Geomorphology, p. 293–314.

Mycielska-Dowgiallo, E., 1977, Channel pattern changes during the last glaciation and Holocene in the northern part of the Sandomirez basin and the middle part of the Vistula valley, Poland, *in* Gregory, K. G., ed., River channel changes: New York, Wiley Interscience, p. 75–87.

Nadler, C. T., and Schumm, S. A., 1981, Metamorphosis of South Platte and Arkansas rivers, eastern Colorado: Physical Geography, v. 2, p. 95–115.

Ohmori, H., 1983, Erosion rates and their relation to vegetation from the viewpoint of world-wide distribution: University of Tokyo, Department of Geography Bulletin, no. 15, p. 77–91.

Richards, K. S., 1979, Channel adjustment to sediment pollution by the China clay industry in Cornwall, England, *in* Rhodes, D. D., and Williams, G. P., eds., Adjustment of the fluvial system: Dubuque, Kendall-Hunt, p. 309–331.

Ritter, D. F., 1982, Complex terrace development in the Nenana valley near Healy, Alaska: Geological Society America, Bulletin, v. 93, p. 345–356.

Saucier, R. T., 1974, Quaternary geology of the lower Mississippi valley: Arkansas Archeological Survey, Research Series 6, 26 p.

——, 1978, Sand dunes and related eolian features of the lower Mississippi alluvial valley: Geoscience and Man, v. 19, p. 23–40.

Saucier, R. T., and Fleetwood, A. R., 1970, Origin and chronologic significance of late Quaternary terraces, Ouachita River, Arkansas and Louisiana: Geological Society of America Bulletin, v. 81, p. 869–890.

Saucier, R. T., and Smith, L. M., 1986, Late Wisconsinan and Holocene evolution of the lower Mississippi valley: Geological Society of America Abstracts With Programs, v. 18, no. 6, p. 739.

Schumm, S. A., 1963, Sinuosity of alluvial rivers on the Great Plains: Geological Society of America Bulletin, v. 74, p. 1089–1100.

——, 1968, River adjustment to altered hydrologic regimen; Murrumbidgee River and paleochannels, Australia: U.S. Geological Survey Professional Paper 598, 65 p.

——, 1969, River metamorphosis: American Society of Civil Engineers Proceedings, Journal Hydraulics Division, v. 95, no. HY1, p. 255–273.

——, 1977, The fluvial system: New York, John Wiley and Sons, 238 p.

——, 1981, Evolution and response of the fluvial system; Sedimentologic implications: Society of Economic Paleontologists and Mineralogists Special Publication 31, p. 19–29.

——, 1984, River morphology and behavior; Problems of extrapolation, *in* Elliott, C. R., ed., River meandering: New York, American Society of Civil Engineers, p. 16–29.

——, 1985, Explanation and extrapolation in geomorphology; Seven reasons for geologic uncertainty: Japanese Geomorphological Union Transactions, v. 6, p. 1–18.

Schumm, S. A., and Khan, H. R., 1972, Experimental study of channel patterns: Geological Society of America Bulletin, v. 83, p. 1755–1770.

Schumm, S. A., Harvey, M. D., and Watson, C. C., 1984, Incised channels; Morphology, dynamics, and control: Littleton, Colorado, Water Resources Publications, 200 p.

Schumm, S. A., Mosley, M. P., and Weaver, W. E., 1987, Experimental fluvial geomorphology: New York, John Wiley and Sons, 410 p.

Shlemon, R. J., 1972, The lower American River area, California; A model of Pleistocene landscape evolution: Association of Pacific Coast Geographers Yearbook, v. 34, p. 61–86.

Smith, N. D., and Smith, D. G., 1984, Williams River; An outstanding example of channel widening and braiding caused by bed-load addition: Geology, v. 12, p. 78–82.

Spaulding, W. G., and Graumlich, L. J., 1986, The last pluvial climatic episodes in the deserts of southwestern North America: Nature, v. 320, p. 441–444.

Starkel, L., 1979, Typology of river valleys in the temperate zone during the last 15,000 years: Acta Universitat Oula, v. 82, p. 9–18.

——, 1983a, The reflection of hydrologic changes in the fluvial environment of the temperate zone during the last 15,000 years, *in* Gregory, K. J., ed., Background to palaeohydrology: New York, Wiley, p. 213–235.

——, 1983b, Climate change and fluvial response, *in* Gardner, R., and Scoging, H., eds., Mega-geomorphology: Oxford University Press, p. 195–211.

Wahl, K. L., 1984, Evolution of the use of channel cross-section properties for estimating streamflow characteristics: U.S. Geological Survey Water Supply Paper 2262, p. 53–66.

Walling, D. E. and Webb, B. W., 1983, Patterns of sediment yield, *in* Gregory, K. G., ed., Background to Paleohydrology: New York, Wiley and Sons, p. 69–100.

Waters, M. R., 1985, Late Quaternary alluvial stratigraphy of Whitewater Draw, Arizona; Implications for regional correlation of fluvial deposits in the American Southwest: Geology, v. 13, p. 705–708.

Weaver, W. E., 1982, Experimental study of alluvial fans [Ph.D. thesis]: Fort Collins, Colorado State University, 450 p.

White, E. M., 1982, Geomorphology of the lower and middle part of the White River Basin, South Dakota: South Dakota Academy of Science Proceedings, v. 61, p. 45–55.

White, E. M., and Hannus, L. A., 1985, Holocene alluviation and erosion in the White River Badlands, South Dakota: South Dakota Academy of Science Proceedings, v. 64, p. 82–94.

Wildman, N. A., 1981, Episodic removal of hydraulic-mining debris, Yuba and Bear river basins, California [M.S. thesis]: Fort Collins, Colorado State University, 107 p.

Williams, G. P., 1984, Paleohydrologic equations for rivers, *in* Costa, J. E., and Fleisher, P. J., eds., Developments and applications of geomorphology: Berlin, Springer-Verlag, p. 343–367.

Wilson, L., 1973, Variations in mean annual sediment yield as a function of mean annual precipitation: American Journal of Science, v. 273B, p. 335–349.

Womack, W. R., and Schumm, S. A., 1977, Terraces of Douglas Creek, northwestern Colorado; An example of episodic erosion: Geology, v. 5, p. 72–76.

MANUSCRIPT ACCEPTED BY THE SOCIETY APRIL 8, 1987

ACKNOWLEDGEMENTS

We thank H. E. Wright, Jr. and J. C. Knox for their careful reviews.

Printed in U.S.A.

The Geology of North America
Vol. K-3, North America and adjacent oceans during the last deglaciation
The Geological Society of America, 1987

Chapter 11

The physical record of lakes in the Great Basin

Larry Benson
U.S. Geological Survey, Box 25046, Denver Federal Center, Denver, Colorado 80225
Robert S. Thompson*
Department of Geological Sciences, Brown University, Providence, Rhode Island 02912

INTRODUCTION

During the late Pleistocene nearly 100 closed basins in the western United States contained lakes (Fig. 1). Most of these were within the Great Basin area of California, Nevada, Utah, and Oregon; others existed in Arizona, New Mexico, and northern Mexico. Today only a few of these basins contain perennial lakes. This paper addresses the chronology and paleoclimatic history associated with fluctuations of lakes in the Great Basin during the period 25 to 9 ka. Specifically: (1) When did the variations occur? (2) Were the lake-level changes synchronous across the region? (3) What types of paleoclimatic changes were responsible for the changes in lake size? (4) Did the paleoclimatic changes implied by lake-level records have a magnitude similar to that inferred from other proxy data, and did these changes occur at the same time in the different systems? (5) Do extreme climatic and hydrologic events in the historic period offer analogs for prehistoric conditions that created the Pleistocene lake systems? (6) What were the effects of the continental ice sheets on the Pleistocene climate of the Great Basin?

Previous investigators (Morrison, 1965; Mifflin and Wheat, 1979; Davis, 1982; Smith and Street-Perrott, 1983; Harrison and Metcalfe, 1985) cataloged lake-level data and presented various interpretations of selected lake-level chronologies. This chapter does not duplicate these works, but instead it focuses on those lake systems that have been intensively studied and that have well-documented absolute-age chronologies. Throughout ensuing sections, calculations and quantitative discussion of processes commonly use Lake Lahontan as an example because of data availability and the authors' greater experience with this lake system. It should be kept in mind that differences in bathymetry and other details of the local setting cause each lake system to respond to climatic change in a somewhat unique manner.

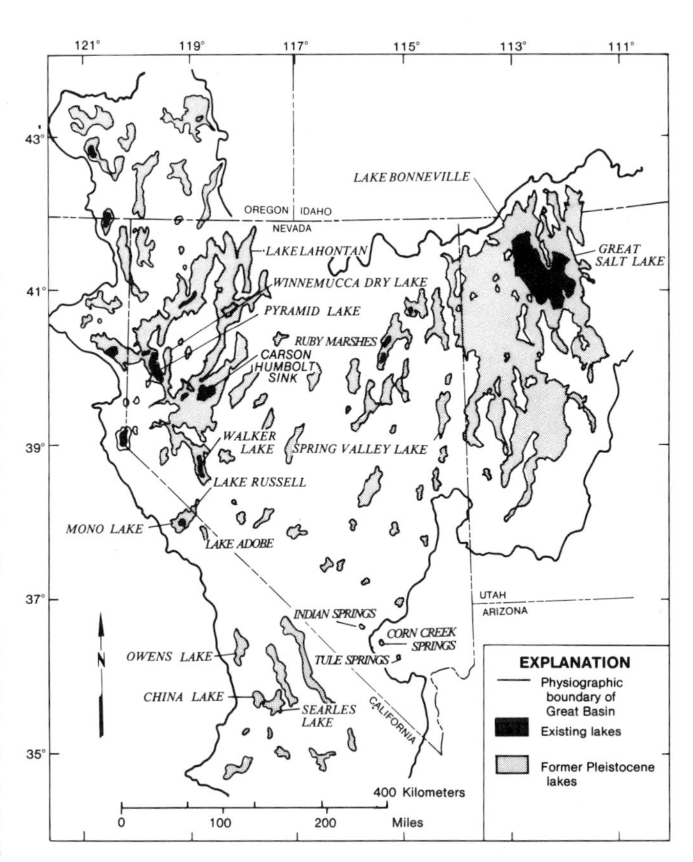

Figure 1. Lakes and marshes of assumed late-Pleistocene age in the Great Basin (after Spaulding and others, 1983).

REGIONAL CLIMATE

Climatic variations in the Great Basin are caused by changes in global atmospheric circulation; in particular, they are the result of changes in the strength of the circumpolar vortex and the wavelength and amplitude of long (Rossby) waves in the mid-latitude westerlies. The pattern of these waves determines the development, movement, and intensity of synoptic-scale features of circulation such as cyclones, anticyclones, fronts, and jet streams; expression of the synoptic-scale circulation at any given point is modified by topography and other characteristics of the regional and local setting.

*Present address: U.S. Geological Survey, Box 25046, Denver Federal Center, Denver, Colorado 80225

Benson, L., and Thompson, R. S., 1987, The physical record of lakes in the Great Basin, *in* Ruddiman, W. F., and Wright, H. E., Jr., North America and adjacent oceans during the last deglaciation: Boulder, Colorado, Geological Society of America, The geology of North America, v. K-3.

Precipitation

Cool-season precipitation from North Pacific sources is dominant throughout the Great Basin (Houghton, 1969). Only the southeastern part of this region receives substantial summer precipitation from subtropical sources. The western part of the Great Basin, more directly in the rain shadow of the Sierra Nevada, receives less annual precipitation than does the eastern portion. Houghton (1969) identified three principal precipitation regimes that occur in the Great Basin, each of which is characterized by a dominance of certain air-mass trajectories. Although these regimes occur throughout the basin, each achieves dominance in different subregions (Fig. 2) and seasons.

From November to April, successive low-pressure centers follow the westerlies from the North Pacific across the northwestern United States (Houghton and others, 1975). These storm tracks shift southward whenever subtropical high-pressure cells contract toward the equator. The development of the thermally induced, stagnant "Great Basin High" during much of the winter season forces storm tracks north of 40°N (Houghton, 1969; Mitchell, 1976). However, this "Great Basin High" dissipates several times during a typical winter, and the path of the migratory cyclones shifts southward into the central Great Basin. Houghton (1969) refers to this pattern as the Pacific component, under which the western Great Basin receives the majority of its precipitation.

Rainfall from April to June and in October and November usually is associated with the development of successive low-pressure cells in the western Great Basin in the lee of the highest parts of the Sierra Nevada. Frequently, these cold cyclones, referred to as "Great Basin Lows" or "Tonopah Lows," entrain moisture from the Pacific Ocean and migrate slowly eastward. The central, eastern, and parts of the southern Great Basin receive most of the annual precipitation in the spring under this regime, which Houghton (1969) refers to as the continental component. This circulation pattern also may bring in moisture from southerly subtropical sources during the warmer months (Houghton, 1969), whereas springtime continental cyclones may obtain moisture from temporary lakes present in valley bottoms.

During July and August the westerlies are weaker, and Pacific storm tracks move far to the north of the Great Basin (Mitchell, 1976). The eastern and southeastern peripheries of the Great Basin have convective storms in summer that bring moisture from the Gulf of California or the Gulf of Mexico or both. Houghton (1969) refers to this precipitation regime as the Gulf component.

Precipitation in the surrounding mountains may be the primary control on lake size in some basins, whereas in other basins the local climate near the lake may be of more importance. The contribution of precipitation on the lake surface to total fluid input ranges from 15 percent in the Lahontan basin (Benson, 1986) to 17 percent in the Lake Russell (Mono Lake) basin (Mason, 1967) and 27 percent in the Bonneville (Great Salt Lake) basin (Stauffer, 1985). In prehistoric times, when lake surfaces were considerably larger, precipitation directly on lake surfaces may have had a larger effect in the creation and maintenance of lake systems.

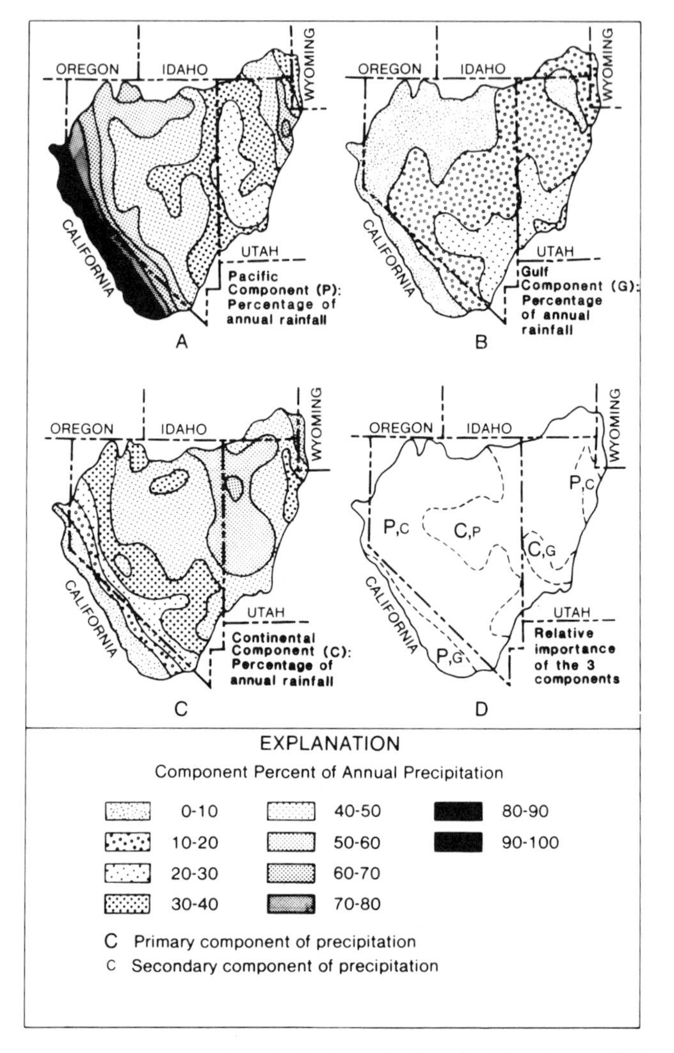

Figure 2. Climatic gradients in the Great Basin (after Houghton, 1979).

Precipitation and Stream Discharge

Precipitation data (1942–83) from Tahoe City in the Sierra Nevada and Fallon in the Lahontan basin, as well as discharge data for the Carson River (measurements made upstream from water diversions for human consumptive use) are illustrated in Figure 3. Precipitation at the mountain station correlates well with the discharge of the nearby Carson River. Precipitation at the basin station generally is much less than that at the mountain station and correlates poorly with both high-altitude precipitation and discharge. Thus, water input to the Lahontan basin primarily is a function of precipitation in the upper reaches of the drainage system.

The dominance of montane climate on lake size also applies to Lake Russell (Mono Lake), which also lies in the rain shadow of the Sierra Nevada. The rain-shadow effect is less important in the Bonneville basin, where the major mountain masses are leeward of the lakes.

Temperature

The Great Basin is an elevated plateau with a continental temperature regime of cold winters and hot summers and a large diurnal range of temperature. In the western Great Basin, winter temperatures are moderated by warm downslope winds off the Sierra Nevada (Houghton and others, 1975). Central and eastern Nevada lie beyond the range of these winds and are much colder in winter. Extreme cold conditions may result from night-time radiative cooling under the stagnant air of Great Basin Highs. Cold conditions occasionally may result from the incursion of Arctic air, but most of this air is shunted east of the Rocky Mountains (Houghton and others, 1975). Although temperature generally decreases with increasing altitude, the steep mountain topography promotes cold-air drainage, which may make the valley bottoms cooler than the lower mountain slopes (Billings, 1954).

Daytime summer temperatures at lower altitudes in the Great Basin frequently exceed 40°C. Valley-bottom altitudes decline to the south, and daytime summer temperatures in the Mojave Desert frequently exceed 44°C.

Evaporation

Similar to the regional trend in temperature discussed before, mean annual lake-evaporation (MAE) in the Great Basin ranges from 1.0 m at the northern boundary to 2.2 m at the southern boundary (Farnsworth and others, 1982). From historical records (1929–1959) of river discharge, precipitation, and lake level, Harding (1965) calculated a MAE of 1.2 m for Pyramid Lake and a MAE of 1.3 m for Winnemucca Lake. Stauffer (1985) used pan-evaporation data from 1919 to 1984 to calculate a MAE of 1.1 for Great Salt Lake. The historic yearly variability in evaporation is small. Pan-evaporation data for Boulder City, Nevada, (1935–1984) and Salt Lake City Airport, Utah, (1919–1984) indicate standard deviations of 6 and 13 percent, respectively, relative to the mean, with maxima 20 and 25 percent greater than the mean, and minima 9 and 19 percent less than the mean.

LAKE-LEVEL FLUCTUATIONS

Variation in lake size occurs as a result of change in the hydrologic budget, which under pristine conditions usually is a function of climatic change. In the absence of written records, prehistoric change in lake levels can be deduced by radiometric dating of materials deposited in or near past shorelines, such as wood, tufa, or gastropods. Variation in the abundance and taxonomic composition of microscopic plants and animals also may be used to identify past lake-level fluctuations in radiometrically dated cores of lacustrine sediments.

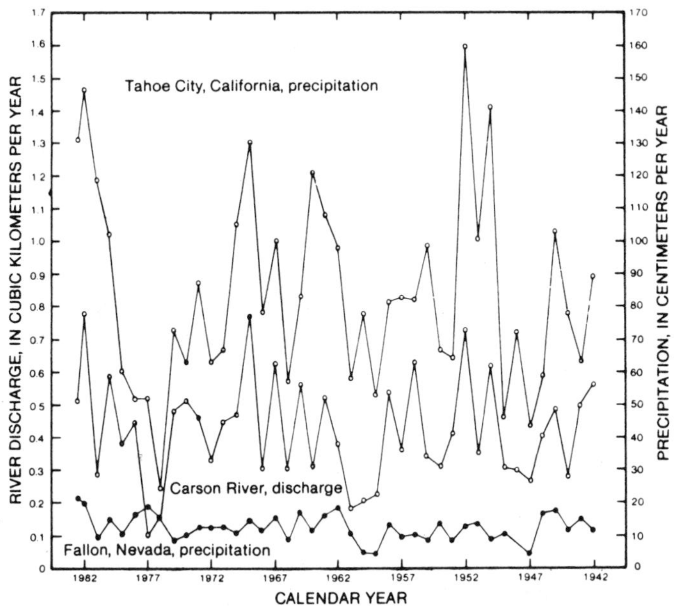

Figure 3. Precipitation records for Fallon, Nevada, and Tahoe City, California, compared with Carson River discharge.

Age-Estimation Methods

Methods used in absolute and relative dating of past lake-level fluctuations include radiocarbon dating, chlorine-36 analysis, tephrochronology, uranium-series dating, uranium trend, thermoluminescence, amino-acid racemization, and salt-balance calculations. Radiocarbon dating, tephrochronology, and salt-balance calculations have been most widely applied to dating sediments pertaining to the last 20,000 years. The application of tephrochronology is reviewed by Davis (1978a) and Sarna-Wojcicki and others (1983), and will not be discussed here.

Radiocarbon Dating. Radiocarbon determinations on samples of limnologic materials from basins other than those discussed in the following text generally have not been proven reliable (Broecker and Walton, 1959; Benson, 1978; Stuiver and Smith, 1979). Contamination potentially affecting these samples results from inclusion of older carbon detritus within a sample, precipitation of secondary carbonate within a sample, recrystallization of metastable carbonate phases (aragonite, high-magnesium calcite, or monohydrocalcite) to stable low-magnesium calcite, remobilization and reprecipitation of inorganic carbon below the sediment-water interface, and introduction into the lake of "old" carbon leached from carbonate rocks by surface or groundwater.

Salt-Balance Calculations. Russell (1885) noted that many modern lakes in the Great Basin were not as rich in dissolved solids as they would be if they were true remnants of the last lake highstands. He postulated that the lakes desiccated and that the precipitated salts were removed by erosion or were buried. Existing lakes were considered to be the result of subsequent rejuvenation. Prior to the development of radiometric methods, attempts were made to use geochemical salt-balance equations to estimate the age of the last desiccation in various lake basins (Russell, 1885; Gilbert, 1890; Gale, 1914, 1915; Huntington, 1914; Van Winkle, 1914; Jones, 1925, 1929; Antevs, 1925, 1938). These estimates varied greatly, depending on the data and assumptions used in the calculations.

Early investigators were not aware of the complicating processes that tend to negate application of the salt-balance approach, nor were hydrologic and water-quality data adequate for the calculations. For a reliable salt-balance estimate of time elapsed since the last desiccation, the following criteria must be met: (1) The volume flux of each fluid source (stream discharge, groundwater discharge, precipitation) must be known as a function of time. (2) Solute concentrations as a function of discharge rate must be known for each fluid source. (3) Only chemically conservative solutes such as the readily soluble halogens (chloride, iodide, and bromide) can be used in salt-balance equations. (4) The system must be closed to other solute sources and sinks (or these must be quantifiable).

Enough is now known about the hydrology and geochemistry of western closed-basin lakes to suggest that it is rarely possible to determine accurately from salt-balance calculations the period since the last desiccation. Historical data indicate that the criteria enumerated above rarely are met (Feth, 1959; Eardley, 1966; Whelan, 1973; Benson, 1978).

The Relation of Lake Level to Basin Bathymetry and Surface Hydrology

Complications involving basin bathymetry and surface hydrology may affect the apparent chronology of lake-level change. The Lahontan basin (as well as other large lake basins) consists of a closed basin comprising several subbasins, separated from each other by sills of varying altitudes. Some subbasins are terminal points for perennial streams; lakes in these subbasins rise and fall with climatic change through time. Subbasins without such streams receive water only when an adjoining subbasin overflows and spills; chronologies from these "streamless" subbasins record only large-scale climatic events.

Catastrophic spillovers and river diversions also affect lake levels. For example, at ~15 ka, a spillover flood from Lake Bonneville poured into the Snake River (Gilbert, 1890; Malde, 1960; Malde and Trimble, 1965) and downcut the outlet at Zenda by 108 m (probably in less than 1 yr; Currey and Oviatt, 1985), and a new pass (Red Rock Pass) was formed 3 km to the south. Changes in drainage patterns in and among subbasins in the Lahontan basin may have affected lake levels in the past (King, 1978; Davis, 1978b, 1982). The shape of the current drainages (Fig. 4) supports this argument; for example, in the Lahontan basin, the Walker River turns sharply from north to southeast before entering Walker Lake; and the Truckee River turns sharply from east to northwest before entering Pyramid Lake. The timing of river "diversions" remains conjectural, although Davis (1982) uses geomorphic criteria to suggest that the Humboldt River flowed into the Quinn River drainage prior to the last lake cycle.

Basin topography also affects lake-level chronology. Lake levels in deep, narrow subbasins sustained by perennial streams respond quickly to change in moisture storage (influx minus evaporation). Such systems (for example, the Lake Russell basin and the Walker Lake subbasin of the Lahontan system) are potentially excellent recorders of high-frequency low-amplitude climatic change on a subregional scale.

Lakes sustained by perennial streams in basins with large surface areas (for example, the Bonneville basin) respond slowly to changes in moisture storage and tend to be excellent recorders of high-amplitude climatic events.

Single-basin lakes not sustained by perennial streams (for example, Spring Valley Lake, Fig. 1) record highly localized changes in the climate and hydrologic balance. The presence of a lake in a similar basin may indicate a change in the surface hydrologic system or a rise of the regional water table. Thus a change in the size (level, surface area, volume) of a lake located in a particular basin (or subbasin) is a complicated function of intrabasin and extrabasin climate and basin topography.

Lake-Level Fluctuations from 25 to 9 ka

The known distribution and extent of Pleistocene Lake systems in the Great Basin are shown in Figure 1. The names and hydrographic characteristics of these paleolakes are summarized in Mifflin and Wheat (1979) and Williams and Bedinger (1984). Shoreline evidence has been found for 56 paleolakes in Nevada: of three shorelines occurring in many basins, two are thought to be of Late Wisconsin (Lahontan) age (Mifflin and Wheat, 1979).

The chronologies of lake-level changes have been studied intensively with radiometric methods in the Bonneville, Lahontan, Russell, and Owens River systems. Radiocarbon dates are available on lake-associated materials from several other lake basins in the Great Basin. In the following section we review chronological evidence from five intensively studied paleolake systems.

Lake Bonneville System. At its highest stage, Lake Bonneville occupied two subbasins (Salt Lake and Sevier) and had a surface area of ~51,300 km^2 (Fig. 5), a volume of ~9,500 km^3, and a maximum depth of 372 m (D. Currey, written communication, 1986). Today, surface discharge constitutes ~66 percent, precipitation ~31 percent, and groundwater ~3 percent of the average influx of water to Great Salt Lake, the major modern lake in the Bonneville basin.

Although numerous workers since Gilbert (1890) have con-

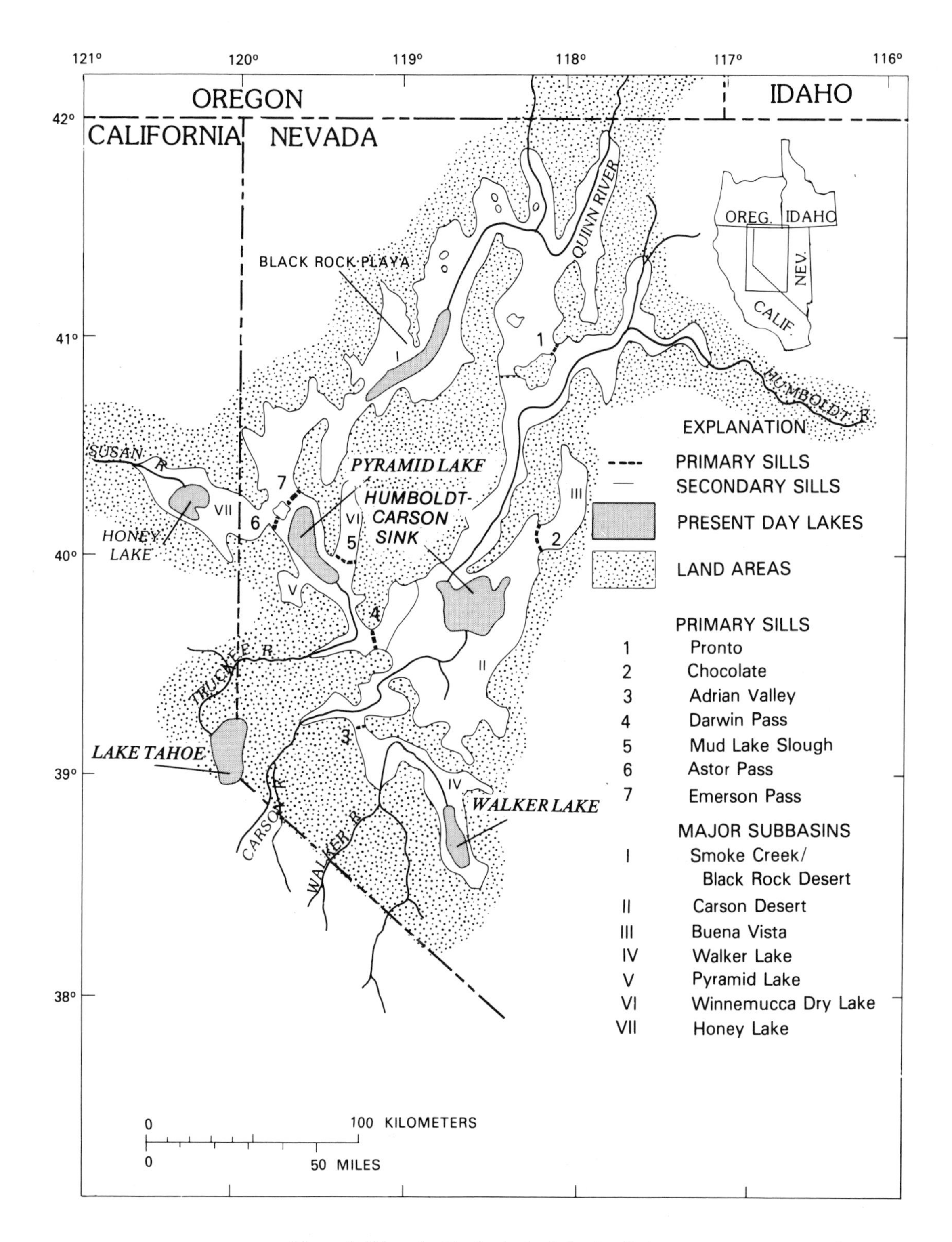

Figure 4. Sills and subbasins in the Lahontan Basin.

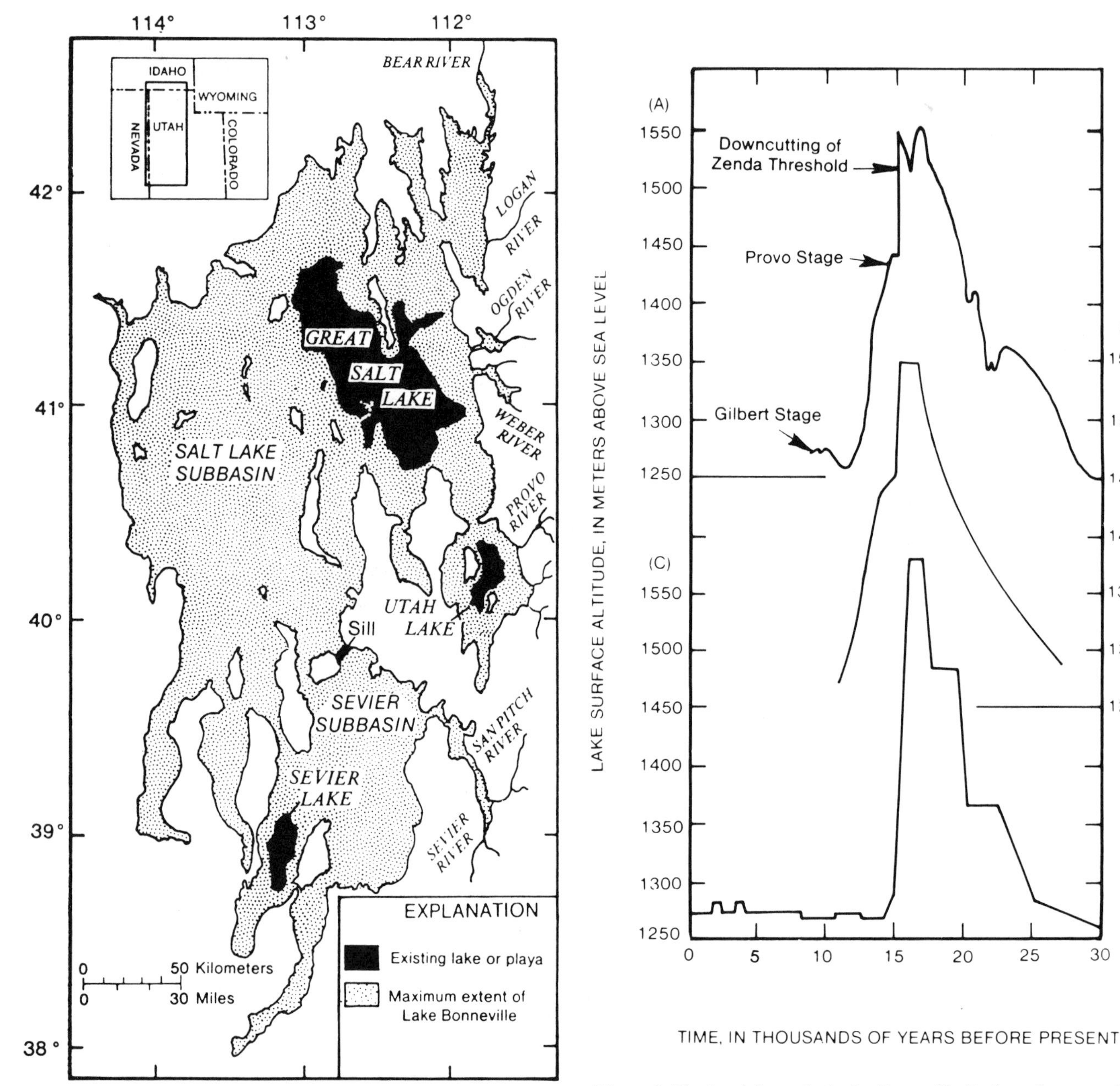

Figure 5. Major sill and subbasins in the Bonneville Basin (modified from Currey and others, 1983).

Figure 6. The last lake cycle in the Bonneville Basin as interpreted by: (A) Currey and Oviatt, 1985; (B) Scott and others, 1983; and (C) Spencer and others, 1984.

tributed to our knowledge of Lake Bonneville, until recently the interpretations of Morrison (1965) and Morrison and Frye (1965) have been cited most frequently. Subsequent stratigraphic and chronologic investigations of exposed deposits in the northeastern Bonneville basin have resulted in a substantially modified interpretation of lake-level fluctuations (Fig. 6; Scott and others, 1983; Spencer and others, 1984; Currey and Oviatt, 1985). Physical tracing between study localities is difficult because of the great distances separating outcrops, and correlation among localities is difficult because deposits of different lake cycles are similar in appearance. Disconformities and abrupt facies changes also inhibit correlations. To overcome these difficulties, Scott and others (1983) relied on amino acid analyses and radiocarbon dating for correlations. Samples from deposits of uncertain age were assigned to a specific lake cycle on the basis of the similarity of their alloisoleucine-to-isoleucine ratios relative to ratios of samples whose relative ages had been determined by other means. The resulting chronology consists of two lake cycles, the Little

Valley cycle (150 to 90 ka) and the Bonneville cycle (26 to 12 ka). Recent work indicates the presence of an additional shallow-lake cycle dating between ~70 to 40 ka (J. Oviatt, written communication, 1987).

By 12 ka, Pleistocene Lake Bonneville ceased to exist, and the remnant body of water was at an extremely low level (Fig. 6; Scott and others, 1983; Spencer and others, 1984; Currey and Oviatt, 1985). Radiocarbon dates on gastropod shells indicate that this extreme lowstand occurred between 12 and 11 ka, and stratigraphic studies firmly place this interval between the Provo and Gilbert stages. Mirabilite ($Na_2SO_4 \cdot 10H_2O$) beds and brine-shrimp pellets were deposited in Great Salt Lake during this period, indicating a very shallow lake. Currey (1980) ascribed desiccation polygons underneath modern Great Salt Lake to a middle Holocene period of aridity, but he now believes (D. R. Currey, written communication, 1986) that these features probably were formed in the earlier desiccation period when the mirabilite beds were deposited. Similarly, the minimum lake level evident in the core stratigraphies of Spencer and others (1984) occurred prior to the deposition of the Mazama tephra (that is, Late Wisconsin or early Holocene), rather than during the middle Holocene.

After 10.9 ka, lake levels rose in the Bonneville basin, and by 10.3 ka, a lake stood at the Gilbert Shoreline, more than 10 m above the historic mean level of Great Salt Lake (1280 m; Currey, 1980; Currey and others, 1983). This lake was intermediate in surface area between Lake Bonneville and Great Salt Lake. The Draper Formation, which Morrison (1965) and Van Horn (1982) attributed to a lake cycle, was reinterpreted by Scott and others (1983) as a valley-fill deposit composed of alluvium and colluvium at its single locality in the Salt Lake subbasin.

The lake receded from the Gilbert Shoreline sometime after 10 ka (Currey and Oviatt, 1985). For the remainder of the Holocene, water levels apparently varied within a small range of altitude. In the Salt Lake subbasin, radiocarbon dates on sheep dung from the base of post-lacustrine sediments in Danger Cave (Jennings, 1957) indicate that lake levels remained below 1314 m since 11.4 ka.

Lake Franklin (Ruby Marshes). The Ruby Marshes are a series of spring-fed marshy pools (~3 m maximum depth) at an altitude of 1818 m on the eastern flank of the Ruby Mountains in northeastern Nevada (Fig. 1). Lake Franklin, which was 35 m deep, occupied the basin during the Late Wisconsin (Mifflin and Wheat, 1979; Thompson, 1984). Analyses of aquatic macrophytes, pollen, acid-resistant algae, diatoms, and ostracodes provide a paleolimnologic record spanning the last 35 kyr (Thompson, 1984; R. Thompson, J. Bradbury, R. Forester, written communication, 1986). The biotic remains record a deep-water period from before 18.5 ± 1.1 ka that lasted to at least 15.4 ± 0.7 ka.

Sediments from immediately above a sample dated at 15.4 ± 0.7 ka change from lacustrine clays to coarser sediments containing small gastropod remains. Thompson (1984) interpreted this sedimentary shift as reflecting a rapid drawdown of the lake surface. Sediment deposition was not reinitiated until the lake rose to slightly above modern levels prior to 9.8 ± 0.4 ka.

Lake Lahontan System. At its highest stage, ~13 ka, Lake Lahontan had a surface area of 22,300 km^2, a volume of 2,020 km^3, and a maximum depth (in the Pyramid Lake subbasin) of 276 m (Benson and Mifflin, 1986). The Lahontan basin consists of seven subbasins separated by sills of varying altitude (Fig. 4). Six rivers terminate in the subbasins, with four (Truckee, Carson, Walker, and Humboldt) contributing 96 percent of the total gaged surface inflow (Benson, 1986). The Truckee, Carson, and Walker rivers have their headwaters in the Sierra Nevada on the western boundary of the Lahontan basin. The Humboldt River drains mountain ranges on the northeast. Subsurface inflow to surface-water bodies in the Lahontan basin is small compared to the total water input as runoff and precipitation (Everett and Rush, 1967; Van Denburgh and others, 1973).

Russell (1885) conducted the first comprehensive study of Pleistocene lake deposits in the Lahontan basin. He identified a "lower lacustral clay" and an "upper lacustral clay" separated by a "medial gravel." Morrison (1965) renamed Russell's (1885) "upper lacustral clay" the Sehoo Formation, the "medial gravel" the Wyemaha Formation, and the "lower lacustral clay" the Eetza Formation. The Sehoo Formation is considered to have been deposited during the last major lake cycle. Broecker and Orr (1958) and Broecker and Kaufman (1965) attempted to assign absolute age estimates to these deposits by radiometric dating; but Morrison and Frye (1965) pointed out that certain radiocarbon dates on tufas were reversed in relation to Morrison's (1964) stratigraphic assignments. Benson (1978, 1981) developed an alternative lake-level chronology for the Pyramid and Walker Lake subbasins, employing a sample-selection procedure that tended to eliminate problems caused by the introduction of secondary carbon into tufa samples. Thompson and others (1986) presented a revised chronology for the last Pleistocene lake cycle in the central Lahontan basin. Their data together with data from Benson and Thompson (1987) form the lake-level chronology of the Lahontan basin illustrated in Figure 7.

The earliest indication of lake level occurs about 40 ka in the Smoke Creek/Black Rock Desert subbasin. At the time a moderate-sized lake connected the Smoke Creek/Black Rock Desert, Honey Lake, Pyramid Lake, and Winnemucca Dry Lake subbasins (Fig. 7, Table 1). This interpretation is supported by the 35 ka Marble Bluff tephra that is present only in the bottoms of the subbasins (Davis, 1978a). By 20 ka the lake level in the western subbasins had risen to about 1265 m, where it remained for 3.5 kyr. Davis (1983) correlated a tephra that crops out at an altitude of 1251 m in the Black Rock Desert with the 23.4-ka Trego Hot Springs tephra. The alternative rise in lake level suggested by this correlation is depicted in Figure 7 as a dotted line. The sill (Darwin pass) that connects the western subbasins with the Carson Desert subbasin also has an altitude of 1265 m, which suggests that lake levels in the western subbasins were stabilized by spill from the Pyramid Lake subbasin to the Carson Desert subbasin until 16.5 ka.

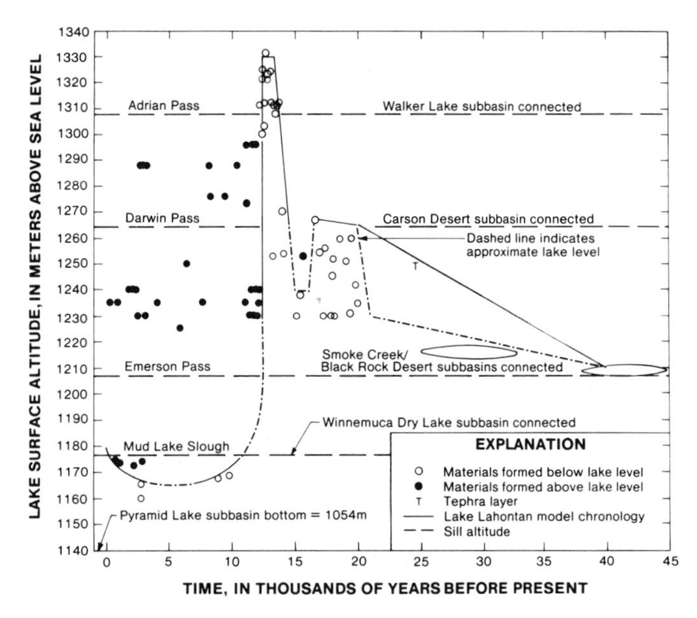

Figure 7. Central Lake Lahontan chronology inferred from data in Thompson and others (1986) and Benson and Thompson (1987).

By 16 ka, lake level in western subbasins may have receded to 1240 m; this recession appears synchronous with a desiccation of Walker Lake (Fig. 8). From about 15 to 13.5 ka, lake level rose rapidly, so that Lake Lahontan was a single body of water by 14 ka. The lake appears to have reached a maximum highstand of 1330 m by 13.5 ka, a condition that persisted until 12.5 ka, at which time lake level plummeted 100 m, during a time interval so short (<500 yr) that the magnitude of the counting error associated with radiocarbon analysis precludes a more precise determination.

No data exist that indicate the level of lakes in the various subbasins between 12 and 10 ka. Radiocarbon dates on nonlithoid tufa in this age range (Broecker and Orr, 1958; Broecker and Kaufman, 1965; Benson, 1978; Benson, 1981) have since been demonstrated to be in error (Thompson and others, 1986). Morrison (1964) suggested that an extreme lowstand occurred in the Carson subbasin following the maximal highstand of Lake Lahontan. Unfortunately, the absence of absolute chronological control on Carson subbasin sediments precludes a clear understanding of the timing of the proposed lowstand.

In the Pyramid Lake subbasin, Born (1972) recovered wood samples dated at 9.7 ± 0.1 ka and 8.8 ± 0.1 ka from altitudes of 1169 and 1168 m from the delta of the ancestral Truckee River; he interpreted these data as indicating that Pyramid Lake stood at an altitude between 1168 and 1219 m. Prior to irrigation, the surface altitude of Pyramid Lake was in the range of 1177 to 1180 m (Born, 1972); thus Born's data indicate that during a portion of the early Holocene, Pyramid Lake was not substantially lower than the historic level and may have been as much as 30 m higher.

TABLE 1. ALTITUDES OF PRIMARY SILLS IN MAJOR SUBBASINS IN THE LAHONTAN BASIN (from Benson and Mifflin, 1986)

Sill name	Sill altitude (m above sea level) Present-day (1985)	Corrected for isostatic rebound and tilting
Adrian Valley	1,308	1,302
Pronto	1,292	1,283
Darwin Pass	1,265	1,253
Chocolate	1,262	1,253
Astor Pass	1,222	1,213
Emerson Pass	1,207	1,195
Mud Lake Slough	1,177	1,177

Evidence exists that an extreme period of aridity did not occur since the last highstand of Lake Lahontan. Persistence of endemic fish (cui-ui and emerald trout) in Pyramid Lake until historic times indicates that the lake did not decrease in size to the point that the fish could not survive (Snyder, 1918). Since 1904, Pyramid Lake has fallen about 15 m, passing through a minimum level (1153 m) in 1967 that was 24 m lower than the 1904 level (1177 m). The emerald trout did not survive this artificial desiccation caused by diversion of water for irrigation, and the cui-ui had to be raised in hatcheries external to the lake, which indicates that Pyramid Lake did not fall much below its historic pre-agricultural lowstand of 1177 m (Harding, 1965) during the last 12.5 kyr.

Geochemical evidence from the Pyramid Lake subbasin also indicates that Pyramid Lake has not desiccated since the last highstand. Benson (1978, 1981) determined that dissolved magnesium, total inorganic carbon, and to a lesser extent, calcium increased downward in pore fluids beneath the sediment-water interface of Pyramid Lake, whereas sodium and chloride decreased downward. Benson (1978, 1981) interpreted these concentration trends as indicating that Pyramid Lake had decreased in size, causing the precipitation of carbonate minerals that continue to undergo dissolution. The absence of sodium- and chloride-concentration gradients of the same sign was interpreted as indicating that Pyramid Lake did not completely desiccate and cause the precipitation of readily soluble chloride salts. The following salt-balance calculation, using chloride, was made to test this interpretation.

The total chloride stored in the Pyramid Lake and Winnemucca Dry Lake subbasins is equal to the mass of chloride dissolved in lake water plus the mass of chloride dissolved in pore fluids located beneath the sediment-water interface. Winnemucca Lake contained 0.26×10^{10} kg of chloride in 1924, and Pyramid Lake contained 5.36×10^{10} kg of chloride in 1976 (Clarke, 1924; Benson, 1984). After correcting for diversion, the mean annual volume of water discharging to Pyramid Lake and Winnemucca

Dry Lake subbasins is estimated to range from 0.75 to 1.00 km^3 yr^{-1}. The lower discharge limit represents the mean annual discharge of the Truckee River measured at the Farad station for the period 1900–1983 (U.S. Geological Survey, 1960, 1963, 1961–1984). The upper discharge limit was calculated with the assumption that Pyramid and Winnemucca lakes had surface areas corresponding to historic lake-surface altitudes of 1175 m and evaporation rates of 1.25 m yr^{-1}.

Instantaneous values of discharge and chloride concentration, available on a monthly basis for 1968–1980 (Desert Research Institute data base), were used to calculate a discharge-weighted, mass-flow rate of chloride. The mass-flow rate (2.13×10^6 kg yr^{-1}) was divided into the total chloride mass dissolved in Pyramid and Winnemucca lakes. The calculation indicates that the Truckee River would have taken 26 kyr to supply the chloride dissolved in these lakes in 1976 and 1924. Accounting for the chloride dissolved in pore-fluids beneath the sediment-water interface of Pyramid Lake in sediments deposited since the last highstand increases the calculated flow period by about 10 percent. Using the upper discharge limit of 1.00 km^3 yr^{-1} decreases the flow period by $\leqslant$33 percent. The part of the calculated flow period in excess of 12.5 kyr is meaningless, because the mass of chloride in excess of that contributed by the Truckee River during the last 12.5 kyr was inherited from Lake Lahontan. In any case, the calculation indicates that Pyramid Lake did not desiccate during the last 2.5 kyr, contrary to the results of earlier calculations (Russell, 1885; Jones, 1925).

Walker Lake is situated in the most southerly subbasin of the Lahontan system and was joined to the central body of Lake Lahontan only when the latter was at its highest levels (>1308 m). Since the initial studies of Russell (1885), it has been postulated that the history of Walker Lake may have been affected by diversion of the Walker River into the Carson River through Adrian Valley (King, 1978; Davis, 1982). King (1978) claims to have found geomorphic evidence for diversion of the Walker River through Adrian Valley. However, the sedimentological record left by the hypothetical diversion remains undated; the diversion could have occurred as the result of Walker lake spilling into the Carson River drainage at 14 ka.

On the basis of chemistry of pore fluids extracted from sediment cores taken from Walker Lake, Benson (1978) deduced that the lake had desiccated in the past. Subsequent studies of the geochemistry, diatom, ostracode, and palynomorph content of these and other cores (L. Benson, J. Bradbury, R. Forester, P. Meyers, R. Thompson, and I. Yang, written communication, 1986) indicate that Walker Lake was dry at ~16, ~5, and ~2 ka (Fig. 8). After this, the lake rose again and, according to the diatom and ostracode data (J. Bradbury, R. Forester, written communication, 1987), reached its highest level in the Holocene just prior to the historic period. Radiocarbon dates on Holocene tufas from the Walker Lake Basin (Fig. 8) are in virtual agreement with this proposed chronology.

Russell (Mono) Lake and Lake Adobe. The Late Wisconsin lake in the Mono Lake Basin (Fig. 9), Lake Russell, was

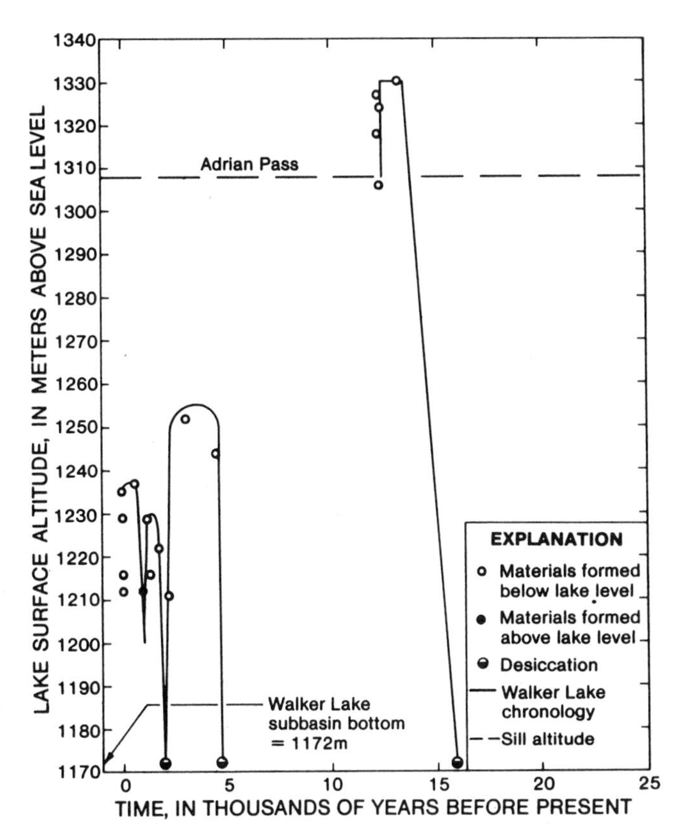

Figure 8. Walker Lake chronology inferred from data in Benson and Thompson (1987).

situated in a relatively small basin and was separated from the Owens River system to the south by a high-altitude sill. Lajoie (1968) made a detailed stratigraphic study of the sediments of the last lake cycle, the chronology of which has been subsequently supplemented by radiocarbon dates on tufa and ostracodes and by tephrochronology (K. Lajoie, written communication, 1986). Smith and Street-Perrott (1983) published a revised version of Lajoie's reconstruction of the history of Lake Russell (Fig. 10). Although K. Lajoie (written communication, 1986) questions some of the interpretations made by these authors, the basic shape of the lake-level curve is correct. The chronology of Lake Russell resembles that of lakes in the Lahontan basin (Fig. 11).

Lake Russell apparently began its main regression from the last major Pleistocene highstand (~2134 m) prior to 13 ka and was at low levels (<1950 m) by ~12 ka (Lajoie, 1968, p. 103). During this regressive phase, "thinolitic" tufa was deposited, and Lajoie (1968, p. 107) states that the deposition of ". . . minor coatings of lithoid tufa on thinolite . . . suggest(s) that the lake may have risen briefly after the main thinolithic phase." This last transgression appears to have reached altitudes as high as 2012 m at what Lajoie estimated to be slightly before 11 ka. Smith and Street-Perrott (1983) did not include the 12 to 11 ka regression-

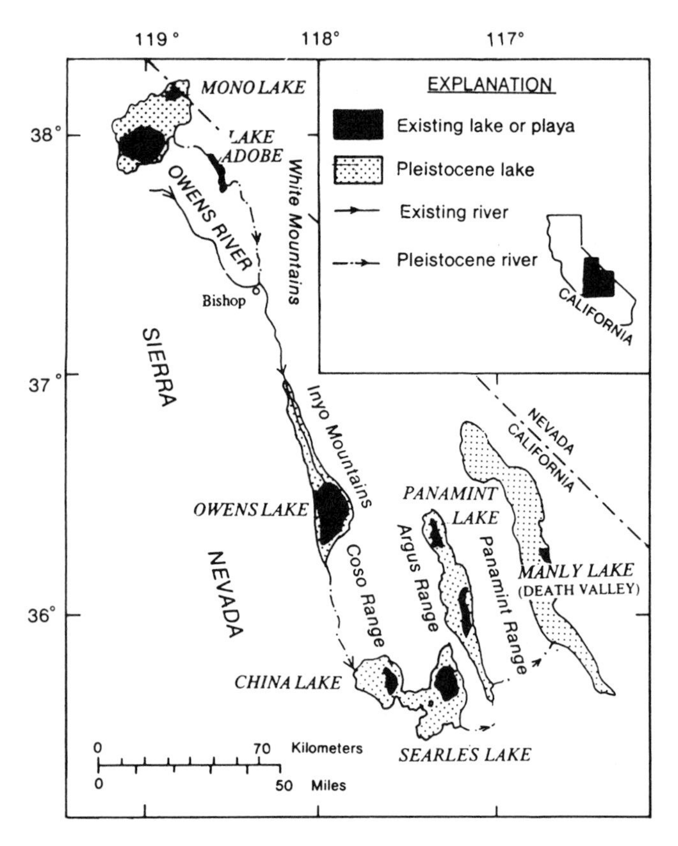

Figure 9. Lakes in the Owens River system (modified from Smith, 1979).

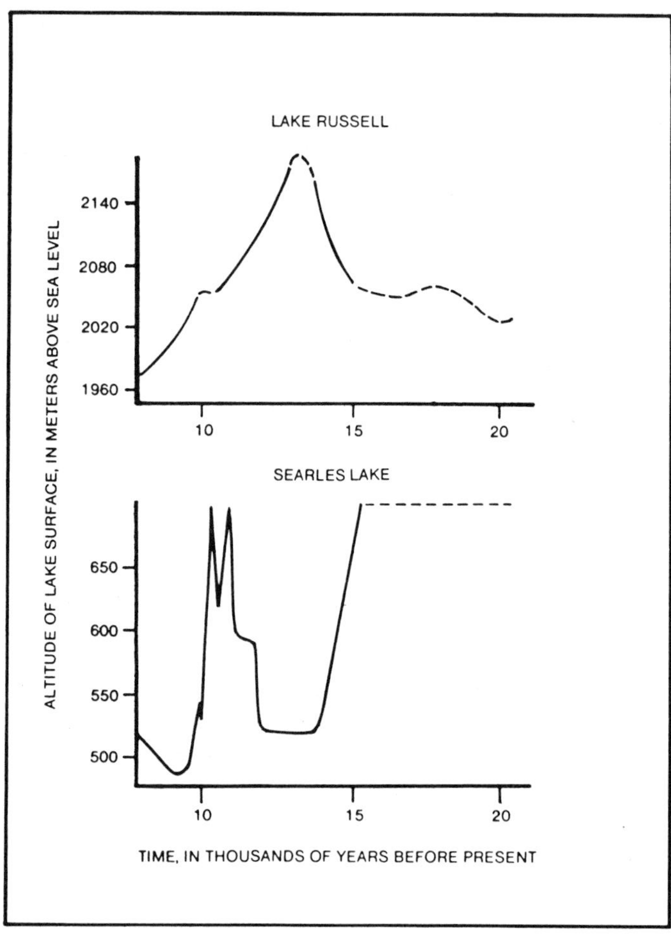

Figure 10. Comparison of the 20 to 7.5-ka chronologies of Lake Russell and Searles Lake (modified from Smith and Street-Perrott, 1983).

transgression sequence in their summary diagram for Lake Russell.

Lake Adobe, which lies south of Mono Lake (Fig. 9), attained a depth of more than 24 m during the late Pleistocene before overflowing into Owens Valley. Batchelder (1970a and b; diagram reproduced in Mehringer, 1977, p. 127) used changes in organic-carbon content, palynomorphs, plant macrofossils, and molluscs from sediment cores to interpret the lake-level history of this basin. He deduced that lake level was low at the base of the core ~11.5 ka and interpreted low organic-carbon concentrations between 11 and 8 ka as indicative of rising lake level.

Owens River System. The Owens River system was a chain of lakes occupying a succession of subbasins along the east side of the Sierra Nevada of California (Fig. 9). Drainage from the Sierra Nevada into the Owens River supplied most of the water in this chain. Although lakes in the Mono Lake basin probably overflowed into the Owens River system during the late Pleistocene, the last Pleistocene lake in this basin (Lake Russell) did not rise high enough to spill (Lajoie, 1968).

A detailed chronology is not yet available for the last Pleistocene lake cycle at Owens Lake, and most of what is known about the Owens River system stems from studies of Searles Lake (Smith, 1962, 1968, 1976, 1979, 1984; Stuiver and Smith, 1979). During the late Pleistocene, Searles Lake was third in a chain of five permanent lakes receiving water from the Owens River (Fig. 9). When Owens Lake filled to a depth of ~60 m, it overflowed into China Lake. When China Lake filled to ~12 m, it overflowed into Searles Lake, and when this lake filled to a depth of ~200 m, it coalesced with China Lake and overflowed into Panamint Lake.

The published chronology for Searles Lake is based solely on radiocarbon analyses of core material (Stuiver and Smith, 1979). Unpublished ^{14}C ages of nearshore materials from the Searles Lake subbasin support the core-based chronology (G. Smith, written communication, 1986). Unfortunately the chronologies for Searles Lake and Lake Russell for 20 to 11 ka are virtually diachronous (Fig. 10). This may be a result of: (1) incorrect interpretation of sedimentary depositional environments; for example, mistaking shallow-water for deep-water sediments; (2) incorrect age assignment to lake sediments and nearshore carbonate materials (tufa); for example, improper adjustment of ^{14}C ages of materials contaminated with secondary carbon, (3) the presence of a distinct climatic boundary that separated the Lake Russell watershed in the central Sierra Nevada from the adjoining Owens River watershed in the central and southern Sierra Nevada.

In the authors' opinion, the existing radiocarbon-based chronologies for Searles Lake and Lake Russell both need to be considered problematic until additional data are published that clearly demonstrate the reliability of samples used for radiocarbon determination.

Marshes in Southern Nevada. No evidence has been found for deep-water lakes of Late Wisconsin or Holocene age in the Mojave Desert part of southern Nevada (Mifflin and Wheat, 1979). Instead, marshes and wet meadows present in the early historic period were more numerous and larger in size. Stratigraphic studies at Tule Springs (Haynes, 1967), Corn Creek Springs, and Indian Springs (Quade, 1983, 1986; Fig. 1) indicate that marshes and associated shallow lakes reached their maximal extent between 30 and 15 ka. A progressive desiccation of the marsh/lake complexes occurred between 13.5 and 8.0 to 7.2 ka. After this time, perennial water was absent, and the sediments deposited during the moister Pleistocene and early Holocene were dissected by erosion. At Corn Creek Springs, the modern water table in the valley center is more than 25 m below the 8.5 ka water level.

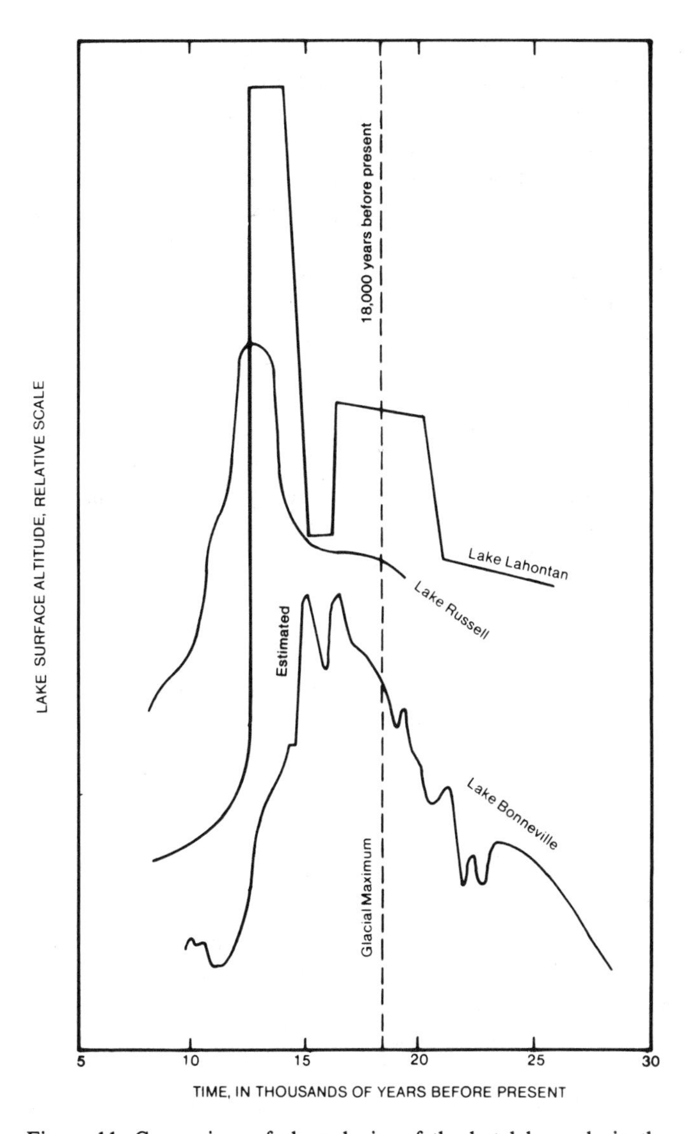

Figure 11. Comparison of chronologies of the last lake cycle in the Lahontan (data from Born, 1972; Benson and Thompson, 1981), Russell (data from Lajoie, 1968; Smith and Street-Perrott, 1983), and Bonneville (Currey and Oviatt, 1985) basins.

PALEOCLIMATIC IMPLICATIONS OF LAKE-LEVEL RECORDS

Numerous workers have attempted to infer the nature of climate change that resulted in high stands of closed-basin lakes in the Great Basin. Galloway (1970) and Brakenridge (1978) contend that times of lake maxima were relatively cold and dry. High lake levels were interpreted to have resulted from increased runoff due to reduced evaporation throughout the basin, combined with regional precipitation rates similar to or even somewhat less than present values. Other researchers argue that calculated temperature reductions for the late Pleistocene were insufficient to cause decreases in evaporation sufficient to maintain lake-level maxima, and therefore that substantial increases in precipitation must have occurred (Leopold, 1951; Antevs, 1952; Snyder and Langbein, 1962). The following section examines the methods used by these researchers to estimate past rates of evaporation, discharge, and precipitation.

Empirical Approaches to Calculation of Evaporation, Runoff, and Precipitation

Calculation of Evaporation. A large number of empirical methods have been developed for estimating evaporation rates from water surfaces, using commonly measured meteorological parameters, such as humidity and air temperature. A number of authors (Leopold, 1951; Antevs, 1952; Broecker and Orr, 1958; Snyder and Langbein, 1962; Galloway, 1970; Reeves, 1973; Brakenridge, 1978; Mifflin and Wheat, 1979) attempted to estimate the amounts of evaporation and precipitation responsible for the maintenance of various paleolake systems of probable Late Wisconsin age. The empirical estimation methods used by these authors to estimate evaporation rates can be generalized as follows: (1) an indicator of the location of past snowline or timberline (cirque-excavation features, relict cryogenic deposits, or nivation hollows) was used to estimate the amount of snowline (or timberline) depression relative to its modern location; (2) location of modern snowline was correlated with some seasonal mean value of temperature (summer 0°C, July 0°C, or −6°C annual isotherm; Leopold, 1951; Brakenridge, 1978). (3) a constant-value temperature lapse-rate, based on ground-level meteorological data or free-air radiosonde data, was used to estimate the amount of air-temperature lowering represented by the decline in snowline; (4) the same amount of air-temperature lowering in the high-altitude parts of the drainage basin was assumed to have occurred at the altitude of the lake surface; (5) the amount of mean-monthly air-temperature lowering was (usually) distributed over the calender year by either imposing a graduated reduction of mean-monthly temperature, with maximum reduction in July and no reduction in January, or by imposing a uniform reduction of monthly air temperature; (6) A correlation between values of mean-annual or mean-monthly air temperature (at lake-surface altitude) and lake-evaporation rate was attempted and applied to the paleolake system.

Early development and application of these empirical procedures by Leopold (1951) and Snyder and Langbein (1962) stimulated consideration of the climatic conditions responsible for paleolake highstands. Although subsequent authors generally did not examine the shortcomings and assumptions of the method, these were clearly stated in these early papers. A recent critique (Benson, 1986) of these empirical procedures concludes that they are flawed and no longer applicable.

Calculation of Runoff, Precipitation, and Air Temperature. The steady-state, mean-annual hydrologic balance for a lake and surrounding closed-basin drainage area can be expressed in terms of total intrabasin discharge to the lake and evaporation from the lake surface.

$$E_l A_l = P_l A_l + D_r + D_g \quad (1)$$

Where E_l = lake evaporation rate; A_l = surface area of the lake; P_l precipitation on the lake; D_r = runoff into the lake; and D_g = groundwater discharge into the lake.

By definition, interbasin groundwater flow and surface outflow do not occur in a closed basin system. For many closed basins the approximation also is made that D_g is insignificant relative to D_r.

Other researchers (for example Kutzbach, 1980) treat the total intrabasin discharge component of equation 1 in terms of precipitation and evapotranspiration processes. This can be expressed as:

$$D_r + D_g = P_{db}A_{db} - ET_{db}A_{db} \quad (2)$$

Where P_{db} = precipitation on the drainage basin; A_{db} = area of the drainage basin; and ET_{db} = evaporation and evapotranspiration in the drainage basin.

The steady state balance equation thus becomes:

$$E_l A_l = P_l A_l + P_{db}A_{db} - ET_{db}A_{db} \quad (3)$$

Equation 1 is preferred when suitable historical data exist for the calculation of D_r. Equation 3 can be applied when the calculation of D_r in terms of P_{db} and ET_{db} is made in terms of a watershed precipitation-runoff model.

Most estimates of the amount of precipitation, discharge, and air-temperature depression necessary to maintain paleolake systems are empirical in nature (Leopold, 1951; Snyder and Langbein, 1962; Galloway, 1970; Reeves, 1973; Mifflin and Wheat, 1979). They can be generalized as follows: (1) groundwater discharge to the lake basin (D_g of Equations 1 and 2) is set to zero; (2) runoff is considered to be a function of precipitation, evapotranspiration (Equation 2), and air temperature; (3) air temperature is computed from the location of the paleosnowline, mean-annual air temperature at the modern snowline, and modern air-temperature lapse rate; (4) temperature-dependent, mean-annual, precipitation-runoff curves such as those of Langbein and others (1949; Fig. 12) are used to estimate values of mean-annual runoff and precipitation necessary to balance evaporative losses from the lake.

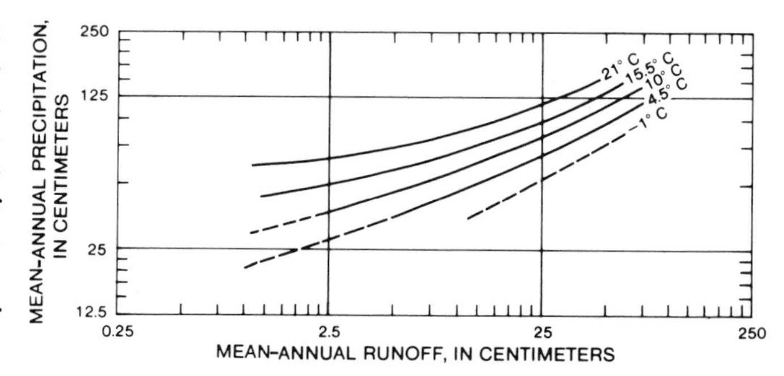

Figure 12. Relation between mean-annual runoff and mean-annual precipitation (modified from Langbein and others, 1949).

Problems with empirical procedures used to estimate air temperature and evaporation rate have been noted previously. Schumm (1965) commented, in respect to the application of mean annual precipitation-temperature-runoff curves (Fig. 12), that ". . . any such estimate (of past runoff) is only a crude approximation at best, and it appears prudent to estimate only the direction and relative magnitude of such a change. . . ." We agree with this statement, and we further note that the curves of Langbein and others (1949) were developed from discharge records from areas not affected materially by diversion or regulation. Large areas in the Great Basin were not equipped with gaging stations at the time of their study; and those areas that were gaged were heavily impacted by diversion and consumptive use. For these reasons the use of curves developed by Langbein and others (1949) does not appear to be appropriate for reconstructing past climates in the Great Basin. Many differences in opinion regarding paleoclimates of the Great Basin probably stem from inaccuracies inherent in the empirical methods used to extract climatic estimates from lake-level records.

Energy-Balance Models

In light of difficulties inherent in the empirical estimates of past evaporation rates discussed above, other approaches need to

TABLE 2. STREAMFLOW STATISTICS FOR RIVERS THAT DISCHARGE TO THE LAHONTAN BASIN

River name	Streamflow-gaging station number	Sample size (complete data sets)	Mean streamflow discharge (km^3 yr^{-1})	1983 streamflow discharge (km^3 yr^{-1})
Carson River	10309000	57	0.352	0.783
	10310000	52	0.102	0.229
Humboldt River	10322500	76	0.344	1.153
	10324500	43	0.033	0.131
	10329000	45	0.022	0.047
	10329500	62	0.030	0.096
			0.580*	0.580*
Quinn River	10353500	34	0.032	0.063+
	10353600	13	0.004	0.019
Susan River	10356500	34	0.085	0.213+
Truckee River	10346000	84	0.725	2.291
Walker River	10293500	31	0.127	0.381+
	10297500	36	0.213	0.548
			0.039*	0.039*

*Estimated consumptive use occurring upstream from streamflow-gaging stations; evapotranspiration rate of 1 m/yr used in estimate.

+Estimate made by regression.

be considered. Covariance, aerodynamic, Dalton, energy-balance, and combined approaches are reviewed and evaluated in Benson (1986) who considered the energy-balance method to be the most appropriate for determination of the sensitivity of evaporation rate to variation in climatic parameters. Each heat term contained in the energy-balance equation can be parameterized in terms of one or more commonly measured climatic variables.

Sensitivity of Evaporation Rate to Climatic Change. Benson (1986) used the Pyramid Lake subbasin as a reference lake-climate system to determine the sensitivity of evaporation rate to change in the amount and type of sky cover, air temperature, water temperature, dew-point temperature (humidity), and solar irradiation at the upper atmosphere. He drew several conclusions from the sensitivity analysis: (1) Evaporation rate decreases strongly with a decrease in the difference in temperature between air and water (T_a–T_o). For example, monthly decreases of 5°C and 10°C in T_a–T_o result in evaporation rate reductions of 0.35 and 0.59 m yr^{-1}. (2) Use of solar-irradiation values for 18 and 12 ka, the times of maximum continental glaciation and lake-level rise, results in negligible changes in the calculated evaporation rate relative to present-day conditions. (3) Relatively large absolute changes in relative humidity (RH) result in small changes in the calculated evaporation rates. For example, an extrapolated monthly increase of RH by 40 percent increases the RH of the driest month (July) to 64 percent, and increases the RH for four of the more humid months to more than 95 percent. However, the calculated mean-annual evaporation rate decreases by only 0.06 m. (4) Changes in the fractional distribution and absolute amount of sky cover can reduce the evaporation rate significantly.

Use of Extreme Historic Values in Simulations of Lake Lahontan Highstands. During the time of their historic highstand, lakes in the Lahontan Basin had a combined surface area of ~1,550 km^2 (King, 1878; Russell, 1885). In contrast, Lake Lahontan existed as a single body of water with a surface area of 22,300 km^2 at 13 ka (Fig. 1). What magnitude of climate change is necessary to account for a lake surface area 14.4 times larger than today's? Although an explicit answer to that question cannot be given, extreme values in historic precipitation, discharge, and evaporation data sets can be used to determine if a changed frequency of present-day hydrologic and meterologic events could account for the increase and maintenance of paleolake Lahontan. An energy-balance approach (Benson, 1986) is used to discuss evaporation-rate change in terms of changes in commonly measured climate parameters (air–water temperature differences, cloudiness, and relative humidity).

Historical discharge records (U.S. Geological Survey, 1960, 1963, 1961–1984) indicate that the annual discharge of rivers to the Lahontan basin achieved a maximum value approximately 2.45× its mean in 1983. Streamflow data for all six rivers that discharge to the Lahontan basin as well as precipitation data from nine basin-floor weather stations were assembled for 1983 (Tables 2 and 3). These data, with water-balance estimates of historical mean-annual evaporation rates (Harding, 1965), were used to estimate the hypothetical surface areas of lakes that would be created in the subbasins as the result of an increase in mean-

TABLE 3. PRECIPITATION STATISTICS FOR CERTAIN WEATHER STATIONS LOCATED ON THE FLOOR OF THE LAHONTAN BASIN, CALIFORNIA AND NEVADA

Weather Station	Record Length (yr)	Sample Size (complete data sets)	Annual Precipitation Statistics (cm)				
			$\bar{x}$	σ	minimum	maximum	1983
Fallon Experimental Station	78	77	12.7	4.2	4.2	23.3	21.5
Gerlach	38	38	15.6	6.8	6.3	36.6	33.0*
Hawthorne Airport	47	37	13.5	4.3	5.8	22.3	17.1
Kings River Valley	28	21	23.5	5.9	16.0	42.7	42.7
Lovelock	91	68	13.7	5.6	2.2	31.2	31.2
Reno Airport	123	123	17.9	6.6	1.3	34.9	33.6
Sand Pass	59	48	16.4	4.8	9.3	26.9	33.0*
Susanville Airport	57	56	38.3	12.5	13.5	63.0	62.0
Winnemucca	114	114	21.3	6.1	8.0	46.7	36.8
Yerrington	71	62	13.2	4.8	3.8	26.9	26.9

*Estimate made by comparison with nearby weather stations.

annual water input equal to 1983 values while leaving the value of the mean-annual evaporation rate unchanged from its historical value. The combined surface area of lakes in the Lahontan basin resulting from this hypothetical situation totals 6,920 km^2 (Fig. 13; Table 4), one third of the surface area that existed during the last highstand. To create the last highstand lake using 1983 values of fluid input (discharge plus on-lake precipitation), the mean annual basinwide evaporation rate must be reduced to about 0.63 m.

The 1983 evaporation rate measured at the Fallon Experimental Station (U.S. Department of Commerce, 1966–1984) was 1.10 m. In general, the variability of recorded evaporation rate is small; for example, the minimum annual evaporation rate measured at the Fallon Experimental Station (U.S. Weather Bureau, 1938–1965; U.S. Department of Commerce, 1966–1984) was 0.87 m, recorded in 1948. A mean annual evaporation rate (0.63 m) sufficiently small to create and maintain the Lake Lahontan highstand, given 1983 values of fluid input, can only be obtained by summing monthly evaporation minima for the historical period of record. Therefore the Lahontan 13-ka highstand in principle can be simulated by combining historical monthly evaporation extremes (minima) with the maximum annual value of discharge to the Lahontan basin (together with 1983 values of basin precipitation) such that a changed frequency of present-day hydrologic and meteorologic events can be hypothesized to account for the growth and maintenance of paleolake Lahonton.

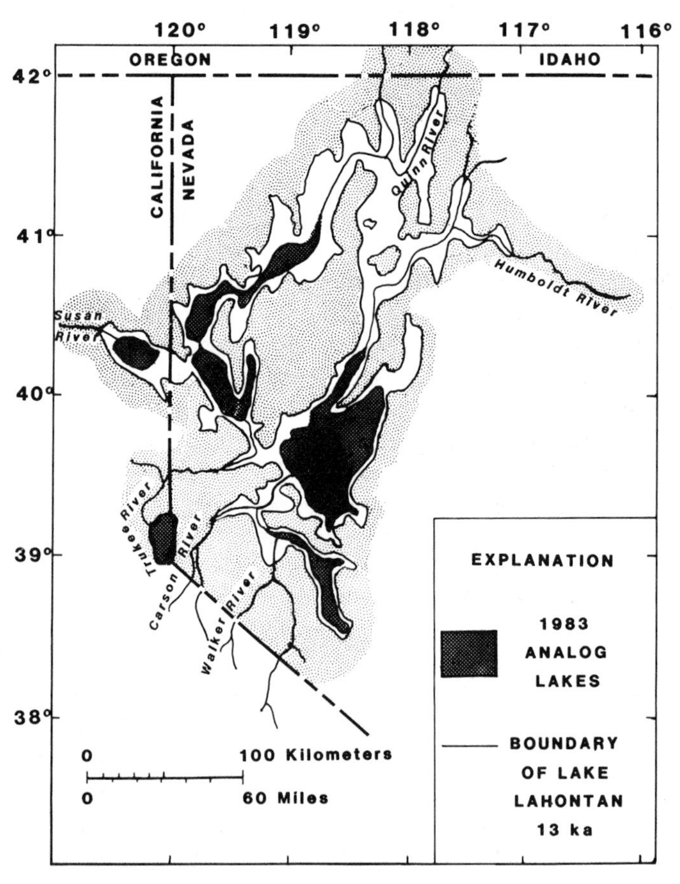

Figure 13. Lake size in the Lahontan Basin that would result if mean-annual discharge was increased to that recorded in 1983.

Implications of Evaporation Reduction. An evaporation rate of 0.63 m yr^{-1} represents a reduction of 0.61 m relative to the present-day mean annual value used in the previous calculations. A recent analysis of the sensitivity of evaporation rate to changes in commonly measured climate parameters (Benson, 1986) indicates that evaporation rate is strongly dependent on the difference between air and water-surface temperatures as well as the type and degree of cloudiness. Recent paleobotanic, geomorphic, and amino-acid studies (Thompson, 1984a, Dohrenwend, 1984; McCoy, 1981) indicate a lowering of Late Wisconsin

TABLE 4. ANALOG CALCULATIONS USING STREAMFLOW DATA FOR 1983 FOR RIVERS THAT DISCHARGE TO THE LAHONTAN BASIN

Subbasin Lake	Area (km^2)	Volume (km^3)	Surface Altitude (m)	Depth (m)
Walker Lake	781	35	1,308	136
Carson Lake	3,215	55	1,204	30
Pyramid-Winnemucca Lake	1,017	68	1,207	153
Black Rock-Smoke Creek Lake	1,560	9	1,188	18
Honey Lake	344	0.65	1,219	2.6

mean-annual air temperature by about 5 to 7°C. Energy-balance calculations (Benson, 1986) indicate that reducing the mean-annual air temperature by 7°C (with sky cover and lake-surface temperature set to present-day values) results in a decreased evaporation rate of 0.64 to 0.72 m yr^{-1}, depending on the manner in which the air-temperature reduction is distributed throughout the annual cycle. Thus a reduction of 7°C in mean-annual air temperature is sufficient, when combined with the 1983 discharge, to achieve the 1330-m Lahontan highstand, if no other factures are considered.

However, the assumption that the mean-annual air temperature decreases without a corresponding proportional decrease in mean-annual water-surface temperature is erroneous; experience dictates that if the air is colder so also is the lake. This fact means that a reduction of 7°C in mean-annual air temperature causes a similar decrease in water temperature and therefore little change in either the air-water temperature difference or in the evaporated rate. This has significant consequences in terms of fixing the cause of evaporation reduction in the Lahontan basin. If, for example, the air and water-surface temperature difference decreased by only 2°C each month rather than by 7°C, significant changes in the amount or type of cloudiness also would be required to reduce the calculated evaporation rate to 0.63 m yr^{-1}, the value required to expand the lake area to its highstand condition (Benson, 1986).

Lake/Climate Feedback Processes

Other physical processes are associated with large lake systems that aid in their growth and maintenance. While the difference between air- and water-surface temperatures may not have decreased significantly during the Late Wisconsin, a lowering of monthly water temperature by 10°C relative to present-day values (Benson and Spencer, 1983; Benson, 1984) would have allowed ice formation during the winter and early spring (November through March). The presence of ice would tend to decrease evaporation by enhancing reflection of short- and long-wave radiation. Additional heat ordinarily used in the evaporation process would be expended during melting of the ice in the spring. Also, presence of ice lasting well into the spring enhances radiation reflection at a time when incoming radiation values are large.

The presence of glaciers in surrounding watersheds also may have contributed to reduction in the lake-evaporation rate. Discharge from glaciers would have occurred at temperatures near 0°C. The transport times of rivers discharging to the Lahontan basin were significantly reduced at 13 ka; and transport would have taken place under lower ambient air temperatures relative to today. These differences would have decreased heat transport to rivers during flow to the lake basin. Colder discharge to a lake would in turn have suppressed evaporation by decreasing the air-water temperature difference.

Large lakes also exert significant effects on weather in their vicinity; for example, lake-induced snow storms often occur when cold Arctic air masses move across relatively warm lakes. These "lake-effect" snow storms have been shown by Braham and Dungey (1984) to cause enhanced snowfall (30 to 120 percent) on the lee side of Lake Michigan.

Eubank and Brough (1980) recently determined that Great Salt Lake affects local temperatures, wind patterns, and precipitation. The largest air-water temperature differences occur in the fall and spring, when major weather patterns are undergoing transition. The "lake effect" appears greatest at these times, indicating thermal instability as the causal mechanism. The probability of significant storminess also increases when the jet stream is located near or just south of the lake and where the winds are from the northwest or southwest. Most heavy snows along the shores of Great Salt Lake occur in narrow strips 40 to 80 km wide located near the jet-stream core. Further precipitation enhancement occurs when moist unstable wind is forced upslope downwind of the lake.

Late Wisconsin Great Basin lake systems probably exerted a significant effect on regional weather. The hypothesis is suggested that the jet stream was forced south of its present-day location by the continental ice sheet (Antevs, 1948; Kutzbach and Wright, 1985), creating the thermal instability necessary to cause precipitation. In addition, precipitation may have been enhanced locally as winds moistened by the lake effect were forced upslope against orographic barriers on lee sides of each lake system, and the increase in storm frequency may have been accompanied by increases in the amount of sky cover and heap-type clouds, causing decreases in evaporation rate. Therefore, after attaining some critical size, Great Basin lakes could have enhanced their own development by altering the regional climate through lake-effect and other feedback processes.

SYNOPTIC CLIMATE AND LAKE-LEVEL FLUCTUATIONS

The sensitivity analysis described above indicates that high lake levels result from reduced air-water temperature difference, increased discharge, increased cloudiness, and lake-climate feedback mechanisms. Given the modern climatic gradients and synoptic climatology of the Great Basin, what are the probable climatic variations that led to past lake-level variations?

The Synoptic Climate of Recent Lake-Level Increases

The regional low lake levels of the 1960s gave way to deeper lakes in the 1980s. In particular, 1983 was an extremely wet year and resulted in significant rises in lake level for Mono, Pyramid, Walker, and Great Salt lakes. In 1983, precipitation was greater than normal for most of the cooler months, with the maximum increase in November in the western region of the Great Basin and in December in the central and eastern regions. Summer rainfall was enhanced in the eastern region; while large as a percentage of the long-term summer mean, it supplied relatively little moisture in absolute amounts.

This example demonstrates that synchronous regional lake-level rises may result from increased cool-season precipitation produced by enhanced cyclonic flow. The Pacific Ocean is the primary source of this moisture under Houghton's (1969) Pacific and Continental regimes. The strength of these two circulation regimes is affected primarily by the position of the polar jet stream. Summer convective storms from the Gulfs of Mexico and California probably are unimportant in generating lake-level rise in the Great Basin, because the high evaporation rate associated with this season reduces the effective input to drainage basins.

The Synoptic Climate of Late Wisconsin Highstands

The path of westerlies bringing moisture-laden air off the Pacific Ocean appears to be an important determinant of lake levels in the Great Basin. The physiography of the Great Basin and adjoining mountain masses along its western periphery channel this flow along certain predictable routes (Bryson and Hare, 1974; Mitchell, 1976; Tang and Reiter, 1984). The distribution of glacial features in the Great Basin indicates that the anchoring effects of physiography during the late Pleistocene also channeled westerly flow along near-modern trajectories (Moran, 1974).

Several authors (Riehl and others, 1954; Horn and Bryson, 1960; Sabbagh and Bryson, 1962; and Pyke, 1972) suggest that the progression of maximum precipitation along the western coast of North America is associated with the southward movement of the mean position of the polar jet stream. Starrett (1949) and Yeh (1950) note the tendency for precipitation maxima to concentrate near the axis of the jetstream; precipitation decreases rapidly south of the axis and less rapidly north of the axis.

Although the overall physiography of the Great Basin has not changed significantly during the past 25 kyr, the boundary conditions for late-Pleistocene climate in western North America were quite different. The presence of a large ice sheet 3 km high across Canada greatly affected the position and intensity of westerly flow, forcing the jetstream southward thoughout the year (Kutzbach, this volume).

Climatic anomalies of opposite sign in the Southwest and Northwest have been inferred from historic climatic records and proxy-climate data for the last 300 yr (Sellers, 1968; LaMarche and Fritts, 1971; Meko and Stockton, 1984). Sellers (1968) ascribes these anomalies to north–south fluctuations in the track of the westerlies, for the Southwest is wet and the Northwest is dry when the jet stream shifts southward. This shift probably prevailed during the Late Wisconsin, because anticyclonic flow over the continental ice sheets forced westerly flow southward (Antevs, 1952; Kutzbach, this volume). In addition, the temperature difference between the ice sheets to the north and the relatively warm sea-surface temperatures to the south should have contributed to the intensification of the summer jet stream.

Paleoclimatic evidence from the Northwest (Barnosky and others, this volume) indicates dry late-Pleistocene conditions, in contrast to relatively moist conditions in the Southwest (Van Devender and others, this volume). These anomalies of opposing signs thus resemble those found in the historic period (Sellers, 1968) and probably reflect past positions of the westerlies. These anomalies have been simulated by model results (Kutzbach, this volume), which indicate that the Northwest in the late-Pleistocene was subject to anticyclonic flow bringing dry continental air into this region from the east (see also Barnosky and others, this volume).

The Synoptic Climate of Post-Wisconsin Lowstands

If the jet-stream hypothesis described above is correct, then regional low lake levels may reflect periods when the mean position of the westerlies lies far north (or south) of the Great Basin. The repositioning of the jet-stream, which reflects a change in the frequency and amplitude of the Rossby Wave pattern, may result from changes in the latitudinal distribution of solar insolation and the removal of the ice-sheet boundary condition.

If the westerlies shifted northward in the middle Holocene (and perhaps during the late-glacial), summer monsoons may be hypothesized to have penetrated farther north. Although available fossil data are not adequate to test this hypothesis, summer rainfall occurring during the season of maximum evaporation would have been less effective than cool-season precipitation in maintaining paleolakes in the Great Basin.

MONTANE GLACIERS AND THE LAST LAKE CYCLE

Late Wisconsin glaciers were present in many of the Great Basin mountain ranges (Blackwelder, 1931); relict periglacial features have been found at high altitudes throughout this region (Dohrenwend, 1984). For more than a century reseachers have attempted to determine if the maximum extent of the glaciers coincided with the highest stands of the Pleistocene lakes.

Stratigraphic Studies

At their maximum extent glaciers and Pleistocene lakes occurred together at only two sites in the Great Basin: the Bonneville basin along the western front of the Wasatch Mountains, and along the western edge of Mono Lake. Antevs (1952) noted the occurrence of moraines beneath deltas and of daltaic sediments in the gap between lateral moraines up to the level of the Bonneville shoreline along the Wasatch front. From this evidence he deduced that the glaciers had reached their maximum levels and had withdrawn prior to the lake reaching this level. Morrison and Frye (1965, p. 25), however, argued that the stratigraphic exposures ". . . demonstrate direct correlations between some of the glacial and lake maxima." These authors further claimed that some of the till was deposited in lake waters.

In contradiction to Morrison and Frye's conclusion, Scott and others (1982, 1983) found soil development on the Pinedale till at the mouths of Bells and Little Cottonwood canyons. This soil was partially eroded and subsequently buried by sediments from the transgression of Lake Bonneville. Through comparison with soils of known Holocene age, Scott and others (1983) estimated that several thousand years had passed between the retreat of the glaciers from their maximal Pinedale extent and the rise of the lake to the Bonneville Shoreline. Scott (cited as a personal communication to Porter and others, 1983, p. 99) estimated that the transgression of Lake Bonneville above this level occurred ~17 ka.

At Mono Lake, Gilbert (1890) also concluded that the glaciers retreated prior to the highest lake levels. He wrote (p. 314) "With one voice these four localities tell us that Mono Lake occupied its maximum level after the glaciers of the Sierra had retreated from their most advanced position. But their testimony goes no farther. . . .If the two sets of phenomena were consequent upon the same series of climatic changes, then the lacustral lagged behind the glacial." Putnam (1950, p. 121) disagreed with this conclusion and argued that the similarities between the patterns of glacial advances and retreats and lake-level rises and declines were similar enough to imply synchroneity. Lajoie (1968) was critical of Putnam's (1950) methods and determined that the pattern of shoreline widths and spacing was the result of variations in the slopes into which these features were eroded, and not to any relationship with glacial phenomena. Moreover, Lajoie (1968, p. 181) determined that, because of poor exposures and complex deltaic stratigraphy, lacustral and glacial events could not be correlated.

Radiometric Studies

As discussed earlier in this paper, Pleistocene lakes in the Great Basin did not achieve their maximum levels until ~15 ka and began their regressions by 14 to 13 ka. Radiocarbon dating of features associated with mountain glaciers in this region indicates that ice retreat probably was well underway before the lakes peaked. Porter and others (1983) determined that Pinedale deglaciation had begun well before 14 ka. At Snowbird Bog in Little Cottonwood Canyon in the Wasatch Mountains, a date of 12.3 ± 0.3 ka places a limiting age on mid-canyon deglaciation well above the Pinedale terminal moraine (Madsen and Currey, 1979). These authors estimated that the glacier had actually retreated above this point by 14 to 13 ka.

On the northern periphery of the Bonneville basin, a radiocarbon date from the basal sediments of a high-altitude lake places ice withdrawal prior to 12 ka (Mehringer and others, 1971; Mehringer, 1977). In the Ruby Mountains of northeastern Nevada, Wayne (1984) obtained a radiocarbon date of 13.0 ± 0.9 ka on basal sediments of a bog formed after the glacier had retreated nearly to its headwall.

The stratigraphic and radiometric data on lake-glacial correlations are somewhat ambiguous, although most of the evidence indicates glacial recessions occurring well before the lakes reached their highest levels. The data, if correct, would appear to indicate increased moisture availability at a time when glaciers in the upper reaches of watersheds were retreating or absent. The hypothesis is suggested that the seasonality of precipitation shifted to warmer months, or alternatively that the winter season became warmer. A shifting of precipitation to warmer months would necessarily have been accompanied by increased cloudiness in order to decrease the normally high, warm-season evaporation rate. This would starve the montane glacier systems in the winter season yet provide sufficient input of water to maintain highstand lakes. The warm-season precipitation is hypothesized to have occurred under a westerly (low-latitude jet stream) regime rather than a monsoonal regime.

SUMMARY

Comparison of chronologies of the last (Late Wisconsin) lake-cycle in the Lahontan and Bonneville basins (Fig. 11) indicates that: (1) lake-level rises occurred gradually in both basins; (2) the gradual rise in lake level appears to have been interrupted by a possible recessional event at ~16 ka; (3) the subsequent decline in lake level occurred rapidly, although perhaps at different times in each basin.

The lowest lake stands recorded in the Bonneville basin occurred immediately after the decline from the pluvial highstand. Following this lowstand, Lake Bonneville rose at about 10.3 ka to the Gilbert shoreline, well above the historic level of Great Salt Lake. A similar sequence of events occurred at the Ruby Marshes, and perhaps at Mono Lake, but in the Lahontan basin, lake-level data for the 12 to 10 ka period are lacking. The events are poorly dated except in the Bonneville basin, and although the stratigraphic sequences are similar across the Great Basin, the events may not have been synchronous. The range of climatic variability, as expressed by changing lake levels, appears to have been much greater between 13 to 10 ka than during the Holocene. The marsh chronologies from the Mojave Desert region of southern Nevada do not reflect the degree of variability indicated by the lake records from farther north. Instead, the marsh data indicate a progressive unidirectional trend in desiccation from 13.5 to 7.5 ka. The delay in apparent desiccation may have resulted from the lag time implicit in the noninstantaneous transport of water from areas of recharge to areas of discharge (marshes).

The last lake cycle is hypothesized to have occurred as the result of two processes: (1) forcing of one branch of the jet stream south by the presence of a large and high continental ice sheet (Kutzbach and Wright, 1985), and (2) by the initiation of lake-climate feedback processes.

Montane glaciers appear to have begun their recession prior to the last lake highstand. It is hypothesized that this was caused by a shift in seasonality, whereby: (1) precipitation generated by the westerlies caused more summer precipitation, or (2) winters became warmer.

REFERENCES CITED

Antevs, E., 1925, On the Pleistocene history of the Great Basin: Carnegie Institute of Washington Publication 352, p. 51–114.

——, 1938, Postpluvial climatic variations in the southwest: Bulletin of the American Metorological Society, v. 19, p. 190–193.

——, 1948, The Great Basin, with emphasis on glacial and post-glacial times; Climatic changes and pre-white man: Bulletin of the University of Utah Biological Series, v. 38, p. 168–191.

——, 1952, Cenozoic climates of the Great Basin: Geologische Rundschau, v. 40, p. 94–108.

Batchelder, G. L., 1970a, Post-glacial fluctuations of lake level in Adobe Valley, Mono County, California: Abstracts of the First Meeting of the American Quaternary Association, p. 7.

——, 1970b, Post-glacial ecology at Black Lake, Mono County, California [Ph.D. thesis]: Tempe, Arizona State University, 181 p.

Benson, L. V., 1978, Fluctuation in the level of pluvial Lake Lahontan during the last 40,000 years: Quaternary Research, v. 9, p. 300–318.

——, 1981, Paleoclimatic significance of lake-level fluctuations in the Lahontan Basin: Quaternary Research, v. 16, p. 390–403.

——, 1984, Hydrochemical data for the Truckee River drainage system, California and Nevada: U.S. Geological Survey Open-File Report 84–440, 35 p.

——, 1986, The sensitivity of evaporation rate to climate change; Results of an energy-balance approach: U.S. Geological Survey Water-Resources Investigations Report 86–4148, 40 p.

Benson, L. V., and Mifflin, M. D., 1986, Reconnaissance bathymetry of basins occupied by Pleistocene Lake Lahontan, Nevada and California: U.S. Geological Survey Water-Resources Investigations Report 85–4262, 14 p.

Benson, L. V., and Spencer, R. J., 1983, A hydrochemical reconnaissance of the Walker River Basin, California and Nevada: U.S. Geological Survey Open-File Report 83–740, 53 p.

Benson, L. V., and Thompson, R. S., 1987, Lake-Level variation in the Lahontan Basin for the past 50,000 years: Quaternary Research (in press).

Billings, W. D., 1954, Temperature inversions in the pinyon-juniper zone of a Nevada mountain range: Butler University Botanical Studies, v. 11, p. 112–118.

Blackwelder, E., 1931, Pleistocene glaciation in the Sierra Nevada and Basin Ranges: Geological Society of America Bulletin, v. 42, p. 865–922.

Born, S. M., 1972, Late Quaternary history, deltaic sedimentation, and mudlump formation at Pyramid Lake, Nevada: Center for Water Resources Research, Desert Research Institute, University of Nevada System, unnumbered publication, 97 p.

Braham, R. R., Jr., and Dungey, M. J., 1984, Quantitative estimates of the effect of Lake Michigan on snowfall: Journal of Climate and Applied Meteorology, v. 23, p. 940–949.

Brakenridge, G. R., 1978, Evidence for a cold, dry full-glacial climate in the American Southwest: Quaternary Research, v. 9, p. 22–40.

Broecker, W. S., and Orr, P. C., 1958, Radiocarbon chronology of Lake Lahontan and Lake Bonneville: Geological Society of America Bulletin, v. 69, p. 1009–1032.

Broecker, W. S., and Kaufman, A., 1965, Radiocarbon chronology of Lake Lahontan and Lake Bonneville II, Great Basin: Geological Society of America Bulletin, v. 76, p. 537–566.

Broecker, W. S., and Walton, A., 1959, The geochemistry of ^{14}C in freshwater systems: Geochimica et Cosmochimica Acta, v. 16, p. 15–38.

Bryson, R. A., and Hare, H. K., 1974, The climates of North America, *in* Bryson, R. A., and Hare, F. K., eds., Climates of North America (Volume 11 of Landsberg, H. E., ed., World survey of climatology [series]): New York, Elsevier, p. 1–47.

Clarke, F. W., 1924, The data of geochemistry, 5th ed.: U.S. Geological Survey Bulletin 770, 841 p.

Currey, D. R., 1980, Coastal geomorphology of Great Salt Lake and vicinity: Utah Geological and Mineralogical Survey Bulletin, v. 116, p. 69–82.

Currey, D. R., and Oviatt, C. G., 1985, Durations, average rates, and probable cause of Lake Bonneville expansions, stillstands, and contractions during the last deep-lake cycle, 32,000 to 10,000 years ago, *in* Kay, P. A., and Diaz, H. F., eds., Problems of and prospects for predicting Great Salt Lake levels: Center for Public Affairs and Administration, University of Utah, p. 1–9.

Currey, D. R., Atwood, G., and Mabey, D. R., 1983, Major levels of Great Salt Lake and Lake Bonneville: Utah Geological and Mineral Survey, Map 73.

Davis, J. O., 1978a, Quaternary tephrochronology of the Lake Lahontan area, Nevada and California: Reno, Nevada Archeological Survey Research Paper 7, 123 p.

——, 1978b, Late Sehoo discharge of the Humboldt River; Stratigraphy archeology at the North Valmy power plant, Humboldt County, Nevada: Geological Society of America Abstracts with Programs, v. 10, p. 386.

——, 1982, Bits and pieces; The last 35,000 years in the Lahontan area, *in* Madsen, D. B., and O'Connell, J. F., eds., Man and environment in the Great Basin: Society for American Archaeology Papers No. 2, p. 53–75.

——, 1983, Level of Lake Lahontan during deposition of the Trego Hot Springs tephra about 22,400 years ago: Quaternary Research, v. 19, p. 312–324.

Dohrenwend, J. C., 1984, Nivation landforms in the western Great Basin and their paleoclimatic significance: Quaternary Research, v. 22, p. 275–288.

Eardley, A. J., 1966, Sediments of Great Salt Lake, *in* Great Salt Lake: Utah Geological Society Guidebook to Geology of Utah, v. 20, p. 105–120.

Eubank, M. E., and Brough, R. C., 1980, The Great Salt Lake and its influence on the weather, *in* Gwynn, J. W., ed., Great Salt Lake; A scientific, historical, and economic overview: Utah Geological and Mineral Survey Bulletin 116, p. 279–283.

Everett, D. E., and Rush, F. E., 1967, A brief appraisal of the Walker Lake area, Mineral, Lyon, and Churchill Counties, Nevada: U.S. Geological Survey Water Resources Reconnaissance Series Report 40, 41 p.

Farnsworth, R. K., Thompson, E. S., and Peck, E. L., 1982, Evaporation atlas for the contiguous 48 United States: U.S. Department of Commerce, National Oceanic and Atmospheric Administration Technical Report NWS 33, 25 p.
Feth, J. H., 1959, Re-evaluation of the salt chronology of several Great Basin Lakes; A discussion: Geological Society of America Bulletin, v. 70, p. 637–640.
Gale, H. S., 1914, Notes on the Quaternary lakes of the Great Basin: U.S. Geological Survey Bulletin 540, 401 p.
—— , 1915, Salines in the Owens, Searles, and Panamint basins, southeastern California: U.S. Geological Survey Bulletin 580, p. 251–323.
Galloway, R. W., 1970, The full-glacial climate in the southwestern United States: Annals of the Association of American Geographers, v. 60, p. 245–256.
Gilbert, G. K., 1890, Lake Bonneville: U.S. Geological Survey Monograph 1, 438 p.
Harding, S. T., 1965, Recent variations in the water supply of the western Great Basin: University of California Archive Series, Report 16, 226 p.
Harrison, S. P., and Metcalfe, S. E., 1985, Spatial variations in lake levels since the last glacial maximum in the Americas north of the equator: Zeitschrift für Gletscherkund und Glazialgeologia, Band 21, p. 1–15.
Haynes, C. V., Jr., 1967, Quaternary geology of the Tule Springs area, Clark County, Nevada: Reno, Nevada State Museum Anthropological Papers No. 13, p. 1–128.
Horn, L. H., and Bryson, R. A., 1960, Harmonic analysis of the annual march of precipitation over the United States: Annals of the Association of American Geographers, v. 50, p. 157–171.
Houghton, J. G., 1969, Characteristics of rainfall in the Great Basin: Reno, Nevada, Desert Research Institute, 205 p.
Houghton, J. G., Sakamoto, C. M., and Glifford, R. O., 1975, Nevada's weather and climate: Nevada Bureau of Mines and Geology Special Publication No. 2, 78 p.
Huntington, E., 1914, The climatic factor as illustrated in arid America: Carnegie Institute of Washington Publication No. 192.
Jennings, J. D., 1957, Danger Cave: University of Utah Anthropological Paper 27, 328 p.
Jones, J. C., 1925, The geologic history of Lake Lahontan: Carnegie Institute of Washington Publication No. 325, p. 3–50.
—— , 1929, Age of Lake Lahontan: Geological Society of America Bulletin, v. 40, p. 533–540.
King, C., 1878, United States geological explorations of the fortieth parallel, systematic geology: Washington, D.C., U.S. Government Printing Office, v. 1, 803 p.
King, G. Q., 1978, The late Quaternary history of Adrian Valley, Lyon County, Nevada [M.S. thesis]: Salt Lake City, University of Utah, Department of Geology, 88 p.
Kutzbach, J. E., 1980, Estimates of past climate at paleolake Chad, North Africa, based on a hydrological and energy-balance model: Quaternary Research, v. 14, p. 210–223.
Kutzbach, J. E., and Wright, H. E., Jr., 1985, Simulation of the climate of 18,000 yr B.P.; Results for the North American/North Atlantic/European Sector: Quaternary Science Reviews, v. 4, p. 147–187.
Lajoie, K. P., 1968, Late Quaternary stratigraphy and geologic history of Mono Basin, eastern California [Ph.D. thesis]: Berkeley, University of California, 271 p.
LaMarche, V. C., Jr., and Fritts, H. C., 1971, Anomaly patterns of climate over the western United States, 1700–1930, derived from principal component analysis of tree-ring data: Monthly Weather Review, v. 99, p. 138–142.
Langbein, W. B., and others, 1949, Annual runoff in the United States: U.S. Geological Survey Circular 52, 13 p.
Leopold, L. H., 1951, Pleistocene climate in New Mexico: American Journal of Science, v. 249, p. 152–168.
Madsen, D. B., and Currey, D. R., 1979, Late Quaternary glacial and vegetation changes, Little Cottonwood Canyon area, Wasatch Mountains, Utah: Quaternary Research, v. 12, p. 254–270.
Malde, H. E., 1960, Evidence in the Snake River plain, Idaho, of a catastrophic flood from Pleistocene Lake Bonneville, *in* Short papers in the geological sciences: U.S. Geological Survey Professional Paper 400–B, p. 295–297.
Malde, H. E., and Trimble, D. E., 1965, Malad Springs to Pocatello: International Association Quaternary Research (INQUA), 7th (1965) Congress, Guidebook, Field Conference E, (Northern and Middle Rocky Mountains), p. 98–103.
Mason, D. T., 1967, Limnology of Mono Lake: Berkeley, University of California publications in Zoology, v. 83, 110 p.
McCoy, W. D., 1981, Quaternary aminostratigraphy of the Bonneville and Lahontan basins, Western U.S., with paleoclimatic implications [Ph.D. thesis]: Boulder, University of Colorado, 603 p.
Mehringer, P. J., Jr., 1977, Great Basin late Quaternary environments and chronology: Reno, Nevada, Desert Research Institute Publications in the Social Sciences, v. 12, p. 113–165.
Mehringer, P. J., Jr., Nash, W. P., and Fuller, R. H., 1971, A Holocene volcanic ash from northwestern Utah: Proceedings of the Utah Academy of Sciences, Arts, and Letters, v. 48, p. 46–51.
Meko, D. M., and Stockton, C. W., 1984, Secular variations in streamflow in the western United States: Journal of Climate and Applied Meteorology, v. 23, p. 889–897.
Mifflin, M. D., and Wheat, M. M., 1979, Pluvial lakes and estimated pluvial climates of Nevada: Nevada Bureau of Mines and Geology Bulletin 94, 57 p.
Mitchell, V. L., 1976, The regionalization of climate in the western United States: Journal of Applied Meteorology, v. 15, p. 920–927.
Moran, J. M., 1974, Possible coincidence of a modern and a glacial-age climatic boundary in the montane West, United States: Arctic and Alpine Research, v. 6, p. 319–321.
Morrison, R. B., 1964, Lake Lahontan; Geology of southern Carson Desert, Nevada: U.S. Geological Survey Professional Paper 401, 156 p.
—— , 1965, Quaternary geology of the Great Basin, *in* Wright, H. E., Jr., and Frey, D. G., eds., The Quaternary of the United States: Princeton, New Jersey, Princeton University Press, p. 265–286.
Morrison, R. B., and Frye, J. C., 1965, Correlation of the middle and late Quaternary successions of the Lake Lahontan, Lake Bonneville, Rocky Mountain, southern Great Plains, and eastern Midwest areas: Reno, Nevada Bureau of Mines Report 9, 45 p.
Porter, S. C., Pierce, K. L., and Hamilton, T. D., 1983, Late Wisconsin mountain glaciation in the western United States, *in* Wright, H. E., Jr., ed., Late Quaternary environments of the United States, v. 1, *in* Porter, S. C., ed., The Late Pleistocene: Minneapolis, University of Minnesota Press, p. 71–114.
Putnam, W. C., 1950, Moraine and shoreline relationships at Mono Lake, California: Geological Society of America Bulletin, v. 61, p. 115–122.
Pyke, C. B., 1972, Some meteorological aspects of the seasonal distribution of precipitation in the western United States and Baja California: University of California Water Resources Center, Contribution No. 139, 205 p.
Quade, J., 1983, Quaternary Geology of the Corn Creek Springs area, Clark County, Nevada [M.S. Thesis], Tucscon, University of Arizona, Tucson, 135 p.
—— , 1986, Late Quaternary environmental changes in the upper Las Vegas Valley, Nevada: Quaternary Research, v. 20, p. 340–357.
Reeves, C. C., Jr., 1973, The full-glacial climate of the southern High Plains, West Texas: Journal of Geology, v. 81, p. 693–704.
Riehl, H., Alaka, M. A., Jordan, C. L., and Renard, R. J., 1954, The jet stream: Meteorological Monograph, v. 2, p. 23–47.
Russell, I. C., 1885, Geological history of Lake Lahontan, a Quaternary lake of northwestern Nevada: U.S. Geological Survey Monograph 11.
Sabbagh, M. E., and Bryson, R. A., 1962, Aspects of the precipitation climatology of Canada investigated by the method of harmonic analysis: Annals of the Association of American Geographers, v. 52, p 426–440.
Sarna-Wojcicki, A. M., Champion, D. E., and Davis, J. O., 1983, Holocene volcanism in the conterminous United States and the role of silicic volcanic ash layers in correlation of latest Pleistocene and Holocene deposits, *in* Wright, H. E., Jr., ed., Late Quaternary environments of the United States,

v. 1, The Holocene: Minneapolis, University of Minnesota Press, p. 52–77.
Schumm, S. A., 1965, Quaternary paleohydrology, *in* Wright, H. E., Jr., and Frey, D. G., eds., The Quaternary of the United States: Princeton, New Jersey, Princeton University Press, p. 783–794.
Scott, W. E., Shroba, R. R., and McCoy, W. D., 1982, Guidebook for the 1982 Friends of the Pleistocene, Rocky Mountain Cell, field trip to Little Valley and Jordan Valley, Utah: U.S. Geological Survey Open-File Report 82-845, 58 p.
Scott, W. E., McCoy, W. D., Shroba, R. R., and Rubin, M., 1983, Reinterpretation of the exposed record of the last two cycles of Lake Bonneville, western United States: Quaternary Research, v. 20, p. 261–285.
Sellers, W. D., 1968, Climatology of monthly precipitation patterns in the western United States, 1931–1966: Monthly Weather Review, v. 96, p. 585–595.
Smith, G. I., 1962, Subsurface stratigraphy of the Late Quaternary deposits, Searles Lake, California; A summary: U.S. Geological Survey Professional Paper 450-C, p. 65–69.
——, 1968, Late Quaternary geologic and climatic history of Searles Lake, southeastern California, *in* Morrison, R. B., and Wright, H. E., Jr., eds., Means of correlation of Quaternary successions: Proceedings of the Seventh Congress of the International Association for Quaternary Research, Salt Lake City, Utah, University of Utah Press, v. 8, p. 293–310.
——, 1976, Paleoclimatic record in the upper Quaternary sediments of Searles Lake, California, U.S.A., *in* Horie, S., ed., Paleolimnology of Lake Biwa and the Japanese Pleistocene, Privately published, Kyoto, Japa, v. 4, p. 577–604.
——, 1979, Subsurface stratigraphy and geochemistry of late Quaternary evaporites, Searles Lake, California: U.S. Geological Survey Professional Paper 1043, 130 p.
——, 1984, Paleohydrologic regimes in the southwestern Great Basin, 0–3.2 m.y. ago, compared with other long records of "global" climate: Quaternary Research, v. 22, p. 1–17.
Smith, G. I., and Street-Perrott, F. A., 1983, Pluvial lakes of the Western United States, *in* Wright, H. E., Jr., ed., Late Quaternary environments of the United States, *in* Porter, S. C., ed., v. 1, The Late Pleistocene: Minneapolis, University of Minnesota Press, p. 190–211.
Snyder, C. T., and Langbein, W. B., 1962, The Pleistocene lake in Spring Valley, Nevada, and its climatic implications: Journal of Geophysical Research, v. 67, p. 2385–2394.
Snyder, J. O., 1918, The fishes of the Lahontan system of Nevada and northeast California: U.S. Bureau of Fisheries Bulletin 35, p. 31–86.
Spencer, R. J., Baedecker, M. J., Eugster, H. P., Forester, R. M., Goldhaber, M. B., Jones, B. F., Kelts, K., McKenzie, J., Madsen, D. B., Rettig, S. L., Rubin, M., and Bowser, C. J., 1984, Great Salt Lake and precursors, Utah; The last 30,000 years: Contributions to Mineralogy and Petrology, v. 86, p. 321–324.
Starrett, L. G., 1949, The relation of precipitation patterns in North America to certain types of jet streams at the 300-millibar level: Journal of Meteorology, v. 6, p. 347–352.
Stauffer, N. E., Jr., 1985, Great Salt Lake water balance model, *in* Kay, P. A., and Diaz, H. F., eds., Problems of and prospects for predicting Salt Lake levels: Salt Lake City, University of Utah, Center for Public Affairs and Administration, p. 168–178.
Stuiver, M., and Smith, G. I., 1979, Radiocarbon ages of stratigraphic units, *in* Subsurface stratigraphy and geochemistry of late Quaternary evaporites, Searles Lake, California: U.S. Geological Survey Professional Paper 1043, p. 68–75.
Tang, M., and Reiter, E. R., 1984, Plateau monsoons of the northern hemisphere; A comparison between North America and Tibet: Monthly Weather Review, v. 112, p. 617–637.
Thompson, R. S., 1984, Late Pleistocene and Holocene environments in the Great Basin [Ph.D. thesis]: Tucson, University of Arizona, 256 p.
Thompson, R. S., Benson, L. V., and Hattori, E. M., 1986, A revised chronology for the last Pleistocene lake cycle in the central Lahontan Basin: Quaternary Research, v. 25, p. 1–9.
U.S. Department of Commerce, National Oceanographic and Atmospheric Administration, 1966–84, Climatological Data–National Summary: Asheville, N.C., Environmental Data Service, National Climatic Center, annual volumes.
U.S. Geological Survey 1960, Compilation of records of surface water of the United States through September 1950; Part 10, The Great Basin: U.S. Geological Survey Water-Supply Paper 1314, 485 p.
U.S. Geological Survey, 1963, Compilations of records of surface water of the United States, October 1950 to September 1960; Part 10, The Great Basin: U.S. Geological Survey Water-Supply Paper 1734, 318 p.
U.S. Geological Survey, 1961–84, Water Resources Data for Nevada, California, Utah and Oregon: U.S. Geological Survey Water-Data Report Series, annual volumes.
U.S. Weather Bureau, 1938–65, Climatological Data–National Summary: Asheville, N.C., Environmental Data Serivce, National Climatic Center, annual volumes.
Van Denburgh, A. S., Lamke, R. D., and Hughes, J. L., 1973, A brief water-resources appraisal of the Truckee River Basin, western Nevada: U.S. Geological Survey Water Resources Reconnaissance Series, Report 57, 122 p.
Van Horn, R., 1982, Surficial geologic map of the Salt Lake City North Quadrangle, Davis and Salt Lake counties, Utah: U.S. Geological Survey Miscellaneous Investigations Series Map I-1404.
Van Winkle, W., 1914, Quality of the surface waters of Oregon: U.S. Geological Survey Water-Supply Paper 363.
Wayne, W. J., 1984, Glacial chronology of the Ruby Mountains–East Humboldt Range, Nevada: Quaternary Research, v. 21, p. 286–303.
Whelan, J. A., 1973, Great Salt Lake, Utah; Chemical and physical variation of the brine, 1966–1972: Utah Geological and Mineral Survey, Water Resource Bulletin 17, 24 p.
Williams, T. R., and Bedinger, M. S., 1984, Selected geologic and hydrologic characteristics of the Basin and Range Province, western United States; Pleistocene lakes and marshes: U.S. Geological Survey Miscellaneous Investigations Survey, Map I-1522-D.
Yeh, T. C., 1950, The circulation of the high troposphere over China in the winter of 1945–46: Tellus, v. 2, p. 173–183.

MANUSCRIPT ACCEPTED BY THE SOCIETY MARCH 16, 1987

ACKNOWLEDGMENTS

The authors wish to acknowledge technical reviews of earlier drafts of this manuscript by D. R. Currey of the University of Utah, and by J. P. Bradbury, J. L. Glenn, and G. I. Smith, all members of the U.S. Geological Survey.

Printed in U.S.A.

The Geology of North America
Vol. K-3, North America and adjacent oceans during the last deglaciation
The Geological Society of America, 1987

Chapter 12

Late Quaternary paleoclimate records from lacustrine ostracodes

Richard M. Forester
U.S. Geological Survey, Mail Stop 919, Box 25046, Denver Federal Center, Denver, Colorado 80225

INTRODUCTION

The fossil record of continental organisms is an important source of paleoclimatic information. Extant taxa provide a way to transfer both qualitative and quantitative climatic information to the fossil record (Delorme and others, 1977; Bartlein and others, 1984). Terrestrial taxa such as insects and especially plants are traditional sources of paleoclimatic information, because their life cycles are directly related to climate (see chapters in The Biological Record on the Continent, this volume). By contrast, aquatic organisms are not traditional sources of paleoclimatic information, because their ecology necessarily is determined in part by hydroenvironmental parameters and because their life cycles must therefore be related to parameters at least one step removed from climate. Thus lacustrine sediments, which typically have the greatest stratigraphic continuity of all nonmarine sediments, are largely sought out only as repositories of terrestrial fossils. Unfortunately this approach ignores the climatic record of the lake itself.

Lakes, especially those in hydrologically closed basins, are a valuable source of paleoclimate records. Paleoshore deposits define paleolake levels that reflect changes in the local paleohydrologic budget. This approach dates from the classic study of Gilbert (1890) and is used today over large geographic areas (Street-Perrott and Harrison, 1985). Lake-level variation is an important tool for interpreting paleohydrography, especially when it is applied to many lakes so as to identify local nonclimatic variability. This form of paleolake-level information, however, depends on the ability to date discrete paleoshore deposits, is incremental in nature, usually omits lower-than-modern levels, and can often only be applied to the last lake cycle. Recent studies (e.g., Lerman, 1978) demonstrate that lakes are complex chemical and physical systems in which parameters such as water temperature and hydrochemistry are often coupled with local climate. When a record of these parameters is preserved in lake sediments, it provides a continuous record of climate. Because hydroenvironmental parameters are important limiting factors for aquatic organisms, their stratigraphic record in lake sediments becomes an important source of paleoclimatic information.

Numerous lacustrine records span the past 18,000–20,000 years as the extensive palynologic literature demonstrates (see chapters by Jacobsen and others, and Barnosky and others, this volume). Nonetheless, most of these records are known only for their terrestrial fossils and not for their paleolimnologic—paleoclimatic story. Consequently, the lacustrine record of late-Pleistocene and Holocene deglaciation in North America is limited to a few recently studied stratigraphic sites, as well as fragmentary records from geographically scattered sites.

This paper examines selected late-Pleistocene and Holocene sites with good stratigraphic records of lacustrine organisms (Fig. 1) in order to provide another perspective of climate change. A secondary purpose is to discuss important relationships between climatic and hydroenvironmental parameters, in order to show how aquatic organisms may record paleoclimate through their environmental sensitivity.

CLIMATE AND HYDROLOGY

Modern-day climate is a product of many complex factors (Bryson and Hare, 1974). The atmosphere is a source or sink for moisture and therefore plays an important role in ground-water and surface-water hydrology of a given basin. The discussion below provides a general basis for evaluating a particular environment and its organisms in terms of climate.

Air temperature and water temperature

Thermally stratified lakes. The epilimnion of a large lake gains and/or loses heat on daily and seasonal cycles (Ragotzkie, 1978). Wind-induced turbulence may move heat downward, especially when the densities of surface and subsurface waters are similar. Lakes in the north-temperate belt of North America typically mix twice a year (Dean, 1981). During the winter the temperature of the entire water column of such dimictic lakes is usually 4°C or lower. Before ice-in and after ice-out, any wind-induced turbulence readily mixes the entire water column, thereby bringing nutrients to the epilimnion and oxygen to the hypolimnion. Warming of the epilimnion produces warmer, less

Forester, R. M., 1987, Late Quaternary paleoclimate records from lacustrine ostracodes, *in* Ruddiman, W. F., and Wright, H. E., Jr., eds., North America and adjacent oceans during the last deglaciation: Boulder, Colorado, Geological Society of America, The Geology of North America, v. K-3.

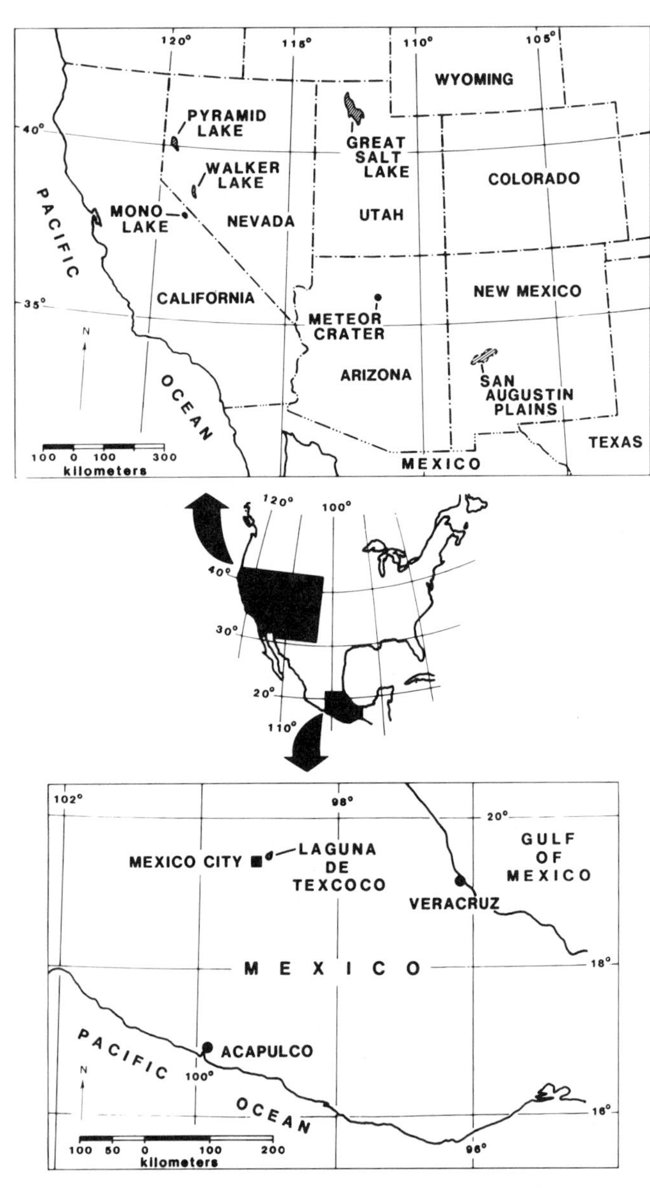

Figure 1. Map showing the location of lakes or lake basins discussed in the text.

dense water that becomes thermally isolated from the hypolimnion. The temperature of the epilimnion increases to some maximum value and then cools to 4°C in the autumn, when the lake mixes again.

In Arctic lakes, the epilimnion may never become warmer than 4°C, so mixing may occur throughout the warm season, whereas coastal temperate lakes may never become colder than 4°C and thus may mix throughout the winter but stratify in summer. These lakes are known as cold and warm monomictic lakes, respectively (Wetzel, 1975). Other stratification types are found in tropical areas. Each mixing style has a latitudinal distribution; thus, knowledge about a lake's mixing schedule provides at least general climatic information (Hutchinson and Loeffler, 1956).

The temperature of the epilimnion, including the littoral zone, may respond to daily or seasonal changes in atmospheric temperature, depending on the volume. Benson (1984), for example, shows that the epilimnion temperature of Pyramid Lake increases from lows in January between 6° and 7°C to highs in July and August between 20° and 25°C. The values approximate the regional monthly average air-temperature maximum in winter and are about 5° to 6°C cooler than the regional monthly air-temperature maximum in summer. Thus, knowledge about the temperature of the epilimnion may provide specific or general information about air temperature over the lake (Fig. 2). Conversely, lakes whose water mass is dominated by river input may exhibit a complex heat budget that is some integration of the river-water heat budget and the atmospheric heat budget. Benson and Thompson (this volume) and Carmack and others (1986) treat this phenomenon in more detail.

The hypolimnion usually remains relatively isothermal as at Pyramid Lake (Benson, 1984), or varies gradually throughout the year as at Walker Lake, where it ranges from about 6°C in winter to 10° or 11°C in summer (Benson and Spencer, 1983). Both Pyramid and Walker lakes are warm monomictic lakes, and their minimum temperature represents the temperature of mixing, which is the winter temperature of the epilimnion. Knowledge about the water temperature of the hypolimnion in any type of northern lake therefore provides little information about seasonal air-temperature variation (Fig. 2).

Shallow lakes, ponds, marshes, seeps, and springs. Many lakes are not deep enough to stratify thermally, and therefore the entire water body loses or gains heat on a daily or other short-term cycle. Maximum and minimum air- and water-temperature data collected weekly from late winter through summer 1986 from a small suburban pond in Lakewood, Colorado, illustrate the relationship between these parameters (Fig. 3). Air temperature varied over a wider range than water temperature, and water temperature usually lagged behind air temperature. Variations in air temperature lasting only a day or so are often not recorded by water temperature. In general, however, when the pond is ice-free, water temperature varies with air temperature. Small lakes may also indirectly record the average winter air temperature through the length of their ice-free season. Small arctic lakes, for example, are only ice-free for 2 to 3 months out of the year (Winter and Woo, 1987), and although they may reach temperatures above 20°C in summer they only do so for short periods of time. In contrast, small lakes in the Denver area are ice-free for at least 9 to 10 months and may remain at temperatures above 20°C for 2 or 3 months.

The temperature of shallow ground-water discharge, e.g., springs or seeps, may approximate mean annual air temperature, but ground water discharging from very deep aquifers may be warmer than mean annual air temperature. Discharging ground water is relatively isothermal throughout the year, but the water temperature of pools or ponds supported by ground water may

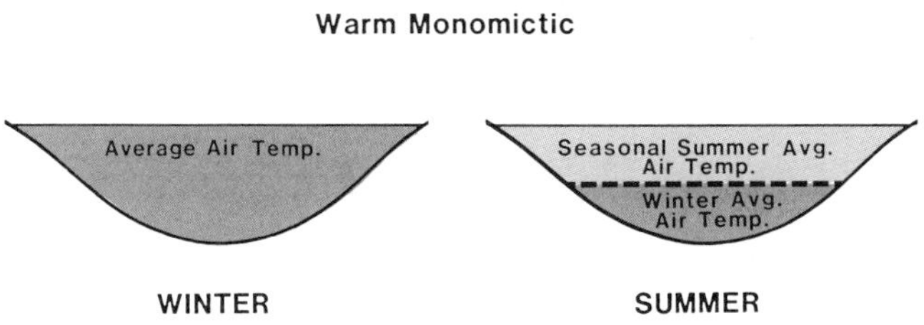

Warm Monomictic

Average Air Temp.

Seasonal Summer Avg. Air Temp.

Winter Avg. Air Temp.

WINTER SUMMER

Figure 2. Schematic diagram showing the generalized seasonal relationship between air and water temperature for a warm monomictic lake.

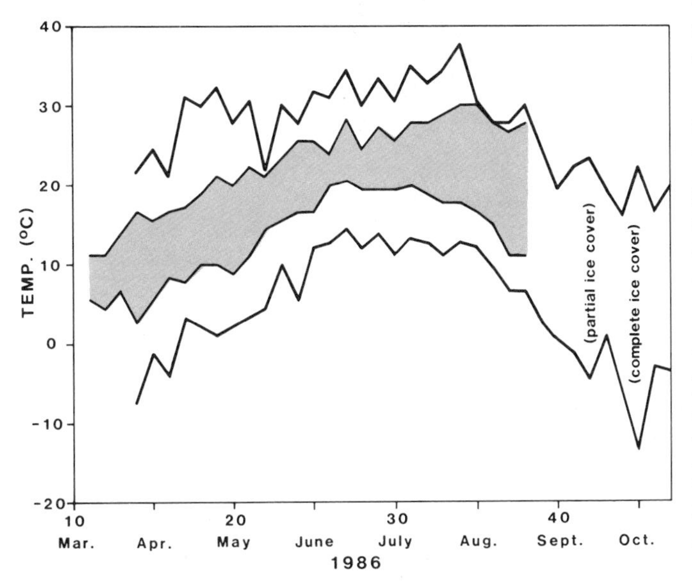

Figure 3. Maximum and minimum weekly air temperature (heavy lines) and water temperature (outlined in blue) for a small pond located in Lakewood, Colorado.

vary with air temperature. Thus, the temperature of ground-water discharge and the aquatic environments it supports can provide information about mean annual air temperature and seasonal air-temperature variability.

Precipitation, evapotranspiration, and hydrochemistry

The annual balance and seasonal distribution of precipitation (P) and evapotranspiration (E) are important climate parameters (Winter and Woo, 1987). Lakes that are chemically coupled to climate can provide information about the P/E ratio through their hydrochemical records. The relationship is complex because the major dissolved-ion composition (solutes) and concentration (salinity) of surface water is a product of both climatic and nonclimatic processes (Gorham, 1961; Jones, 1965; Hardie, 1968; Garrels and McKenzie, 1967; Jones and Bowser, 1978; Eugster and Hardie, 1978; Eugster and Jones, 1979; Jones and Weir, 1983). Inflow from rock-water reactions (weathering), selective solution of precipitated minerals, atmospheric precipitation, CO_2 solution, and solution of atmospheric dust are primary solute sources. Selective mineral precipitation due to evaporative concentration or mixing of waters with different salinities or compositions are the primary solute sinks, along with biologic processes, CO_2 outgassing, exchange reactions, and outflow (Fig. 4).

Eugster and Jones (1979) describe how solute composition evolves during evaporative concentration. The relationship of this process to climate is described by considering the expected hydrochemical behavior under different annual moisture budgets of three lakes whose chemistry is coupled to the atmosphere and whose geologic setting is the same.

Example 1 is a lake located in an area with a very wet climate and insignificant evapotranspiration. Dissolved ions entering the lake via atmospheric precipitation, surface runoff, and ground-water discharge have a short residence time and leave via surface and subsurface discharge. The dominant ions are typically Ca^{2+} and HCO_3^- derived from the solution of carbonates and the dissociation of carbonic acid (Kelts and Hsu, 1978; Dean, 1981). On very warm days, or if photosynthesis is common enough, CO_2 may be lost from the water column, increasing the pH and causing the precipitation of calcite (Dean, 1981). The other major ions (Na^+, K^+ Mg^{2+}, SO_4^{2-} and Cl^-) are usually transient and are present at low concentrations, so salinity is low. Lakes having these hydrochemical properties are known from northeastern Minnesota, for example (Gorham and others, 1983), and more generally from the areas having a very high P/E ratio (Winter and Woo, 1987). The P/E ratio may be high because total precipitation is very high or because total evapotranspiration is very low owing to cold air temperature and/or ice cover (with zero evaporation) through part of the year. Lakes dominated by through-flowing streams or ground water also tend to exhibit these characteristics, even though they may be located in areas with a dry climate.

Example 2 is a lake in an area where $P \geqslant E$. Surface or subsurface outflow is less important or insignificant, but evaporation is important, so salinity is typically higher and major ions reside longer than in example 1. During the summer the epilimnion routinely precipitates calcite, which may accumulate in the sediments, or in the case of deep lakes it may dissolve in the hypolimnion (Dean, 1981). Ca^{2+} and HCO_3^- are precipitated from the lake in equivalent proportions, but other ions, Mg^{2+} for example, remain in solution, so that the Mg^{2+}/Ca^{2+} ratio increases, as does salinity, which may go through some sort of seasonal cycle. If the Mg^{2+}/Ca^{2+} ratio increases sufficiently, other carbonate minerals may precipitate, potentially recording these chemical changes by depositing a distinctive carbonate mineral stratigraphy (Spencer and others, 1984). In areas of the upper Midwest where annual $P \geqslant E$, the lakes typically have waters dominated by Ca^{2+}+(Mg^{2+})-HCO_3^-, with other ions such as Na^+ or SO_4^{2-} being common (Gorham and others, 1983). The differ-

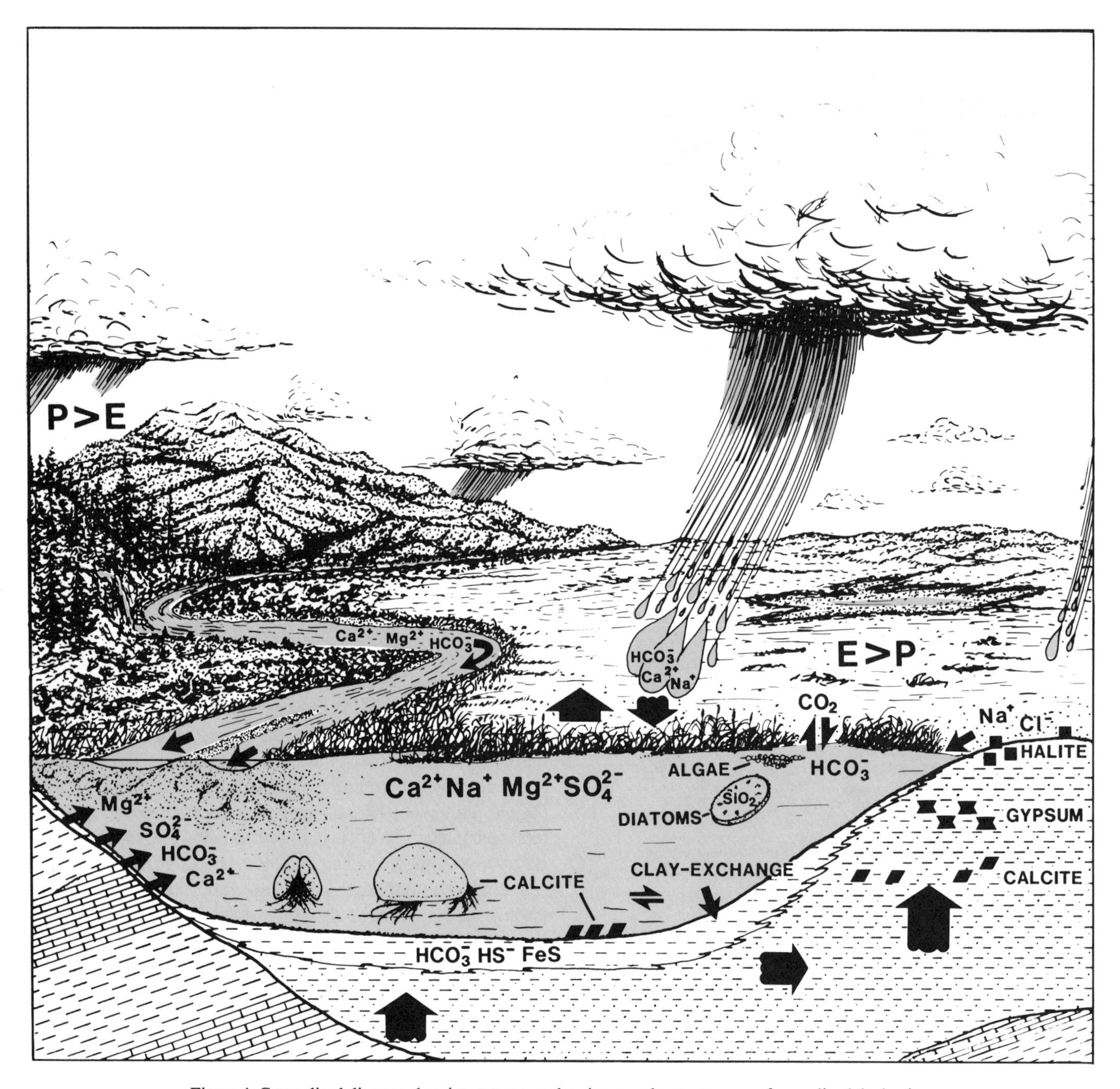

Figure 4. Generalized diagram showing common solute input and output sources for a saline lake having Ca-enriched saline water and no surface outlet. The water chemistry is a product of regional and local climate, together with rock-water reactions and other processes (see text).

ences between the P/E ratios in examples 1 and 2 therefore account for the distinctive difference in lake chemistry.

Example 3 is a lake in an area where P <<E. Water enters the lake via atmospheric precipitation or surface and subsurface discharge but generally leaves via evaporation (see examples in Winter and Woo, 1987). Dissolved ions usually accumulate in the lake, so that salinity must increase over time, although ions may be lost: 1) to salt precipitation, especially near shore, 2) to deflation, 3) to salt storage in pore fluids (Jones and Van Denburgh, 1966), 4) through discharge to ground water, and 5) through periodic surface outflow when atmospheric precipitation is especially heavy. The water column may be saturated or even supersaturated with respect to calcite or other carbonate minerals, especially on a seasonal basis, and precipitated carbonate miner-

als are routinely stored in the sediment. Carbonate mineral production that is common in example 2 continues in lakes along the evaporation gradient until eventually a state is reached where either Ca^{2+} or HCO_3^-, depending upon the initial ratio of the two, is depleted from the water column (Eugster and Jones, 1979), whereupon the other ion's concentration increases. Thus the water may now be described as depleted in Ca^{2+} or HCO_3^- and is now dominated by a combination of ions other than Ca^{2+}+(Mg^{2+})–HCO_3^- (Fig. 5). Lakes having these compositional characteristics are common on the North American prairies (Gorham and others, 1983) or in other semi-arid to arid regions such as the Great Basin. Because the P/E ratio is lower than for examples 1 and 2, the composition of the water column is significantly different.

The chemical state at which either a cation or anion is depleted from solution is known as a solute branchpoint, and in between examples 2 and 3 the calcite branchpoint is reached. The calcite branchpoint is especially important climatically because it is the only solute branchpoint that occurs at relatively low salinity (roughly 1 to 3 ppt), and thus it marks a major change in water chemistry as evapotranspiration becomes important or dominant. In this sense the calcite branchpoint defines an aquatic ecotone like the prairie/forest transition. Thus, knowledge about the solute composition of lake water provides a means of interpreting the local moisture budget, especially within a range where P>> E to P<E (Fig. 5).

The same process described in the three examples above operates within the capillary fringe surrounding the lake, so that, on average, once climate becomes dry, an efflorescent crust of highly soluble minerals such as halite (NaCl) or thenardite (Na_2SO_4) will form. Any atmospheric precipitation or surface runoff quickly dissolves the more soluble minerals, and the solutions are added to the lake or local ground water, whereas the less soluble minerals, such as the alkaline-earth carbonates and gypsum, tend to remain in the sediment (Eugster and Jones, 1979; Fig. 4). This mechanism permits lake water to evolve chemically to end-member solutions, even when the lake never becomes volumetrically reduced enough for those compositions to evolve within the lake. Efflorescent crusts usually become more common as climate becomes drier, as long as ground water remains near the earth's surface. In very arid areas where the lakes (dry playas) are supported principally by infrequent precipitation and overland flow, the limited salts that precipitate from these dilute waters are often lost to deflation or dispersal within the sediment. Because salts do not accumulate, the water on dry playas is typically fresh but variable in composition. Therefore, brine lakes usually do not exist in extremely arid areas except when supported by extensive ground-water discharge.

Consider a pair of lakes (example 4) that are identical in every respect except that the surface outlet of one lake is lower than that of the other (Fig. 6). If the input of water is sufficient to permit the first lake to discharge continuously but the second to discharge only intermittently, then the second will retain more salts and may be more advanced in its state of solute evolution.

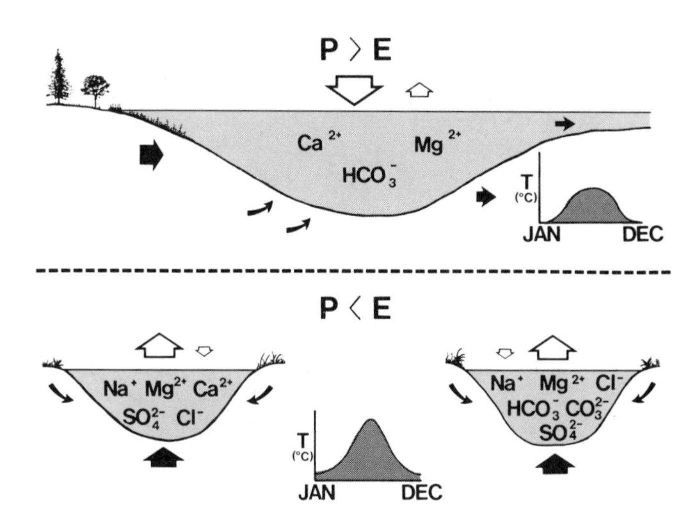

Figure 5. Schematic diagram illustrating an ideal relationship in which water chemistry is completely coupled to the local balance between precipitation and evaporation.

The chemistry of the two lakes when viewed from a purely climatic perspective would suggest that the first lake was located in an area with a wetter climate than the second. Conversely, if the salinity of both lakes increased during a very dry event, the one with the lower outlet would flush salts sooner with a shift to a wetter climate, and hence it would potentially record the climate change in its water column before the lake with a higher outlet. Both lakes, if studied independently, might preserve the same relative climate record, but if their records were compared, the magnitude and timing of climate records would differ. Thus, in a region having many lakes, local circumstances will also influence the relationship between chemistry and climate, resulting in differences in their climate records. These examples represent ideal situations, just as the solute evolutionary process described by Eugster and Jones (1979) is based largely on the importance of precipitate mass action and inference of geochemical process from conservative solutions of the major ions. The actual processes in nature are usually more complex and complicate the interpretation of paleoclimate from lake records.

The composition of lake water is also strongly dependent on local rock type. Carbonate-enriched waters are often associated with volcanic rocks, and carbonate-depleted waters with sedimentary, plutonic, and metamorphic rocks (Jones, 1966; Bodine and Jones, 1986). The prominent changes in water chemistry between the prairie lakes and the Minnesota forest lakes reported by Gorham and others (1983) relate in part to a change in lithology of the glacial drift and in part to differences in climate. Large western lakes having huge catchments, for example, may receive surface drainage via rivers from many different rock types (see, for example, Spencer and others, 1985a, b). Variation in the discharge rate of such rivers in the catchment over time might

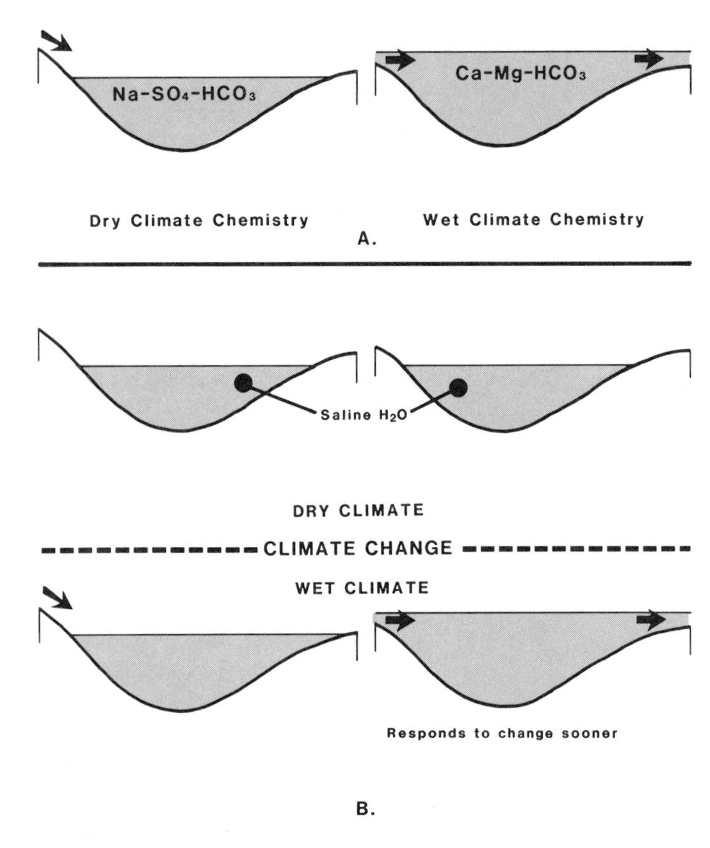

Figure 6. Schematic diagram showing how different lakes in the same region may respond to climate in different ways. A) Two lakes located in a region with a dry climate, but one lake is capable of draining, resulting in chemistry that is more typical of a wetter climate. B) Top: climate becomes drier than for A so that both lakes are now saline. Bottom: climate change towards wetter conditions results in the relation that the lake draining first records the change sooner than the other lake.

result in differing compositions of the water column in the lake, which might be erroneously interpreted as differences in climate.

The relationship between local climate and a lake's limnology illustrates how knowledge about the temperature and chemistry of the water column may provide very detailed climatic information or virtually none. Discrepancies between the character and timing of climate records from lakes in adjacent basins may reflect nothing more than the way particular lakes record a given process. Extracting climate information from lake sediments therefore involves substantially more than the logistics of coring or selection of an outcrop of lake deposits. Caution must always be applied to the climatic interpretation of any lake record, especially because most endogenic components left in lake sediments are records of limnologic processes occurring within the water column. When these limnologic processes are coupled with climate, their records may be translated into a climate story.

THE RESOLUTION OF LACUSTRINE SEDIMENT RECORDS

Seasonal changes in air temperature, precipitation, and evapotranspiration may be recorded with varying resolution by the water temperature and chemistry of the water column. Unfortunately, sediment-accumulation rates in most lakes are too low to record the events of a single year, let alone seasons. Small lakes or ponds that are potentially the most climatically sensitive usually have the lowest sediment accumulation rates. The best sediment-time resolution comes from rhythmic sediments in some dimictic lakes, where each layer integrates about 6 months (Anderson and others, 1985). Otherwise, a lake's sediment-resolution scale reduces the daily to yearly sensitivity of its water column to decadal or coarser temporal scales.

AQUATIC ORGANISMS AND THEIR ENVIRONMENT

Aquatic organisms, through their ecologic response to the environment, provide an important paleohydroenvironmental record, which is especially sensitive to dilute saline or freshwater environments (Forester, 1986). Endogenic minerals also provide a primary record of a lake's physical and chemical properties, but they usually provide the most information for very high salinity waters, where numerous solute- and temperature-sensitive evaporite minerals can precipitate (Bodine and Jones, 1986). The study of the mineralogy and pore-water chemistry of cores from the Great Salt Lake by Spencer and others (1984, 1985a, b) is a recent example of these studies.

Although lakes also contain numerous aquatic organisms, few of these organisms can be commonly recovered as fossils in statistically useful numbers. A typical microfossil record consists primarily of ostracodes and diatoms, although Cladocera, charophytes, chrysomonad cysts, thecamoebans, and occasionally mollusks may be common. Ostracode valves, which are composed of low-magnesium calcite, are readily preserved in all alkaline environments and occasionally in some acidic environments. Diatom valves, which are composed of opaline silica, are readily preserved in acidic environments and usually in all but the most alkaline environments. Ostracode and diatom species are environmentally and geographically diverse and are known from the arctic to the tropics, from deserts to high-mountain lakes, and from very dilute to highly saline waters. Ostracodes and diatoms form the primary organism record of the physical and chemical properties of the water column and from that a record of paleoclimate.

Ostracodes, although common in virtually all nonmarine aquatic environments, have not been studied extensively, and their paleoenvironmental interpretive utility is not widely appreciated. In contrast, diatom biology, ecology, and systematics are the subject of an extensive literature, and, although much remains to be understood, diatom paleoenvironmental utility is widely accepted. Consequently, the forthcoming discussion about the relationship between organisms and their aquatic environment will focus on ostracodes rather than diatoms.

The poor reputation of ostracodes for various Quaternary studies apparently lies in part with the belief that the carapace is difficult to identify to species and that most species are environmentally insensitive. Both views are unfounded: ostracodes are no more or less taxonomically complex than any other group of organisms, and recent studies show them to be highly suited to detailed paleoenvironmental interpretations. Delorme (1969) was the first to recognize that ecology and biogeography of nonmarine ostracodes were so habitat-specific that they could be used to make quantitative paleoenvironmental interpretations. He established a modern quantitative data base for Canadian ostracodes and has used this information to interpret past hydroenvironments and climate (e.g., Delorme and others, 1977). In addition, ostracode valves are now being studied geochemically for a variety of parameters, including Mg/Ca, Sr/Ca ratios (Chivas and others, 1983), amino acid epimerization (McCoy and Forester, 1982), the stable isotopes C^{13} and O^{18}, Sr isotopes, C^{14}, and uranium series dating. These parameters provide information about water temperature, salinity, and age.

Relative to most microorganisms, ostracode species are long-lived animals. An ostracode life cycle may be as short as 3 to 5 weeks or as long as a year or more (Delorme, 1978). Modern ostracode species are often latitudinally limited in their distribution, suggesting that not only water temperature per se but its seasonal variation are important limiting factors (Delorme, 1971, references therein; Forester, 1985). Delorme and Zoltai (1984) show that the distribution of certain shallow-water species may be defined in terms of both water- and air-temperature ranges. Studies of the living ostracodes from a Lakewood, Colorado, pond suggest that hatching and growth to reproductive maturity are temperature-dependent (Forester, unpublished observations). Martens and others (1985) show that *Mytilocypris henricae* (Chapman) populations take longer (4 to 5 months) to reach maturity in winter than in summer (2 to 2.5 months) and also note that eggs will not hatch below certain temperatures. McLay (1978a, b) shows how water temperature plays a vital role in several North American ostracode species life cycles, including egg hatching. Most species therefore must have lower and upper survival temperatures and life-cycle temperatures as well as particular egg-hatching temperatures. Knowledge of a species' temperature requirements should provide at least seasonal temperature ranges for a water body, whereas such knowledge applied to many species may provide detailed temperature information. Moreover, if ostracodes have survival and life-cycle temperature limitations, they must also have optimal temperature ranges within which productivity is maximized, so that abundance of taxa within a sample should provide additional temperature information. These biological-scale responses to water temperature are integrated over a species' biogeographic range on a geologic time scale, because species are known to expand and contract their biogeographic ranges with climate change (Delorme and Zoltai, 1984; Forester, 1985). Thus an ostracode species living in waters that are thermally coupled with the atmosphere will provide general information about air temperature, whereas an assemblage of species in the same situation may provide detailed air-temperature information. Ostracodes from the bottom waters of deep lakes provide little or no information about seasonal air-temperature variability.

Hydrochemistry also plays a vital role in ostracode ecology. Delorme (1969) showed that ostracode hydrochemical sensitivity is great enough to describe particular environments in terms of salinity and solute composition. Forester (1983) and Forester and Brouwers (1985) suggested that solute composition is more important than salinity for some marine and nonmarine ostracode species. Forester (1986) suggested that ostracode species delineate species-specific areas on anion trilinear diagrams and that an ostracode's upper salinity tolerance might be anion-specific. In general, freshwater ostracodes are usually restricted to waters dominated by Ca^{2+}+(Mg^{2+})–HCO_3^-, whereas saline-water ostracodes are usually restricted to or prefer Ca^{2+} or carbonate-enriched waters. Even when the salinity of a water body is low (<3 ppt), it will contain "saline"-water ostracodes if it is not dominated by Ca^{2+}+(Mg^{2+})–HCO_3^-. Preliminary observations suggest that the changing chemical properties of surface water described with examples 1, 2, and 3 above have a corresponding change in the occurrence or abundance of particular ostracode species (Forester, unpublished data). Ostracodes therefore seem to be sensitive to the hydrochemical parameters that vary with climate. Ostracode-species assemblages from numerous samples of a core should reflect changes in the chemistry of a lake over time and in stratigraphic order. When the chemistry of the water column is coupled to climate through evaporation, ostracodes may then provide information about the local or regional P/E ratio from hydrochemical reconstructions.

Ostracodes from climatically sensitive settings offer the potential of distinguishing between air temperature and moisture variability, in part because these properties are expressed as quite different parameters, i.e., temperature versus chemistry in the water column. Unfortunately, although ostracodes have obvious paleoenvironmental and paleoclimatic potential, quantified ostracode-ecologic studies are in their earliest phases of documentation and are being conducted by relatively few investigators.

LATE-PLEISTOCENE AND HOLOCENE LACUSTRINE RECORDS

The western region of the United States south to central Mexico contains numerous basins, some of which have shallow ephemeral lakes or springs. During the last glaciation, however, many if not most of these basins contained lakes, some of which were extremely large (see Benson and Thompson, this volume). Ancient Lakes Bonneville and Lahontan are probably the best known "pluvial" lakes, and the sediments and especially the shore-line geomorphology of these lakes have been extensively studied (Scott and others, 1983; Benson, 1978; Thompson and others, 1986). The chain of lakes in Owens Valley, including

Owens, China, Searles, Panamint, and Death Valley lakes have also been studied in detail (Smith, 1979). Smith and Street-Perrott (1983) provide an excellent review of existing studies on deposits of late-Pleistocene age in western basins. These studies generally do not include aquatic microfossils. Studies of the endogenic minerals and sediment pore fluids in a few basins (Benson, 1978; Smith, 1979; Spencer and others, 1985a, b) provide hydrochemical data that may be interpreted in terms of lake paleohydrography and paleoclimate. Many other studies have focused on geomorphologic, stratigraphic, sedimentologic, and chronologic aspects of the sedimentary deposits (Smith and Street-Perrott, 1983).

The ostracode record from cores taken in the Great Salt Lake (Spencer and others, 1984; Forester, additional unpublished data); from core, auger, and outcrop samples taken from the San Agustin Plains dry playa (Markgraf and others, 1984; Forester and Markgraf, 1984); and from outcrops in Meteor Crater, Arizona, and Mono Lake, California (Forester, unpublished reports) form the primary data for the discussion that follows. The diatom record from Laguna de Texcoco (Bradbury, 1971), together with preliminary ostracode data, is also discussed.

Great Salt Lake/Lake Bonneville

Great Salt Lake occupies one of numerous hydrologically closed basins in northwestern Utah (Fig. 1). The lakes in these basins could form a single water mass at moderate to low level if they expanded, but they could only drain if lake level reached a high-altitude surface outlet (Currey and others, 1984). The paleolake-level history of this basin is based largely on studies of elevated shoreline deposits and associated lacustrine sediments. Smith and Street-Perrott (1983) summarize these studies and draw attention to discrepancies between the lake-level chronologies among the studies. Subsequently Spencer and others (1984) studied several cores from the south arm of Great Salt Lake and proposed a new lake-level history that, with some modifications, is in good agreement with a very detailed chronologic study of shore-zone deposits (Currey and Oviatt, 1985).

The cores (Spencer and others, 1984) contain a distinctive stratigraphy (Fig. 7). Both units II and IV contain brine-shrimp egg capsules and pellets, whereas unit III contains ostracodes. Modern-day Great Salt Lake is well known for its brine shrimp population, but it does not contain ostracodes, because for chloride-dominated waters the salinity of the lake is too high for the survival of any North American ostracode (Forester, 1986). Thus, sediments forming units II and IV were presumably deposited in a high-salinity chloride-rich lake at relatively low level. The absence of brine shrimp and the presence of ostracodes in unit III, however, implies lower salinity and a higher lake level than for units II and IV.

The adult ostracodes found in various samples from unit III are listed in Figure 7 with their absolute abundances. The stratigraphically lower ostracode assemblages from the core are dominated by limnocytherid ostracodes, whereas the stratigraphically higher assemblages are dominated by candonid ostracodes. Comparison of these assemblages with those from shoreline deposits and basin-margin outcrops shows that the limnocytherid-dominated assemblages occur in the lower-elevation deposits, whereas the candonid-dominated assemblages occur in the higher. The Bonneville shoreline deposits, which are at the highest elevation, contain two candonid species not known from any of the lower-shoreline deposits. These species define subunit IIIb in the core, whereas the other unit III subunits are either defined by the occurrence of particular ostracode species or by species abundances together with species combinations. Correlation of an ostracode assemblage from a core sample to a particular shoreline, of course, does not mean that the water depth was exactly equal to the altitude of the shoreline deposit, but rather it implies that water depth fell into some range of values that include a particular shoreline altitude. Ostracodes have not been collected from an altitudinally diverse suite of samples, and thus the range of water depth associated with the unit III subdivisions is not defined (Fig. 8a).

The chronology of the paleoenvironmental events may be obtained in two ways: (1) from datable materials within the core, such as volcanic ashes and C^{14} determinations on wood or other carbon fractions, and (2) from dates on shoreline features that can be correlated with the core (Currey and Oviatt, 1985). These two types of chronologic information yield similar but not identical lake-level histories. The shoreline studies show rising lake level earlier than do core studies, whereas the opposite is true for falling lake level. These differences are most probably related to the lag times needed for the lake to be habitable by ostracodes and to lag times needed for the lake to reach a new lower level following climate change. The rise of Great Salt Lake above modern levels occurred about 26 ka (Currey and Oviatt, 1985), whereas the transition from a brine shrimp lake to an ostracode lake occurred about 25 ka. The rapid rise of lake level toward the Provo and Bonneville levels appears to have occurred at about 20 ka, the maximum lake is at about 17–18 ka, and the drop toward the low stand began about 14.5–15 ka (Currey and Oviatt, 1985; Spencer and others, 1984; Fig. 8a).

The lake-level data from Currey and Oviatt (1985) shows that, when ostracodes lived at core site C in the central area of the lake (Spencer and others, 1984), the water depth ranged from about 50 m to over 300 m deep. Thus, the ostracodes were living in the hypolimnion, which means that their record of temperature was for water that was not related to seasonal variations in air temperature. The ostracodes found throughout unit III appear to be cold-water taxa. None of the species is known to require warm water, confirming that the water was cold throughout the year.

The ostracode *Limnocythere staplini* Gutentag and Benson, 1962, which typically lives in carbonate-depleted saline waters (Forester, 1986), is the first species to appear. It tolerates salinities up to about 40 ppt in Cl^--dominated water, and up to 200 ppt in SO_4^{2-}-dominated water, but tolerates only very low salinities in carbonate-enriched water (Delorme, 1969). Modern-day Great Salt Lake chemistry, and presumably the chemistry of all pre-

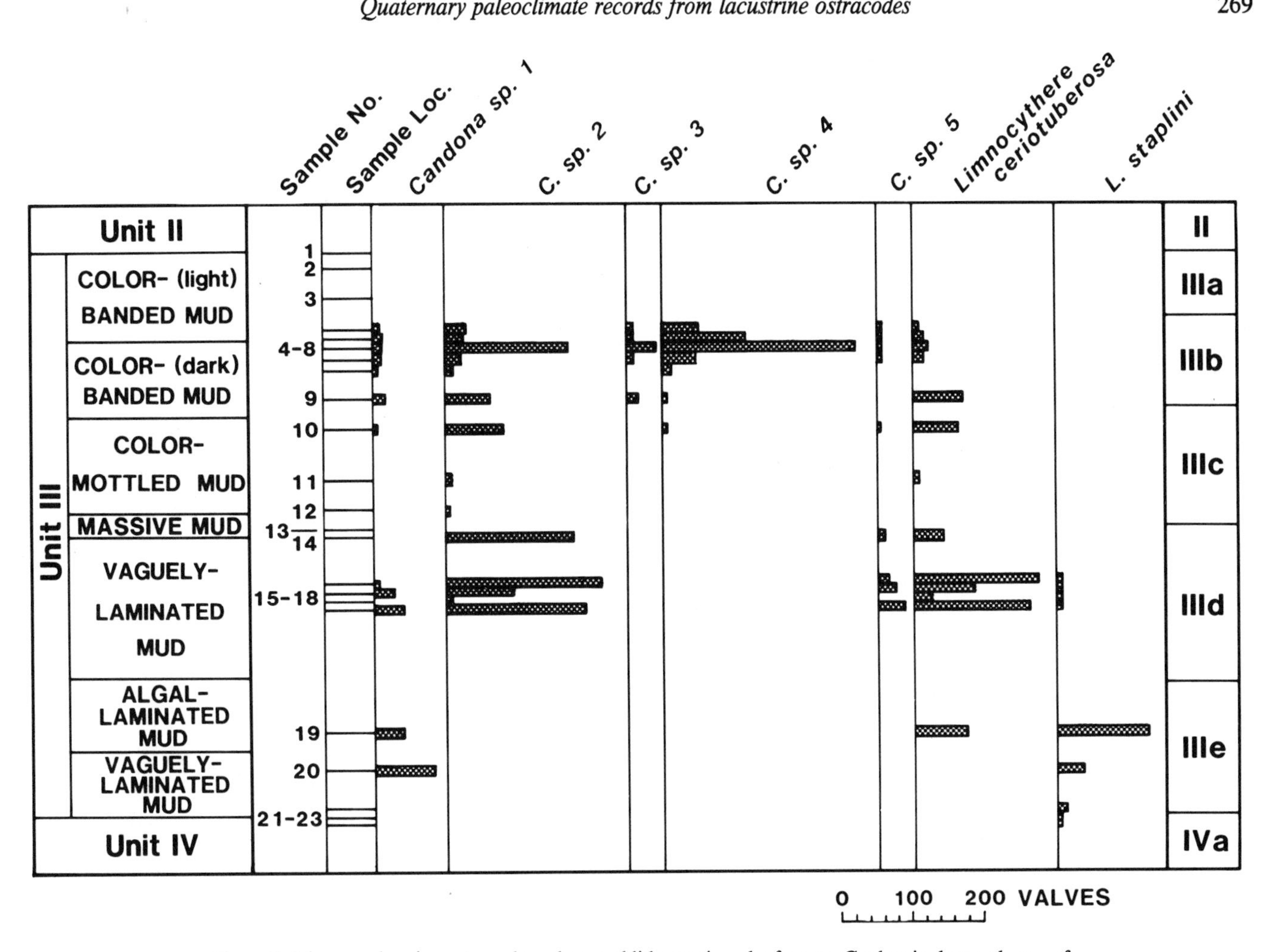

Figure 7. Diagram showing ostracode and general lithostratigraphy for core C taken in the south arm of Great Salt Lake. Units II, III, and IV are defined by lithology, mineralogy, and paleontology. See text or Spencer and others (1984) for more details.

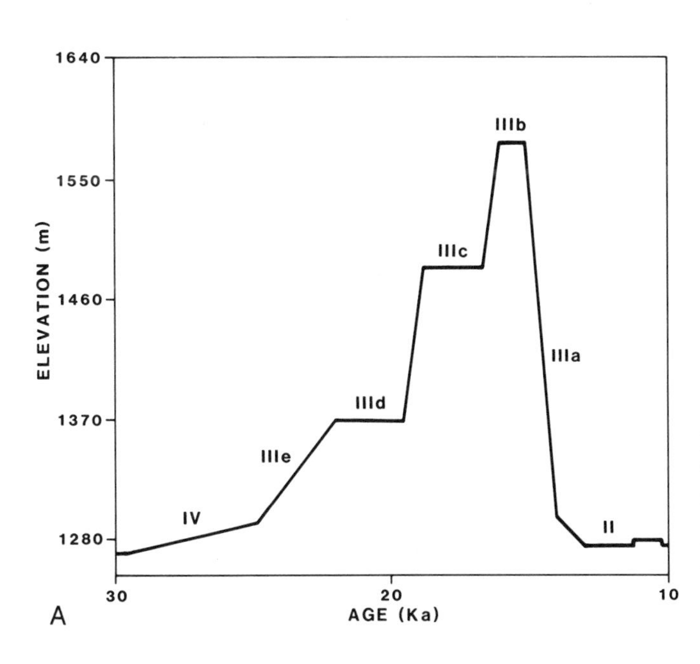

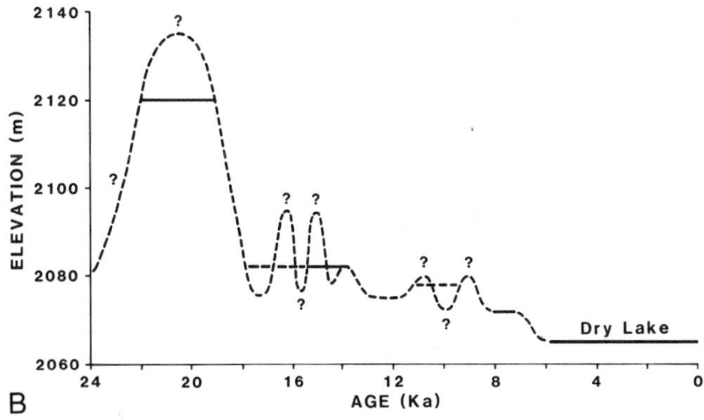

Figure 8. Generalized diagram showing lake-level change versus time for two lakes in the eastern Great Basin. A) Great Salt Lake, Utah, modified from Spencer and others (1984) see text. Roman numerals correspond to units shown in Figure 7. B) San Agustin Basin, New Mexico, from data in Markgraf and others (1983) and Forester and Markgraf (1984). Lake level is poorly constrained in this basin and is based on Pleistocene tufa elevations, a Holocene marsh deposit (solid lines), and paleoenvironmental inferences from ostracodes in lake sediments (dashed lines), see Figure 9 and text.

vious low-stand lakes, is suited to the occurrence of *L. staplini,* but high Cl^--dominated salinity serves as a barrier. The appearance of *L. staplini* and the disappeareance of the brine shrimp therefore suggest decreased salinity.

Limnocythere ceriotuberosa Delorme, 1971, occurs first with *L. staplini* near the top of unit IIIE and then as the only limnocytherid through most of the remainder of unit III (Fig. 7 and unpublished data). It typically lives in carbonate-enriched water that ranges in salinity from about 500 ppm to 25 ppt, although apparently tolerating higher salinity in Cl^--dominated water (Forester 1986). Its occurrence with *L. staplini* suggests rising alkalinity, whereas the absence of *L. staplini* suggests yet higher alkalinity. An alkalinity of approximately 15 equivalent-percent at salinities above 3 ppt is the approximate upper tolerance level for *L. staplini* (Forester, 1986). Spencer and others (1985 a, b) show that all river water entering the Great Salt Lake, but especially the Bear River, are relatively rich in HCO_3^-, as is atmospheric precipitation. Thus, water entering the lake would add primarily HCO_3^- to the existing Cl^- and SO_4^{2-}, and thus total alkalinity would rise as salinity fell. The rise in alkalinity recorded by the ostracodes is therefore an expression of chemical changes in the water column as the lake becomes deeper, just as the stepped shoreline terraces are a physical record of rising lake level.

The candonids become common in the middle of unit IIId and reach maximum species diversity in unit IIIb (Fig. 7). The relatively low abundance of all taxa in unit IIIc is believed to be due to low dissolved-oxygen levels on the lake floor and to frequent turbidity flows (e.g., massive mud Fig. 7). None of these candonid species is known from modern environments, but their occurrence in the high-elevation shoreline deposits suggests that, like many modern candonid species, these preferred low salinity and their occurrence at core site C shows they were abundant in the profundal zone. The increase in candonid species diversity at the base of unit IIIb coincides with a transition from aragonite to low Mg-calcite precipitation and a gradual shift to light oxygen isotopes, all implying a reduction in salinity from unit IIIc. This salinity reduction probably coincides with the lake reaching its topographic outlet and draining (Spencer and others, 1984).

The abundance of the candonid species decrease near the top of subunit IIIb, and *Limnocythere cerituberosa* occurs in most samples. Moveover, the carbonate mineralogy shifts from low-Mg calcite to high-Mg calcite, and the stable oxygen isotopes become heavier (Spencer and others, 1984). These changes suggest that the salinity of the lake increased. Following these events, the ostracode assemblage (unit IIIa) contains predominately mud-coated ostracodes, with a few specimens of limnocytherids that are not mud-coated. The apparent increase in salinity in the upper part of unit IIIb probably represents the beginnings of the evaporative concentration of solutes in the lake after it fell to the Provo Shoreline, which followed the rapid downcutting of the outlet at Red Rock Pass. The mud-coated ostracodes were redeposited from exposed lake-floor sediments into the shrinking lake as it dropped from the Provo shoreline. Samples examined since the study by Spencer and others (1984) show that the limnocytherid transition from *L. staplini* to *L. ceriotuberosa* described previously is reversed in unit IIIa, suggesting that the lake changed from carbonate-enriched water to carbonate-depleted saline water.

Spencer and others (1985b) describe the geochemical evidence for a rapid and continuous fall in lake level following the rise to the Bonneville Shoreline, which is supported by the dated shoreline evidence of Currey and Oviatt (1985). The transition from ostracodes to brine shrimp during the fall of the lake is sharp, implying that the ostracode salinity threshold was exceeded and that the lake returned to high salinity. Thus the lake went from a maximum level supported by a large input of water and/or greatly reduced evaporation to a system in which evaporation greatly exceeded input. In my view, the most significant aspect of this record is not the timing of the return of lake level to modern or even brine-shrimp values, but rather the abrupt change from a very wet and/or cold climate to a dry and/or warm climate. The fall of the lake from the Provo shoreline represents a lag behind this change in the hydrologic budget, which presumably was produced by a major change in climate at 14.5 to 15 ka.

San Agustin Plains dry playa

The San Agustin Plains in west-central New Mexico (Fig. 1) is made up of three subbasins and is known to contain at least a 1.6-m.y. sedimentary record (Markgraf and others, 1983). The paleoenvironmental history indicates that there was a high lake level at about 19–22 ka, an intermediate lake level from about 14–18 ka, a low lake level at about 6–12 ka, and a dry playa from about 5 ka (Markgraf and others, 1984; Forester and Markgraf, 1984).

The principal ostracode record from the San Agustin Plains consists of *Limnocythere ceriotuberosa* and *L. bradburyi* Forester, 1985. *L. ceriotuberosa* lives in deep and shallow lakes, including ephemeral lakes throughout the Great Basin and north into the Canadian prairies. *L. bradburyi* lives in several shallow, often ephemeral lakes on the central Mexican Plateau (Forester, 1985) and in an ephemeral lake in southern New Mexico (Robert Angell, verbal communication, 1985). The northern occurrence of *L. bradburyi* today is just south of the frost line, whereas *L. ceriotuberosa* occurs only north of the frost line. The modern biogeography of these two species therefore suggests that seasonal variations in water temperature determine their occurrences within chemically suitable waters.

Limnocythere ceriotuberosa lives both in the bottom waters of deep lakes (e.g., Walker Lake, Nevada), which remain cold throughout the year, and in shallow ephemeral lakes that have warm water during part of its life cycle. Thus, water temperature, at least within the range of about 0° to 25°C, is within the tolerance limits of this species, so that its absence from lakes south of the frostline, where water temperatures are within that range, must be due to other factors. McLay (1978a, b) concluded, on the basis of both experimental and field data, that ostracode eggs often require particular conditions such as certain temperatures or desiccation to hatch. The occurrence of *L. ceriotuberosa* north of the frost line may indicate that its eggs must experience cold temperatures before they will hatch, whereas *L. bradburyi* eggs might find freezing conditions lethal. *L. bradburyi* might be able to survive and reproduce in the bottom waters of a permanent northern lake, but would require warm temperatures during the year in order to reach maturity.

A composite stratigraphy of *Limnocythere ceriotuberosa* and *L. bradburyi* in various kinds of sediment samples from the San Agustin Plains is shown in Figure 9. The stratigraphically lowest assemblage (Fig. 9; assemblage III of Forester and Markgraf, 1984) is largely composed of *L. ceriotuberosa* and is believed to range in age from 19 to 22 ka. Limited shoreline chronology (Forester and Markgraf, 1984) suggest that the San Agustin lake was deep enough to enter the adjoining Ake Basin at about 22 ka (Fig. 8b). Thus the *L. ceriotuberosa* assemblage existed in deep water, and the general absence of *L. bradburyi* implies that the hypolimnion of the lake remained cold throughout the year. The presence of deep cold water in this basin requires a substantial change in climate in order to support such a lake at this latitude. The timing of this event suggests the climatic event is coincident with the rise of the Great Salt Lake toward the Provo Shoreline.

Limnocythere ceriotuberosa and *L. bradburyi* co-occur in large numbers in the assemblage dating from about 18 to at least 14 ka. The appearance of *L. bradburyi* implies an effective warming, resulting either from increased air temperature or from decreased water depth and thereby more mixing in summer. The date of 14 ka reported by Forester and Markgraf (1984) for shoreline tufa suggests that the lake was relatively shallow at this time (Fig. 8b).

Limnocythere bradburyi dominates the lower part of the next ostracode assemblage (Forester and Markgraf, 1984), which occurs in sediments that are entirely Holocene in age, although additional data from a poorly curated core suggests that this assemblage could be as old as 12 ka. The commonness of *L. bradburyi* and the rarity of *L. ceriotuberosa* implies that the bottom waters were warm throughout the year or, if the lake were dry in winter, winter air temperatures were warm. In either event, the dominance of *L. bradburyi* suggests that winters were much warmer than today or at 14–22 ka. Today a warm winter in this area is usually due to infrequent arctic air incursions. A warm winter may also be a dry winter, because cold arctic air is usually needed to lift westerly maritime air in order to produce substantial quantities of snow or rain. A dry-warm winter would imply the lake was maintained by a regular (relative to today) summer rain period. The increase in abundance of *L. ceriotuberosa* stratigraphically upward implies colder winters and probably winter moisture derived from frontal activity. Then, about 5 ka, the available precipitation became so low that the lake changed to its present dry playa (Fig. 8b).

The predominance of limnocytherids throughout this record, together with the rarity of candonids (the only species present is known to live largely in saline water), suggest that the water was saline. Compositionally, both species live in hydrochemically similar waters that are carbonate enriched and dominated by Na^{+}+(Mg^{2+})-HCO_3^{-} or Cl^{-} ions (Fig. 10). These compositions, together with the saline water, suggest that evaporation was important at least on a seasonal basis and that the lake remained topographically closed, which is consistent with the lake-level history (Fig. 8b). The apparent absence of salts in the lake sediments is probably due to ground-water leakage at depth (Blodgett and Titus, 1973).

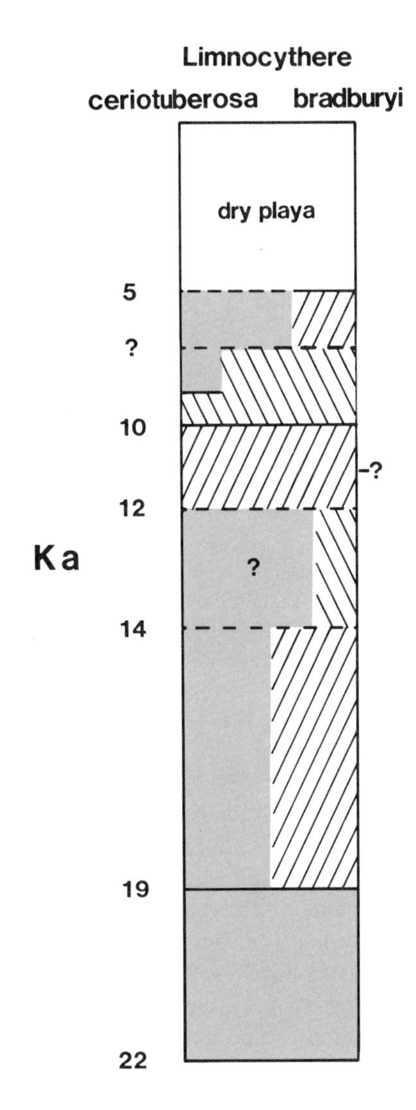

Figure 9. Composite ostracode stratigraphy for the San Agustin Plains, derived from core, auger, and pit-wall samples showing approximate relative abundance of two key species plotted against time. See text or Markgraf and others (1984) for more details.

Meteor Crater

Meteor Crater is a well-known impact structure in central Arizona (Shoemaker and Kieffer, 1974; Fig. 1). Samples col-

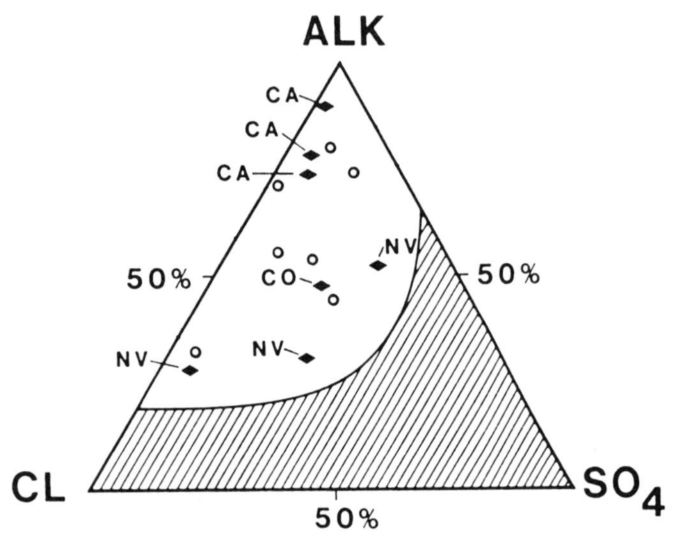

Figure 10. Anion trilinear diagram showing solute composition for lakes where *Limnocythere ceriotuberosa* (diamonds) or *L. bradburyi* (circles) are living. CA= California, CO= Colorado, NV= Nevada, and all circles are lakes in Mexico.

lected in 1958 from the walls of exploration shafts (Shoemaker, verbal communication, 1983) in crater-floor sediments contain a diverse ostracode and diatom assemblage, and one sample contains benthic foraminifera. The ostracodes (19 species) include forms that typically live in either lacustrine or ground-water-discharge environments and show that the first permanent water to enter the crater formed a shallow, slightly saline lake having a dominant dissolved-ion composition of Na^{+}+(Mg^{2+})-Cl^{-}. This lake probably had saline seeps and springs along its margins. The saline lake subsequently freshened, resulting in a shallow freshwater lake containing ostracodes that require or tolerate cold water in order to reproduce. Freshwater seeps and springs surrounded this lake. The next ostracode assemblage is dominated by freshwater spring and seep species, suggesting that the shallow lake has been reduced to a marsh. The fresh water was compositionally between one dominated by Ca^{2+}+(Mg^{2+})-HCO_3^-+(Cl^-)+SO_4^{2-}) and one Ca-enriched but dominated by Na^++(Mg^{2+})-Cl^-+(HCO_3^-)+(SO_4^{2-}). The ostracode-bearing sediments are overlain by red silts and sands similar to the modern dry-playa sediments, implying that ground-water discharge no longer occurred on the crater floor.

Today ground water is 65 m below the surface of the dry playa; it must have risen at least 30 m in order to cover the crater paleofloor (Shoemaker and Kieffer, 1974). The appearance of saline water, followed by its transition to fresh water, implies a progressive increase in effective moisture. The presence of lacustrine freshwater ostracodes requiring cold water implies at least seasonally cold air temperatures, whereas the presence of cold-requiring or cold-tolerant spring and seep taxa implies cold mean-annual air temperature.

The stratigraphic succession of environments is similar to the environmental stratigraphy in Great Salt Lake and San Agustin Basin (Fig. 8). Unfortunately the lacustrine sediments in Meteor Crater are not precisely dated, although a volcanic ash at the base of the freshwater lake beds is thought to be about 16 ka (Shoemaker and Kieffer, 1974; Shoemaker, verbal communication, 1983). Amino acid data on aquatic snails indicate the lake beds are late Pleistocene in age (McCoy, verbal communication, 1984). If the age of the volcanic ash is estimated correctly, then the shallow freshwater lake is correlative with the highest lake level (Bonneville) in Great Salt Lake basin.

Mono Lake

Mono Lake is a large saline lake located in east-central California (Fig. 1). Lajoie (1968) conducted a detailed study of late Pleistocene and Holocene deposits surrounding the lake and from that study developed a lake-level history for this system (Fig. 11). The Wilson Creek beds were studied by Lajoie (1968; see also Lajoie and others, 1982) in detail and contain a relatively continuous sedimentary record, which was dated by tephrochronology and C^{14}. The entire Wilson Creek section contains abundant ostracodes, which often make up the principal sand fraction. Ostracodes were examined from a number of Lajoie's original samples of these beds to determine how the ostracode paleoenvironmental history compared to the lake-level interpretations. The resulting interpretations based on the ostracodes were used to arrange the samples in a relative order of salinity, assuming that water depth and salinity are directly related in this lake. Actual water depth is known for some samples, because they were collected from paleoshore deposits. This information, when plotted against time, provides a relative lake-level curve that is quite similar to the one constructed by Lajoie (Fig. 11). Lake level (relative to today) appears to have been intermediate about 15–22 ka. Lake level then rose to a high stand about 12–14 ka, on the basis of both shoreline deposits and ostracode data (Fig. 11). The lake then fell rapidly, reaching near-modern levels in the early Holocene. Lake-level history at Mono Lake is similar to the Lake Lahontan history recently discussed by Thompson and others (1986) and by Benson and Thompson (this volume). These lake-level histories are distinctly out of phase with those of the eastern Great Basin lakes discussed previously (Fig. 8).

Laguna de Texcoco

Laguna de Texcoco is one of several lake basins located in the Valley of Mexico (Fig. 1) whose paleolimnologic history was established with diatoms from cores believed to have a record dating to 100 ka (Bradbury, 1971). Bradbury (1971) suggested that a large freshwater lake existed in the distant past, but not since 20 ka. The *Nitzschia frustulum* diatom assemblage at 27.2–12 ka implies that Laguna de Texcoco contained a shallow, saline water body when the large pluvial lakes existed to the north.

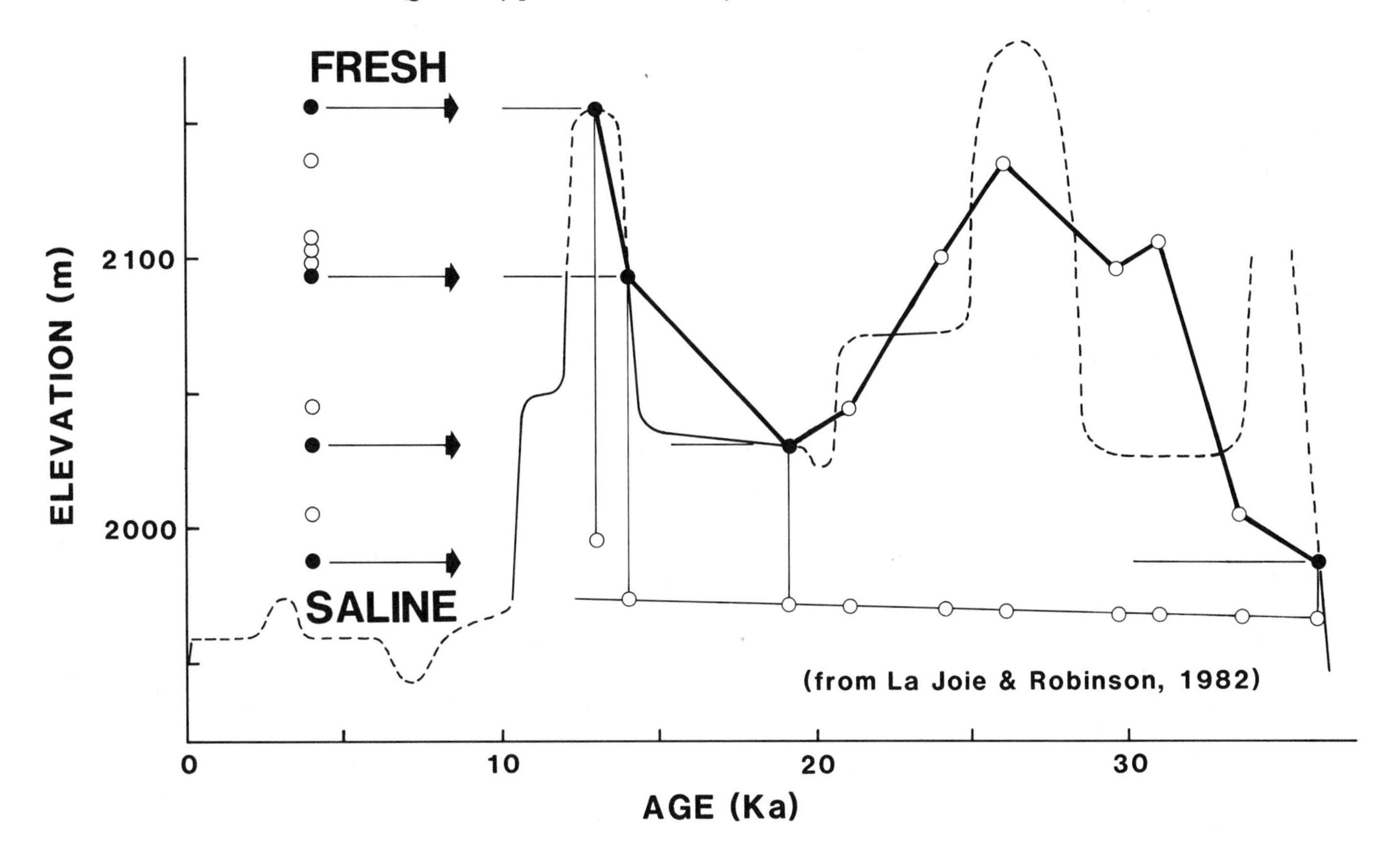

Figure 11. Lake-level history of Mono Lake, California, inferred from unpublished sedimentologic and stratigraphic studies by Lajoie, shown as dashed and solid lines. Ostracode data are from samples taken by Lajoie from paleoshorelines (solid circles) and from open lacustrine sediments (open circles). Samples are arranged from least saline (left-top) to most saline (left-bottom). Paleoshoreline samples provide absolute water depth, whereas water depths from open lacustrine samples are inferred from salinity interpretation.

A preliminary examination of ostracodes in six samples from a core taken by Bradbury from the center of the basin in central Mexico City contains *Limnocythere sappaensis* in abundance, together with occasional ground-water–discharge taxa in an interval that dates from 18.7–11.7 ka. These ostracodes imply that the lake was very saline and shallow, possibly even ephemeral, during this interval and that springs and seeps existed near the core site. A sample taken just 5 cm below the sample dated at 18.7 ka contains planktonic diatoms (Bradbury, verbal communication, 1986) together with *L. bradburyi.* These taxa suggest somewhat deeper and less saline water existed in the central area of the basin, but it was not deep enough to leave open lacustrine sediments on the edge of the basin at the core site studied by Bradbury (1971). The presence of only warmth-requiring or warmth-tolerant ostracodes from shallow-water deposits suggests air temperature was warm at this site, whereas the presence of only salt-loving taxa suggests the P/E ratio was less than 1.0.

DISCUSSION

The paleoenvironmental interpretations of the records from several lakes in the Great Basin and central Mexico suggest that each lake had a distinctive and relatively unique lake-level history since 20 ka. The lake-level history of each basin is thought to be at least an integrated hydrologic summary of climate and to have been hydrochemically, and in some cases thermally, coupled with climate. The histories of the lakes in the eastern Great Basin (Great Salt, San Agustin, and Meteor Crater; Fig. 1) indicate that all reached their deepest late-Pleistocene phase prior to 16 ka and then fell quickly to low stands or record climate change around 14.5–15 ka (Fig. 8). The lakes in the western Great Basin (Mono, and those forming Lahontan) had an intermediate or low stand prior to 16 ka and then rose to a maximum around 12–14 ka, followed by a rapid fall to modern low stands (see discussion of LaJoie, 1968; Thompson and others, 1986; Benson and Thompson, this volume; Fig. 11). Laguna de Texcoco (Bradbury, 1971) appears to be completely out of phase with the northern lakes and records no significant rise since 20 ka, although the existence of planktonic diatoms and *Limnocythere bradburyi* around 19–20 ka may imply a weak alliance with the eastern Great Basin lakes.

The chronology of the lake-level histories is obviously crucial to any discussion that seeks regional paleoenvironmental patterns. Dating lacustrine events occurring since 20 ka in Great Basin lakes is largely based on C^{14}, which is subject to a variety of geochemical problems (see, for example, Riggs, 1984). The chronology of lacustrine events dated in only a few places is necessarily suspect and subject to verification by additional studies. The chronologies developed for lake-level histories in the Great Salt Lake and Pyramid Lake basins are based upon numerous dates from many localities (see Currey and Oviatt, 1985; Thompson and others, 1986; references therein). The lake-level history of Mono Lake has also been documented by numerous dates (LaJoie, 1968; LaJoie and others, 1982) and is chronologically similar to that for Pyramid Lake. The chronologic studies in these basins thus provide the principal basis for suggesting that the timing of maximum lake level in the eastern and western Great Basin was not synchronous.

The proximity of the eastern and western Great Basin implies that air temperature should have been similar, so that the lake-level differences are probably not due to substantial differences in evaporation but rather in precipitation. Today, snow from winter storms is usually stored in the mountains surrounding the lakes and is available in the spring or summer as cold runoff (Benson and Thompson, this volume). If winter storm activity were greatly increased during the late Pleistocene, then available runoff would be increased. Continued storm activity during the summer months due to the interaction of maritime and arctic air might sustain the runoff and increase cloud cover, which together with cold water, would greatly reduce evaporation from the lake (Benson, 1986). The amount of precipitation and runoff, together with the reduction in evaporation necessary to produce the Lake Lahontan high stand, for example, is believed to have greatly exceeded the modern extremes (Benson and Thompson, this volume).

The asynchroneity between maximum-lake-level histories in the eastern and western Great Basin suggests that the input of water and its preservation through presumably lower evaporation must have varied regionally over time. Kutzbach (this volume, references therein) describes how atmospheric circulation would vary in response to a continental glacier and its retreat. Those circulation models show how the position of the jet stream varies in response to a large ice mass. Because the jet stream today serves as a steering current for large storms advancing eastward from the Pacific Ocean (maritime air), it must have controlled the position of storm tracks in the past as well. Our modern climatic pattern usually involves a strong jet during the winter and a meandering weak jet during the summer (Lamb, 1972; Bryson and Hare, 1974), resulting in strong zonal and meridional flow of air masses respectively. The presence of the continental ice sheet might maintain a year-round winter zonal circulation pattern and might also limit the path of the jet stream, which would focus storms onto a particular area. Because the edge of the Arctic air mass would be fixed south of the ice margin, the storms would intensify along a front and dump large quantities of snow or rain, depending on season and elevation. If the ice mass advanced or retreated differentially, the average position of the jet should also change, focusing storms in different areas. Smith (1987) suggests that moisture-laden air might be drawn from the tropics and dispersed over land, thus supplying a substantial increase in available water vapor. Tropical storms should move rapidly to land due to the intensified strength of the jet and then would lose their moisture in large quantities at the first opportunity, which would typically involve rising over mountains or cold air.

Summaries of the glacial geology (Mickelson and others, 1983; Waitt and Thorson, 1983; and Porter and others, 1983) show that the Laurentide and Cordilleran ice masses expanded and retreated at different rates, as did glaciers in the western mountains. If indeed these glaciers exerted sufficient control over circulation, then their advance and retreat may account for the large western lakes as well as the time lag behind the high stands of the lakes in the eastern and western Great Basin. The Cordilleran ice sheet was apparently in retreat by 14 ka, whereas the western mountain glaciers had disappeared (Waitt and Thorson, 1983; Porter and others, 1983). The timing of the latter changes in ice volume is coincident with the fall of Bonneville and the rise of Lahontan.

Lacustrine records of air temperature derived from ostracodes are limited to the southern basins (San Agustin, Meteor Crater, and Laguna de Texcoco) because only those basins seem to have remained shallow enough to have seasonal variability of water and air temperature coupled with each other. The ostracodes suggest that temperatures were much colder than today prior to 18 ka in the southern Great Basin, implying that arctic-air incursions were more frequent than today, and perhaps as a consequence, winters were also longer. Between about 18 and 14 ka, the southern Great Basin shows marked seasonality, probably characterized by cold winters and warm summers. From the early Holocene to perhaps 12 ka, the region appears to have been warmer than today, with mild winters that probably had few incursions of arctic air. This warm period gradually changed to a more seasonal pattern during the Holocene, reaching the modern state at an unknown time in the past.

The late-Pleistocene lake-level and lake-temperature histories seem to suggest that both increased precipitation and colder temperatures were common throughout the Great Basin prior to 15 ka, although storm activity was more frequent or intense in the eastern Great Basin. Climate change, perhaps due to changes in ice volume especially of the western glaciers, appears to result in a westward shift of storms—and perhaps a northward shift in the jet stream as well—around 14.5–15 ka, resulting in the rise of western and the fall of eastern Great Basin lakes.

The timing of the climate change that resulted in the termination of the large Great Basin lakes is similar to the marine O^{18}

record, in which the first signal of glacial meltwater occurs about 14 ka, with a possible second step around 9–10 ka (Mix and Ruddiman, 1985; Mix, this volume). Obviously, caution needs to be applied in interpreting the similarity in these dates, because the possibility of C^{14} errors is large and because apparent similarities in the timing of events in the Great Basin may have little to do with events in the Atlantic Ocean. Nonetheless, the expected sequence of events should start with climatic change and be followed by glacial melting and a record of that melting in the oceans. The record of a climate change at about 14.5–15 ka in Great Salt Lake is indeed slightly earlier than the 14-ka marine isotopic record. If the dates are accurate, and if one event is related to another, the timing of the two events suggests that the effects of the change in climate were abrupt, resulting in a geologically rapid response from the glaciers. Such rapid changes are usually attributed to substantial changes in atmospheric circulation. If deglaciation and lake levels are related, the mechanism(s) that resulted in these changes should have been operative prior to the record of change (14.5–15 ka) at Great Salt Lake. Unfortunately, although lakes may contain crucial climate records pertaining to the mechanisms of deglaciation, their records have not been studied in sufficient numbers or details from a geographic grid to favor one solution over another.

REFERENCES

Anderson, R. Y., Dean, W. E., Bradbury, J. P., and Love, D., 1985, Meromictic lakes and varved lake sediments in North America: U.S. Geological Survey Bulletin 1607, p. 1–19.

Bartlein, P. J., Webb, T., III, and Fleri, E., 1984, Holocene climatic change in the northern midwest; Pollen-derived estimates: Quaternary Research, v. 22, p. 361–374.

Benson, L. V., 1978, Fluctuation in the level of pluvial Lake Lahontan during the last 40,000 years: Quaternary Research, v. 9, p. 300–318.

—— , 1984, Hydrochemical data for the Truckee River drainage system, California and Nevada: U.S. Geological Survey Open-File Report 84–440, 35 p.

—— , 1986, The sensitivity of evaporation rate to climate change; Results of an energy-balance approach: U.S. Geological Survey Water-Resources Investigation Report 86–4148, 40 p.

Benson, L. V., and Spencer, R. J., 1983, A hydrochemical reconnaissance study of the Walker River Basin, California and Nevada: U.S. Geological Survey Open-File Report 83-740, 53 p.

Blodgett, D. D., and Titus, F. B., 1973, Hydrogeology of the San Agustin Plains, New Mexico: New Mexico Bureau of Mineral Resources Open File Report 51, 54 p.

Bodine, M., and Jones, B. F., 1986, The salt norm; A quantitative chemical-mineralogical characterization of natural waters: U.S. Geological Survey Water Resources Investigation Report, 86–4086, 130 p.

Bradbury, J. P., 1971, Paleolimnology of Lake Texcoco, Mexico, evidence from diatoms: Limnology and Oceanography, v. 16, p. 180–200.

Bryson, R. A., and Hare, F. K., 1974, Climates of North America: New York, Elsevier, 420 p.

Carmack, E. C., Wiegand, R. C., Daley, R. J., Gray, C.B.J., Jasper, S., and Pharo, C. H., 1986, Mechanisms influencing the circulation and distribution of water mass in a medium residence-time lake: Limnology and Oceanography, v. 31, p. 249–265.

Chivas, A. R., DeDeckker, P., and Shelley, J.M.G., 1983, Magnesium, strontium, and barium partitioning in nonmarine ostracode shells and their use in paleoenvironmental reconstructions; A preliminary study, *in* Maddocks, R. F., ed., Proceedings Eighth International Symposium on Ostracoda, University of Houston, Geosciences 1982, Applications of Ostracoda: Houston, Texas, Department of Geosciences, University of Houston, p. 238–249.

Currey, D. E., and Oviatt, C. G., 1985, Durations, average rates and probable causes of Lake Bonneville expansions, stillstands, and contractions during the last deep-lake cycle, 32,000 to 10,000 years ago, *in* Kay, P. A., and Diaz, H. F., eds., Proceedings of Problems of and Prospects for Predictions Great Salt Lake Levels, Salt Lake City: Salt Lake City, University of Utah, Center for Public Affairs and Administration, p. 9–24.

Currey, D. E., Atwood, G., and Mabey, D. R., 1984, Major levels of Great Salt Lake and Lake Bonneville: State of Utah Department of Natural Resources Utah Geological Survey Map 73, scale 1:750,000.

Dean, W. E., 1981, Carbonate minerals and organic matter in sediments of modern North temperate hard-water lakes: Society of Economic Paleontologists and Mineralogists Special Publication No. 31, p. 213–232.

Delorme, L. D., 1969, Ostracodes as Quaternary paleoecological indicators: Canadian Journal of Earth Sciences, v. 6, p. 1471–1476.

—— , 1971, Freshwater ostracodes of Canada; Part V, Families Limnocytheridae, Loxoconchidae: Canadian Journal of Zoology, v. 49, p. 43–64.

—— , 1978, Distribution of freshwater ostracodes in Lake Erie: Journal Great Lakes Research, International Association Great Lakes Research, v. 4, p. 216–220.

Delorme, L. D., and Zoltai, S. C., 1984, Distribution of an Arctic ostracod fauna in space and time: Quaternary Research, v. 21, p. 65–73.

Delorme, L. D., Zoltai, S. C., and Kalas, L. L., 1977, Freshwater shelled invertebrate indicators of paleoclimate in northwestern Canada during late glacial times: Canadian Journal of Earth Sciences, v. 14, p. 2029–2046.

Eugster, H. P., and Hardie, L. A., 1978, Saline lakes, *in* Lerman, A., ed., Lakes; Chemistry, geology, and physics: New York, Springer-Verlag, p. 237–293.

Eugster, H. P., and Jones, B. F., 1979, Behavior of major solutes during closed-basin brine evolution: American Journal of Science, v. 279, p. 609–631.

Forester, R. M., 1983, Relationship of two lacustrine ostracode species to solute composition and salinity; Implications for paleohydrochemistry: Geology, v. 11, p. 435–438.

—— , 1985, *Limnocythere bradburyi* n. sp.; A modern ostracode from Central Mexico and a possible Quaternary paleoclimate indicator: Journal of Paleontology, v. 59, p. 8–20.

—— , 1986, Determination of the dissolved anion composition of ancient lakes from fossil ostracodes: Geology, v. 14, p. 796–798.

Forester, R. M., and Brouwers, E. B., 1985, Hydrochemical parameters governing the occurrence of estuarine and marginal estuarine ostracodes; An example from south-central Alaska: Journal of Paleontology, v. 59, p. 344–369.

Forester, R. M., and Markgraf, V., 1984, Late Pleistocene and Holocene seasonal climatic records from lacustrine ostracode assemblages and regional (pollen) vegetational patterns in southwestern, U.S.A.: American Quaternary Association Eighth Biennial Meeting, Boulder, Colorado, Program and Abstracts, p. 43–45.

Garrels, R. M., and McKenzie, F. T., 1967, Origin of the chemical composition of some springs and lakes, *in* Equilibrium concepts in natural water systems: American Chemical Society Advances in Chemistry, v. 67, p. 222–242.

Gilbert, G. K., 1890, Lake Bonneville: U.S. Geological Survey Monograph 1, 438 p.

Gorham, E., 1961, Factors influencing supply of major ions to inland waters, with special reference to the atmosphere: Geologic Society of America Bulletin, v. 72, p. 795–840.

Gorham, E., Dean, W. E., and Sanger, T. E., 1983, The chemical composition of lakes in the north-central United States: Limnology and Oceanography, v. 28, p. 287–301.

Gutentag, E. D., and Benson, R. H., 1962, Neogene (Plio-Pleistocene) freshwater ostracodes from the central High Plains: Kansas Geological Survey Bulletin 157, pt. 4, 60 p.

Hardie, L. A., 1968, The origin of the recent non-marine evaporite deposit of Saline Valley, Inyo County, California: Geochimica et Cosmochimica Acta, v. 32, p. 1279–1301.

Hutchinson, G. E., and Loffler, H., 1956, The thermal classification of lakes: Proceedings of the National Academy of Sciences, v. 42, p. 84–86.

Jones, B. F., 1965, The hydrology and mineralogy of Deep Springs Lake, Inyo County, California: U.S. Geological Survey Professional Paper 502–A, 56 p.

——, 1966, Geochemical evolution of closed basin water in the western Great Basin, *in* Rau, J. L., ed., Second Symposium on Salt: Northern Ohio Geological Society, v. 1, p. 181–200.

Jones, B. F., and Bowser, C. J., 1978, The mineralogy and related chemistry of lake sediments, *in* Lakes; Lerman, A., ed., Chemistry, geology, and physics: New York, Springer-Verlag, p. 179–235.

Jones, B. F., and van Denburgh, A. S., 1966, Geochemical influences on the chemical character of closed lakes: International Association of Scientific Hydrology Publication 70, p. 435–446.

Jones, B. F., and Weir, A. H., 1983, Clay minerals of Lake Albert, an alkaline, saline lake: Clays and Clay Minerals, v. 31, p. 161–172.

Kelts, K., and Hsü, K. J., 1978, Freshwater carbonate sedimentation, *in* Lerman, A., ed., Lakes; Chemistry, geology, and physics: New York, Springer-Verlag, p. 295–324.

Lajoie, K. R., 1968, Quaternary stratigraphy and geologic history of Mono Basin, eastern California [Ph.D. thesis]: Berkeley, University of California, 271 p.

Lajoie, K. R., Robinson, S. W., Forester, R. M., and Bradbury, J. P., 1982, Rapid climatic cycles recorded in closed-basin lakes: American Quaternary Association Seventh Biennial Conference, Program and Abstracts: Seattle, University of Washington, p. 53.

Lamb, H. H., 1972, Climate; Present, past, and future: v. 1, Fundamentals and climate now: London, Methuen and Co., 613 p.

Lerman, A., 1978, Lakes; Chemistry, geology, and physics: New York, Springer-Verlag, 363 p.

Markgraf, V., Bradbury, J. P., Forester, R. M., McCoy, W., Singh, G., and Sternberg, R., 1983, Paleoenvironmental reassessment of the 1.6-million-year-old record from San Agustin Basin, New Mexico: New Mexico Geological Society Guidebook, 34th Field Conference, Socorro Region II, p. 291–297.

Markgraf, V., Bradbury, J. P., Forester, R. M., Singh, G., and Sternberg, R. S., 1984, San Agustin Plains, New Mexico; Age and paleoenvironmental potential reassessed: Quaternary Research, v. 22, p. 336–343.

Martens, K., DeDeckker, P., and Marples, T. G., 1985, Life history of *Mytilocypris henricae* (Chapman) (Crustacea: Ostracoda) in Lake Bathurst, New South Wales: Australian Journal of Marine and Freshwater Research, v. 36, p. 807–819.

McCoy, W. D., and Forester, R. M., 1982, Epimerization of isoleucine in Quaternary nonmarine ostrocodes: The Geological Society of America Abstracts with Program, v. 14, p. 560.

McLay, C. L., 1978a, Comparative observations on the ecology of four species of ostracods living in a temporary freshwater puddle: Canadian Journal of Zoology, v. 56, p. 663–675.

——, 1978b, The population biology of *Cyprinotus carolinensis* and *Herpetocypris reptans:* Canadian Journal of Zoology, v. 56, p. 1170–1179.

Mickelson, D. M., Clayton, L., Fullerton, D. S., and Borns, H. W., Jr., 1983, The Late Wisconsin glacial record of the Laurentide Ice Sheet in the United States, *in* Wright, H. E., Jr., ed., Late Quaternary environments of the United States, *in* Porter, S. C., ed., v. 1, The Late Pleistocene, Minneapolis, University of Minnesota Press, v. 1, p. 3–37.

Mix, A. C., and Ruddiman, W. F., 1985, Structure and timing of the last deglaciation; Oxygen-isotope evidence: Quaternary Science Reviews, v. 4, p. 59–108.

Porter, S. C., Pierce, K. L., and Hamilton, T. D., 1983, Late Wisconsin mountain glaciation in the western United States, *in* Wright, H. E., Jr., ed., Late Quaternary environments of the United States, *in* Porter, S. C., ed., v. 1, The Late Pleistocene: Minneapolis, University of Minnesota Press, v. 1, p. 54–71.

Ragotzkie, R. A., 1978, Heat budgets of lakes, *in* Lerman, A., ed., Lakes; Chemistry, geology, and physics: New York, Springer-Verlag, p. 1–20.

Riggs, A. C., 1984, Major Carbon-14 deficiency in modern snail shells from southern Nevada Springs: Science, v. 224, p. 58–61.

Scott, W. E., McCoy, W. D., Shroba, R. R., and Meyer, R., 1983, Reinterpretation of the exposed record of the last two cycles of Lake Bonneville, western United States: Quaternary Research, v. 20, p. 261–285.

Shoemaker, E. M., and Kieffer, S. W., 1974, Guide to the Astronaut Trail at Meteor Crater, *in* Shoemaker, E. M., and Kieffer, S. W., eds., Guidebook to the geology of Meteor Crater Arizona: 37th Annual Meeting of the Meteoritical Society, August 7, 1974, p. 34–62.

Smith, G. I., 1979, Subsurface stratigraphy and geochemistry of late Quaternary evaporites, Searles Lake, California: U.S. Geological Survey Professional Paper 1043, 129 p.

——, 1987, Continental paleoclimate records and their significance, *in* Morrison, R. B., ed., Quaternary nonglacial geology; Conterminous U.S.: Boulder, Colorado, Geological Society of America, The Geology of North America, v. K-2 (in press).

Smith, G. I., and Street–Perrott, F. A., 1983, Pluvial lakes of the western United States, *in* Wright, H. E., Jr., Late Quaternary environments of the United States, *in* Porter, S. C., ed., v. 1, The Late Pleistocene: Minneapolis, University of Minnesota Press, p. 190–214.

Spencer, R. J., and 11 others, 1984, Great Salt Lake, and precursors, Utah; The last 30,000 years: Contributions to Mineralogy and Petrology, v. 86, p. 321–334.

Spencer, R. J., Eugster, H. P., Jones, B. F., and Rettig, S. L., 1985a, Geochemistry of Great Salt Lake, Utah I; Hydrochemistry since 1850: Geochimica et Cosmochimica Acta, v. 49, p. 727–738.

——, 1985b, Geochemistry of Great Salt Lake, Utah; II, Pleistocene–Holocene evolution: Geochimica et Cosmochimica Acta, v. 49, p. 739–748.

Street–Perrott, F. A., and Harrison, S. P., 1985, Lake levels and climate reconstructions, *in* Hecht, A. D., ed., Paleoclimate analysis and modeling: New York, John Wiley and Sons, p. 291–340.

Thompson, R. S., Benson, L. V., and Hattori, E. M., 1986, A revised chronology for the last Pleistocene lake cycle in the central Lahontan Basin: Quaternary Research, v. 25, p. 1–9.

Waitt, R. B., and Thorson, R. M., 1983, The Cordilleran ice sheet in Washington, Idaho, and Montana, *in* Wright, H. E., Jr., ed., Late Quaternary environments of the United States, *in* Porter, S. C., ed., v. 1, The Late Pleistocene: Minneapolis, University of Minnesota Press, p. 53–70.

Wetzel, R. G., 1975, Limnology: Philadelphia, W. B. Saunders, 743 p.

Winter, T. C., and Woo, Ming-ko, 1987, Lakes and wetlands, *in* Moss, M. L., Wolman, M. G., and Riggs, H. C., eds., Surface water hydrology of North America: Boulder, Colorado, Geological Society of America, The Geology of North America, v. 0-1 (in press).

Manuscript Accepted by the Society March 16, 1987

ACKNOWLEDGMENTS

I would like to thank David P. Adam, L. Denis Delorme, and Blair F. Jones for reviewing earlier drafts of this paper and making valuable suggestions for improvement in style and content. I would also like to thank J. Platt Bradbury, Donald R. Curry, Jack Oviatt, Ronald J. Spencer, Vera Markgraf, Ken R. Lajoie, Eugene M. Shoemaker, and Larry V. Benson for valuable discussions.

Printed in U.S.A.

The Geology of North America
Vol. K-3, North America and adjacent oceans during the last deglaciation
The Geological Society of America, 1987

Chapter 13

Patterns and rates of vegetation change during the deglaciation of eastern North America

George L. Jacobson, Jr.
Department of Botany and Plant Pathology and Institute for Quaternary Studies, University of Maine, Orono, Maine 04469
Thompson Webb III
Department of Geological Sciences, Brown University, Providence, Rhode Island 02912
Eric C. Grimm
Limnological Research Center, University of Minnesota, Minneapolis, Minnesota 55455

INTRODUCTION

Environmental changes during deglaciation are among the most profound and rapid of any during the Quaternary. Palynological evidence reveals the changing patterns of vegetation on late Quaternary landscapes, and provides abundant evidence for large shifts during the late Wisconsin termination in particular. Changes in the oxygen-isotope ratios in marine microfossils show that the large, rapid termination of the most recent ice age is similar to those during previous cycles back to at least 400 ka (Broecker and van Donk, 1970). The abundant evidence for changes from 18 to 6 ka (thousands of radiocarbon years B.P.) therefore provides clues to the possible behavior of terrestrial and marine ecosystems during earlier times.

Palynological evidence for major changes in the North American vegetation began with the contributions of pioneers like Deevey (1939, 1949, 1951), Sears (1942), Leopold (1956), and Fries (1962) and has grown dramatically since Wright and others (1963) and McAndrews (1966) demonstrated that the prairie-forest border in Minnesota had moved considerable distances in apparent response to Holocene climatic changes. More recently, increases in both the amount of data and in the use of computers to compile and evaluate these data has permitted mapping of population expansions (Davis, 1981a) and population densities (Bernabo and Webb, 1977; Webb, 1987). These maps reveal broad-scale patterns and complex behavior in the vegetational changes across eastern North America during the past 18,000 yr.

Here we use three separate approaches to the reconstruction of vegetation during the most recent deglaciation. The first involves mapping the changing patterns of distribution and abundance of selected plant taxa across eastern North America. The second approach uses the combined evidence from several taxa to illustrate how the different taxa interacted through time and to reveal patterns in the changing composition of the vegetation. The third is a numerical approach designed to reveal the rate of change in the vegetation per unit time, and thus to show whether periods of relative constancy have existed between periods of especially rapid change, or whether changes have been gradual throughout the past 18,000 yr.

Each of these approaches produces evidence of vegetational changes that are not only climatically controlled but also ecologically mediated by local events such as fire and other disturbance. By using the three approaches together, we attempt to differentiate the nature, timing, and causes of these changes.

Fossil pollen stratigraphies provide the primary data for our study. These stratigraphies come from among the several hundred sites in eastern North America that have been studied over the past 25 years by modern palynological methods and radiometric dating. Most of these data are now in a computer-based repository at Brown University and thus available for systematic study.

PLANT POPULATIONS

Mapped data

Radiocarbon-dated pollen profiles covering the last 10,000 yr have been completed for over 250 sites in eastern North America. Approximately 70 sites date back to 12 ka, and 15 to 18 sites to 18 ka (Fig. 1). Names and locations of sites are presented elsewhere: Webb and others (1983) for the Midwest, Gaudreau and Webb (1985) for the northeastern United States and southern Quebec, and Bartlein and Webb (1985) for all sites with data for 6 ka. Lines of equal pollen percentages (isopoll lines) were drawn on maps of the data from sites available for different times, starting at 18 ka (Plate 1). For this analysis we present maps for

Jacobson, G. L., Jr., Webb, T., III, and Grimm, E. C., 1987, Patterns and rates of vegetation change during the deglaciation of eastern North America, *in* Ruddiman, W. F., and Wright, H. E., Jr., eds., North America and adjacent oceans during the last deglaciation: Boulder, Colorado, Geological Society of America, The Geology of North America, v. K-3.

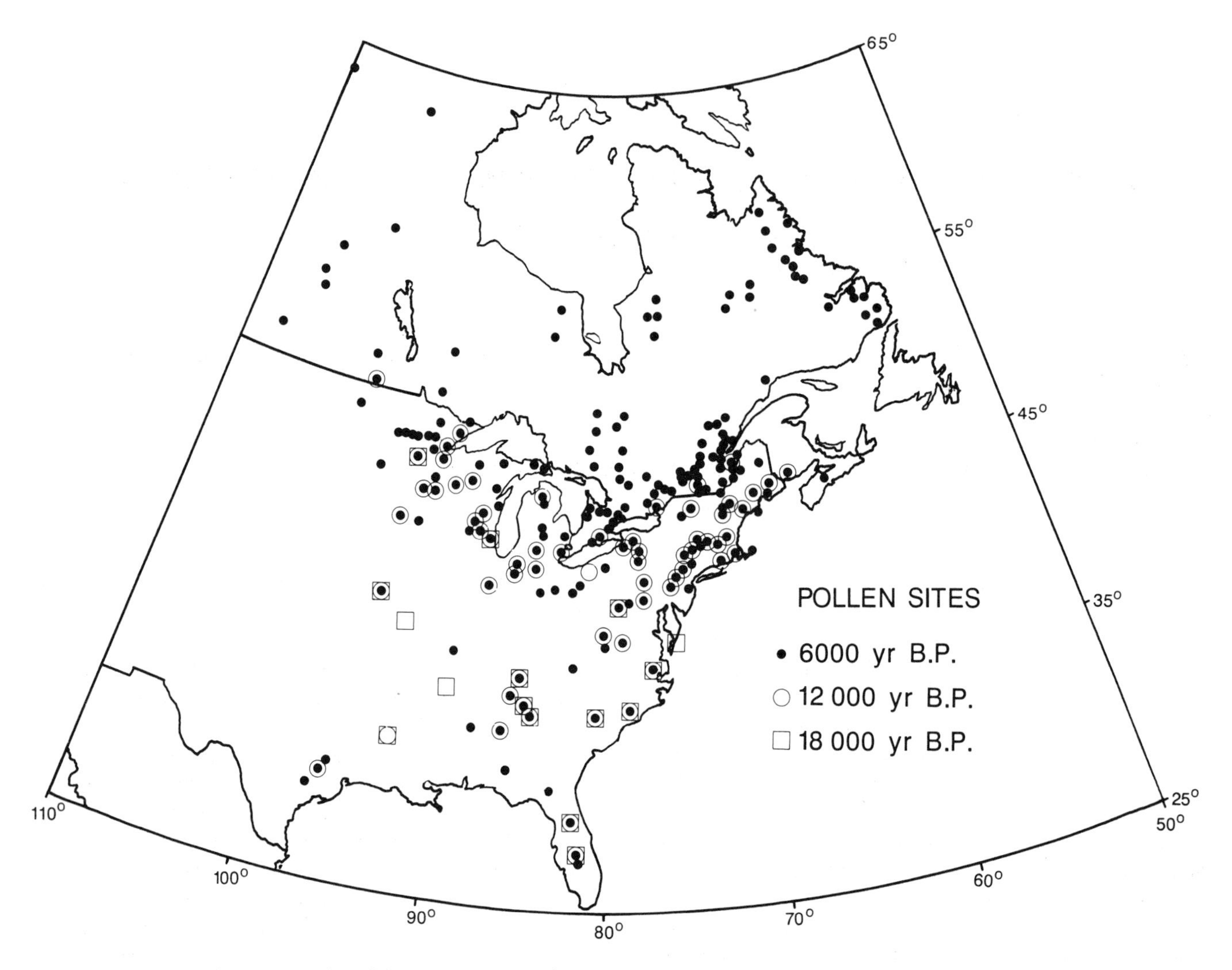

Figure 1. Location of sites with pollen data for 6 ka (black dots). Open squares mark sites with data for 12 ka, and open circles mark sites with data for 18 ka.

18, 14, 12, 10, 8, and 6 ka, along with a map for 0.5 ka that shows "modern" pollen abundances as they existed before European immigrants disturbed the vegetation.

For each of the maps, isopoll contours vary from levels that indicate low abundance of that taxon up to levels that reveal the location of major population centers. Actual percentages for the contoured lines differ from one plant group to another, because the amount of pollen produced and its dispersal distance varies widely from taxon to taxon (Delcourt and others, 1984; Bradshaw and Webb, 1985; Prentice, 1985). Some smoothing of the plotted data inevitably accompanies the contouring, and we recognize that the maps are not accurate in every detail. Furthermore, palynological studies have not differentiated between the several component species for many of the families and genera mapped. We may assume, therefore, that the maps fail to reveal some important geographic patterns within the taxa shown. Our goal, however, is to show how the population centers for important plants shifted position during the most recent deglaciation of eastern North America.

The modern vegetation of this large region (Plate 1) grades from tundra in the far north of Canada through boreal forest dominated by spruce (*Picea*) trees and other conifers to Great Lakes forest of mixed deciduous and coniferous trees in the Great Lakes states, New England, and the Maritime Provinces of Canada. A diverse forest of deciduous trees including oak (*Quercus*), hickory (*Carya*), ash (*Fraxinus*), elm (*Ulmus*), and ironwood (*Ostrya/Carpinus*) lies south of the Great Lakes, and a forest of southern conifers, mostly pine (*Pinus*), grows in the Southeast and along the Gulf of Mexico. This population of southern pines is distinct from the northern pine populations of the Great Lakes forest and northward. Prairie covers the Great Plains from Texas north through the Dakotas and western Minnesota into Canada. These biome-scale patterns on the modern landscapes are revealed nicely in selected isopoll maps for 0.5 ka (Plate 1).

As examples of how individual plant taxa changed in location and abundance during deglaciation, on Plate 1 we present isopoll maps for sedge (Cyperaceae), spruce, birch (*Betula*), forbs (nongraminoid herbs), oak, hickory, ash, pine, and hemlock (*Tsuga*). Ice positions follow the mapped summaries of Denton and Hughes (1981).

Sedge. (Plate 1). The sedges, which today are abundant in regions of prairie and tundra, were widespread at 18 ka across a broad region of eastern North America south of the Laurentide ice sheet. The geographic extent of this family of grasslike plants remained broad scale until sometime between 12 and 10 ka, when the widespread populations in the east decreased to two centers, one in southern Labrador and the other in the southern Appalachians; the only broad-scale distribution of sedges was in the Great Plains, where prairie grows today. The high values of sedge pollen along the Labrador coast by about 10 ka indicate the development of modern tundra (Lamb, 1980), and western sedge populations increased to mark the mid-Holocene maximum extent of prairie. The eastward extension of this taxon at 8 ka and 6 ka is consistent with the expansion of prairie eastward into Indiana and western Ohio (Wright, 1968).

Spruce. (Plate 1). During the full-glacial conditions of 18 ka, spruce was abundant throughout a broad region south of the Laurentide ice sheet, similar in extent to the sedges, except that the population center for spruce was primarily south of the western Great Lakes, with lower abundances south of the ice margin in the east. Between 18 and 12 ka the population increased steadily toward the east but still retained its presence across a broad region south of the ice. By 12 ka spruce was present along the entire southern edge of the wasting ice, except for northern New England and nearby Canada, where spruce trees had yet to colonize the recently deglaciated landscapes. By 10 ka the extensive populations of spruce had begun to collapse, with relatively high abundances restricted to a narrow band immediately south of the ice. The decline in these populations continued until at least 8 ka, when only a few isolated centers of abundance remained, scattered across central Canada; this was the low point for spruce population numbers for the entire time from 18 ka to the present. The modern distribution of spruce in the boreal forest had emerged by 6 ka, as all but the last vestiges of the ice sheet disappeared.

Birch. (Plate 1). Today birch trees are an important component in mixed forests of the Great Lakes region and New England, but at 18 ka the birches were rare south of the ice and continued so until 12 ka. Only around 12 ka did birch trees and shrubs emerge in notable abundance, and then only in the northeast. By 10 ka they grew in significant abundances across most of the area south of the ice sheet. As the ice became restricted to smaller areas of eastern Canada, birch populations increased rapidly in the west, and by 8 ka they covered a broad band south of the ice and north of the Great Lakes. This extensive distribution subsequently split up, and by 6 ka the most abundant populations of birches were largely restricted to eastern Canada.

Forbs. (Plate 1). Forbs include taxa that are important components of prairie vegetation and are well represented in the pollen record; these include sage (*Artemisia*), ragweed (*Ambrosia*), other members of the Sunflower family (Asteraceae), Goose Foot family (Chenopodiaceae), and Amaranth family (Amaranthaceae). During full-glacial conditions these plants were present in two separate locations, one in Texas and the second just south of the ice in the western Great Lakes region. Here, sage pollen was dominant among the forb types. The two population centers had merged by 14 ka, primarily as a result of expansion of the southern population. Between 14 and 12 ka the northern population (mostly sage) expanded eastward along the immediate southern margin of the ice, and in southern Quebec sage populations continued to grow as late as 10 ka. Also at 12 ka, the southern population expanded eastward along the Gulf coast and especially into Florida, where a minor center persisted at least through 6 ka (Watts, 1969). By 10 ka the major center of distribution (dominated by members of the Goose Foot/Amaranth families) was restricted to the Great Plains. This population had expanded dramatically across the Midwest by 8 ka and remained in that configuration through 6 ka, when prairie grew farther to the east than it does today.

Oak. (Plate 1). The oaks have undergone dramatic changes in abundance and distribution since 18 ka, when they were restricted primarily to scattered areas in the South from the Atlantic coast across the region north of the Gulf of Mexico. By 14 ka these trees had become more abundant in a band from Florida across most of the Gulf coast. A major population increase occurred by 12 ka, and the center of distribution moved slightly to the north. After 12 ka this northward shift accelerated, and oak trees became abundant as far north as Lake Michigan by 10 ka. This pattern of distribution and abundance continued up through 6 ka, although by then the populations in Florida had begun to decline.

Hickory. (Plate 1). During full-glacial conditions hickory was consistently present in the lowlands bordering the Gulf of Mexico. By 14 ka the major population center was in the northern panhandle of Florida, from which it spread to form increasingly large southeastern populations at 12 and 10 ka. Between 10 and 6 ka the population center shifted northward and westward, so that in the mid-Holocene the greatest concentrations of hickory were generally along the Mississippi drainage from the Gulf of Mexico to the southern Great Lakes. Hickory trees arrived in the Midwest about 10 ka as oak populations expanded there, but hickory trees arrived in New England 2,000 to 3,000 years after the oak populations first increased in that region.

Ash. (Plate 1). During full-glacial conditions ash populations were apparently restricted to small microhabitats south of the ice in the Midwest. By 14 ka these trees were abundant immediately south of the Great Lakes, and by 12 ka this population had expanded in this same region to levels that have not been exceeded since that time. Even by 10 ka the ash component in the vegetation was decreasing, and it continued to do so up to 6 ka. Throughout all of these major changes in abundance, the geographic center of the ash population remained relatively fixed, especially in comparison with the taxa mentioned above. Evidence from a number of sites where ash-pollen types were separated (e.g., Cushing, 1967) indicates that this was primarily black ash (*F. nigra*) from 12 ka to 8 ka, but that the modern center for ash pollen coincides with the distribution of white and green ash (*F. americana* and *F. pennsylvanica*).

Pine. (Plate 1). Major populations of pine in full-glacial times covered the southern Atlantic Coastal Plain and an area several hundred kilometers to the west. The data from Tulane Lake, Florida (W. A. Watts, personal communication, 1986) indicate that southern pines grew in southern Florida, and plant macrofossils from northern Georgia indicate that northern pines (e.g., *Pinus banksiana*) grew there (Watts, 1970). Between 18 and 12 ka the pines dominated a decreasing area of the Southeast and became restricted to a north-south band along the Atlantic Coastal Plain east of the Appalachian Mountains. A major shift in the distribution of these conifers came between 12 and 10 ka, as the north-south–trending population ceased to exist and the center of abundance was oriented in an east-west band south of the wasting ice; this distribution subsequently expanded to form the northern pine forests of today. By 8 ka the first major population expansions for southern pines were occurring in Florida and adjacent areas, and by 6 ka the population center for southern pines had become well established, although not quite so extensive as it exists today.

Hemlock. (Plate 1). From an apparently small center of distribution in the central Appalachians at 12 and 10 ka, hemlock populations expanded to the northeast in the early Holocene. These trees became abundant from the eastern Great Lakes region through New England by 8 ka and expanded northward and westward to reach their Holocene maximum in abundance at about 6 ka.

VEGETATION

The distinctive behavior of the individual taxa mentioned above produced ever-changing patterns of vegetation during deglaciation (Davis, 1983; Webb, 1987). To take advantage of the wealth of information available in the isopoll maps, we prepared a series of colored maps that illustrate how the individual populations combined to form a changing mosaic of vegetation (Plate 2). On a base map similar to that used above, we show three maps for each time from 18, 14, 12, 10, 8, and 6 ka, along with composites showing the modern (0.5 ka) distribution of these taxa in the pollen record. The maps have been designed with a combination of primary colors for key taxa in such a way that the secondary colors produced by overlapping primary colors can reveal the changing nature of the composition and distribution of the vegetation. For example, the woodland of scattered spruce in tundra that constitutes the transition from boreal forest to tundra is shown by the orange pattern, which is a secondary color resulting from the overlapping of sedge (yellow) and spruce (red) in Series A.

In one set of maps (Plate 2, Series A) we show the major pollen types in the modern tundra, prairie, and boreal forest (spruce, birch, sedge, and forbs); in the second map for each time (Plate 2, Series B) we show the major pollen types in the mixed forest and deciduous forest (pine, oak, hickory, hemlock, and beech [*Fagus*]); in the third set (Plate 2, Series C) we show three pollen types (ash, elm, and ironwood) that had more extensive distributions from 14 ka to 10 ka than they do today. Each taxon is shown only when it was so abundant as to have been an important component in the vegetation. Isopoll values above which the taxa were mapped are 20% spruce, 20% birch, 5% sedge, 10% forbs, 20% pine, 20% oak, 3% hickory, 5% hemlock, 5% beech, 3% ash, 5% elm, and 3% ironwood.

Modern pollen and vegetation. Correspondence of the composite maps for 0.5 ka (Plate 2) with the map of modern vegetation (Plate 1) is good, although the 12 pollen types together provide considerable detail that is not evident in the generalized map of vegetation. Several of the pollen types appear to be especially good indicators of biome-scale vegetation. Sedge, for example, closely matches tundra in the north of Canada, and along with forbs follows prairie in the Great Plains (Plate 2, Series A). Spruce provides a good indication of boreal forest, although pine and birch are also components of that forest in the area south and west of Hudson Bay. The mapped patterns of oak and history (Series B) generally follow the deciduous forest south of the Great Lakes and the northern portion of the southeastern pine-oak forest; the southern population of pines matches the latter forest even more closely.

Full-glacial conditions. Patterns of vegetation at 18 ka must have borne little resemblance to anything that exists in southern North America today (Whitehead, 1973; Watts, 1983). Pine forests were limited to the Southeast, with fossil evidence indicating that boreal species such as jack pine (*Pinus banksiana*) were present as far south as North Carolina (Whitehead, 1981) and Georgia (Watts, 1970). Tundra elements, including sedges, were abundant across a broad band south of the ice, but important differences in the vegetation existed across that region. Spruce trees grew with the sedges in the Midwest, forming a woodland of scattered trees with an open tundralike understory that continued essentially to the margin of the ice sheet, even between ice lobes in Minnesota (Birks, 1976). South of the ice in the East, however, trees apparently were not present, and the vegetation was more open and tundralike (Davis and Jacobson, 1985). Oak was present in significant amounts only in a small area within the pine-dominated region of the Southeast, and hickory was restricted to an area along the Gulf of Mexico from Texas to west Florida. Other deciduous taxa were conspicuously absent in any large quantities, although several, including hickory, may have been present in isolated small populations where suitable microhabitats existed throughout the South (Delcourt and others, 1980). We have little knowledge of the vegetation on the exposed continental shelf along the Atlantic coast during full-glacial time, but pines and probably some deciduous trees may have grown there while sea level remained low.

Deglaciation. By 14 ka the ice sheet had begun to thin (Hughes, this volume), even though in the Midwest the Des Moines and James lobes were at their southernmost late Wisconsin position (Clayton and Moran, 1982). Southern New England was free of ice, and areas of higher elevation had emerged in the north, leaving only the lowlands and most of Maine still covered (Davis and Jacobson, 1985).

Several notable developments had occurred in the vegetation by that time. Tundra extended from the recently deglaciated areas of New England across a broad band south of the ice. Forbs and sedges occurred across the Great Plains from Texas north to the ice lobes in the Midwest, perhaps signaling the expansion of the modern prairie, at least in the southern Great Plains. Spruce populations remained present, primarily still in woodlands (orange on Plate 2, Series A), south of the ice in the midcontinent. Spruce trees were not present, however, in the deglaciated regions of New England, where tundralike vegetation remained for several thousand years in places such as Moulton Pond, Maine (Davis and others, 1975).

By 14 ka pine forests were restricted to the Southeast, but they extended northward on the east side of the Appalachians as far as New Jersey. Oak-hickory forests (green on Series B) by now covered the Coastal Plain surrounding the Gulf of Mexico, both taxa having expanded their centers of abundance dramatically since 18 ka.

In some ways the most striking development of the period was the emergence of black ash as a major component of the spruce woodland immediately south of the ice sheet in the Midwest. Ironwood, another hardwood of some importance at a slightly later time, was abundant at just a few scattered localities from the midwestern ice margin down to the Gulf coast.

As the ice sheet continued to melt during late-glacial time, large regions were opened up for colonization by vegetation. Calving ice in the Northeast rapidly opened the St. Lawrence lowland and cut off the supply of ice to the south, so that by 12 ka only a small remnant ice mass remained in Maine and adjacent Quebec (Davis and Jacobson, 1985). At the same time, however, the Laurentide ice sheet still extended south across the Great Lakes. This rapid retreat of ice in the Northeast exposed an area that was entirely bounded on the west by ice.

For the most part, the vegetational shifts between 14 and 12 ka were more changes in detail than rearrangements of major vegetation patterns on the landscape. Treeless boreal vegetation continued to be important across a broad band south of the ice sheet, having extended up through the newly opened regions in New England and southeastern Canada. In some areas, plant cover was apparently not sufficient to prevent soil erosion (Anderson and others, 1984). Sedges and forbs, along with grasses, formed a prairielike landscape in the Great Plains, presumably with more open boreal qualities in the area just south of the ice sheet.

Spruce woodland remained across the same broad region south of the ice, although it was only slowly advancing eastward into southern New England. Pine forests had shifted slightly eastward and considerably farther northward to cover a relatively narrow area east of the Appalachians.

In the Southeast, oak trees became more abundant in the forests, but the population center for oaks did not cover an area much larger than at 14 ka. Oak along with hickory and, in a few small areas, beech formed an increasingly rich deciduous forest along the Gulf coast (Series B).

The most notable additions to the forests and woodlands of that time were ash and ironwood, which grew abundantly over largely overlapping areas of the Midwest and in larger regions than at any time from 18 ka to the present. The widespread appearance of ironwood in particular further supports the notion that a broad woodland of open-grown vegetation existed south of the ice sheet; this tree flowers and produces abundant pollen only when growing in well-lighted conditions. About this time, enough hemlock pollen was recorded at Cranberry Glades, Virginia (Watts, 1979) to indicate existence of a major population center in a small region of the central Appalachians. Birch shrubs and trees became abundant for the first time in forests along the coastal lowlands of southern New England.

Transition to interglacial conditions. The greatest reorganization of the vegetation in the past 18 ka took place between 12 and 10 ka. During this interval the ice had melted from essentially all of the northeastern United States and south-central and southeastern Canada. Tree populations expanded throughout the southeastern and south-central regions of the continent, and the broad band of spruce woodland dominated by sedges and scattered spruce trees was almost entirely gone. Sedges were abundant enough to indicate the presence of open tundralike vegetation only in southern Ontario, southern Quebec, and a small region of the central Appalachians, presumably in a high-elevation remnant of the plant cover that had been so common previously (Series A). Richard (1987) concluded that the treeless vegetation in central Quebec between 11 and 10 ka was more similar to modern tundra than were treeless areas farther south.

Pine forests shifted location from along the Atlantic coast to an east-west band across the Great Lakes and along the southern margin of the wasting ice (Series B). This increase in pine forests across the midcontinent came at the expense of the spruce woodland and to some extent the ash woodlands, which had been so abundant at 12 ka. Spruce remained an important component of the forests along a narrow band just south of the ice.

Oak-history forests (Series B) continued the expansion of the deciduous vegetation that had begun along the Gulf coast several thousand years before; by 10 ka they dominated the Southeast and also an area south of the Great Lakes. Several other hardwoods also played a large role in the vegetation of this time. Elm in particular expanded dramatically during the previous two thousand years, and it was abundant over a major area of the midcontinent (Series C). Ash was actually slightly less abundant than it had been at 12 ka in that area, but together with elm and ironwood it formed for the first time a mixed deciduous forest similar to that which occurred just south of the Great Lakes from that time to the present (Series C). Hemlock and ironwood remained in configurations almost identical to those of the earlier time.

The landscapes at the start of the Holocene were covered by conifer forests immediately south of the retreating ice sheet, except along the eastern margin in Canada, where birch populations dominated. Newly formed deciduous forests grew in a large area south of the band of pine and spruce. Over much of the Midwest the vegetation consisted of a mixture of oak, hickory, ash, elm, and ironwood. Prairie covered the Great Plains, apparently with considerable elm and ash growing in scattered localities, probably along rivers in the central plains.

Early Holocene. The changes in vegetation between 10 and 8 ka were nearly as dramatic as those in the previous interval. The final remnant of the Laurentide ice sheet persisted in an area surrounding Hudson Bay; the ice sheet by then probably was thin and largely stagnant (Hughes, this volume). Sedges and shrub birches along the retreating ice margin in the Quebec/Labrador plateau indicate the presence of tundra there, but apparently not elsewhere along the southern margin of the ice. Dense populations of spruce were almost nonexistent, save for a few isolated patches immediately south of the ice.

Pine forests continued in an east-west orientation from the Great Lakes northward, but by this time they contained large amounts of birch, which had spread from southeastern Canada and northern New England westward for several thousand kilometers. The extent of both pine and birch at this time suggests that conditions were dry over much of this region, and that fire frequency may have been higher than in earlier or later times. Forests of oak and hickory continued to dominate much of southeastern North America. In the Midwest, however, several previously important hardwoods had much-reduced roles in the vegetation. The major populations of ash, elm, and ironwood by this time were restricted to an area just south of the Great Lakes and were far less abundant and extensive than in late-glacial time.

Regions with abundant sedges and forbs were larger than previously, owing to the eastward expansion of prairie. This advance occurred in the northern plains, where it was first documented in detail by McAndrews (1966), as well as in the region south of the Great Lakes in the "prairie peninsula" (Wright 1968, 1970; Webb and others, 1983b).

Hemlock trees expanded their range from the eastern Great Lakes region to New England for the first time between 10 and 8 ka (see also Davis, 1981a). From the area of their limited previous occurrence in the central Appalachians, these populations expanded northward along the mountains and then eastward into southern New England. Beech, on the other hand, decreased as a component of forests in the southeast during this same time interval, so that its earlier extent in the mid-Atlantic region was now restricted to only a small area along the coast (Bennett, 1985). By then a similar small population of beech had also emerged in Connecticut, where it later became the center of expansion for beech into the forests of New England and the northern Appalachians (Dexter and others, 1987).

Over the next 2,000 years the last remaining ice in northeastern Canada melted completely (Hughes, this volume; King, 1985), and the configuration of vegetation in eastern North America began to approach modern conditions (Davis, 1983). The boreal forest of spruce trees, in combination with pine trees in the west and birch trees in the east, became established in a band across southern Canada (Ritchie, 1976; Webb, 1987). Spruce forests in particular expanded during this interval, and the extensive birch populations (probably mostly white birch, *B. papyrifera*) of 8 ka were less extensive. By 6 ka the population center for birch became restricted to northern New England, the Canadian Maritime Provinces, Quebec, and Labrador. Pine became a more important component in the forests of two regions: one the area between the Great Lakes and Hudson Bay, and the other in coastal lowlands along the Gulf of Mexico and the Atlantic Ocean adjacent to and including Florida. The northern pine forests have never reached into Labrador.

In the northeastern United States, hemlock continued to become more numerous in the mixed forests from the Great Lakes eastward across New England and into the Canadian Maritimes. For the first time, beech populations expanded within the forests of the northeast. From New England and southern Quebec to the Great Lakes it formed an association with hemlock characteristic of modern forests. Yellow birch (*Betula lutea*) probably became more extensive in this area, expanding with hemlock and beech during the mid-Holocene. In the sites where macrofossil evidence is available (Jacobson and others , unpublished) fossils for yellow birch consistently appear only after those for hemlock are first recorded. Thus the modern hemlock-beech-birch forests of the east formed during the mid-Holocene, largely in place of the white pine-dominated forests of earlier times (Webb and others, 1983a).

South of the Great Lakes the populations of elm, ash, and ironwood trees remained similar in extent to what they were at 8 ka (Plate 2, Series C). Although far less extensive than they were at 12 and 10 ka, these three taxa had become characteristic components of the mixed hardwood forests in this region and essentially nowhere else. Oak and hickory were also present in the forests south of the Great Lakes but were more abundant in the areas more to the south and southeast, forming the dominant vegetation in southeastern North America (Plate 2, Series B). Development of the deciduous forest has been discussed in more detail by Davis (1981a, 1983), Watts (1980), Watts and Stuiver (1980), Webb and others (1983a, b), Delcourt and Delcourt (1985), Gaudreau and Webb (1985), among others. These recent studies have shown that the history of the rich deciduous forest has been much more dynamic than earlier thought (e.g., Braun, 1947).

Prairie remained somewhat east of the modern prairie-forest ecotone. Relatively xeric conditions continued in the midcontinent, although only shortly later forests began to expand westward to modern locations (Webb and others, 1983b).

Rates of Vegetation Change

Method and Data. The mapped patterns of plant populations (Plate 1) and vegetation (Plate 2) illustrate the appearance and disappearance of vegetation regions and ecotones as each taxon responded individualistically to the long-term major climatic changes during the most recent deglaciation. The 2,000-yr intervals between maps, however, are too imprecise to reveal much about the rapidity of changes or the site-specific variation of those changes. Jacobson and Grimm (1986) developed a method for measuring rates of vegetation change in continuous pollen-stratigraphic sequences. In these sequences, they calculated the multivariate distances between adjacent pollen spectra interpolated to represent even time intervals. These distances per unit time (e.g., 100 yr) provide a quantitative measure of the rate of palynological change, which is a proxy for the amount of vegetational change. A plot of the increments of change over long periods of time shows the rate of vegetational change.

With an approach similar to that of Jacobson and Grimm (1986), we have graphed the rate of palynological change for 18 sites from across eastern North America. For each site, we first smoothed the individual pollen curves in a multidimensional taxon space, which results in a smoothed curve in a hyperspace defined by the original taxa used in the analysis. We used the 25 most abundant taxa reliably identified by the different palynologists: *Abies, Acer saccharum, Alnus, Ambrosia, Artemisia, Betula, Carya, Castanea,* Chenopodiaceae/Amaranthaceae, *Corylus,* Cupressaceae, Cyperaceae, Ericaceae, *Fagus, Fraxinus, Larix, Ostrya/Carpinus,* Poaceae, *Picea, Pinus, Populus, Quercus, Tsuga,* Tubuliflorae (Asteraceae subfamily Tubuliflorae undifferentiated), and *Ulmus.*

For smoothing the pollen curve in multidimensional space, we developed a somewhat *ad hoc* method to accommodate unequal spacing of samples in time. The method requires that a minimum number of samples be used for each smoothing interval. If these samples span less than a given time interval, more samples are included, up to some maximum, until the given time interval is exceeded. For a sample interval of 200 years, the smoothing method proved satisfactory with a minimum of five samples; when necessary, however, more samples were included until a time interval of 300 years or a maximum of nine samples was exceeded. Positions along the smoothed curve were interpolated for constant time intervals, and chord distances were calculated between these points. The chord distance is a Euclidian distance of the square roots of the pollen percentages. The square-root transformation increases the signal and reduces the noise among the percentages (Prentice, 1980; Overpeck and others, 1985).

From among the many pollen stratigraphies available from eastern North America, we selected 18 for our analysis. The sites represented different geographic regions of the larger area. Few of the many possible sites had the two characteristics necessary for successful analysis of rates of change: reliable dating control and sample intervals close in time, ideally at least one sample per century. After examining the data for sites in the Brown University data bank, we chose 18 that were adequate for this study. (Data for two neighboring sites in Minnesota [Billy's Lake and Wolf Creek] cover complementary time periods and are presented together in the results.)

As an example, the rate-of-change graph for Gould Pond, Maine (Jacobson and others, unpublished; Grimm and Jacobson, 1985) has times of rapid change as well as extended periods of relatively little change (Fig. 2), when the pollen spectra indicate that forests of central Maine remained in one consistent configuration. Times of relative vegetational constancy were in the early Holocene when white pine–hardwood forest prevailed, and in the late Holocene when hemlock-hardwood forest dominated. Dramatic change accompanied important ecological events such as the mid-Holocene decline in hemlock abundance and the late-glacial decline in the abundance of spruce trees.

Results. Plots of rates of change at the 18 sites show considerable variability across eastern North America (Fig. 3). Most sites had two or three times of especially rapid change during the period 18 to 6 ka, although some plots have much more overall change than others. Medicine Lake, for example, had rapid change around 10 ka but little for the next 4,000 years, whereas Lake West Okoboji had three or four times of dramatic change in the same time interval, when the forests and subsequent prairies of northwestern Iowa changed markedly in composition.

The period around 10 ka appears to have been a time of rapid vegetational change near most of the sites (Fig. 3). Several sites also recorded rapid change at one or more times between 14 and 12 ka, but changes of similar magnitude occurred at some other times at most of the sites.

In order to separate local or site-specific events from others that might have resulted from continent-scale shifts in atmospheric circulation or other such phenomena, we examined the rates of change for several sites collectively, first by region and then for all 18 sites. To accomplish the separation we calculated the mean rate of change for each 200 years at the group of sites in question.

At sites near the prairie-forest ecotone (Fig. 4A), change was consistently rapid around 10 ka, from 13.5 to 14.0 ka, and possibly at 12.3 ka. Otherwise, much of the variability evident at the individual sites averages out in the mean plots. Thus, individual sites often show major vegetational changes during the Holocene, but those changes are local or time transgressive.

Similarly the plot (Fig. 4B) for six sites from the eastern Great Lakes region to New England (the area in which hemlock today is an important component of the forest) shows a major peak at 10 ka, as well as two other significant peaks, one at 12.3 ka, the other at 13.5 ka. These sites also show a mid-Holocene peak at the time of the regional demise of hemlock (Webb, 1982). Davis (1981b) and Allison and others (1986) attribute the hemlock decline to a pathogen that specifically affected these conifers, in a fashion similar to that represented in the chestnut blight and Dutch elm disease of the present century.

The two regions thus show similar times of rapid, synchronous change in the late-glacial and the earliest Holocene. A plot of the mean rate of change for all 18 sites (Fig. 4C) shows three times of regionally synchronous change: at 10 ka, which is the

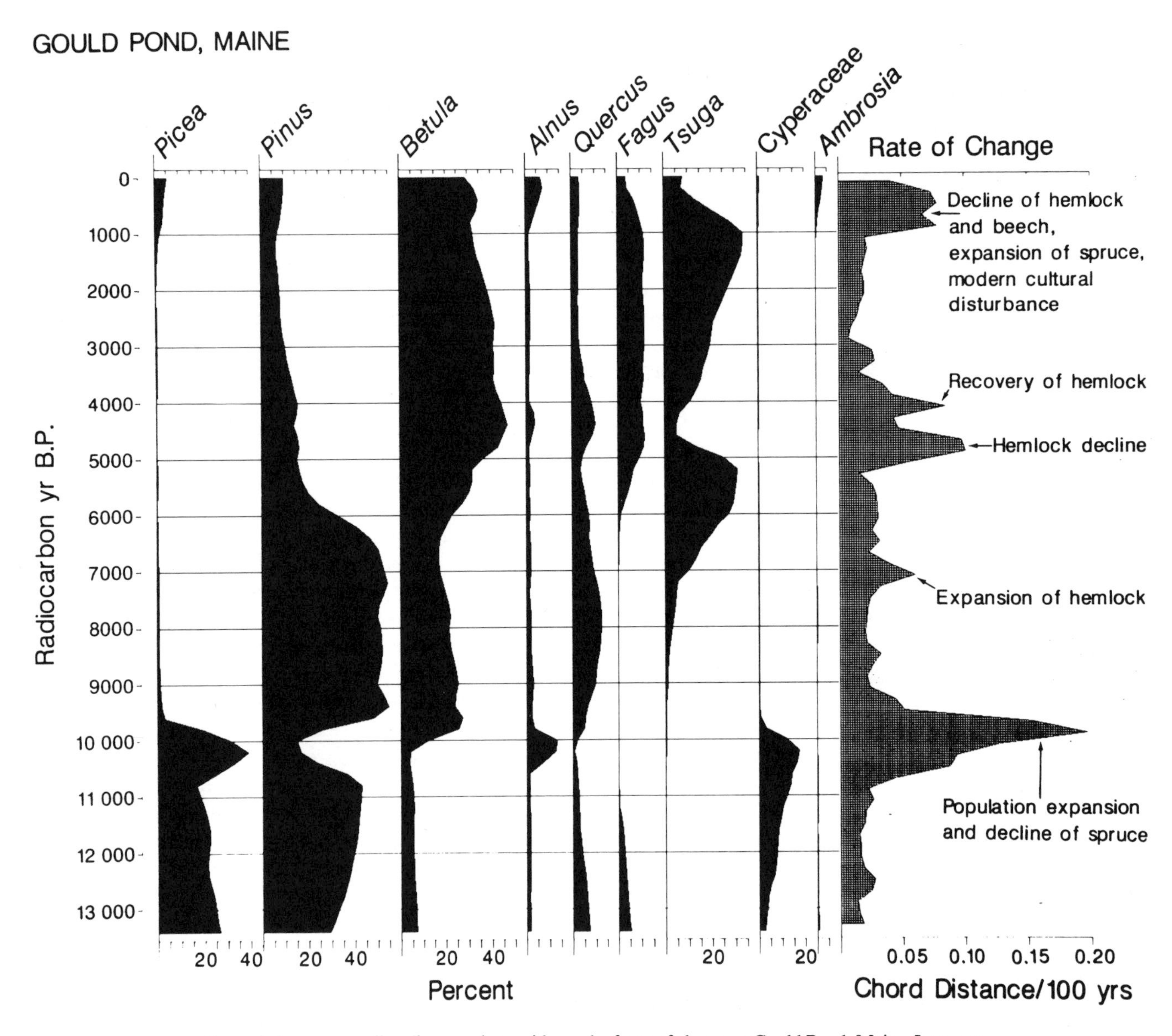

Figure 2. Summary pollen diagram along with graph of rate of change at Gould Pond, Maine. Important vegetational changes are labelled.

most pronounced, and at 12.3 ka, and 13.5 ka. In fact, in the records of the southern sites (e.g., Anderson Pond and Lake Tulane) these earlier peaks represent the times of most rapid change. These earlier times, however, also saw rapid change at northern sites such as Rogers Lake, Tannersville, and Wolf Creek, among others (Fig. 3).

The plots of mean rates of change for the two regions and for eastern North America thus show consistent patterns in the timing of rapid changes in vegetation during deglaciation. The nature of the changes differed from site to site and region to region. The vegetational transitions of the late glacial were variable across the continent; as oak expanded in the Southeast, pine spread across the north, and ash, elm, and other hardwoods became abundant south of the Great Lakes (Plates 1 and 2).

CONCLUSIONS

The maps of pollen data illustrate the varied botanical and ecological history of this large area. Species assemblages and abundances changed significantly within every 2,000-yr period. The change was continuous but varied in rate. Each site has several periods of especially fast reorganization of the vegetation, separated by periods of relative constancy.

Of particular interest during the most recent deglaciation of

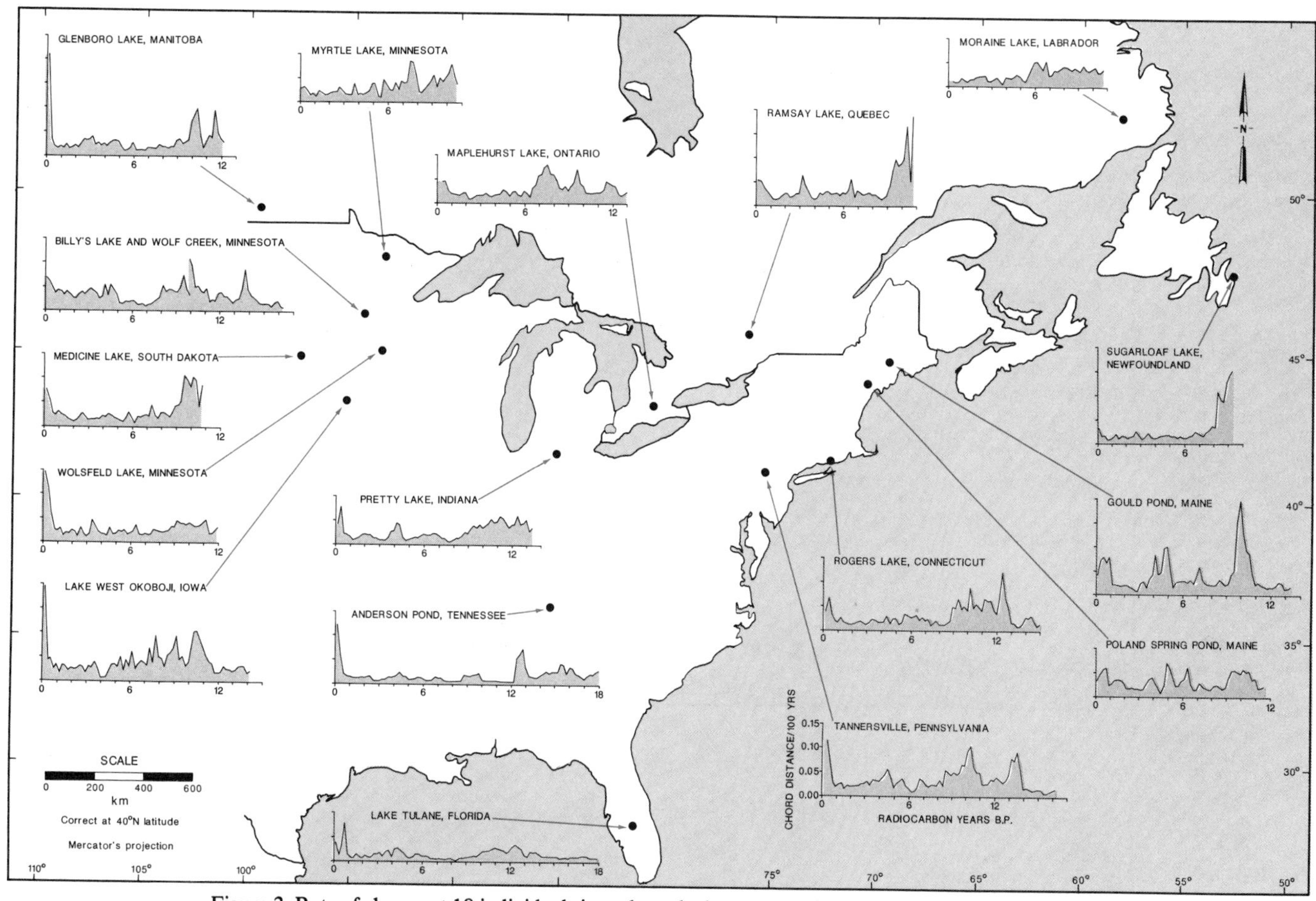

Figure 3. Rate of change at 18 individual sites, along the base map showing their locations. Citations for sites are: Glenboro Lake (Ritchie, 1976); Myrtle Lake (Janssen, 1968); Maplehurst Lake (Mott and Farley-Gill, 1978); Ramsay Lake (Mott and Farley-Gill, 1981); Moraine Lake (Engstrom and Hansen, 1985); Sugarloaf Pond (Macpherson, 1982); Gould Pond and Poland Spring Pond (Jacobson and others, unpublished); Rogers Lake (Davis, 1969); Tannersville (Watts, 1979); Lake Tulane (W. A. Watts, personal communication, 1986); Anderson Pond (Delcourt, 1979); Pretty Lake (Williams, 1974); West Okoboji Lake (Van Zant, 1979); Wolsfeld Lake (Grimm, 1983); Medicine Lake (Radle, 1981); Billy's Lake (Jacobson and Grimm, 1986); and Wolf Creek (Birks, 1976).

North America are the several times of rapid synchronous change centered at 13.5, 12.3, and 10.0 ka. These are the only occasions when sites all across the study region show such synchroneity and are thus the times most likely to have been subject to broad changes in atmospheric circulation. As the first strong evidence of synchroneity in terrestrial events during the late glacial, these results constitute a useful independent estimate of the responses of terrestrial vegetation to changing climate during deglaciation.

The linkage is indirect between the late-glacial vegetational patterns of eastern North America and either the height and area of the ice sheet or the North Atlantic sea-surface conditions; the connection requires the use of climatic models (Manabe and Broccoli, 1985; Kutzbach and Wright, 1985; Kutzbach and Guetter, 1986; Kutzbach, this volume) and biological response functions (Webb and others, this volume). Duplessy and others (1981), Mix and Ruddiman (1985), and Mix (this volume) present evidence for steplike changes in the marine stratigraphic record of oxygen isotopes. Such changes in the isotope stratigraphies may imply that deglaciation occurred in several stages, rather than at a constant rate.

Recent estimates suggest that the most abrupt changes in ice volume occurred between 14 and 12 ka, and again at 10 ka (Mix and Ruddiman, 1985; Mix, this volume). Our results for rates of vegetation change indicate that on land there may have been two distinct but short-lived times of rapid change between 14 and 12 ka, and that the most dramatic changes occurred at 10 ka. This is especially so in the north-temperate latitudes, where pine populations shifted their orientation from north-south along the east coast to east-west south of the Great Lakes. Mix and Ruddiman (1985) also report a possible final step in deglaciation at about

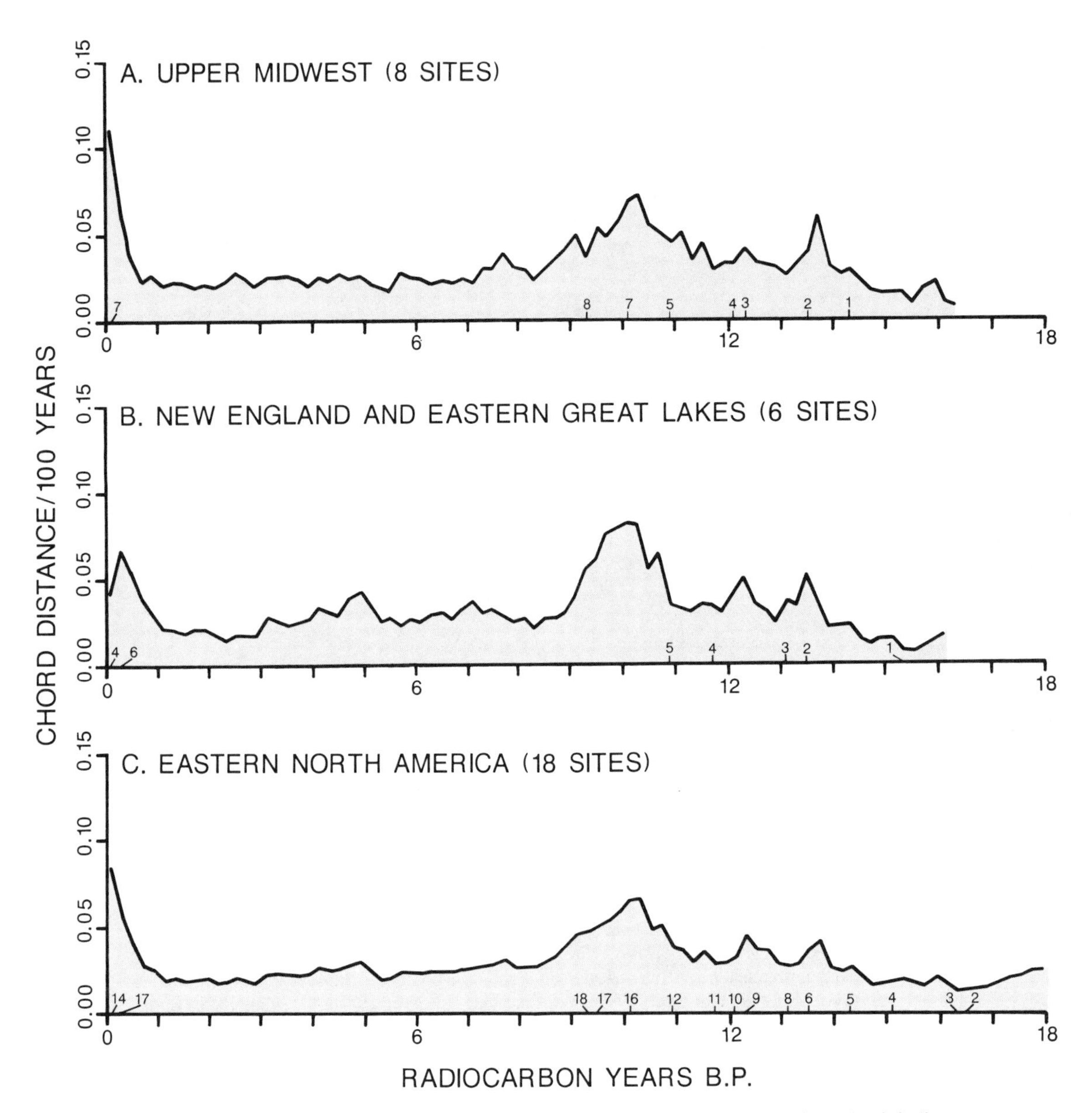

Figure 4. Average rates of change for different regions. Small number above the horizontal axis indicates number of sites in average at that point and to the right. A, Upper Midwest; B, New England and eastern Great Lakes; C, Eastern North America.

8–6 ka (their termination 1c). Our terrestrial evidence does not show regionally synchronous change at that time (see Fig. 4). The pollen record for Moraine Lake, Labrador, the one site among our 18 that was closest to the last existing Laurentide ice, registered an abrupt change at 6 ka (Fig. 2), so we cannot rule out some localized climatic adjustments in far eastern Canada. The pollen data from Sugarloaf Lake, Newfoundland, however, show no detectable changes at 6 ka.

The various results presented in this chapter provide different perspectives on the character and rates of vegetation change during the deglaciation in eastern North America. The isopoll maps show distinctly individualistic behavior of taxa in time. These separate changes in individual taxa led to continuous modifications in the composition and location of the different regions of vegetation. Rates of change measured in pollen records at individual sites are variable, demonstrating that the vegetational transitions were more abrupt at some times than at others. Taken together, the mean rates of change for eastern North America reveal several times when changes were synchronous and rapid everywhere. These were likely caused by significant shifts in patterns of atmospheric circulation on a continental scale (Bryson and Wendland, 1967; Kutzbach, this volume).

REFERENCES CITED

Allison, T. D., Moeller, R. E., and Davis, M. B., 1986, Pollen in laminated sediments provides evidence for a mid-Holocene forest pathogen outbreak: Ecology, v. 67, p. 1101–1105.

Anderson, R. S., Homola, R. L., Davis, R. B., and Jacobson, G. L., Jr., 1984, Fossil remains of the mycorrhizal fungus *Glomus fasciculatum* complex in postglacial lake sediments from Maine: Canadian Journal of Botany, v. 62, p. 2325–2328.

Bartlein, P. J., and Webb, T., III, 1985, Mean July temperature for eastern North America at 6,000 yr. B.P.; Regression equations for estimates based on fossil-pollen data: Syllogeus, v. 55, p. 301–342.

Bennett, K. D., 1985, The spread of *Fagus grandiofolia* across eastern North America during the last 18,000 years: Journal of Biogeography, v. 12, p. 147–164.

Bernabo, J. C., and Webb, T., III, 1977, Changing patterns in the Holocene pollen record from northeastern North America; A mapped summary: Quaternary Research, v. 8, p. 64–96.

Birks, H.J.B., 1976, Late-Wisconsinan vegetational history at Wolf Creek, central Minnesota: Ecological Monographs, v. 46, p. 395–429.

Braun, E. L., 1947, Development of the deciduous forest of eastern North America: Ecological Monographs, v. 17, p. 211–219.

Bradshaw, R.H.W., and Webb, T., III., 1985, Relationships between contemporary pollen and vegetation data from Wisconsin and Michigan, USA: Ecology, v. 66, p. 721–737.

Broecker, W. S., and van Donk, J., 1970, Insolation changes, ice volumes, and the ^{18}O record in deep-sea cores: Reviews of Geophysics and Space Physics, v. 8, p. 169–198.

Bryson, R. A., and Wendland, W. M., 1967, Tentative climatic patterns for some late-Glacial and postglacial episodes in central North America, *in* Mayer-Oakes, W. J., ed., Life, land, and water: Winnipeg, University of Manitoba Press, p. 271–289.

Clayton, L., and Moran, S. R., 1982, Chronology of late-Wisconsinan glaciation in middle North America: Quaternary Science Reviews, v. 1, p. 55–82.

Cushing, E. J., 1967, Late-Wisconsin pollen stratigraphy and the glacial sequence in Minnesota, *in* Cushing, E. J., and Wright, H. E., Jr., eds., Quaternary Paleoecology: New Haven, Yale University Press, p. 59–88.

Davis, M. B., 1969, Climatic changes in southern Connecticut recorded by pollen deposition at Rogers Lake: Ecology, v. 50, p. 409–422.

—— , 1981a, Quaternary history and the stability of forest communities, *in* West, D. C., Shugart, H. H., and Botkin, D. B., eds., Forest succession, concepts, and application: New York, Springer-Verlag, p. 132–153.

—— , 1981b, Outbreaks of forest pathogens in Quaternary history, *in* Proceedings, Fourth International Palynological Conference (1976–77), Volume 3: Lucknow, India, p. 216–227.

—— , 1983, Quaternary history of deciduous forests of eastern North America and Europe: Annals of the Missouri Botanical Garden, v. 70, p. 550–563.

Davis, R. B., and Jacobson, G. L., Jr., 1985, Late-glacial and early Holocene landscapes in northern New England and adjacent areas of Canada: Quaternary Research, v. 23, p. 341–368.

Davis, R. B., Bradstreet, R. E., Stuckenrath, R., Jr., and Borns, H. W., Jr., 1975, Vegetation and associated environments during the past 14,000 years near Moulton Pond, Maine: Quaternary Research, v. 5, p. 436–465.

Deevey, E. S., Jr., 1939, Studies on Connecticut lake sediments, Part I; A postglacial climatic chronology for southern New England: American Journal of Science, v. 237, p. 691–723.

—— , 1949, Biogeography of the Pleistocene; Part I; Europe and North America: Geological Society of America Bulletin, v. 60, p. 1315–1416.

—— , 1951, Late-glacial and postglacial pollen diagrams from Maine: American Journal of Science, v. 249, p. 177–207.

Delcourt, H. R., 1979, Late Quaternary vegetation history of the eastern highland rim and adjacent Cumberland Plateau of Tennessee: Ecological Monographs, v. 49, p. 255–280.

Delcourt, H. R., and Delcourt, P. A., 1985, Quaternary palynology and vegetational history of the southeastern United States, *in* Bryant, V. M., and Holloway, R. G., eds., Pollen records of late-Quaternary North American sediments: Dallas, American Association of Stratigraphic Palynologists, p. 1–37.

Delcourt, P. A., Delcourt, H. R., Brister, R. C., and Lackey, L. E., 1980, Quaternary vegtation history of the Mississippi Embayment: Quaternary Research, v. 13, p. 111–132.

Delcourt, P. A., Delcourt, H. R., and Webb, T., III., 1984, Atlas of mapped distributions of dominance and modern pollen percentages for important tree taxa of eastern North America: American Association of Stratigraphic Palynologists Contributions Series no. 14, 131 p.

Denton, G. H., and Hughes, T. J., 1981, The last great ice sheets: New York, Wiley Interscience, 484 p.

Dexter, F., Banks, H. T., and Webb, T., III, 1987, Modeling Holocene changes in the location and abundance of beech populations in eastern North America: Review of Palaeobotany and Palynology, v. 50, p. 273–292.

Duplessy, J. C., Delibrias, G., Turon, J. L., Pujol, C., and Duprat, J., 1981, Deglacial warming of the northeastern Atlantic Ocean; Correlation with the paleoclimate evolution of the European continent: Palaeogeography, Palaeoclimatology, Palaeoecology, v. 35, p. 121–144.

Engstrom, D. R., and Hansen, B.C.S., 1985, Postglacial vegetational change and soil development in southeastern Labrador as inferred from pollen and chemical stratigraphy: Canadian Journal of Botany, v. 63, p. 543–561.

Fries, M., 1962, Pollen profiles of late Pleistocene and Recent sediments at Weber Lake, northeastern Minnesota: Ecology, v. 43, p. 295–308.

Gaudreau, D. C., and Webb, T., III., 1985, Late-Quaternary pollen stratigraphy and isochrone maps for the northeastern United States, *in* Bryant, V. M., Jr., and Holloway, R. G., eds., Pollen records of late Quaternary North American sediments: Dallas, American Association of Stratigraphic Palynologists Foundation, p. 247–280.

Grimm, E. C., 1983, Chronology and dynamics of vegetation change in the prairie-woodland region of southern Minnesota: New Phytologist, v. 93, p. 311–350.

Grimm, E. C., and Jacobson, G. L., Jr., 1985, Rates of vegetation change; A comparison between Minnesota and Maine: Ecological Society of America Bulletin, v. 66, p. 183.

Jacobson, G. L., Jr., and Grimm, E. C., 1986, A numerical analysis of Holocene forest and prairie vegetation in central Minnesota: Ecology, v. 67, p. 958–966.

Janssen, C. R., 1968, Myrtle Lake; A late- and postglacial pollen diagram from northern Minnesota: Canadian Journal of Botany, v. 46, p. 1397–1408.

King, G. A., 1985, A standard method for evaluating radiocarbon dates of local deglaciation; Application to the deglaciation history of southern Labrador and adjacent Quebec: Geographie Physique et Quaternaire, v. 39, p. 163–182.

Kutzbach, J. E., and Guetter, P. J., 1986, The influence of changing orbital parameters and surface boundary conditions on climate simulations for the past 18,000 years: Journal of the Atmospheric Sciences, v. 43, p. 1726–1759.

Kutzbach, J. E., and Wright, H. E., Jr., 1985, Simulation of the climate of 18,000 yr B.P.; Results for the North American/North Atlantic/European sector and comparison with the geologic record: Quaternary Science Reviews, v. 4, p. 147–187.

Lamb, H. F., 1980, Late Quaternary vegetational history of southeastern Labrador: Arctic and Alpine Research, v. 12, p. 117–135.

Leopold, E. B., 1956, Two late-glacial deposits in southern Connecticut: Proceedings of the National Academy of Sciences, v. 52, p. 863–867.

Macpherson, J. B., 1982, Postglacial vegetational history of the eastern Avalon Peninsula, Newfoundland, and Holocene climatic change along the eastern Canadian seaboard: Geographie Physique et Quaternaire, v. 36, p. 175–196.

Manabe, S., and Broccoli, A. J., 1985, The influence of continental ice sheets on the climate of an ice age: Journal of Geophysical Research, v. 90,

p. 2167–2190.
McAndrews, J. H., 1966, Postglacial history of prairie, savanna, and forest in northwestern Minnesota: Memoirs of the Torrey Botanical Club, v. 22, p. 1–72.
Mix, A. C., and Ruddiman, W. F., 1985, Structure and timing of the last deglaciation; Oxygen-isotope evidence: Quaternary Science Reviews, v. 4, p. 59–108.
Mott, R. J., and Farley-Gill, L. D., 1978, A late-Quaternary pollen profile from Woodstock, Ontario: Canadian Journal of Earth Sciences, v. 15, p. 1101–1111.
—— , 1981, Two late Quaternary pollen profiles from Gatineau Park, Quebec: Geological Survey of Canada Paper 80-31, p. 1–10.
Overpeck, J. T., Webb, T., III, and Prentice, I. C., 1985, Quantitative interpretation of fossil pollen spectra; Dissimilarity coefficients and the method of modern analogs: Quaternary Research, v. 23, p. 87–108.
Prentice, I. C., 1980, Multidimensional scaling as a research tool in Quaternary palynology; A review of theory and methods: Review of Palaeobotany and Palynology, v. 31, p. 71–104.
—— , 1985, Pollen representation, source area, and basin size; Toward a unified theory of pollen analysis: Quaternary Research, v. 23, p. 76–86.
Radle, N. J., 1981, Vegetation history and lake-level changes at a saline lake in northeastern South Dakota [M.S. thesis]: Minneapolis, University of Minnesota, 126 p.
Richard, P.J.H., 1987, Patterns of post-Wisconsin plant colonization in Quebec-Labrador, in Matthews, J. V., Jr., and others, Quaternary Environments of Canada as Documented by Paleobotanical Case Histories, *in* Fulton, R. J., Heginbottom, J. A., and Funder, S., eds., Quaternary Geology of Canada and Greenland: Ottawa, Ontario, Geological Survey of Canada, The Geology of North America, in press.
Ritchie, J. C., 1976, The late-Quaternary vegetational history of the western interior of Canada: Canadian Journal of Botany, v. 54, p. 1793–1818.
Sears, P. B., 1942, Postglacial migration of five forest genera: American Journal of Botany, v. 29, p. 684–691.
Van Zant, K. L., 1979, Late glacial and postglacial pollen and plant macrofossils from Lake West Okoboji, northwestern Iowa: Quaternary Research, v. 12, p. 358–380.
Watts, W. A., 1969, A pollen diagram from Mud Lake, Marion County, north-central Florida: Geological Society of America Bulletin, v. 80, p. 631–642.
—— , 1970, The full-glacial vegetation of northwestern Georgia: Ecology, v. 51, p. 19–33.
—— , 1979, Late Quaternary vegetation of central Appalachia and the New Jersey Coastal Plain: Ecological Monographs, v. 49, p. 427–469
—— , 1980, The late Quaternary vegetation history of southeastern United States: Annual Reviews of Ecology and Systematics, v. 11, p. 387–409.
—— , 1983, Vegetation history of eastern United States 25,000 to 10,000 years ago, *in* Porter, S., ed., The late Quaternary of the United States, v. 1, The late Pleistocene: Minneapolis, University of Minnesota Press, p. 294–310.
Watts, W. A., and Stuiver, M., 1980, Late Wisconsin climate of northern Florida and the origin of species-rich deciduous forest: Science, v. 210, p. 325–327.
Webb, T., III., 1982, Temporal resolution in Holocene pollen data, *in* Proceedings, Third North American Paleontological Convention, v. 2, Montreal, p. 569–572.
—— , 1987, The appearance and disappearance of major vegetational assemblages; Long-term vegetational dynamics in eastern North America: Vegetatio, v. 69, p. 177–187.
Webb, T., III, Richard, P.J.H., and Mott, R. J., 1983a, A mapped history of Holocene vegetation in southern Quebec: Syllogeus, v. 49, p. 273–336.
Webb, T., III, Cushing, E. J., and Wright, H. E., Jr., 1983b, Holocene changes in the vegetation of the Midwest, *in* Wright, H. E., Jr., ed., Late-Quaternary environments of the United States, v. 2, The Holocene: Minneapolis, University of Minnesota Press, p. 142–165.
Whitehead, D. R., 1973, Late-Wisconsin vegetational changes in unglaciated eastern North America: Quaternary Research, v. 3, p. 621–631.
—— , 1981, Late-Pleistocene vegetational changes in northeastern North Carolina: Ecological Monographs, v. 51, p. 451–471.
Williams, A. S., 1974, Late-glacial–postglacial vegetational history of the Pretty Lake region, northeastern Indiana; Hydrologic and biological studies of Pretty Lake, Indiana: Washington, D.C., U.S. Geological Survey Professional Paper 686, p. D1–D23.
Wright, H. E., Jr., 1968, History of the prairie peninsula, *in* Bergstrom, R. E., ed., The Quaternary of Illinois: Urbana, College of Agriculture, University of Illinois, Special Publication 14, p. 78–88.
—— , 1970, A vegetational history of the central Plains, *in* Dort, W., Jr., and Jones, J. K., Jr., eds., Pleistocene and Recent environments of the central Great Plains: Lawrence, University of Kansas Press, p. 157–172.
Wright, H. E., Jr., Winter, T. C., and Patten, H. L., 1963, Two pollen diagrams from southeastern Minnesota; Problems in the late- and postglacial vegetational history: Geological Society of America Bulletin, v. 74, p. 1371–1396.

MANUSCRIPT ACCEPTED BY THE SOCIETY FEBRUARY 2, 1987

ACKNOWLEDGMENTS

W. A. Watts kindly provided unpublished pollen data for Lake Tulane, Florida. Otherwise, the pollen maps were produced for the data base available as of November 1983, and unpublished data were generously provided by R. E. Bailey, R. G. Baker, E. J. Cushing, R. P. Futyma, D. C. Gaudreau, R. O. Kapp, G. A. King, H. F. Lamb, J. H. McAndrews, R. J. Nickmann, J. G. Ogden III, W. A. Patterson III, P.J.H. Richard, L.C.K. Shane, S. K. Short, K. M. Trent, K. L. Van Zant, and D. R. Whitehead. Data for several sites in Maine (cited in the text as Jacobson and others, unpublished) were gathered by R. S. Anderson and M. Tolonen, who worked on a project in association with G. L. Jacobson and R. B. Davis. Grants from NSF Climate Dynamics Program to COHMAP (Cooperative Holocene Mapping Project) and from DOE Carbon Dioxide Research Program (DE-FG02G-85ER60304) supported the contribution of T. Webb. We thank J. Allard and H. A. Jacobson for assistance in preparing figures and plates. Helpful suggestions for improving the manuscript were made by L. J. Maher, Jr., D. R. Whitehead, W. F. Ruddiman, and H. E. Wright, Jr.

Printed in U.S.A.

The Geology of North America
Vol. K-3, North America and adjacent oceans during the last deglaciation
The Geological Society of America, 1987

Chapter 14

The northwestern U.S. during deglaciation; Vegetational history and paleoclimatic implications

Cathy W. Barnosky
Carnegie Museum of Natural History, 4400 Forbes Avenue, Pittsburgh, Pennsylvania 15213
Patricia M. Anderson
College of Forest Resources, University of Washington, Seattle, Washington 98195
Patrick J. Bartlein
Department of Geography, University of Oregon, Eugene, Oregon 97403

INTRODUCTION

To understand the way late Quaternary environmental change has shaped vegetation requires a network of detailed paleoecologic records. In the northwestern conterminous U.S. (hereafter referred to as the northwestern U.S.) and Alaska, this is especially true, inasmuch as the habitats and biota of a mountainous area are complex. Superimposed on the geographic and ecologic diversity of these regions is a climate greatly modified by topography and highly variable through geologic time. Modernization of the vegetation in the late Cenozoic was a response to the gradual cooling of climate and to the shorter climatic oscillations that culminated with the glacial and interglacial cycles of the Quaternary. The effect of a single cycle on plant communities in the course of long-term climatic deterioration—indeed even the impact of the last glacial/interglacial cycle—is not well known.

The response of vegetation to late Pleistocene glaciation and the climatic and environmental changes that ushered in the postglacial period are the subject of this chapter. In the last few years several review papers have summarized the late Quaternary vegetational history of northwestern North America, and it is not our goal to duplicate that material here (see Baker, 1983; Heusser, 1983a, 1985; Ager, 1983; Ager and Brubaker, 1985; Mehringer, 1985). Instead, this chapter focuses on areas where the network of pollen sites is sufficiently dense to analyze spatial and temporal patterns in vegetation and environment during the period from ca. 20 to 7 ka. For the conterminous northwestern U.S. the discussion is confined to vegetation records from four areas: (1) Washington and southwestern British Columbia in the Pacific Northwest; (2) the Yellowstone Plateau, the Snake River Plain, and adjacent mountain ranges in the northern Rocky Mountains; (3) the San Juan Mountains and the Front Range of the Colorado Rockies; and (4) the northern Great Plains and intermontane basins of Wyoming and Montana (Fig. 1). In Alaska, sites from the northern, central, and southern parts are considered, as are sites from far northwestern Canada.

The chapter is organized as follows: the late Pleistocene and early Holocene vegetation and climate of the northwestern U.S. are presented by region, followed by a discussion of the fossil-plant records from Alaska. The final section provides a summary of both areas and compares the fossil records with climate reconstructions developed from paleoclimate model experiments of atmospheric circulation.

THE NORTHWESTERN U.S.

The western Cordillera owes much of its rugged scenery to the effects of ice-sheet and alpine glaciation in the Pleistocene. The Cordilleran ice sheet, which had its source in British Columbia, extended southward at its maximum to fill valleys and bury low hills from northern Washington eastward to the Continental Divide of northwestern Montana (see Booth, this volume; Fig. 2). At the same time, alpine glaciers were active on the highest parts of the Coast Range, Cascade Range, and Rocky Mountains. Most mountain glaciers were confined to preexisting valleys, leaving lower slopes and intermontane basins free of ice. Large ice fields, however, did form in the Cascade Range, Colorado Front Range, Wind River Range, Yellowstone Plateau, and the San Juan Mountains. East of the Continental Divide in northern Montana, Cordilleran ice combined with local glaciers to form broad piedmont lobes that extended east onto the Great Plains. The distribution of late Pleistocene glaciers reflected a combination of topography and distance from prevailing westerly sources of moisture, with the most extensive ice cover along the Pacific coast and on the highest ranges (Porter and others, 1983).

The modern vegetation patterns in the western U.S. are

Barnosky, C. W., Anderson, P. M., Bartlein, P. J., 1987, The northwestern U.S. during deglaciation; Vegetational history and paleoclimatic implications, *in* Ruddiman, W. F., and Wright, H. E., Jr., eds., North America and adjacent oceans during the last deglaciation: Boulder, Colorado, Geological Society of America, The Geology of North America, v. K-3.

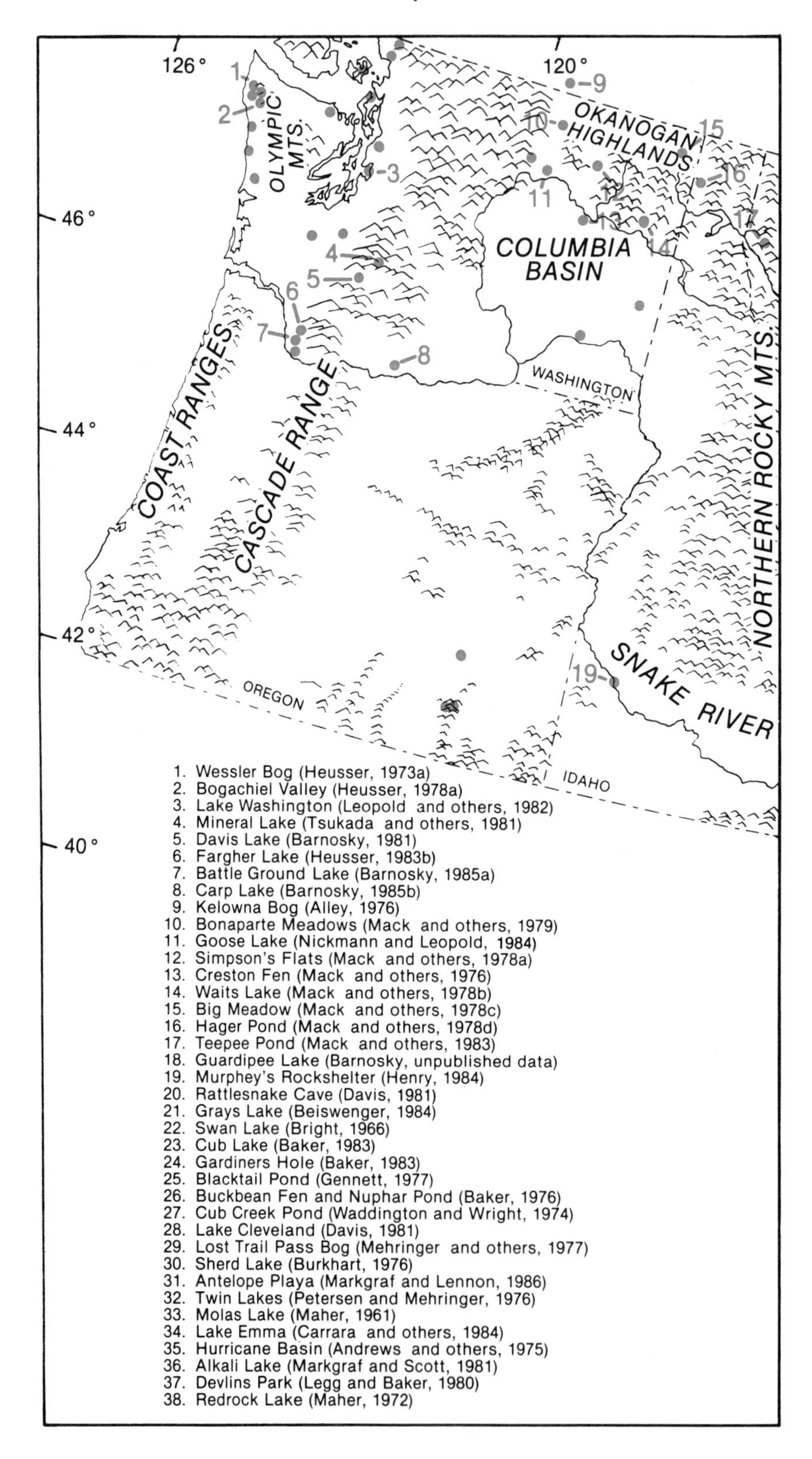

Figure 1 (this and facing page). Location of radiocarbon-dated pollen sites in the northwestern U.S. that are considered in this discussion. Numbered sites are specifically cited in text. For a more complete list of pollen sites see Baker (1983), Heusser (1985), and Mehringer (1985).

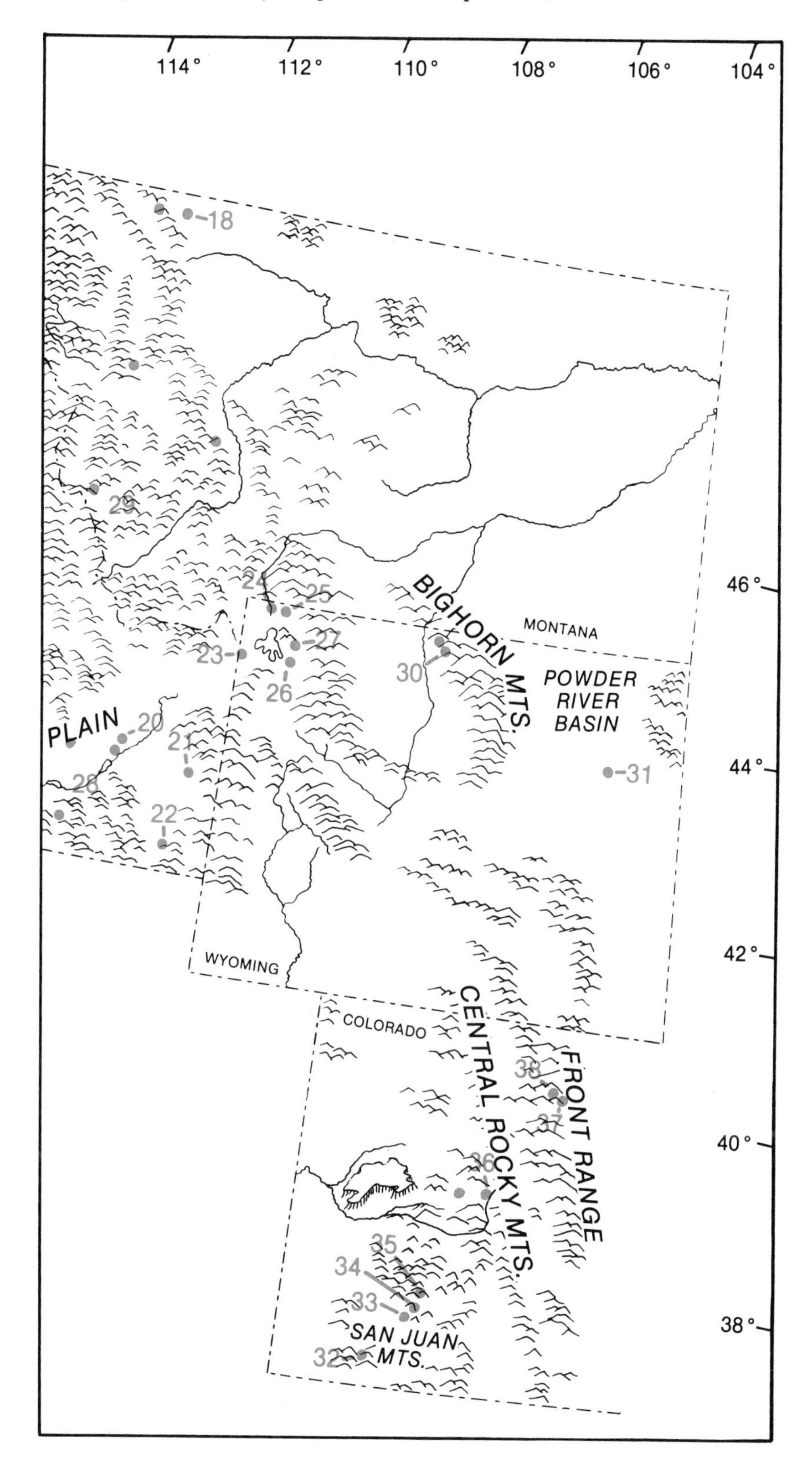
114°
112°
110°
108°
106°
104°
46°
44°
42°
40°
38°
18
29
24
25
23
27
26
30
BIGHORN MTS.
MONTANA
POWDER RIVER BASIN
PLAIN
20
21
28
22
31
WYOMING
COLORADO
CENTRAL ROCKY MTS.
FRONT RANGE
38
37
36
35
34
33
SAN JUAN MTS.
32

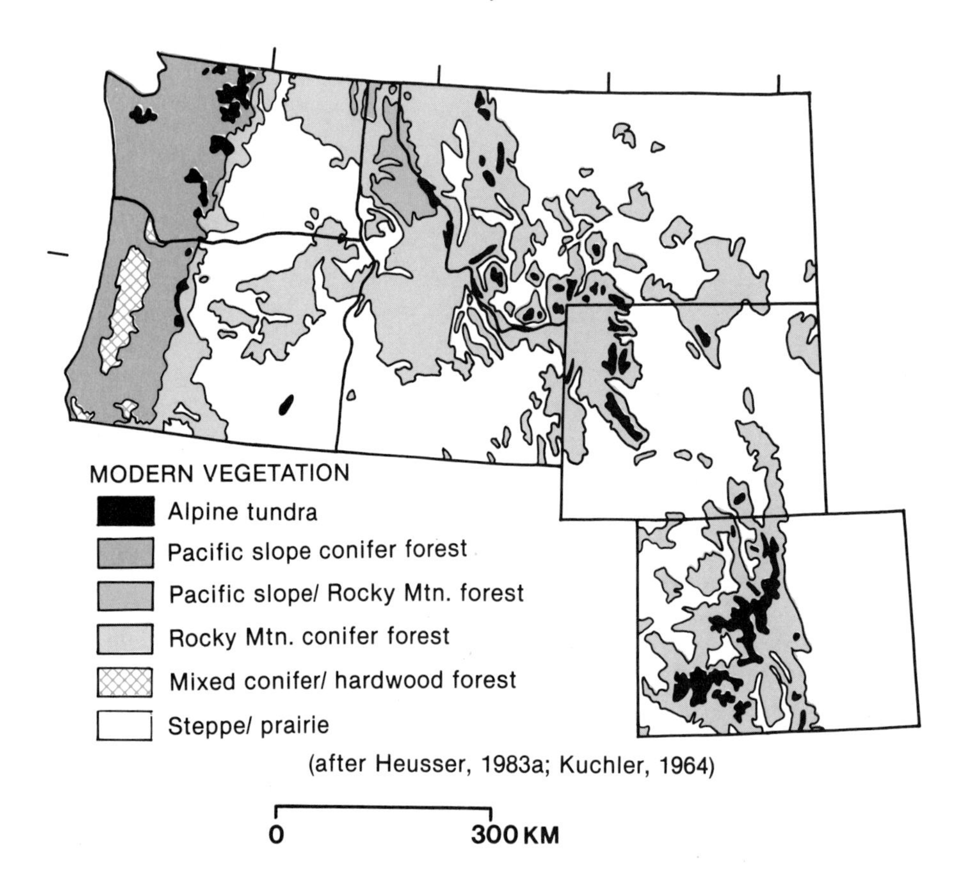

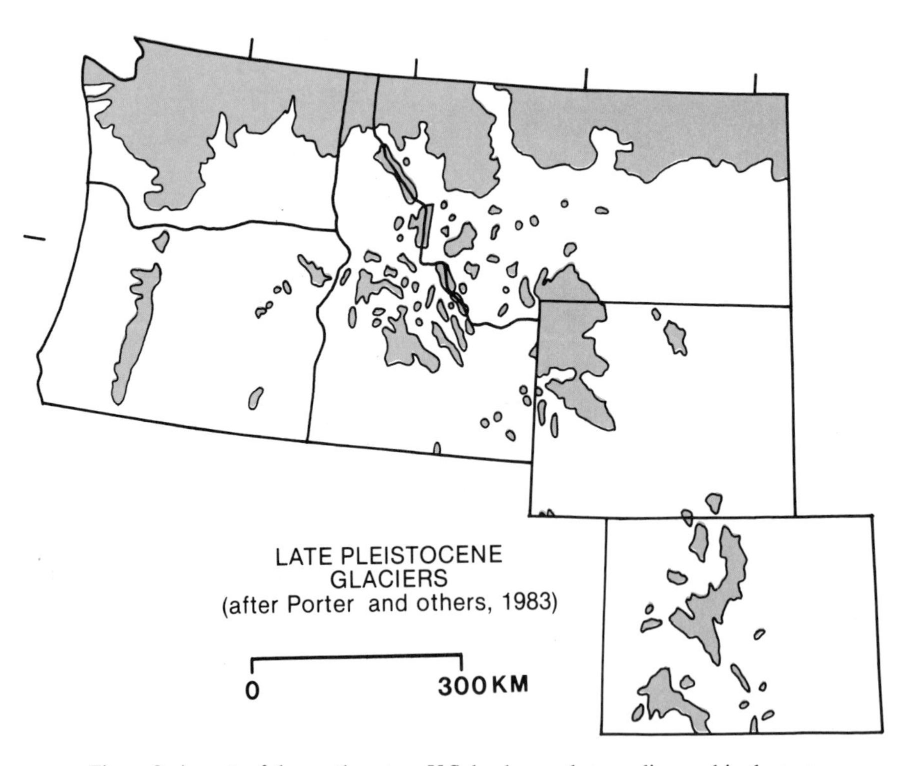

Figure 2. Aspects of the northwestern U.S. landscape that are discussed in the text.

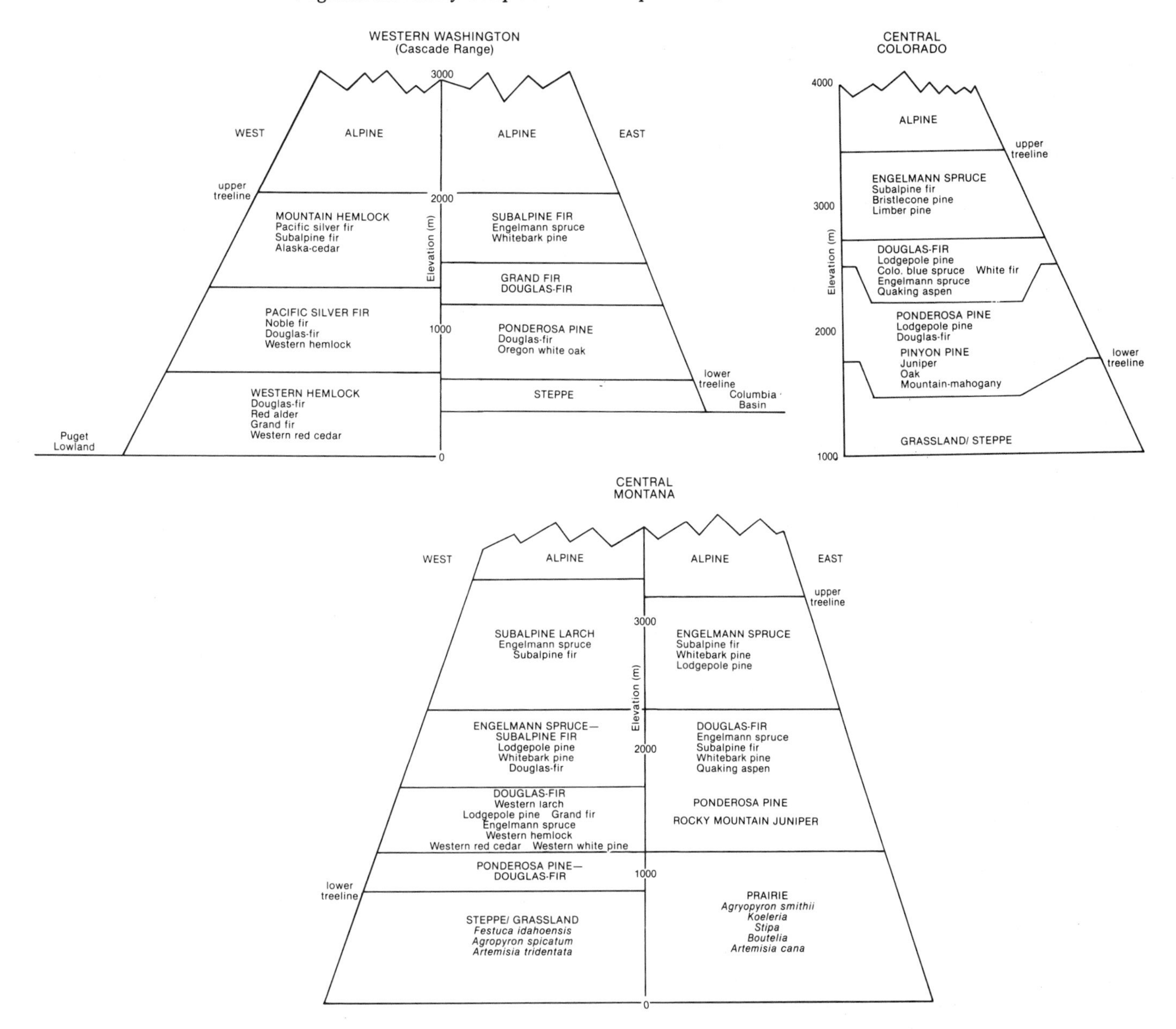

Figure 3. Generalized zonation of modern vegetation with elevation in three areas of the northwestern U.S. (data from numerous sources).

broadly zonal with elevation (Daubenmire, 1943; Figs. 2 and 3). In general, the basins support grassland or sagebrush steppe as a result of aridity, while at higher elevations the mountain ranges intercept westerly moisture and sustain conifer forests. Exceptions are the Pacific coast and Puget Trough, where sufficient precipitation supports forest at low elevations as well. The zonation of these forests reflects increasing precipitation with elevation. Upper altitudinal limits are determined by summer growing-season temperature, although drought resulting from high winds and low soil temperatures is also a factor. Only on the highest peaks and ridges does alpine tundra occur. Lower altitudinal limits are determined by summer drought or locally by fine-textured soils on lower slopes and in basins. Species near their altitudinal limit are restricted to the most favorable habitats. For example, south-facing slopes are preferred at the upper altitudinal limit, whereas mesic substrates are favored at the lower limit. Soil conditions at lower treeline or localized environmental variations with elevation may disrupt the regional zonation and in places eliminate a vegetation type altogether (Daubenmire, 1943).

It is not surprising that western plant distributions coincide with major climatic boundaries. Several species, for example, have a range limit between Lat. 40° and 44°N near the winter boundary of Pacific air masses. Similarly, maritime species grow far east of their coastal range into northwestern Montana, along the southern limit of Pacific air masses in summer. Plant distributions also reflect the west-to-east gradient in effective moisture created by the barriers to westerly air masses—successively, the Coast Ranges, Cascade Range, and Rocky Mountains. The orographic effect is most obvious on opposite sides of the Cascade Range, where a decline in precipitation by an order of magnitude produces a shift from conifer rainforests to sagebrush steppe within 150 km across the range.

The Pacific Northwest: Western Washington and Southwestern British Columbia

Nowhere in northwestern North America is the coverage of pollen records more extensive than in the Pacific Northwest, particularly in western Washington and southwestern British Columbia. Pollen sites from the Pacific Coast, Puget Trough, Fraser Lowland, and southwestern Columbia Basin provide a vegetational history back to full-glacial time.

The late Pleistocene Fraser glaciation (ca. 25–10 ka) includes several stades and interstades (see Booth, this volume and references therein). During alpine glaciation between ca. 22 and 19 ka, ice caps formed in the Cascade Range and Olympic Mountains, but glaciers were largely absent from the lowlands. The alpine glaciers, however, had retreated from their maximum position before ca. 18 ka when Cordilleran ice lobes advanced into the northern and central Puget Trough and the present-day Strait of Juan de Fuca. Both ice lobes retreated from a maximum position between 15 and 14 ka, but their subsequent recession northward was interrupted by a readvance of the ice sheet to the International Boundary ca. 11–10 ka (Sumas Stade).

The three lowland areas—the Pacific Coast, the Puget Trough, and the southwestern Columbia Basin—register a distinctive vegetational history during deglaciation (Fig. 4). Between 20 and 16.8 ka the Pacific coast of the Olympic Peninsula was covered by parkland communities. Tundra, if present, was apparently restricted to areas marginal to alpine glaciers, inasmuch as pollen data farther from ice margins record the presence of spruce, pine, mountain hemlock, and western hemlock (see Table 1 for scientific names). A modern analogue for the glacial vegetation is the present-day subalpine parkland of the Olympic Mountains (Heusser, 1977), which implies a cold, humid climate in the past. The first indication of warming occurs in the Bogachiel Valley (Heusser, 1978a), where present-day lowland taxa, including spruce (cf. Sitka spruce), replaced grass and other herbs at 16.8 ka. This amelioration apparently began while the Juan de Fuca lobe was still advancing westward between 17 and 14.4 ka (Waitt and Thorson, 1983). The fact that the ice body to the north had little effect on the coastal vegetation is not surprising in light of the moderating effect of the Pacific Ocean today.

With the retreat of the Juan de Fuca lobe, the deglaciated terrain was invaded first by lodgepole pine and at some sites by Sitka alder, but as early as 13 ka, red alder, mountain hemlock, western hemlock, and spruce were also present (Heusser, 1973a). At 10 ka the forest consisted of western hemlock, spruce, Douglas-fir, and alder. The change in coastal vegetation from a parkland of mixed montane and lowland species to a closed forest of lowland taxa between 16.8 and 10 ka is evidence of gradual warming. Heusser (1973a) postulates a climatic reversal toward cooler conditions at ca. 11 ka from an increase in *Sphagnum* and mountain hemlock at Wessler Bog, but this episode is poorly dated and has not been recognized elsewhere. After 8 ka modest increases in Douglas-fir, alder, and bracken fern percentages indicate a climate warmer and drier than that of the present.

The Puget Trough records show a somewhat different vegetational sequence: between 19 and 17 to 16 ka, the lowland supported tundra parkland, in which the major taxa were grass, sedge, *Artemisia,* spruce, pine, and herbs diagnostic of tundra. Plant macrofossils of Engelmann spruce (or possibly white spruce) and lodgepole pine are locally found (Barnosky, 1981, 1985a). The present range of Engelmann spruce points to the subalpine parkland of the northern Rocky Mountains as a possible modern analogue for the full-glacial environment. Thus treeline was not only lower at that time, but there was an incursion of xerophytic subalpine taxa (see also Janssens and Barnosky, 1985). Stronger easterly winds have also been inferred from the presence of Tertiary diatoms blown from eastern Washington, as well as from the mineralogy of glacial sediments (Barnosky, 1983).

The paleoecologic record provides some evidence of warmer summers between ca. 17 and 15 ka. At Davis and Mineral lakes an expansion of lodgepole pine has been attributed to higher temperatures (Barnosky, 1981; Tsukada and others, 1981) and to the availability of suitable substrates as a result of ice retreat in the Cascade Range (Barnosky, 1985a). At Battle Ground Lake, the occurrence of Douglas-fir and Sitka-spruce macrofossils at 17 to 16 ka also argues for summer warming, although the pollen record shows no change there or at nearby Fargher Lake (Heusser, 1983a). A fossil insect assemblage from the central Puget Trough, however, suggests temperate dry conditions at 16.4 ka, not unlike those of prairie regions in western Washington today (Nelson and Coope, 1982), and thus supports the interpretation from plant macrofossils.

From 15 to 12.5 ka, Sitka and Engelmann spruce, pine, mountain hemlock, and fir were present. The nearest modern analogue is the high-altitude open forests of the western Cascade Range (Barnosky, 1981) rather than the northern Rocky Mountains, suggesting an increase in effective moisture. Just as on the Pacific Coast, lodgepole pine invaded the deglaciated northern and central Puget Trough, although lignin evidence from Lake Washington suggests its distribution was uneven (Leopold and others, 1982; Fig. 5). Pollen of western hemlock provides the first paleobotanical evidence of warming as early as 12.5 ka, and by 12 ka, summer temperatures may have been equal to the present, judging from the occurrence of temperate aquatic taxa on the northeastern Olympic Peninsula (Petersen and others, 1983). Between 12.5 and 10.5 ka, both montane species (mountain hemlock, fir, and Sitka alder) and lowland species (western hemlock, Sitka spruce, Douglas-fir, grand fir, and red alder) grew in the

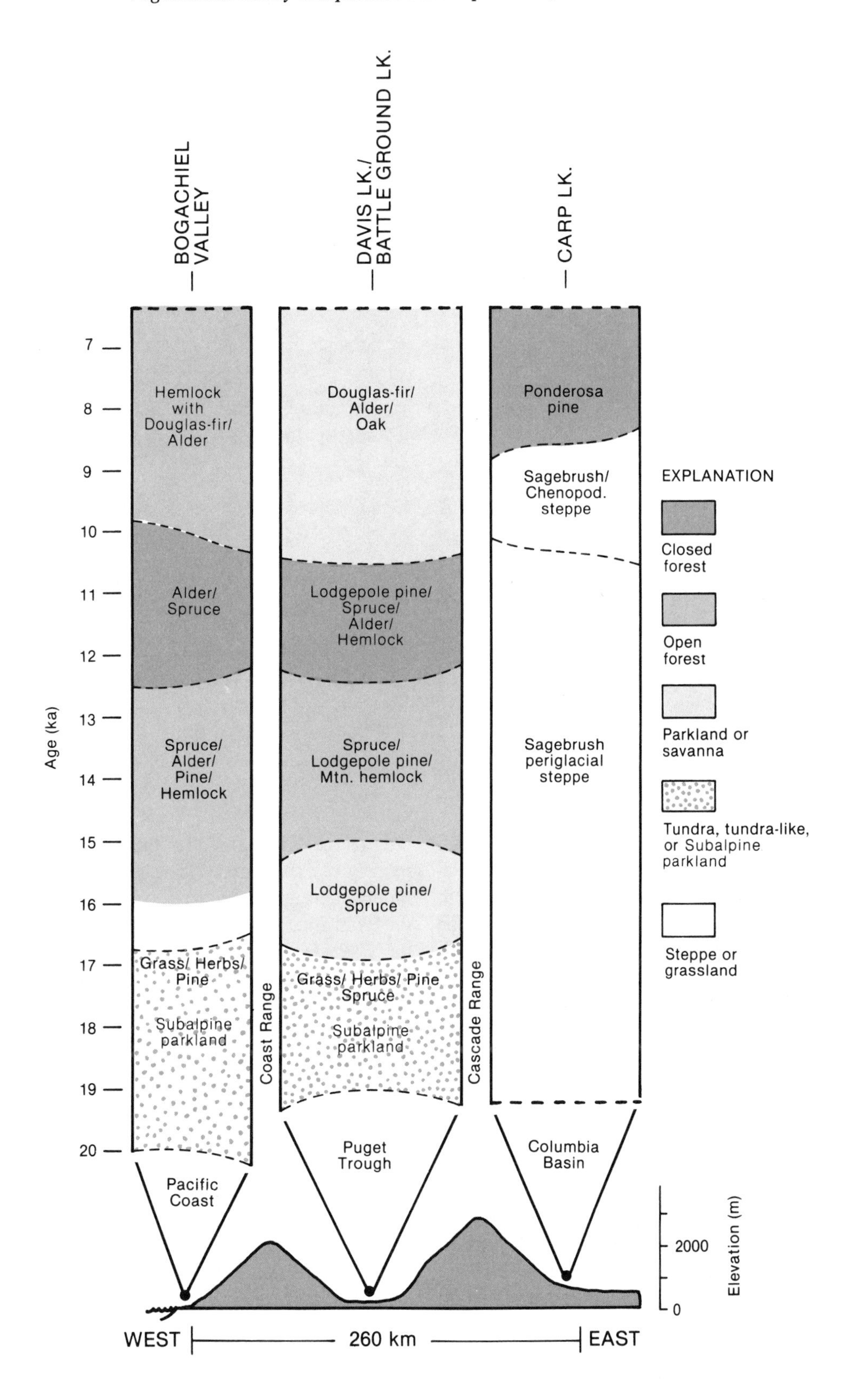

Figure 4. Vegetational changes through time inferred from a transect of pollen sites in the unglaciated part of western Washington.

TABLE 1. SCIENTIFIC NAMES FOR PLANTS MENTIONED IN TEXT*

Common Name	Scientific Name	Common Name	Scientific Name
Alder, red	*Alnus rubra*	Larch, subalpine	*Larix lyalli*
Sitka	*A. sitchensis*	western	*L. occidentalis*
Birch, paper	*Betula papyrifera*	Mountain-mahogany	*Cercocarpus montanus*
Bracken fern	*Pteridium aquilinum*	Oak, Oregon white	*Quercus garryana*
Cedar, western red	*Thuja plicata*	Pine, bristlecone	*Pinus aristata*
Alaska-	*Chamaecyparis nootkatensis*	limber	*P. flexilis*
Chenopods/Amaranths	Chenopodiineae	lodgepole	*P. contorta*
Crowberry	*Empetrum nigrum*	pinyon	*P. edulis*
Douglas-fir	*Pseudotsuga menziesii*	ponderosa	*P. ponderosa*
Fir, grand	*Abies grandis*	western white	*P. monticola*
Pacific Silver	*A. amabilis*	whitebark	*P. albicaulis*
subalpine	*A. lasiocarpa*	Poplar, balsam	*Populus balsamifera*
white	*A. concolor*	Sagebrush/wormwood	*Artemisia*
Grass	Gramineae	Sedge	Cyperaceae
Hemlock, mountain	*Tsuga mertensiana*	Spruce, black	*Picea mariana*
western	*T. heterophylla*	Engelman	*P. engelmannii*
Juniper, common	*Juniperus communis*	Sitka	*P. sitchensis*
Rocky Mtn.	*J. scopulorum*	white	*P. glauca*
western	*J. occidentalis*	Willow	*Salix*

*Nomenclature follows Hitchcock and Cronquist (1973), Hulten (1968), and Weber (1976).

unglaciated southern Puget Trough. In the northern Puget Trough and Fraser Lowland, western hemlock dominates the pollen record between 11 and 10 ka (Mathewes, 1985). It is tempting to consider the high percentages as evidence of cooling during the Sumas readvance, but an alternative explanation is that hemlock thrived from the combination of moist substrates created during deglaciation and increasing summer warmth. As temperatures continued to increase, summer drought became limiting and hemlock and other mesophytic taxa were replaced by Douglas-fir and alder. By 10.5 ka, Puget Trough and Fraser Lowland forests must have consisted almost exclusively of Douglas-fir, with bracken fern as an important fire-adapted ground cover and red alder in riparian settings. Prairie regions in the central Puget Trough also expanded as a result of drought. Oak savanna, now reaching its northern limit in the Willamette Valley of northwestern Oregon, extended its range northward into the southern Puget Trough.

Only a single site, Carp Lake in the southwestern Columbia Basin, provides a record of both glacial and postglacial vegetation east of the Cascade Range (Barnosky, 1985b). The temperate steppe of the Columbia Basin today was apparently replaced from 23.5 to ca. 10 ka by a more sparsely vegetated periglacial steppe. The high values of grass and *Artemisia* pollen, low rates of pollen accumulation (<1000 grains/cm^2/yr), the occurrence of alpine herbs, and the absence of temperate aquatic taxa (which are present before and after this period) indicate cold and extremely dry conditions. There is no evidence that the range of montane trees shifted downslope to the Columbia Basin. Instead, the forest zone was probably compressed at middle elevations in the Cascade Range or displaced to undiscovered refugia in Oregon. The sedimentologic record at Carp Lake indicates lower lake level and increased erosion within the catchment during this period. At ca. 13 ka, warming is inferred from continued shallowing of the lake and from increased spruce pollen, attributed to the spread of spruce forest in the Cascade Range. The Columbia Basin itself was still covered by steppe vegetation, however. Further warming between ca. 10 and 8.5 ka is indicated by Chenopodiineae and temperate aquatic taxa in the pollen record and by sedimentologic evidence for very low water levels. Steppe vegetation is assumed to have changed from periglacial to temperate by ca. 10 ka, although the main pollen taxa (*Artemisia* and grass) lack the taxonomic resolution to make a clear distinction between these vegetation types. A sharp rise in pine percentages at 8.5 ka records the establishment of ponderosa pine forest, which persists to the present day at the site. The appearance of pine coincides with evidence of higher lake levels, suggesting the onset of a climate cooler and moister than before.

A comparison of lowland records from the Pacific Coast, the Puget Trough, and the Columbia Basin gives insight into the

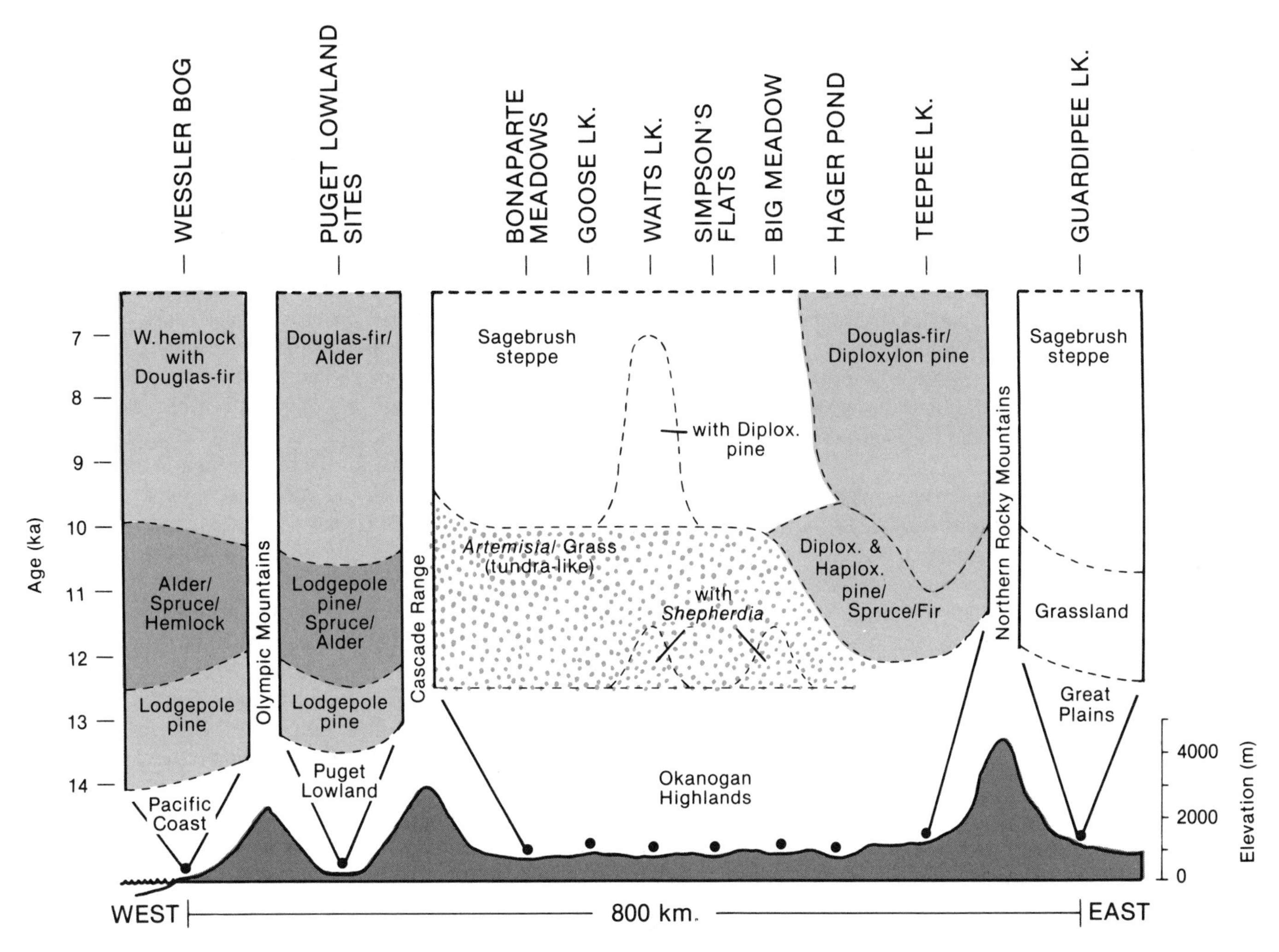

Figure 5. Vegetational changes through time inferred from a transect of pollen sites along the former southern edge of the Cordilleran ice sheet. See Figure 4 for explanation.

regional climate during the deglaciation. During alpine glaciation, cool humid conditions are recorded west of the Olympic Mountains. Quantitative climatic reconstructions suggest that temperatures were lower than today's by 5°C, although precipitation was unchanged (Heusser and others, 1980). In the Puget Trough, both growing season temperature and annual precipitation were lower, as indicated by the presence of xerophytic subalpine species, but adequate transfer functions have not yet been developed to quantify the glacial-age climate. The driest cold interval occurred between 20 and 17 ka, when mesophytic species were nearly absent from the Puget Trough as well as from the Fraser Lowland (Hicock and others, 1982). The Columbia Basin, covered by periglacial steppe, was under the influence of a cold, dry, and probably windy climate throughout the Fraser glaciation.

The paleoecologic record suggests that moisture was confined to west of the Olympic Mountains, and storm systems did not penetrate far inland, especially during the period from ca. 20 to 17 ka. Lower sea-surface temperatures at that time would have significantly reduced the amount of moisture carried onshore, and several features of the local landscape would have further enhanced aridity. Lowered sea level created a coastal plain that extended some 50 km farther west than that of today, and the orographic barrier posed by the Olympic Mountains, combined with the greater contrasts in land and sea-surface temperatures, may have caused much of the precipitation to be shed on the coastal plain and the Olympic Mountains, leaving little for lowlands to the east. Furthermore, the absence of Puget Sound meant the loss of an important moderating influence on climate; the Puget Trough would have more resembled interior montane basins to the east than the coastal plain to the west.

A shift from cold and dry to cool and humid conditions occurred in the southern Puget Trough ca. 15 ka. This change was concurrent with the initiation of a gradual warming trend on the Pacific slope and suggests widespread climatic change. To some extent, persistent coolness in the Puget Trough may have been the result of the advancing Puget lobe, although conditions were mild enough to allow trees to grow within 60 km of the ice margin. To the east, Carp Lake, which lay a greater distance from the ice sheet, shows little or no change in vegetation during this period.

Warming is registered in the vegetation first along the Pacific coast at 16.8 ka, then in Puget Sound at 12.5 ka, and finally in the Columbia Basin between ca. 13 and 10 ka. The timing corresponds to some degree to the chronology of deglaciation in each area. Along the coast, alpine glaciers had retreated by 18 ka, and the subsequent effect of Cordilleran ice in the Strait of Juan de Fuca was negligible. Thus vegetation change occurred early. In the Puget Trough and Columbia Basin the ice sheet retreated after 15 ka, and the pollen response to warmer conditions came 2,000 yrs or more after that.

The early Holocene vegetation throughout western Washington and southwestern British Columbia suggests increased summer drought. This interpretation, however, would seem to conflict with evidence for early Holocene glaciation in the North Cascade Range, which must have resulted from either decreased temperatures or increased precipitation at high elevations (Beget, 1984; Waitt and others, 1982). The paradoxical situation may have been caused by a steepening of the temperature lapse rate during a period of aridity. A steeper lapse rate would have produced cooler conditions at high altitudes while low areas remained warm.

The Pacific Northwest: The Okanogan Highlands and Adjacent Northern Rocky Mountains

No long records have been recovered from immediately south of the former Cordilleran ice limit, but an extensive network of well-dated pollen sites is located in the Okanogan Highlands of eastern Washington and in the Rocky Mountains of northern Idaho and northwestern Montana (Fig. 5).

The Okanogan sites lie in valleys once occupied by the Okanogan, Sanpoil, Colville, and Pend Oreille lobes of the ice sheet. The northern Rocky Mountain records are from lakes in areas formerly covered by the Purcell Trench lobe in Idaho and the West Kootenai glacier of northwestern Montana. The retreat of eastern Washington lobes was broadly synchronous with deglaciation of the Puget lobe, and by ca. 11 ka, ice termini lay at the International Boundary (Waitt and Thorson, 1983). The chronology of deglaciation and flood history in the region rests largely on tephrochronology, and in particular on the occurrence of Glacier Peak G ash (dated at 11.2 ka; Mehringer and others, 1984) in glacial deposits and lake sediments.

In contrast to differences in modern vegetation from site to site, the late-glacial vegetation was fairly uniform. The pollen record prior to ca. 10 ka is characterized by high percentages of *Artemisia* and grass, along with low amounts of conifers and herbs. The landscape at this time was apparently covered by open grassland with isolated pockets of forest species, including haploxylon pine (either whitebark or western white pine), spruce, and fir. Sites in large valleys show especially high values of *Artemisia,* implying that it was an important colonizer of stagnant-ice terrain and outwash deposits (Mack and others, 1978a). High pollen percentages of sedge, willow, birch, and alder indicate wet, disturbed ground around most sites, and many sites show a surprising amount of *Shepherdia canadensis* (buffaloberry), a shrub of wide distribution that commonly occurs in pioneer communities.

These early communities in the Okanogan region have been difficult to interpret, in part because the major taxa (e.g., grass, *Artemisia,* and pine) have broad ecological tolerances and pollen that can be transported long distances. For example, the *Artemisia* pollen could be from one of several steppe shrubs—the sagebrushes—or from arctic-alpine species. Although no modern analogues have been found for the combination of high percentages of *Artemisia* and grass and low values of haploxylon pine, fir, and spruce, the assemblage has been called tundralike (Mack and others, 1978b). In the case of the Okanogan region, true tundra is ruled out by the presence of *Typha* (cattail) pollen.

An early and long-standing hypothesis that the Columbia Basin was covered by lodgepole pine parkland during the last glaciation (Hansen, 1947) can be dismissed because of the low percentages of diploxylon pine pollen at most late-glacial sites in eastern Washington and the evidence for periglacial steppe at Carp Lake. Vertebrate fossil data from the central Columbia Basin also point to an open rather than forested landscape during full-glacial time (Gustafson, 1972; Martin and others, 1982). Tree pollen is more abundant in late-glacial spectra at sites in the eastern and western Okanogan Highlands than in the central part, perhaps because glacial refugia were located in the bordering highlands rather than in the basin to the south. Mack and others (1976) postulate that conifers grew in the loess-covered hills of the northern Columbia Basin, although paleoecologic data are lacking from that region. High amounts of diploxylon pine pollen featured in late-glacial records from Williams Fen (Nickmann, 1979), Simpson's Flats, and Waits Lake (Mack and others, 1978a, 1978b) suggest that lodgepole pine was able to grow on coarse mineral substrates and colonize areas affected by glacial meltwater and Scabland flooding.

That haploxylon pine, fir, and spruce pollen are present in nearly all late-glacial records from the Okanogan region argues for a climate cooler and wetter than that of today, and it is noteworthy that these particular species are absent from Carp Lake, where dry conditions are inferred. Periglacial mounds and patterned ground in the northern Columbia Basin have also been cited as additional evidence of cold conditions at this time; although these features are undated and many are of questionable periglacial origin.

In northern Idaho and Montana lodgepole pine, western white pine or whitebark pine, fir, and spruce were present in an open forest by ca. 10 ka. The two records, Hager Lake and Teepee Lake (Mack and others, 1978d, 1983), are too young to contain an earlier *Artemisia* and grass period. The assemblages are typical of modern open subalpine forests east of the Continental Divide in southwestern Montana, and, like late-glacial assemblages from northeastern Washington, they imply a climate cooler and more humid than that of the area today.

Nearly all sites show a period of increased summer drought beginning between ca. 10.6 and 9 ka, with the exception of Kelowna Bog in southern British Columbia (Alley, 1976) and Hager Pond where it is not registered until 8.3 ka. Drought is inferred from increasing percentages of grass, *Artemisia,* and diploxylon pine (either lodgepole pine or ponderosa pine), at the expense of more-mesophytic conifer taxa (haploxylon pine, fir, spruce). The resemblance of these assemblages with modern pollen spectra from steppe vegetation in the Columbia Basin (Mack and Bryant, 1974) suggests that the forest/steppe ecotone had shifted northward at least 100 km in the early and middle Holocene.

The sedimentary records at a number of Okanogan sites indicate that lake levels in the early Holocene were lower than at present. For example, unconformities occurred ca. 9.7 ka at Big Meadow (Mack and others, 1978c), between 9 and 6.7 ka at Simpson's Flats and Bonaparte Meadows (Mack and others, 1979), and possibly ca. 6.7 ka at Waits Lake. Both the sedimentologic and pollen data suggest that the timing of maximum drought was highly variable from place to place. At Big Meadow, Simpson's Flats, Bonaparte Meadows, Goose Lake (Nickmann and Leopold, 1984), and Kelowna Bog, aridity was greatest in the early Holocene between ca. 10 and 7 ka. At Carp and Goose Lakes it was over as early as 8.5 to 8 ka, whereas at Creston Fen (Mack and others, 1976), Williams Fen, Hager Pond, and Teepee Lake maximum drought did not commence until after 7.5 ka.

The Snake River Plain, Yellowstone Plateau, and Adjacent Regions

Several postglacial pollen records are available from southwestern Montana, northwestern Wyoming, and southern Idaho (Fig. 6). During the Pinedale glaciation (between ca. 25 and 9 ka; Porter and others, 1983), much of the area was covered by the Yellowstone ice cap, which was the source for outlet glaciers flowing into south-central Montana, northwest Wyoming, and southeast Idaho. Concurrently, local valley glaciers in adjacent mountain ranges were expanded as well. Wastage of the Yellowstone ice cap was largely completed by 14 to 13 ka, and, with few exceptions, glaciers receded upslope above their Neoglacial positions by 10 ka (Porter and others, 1983).

Pollen records from the sagebrush steppe of the Snake River Plain are available at four sites—Murphey's Rockshelter (Henry, 1984), Rattlesnake Cave (Davis and others, 1986), Grays Lake (Beiswenger, 1984), and Swan Lake (Bright, 1966). The first three sites show steppe vegetation throughout the Holocene, although the time of maximum drought implied by the pollen data varied. A 10,350-yr pollen record from Murphey's Rockshelter on the western Snake River Plain suggests widespread grassland and a climate cooler and moister than today's prior to 9.9 ka. Increasing percentages of Chenopodiineae pollen between 9.9 and 6.3 ka define a period of drought with sparsely vegetated chenopod steppe. At Rattlesnake Cave the increase in Chenopodiineae pollen occurred ca. 8.3 ka. Preliminary data from Grays Lake at the northeastern edge of the Snake River Plain show high values of *Artemisia* and small amounts of pine during the last glaciation. At 12 ka, pollen of pine and *Artemisia* increase in abundance and are joined by low frequencies of spruce. Increased drought begins ca. 10 ka, judging from the high percentages of pollen of juniper-type, grass, *Artemisia,* and Chenopodiineae, with the driest interval at ca. 7.3 ka (Beiswenger, 1984).

Swan Lake, which lies in sagebrush steppe in a narrow valley at the southeastern margin of the Snake River Plain, shows a different sequence (Bright, 1966). A cold humid climate for the period from 12 to 11.4 ka is inferred from high pollen percentages of haploxylon pine (limber or whitebark pine) and spruce. Unlike the situation at other basin sites, conifers apparently grew near Swan Lake in the late Pleistocene before their replacement by steppe with small numbers of low-elevation conifers. At 11.4 ka, a warming trend is registered by increased percentages of *Artemisia* and Chenopodiineae pollen. Just as at Grays Lake, the contribution of spruce pollen declined, while that of lodgepole pine and haploxylon pine (either limber pine or whitebark pine) increased. Between 10.8 and 10.3 ka, increasing warmth and aridity are implied by the high values of lodgepole pine, Douglas-fir, and *Artemisia.* By 10.3 ka, steppe had expanded to its present altitudinal range and the forest was similar to that of today (Bright, 1966). A shift toward more xerophytic communities is documented by the high values of Chenopodiineae pollen between 8.4 and 3.1 ka. Thus the timing of drought was later than at other sites on the Snake River Plain.

Three sites provide pollen data at the lower forest border—Blacktail Pond (Gennett and Baker, 1986), Cub Lake, and Gardiners Hole (Baker, 1983). Prior to 11.8 ka the area was covered by tundralike vegetation dominated by *Artemisia* and grass. Pine and spruce, represented by low pollen percentages, probably grew in more mesic areas. At Gardiners Hole and Blacktail Pond, spruce increased in abundance at 11.8 ka, presumably because of increased summer temperature and effective moisture. It soon was joined in large numbers by whitebark pine, limber pine, and lodgepole pine to form an open forest. With further warming at ca. 11.3 ka open forest was replaced by closed forest dominated by lodgepole pine. At Cub Lake, however, the development of lodgepole pine forest was delayed until ca. 8 ka. The driest interval at all three sites is identified by high pollen percentages of Douglas-fir and *Artemisia* as well as lodgepole pine, beginning between 8 and 7 ka.

Records from the subalpine forest of the northern Rocky Mountains are remarkably similar from one range to another (Fig. 6). In Yellowstone Park, Cub Creek Pond (Waddington and Wright, 1974) with a basal date of 14.3 ka is the oldest site, although the date has been questioned on the grounds that it brackets the Glacier Peak B ash (Westgate and Evans, 1978), ca. 11.2 ka. Nuphar Pond and Buckbean Fen (Baker, 1976) show a

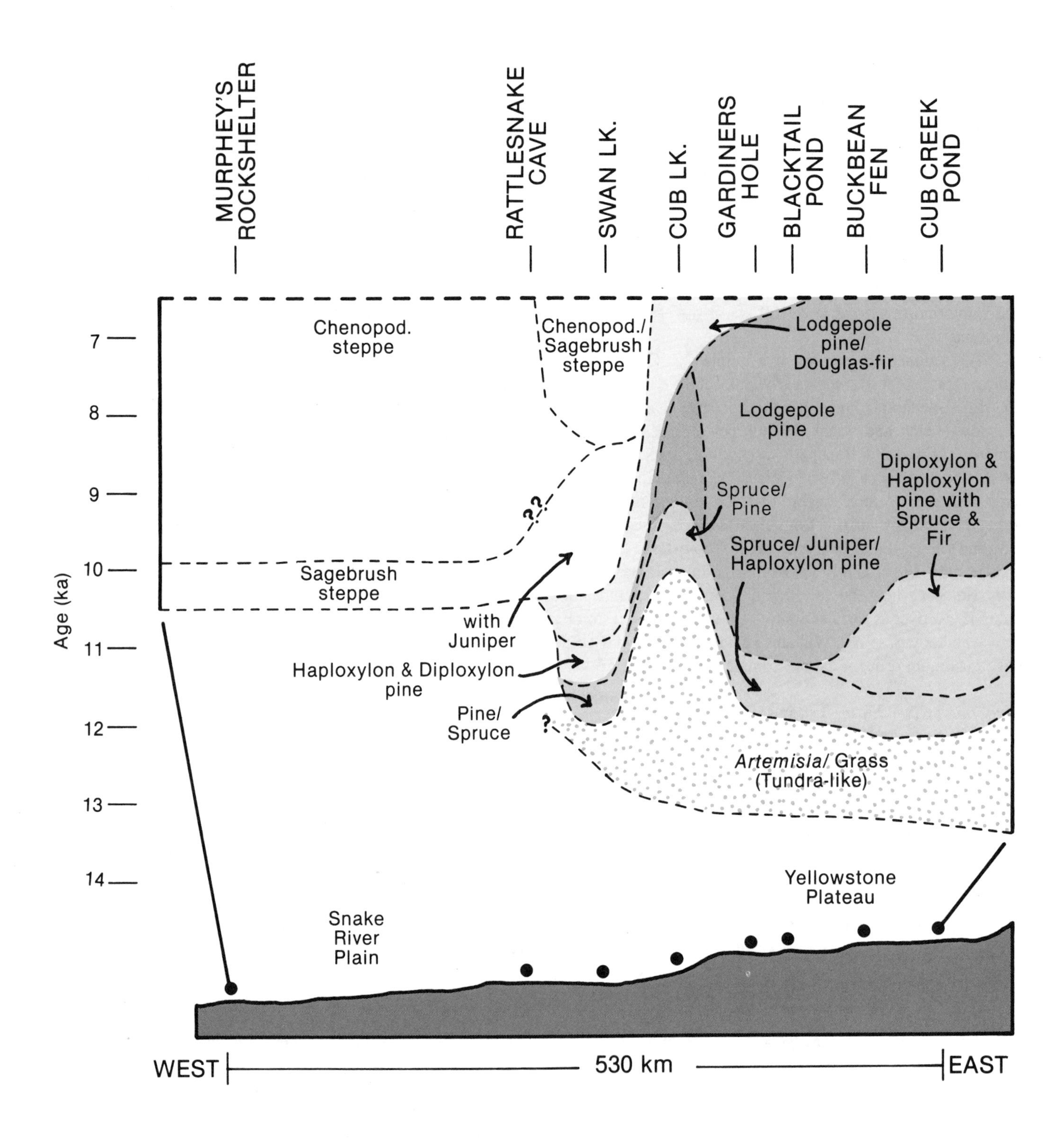

Figure 6 (this and facing page). Vegetational changes through time inferred from pollen sites in the Snake River Plain, Yellowstone Plateau, and adjacent mountains. See Figure 4 for explanation.

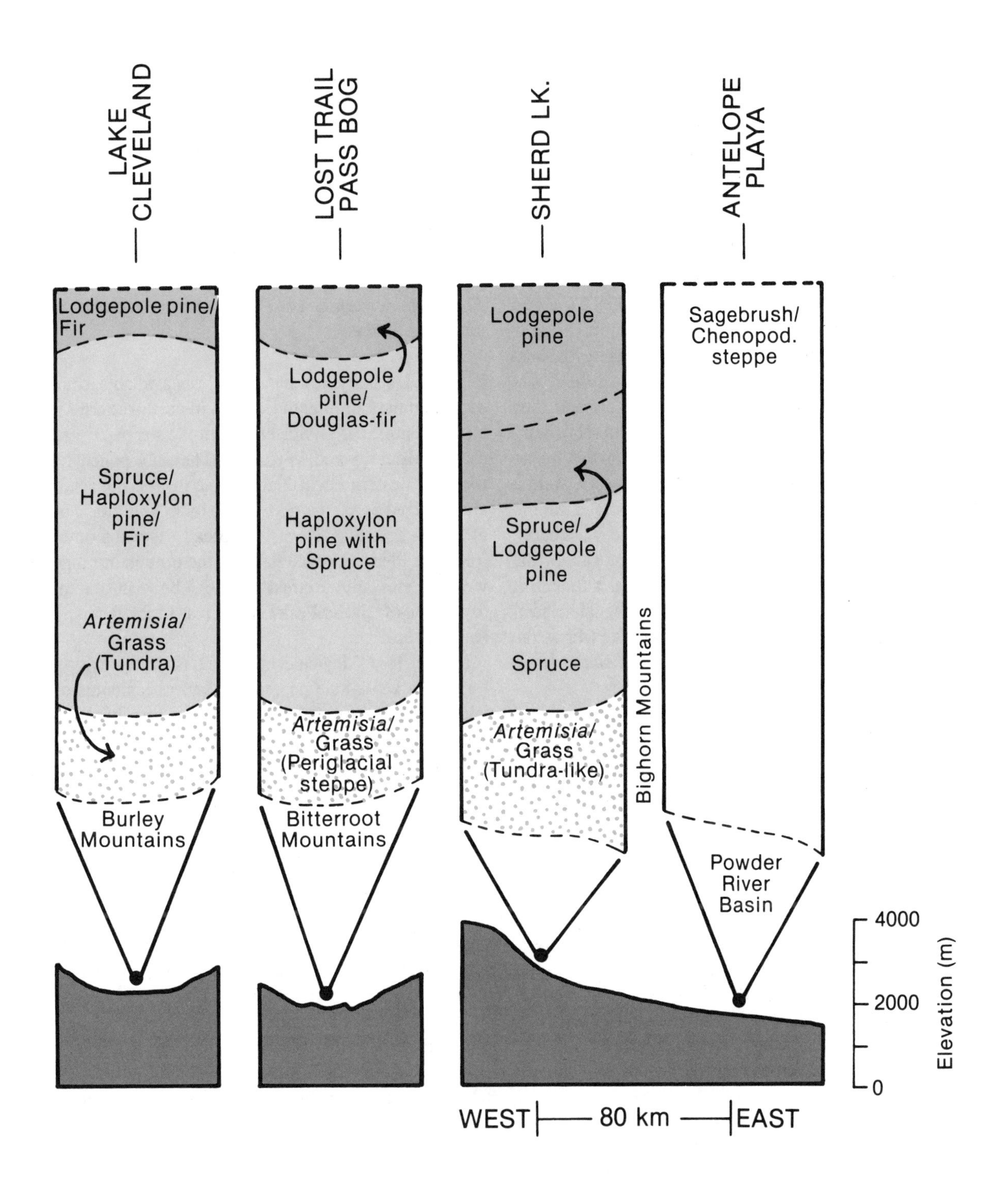
LAKE CLEVELAND
LOST TRAIL PASS BOG
SHERD LK.
ANTELOPE PLAYA
Lodgepole pine/ Fir
Spruce/ Haploxylon pine/ Fir
Artemisia/ Grass (Tundra)
Burley Mountains
Lodgepole pine/ Douglas-fir
Haploxylon pine with Spruce
Artemisia/ Grass (Periglacial steppe)
Bitterroot Mountains
Lodgepole pine
Spruce/ Lodgepole pine
Spruce
Artemisia/ Grass (Tundra-like)
Bighorn Mountains
Sagebrush/ Chenopod. steppe
Powder River Basin
4000
2000
0
Elevation (m)
WEST
80 km
EAST

basal age of about 13.5 ka. In the Albion Mountains of east-central Idaho a record from Lake Cleveland begins at 12.6 ka (Davis and others, 1986), about the same time as Sherd Lake in the Bighorn Mountains (Burkhart, 1976). Lost Trail Pass Bog in the Bitterroot Mountains of southwestern Montana has a basal date of ca. 11.5 ka (Mehringer and others, 1977). The oldest sites reveal that prior to 11.5 ka the subalpine region was covered by *Artemisia*-dominated vegetation. *Artemisia* values in the late-glacial spectra are higher than any in modern surface samples from alpine or tundra vegetation and are attributed largely to sagebrush taxa.

At Lost Trail Pass Bog the pollen data are interpreted as steppe, implying cold and also dry conditions, whereas the subsequent invasion of trees is thought to suggest warming and increased moisture (Mehringer and others, 1977). Sites in Yellowstone, the Bighorn Mountains, and Idaho feature the same pollen assemblages, but in this case the environmental reconstruction is for alpine tundra and cold humid conditions. If the reconstruction of alpine tundra rather than steppe is correct, then increased temperatures would have permitted the subsequent invasion of trees. As at lower elevations, spruce was the first tree to colonize the postglacial landscape at 11.6 ka. Shortly afterward it was joined by lodgepole pine and limber pine in the Albion Mountains and by lodgepole pine, Douglas-fir, and whitebark/limber pine in Yellowstone. Subalpine fir was not part of the late-glacial vegetation around Lost Trail Pass Bog, although it was present in other regions. Pine is absent from the late-glacial interval in the Bighorn Mountains, leaving spruce and fir as the major conifers. Douglas-fir, juniper, birch, and *Populus* were present in Yellowstone but not at other sites.

The early Holocene was a period of maximum warmth at most high-elevation sites. In Yellowstone, evidence of conditions warmer and possibly drier than today are found in the high percentages of lodgepole pine and trace amounts of Douglas-fir; both taxa are common today at lower elevations. This period occurs from 10.2 to 5 ka at Buckbean Fen and between 9 and 4.5 ka at Cub Creek Pond. Corroboration comes in the form of a major hiatus between ca. 9 and 6 ka in the sedimentary record at Buckbean Fen, when the basin apparently dried. At Lake Cleveland, early Holocene warming is also inferred, from macrofossils of aquatic plants and needles of the xerophytic limber pine (Davis and others, 1986). At Lost Trail Pass Bog, maximum warmth occurred later, after 7 ka, as inferred by higher percentages and pollen accumulation rates of Douglas-fir and lodgepole pine. The Bighorn sites show no change in the pollen record to indicate a postglacial episode of more xerothermic conditions.

Several conclusions can be drawn from these data. The Snake River Plain was largely treeless during glacial time, although subalpine species grew in protected mesic habitats. At ca. 12 ka an increase in temperature and possibly precipitation allowed conifers to grow at all elevations, and by 11.6 ka many species grew in moderate abundance at and above their present altitudinal range. For example, Douglas-fir apparently then grew around Yellowstone Lake, as did spruce in the Albion Mountains. With the exception of Swan Lake, sites on the Snake River Plain record greatest drought in the early Holocene. At slightly higher elevation, however, the open-forested slopes experienced greatest drought in the middle and late Holocene after 7 ka. An explanation for the asynchrony in such close proximity is not obvious. High-elevation sites provide an equally confusing picture of past fluctuations in summer temperature, and thus the variability is not strictly due to elevation alone. In Yellowstone, maximum warmth occurred between 10 and 4 ka. In the Burley Mountains it began ca. 11 ka and was over by 7 ka. The record from the Bitterroot Mountains clearly shows middle Holocene warming beginning 7 ka, whereas no thermal maximum is evident from pollen data in the Bighorn Mountains.

The northwestern Great Plains and northeastern montane basins

Until recently, few data have been available from the eastern margin of the Cordilleran ice sheet in northwestern Montana east of the Continental Divide and from the steppe-covered basins of central Montana and Wyoming. The only pollen records from the northwestern Great Plains come from postglacial sites in the area covered by the Two Medicine lobe. This lobe was the largest of the piedmont glaciers that spread eastward onto the Plains from late Pleistocene ice fields in the mountain ranges of northwestern Montana. Retreat of the lobe occurred prior to the deposition of Glacier Peak layer G (ca. 11.2 ka) and St. Helens Jy (ca. 11.4 ka).

The Two Medicine lobe area is dotted with small lakes formed in kettle-hole depressions and as remnants of proglacial lakes. At Guardipee Lake (C. Barnosky, unpublished data) in the prairie, the late-glacial pollen record is dominated by pine (largely diploxylon-type and attributed to distant lodgepole pine populations), grass, *Artemisia,* and Chenopodiineae. Values of spruce are low, unlike the high percentages in late-glacial assemblages at Marias Pass, the former ice divide some 70 km to the west (Carrara and others, 1986) and on the central Great Plains, where a spruce forest containing some hardwood elements grew (Webb and others, 1983). Instead, the Guardipee Lake record suggests that the late-glacial landscape near the mountain front was essentially treeless and that the montane forest zone was less extensive than the modern belt. The treeless landscape may have been a response to prolonged and locally severe conditions near the confluence of Cordilleran and Laurentide ice. In this case, cold windy conditions within the periglacial zone fostered the open landscape, an interpretation consistent with evidence of permafrost (Mears, 1981; Péwé, 1983). Alternatively, the grassland period may have developed in response to early postglacial warming in the northwestern Great Plains. Support for early warming comes from the rapid retreat of the Two Medicine lobe after 12 ka (Carrara and others, 1986). In addition, the minor taxa present at Guardipee include *Ambrosia*-type, *Shepherdia canadensis,* and other herbs, all of which have wide ranges but none diagnostic of tundra.

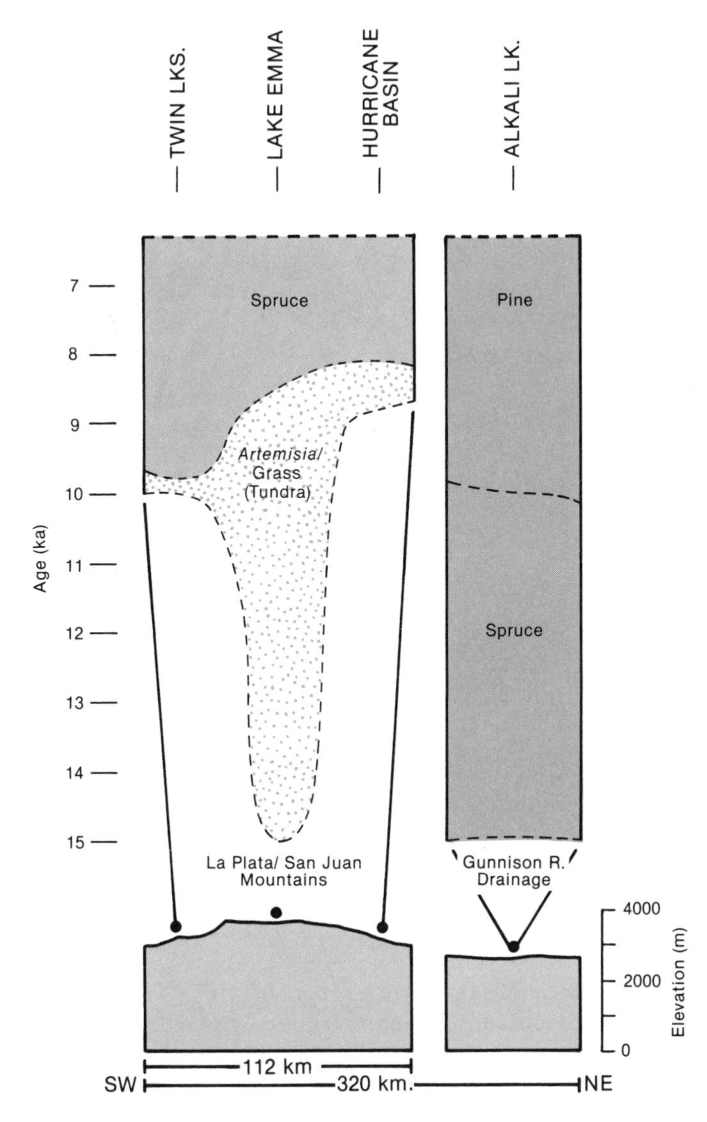

Figure 7. Vegetational changes through time inferred from pollen sites in the Colorado Rocky Mountains. See Figure 4 for explanation.

The early Holocene pollen record at Guardipee Lake features high percentages of *Artemisia* and pine relative to grass and other herbs. Conditions were more arid than in late-glacial time, and the lake was probably dry intermittently.

A 13,000-yr record from Antelope Playa in the Powder River Basin of eastern Wyoming, between the Black Hills and the Bighorn Mountains, also indicates steppe vegetation with virtually no trees throughout the late Quaternary (Markgraf and Lennon, 1986). Unlike other sites in the Rocky Mountain West, increased accumulation rates of *Artemisia* and grass pollen are interpreted as evidence of aridity after 5 ka. The absence of adult ostracods in the late Holocene supports this conclusion by implying seasonal drying of the playa. It seems unwise to place too much reliance on pollen accumulation rates from this site, because the basin has had irregular sedimentation throughout its history. If percentage values from Antelope Playa alone are considered, the late-glacial and early Holocene were drier than subsequently, just as at Guardipee Lake.

The Colorado Rocky Mountains

The vegetational history of the Colorado Rockies is known primarily from sites in the Colorado Front Range and the San Juan Mountains (Fig. 7). Whereas pollen data from the subalpine and alpine sites are reasonably similar for the late Pleistocene, the interpretation of the Holocene record after ca. 7 ka varies considerably. Reconciliation of the discrepancies has been discussed in recent reviews (Nichols, 1982; Baker, 1983).

The oldest pollen record in the Front Range comes from Devlins Park in the subalpine forest (Legg and Baker, 1980). Pollen recovered from lacustrine silts dated between 22 and 12 ka show high values of *Artemisia* and pine, with lesser amounts of grass, Chenopodiineae, juniper-type, and spruce. Alpine tundra vegetation and a treeline 500 m lower than at present are inferred from the combination of low pollen-accumulation rates and diagnostic herb pollen. Climate was probably both cooler and wetter than today's, with a reduction in the present growing season by at least 5 days (Legg and Baker, 1980).

The San Juan Mountains apparently were rid of some of the largest Pinedale ice fields very early. A study of Lake Emma, which lies in a south-facing cirque presently above forest and krumholz vegetation, suggests deglaciation by ca. 14.9 ka (Carrara and others, 1984). Carrara and others (1984) postulate that much of the San Juan Mountains was situated just above the late Pleistocene glaciation threshold, so that a slight increase in temperatures created a large ablation area. Other paleoecologic records in the region strengthen the argument for early deglaciation: the basal date of 15.5 ka at nearby Molas Lake, originally rejected by Maher (1961) as too old, is supported by the Lake Emma findings. At Hurricane Basin, north-facing cirques were ice free by 9 ka, and south-facing cirques may have been deglaciated at least 2,000 yrs before that (Andrews and others, 1975). Preliminary data from the Crested Butte area suggest deglaciation as early as 15 ka (Fall, 1984). In the Chuska Mountains of northeastern New Mexico, a change from spruce to pine forest at ca. 13.5 ka also indicates early amelioration (Wright and others, 1973).

From 15 to 10 ka high elevations were generally treeless, dominated by *Artemisia,* grass, and other herbs. Whether the vegetation was shrub steppe (Fall, 1984) or alpine tundra (Carrara and others, 1984; Andrews and others, 1975; Legg and Baker, 1980) is disputed, and once again the discrimination depends largely on the interpretation of *Artemisia* and the presence of tundra-indicator species in the pollen record. At Lake Emma the first 5,000 yrs featured sparse tundra vegetation, although spruce forest may have grown near the site within 500 yrs of deglaciation. Other late-glacial records show increasing spruce pollen prior to 10 ka, suggestive of slowly but steadily rising treeline.

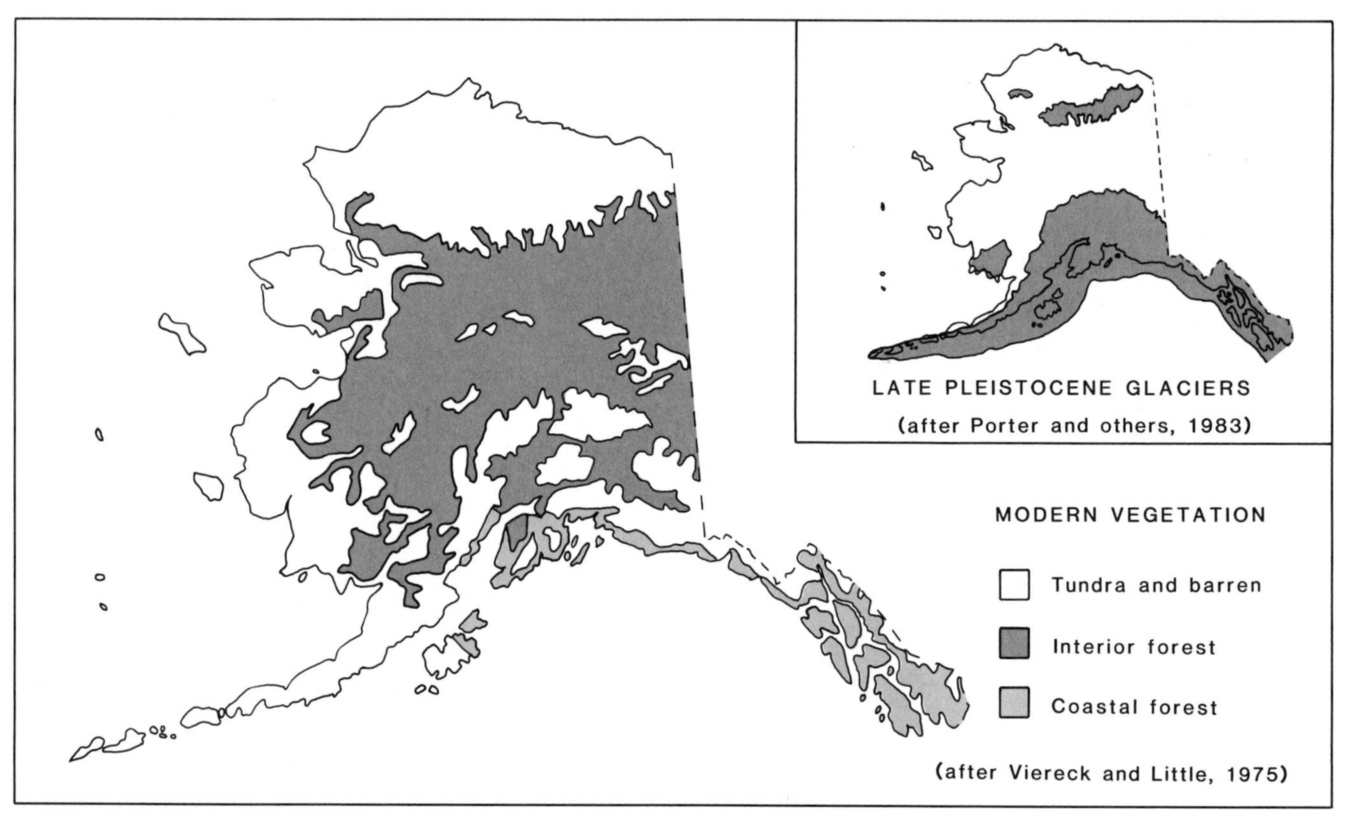

Figure 8. Aspects of the Alaskan landscape discussed in the text.

The elevation of upper treeline in most Colorado studies is calculated from spruce/pine ratios, which are compared to Maher's (1963) analysis of spruce/pine ratios from surface samples. Radiocarbon dates on wood fragments found above treeline provide supporting evidence for periods of higher treeline in the past. At Crested Butte, spruce forest grew at the elevation of the present subalpine forest as early as 11 ka (Fall, 1984). A 9,800-yr record from the La Plata Mountains suggests that subalpine forests there were not established until 8.3 ka, but that treeline lay at elevations higher than today until 6.7 ka (Petersen and Mehringer, 1976). Similarly, in the San Juan Mountains, the period of highest treeline began between 9 and 8 ka (Carrara and others, 1984; Andrews and others, 1975; Maher, 1961). Wood fragments above present treeline in the Front Range are dated as early as ca. 9.2 ka (Benedict, 1973). Thus, all these records suggest conditions warmer than present in the early to middle Holocene. Only Maher (1972) diverges from this interpretation in his study of Redrock Lake in the Front Range. The period from 10 to 7.6 ka is thought to be cooler and moister, whereas the climate from 7.6 to 6.7 ka was like that of the present day.

Alkali Lake in the *Artemisia* steppe below the forest zone presents a 15,000-yr record from low elevations (Markgraf and Scott, 1981). Prior to 10 ka, cool moist conditions were widespread, and a subalpine spruce forest grew at the site. A lowering of upper treeline by 300 to 600 m is inferred. With warming, a forest predominantly of limber pine developed and grew at the site for the next 6,000 yrs with little modification. Its protracted occurrence suggests conditions more humid than today's, which Markgraf and Scott (1981) attribute to enhanced summer monsoons as a result of regional warming in the middle Holocene.

ALASKA

The modern landscape of Alaska is varied, with wet coastal meadows and rolling tundra in the north, great expanses of spruce forest in the valleys and plateaus of the interior, and almost impenetrable spruce-hemlock forests along the southeastern Pacific coast. Glaciers cap the highest peaks of southern Alaska and the Brooks Range, whereas barren slopes of the lower mountains attest to the fact that much of Alaska still remains a land of low temperatures and forbidding environments.

The modern vegetation of Alaska is floristically simple and can be divided into three formations: tundra, interior boreal forest, and coastal forest (Fig. 8). Tundra occurs along the coasts of northern and western Alaska and at high elevations throughout the interior (Fig. 9). The boreal forest dominates low to middle elevations in the interior, primarily in the intermontane regions.

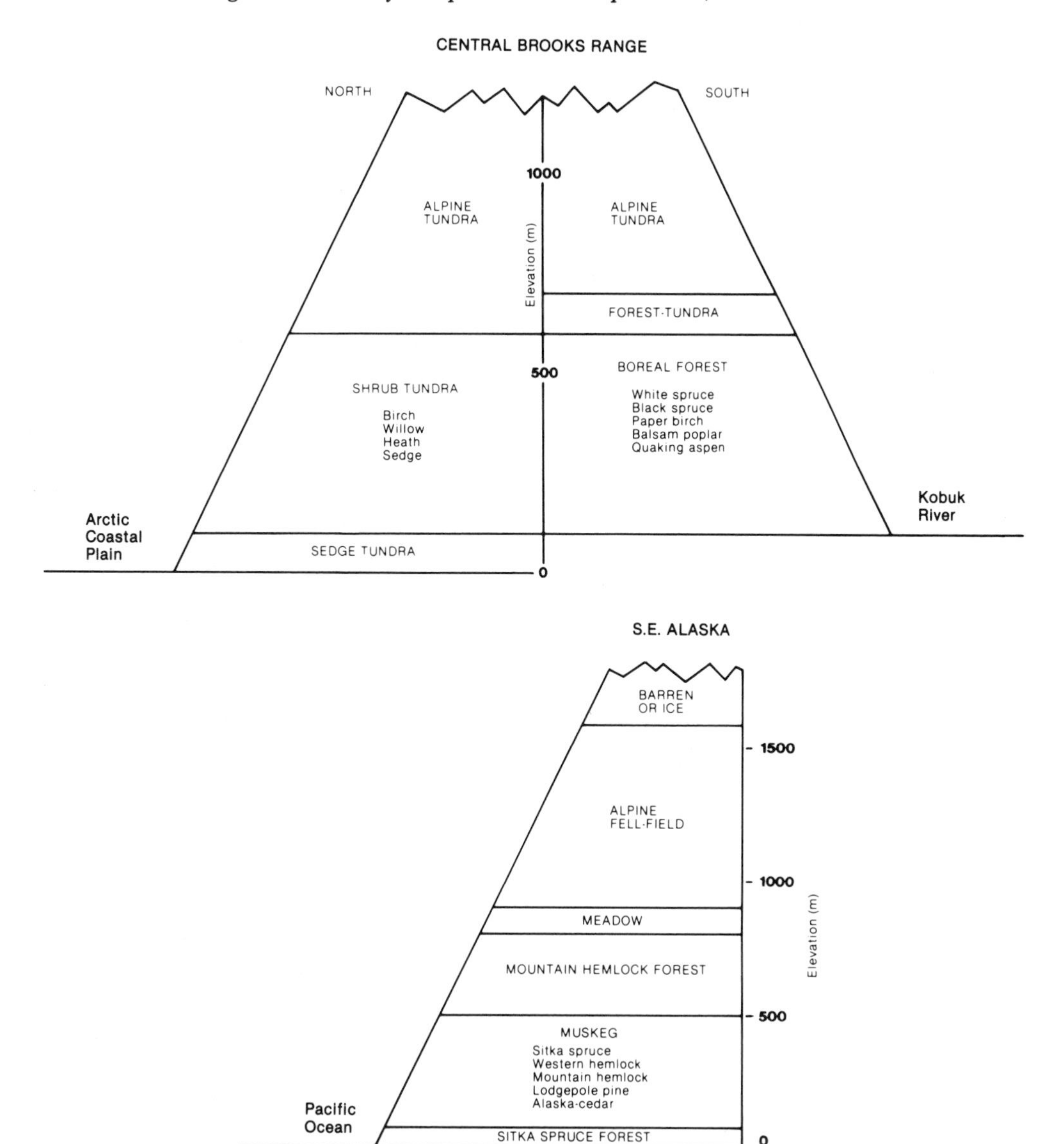

Figure 9. Generalized zonation of modern vegetation in two areas of Alaska (data from numerous sources).

The coastal forest is most extensive in the panhandle of southeastern Alaska and along a narrow strip on the northern coast of the Gulf of Alaska (see Viereck and Little, 1975; and Hultén, 1968).

The time under consideration in this paper spans portions of the Duvanny Yar glacial interval (30 to 14 ka), the late-glacial and early Holocene birch interval (14 to 8.5 ka), and the middle Holocene (8.5 to 6 ka; Hopkins, 1982). These three periods roughly correspond to the herb pollen zone, the birch pollen zone, and the alder (or alder-spruce) pollen zone defined in northern and central Alaska (Livingstone, 1955). In southern Alaska most pollen records span only the last 10,000 yrs, so the stratigraphic zones do not correspond so simply with the above intervals.

Full-glacial vegetation: Grassland, barrens, or tundra mosaic?

The controversy over the nature of the Alaskan landscape during Duvanny Yar time centers on a major paradox. Faunal evidence indicates greater concentration and diversity of large grazing mammals during the late Pleistocene than currently present in Alaska. Some pollen data, however, suggest a floristically depauperate barren vegetation that palynologists assert could not support a sizable population of large mammals. Although it is beyond the scope of this paper to summarize the debate in detail (see Hopkins and others, 1982), three alternative descriptions of the full-glacial landscape of eastern Beringia have been offered: (1) continuous steppe tundra, arctic steppe, or grassland (Guthrie, 1985); (2) discontinuous fell field or polar desert (Cwynar, 1982); and (3) mosaic of open ground and tundra (Schweger, 1982; see also Hopkins and others, 1982; Ritchie, 1984).

Proponents of the first model cite the abundant fossils of bison (*Bison priscus*), horse (*Equus*), and mammoth (*Mammuthus primigenius*); the high percentages of grass, sedge, and wormwood pollen; and the local presence of invertebrate fossils characteristic of steppe. This argument is based on the size and diversity of animal populations during Duvanny Yar time. Certainly if large or numerous herds roamed concurrently throughout eastern Beringia, a productive grassland or steppe was necessary to satisfy the different types of feeders (e.g., browsers versus grazers).

Paleoecologists who think the full-glacial vegetation was discontinuous and more similar to a polar desert offer the following evidence to support their model: (1) low rates of pollen accumulation more comparable to modern fell fields than to grasslands, (2) highly inorganic lake sediments implying abundant inwash of sediment from sparsely vegetated slopes, and (3) minor pollen types that have a stronger affinity to arctic/alpine tundra than to steppe. Proponents of this model doubt that particularly large herds were present in Beringia during Duvanny Yar time.

We favor the third model, which states that the landscape was covered by a mosaic of tundra types whose distribution was primarily a function of altitude and available moisture. As such, low-lying areas, in particular river valleys, supported mesic meadowlike tundra, whereas slopes and ridgetops were vegetated by a more xeric tundra that became discontinuous with increasing altitude. Mesic tundra was dominated by moisture-dependent graminoids and forbs with local occurrences of shrub thickets, while xeric tundra was dominated by wormwood (e.g., *Artemisia arctica* ssp. *arctica* and *A. arctica* ssp. *comata*) and herbaceous species that could survive in harsh environments. These plants probably provided a discontinuous ground cover similar to that found in modern fell fields. The mosaic model assumes that the mesic tundra provided sufficient biomass to support small herds of ungulates.

Northern Alaska

Much of northern Alaska remained unglaciated during the late Pleistocene (Fig. 8). Ice cover was greatest in the Brooks Range between 25 and 11.5 ka, but even these glaciers were not as large as those of earlier ice advances. In general, alpine glaciers were smallest in the north and west, and nowhere did they extend beyond the mountain flanks (Porter and others, 1983). Two glacial maxima occurred at 24 ka and 17 ka with a mild interstade between 22 and 19.5 ka. A slight readvance took place at 12.8 ka.

The earliest palynological research in northern Alaska focused on the central Brooks Range (Livingstone, 1955, 1957), the Arctic Slope (Colinvaux, 1964b), and Seward Peninsula (Colinvaux, 1964a). The addition of new data from the Brooks Range (Schweger, 1982; Brubaker and others, 1983; Edwards and others, 1985; Anderson, 1985; Bergstrom, 1984) permits a detailed analysis of the vegetational history of that area. Research has also continued on the Seward Peninsula and the Arctic Slope (Matthews, 1974b; Shackleton, 1982; Nelson, 1982; Wilson, 1984), although the results will not be discussed here.

Brooks Range. Pollen records from Kaiyak and Squirrel lakes in northwestern Alaska date to ca. 37 and 28 ka, respectively (see Fig. 10 for site locations). The data imply that mesic tundra vegetation grew in the river valley bottoms prior to 14 ka and was probably rich enough to support small populations of late Pleistocene megafauna (Anderson, 1985). Records from Ranger and Rebel lakes at higher elevations in the central Brooks Range are more difficult to interpret because of their poor radiocarbon chronology, but they suggest the presence of xeric tundra or fell-field vegetation (Brubaker and others, 1983; Edwards and others, 1985; Fig. 11A). Pollen data from Sands of Time Lake and Ped Pond in northeastern Alaska also record a sparse vegetation cover even though these sites lie at relatively low elevations (Edwards and Brubaker, 1984, 1986). These records seem to argue against vegetational patterns defined solely by altitude and give greater emphasis to the role of effective moisture in determining the vegetation (Schweger, 1982). The Brooks Range sites, when considered together, suggest a broad geographic trend toward drier, more severe conditions in northeastern Alaska concurrent with more mesic conditions at least at the lower elevations in the northwest (Anderson, 1985). In northwestern Canada, data from Hanging Lake and Lateral Pond show xeric herb tundra at middle to high elevations, consistent with the pattern in northeastern Alaska (Ritchie, 1984).

Pollen records of the last 14,000 yrs show significant variability in the Brooks Range. The shift from xeric herb tundra to birch shrub tundra occurred earliest in northwestern Alaska at ca. 14 ka (Kaiyak and Squirrel lakes). Sites from the central Brooks Range (e.g., Ruppert Lake, Brubaker and others, 1983; Screaming Yellowlegs Pond, Edwards and others, 1985; Toolik Lake, Bergstrom, 1984) and the Yukon lowlands (Sands of Time Lake) record this vegetation change 1,000 to 2,000 yrs later. The asynchrony may relate to several factors: (1) local differences in the timing of deglaciation, (2) distance from ice-age refugia and thus seed source, or (3) a more favorable climate for birch growth earlier in southwestern portions of the Brooks Range.

Sands of Time Lake and some sites in the central Brooks Range show increased percentages of poplar pollen between 11 and 9 ka. In addition, macrofossils document a northward extension of poplar onto the Arctic Slope about 9.5 ka (Hopkins and others, 1981; Nelson, 1982). Poplar pollen is conspicuously absent in Arctic Slope sediments of this age, but the pollen records are few and from sediments that may not preserve this fragile grain. The expansion of poplar (thought to be balsam poplar) is attributed to increased spring or summer temperature and/or the increased availability of disturbed ground resulting from spring floods (Brubaker and others, 1983; Edwards and Dunwiddie, 1985). The decline in poplar communities after 9 ka has been explained by decreased spring/summer temperatures, decreased snow pack that reduced spring flooding, and development of organic-rich soils unsuited for poplar growth (Brubaker and others, 1983; Cwynar, 1982). Concurrent range extensions of beaver, cattail, and poplar as well as melting of ice wedges near Kotzebue Sound (McCulloch and Hopkins, 1966) support the hypothesis that early Holocene summer temperatures were higher than at present. Downcutting of streams and range extensions of the beetle *Agonum quadripunctatum* and the ostracod *Candona acuta* on the Arctic Slope also suggest a warmer climate between ca. 12 and 8 ka (Carter and others, 1984).

Alder appeared first in the Kotzebue Sound drainage at ca. 9 ka, but dates of 8 ka at Sands of Time Lake and 9.5 ka at Toolik Lake indicate its relatively early arrival farther east as well. Alder is present in the central Brooks Range by ca. 7 ka. The overall pattern in the Brooks Range, along with a date of 8.5 ka for alder arrival at Hanging Lake in the northern Yukon (Cwynar, 1982), suggests several small refugia in eastern Beringia, although this has not yet been confirmed by plant macrofossils (Hopkins and others, 1981).

The earliest postglacial record of spruce in northern Alaska comes from Sands of Time Lake at ca. 8.7 ka. Pollen data from Screaming Yellowlegs Pond indicate that spruce arrived in the east-central Brooks Range between 8.5 and 8 ka. Farther to the west, data from Ruppert Lake are more equivocal, because increased percentages of spruce pollen between 8.5 and 8 ka do not coincide with high pollen accumulation rates. Most likely, spruce pollen was blowing in from distant populations farther east. Overall, the vegetational history delineates an east-to-west migration of the boreal forest across the southern flanks of the Brooks Range (Anderson, 1985). Early Holocene environments differed markedly between northwestern and northeastern Alaska, the former supporting birch-alder shrub tundra and the latter a spruce parkland.

Central Alaska

The Alaska Range was covered by a late Pleistocene ice cap that reached its maximum ca. 25 ka (Porter and others, 1983). The northern valley glaciers of the Alaska Range never extended far beyond the mountain flanks. Deglaciation occurred between ca. 14 and 9.6 ka, with a slight readvance in the McKinley area between 12.8 and 11.8 ka.

Palynological research in eastern portions of central Alaska has concentrated on the Fairbanks area, Tanana Valley, and northern foothills of the Alaska Range (Ager, 1983; Ager and Brubaker, 1985; Matthews, 1974a). In the western portion, the Bering Sea islands (Colinvaux, 1967, 1981) and the Yukon Delta–Norton Sound region (Ager, 1982, 1983) are the best studied.

Tanana Valley and the northern foothills of the Alaska Range. Many sites have been described from the Tanana Valley of east-central Alaska, but few date to full-glacial time. In the Isabella Basin, late Pleistocene colluvium yielded pollen as well as fossil plants, insects, and mammals indicative of steppe tundra (Matthews, 1974a). However, the lack of dating control and a possible hiatus cast doubt on this assemblage being late Wisconsinan in age (Ager and Brubaker, 1985). Two well-dated pollen records from Birch and Harding lakes in the Tanana Valley are of little help in reconstructing the full-glacial landscape (Fig. 11B). Birch Lake has a thin herb zone and no basal radiocarbon date. Rates of pollen accumulation for the herb zone are extrapolated down-core from a 14.7 ka date, and Ager (1983) interprets the vegetation as a xeric steppe tundra. The Harding Lake record extends to the middle Wisconsin but apparently lacks or has an incomplete full-glacial sequence (Ager and Brubaker, 1985).

From these scanty data it appears that central Alaska was covered by a herb-dominated tundra or steppe tundra, in which willow was the only common shrub. Poplar and perhaps tamarack may have survived in isolated stands in eastern Beringia, but there is no evidence for a spruce refugium in east-central Alaska as was once suggested (see Hopkins and others, 1981).

The subsequent vegetation history of the Tanana Valley is well documented (Ager, 1983; Ager and Brubaker, 1985). At 14 ka, birch-dominated shrub tundra spread rapidly through the valley. Shrub tundra was replaced by a willow-poplar scrub forest at 11 ka. At 9.5 ka, spruce first appeared and evidently spread rapidly to the east (based on a date of 8.7 ka at Antifreeze Pond in the Yukon Territory; Rampton, 1971), south (9.1 ka in the Gulkana uplands; Schweger, 1981), and north (8.7 ka in the Yukon Flats; Edwards and Brubaker, 1984), and more slowly to the west (7.5 ka at Eightmile Lake and 5.7 ka at Lake Minchumina; Ager, 1983). Alder was present in the Tanana Valley after ca. 8.4 ka.

The northern foothills of the Alaska Range feature a somewhat different vegetation history (Ager, 1983). At Eightmile Lake, the herb zone has more sedge and less wormwood pollen than the Tanana sites. The birch zone, dating between 13 and 7.5 ka, contains a poplar-willow pollen subzone beginning at 10 ka. Spruce and alder pollen rise simultaneously at 7.5 ka, unlike the Tanana Valley records, where spruce precedes alder by about 1,100 yrs. The relatively late appearance of alder in east-central Alaska suggests a migration into interior Alaska from either the western or southern coast.

The earliest appearance of spruce in eastern Beringia is at ca. 10 ka from the Tuktoyaktuk Peninsula area of northwestern Canada (Ritchie, 1984). Although the pattern of spruce migration is broadly east to west, as suggested by Hopkins and others (1981), spruce woodlands in Alaska actually seem to radiate from the Tanana Valley to the south, northeast, and northwest. The available pollen data suggest that if spruce migrated into Alaska from northwestern Canada, it did so quickly (ca. 775 km over 1,000 yrs) and did not follow a route along the Porcupine River (Edwards and Brubaker, 1986; Anderson, unpublished data).

Yukon Delta–Norton Sound. Recent work in the Yukon Delta–Norton Sound area indicates that despite some local varia-

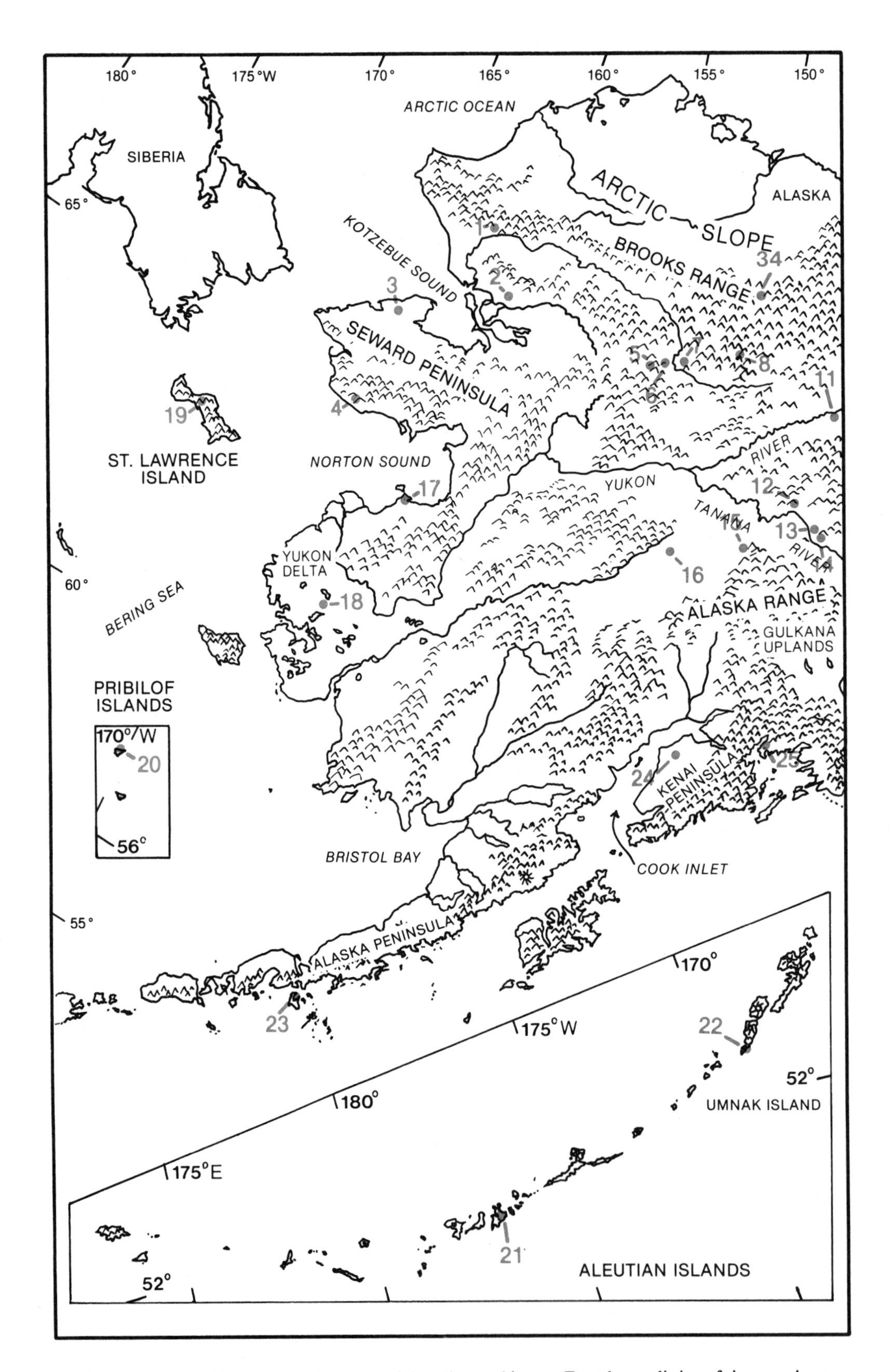

Figure 10 (this and facing page). Location of sites discussed in text. For a longer listing of sites, see Ager and Brubaker (1985).

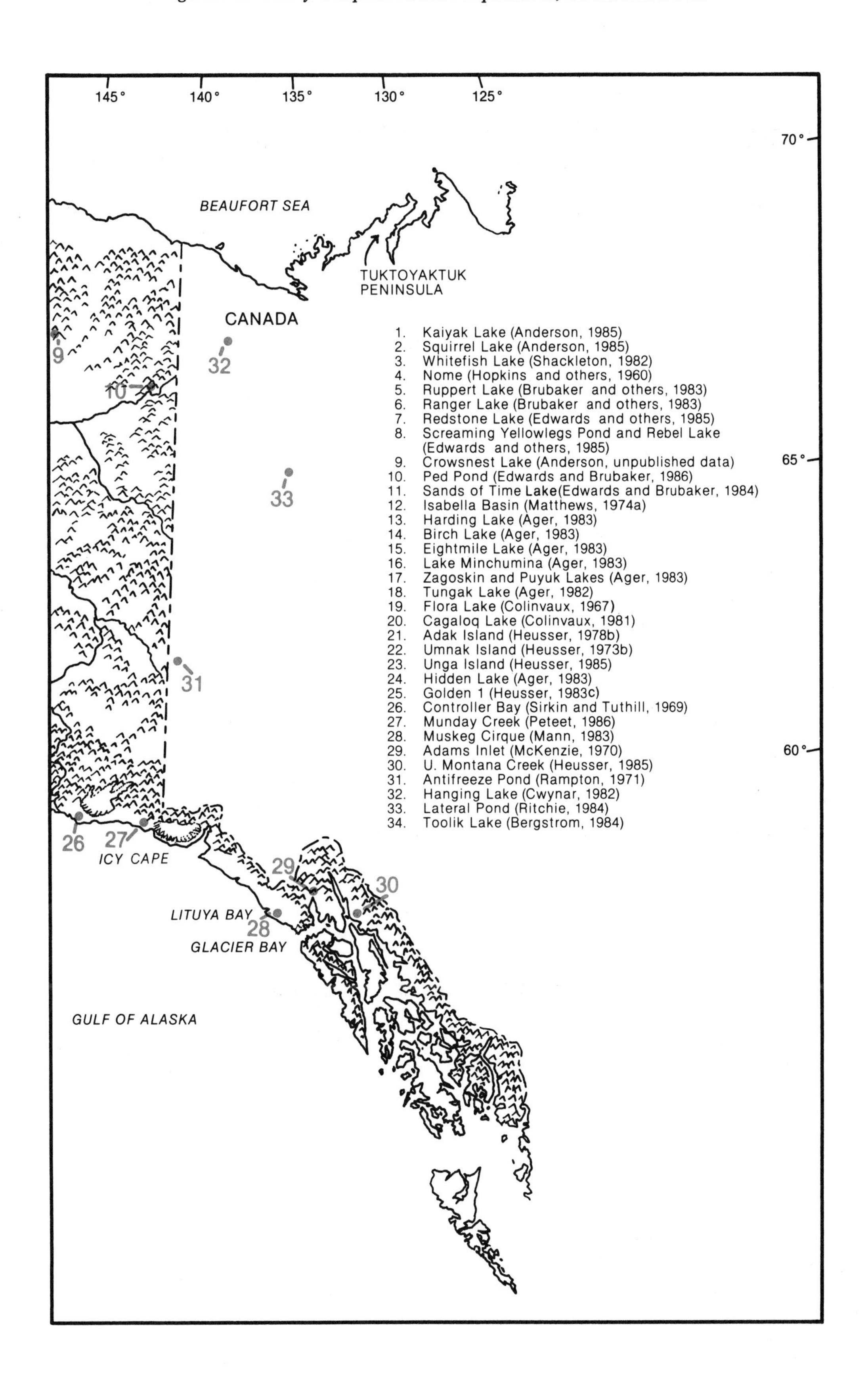
145°
140°
135°
130°
125°
70°
65°
60°
BEAUFORT SEA
TUKTOYAKTUK PENINSULA
CANADA
ICY CAPE
LITUYA BAY
GLACIER BAY
GULF OF ALASKA
9
10
26
27
28
29
30
31
32
33
1. Kaiyak Lake (Anderson, 1985)
2. Squirrel Lake (Anderson, 1985)
3. Whitefish Lake (Shackleton, 1982)
4. Nome (Hopkins and others, 1960)
5. Ruppert Lake (Brubaker and others, 1983)
6. Ranger Lake (Brubaker and others, 1983)
7. Redstone Lake (Edwards and others, 1985)
8. Screaming Yellowlegs Pond and Rebel Lake (Edwards and others, 1985)
9. Crowsnest Lake (Anderson, unpublished data)
10. Ped Pond (Edwards and Brubaker, 1986)
11. Sands of Time Lake(Edwards and Brubaker, 1984)
12. Isabella Basin (Matthews, 1974a)
13. Harding Lake (Ager, 1983)
14. Birch Lake (Ager, 1983)
15. Eightmile Lake (Ager, 1983)
16. Lake Minchumina (Ager, 1983)
17. Zagoskin and Puyuk Lakes (Ager, 1983)
18. Tungak Lake (Ager, 1982)
19. Flora Lake (Colinvaux, 1967)
20. Cagaloq Lake (Colinvaux, 1981)
21. Adak Island (Heusser, 1978b)
22. Umnak Island (Heusser, 1973b)
23. Unga Island (Heusser, 1985)
24. Hidden Lake (Ager, 1983)
25. Golden 1 (Heusser, 1983c)
26. Controller Bay (Sirkin and Tuthill, 1969)
27. Munday Creek (Peteet, 1986)
28. Muskeg Cirque (Mann, 1983)
29. Adams Inlet (McKenzie, 1970)
30. U. Montana Creek (Heusser, 1985)
31. Antifreeze Pond (Rampton, 1971)
32. Hanging Lake (Cwynar, 1982)
33. Lateral Pond (Ritchie, 1984)
34. Toolik Lake (Bergstrom, 1984)

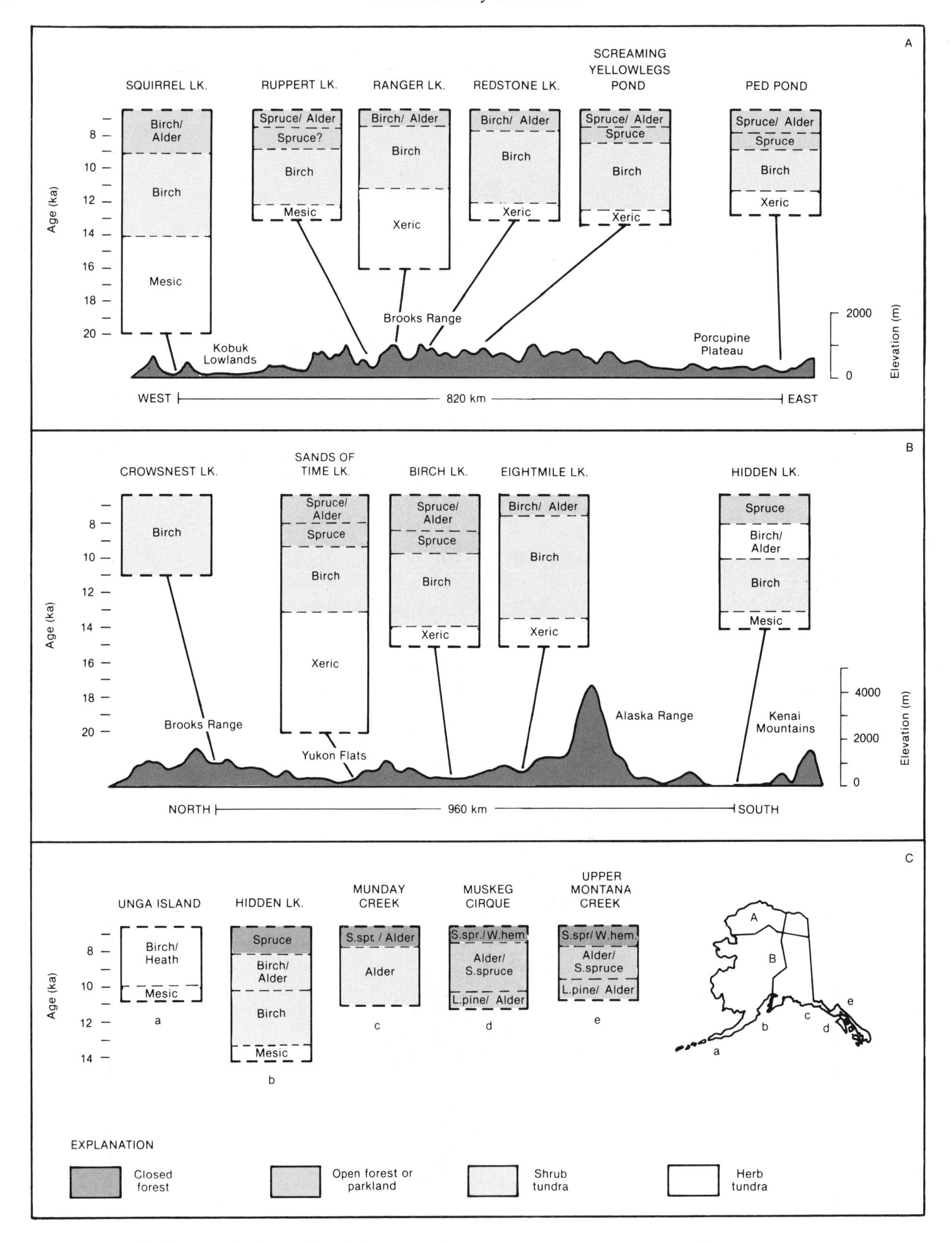

Figure 11. Vegetational changes through time inferred from three transects of pollen sites in Alaska. (Abbreviations: L. pine = lodgepole pine; S. spruce = Sitka spruce; W. hem. = western hemlock).

tion among sites in the Yukon Delta area, a consistent picture of the regional vegetation exists (Ager, 1982, 1983). The full-glacial vegetation was a xeric tundra, which may have been discontinuous, but small amounts of birch and heath pollen suggest that mesic habitats were also present. Birch shrub tundra is recorded after 14 ka and probably was accompanied by an expansion of poplar and willow between 14 and 10 ka. Alder may have been present as early as 12 to 11 ka, but was not important in the vegetation until after 7.5 ka.

The history of alder in western Alaska is puzzling. If alder expansion in Kotzebue Sound was related to increased effective moisture resulting from the flooding of the land bridge, dates of ca. 9 ka would also be expected for the Yukon Delta–Norton Sound region. If the recolonization was controlled by climate alone, however, the late arrival of alder in the Yukon Delta–Norton Sound area argues against a southern seed source. The Kotzebue Sound drainage seems a likely candidate for a refugium, although it is possible that alder was also present in small numbers in the Yukon Delta region and had to await appropriate climatic conditions before it could spread. In the latter case, its migration would have been controlled by a combination of temperature and precipitation.

Bering Strait. The islands of the Bering Sea have yielded exceptionally long records, as might be expected given their location on the unglaciated and exposed portions of the Bering platform. On the Pribiloff Islands, a pollen sequence at Cagaloq Lake spans the last 30,000 yrs (Colinvaux, 1981). The record from Flora Lake on St. Lawrence Island extends back to the Sangamon interglaciation (Colinvaux, 1967). The great age of the sites combined with problematic or few radiocarbon dates, however, make interpretation of the pollen stratigraphy difficult.

The herb zone is dominated by wormwood, grass, and sedge, a typical full-glacial pollen assemblage indicative of xeric tundra. Layers of aeolian sand at Cagaloq Lake are further evidence of an arid climate. Maritime herb tundra developed ca. 11 ka when the Pribiloff Islands were a single land mass located several hundred kilometers off the coast.

Moderate rates of pollen accumulation of spruce, birch, and alder at Cagaloq Lake are used as evidence of a local source for these plants during the full-glacial (Colinvaux, 1981). The Yukon and Kuskokwim deltas have also been suggested as a source for the pollen, but no such forests have been located (see Hopkins and others, 1981).

Southern Alaska

In late Pleistocene time an extensive ice complex covered much of the current land and continental shelf near the Alaska Peninsula and Aleutian Islands (Fig. 11C). Although ice streams, piedmont glaciers, and intermontane ice fields were part of the larger Cordilleran ice sheet, they were not always in phase with extraregional glacial fluctuations. Deglaciation of the Aleutian Islands and Alaska Peninsula occurred between 12 and 10 ka, the Prince William Sound area between 13.5 and 10 ka, and the Alaskan panhandle between 13 and 9.5 ka (Hamilton and Thorson, 1983).

Aleutian and Shumagin islands. Of the few pollen records from southwestern Alaska, the best sequences come from Umnak (Heusser, 1973b), Adak (Heusser, 1978b), and Unga islands (Heusser, 1985). These records begin between 11 and 10 ka, probably shortly after deglaciation. The Adak Island record shows an early phase of sedge, willow, crowberry, and *Lycopodium* at ca. 10 ka, which is replaced at ca. 3 ka by high percentages of sedge and crowberry pollen. The record from Umnak Island features subtle changes in percentages of grass, sedge, and willow pollen. The period from ca. 10 to 8.5 ka is dominated by sedge pollen and is interpreted as a time of cool moist conditions. Between 8.5 and 7.5 ka the climate became warmer, drier, and less windy than before, as evidenced by increased grass and willow pollen. On Unga Island, herbs dominated the vegetation until 9.5 ka when shrub birch and crowberry colonized the area. Alder first grew on the island between 5 and 6 ka. The Unga pollen record suggests a general warming trend between 10 and 5 ka, but paleoclimatic interpretations are complicated by the possible effects of other ecological disturbances, including volcanic eruptions.

Kenai Peninsula to Icy Cape. The oldest pollen records from south-central Alaska begin between ca. 14 and 10 ka, but most are poorly dated and span only the late Holocene (see Ager, 1983; Heusser, 1985). Four exceptions are Hidden Lake on the Kenai Peninsula (Ager, 1983), Golden 1 near Prince William Sound (Heusser, 1983c), Munday Creek near Icy Cape (Peteet, 1986), and peat exposures between Controller Bay and Icy Cape (Fig. 11C; Heusser, 1985; Sirkin and Tuthill, 1969). The Hidden Lake record spans approximately the last 14,000 yrs, whereas the oldest Controller Bay and Icy Cape sections date to 10 and 9 ka. Ager (1983) dismisses a basal date of ca. 14.4 ka on one profile from Controller Bay because the associated pollen assemblages are rich in alder pollen and atypical of late Pleistocene spectra from this area. In addition, the marine clays that were dated may have contained old carbon, thereby producing an erroneous date.

Shortly after deglaciation, mesic tundra or mixed herb-shrub communities colonized the lowlands of the Kenai Peninsula. Poplar invaded the area ca. 10.5 ka, probably growing in scattered stands. Willow too became important, and a mixed poplar-willow scrub forest persisted until ca. 8.5 ka. Alder first invaded the area at ca. 9.5 ka, apparently from the southern coast, and it then spread rapidly to the north and west. Spruce (probably white spruce), on the other hand, migrated rapidly from interior Alaska southward to the Kenai Peninsula. Dates for the arrival of spruce are 9.1 ka in the Gulkana uplands, 8.7 ka in the northern Chugach Mountains, 8 ka in upper Cook Inlet, and 7.8 ka in the central Kenai Peninsula. Coastal forests of Sitka spruce and mountain hemlock developed between ca. 4 to 3 ka as part of a gradual northward spread along the southeastern Alaskan coast.

Golden 1, the oldest record from Prince William Sound, indicates the presence of mesic sedge-dominated tundra with local thickets of willow and alder at 10 ka. Alder increased in abundance until 8.3 ka, when sedge and crowberry became more abundant. At this time alder communities either decreased in size or produced less pollen. Sitka spruce, mountain hemlock, and western hemlock appeared ca. 2.7 ka indicating the development of a conifer woodland at low elevations.

Between 11 and 9 ka mesic sedge-heath tundra covered much of the Icy Cape area. Alder was present by 9 ka and remained dominant until ca. 7.6 ka when Sitka spruce invaded the area. Western hemlock and mountain hemlock first appeared at ca. 3.8 and 3.5 ka, respectively.

Alaska panhandle. The oldest sequence in the panhandle of southeastern Alaska comes from Montana Creek near Juneau (Heusser, 1985) with a date of ca. 10.8 ka. The initial vegetation was lodgepole pine parkland with alder and sedge. Sitka spruce, western hemlock, and mountain hemlock migrated into the area in that order during the early to middle Holocene. A similar but better-dated pollen record from Muskeg Cirque near Lituya Bay (Mann, 1983) shows a decline in pine pollen and an increase in spruce (probably Sitka spruce) and alder at 10.5 ka. Pollen of western hemlock increased at 7.1 ka and that of mountain hemlock at 4.9 ka. Exposures in Adams Inlet in Glacier Bay provide further evidence of a lodgepole pine parkland ca. 11 ka (Ager, 1983, and references therein). A spruce cone dated to ca. 11.2 ka implies that Sitka spruce was present in southeastern Alaska early on. The low pollen percentages, however, suggest that spruce was rare and grew only in scattered stands.

In summary, after local deglaciation, most of southern Alaska was vegetated by sedge-heath-willow tundra except near Lituya Bay, where a lodgepole pine parkland grew. Shrub tundra characterized the Aleutian vegetation throughout the Holocene. In areas of south-central and southeastern Alaska, however, alder was added to the shrub component of the vegetation between ca. 10 and 9 ka. The dominance of alder during the early Holocene is thought to indicate the occurrence of a warm relatively dry interval (Peteet, 1986). The first appearance of Sitka spruce at Lituya Bay is dated to ca. 10.5 ka, at Icy Cape to 7.6 ka, and on the Kenai Peninsula and Prince William Sound to between 4 and 3 ka, suggesting a northwestward migration of this species along the coast. Western hemlock did not appear in southeastern Alaska until 7.1 ka (Lituya Bay), and farther north at Icy Cape not until ca. 3.8 ka. The history of mountain hemlock shows a similar migration history, with arrival dates of 4.9 ka at Lituya Bay, 3.5 ka on Icy Cape, and 2.7 ka at Prince William Sound. The northwestern spread of these conifer species has been related to increased storm frequencies and precipitation (Heusser, 1983c; Peteet, 1986).

SUMMARY AND PALEOCLIMATIC IMPLICATIONS

Widespread patterns in the vegetational history

Vegetation during the full- and late-glacial periods. Traditional opinion has held that during glacial periods, montane tree species of the northwestern U.S. were shifted to lower elevations, where they occupied the basins and lower slopes. Estimates of the altitudinal displacement have varied. In the Olympic Mountains, treeline shifted 1,000 m downslope between ca. 20 and 16 ka coincident with alpine glaciation (Heusser, 1977). On the Yellowstone Plateau, 1,200 m is the best estimate, based on the postglacial occurrence of subalpine and tundralike vegetation at the lowest-elevation site, Cub Lake. The Bighorn records suggest a treeline depression of 650 m, but this is a minimum figure from sites at midslope (Baker, 1983). Swan Lake on the Snake River Plain suggests a lowering of treeline by as much as 1,100 m (Baker, 1983). Alkali Lake in Colorado features subalpine forest at 15 ka, which argues for a treeline depression of as much as 600 m (Markgraf and Scott, 1981).

Although some studies propose a downslope shift and telescoping of entire vegetation zones (Wright and others, 1973), more precise characterization of past communities, provided in large part by macrofossil data, suggests that zonal displacement was the exception rather than the rule. Compared with present-day communities, the glacial vegetation at many sites featured mixtures of lowland and subalpine species that are not found or are very rare on the modern landscape. Certain tree species (e.g., bristlecone pine in the Southwest, Engelmann spruce and lodgepole pine in the Northwest) had broader altitudinal and latitudinal ranges in glacial time than they have today, whereas others were much more restricted.

The nature of nonarboreal communities at high elevations during the glacial maximum is unclear, largely because records are sparse. In the absence of direct data, inferences are drawn from the earliest pollen spectra at deglaciated sites, even though these assemblages represent only the component of the late Pleistocene vegetation that could adapt to new substrates. The low rates of pollen accumulation in early postglacial assemblages, the occurrence of herb pollen typical of the alpine zone today, and the minerogenic nature of late-glacial sediments suggest tundra-like vegetation; but, considering the poor taxonomic resolution of the main pollen contributors (grasses, *Artemisia,* sedges), the possibility of steppe in some high-elevation areas cannot be dismissed.

With the exception of western Washington, it appears that basins in glacial time were covered not by forest, but by communities most closely resembling periglacial steppe (Table 2). Apparently the environment was both too dry and too cold to support widespread forest. In the Puget Trough of western Washington, conditions were less extreme but nevertheless colder and drier than at present, based on the occurrence of Engelmann spruce and lodgepole pine parkland. Only along the Pacific coast, where moisture was not limiting, were mesophytic subalpine trees able to grow in moderate abundance. Greater continentality inferred from paleoecologic data is consistent with the evidence in many interior basins of late Pleistocene patterned ground, ice and sand wedges, and other periglacial features (Mears, 1981; Péwé, 1983).

With forests much contracted from their present distribu-

TABLE 2. SUMMARY OF THE FOSSIL RECORD AND PALEOCLIMATIC INFERENCES BASED ON IT FOR NORTHWESTERN NORTH AMERICA

Northwest United States	Alaska
10 to 8 ka W. Washington: spread of Douglas fir; open forest/ savanna (drier than earlier; maximum summer drought) E. Washington: spread of temperate steppe (drier than earlier; maximum summer drought) N. Rocky Mountains: Douglas fir/pine forest (drier than earlier; maximum summer drought) N. Great Plains: spread of *Artemisia* (drier than earlier; maximum summer drought) Snake River Plain: spread of Chenopods (drier than earlier; maximum summer drought) San Juan Mountains: pine forests at low elevations; maximum elevation of upper treeline (warmer and wetter than earlier)	**9 to 7.5 ka** North: alder-birch shrub tundra, west; spruce parkland, east (warmer and wetter than present) Central: spruce parkland; birch shrub tundra (warmer and wetter than present) N.W. Canada: spruce parkland; birch shrub tundra (warmer and wetter than present) South: birch-alder shrub tundra, central; spruce parkland, east (warmer and drier than present)
12.5 to 10 ka W. Washington: spread of temperate taxa (warmer than earlier) E. Washington: deglaciation; tundralike communities; spread of spruce (warmer than earlier) N. Rocky Mountains: spread of subalpine forest (warmer than earlier) Snake River Plain, Powder River Basin, N. Great Plains: steppe/grassland (warmer than earlier)	**11 to 9 ka** North: birch shrub tundra with poplar (warmer than earlier) Central: willow-poplar scrub (warmer than earlier) N.W. Canada: birch shrub tundra with poplar (warmer than earlier; warmer than present) South: willow-poplar scrub, central; forest/ parkland, east (warmer than earlier; warmer than present)
16 to 12.5 ka W. Washington: deglaciation; mesophytic subalpine parkland (dry, but moister than earlier) E. Washington: periglacial steppe (still cold and dry) N. Rocky Mountains: deglaciation; tundralike (warmer than earlier) San Juan Mountains: deglaciation; spruce forest at low elevations; tundra at high elevations (warmer than earlier)	**14 to 11 ka** North: birch shrub tundra (warmer than earlier) Central: birch shrub tundra (warmer than earlier) N.W. Canada: birch shrub tundra (warmer and wetter than earlier)
18 to 16 ka W. Washington: subalpine parkland (drier and colder than present). E. Washington: periglacial steppe (drier and colder than present) Rocky Mountains: periglacial steppe; tundra (drier and colder than present)	**18 to 14 ka** North: xeric tundra; more mesic in west (colder and drier than present; possibly warmer summers) Central: xeric tundra; more mesic in west (colder and drier than present; possibly warmer summers) N.W. Canada: xeric tundra (colder and drier than present) South: glaciated (colder than present)

tion, subalpine trees apparently grew along the bases of mountains, where the slightly greater moisture, yet tolerable temperatures, provided them refuge from the otherwise harsh environment below and above. Trees may also have grown on unglaciated hillslopes above Pleistocene ice fields and valley glaciers, although the estimated amount of snowline and treeline depression argues against this possibility. The forest zone was apparently discontinuous on many mountain ranges. In the eastern Cascade Range, for example, upper and lower treeline may have merged to eliminate forest altogether, resulting in a broad expanse of alpine tundra and periglacial steppe that extended from high elevations down to the Columbia Basin.

Although the full-glacial vegetation of eastern Beringia has been interpreted both as a uniform steppe or herb tundra and as a mosaic of tundra types, regional analysis of the data shows that there was also an east-to-west gradient. While the growing season was sufficiently cold and dry to support only herb-dominated tundra, the climate was apparently more severe in the east than in the west. In the east, plant growth was limited even at low elevations, whereas in the west, low elevations supported a comparatively lush meadowlike mesic tundra. To some extent, the regional variability is probably related to the relative proximity of Laurentide ice and piedmont glaciers. Eastern Alaska lay closest to the ice sheet and thus experienced colder conditions than the west. The east also was affected more by rainshadows of the Alaska and Brooks ranges. Widespread and thick aeolian deposits, active dune fields, and the sparse vegetation all attest to the aridity of the eastern Beringian climate. Snow cover probably was slight if not absent in northern Alaska and the Yukon Territory, and summer precipitation was less than today.

Postglacial migration of important plant species. The data from the northwestern U.S. and Alaska shed some light on questions of glacial-age refugia and postglacial migration. In the northwestern U.S. there is little evidence to suggest that temperate species were displaced any great distance in glacial time, but in truth the late Pleistocene ranges are unknown in most cases. The spread of western red cedar from southwestern Washington to northwestern British Columbia in the early and middle Holocene implies its survival south of the ice boundary (Hebda and Mathewes, 1984). Likewise, Tsukada (1982) proposes a glacial refugium for coastal populations of Douglas-fir in the southern Puget Trough and Willamette Valley, but this hypothesis is not supported by pollen data from southwestern Washington. Furthermore, the chronology of its appearance in the Puget Trough and Fraser Lowland does not clearly reveal dispersal from the south (Barnosky, 1985a). A refugium for the northern interior populations of Douglas-fir has been proposed in the southern Rocky Mountains on the basis of its late appearance in northern Idaho at 10 ka (Tsukada, 1982). Alternatively, Douglas-fir may have migrated from the Southwest, where it had a broad range in full- and late-glacial time (Spaulding and others, 1983; Wells, 1983). The early postglacial occurrence of Douglas-fir throughout the northwestern U.S. implies, however, that the species was probably better suited to the rigors of a glacial climate than originally thought and survived locally in small populations.

In southern Alaska, white spruce apparently was the only tree species to spread from a northern source. The other species probably migrated from southern Alaska and/or British Columbia, although the location of full-glacial refugia has been debated since the 1930s (see Mann, 1983). Refugia have been proposed on the Aleutian Islands, the exposed Bering shelf, the southern coast of Alaska, Kodiak Island, unspecified areas of the north Pacific coast, the Queen Charlotte Islands, and Lituya Bay. Thus far the strongest case comes from the Queen Charlotte Islands (Warner and others, 1982), where laminated sands and clays dated to 16 ka contain macrofossils of willow, spruce, and various herbaceous and aquatic plants. Mann (1983), however, argues that the postglacial spread of taxa in the Lituya area is part of a migration northward from western Washington. Certainly, the absence of western hemlock or mountain hemlock pollen in late Pleistocene and early Holocene records from southern Alaska is consistent with migration from south of the ice sheet.

The widespread and nearly synchronous appearance of birch shrub tundra in northern and central Alaska as well as on the Kenai Peninsula implies that shrub birch survived the full-glacial interval in small scattered populations. Climatic amelioration in particular must have been critical in promoting its rapid spread. Rise in sea level, flooding of the land bridge, local deglaciation, and increased solar radiation between 14 and 13 ka probably increased summer temperature and precipitation and likely triggered this vegetational change.

In eastern Beringia the location of a full-glacial refugium for spruce has not yet been discovered, although the Tuktoyaktuk Peninsula of northwestern Canada has been proposed (Ritchie, 1984). Hopkins and others (1981) suggest that spruce migrated into Alaska in the early Holocene along the Porcupine and Yukon valleys. However, the fact that spruce appears in the Tanana Valley at 9.5 ka before it is present in the Porcupine River area to the east makes this an unlikely route of westward migration. Between 9.5 and 8.5 ka, spruce spread quickly in all directions from the Tanana Valley, appearing in the Yukon lowlands by 8.5 ka, the central Brooks Range between 7.5 and 6 ka, and the lower Kobuk valley of the western Brooks Range by 4 ka.

The spread of alder follows a more complicated pattern. Alder was present in the Kotzebue Sound area between 10 and 9 ka and may have been a minor component of the vegetation in the Yukon Delta area as early as 14 ka. It expanded rapidly to achieve its modern distribution first in Alaska between 8 and 7 ka and then in northwestern Canada between 6 and 5 ka. This expansion may represent dispersal from coastal areas to the interior or expansion from isolated populations throughout eastern Beringia. The pattern of alder migration appears to have been largely controlled by increased effective moisture (P. Anderson, unpublished data).

Vegetation during the early Holocene. In both the northwestern U.S. and Alaska, the early Holocene was a time of increased summer temperatures and summer drought, although the timing of severe drought differed from region to region and apparently also with altitude. In western Washington, lowland records registered greatest warmth and aridity between ca. 10.5 and 7 ka. So do many pollen and lake-level records from eastern Washington. Two records from the Columbia Basin place the end of this period as early as ca. 8 ka, although in most cases it culminated ca. 6 ka. In a few areas of eastern Washington, northern Idaho, and western Montana, the xerothermic period is delayed until the middle Holocene (between ca. 8.5 and 3.5 ka). High-elevation sites in Yellowstone and southeastern Idaho imply increased summer temperatures and possible drought conditions between ca. 9 and 4 ka, whereas sites from the lower forest there show maximum dryness from 7 to ca. 1.5 ka. The records from the Snake River Plain show drought between 10 and 6 ka, with the exception of Swan Lake, where it occurs in the middle Holocene. Data from the Colorado Rockies indicate early to middle Holocene warming at high elevations, when treeline moved upslope by as much as 100 m. Concurrently, at low elevations, summer precipitation increased in response to more active monsoonal circulation brought on with warming. The northern Great Plains apparently experienced increased drought in the early postglacial after 11.2 ka, although the interval is poorly dated. Data from the Laramie Basin in Wyoming suggest that it may have been over by 5.4 ka.

The vegetational sequence in south-central and southeastern Alaska between 9 and 6 ka shows the replacement of early shrub tundra communities by coastal forest. In southeastern Alaska, conditions between 10 and 7.6 ka were warmer and drier than today's (Peteet, 1986; Heusser and others, 1985). Farther west on the Kenai Peninsula a similar climate prevailed between 10.5 and 8.5 ka. Evidence from the Aleutian Islands, however, argues for a cool, moist period between 10 and 8.5 ka followed by warmer, drier conditions in the middle Holocene.

In northwestern North America, summer temperatures between 11 and 9 ka were significantly warmer than at present. This thermal maximum is best documented in northwestern Canada by pollen data (Ritchie and others, 1983), and in Alaska by macrofossil and geomorphic data (McCulloch and Hopkins, 1966; Hopkins and others, 1981; Carter and others, 1984). The Alaskan pollen records, however, give a somewhat ambiguous indication of warming. Birch continued to dominate with little change from before, although the fact that poplar was also abundant in many areas argues for greater warmth.

The expansion of both spruce and alder at ca. 9 ka suggests continued warm conditions following the postglacial thermal maximum between ca. 11 and 9 ka. Alder and spruce were not important in pollen records until after 9 ka, either because their migration lagged behind the climatic change or summers between 11 and 9 ka were too dry. Ecophysiological studies on white spruce (see Brubaker and others, 1983) and response-surface studies of modern pollen and climate (P. Bartlein, unpublished data) show that modern spruce and alder are sensitive to changes in effective moisture as well as to temperature.

Paleoclimatic Interpretations Provided by Model Simulations

Although the details of the paleoecological record and its interpretation vary from region to region, the patterns show sufficient spatial coherency to suggest that vegetation responded to climatic changes that affected all of northwestern North America. Climate simulation models (Kutzbach, this volume; Kutzbach and Guetter, 1986) suggest some reasons for the similarities. Models such as the National Center for Atmospheric Research (NCAR) Community Climate Model (CCM) described by Kutzbach do not reconstruct climate *per se,* but instead are used to show physically consistent patterns in the response of climate to prescribed changes in a set of controls or boundary conditions. Simulations can provide information on a range of climatic variables such as atmospheric circulation, components of the energy balance, and the seasonal values of standard climate variables. They have utility both in explaining the fossil record and in formulating hypotheses testable with the paleoecologic data.

The accuracy of comparisons between the paleoclimatic inferences drawn from the fossil record and the model simulations is constrained by (1) the qualitative nature of the paleoclimatic inferences, (2) the precision of the fossil record to define times of change, (3) the choice of specific boundary conditions for each simulation, and (4) the inherent inability of the model to properly simulate the present climate. Furthermore, the CCM has a coarse spatial resolution of approximately 4.4° latitude by 7.5° longitude. In western North America, a grid of this size represents the entire western cordillera as a broad dome, approximately 1.5 km at its highest elevation. Smaller-scale topographic features that may have regional climatic significance, such as the Cascade Range and Alaska Range, thus fall below the scale of resolution of the model. Despite these uncertainties and limitations, the broad-scale features of the inferred paleoclimate still compare favorably with the model results.

For the interval from 18 to 6 ka, the experiments incorporated important changes in the boundary conditions of the climate system, including amplification of the seasonal cycle of solar radiation, decreases in ice-sheet size and sea-ice extent, and increases in sea-surface temperature. The principal features of the model's response to these changing boundary conditions are illustrated by the experiments for 18 ka (representing full-glacial conditions—large ice sheet and seasonal cycle of insolation like that at present), 12 ka (late-glacial), and 9 ka (early Holocene—small ice sheet with maximum amplification of the seasonal cycle of solar radiation; Fig. 12). The following discussion of the simulations focuses on the pattern of circulation at the surface and aloft (Fig. 12), and on surface temperature, precipitation, effective moisture (precipitation minus evaporation [P–E]), net radiation, and wind speed and direction averaged over several model grid points in the northwestern U.S. and Alaska (Fig. 13; see Fig. 12 in Kutzbach, this volume, for definition of regions). For comparison, the fossil record and its qualitative paleoclimatic interpretation have been summarized by regionally significant episodes (in Table 2).

At the global scale, the model's response to the insolation variations was greatest in tropical regions, where it was expressed by stronger monsoonal circulation and greater precipitation around 9 ka (Kutzbach and Street-Perrott, 1985). The response to the changes in the surface boundary conditions during the interval 18 to 9 ka was greatest in the middle and high latitudes of the northern hemisphere (Kutzbach and Guetter, 1986; Kutzbach and Wright, 1985). The response to the insolation variations was not unimportant, however.

The 18-ka model simulation shows that atmospheric circulation around North America was greatly influenced by the presence of the Laurentide ice sheet. Because of its height, the ice sheet at its maximum split the jet stream. One branch passed north of the ice sheet, while the second crossed the west coast at Lat. 45°N, farther south than at present (Kutzbach and Guetter, 1986). As a result of low surface temperature over the ice sheet, a "glacial anticyclone" developed, which had maximum expression in January (Fig. 12A).

In the northwestern U.S., easterly winds were stronger and temperatures significantly lower than today's, as were annual precipitation and P–E (Fig. 13). In the fossil record, these cold dry conditions are evidenced by widespread periglacial steppe and tundra over much of the northwestern U.S., and xerophytic subalpine parkland west of the Cascade Range in Washington (Table 2).

In Alaska at 18 ka, the anticyclonic circulation produced stronger southeasterly winds in January than at present (Figs. 12A, 13); although these winds may have not been dune forming

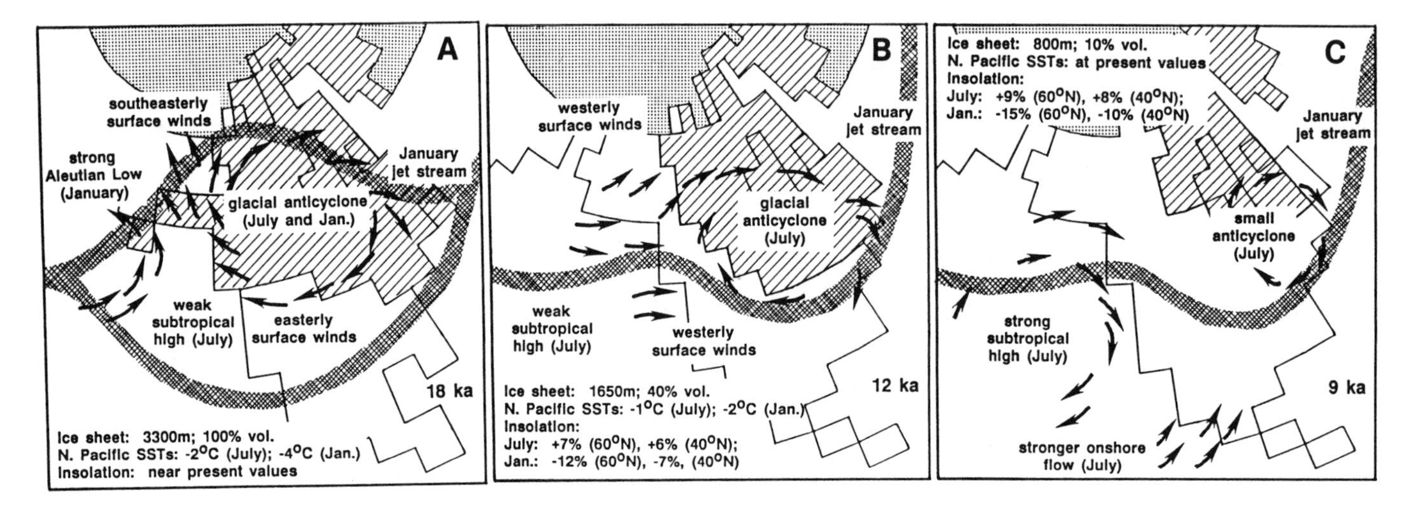

Figure 12. Schematic summary of NCAR Community Climate Model simulations for (A) 18 ka, (B) 12 ka, and (C) 9 ka. Based on Kutzbach (this volume), Kutzbach and Guetter (1986), and unpublished data (Kutzbach, personal communication, 1986).

(for further discussion, see Kutzbach and Wright, 1985). Adiabatic warming as this air descended from the broad dome of ice, coupled with general southerly flow aloft associated with the northern branch of the split jet, resulted in the simulation of January temperatures warmer than at present (see also Barry, 1983). In July, southerly flow was weaker than in January, and simulated temperatures were colder than at present. The dominance of herb tundra at 18 ka corroborates the model results for cooler summers, but because Arctic vegetation responds mostly to temperatures during the short growing season, there is little independent evidence for warmer winters at the glacial maximum. Mild winters may have been critical in permitting megafauna to subsist, however, perhaps even in sparsely vegetated landscapes. July precipitation was significantly lower in the model at 18 ka than at present, but western Alaska was less dry than eastern (model results not shown), which may explain the development of the more mesic herb tundra in the west at that time (Table 2).

Except for increased insolation, the boundary conditions at 15 ka were unchanged from their 18 ka values. Consequently there was little appreciable difference in the response of the model (Fig. 13). In the Pacific Northwest, however, some amelioration is implied in western Washington at 16 ka when parkland changed from xerophytic to mesophytic. While there is little change in temperature, simulated net radiation increased to about 110 percent of its present value in July, in response to the increasing insolation. The culmination and retreat of the Puget and Juan de Fuca lobes also occurred shortly after 15 ka, as did the retreat of glaciers on the Yellowstone Plateau and San Juan Mountains. In Alaska, herb tundra remained essentially unchanged until after 15 ka (Table 2). The patterns of changes in the fossil record are in the same direction as those implied by the model simulations at 12 ka, and it is likely that the assumption of little change in the surface boundary conditions in the model until after 15 ka is unrealistic.

Between 13 ka and 12 ka, temperate taxa first appeared in the subalpine forests of western Washington, and forests spread into areas of the Rocky Mountains that were previously treeless (Table 2). The CCM simulation for 12 ka provides a framework for interpreting these late-glacial events. The North American ice sheet in the model was reduced to half of its 18 ka height and 40 percent of its volume, and consequently its effect on atmospheric circulation was diminished (Kutzbach and Guetter, 1986). Circulation became more zonal over North America as a whole, and the anticyclonic circulation at the surface contracted (Fig. 12B). As a result, the easterly winds in January that prevailed earlier in the northwestern U.S. were diminished in the model at 12 ka, and January temperature increased. July temperature apparently also increased significantly in response to the increased insolation (6 percent greater than today at Lat. 40°N) and the attendant increase in net radiation at the surface. The vegetation shift from glacial to postglacial patterns thus seems related both to the collapse of the ice sheet and to increased July insolation.

A key shift in the vegetation of Alaska occurred at 14 ka with the development of birch shrub tundra (Table 2). Simulated July temperature at 12 ka approached its modern value, as did precipitation and effective moisture (Fig. 13). Just as in the northwestern U.S., these changes probably reflected circulation changes brought on by the collapse of the ice sheet and by increased summer isolation (7 percent greater than today at Lat. 60°N in July). Simulated January temperature decreased significantly at 12 ka in Alaska, in part because of reduced insolation in winter (12 percent less than today at Lat. 60°N in January), and in part because the descending southeasterly flow from the ice

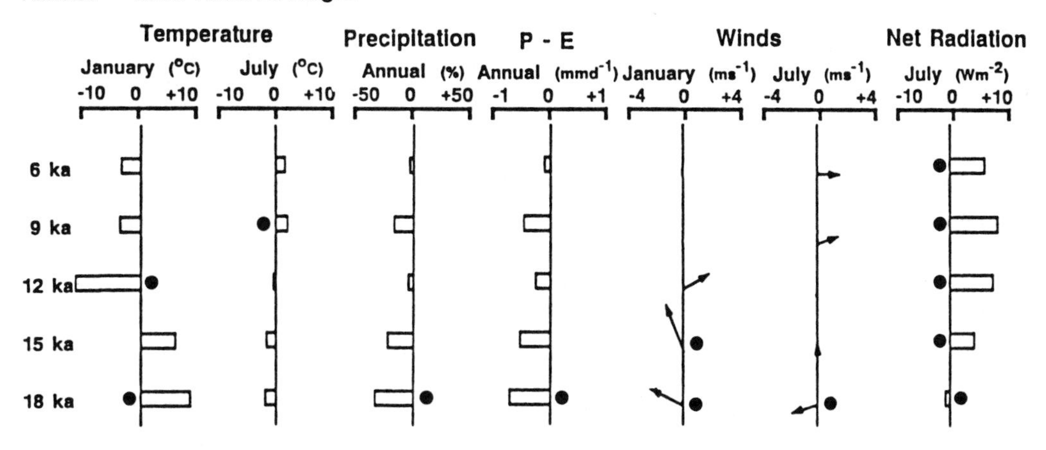

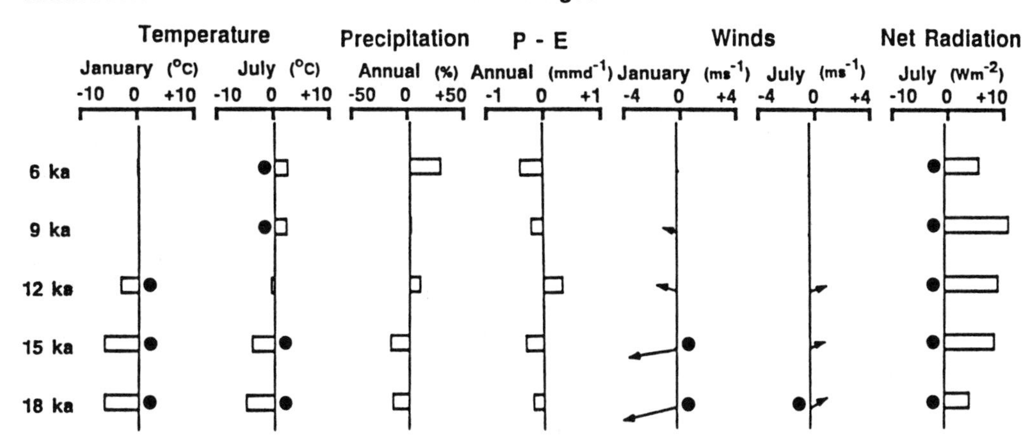

Figure 13. Climate anomalies from 18 to 6 ka simulated by the NCAR Community Climate Model. (Abbreviations: mmd^{-1} = millimeters per day; ms^{-1} = meters per second; Wm^{-2} = Watts per square meter; P-E = precipitation minus evaporation). Values of the climatic variables are averaged over several model gridpoints. Anomalies are calculated as the difference between the paleoclimatic experiment and the modern control simulations. The regions to which these data apply are shown in Fig. 12 of Kutzbach (this volume). Wind anomalies are shown as arrows, with north toward the top of the page. Significant anomalies, relative to the model's natural variability, are indicated by dots. See Kutzbach and Guetter (1986) for further explanation of significance determinations.

sheet was replaced by stronger flow from the North Pacific (still prescribed to be about 2°C cooler than at present). Again, there is little direct evidence to corroborate the January temperatures suggested by the model, but it seems possible that decreased winter temperature triggered the disruption of Guthrie's (1982) tundra mosaic and led to the demise of the megafauna. A similar shift toward long cold winters in Ireland has been used to explain extinction there (Barnosky, 1986). The simulated conditions for July are consistent with the replacement of herb tundra by shrub birch tundra in many parts of Alaska after 14 ka.

At 9 ka, the ice sheet in the model was set at one quarter of its full-glacial elevation and 10 percent of its volume. Insolation in July at 9 ka was 9 percent greater than its present value at Lat. 60°N, and 8 percent greater at Lat. 40°N. As a result, the glacial anticyclone in the model was further reduced in size, its influence essentially confined to northeastern North America in January (although weaker-than-present westerly winds continued in the northwestern U.S.—Figs. 12C and 13). Summer insolation increased and warmed the interior of the continent, while lower surface pressure existed over the southwestern U.S. (Kutzbach and Guetter, 1986). Over northwestern North America and the adjacent North Pacific, net radiation increased in summer and the subtropical high-pressure system expanded (Figs. 12C and 13).

In both the northwestern U.S. and Alaska, simulated July temperature during the early Holocene was higher and effective moisture lower than before or at present (Fig. 13). In the northwestern U.S. the establishment of more xerothermic vegetation, as well as an upward shift of both upper and lower treeline suggest warmer and drier summers during the interval from 10 to 8 ka. In Alaska, increased summer warmth is implied by changes in the composition of the birch shrub tundra between 11 and 9 ka and by the subsequent spread of spruce woodland (Table 2; see also Ritchie and others, 1983; Heusser and others, 1985).

After 9 ka, the surface boundary conditions in the model were set at their modern values, while the amplification of the seasonal cycle of solar radiation remained fairly large (Fig. 13). The simulations for 6 ka thus illustrate the response of climate to the altered radiation regime alone (Kutzbach and Guetter, 1986), whereas the vegetation changes after 9 ka reflect the completion of the transition from glacial to interglacial patterns. In the northwestern U.S., both simulated July temperature and net radiation remained greater than present, while precipitation increased (Fig. 13). Vegetation changes at this time reflect a consistent decrease in summer drought. Simulated precipitation and effective moisture increased to near their modern values in western Alaska but remained lower than present in eastern Alaska (not shown). Increased moisture in western Alaska can be inferred from the range expansion of alder there, and while summer temperatures were probably no longer warmer than those at present, there was sufficient warmth to favor the continued expansion of spruce parkland.

In conclusion, when compared with the features of the simulated climate for northwestern North America, the fossil record reveals a coherent pattern of vegetation change during the last deglaciation. Following a "glacial" interval lasting to ca. 12 ka, the vegetation was reorganized into an "interglacial" pattern, with maximum expression at 9 ka. Circulation variations were closely tied to the size of the ice sheet. From 18 to 12 ka, the Laurentide ice sheet exerted considerable influence on atmospheric circulation over northwestern North America. It split the jet stream, shifting storm tracks southward and generating anticyclonic circulation at the surface that brought easterly winds to the northwestern U.S. and southerly winds to Alaska. At 12 ka, amplification of the seasonal cycle of insolation was already significant and the influence of the ice sheet on circulation was diminishing, so that by 9 ka, greater summertime net radiation and elevated temperatures prevailed. As in eastern North America (Webb and others, this volume), the broad-scale patterns in the vegetational history of northwestern North America reflected the replacement of one climatic control by another of equally large scale.

The response of plant communities to these events should be viewed as part of the long-term development of vegetation in northwestern North America. In this process, the combination of the increasing continentality in the late Cenozoic, the glacial/interglacial oscillations of the Quaternary, and even seasonal variations have all been important. Because ice-sheet size and variations in insolation are critical in determining broad-scale climate, the vegetation of northwestern North America has had to adjust to a continuously varying set of conditions. When the seasonal cycle of insolation was amplified there was relative warmth and least effective moisture. When the ice sheet was large the climate was relatively cool and dry as a result of the glacial anticyclone. Effective moisture was greatest when the seasonal cycle of insolation was less than that of today, because summer evaporation would have been low. The inexorable interplay of these factors has shaped a vegetation and flora that Grichuk (1984) terms "orthoselective"—the extremes of climate during each glacial/interglacial oscillation have resulted in the extinction of some species; whereas the ranges of more robust species, the survivors, have been comparatively little affected. During the last ice age these survivors presumably occupied favorable microhabitats scattered throughout the mountainous landscape. Microhabitats and the populations they sustained grew with regional climatic change to produce the present composition and distribution of forests in northwestern North America.

REFERENCES CITED

Ager, T. A., 1982, Vegetational history of western Alaska during the Wisconsin glacial interval and the Holocene, *in* Hopkins, D. M., Mathews, J. V., Jr., Schweger, C. E., and Young, S. B., eds., Paleoecology of Beringia: New York, Academic Press, p. 75–93.

——, 1983, Holocene vegetational history of Alaska, *in* Wright, H. E., Jr., ed., Late-Quaternary environments of the United States, v. 2: Minneapolis, University of Minnesota Press, p. 128–140.

Ager, T. A., and Brubaker, L. B., 1985, Quaternary palynology and vegetational history of Alaska, *in* Bryant, V. M., Jr., and Holloway, R. G., eds., Pollen records of late-Quaternary North American sediments: Dallas, American Association of Stratigraphic Palynologists Foundation, p. 353–384.

Alley, N. F., 1976, The palynology and paleoclimatic significance of a dated core of Holocene peat, Okanogan valley, southern British Columbia: Canadian Journal of Earth Sciences, v. 13, p. 1131–1141.

Anderson, P. M., 1985, Late Quaternary vegetational change in the Kotzebue Sound area, northwestern Alaska: Quaternary Research, v. 24, p. 307–321.

Andrews, J. T., Carrara, P. E., King, F. B., and Stuckenrath, R., 1975, Holocene environmental changes in the alpine zone, northern San Juan Mountains, Colorado; Evidence from bog stratigraphy and palynology: Quaternary Research, v. 5, p. 173–197.

Baker, R. G., 1976, Late Quaternary vegetation history of the Yellowstone Basin, Wyoming: U.S. Geological Survey Professional Paper 729-E, 48 p.

——, 1983, Holocene vegetation history of the western United States, *in* H. E. Wright, Jr., ed., Late Quaternary environments of the United States, v. 2:

Minneapolis, University of Minnesota Press, p. 109–127.

Barnosky, A. D., 1986, "Big Game" extinction caused by late Pleistocene climatic change: Irish elk *(Megaloceros giganteus)* in Ireland: Quaternary Research, v. 25, p. 128–135.

Barnosky, C. W., 1981, A record of late Quaternary vegetation from Davis Lake, southern Puget Lowland, Washington: Quaternary Research, v. 16, p. 221–239.

—— , 1983, Late Quaternary vegetational and climatic history of southwestern Washington [Ph.D. thesis]: Seattle, University of Washington, 201 p.

—— , 1985a, Late-Quaternary vegetation near Battle Ground Lake, southern Puget Trough, Washington: Geological Society of America Bulletin, v. 96, p. 263–271.

—— , 1985b, Late Quaternary vegetation in the southwestern Columbia Basin, Washington: Quaternary Research, v. 23, p. 109–122.

Barry, R. G., 1983, Late Pleistocene climatology, *in* Porter, S. C., ed., Late Quaternary environments of the United States: Minneapolis, University of Minnesota Press, v. 1, p. 390–403.

Beget, J. E., 1984, Tephrochronology of late Wisconsin and Holocene glacier fluctuations near Glacier Peak, North Cascade Range, Washington: Quaternary Research, v. 21, p. 304–316.

Beiswenger, J. M., 1984, Late Quaternary vegetational history of Grays Lake basin, southeastern Idaho: American Quaternary Association Conference, 8th, Boulder, Abstracts, p. 8.

Benedict, J. B., 1973, Chronology of cirque glaciation, Colorado Front Range: Quaternary Research, v. 3, p. 584–599.

Bergstrom, M. F., 1984, Late Wisconsin and Holocene history of a deep arctic lake, north-central Brooks Range, Alaska [M.S. thesis]: Columbus, Ohio State University, 112 p.

Bright, R. C., 1966, Pollen and seed stratigraphy of Swan Lake, southeastern Idaho; Its relation to regional vegetational history and to Lake Bonneville history: Tebiwa, v. 9, p. 1–47.

Brubaker, L. B., Garfinkel, H. L., and Edwards, M. E., 1983, A late-Wisconsin and Holocene vegetation history from the central Brooks Range; Implications for Alaskan paleoecology: Quaternary Research, v. 20, p. 194–214.

Burkhart, M. R., 1976, Biostratigraphy and late Quaternary vegetation history of the Big Horn Mountains, Wyoming [Ph.D. thesis]: Iowa City, University of Iowa, 70 p.

Carrara, P. E., Mode, W. N., Meyer, R., and Robinson, S. W., 1984, Deglaciation and postglacial timberline in the San Juan Mountains, Colorado: Quaternary Research, v. 21, p. 42–56.

Carrara, P. E., Wilcox, R. E., and Short, S. K., 1986, Deglaciation of the mountainous region of northwestern Montana east of the Continental Divide: Arctic and Alpine Research, v. 18, p. 317–325.

Carter, L. D., Nelson, R. E., and Galloway, J. P., 1984, Evidence for early Holocene increased precipitation and summer warmth in arctic Alaska: Geological Society of America Cordilleran Section Meeting, 80th, Anchorage, Abstracts, p. 274.

Colinvaux, P. A., 1964a, The environment of the Bering land bridge: Ecological Monographs, v. 34, p. 297–329.

—— , 1964b, Origin of ice ages; Pollen evidence from arctic Alaska: Science, v. 145, p. 707–708.

—— , 1967, A long pollen record from St. Lawrence Island, Bering Sea, Alaska: Palaeogeography, Palaeoclimatology, and Palaeoecology, v. 3, p. 29–43.

—— , 1981, Historical ecology in Beringia; The south land bridge coast at St. Paul Island: Quaternary Research, v. 16, p. 18–36.

Cwynar, L. C., 1982, A late-Quaternary vegetation history from Hanging Lake, northern Yukon: Ecological Monographs, v. 52, p. 1–24.

Daubenmire, R. F., 1943, Vegetational zonation in the Rocky Mountains: Botanical Review, v. 9, p. 326–393.

Davis, O. K., 1981, Vegetation migration in southern Idaho during the late Quaternary and Holocene [Ph.D. thesis]: Minneapolis, University of Minnesota, 252 p.

Davis, O. K., Sheppard, J. C., and Robertson, S., 1986, Contrasting climatic histories for the Snake River Plain, Idaho, resulting from multiple thermal maxima: Quaternary Research, v. 26, p. 321–339.

Edwards, M. E., and Brubaker, L. B., 1984, A 23,000 year pollen record from northern interior Alaska: American Quaternary Association Conference, 8th, Boulder, Abstracts, p. 35.

—— , 1986, Late-Quaternary environmental history of the Fishhook Bend area, Porcupine River, Alaska: Canadian Journal of Earth Sciences, v. 23, p. 1765–1773.

Edwards, M. E., and Dunwiddie, P. W., 1985, Dendrochronological and palynological observations on *Populus balsamifera* in northern Alaska, U.S.A.: Arctic and Alpine Research, v. 17, p. 271–278.

Edwards, M. E., Anderson, P. M., Garfinkel, H. L., and Brubaker, L. B., 1985, Late Wisconsin and Holocene vegetation history of the upper Koyukuk region, Brooks Range, Alaska: Canadian Journal of Botany, v. 63, p. 616–626.

Fall, P. L., 1984, Development of modern alpine tundra and subalpine forests during the late Quaternary and their climatic implications: American Quaternary Association Conference, 8th, Boulder, Abstracts, p. 41.

Gennett, J. A., 1977, Palynology and paleoecology of sediments from Blacktail Pond, northern Yellowstone Park, Wyoming [M.S. thesis]: Iowa City, University of Iowa, 74 p.

Gennett, J. A., and Baker, R. G., 1986, A late Quaternary pollen sequence from Blacktail Pond, Yellowstone National Park, Wyoming, U.S.A.: Palynology, v. 10, p. 61–71.

Grichuk, V. P., 1984, Late Pleistocene vegetation history, *in* Velichko, A. A., Wright, H. E., Jr., and Barnosky, C. W., eds., Late Quaternary environments of the Soviet Union: Minneapolis, University of Minnesota Press, p. 155–178.

Gustafson, C. E., 1972, Faunal remains from the Marmes Rockshelter and related archaeological sites in the Columbia Basin [Ph.D. thesis]: Pullman, Washington State University, 151 p.

Guthrie, R. D., 1982, Mammals of the mammoth steppe as paleoenvironmental indicators, *in* Hopkins, D. M., Matthews, J. V., Jr., Schweger, C. E., and Young, S. B., eds., Paleoecology of Beringia: New York, Academic Press, p. 307–326.

—— , 1985, Woolly arguments against the mammoth steppe—a new look at the palynological data: The Quarterly Review of Archaeology, v. 6, p. 9–16.

Hamilton, T. D., and Thorson, R. M., 1983, The Cordilleran ice sheet in Alaska, *in* Porter, S. C., ed., Late-Quaternary environments of the United States, v. 1: Minneapolis, University of Minnesota Press, p. 38–52.

Hansen, H. P., 1947, Postglacial forest succession, climate and chronology in the Pacific Northwest: American Philosophical Society, v. 37, p. 1–130.

Hebda, R. J., and Mathewes, R. W., 1984, Holocene history of cedar and native Indian cultures of the North American Pacific coast: Science, v. 225, p. 711–713.

Henry, C., 1984, Holocene paleoecology of the western Snake River Plain, Idaho [M.S. thesis]: Ann Arbor, University of Michigan, 171 p.

Heusser, C. J., 1973a, Environmental sequence following the Fraser advance of the Juan de Fuca lobe, Washington: Quaternary Research, v. 3, p. 284–304.

—— , 1973b, Postglacial vegetation on Umnak Island, Aleutian Islands, Alaska: Review of Palaeobotany and Palynology, v. 15, p. 277–285.

—— , 1977, Quaternary paleoecology of the Pacific slope of Washington: Quaternary Research, v. 8, p. 282–306.

—— , 1978a, Palynology of Quaternary deposits of the lower Bogachiel River area, Olympic Peninsula, Washington: Canadian Journal of Earth Sciences, v. 15, p. 1568–1578.

—— , 1978b, Postglacial vegetation on Adak Island, Aleutian Islands, Alaska: Bulletin of the Torrey Botanical Club, v. 105, p. 18–23.

—— , 1983a, Vegetational history of the northwestern United States, including Alaska, *in* Porter, S. C., ed., Late-Quaternary environments of the United States, v. 1: Minneapolis, University of Minnesota Press, p. 239–258.

—— , 1983b, Pollen diagrams from the Shumagin Islands and adjacent Alaska Peninsula, southwestern Alaska: Boreas, v. 12, p. 279–295.

—— , 1983c, Holocene vegetation history of the Prince William Sound region, south-central Alaska: Quaternary Research, v. 19, p. 337–355.

——, 1985, Quaternary pollen records from the interior Pacific Northwest coast; Aleutians to the Oregon-California boundary, *in* Bryant, V. M., Jr., and Holloway, R. G., eds., Pollen records of late-Quaternary North American sediments: Dallas, American Association of Stratigraphic Palynologists Foundation, p. 141–165.

Heusser, C. J., Heusser, L. E., and Streeter, S. S., 1980, Quaternary temperatures and precipitation for the northwest coast of North America: Nature, v. 286, p. 702–704.

Heusser, C. J., Heusser, L. E., and Peteet, D. M., 1985, Late-Quaternary climatic change on the American North Pacific coast: Nature, v. 315, p. 485–487.

Hicock, S. R., Hebda, R. J., and Armstrong, J. E., 1982, Lag of the Fraser glacial maximum in the Pacific Northwest; Pollen and macrofossil evidence from western Fraser Lowland, British Columbia: Canadian Journal of Earth Sciences, v. 19, p. 2288–2296.

Hitchcock, C. L., and Cronquist, A., 1973, Flora of the Pacific Northwest: Seattle, University of Washington Press, 730 p.

Hopkins, D. M., 1982, Aspects of the paleogeography of Beringia during the late Pleistocene, *in* Hopkins, D. M., Matthews, J. V., Jr., Schweger, C. E., and Young, S. B., eds., Paleoecology of Beringia: New York, Academic Press, p. 3–28.

Hopkins, D. M., MacNeil, F. S., and Leopold, E. B., 1960, The Coastal Plain at Nome, Alaska; A late Cenozoic type section for the Bering Strait region: Report of the 21st International Geological Congress, Part 4, Copenhagen, p. 44–57.

Hopkins, D. M., Smith, P. A., and Matthews, J. V., Jr., 1981, Dated wood from Alaska and the Yukon; Implications for forest refugia in Beringia: Quaternary Research, v. 15, p. 217–249.

Hopkins, D. M., Matthews, J. V., Jr., Schweger, C. E., and Young, S. B., 1982, Paleoecology of Beringia: New York, Academic Press, 489 p.

Hultén, E., 1968, Flora of Alaska and neighboring territories: Stanford, University of Stanford Press, 1,008 p.

Janssens, J. A., and Barnosky, C. W., 1985, Late Pleistocene and early Holocene bryophytes from Battle Ground Lake, Washington, U.S.A.: Review of Palaeobotany and Palynology, v. 46, p. 97–116.

Küchler, A. W., 1964, Potential natural vegetation of the conterminous United States (map): American Geographical Society Special Publication, v. 36, 39 p., map scale 1:3,168,000.

Kutzbach, J. E., and Guetter, P. J., 1986, The influence of changing orbital parameters and surface boundary conditions on climate simulations for the past 18,000 years: Journal of the Atmospheric Sciences, v. 43, p. 1726–1759.

Kutzbach, J. E., and Street-Perrott, F. A., 1985, Milankovitch forcing of fluctuations in the level of tropical lakes from 18 to 0 kyr B.P.: Nature, v. 317, p. 130–134.

Kutzbach, J. E., and Wright, H. E., Jr., 1985, Simulation of the climate of 18,000 yr B.P.: Results for the North American/North Atlantic/European sector: Quaternary Science Reviews, v. 4, p. 147–187.

Legg, T. E., and Baker, R. G., 1980, Palynology of Pinedale sediments, Devlins Park, Boulder County, Colorado: Arctic and Alpine Research, v. 12, p. 319–333.

Leopold, E. B., Nickmann, R. J., Hedges, J. I., and Ertel, J. R., 1982, Pollen and lignin records of late Quaternary vegetation, Lake Washington: Science, v. 218, p. 1305–1307.

Livingstone, D. A., 1955, Some pollen profiles from arctic Alaska: Ecology, v. 36, p. 587–600.

——, 1957, Pollen analysis of a valley fill near Umiat, Alaska: American Journal of Science, v. 255, p. 254–260.

Mack, R. N., and Bryant, V. M., Jr., 1974, Modern pollen spectra from the Columbia Basin, Washington: Northwest Science, v. 48, p. 183–194.

Mack, R. N., Bryant, V. M., Jr., and Fryxell, R., 1976, Pollen sequence from the Columbia Basin, Washington; reappraisal of postglacial vegetation: American Midland Naturalist, v. 95, p. 390–397.

Mack, R. N., Rutter, N. W., and Valastro, S., 1978a, Late Quaternary pollen record from the Sanpoil River valley, Washington: Canadian Journal of Botany, v. 56, p. 1642–1650.

Mack, R. N., Rutter, N. W., Valastro, S., and Bryant, V. M., Jr., 1978b, Late Quaternary vegetation history at Waits Lake, Colville River valley, Washington: Botanical Gazette, v. 139, p. 499–506.

Mack, R. N., Rutter, N. W., Bryant, V. M., Jr., and Valastro, S., 1978c, Late Quaternary pollen record from Big Meadow, Pend Oreille County, Washington: Ecology, v. 59, p. 956–965.

——, 1978d, Reexamination of postglacial vegetation history in northern Idaho: Hager Pond, Bonner County: Quaternary Research, v. 10, p. 241–255.

Mack, R. N., Rutter, N. W., and Valastro, S., 1979, Holocene vegetation history of the Okanogan valley, Washington: Quaternary Research, v. 12, p. 212–225.

——, 1983, Holocene vegetational history of the Kootenai River valley, Montana: Quaternary Research, v. 20, p. 177–193.

Maher, L. J., Jr., 1961, Pollen analysis and post-glacial vegetation history in the Animas Valley [Ph.D. thesis]: Minneapolis, University of Minnesota, 85 p.

——, 1963, Pollen analysis of surface materials from the southern San Juan Mountains, Colorado: Geological Society of America Bulletin, v. 74, p. 1485–1504.

——, 1972, Absolute pollen diagram of Redrock Lake, Boulder County, Colorado: Quaternary Research, v. 2, p. 531–553.

Mann, D. H., 1983, The Quaternary history of the Lituya glacial refugium, Alaska [Ph.D. thesis]: Seattle, University of Washington, 268 p.

Markgraf, V., and Lennon, T., 1986, Paleoenvironmental history of the last 13,000 years of the eastern Powder River Basin, Wyoming, and its implication for prehistoric cultural patterns: Plains Anthropologist, v. 31, p. 1–12.

Markgraf, V., and Scott, L., 1981, Lower timberline in central Colorado during the past 15,000 yr: Geology, v. 9, p. 231–234.

Martin, J. E., Barnosky, A. D., and Barnosky, C. W., 1982, Fauna and flora associated with the West Richland mammoth from the Pleistocene Touchet Beds in south-central Washington: Research Report of the Thomas Burke Memorial Washington State Museum, no. 3, 61 p.

Mathewes, R. W., 1985, Paleobotanical evidence for climatic change in southern British Columbia during late-glacial and Holocene time, *in* Harington, C. R., ed., Climatic change in Canada 5; Critical periods in the Quaternary climatic history of northern North America: Syllogeus, v. 55, p. 397–422.

Matthews, J. V., Jr., 1974a, Wisconsin environment of interior Alaska; Pollen and macrofossil analysis of a 27-meter core from Isabella basin (Fairbanks, Alaska): Canadian Journal of Earth Sciences, v. 11, p. 828–841.

——, 1974b, Quaternary environments at Cape Deceit (Seward Peninsula, Alaska); Evolution of a tundra ecosystem: Geological Society of America Bulletin, v. 85, p. 1353–1384.

McCulloch, D. S., and Hopkins, D. M., 1966, Evidence for an early Recent warm interval in northwestern Alaska: Geological Society of America Bulletin, v. 77, p. 1089–1108.

McKenzie, G. D., 1970, Glacial geology of Adams Inlet, southeastern Alaska: Institute of Polar Studies Report, v. 25, 121 p.

Mears, B., Jr., 1981, Periglacial wedges and the late Pleistocene environment of Wyoming's intermontane basins: Quaternary Research, v. 15, p. 171–198.

Mehringer, P. J., Jr., 1985, Late-Quaternary pollen records from the interior Pacific Northwest and Northern Great Basin of the United States, *in* Bryant, V. M., Jr., and Holloway, R. G., eds., Pollen records of Late-Quaternary North American sediments: Dallas, American Association of Stratigraphic Palynologists Foundation, p. 167–189.

Mehringer, P. J., Jr., Arno, S. F., and Petersen, K. L., 1977, Postglacial history of Lost Trail Pass Bog, Bitterroot Mountains, Montana: Arctic and Alpine Research, v. 9, p. 345–368.

Mehringer, P. J., Jr., Sheppard, J. C., and Foit, F. F., Jr., 1984, The age of Glacier Peak tephra in west-central Montana: Quaternary Research, v. 21, p. 36–41.

Nelson, R. E., 1982, Late Quaternary environments of the western Arctic Slope, Alaska [Ph.D. thesis]: Seattle, University of Washington, 141 p.

Nelson, R. E., and Coope, G. R., 1982, A late-Pleistocene insect fauna from Seattle, Washington: American Quaternary Association Conference, 7th, Seattle, Abstracts, p. 146.

Nichols, H., 1982, Review of late Quaternary history of vegetation and climate in the mountains of Colorado, *in* Halfpenny, J. C., ed., Ecological Studies in the Colorado Alpine; A festschrift for John W. Marr: Institute of Arctic and Alpine Research Occasional Paper No. 37, University of Colorado, Boulder, p. 27–33.

Nickmann, R. J., 1979, The palynology of Williams Lake fen, Spokane County, Washington [M.S. thesis]: Cheney, Eastern Washington State University, 57 p.

Nickmann, R. J., and Leopold, E., 1984, A postglacial pollen record from Goose Lake, Okanogan County, Washington; Evidence for an early Holocene cooling: Chief Joseph Summary Report, Office of Public Archeology, University of Washington, Seattle, p. 131–148.

Peteet, D. M., 1986, Modern pollen rain and vegetational history of the Malaspina glacier district, Alaska: Quaternary Research, v. 25, p. 100–120.

Petersen, K. L., and Mehringer, P. J., Jr., 1976, Postglacial timberline fluctuations, La Plata Mountains, southwestern Colorado: Arctic and Alpine Research, v. 8, p. 275–288.

Petersen, K. L., Mehringer, P. J., and Gustafson, C. E., 1983, Late-glacial vegetation and climate at the Manis Mastodon site, Olympic Peninsula, Washington: Quaternary Research, v. 20, p. 215–231.

Péwé, T. L., 1983, The periglacial environment in North America during Wisconsin time, *in* Porter, S. C., ed., Late-Quaternary environments of the United States, v. 1: Minneapolis, University of Minnesota Press, p. 157–189.

Porter, S. C., Pierce, K. L., and Hamilton, T. D., 1983, Late Wisconsin mountain glaciation in the western United States, *in* Porter, S. C., ed., Late-Quaternary environments of the United States, v. 1: Minneapolis, University of Minnesota, p. 71–111.

Rampton, V. N., 1971, Late Quaternary vegetational and climatic history of the Snang-Klutlan area, southwestern Yukon Territory, Canada: Geological Society of America Bulletin, v. 82, p. 959–978.

Ritchie, J. C., 1984, Past and present vegetation of the far northwest of Canada: Toronto, University of Toronto Press, 251 p.

Ritchie, J. C., Cwynar, L. C., and Spear, R. W., 1983, Evidence from northwest Canada for an early Holocene Milankovitch thermal maximum: Nature, v. 305, p. 126–128.

Schweger, C. E., 1981, Chronology of late-glacial events from the Tangle Lakes, Alaska Range: Arctic Anthropology, v. 18, p. 97–101.

—— , 1982, Late Pleistocene vegetation of eastern Beringia; Pollen analysis of dated alluvium, *in* Hopkins, D. M., Matthews, J. V., Jr., Schweger, C. E., and Young, S. B., eds., Paleoecology of Beringia: New York, Academic Press, p. 95–112.

Shackleton, J., 1982, Environmental histories from Whitefish and Imuruk lakes, Seward Peninsula, Alaska: Ohio State University Institute of Polar Studies, Report 76, 50 p.

Sirkin, L. A., and Tuthill, S. J., 1969, Late Pleistocene palynology and stratigraphy of Controller Bay region, Gulf of Alaska: Études sur le Quaternaire dans le Monde, International Association for Quaternary Research Congress, 8th, p. 197–208.

Spaulding, W. G., Leopold, E. B., and Van Devender, T. R., 1983, Late Wisconsinan paleoecology of the American Southwest, *in* Porter, S. C., ed., Late Quaternary environments of the United States, v. 1: Minneapolis, University of Minnesota Press, p. 259–293.

Tsukada, M., 1982, *Pseudotsuga menziesii* (Mirb.) Franco; Its pollen dispersal and late Quaternary history in the Pacific Northwest: Japanese Journal of Ecology, v. 32, p. 159–187.

Tsukada, M., Sugita, S., and Hibbert, D. M., 1981, Paleoecology of the Pacific Northwest; I. Late Quaternary vegetation and climate: Verhandlungen der Internationalen Vereingung für theoretische und angewandte Limnologie, v. 21, p. 730–737.

Viereck, L. A., and Little, E. L., Jr., 1975, Atlas of the United States trees, v. 2, Alaskan trees and common shrubs: U.S. Department of Agriculture Forest Service Miscellaneous Publication 1293, 127 p.

Waddington, J.C.B., and Wright, H. E., Jr., 1974, Late Quaternary vegetational changes on the east side of Yellowstone Park, Wyoming: Quaternary Research, v. 4, p. 175–184.

Waitt, R. B., Jr., and Thorson, R. M., 1983, The Cordilleran ice sheet in Washington, Idaho, and Montana, *in* Porter, S. C., ed., Late Quaternary environments of the United States, v. 1: Minneapolis, University of Minnesota Press, p. 53–70.

Waitt, R. B. Jr., Yount, J. C., and Davis, P. T., 1982, Regional significance of an early Holocene moraine in the Enchantment Lakes basin, North Cascade Range, Washington: Quaternary Research, v. 17, p. 191–210.

Warner, B. G., Mathewes, R. W., and Clague, J. J., 1982, Ice-free conditions on the Queen Charlotte Islands, British Columbia, at the height of late Wisconsin glaciation: Science, v. 218, p. 675–677.

Webb, T., III, Wright, H. E., Jr., and Cushing, E. J., 1983, Holocene changes in the vegetation of the Midwest, *in* Wright, H. E., Jr., ed., Late-Quaternary environments of the United States, v. 2: Minneapolis, University of Minnesota Press, p. 142–165.

Weber, W. A., 1976, Rocky Mountain flora: Boulder, Colorado Associated University Press, 479 p.

Wells, P. V., 1983, Paleobiogeography of montane islands in the Great Basin since the last glaciopluvial: Ecological Monographs, v. 53, p. 341–382.

Westgate, J. A., and Evans, M. E., 1978, Compositional variability of Glacier Peak tephra and its stratigraphic significance: Canadian Journal of Earth Sciences, v. 15, p. 1554–1567.

Wilson, M. J., 1984, Modern and Holocene environments of the North Slope of Alaska [M.A. thesis]: Boulder, University of Colorado, 253 p.

Wright, H. E., Jr., Bent, A. M., Hansen, B. S., and Maher, L. J., Jr., 1973, Present and past vegetation of the Chuska Mountains, northwest New Mexico: Geological Society of America Bulletin, v. 84, p. 1155–1180.

Manuscript Accepted by the Society February 2, 1987

ACKNOWLEDGMENTS

This paper benefited greatly from discussions with John E. Kutzbach, Thompson Webb III, Herbert E. Wright, Jr., Linda B. Brubaker, and James C. Ritchie. Richard G. Baker, David M. Hopkins, and Stanton A. Cook provided thoughtful reviews. We also extend our appreciation to Peter Guetter for furnishing model output, to Nancy Perkins, Gail Richards, and Richard Reanier for preparing illustrations, and to Elizabeth Hill for typing the manuscript. Funding for this study came from the National Science Foundation under these grants: Ecology BSR-8311473, Polar Programs DPP-8403598, DPP-810686, DPP-8413620, and Climate Dynamics ATM-8118870.

Printed in U.S.A.

The Geology of North America
Vol. K-3, North America and adjacent oceans during the last deglaciation
The Geological Society of America, 1987

Chapter 15

Vegetation history of the deserts of southwestern North America; The nature and timing of the Late Wisconsin–Holocene transition

Thomas R. Van Devender
Arizona-Sonora Desert Museum, 2021 N. Kinney Road, Tucson, Arizona 85743
Robert S. Thompson
U.S. Geological Survey, MS 919, Federal Center, Denver, Colorado 80225
Julio L. Betancourt
U.S. Geological Survey, 300 W. Congress, FB Box 44, Tucson, Arizona 85701

INTRODUCTION

The Southwest is a region of mountains and basins in the southwestern United States and northern Mexico. Four major deserts are found in rainshadows of the Sierra Madre Oriental and Occidental, which enclose the Mexican Plateau, and of the Rocky Mountains to the east and the Sierra Nevada/Transverse Ranges to the west. The climate of these now-arid regions during the Wisconsin was greatly modified, although North America's continental glaciers were distant. The global thermal regimes and general circulation of the atmosphere were evident in advances and retreats of alpine glaciers, fluctuating lake levels, and changing vegetation. The climate of arid North America is especially sensitive to changes in the general circulation of the atmosphere because important climatic variables, including rainfall and freezing temperatures, reflect air masses imported from both north-temperate and subtropical regions. In this chapter we summarize the evidence for vegetational and climatic history for the last 22,000 yr in the Southwest, primarily from the fossil middens of packrats (*Neotoma* sp.)

Antiquity of southwestern vegetation

The North American desert floras are rich in endemic plants. Many of them have unusual or even bizarre adaptations to aridity, including the succulence of cacti, the chlorophyllous bark of microphyllous leguminous trees, and the drought deciduous stems of shrubs. The evolution of these desert plants from subtropical and temperate relatives probably began in the late Miocene 5–8 million years ago (Axelrod, 1979); surely desertscrub communities have occupied our modern desert regions for a very long time.

However, early palynological studies in the Southwest in such places as San Agustin Plains, New Mexico (Clisby and Sears, 1956), Willcox Playa, Arizona (Martin, 1963), and Tule Springs, Nevada (Mehringer, 1967), suggest that during the late Wisconsin glacial period, forest and woodland communities grew as much as 1,200 m lower than today, replacing desertscrub communities over most of the elevational gradient (Martin and Mehringer, 1965). Since 1964, taxonomic and ecologic resolution from the study of plants preserved in packrat middens has allowed detailed vegetation reconstructions from the deserts themselves. Macrofossil studies show that plant species responded to climatic change in an individualistic manner (Spaulding and others, 1983), and that associations of species in modern communities are primarily manifestations of past and recent climatic regimes. Species in desertscrub communities weathered the ice ages in different geographical or elevational ranges, on different slopes, or in woodland communities. The geographic extent of late Wisconsin woodland and desertscrub communities and the timing of the development of the modern vegetation are presented in the regional summaries that follow.

Paleoclimatic hypotheses

Reconstructions of Wisconsin climates vary. Galloway (1970, 1983), Brakenridge (1978), and Dohrenwend (1984) inferred from past distributions of glacial and periglacial features in the highlands that full-glacial mean annual temperatures were from 7 to 11°C colder than those of today, with little or no increase in mean annual precipitation. Others interpreted high lake levels as evidence for increased rainfall, with only a moderate decrease in mean annual precipitation (Antevs, 1948; Mifflin and Wheat, 1979; Snyder and Langbein, 1962). Benson (1981)

Van Devender, T. R., Thompson, R. S., and Betancourt, J. L., 1987, Vegetation history of the deserts of southwestern North America; The nature and timing of the Late Wisconsin-Holocene transition, *in* Ruddiman, W. F., and Wright, H. E., Jr., eds., North America and adjacent oceans during the last deglaciation: Boulder, Colorado, Geological Society of America, The Geology of North America, v. K-3.

concluded that a reduction in mean annual temperature coupled with an increase in cloud cover maintained pluvial Lake Lahontan. Van Devender and Spaulding (1979) thought plant macrofossils from late Wisconsin packrat middens reflected increased precipitation, cool summers, and mild winters. Spaulding (1985a) and Spaulding and others (1984) later interpreted midden plant remains from southern Nevada as indicating both lower temperatures (7°C below modern mean annual temperature) and increased winter rainfall. Wells (1979) saw the packrat midden data as reflecting increased summer rainfall, with little reduction in temperature.

The notions on Holocene climates are equally diverse. Antevs (1948), partly on the basis of pluvial lake studies in the Great Basin, argued that his Altithermal Period (our middle Holocene and the Atlantic Period of Europe) was characterized by hot and dry conditions across the entire western United States from 7.5 to 4.0 ka. In contrast, Martin (1963) believed that this was a period of plentiful summer precipitation for the monsoonal portions of the Southwest.

Variations in the earth's orbital parameters have caused changes in the distribution of solar radiation on earth throughout the Quaternary. In the early and middle Holocene, these variations may have raised summer temperatures and intensified monsoons in many parts of the world (Kutzbach and Otto-Bliesner, 1982), with related effects on circulation patterns and plant distributions in others (Wright, 1981). Spaulding and Graumlich (1986) and Cole (1985) suggest that maximum solar radiation in the latest Wisconsin and early Holocene enhanced monsoonal precipitation in the Mohave Desert in Nevada and Arizona. Herein we review the late Quaternary vegetational records from the Southwest and evaluate the various hypotheses.

METHODS

Packrat midden analysis

Packrats or woodrats are medium-sized cricetid rodents in the genus *Neotoma*. Approximately 21 species are found from British Columbia to Central America in many habitats from sea level to at least 3,350 m elevation. All species collect various plant materials for food or construction of a house or den (Finley, 1958). When packrats live in dry rockshelters, portions of their houses can become indurated by urine into hard, dark organic deposits termed middens. These deposits are a fortuitous source of well-preserved, perishable organic remains from the North American deserts, woodlands, and forests. Abundant plant macrofossils provide an excellent sample of the local vegetation within 30 to 50 m of the rockshelter. Pollen is abundant (up to a million grains/gram of matrix) but yields less detailed information than the plant macrofossils because of lower taxonomic resolution and uncertainty as to source area (Thompson, 1985). Bones and arthropod exoskeletons are present but less common than the plant remains.

Packrat middens can be preserved beyond the limits of radiocarbon dating in rockshelters if they remain dry. They have been found in many rock types, although limestone in winter-dry climates like southern Nevada may be ideal for midden preservation. Middens must be collected with care to separate stratigraphic units and to clean outer surfaces of contaminants. Most workers disaggregate the samples in water to separate and clean the fossils, although Wells (1976) prefers to examine the plants on the midden surface. A great many more plant macrofossils, plant taxa, and animal remains are recovered with washing. After drying, the plant materials are sorted and analyzed with various semi-quantitative and quantitative methods (Van Devender, 1973; Spaulding, 1981, 1985a; Cole, 1981; Betancourt, 1984). The diagrams presented in this chapter utilize an internal relative-abundance scale ranging from abundant to rare (5–1). A typical sample, weighing 800 to 1,000 gm initially and 75 to 150 gm after washing, may contain more than 1,000 identifiable plant specimens. Relative abundances are assigned as follows: 1 specimen = 1, 2 to 29 specimens = 2, 30 to 99 specimens = 3, 100 to 200 specimens = 4, more than 200 = 5. The ranges of relative-abundance classes for smaller samples are smaller.

Radiocarbon dating

More than 1,100 radiocarbon ages have been determined on packrat midden materials, many on individual species (Webb, 1985; Webb and Betancourt, 1987). In recent years more middens were dated on packrat fecal pellets and midden debris to conserve identifiable plant materials for other uses, including taxonomic study and stable-isotope or biochemical analyses. Since 1984, routine dating by tandem accelerator mass spectrometer (TAMS) of samples as small as ten milligrams has become a powerful technique to detect contamination, verify unusual community associations, document Holocene dispersal of plants, and refine biochronological boundaries marked by the disappearance or appearance of climatically sensitive species (Van Devender and others, 1985).

Most middens have a single radiocarbon date that can be associated with all of the species in the assemblage if the sample was from a discrete stratigraphic context and outer layers were removed. Badly mixed samples are discarded. Concordant multiple dates from a sample are averaged (Long and Rippeteau, 1974).

Packrat midden chronologies

With the discovery that ancient packrat middens could provide detailed records of local vegetation (Wells and Jorgensen, 1964), surveys of the history of vegetation and climate of the North American deserts were initiated. Researchers found Wisconsin-age pinyon pines, junipers, shrub oaks, and montane conifers in desert lowlands (Wells, 1966; Wells and Berger, 1967; Mehringer and Ferguson, 1969; Van Devender, 1973; Van Devender and Spaulding, 1979; Spaulding and others, 1983). Xeric woodland communities lingered in the deserts for several thousand years into the early Holocene (Van Devender, 1977). Indeed, the locations of ice-age desert communities devoid of woodland trees and shrubs are only now coming to light (Cole, 1986; Spaulding, 1983).

Since 1975 packrat midden analysis has shifted toward studies where plant macrofossil sequences are developed into detailed local chronologies of vegetation and climate (Spaulding, 1981; Cole, 1981, 1982; Betancourt and Van Devender, 1981; Thompson, 1984; Van Devender and others, 1984). Radiocarbon-dated midden assemblages are discrete points in time unlike the continuous stratigraphic data available from pollen studies of sedimentary basins. However, we are confident that midden chronologies from specific localities adequately record the major details of local vegetational history.

Reconstructing paleoclimates

Changes in the geographic distribution of a plant can be compared to maps of climatic parameters as indicators of both the direction and the magnitude of climatic change. The modern geographic ranges of most of the important trees, shrubs, and succulents are well known.

The climate of a modern analogue site, where the present vegetation is similar to a fossil assemblage, can give some idea of the paleoclimate represented by the assemblage (Spaulding, 1985a). However, the modern analogues usually differ from the assemblages in some significant aspect, and glacial climatic regimes were inherently different from those of today. The perils of the analogue approach to pollen analysis discussed by Bryson (1985) apply to packrat midden analysis as well. In general each plant species should reflect climate independently, and a reconstruction should be a compilation of these inferences.

Several authors have calculated the amount of elevational depression of a plant species relative to its modern limits. Using the modern environmental lapse rates for temperature and precipitation, they have estimated the corresponding climatic differences (Van Devender, 1973; Spaulding, 1981; Betancourt, 1984). This approach assumes similarity between present and past climatic gradients with elevation, which is also unlikely.

The best paleoclimatic inferences are those based on phenological responses or physiological tolerances of individual species. The phenologies of most southwestern plants are fairly well known. Unfortunately, the physiological tolerances of many key plants have not been studied. Qualitative estimates of paleoclimates from packrat midden assemblages are quite robust, because of identification to the species level and strong differentiation of climate along geographic and elevational gradients.

Chronology of vegetation change

As midden chronologies developed, it became apparent that the timing of disappearances or appearances of trees, shrubs, or long-lived succulents was remarkably similar across the Southwest. Van Devender and Spaulding (1979) proposed a chronology for the last 22,000 yr, placing the end of the Late Wisconsin (LW) at about 11.0 ka, the early Holocene (EH) from 11.0 to ca. 8.0 ka, the middle Holocene (MH) from 8.0 to ca. 4.0 ka and the late Holocene (LH) from ca. 4.0 ka to present. These bioclimatic divisions, their boundaries, and comparisons between full-glacial (22.0 to 14.0 ka) and late-glacial (14.0 to 11.0 ka) vegetation are evaluated below.

The disappearance of a plant in a midden sequence is a better indicator of climatic change than the appearance of a plant because population mortality after a major climatic change can be relatively rapid. Although relict populations can persist after a climatic change, few examples have come to light in the packrat midden record. The disappearance of various pinyons and their mesic woodland associates from the Chihuahuan, Sonoran, and Mohave Desert lowlands at about 11.0 ka is a good marker for the end of the Wisconsin (Van Devender and Spaulding, 1979; Lanner and Van Devender, 1981; Van Devender, 1986a). Although the first appearance of a plant in a packrat-midden sequence can indicate a climatic change, the timing is subject to the different dispersal capabilities and migrational distances for the individual species. Although dispersal along an elevational gradient after a climatic change could be rapid, dispersal along latitudinal gradients may lag because of differential migration rates. In either case, immigrating plants may be delayed due to competition if plants with similar ecological adaptations occupy an area first. Cole (1985) concluded that lags in departures and arrivals resulted in lower species richness in Grand Canyon plant communities at the Late Wisconsin–Holocene transition. His intriguing hypothesis of vegetational inertia during a climatic change needs to be evaluated with more extensive data sets from various ecological settings. The apparent synchroneity in vegetation changes in widely separated areas suggests that vegetational inertia has not been overly important and that plant distributions generally kept pace with changes in regional climate.

The vegetation of the deserts is quite sensitive to changes in the general circulation of the atmosphere (Neilson, 1986). Except for summer temperatures throughout the Southwest and some winter temperatures in the Great Basin, which are controlled by local insolation and cooling, the major features of climate are imported from other regions. In the Great Basin, low winter temperatures frequently result from re-radiational cooling under clear skies in a regional high-pressure zone (Houghton and others, 1975). Also, freezing air may penetrate the Southwest during incursions of Arctic air through the Great Plains, Rocky Mountains, or Great Basin. Winter frontal storms originate in the Pacific Ocean, and the timing, intensity, and latitude of the winter storm tracks depends on the development and position of the Aleutian Low. The monsoons from the Gulf of Mexico and the Gulf of California bring subtropical moisture to the Southwest in summer. Their intensity, timing, and penetration depends on the development of the Bermuda High (Bryson and Lowry, 1955; Hales, 1974). Late-summer tropical storms or hurricanes that result in major precipitation events in the deserts are controlled by the sea-surface temperatures in the Pacific Ocean west of Baja

California (Douglas, 1976; Huning, 1978). The potential for sensitive paleoclimatic reconstructions is excellent because of the interplay of these imported climatic features from both northern temperate and southern subtropical circulation systems.

Regional vegetation closely reflects major climatic gradients. Winter-rainfall chaparral and Mohave desertscrub in California give way to Sonoran and Chihuahuan desertscrub and desert-grassland as summer rainfall increases and temperatures fall with increasing elevation eastward to Arizona, New Mexico, and Texas (Neilson, 1987). Summer rainfall also increases eastward from the Great Basin in Nevada toward the Rocky Mountains in Utah, Colorado, and New Mexico. The importance of winter rainfall decreases from the Great Basin and Colorado Plateau south into the Sonoran and Chihuahuan deserts and the northern edge of the Neotropics. Winter temperatures increase to the south, where catastrophic freezes are less frequent or prolonged. The Sonoran Desert merges into frost-free Sinaloan thornscrub in central Sonora. In the Chihuahuan Desert, subtropical components in the vegetation are common at lower latitudes, although incursions of Arctic air (blue northers or nortes) can bring freezing temperatures throughout the Mexican Plateau as far south as Mexico City. Local climates are dramatically affected by topography. Not only does the amount of rainfall throughout the year increase with elevation as temperature decreases, but at the southerly latitudes the importance of orographic summer rainfall increases with elevation as well.

Regional vegetation is a complex reflection of all of these climatic gradients. Disappearances and appearances of ecologically important plants in packrat midden sequences imply adjustments in their elevational or geographical ranges as climatic parameters changed. The modern vegetation differs substantially from that of the late Wisconsin vegetation in virtually the entire Southwest. Possible exceptions would be edaphic communities dominated by halophytes on the Colorado Plateau or near the head of the Gulf of California. The modern plant communities have been in place less than 8,900 yr, and mostly less than 5,000 yr. Major vegetation changes recorded in midden sequences probably reflect major shifts in regional circulation patterns of western North America during deglaciation, i.e., the winter frontal storm tracks, the intensity of the summer monsoon, freeze frequency, and the amount of summer insolation.

Conventions

The temporal coverage in the present chapter is from the late Wisconsin full-glacial (22 ka) to the middle Holocene (6 ka). Middle Wisconsin and Holocene records younger than 6 ka in longer time series are also discussed to establish a frame of reference.

Ages are abbreviated to thousands of years (ka). Standard deviations and laboratory numbers are cited for individual radiocarbon dates in yr B.P. (radiocarbon years before 1950). Additional information on midden radiocarbon dates can be sought in the literature cited.

Common names are used for plants throughout the text and in chronological diagrams. Scientific names are presented the first time a common name is used. Common and scientific names for plants are presented in Appendix I.

ENVIRONMENTAL SETTING

The four major deserts of North America are found in the Basin and Range Province (Great Basin, Mohave, and Sonoran deserts), on the Colorado Plateaus (Great Basin Desert), and on the Mexican Plateau (Chihuahuan Desert). The term Southwest is used here to refer to the major deserts of North America, although some of the areas in the Great Basin, California, and Texas have not been included in other definitions of the Southwest.

Chihuahuan Desert

The Chihuahuan Desert is an interior continental desert north of the highlands of the Mexican Plateau, west of the Sierra Madre Oriental, east of the Sierra Madre Occidental, and south of the Rocky Mountains (Fig. 1; Morafka, 1977; Johnston, 1977). It extends from northern Zacatecas (25°N) to New Mexico and Arizona (33°N). Most mountains of the Chihuahuan Desert are of Paleozoic or Cretaceous limestones and are aligned north–northeast. The Bolson de Mapimi in Coahuila, Durango, and Chihuahua is a region of internally drained basins at 1,075 to 2,000 m elevation with now-dry playa lakes and limestone sierras. The lowlands along the Rio Grande in Texas (600 to 1,675 m) form the Trans-Pecos subdivision of the Chihuahuan Desert, which is bound by the woodlands of the Edwards Plateau on the east, by the grasslands of the Great Plains, and by the grasslands and woodlands of the Davis Mountains on the north. The northern Chihuahuan Desert merges with the Rocky Mountains in the Guadalupe-Sacramento Mountains.

The vegetation of the Chihuahuan Desert is a xeric desertscrub, commonly dominated by creosote bush (*Larrea divaricata*). Succulents, including lechuguilla (*Agave lechugilla*), and subtropical shrubs, increase in importance southward toward the northern edge of the Neotropics. Grasses gain importance to the north and west. Chihuahuan desertscrub gives way to subtropical scrub southward in the semi-arid Saladan area and eastward into Tamaulipan thornscrub on the coastal plain of the Gulf of Mexico. In the west and north, desertscrub grades into grasslands of Chihuahua and the Great Plains. More mesic habitat on mountains within and surrounding the Chihuahuan Desert support pine-oak woodlands. The climate of the Chihuahuan Desert is dominated by summer monsoonal rainfall and by temperature regimes that become subtropical in the south. Incursions of Arctic air masses, however, can bring devastating freezes throughout the Mexican Plateau.

Colorado Plateau

The Colorado Plateau is a well-marked physiographic province that covers almost 400,000 km^2 in northwestern New

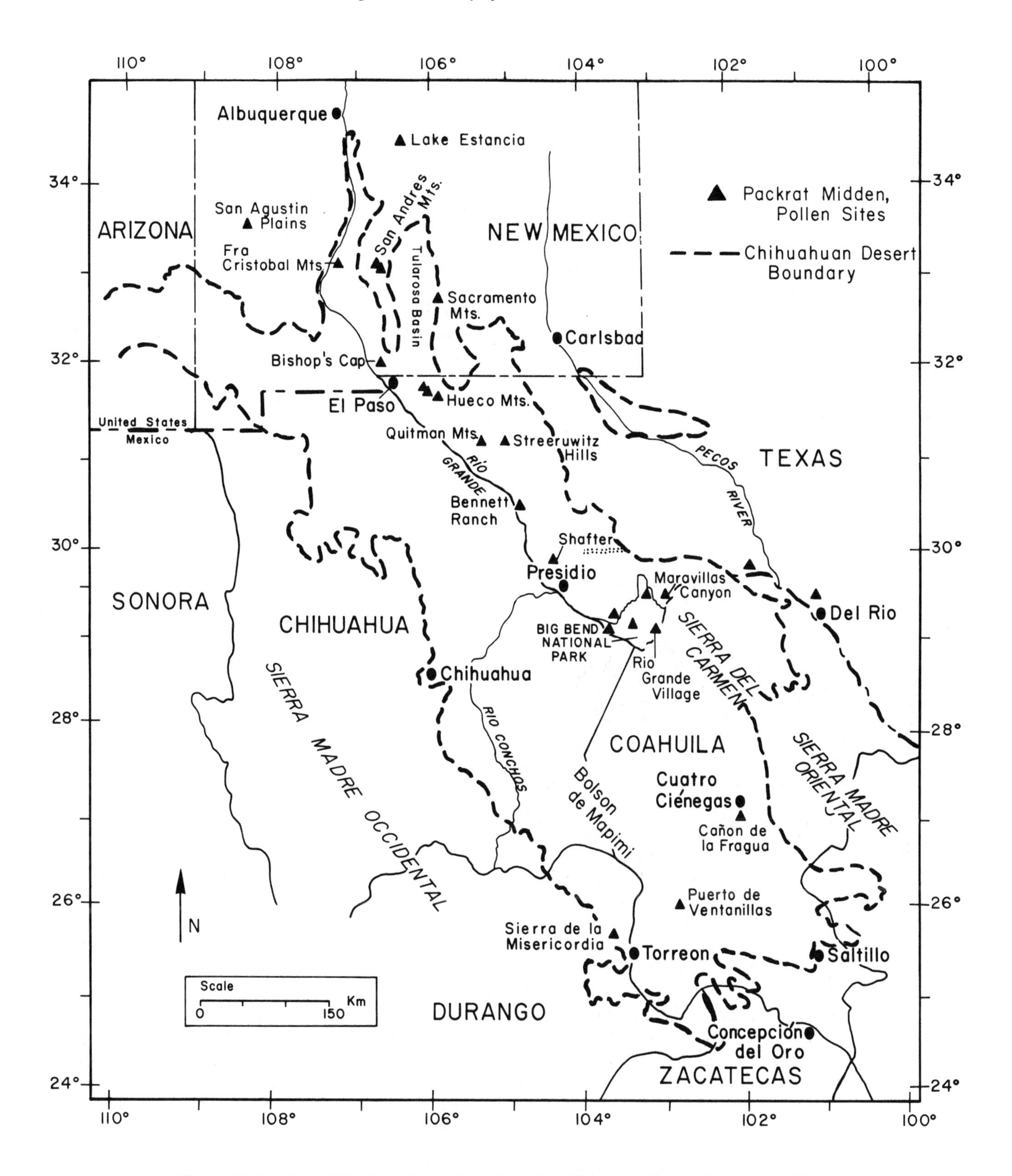

Figure 1. Map of the Chihuahuan Desert. Boundary of the Chihuahuan Desert after Schmidt (1979). Triangles mark packrat midden or pollen sites discussed in text and references cited.

Mexico, northeastern Arizona, western Colorado, and southeastern Utah (Fig. 2). The central and southern Rocky Mountains bound the region to the north and east. The Mogollon Rim and the High Plateaus of Utah form the boundary with the Basin and Range Province. In the region's interior, great sedimentary systems of alternating resistant and easily erodible rocks dip gently to the northeast, opposite to the average drainage direction. Igneous rocks occur in the interior as laccolithic ranges, including the Navajo, Henry, La Sal, Abajo, and Carrizo mountains; volcanic rocks are more common along the periphery in the southern Rocky Mountains, the Datil-Mogollon Volcanic Plateau, the San Francisco Mountains, and the lava-capped High Plateaus. Physiographically, the major landforms are high plateaus on the upfolds, hogbacks on their flanks, lower plateaus between the upfolds, laccolithic mountains, and an intricate set of canyons, most spectacular in the Canyon Lands and Grand Canyon sections (Hunt, 1956).

The vegetation of the Colorado Plateau is an elevational zonation with alpine tundra restricted to above 3,480 m on the loftiest peaks, subalpine forest dominated by Engelmann spruce (*Picea engelmannii*) and subalpine fir (*Abies lasiocarpa*), with sparse Great Basin bristlecone pine (*Pinus longaeva*) and limber pine (*P. flexilis*) at 3,480 to 2,900 m. Below the subalpine zone lies a mixed-conifer forest (2,900 to 2,600 m) of Douglas-fir (*Pseudotsuga menziesii*), white fir (*Abies concolor*), blue spruce (*Picea pungens*), and quaking aspen (*Populus tremuloides*). Ponderosa pine (*Pinus ponderosa*) forest is best developed from 2,500 to 2,100 m, most impressively on the Mogollon Rim in Arizona. In the northwest part of the Plateau, ponderosa pine is replaced by a deciduous woodland of Gambel oak (*Quercus gambelii*) and bigtooth maple (*Acer grandidentatum*). Pinyon–juniper woodland with Colorado pinyon (*Pinus edulis*) and oneseed juniper (*Juniperus monosperma*) in the east and Utah juniper (*J. osteosperma*) in the west occupy extensive areas at 2,100 to 1,600 m. In moist alcoves throughout the otherwise desolate canyon country, perennial springs support low-elevation stands of trees more characteristic of mesic highland forests.

The desertscrub communities above 1,600 m are an extension of the Great Basin Desert with big sagebrush (*Artemisia tridentata*) desertscrub on relatively deep, noncalcareous soils (Turner, 1982). Desertscrub and grassland communities below 1600 m are dominated by shadscale (*Atriplex confertifolia*), mat saltbush (*A. corrugata*), black brush (*Coleogyne ramosissima*), or grasses. Mohave Desert communities enter the Colorado Plateau below 1,200 m in southwesternmost Utah and within the Grand Canyon in Arizona.

The present climate of the Colorado Plateau is strongly affected by topography, with the interior occupying a major rainshadow in the lee of highlands to the south, east, and west. Mean annual precipitation is less than 220 mm at most stations below 1,800 m elevation. Precipitation is commonly half that of stations at equivalent altitude below the Mogollon Rim (Fig. 2). Snowfall can vary from zero in the bottom of the Grand Canyon to a few meters in the uplands. Warm-season (April–September) rainfall accounts for 35 to 65 percent of the annual total; the importance of winter rainfall increases to the northwest and locally toward higher elevations.

Sonoran Desert

The Sonoran Desert is the subtropical arid area centered around the head of the Gulf of California in western Sonora, southwestern Arizona, southeastern California, and much of Baja California (Shreve, 1964; Fig. 3). Desert vegetation is found at elevations ranging from sea level at the Gulf of California to 278 m at Needles along the Colorado River and up to about 1,000 m to the east and north in Arizona (Turner and Brown, 1982). The landscape is a basin-and-range mosaic with granitic mountain ranges formed in the Cretaceous Laramide Orogeny and ranges composed of Middle Tertiary rhyolites and basalts.

Subdivisions of the Sonoran Desert were proposed by Shreve (1964) and refined by Turner and Brown (1982) on the basis of vegetation and climate. Trees, especially foothills palo verde (*Cercidium microphyllum*) and ironwood (*Olneya tesota*), and arborescent cacti, including saguaro (*Carnegiea gigantea*), are common throughout the Sonoran Desert and reflect strong floristic affinities with subtropical thornscrub to the south. All portions of the Sonoran Desert may occasionally experience freezing temperatures, although the duration is rarely more than a single night. In general the frequency and intensity of winter freezes decreases southward. Rainfall ranges from a biseasonal regime with strong summer monsoons in Sonora and Arizona to a winter rainfall regime in Baja California.

Mohave Desert

The Mohave Desert covers portions of southeastern California, southern Nevada, northwestern Arizona, and a small corner of southwestern Utah (Fig. 4). The Mohave Desert has basin-and-range physiography, with internal drainage in the western and southern portions and drainage to the Colorado River in the east. The mountains of the Mohave Desert are composed of granites, rhyolite, and basalt, although calcareous rocks predominate north of 35–36° latitude from the Grand Canyon and southern Nevada into the Great Basin.

The elevations of the valley bottoms in the Mohave Desert range from below sea level in Death Valley to approximately 1,000 m. Creosote bush, Joshua tree (*Yucca brevifolia*), and Mohave yucca (*Y. schidigera*) are characteristic. The mountain ranges imbedded in the "sea" of creosote bush have a vegetation zonation similar to the Great Basin (Billings, 1951) and the Colorado Plateau. Black brush desertscrub occurs on the bajadas, with big sagebrush on the higher bajadas and lower mountain slopes. Pinyon–juniper woodlands with singleleaf pinyon (*Pinus monophylla*) and Utah juniper grow on the mountain slopes, while the summits of the higher ranges in the northern sector support Great Basin bristlecone pine.

The climate is characterized by a dominance of winter rain-

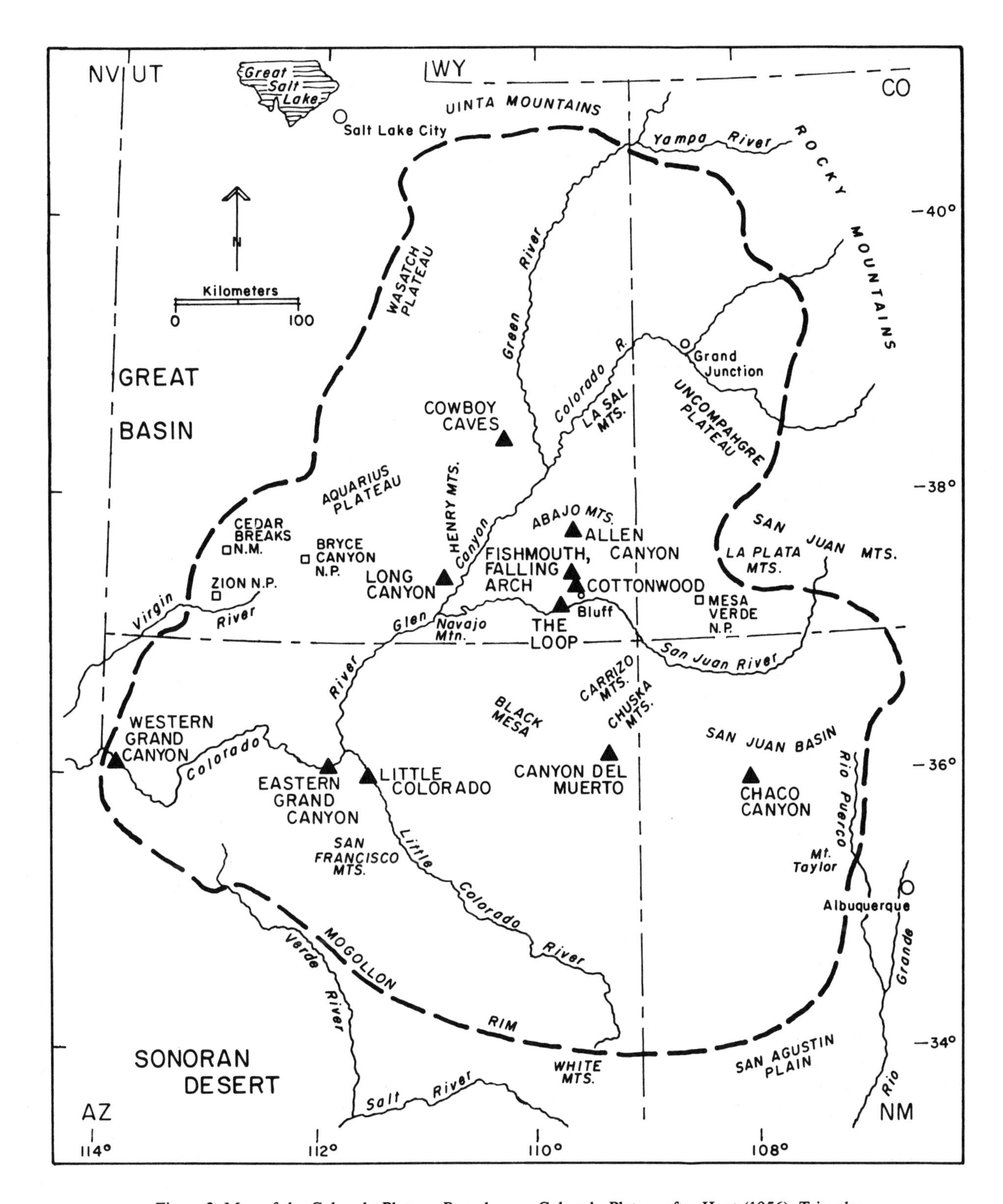

Figure 2. Map of the Colorado Plateau. Boundary or Colorado Plateau after Hunt (1956). Triangles mark packrat midden or pollen sites discussed in text and references cited.

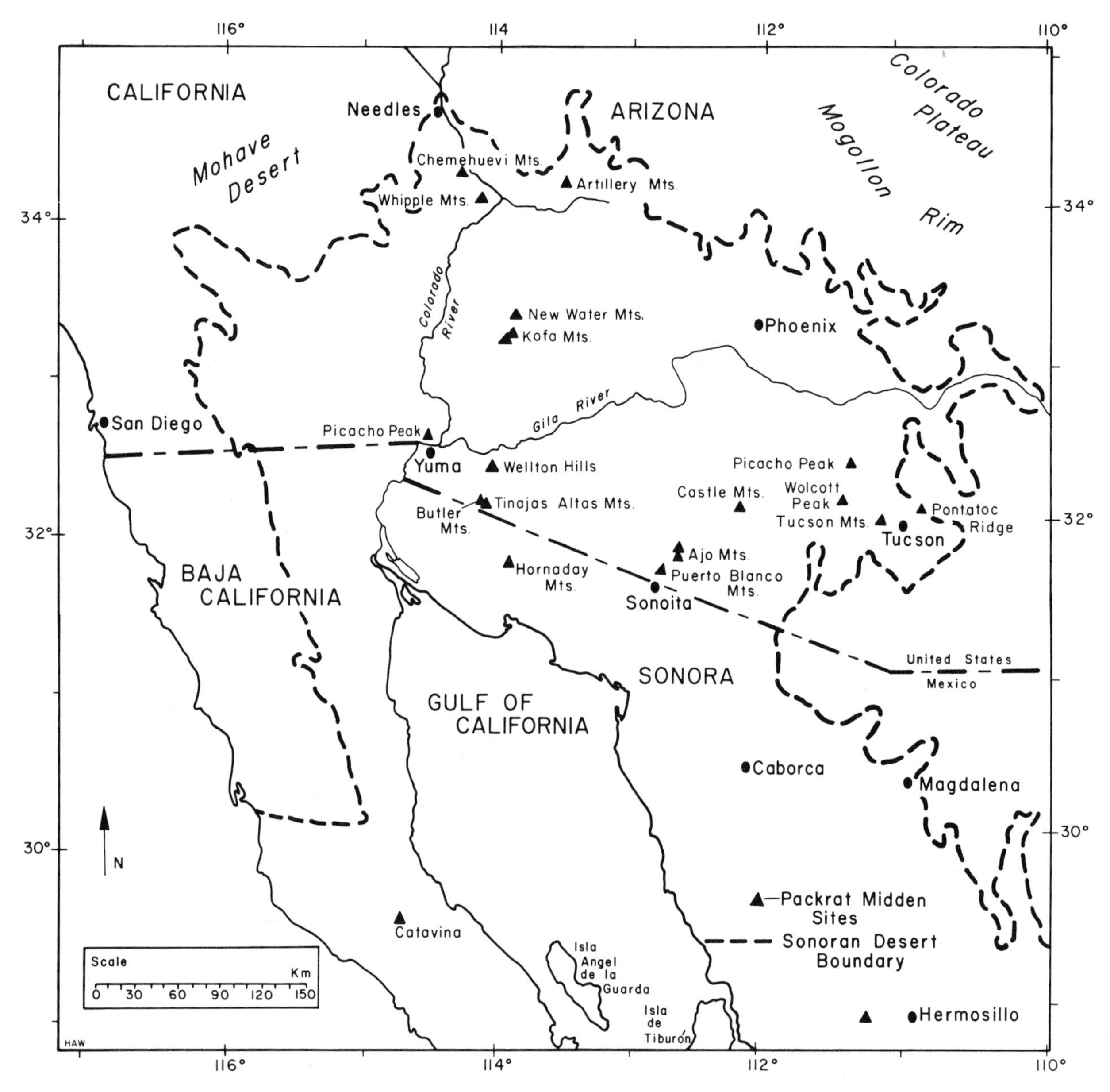

Figure 3. Map of the Sonoran Desert as defined by Shreve (1964). Triangles mark packrat midden sites discussed in text and references cited.

fall with hot, dry summers. In contrast with the Great Basin to the north and the Sierra Nevada to the west, which have their driest periods in July and August, much of the Mohave Desert experiences its greatest drought in late May and June.

Great Basin Desert

The Great Basin is an area of basin-and-range topography in eastern California, central and northern Nevada, southeastern Oregon, and western Utah (Fig. 4). This is a region of internally drained basins bound by limestone mountain ranges in the south and volcanic or metamorphic ranges to the north.

The vegetation of the Great Basin is an elevational series with big sagebrush and/or shadscale desertscrub in the valley bottoms, pinyon-juniper woodland on the lower mountain slopes, and montane or subalpine conifer forests with ponderosa pine, white fir, Douglas-fir, Great Basin bristlecone pine, and limber pine at higher elevations. Big sagebrush and its relatives can be a dominant or associate in diverse communities from 1,200 m in valley bottoms to lofty peaks about 4,000 m. In the extreme northwestern Great Basin, big sagebrush desertscrub covers the entire elevational range. The Great Basin Desert extends to adjacent areas in the Colorado River drainage outside the physiographic Great Basin (Turner, 1982).

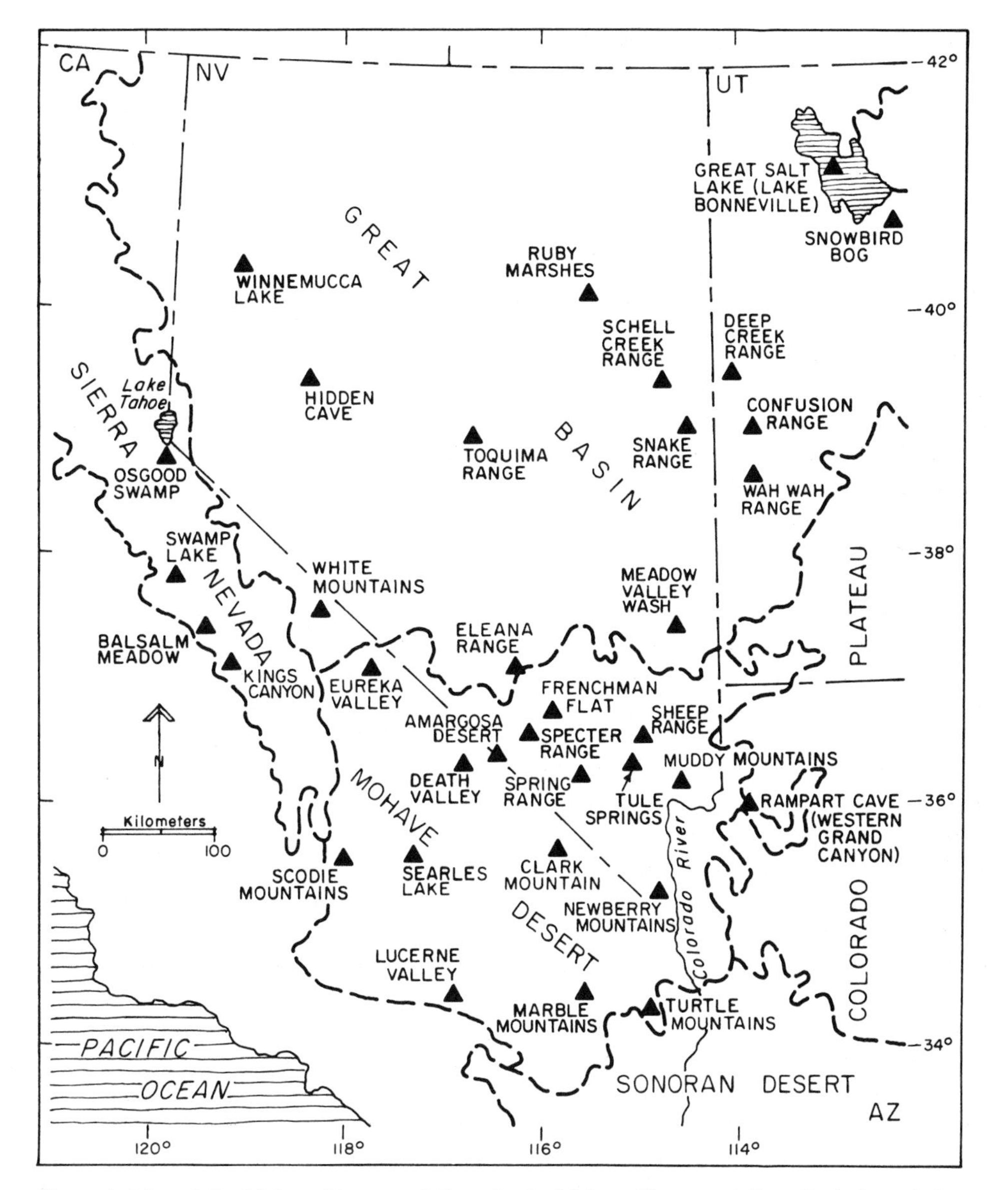

Figure 4. Map of the Mohave Desert and Great basin. Mohave Desert and Great Basin boundaries modified from Küchler (1966). Triangles mark packrat midden or pollen sites discussed in text and references cited.

The Great Basin lies in the rainshadows of the Sierra Nevada to the west, and, to a lesser degree, of the Rocky Mountains to the east. The climate is continental, with cold winters, warm summers, and predominantly winter rainfall. Significant summer rainfall is recorded only in the eastern and southeastern Great Basin (Houghton, 1969).

DISCUSSION AND SUMMARY

Ancient packrat middens have yielded detailed records of late Quaternary vegetational history of rocky slopes throughout the Southwest. Middens have been studied in all of the North American deserts along gradients from the southern Chihuahuan Desert in Durango (25°52′N; Fig. 1) to the Colorado Plateau in Utah (37°47′N; Fig. 2), from the Sonoran Desert in Sonora (29°N; Fig. 3) to the Great Basin (39°N; Fig. 4), and from the eastern Chihuahuan Desert in Coahuila (102°10′W) to the Sierra Nevada in California (118°48′W). The midden sites are from an elevational range of 240 to 2,680 m. The glacial climates of the Late Wisconsin had a profound impact on vegetation of the entire area, as woodland and forest trees expanded widely into lowland habitats and desertscrub dominants retreated. Desertscrub communities lacking woodland dominants were found only in the driest parts of the desert core: the Lower Colorado River Valley below 300 m elevation (Cole, 1986) and the Amargosa Desert in the Sierra Nevada rainshadow (Spaulding, 1983, 1985b).

Paleovegetation

With some exceptions, the late Wisconsin to Holocene vegetational sequence is unidirectional, culminating in relatively modern plant communities. The developmental sequences are very similar to the modern vegetation along north-to-south latitudinal and high-to-low elevational gradients. In the Hueco Mountains, Texas, for example, a four-step vegetation sequence (a Late Wisconsin pinyon-juniper-oak woodland, an early Holocene oak-juniper woodland, a middle Holocene desert-grassland, and a late Holocene Chihuahuan desertscrub) is similar both to the general regional vegetation from north-central New Mexico south along the Rio Grande into Trans-Pecos Texas and to the elevational zonation on nearby mountains. The number of vegetation types in developmental sequences were as few as two in areas where the difference between Late Wisconsin and modern climates was minimal and where the summer monsoon has not been important, e.g., the lowest elevations along the Colorado River in Arizona and California.

The vegetation sequence in the Snake Range in the Great Basin was a late Wisconsin Great Basin bristlecone pine forest, an early Holocene limber pine woodland, a middle Holocene Utah juniper–big sagebrush woodland, and a middle to late Holocene pinyon-juniper woodland. A major difference from other southwestern sequences is that the most xeric plant community was apparently in the middle Holocene (7.4 to 6.3 ka) and not the late Holocene.

Chihuahuan Desert. A Late Wisconsin pinyon-juniper-oak woodland covered rocky slopes at 600 to 1,675 m, the entire elevational gradient of the Chihuahuan Desert (Van Devender, 1986a, 1986b, 1987b). The importance of succulents and other desertscrub plants in the woodland increased to the south. Three sites in the Bolson de Mapimi, Mexico (25°52′–26°39′N), the large area of internal drainage basins on the Mexican Plateau, recorded woodland with papershell pinyon (*Pinus remota*) and juniper (*Juniperus* sp.), or juniper alone, with many succulents about 12.0 ka near the northern edge of the Neotropics (Van Devender, 1987b; Van Devender and Burgess, 1985). The expansion of papershell pinyon and juniper at Cañon de la Fragua, Coahuila, refutes the conclusions from a pollen study from nearby marshes that there was little change in the Chihuahuan desertscrub on the slopes of the Cuatro Cienegas Basin during the Late Wisconsin (Meyer, 1973). Unique endemic Chihuahuan Desert plants survived the glacial climates as members of equable xeric woodland communities. Xeric desertscrub communities formed in the early Holocene soon after the end of the Late Wisconsin.

Extensive midden sequences from two study areas (Maravillas Canyon and the Rio Grande Village area; 29°11–33′N; Fig. 5) and isolated records from six other sites provide an excellent vegetational history for the last 40,000 yr in the Big Bend of Texas. A Late Wisconsin pinyon-juniper-oak woodland grew on limestone slopes that now support succulent Chihuahuan desertscrub from 22.0 to 11.0 ka, with little difference between full- and late-glacial. At the lowest-elevation site (Rio Grande Village, 610 m), several important Chihuahuan Desert plants including lechuguilla and allthorn (*Koeberlinia spinosa*) were important components in the woodland. A xeric oak-juniper woodland without pinyon persisted in the early Holocene until after 9,870 ± 150 yr B.P. (A-4237). After the disappearance of juniper and oak, silver wolfberry (*Lycium puberulum*) and Chihuahuan crucifixion thorn (*Castela stewarti*) continued to be important in the vegetation until the formation of a more xeric desertscrub after 8,560 ± 380 yr B.P. (A-2964). The modern composition of the plant community was reached after 4,330 ± 110 yr B.P. (A-3175).

A similar vegetation sequence was recorded for the northern Chihuahuan Desert (Hueco Mountains, Texas; 31°50′N; Fig. 6), with pinyon-juniper-oak woodland growing on limestone slopes at 1,270 to 1,495 m elevation for at least 32,200 yr in the middle and Late Wisconsin (42.0 to 10.8 ka). The woodland contained both papershell and Colorado pinyons (*Pinus remota, P. edulis*) without Chihuahuan Desert succulents. The early Holocene oak-juniper woodland was well developed until 8.4 to 8.1 ka. A middle Holocene desert-grassland gave way to Chihuahuan desertscrub by about 4,200 ± 600 yr B.P. (AA-381) as more subtropical plants, including lechuguilla and ocotillo (*Fouquieria splendens*), migrated into the area from the Big Bend. The desertscrub corridor across the Continental Divide in southern New Mexico connecting the Chihuahuan and Sonoran deserts was established at that time.

Several sites in south-central New Mexico (Sacramento and San Andres mountains; 32°45′–33°11′N) record the upper limit of the Late Wisconsin pinyon-juniper-oak woodland and the lower edge of a mixed-conifer forest with Douglas-fir, blue spruce, Rocky Mountain juniper (*Juniperus scopulorum*), and ponderosa pine at about 1,585 m (Van Devender and Toolin, 1983; Van Devender and others, 1984). The modern pine forest was apparently absent from the late-glacial vegetation zonation.

Colorado Plateau. The Late Wisconsin vegetation at 2,000 to 1,450 m on the Colorado Plateau was characterized by montane conifers (Figs. 7, 8, and 9). More mesic areas supported mixed-conifer forest with Douglas-fir, blue spruce, limber pine, common juniper (*Juniperus communis*), and Rocky Mountain juniper. In drier areas, this forest was restricted to moist canyons and alcoves. Perennial springs probably occurred in moist "window box" alcoves, including those dry today, accentuating the contrast between the vegetation of open slopes and riparian habitats. On sandstone, blue spruce and water birch (*Betula occidentalis*) followed perennial streams down to 1,300 m, where they were limited by warm-season temperatures approximately 6°C cooler than today (Betancourt, 1984).

More-open slopes probably supported limber pine–Rocky Mountain juniper woodlands physiognomically similar to mod-

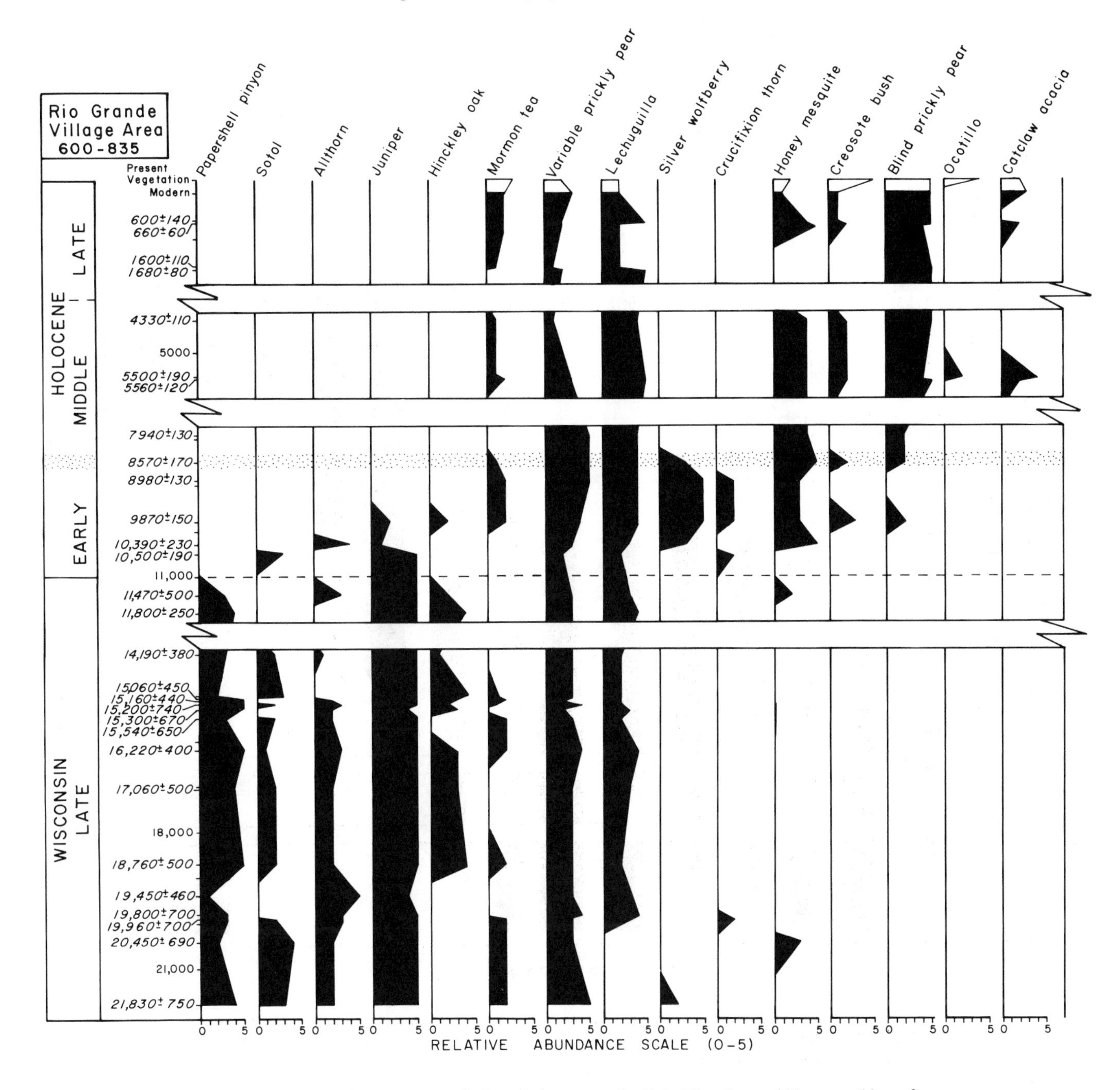

Figure 5. Chronological summary of selected plant macrofossils in 29 packrat midden assemblages from the Rio Grande Village area, Big Bend National Park, Texas. Relative abundance scale: 1=rare, 2=uncommon, 3=common, 4=very common, 5=abundant.

ern pinyon-juniper woodlands. Modern analogues to this Late Wisconsin woodland occur today in the foothills of the central Rocky Mountains (Idaho, Montana, Wyoming). Utah juniper woodland was zoned directly below mixed-conifer forest in the eastern Grand Canyon but restricted by higher base levels farther upstream in Utah.

During the Late Wisconsin, Great Basin desertscrub dominated by big sagebrush probably covered fine-grained soils in the valleys. Alkaline flats supported salt-tolerant plants that today extend to the Canadian border, yet are commonly thought of as xerophytes. These include greasewood (*Sarcobatus vermiculatus*), shadscale, mat saltbrush, fourwing saltbush (*Atriplex canescens*), spiny hop sage (*Grayia spinosa*), and winterfat (*Ceratoides lanata*). Some of these plant communities may have been little

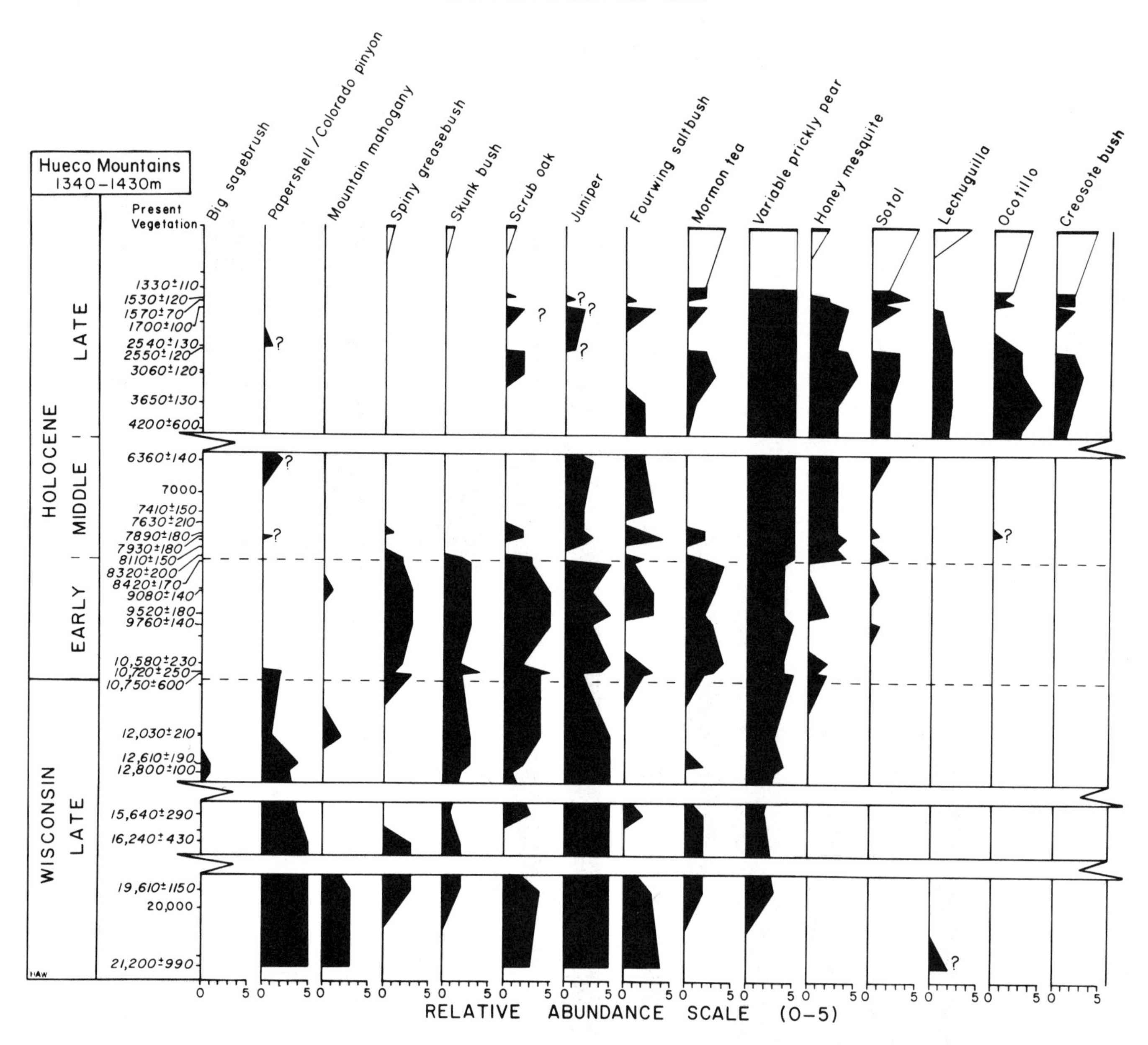

Figure 6. Chronological summary of selected plan macrofossils in 29 packrat midden assemblages from the Hueco Mountains, Texas. Relative abundance scale: 1=rare, 2=uncommon, 3=common, 4=very common, 5=abundant; ?=probable contaminant.

different during the Late Wisconsin. Big sagebrush and halophytic desertscrub probably expanded in these habitats, with black brush restricted to the western Grand Canyon juniper woodlands.

The flora of the Colorado Plateau is rich in endemics, many of which are restricted to the hypersaline Mancos Shale. Such genera as *Astragalus, Eriogonum, Pediocactus,* and *Sclerocactus* have undergone majot evolutionary radiations on the Colorado Plateau. Many of these "desert" endemics probably grew at similar elevations as the mixed-conifer forest in the Late Wisconsin (Betancourt, 1984).

Above 2,000 m, the Late Wisconsin mixed-conifer forest merged into subalpine or spruce-fir forest as Engelmann spruce and subalpine fir assumed dominance. Upper treeline was probably at ca. 2,900 m, only 575 m lower than today (Carrara and others, 1984; Jacobs, 1983; Richmond, 1962; Merrill and Péwé, 1977). Subalpine forest expanded its current distribution on igneous peaks or lava-capped plateaus onto the vast sandstone country in the lowlands. The elevational lowering meant considerable increase in area over the modern spruce-fir forest, closing the distance between now isolated boreal habitats. Spruce-fir forest

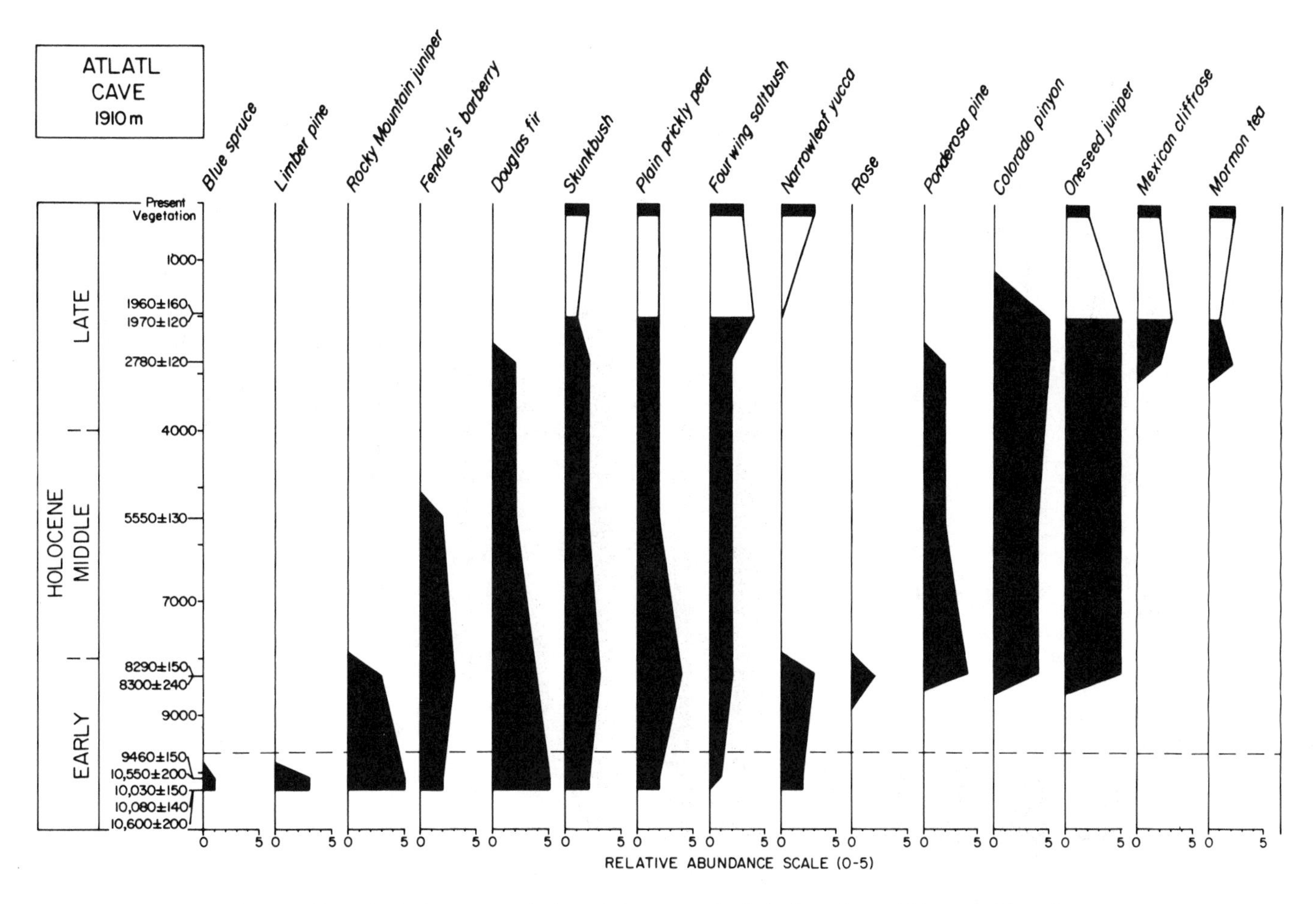

Figure 7. Chronological summary of selected plant macrofossils in six packrat midden assemblages from Atlatl Cave, New Mexico. Relative abundance scale: 1=rare, 2=uncommon, 3=common, 4=very common, 5=abundant.

replaced the extensive modern ponderosa pine forest on the Mongollon Rim.

Treeline lowering by ca. 600 m greatly expanded the potential habitat for alpine tundra on isolated peaks but was not enough to establish continuous connections. Dispersal of alpine plants between these isolated peaks was either long-distance via birds or wind, or through Wisconsin forests. In the absence of tundra plants, big sagebrush and bunch grasses probably moved into available habitats on isolated peaks and plateaus.

At Cottonwood Cave, Utah (37°19′N, 1390 m), dominants in the Late Wisconsin mixed-conifer forest shifted from blue spruce in a 15,660 ± 320 yr B.P. (A-4469) full-glacial sample to Douglas-fir, limber pine, and Rocky Mountain juniper by 12,690 ± 230 yr B.P. (A-4307) in the late-glacial. At the end of the Late Wisconsin, montane conifers declined again, coincident with vegetational changes elsewhere in the Southwest. Engelmann spruce and subalpine fir were replaced by blue spruce and ponderosa pine between 11.3 and 10.0 ka, with further reduction of the montane flora at 10.0 to 7.2 ka. At 1,585 m, blue spruce and common juniper dropped out between 12.8 and 10.4 ka, and limber pine and Rocky Mountain juniper were replaced by Utah juniper at 10.4 to 9.7 ka. Montane conifers such as limber pine and blue spruce also persisted in the Chaco Canyon lowlands (1,910 m) until 10 ka. Douglas fir persisted into the middle Holocene in Canyon Lands and the San Juan Basin. Ponderosa pine appears to have expanded its lower limit in the early and middle Holocene. In the middle and lower portions of its elevational range, pinyon-juniper woodland was established by the middle Holocene. Colorado pinyon did not expand into its present upper elevational range until the late Holocene, when the modern plant community composition and vegetational zonation were established.

Sonoran Desert. In the subtropical Sonoran Desert, a pinyon–juniper–oak woodland with singleleaf pinyon grew in the modern Arizona Upland Subdivision from 1,555 to ca. 60 m elevation during the Late Wisconsin. The lowest record for singleleaf pinyon was 460 m in the Tinajas Altas Mountains, Arizona (Fig. 10). Few Sonoran Desert plants persisted in Late Wisconsin

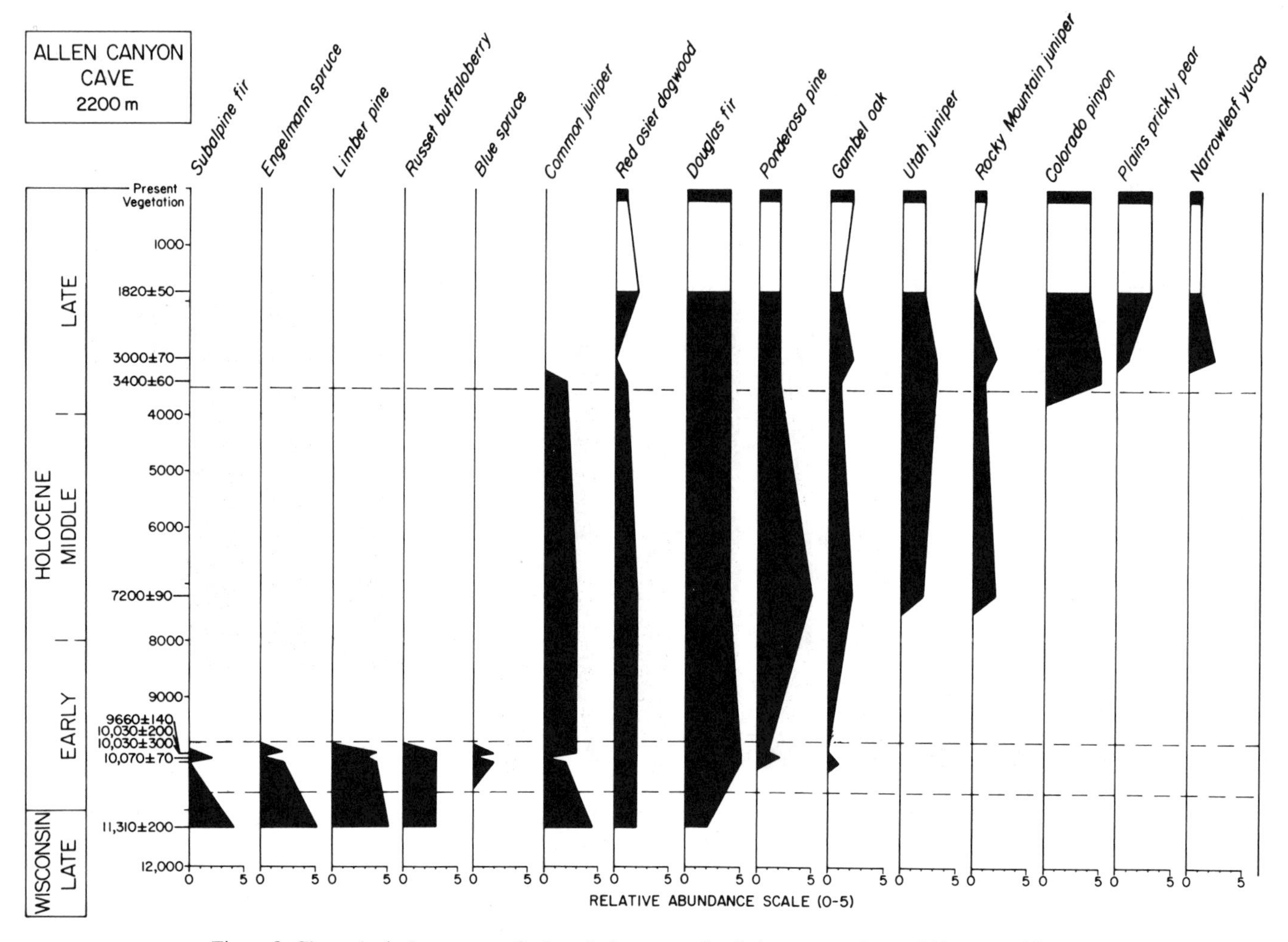

Figure 8. Chronological summary of selected plant macrofossils in seven packrat midden assemblages from Allen Canyon Cave, Utah. Relative abundance scale: 1=rare, 2=uncommon, 3=common, 4=very common, 5=abundant.

woodlands, although Mohave Desert plants, especially Joshua tree, expanded widely. Below 600 m, a xeric juniper woodland with California juniper (*Juniperus californica*), shrub live oak (*Quercus turbinella*), Joshua tree, Whipple yucca (*Yucca whipplei*), and Bigelow beargrass (*Nolina bigelovii*) was present down to ca. 305 m (Van Devender, 1987c). A radiocarbon date of 18,700 ± 1,050 yr B.P. (AA-536; Van Devender, 1987c) on creosote bush from a California juniper–Joshua tree assemblage from 330-m elevation in the Tinajas Altas Mountains establishes its presence in the full-glacial. This record refutes the proposal of Wells and Hunziker (1976) that creosote bush immigrated to the Chihuahuan Desert from Argentina after 11.5 ka. The lowest Wisconsin record for California juniper was 240 m in a creosote bush–white bursage (*Ambrosia dumosa*) desertscrub assemblage in the Butler Mountains, Arizona.

Xeric Mohave desertscrub with creosote bush, black brush, Joshua tree, and Whipple yucca was present at 240 to 300 m from 13.4 ca. 11.0 ka in the Late Wisconsin in the Picacho Peak area, California, near Yuma, Arizona (Fig. 3; Cole, 1986). Thus, desertscrub without woodland plants in the Sonoran Desert was restricted to below 300 m elevation in the xeric trough of the Lower Colorado River Valley from perhaps 34°N south into the lowlands surrounding the head of the Gulf of California, including the land exposed by ca. 100 m lowering of sea level. This area has probably been a core North American desert for much of the Quaternary, mostly resembling the modern Mohave Desert, with subtropical Sonoran Desert plants moving north along the Colorado River only during warmer portions of interglacials.

The end of the Late Wisconsin in the Sonoran Desert is marked by the disappearance of singleleaf pinyon and other mesic woodland plants from the 600 to 1,555 m elevational

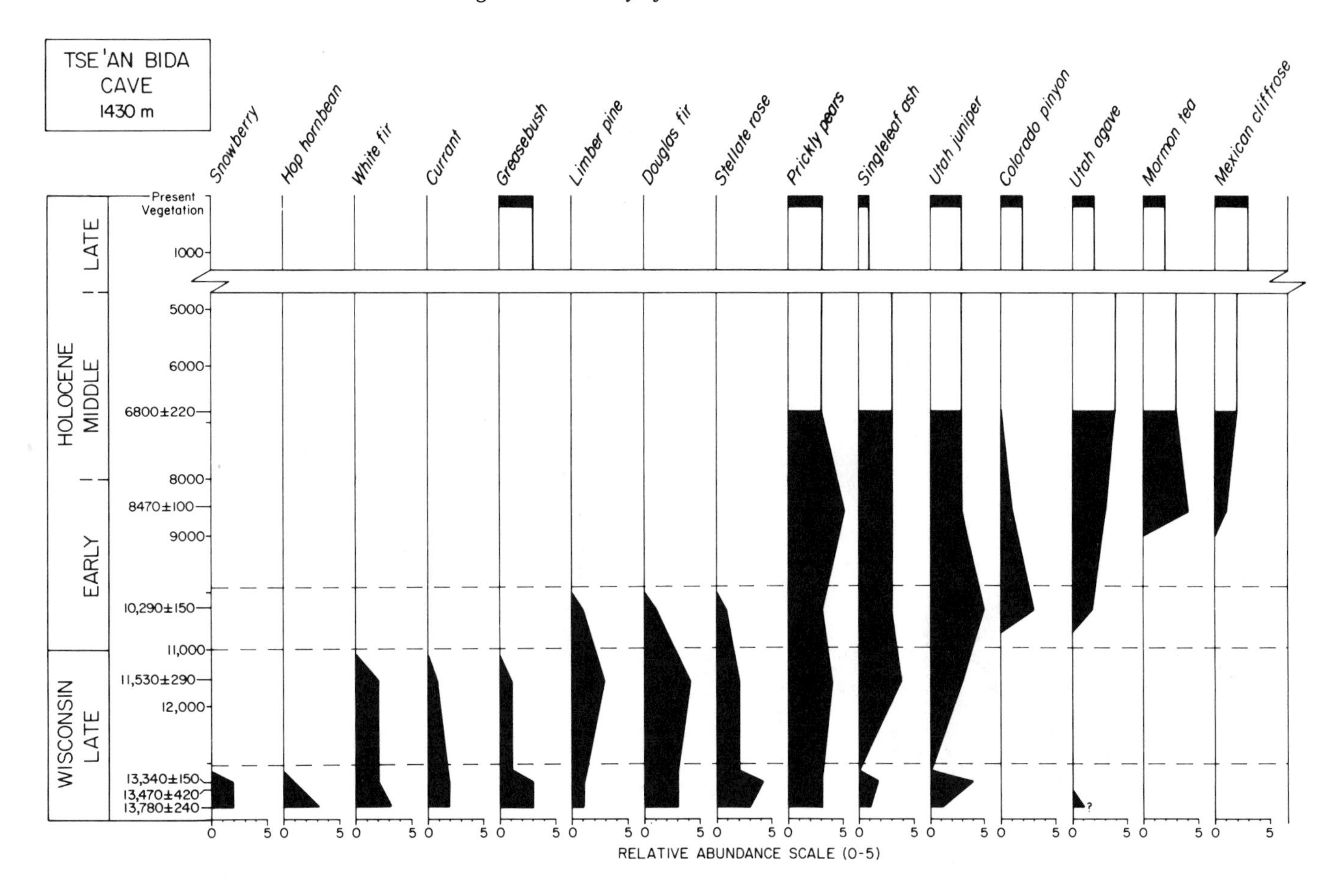

Figure 9. Chronological summary of selected plant macrofossils in seven packrat midden assemblages from Tse'an Bida Cave, Grand Canyon National Park. Data from Cole (1981). Relative abundance scale: 1=rare, 2=uncommon, 3=common, 4=very common, 5=abundant.

range about 11.0 ka (Fig. 10 and 11). A xeric juniper–shrub live oak woodland/chaparral continued into the Holocene, as did California juniper woodlands at lower elevations. Early Holocene plant communities with juniper were present over an elevational range of ca. 1,315 m in the Sonoran Desert. Juniper grew in a lowland Sonoran Desert site near Catavina in central Baja California, Mexico (Fig. 3), that now supports the unusual boojum tree (*Fouquieria columnaris*; Wells, 1976). Several junipers occupied the Sonoran Desert, with Utah, oneseed, redberry (*Juniperus erythrocarpa*), and even Rocky Mountain junipers in the higher areas and California, and possibly Utah juniper at the lowest elevations (Van Devender, 1987c). The vegetation ranged from juniper–shrub live oak woodland/chaparral to creosote bush desertscrub with California juniper. The disappearance of these junipers from the Sonoran Desert at about 8.9 ka (Puerto Blanco and Tinajas Altas mountains in Arizona, Whipple Mountains in California; Van Devender, 1987a, 1987c) marks the end of the early Holocene. A 7,870 ± 750 yr B.P. (A-1284) date on juniper from the New Water Mountains, Arizona (Van Devender, 1977) appears to represent survival of a local population without regional significance.

In the Puerto Blanco Mountains, Arizona (Fig. 3, 535 to 605 m), California juniper was uncommon in early Holocene midden assemblages, while important Sonoran desertscrub dominants, including saguaro and brittle bush, had returned from their glacial refugia to the south (Van Devender, 1987a). In lower areas such as the Picacho Peak area, California (240 to 300 m), and the Hornaday Mountains, Sonora (Fig. 3; 240 m), juniper was not present in the early Holocene, and desertscrub vegetation was well established. Mohave Desert plants declined or disappeared in the early Holocene in the eastern Sonoran Desert but persisted near the Colorado River.

After ca. 8.9 ka, desertscrub vegetation was established in the Sonoran Desert. However, the communities in the Puerto Blanco and Tinajas Altas mountains were very different from those of today. Catclaw acacia (*Acacia greggii*), velvet mesquite

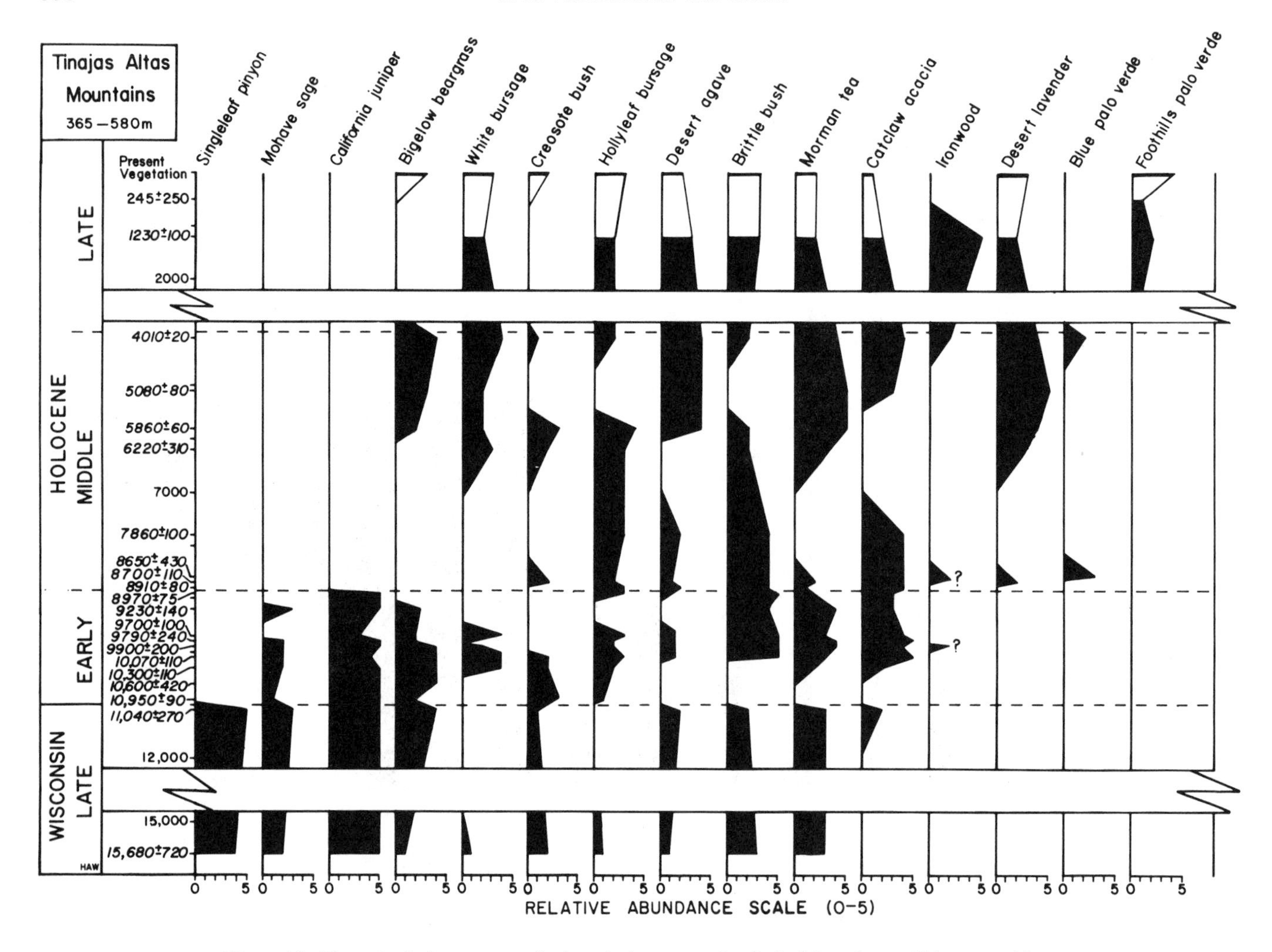

Figure 10. Chronological summary of selected plant macrofossils in 20 packrat midden assemblages from Tinajas Atlas Mountains, Arizona. Relative abundance scale: 1=rare, 2=uncommon, 3=common, 4=very common, 5=abundant; ?=probable contaminant.

(*Prosopis velutina*), and blue palo verde (*Cercidium floridum*), now retricted to riparian wash habitats, grew on hot, dry south-facing slopes, suggesting summer rainfall much greater than today. The absence of frost-sensitive subtropical Sonoran plants also suggests that winter freezes were more frequent than today. To the west the middle Holocene vegetation of the Picacho Peak area, California, was essentially modern, while that of the Whipple Mountains suggested a slight increase in summer rainfall. The northwestern boundary between the Sonoran and Mohave deserts has been in essentially the same position for the last 8,900 yr.

After about 4.0 ka, relatively mdoern vegetation and climate were established throughout the Sonoran Desert as summer rainfall and winter freeze frequency declined to near present levels. In the late Holocene, subtropical vegetation reached its northernmost extent since the last interglacial.

Mohave Desert. The only area in the Mohave Desert that probably supported a creosote bush desertscrub vegetation during the Late Wisconsin was below about 300 m in the Colorado River Valley. The lowest record for juniper was 258 m in the Chemehuevi Mountains, California (34°32′N; Fig. 4; Wells, 1983). The 510-m record for singleleaf pinyon in the Whipple Mountains, California (34°14′N), on the Mohave–Sonoran Desert boundary, suggests near proximity to the lower edge of pinyon–juniper woodland. Late Wisconsin pinyon–juniper woodland was at 730 to 850 m in both the Turtle Mountains, California (34°24′N; Wells and Berger, 1967), and the Newberry Mountains, Nevada (35°15′N; Leskinen, 1975). Low-elevation juniper woodland extended up the Colorado River into the western Grand Canyon (425 to 675 m, 36°06′N; Phillips, 1977; Fig. 12) and southern Nevada (530 to 670 m; Wells and Berger, 1967; Wells, 1983). A single midden from 635 m records the lower limit of pinyon in the Grand Canyon (Phillips, 1977). In the southern

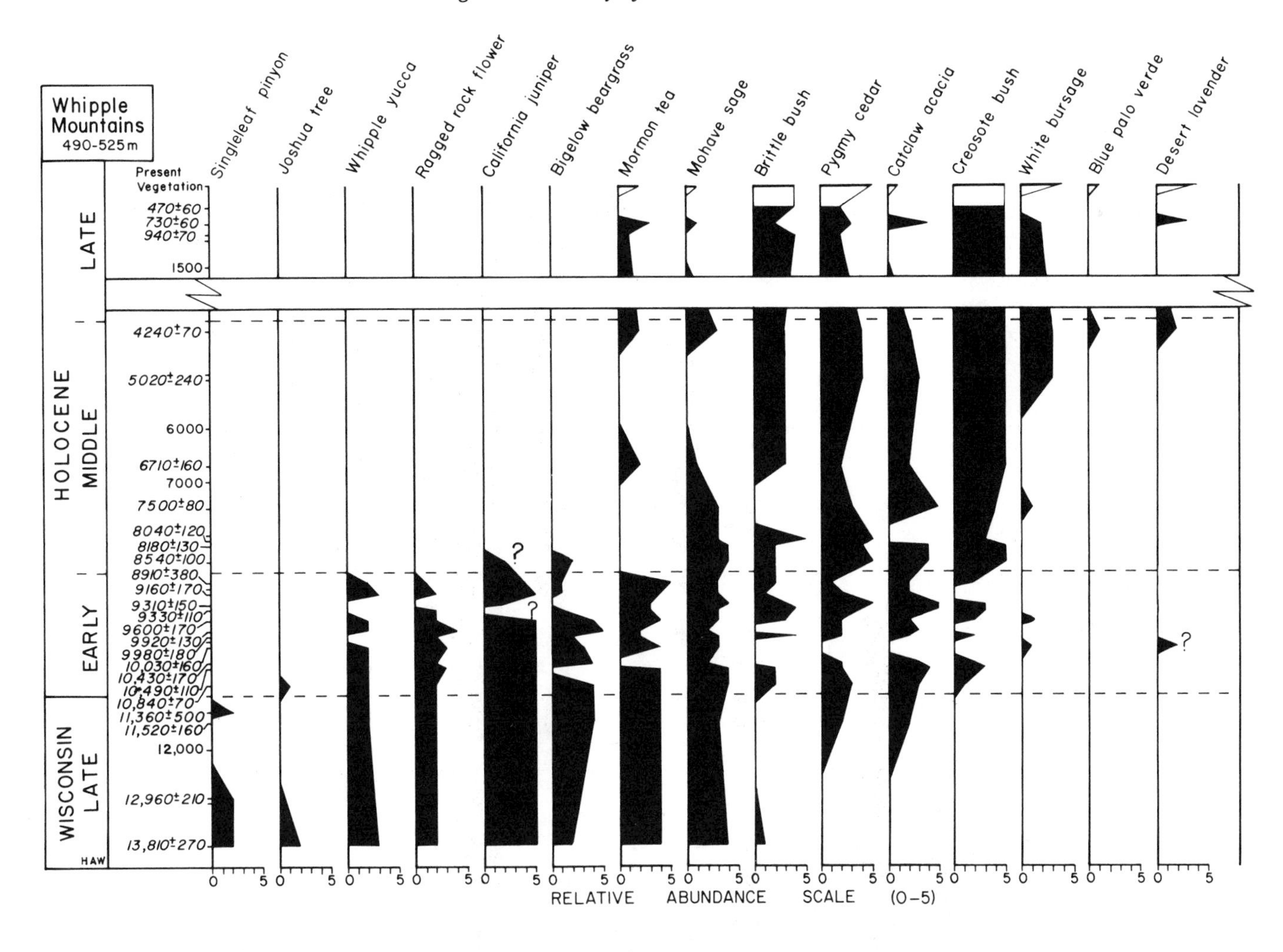

Figure 11. Chronological summary of selected plant macrofossils in 25 packrat midden assemblages from Whipple Mountains, California. Relative abundance scale: 1=rare, 2=uncommon, 3=common, 4=very common, 5=abundant; ?=questionable record.

Mohave Desert, juniper woodland was at 972 m in the Lucerne Valley (34°29′N), with pinyon-juniper woodland at 1,220 m on nearby Ord Mountain (King, 1976). The lower limits of woodland were highest in the westernmost Mohave Desert, as they are today, tracking increasing aridity into the rainshadows of the Transverse Range north of the Los Angeles Basin and the southern Sierra Nevada (Spaulding, 1987). Middens from 1,130 m in the Scodie Mountains (35°36′N) in the westernmost Mohave Desert extend the late Wisconsin pinyon-juniper woodland to the base of the Sierra Nevada. Joshua tree was a common component in these Late Wisconsin pinyon-juniper woodlands as it is today on the north slopes of the San Bernardino Mountains, California.

In the northern Mohave Desert, Late Wisconsin juniper woodland extended from the low-elevation Colorado River sites up to 1,100 m in the Sheep Range (36°30′N), Nevada, with sparse singleleaf pinyon up to about 1,700 m (Spaulding, 1981). Juniper woodland without pinyon was at 1,100 to 1,280 m in the Frenchman Flat area (36°42′N; Wells and Berger, 1967). Juniper woodland had an elevational range of at least 855 m, with singleleaf pinyon rare or absent as the vegetation zonation became even drier for a given elevation. The only late Wisconsin record for pinyon-juniper woodland west of the Sheep Range is from 1,100 to 1,190 m in the Specter Range, Nevada (36°40′N; Spaulding, 1985a). Juniper woodland was growing at 790 to 800 m in Owl Canyon (36°24′N) and 960 m in the Last Chance Range, Nevada (36°17′N; Spaulding, 1985a, 1985b), and 425 to 1,280 m in Death Valley, California (36°28′N; Wells and Woodcock, 1985). A midden assemblage from the Eureka Valley, California, records a Utah juniper woodland with sparse limber pine at yet a higher elevation (1,430 m) in the northern Mohave Desert (Spaulding, 1980). Shrub assemblages with little

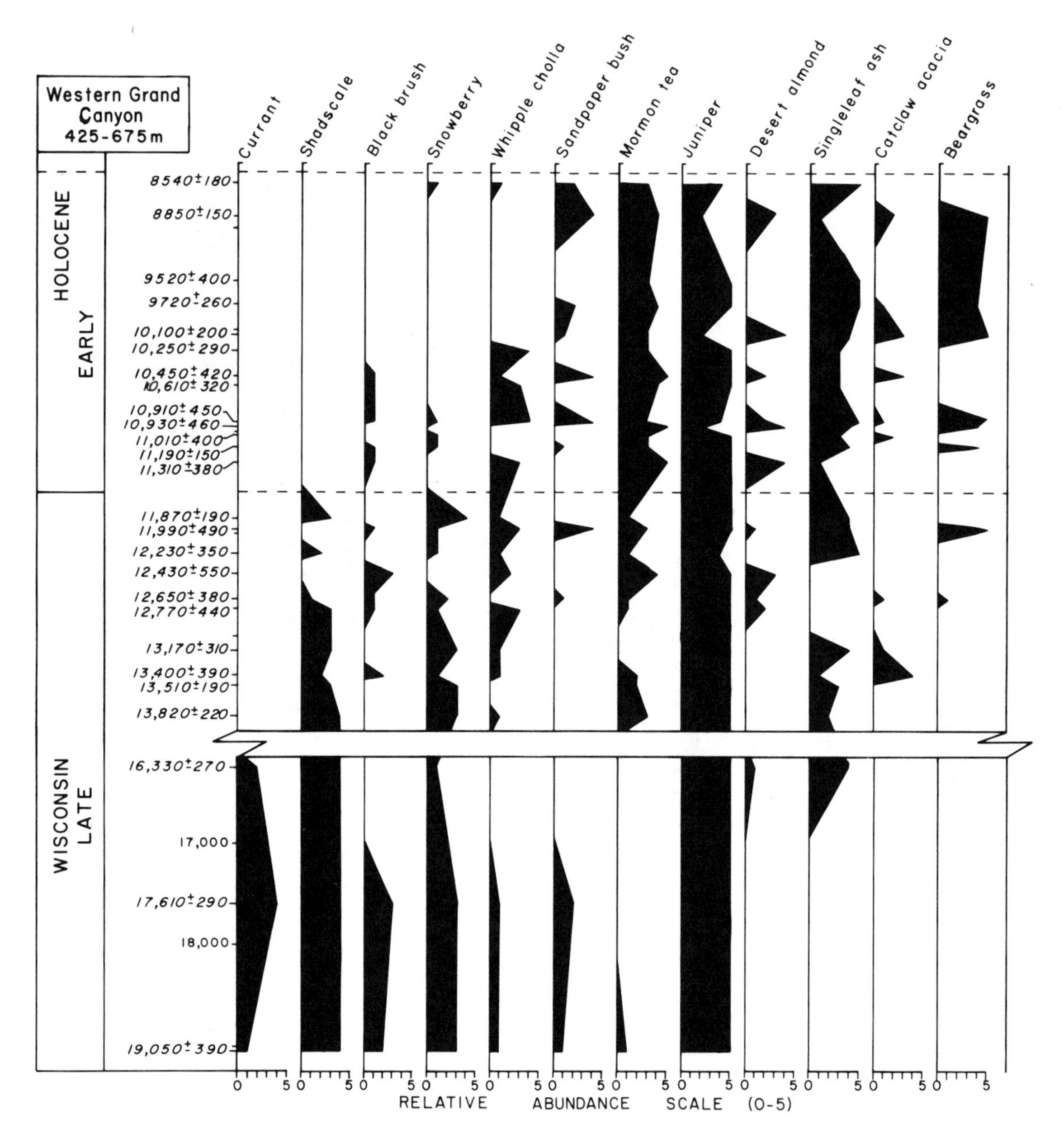

Figure 12. Chronological summary of selected plant macrofossils in 26 packrat midden assemblages from Rampart Cave area, western Grand Canyon, Arizona. Data from Phillips (1977). Relative abundance scale: 1=rare, 2=uncommon, 3=common, 4=very common, 5=abundant.

or no juniper from 900 to 930 m from Point of Rocks, Nevada (36°34′N; Spaulding, 1985a), mark the extension of the regional gradient of decreasing rainfall to the northwestern Mohave Desert in the lee of the Sierra Nevada (Spaulding, 1983). All of the regional differences in vegetational zonation and climate observed in the Late Wisconsin parallel modern environmental gradients.

Study sites in the northern Mohave Desert record mixed-conifer forest zoned above late Wisconsin woodlands. Assemblages with Great Basin bristlecone pine, limber pine, and white fir were found at 1,900 to 2,100 m on Clark Mountain, California (35°33′N; Mehringer and Ferguson, 1969), at 1,850 m on Potosi Mountain (35°39′N; Thompson and Mead, 1982), and at ca. 1,700 to 2,400 m in the Sheep Range, Nevada (Spaulding, 1981). Isolated stands of Great Basin bristlecone pine occur today in the Inyo, White, and Funeral mountains of California.

During the Late Wisconsin, the big sagebrush and shadscale desertscrub communities of fine-grained soils in the valleys expanded southward from the Great Basin into the modern Mohave Desert. The elevational and geographic ranges of big sagebrush, shadscale, and other Great Basin desertscrub plants were much greater than now. The upper elevational limits of shadscale were actually 320 m higher than today in the Sheep Range, resulting in the rare association of shadscale with Great Basin bristlecone pine (Spaulding, 1981; Spaulding and others, 1983). Black brush and creosote bush desertscrub communities were displaced to lower elevations and latitudes.

A shift to more xeric vegetation between 12.0 and 11.0 ka marks the end of the Late Wisconsin. Singleleaf pinyon and its more mesic associates retreated from the lowlands of the southern Mohave Desert, leaving juniper woodland over an 815-m range from 305 m in the Colorado River Valley up to at least 1,220 m. Juniper woodlands below the late Wisconsin range of singleleaf pinyon continued into the early Holocene with relatively minor changes in composition. The final disappearance of juniper from the deserts was after 9,500 ± 240 yr B.P. (A-1017) at 730 to 850 m in the Newberry Mountains, Nevada (Leskinen, 1975), after 8,560 ± 260 yr (B.P. (A-1469) at 425 to 675 m in the western Grand Canyon, Arizona (Phillips, 1977), after 9,900 ± 910 yr B.P. (UCLA-1660) at 530 to 554 m in southern Nevada (Wells, 1983), after 7,800 ± 150 yr B.P. (UCLA-560) at 1,280 m in the Frenchman Flat area, Nevada (Wells and Berger, 1967), after 8,905 ± 265 yr B.P. (GX-6188) at 840 to 860 m in the Marble Mountains, California (34°40′N; Spaulding, 1980), and after 7,800 ± 350 yr B.P. (UCR-249) in the Lucerne Valley, California (King, 1976). Three samples from 475 to 900 m in the Marble Mountains contained xeric shrubs without juniper at 10.5 to 9.5 ka, although the modern Mohave desertscrub dominants arrived later (Spaulding, 1980, 1983). In all of these areas the shift to a more xeric plant community between 8.9 and 7.9 ka marks the establishment of relatively modern vegetation. Vegetation of the middle and late Holocene was similar throughout most of the Mohave Desert, with some minor fluctuations.

In the Sierran rainshadow areas of the northwestern Mohave Desert, the early Holocene vegetation transition was more variable. Utah juniper was gone by 11,210 ± 380 yr B.P. (GX-7806) at 425 m in Death Valley, California (Wells and Woodcock, 1985) and by 10,260 ± 520 yr B.P. (A-2402) at 790 to 800 m in Owl Canyon, Nevada (Spaulding, 1985a), but survived until after 9,280 ± 210 yr B.P. (A-2414) in the Last Chance Range, Nevada (Spaulding, 1985a). At Point of Rocks, Nevada, xeric Late Wisconsin desertscrub at 900 to 930 m gave way to a relatively modern Joshua tree community by 9.6 ka, with modern desertscrub developing after 9.3 ka (Spaulding, 1985b). The nearby Skeleton Hills (36°38′N) recorded white bursage and pygmy cedar (*Peucephyllum schottii*) at 9.2 to 8.8 ka (Spaulding, 1985b). At 425 m in Death Valley, white bursage arrived by 10,230 ± 320 (GX-7812), with creosote bush arriving later (Wells and Woodcock, 1985). At 1,430 to 1,635 m in the Eureka Valley, California (37°20′N), Utah juniper was gone by 8,300 ± 250 yr B.P. (GX-6231), but creosote bush reached its northern limits after 5.0 ka (Spaulding, 1980).

Few sites record the upward or northward migration of singleleaf pinyon in the Mohave Desert. It arrived by 12,460 ± 190 yr B.P. (I-3690) in a Late Wisconsin mixed-conifer forest site at 1,910 m on Clark Mountain, California (Mehringer and Ferguson, 1969). Singleleaf pinyon appeared at 1,550 m in the Spotted Range, Nevada, by 9,450 ± 90 yr B.P. (UCLA ?; Wells and Berger, 1967) as a migrant from the south or from sparce populations on larger mountain ranges such as the Sheep Range. Well-developed pinyon-juniper woodland was established at 1,100 to 1,700 m within its modern elevational range by 11,500 ± 150 yr B.P. (A-1774) in the Sheep Range (Spaulding, 1981). Singleleaf pinyon apparently moved into its modern elevational range without intermediate populations in the elevations that supported late Wisconsin juniper woodlands.

Great Basin. Late Wisconsin forests dominated by subalpine conifers expanded throughout the mountains of the eastern Great Basin. Limber pine and Douglas-fir were present at 1,350 to 1,615 m in the Meadow Valley Wash, southeastern Nevada (Madsen, 1973; Wells, 1983). At 1,600 to 2,340 m in the Confusion and Wah Wah ranges, western Utah, forest assemblages included Great Basin bristlecone pine, limber pine, Douglas fir, and common and Rocky Mountain junipers (*Juniperus communis, J. scopulorum*; Wells, 1983). In the Snake Range of eastern Nevada, Great Basin bristlecone pine, limber pine, Engelmann spruce, and common juniper were present in late Wisconsin assemblages at 1,640 to 2,680 m (Thompson, 1979, 1984; Wells, 1983; Fig. 13). Pinyon-juniper woodland and ponderosa pine forest were absent from the Late Wisconsin zonation of the Great Basin.

Farther west the late Wisconsin vegetation reflected aridity associated with the Sierra Nevada rainshadow. Assemblages from 1,810 m in the Eleana Range, Nevada (37°07′N), on the Great Basin–Mohave Desert transition were dominated by limber pine (Spaulding, 1985a). As on the Colorado Plateau, limber pine probably formed an open woodland of small trees in areas that now support pinyon-juniper woodland. Today, limber pine forms a woodland with Utah juniper in some of the isolated ranges in central Nevada and on the northern periphery of the Great Basin in northeastern Nevada. Apparently Late Wisconsin limber pine woodlands in the drier parts of the Great Basin lacked any associated junipers. Late Wisconsin middens from 1,230 to 1,296 m from Winnemucca Lake, Nevada (40°12′N) in the pluvial Lake Lahontan basin were dominated by western juniper (*Juniperus occidentalis*), in association with xeric desert shrubs (Thompson and others, 1986). These sparse results parallel the modern climatic gradient of increasing aridity to the west and northwest in the Great Basin (Houghton, 1969). Future work in western and northwestern Nevada may reveal Late Wisconsin landscapes covered with big sagebrush desertscrub without trees or with sparse western juniper or limber pine.

Big sagebrush is an adaptable shrub of cold-temperate climates that can grow anywhere along the elevational gradient with or without most other plant communities. In the Late Wisconsin, big sagebrush desertscrub expanded into the montane

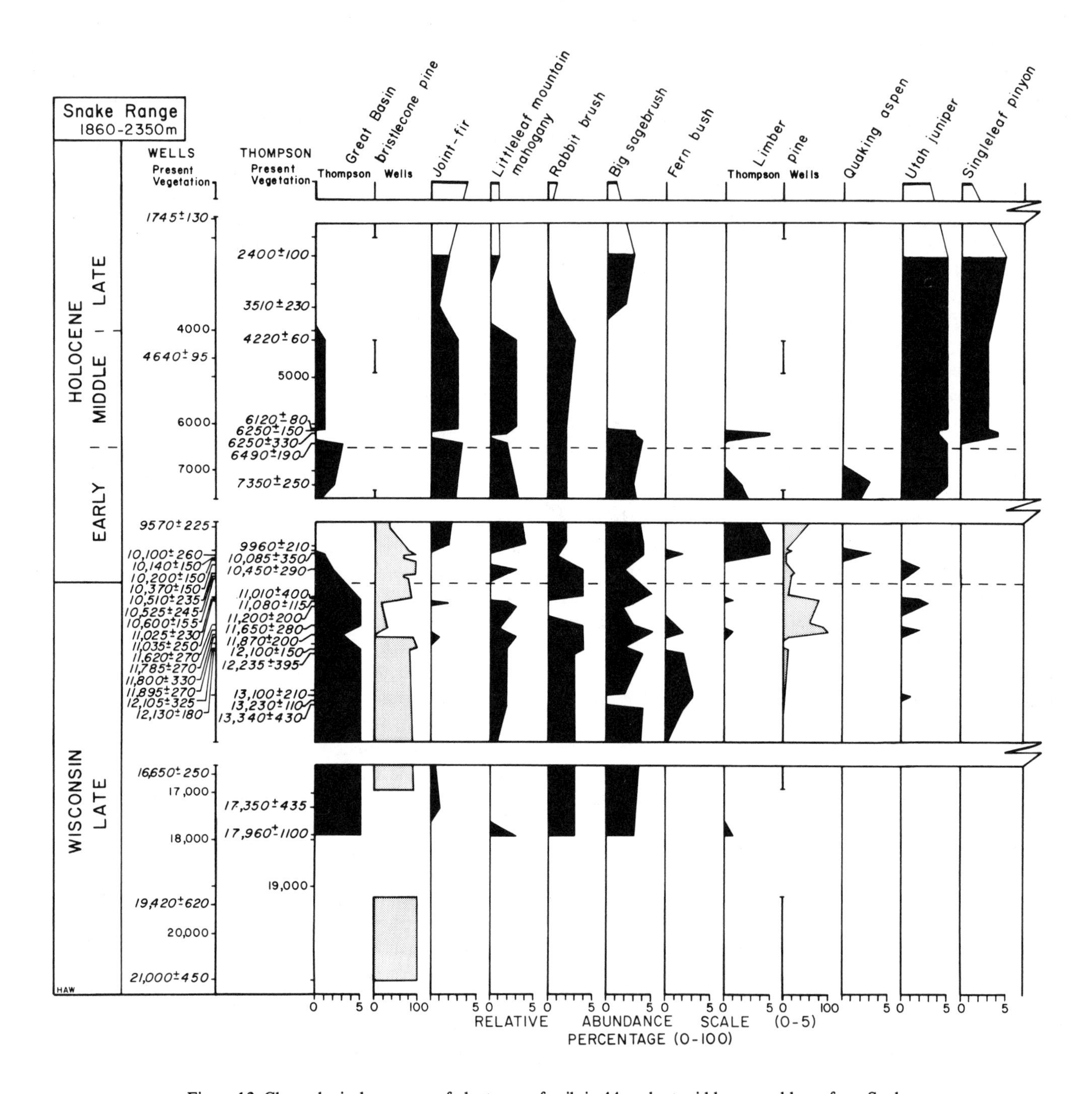

Figure 13. Chronological summary of plant macrofossils in 44 packrat midden assemblages from Snake Range, Nevada. Although the elevational range (1860 to 2350 m) is great, all sites are in pinyon–juniper woodland. Persistence of minor quantities of Great Basin bristlecone pine and other subalpine elements into the Holocene probably reflects the relatively mesic setting of the sites. Thompson data in black uses relative abundance scale: 1=rare, 2=uncommon, 3=common, 4=very common, 5=abundant. Stippled percentages from Wells (1983).

forest at higher elevations and replaced shadscale desertscrub at lower elevations. It was the associate of subalpine conifers growing as much as 1,000 m lower than today, and it was the dominant shrub on fine-grained soils in valleys.

The shift from Late Wisconsin vegetation to modern regimes in the Holocene occurred in different fashions in different portions of the Great Basin. The transition from Late Wisconsin conifer forests or woodlands began in most areas by 11.0 ka. At the Eleana Range, Nevada, in the southern Great Basin, a relatively modern pinyon-juniper woodland was established by 11,700 ± 85 yr B.P. (USGS-914; Spaulding, 1985a). The disappearance of western juniper from Winnemucca Lake in the western Great Basin by 11.0 ka probably marks the development of Great Basin desertscrub in the area. Most other sites have more stages in the Holocene vegetation developmental sequence. The early Holocene was a transitional period in the ranges of western Utah and eastern Nevada, when subalpine conifers of Wisconsin forests were gradually replaced by more xeric woodland and forest trees. By 8.9 to 7.4 ka in Meadow Valley Wash, Nevada (37°31′N; Madsen, 1973; Wells, 1983), the Confusion and Wah Wah ranges, Utah (38°30′-39°05′N; Wells, 1983), and the Snake Range, Nevada (Thompson, 1984), juniper woodlands with Utah and Rocky Mountain junipers were established coincident with the development of desertscrub communities in the southern deserts. However, the late arrival of singleleaf pinyon (Thompson and Hattori, 1983) and the replacement of big sagebrush desertscrub by shadscale desertscrub at 6.6 to 6.0 ka are unique events not easily correlated with developments elsewhere in the Southwest. Records of singleleaf pinyon from the Eleana and Sheep ranges by 11.5 ka suggest that its migration into the interior of the Great Basin was delayed until the development of warmer climates or increased summer precipitation (Thompson and Hattori, 1983) or the arrival of Archaic peoples (Mehringer, 1986). Thus, two distinct middle Holocene vegetation types characterize the Great Basin. The arrival of joint-fir (*Ephedra viridis*) and the local retreat of Rocky Mountain juniper at about 2,380 ± 100 yr B.P. (A-2435) in the Toquima Range, Nevada (39°N), mark a shift to a drier woodland in the late Holocene (Thompson and Kautz, 1983).

Chronology

The chronology for the Southwest based on the vegetational history proposed by Van Devender and Spaulding (1979) was found to be basically accurate with examination of the current packrat midden data set. Refinements of vegetation boundaries and regional differences are discussed below.

Middle Wisconsin/Late Wisconsin. Fifteen midden sequences from the region span middle to Late Wisconsin, showing little difference in vegetation. The pinyon-juniper-oak woodland in the Hueco Mountains, Texas, was similar for 42.0 to 10.8 ka (Van Devender and others, 1987). In Owl Canyon, Nevada, Utah juniper woodland persisted from 32.9 to 13.2 ka (Spaulding, 1985a). Middle Wisconsin aridity ended between 24.1 and 21.8 ka at Rio Grande Village, Texas (Van Devender, 1987b), and between 20.2 and 19.3 ka in the Specter Range, Nevada (Spaulding, 1985a).

Late Wisconsin full-glacial/late-glacial. A total of 22 packrat midden sites from across the Southwest show little or no difference between the vegetation of the full-glacial (22 to 14 ka) and the late-glacial (14 to 11 ka). A few sites have evidence of a modest warming with more xeric vegetation after 13.0 to 12.0 ka: Cottonwood Canyon in Utah, Pontatoc Ridge (Van Devender, 1987c) and the western Grand Canyon in Arizona (Phillips, 1977; Fig. 12), and the Turtle Mountains in California (Wells and Berger, 1967). On the basis of the terrestrial vegetational record, the division of the Late Wisconsin into two bioclimatic units is unwarranted for the Southwest. A 13- to 12-ka vegetational record from most areas in the Southwest would probably represent the local vegetation for the entire Late Wisconsin. This is probably due to the continued effects of the continental ice sheets on the circulation patterns of western North America, producing pluvial regimes. The modest record of vegetational change in the late-glacial could be from sites or regions where moisture was not so limiting that background temperature changes could be influential.

Late Wisconsin/early Holocene. A total of 48 sites have midden assemblages that allow inferences about the major vegetational changes between 11.7 and 10.7 ka at the end of the Late Wisconsin. In the Chihuahuan and Sonoran deserts, the beginning of the Holocene vegetation was very close to 11.0 ka. The best sites are Rio Grande Village (11.5/10.5 ka), Maravillas Canyon (11.2/10.3 ka), and Hueco Mountains (11.0/10.9 ka) in Texas (Van Devender, 1987b), San Andres Mountains in New Mexico (11.0/10.9 ka; Fig. 14; Van Devender and Toolin, 1983), Tinajas Altas Mountains in Arizona (11.04/10.95 ka; Fig. 10; Van Devender, 1987c), and Whipple Mountains in California (11.4/10.9 ka; Fig. 11).

On the Colorado Plateau, a combination of several sites (Allen Canyon, Utah, of Betancourt, 1984; Hance Canyon, Arizona, of Cole, 1981) suggests a LW/EH vegetation boundary at 11.3/10.7 ka. One site (Fishmouth Cave, Utah) exhibited a secondary vegetational change at 10.4/9.7 ka (Betancourt, 1984), which may become better defined as work progresses in the region.

Midden sequences in the Mohave Desert have poor resolution across the Late Wisconsin/early Holocene boundary. The vegetational change was after 11.4 ka in some records and before 11.7 ka in others. Three sites indicate significant vegetational changes before 11.0 ka: Death Valley in California (before 11.2 ka; Wells and Woodcock, 1985), and Point of Rocks (before 11.7 ka; Spaulding, 1985a) and the Sheep Range in Nevada (before 11.6 ka; Spaulding, 1981). A nearby Sheep Range record yielded a younger boundary (11.4/10.0 ka; Spaulding, 1981).

The vegetational change at the end of the Late Wisconsin in the Great Basin was after 11.0 and before 10.5 ka. The best midden sequences are from the Snake Range, Nevada (11.0/10.5 ka; Fig. 13; Thompson, 1984, 1987), and the Wah Wah Range,

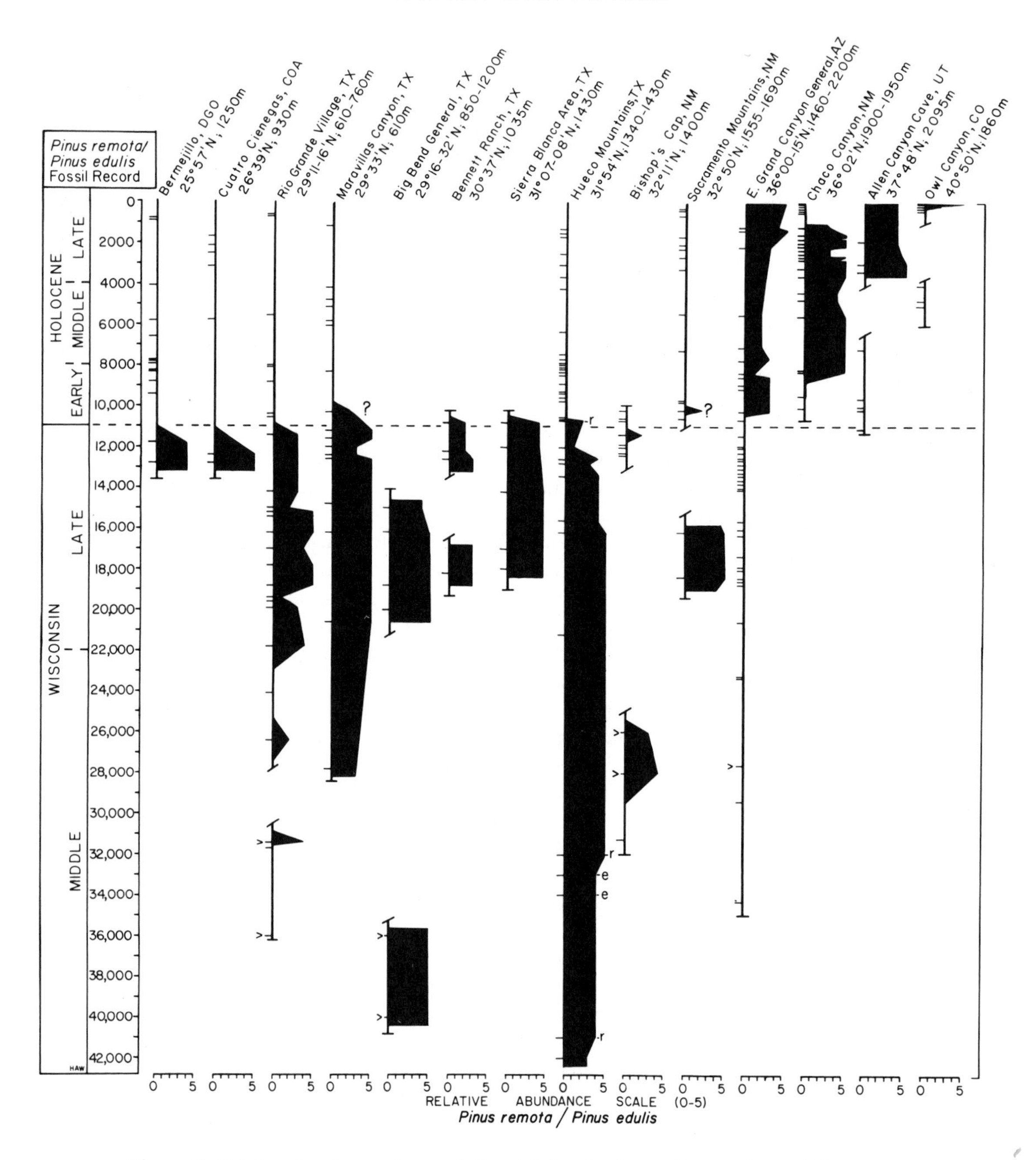

Figure 14. Relative abundances of papershell and Colorado pinyons from packrat midden sequences along a latitudinal gradient from Durango, Mexico, to northern Colorado. Data from references cited in text. Length of baseline indicates midden coverage for each area. Radiocarbon dates marked on baseline. >=infinite date, ?=questionable record. From Van Devender (1986c).

Utah (11.9/10.3 ka; Wells, 1983). The Eleana Range, Nevada, yielded a record of an early vegetational change (before 11.7 ka; Spaulding, 1985a).

Most midden chronologies from arid North America reflect a major vegetational change marking the end of the Wisconsin at ca. 11.0 ka. Verification or replication of the early dates in the northern Mohave Desert and southern Great Basin would lead to revision or amplification of this time boundary. In light of these data, the 10.0-ka date proposed for the end of the Wisconsin by the INQUA Holocene Commission (Olausson, 1982), is inappropriate for the Southwest.

Early Holocene/middle Holocene. The early Holocene was a transitional period between Late Wisconsin and more modern environments of the middle and late Holocene. A total of

38 packrat midden sites have yielded assemblages that reflect a vegetational shift at the beginning of the middle Holocene at 9.9/7.8 ka. In the Chihuahuan Desert this was at 9.0/7.8 ka, with the best records at Rio Grande Village (9.0/8.6 ka; Fig. 5; Van Devender, 1986a, 1986b), Maravillas Canyon (before 8.2 ka; Van Devender, 1987c), and the Hueco Mountains (8.3–8.1 ka; Fig. 6; Van Devender and others, 1987) in Texas, and the Sacramento Mountains (9.6/7.4 ka; Van Devender and others, 1984) and the San Andres Mountains (after 9.1 ka; Van Devender and Toolin, 1983) in New Mexico. The Fra Cristobal Mountains, New Mexico, yielded a relatively late vegetational change (after 7.8 ka; Van Devender, 1987b). The Hueco and Fra Cristobal mountains sites are in the upper elevational range of the early Holocene oak-juniper woodland (1,340 to 1,430 m, 1,675 to 1,740 m), where woodland plants may have persisted as local relicts at favorable sites.

On the Colorado Plateau, chronological coverage is not good through the early/middle Holocene vegetational change. Montane conifers persisted at surprisingly low elevations (1,585 m) until 10 ka. Utah juniper arrived by 9.7 ka at Fishmouth Cave and by 9.5 ka and 9.2 ka at nearby sites, suggesting a major vegetational change (Betancourt, 1984). At Chaco Canyon the transition from montane conifers to a pinyon-juniper woodland was at 10.0 to 8.3 ka (Fig. 7; Betancourt and others, 1983). At Nankoweap/Chuar Valley in the eastern Grand Canyon there was a vegetational change at 9.1/7.9 ka (Cole, 1981, 1982).

In the Sonoran Desert the vegetational change at the end of the early Holocene was at about 8.9 ka. The best records are from the Puerto Blanco Mountains (9.1/8.8 ka; Van Devender, 1987a), the Tinajas Atlas Mountains (8.97/8.91 ka; Fig. 10; Van Devender, 1987c), and the Wellton Hills (before 8.8 ka; Van Devender, 1973) in Arizona, and the Whipple Mountains in California (8.9/8.5 ka; Fig. 11; Van Devender, 1987c). A late record (7.9 ka) for woodland plants in the desert from the New Water Mountains, Arizona (Van Devender, 1977), may be due to survival of a relict population at a relatively high elevation (605 to 615 m) for xeric juniper woodland.

In the Mohave Desert the beginning of the middle Holocene is not clearly marked in the vegetation but appears to have been between 8.5 and 8.4 ka. The best records are from the western Grand Canyon, Arizona (after 8.5 ka; Fig. 12; Phillips, 1977), the Marble Mountains Locality C, California (8.9/7.9 ka; Spaulding, 1980), and the Skeleton Hills (8.8/8.2 ka; Spaulding, 1985b), Nevada. Older records for this vegetational change are available from the Marble Mountains Locality B (near 9.5 ka; Spaulding, 1980) and Owl Canyon, Nevada (9.6–9.3 ka; Spauding, 1985a). Significantly younger records are from Lucerne Valley, California (after 7.8 ka; King, 1976), and Frenchman Flats, Nevada (after 7.8 ka; Wells and Berger, 1967).

In the Great Basin, recognition of a vegetational change at the beginning of the middle Holocene is complicated by differences in radiocarbon dating of midden and pollen sequences and by the late arrival of singleleaf pinyon. There is a vegetational change between 8.9 ka and 8.6 ka in the Snake Range, Nevada (10.0/7.4 ka; Fig. 13; Thompson, 1984, 1987). In pollen profiles the vegetation changes slightly later, e.g., Snowbird Bog (ca. 8.0 ka; Madsen and Currey, 1979) and the Raft River Mountains (ca. 7.5 ka; Mehringer, 1977) in Utah, and the Ruby Marshes in Nevada (8.5/8.0 ka; Thompson, 1984). A major secondary change in the Middle Holocene of the Great Basin (ca. 6.2 ka) was primarily marked by the arrival of singleleaf pinyon from its Late Wisconsin range in the Mohave and Sonoran deserts. The best records of this event are from the Snake River (6.3/6.1 ka; Fig. 13; Thompson, 1984, 1987) and Gatecliff Rockshelter in the Toquima Range, Nevada (ca. 6.0 ka; Thompson and Kautz, 1983). At Hidden Cave (ca. 6.5 ka; Wigand and Mehringer, 1985) and the Ruby Marshes (ca. 6.7 ka; Thompson, 1984) in Nevada, big sagebrush communities gave way to shadscale desertscrub. This vegetational change is unique for the Southwest and needs further study.

In summary, the vegetational change marking the early/middle Holocene transition was somewhat more gradual than the vegetational change at the end of the Wisconsin. Favored sites in the Chihuahuan, Sonoran, and Mohave deserts supported woodland trees until 7.8 ka, leading to the inference of a widespread vegetational change at ca. 8.0 ka (Van Devender, 1977; Van Devender and Spaulding, 1979). However, at most sites across the Southwest, the final disappearance of Wisconsin woodland or forest plants from desert sites was at 8.9 to 8.4 ka, suggesting an earlier change. This is further complicated by a few earlier dates in the Mohave Desert and Colorado Plateau and by a secondary vegetational change at about 6.1 ka in the Great Basin.

Middle Holocene/late Holocene. The final modernization of the vegetation at the beginning of the late Holocene is expressed at a total of 13 sites. In the Chihuahuan Desert this boundary is a major vegetational change at 4.4/4.0 ka in areas where middle Holocene desert–grassland gives way to subtropical desertscrub (Van Devender, 1987c). The Hueco Mountains, Texas, and the Sacramento Mountains, New Mexico, in the northern Chihuahuan desert have the most marked vegetational changes. The boundary is not as well marked in other sequences from desert-grassland at higher elevations or desertscrub from lower areas.

On the Colorado Plateau there was little difference between the middle and late Holocene vegetation. Perhaps the most dramatic change is recorded at Allen Canyon Cave at 7.2 ka, when ponderosa pine was dominant, and 3.4 ka, when Colorado pinyon and other xerophytes make their first appearance (Betancourt, 1984; 1987). The disappearance of Colorado pinyon at Chaco Canyon, New Mexico, at 1.2 to 1.0 ka is a major vegetational change that was apparently due to prehistoric fuel harvesting (Betancourt and others, 1983). Ponderosa pine and Douglas fir last appeared at 2.4 to 2.2 ka.

In the Sonoran Desert, the beginning of the late Holocene marks a major vegetational change at ca. 4.0 ka in the Arizona Upland subdivision, where a cool, frost-limited desertscrub is replaced by the modern subtropical desertscrub (Van Devender, 1987c). This change is best exhibited in the Puerto Blanco Mountains (5.2/3.5 ka; Van Devender, 1987a) and Tinajas Altas Mountains (after 4.0 ka; Van Devender, 1987c), Arizona. In warmer, drier areas at low elevation in the Lower Colorado River valley, the vegetational change is not well marked (Cole, 1986).

In most of the Mohave Desert, the middle/late Holocene vegetational boundary is not well marked, with essentially modern vegetation established in the middle Holocene. One exception is the Eureka Valley, California, in the northern Mohave Desert, where creosote bush arrived between 5.4 and 3.9 ka (Spaulding, 1980).

In the Great Basin the vegetation of the late Holocene is similar to that of the second half of the middle Holocene. There are a few indications of vegetational/climatic changes close to 4.0 ka. The vegetation was modern at Raft River, Utah, by 3.5 ka (Mehringer, 1977). The Ruby Marshes were again wet after a dry period at 6.7 to 4.7 ka, when big sagebrush returned (Thompson, 1984). At Gatecliff Rockshelter, Nevada, the modern vegetation developed after 2.4 ka, with a decline in Rocky Mountain juniper and an increase in joint-fir (Thompson and Kautz, 1983).

In summary, the final modernization of the vegetation of the Southwest was complete by about 5.0 ka, as desert-grassland and cool desertscrub were replaced by subtropical desertscrub communities in the Chihuahuan and Sonoran deserts (Van Devender, 1987b, 1987c), and creosote bush moved to its modern northern limits in the Mohave Desert (Spaulding, 1980, 1983, 1987). Otherwise, relatively modern vegetation was established in the middle Holocene by 8.9 to 8.4 ka in most areas or 6.1 ka in the Great Basin. In areas on the Colorado Plateau and in the central Great Basin, a final change occurred after 2.4 ka. The activities of prehistoric and historic peoples in the Southwest resulted in further vegetational changes.

Paleoclimates

The packrat midden record provides an excellent proxy record for the climatic history of the Southwest and the evaluation of various paleoclimatic hypotheses.

Late Wisconsin. Compared to today the principal features of the Late Wisconsin climate were the strong summer cooling and the shift to winter precipitation from Pacific frontal storms, with greatly diminished monsoonal flow. Evidence cited for colder winter temperatures (e.g., elevational lowering or southerly displacement of the ranges of cold-limited plants) can be interpreted as well as reflecting cooler summers and/or shifts in rainfall seasonality (e.g., the absence of lechuguilla in the northern Chihuahuan Desert and saguaro in the Sonoran Desert). The presence of cold-tolerant plants such as Great Basin bristlecone pine, limber pine, and big sagebrush in the Late Wisconsin can only reflect summer temperatures. In contrast, records supporting equable climates with mild winters and cool summers are more convincing, e.g., lechuguilla and Chihuahuan crucifixion thorn in pinyon-juniper-oak woodlands at Rio Grande Village and Livingston Hills, Texas (Van Devender and others, 1985), and shadscale moving to higher elevations in mixed-conifer forest in the Sheep Range, Nevada (reinterpreted from Spaulding, 1981; Spaulding and others, 1983)

The most likely explanation of mild winter temperatures in the Late Wisconsin is the blocking of Arctic air masses by the continental glaciers (Bryson and Wendland, 1967). Equable climates would release the lower and southerly range limits of woodland and forest plants and animals but not necessarily extirpate the desertscrub residents, resulting in occasional ecological mixtures that would not be found or would be very unusual today. The hypotheses presented by various authors for lower mean annual temperatures with temperatures equally cooled throughout the year during the Late Wisconsin are not supported.

The arguments by Kutzbach and Otto-Bleisner (1982) and Spaulding and Graumlich (1986) for a gradual increase in summer insolation from the full-glacial, culminating in warmer than modern summer temperatures by 12 ka, are not supported (Van Devender, 1987a, 1987c). The series of seven middens from the Eleana Range in Nevada cited in Spaulding and Graumlich (1985) as indicating decreasing effective moisture beginning by 15 ka, actually contain quite similar limber pine macrofossil assemblages from 17.1 to 13.2 ka (Spaulding, 1983). Lower abundances of some plants in the youngest sample are not greatly different from values in some of the older samples. The pinyon-juniper woodland assemblage in the 11.7-ka sample was very similar to a 10.6-ka sample, suggesting that the 11.7-ka date could be too old and that early Holocene climates may have developed later. Apparently the effects of the continental glaciers on regional circulation regimes, especially the patterns of precipitation for the Southwest, were much more important than changes in insolation even as far south as 26°N.

Late Wisconsin precipitation gradients were similar to those of today, with moisture increasing with elevation and a shift from winter to summer rainfall at lower latitudes. The subtropical high-pressure cells were very weak; summer rainfall was completely absent from the Mohave and Great Basin deserts. Evidence for modest summer rainfall was available from east of 112°42′W (Ajo Mountains, Arizona, of Van Devender, 1987a) and south of 32–33°N (Pontatoc Ridge, Arizona, of Van Devender, 1987c; Sacramento Mountains, New Mexico, of Van Devender and others, 1984; Hueco Mountains, Texas, of Van Devender, 1987b). The summer rainfall climates of the Chihuahuan Desert and the eastern portion of the Sonoran Desert were greatly reduced in area and displaced into Mexico. The arguments of Wells (1979) for increased summer rainfall in the late Wisconsin, based on the expansion of shrub live oak, a dominant in winter rainfall chaparral, are not supported.

Late Wisconsin precipitation appears to have been greater than today throughout the Southwest and can be described as pluvial. However, the relative increase in precipitation was not uniform across the Southwest. Most areas received at least 60 to 100 percent more rainfall than today. The major exception to this was in the northern Mohave Desert and the Great Basin, where the intense rainshadow of the Sierra Nevada limited increases to

15 to 40 percent over today (Dohrenwend, 1984; Spaulding, 1985a; Spaulding and others, 1984; Thompson, 1984). The rainfall gradients between this area east to the Rocky Mountains and south to the Mohave Desert were much steeper than today. In contrast, the winter-summer rainfall gradient from the Mohave Desert east to the Chihuahuan Desert was much weaker than today.

Galloway (1970, 1983) and Brakenridge (1978) suggested that high lake levels and lowering of periglacial conditions on mountain tops during the Late Wisconsin could be explained by decreases of 11–12°C and 7–8°C for each month of the year, with no increase in precipitation. Pluvial lake hydrologic budgets can be balanced by adjusting either temperature or precipitation (Snyder and Langbein, 1962). The undated deposits exhibiting periglacial features in southwestern mountains were assumed to represent full-glacial time and to be solely the result of lowered temperatures, although precipitation regimes were probably important. The cold-dry climate models called for a 1,300- to 1,400-m timberline depression and a 1,000-m snowline depression. If an environmental lapse rate of 6°C per 1,000 m is applied to the data used by Galloway and Brakenridge, the vegetation zones would have been lowered 1,750 m and 1,250 m. These estimates are far in excess of the actual vegetational lowering recorded in the fossil packrat midden record. Cooling of lesser magnitude would be insufficient to fill pluvial lakes without increased precipitation. The Late Wisconsin biotic record does not provide support for the cold-dry paleoclimatic model.

Early Holocene. The climate of the early Holocene was transitional between that of the late Wisconsin and more modern regimes. Winter rainfall continued to be greater than today in most of the Southwest. The southern edge of the winter storm track moved northward in the southern Chihuahuan Desert, resulting in an early (by 9.4 ka) onset of xeric conditions in the Bolson de Mapimi. An increase in summer temperatures was reflected in a modest increase in the development of the Bermuda High and the summer monsoon. In the Sacramento Mountains, New Mexico, in the northern Chihuahuan Desert, the summer rainfall increased to perhaps 30 to 40 percent of the annual total (60 to 70 percent today; Van Devender and others, 1984). In the Sonoran Desert (Puerto Blanco Mountains, Arizona) summer rainfall increased to about 30 percent (40 percent today; Van Devender, 1987a). On the northwestern edge of the Sonoran Desert (Whipple Mountains, California; Fig. 11) summer rainfall was not important (27 percent today) in the early Holocene. Monsoonal expansion farther west or north into the Mohave Desert was unlikely to have maintained early Holocene high water levels in Searles Lake, California. In the early Holocene the abundance of juniper at low elevations in the Lower Colorado River Valley increased from east to west at lower elevations toward the source of winter precipitation (Van Devender, 1987a).

The early Holocene summer monsoon maximum proposed by Spaulding and Graumlich (1986) for the Mohave and Sonoran deserts warrants discussion. The evidence for summer rainfall was the increased abundances of undesignated succulents and grasses in packrat middens from the Eleana Range, Point of Rocks, and Skeleton Hills in Nevada. The succulents in these assemblages are Utah agave, beavertail cactus (*Opuntia basilaris*), Joshua tree, and Mohave prickly pear (*O.* cf. *erinacea*; Spaulding, 1983, 1985a, 1985b). All of these are Mohave Desert plants that occur mostly in areas where winter rainfall is dominant, including areas where summer rainfall is unimportant or absent. These succulents are little dependent on summer rainfall. Much of the increase in grasses in the samples was for unidentified Poaceae, which could be species that utilize summer or winter precipitation. Samples from Point of Rocks site 2, dated at 9.3 to 9.6 ka, contained remains of six weeks threeawn (*Aristida* cf. *adscensionis*) and fluff grass (*Erioneuron pulchellum*), an annual and a small perennial grass that are widespread in summer-rainfall habitats in the Sonoran and Chihuahuan deserts. However, these opportunistic grasses can also flower in response to winter-spring moisture if temperatures are warm. In the Mohave Desert they are mostly active from February to June (Munz, 1974), although they respond readily to occasional tropical storms from late August to early October (A. Sanders, personal communication, 1987). Indian rice grass(*Oryzopsis hymenoides*), an important perennial in winter-rainfall areas in the Mohave and Great Basin deserts, was also in the early Holocene assemblages. These data are unconvincing as evidence of a summer-rainfall maximum in the early Holocene. The expansion of ponderosa pine and Whipple cholla into the eastern Grand Canyon in Arizona in the early Holocene could be due to a combination of increasing summer temperatures and rainfall and maintenance of Late Wisconsin winter rainfall instead of a monsoonal maximum (Cole, 1985). The evidence for a summer monsoonal maximum in the early Holocene of the Southwest is equivocal at best.

Middle Holocene. With the final demise of the Late Wisconsin winter-rainfall regime after 8.9 ka, a relatively modern climatic regime was established. Reduction of winter rainfall in the Mohave Desert and Great Basin resulted in drought. However, dramatically increased summer temperatures produced maximum summer monsoonal rainfall in the Chihuahuan and Sonoran deserts (Van Devender, 1987a, 1987c; Van Devender and Spaulding, 1979; Van Devender and others, 1984) and on the Colorado Plateau (Betancourt, 1984; Betancourt and others, 1983). Permanent water remained in now-dry playa lakes in New Mexico (Markgraf and others, 1984), although the seasonality and source area of the moisture shifted dramatically.

Summer rainfall was considerably greater than today in the Sonoran Desert (Puerto Blanco and Tinajas Altas mountains, Arizona), but only slightly so near the Mohave Desert boundary (Whipple Mountains, California; Van Devender, 1987a, 1987c). The enhanced summer monsoonal hypothesis of Martin (1963) has been substantiated while Antev's (1948) Altithermal drought has been restricted to the Mohave Desert, Great Basin (Mehringer, 1986), and possibly coastal southern California. The middle Holocene warm period (8.9 to 4.0 ka) was longer than Antev's Altithermal (7.0 to 4.5 ka). Although the middle Holocene was probably the time of maximum westward penetration of the summer monsoon, and possibly of maximum winter drought, warm-season rainfall was probably only significantly greater than today in the mountains of the eastern Mohave Desert and west of the Lower Colorado River valley, and not in the desert lowlands. The only important consequence of this increase in summer rainfall in this area may have been that conditions for seedling establishment of plants were met more often.

Although summer temperatures were hot, the frequency of severe winter freezes due to southerly incursions of Arctic air was apparently much greater than today. Intense freezes may have prevented subtropical desertscrub plants, including lechuguilla, from invading the northern Chihuahuan Desert and favored the development of grassland. Winter freezes apparently prevented more subtropical Sonoran desertscrub plants such as the organ pipe cactus (*Stenocereus thurberi*) and Mexican jumping bean (*Sapium biloculare*) from reaching the Puerto Blanco Mountains, Arizona (Van Devender, 1987a). Freezes may well have limited the northern range of creosote bush in the northern Mohave Desert, preventing its migration into the Eureka Valley, California (Spaulding, 1980).

Late Holocene. By 4.0 ka the modern climatic regime was established, as reduction in the frequency of winter freezes and adequate summer rainfall combined to produce the most subtropical climates in the Holocene for North America (Van Devender, 1987a). In the northern Chihuahuan Desert and the Sonoran Desert, more subtropical elements of the flora migrated to their present northern ranges. In the Mohave Desert, Colorado Plateau, and the Great Basin, where winter freezes are more important in the modern climate, little difference was seen between middle and late Holocene climates. The deserts of North America reached their greatest area in the late Holocene. In the Sonoran Desert (Puerto Blanco Mountains, Arizona) the last ca. 1,000 yr have been even drier, although there was a short, cool, wet period about 990 ± 50 yr B.P. (A-3972; Van Devender, 1987a). A similar shift to a more xeric vegetation type was seen in the central Great Basin (Toquima Range) after 2.4 ka (Thompson, 1984, 1987).

Biogeography

The fossil packrat midden record provides an excellent view of the phytogeography of the Southwest for species, communities, and vegetation zonation during the last glacial/interglacial cycle. The views of Gleason (1939), that plant communities are composed of individual species with their own physiological tolerances and geographical ranges limited by environmental factors, are overwhelmingly supported (Cole, 1982, 1985; Spaulding and others, 1983; Van Devender, 1987a). Community composition has changed continuously for the last 11,000 yr as Wisconsin plants retreated from desert lowlands and as woodland, grassland, and desertscrub plants migrated into new areas. Relatively modern compositions of many southwestern plant communities were only attained in the last 4,000 to 5,000 yr at the beginning of the late Holocene. Community composition continued to vary subsequently, reflecting lesser climatic fluctuations in modern regimes. Davis (1986) and Webb (1986) suggested that differential responses of plant species to climatic changes and continuous variation in climate on several time scales result in dynamic plant communities that are in flux as they track changing climates. Rarely do these communities reach stable equilibrium in the sense often used by modern ecologists. Similar processes occur in southwestern plant communities, with the exception of simple communities in extremely harsh climates or substrates.

Late Wisconsin. Under Late Wisconsin climates, the distributions of virtually all plant species and most plant communities were very different from today. The geographic distributions of Great Basin and Rocky Mountain plants such as big sagebrush, shadscale, limber pine, Engelmann and blue spruces, and Rocky Mountain juniper were greatly expanded. Other plants such as ponderosa pine, Colorado pinyon, and saguaro were displaced to the south into greatly reduced ranges.

The only areas in the Southwest where plant communities were unchanged by Late Wisconsin climates were those severely limited by climatic extremes or inhospitable substrates. Big sagebrush desertscrub in the fine-grained valley soils and the shadscale desertscrub of saline lake margins in the Great Basin were probably very similar to those of today. Some of the halophytic shrub communities on the hypersaline Mancos Shale on the Colorado Plateau have probably been stable as well. Similarly the halophytic shrub communities and the adjacent creosote bush-white bursage desertscrub communities of the Gran Desierto surrounding the head of the Gulf of California in Sonora were probably little changed.

In areas of continental glaciation, the Holocene vegetational sequence begins with pioneers on bare, unweathered substrate and ends with the modern vegetation. The only areas in the Southwest that lacked vegetation in the late Wisconsin were small areas covered by montane glaciers and pluvial lakes. In most areas, the vegetation changed from one type to another through time.

The subalpine or boreal forest was greatly expanded during the Late Wisconsin. The lowering of treeline based on the pollen record is modest. A pollen record from Hay Lake at 2,780 m in the White Mountains of Arizona recorded a subalpine forest throughout the late Wisconsin (Jacobs, 1983). Upper treeline was probably near 2,900 m for the Colorado Plateau in general. This cool, mesic forest, dominated by Engelmann spruce, subalpine fir, and limber pine at higher elevations and Douglas-fir at lower elevations, probably covered a 1,600-m elevational range on mesic slopes. Apparently the modern pine forest zone dominated by ponderosa pine was absent from the vegetational zonation at middle latitudes (Cole, 1982; Betancourt, 1984; Van Devender and Toolin, 1983); mixed-conifer forest directly contacted woodlands. Late Wisconsin corridors between forested areas would have been in mixed-conifer forest. Isolated populations of subalpine forest trees such as Engelmann spruce and subalpine fir in the Graham Mountains, Arizona, were not connected, although flight distances by jays and nutcrackers (Corvidae) for seed dispersal would have been much reduced, and riparian gallery forest corridors in valleys would have been well developed. The Mogollon Rim was probably a moist forest corridor connecting the Rocky Mountains of south-central Colorado with the White and San Francisco mountains of Arizona. Rocky Mountain bristlecone pine (*Pinus aristata*) dispersed along the Mogollon Rim to establish the isolated population on the San Francisco Mountains, but not to the interior highlands of the Colorado Plateau.

In the Great Basin, subalpine conifers, including Great Basin bristlecone pine, limber pine, and Douglas-fir, expanded from the Rocky Mountains into eastern Nevada in the late Wisconsin. Limber pine apparently extended farther into the isolated ranges of the central Great Basin. The only record of a Sierra Nevada plant expanding its range to the east in the late Wisconsin is for western juniper at Winnemucca Lake, Nevada, in the western Great Basin (Thompson and others, 1986).

In both the Chihuahuan and Sonoran deserts, pinyon-juniper-oak woodlands with papershell, Colorado, or singleleaf pinyon were present from about 600- to 1,525-m elevation. Montane conifers now isolated on the Chisos, Davis, Guadalupe, Grahams, Santa Catalina, and other mountains probably reached these areas through long-distance transport and not by direct vegetational corridors. Pinyon-juniper woodlands were displaced southward out of the Great Basin and the Colorado Plateau, where mixed-conifer forests contacted Great Basin desertscrub directly. In the modern Sonoran Desert, xeric juniper woodlands extended down to about 300 m, with Mohave desertscrub at lower elevations along the Colorado River. A simplification of plant zonation similar to that of the Late Wisconsin without pine forest in middle elevations, or pinyon-juniper woodlands in the Great Basin, occurs today at higher latitudes in northern Nevada (Thompson, 1984; Spaulding, 1985). The replacement of pinyon-juniper woodland by limber pine–Rocky Mountain juniper woodland in drier areas on the Colorado Plateau and limber pine woodland in the Great Basin is an interesting convergence in physiognomic structure by distantly related trees.

Holocene. The Holocene was a very dynamic time in the Southwest. As summers warmed and the summer monsoon began to develop in the early and middle Holocene, woodlands retreated from southern deserts and migrated with the newly formed ponderosa pine forest north to the Rocky Mountains, onto the Colorado Plateau, through the northern Mohave Desert, and into the Great Basin. Colorado pinyon apparently expanded its late Wisconsin latitudinal range of 2–3° to 10° (Van Devender and others, 1984). Ponderosa pine probably dispersed from relict populations in southern Arizona and southern New Mexico through the Rockies to Canada and recontacted the Sierra Nevada populations. Over most of the Pleistocene the Rocky Mountains have supported depauperate spruce-fir forest, tundra, and ice. Many trees, such as white fir, Colorado pinyon, oneseed juniper, Utah juniper, and ponderosa pine, traditionally thought of as characteristic of the Rocky Mountains, migrated upward and northward from the Colorado Plateau and highlands below the Mogollon Rim (Betancourt, 1987).

In contrast, the range of papershell pinyon shrank into modern relict populations southeast of the Chihuahuan Desert in Texas and on mountaintops in northeastern Mexico. A suite of Mexican pine–oak woodland trees, especially the evergreen oaks and Mexican pinyon (*Pinus cembroides*), probably dispersed into Texas from the Sierra Madre Oriental, and into Arizona and New Mexico from the Sierra Madre Occidental as northern forest and woodland trees retreated.

The middle Holocene, with maximum summer rainfall and frequent winter freezes, favored the development of grassland. Unbroken expanses of grassland probably extended from central Canada through the Great Plains to the Mexican Plateau and west to Arizona (Van Devender and others, 1984). Warmth-loving perennial C-4 grasses probably expanded into the Great Plains from the Mexican Plateau and displaced cold-adapted C-3 grasses in the central Great Plains.

In the late Holocene the deserts of North America reached their maximum extent at the expense of grasslands. The ranges of dominants such as creosote bush, lechuguilla, and organ pipe cactus expanded greatly. Desertscrub corridors opened across the Continental Divide in southeastern Arizona, connecting the Chihuahuan and Sonoran deserts for the first time since the last interglacial. Subtropical desertscrubs, with thornscrub on their southern peripheries, reached their maximum development for the Holocene as the northern edge of the Neotropics shifted into its modern position.

APPENDIX I. COMMON AND SCIENTIFIC NAMES OF PLANTS MENTIONED IN TEXT AND FIGURES.

Allthorn	*Koeberlinia spinosa*
Beargrass	*Nolina* sp.
Beavertail cactus	*Opuntia basilaris*
Bigelow beargrass	*Nolina bigelovii*
Big sagebrush	*Artesmisia tridentata*
Bigtooth maple	*Acer grandidentatum*
Black brush	*Coleogyne ramosissima*
Blind prickly pear	*Opuntia rufida*
Blue palo verde	*Cercidium floridum*
Blue spruce	*Picea pungens*
Boojum tree	*Fouquieria columnaris*
Brittle bush	*Encelia farinosa*
California juniper	*Junperus californica*
Catclaw acacia	*Acacia greggii*
Chihuahuan cricifixion thorn	*Castela stewarti*
Colorado pinyon	*Pinus edulis*
Common juniper	*Juniperus communis*
Creosote bush	*Larrea divaricata*
Currant	*Ribes sp.*
Desert agave	*Agave deserti*
Desert almond	*Prunus fasciculata*
Desert lavender	*Hyptis emoryi*

Douglas fir	*Pseudotsuga menziesii*	Prickly pears	*Opuntia* spp.
Engelmann spruce	*Picea engelmannii*	Pygmy cedar	*Peucephyllum schottii*
Fendler barberry	*Berberis fendleri*	Quaking aspen	*Populus tremuloides*
Fern bush	*Chamaebatiaria millefolium*	Rabbit bush	*Chrysothamnus* sp.
Fluff grass	*Erioneuron pulchellum*	Ragged rock flower	*Crossosoma bigelovii*
Foothills palo verde	*Cercidium microphyllum*	Redberry juniper	*Juniperus erythrocarpa*
Fourwing saltbush	*Atriplex canescens*	Red osier dogwood	*Cornus stolonifera*
Gambel oak	*Quercus gambelii*	Rocky Mountain bristlecone pine	*Pinus aristata*
Greasebush	*Forsellesia nevadensis*	Rocky Mountain juniper	*Juniperus scopulorum*
Greasewood	*Sarcobatus vermiculatus*	Rose	*Rosa* sp.
Great Basin bristlecone pine	*Pinus longaeva*	Russet buffaloberry	*Shepherdia canadensis*
Hinckley oak	*Quercus hinckleyi*	Saguaro	*Carnegiea gigantea*
Hollyleaf bursage	*Ambrosia ilicifolia*	Sandpaper bush	*Mortonia scabrella*
Honey mesquite	*Prosopis glandulosa*	Scrub oak	*Quercus pungens*
Hop hornbean	*Ostrya knowltoni*	Shadscale	*Atriplex confertifolia*
Indian ricegrass	*Oryzopsis hymenoides*	Shrub live oak	*Quercus turbinella*
Ironwood	*Olneya tesota*	Silver wolfberry	*Lycium puberulum*
Joint-fir	*Ephedra viridis*	Singleleaf ash	*Fraxinus anomala*
Joshua tree	*Yucca brevifolia*	Singleleaf pinyon	*Pinus monophylla*
Juniper	*Juniperus* sp.	Six weeks threeawn	*Aristida adscensionis*
Lechuguilla	*Agave lechuguilla*	Skunk bush	*Rhus trilobata*
Limber pine	*Pinus flexilis*	Snowberry	*Symphoricarpos* sp.
Littleleaf mountain mahogany	*Cercocarpus intricatus*	Sotol	*Dasylirion* sp.
Mat saltbush	*Atriplex corrugata*	Spiny greasebush	*Forsellesia spinescens*
Mexican cliffrose	*Cowania mexicana*	Spiny hop sage	*Grayia spinosa*
Mexican jumping bean	*Sapium biloculare*	Stellate rose	*Rosa stellata*
Mexican pinyon	*Pinus cembroides*	Subalpine fir	*Abies lasiocarpa*
Mohave prickly pear	*Opuntia erinacea*	Utah agave	*Agave utahensis*
Mohave sage	*Salvia mohavensis*	Utah juniper	*Juniperus osteosperma*
Mohave yucca	*Yucca schidigera*	Variable prickly pear	*Opuntia phaeacantha*-type
Mormon tea	*Ephedra nevadensis*	Velvet mesquite	*Prosopis velutina*
Mountain mahogany	*Cercocarpus montanus*	Water birch	*Betula occidentalis*
Narrowleaf yucca	*Yucca angustissima*	Western juniper	*Juniperus occidentalis*
Ocotillo	*Fouquieria splendens*	Whipple cholla	*Opuntia whipplei*
Oneseed juniper	*Juniperus monosperma*	Whipple yucca	*Yucca whipplei*
Organ pipe cactus	*Stenocereus thurberi*	White bursage	*Ambrosia dumosa*
Papershell pinyon	*Pinus remota*	White fir	*Abies concolor*
Plains prickly pear	*Opuntia polyacantha*	Winter fat	*Ceratoides lanata*
Ponderosa pine	*Pinus ponderosa*		

REFERENCES CITED

Antevs, E., 1948, The Great Basin, with emphasis on glacial and post-glacial times; Climatic changes and pre-white man: Bulletin of the University of Utah, v. 38, p. 168–191.

Axelrod, D. I., 1979, Age and origin of Sonoran Desert vegetation: California Academy of Sciences Occasional Papers, v. 132, p. 1–74.

Benson, L. V., 1981, Paleoclimatic significance of lake-level fluctuations in the Lahontan Basin: Quaternary Research, v. 16, p. 390–403.

Betancourt, J. L., 1984, Late Quaternary plant zonation and climate in southeastern Utah: Great Basin Naturalist, v. 44, p. 1–35.

——, 1987, Late Quaternary phytogeography of the Colorado Plateau, *in* Martin, P. S., Van Devender, T. R., and Betancourt, J. L., eds., Fossil packrat middens; The last 40,000 years of biotic change in the arid West: Tucson, University of Arizona Press (in press).

Betancourt, J. L., and Van Devender, T. R., 1981, Holocene vegetation in Chaco Canyon, New Mexico: Science, v. 214, p. 656–658.

Betancourt, J. L., Martin, P. S., and Van Devender, T. R., 1983, Fossil packrat middens from Chaco Canyon, New Mexico; Cultural and ecological significance: 1983 American Geomorphological Field Group Guidebook, p. 207–217.

Billings, W. D., 1951, Vegetational zonation in the Great Basin of western North America; Compt. Rend. du Colloque sur les Bases Ecologiques de la Regeneration de la Vegetation des Zones Arides: Paris, Union International Societe Biol., p. 101–122.

Brakenridge, G. R., 1978, Evidence for a cold, dry full-glacial climate in the American Southwest: Quaternary Research, v. 9, p. 22–40.

Bryson, R. A., 1985, On climatic analogs in paleoclimatic reconstructions: Quaternary Research, v. 23, p. 275–286.

Bryson, R. A., and Lowry, W. D., 1955, The synoptic climatology of the Arizona summer precipitation singularity: American Meteorological Society Bulletin, v. 36, p. 329–339.

Bryson, R. A., and Wendland, W., 1967, Tentative climatic patterns for some late glacial and post-glacial episodes in central North America, *in* Mayer–Oakes, S. J., ed., Life, land, and water: Winnepeg, University of Manitoba Press, p. 271–298.

Carrara, P. E., Mode, W. N., Meyer, R., and Robinson, S. W., 1984, Deglaciation and post-glacial timberline in the San Juan Mountains, Colorado: Quaternary Research, v. 21, p. 42–55.

Clisby, K. H., and Sears, P. B., 1956, San Augustin Plains; Pleistocene climatic changes: Science, v. 124, p. 537–539.

Cole, K. L., 1981, Late Quaternary environment in the eastern Grand Canyon; Vegetational gradients over the last 25,000 years [Ph.D. thesis]: Tucson, University of Arizona, 170 p.

——, 1982, Late Quaternary zonation of vegetation in the eastern Grand Canyon: Science, v. 217, p. 1141–1145.

——, 1985, Past rates of change, species richness, and a model of vegetational inertia in the Grand Canyon, Arizona: The American Naturalist, v. 125,

p. 289–303.
—— , 1986, The Lower Colorado Valley; A Pleistocene desert: Quaternary Research, v. 25, p. 392–400.
Davis, M. B., 1986, Climatic instability, time lags, and community disequilibrium, *in* Diamond, J., and Case, T. J., eds., Community ecology: New York, Harper and Row Publishers, p. 269–284.
Dohrenwend, J. C., 1984, Nivation landforms in the western Great Basin and their paleoclimatic significance: Quaternary Research, v. 22, p. 275–288.
Douglas, A. V., 1976, Past air-sea interactions over the eastern North Pacific Ocean as revealed by tree-ring data [Ph.D. thesis]: Tucson, University of Arizona, 196 p.
Finley, R. B., Jr., 1958, The wood rats of Colorado; Distribution and ecology: University of Kansas Publications, Museum of Natural History, v. 10, p. 213–552.
Galloway, R. W., 170, The full-glacial climate in the southwestern United States: Annals of the Association of American Geographers, v. 60, p. 245–256.
—— , 1983, Full-glacial southwestern United States; Mild and wet or cold and dry?: Quaternary Research, v. 19, p. 236–248.
Gleason, H. A., 1939, The individualistic concept of the plant association: American Midland Naturalist, v. 21, p. 92–110.
Hales, J. E., Jr., 1974, Southwestern United States summer monsoon source—Gulf of Mexico or Pacific Ocean?: Journal of Applied Meteorology, v. 13, p. 331–342.
Houghton, J. G., 1969, Characteristics of rainfall in the Great Basin: Reno, Nevada, Desert Research Institute, 205 p.
Houghton, J. G., Sakamoto, C. M., and Gifford, R. O., 1975, Nevada's weather and climate: Nevada Bureau of Mines Special Publication 2, 78 p.
Huning, J. R., 1978, A characterization of the climate of the California desert: Riverside Desert Planning Staff, California Bureau of Land Management unpublished report, 220 p.
Hunt, C. B., 1956, Cenozoic geology of the Colorado Plateau: U.S. Geological Survey Professional Paper 279, 95 p.
Jacobs, B. F., 1983, Past vegetation and climate of the Mogollon Rim Area, Arizona [Ph.D. thesis]: Tucson, University of Arizona, 166 p.
Johnston, M. C., 1977, Brief resumé of botanical, including vegetational, features of the Chihuahuan Desert region with special emphasis on the uniqueness, *in* Wauer, R. H., and Riskind, D. H., eds., Transactions of the symposium on the biological resources of the Chihuahuan Desert region, U.S. and Mexico, Alpine, Texas, 1974: Washington, D.C., U.S. Department of the Interior, National Park Service Transactions and Proceedings Series 3, p. 335–362.
King, T. J., 1976, Late Pleistocene–early Holocene history of coniferous woodlands in the Lucerne Valley region, Mojave Desert, California: Great Basin Naturalist, v. 36, p. 227–238.
Küchler, A. W., 1966, Potential natural vegetation map: U.S. Geological Survey, scale 1:7,500,000.
Kutzbach, J. E., and Otto-Bleisner, B. L., 1982, The sensitivity of the African-Asian monsoonal climate to orbital parameters changes for 9,000 years B.P. in a low-resolution general circulation model: Journal of the Atmospheric Sciences, v. 39, p. 1177–1188.
Lanner, R. M., and Van Devender, T. R., 1981, Late Pleistocene piñon pines in the Chihuahuan Desert: Quaternary Research, v. 15, p. 278–290.
Leskinen, P. H., 1975, Occurrence of oaks in late Pleistocene vegetation in the Mojave Desert of Nevada: Madroño, v. 23, p. 234–235.
Long, A., and Rippeteau, B., 1974, Testing contemporaneity and averaging radiocarbon dates: American Antiquity, v. 39, p. 205–215.
Madsen, D. B., 1973, Late Quaternary paleoecology in the southeastern Great Basin [Ph.D. thesis]: Columbia, University of Missouri, 125 p.
Madsen, D. B., and Currey, D. R., 1979, Late Quaternary glacial and vegetation changes, Little Cottonwood Canyon area, Wasatch Mountains, Utah: Quaternary Research, v. 12, p. 254–270.
Markgraf, V., Bradbury, J. P., Forester, R. M., Singh, G., and Sternberg, R. S., 1984, San Agustin Plains, New Mexico; Age and paleoenvironmental potential reassessed: Quarternary Research, v. 22, p. 336–343.
Martin, P. S., 1963, The last 10,000 years: Tucson, University of Arizona Press, 87 p.
Martin, P. S., and Mehringer, P. J., Jr., 1965, Pleistocene pollen analysis and biogeography of the Southwest, *in* Wright, H. E., Jr., and Frey, D. G., eds., The Quaternary of the United States: Princeton, New Jersey, Princeton University Press, p. 433–451.
Mehringer, P. J., Jr., 1967, Pollen analysis of the Tule Springs area, Nevada: Nevada State Anthropological Papers, v. 13, p. 129–200.
—— , 1977, Great Basin late Quaternary environments and chronology: Desert Research Institute Publications in the Social Sciences, v. 12, p. 113–167.
—— , 1986, Prehistoric environments, *in* d'Azcvedo, W. L., ed., Handbook of North American Indians, v. 11, Great Basin: Washington, D.C., Smithsonian Institution, p. 31–50.
Mehringer, P. J., Jr., and Ferguson, C. W., 1969, Pluvial occurrence of bristlecone pine, *Pinus aristata,* in a Mohave Desert mountain range: Journal of the Arizona Academy of Science, v. 5, p. 284–292.
Merrill, R. K., and Péwé, T. L., 1977, Late Cenozoic geology of the White Mountains, Arizona: Arizona Bureau of Geology and Mineral Technology Special Paper 1, 65 p.
Meyer, E. R., 1973, Late Quaternary paleoecology of the Cuatro Cienegas Basin, Coahuila, Mexico: Ecology, v. 54, p. 982–995.
Mifflin, M. D., and Wheat, M. M., 1979, Pluvial lakes and estimated pluvial climates of Nevada: Nevada Bureau of Mines and Geology Bulletin, v. 94, 57 p.
Morafka, D. J., 1977, A biogeographical analysis of the Chihuahuan Desert through its herpetofauna: The Hague, W. Junk, 313 p.
Munz, P. A., 1974, A flora of southern California: Berkeley, University of California Press, 1086 p.
Neilson, R. P., 1986, High-resolution climatic analysis and Southwest biogeography: Science, v. 232, p. 27–34.
—— , 1987, Biotic regionalization and climatic controls in western North America: Vegetatio (in press).
Olausson, E., 1982, The Pleistocene/Holocene boundary in southwestern Sweden: Sveriges Geologiska Undersokning, Serie C, NR 74.
Phillips, A. M., 1977, Packrats, plants, and the Pleistocene of the lower Grand Canyon, Arizona [Ph.D. thesis]: Tucson, University of Arizona, 123 p.
Richmond, G. M., 1962, Quaternary stratigraphy of the La Sal Mountains, Utah: U.S. Geological Survey Professional Paper 324, 135 p.
Schmidt, R. H., Jr., 1979, A climatic delineation of the 'real' Chihuahuan Desert: Journal of Arid Environments, v. 2, p. 243–250.
Shreve, F., 1964, Vegetation of the Sonoran Desert, *in* Shreve, F., and Wiggins, I. L., Vegetation and flora of the Sonoran Desert (2 volumes): Stanford University Press, v. 1, pt. 1, p. 6–186.
Snyder, C. T., and W. B., Langbein, 1962, The Pleistocene lake in Spring Valley, Nevada, and its climatic implications: Journal of Geophysical Research, v. 67, p. 2385–2394.
Spaulding, W. G., 1980, The presettlement vegetation of the California Desert: Desert Planning Staff, California Bureau of Land Management, 97 p.
—— , 1981, The late Quaternary vegetation of a southern Nevada Mountain Range [Ph.D. thesis]: Tucson, University of Arizona, 271 p.
—— , 1983, Late Wisconsin macrofossil records of desert vegetation in the American Southwest: Quaternary Research, v. 19, p. 256–264.
—— , 1985a, Vegetation and climates of th last 45,000 years in the vicinity of the Nevada Test Site, south-central Nevada: U.S. Geological Survey Professional Paper 1329, 83 p.
—— , 1985b, Ice-age desert in the southern Great Basin: Current Research in the Pleistocene, v. 2, p. 83–85.
—— , 1987, Vegetational development of the Mojave Desert; The last glacial maximum to the present, *in* Martin, P. S., Van Devender, T. R., and Betancourt, J. L., eds., Fossil packrat middens, The last 40,000 years of biotic changes in the arid West: Tucson, University of Arizona Press (in press).
Spaulding, W. G., and Graumlich, L. J., 1986, The last pluvial climate episodes in the deserts of southwestern North America: Nature, v. 320, p. 441–444.
Spaulding, W. G., Leopold, E. B., and Van Devender, T. R., 1983, Late Wiscon-

sin paleoecology of the American Southwest, *in* Porter, S. C., ed., The late Pleistocene of the United States: University of Minnesota Press, p. 259–293.
Spaulding, W. G., Robinson, S. W., and Paillet, F. L., 1984, Preliminary assessment of climatic change during late Wisconsin time, southern Great Basin and vicinity, Arizona, California, and Nevada: U.S. Geological Survey Water-Resources Investigations Report 84–4328, 40 p.
Thompson, R. S., 1979, Late Pleistocene and Holocene packrat middens from Smith Creek Canyon, White Pine County, Nevada: Nevada State Museum Anthropological Papers, v. 17, p. 362–380.
—— , 1984, Late Pleistocene and Holocene environments in the Great Basin [Ph.D. thesis]: Tucson, University of Arizona, 256 p.
—— , 1985, Palynology and *Neotoma* middens, *in* Fine-Jacobs, B. L., Fall, P. L., and Davis, O. K., eds., Late Quaternary vegetation and climates of the American Southwest: American Association of Stratigraphic Palynologists Contributions Series, v. 16, p. 89–112.
—— , 1987, Late Quaternary vegetation and climate in the Great Basin; The packrat midden evidence, *in* Martin, P. S., Van Devender, T. R., and Betancourt, J. L., eds., Fossil packrat middens; The last 40,000 years of biotic change in the arid West: Tucson, University of Arizona Press (in press).
Thompson, R. S., and Hattori, E. M., 1983, Packrat, *Neotoma* middens from Gatecliff Shelter and Holocene migrations of woodland plants: Anthropological Papers of the American Museum of Natural History, v. 59, p. 157–167.
Thompson, R. S., and Kautz, R. R., 1983, Pollen analysis: Anthropological Papers of the American Museum of Natural History, v. 59, p. 136–151.
Thompson, R. S., and Mead, J. I., 1982, Late Quaternary environments and biogeography in the Great Basin: Quaternary Research, v. 17, p. 39–55.
Thompson, R. S., Benson, L., and Hattori, E. M., 1986, A revised chronology for the last Pleistocene lake cycle in the central Lahontan Basin: Quaternary Research, v. 25, p. 1–9.
Turner, R. M., 1982, Great Basin Desertscrub, *in* Brown, D. E., ed., Biotic communities of the American Southwest; United States and Mexico: Desert Plants, v. 4, p. 145–155.
Turner, R. M., and Brown, D. E., 1982, Sonoran Desertscrub, *in* Brown, D. E., ed., Biotic communities of the American Southwest; United States and Mexico: Desert Plants, v. 4, p. 181–221.
Van Devender, T. R., 1973, Late Pleistocene plants and animals of the Sonoran Desert; A survey of ancient packrat middens in southwestern Arizona [Ph.D. thesis]: Tucson, University of Arizona, 179 p.
—— , 1977, Holocene woodlands in the southwestern Deserts: Science, v. 198, p. 189–192.
—— , 1986a, Climatic cadences and the composition of Chihuahuan Desert communities; The late Pleistocene packrat midden record, *in* Diamond, J., and Case, T. J., eds., Community ecology: New York, Harper and Row, p. 285–299.
—— , 1986b, Pleistocene climates and endemism in the Chihuahuan Desert flora, *in* Barlow, J. C., Powell, A. M., and Timmermann, B. N., eds., Second Symposium on the Resources of the Chihuahuan Desert: Alpine, Texas, Chihuahuan Desert Research Institute, p. 1–19.
—— , 1986c, Late Quaternary history of pinyon-juniper-oak woodlands dominated by *Pinus remota* and *Pinus edulis*, *in* Everett, R. L., ed., Proceedings of pinyon-juniper conference: U.S. Department of Agriculture, Forest Service, General Technical Report Int-215, 99–103.
—— , 1987a, Holocene vegetation and climate in the Puerto Blanco Mountains, southwestern Arizona: Quaternary Research, v. 27, p. 51–72.
—— , 1987b, Late Quaternary vegetation and climate in the Chihuahuan Desert, United States and Mexico, *in* Martin, P. S., Van Devender, T. R., and Betancourt, J. L., eds., The last 40,000 years of biotic change in the arid West: Tucson, University of Arizona Press, (in press).
—— , 1987c, Late Quaternary vegetation and climate in the Sonoran Desert, United States and Mexico, *in* Martin, P. S., Van Devender, T. R., and Betancourt, J. L., eds., The last 40,000 years of biotic change in the arid West: Tucson, University of Arizona Press (in press).
Van Devender, T. R., and Burgess, T. L., 1985, Late Pleistocene woodlands in the Bolson de Mapimi; A refugium for the Chihuahuan Desert biota?: Quaternary Research, v. 24, p. 346–353.
Van Devender, T. R., and Spaulding, W. G., 1979, Development of vegetation and climate in the southwestern United States: Science, v. 204, p. 701–710.
Van Devender, T. R., and Toolin, L. J., 1983, Late Quaternary vegetation of the San Andres Mountains, Sierra County, New Mexico, *in* Eidenbach, P. L., ed., The prehistory of Rhodes Canyon, survey and mitigation: Tularosa, New Mexico, Human Systems Research, Inc., p. 33–54.
Van Devender, T. R., Betancourt, J. L., and Wimberly, M., 1984, Biogeographic implications of a packrat midden sequence from the Sacramento Mountains, south-central New Mexico: Quaternary Research, v. 22, p. 344–360.
Van Devender, T. R., Martin, P. S., Thompson, R. S., Cole, K. L., Jull, A.J.T., Long, A., Toolin, L. J., and Donahue, D. J., 1985, Fossil packrat middens and the tandem accelerator mass spectrometer: Nature, v. 317, p. 610–613.
Van Devender, T. R., Bradley, G. L., and Harris, A. H., 1987, Late Quaternary mammals from the Hueco Mountains, El Paso, and Hudspeth counties, Texas: Southwestern Naturalist, v. 32, p. 179–195.
Webb, R. H., 1985, Spatial and temporal distribution of radiocarbon ages on rodent middens from the southwestern United States: Radiocarbon, v. 28, p. 1–8.
Webb, R. H., and Betancourt, J. L., 1987, Bias in the spatial and temporal distribution of radiocarbon ages from packrat middens, *in* Martin, P. S., Van Devender, T. R., and Betancourt, J. L., eds., Fossil packrat middens; The last 40,000 years of biotic change in the arid West: Tucson, University of Arizona Press (in press).
Webb, T., III, 1986, Is vegetation in equilibrium with climate? How to interpret late-Quaternary pollen data: Vegetatio, v. 67, p. 75–91.
Wells, P. V., 1966, Late Pleistocene vegetation and degree of pluvial climatic change in the Chihuahuan Desert: Science, v. 153, p. 970–975.
—— , 1976, Macrofossil analysis of wood rat, *Neotoma* middens as a key to the Quaternary vegetational history of arid America: Quaternary Research, v. 6, p. 223–248.
—— , 1979, An equable glaciopluvial in the West; Pleniglacial evidence of increased precipitation on a gradient from the Great Basin to the Sonoran and Chihuahuan deserts: Quaternary Research, v. 12, p. 311–325.
—— , 1983, Paleobiogeography of montane islands in the Great Basin since the last glaciopluvial: Ecological Monographs, v. 53, p. 341–382.
Wells, P. V., and Berger, R., 1967, Late Pleistocene history of coniferous woodlands in the Mohave Desert: Science, v. 155, p. 1640–1647.
Wells, P. V., and Hunziker, J. H., 1976, Origin of the creosote bush deserts of southwestern North America: Annals of the Missouri Botanical Garden, v. 63, p. 843–861.
Wells, P. V., and Jorgensen, C. D., 1964, Pleistocene wood rat middens and climatic change in Mohave Desert; A record of juniper woodlands: Science, v. 143, p. 1171–1174.
Wells, P. V., and Woodcock, D., 1985, Full-glacial vegetation of Death Valley, California; Juniper woodland opening to Yucca semidesert: Madroño, v. 32, p. 11–23.
Wigand, P. E., and Mehringer, P. J., Jr., 1985, Pollen and seed analysis: Anthropological Papers of the American Museum of Natural History, v. 61, p. 108–124.
Wright, H. E., Jr., 1981, Vegetation east of the Rocky Mountains 18,000 years ago: Quaternary Research, v. 15, p. 113–125.

MANUSCRIPT ACCEPTED BY THE SOCIETY MARCH 16, 1987

ACKNOWLEDGMENTS

We thank Paul S. Martin for 15 years of encouragement and support. The National Science Foundation, the U.S. Geological Survey, the Bureau of Land Management, the National Park Service, Southwest Parks and Monuments Association, Grand Canyon Natural History Association, the Department of Energy, Fort Bliss Army Base, Holloman Air Force Base, the National Geographic Society, and Human Systems Research, Inc. have provided funds for packrat midden research. The careful reviews of Vera Markgraf, J. Platt Bradbury, Peter J. Mehringer, Jr., W. Geoffrey Spaulding and Herbert E. Wright, Jr., improved the manuscript. Helen Wilson and Charles Sternberg drafted the figures. Jean Morgan typed the manuscript.

Printed in U.S.A.

The Geology of North America
Vol. K-3, North America and adjacent oceans during the last deglaciation
The Geological Society of America, 1987

Chapter 16

Late Wisconsin and early Holocene paleoenvironments of east-central North America based on assemblages of fossil Coleoptera

Alan V. Morgan
Quaternary Sciences Institute, Department of Earth Sciences, University of Waterloo, Waterloo, Ontario N2L 3G1, Canada

INTRODUCTION

The examination of fossil assemblages of Coleoptera (beetles) in North America is a relatively new but rapidly growing research area. This chapter is an attempt to use suites of fossil Coleoptera to establish the nature and timing of different paleoenvironments at the margins of the Laurentide Ice Sheet as it advanced and retreated from the Great Lakes region. Some of the work is synthesized from previous publications, but some important data are from unpublished sites. Before the various sites illustrated in Figure 1 are discussed, certain assumptions must be restated that are implicit in the study of Quaternary insect assemblages.

The first of these is that beetle species can be identified by characters that are not always used by neotaxonomists. The latter frequently describe species by using portions of the exoskeleton that are destroyed or scattered following the death of the organism. In many cases a specific determination cannot be made with the available fossil components, but even narrowing the identification to a species group often is sufficient to provide valuable ecological data.

The second assumption is that species have not evolved during the Pleistocene. Early in the century a general belief prevailed that intense speciation occurred during periods of glaciation. This was one of the greatest barriers to the development of paleoentomology, and it was not until the work of Henriksen (1933), Lindroth (1948), and Coope (1959, 1968; Coope and others, 1961) that this belief was shown to be fallacious. Beetles recovered from Quaternary deposits can be matched precisely with modern species, although they may occur currently thousands of kilometers away from the fossil locality. Examples of major late Pleistocene geographic shifts are demonstrated in papers by Coope (1973) and Hammond and others (1979).

Some general statements can be added to the two basic tenets stated above. Paleoenvironments or paleoclimates are not established by use of individual species, but rather of species assemblages. Inherent is the assumption that we know the full modern distribution of a species (or at least the range limits), and that adequate ecological data are available. We assume that the ecology of the fossil species is the same as that of the modern species, and that little or no physiological change has occurred during this period. Publications commenting on some of the points mentioned above include two excellent summaries by Coope (1970, 1979). North American reviews are provided by Ashworth (1979), Matthews (1977, 1980), Morgan and Morgan (1980a, 1980b), and Morgan and others (1983a).

The problems of paleoclimatic reconstruction

Recent research has reinforced and refined some of the methodologies of paleoclimatic reconstruction from fossil insect assemblages outlined by Morgan and Morgan (1981). Eurythermic species tend to be noted but ignored in paleoclimatic reconstruction because there are problems in determining limiting climatic parameters. Stenotherms can be defined from the knowledge of different distributions of species within certain families (especially the Carabidae and the Cicindelidae), which are particularly important in the reconstruction of paleoenvironments on recently deglaciated landscapes. Distribution patterns of modern Coleoptera in northern North America are broadly transcontinental and tend to run parallel to the July and annual isotherms that cross this portion of the continent. There is some confusion in this simplistic picture because of the montane belts of western and eastern North America. Nevertheless, species distributions can be summarized in a general manner (Fig. 2). However, it would be wrong to assume that all species fit exactly into the obvious patterns outlined in this figure, for a great many tend to transgress boundaries of categories. Some species depend upon host plants, others upon decomposition products (fungus, dung, or carrion), while still more may be limited by substrate type, humidity levels and precipitation, or even winter temperatures.

Morgan, A. V., 1987, Late Wisconsin and early Holocene paleoenvironments of east-central North America based on assemblages of fossil Coleoptera, *in* Ruddiman, W. F., and Wright, H. E., Jr., eds., North America and adjacent oceans during the last deglaciation: Boulder, Colorado, Geological Society of America, The geology of North America, v. K-3.

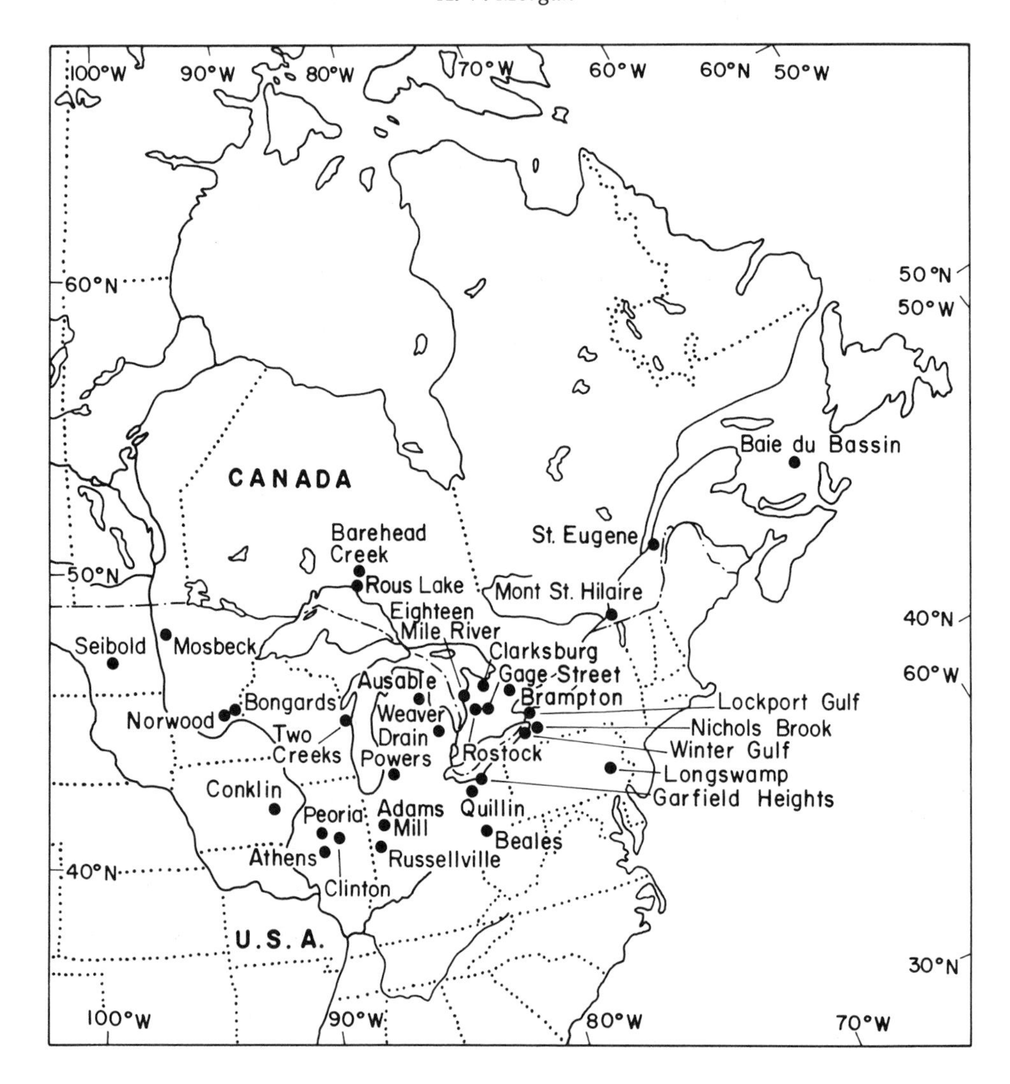

Figure 1. Location of sites mentioned in the text.

Winter temperatures. Paleoentomologists have assumed that winter temperatures are relatively unimportant to overwintering Coleoptera, and this is true to some extent. Once a beetle has gone into hibernation, the temperature of the external environment has been largely disregarded. Beetles overwinter by two related but different strategies. The first is by supercooling, in the case of species that are frost susceptible and possibly limited to winter lows of –30° to –35°C. The second is by freezing, in the case of frost-tolerant forms. The second category is more energy efficient and is probably used by most boreal species. Certain beetles are extremely well adapted to very low winter temperatures, building up glycerol levels in the haemolymph to withstand winter lows that may be below –60°C. These species cannot easily survive climatic amelioration in the middle of the hibernation period, however, because the glycerol content drops very rapidly (often by 50 percent in less than 24 hours). Any subsequent return to even moderately cold conditions causes the death of the individual by freezing and cellular disruption (Baust and Miller, 1970).

Other frost-tolerant species have low supercooling potential and freeze at relatively high temperatures (around –4° to –8°C). Some carabids can only survive if buffered from external temperatures beneath thick snow cover, and are often killed by winter lows as high as –10°C. Still others are capable of surviving in anoxybiotic conditions, such as those found beneath an ice cover (Thiele, 1977). Descriptions of overwintering Coleoptera are provided by Miller (1969), Baust and Miller (1970). Ring and Tesar (1980), and references therein.

If we knew more about the cold-hardiness and overwintering strategies of species found in the fossil record it might be possible to comment on the nature of past winter temperatures and precipitation. Enough data exist to demonstrate that this is an

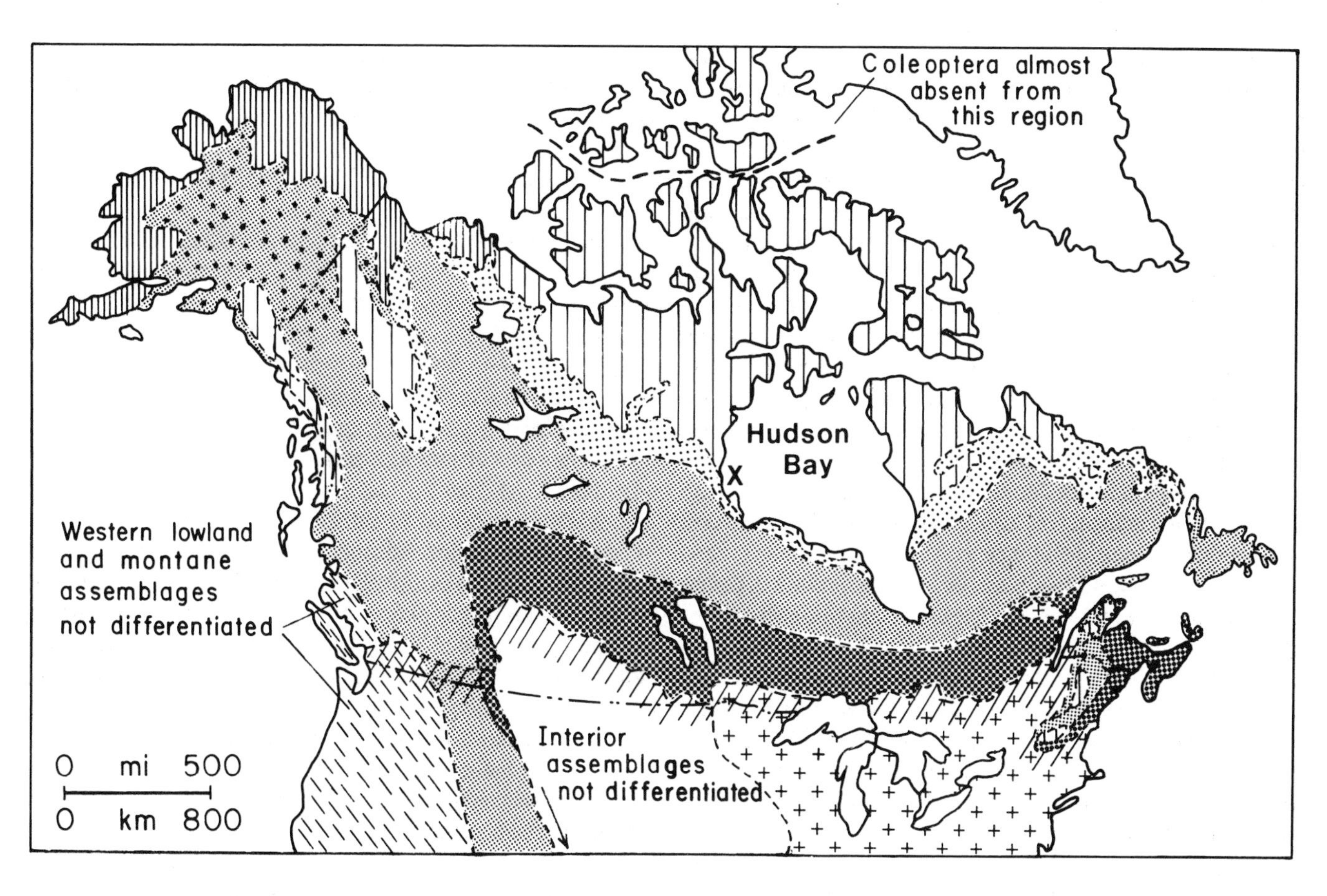

Figure 2. Approximate distributional zones of Coleoptera in northern North America. **Vertical shading:** faunas found north of the arctic treeline or altitudinally above timber limit in western and eastern North America. Some species belonging to this category are found only west of Hudson Bay and north to the mainland coast. Their known southestern ranges terminate near X. Close vertical shading indicates the approximate range of open-ground faunas centered in Alaska. Most high-altitude faunas of the western Cordillera are not indicated because of the complexity of the topography. Eastern high-altitude faunas are confined to the northern end of the Appalachian Mountains, particularly in northern New England, Gaspé, and unmarked areas of Newfoundland and Labrador. **Stipple shading:** faunas south of arctic treeline and below timber limit, typical of the boreal zone of North America. Light stipple represents treeline species in assemblages associated with areas of thin conifers. Medium stipple is more typical of north to central boreal species in transcontinental distribution to Alaska. The arrow indicates a western montane extension (to New Mexico) and is a generalized indication of this category southward at high elevations. Certain species of this group can be found through the Chiricahua Mountains of Arizona and into northern Mexico. Dense stipple indicates central to southern boreal distributions west to Alberta and Montana. The southern limit of this zone east of Manitoba is transitional with faunas of deciduous woodlands, and from Manitoba westward, with faunas of the aspen parkland and dry grassland. Very coarse irregular stipple in Alaska and the Yukon indicates the approximate extension of many Alaskan boreal forms. **Crosses** indicate the distribution of species found today south of the boreal forest but confined to the eastern third of the continent. **Diagonal shading** demarcates the northern limit of transcontinental species noted in many assemblages. These are found at low elevations in the mountains of western North America. See figure for additional remarks.

interesting line to follow, but the vast number of species involved means that paleoentomologists probably will be without critical information for many years to come.

Mean annual temperatures. The estimation of mean annual temperatures is problematic without firm data on the overwintering habits of the species represented. In all the cases cited below, mean annual temperature estimates are based upon the current distribution of beetles. In some cases, especially where these species are confined to the extreme northwest of the continent, the estimates may be slightly low. However, I have tried to compensate for the substantial latitudinal differences between arctic sites today and the position of midlatitude fossil sites. The mean annual temperatures at treeline in the Northwest Territories (ca. −10°C) are quite different from the treeline temperatures in Quebec (ca. −6°C), even though the mean July temperatures (ca. 11°C) are quite similar (Anonymous, 1984). One other unknown factor is the regional climate that could be induced by the proximity of the Laurentide Ice Sheet. This could have reduced both July and mean annual temperatures and had considerable control on humidity and precipitation patterns.

Colonization. Carabid species are important for reconstructing environments after ice retreat, for many are active predators and able to fly. Some are well adapted to living on barren substrates (e.g., they have long legs and heat-absorbing surfaces), but most importantly they have the ability to live among pioneering plant species in newly exposed landscapes. Environments near glacier margins are notoriously unstable. They appear barren or sterile, yet there is a multitude of plant and animal life on or within different substrates. Lower plant forms (algae, lichens, and mosses) provide food for Collembola (spring tails), various families of Diptera (flies), herbivorous arthropods (moss mites), predaceous arthropods (spiders), and Coleoptera, particularly species of *Bembidion, Elaphrus, Pterostichus,* and *Nebria.* These rapidly moving carnivores are able to run or fly swiftly along unstable fluvial margins and can run or swim across water surfaces. They are frequently active at night, and some species can move beneath the ablating margins of glacial ice (personal observation).

Cincindelids (tiger beetles) are also active hunters, moving after prey on bare expanses of heated sand. Although the modern arctic would seem to be an ideal place for this family (large expanses of outwash sands, beach sediments, and exposed eskers), cicindelids are limited to the central-northern boreal zone and southward. A review of Quaternary fossil cicindelids is given in Ngano and others (1982).

Certain aquatic and semiaquatic species can live on recently deglaciated substrates. Many members of the Dytiscidae and Hydrophilidae are excellent fliers, and they inhabit arctic and subarctic water bodies today. For reasons that are often unclear, certain species inhabit different types of aquatic habitats (small tundra ponds, large lakes, small streams, or densely vegetated margins of larger water bodies). Their distributions may also be limited to sites north of treeline, at treeline, south of treeline, or combinations thereof. Because treeline is a relatively well-defined climatic line, it is possible to utilize the species distributions in stenothermic categories, particularly when utilizing July temperatures. Many representatives of the families mentioned above, together with species of other families, can be placed in stenothermic groups. These may even include species that depend upon host plants, but whose ranges terminate according to temperature, rather than the range of the host plant (Morgan, 1973).

PREGLACIAL SITES AND ENVIRONMENTS

Preglacial environments are described from assemblages at sites beneath till units that are regarded as correlative with the maximum Wisconsin Laurentide advance. These assemblages provide some clues to the species inhabiting the region prior to the main ice advance. Names of individual genera or species are mentioned only where they are important to the understanding of distributional shifts, or where they relate specifically to climatic interpretations. Full species lists are given in Morgan and Morgan (1980a).

Early and mid-Wisconsin sites in southern Ontario have produced insect faunas that indicate a variety of environments ranging from boreal to treeline (Morgan, 1972; Morgan and Morgan, 1980a). They differ from Sangamon assemblages, which contain species found living in the lower Great Lakes basins today (Morgan and Morgan, 1980a; Williams and Morgan, 1977; Pilny and Morgan, 1987).

Figure 3 illustrates specific sites mentioned in the next two text sections. These localities, with the exception of the Athens and Conklin Quarry sites, are covered by Late Wisconsin till of the Huron and Michigan lobes.

A site at Clarksburg, Ontario (>36,000 yr B.P.), contains a well-preserved insect fauna that is sparse in numbers and species diversity. The presence of *Chrysolina subsulcata* and *C. magniceps, Elaphrus parviceps,* and species of the subgenus *Cryobius* (*P. pinguedinius, P. ventricosus, P. brevicornis*) coupled with the total absence of phytophagous scolytids and other genera normally associated with the boreal forest, indicate that this is a true tundra assemblage, the coldest so far recovered in Ontario. Some of the species only live today in the extreme northwest of the continent. The mean July temperature at the site must have been below 10°C, and the mean annual temperature perhaps as low as −8°C. The insect assemblage favors an earlier Wisconsin age for the site, although the stratigraphy suggests a time just before the Late Wisconsin maximum (Warner and others, in preparation).

South of Lake Erie the youngest site predating the last glaciation is at Garfield Heights, Ohio (Coope, 1968; White, 1968). In 1981 the site was resampled and insects recovered from two small concentrations of organic detritus (presumably dated to 28 to 24 ka). None of the species at this site is confined to the tundra today, although several do range marginally beyond the treeline onto the tundra. Several species require dry, open-ground habitats. Substrates were probably covered with mosses and lichens, together with grasses and other herbs. Plant macrofossils recovered during insect separation include *Dryas integrifolia, Vaccinium uliginosum,* and scattered *Picea* needles. This associa-

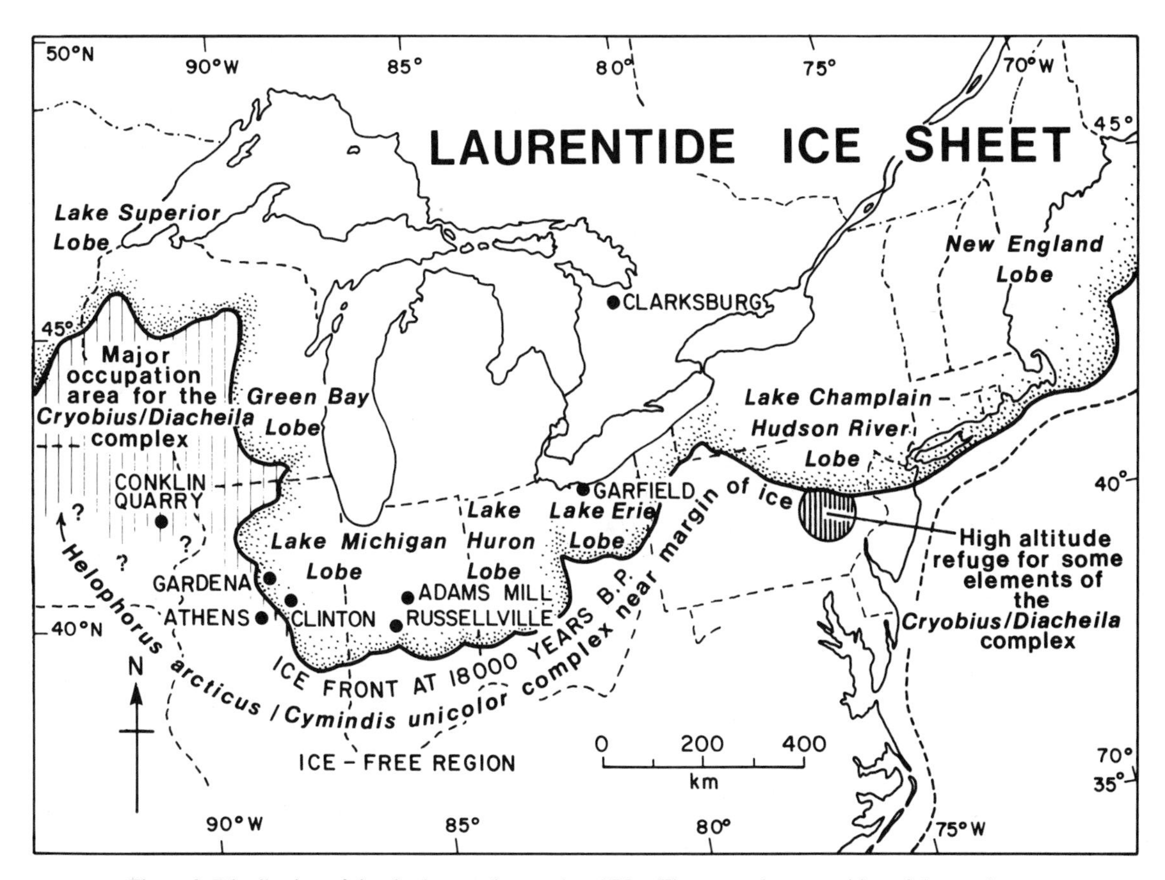

Figure 3. Distribution of sites in the text that predate 17 ka. The approximate position of the maximum extent of Laurentide ice is marked, and the major ice lobes are noted. Vertical shading denotes probable centers of residence for the *Cryobius/Diacheila* complex. Non-marked refugia may also have been present in parts of the Alleghenies, or in enclaves along the ice margin. The Appalachian distribution is generalized.

tion is identical to that reported by Berti (1975) for this site. The presence of *Picea* needles, together with the log mentioned by Coope (1968), suggests that at least isolated conifers were present, but no insects associated with trees were found. Several carabid species (*Miscodera arctica, Bembidion grapei, Pterostichus adstrictus*) and the predatory bug *Saldula saltatoria,* usually are found today on open sandy to organic substrates, or in vegetation near water, but there is an absence of totally aquatic beetle species. All of the insect species identified are found in eastern Canada today. The past mean July temperature for the site is estimated at below 15°C, and the mean annual temperature at ca. 0°C.

Several sites south and west of Garfield provide some information on the environments outside the advancing ice front. An insect assemblage recovered from the Adams Mill section (Fig. 1), dated at ca. 22 ka, indicates deposition on a moist substrate in a boreal environment (Morgan and others, 1983b). The presence of water is indicated by caddisflies, dytiscids, and hydrophilids, while scolytids and wood fragments show that trees were in the vicinity. Several carabid, staphylinid, and byrrhid species suggest pond margins with a rich growth of sedges and mosses. July temperatures are estimated at ca. 16°C, and the mean annual temperature 0° to –1°C.

The Russellville site (ca. 21 ka) southwest of Adams Mill contains species reflecting a marginally cooler boreal environment (Morgan and others, 1983b). Scolytids dependent upon *Picea* were recovered, and numerous logs can be seen in the overlying till. The July temperature at this site is estimated at ca. 14° to 16°C, and the mean annual temperature at ca. -1°C.

Organic silts exposed at Athens, Illinois (Fig. 1), dated at 27 to 22.5 ka, are believed to be equivalent to the lowest portion of the Peoria Loess (Follmer and others, 1979). Insects recovered are typically boreal, with several species of scolytids as well as aquatic genera. No apparent climatic change is indicated within the sequence, and the assemblage probably represents midboreal conditions (July mean temperature of 15° to 17°C; mean annual temperature of ca. 0°C).

Insects were recovered from two different horizons at the

Gardena section near Farm Creek, Peoria, Illinois (Fig. 1). The lower portion of the Morton Loess (with wood dated ca. 25 ka; Follmer and others, 1979) contains a poorly preserved but numerically rich insect assemblage, similar to that recovered from the Athens site. It is typically boreal and contains a high percentage of scolytid beetles and a number of ants, as well as taxa suggesting waterside conditions. The presence of large numbers of dytiscids (water beetles) indicates standing water. The climatic regime is probably best matched to midboreal conditions (mean July temperature ca. 15° or 16°C).

At least two insect assemblages west of Russellville are assumed to be approximately contemporaneous with the maximum extent of the late Wisconsin ice advance into Illinois. Since both localities are till covered, technically they are preglacial. However, the radiocarbon dates and the proximity of the sites to the glacial margin suggest that these assemblages represent the faunas existing close to the ice margin in this region. Both are considered in the next section.

FULL-GLACIAL SITES AND ENVIRONMENTS

A deep excavation near Clinton, Illinois (Fig. 1), exposed a thin organic horizon of compacted mosses (dated at ca. 20.5 ka) under till of the Wedron Formation. The insect fauna from the site contained large numbers of the staphylinid *Olophrum rotundicolle* and the hydrophylid *Helophorus sempervarians.* Both species are normally residents of the boreal zone of North America. The carabids recovered from the site include *Diacheila polita, Elaphrus lapponicus,* and *Agonum exaratum.* All three are northern species. *Diacheila polita* is confined to the northwest of the continent, and *Agonum exaratum* extends from western Alaska into the Yukon, the barren land of Keewatin (Northwest Territories), and eastward to Churchill, Manitoba. The distribution and habitats of *Diacheila polita* and *Elaphrus lapponicus* are discussed later, but all three species may be regarded as open-ground residents. No insects dependent on trees were found at the site.

The uppermost samples (ca. 19.5 ka) at the Gardena section, Peoria, were taken from a compressed moss and silt horizon (2 m above the lower samples) immediately below the Delavon Till and above the uppermost Morton Loess (Follmer and others, 1979). The stratigraphic and lithologic relationships closely resemble those seen at the Clinton site.

The upper fauna is quite different from the assemblage 2 m below at Gardena but is almost identical to that at the Clinton site. The predominant staphylinids are *Olophrum rotundicolle* and *Acidota quadrata,* and the most interesting carabids are *Diacheila polita* and *Elaphrus lapponicus.* Several points can be made. *E. lapponicus* is a hygrophilous species that Lindroth (1961) reports from cold-water areas where the vegetation consists mainly of mosses. It is rarely found above timber line and is not a true resident of the tundra. Goulet (1983) states that the preferred substrate is of neutral pH where *Paludella* mosses and other short vegetation such as *Marchantia* grow. Scattered small conifers are present at some sites. The typical habitat is near springs, brooks, and ponds. The distribution reported by Goulet (1983) extends north to treeline and across Canada from southern Labrador to the Anderson and Mackenzie River Deltas (Northwest Territories) and into the Fairbanks region of Alaska. The northern limit of distribution coincides closely with the 12°C July isotherm.

Diacheila polita is described as inhabiting peaty soil on the open tundra as well as the margins of *Carex* pools on moist, soft ground (Lindroth, 1961). The species is also reported from rather dry places with *Betula nana, Vaccinium uliginosum, Rubus,* carices, and mosses (Lindroth 1961) and in the same area on a coarse sand substrate with grasses, *Alnus,* and *Salix* (Williams and others, 1981). The distribution of this beetle is confined to Alaska, the Yukon, and extreme northwestern Northwest Territories, where it is found almost exclusively in the tundra. The species reaches below treeline in several areas, and there is one unpublished collecting record near Fairbanks (J. V. Matthews, personal communication, 1987). Other carabids in this fossil assemblage have distributions that range through the boreal zone onto the tundra or have ranges that overlap at treeline. Most of them are hygrophilous and often are associated with peaty, moist pool margins mantled with mosses. There are no scolytids in the insect assemblage. I estimate the mean July temperature at both the Clinton and Gardena (top) sections to be 11° to 12°C. The mean annual temperature could have been lower than at any of the sites described earlier (except for Clarksburg) and may well have been as low as −7° or −8°C.

An interesting insect assemblage has been reported from a fossiliferous unoxidized loess near Iowa City by Schwert and others (1981). The Conklin Quarry locality is dated between 18.1 and 16.7 ka from wood fragments in the sediments (Baker and others, 1986). The insect fauna (like that of the Clinton and Peoria sites) is a mixture of boreal, treeline, and tundra species. Some of the species inhabit dry, sandy to gravelly substrates, while others prefer moist habitats, often with a rich vegetation cover, including *Alnus* and *Salix.* Despite the mixture of species at the site, the faunal elements overlap at treeline. Schwert and others (1981) suggest a July temperature of 10° to 12°C for the site at the time of deposition.

LATE-GLACIAL SITES AND ENVIRONMENTS

Late-glacial sites contain organic sequences on top of till within or on top of ice-contact sediments that postdate the retreat of ice from its maximum and predate 10 ka (Fig. 4). Localities are extremely common throughout the lower Great Lakes region, although the detail of investigation varies considerably from site to site. Many of the fossil assemblages come from kettle infillings, paleostream valleys, marl deposits, or sites of what were probably ephemeral ponds. In most cases the time represented by a particular site is generally short (several hundred radiocarbon years). Here I attempt to divide the different assemblages into 1,000-yr intervals, starting with the oldest records.

15 to 14 ka. Two sites fall into this time range. The older is

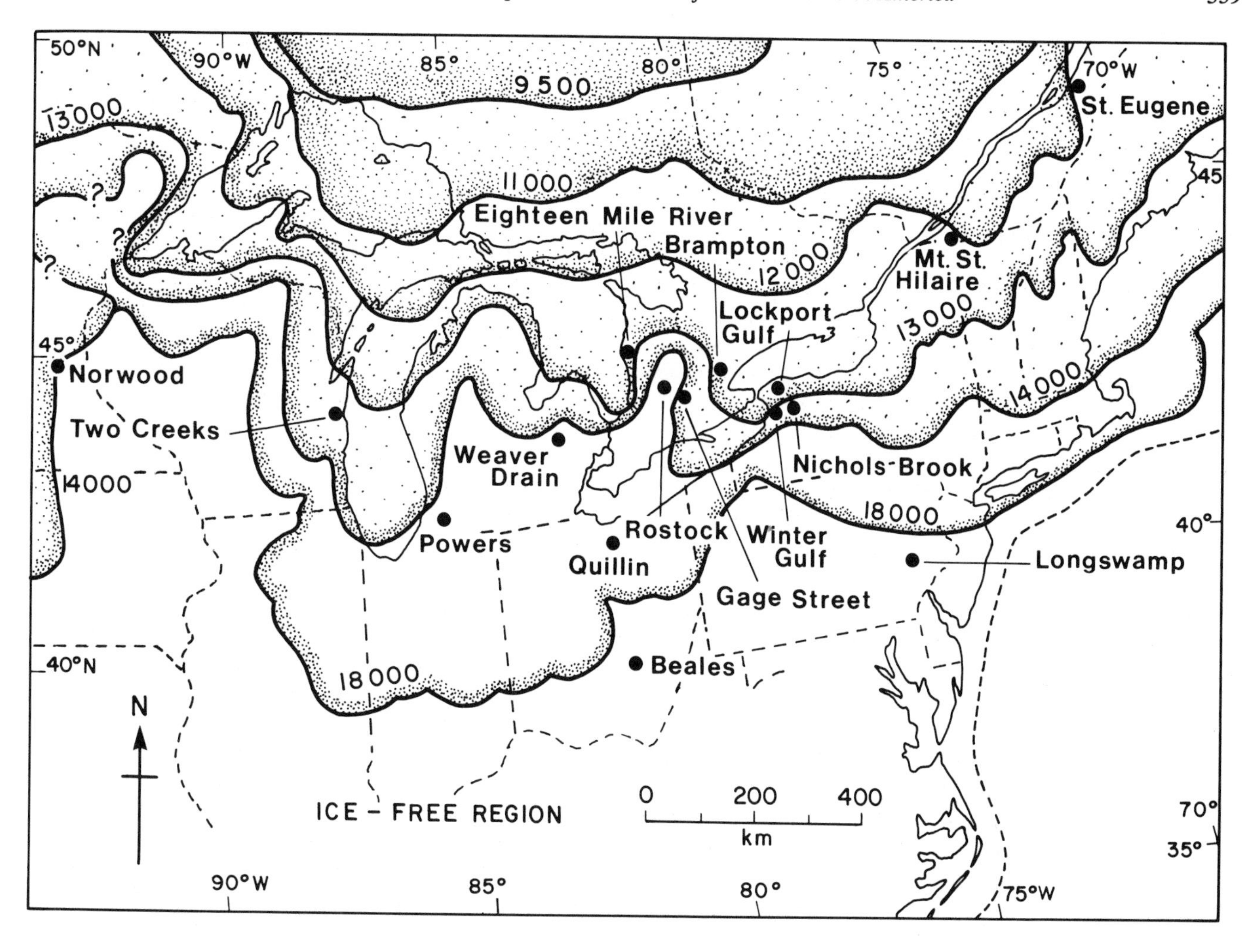

Figure 4. Distribution of sites mentioned in text that postdate 16 ka and predate 10 ka. Approximate ice fronts are shown after Prest (1969) and Mickelson and others (1983). For sites not included on this diagram see Figure 1.

probably Longswamp (near Mertztown, Pennsylvania), about 60 km outside the Laurentide ice margin. A summary of the geological setting, and details of the plant macrofossils and pollen from the site are given in Watts (1979). Problems with the radiocarbon chronology at this site (compared with others close by in Appalachia) led Watts (1979) to conclude that the basal date should be as much as 15 ka or older. On this assumption the oldest insects from the site (ca. 15 to 14 ka), although generally agreeing with the environmental interpretation provided by the plants, give contradictory evidence concerning the climate. The plants suggest tundra conditions, with implied July averages of less than 11°C, while the basal insects reveal fairly dry, poorly vegetated uplands with wet depressions. No scolytids were found, but the beetles (*Notiophilus semistriatus, Stenus comma, Tachinus elongatus*) are typical residents of open ground in the boreal zone and suggest July temperatures of 13°C or 14°C.

The Quillin (Lodi) site south of Cleveland, Ohio, has a basal radiocarbon age from a "litter layer" of 14,500 ± 150 yr B.P. (ISGS-402). Fossil insects recovered from this level indicate an abundance of boreal species that occupy habitats ranging from flowing water to vegetated pond margins. Large numbers of scolytids indicate that the plant and animal communities jointly invaded a recently deglaciated landscape. The mean July temperature was probably about 15°C and the mean annual temperature about –2° to –3°C.

14 to 13 ka. Fossil insect assemblages associated with continuing deposition at Longswamp and Quillin indicate temperatures similar to those cited above. At Longswamp, the presence of staphylinid species (*Tachinus elongatus, Tachyporus mexicanus*), only recorded today as far north as the northern and central boreal forest, suggest slightly warmer July temperatures (perhaps 14° to 15°C). These insects are found with a plant assemblage attributed to tundra, for example, *Dryas integrifolia, Empetrum* cf. *nigrum,* and *Vaccinium uliginosum* (Watts, 1979).

Temperatures at the Quillin site are presumed to be the same as those mentioned earlier (ca. 15°C), but some interesting comparisons can be made with the Weaver Drain site, 240 km northwest in southeastern Michigan (Fig. 1), where bands of organic detritus within silts were found in an abandoned Lake Maumee overflow channel. Plants identified as *Dryas integrifolia, Vaccinium uliginosum,* and *Salix herbacea* were dated at 13,770 ± 210 yr B.P. (I-4899; Burgis, 1970). Coleoptera from the site reflect open-ground conditions. A few species are found today extending north to treeline or beyond (*Pterostichus haematopus, Amara glacialis, Cymindis unicolor*) and others inhabit open areas in the northern half of the boreal forest. One scolytid was present in the sample, suggesting that although trees were not present locally, both trees and these insects were starting to colonize southeastern Michigan. The beetles indicate July temperatures of ca. 11°C and mean annual temperatures of –4° to –3°C.

Contrasting with the Weaver Drain insects is a very small assemblage recovered from the Beales site 60 km beyond the ice limit in southern Ohio. The site is 190 km south of Lodi and has a radiocarbon age of 13,800 yr B.P. The insects are independent of trees, but species such as *Tachinus canadensis* are typical residents of mixed deciduous/coniferous woodlands today. July temperatures may have been close to 17°C and the mean annual temperature perhaps 2°C.

13 to 12 ka. The Longswamp site shows interesting changes in Coleoptera just before major changes in the pollen and plant macrofossils (Watts, 1979). Uncertainties in the radiocarbon dates at this site preclude an exact timing of events, but at approximately 12.5 ka and just at the transition from pollen zone LS-1 (the tundra zone) to LS-2a (characterized by high percentages of *Betula* pollen and *Betula glandulosa* macrofossils), there is a large influx of scolytids (*Scolytus piceae, Polygraphus rufipennis, Cryphalus ruficollis, Dryocoetes affaber, Ploeotribus piceae,* etc.). Most of these species are hosted by various conifers, and particularly by *Picea.* The insect evidence thus suggests the presence of spruce trees close to the site. This is substantiated by increasing percentages of *Picea* pollen, although tundra plant macrofossils were recorded at this level. Because of the problems of dating at Longswamp the transitional events noted above may be as old as 13.5 ka (Watts, 1979).

Three insect sites in southern Ontario and two in New York State provide information about the late-glacial environment as the ice retreated into southern Canada. The oldest is believed to be the Rostock site, estimated at 13 ka. The insects recovered from the lowest levels of this marl-filled morainic depression are residents of the northern boreal zone today. Scolytids are completely absent, and the aquatic beetles now inhabit cold, clear water bodies. Most of the ground beetles are found on various substrates poor in vegetation and frequently at the margins of snow fields or near water emanating from glaciers. Many of the carabids have western and northern Canadian distributions, and some are found on alkaline, saline substrates (Pilny and others, 1987). Temperatures are difficult to estimate because of the extreme disparities in the modern distributions of species recovered from the basal assemblages. I suggest mean July temperatures of 12° to 14°C, and a mean annual temperature of 0°C.

Two insect faunas near Buffalo provide environmental data at locations away from the ice front. The Winter Gulf site was deposited between ca. 12.7 and 12.5 ka (Schwert and Morgan, 1980). The coleopteran assemblage implies an open mire with trees. A July mean temperature slightly above 16°C and a mean annual temperature of approximately 2°C are suggested.

The Nichols Brook site, about 35 km east of Winter Gulf, contains basal deposits believed to be contemporaneous with the youngest materials at the Winter Gulf site. The oldest beetle assemblage is quite similar to the fauna from Winter Gulf and reflects conditions found today in the central boreal forest of Ontario. July and annual mean temperatures may be the same as for Winter Gulf (Fritz and others, 1987).

A marl sequence at Gage Street in Kitchener, Ontario, has been intensively studied for insects, pollen, and plant macrofossils (Schwert and others, 1985) and stable isotopes (Fritz and others, 1987). Deposition is believed to have commenced about 12.8 ka, although there are hardwater-contamination problems with the basal radiocarbon dates. The oldest insects recovered from the site include open-ground carabids, aquatic species that inhabit shallow oligotrophic water bodies, and scolytids. No obligate tundra species occur at the site, and the fauna is dominantly boreal. July temperatures are estimated between 15° and 17°C.

A kettle depression at Brampton, Ontario, infilled with basal organic silts overlain by peat, provides an interesting comparison with the above sites. Like Rostock and Gage Street, the oldest Brampton fauna (estimated at >12.5 ka) indicates a clear, cold water body receiving detritus from an open landscape. As in the two other Ontario sites, the oldest plant assemblages include such "tundra" species as *Dryas integrifolia, Vaccinium uliginosum,* and *Salix herbacea.* The insects are boreal; some, like *Cymindis unicolor* and *Helophorus arcticus* are found today at treeline, but others do not reside north of the central boreal forest. A description of one such species, *Cicindela limbalis,* is given in Morgan and Freitag (1982). July temperatures for the basal portion of the Brampton section are estimated at 15°C, and the mean annual temperature at 0°C.

The Norwood site in Minnesota has a geologic setting similar to that of the Brampton site. Ashworth and others (1981) describe the fauna and flora of a late-glacial community that colonized a stagnant-ice landscape. The first well-recorded insects come from the top of a silt dated prior to 12,400 ± 60 yr B.P. (QL-1083). This assemblage contains many species present in the Brampton locality (*Cymindis unicolor, Helophorus arcticus,* and a tiger beetle, *Cicindela* cf. *sexguttata*). Unlike the basal Brampton assemblage, a few scolytid species were found associated with the more open ground forms mentioned above. The assemblage is assumed to represent the transition from an unstable open environment to a stable coniferous forest. As in the case of the Rostock site, it is difficult to determine the climate because of the widespread distribution of species contained within the assemblage. Ashworth and others (1981, p. 78) suggest, ". . . the temperature regime probably reflects that of the boreal forest rather than that of the tundra-forest transition zone." They cite the absence of obligate tundra forms and the limited number of northern stenotherms to support this suggestion.

12 to 11 ka. Insects at the Longswamp site are restricted in numbers and reflect boreal conditions. At Rostock the faunal assemblage indicates a possible deepening of the water body, arrival of varied vegetation on the surrounding landscape, and an increase in aquatic plants in the littoral area. The character of the species assemblage is confused, with subarctic, boreal, and even temperate taxa present. I conclude that July temperatures could be quite high, possibly in the 16° to 18° range.

The Nichols Brook site also indicates ameliorating July temperatures (estimated at 18°C; Fritz and others, 1987), and this regime is also suggested by both the Gage Street and Brampton insects (ca. 18°C). All three sites (like Rostock) show the first appearance of insects whose northern limits only reach the southern limit of the boreal zone. Plant assemblages are typically boreal and are dominated by *Picea* (both pollen and plant macrofossils).

Two faunas from Wisconsin and Michigan show slightly different temperatures. The older of these (Morgan and Morgan, 1979) is from the Two Creeks Forest Bed near Manitowoc, Wisconsin, dated many times at ca. 11.8 to 12 ka. The assemblage reflects an open boreal environment, with July temperatures of ca. 15° to 16°C. The Powers site in southwestern Michigan, dated on bone collagen from a mastodon rib at 11,220 ± 310 yr B.P. (Beta-9482) has produced a boreal insect assemblage. If the date is correct the estimated July temperature (17°C) was somewhat cooler than that in Ontario at the same time. The site has a mixture of species, similar to those seen in other late-glacial assemblages, but it also contains Coleoptera that now have their northern limit in southern Ontario and southwestern Quebec.

11 to 10 ka. Estimated July temperatures at the Nichols Brook, Gage Street, and Brampton sites continue to rise during this period. At 11 ka the Ontario regional July temperature away from the glacially fed proto–Great Lakes was ca. 18°C, increasing to perhaps 20°C by 10 ka. The mean annual temperature is suggested to be 4°C at 11 ka and 5°C at 10 ka (Edwards and others, 1985).

Sites on the shores of the Great lakes or Champlain Sea, as well as one in the lee of the Niagara Escarpment, indicate cooler temperatures than those cited above. The oldest of these is the St. Eugene site east of Quebec City on the southern shore of the Champlain Sea. A thin organic horizon within sandy shore gravels is dated at 11,050 ± 130 yr B.P. (QU-448) (Mott and others, 1981). The assemblage includes treeline and northern boreal insects. *Amara glacialis* and *Pterostichus haematopus,* seen at Weaver Drain, are associated here with *Cymindis unicolor* and *Helophorus arcticus,* found at the Brampton site. Other significant northern Coleoptera include *Amara alpina, Elaphrus lapponicus, E. parviceps,* and *Bembidion* spp. The lack of *Cryobius* in the site is noted by Mott and others (1981). The total absence of scolytids indicates that trees were remote from this site, which also has some of the "tundra" plants reported from sites in the United States and southwestern Ontario (*Dryas integrifolia, Salix herbacea*). Mott and others (1981) postulate a July mean of 17° to 18°C and a mean annual temperature of ca. -1°C.

Contrasting with the St. Eugene site are the assemblages from the basal samples of Lockport Gulf in western New York. The oldest materials have been dated at 10,920 ± 160 yr B.P. (1-5841; Miller and Morgan, 1982). The site contains a predominantly boreal insect assemblage, and the authors found little change in the temperatures estimated for the period from ca. 11 to 9 ka (a July mean in the range of 16° to 18°C). They presume that the fauna of the site reflects local conditions rather than the regional climate, for the site is contained within a deep, north-facing valley incised into the Niagara Escarpment. As such it could retain winter snows and be sheltered from warm southerly winds and sunlight.

By comparison, the Nichols Brook site 75 km south of Lockport suggests a July mean of 18° to 19°C at 11 ka, rising to 20°C at 10 ka. Similar temperatures are postulated for both the Gage Street and Brampton sites (Edwards and others, 1985; Fritz and others, 1987), and these are believed to be most representative of the regional climate.

The Eighteen Mile River locality (Ashworth, 1977) was marginal to Lake Algonquin at 10,600 ± 160 yr B.P. (GSC-1127), and the fauna recovered from the site is composed of mixed boreal elements, with no modern analog. Ashworth (1977) suggests that the assemblage lived in a cold microenvironment surrounded by regionally warmer conditions. The insects living in the valley reflect mean July temperatures as low as 12° to 13°C, but these values shoulud not be regarded as representative of regional climate. Other sites in the region suggest July temperature values of 18° to 19°C at this time (Morgan and others, 1982).

The last insect assemblage to fall into this interval is the Mont St. Hilaire locality south of Montreal. The organic debris at this site is believed to have been deposited by a stream flowing from this steep-sided hill into the Champlain Sea. A radiocarbon date on the debris indicates an age of 10,100 ± 150 yr B.P. (GSC-2200; Mott and others, 1981). On the basis of relatively thermophilous insects, the authors suggest a mean July temperature of ca. 19° to 20°C.

POST-GLACIAL SITES AND ENVIRONMENTS

All of the sites postdating 10 ka are outside the influence of the waning Laurentide Ice Sheet, and many of the insect assemblages remain unpublished. Continuing deposition at several southern sites and a number of more northerly localities permits some interpretations for earliest Holocene time (10–9 ka), but thereafter the limited number of analyses precludes this type of determination. Accordingly, the later descriptions are limited to two larger time intervals (9–5 ka and post 5 ka). All the sites mentioned in this section are located on Figure 1.

10 to 9 ka. Deposition of sediments containing insects continued in the Nichols Brook, Lockport Gulf, and Gage Street sites in northern New York and southern Ontario. At Nichols Brook the assemblages indicate the presence of conifers, but obligatory *Picea* feeders are absent, and the scolytids may have been feeding on *Pinus* species. Carabids found at this level are usually associated with open fields or deciduous woodlands. Species that only just reach southernmost Ontario today from the south are seen in the early Holocene levels in both the Gage Street and Brampton sites. July temperatures were close to 21°C, the July mean for the region today.

Three sites at widely spaced localities provide paleoentomological data about more northerly assemblages in earliest Holocene time. These are Baie du Bassin in Quebec, the Mosbeck site in Minnesota, and Seibold in North Dakota.

The Baie du Bassin site on Amherst Island (Magdalen Islands), dated at 10,000 ± 130 yr B.P. (BGS-313; Prest and others, 1976), has produced a small insect fauna—one that could be found near poorly drained sites on the Magdalen Islands today. The presence of scolytids and *Picea* needles indicates that trees were living close to the site. The authors do not include climatic interpretations in the paper.

The Mosbeck site in northwestern Minnesota has a basal date of 9,940 ± 160 yr B.P. (1-3880; Ashworth and others, 1972). The insects indicate a wetland environment bordered by boreal spruce forest. The authors suggest climate and vegetation similar to southeastern Manitoba today. Although no climatic statistics are provided, the postulated July and annual mean temperatures might be 17° to 18°C and 0°C respectively. Such temperatures could also have been present in the Seibold site on the Missouri Coteau of North Dakota, which is dated at 9,750 ± 140 yr B.P. and has an insect assemblage similar to that found today in the south-central boreal zone (Ashworth and Brophy, 1972).

9-5 ka. Excavations in the Rous Lake sand pit near Marathon, Ontario, revealed detrital organics dated at 8,310 ± 100 (WAT-1508), exposed in deltaic sands. The insects indicate a boreal assemblage that might be found in the Lake Superior region today, except for western species that are ecologically compatible with the rest of the fauna but are not found nearby at present. A mean July temperature of 15° to 16°C and a mean annual temperature of ca. 2°C is suggested (Bajc and others, in preparation).

At the Gage Street site the insect assemblages terminate at ca. 6.9 ka. Schwert and others (1985) state that the presence of stenotherms in levels from 8.6 to 7.9 ka indicates July temperatures of 22° to 23°C, with temperatures increasing from 23° to 25°C. The youngest levels of the Gage Street site were deposited at the same time as the Barehead Creek site near Marathon, Ontario. Fossil insects from this boreal assemblage, dated at 6,950 ± 150 yr B.P. (WAT-1181), suggest July temperatures of ca. 17° to 18°C, only marginally above those in the area today. Another site on the Goulais River near Sault Ste. Marie, Ontario, dated at approximately 6.4 ka, has a small insect assemblage that could be found in the same area today. July temperatures were probably ca. 17°C, and the mean annual temperature ca. 4°C (Warner and others, 1987).

5 ka and younger. The only mid-Holocene site so far investigated in detail in the lower Great Lakes region is the Au Sable River assemblage (Morgan and others, 1985). The insect fauna represents communities that lived in upland forests and fluvial and marsh habitats along a river draining to the main Nipissing shore. Dates on a log in the lower part of the sequence are 4,130 ± 90 yr B.P. (I-8736) and 4,020 ± 30 yr B.P. (WAT-72). The assemblage of insects is one that could be found today in northern and central Michigan, where July temperatures are 20°C. Evidence from the beetle ecologies and the plant macrofossils also indicates that the animal and plant communities were ecologically compatible (Morgan and others, 1985).

Schwert and Ashworth (1985) described a late Holocene insect fauna from Bongards, Minnesota, with a basal date of 3,500 ± 150 yr B.P. (GX-6511). The assemblage is one that might reside in south-central Minnesota today, and the authors infer no climatic change in the region over the last 3,500 years.

DISCUSSION

Undoubtedly many factors influence insect communities, and it should not be assumed that the principal factor is the July temperature. Controls that influence the modern distributions of Coleoptera may include competition from other insect or arthropod groups, a wide variety of macro- or micro-climatic conditions, substrate type, host plants, and colonization strategies. The fossil record nevertheless reveals that late Holocene communities formed ecological units that match those living in the same area today. Presumably the environmental parameters (including the present climate at that geographic location) would have been similar to the environment (and climate) experienced at the time that the fossil assemblage lived.

Communities in mid- and early Holocene time are still "in balance" (i.e., found as an assemblage in close geographic proximity today) and it is not until earliest Holocene time that we find "peculiarities" in the insect assemblages. At that time the climatic and/or ecological conditions were such that individual species (which today are often separated by large geographic distances) were able to live as part of a single assemblage. Some of these fossil communities were extremely long-lived as distinct assemblages during Wisconsin time, although they do not exist as ecological communities today. The different distributions of beetle species in North America are therefore a function of selective filters that have produced the present ranges on the continent. The rapidly changing environments of late-glacial and earliest Holocene time were largely responsible for breaking up the early and middle Wisconsin assemblages that had migrated into more southerly latitudes during the onset of Wisconsin glaciation. Modern arctic and subarctic species, which were able to survive the maximum glaciation in the northwestern (Beringian) refugium, formed a separate population, which was able to recolonize large areas of the Yukon and Northwest Territories as well as part of the northern Cordillera (Morgan and Morgan, 1980a, 1982; Morgan and others, 1984, 1986).

The earliest faunas predating the ice advance across southern Ontario and into the United States contain genera (*Cryobius, Chrysolina*) that are commonly found associated with open-ground or treeline beetles (*Diacheila polita, Elaphrus lapponicus, E. parviceps, Helophorus arcticus*). These assemblages are seen in early Wisconsin sites in southern Ontario and the St. Lawrence Valley (Morgan, 1972; Morgan and Morgan, 1980a; Williams and others, 1981; Matthews, unpublished data) and appear to be long-term residents of the region. Away from the ice front was an open-ground boreal assemblage of Coleoptera, which also contains scolytids and thus indicates the presence of trees.

Northern elements living in harmony in eastern and central Canada in early Wisconsin time migrated together toward the continental interior at or near the time of maximum glaciation. The assemblage can be divided into two complexes. The *Diacheila/Cryobius* complex is defined (in its simplest terms) as consisting of species today centered in the cold, relatively dry northern and western parts of the continent, and whose direct ancestors probably survived in the Beringian refugium. The portion of this complex that had migrated south and west from eastern Canada was eliminated from lowland areas south of the Laurentide Ice Sheet commencing at ca. 16.5 ka and eventually from high-altitude eastern Appalachian sites as late (or later?) than 11 ka. Species of this complex are found living today in the region west of Hudson Bay and largely, but not exclusively, north of treeline. Not all members of either *Cryobius* or *Diacheila* belong to this complex (e.g., *P. arcticola, P. pinguedinius,* and *P. brevicornis* would be exceptions, as well as *D. arctica*). *Pterostichus caribou, P. mandibularoides, Blethisa catenaria,* and *Helophorus splendidus* would, however, be included.

The *Cymindis unicolor/H. arcticus* assemblage is represented by cold stenotherms now found across the continent in northern latitudes. The ancestral forms of this assemblage principally survived south of the ice and successfully colonized the eastern arctic and subartic and also boreal North America. A second ancestral pool for certain species included in this assemblage also survived in the Beringian refugium. A suggestion has been made that a third refugial area (on the eastern margin of the ice) functioned for a limited number of species (Morgan and Morgan, 1980a; Morgan and others, 1984). *C. unicolor* and *H. arcticus* were chosen as two representatives of species that live beyond or at treeline, or in open areas of the northern boreal forest. Many other species (*Elaphrus lapponicus, Pterostichus haematopus, Amara glacialis, Notiophilus borealis, Nebria nivalis,* etc.) would also belong to this complex.

At or near the time of maximum glaciation, both complexes moved south and west toward the continental interior. The *Cryobius/Diacheila* complex has not been found in sites immediately south of southern Ontario (i.e., Pennsylvania/Ohio/Indiana) in the 22 to 21 ka period, but it was resident at the Clinton and Gardena sites at 21 to 19 ka (Morgan and Morgan, 1986) and in Iowa at 17 ka (Baker and others, 1986). In contrast, the assemblages found in central Indiana shortly before the ice advance, and in central Illinois prior to 22.5 ka, are commonly seen in boreal communities of eastern Canada today and thus are adapted to moister conditions.

The ice-marginal communities of the late Wisconsin maximum (ca. 20 to 18 ka) contain a peculiar mixture of species characteristic today of northwestern and transcontinental treeline areas and of open ground within the northern boreal forest. The presence of typical boreal species suggests that trees were suppressed because of other environmental factors, such as insufficient moisture or strong winds, for the *Cryobius/Diacheila* complex is adapted to cold/dry conditions. In this way unstable substrates or continuous wind abrasion of plants may have precluded the establishment of a tree cover dense enough to be commonly recorded in the plant-macrofossil or pollen record. Areas of Krummholz conifers should have existed in many sheltered sites.

We have not yet analyzed sites south of the late Wisconsin maximum in Illinois that are contemporaneous with the Clinton and Gardena localities (i.e., 21 to 19 ka). However, the lower Gardena section and the Athens Quarry site provide some idea of the environmental conditions prevailing prior to the arrival of the Lake Michigan lobe. Both sites indicate boreal insect assemblages from 25 to 22 ka. Even in the youngest samples, taken from the Athens site (ca. 22.5 ka), there are reasonable numbers of scolytids, indicating *Picea* trees in the vicinity. The *Cryobius/Diacheila* community is not present at this time, although it may have been farther north, closer to the southward-moving ice margin. An open coniferous environment with permanent water bodies here and in Indiana and Ohio is suggested. The cessation of organic deposition at the Athens site after ca. 22 ka may have been due to the onset of drier conditions with the approaching ice front, as the organic horizons are covered with loess (Follmer and others, 1979).

The *Cryobius/Diacheila* assemblage appears to have been well represented in the area west of the Lake Michigan lobe after 20 ka (Fig. 3), and it may have occupied a large expanse of ice-free ground across northwestern Illinois, eastern and northeastern Iowa, southwestern Wisconsin, and southeastern Minnesota (Mickelson and others, 1983). Like the interlobate region in southwestern Ontario some 4,000 yr later, the climate in this region was strongly modified by the proximity of the ice. The area may also have been under the influence of cold and dry arctic air west of the Lake Michigan lobe, and arctic and subarctic species of Coleoptera could have survived the low winter temperatures and dry conditions prevailing in the region.

Between 17 and 15 ka an important climatic change caused the demise of the *Cryobius/Diacheila* assemblage along the southern margin of the Laurentide ice in the area west of the junction of the Lake Michigan and Lake Huron lobes. Other open-ground carabid species (of the *H. arcticus/C. unicolor* complex) survived the transition, migrating northward with the retreating ice margin. Although the lowland communities of the

Cryobius/Diacheila complex seem to have been eliminated, elements of the complex probably survived close to the ice margin at higher altitudes in Appalachia (Morgan and Morgan, 1980a). It is difficult to assess the ecological parameters that preferentially eliminated the *Cryobius/Diacheila* complex from lowland areas in the west. Perhaps species in this complex, although being able to survive extremely cold and dry winter conditions, needed cooler and drier summers than do the open-ground carabids of the boreal zone today. Such conditions prevailed with altitude modification in the Appalachians and allowed the survival of elements of the complex in this region almost to the close of Wisconsin time. Certainly there is little difference in temperature preferences for the two complexes that overlap at modern treeline in western Canada and Alaska. Contributing causes for the demise of the *Cryobius/Diacheila* complex may have involved increasing insolation in midlatitudes, or thermal stress (Morgan and Morgan, 1982), or the onset of wet conditions at the ice margin around 16.5 to 15.5 ka.

The most marked change perhaps involved increased winter rainfall. Rapidly increasing summer temperatures combined with winter precipitation as rain would have caused rapid melting of the Lake Michigan and Lake Huron lobes. Trees situated near the eastern margin of the ice could have rapidly colonized the exposed land and even the stagnant ice itself. The associated insect fauna would have moved with the trees. If ablation rates exceeded the colonization potential of the trees, "tundra" plant assemblages accompanied by open-ground beetles could have moved into deglaciated areas. These open-ground regions were highly unstable in the climatic, ecologic, and physical sense. Moisture, either as water within the ice or as snowfall on the higher ice surfaces, might have been responsible for numerous minor, and some major, ice readvances within an overall continuous retreat. In the area west of the Lake Michigan lobe, increasing humidity and temperatures would allow tree growth, although the region may well have remained outside the principal storm tracks affecting the southernmost portion of the ice. A general increase in temperatures might also have created unfavorable conditions for the summer-active Coleoptera larvae, contributing still more stress to the cold-adapted forms.

The climatic gradients away from the ice margin would have been steep. At ca. 14 ka, mean July temperatures in southern Michigan at the ice front (10° to 11°C) were 6° to 7°C cooler than those in a zone 400 to 450 km farther south. In adjacent southern Ontario, in the interlobate region between the Lake Huron/Erie and Ontario lobes, the presence of ice-wedge polygons substantiates a mean annual temperature of −3° to −4°C (Morgan, 1982). It is not surprising that this region should harbor "tundra" plant assemblages and treeline open-ground beetles. However, the absence of periglacial landforms on tills exposed after 13 ka suggests that permafrost conditions were short-lived. After 13 ka the mean annual temperature rose above 0°C and was never again this low. Permafrost may have been only a localized phenomenon associated with the cold air flowing off the ice surface, as seen earlier in the region northwest of Illinois. In this way the onset of lower temperatures did not cause the glacial advance but rather was a result of it.

Once the action of warm moist winds removed the southern portion of the Lake Huron lobe, conditions inside the former Ontario Island would have ameliorated quickly. The subsequent removal of the Lake Ontario lobe would have allowed warm air to encroach on a "straighter" ice margin, again creating rapid ice retreat (Fig. 5). July temperatures in southern Ontario rose from ca. 12°C at or slightly before 13 ka, to 20°C by 10 ka. In the same period the mean annual temperature went from ca. −3°C just before 13 ka to 2°C about 12.5 ka and then to about 5°C at 10 ka. At the close of Wisconsin time the January means may have been about 5°C below modern values, although the July temperatures would have been close to the current July mean in southwestern Ontario. Precipitation during this period would have been principally as increased winter snowfall. The presence of western Canadian (prairie) beetles suggests that summers were warm and dry. Insect assemblages, particularly those from the Rostock locality, suggest strong evaporative conditions during the early stages of deposition between 13 and 12 ka. Farther east the period of rapid and continuously ameliorating change between 12.5 and 11 ka probably provided the extra ecological stress that eliminated all but the last few members of the *Cryobius/Diacheila* complex in the highest areas of the northern Appalachians.

Shortly after ice retreat the animal and plant communities appear to be out of phase, and the tundra plant assemblages may not reflect thermal conditions so much as pioneering communities (Morgan and Morgan, 1980a). Local faunas along the shores of Lake Algonquin and in the lee of the Niagara Escarpment were modified by cool winds and prolonged winter snowfall. Similar "lake effects" might have been experienced to varying degrees in the Two Creeks and Powers sites between 12 and 11 ka. The insect fauna at the St. Eugene site (ca. 11 ka) differs substantially from that recorded at the Mont St. Hilaire site less than 1,000 yr later. The northern and western species present at the former site are not recorded in the Montreal region by 10.1 ka. This is hardly surprising because by 10 ka the northern and western Canadian insects had also apparently vanished from southern Ontario.

Some of these western Canadian elements within boreal assemblages are recorded from the area north of Lake Superior after 10 ka. Accordingly, it is possible merely to attribute the faunal shift to a direct migration, although the reason for the local extinction in southwestern Ontario is not apparent. Perhaps humidity increased, midwinter thaws might have brought the insects out of hibernation, predators may have increased in the area, or fungi or bacteria may have attacked the overwintering insects. From these data it should be possible to predict that about 11 ka the northern species present east of Quebec City migrated to higher altitudes in the northern Appalachians (particularly New England, Maine, and Gaspé), as well as to higher latitudes in Quebec/Labrador. Many species in these assemblages were able to survive the Holocene warming trends, and a few have survived in these areas to the present.

By 7.5 ka the insect assemblages in the region between

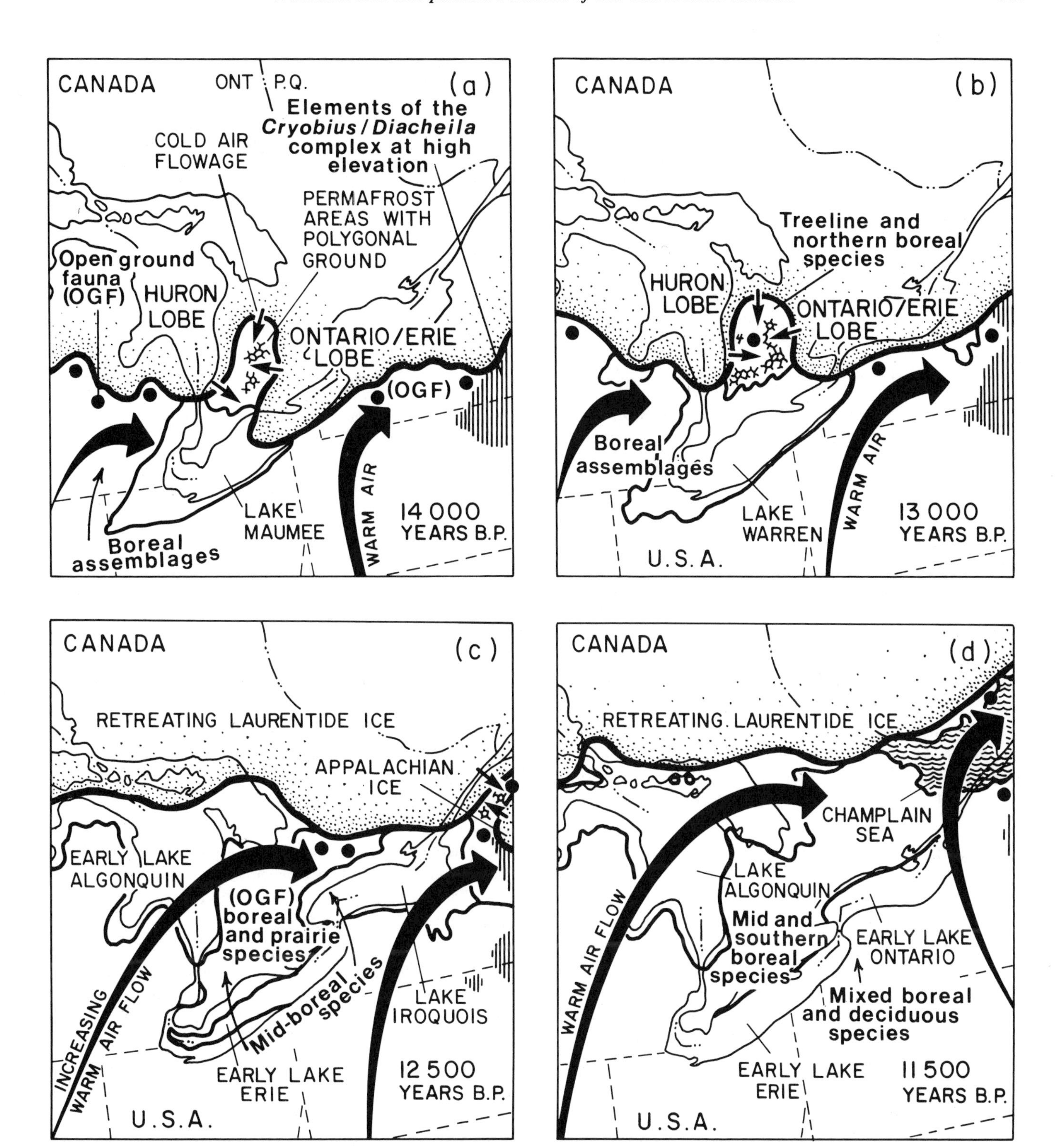

Figure 5. Approximate ice-front positions and proglacial lakes in the lower Great Lakes region from 14 to 11.5 ka. This period saw the disappearance of permafrost (5a and 5b) from southwestern Ontario, the rapid retreat of ice (compare 5b, 5c, and 5d), and the invasion of a thermophilous beetle fauna. The Champlain Sea incursion of the St. Lawrence and Ottawa Valleys is shown in 5d. Glacial lake positions and ice fronts from Prest (1969).

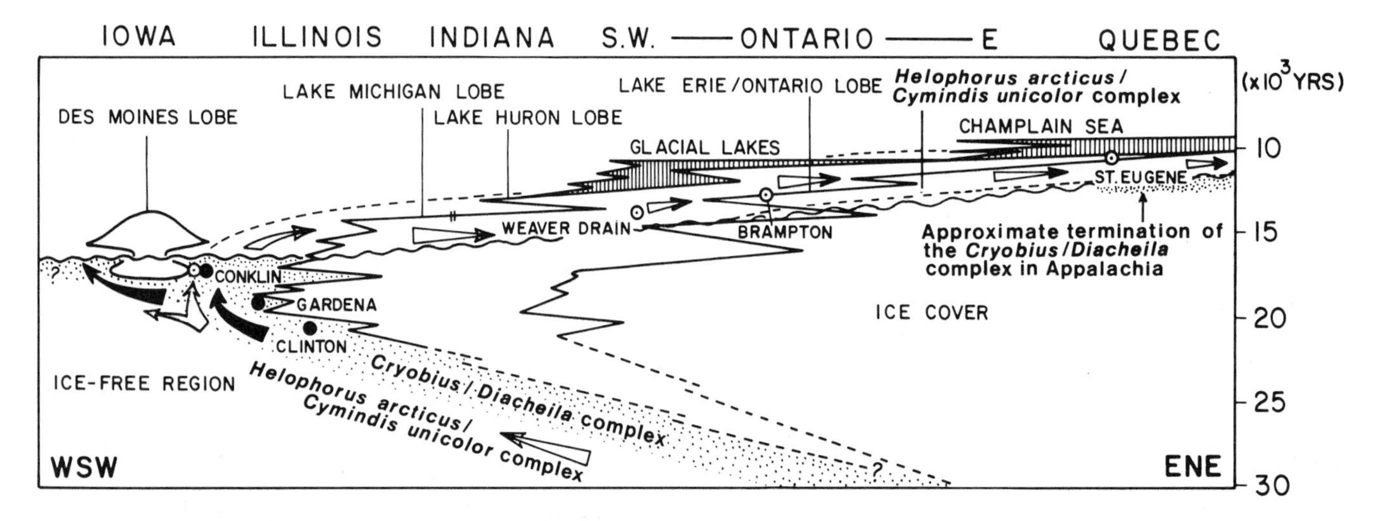

Figure 6. Schematic section illustrating approximate ice fronts and the movement of two different insect assemblages (described in the text) through the Great Lakes region.

southern Lake Huron and Lake Ontario were similar to those found in southernmost Ontario today. Summer temperatures were above the current July means, winter temperatures may have been slightly below, and the mean annual temperature was practically the same as today. Wood cellulose analyses show increases in the growing-season humidity levels (Edwards and others, 1985), and these may have been close to current values by 7 ka. Certainly the insect assemblages of the Marathon district about 7 ka and those of central Ontario, Michigan, and Minnesota, at 6.5, 4, and 3.5 ka respectively, are the same as those presently found in these regions (Morgan and others, 1985; Schwert and Ashworth, 1985; Warner and others, 1987).

SUMMARY

The data presented above are summarized in a series of diagrams to illustrate the insect recolonization following retreat of the ice from its maximum at 18 ka. Figure 6 illustrates the approximate ice fronts and the movements of two different types of insect assemblages through the Great Lakes region. The *Cryobius/Diacheila* complex (today centered in the northern and western portions of North America) moved westward in front of advancing Laurentide ice. It occurred in eastern Canada in the early Wisconsin and terminated west of the Lake Michigan and Erie Lobes ca. 16.5 to 15.5 ka instead of following the retreating ice Populations containing elements of this complex persisted in the higher areas of Appalachia, particularly in those regions close to the ice front. The termination of the complex on high ground in the east has been envisaged as diachronous, possibly lasting as late as 11 ka. I assume that the primary factor eliminating the former complex was a combination of increasing insolation, rapidly ameliorating summer temperatures leading to thermal stress, and the arrival of moist air at the southern margin of the Laurentide Ice Sheet about 16.5 ka. This may have been in the form of winter rain (which brought overwintering adult insects out of hibernation and/or made them particularly susceptible to predators and disease) and increased summer temperatures and humidity (creating stress for the summer-active larvae).

The *Helophorus arcticus/Cymindis unicolor* complex, which was present in eastern Canada in the early Wisconsin and has both northeastern and northern transcontinental distributions today, advanced south and west with the ice. The complex then followed the retreating ice front north and east and was present in a number of lowland sites postdating 15 ka.

A set of diagrams (Fig. 7) illustrates the northward movement of insect and plant communities from 18 to about 10 ka. I have ignored topography by concentrating on migration in lowland (<350 m elevation); the ice profile is purely diagrammatic, and the time is approximate. The modern section shows the present general distribution of plants from Cape Henrietta Maria south into the deciduous woodlands of southern Ontario. Insect distributions along the same transect are represented by (A) open-ground/treeline species, some of which range onto true tundra; (B) boreal species whose distribution may or may not be controlled by the presence of trees; (C) Coleoptera found in the area occupied by mixed deciduous and coniferous woodland today; and (D) Coleoptera found in the area of deciduous woodland. The insects of (C) and (D) do not necessarily depend upon trees.

The 18-ka diagram illustrates the narrow extent of (A) faunas associated with the maximum extent of Laurentide ice. Note also the compression of (B) and (C) faunal zones. Although the zone occupied by (A) in the Peoria region is narrow, the potential occupation area west of the Lake Michigan lobe would have extended far north as the point marked (x). Arctic air flowing off the ice provided a narrow area for species of the *Cryobius/Dia-*

cheila complex (black spot) adapted to cold, dry conditions at the ice margin in central Illinois, but a much larger area would have existed for this complex in the ice-free regions to the west (Fig. 3).

The transect for the 14-ka diagram is moved eastward, and it represents a line from Cape Henrietta Maria (extreme northeastern Ontario) south to southern Ontario and then across Lake Erie into the Cleveland area. The area south of the Georgian Bay lobe and between the Lake Huron and Ontario/Erie lobes had extensive permafrost caused by cold-air drainage. Boreal Coleoptera (B) were well established south of glacial Lake Maumee and were also starting to move into the area west of the Maumee outlet in Michigan (Fig. 5a). By at least 13 ka the interlobate and ice-marginal areas were occupied by open-ground and treeline beetles. Coleoptera of the *Helophorus arcticus/Cymindis unicolor* complex were capable of surviving in these regions.

By 12.5 ka the ice front had receded to the southern margin of Lake Huron (Georgian Bay region). The area between the lakes was an expanse of open ground, mostly colonized by "tundra" plants but with conifers invading the extreme south. The insects range from treeline open-ground species (A) to beetles residing either on the prairies or in open-ground areas of the central and southern boreal zone today (B). By 11.5 ka, northern open-ground species (A) were being forced to the cool marginal areas of glacial Lake Algonquin, and typical boreal species (B) occupied much of southern Ontario.

At 10 ka the northern species had vanished from southern Ontario. Boreal species (B) were well established both in southern Ontario and north of Lake Huron; mixed woodland species (C) and species that only just reach southern Ontario today (D) were present in the south of the province and in northern New York. Mean July temperatures in southern Ontario were close to 20°C (the present value).

Figure 8 provides a general time/space diagram summarizing some of the main events presented in the text. The solid line indicates diagrammatic frontal positions as the ice advanced south to Illinois and then retreated north. The early period is represented by a cool dry environment outside the ice margin to the most southerly position at ca. 18 ka. At the ice margin winds flowing off the ice surface produced the local cold and dry conditions that supported the *Cryobius/Diacheila* complex in lowland areas. Sometime after 17 ka, and probably centered around 16.5 to 15.5 ka, this cold-adapted insect assemblage disappeared from the southern margins of the Laurentide Ice Sheet. Sites both older and younger than this date contain northern species of eastern or transcontinental distribution today (represented by the *Helophorus arcticus/Cymindis unicolor* complex). These species, together with many boreal Coleoptera, indicate differences in survival mechanisms between members of the two complexes. The regional demise of the *Diacheila/Cryobius* complex might be related to warmer conditions (increasng insolation and July temperatures, possibly accompanied by higher humidity levels and associated with winter rain), on the southern midcontinental section of the Laurentide Ice Sheet. The ice underwent rapid

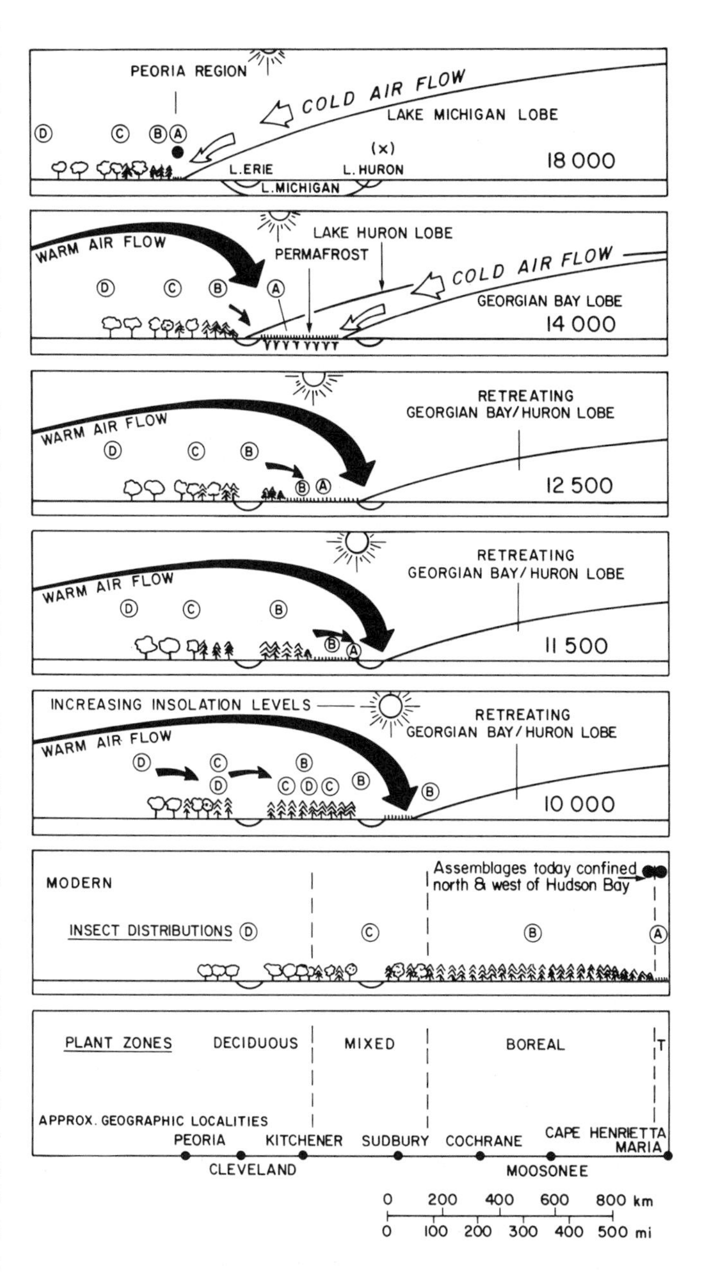

Figure 7. A series of diagrams to illustrate the movement of the two major insect complexes and the recolonization following ice retreat from 18 to 10 ka. See text for details.

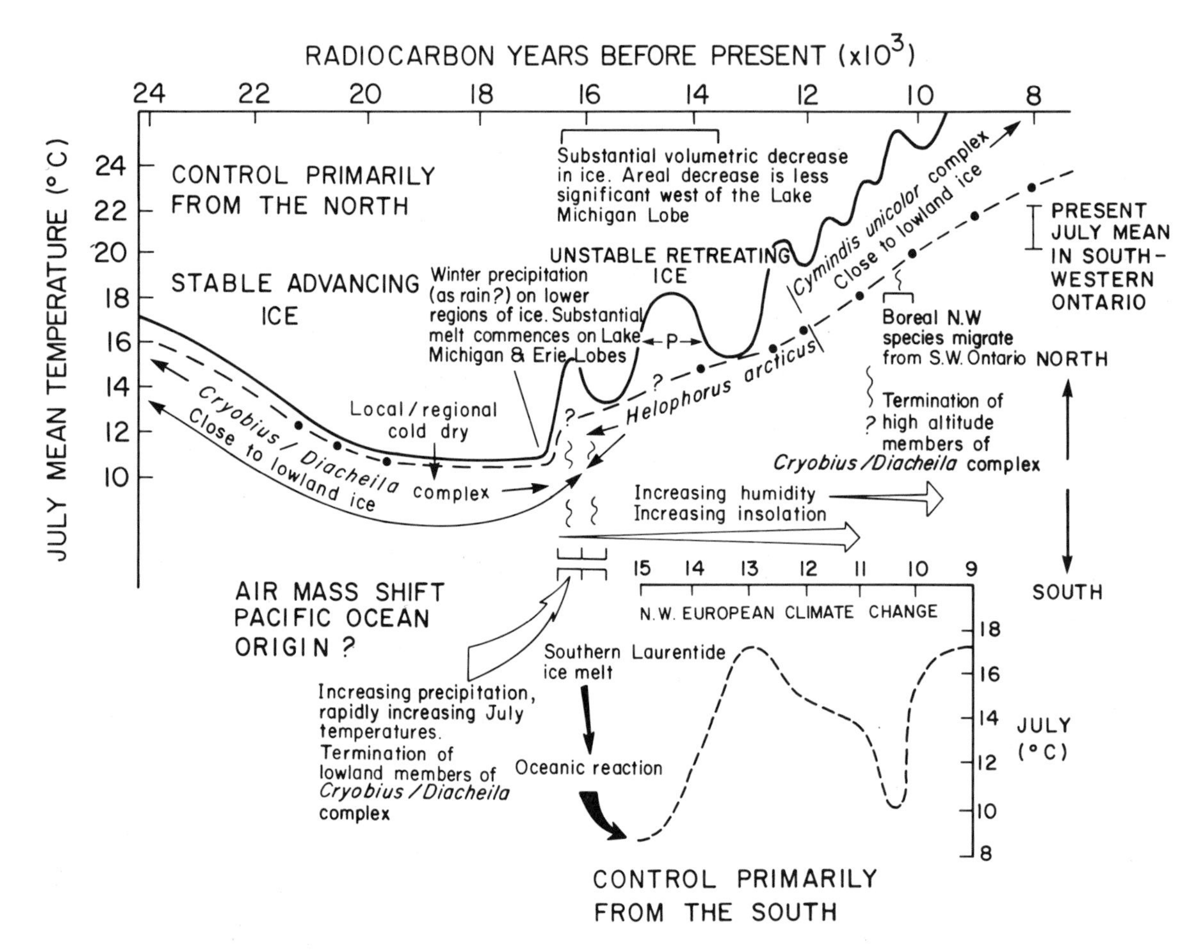

Figure 8. A general time/space diagram summarizing some of the main events described in the text. The mean July temperature curve is generalized for sites away from the influence of the ice.

ablation, stagnation, and retreat. Increased meltwater may have contributed to atmospheric moisture and rising humidities. Elements of the *Cryobius/Diacheila* complex survived at higher elevations in the east, probably on north-facing slopes, and particularly in areas close to the ice in Appalachia.

In east-central North America, rapid northward ice retreat was broken by many short-lived minor and some major ice advances. Insect assemblages in areas distant from the ice show no climatic reversals but a continuously ameliorating climatic trend. The Dryas/Allerød fluctuations of western Europe, recorded especially by insects in southern Britain (Coope and others, 1971; and lower right of Fig. 8), are not reflected in the insect assemblages of the lower Great Lakes region, although I do not preclude the possibility that they might be seen in assemblages from eastern Canada.

Faunas beside the ice margin or in reentrants in the ice front indicate much cooler conditions than the faunas living well beyond the ice margin at the same time. Mean annual temperatures inside these interlobate areas were low enough to produce permafrost (P), but climatic conditions were sufficiently different from the preglacial conditions so that the *Cryobius/Diacheila* complex did not occur. Ice readvance seems to be a product of instability within the ice sheet rather than climatic deterioration along the ice margin. Again it seems that the major contributor to this pattern of retreat was winter precipitation (accompanied by midwinter thaws), because many species outside the ice margin up to 10.5 or 10 ka are forms relatively well adapted to prairie or northwestern conditions. The present July mean temperature in southwestern Ontario (20° to 22°C) was reached at 10 ka, when the western and northern species emigrated from the area. By ca. 8.5 ka, the July mean temperature was above the current value experienced in southwestern Ontario. The modern fauna seems to have become established in this region about 7 ka.

Undoubtedly further investigations of insect assemblages will help to refine the hypotheses suggested above. The present initial synthesis should provide some different insights on the vexing problem of the demise of the Laurentide Ice Sheet in central North America.

REFERENCES CITED

Anonymous, 1984, Climatic Atlas Climatique, Canada; Temperature and degree days: Ottawa, Ontario, Canadian Government Publishing Centre, Environment Canada, Atmospheric Environment Map Series 1, scale approximately 1:23,760,000.

Ashworth, A. C., 1977, A late Wisconsinan coleopterous assemblage from southern Ontario and its environmental significance: Canadian Journal of Earth Sciences, v. 14, p. 1625–1634.

——, 1979, Quaternary Coleoptera studies in North America; Past and present, *in* Erwin, T. L., Ball, G. E., and Whitehead, D. R., eds., Carabid beetles; Their evolution, natural history, and classification: The Hague, Dr. W. Junk bv Publishers, p. 395–406.

Ashworth, A. C., and Brophy, J. A., 1972, A late Quaternary fossil beetle assemblage from the Missouri Coteau, North Dakota: Geological Society of America Bulletin, v. 83, p. 2981–2988.

Ashworth, A. C. Clayton, L., and Bickley, W. B., 1972, The Mosbeck site; A paleoenvironmental interpretation of the late Quaternary history of Lake Agassiz based on fossil insect and mollusk remains: Quaternary Research, v. 2, p. 176–188.

Ashworth, A. C., Schwert, D. P., Watts, W. A., and Wright, H. E., Jr., 1981, Plant and insect fossils at Norwood in south-central Minnesota: A record of late-glacial succession: Quaternary Research, v. 16, p. 66–79.

Baker, R. G., and 6 others, 1986, A full-glacial biota from southeastern Iowa: Journal of Quaternary Sciences, v. 1, p. 91–107.

Baust, J. G., and Miller, L. K., 1970, Seasonal variations in glycerol content and its influence in cold hardiness in the Alaskan carabid beetle *Pterostichus brevicornis:* Journal of Insect Physiology, v. 16, p. 979–990.

Berti, A. A., 1975, Paleobotany of Wisconsinan interstadials, eastern Great Lakes region, North America: Quaternary Research, v. 5, p. 591–619.

Burgis, W. A., 1970, The Imlay Outlet of glacial Lake Maumee, Imlay City, Michigan [M.S. thesis]: Ann Arbor, University of Michigan, 74 p.

Coope, G. R., 1959, A late Pleistocene fauna from Chelford, Cheshire: Proceedings of the Royal Society of London, series B, v. 151, p. 70–86.

——, 1968, An insect fauna from mid-Weichselian deposits at Brandon, Warwickshire: Philosophical Transactions of the Royal Society of London, series B, v. 254, p. 425–456.

——, 1970, Interpretations of Quaternary insect fossils: Annual Review of Entomology, v. 15, p. 97–120.

——, 1973, Tibetan species of dung beetle from late Pleistocene deposits in England: Nature, v. 245, p. 335–336.

——, 1979, Late Cenozoic fossil Coleoptera; Evolution, biogeography, and ecology: Annual Review of Ecology and Systematics, v. 10, p. 247–267.

Coope, G. R., Morgan, A., and Osborne, P. J., 1971, Fossil Coleoptera as indicators of climatic fluctuation during the last glaciation in Britian: Palaeogeography, Palaeoclimatology, Palaeoecology, v. 10, p. 87–101.

Coope, G. R., Shotton, F. W. and Strachan, I., 1961, A Late Pleistocene fauna and flora from Upton Warren, Worcestershire: Philosophical Transactions of the Royal Society, London, series B, v. 244, p. 379–421.

Edwards, T.W.D., Aravena, R. O., Fritz, P., and Morgan, A. V., 1985, Interpreting paleoclimate from ^{18}O and ^{2}H in plant cellulose; Comparison with evidence from fossil insects and relict permafrost in southwestern Ontario: Canadian Journal of Earth Sciences, v. 22, p. 1720–1726.

Follmer, L. R., McKay, E. D., and Lineback, J. A., 1979, Wisconsinan, Sangamonian, and Illinoian stratigraphy in central Illinois: Midwest Friends of the Pleistocene 26th Field Conference, Illinois State Geological Survey Guidebook 13, 93 p.

Fritz, P., Morgan, A. V., Eicher, U., and McAndrews, J. H., 1987, Stable isotope, fossil Coleoptera and pollen stratigraphy in late Quaternary sediments from Ontario and New York State: Palaeogeography, Palaeoclimatology, Palaeoecology, v. 58, p. 183–202.

Goulet, H., 1983, The genera of holarctic Elaphrini and species of *Elaphrus* Fabricus (Coleoptera:Carabidae); Classification, phylogeny, and zoogeography: Questiones Entomologicae, v. 19, p. 219–482.

Hammond, P., Morgan, A., and Morgan, A. V., 1979, On the *gibbulus* group of *Anotylus,* and fossil occurrences of *Anotylus gibbulus* (Staphylinidae): Systematic Entomology, v. 4, p. 215–221.

Henriksen, K. L., 1933, Undersøgelser over Danmark Skanes kvartaere insektfauna: Videnskabelige Meddedelser fra Dansk Naturhistorik Forening, v. 96, p. 77–355.

Lindroth, C. H., 1948, Interglacial insect remains from Sweden: Arsbok Sveriges Geologiske Undersökning, ser. C, p. 1–7.

——, 1961, The ground-beetles (Carabidae excl. Cicindelinae) of Canada and Alaska, Part 2: Opuscula Entomologica Supplementum 20, p. 1–200.

Matthews, J. V., Jr., 1977, Coleoptera fossils; Their potential value for dating and correlation of late Cenozoic sediments: Canadian Journal of Earth Sciences, v. 14, p. 2339–2347.

——, 1980, Tertiary land bridges and their climate; Backdrop for the development of the present Canadian fauna: Canadian Entomologist, v. 112, p. 1089–1103.

Mickelson, D. M., Clayton, L., Fullerton, D. S., and Borns, H. W., Jr., 1983, The Late Wisconsin glacial record of the Laurentide Ice Sheet in the United States, *in* Porter, S., and Wright, H. E., Jr., eds., Late Quaternary environments of the United States; Volume 1, The Late Pleistocene: Minneapolis, University of Minnesota Press, p. 3–37.

Miller, L. K., 1969, Freezing tolerance in an adult insect: Science, v. 166, p. 105–106.

Miller, R. F., and Morgan, A. V., 1982, A postglacial coleopterous assemblage from Lockport Gulf, New York: Quaternary Research, v. 17, p. 258–274.

Morgan, A., 1972, The fossil occurrence of *Helophorus arcticus* Brown (Coleoptera: Hydrophilidae) in Pleistocene deposits of the Scarborough Bluffs, Ontario: Canadian Journal of Zoology, v. 50, p. 555–558.

——, 1973, Late Pleistocene environmental changes indicated by fossil insect faunas of the English Midlands: Boreas, v. 2, p. 173–212.

Morgan, A., and Morgan, A. V., 1982, *Diacheila polita* Fald. (Coleoptera: Carabidae) and *Helophorus arcticus* Brown (Coleoptera: Hydrophilidae) in late Wisconsinan deposits: Joint Meeting of Entomological Societies of America, Canada, and Ontario, Toronto, Ontario, programme volume, p. 101.

Morgan, A., Morgan, A. V., and Elias, S. A., 1985, Fossil insects and paleoecology of the Au Sable River, Michigan: Ecology, v. 6, p. 1817–1828.

Morgan, A. V., 1982, Distribution and probable age of relict permafrost features in southwestern Ontario, *in* French, H. M., ed., Proceedings of the 4th Canadian Permafrost Conference; The Roger J. E. Brown Memorial Volume: National Research Council of Canada, p. 91–100.

Morgan, A. V., and Freitag, R., 1982, The occurrence of *Cicindela limbalis* Klug (Coleoptera: Cicindelidae) in a late-glacial site at Brampton, Ontario: Coleopterists Bulletin, v. 36, p. 105–108.

Morgan, A. V., and Morgan, A., 1979, The fossil Coleoptera of the Two Creeks Forest Bed, Wisconsin: Quaternary Research, v. 12, p. 226–240.

——, 1980a, Faunal assemblages and distributional shifts of Coleoptera during the late Pleistocene in Canada and the northern United States: The Canadian Entomologist, v. 112, p. 1105–1128.

——, 1980b, Beetle bits; The science of paleoentomology: Geoscience Canada, v. 27, p. 22–29.

——, 1981, Paleoentomological methods of reconstructing paleoclimate with reference to interglacial and interstadial insect faunas of southern Ontario, *in* Mahaney, W. C., ed., Quaternary paleoclimate: Norwich, England, Geo Abstracts Limited, p. 173–192.

——, 1986, A preliminary note on fossil insect faunas from central Illinois; Quaternary records of central and northern Illinois: American Quaternary Association Field Guide, 9th Biennial Meeting, Champaign, Illinois State Geological Survey, 84 pages.

Morgan, A. V., Miller, R. F., and Morgan, A., 1982, Paleoenvironmental reconstruction of southwestern Ontario between 11,000 and 10,000 yr B.P. using fossil insects as indicators:Montreal, Quebec, 3rd North American Paleontology Convention Proceedings, v. 2, p. 381–386.

Morgan, A. V., Morgan, A., Ashworth, A. C., and Matthews, J. V., Jr., 1983a, Late Wisconsin fossil beetles in North America, *in* Porter, S., and Wright, H. E., Jr., eds., Late Quaternary environments of the United States; Volume 1, The Late Pleistocene: Minneapolis, University of Minnesota Press, p. 354–363.

Morgan, A. V., Morgan, A., and Miller, R. F., 1983b, A preliminary report on two fossil insect assemblages from west-central Indiana, *in* Bleuer, N. K., Melhorn, W. N., and Pavey, R. R., eds., Interlobate stratigraphy of the Wabash Valley, Indiana: Lafayette, Indiana, Purdue University Press, 136 p.

Morgan, A. V., Morgan, A., and Miller, R. F., 1984, Range extension and fossil occurrences of *Holoboreaphilus nordenskioeldi* (Mäklin) (Coleoptera: Staphylinidae) in North America: Canadian Journal of Zoology, v. 62, p. 463–467.

Morgan, A. V., Morgan, A., Nelson, R. E., and Pilny, J. J., 1986, Current status of knowledge on the past and present distribution of the genus *Blethisa* (Coleoptera: Carabidae) in North America: Coleopterists Bulletin, v. 40, p. 105–115.

Mott, R. J., Matthews, J. V., Jr., and Anderson, T., 1981, Late-glacial paleoenvironments bordering the Champlain Sea based on pollen and macrofossil evidence, *in* Mahaney, W. C., ed., Quaternary paleoclimate: Norwich, England, Geo Abstracts Limited, p. 129–171.

Ngano, C. D., Miller, S. E., and Morgan, A. V., 1982, Fossil tiger beetles (Coleoptera: Cicindelidae); Review and new Quaternary records: Psyche, v. 89, p. 339–346.

Pilny, J. J., and Morgan, A. V., 1987, Paleontomology and paleoecology of a possible Sangamon age site near Innerkip, Ontario: Quaternary Research, v. 28, p. 157–174.

Pilny, J. J., Morgan, A. V., and Morgan, A., 1987, Paleoclimatic implications of a late Wisconsin insect assemblage from Rostock, southwestern Ontario: Canadian Journal of Earth Sciences, v. 24, p. 617–630.

Prest, V. K., 1969, Retreat of Wisconsin and recent ice in North America: Geological Survey of Canada Map 1257A, scale approximately 1:20,000,000.

Prest, V. K., Terasmae, J., Matthews, J. V., Jr., and Lichti-Federovich, S., 1976, Late Quaternary history of Magdalen Islands, Quebec: Maritime Sediments, v. 12, p. 35–59.

Ring, R. A., and Tesar, D., 1980, Cold-hardiness of the Arctic beetle, *Pytho americanus* Kirby Coleoptera, Pythidae (Salpingidae): Journal of Insect Physiology, v. 26, p. 763–774.

Schwert, D. P., and Ashworth, A. C., 1985, Fossil evidence for late Holocene faunal stability in southern Minnesota (Coleoptera): The Coleopterists Bulletin, v. 39, p. 67–79.

Schwert, D. P., and Morgan, A. V., 1980, Paleoenvironmental implications of a late-glacial insect assemblage from northwestern New York: Quaternary Research, v. 13, p. 93–110.

Schwert, D. P., Ashworth, A. C., and Baker, R. G., 1981, An arctic insect assemblage from the late Wisconsinan of midcontinental North America: Geological Society of America Abstracts with Programs, v. 13, p. 550.

Schwert, D. P., Anderson, T. W., Morgan, A., Morgan, A. V., and Karrow, P. F., 1985, Changes in late Quaternary vegetation and insect communities in southwestern Ontario: Quaternary Research, v. 23, p. 205–226.

Thiele, H.-U., 1977, Carabid beetles in their environments; A study on habitat selection by adaptations in physiology and behaviour: New York, Springer-Verlag, p. 1–369.

Warner, B. G., Karrow, P. F., Morgan, A. V., and Morgan, A., 1987, Plant and insect fossils from Nipissing sediments along the Goulais River, southeastern Lake Superior: Canadian Journal of Earth Sciences (in press).

Watts, W. A., 1979, Late Quaternary vegetation in central Appalachia and the New Jersey coastal plains: Ecological Monographs 49, p. 427–469.

White, G. W., 1968, Age and correlation of Pleistocene deposits at Garfield Heights (Cleveland), Ohio: Geological Society of America Bulletin, v. 79, p. 749–755.

Williams, N. E., and Morgan, A. V., 1977, Fossil caddisflies (Insecta: Trichoptera) from the Don Formation, Toronto, Ontario, and their use in paleoecology: Canadian Journal of Zoology, v. 55, p. 519–527.

Williams, N. E., Westgate, J. A., Williams, D. D., Morgan, A., and Morgan, A. V., 1981, Invertebrate fossils (Insecta: Trichoptera, Diptera, Coleoptera) from the Pleistocene Scarborough Formation at Toronto, Ontario, and their paleoenvironmental significance: Quaternary Research, v. 16, p. 146–166.

MANUSCRIPT ACCEPTED BY THE SOCIETY APRIL 10, 1987
QUATERNARY ENTOMOLOGY LABORATORY CONTRIBUTION 100

ACKNOWLEDGMENTS

I acknowledge the assistance of numerous geologists, entomologists, and Quaternary co-workers who have provided help and information used in the development of this chapter. Colleagues and students in the Department of Earth Sciences at the University of Waterloo are particularly worthy of mention, although I take sole responsibility for any contentious statements made in the text. I appreciate the data provided by G. R. Coope and used in Figure 8 and the comments provided by H. E. Wright in the preparation of various drafts. Similarly, I am grateful for the review provided by J. V. Matthews and A. C. Ashworth, and I apologize for points on which I prefer to differ. I acknowledge the vital financial role made in numerous research grants from the Canadian Natural Sciences and Engineering Research Council for individual projects reported in this chapter and for the final synthesis of these results. Finally, I express special thanks to my wife, Anne, who has shared in so many aspects of this work during the last decade.

Printed in U.S.A.

The Geology of North America
Vol. K-3, North America and adjacent oceans during the last deglaciation
The Geological Society of America, 1987

Chapter 17

Environmental fluctuations and evolution of mammalian faunas during the last deglaciation in North America

Russell W. Graham
Illinois State Museum, Springfield, Illinois 62706
Jim I. Mead
Department of Geology and The Museum of Northern Arizona, Northern Arizona University, Flagstaff, Arizona 86011

INTRODUCTION

Environmental changes associated with the last deglaciation (Termination 1 in the marine record) had profound effects on the evolution of biotic communities in North America. Vertebrate species, especially mammals, are particularly sensitive proxies for these changes, so they provide excellent documentation of the climatic fluctuations of the late Quaternary. However, vertebrate communities are not tightly bound aggregates of species; rather, they are collections of species randomly distributed along environmental gradients (Whitakker, 1970). Consequently, environmental change will not illicit a response from the entire community, but instead each species will respond according to its own tolerances. The species composition of a community may therefore be constantly in flux, and significant environmental fluctuations may cause major biotic reorganizations. Modern communities are not direct analogues for past ones, but changes in the distribution, abundance, and clinal variation of individual species can provide invaluable information about past environments.

The composition of Pleistocene mammalian communities, even if extinct taxa are excluded, was not the same as modern ones (Graham, 1985b). In fact, many late Pleistocene mammalian communities were composed of species that today are geographically allopatric and appear to be ecologically incompatible. Figure 1 shows the modern distribution of species occurring at the same level in Baker Bluff Cave, Sullivan County, Tennessee (Guilday and others, 1978). These communities have been referred to as disharmonious (Lundelius and others, 1983), but this does not mean that the organisms were not in harmony with prevailing Pleistocene environments. Quite the contrary, disharmonious biotas were apparently maintained by equable climates that have no modern analogues.

By equable climates we mean ones with reduced seasonal temperature extremes as originally defined by Hibbard (1960). This definition does not equate with maritime climates as implied by Rhodes (1984, p. 33), nor "oceanic," as discussed by Guthrie (1982, p. 323), although maritime climates can be equable. Because mean annual temperature has negligible influence on equability, it is possible to have both cold and warm equable climates. Also, mean annual temperatures for equable climates may be the same as means for cold or warm continental nonequable climates (Fig. 2).

Because moisture and temperature are integrally related through effective moisture (net annual precipitation minus net annual evapotranspiration), the cool equable climates of the North American late Pleistocene may have been moister than those of today. Also, with this definition an even seasonal distribution of precipitation is not inherent in equability, as inferred by Guthrie (1982). Thus, equable climates, like those of the Serengeti Plain today, may have strong seasonal precipitation patterns.

The biotic effects of the equable and continental climates are significantly different. The distributions of organisms are obviously controlled by complex interactions among a host of biotic and environmental variables. However, with regard to climatic factors, it is the extremes and not the means that are the principal limiting factors. For instance, winter low temperatures might control the northern limit of a species range, whereas summer high temperatures might restrict the southern distribution of a species.

If the climatic extremes are relaxed, as in an equable climate, then southern and northern species would tend to move together along an environmental gradient. The broad overlap in species distributions would tend to obscure the definition of ecotones, although there would still be northern and southern taxa (Graham, 1979). With nonequable climates the species would be dispersed along the environmental gradient, and ecotones could be more clearly defined.

Finally, the transition from a cold equable climate (Pleistocene) to a warm continental climate (Holocene) would have differential effects on the climatic extremes and thus the distribution of plants and animals. In this transition the greatest amount

Graham, R. W., and Mead, J. I., 1987, Environmental fluctuations and evolution of mammalian faunas during the last deglaciation in North America, *in* Ruddiman, W. F., and Wright, H. E., Jr., eds., North America and adjacent oceans during the last deglaciation: Boulder, Colorado, Geological Society of America, The Geology of North America, v. K-3.

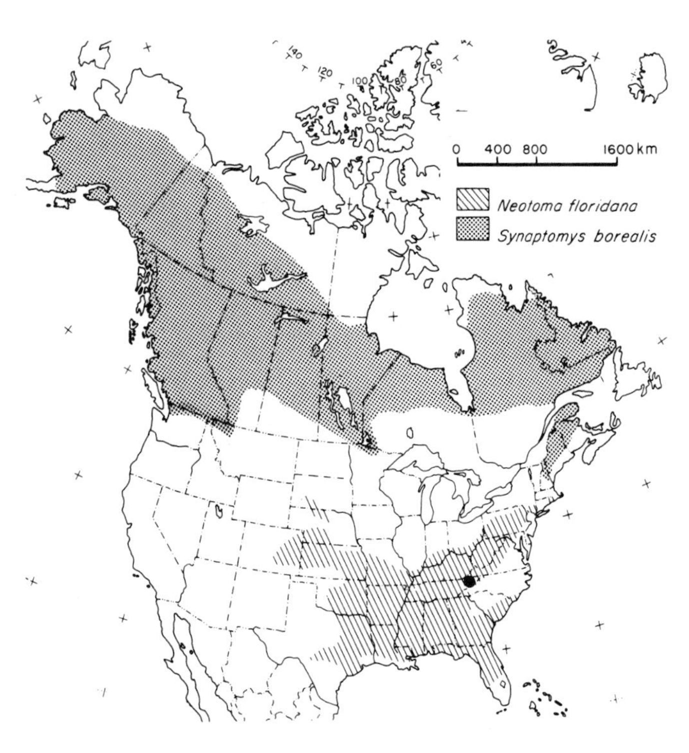

Figure 1. Modern distribution of mammalian species found at the same horizon illustrating disharmonious pairs in the Baker Bluff Cave local fauna, Sullivan County, Tennessee. (Modified after Graham, 1985b.)

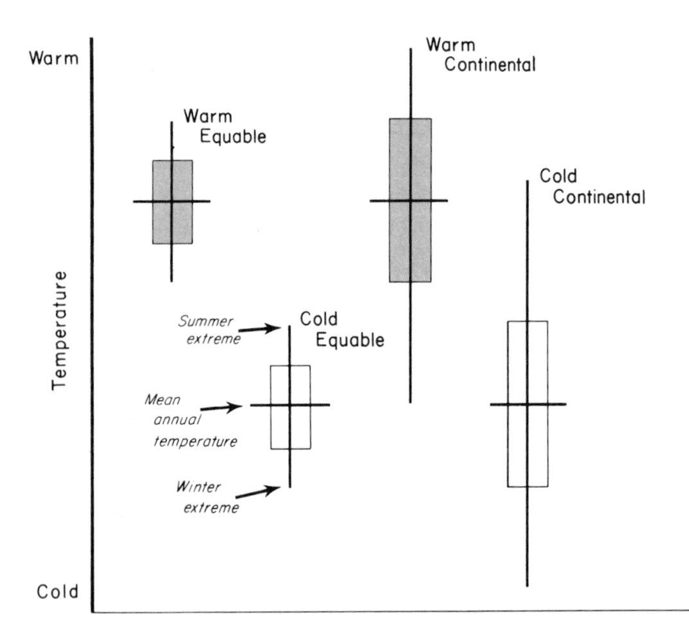

Figure 2. Schematic diagram comparing temperature variables for hypothetical warm and cold equable climates as well as warm and cold continental climates.

of change would be reflected in the summer extremes (Fig. 2). Therefore the southern limits of species would show the greatest amount of range adjustment, whereas the northern limits would be essentially unaltered. This general pattern seems to fit with the observed changes in Pleistocene and Holocene communities.

During glacial stages, species were physically forced southward by glacial ice in North America. However, resident species south of the ice sheet were not displaced en masse, because winter extremes were not accentuated. Boreal and arctic taxa were integrated with preexisting temperate communities under the more equable climatic conditions. Because climatic extremes were not so severe, other limiting factors may have been more important (E. L. Lundelius, personal communication, 1985). Thus microenvironmental differentiation undoubtedly played an important role in supporting the late Pleistocene communities (Graham, 1985a, 1985b; Guthrie, 1982, 1984; Rhodes, 1984). This amalgamation of ecotypes produced a much finer grained environmental mosaic, with increased diversity for certain taxa (Graham, 1976a).

For instance, the number of shrew and vole species from the Midwest was higher during the late Pleistocene than in either the Holocene or modern times (Figs. 3 and 4). These differences are probably not the result of some artificial sampling or taphonomic biases, for these effects should be similarly reflected in the Pleistocene and Holocene fossil samples (Graham, 1985b). However, cave sites tend to have higher diversities than open sites. Thus the dramatic decrease in shrew and vole diversity at the end of the Pleistocene (ca. 10 ka) is probably the result of more continental climates during the Holocene (Graham, 1976a). The fine-grained environmental mosaic of the late Pleistocene was probably important in supporting megaherbivore diversity as well (Guthrie, 1982).

Increasing continentality caused species to disperse along environmental gradients, and the southern limits of arctic and boreal species were significantly restricted. With the retreat of glacial ice, species began to move north and colonize the newly opened landscapes. Little fossil evidence is available to document the actual mechanisms of recolonization by the vertebrate fauna. However, modern distributions and diversity patterns may provide insight into this process.

Species numbers for arbitrarily defined areas (state boundaries) can be calculated from standard distributional maps (Hall, 1981). These data can then be plotted on a species-area curve as defined by MacArthur and Wilson (1967). If glaciation and recolonization significantly affected equilibrium of species numbers, then states of equal area with similar glacial histories (i.e., unglaciated, partially glaciated, completely glaciated, or under alpine glaciation) should cluster together on the species-area curves. In other words, glaciated states should have the lowest species numbers, and unglaciated states should have the highest. This is not the case (Fig. 5). Therefore, it appears that the number of mammalian species recolonizing glaciated terrain reached equilibrium soon after deglaciation (in some cases before 8 ka), but the species composition of communities (or assortment equilibrium) may have continued to change (Graham, 1985a). This is

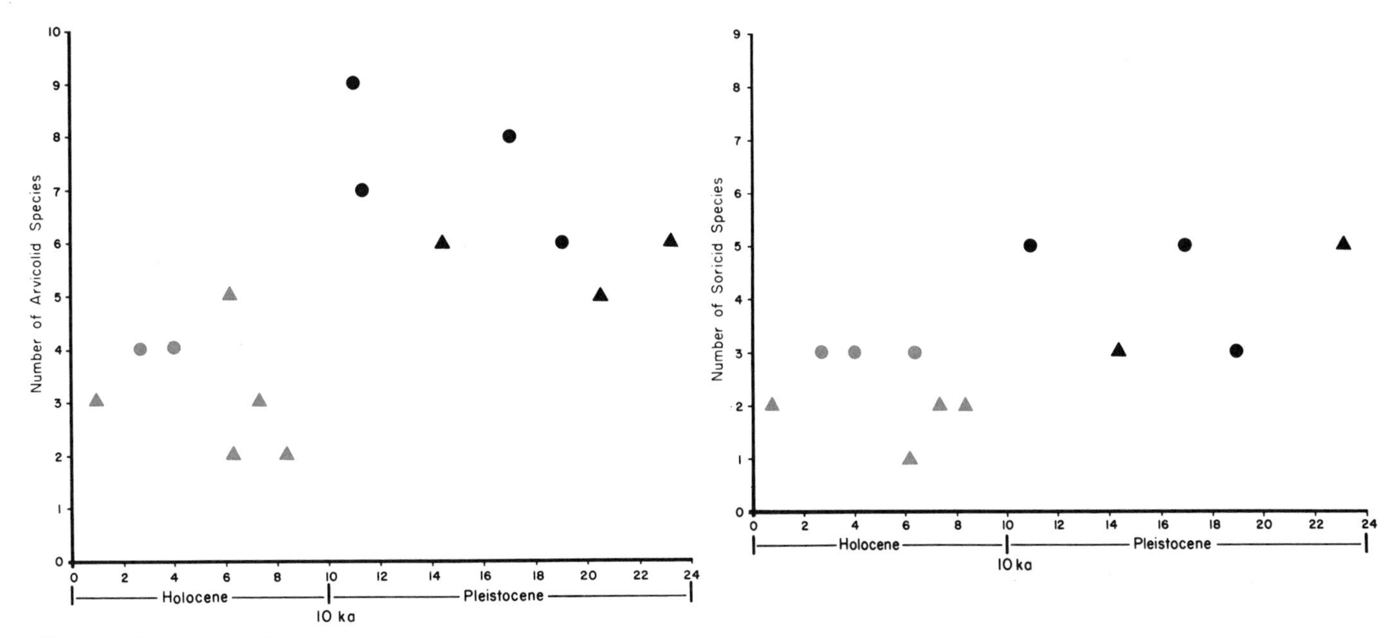

Figure 3. Comparison of the number of vole (Arvicolid) species (ordinate) from fossil sites of various absolute ages (abscissa) in the midwestern United States. Δ open sites, ● cave sites. (Modified after Graham, 1985b.)

Figure 4. Comparison of the number of shrew (Soricid) species (ordinate) from fossil sites of various absolute ages (abscissa) in the midwestern United States. Δ open sites, ● cave sites. (Modified after Graham, 1985b.)

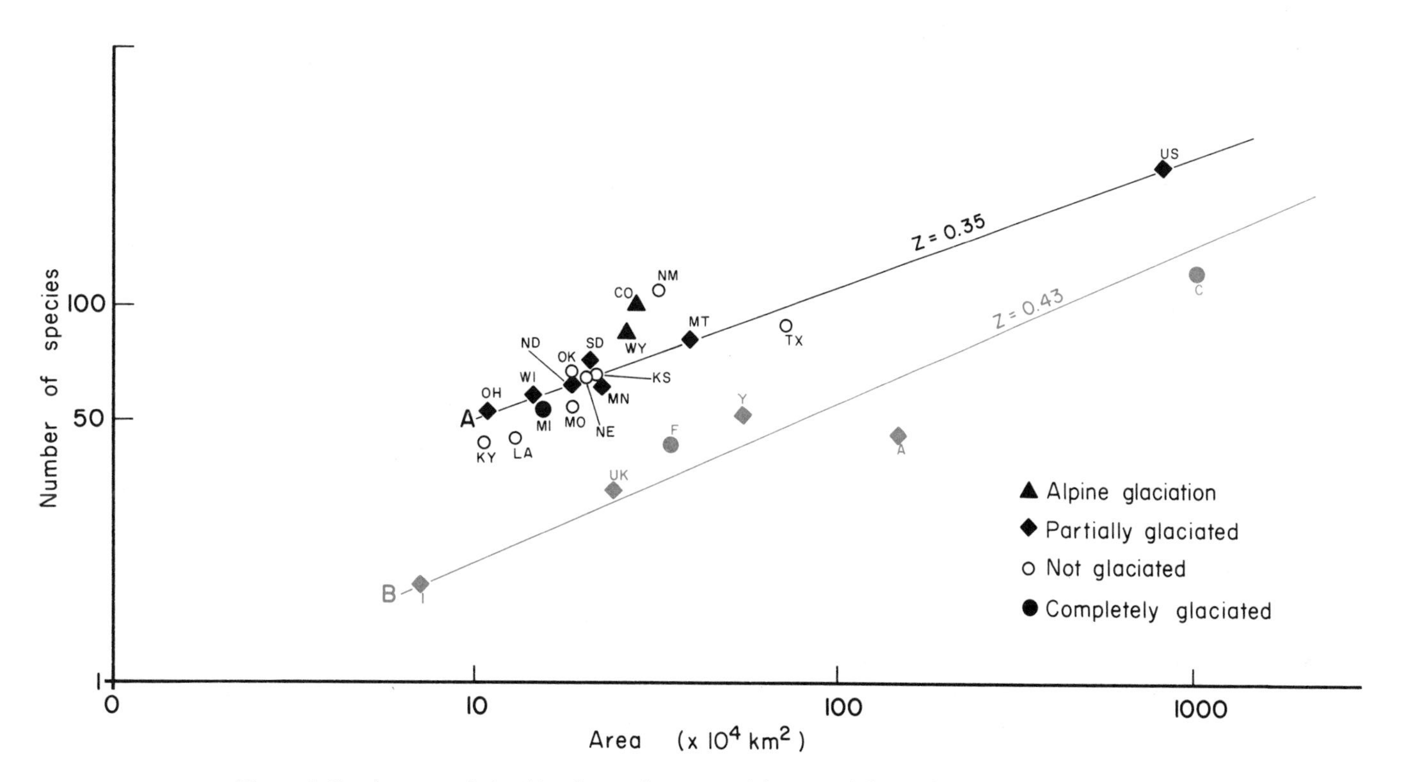

Figure 5. Species-area relationships for modern terrestrial mammal faunas between 26° and 49° N. latitude (line A) and 49° and 70° N. latitude (line B). Location abbreviations: A, Alaska; CO, Colorado; KS, Kansas; KY, Kentucky; LA, Louisiana; MI, Michigan; MN, Minnesota; MO, Missouri; MT, Montana; NE, Nebraska; NM, New Mexico; ND, North Dakota; OH, Ohio; OK, Oklahoma; SD, South Dakota; TX, Texas; WI, Wisconsin; WY, Wyoming; I, Ireland; UK, United Kingdom; F, Finland; Y, Yukon Territory; C, Canada (entire mainland); US, United States (entire contiguous 48) (Modified after Graham, 1985a.)

consistent with diversity patterns for vegetational recolonization of the Klutlan Glacier area in the Yukon Territory (Birks, 1980). Finally, it demonstrates that the latitudinal diversity gradients are not the result of historical events (Graham, 1985a).

FULL-GLACIAL FAUNAS

Full-glacial (20 to 17 ka) local faunas (l.f.) are not abundant, and they are widely dispersed in the U.S. In the East, Baker Bluff and Guy Wilson caves in Sullivan County, Tennessee, are the only dated full-glacial local faunas (Fig. 6, Table 1), although some undated local faunas are assigned to this interval. Baker Bluff Cave, a stratified local fauna, ranges in age from 19 to 10 ka (Guilday and others, 1978). It contains two bird and ten mammal species that no longer occur in the southern Appalachian area, nine mammal species that exist at higher elevations in the region, four that exhibit significant chronoclinal shifts in body size, and six that are extinct. Such latitudinal and altitudinal changes in geographic ranges, shifts in body size, and extinctions are characteristic of most late Pleistocene mammalian sequences in North America.

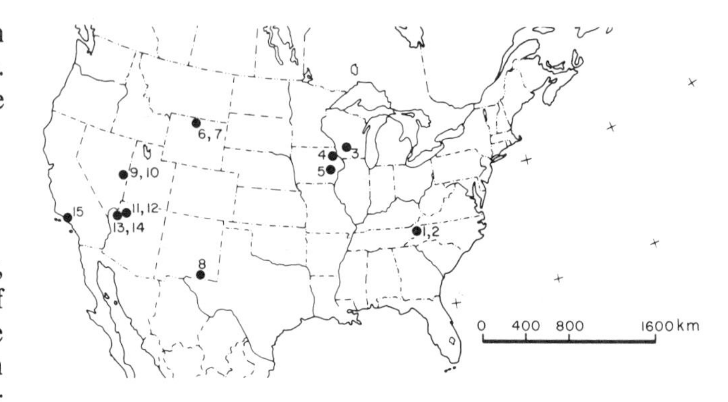

Figure 6. Location of full-glacial (20 to 17 ka) vertebrate faunas. See Table 1 for identification of sites.

Guilday and others (1978) believe that the Baker Bluff Cave l.f. records a transition from cool-temperature deciduous forest/coniferous woodland to boreal coniferous parkland. This interpretation is supported by the presence of the arctic shrew, longtail shrew, and pygmy shrew in upper levels (see Appendix I for scientific names of taxa). Numbers of boreal red squirrels and northern flying squirrels also increase upward, whereas temperate eastern gray squirrels, southern flying squirrels, and eastern chipmunks were more abundant in the lower levels. The progressive increase in abundance of the least shrew, least chipmunk, thirteenline ground squirrel, and meadow vole in the upper levels reflects an opening of the forest.

This interpretation of a full-glacial temperate deciduous forest is not consistent with vegetational reconstructions based on pollen sequences. Watts (1983) believes that conifer forests with jackpine and spruce must have been dominant, but that some broad-leaved trees and herbs may have also been important. He suggests a full-glacial climate similar to the modern climate of coastal Maine, which is characterized by short growing seasons and severe winters. The differences between faunal and palynological interpretations could result from invalid dates for Baker Bluff Cave. However, they could also be due to variance in regional (pollen) and local (vertebrates) sampling. Thus the Baker Bluff Cave full-glacial faunule may represent a localized deciduous environment within the more boreal landscape.

Guy Wilson Cave, approximately 74 km to the east, contains extinct species as well as extirpated boreal species (Guilday and others, 1975). The record of caribou represents the most southerly distribution of this boreal species in North America (Kurtén and Anderson, 1980) attesting to cooler summers. Caribou also occur at Baker Bluff Cave, but the remains were apparently mixed with younger sediments (Guilday and others, 1978).

New Paris No. 4, Bedford County, Pennsylvania, is a vertical pit cave with 9 m of sediment containing a rich and diverse late Pleistocene vertebrate fauna (Guilday and others, 1964). Charcoal from the 4.6-m level has been dated at 11.3 ka, but it is difficult to determine the actual amount of time represented by the entire faunal sequence. Changes in faunal composition suggest that most of the late Wisconsin may be represented and that lower levels may represent full-glacial environments. The Hudson Bay collared lemming and the arctic shrew do not occur above the 6.1-m level, and the water shrew only occurs at the 8.5-m level. In fact the fauna from the lowest 1-m level (8 to 9 m) contains a strong boreal and arctic component (Guilday and others, 1964, Table 40). Openness of the full-glacial environments is suggested by the presence of the thirteenline ground squirrel in the lower levels. The garter snake and the Hudson Bay toad, which are among the most-northerly distributed reptile and amphibian species today, are the only herpetological species in the lower levels at New Paris No. 4.

Three sites in southwestern Wisconsin and eastern Iowa were close to the ice front during full-glacial times. The ca. 17 ka Moscow Fissure l.f. (Table 1) in the Driftless Area of southwestern Wisconsin contains one tundra species (collared lemming), and several boreal species (yellowcheek vole and heather vole) that reflect a climate cooler than at present (Foley, 1984). The Moscow Fissure also contains northern pocket gopher and western jumping mouse, whose modern distributions are several hundred kilometers west of the site (Foley, 1984). These species reflect open environments but not necessarily a more arid climate, for today both species inhabit open high-altitude boreal forests of the western United States.

The full-glacial climate was not severe enough to eliminate the garter snake, fox snake, and milk snake. The limiting factors for the oviparous (egg-laying) snakes such as fox and milk snakes may not be low winter temperatures, which can be avoided by hibernation. Rather, low summer temperatures limit mating, egg-laying, incubation, hatching, and post-hatching activities (Foley,

TABLE 1. RADIOCARBON DATES FOR FULL GLACIAL FAUNAS

Site Name (County and State)	Youngest Date* yr B.P.	Material Dated and Lab No.	Oldest Date yr B.P.	Material Dated and Lab No.	Major References
1. Baker Bluff Cave 9-10 ft level (Sullivan, TN)	19,100±850	bone apatite, GX-3495			Guilday and others, 1978
2. Guy Wilson Cave (Sullivan, TN)	19,700±600	*Platygonus* bone, I-4163			Guilday and others, 1978
3. Moscow Fissure (Iowa, WI)	17,050±1500	snake vertebra, bone collagen, I-10, 153			Foley, 1984
4. Elkader (Clayton, IA)	20,530±130	*Picea* wood, Beta-2748			Woodman, 1982
5. Conklin Quarry (Johnson, IA)	17,170±205	wood, DIC-1240			Baker and others, 1986
6. Natural Trap Cave Stratum 3 (Bighorn, WY)	17,620±1490	*Equus* bone collagen, DI-690	20,250±275	*Ovis* bone collagen, DI-1687	Martin and Gilbert, 1978
7. Prospects Shelter Middle Stratum IV (Bighorn, WY)	17,500±4.5%	thermo-luminescence			Chomko and Gilbert, 1987
8. Musk Ox Cave (Eddy, NM)	18,140±200	bone, not given	25,500±1100	bone, not given	Harris, 1985
9. Streamview No. 2 (White Pine, NV)	17,350±435	*Pinus longaeva*, GX-5866			Thompson and Mead, 1982
10. Ladder Cave (White Pine, NV)	17,960±110	twigs and *Nothrotheriops* dung, A-2092			Mead and others, 1982
11. Stanton's Cave (Coconino, AZ)	19,320±380	*Oreamnos harringtoni* horn sheath, AA-1850 TAMS			Mead and others, 1986
12. Tse'an Kaeton Cave (Coconino, AZ)	17,500±300	*Oreamnos harringtoni* dung, A-2723			Mead and others, 1986
13. Rampart Cave Locality 7204 (Mohave, AZ)	18,430±300	*Oreamnos* dung, A-1278	19,980±210	*Fraxinus anomala* twigs, AA-1840 TAMS	Mead, 1981
14. Vulture Cave (Mohave, AZ)	17,030±760	*Juniperus*, A-1768	19,050±390	*Juniperus*, A-1606	Mead and Phillips, 1981
15. Rancho La Brea Pit 3 (Los Angeles, CA)	19,300±395	*Smilodon* bone collagen (amino acid), UCLA-1292K	20,500±900	*Smilodon* bone collagen (amino acid), UCLA-1292J	Marcus and Berger, 1984

*If there is only one date from the site or unit, it is listed as the "youngest."

1984, p. 28). Adiabatic warming of air masses descending the ice sheet may have helped to maintain the distribution of some amphibian and reptile species near the ice front.

The Elkader local biota (l.b.) from Clayton County, Iowa, dated on spruce wood at 20.53 ka, is one of the most tundralike faunas known from the Pleistocene of the U.S. (Woodman, 1982), with collared lemming, arctic ground squirrel, and singing vole. However, it also contains other species like meadow vole and yellowcheek vole that do not inhabit the tundra today. The mammalian, molluscan, and vegetational remains from Elkader reflect an environment similar to the transition between modern boreal forest and tundra (Woodman, 1982).

The Conklin Quarry l.b., Johnson County, Iowa, at the base of Wisconsin loess, dates to 17.17 ka (Baker and others, 1986). Vertebrates are rare, but the biota contains rich assemblages of insects, terrestrial and aquatic molluscs, pollen, and plant macrofossils. The mammalian fauna includes meadow vole, yellowcheek vole, possibly singing vole, collared lemming, redback vole, and heather vole. The mammals as well as the rest of the biota reflect a tundra/tree-line environment.

In the western United States, Natural Trap Cave and Prospects Shelter (Chomko and Gilbert, 1987) preserve similar full-glacial mammalian local faunas, including species also found in the midwestern and eastern full-glacial faunas (i.e., collared lemming, heather vole, yellowcheek vole, meadow vole, and northern bog lemming). However, these local faunas also have western species (sagebrush vole, yellowbelly or hoary marmot, and pika) that are not found in the midwestern or eastern local faunas but occur in the Bighorn Mountains today. The full-glacial faunas from the Bighorn Mountains reflect open environments, but they are not similar to modern arctic tundra faunas. Some authors (Chomko and Gilbert, 1987; Walker, 1987) suggest that these local faunas reflect environments similar to the Pleistocene arctic steppe of the Alaskan refugium (Matthews, 1982) or the modern high-alpine tundra of the Bighorns. The arctic steppe of Alaska contained brown lemming, wooly mammoth, and saiga, which are not found in the full-glacial faunas of the Bighorn Mountains or other areas in the U.S. Furthermore, collared lemming and yellowcheek vole do not occur in any alpine tundra environments in the United States but are abundant in the full-glacial faunas of the Bighorns. We therefore believe that the full-glacial faunas of the northern plains are more like other full-glacial faunas south of the ice sheet than like the Arctic steppe faunas of Alaska. However, the northern plains faunas contain more grassland species than the full-glacial faunas farther east. This suggests that parts of the northern plains were characterized by cool, moderately dry climates with scattered woodlands.

The southwestern U.S. had full-glacial faunas that reflect moister and cooler climates than at present. Boreomontane species that today occur in alpine environments ranged to much lower elevations during full-glacial times. For example, Musk Ox Cave, Eddy County, New Mexico, contains boreomontane species like yellowbelly marmot, meadow vole, masked shrew, and water shrew as well as the eastern least shrew (Harris, 1985). At the same time more xeric species like gray shrew, silky pocket mouse, and southern grasshopper mouse were also present.

The well-dated faunas from the Grand Canyon area of Arizona and southern Great Basin of Nevada are derived from packrat middens and are therefore rather small. Their taphonomic pathways are not comparable to many other full-glacial faunas, and the extreme relief of the canyons may actually mute regional environmental fluctuations. However, full-glacial middens from Rampart Cave (Mead, 1981) and Vulture Cave (Mead and Phillips, 1981) contain boreomontane species like voles and yellowbelly marmot.

These faunal displacements to lower elevations are consistent with paleobotanical data. An open juniper woodland dominated the lower elevations within the Grand Canyon, a community now 1,000 m higher than the vegetation at Rampart and Vulture caves today (Mead and Phillips, 1981; Phillips, 1984). Limber pine and Douglas fir grew below the canyon rims at the higher and eastern portions of the Grand Canyon (Cole, 1982).

Harrington's mountain goat was common throughout the Grand Canyon during full-glacial times, as indicated by radiocarbon dates on its dung and horn-core sheaths (Mead and others, 1986). The Shasta ground sloth was absent from the Grand Canyon during full-glacial times, but it was present both before and after this time.

Ladder Cave, Streamview Shelter, and Smith Creek Canyon in eastern Nevada have yielded full-glacial packrat middens with records for pika and longtail vole. The heather vole is also recorded, but its chronologic placement is not certain (Mead and others, 1982). Grayson (1982) has shown that pika, yellowbelly marmot, heather vole, and pine marten were much more widespread in other parts of the Great Basin during the late Wisconsin. Unfortunately, few of these sites date specifically to full-glacial times (Grayson, 1982).

Part of the well-known fauna from the tar pits at Rancho La Brea may also date to full-glacial times, as pits 3, 4, 16, and 2051 have produced radiocarbon dates within this time interval (Marcus and Berger, 1984, Table 8.1). The dates seem secure, but many of the pits contain temporally mixed faunal assemblages. The absolutely dated megafauna was widely distributed throughout the western United States and apparently not restricted to any specific habitat type. Therefore, paleoenvironmental interpretations are limited to regional reconstructions of savanna.

LATE-GLACIAL FAUNAS

Faunas from 15 to 13 ka

In the eastern U.S. there are only two local faunas that have absolute dates within the late-glacial interval (Fig. 7, Table 2). Saltville in Smyth County, Virginia, is a stratified site with lacustrine (W2) and fluvial (W3) units dating between 13.13 and 14.48 ka respectively, according to McDonald (1986). Older radiocarbon dates of 17,710 ± 130 yr B.P. (Beta-5724) and 19,080

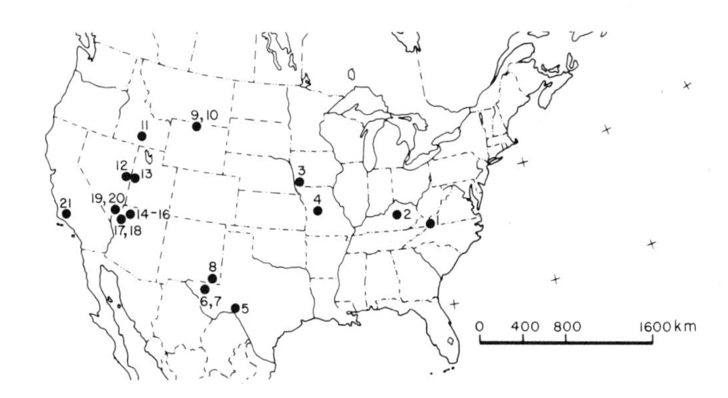

Figure 7. Location of late-glacial (15 to 13 ka) vertebrate faunas. See Table 2 for identification of sites.

± 650 yr B.P. (A-2987), both derived from bulk sediment samples from W2 and W3 respectively, have been rejected by McDonald (1986) as too old. Recent investigations at Saltville have not been fully reported, but McDonald (1986) suggests that between 15 and 13 ka the local vegetation was sedge meadows, alder swamps, and herbs on the valley floor within a regional coniferous forest. This environment supported a diverse megafauna (Ray and others, 1967).

The Welsh Cave l.f. in Woodford County, Kentucky, is radiocarbon dated at 12.95 ka on flathead peccary bones (Table 2). Guilday and others (1971) believe that this date applies to the majority of the faunal remains recovered from the cave and that the fauna from the primary Pleistocene deposits has a similar age. The occurrence of both boreal forest and grassland species appears to represent a parkland environment supported by a relatively cool and moist climate.

In the Midwest, Boney Spring in Benton County, Missouri, contains fish, amphibians, and reptiles that occur in the area today, as do all the small mammals except the meadow vole and the woodland jumping mouse. The harmonious nature of the Boney Spring l.f. is unlike most other late Pleistocene faunas from Missouri or other parts of North America. This difference may be the result of taphonomic conditions inherent in spring deposits (Saunders, 1977). Thus the small-mammal fauna from Boney Spring may not fully reflect all of the environments that were in the area but instead documents local woodland environments around the spring. Horse and perhaps Harlan's ground sloth suggest the presence of open habitats. Pollen and plant macrofossils suggest a mixture of spruce and deciduous elements following the full-glacial boreal grasslands (King, 1973, p. 562). Cool and moist environments are implied.

The Craigmile l.f. from Mills County, western Iowa, dates ca. 14.8 ka (Fig. 7, Table 2). It is stratigraphically superimposed upon the Waubonsie l.f., which is dated at ca. 22.3 ka (Rhodes, 1984). Craigmile contains mammalian species that today inhabit the deciduous forest, boreal forest, and grasslands (Rhodes, 1984). A greater abundance of arboreal species in the Craigmile l.f. reflects a more mesic environment than present during the later accumulation of the Waubonsie local fauna.

To the west, Natural Trap Cave and Prospects Shelter (Fig. 7) provide stratified faunal records that show 12 genera of megafauna disappearing simultaneously around 15.5 ka (Chomko and Gilbert, 1987). Mammoth appears for the first time in the area at about 14 ka, and at 13.5 ka boreal and arctic small mammals increase, such as collared lemming, northern bog lemming, and pika (Chomko and Gilbert, 1987). A similar fauna is present at approximately 14 ka in the Pryor Mountains about 50 km northwest of Little Mountain (Bonnichsen, 1985). At about the same time, grassland species such as the northern pocket gopher, pocket mouse, least chipmunk, and prairie vole decrease in frequency.

In the southern plains during late-glacial times, some mammalian faunas were characterized by a mosaic of species that today inhabit boreomontane conifer forests, eastern deciduous forests, and grassland. However, the boreal species were quite different from those of the northern plains, in that collared lemming, yellowcheek vole, and northern bog lemming did not extend that far south. Instead, montane species like yellowbelly marmot, meadow vole, and vagrant shrew, which occur in Dry Cave (Animal Fair locality), extended their ranges to lower elevations (Harris, 1985).

The late-glacial fauna from Cueva Quebrada, Val Verde County, Texas (Fig. 7), which is farther south than Dry Cave, does not contain any boreomontane species, although they occur in other late Pleistocene sites farther south, like San Josecito Cave (Jakaway, 1958). The pygmy mouse is the only extralimital species in the Cueva Quebrada local fauna. Lundelius (1984) feels that it indicates a more humid climate and brushy environment during late-glacial times. The extinct herbivore fauna is dominated by grazers (e.g., bison, horse, antelope), which suggest extensive open areas. No woodland animals are present.

In the southwestern United States, the late-glacial record is far more extensive than the full-glacial one. In the Grand Canyon, a mosaic of extralocal woodland species was intermingled with desert plants that are still prominent in the region today. Harrington's mountain goat was present, and the Shasta ground sloth still inhabited the lower western end of the Grand Canyon, which was an open juniper woodland with single-leaf ash (Mead and Phillips, 1981; Phillips, 1984).

The chuckwalla and desert spiny lizard first appear in the late Wisconsin record at this time, and the desert tortoise and collared lizard reappear. The northern distributions of the chuckwalla and desert tortoise may be limited by winter temperature extremes, so late-glacial winter temperatures must have warmed and reached a level no colder than today. Cooler summer temperatures are suggested by boreal mammals at lower elevations.

Few of the numerous late Pleistocene sites in the Great Basin are accurately dated (Grayson, 1982) but late-glacial faunas also

TABLE 2. RADIOCARBON DATES FOR LATE GLACIAL FAUNAS FROM 13 TO 15 KA

Site Name (County and State)	Youngest Date* yr B.P.	Material Dated and Lab No.	Oldest Date yr B.P.	Material Dated and Lab No.	Major References
1. Saltville Units W2 and W3 (Smyth, VA)	13,130±330	woody stem, A-2985	14,480±300	woody stem, Beta-5724	McDonald, 1986
2. Welsh Cave (Woodford, KY)	12,950±550	*Platygonus* bone collagen, I-2982			Guilday and others, 1971
3. Craigmile (Mills, IA)	14,430±1030	charcoal, I-7496	14,830+1060 -1220	charcoal, DIC-1688	Rhodes, 1984
4. Boney Spring (Benton, MO)	13,550±400	organic debris, A-1079	13,700±600	organic debris, M-2211	Saunders, 1977
5. Cueva Quebrada (Val Verde, TX)	13,920±210	wood, TX-880	14,300±220	wood, TX-881	Lundelius, 1984
6. Dust Cave (Culberson, TX)	13,000±730	*Picea* needles, A-1539			Van Devender and others, 1979
7. Upper Sloth Cave (Culberson, TX)	10,750±140	*Nothrotheriops shastensis* dung, A-1583	13,060±280	*Picea* needles, A-1549	Logan and Black, 1979 Van Devender and others, 1979
8. Dry Cave Locality 22 (Eddy, NM)	15,030±210	bone collagen, I-6201			Harris, 1985
Dry Cave Harris's Pocket	14,470±250	*Neotoma* dung, I-3365			Harris, 1977
9. Natural Trap Cave Unit 2 (Bighorn, WY)	12,777±900	bone collagen, not given	14,670+670 -730	bone collagen, DIC-689	Chomko and Gilbert, 1987
10. Prospects Shelter stratum IV (Bighorn, WY)	13,500±4.8%	thermo-luminescence of bone	17,500±4.5%	thermo-luminescence of bone	Chomko and Gilbert, 1987
11. Wilson Butte Cave (Jerome, ID)	14,500±500	bone, M-1409	15,000±800	bone, not given	Harris, 1985
12. Smith Creek Cave No. 5 (White Pine, NV)	13,340±430	*Neotoma* dung, A-2094			Mead and others, 1982 Thompson and Mead, 1982
13. Garrison No. 2 (Millard, UT)	13,480±250	*Pinus flexilis*, A-3213			Mead and others, 1982 Thompson and Mead, 1982

TABLE 2. (CONTINUED)

Site Name (County and State)	Youngest Date* yr B.P.	Material Dated and Lab No.	Oldest Date yr B.P.	Material Dated and Lab No.	Major References
14. Stanton's Cave (Coconino, AZ)	13,120±130	*Oreamnos harringtoni* horn sheath, AA-1849	13,770±500	*Oreamnos harringtoni* horn sheath, AA-1132	Mead and others, 1986
15. Tse'an Bida Cave (Coconino, AZ)	13,100±700	*Oreamnos harringtoni* dung, RL-1133			Mead, 1981 Mead and others, 1986
16. Tse'an Kaetan Cave (Coconino, AZ)	14,220±320	*Oreamnos harringtoni* dung, A-2835			Mead and others, 1986
17. Vulture Cave (Mohave, AZ)	13,820±220	*Juniperus*, A-1564	15,130±210	*Juniperus*, A-1607	Mead and Phillips, 1981
18. Rampart Cave (Mohave, AZ)	13,140±320	*Nothrotheriops shastensis* dung, A-1207	13,430±130	*Oreamnos harringtoni* horn sheath, AA-1839	Long and others, 1974 Mead and others, 1986
19. Quade No. 1 (Clark, NV)	14,300±800	*Ochotona* dung, AA-782			Van Devender and others, 1985
20. Tule Springs E1 (Clark, NV)	13,000±200	carbonized wood, UCLA-552	13,100±200	carbonized wood, not given	Haynes, 1967 Mawby, 1967
21. Rancho La Brea Pit 3 (Los Angeles, CA)	13,035±275	*Smilodon* bone collagen (amino acid), QC-279	15,200±150	cypress wood, QC-422A	Marcus and Berger, 1984

* If there is only one date from the site or unit, it is listed as the "youngest."

document altitudinal shifts of boreomontane species to valley bottoms. At Quade No. 1, pika apparently inhabited a xeric juniper woodland environment (Van Devender and others, 1985), and at Garrison No. 2, pikas were within a few kilometers of the Lake Bonneville shoreline. At Smith Creek Cave No. 5, pika may have lived along canyon slopes vegetated with bristlecone pine and sagebrush (Thompson and Mead, 1982). The late-glacial fauna (Unit E1) at Tule Springs provides an extralimital record of the pygmy rabbit, which presently inhabits thick sagebrush flats (Grayson, 1982).

Faunas from 12 to 11 ka

Although its origins are not clear, the Clovis Cultural Complex appears to be restricted to a fairly narrow temporal interval 11.5 to 11 ka (Haynes, 1980). Vertebrate faunas associated with Clovis archeological sites can therefore be closely dated even if radiocarbon dates are not available for a site. These faunas also share similar taphonomic pathways. Stratified Clovis sites with faunal remains are restricted to the U.S. west of the Mississippi River (Fig. 8, Table 3), with the possible exception of the Shawnee-Minisink site in Pennsylvania (McNett and others, 1977).

The micromammal faunas from most of the Clovis sites are represented by small samples, but extralimital species (Table 4) document cooler and moister climates than today even though modern community patterns were beginning to emerge.

Environmental gradients persisted (i.e., colder to the north and drier to the west). For instance, southern Clovis sites (i.e., San Pedro Valley, Blackwater Draw Locality No. 1, and Domebo) contain northern species but not extreme boreal species like heather vole, boreal redback vole, or pygmy shrew, which occur in the Clovis sites of the northern plains (i.e., Agate Basin and Lange/Ferguson). Western Clovis faunas include bison, pocket mouse, and northern grasshopper mouse, which prefer open habitats, whereas Kimmswick, to the east, has woodland species like tree squirrels, whitetail deer, and woodchuck. The westerly decreasing moisture gradient is also reflected by the extinct taxa from these sites. For instance, the western sites are characterized by horse, flathead peccary, camel, antelope, and mammoth, which are considered to be predominantly grazers, although most were capable of mixed feeding. At Kimmswick, browsers such as longnose peccary, Harlan's ground sloth, and American mastodon were dominant forms.

Faunas that date to the Clovis interval but are not associated with diagnostic Clovis artifacts (Fig. 9, Table 5) provide a similar paleoenvironmental reconstruction. In the Grand Canyon area, low-elevation faunas still contained extralimital boreomontane species like voles, porcupine, and yellowbelly marmot (Mead and Phillips, 1981). The plant communities surrounding Rampart and Vulture caves were composed of a mixture of woodland species found today above the rims of the Grand Canyon, high-desert species found today at or just below the rims or as relict populations at low elevations, and low-desert species of which the modern flora is exclusively composed (Phillips, 1984).

In the Glen Canyon area of southern Utah, low-elevation xeric woodlands contain desert mammals like kangaroo rat and pocket mouse as well as reptiles like the chuckwalla and western shovelnose snake. However, there are no records of boreomontane or mesic species (Mead and others, 1983). Pollen analysis suggests that the regional environment was a sagebrush steppe with riparian woodlands (Davis and others, 1984).

Several extinct mammalian herbivores also persisted in the Grand Canyon area at this time. The average of the six youngest radiocarbon dates obtained on Harrington's mountain goat is 11,160 ± 125 yr B.P. (Mead and others, 1986). This date is not significantly older than the average (11,018 ± 50 yr B.P.) for the youngest dates on Shasta ground sloth (Martin, 1984). The youngest date on mammoth dung from Bechan Cave in southern Utah is 11,670 ± 300 yr B.P. (Davis and others, 1984).

A study by Hansen (1978) suggests that the Shasta ground sloth primarily ate desert plant species, most of which still occur near Rampart Cave. Analysis of mammoth dung from Bechan Cave indicates that mammoths predominantly fed on graminoids (grasses, sedges, and rushes), but they also ate saltbush, cactus, and sagebrush (Davis and others, 1984). Some investigators (Davis and others, 1984; Mead and others, 1986; Thompson and others, 1980) believe that the studies of dung from mammoth, Harrington's mountain goat, and Shasta ground sloth do not indicate any stress in their diets. However, dietary stress is difficult to identify, because it depends on the densities of both plant and

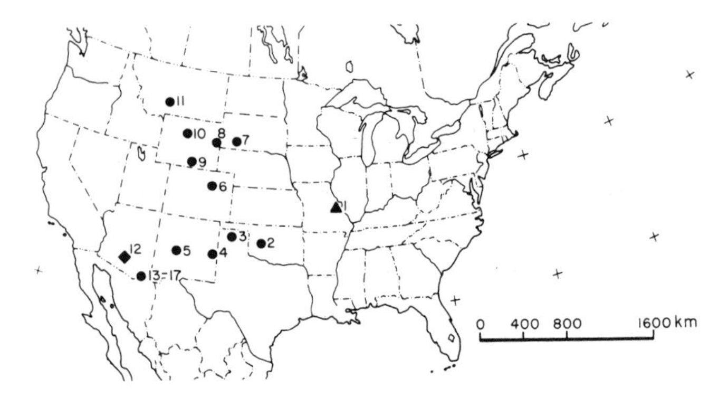

Figure 8. Location of faunas associated with artifacts of the Clovis Cultural Complex. ● mammoth, ▲ American mastodon, ♦ uncertain faunal association. See Table 3 for identification of sites.

Figure 9. Location of noncultural late-glacial (12 to 11 ka) vertebrate faunas. See Table 5 for identification of sites.

TABLE 3. FAUNAL SITES ASSOCIATED WITH ARTIFACTS OF THE CLOVIS CULTURAL COMPLEX

Site Name (County and State)	Youngest Date* yr B.P.	Material Dated and Lab No.	Oldest Date yr B.P.	Material Dated and Lab No.	Major References
1. Kimmswick (Jefferson, MO)	No Radiometric Dates				Graham and others, 1981
2. Domebo (Caddo, OK)	11,045±647	wood & bone organics, SM-695	11,220±500	wood & bone organics SI-172	Leonhardy and Anderson, 1966
3. Miami (Roberts, TX)	No Radiometric Dates				Haynes, 1970
4. Blackwater Draw Locality No. 1 Brown Sand Wedge (Roosevelt, NM)	11,040±500	carbonized plant remains, A-490	11,630±400	carbonized plant remains, A-491	Hester, 1972
5. Mockingbird Gap (Socorro, NM)	No Radiometric Dates				Haynes, 1970
6. Dent (Weld, CO)	11,200±500	*Mammuthus* bone, I-622			Haynes, 1964
7. Lange/Ferguson (Shannon, SD)	10,670±300	sedimentary organic material, I-11, 710	10,730±530	*Mammuthus* bone collagen, I-13, 104	Hannus, 1982
8. Sheaman (Niobrara, WY)	10,030±280	*Bison* bone, RL-1263			Frison, 1982
9. Union Pacific (Carbon, WY)	11,280±350	*Mammuthus* bone, I-449			Haynes, 1970
10. Colby (Washakie, WY)	11,200±200	*Mammuthus* bone, RL-392			Frison and others, 1978
11. Anzick (Park, MT)	No Radiometric Dates				Lahren and Bonnichsen, 1974
12. Ventana Cave (Pima, AZ)	11,300±1200	charcoal, not given			Haynes, 1964
13. Lehner Ranch (Cochise, AZ)	11,000 (average of 21 dates)				Haynes, 1980
14. Murray Springs (Cochise, AZ)	11,230±340 (average of 10 dates)				Haynes, 1980
15. Naco (Cochise, AZ)	9250±300	solid carbon, A-9 + 10			Haynes, 1964
16. Leikem (Cochise, AZ)	No Radiometric Dates				Haynes, 1970
17. Escapule (Cochise, AZ)	No Radiometric Dates				Haynes, 1970

* If there is only one date from the site or unit, it is listed as the "youngest."

TABLE 4. MAMMALS FROM CLOVIS SITES IN THE SOUTHWEST, GREAT PLAINS, AND MIDWESTERN UNITED STATES*

	SITES					
	SW - - - - - - -	PLAINS - - - - - - - -				NE
		South		North		
	SPV	BWD	DOM	AGB	LFM	KIM
MAMMALIA						
Insectivora						
Sorex cinereus/haydeni	o	o	o	o	●	o
Sorex hoyi	o	o	o	o	●	o
Sorex sp.	o	o	o	o	●	o
Blarina brevicauda	o	o	o	o	●	o
Blarina sp.	o	o	o	o	o	●
Edentata						
Glossotherium harlani	o	o	o	o	o	●
Carnivora						
Mustela vison	o	o	o	o	o	●
Canis dirus	●	●	o	o	o	o
Canis lupus	o	●	o	o	o	o
Canis latrans	●	●	o	o	o	o
Canis sp.	o	o	o	●	o	●
Vulpes velox	o	●	o	o	o	o
Ursus americanus	●	o	o	o	o	o
Smilodon populator	o	●	o	o	o	o
Rodentia						
Marmota monax	o	o	o	o	o	●
Spermophilus tridecemlineatus	o	o	o	o	o	●
Spermophilus franklinii	o	o	o	o	o	●
Spermophilus sp.	o	o	o	o	●	●
Sciurus cf. *niger*	o	o	o	o	o	●
Sciurus sp.	o	o	o	o	o	●
Tamias sp.	o	o	o	o	●	o
Thomomys talpoides	o	o	o	o	●	o
Thomomys sp.	●	o	o	o	o	o
Geomys cf. *bursarius*	o	o	●	o	o	●
Perognathus hispidus	o	o	●	o	o	o
Perognathus sp.	●	o	o	o	o	o
Reithrodontomys sp.	o	o	o	o	●	o
Peromyscus leucopus	o	o	o	o	●	o
Peromyscus sp.	o	o	o	o	o	●
Onychomys leucogaster	o	o	o	o	●	o
Sigmodon hispidus	o	o	●	o	o	o
Neotoma sp.	●	o	o	o	o	o
Clethrionomys gapperi	o	o	o	o	●	o
Phenacomys intermedius	o	o	o	●	o	o
Microtus pennsylvanicus	o	o	o	●	●	●

TABLE 4. (CONTINUED)

	SITES					
	SW - - - - - - -	PLAINS - - - - - - - - -				NE
		South		North		
	SPV	BWD	DOM	AGB	LFM	KIM
Microtus longicaudus	o	o	o	•	o	o
Microtus ochrogaster/pinetorum	o	o	•	o	o	•
Microtus sp.	•	o	o	o	o	•
Ondatra zibethicus	•	•	•	o	•	•
Synaptomys cooperi	o	o	•	o	o	•
Zapus princeps	o	o	o	o	•	o
Lagomorpha						
Sylvilagus nuttallii/audubonii	o	o	o	•	o	o
Sylvilagus sp.	•	o	o	o	o	•
Lepus sp.	•	o	o	o	o	o
Perissodactyla						
Equus conversidens	o	•	o	o	o	o
Equus niobrarensis	o	•	o	o	o	o
Equus scotti	o	•	o	o	o	o
Equus sp.	•	o	o	o	o	o
Tapirus sp.	•	o	o	o	o	o
Artiodactyla						
Platygonus sp.	•	•	o	o	o	o
Mylohyus nasutus	o	o	o	o	o	•
Camelops cf. *huerfanensis*	•	o	o	o	o	o
Camelops sp.	•	•	o	•	o	o
Paleolama macrocephala	o	•	o	o	o	o
Odocoileus virginianus	o	o	o	o	o	•
Cervid	o	o	o	o	•	o
Antilocapra americana	o	o	o	•	o	o
Antilocaprid	o	•	o	o	o	o
Bison antiquus	•	•	•	•	o	o
Bison sp.	•	o	o	o	•	o
Proboscidea						
Mammuthus columbi	•	o	•	o	o	o
Mammuthus sp.	•	•	o	•	•	o
Mammut americanum	o	o	o	o	o	•

* Key to column heads:
SPV = San Pedro Valley sites: Lehner and Murray Springs (Saunders, 1987)
BWD = Blackwater Draw Locality No. 1 (Lundelius, 1972)
DMB = Domebo (Slaughter, 1966)
AGB = Agate Basin (Walker, 1982)
LFM = Lange/Ferguson (Martin, 1987)
KIM = Kimmswick (this report)

TABLE 5. RADIOCARBON DATES FOR LATE GLACIAL FAUNAS FROM 11 TO 12 KA

	Site Name (County and State)	Youngest Date* yr B.P.	Material Dated and Lab No.	Oldest Date yr B.P.	Material Dated and Lab No.	Major References
1.	Little Salt Spring (Sarasota, FL)	12,030±200	wood spear shaft, TX-2636			Clausen and others, 1979
2.	Baker Bluff Cave 6-7 ft level (Sullivan, TN)	11,640	bone collagen, GX-3370b			Guilday and others, 1978
3.	New Paris No. 4 (Bedford, PA)	11,300±1000	charcoal, not given			Guilday and others, 1964
4.	Lindsay Mammoth (Dawson, MT)	10,700±290	bone, WSU-652	11,925±350	bone, S-918	Frison, 1978
5.	Agate Basin pre-Folsom strata (Niobrara, WY)	11,450±110	charcoal, SI-3734	11,840±130	charcoal, I-10899	Frison, 1982
6.	Dutton Gleysol (Yuma, CO)	11,710±150	*Mammuthus* bone collagen, SI-2877			Graham, 1981
7.	Lamb Spring Unit 1 (Douglas, CO)	11,735±95	*Mammuthus* bone collagen, SI-4850			Stanford, 1983
8.	Robert (Meade, KS)	11,100±390	land snail shell, SM-762			Schultz, 1967
9.	Lubbock Lake Stratum 1 (Lubbock, TX)	11,140±80	not given	12,650±250	clam shells, LL-308	C. Johnson, 1974
10.	Ben Franklin (Fannin & Delta, TX)	9550±377	charcoal, SM-532	11,135±450	clam shells, SM-533	Slaughter and Hoover, 1963
11.	Cave Without A Name (Kendall, TX)	10,900±190	bone, TX-250			Lundelius, 1967
12.	Lower Sloth Cave (Culberson, TX)	11,590±230	*Nothrotheriops* dung, A-1519			Harris, 1985
13.	Upper Sloth Cave (Culberson, TX)	10,750±140	artiodactyl dung, A-1583	11,760±610	*Nothrotheriops* dung, A-1519	Harris, 1985

TABLE 5. (CONTINUED)

Site Name (County and State)	Youngest Date* yr B.P.	Material Dated and Lab No.	Oldest Date yr B.P.	Material Dated and Lab No.	Major References
14. Williams Cave (Culberson, TX)	11,140±320	*Nothrotheriops* dung, A-1589	12,100±210	*Nothrotheriops* dung, A-1563	Harris, 1985
15. Hermit's Cave (Eddy, NM)	11,850±350	wood, W-498	12,900±350	charcoal, W-495	Harris, 1985
16. Dry Cave Stalag 17 (Eddy, NM)	11,880±250	charcoal, I-5987			Harris, 1985
17. Shelter Cave (Dona Ana, NM)	11,330±370	*Nothrotheriops* dung, not given			Harris, 1985
18. Falling Arches No. 1 (San Bernardino, CA)	11,650±190	*Juniperus* twigs, A-1548			Van Devender and Mead, 1978
19. Rancho La Brea Pit 61-67 (Los Angeles, CA)	11,130±275	*Smilodon* bone collagen, QC-413	12,200±200	*Smilodon* bone collagen, UCLA-1292Y	Marcus and Berger, 1984
20. Rancho La Brea Pit 81 (Los Angeles, CA)	10,940±510	*Equus* bone collagen, QC-405			Marcus and Berger, 1984
21. Rampart Cave (Mohave, AZ)	11,000±140	*Nothrotheriops* dung, A-1066	11,480±200	*Nothrotheriops* dung, A-1041	Long and others, 1974
22. Gypsum Cave (Clark, NV)	11,360±260	*Nothrotheriops* dung, A-1202	11,690±250	*Nothrotheriops* dung, IJ-452	Long and others, 1974
23. Jaguar Cave (Lemhi, ID)	10,370±350	charcoal, not given	11,580±250	charcoal, GX-395	Kurten and Anderson, 1981
24. Manis (Clallum, WA)	11,100±150	micro-organic matter, WSU-2208	12,100±310	*Salix* seeds/wood, WSU-1866, 1867	Gustafson and others, 1979

* If there is only one date from the site or unit considered, it is listed as "youngest."

animal species, and these data are not readily available for evaluation.

In the central Rocky Mountains of eastern Idaho, the Jaguar Cave l.f. has been dated at 11.58 ka from a hearth (Sadek-Kooros, 1966). The fauna contains boreomontane and arctic species like pika, collared lemming, and caribou as well as grassland species like horse, pronghorn, and bison (Guilday and Adam, 1967; Kurtén and Anderson, 1972). This local fauna suggests an environmental mosaic with extensive open areas and scattered pockets of trees and low shrubs. The pika may indicate exposed talus slopes.

Lubbock Lake, a stratified site in northern Texas, has one of the best-dated faunal sequences from the southern plains. The oldest fossiliferous deposits (Stratum 1), dated to ca. 11 ka, contain a diverse extinct megafauna (E. Johnson, 1974). The microfauna from these horizons is sparse, but it does not contain boreomontane species like water shrew, masked shrew, and meadow vole that are found in other late Pleistocene faunas from northern Texas and eastern New Mexico (e.g., Dalquest, 1965; Slaughter, 1975; Slaughter and Hoover, 1963). It is possible that the absence of these taxa from Lubbock Lake could be attributed to sampling problems. However, the Lubbock Lake Clovis faunule is similar to that from the Domebo Clovis site in southwestern Oklahoma (Slaughter, 1966). The prairie vole and muskrat, the only extralimital species (E. Johnson, 1974), would indicate that the climates at Lubbock were only slightly cooler and moister than today.

The Folsom-age faunule (ca. 10.5 ka) from Lubbock Lake (Stratum 2A) is sparse but more diverse than the Clovis-age faunule. It also contains only two extralimital species, the prairie vole and muskrat (E. Johnson, 1974). However, the Folsom faunule contains more xeric species than the Clovis faunule (e.g., pocket mouse and blacktail jackrabbit). The Folsom faunule has only one extinct taxon, the diminuitive pronghorn, but neither its provenience nor its identification are discussed by E. Johnson (1974).

In the eastern U.S. three dated sites fit within the Clovis time period (Table 5). At approximately the 5-m level, in New Paris No. 4, boreomontane species like heather vole and northern bog lemming are lost from the record. At the same time, the timber rattlesnake, which prefers deciduous woodlands, significantly increases in abundance. However, extralimital species like the yellowcheek vole, yellownose vole, and perhaps redback vole occur within and above the 4.6-m level. These changes suggest that Clovis-age climates were warmer than the preceding full- and early late-glacial climates but still cooler than modern climates. The Clovis-age environments were probably a moderately closed coniferous-deciduous forest mosaic with deciduous elements dominant.

The Clovis-age horizons (11.64 ka) at Baker Bluff Cave mark the abrupt appearance of boreal species like arctic shrew and pygmy shrew and an increase in other boreal species like northern flying squirrel, red squirrel, and redback vole, which were present in earlier levels. However, temperate species like gray squirrel, southern flying squirrel, southern bog lemming, and prairie or pine vole were not eliminated from the record, although their abundances were reduced. In fact, the appearance of one temperate species, the least shrew, is coincident with the appearance of the boreal species.

In addition to the boreal taxa, open-habitat species like thirteenline ground squirrel, least chipmunk, and meadow vole also increased in abundance during Clovis times. The Clovis-age herpetological fauna from Baker Bluff Cave is more diverse than the full-glacial ones from there or New Paris No. 4. Guilday and others (1978) believe that these faunal changes reflect an environmental shift during Clovis times from temperate closed forest to more open boreal forest, an environmental change opposite to that reflected at New Paris No. 4.

At Little Salt Spring, Sarasota County, Florida, eleven vertebrate species have been recovered from a ledge 26 m below the modern water surface, associated with an extinct giant tortoise that was presumably killed by humans with a sharpened wooden stake radiocarbon-dated at 12.03 ka (Holman and Clausen, 1984). Holman and Clausen (1984) believe that the Little Salt Spring l.f. is divisible into one group of species that inhabited the pond or surrounding mesophytic habitats and a second group adapted to the upland xerophytic habitats described for the regional palynological record (Watts, 1983).

LATE PLEISTOCENE EXTINCTION

The end of the Pleistocene is marked by a major extinction event, which primarily affected the terrestrial vertebrate communities, especially the large-mammal herbivore and carnivore guilds. Most of the extinction occurred between 10 and 12 ka in North America (Mead and Meltzer, 1984; Kurtén and Anderson, 1980). The best absolute chronologies for individual taxa are restricted to the southwestern United States. Certain taxa, like the American mastodon, may have existed beyond 10 ka in other areas of North America (King and Saunders, 1984).

Many theories have been proposed to explain this extinction event, and the recent book edited by Martin and Klein (1984) provides the most comprehensive statements on these theories to date. The theories can generally be categorized within two schools of thought, over-predation by human hunters and climatic change. The latter hypothesis provides a broad umbrella for many diverse ideas. In North America the extinction event is essentially coincident with the appearance of the Clovis culture, and Martin (1984) has stressed this correlation in arguing for prehistoric overkill. Significant climatic changes also occur during this interval of time, however. Therefore it is difficult if not impossible to differentiate these two hypotheses on the basis of temporal correlations alone in North America. Both schools of thought concede that the extinction of the large carnivores is probably the consequence of extinction of their prey, the large herbivores. Grayson (1984) provides a good review of the strengths and weaknesses of these two basic hypotheses, so they are not repeated here.

Reorganization of plant and animal communities at the end of the Pleistocene may have contributed to the extinction event (Graham, 1986; Graham and Lundelius, 1984; Guthrie, 1984). The ultimate environmental factor in the extinction is probably habitat destruction, which had several proximate consequences. With the rapid and individualistic reorganization of plant communities, the foraging strategies of herbivores must have been disrupted, and competition may have been heightened during the change to new feeding systems. Also, the formation of new vegetational associations may have acted as ecological barriers in limiting the ability of animals to migrate over large areas. As in the Serengeti today (MacNaughton, 1983), this would limit population movement out of areas adversely affected by local climatic changes.

The overall effective nutritional value of vegetation may have been reduced through selection for plant communities dominated by species with allochemic protection mechanisms throughout their life histories. For instance, in the Holocene the dominance of spruce, which contains noxious terpenes, and the reduction in grasses, herbs, and forbs through closure of the boreal forests would have reduced the effective nutritional value of this vegetational type. As a consequence, the modern boreal forest does not support a diverse mammalian community.

Implicit in all of the climatic models is the idea that the environmental changes at the end of the Pleistocene were unique. Jacobson and others (this volume) demonstrate that the magnitude of the vegetational change during the last deglaciation was greatest between 10 and 12 ka. This also appears to be the case for the mammalian communities (Graham and Lundelius, 1984). Graham (1986) argues that the environmental changes of the last deglaciation (Wisconsin to Holocene) were significantly different from the previous deglaciation event (Illinoian to Sangamonian). Specifically, the climatic changes of the last deglaciation appear to be more continental. Further support for this contention is provided by a recent study of an Illinoian-Sangamonian sequence of vegetation and vertebrates from central Illinois (King and Saunders, 1986).

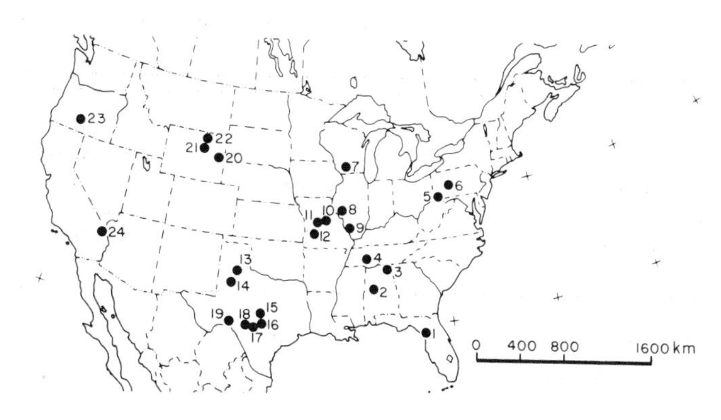

Figure 10. Location of early Holocene (10 to 8 ka) vertebrate faunas. See Table 6 for identification of sites.

EARLY HOLOCENE FAUNAS

Eastern United States

In general, early Holocene (10 to 8 ka) faunas throughout the United States (Fig. 10, Table 6) can be characterized by the absence of diverse extinct mammalian taxa and by a more modern aspect of the extant component. However, the early Holocene faunas were still quite different from the modern faunas of the region, for the dissolution of disharmonious faunas is transgressive, both temporally and spatially, although most had disappeared by the end of the Pleistocene (ca. 10 ka).

The Devil's Den l.f., recovered from a sinkhole cave in Levy County, Florida, has frequently been considered to date from the latest Pleistocene or the earliest Holocene, for it contains extant but northern extralimital taxa like gray myotis, muskrat, and meadow vole, along with a diverse extinct megafauna (Martin and Webb, 1974; Semken, 1983; Kurtén and Anderson, 1980). An unpublished geological report by H. K. Brooks suggests dates between 7 and 8 ka for the site (Martin and Webb, 1974), presumably based upon radiocarbon analysis of bone (Holman, 1978a; Kurtén and Anderson, 1980, p. 56). It is not clear whether the dates were derived from the collagen or apatite fraction of the bone, and standard deviations were not reported. Radiocarbon dates on bones from caves must be considered as minimum ages at best (Land and others, 1980). Furthermore, Martin and Webb (1974, p. 135) note that the extinct taxa in the upper blackened surface layer may be redeposited. None of the published accounts (Martin and Webb, 1974; Holman, 1978a) indicate the stratigraphic context of the radiocarbon dates. We therefore believe that the Devil's Den l.f. must be dated more precisely before it can be regarded as early Holocene.

At Russell Cave, Jackson County, northern Alabama, an early Holocene component (Unit G), dated 7.565 to 8.5 ka, has produced a vertebrate fauna essentially the same as the modern fauna, with two exceptions (Weigel and others, 1974). Two teeth, perhaps representing a single individual, of the extinct longnose peccary have been recovered from the base of the lowest level (Unit G). This species has also been found in Holocene strata IV and V of Cheek Bend Cave, Tennessee (Klippel and Parmalee, 1982). However, Klippel and Parmalee (1982, p. 223) caution that "the presence of these teeth in post-Pleistocene strata can only be viewed as fortuitous until—and if—additional elements are recovered during future excavations and their placement in the cave [Cheek Bend] stratigraphy can be established with absolute certainty." Similarly, the Russell Cave specimens of the longnose peccary may result from redeposition from remnant Pleistocene strata. Neither the ages nor the paleontological contents are known for the basal stratigraphic units. Furthermore, because of the uneven topography of the basal breakdown, it would be easy for earlier material to move downslope and be redeposited in Unit G.

TABLE 6. RADIOCARBON DATES FOR EARLY HOLOCENE FAUNAS

Site Name (County and State)	Youngest Date* yr B.P.	Material Dated and Lab No.	Oldest Date yr B.P.	Material Dated and Lab No.	Major References
1. Devil's Den (Levy, FL)	ca. 8000				Holman, 1978a
2. Stanfield-Worley Bluff Shelter (Tuscaloosa, AL)	8920±400	charcoal, M-1153	9640±450	charcoal, M-1152	Parmalee, 1963
3. Russell Cave (Jackson, AL)	7565±250	charcoal, not given	8500±320	charcoal, not given	Griffin, 1974
4. Cheek Bend Cave (Maury, TN)	7505±440	charcoal, GX-7855			Klippel and Parmalee, 1982
5. Meadowcroft Rockshelter (Washington, PA)	8010±110	charcoal, SI-2064	11,300±700	charcoal, SI-2491	Adovasio and others, 1984
6. Hosterman Pit (Centre, PA)	9240±1000	charcoal, M-1291			Guilday, 1967
7. Raddatz Rockshelter (Sauk, WI)	5200±400	not given M-813	11,600±600	charred wood, M-812	Wittry, 1959
8. Koster (Greene, IL)	8130±75	wood charcoal, ISGS-337	8480±110	wood charcoal, ISGS-236	Brown and Vierra, 1983
9. Modoc Rock Shelter (Randolph, IL)	8010±140	carbonized plant, ISGS-830	9320±230	carbonized plant, ISGS-820	Styles and others, 1983
10. Graham Cave Zone IV (Callaway, MO)	9290±300	charcoal, M-1889	9470±400	charcoal, M-1928	McMillan and Klippel, 1981
11. Brynjulfson Cave #1 (Boone, MO)	9440±760	bone, ISGS-D70			Parmalee and Oesch, 1972
12. Rodgers Rockshelter (Benton, MO)	8100±300	carbonized wood, A-868A	10,530±650	carbonized wood, ISGS-48	McMillan and Klippel, 1981
13. Rex Rodgers (Briscoe, TX)	8985±83	bone apatite, SMU-274			Speer, 1978

TABLE 6. (CONTINUED)

14.	Lubbock Lake Plainview (Lubbock, TX)	9985±100	humate, SMU-720			Holliday and Johnson, 1981
	Lubbock Lake Firstview (Lubbock, TX)	8095±230	humate, SMU-827	8655±90	humate, SMU-4177	Johnson and Holliday, 1981
15.	Miller's Cave (Llano, TX)	7200±300	bone, UTA-326			Patton, 1963
16.	Friesenhahn Cave (Bexar, TX)	8010±430	bone apatite, TX-2392	9640±440	bone apatite, TX-2394	Graham, 1976b
17.	Klein Cave (Kerr, TX)	7685±645	bone, SM-1196			Roth, 1972
18.	Schulze Cave (Edwards, TX)	9310±310	bone, I-2741A	9680±700	bone, SM-807	Dalquest and others, 1969
19.	Baker Cave (Val Verde, TX)	9020±150	charcoal, TX-2466	9180±220	charcoal, RL-828	Hester, 1983
20.	Agate Basin Folsom (Niobrara, WY)	10,665±85	charcoal, SI-3732			Walker, 1982
	Agate Basin Hell Gap (Niobrara, WY)	10,445±110	charcoal, SI-4430			Walker, 1982
21.	Bush Shelter (Washakie, WY)	9000±240	charcoal, RL-1407			Walker, 1987
22.	Medicine Lodge Creek (Bighorn, WY)	9590±180	bone, RL-393			Frison, 1978
23.	Connley Caves Stratum 3 (Lake, OR)	7200	not given	9500	not given	Grayson, 1979
	Connley Caves Stratum 4 (Lake, OR)	9500	not given	11,200	not given	Grayson, 1979
24.	Tunnel Ridge No. 2 (San Bernardino, CA)	10,330±300	*Juniperus* wood, A-1470			Mead and others, 1983

* If there is only one date from the site or unit, it is listed as the "youngest."

Extralimital records for the porcupine have been documented from the early Holocene deposits at both Russell Cave and the Stanfield-Worley Bluff Shelter in Alabama (Parmalee, 1962, 1963). The porcupine had a much wider geographic distribution throughout the late Pleistocene and most of the Holocene prior to 1 ka (Semken, 1983). Consequently, its early Holocene distribution may not signify any major environmental change. Faunal remains from the Stanfield-Worley site are sparse, but the species composition and frequencies of the whitetail deer and gray squirrel are similar to those of the modern fauna (Parmalee, 1962).

Charcoal associated with faunal remains from Hosterman Pit, Centre County, Pennsylvania, has been dated at 9.24 ka by Guilday (1967). Unfortunately, he does not clearly report the temporal range that may be represented by the assemblage, but he does state ". . . there is every indication that the charcoal/bone association was a primary one and that the fauna can therefore be dated at about 9,000 years old" (Guilday, 1967, p. 231–232). Species composition is modern in all aspects, without any extinct taxa (Guilday, 1967, p. 232, Table 1). Other early Holocene sites from this region also suggest that the fauna of the northeastern U.S. had essentially reached its modern composition by at least 9 ka, as suggested by Guilday (1967). The early Holocene component (8.92 ka) at Sheep Rock Shelter, Huntingdon County, Pennsylvania, contains a fauna similar to today's and has no extinct taxa (Guilday and Parmalee, 1965).

At Meadowcroft Rockshelter, Washington County, Pennsylvania, a cultural horizon (middle Stratum IIa) radiocarbon dated at 11.3 ka, has produced a vertebrate fauna characterized as "modern" temperate biota (Guilday and Parmalee, 1982). In fact, the entire vertebrate sequence at Meadowcroft is essentially modern (Mead, 1980). The accuracy of the radiocarbon dates has been questioned, especially for the older deposits in lower Stratum IIa (Haynes, 1980), but counterarguments have been offered to support the validity of the dates (Adovasio and others, 1980; Stuckenrath and others, 1982). The faunule from Stratum IIa does not fit well with late Pleistocene paleoenvironmental reconstructions for the mid-Appalachian area based on other vertebrate and paleobotanical sites (Guilday and Parmalee, 1982; Mead, 1980). Furthermore, Guilday and Parmalee (1982, p. 173) state, "It is not reasonable to view the Upper Ohio Valley as a temperate biotic refugium during late glacial times, i.e., prior to ca. 12,000 years ago. . . ." Therefore, until the Meadowcroft record can be corroborated by other sites or new dating methods, all available evidence suggests that the fauna of the northeastern, and perhaps eastern, U.S. reached its modern composition sometime between 11,000 and 9,000 years ago.

Midwest and Prairie Peninsula

In the midwestern U.S. several rockshelters preserve long Holocene cultural and faunal sequences (Fig. 10, Table 6). The Raddatz Rockshelter, Sauk County, Wisconsin, has a deeply stratified faunal sequence with two radiocarbon dates, 11.602 and 5.191 ka, from the lowest and middle levels respectively (Cleland, 1966). Early Archaic artifacts in Stratum I suggest an early Holocene age; the faunal composition reflects predominantly deciduous forest mixed with conifers. Cleland (1966) proposes a cool moist environment in southern Wisconsin during the early Holocene.

Recent investigations, especially new radiocarbon dates (Styles and others, 1981), demonstrate relatively complete stratified early and middle Holocene sequences at Modoc Rock Shelter, Randolph County, Illinois. Level 15 at 9 ka contains two extralimital species—rice rat and Franklin's ground squirrel—which are not currently sympatric. Today, the rice rat occurs less than 90 km south of Modoc, and Franklin's ground squirrel less than 40 km north. The co-occurrence of these two taxa may suggest that winters were slightly warmer and summers slightly cooler than today. However, the distance between the ranges of these species is so slight that their co-occurrence does not reflect Pleistocene climatic conditions. In fact, the presence of these two species may indicate a climate with greater effective moisture.

The rice rat occurs in both early and middle Holocene deposits, but Franklin's ground squirrel is restricted to level 15. The thirteenline ground squirrel in level 11 (ca. 8.3 ka), is found slightly north of Modoc today, but it prefers short-grass areas, whereas Franklin's ground squirrel inhabits moister areas with taller grass. This change in ground squirrel species may reflect decreasing effective moisture in the early Holocene.

The mammalian fauna from Horizon 11 at the Koster site, Greene County, Illinois, dates between 8.13 and 8.48 ka (Brown and Vierra, 1983). It is similar to the historic fauna of the area (Neusius, 1982), so environmental conditions must have been similar to those of today. However, as Neusius (1982) cautions, the analysis of the Horizon 11 faunule is only preliminary, and future studies, especially of the microvertebrate fraction, may alter the interpretations. The presence of the badger and plains pocket gopher in the Horizon 11 faunule suggests that even during the early Holocene the environments at Koster were more open than at Modoc to the south.

In central Missouri Graham Cave Zone IV (9.29 to 9.47 ka) has a fauna of predominantly forest species, and less than 35 percent of the species prefer forest edge (McMillan and Klippel, 1981, p. 236). It thus reflects a relatively closed forest. Moister and perhaps cooler early Holocene climates at Graham Cave are also suggested by the smaller size of gray squirrels, which today is highly correlated with moister environments (Purdue, 1980, p. 254, Fig. 5).

A radiocarbon date of 9.44 ka has been derived from bone from Brynjulfson Cave No. 1, Boone County, Missouri (Parmalee and Oesch, 1972). However, bone from this cave has also been dated at more than 27 ka (Coleman and Liu, 1975). The fauna is a mixture of extinct, extirpated, and extant species as well as introduced European taxa (Parmalee and Oesch, 1972). Without stratigraphic information, it is difficult to interpret the paleoenvironmental implications of this fauna, and neither Brynjulfson Cave Nos. 1 nor 2 should be considered as evidence that the

Ozark Highland was a Holocene refugium for extinct Pleistocene megafauna.

Farther west at Rodgers Shelter, Benton County, Missouri, the climate was beginning to show signs of desiccation by the early Holocene. Sediments dated 8.6 to 9.4 ka document a period of maximum upland erosion and maximum aggradation by the Pomme de Terre River (Ahler, 1976). McMillan and Klippel (1981) believe that these deposits represent the transition of the uplands from forest to grasslands. The first appearance of grassland species like bison, pronghorn, and prairie chicken between 8.6 and 8 ka supports this contention.

In western Iowa, pollen and plant macrofossils from Lake West Okoboji indicate that at ca. 11 ka the environment was a closed deciduous forest dominated by elm and oak (Van Zant, 1979). This forest became progressively more open as the climate became warmer and drier, and by 9.075 ka prairie was established over much of the uplands (Van Zant, 1979). The faunal sequence from the Cherokee site, northwestern Iowa, agrees well with the vegetational record from Okoboji. The Cultural Horizon IIIa faunule (ca. 8.4 ka) suggests environments drier than today but moister than the following middle Holocene (Semken, 1980).

Southern Plains

Following Slaughter (1967), Semken (1983, p. 200) has stated that "... aspects of typical Pleistocene conditions appear to have persisted some 2,000 years later on the southern Plains than the 10 ka date frequently assigned to the boundary." We agree that early Holocene climates and environments were different from those of later parts of the Holocene. However, we also believe that the early Holocene environments were distinctly different from those of the Pleistocene, and that Pleistocene conditions, especially the existence of disharmonious faunas and extinct megafauna, had essentially been terminated by 10 ka. Early Holocene climates of the southern plains, like those of the midwestern U.S., were warmer than the Pleistocene climates, but they were moister than later stages.

These climatic differences are indicated by the regional biota. Stratum 2B at Lubbock Lake with Plainview cultural materials has been dated ca. 10 ka (Johnson and Holliday, 1980; Holliday and Johnson, 1981). No extinct taxa are known from these deposits, but the prairie vole, muskrat, meadow vole, and southern bog lemming are extralimital. All prefer low moist open environments with a heavy growth of vegetation. Also, because all four occur farther north today, they suggest that 10-ka climates at Lubbock Lake were cooler as well as moister than present.

A Lubbock Lake Firstview feature (FA6-3) with faunal remains has been dated between 8.655 and 8.095 ka (Johnson and Holliday, 1981). By this time the meadow vole, prairie vole, southern bog lemming, and muskrat no longer occurred at Lubbock Lake, as evidenced by the Firstview faunule. Thus modern environments and climates seem to have been established there by 8.655 ka at the latest (Johnson and Holliday, 1981), and perhaps as early as 10 ka.

A limited microfauna was recovered from the Paleoindian bison kill at the Rex Rodgers site, approximately 110 km north of Lubbock Lake (Speer, 1978). One radiocarbon date of 8.985 ka was derived from bone apatite. As Speer notes, bone apatite dates are susceptible to contamination by isotopic exchange, but on the basis of experimentally derived fractionation standards she considers a corrected date of 9,118 ± 83 yr B.P. to be the most reliable. However, the five projectile points from the Rex Rodgers site are like variants of the early Paleoindian traditions (e.g., Folsom or Plainview), which date ca. 10 ka (Holliday and Johnson, 1981). We believe that the date for the Rex Rodgers site is probably closer to 10 than 9 ka.

The Rex Rodgers microfauna contained three extralimital species, the shorttail shrew, meadow vole, and water snake. All of these species occur more than 320 km north and east of the site today (Schultz, 1978; Holman, 1978b). Thus, the Rex Rodgers l.f. lends credence to moister and cooler climates in northern Texas at 10 ka. Furthermore, there were no extinct taxa recovered.

Semken (1983) suggests that faunas from central Texas sites like Schultze Cave (Dalquest and others, 1969), Friesenhahn Cave (Graham, 1976), and Klein Cave (Roth, 1972) reflect the continuation of Pleistocene-like "pluvial" climates into the early Holocene. The radiocarbon dates from all of these sites were derived from bone, and, as Semken (1983) indicates, they may not be accurate (Land and others, 1980). The faunas from Schultze and Klein caves contain more boreomontane species than the early Holocene faunules (e.g., Folsom and Plainview) at Lubbock Lake. At Friesenhahn Cave the absence of boreal species and the occurrence of southern bog lemming, shorttail shrew, and eastern chipmunk reflect moister climates, but not necessarily the coldness of the Pleistocene.

Finally, Semken (1983) suggests that the Travertine faunule from Miller's Cave, Llano County, Texas (Patton, 1963), dated ca. 7.2 ka, marks the shift to a nonpluvial climate, although temperatures were similar to areas north of the site today. The presence of extralimital species like the shorttail shrew, prairie vole, southern bog lemming, and muskrat suggest that climates were moister as well as cooler than present. All of these extralimital species occur in other Texas faunas that date ca. 10 ka or older (i.e., Lubbock Lake, Rex Rodgers). In addition, the Travertine faunule contains an extinct taxon, the beautiful armadillo. We do not believe that the bone date of 7.2 ka for the Travertine faunule is accurate. Therefore the temporal placement of this faunule is uncertain, but as suggested by Patton (1963), the fauna appears to be Late Wisconsin in age.

We thus find little evidence that extinct taxa survived beyond 10 ka on the southern plains. Also, most of the disharmonious faunas characteristic of the late Pleistocene are apparently gone by this time. However, certain species like the meadow vole, southern bog lemming, prairie vole, shorttail shrew, and muskrat may have persisted into the early Holocene. These species reflect moister and somewhat cooler environments than today, but they do not necessarily indicate that climates were as cool as those of the Pleistocene. By at least ca. 8.6 ka, the fauna of the southern plains began to take on a modern aspect. The persistence of moist and cool climatic conditions into the early Holocene on the southern plains is similar to paleoenvironmental changes in the Midwest but significantly different from those of the eastern U.S.

Northern Plains and West

All of the well-dated early Holocene faunal sites of the northern plains are from Wyoming. The Medicine Lodge Creek site on the western slope of the Bighorn Mountains is a deeply stratified occupation (Frison, 1978). One cultural level, near the bottom of the excavated deposits and dated at 9.59 ka, contained an extensive small-mammal faunule. The presence of heather vole, boreal redback vole, and pika may suggest that boreal forests extended down Medicine Lodge Creek valley to at least 1,463 m (Walker, 1987). These midden deposits may have been partially derived from the foraging activities of prehistoric human residents at higher elevations (Walker, 1975). However, it is also possible that these species reflect favorable environmental conditions at lower elevations, which may be a function of moister and cooler climates. This latter scenario is supported by other early Holocene faunas from Wyoming.

Bush Shelter in the southern Bighorn Mountains contains a stratified fauna that documents late Pleistocene as well as early and middle Holocene environments. Only one radiocarbon date (9 ka) is available from the middle of the sequence (Walker, 1987). Walker (1987) provides the following discussion of the late Pleistocene and early Holocene faunal sequence: "One meter below the 9 ka B.P. level, *Dicrostonyx* [collared lemming] occurs in larger numbers than either *Ochotona* [pika], *Phenacomys* [heather vole], *Clethrionomys* [redback vole], or other microtines. Thirty-five centimeters below the 9 ka B.P. level, *Phenacomys* and *Clethrionomys* increase in abundance and *Dicrostonyx* has almost disappeared. At the 9 ka B.P. level, *Dicrostonyx* disappears and *Phenacomys* becomes rare. Composition of modern microtine guilds is beginning to form."

A similar stratified faunal record is preserved at the Agate Basin site, Niobrara County, eastern Wyoming (Walker, 1982). Comparison of Folsom (10.7 ka) and Hell Gap (10.4 ka) faunules demonstrates the loss of one boreomontane species, the boreal redback vole, and the addition of one steppe species, the Wyoming pocket mouse. After the Hell Gap occupation, three boreomontane micromammals—longtail vole, meadow vole, and heather vole—were lost, as well as one steppe species, the sagebrush vole (Walker, 1982). These faunal sequences from Wyoming clearly illustrate the individualistic response of species to changing climates and environments.

These paleoenvironmental reconstructions appear to contradict a sediment column with pollen, ostracod, and plant macrofossils from Antelope Playa in the eastern Powder River Basin, Wyoming. Markgraf and Lennon (1986, p. 8) state, "Apparently, sagebrush-grassland dominated in this region in late Pleistocene and Holocene times, and there was no woodland in the lowlands, at least not as late as 13,000 years B.P. The only environmental change that can be detected occurs at 5,000 years B.P." One of the differences between the palynological record and the faunal interpretations could relate to proximity of montane environments. The faunal sites are close to or in mountainous areas, whereas Antelope Playa is located in a large low-relief basin.

The early Holocene faunal history for the western U.S. is meager. Grayson (1982, p. 86) states "Archeological excavations have yielded vertebrate faunas from virtually every part of the Great Basin, though most of these faunas represent only temporal spot samples and have been collected with one-quarter-inch screens. Ancient packrat middens have also provided important data on these mammals. Still rare, however, are studies of well stratified, deep sequences of vertebrates collected with techniques designed to retrieve small mammal elements." This assessment is also broadly applicable to the southwestern U.S.

In the Great Basin, boreomontane species persisted at lower elevations or in mountains not inhabited by these species today. At Connley Caves in south-central Oregon, pika occurs in deposits dated between 11 and 7 ka, but after the fall of the Mazama ash (ca. 7 ka) pikas were no longer incorporated into the sediments of these shelters (Grayson, 1982). Ziegler (1963) identified pikas in sediments deposited prior to 9.5 ka at Deer Creek Cave, northeastern Nevada. Pikas have also been recovered at Streamview No. 1. Other species like the golden-mantled ground squirrel and pygmy cottontail show a similar distribution during the late Quaternary (Grayson, 1982).

In the southwestern U.S., boreomontane species may have retreated to higher elevations before the end of the Pleistocene (ca. 10 ka). The Tunnel Ridge No. 2 packrat midden from the lower Colorado River valley, California, contains a diverse vertebrate fauna dated at 10.33 ka on juniper wood (Mead and others, 1983). No boreomontane species occur in this fauna; instead it is primarily a desert-adapted fauna with redspot toad, kangaroo rat, pocket mouse, and leopard lizard (Mead and others, 1983). The vegetational record indicates a juniper-beargrass woodland (Van Devender and Mead, 1978; Van Devender and Spaulding, 1979). Discontinuous vegetational ground cover is suggested by the presence of the leopard lizard, which requires barren ground for locomotion.

MIDDLE HOLOCENE FAUNAS

The species composition of the middle Holocene (7 to 5 ka) faunas from the eastern United States (Fig. 11, Table 7) is essentially the same as modern ones. As indicated by Guilday (1967, p. 232), "All Indian archaeofaunas from the East within the time span of the last 6,000 years contain faunas which are essentially modern." Little environmental change is indicated. However, Guilday (1965) also suggested that the presence of fox squirrel and box-turtle at the Lamoka Lake site, New York, may indicate an opening of the forest as a result of warmer climates between 5.5 and 3.5 ka.

Figure 11. Location of middle Holocene (7 to 5 ka) vertebrate faunas. See Table 7 for identification of sites.

Although there are questions about the absolute chronology at Meadowcroft (Haynes, 1980; Mead, 1980), the fauna from eleven distinct strata that postdate 11.3 ka shows little evidence for climatic change (Guilday and Parmalee, 1982). Furthermore, clinal variation of two mammalian taxa from Meadowcroft support an hypothesis for stable Holocene climates. Adovasio and others (1984, p. 358) state, "Measurements of southern flying squirrel humeri and southern bog lemming first lower molars were compared at different strata in the Meadowcroft deposit down to and including Stratum IIa to see if there was any indication of time-related size changes that might suggest climatic change. In both cases the results were negative, and the sample parameters agree with those of Recent material at all levels inside the dripline, suggesting temperate conditions throughout."

The middle Holocene faunule from Sheep Rock shows no significant environmental fluctuations (Guilday and Parmalee, 1965). Major changes in the ratios of eastern cottontail rabbits and southern flying squirrels, which may reflect an opening of the forest, occurred during the Woodland occupation (0.46 to 0.50 ka). This change probably resulted from land clearing for horticulture by the human occupants (Guilday and Parmalee, 1965). Layer F (5.980 to 6.310 ka) of Russell Cave, northern Alabama, contains a faunule that is the same as the modern fauna of the area, except for the occurrence of the porcupine. The Stanfield-Worley Bluff Shelter in Alabama also has a "modern" middle Holocene faunule with porcupine.

At Cheek Bend Cave, Maury County, Tennessee, a stratified insectivore sequence appears to document environmental change during the middle Holocene. Significant variation in the frequency distribution of shorttail shrew and least shrew occur within the Holocene strata (Klippel and Parmalee, 1982). Generally the shorttail shrew prefers moist forested environments, whereas the least shrew usually inhabits drier and more open environments. Stratum V, which is a middle Holocene deposit, contained greater numbers of least shrews, while shorttail shrews were predominant in strata below and above. Klippel and Parmalee (1982, p. 456) believe these frequency shifts reflect decreased summer rainfall and/or increased summer temperatures, which would transform the forested margins of present-day cedar glades into more open habitat. This is congruent with pollen records from the Western Highland Rim in middle Tennessee, which suggest that middle Holocene vegetation was much like the Ozark Mountains of eastern Missouri today (Delcourt, 1980).

The effects of the middle Holocene warming were even more pronounced in the midwestern United States. Environmental changes were a continuation of climatic warming and drying trends initiated in the western Prairie Peninsula during the early Holocene. This climatic change was differentially reflected along east-west environmental gradients, with the severest aridity in the west and only slight or no effects in the eastern U.S.

For instance, Cleland (1966) believes that the faunal sequence from Raddatz Rockshelter reflects environmental fluctuations along the northern border of the Prairie Peninsula. By scoring habitat preferences for the various species in the stratigraphic sequence, Cleland (1966, p. 104) concluded that at Level 11 there is a decline in deciduous forest habitat and a sudden appearance of grasslands. Although not securely dated, these changes presumably reflect the onset of the middle Holocene warming.

However, the magnitude of this change is questionable. Species considered diagnostic of grasslands by Cleland (1966, p. 102, Table 8) actually have a much wider habitat preference. For instance, elk, turkey vulture, and redtail hawk are widespread species that occurred throughout most of the U.S. historically. Furthermore, the remains of turkey vulture and redtail hawk could have been brought to the site from long distances by the human inhabitants for a variety of cultural reasons. Therefore we believe that the faunal record from Raddatz Rockshelter, as currently known, may not reflect any significant environmental fluctuations.

Modoc Rock Shelter, at the southern boundary of the Prairie Peninsula, has a Middle Archaic faunule that may reflect subtle middle Holocene climatic change. Two extralimital species, spotted skunk and hispid cotton rat, occur in the Middle Archaic horizons (Styles and others, 1981). The presence of these two species suggests that the environments around Modoc were more open in the middle Holocene than they are today. However, they do not necessarily reflect drier environments. Today, both these species inhabit mesic environments in the Southeast. In fact, even in the western limits of its distribution, the cotton rat prefers tall grass and moist environments. The southern distribution of these species today may indicate that middle Holocene climates were warmer.

Between 8 and 7.6 ka sediments from Graham Cave, central Missouri, show an abrupt increase in eolian silts, which presumably reflect warmer and drier climates (McMillan and Klippel, 1981). Subsequent sedimentation is different from the earlier Holocene. The middle Holocene faunule analyzed by Klippel (1971) does not document any significant range shifts for verte-

TABLE 7. RADIOCARBON DATES FOR MIDDLE HOLOCENE FAUNAS

Site Name (County and State)	Youngest Date* yr B.P.	Material Dated and Lab No.	Oldest Date yr B.P.	Material Dated and Lab No.	Major References
1. Lamoka Lake (Schuyler, NY)	4401±250	charcoal, M-912	5383±250	charcoal, C-367	Ritchie, 1969
2. Sheep Rock Shelter (Huntington, PA)	3220±160	charcoal, M-2085	7050±250	charcoal, M-1908	Michels and Dutt, 1968
3. Meadowcroft Rockshelter Stratum IIb (Washington, PA)	3210±95	charcoal, SI-1681	6670±140	charcoal, SI-2055	Adovasio and others, 1984
4. Cheek Bend Cave Stratum V-VI (Maury, TN)	4655±75	charcoal, UGa-2775	7505±440	charcoal, GX-7855	Klippel and Parmalee, 1982
5. Russell Cave Layer F (Jackson, AL)	5980±200	charcoal, not given	6310±140	charcoal, not given	Griffin, 1974
6. Raddatz Rockshelter (Sauk, WI)	No dates for middle Holocene units.				
7. Modoc Rock Shelter (Randolph, IL)	4720±300	charcoal, M-483	5955±235	charcoal, C-899	Styles and others, 1981
8. Graham Cave Zone II (Callaway, MO)	7610±140	charcoal, I-5218	7630±120	charcoal, I-5217	McMillan and Klippel, 1981
9. Rodgers Shelter (Benton, MO)	5100±400	charcoal, M-2332	7010±16	charcoal, GAK-1171	McMillan and Klippel, 1981
10. Cherokee Sewer Horizon II (Cherokee, IA)	6800±190	charcoal, UCR-491	8500±200	charcoal, UCLA-1877A	Hoyer, 1980
Cherokee Sewer Horizon I (Cherokee, IA)	6300±90	charcoal, UCLA-1877B	6500±200	carbonized wood, UCR-492	Hoyer, 1980
11. Mud Creek (Cedar, IA)	6220±110	wood, I-6228			Kramer, 1972
12. Wunderlich (Comal, TX)	4170±200	shell, TX-15	5405±300	charcoal, TX-16	Lundelius, 1967
13. Snyder (Butler, KS)	No absolute dates for site.				
14. Coffey (Pottawatomie, KS)	4840±95	wood, N-1549	5850±135	wood, N-1550	Schmits, 1978
15. Dead Indian Creek (Park, WY)	3800±110	charcoal, RL-321	4430±250	charcoal, W-2599	Walker, 1987
16. Connley Caves Stratum 2 (Lake, OR)	3400	not given	4400	not given	Grayson, 1979
17. Gatecliff Shelter (Nye, NV)	5000±80	charcoal, UCLA-1926e	7080±675	charcoal, QC-291	Thomas, 1982

* If there is only one date from the site or unit, it is listed as the "youngest."

brate species. Changes in the relative proportions of deer, eastern cottontail rabbits, and tree squirrels suggest an opening of the forest during the middle Holocene (McMillan and Klippel, 1981). The warmer and drier climate of the middle Holocene is also documented by clinal variation in the size of squirrels and rabbits from Graham Cave (Purdue, 1980).

Farther west the middle Holocene climate had a more profound effect on the vertebrate communities. At Rodgers Shelter, four new grassland species, badger, pocket mouse, jackrabbit, and spotted skunk were added to the Middle Archaic faunule (7 to 6.3 ka; McMillan and Klippel, 1981). These species complement the grassland species of bison, pronghorn, and prairie chicken that were already present in the early Holocene. These changes suggest increasing aridity during the middle Holocene, which is also supported by Purdue's (1980) analysis of clinal variation of squirrels and rabbits.

Stratified faunules from the cultural horizons at the Cherokee site seem to document increasing aridity during the middle Holocene as well. The Cultural Horizon II faunule (ca. 7.3 ka) contains species like hoary tree bat, big brown bat, and red squirrel, which require large trees for roosting and nesting (Semken, 1980). These species are not present in the Cultural Horizon I faunule (ca. 6.35 ka). Both middle Holocene faunules contain grassland species like plains pocket gopher, hispid pocket mouse, western harvest mouse, and prairie vole. However, the Horizon I faunule has a higher frequency of grassland or prairie species (Semken, 1980, Fig. 3.4). This is created by the appearance in Horizon I of other grassland species like Franklin's ground squirrel and northern grasshopper mouse and the disappearance of species like red squirrel, southern bog lemming, and meadow jumping mouse, which occurred in Horizons II and III (Semken, 1980, Table 3.2).

In eastern Iowa, the Mud Creek l.f. was collected from several outcrops, and the age of the entire biota depends upon stratigraphic correlations with dated sequences (ca. 6.2 ka; Kramer, 1972). Because the biota contains plant and mammal species that occur farther east today, greater effective moisture is suggested for eastern Iowa at 6.2 ka. This may represent the moist phase during the middle Holocene that has been detected in varve sequences from Minnesota (Dean and others, 1984). It is also possible, as suggested by Semken (1983), that the middle Holocene climatic gradient in Iowa was more pronounced than it is today.

Paleoenvironmental data for the middle Holocene of the southern plains appears to demonstrate a trend for increased aridity throughout this time. In central Texas early Holocene species adapted to moist habitats (eastern chipmunk, southern bog lemming, and shorttail shrew) are absent from middle Holocene faunas. At the Wunderlich site on the eastern edge of the Edwards Plateau, pronghorn expanded its range eastward at 5.4 ka, but aquatic habitats persisted, as indicated by the presence of the river otter (Lundelius, 1967). Also, it appears that the shorttail shrew progressively withdrew eastward from the Edwards Plateau. Lundelius (1967) believes that this reflects a warming and drying of the climate from west to east across the Edwards Plateau.

In Val Verde County, Texas, several deeply stratified rockshelters along the Rio Grande drainage preserve a Holocene biotic and cultural record. Collectively, the vertebrate sequences from these rockshelters suggest that the middle Holocene environments had greater effective moisture with more vegetational cover and better developed soils than today (Graham, 1987). However, the middle Holocene environments were not as moist as those of the early Holocene or late Pleistocene (Graham, 1987), thus reflecting a trend for general desiccation throughout the Holocene. This trend is further supported by palynological (Bryant, 1966), ethnobotanical (Williams-Dean, 1978), and geomorphological (Patton and Dibble, 1982) data.

A dramatic increase in the abundance of aquatic resources (i.e., fish, turtle, muskrat, beaver) in these rockshelter deposits around 7 ka is of particular interest. A similar florescence of aquatic resources appears in the Prairie Peninsula at about the same time (Styles, 1986). The increased utilization of aquatic resources may be the result of human populations moving to major river valleys in response to increasing aridity (Styles, 1986). It may also result from the increase in the size of human populations (demographic pressure) or development of new technologies (Styles, 1986; Lord, 1984). In reality, all three factors are interrelated.

Although there are no middle Holocene faunas from northern Texas, pedological and geomorphological relationships suggest a similar environmental trend for increasing desiccation. From 8.5 to 5.5 ka, soils developed in draws (ephemeral drainage ways) formed in lacustrine sediments and marshy environments, but between 5.5 and 4.5 ka, eolian sediments filled the draws indicating drier conditions (Holliday, 1985). Pedologic and geomorphologic data from playas and dune fields also reflect drier environments in the middle Holocene.

Faunas from two Middle Archaic sites in Kansas preserve a middle Holocene paleoenvironmental record. The vertebrate fauna from the Coffey site dates between 5.163 and 5.270 ka (Davis, 1987). This fauna records a grassland environment with riparian woodlands similar to today (Schmits, 1978). The Snyder site in Butler County is undated but appears to be culturally contemporaneous with the Coffey site. The fauna from this site also documents grassland and riparian environments, but it includes pronghorn and jackrabbit, which are missing from the Coffey site to the northeast (Davis, 1987). If sampling is not a problem, then these differences might reflect an environmental gradient, with greater aridity to the south and west.

The middle Holocene faunal records for the northern plains and western U.S. are meager. Walker (1987) provides a preliminary discussion of the Dead Indian Creek l.f. (4.18 to 4.43 ka). He concludes that it represents modern environmental conditions, in agreement with the paleoenvironmental record from Antelope Playa, eastern Powder River Basin (Markgraf and Lennon, 1986).

In the Great Basin and northwestern U.S. the middle Holocene also appears to be a time of climatic warming and decreasing effective moisture. At the Connely Caves, south-central Oregon, pika is eliminated from the record by 7 ka, and the frequency of pygmy cottontails and jackrabbits is reduced significantly (Grayson, 1982). The relative abundance of pygmy cottontails was also reduced at about the same time at the Wasden site, southern Idaho (Butler, 1972). At Gatecliff Shelter, central Nevada, the abundance of the pygmy cottontail was reduced after 5 ka, and hares showed a marked decrease ca. 5.3 ka (Grayson, 1982). The heather vole, a boreomontane species, was also eliminated from the Gatecliff faunal sequence at ca. 5.3 ka (Grayson, 1981).

CONCLUSIONS

Vertebrate faunal remains from paleontological and archeological sites serve as sensitive proxies for paleoenvironmental changes during the late Quaternary. Therefore vertebrate faunas, especially mammals, can be employed in paleoenvironmental and paleoclimatic reconstructions. However, most Pleistocene faunas contain combinations of extant species that do not occur together today. This limits the application of direct analogs in paleoenvironmental reconstructions. These nonanalog (disharmonious) associations were apparently formed by the individualistic response of species to environmental fluctuations. Paleoecological analyses of the constituent species of late Pleistocene communities suggest that they were supported by a fine-grained environmental mosaic within a cold equable climate.

Full-glacial faunas throughout most of North America contain boreomontane and arctic species that extended to lower latitudes and altitudes than they do today. Environmental gradients existed, and although boreomontane species distributions were lowered, the number of boreal species declined with decreasing elevation and latitude. Faunas near the ice front contained species like arctic shrews, lemmings, voles, and ground squirrels, which today inhabit arctic tundra. Several full-glacial faunas from Iowa and Wisconsin are the most tundralike of any south of the ice sheet. However, these small mammal guilds are not directly analogous to those of the modern tundra, for full-glacial guilds include species that do not inhabit the tundra or northern boreal forest today (e.g., prairie vole, sagebrush vole, thirteenline ground squirrel, eastern chipmunk). The presence of these species in ice-marginal faunas suggests that winter temperature extremes were not as severe as those in the modern arctic; in fact they probably were not any more severe than those currently in the northern U.S. Furthermore, summer temperatures were cold enough to support the arctic species but not cold enough to eliminate all amphibian and reptile species.

These equable climatic conditions are probably the result of several factors. Adiabatic warming of air descending from the ice mass would have moderated cold temperatures, at least locally. Also, as indicated by climatic simulation models (Kutzbach and Wright, 1985), the jet stream was split into northern and southern branches by the continental glaciers. Thus, the ice would have effectively blocked the southerly flow of arctic highs in the winter, and the southern branch of the jet stream would have made January temperatures warmer than present (Kutzbach and Wright, 1985). Conversely, the simulation model would suggest that in July southerly flow was weakened and temperatures would be cooler (Kutzbach and Wright, 1985).

The composition of the ice-marginal faunas also varied along east-west moisture gradients. The number of deciduous forest species decreased to the west as the frequency of grassland species increased. This gradient for decreasing moisture and more open environments towards the west is also reflected in faunas from lower latitudes (e.g., Florida - Texas - Arizona). It is also apparent in the distribution of extinct species as well. Woodland-adapted species like American mastodon, stagmoose, and giant beaver occur in the eastern half of the U.S., whereas camel, horse, and mammoth were more prevalent in the west during the late Wisconsin (Graham, 1979; Martin and Neuner, 1978; Martin and others, 1985).

Early late-glacial (15 to 13 ka) faunas are also disharmonious and generally reflect cold equable climates. However, many of the stratified faunas tend to exhibit a slight warming trend. At New Paris No. 4 in Pennsylvania, arctic species (collared lemming) and some boreomontane taxa (heather vole and northern bog lemming) were already lost by this time. In the upper Midwest the collared lemming is not known and may have already migrated north. Yet the late-glacial faunas still contain other boreal species like yellowcheek vole, heather vole, and boreal redback vole that occur farther north today. Faunas from the southern plains, Southwest, and Great Basin show similar trends, although some of the boreomontane species involved differ (e.g., pika, yellowbelly marmot, vagrant shrew).

Early late-glacial faunas from Wyoming and Montana may indicate fairly rigorous environments. At Natural Trap Cave, 12 genera of megafauna disappear simultaneously around 15.5 ka and at 13.5 ka boreomontane and arctic species increase while grassland species decrease. A 14-ka fauna from the Pryor Mountains, Montana, is similar to those from the Bighorn Mountains. Boreal and arctic faunas persisted into the later parts of the late-glacial in the northwestern United States, and diverse megafaunas were also present.

Climatic and environmental fluctuations at the end of the Pleistocene (ca. 10 ka) were significantly greater than those of the late Pleistocene or entire Holocene. Milankovitch forcing suggests that this is a time of maximum seasonal contrast (Kutzbach and Guetter, 1986). Faunal evidence supports such a contention and indicates change from a cold equable Pleistocene climate to a warmer more continental one. The magnitude of these changes is reflected in the response of the vertebrate communities. Enhancement of seasonal extremes (colder January and warmer July) caused species to migrate along environmental gradients in different directions, at different times, and with different rates. Modern community patterns began to emerge as the disharmonious biotas of the late Pleistocene were individualistically disassembled. Extinction of 32 genera of large mammals in North America may be one of the major consequences of this environmental change.

These environmental fluctuations were geologically rapid but time-transgressive. Thus some early Holocene faunas were still different from modern ones. For the southern plains, early Holocene climates were still cool and moist, as suggested by Semken (1983), but they were not the same as Pleistocene climates or environments. Similarly, some boreomontane species remained at lower latitudes and elevations in the Great Basin and northern plains during the early Holocene. However, in the eastern U.S., modern climates and mammalian community composition were reached by 10 ka. These reconstructions are generally consistent with climatic simulation models that suggest that early to middle Holocene climates for the midcontinent were "warm and dry" (Kutzbach and Guetter, 1986). The cooler and moister climates of the early Holocene (ca. 9 to 10 ka) of the Plains area may have resulted from a lingering ice sheet. However, shortly after 9 ka, Plains faunas indicate drier and warmer climatic conditions.

The environmental changes of the middle Holocene are also time-transgressive and exhibit regional differences. In the eastern U.S. there is little faunal evidence for environmental change, although forests may have been slightly more open. In the Great Plains and central Midwest there is abundant evidence for a major period of warming and increasing aridity. These environmental changes are responsible for the eastward expansion of the Prairie Peninsula, with maximum expansion and greatest aridity at ca. 7 ka. This is consistent with the climatic simulation models for the continental interiors during this time (Kutzbach and Guetter, 1986).

Drier and warmer climates allowed grassland vertebrate species like prairie chicken, bison, antelope, badger, and pocket mouse to invade western Missouri, as indicated by the Rodgers Shelter local fauna. Farther east at Graham Cave in central Missouri these grassland species were not present, but fluctuations in relative abundance and shifts in clinal variation of deer, rabbits, and tree squirrels suggest an opening of the forest. Finally, at Modoc Rock Shelter in southwestern Illinois, southern rather than western species expanded into the area. The occurrence of rice rat, hispid cotton rat, and wood rat may indicate climatic warming. However, the rice rat and hispid cotton rat do not necessarily reflect more arid climatic conditions, because they both prefer moist grassy environments in the southern U.S. today.

Warm and arid climatic conditions are also suggested by the middle Holocene faunas from the southern plains, desert Southwest, and Great Basin. Climatic conditions in these areas for the late Holocene remained arid, unlike the Prairie Peninsula which exhibited a reversal to more mesic conditions. The warming of middle Holocene climates continued to isolate boreomontane species on mountain tops in the western U.S. Extinction without replacement in these island habitats has left small-mammal communities in nonequilibrium (Brown, 1971).

The environmental changes of the late Quaternary have played a fundamental role in shaping the species composition, diversity, and structure of modern mammalian communities. Although not discussed in this contribution, environmental changes during the late Holocene and Historic times also had many effects on vertebrate communities (Semken, 1983). Future environmental fluctuations, including both climatic and cultural events, are almost assured, and undoubtedly vertebrate communities will continue to evolve.

Appendix I. Common and scientific names of taxa mentioned in text. Common names are listed in alphabetical order rather than in the standard phylogenetic system.

COMMON NAME	SCIENTIFIC NAME
American mastodon	*Mammut americanum*
antelope	Antilocaprid undifferentiated
arctic ground squirrel	*Spermophilus parryii*
arctic shrew	*Sorex arcticus*
badger	*Taxidea taxus*
beautiful armadillo	*Dasypus bellus*
beaver	*Castor canadensis*
big brown bat	*Eptesicus fuscus*
bison	*Bison* sp.
blacktail jackrabbit	*Lepus californicus*
boreal redback vole	*Clethrionomys gapperi*
box turtle	*Terrapene carolina*
brown lemming	*Lemmus sibiricus*
camel	*Camelops* sp.
caribou	*Rangifer tarandus*
chuckwalla	*Sauromalus obesus*
collared lemming	*Dicrostonyx torquatus*
collared lizard	*Crotaphytus collaris*
desert spiny lizard	*Sceloporus magister*
desert tortoise	*Gopherus agassizi*
eastern chipmunk	*Tamias striatus*
eastern cottontail rabbit	*Sylvilagus floridanus*
eastern gray squirrel	*Sciurus carolinensis*
elk	*Cervus elaphus*
flathead peccary	*Platygonus compressus*
fox snake	*Elaphe vulpina*
fox squirrel	*Sciurus niger*
Franklin's ground squirrel	*Spermophilus franklinii*
garter snake	*Thamnophis* sp.
giant beaver	*Castoroides ohioensis*
giant tortoise	*Geochelone crassiscutata*
golden mantle ground squirrel	*Spermophilus lateralis*
gray fox	*Urocyon cinereoargenteus*
gray myotis bat	*Myotis grisescens*
gray shrew	*Notiosorex crawfordi*
gray squirrel	*Sciurus carolinensis*
Harlan's ground sloth	*Glossotherium harlani*
Harrington's mountain goat	*Oreamnos harringtoni*

heather vole	Phenacomys intermedius	pygmy shrew	*Sorex hoyi*
hispid cotton rat	Sigmodon hispidus	redback vole	*Clethrionomys* sp.
hispid pocket mouse	Perognathus hispidus	redspot toad	*Bufo punctatus*
hoary marmot	Marmota caligata	red squirrel	*Tamiasciurus hudsonicus*
hoary tree bat	Lasiurus cinereus	redtail hawk	*Buteo jamaicensis*
horse	Equus sp.	rice rat	*Oryzomys palustris*
Hudson Bay collared lemming	*Dicrostonyx husonius*	river otter	*Lutra canadensis*
Hudson Bay toad	*Bufo americanus copei*	sagebrush vole	*Lagurus curtatus*
jackrabbit	*Lepus* sp.	Shasta ground sloth	*Nothrotheriops shastensis*
kangaroo rat	*Dipodomys* sp.	shorttail shrew	*Blarina brevicauda*
least chipmunk	*Tamias minimus*	saiga	*Saiga tartaracus*
least shrew	*Cryptotis parva*	silky pocket mouse	*Perognathus flavus*
leopard lizard	*Crotaphytus wislizeni*	singing vole	*Microtus miurus*
longnose peccary	*Mylohyus nasutus*	southern bog lemming	*Synaptomys cooperi*
longtail shrew	*Sorex dispar*	southern flying squirrel	*Glaucomys volans*
longtail vole	*Microtus longicaudus*	southern grasshopper mouse	*Onychomys torridus*
mammoth	*Mammuthus* sp.	spotted skunk	*Spilogale putorius*
masked shrew	*Sorex cinereus*	stag-moose	*Cervalces scotti*
meadow jumping mouse	*Zapus hudsonius*	thirteenline ground squirrel	*Spermophilus tridecemlineatus*
meadow vole	*Microtus pennsylvanicus*	timber rattlesnake	*Crotallus horridus*
milk snake	*Lampropeltis triangulum*	toad	*Bufo* sp.
muskrat	*Ondatra zibethicus*	tree squirrels	*Sciurus* sp.
northern bog lemming	*Synaptomys borealis*	turkey vulture	*Cathartes aura*
northern flying squirrel	*Glaucomys sabrinus*	vagrant shrew	*Sorex vagrans*
northern grasshopper mouse	*Onychomys leucogaster*	voles	*Microtus* sp.
northern pocket gopher	*Thomomys talpoides*	water shrew	*Sorex palustris*
packrat	*Neotoma* sp.	water snake	*Natrix sipedon*
pika	*Ochotona princeps*	western harvest mouse	*Reithrodontomys megalotis*
pine marten	*Martes americana*	western jumping mouse	*Zapus princeps*
pine vole	*Microtus pinetorum*	western shovelnose snake	*Cionactis occipitalis*
plains pocket gopher	*Geomys bursarius*	whitetail deer	*Odocoileus virginianus*
pocket mouse	*Perognathus* sp.	woodchuck	*Marmota monax*
porcupine	*Erethizon dorsatum*	woodland jumping mouse	*Napeozapus insignis*
prairie chicken	*Tympanuchus cupido*	wooly mammoth	*Mammuthus primigenius*
prairie vole	*Microtus ochrogaster*	Wyoming pocket mouse	*Perognathus fasciatus*
pronghorn	*Antilocapra americana*	yellowbelly marmot	*Marmota flaviventris*
pygmy mouse	*Baiomys taylori*	yellowcheek vole	*Microtus xanthognathus*
pygmy rabbit	*Brachylagus idahoensis*	yellownose vole	*Microtus chrotorrhinus*

REFERENCES CITED

Adovasio, J. M., Gunn, J. D., Donahue, J., Stuckenrath, R., Guilday, J., and Volman, K., 1980, Yes, Virginia, it really is that old; A reply to Haynes and Mead: American Antiquity, v. 45, p. 588–595.

Adovasio, J. M., Donahue, J., Carlisle, R. C., Cushman, K., Stuckenrath, R., and Wiegman, P., 1984, Meadowcroft Rockshelter and the Pleistocene/Holocene transition in southwestern Pennsylvania, *in* Genoways, H. H., and Dawson, M. R., eds., Contributions in Quaternary vertebrate paleontology; A volume in memorial to John E. Guilday: Carnegie Museum of Natural History Special Publication 8, p. 347–369.

Ahler, S. A., 1976, Sedimentary processes at Rodgers Shelter, *in* Wood, W. R., and McMillan, R. B., eds., Prehistoric man and his environments; A case study in the Ozark Highland: New York, Academic Press, Incorporated, p. 123–139.

Baker, R. G., Rhodes, R. S. II, Schwert, D. P., Ashworth, A. C., Frest, T. J., Hallberg, G. R., and Janssens, J. A., 1986, A full-glacial biota from southeastern Iowa, U.S.A.: Journal of Quaternary Science, v. 1, no. 2, p. 91–107.

Birks, H.J.B., 1980, The present flora and vegetation of the moraines of the Klutlan Glacier, Yukon Territory, Canada; A study in plant succession: Quaternary Research, v. 14, p. 60–86.

Bonnichsen, R., 1985, High-altitude adaptation at False Cougar Cave: National Geographic Research Reports, v. 20, p. 27–40.

Brown, J. A., and Vierra, R. K., 1983, What happened in the Middle Archaic?; Introduction to an ecological approach to Koster site archaeology, *in* Phillips, J. L., and Brown, J. A., eds., Archaic hunters and gatherers in the American Midwest: New York, Academic Press, p. 165–195.

Brown, J. H., 1971, Mammals and mountaintops; Nonequilibrium insular biogeography: American Naturalist, v. 105, p. 467–478.

Bryant, V. M., Jr., 1966, Pollen analysis of the Devil's Mouth site; A preliminary study of the paleoecology of the Amistad Reservoir Area: Washington, D.C., Final Report to the National Science Foundation (unpublished), p. 129–164.

Butler, B. R., 1972, The Holocene or postglacial ecological crisis on the eastern Snake River Plain: Tebiwa, v. 15, p. 49–63.

Chomko, S. A., and Gilbert, B. M., 1987, The late Pleistocene/Holocene faunal record in the northern Bighorn Mountains, Wyoming, *in* Graham, R. W., Semken, H. A., Jr., and Graham, M. A., eds., Late Quaternary mammalian biogeogeography and environments of the Great Plains and Prairies: Springfield, Illinois State Museum Scientific Papers no. 22 (in press).

Clausen, C. J., Cohen, A. D., Emiliani, C., Holman, J. A., and Stip, J. J., 1979, Little Salt Spring, Florida; A unique underwater site: Science, v. 203, p. 609–614.

Cleland, C. E., 1966, The prehistoric animal ecology and ethnozoology of the Upper Great Lakes region: Ann Arbor, University of Michigan, Museum of

Anthropology, Anthropological Papers, v. 29, 294 p.

Cole, K., 1982, Late Quaternary zonation of vegetation in the eastern Grand Canyon: Science, v. 217, p. 1141–1145.

Coleman, D. D., and Liu, C. L., 1975, Illinois State Geological Survey radiocarbon dates VI: Radiocarbon, v. 17, p. 160–173.

Dalquest, W. W., 1965, New Pleistocene formation and local fauna from Hardeman County, Texas: Journal of Paleontology, v. 39, p. 63–79.

Dalquest, W. W., Roth, E., and Judd, F., 1969, The mammal fauna of Schulze Cave, Edwards County, Texas: Bulletin of the Florida State Museum, v. 13, p. 206–276.

Davis, L. C., 1987, Late Pleistocene/Holocene environmental changes in the central Plains of the United States; The mammalian record, *in* Graham, R. W., Semken, H. A., Jr., and Graham, M. A., eds., Late Quaternary mammalian biogeography and environments of the Great Plains and prairies: Springfield, Illinois State Museum Scientific Paper 22 (in press).

Davis, O. K., Agenbroad, L., Martin, P. S., and Mead, J. I., 1984, The Pleistocene dung blanket of Bechan Cave, *in* Genoways, H. H., and Dawson, M. R., eds., Contributions in Quaternary vertebrate paleontology; A volume in memorial to John E. Guilday: Carnegie Museum of Natural History Special Publication 8, p. 267–282.

Dean, W. E., Bradbury, J. P., Anderson, R. Y., and Barnosky, C. W., 1984, The variability of Holocene climatic change; Evidence from varved lake sediments: Science, v. 226, p. 1191–1194.

Delcourt, H. R., 1980, Late Quaternary vegetation history of the eastern highland rim and adjacent Cumberland Plateau of Tennessee: Ecological Monographs, v. 49, p. 255–280.

Foley, R. L., 1984, Late Pleistocene (Woodfordian) vertebrates from the Driftless Area of southwestern Wisconsin; The Moscow Fissure local fauna: Illinois State Museum Reports of Investigations, v. 39, p. 1–50.

Frison, G. C., 1978, Prehistoric hunters of the High Plains: New York, Academic Press, 457 p.

——, 1982, Radiocarbon dates, *in* Frison, G. C., and Stanford, D. J., eds., The Agate Basin site; A record of the Paleoindian occupation of the northwest High Plains: New York, Academic Press, p. 178–180.

Frison, G. C., Walker, D. N., Webb, S. D., and Zeimens, G. M., 1978, Paleoindian procurement of *Camelops* on the northwestern Plains: Quaternary Research, v. 10, p. 385–400.

Graham, R. W., 1976a, Late Wisconsin mammal faunas and environmental gradients of the eastern United States: Paleobiology, v. 2, p. 343–350.

——, 1976b, Pleistocene and Holocene mammals, taphonomy, and paleoecology of the Friesenhahn Cave local fauna, Bexar County, Texas [Ph.D. thesis]: Austin, University of Texas, 232 p.

——, 1979, Paleoclimates and late Pleistocene faunal provinces in North America, *in* Humphrey, R. L., and Stanford, D., eds., Pre-Llano cultures of the Americas; Paradoxes and possibilities: The Anthropological Society of Washington, p. 49–69.

——, 1981, Preliminary report on late Pleistocene vertebrates from the Selby and Dutton archeological/paleontological sites, Yuma County, Colorado: University of Wyoming, Contributions to Geology, v. 20, no. 1, p. 33–56.

——, 1985a, Diversity and community structure of the late Pleistocene mammal fauna of North America: Acta Zoologica Fennica, v. 170, p. 181–192.

——, 1985b, Response of mammalian communities to environmental changes during the late Quaternary, *in* Diamond, J., and Case, T. J., eds., Community ecology: New York, Harper and Row Publishers, p. 300–313.

——, 1986, Plant-animal interactions and Pleistocene extinctions, *in* Elliott, D. K., ed., Dynamics of extinctions: New York, John Wiley and Sons, p. 131–154.

——, 1987, The late Pleistocene and Holocene mammalian fauna of the southwestern Plains of the United States, *in* Graham, R. W., Semkens, H. A., Jr., and Graham, M. A., eds., Late Quaternary mammalian biogeography and environments of the Great Plains and prairies: Springfield, Illinois State Museum Scientific Paper 22 (in press).

Graham, R. W., Haynes, C. V., Johnson, D. V., and Kay, M., 1981, Kimmswick; A Clovis-mastodon association in eastern Missouri: Science, v. 213, p. 1115–1117.

Graham, R. W., and Lundelius, E. L., Jr., 1984, Coevolutionary disequilibrium and Pleistocene extinctions, *in* Martin, P. S., and Klein, R. G., eds., Quaternary extinctions; A prehistoric revolution: Tucson, University of Arizona Press, p. 223–249.

Grayson, D. K., 1979, Mount Mazama, climatic change, and Fort Rock Basin archaeofaunas, *in* Sheets, P. D., and Grayson, D. K., eds., Volcanic activity and human ecology: New York, Academic Press, p. 427–457.

——, 1981, A mid-Holocene record for the heather vole, *Phenacomys* cf. *intermedius,* in the central Great Basin and its biogeographic significance: Journal of Mammalogy , v. 62, p. 115–121.

——, 1982, Toward a history of Great Basin mammals during the past 15,000 years, *in* Madsen, D. B., and O'Connell, J. F., eds., Man and environment in the Great Basin: Society of American Archaeology Papers 2, p. 82–101.

——, 1984, Nineteenth-century explanations of Pleistocene extinctions; A review and analysis, *in* Martin, P. S., and Klein, R. G., eds., Quaternary extinctions; A prehistoric revolution: Tucson, University of Arizona Press, p. 5–39.

Griffin, J. W., 1974, Part I, The site and excavations, *in* Griffin, J. W., ed., Investigations in Russell Cave: Washington, D.C., U.S. Department of Interior, National Park Service, p. 1–15.

Guilday, J. E., 1965, Bone refuse from the Lamoka Lake site, *in* Ritchie, W. A., ed., The archeology of New York State: Garden City, Natural History Press, p. 56–58.

——, 1967, The climatic significance of the Hosterman's Pit local fauna, Centre County, Pennsylvania: American Antiquity, v. 32, p. 231–232.

Guilday, J. E., and Adam, E. K., 1967, Small mammal remains from Jaguar Cave, Lemhi County, Idaho: Tebiwa, v. 10, p. 26–36.

Guilday, J. E., and Parmalee, P. W., 1965, Animal remains from the Sheep Rock Shelter (36 HU 1), Huntington County, Pennsylvania: Pennsylvania Archaeologist, v. 35, no. 1, p. 34–49.

——, 1982, Vertebrate faunal remains from Meadowcroft Rockshelter, Washington County, Pennsylvania; Summary and interpretation, *in* Carlisle, R. C., and Adovasio, J. M., eds., Meadowcroft; Collected papers on the archaeology of Meadowcroft Rockshelter and the Cross Creek drainage: Minneapolis, Society of American Archaeology, 47th Annual Meeting, p. 163–174.

Guilday, J. E., Martin, P. S., and McCrady, A. D., 1964, New Paris No. 4; A Pleistocene cave deposit in Bedford County, Pennsylvania: Bulletin of the National Speleological Society, v. 26, p. 121–194.

Guilday, J. E., Hamilton, H. W., and McCrady, A. D., 1971, The Welsh Cave peccaries (*Platygonus*) and associated fauna, Kentucky Pleistocene: Annals of the Carnegie Museum, v. 43, p. 249–320.

Guilday, J. E., Hamilton, H. W., and Parmalee, P. W., 1975, Caribou (*Rangifer tarandus* L.) from the Pleistocene of Tennessee: Journals of the Tennessee Academy of Science, v. 50, p. 109–112.

Guilday, J. E., Hamilton, H. W., Anderson, E., and Parmalee, P. W., 1978, The Baker Bluff Cave deposit, Tennessee, and the late Pleistocene faunal gradient: Bulletin of the Carnegie Museum of Natural History, v. 11, p. 1–67.

Gustafson, C. E., Gilbow, D., and Daugherty, R. D., 1979, The Manis mastodon site; Early man on the Olympic Peninsula: Canadian Journal of Archeology, v. 3, p. 157–164.

Guthrie, R. D., 1982, Mammals of the mammoth steppe as paleoenvironmental indicators, *in* Hopkins, D. M., Matthews, J. V., Jr., Schweger, C. E., and Young, S. B., eds., Paleoecology of Beringia: New York, Academic Press, p. 307–326.

——, 1984, Mosaics, allelochemics, and nutrients, *in* Martin, P. S., and Klein, R. G., eds., Quaternary extinctions; A prehistoric revolution: Tucson, University of Arizona Press, p. 259–298.

Hall, E. R., 1981, The mammals of North America: New York, John Wiley and Sons, 2 vols., 1181 p.

Hannus, L. A., 1982, Evidence of mammoth butchering at the Lange/Ferguson (39SH33) Clovis kill site; Society of American Archaeology, 47th Annual Meeting, Programs and Abstracts, p. 53.

Hansen, R. M., 1978, Shasta ground sloth food habits, Rampart Cave, Arizona:

Paleobiology, v. 4, p. 302–319.
Harris, A. H., 1977, Wisconsin age environments in the northern Chihuahuan desert; Evidence from the higher vertebrates, *in* Wauer, R. H., and Riskind, D. H., eds., Transactions of the symposium on the biological resources of the Chihuahuan Desert region, United States and Mexico: Washington, D.C., National Park Service Transactions and Proceedings Series 3, p. 23–52.
—— , 1985, Late Pleistocene vertebrate paleoecology of the west: Austin, University of Texas Press, 293 p.
Haynes, C. V., 1964, Fluted projectile points; Their age and dispersion: Science, v. 145, p. 1408–1413.
—— , 1967, Carbon-14 dates and early man in the New World, *in* Martin, P. S., and Wright, H. E., Jr., eds., Pleistocene extinctions; The search for a cause: New Haven, Yale University Press, p. 267–286.
—— , 1970, Geochronology of man-mammoth sites and their bearing on the origin of the Llano Complex, *in* Dort, W., Jr., and Jones, J. K., Jr., eds., Pleistocene and Recent environments of the central Great Plains: Lawrence, University of Kansas Special Publication 3, p. 77–92.
—— , 1980, Paleoindian charcoal from Meadowcroft Rockshelter; Is contamination a problem: American Antiquity, v. 45, no. 3, p. 582–587.
Hester, J. J., 1972, Blackwater locality No. 1; A stratified early man site in eastern New Mexico: Fort Burgwin Research Center Publication, v. 8, p. 1–239.
Hester, T. R., 1983, Late paleo-indian occupations at Baker Cave, southwestern Texas: Bulletin of the Texas Archeological Society, v. 53, p. 101–119.
Hibbard, C. W., 1960, Pliocene and Pleistocene climates in North America: Annual Report of the Michigan Academy of Science, Arts, and Letters, v. 62, p. 5–30.
Holliday, V. T., 1985, Holocene soil-geomorphological relations in a semi-arid environment; The southern High Plains of Texas, *in* Boardman, J., ed., Soils and Quaternary landscape evolution: New York, John Wiley and Sons Limited, p. 325–357.
Holliday, V. T., and Johnson, E., 1981, An update on the Plainview occupation at the Lubbock Lake site: Plains Anthropologist, v. 26, p. 251–253.
Holman, J. A., 1978a, The late Pleistocene herpetofauna of Devil's Den Sinkhole, Levi County, Florida: Herpetologica, v. 34, no. 2, p. 228–237.
—— , 1978b, Supplementary data on the Rex Rodgers site; Amphibian and reptile remains, *in* Hughes, J. T., and Willey, P. S., eds., Salvage archeology at Mackenzie Reservoir: Austin, Texas Historical Commission Archeological Survey Report no. 24, p. 107.
Holman, J. A., and Clausen, C. J., 1984, Fossil vertebrates associated with paleo-indian artifact at Little Salt Spring, Florida: Journal of Vertebrate Paleontology, v. 4, no. 1, p. 146–154.
Hoyer, B. E., 1980, The geology of the Cherokee Sewer site, *in* Anderson, D. C., and Semken, H. A., Jr., eds., The Cherokee Sewer excavations; Holocene ecology and human adaptations in northwestern Iowa: New York, Academic Press, p. 21–66.
Jakaway, G. E., 1958, Pleistocene Lagomorpha and Rodentia from the San Josecito Cave, Nuevo Leon, Mexico: Transactions of the Kansas Academy of Science, v. 61, no. 3, p. 313–327.
Johnson, C., 1974, Geologic investigations at the Lubbock Lake site, *in* Black, C. C., ed., History and prehistory of the Lubbock Lake site: Lubbock: West Texas Museum Association, The Museum Journal, v. 15, p. 79–105.
Johnson, E., 1974, Zooarcheology and the Lubbock Lake site, *in* Black, C. C., ed., History and prehistory of the Lubbock Lake site: Lubbock: West Texas Museum Association, The Museum Journal, v. 15, p. 107–122.
Johnson, E., and Holliday, V. T., 1980, A Plainview kill/butchering locale on the Llano Estacado; The Lubbock Lake site: Plains Anthropologist, v. 25, no. 88, p. 89–111.
—— , 1981, Late paleo-indian activity at the Lubbock Lake site: Plains Anthropologist, v. 26, p. 173–193.
King, J. E., 1973, Late Pleistocene palynology and biogeography of the western Missouri Ozarks: Ecological Monographs, v. 43, no. 4, p. 539–565.
King, J. E., and Saunders, J. J., 1984, Environmental insularity and the extinction of the American mastodont, *in* Martin, P. S., and Klein, R. G., eds., Quaternary extinctions; A prehistoric revolution: Tucson, University of Arizona Press, p. 315–344.
—— , 1986, *Geochelone* in Illinois and the Illinoian–Sangamonian vegetation of the type region: Quaternary Research, v. 25, p. 89–99.
Klippel, W. E., 1971, Prehistory and environmental change along the southern border of the Prairie Peninsula during the Archaic Period [Ph.D. thesis]: Columbia, University of Missouri, 188 p.
Klippel, W. E., and Parmalee, P. W., 1982, Diachronic variation in insectivores from Cheek Bend Cave and environmental change in the midsouth: Paleobiology, v. 8, no. 4, p. 447–458.
Kramer, T. L., 1972, The paleoecology of the postglacial Mud Creek biota, Cedar and Scott counties, Iowa [M.S. thesis]: Iowa City, University of Iowa, 69 p.
Kurtén, B., and Anderson, E., 1972, The sediments and fauna from Jaguar Cave; II, The fauna: Tebiwa, v. 15, p. 21–45.
—— , 1980, Pleistocene mammals of North America: New York, Columbia University Press, 442 p.
Kutzbach, J. E., and Wright, H. E., Jr., 1985, Simulation of the climate of 18,000 yr B.P.; Results of the North American/North Atlantic/European sector: Quaternary Science Reviews, v. 4, p. 147–187.
Kutzbach, J. E., and Guetter, P. J., 1986, The influence of changing orbital parameters and surface boundary conditions on climatic simulations for the past 18,000 years: Journal of the Atmospheric Sciences, v. 43, p. 1725–1759.
Land, L. S., Lundelius, E. L., Jr., and Valastro, S., 1980, Isotopic ecology of deer bones: Palaeogeography, Palaeoclimatology, Palaeoecology, v. 32, p. 143–152.
Lahren, L. A., and Bonnichsen, R., 1974, Bone foreshafts from a Clovis burial in southwestern Montana: Science, v. 186, p. 147–150.
Leonhardy, F. C., and Anderson, A. D., 1966, The archaeology of the Domebo site, *in* Leonhardy, F. C., ed., Domebo; A paleo-indian mammoth kill in the prairie-plains: Lawton, Contribution of the Museum of the Great Plains, no. 1, p. 14–26.
Logan, L. E., and Black, C. C., 1979, The Quaternary vertebrate fauna of Upper Sloth Cave, Guadalupe Mountains National Park, Texas, *in* Genoways, H. H., and Baker, R. J., eds., Biological investigations in the Guadalupe Mountains National Park, Texas: Washington, D.C., National Park Service, Proceedings and Transactions, Series no. 4, p. 141–158..
Long, A., Hansen, R. M., and Martin, P. S., 1974, Extinction of the Shasta ground sloth: Geological Society of America Bulletin, v. 85, p. 1843–1848.
Lord, K. L., 1984, The zoogeography of Hinds Cave, Val Verde County, Texas: College Station, Texas A & M University, Department of Anthropology, 296 p.
Lundelius, E. L., Jr., 1967, Late Pleistocene and Holocene faunal history of central Texas, *in* Martin, P. S., and Wright, H. E., Jr., eds., Pleistocene extinctions; A search for a cause: New Haven, Yale University Press, p. 287–319.
—— , 1972, Vertebrate remains from the Gray Sand, *in* Hester, J. J., ed., Blackwater Draw locality no. 1; A stratified early man site in eastern New Mexico: Taos, Fort Burgwin Research Center Publication 8, p. 148–163.
—— , 1984, A Late Pleistocene mammalian fauna from Cueva Quebrada, Val Verde County, Texas, *in* Genoways, H. H., and Dawson, M. R., eds., Contributions in Quaternary vertebrate paleontology; A volume in memorial to John E. Guilday: Pittsburgh, Carnegie Museum of Natural History Special Publication 8, p. 456–481.
Lundelius, E. L., Jr., Graham, R. W., Anderson, E., Guilday, J., Holman, J. A., Steadman, D. W., and Webb, S. D., 1983, Terrestrial vertebrate faunas, *in* Porter, S. C., ed., Late Quaternary environments of the United States; Volume 1, The late Pleistocene: Minneapolis, University of Minnesota Press, p. 311–353.
MacArthur, R. H., and Wilson, E. O., 1967, The theory of island biogeography: Princeton, Princeton University Press, 203 p.
MacNaughton, S. J., 1983, Serengeti grassland ecology; The role of composite environmental factors and contingency in community organization: Ecological Monographs, v. 53, no. 3, p. 291–320.
Marcus, L. F., and Berger, R., 1984, The significance of radiocarbon dates for Rancho La Brea, *in* Martin, P. S., and Klein, R. G., eds., Quaternary extinc-

tions; A prehistoric revolution: Tucson, University of Arizona Press, p. 159–183.

Markgraf, V., and Lennon, T., 1986, Paleoenvironmental history of the last 13,000 years of the eastern Powder River Basin, Wyoming, and its implications for prehistoric cultural patterns: Plains Anthropologist, v. 31, p. 1–12.

Martin, J. E., 1987, Paleoenvironment of the Lange/Ferguson Clovis kill site in the Badlands of South Dakota, *in* Graham, R. W., Semken, H. A., Jr., and Graham, M. A., eds., Late Quaternary mammalian biogeography and environments of the Great Plains and prairies: Springfield, Illinois State Museum Scientific Paper 22 (in press).

Martin, L. D., and Gilbert, B. M., 1978, Excavations at Natural Trap Caves: Transactions of the Nebraska Academy of Science, v. 6, p. 107–116.

Martin, L. D., and Neuner, A. M., 1978, The end of the Pleistocene in North America: Transactions of the Nebraska Academy of Science, v. 6, p. 117–126.

Martin, L. D., Rogers, R. A., and Neuner, A. M., 1985, The effect of the end of the Pleistocene on man in North America, *in* Mead, J. I., and Meltzer, D. J., eds., Environments and extinctions; Man in late glacial North America: Orono, Maine, Center for the Study of Early Man, p. 15–30.

Martin, P. S., 1984, Prehistoric overkill; The global model, *in* Martin, P. S., and Klein, R. G., eds., Quaternary extinctions; A prehistoric revolution: Tucson, University of Arizona Press, p. 354–403.

Martin, P. S., and Klein, R. G., 1984, Quaternary extinctions; A prehistoric revolution: Tucson, University of Arizona Press, 892 p.

Martin, R. A., and Webb, S. D., 1974, Late Pleistocene mammals from the Devil's Den fauna, Levi County, *in* Webb, S. D., ed., Pleistocene mammals of Florida: Gainesville, University Presses of Florida, p. 114–145.

Matthews, J. V., Jr., 1982, East Beringia during late Wisconsin time; A review of the biotic evidence, *in* Hopkins, D. M., Matthews, J. V., Jr., Schweger, C. E., and Young, S. B., ed., Paleoecology of Beringia: New York, Academic Press, p. 127–152.

Mawby, J. E., 1967, Fossil vertebrates of the Tule Springs site, Nevada: Nevada State Museum Anthropological Papers, v. 13, p. 105–128.

McDonald, J. N., 1986, Valley-bottom stratigraphy of Saltville Valley, Virginia, and its paleoecological implications: National Geographic Research Reports, v. 21, p. 291–296.

McMillan, R. B., and Klippel, W. E., 1981, Post-glacial environmental change and hunting societies of the southern Prairie Peninsula: Journal of Archaeological Sciences, v. 8, p. 215–245.

McNett, C. W., Jr., McMillan, B. A., and Marshall, S. B., 1977, The Shawnee-Minisink site, *in* Newman, W. S., and Salwen, B., eds., Amerinds and their paleoenvironments in northeastern North America: Annals of the New York Academy of Sciences, v. 288, p. 282–296.

Mead, J. I., 1980, Is it really that old? A comment about the Meadowcroft Rockshelter "Overview": American Antiquity, v. 45, p. 579–582.

—— , 1981, The last 30,000 years of faunal history within the Grand Canyon, Arizona: Quaternary Research, v. 15, p. 311–326.

Mead, J. I., and Meltzer, D. J., 1984, North American late Quaternary extinctions and the radiocarbon record, *in* Martin, P. S., and Klein, R. G., eds., Quaternary extinctions; A prehistoric revolution: Tucson, University of Arizona Press, p. 440–450.

Mead, J. I., and Phillips, A. M., III, 1981, The late Pleistocene and Holocene fauna and flora of Vulture Cave, Grand Canyon, Arizona: Southwestern Naturalist, v. 26, p. 257–288.

Mead, J. I., Thompson, R. S., and Van Devender, T. R., 1982, Late Wisconsinan and Holocene fauna from Smith Creek Canyon, Snake River Range, Nevada: Transactions of the San Diego Society of Natural History, v. 20, p. 1–26.

Mead, J. I. Van Devender, T. R., and Cole, K., 1983, Late Quaternary small mammals from Sonora Desert packrat middens, Arizona and California: Journal of Mammalogy, v. 64, p. 173–180.

Mead, J. I., and 7 others, 1986, Extinction of Harrington's mountain goat: Proceedings of the National Academy of Science, v. 83, p. 836–839.

Michels, J. W., and Dutt, J. S., 1968, Archaeological investigations of Sheep Rock Shelter, Huntington County, Pennsylvania: University Park, Pennsylvania State University, Occasional Papers in Anthropology, v. 3, 505 p.

Neusius, S. W., 1982, Early-Middle Archaic subsistence strategies; Changes in faunal exploitation at Koster site [Ph.D. thesis]: Evanston, Northwestern University, 373 p.

Parmalee, P. W., 1962, Faunal remains from the Stanfield-Worley Bluff Shelter, Colbert County, Alabama: Journal of Alabama Archaeology, v. 8, p. 112–114.

—— , 1963, A prehistoric occurrence of porcupine in Alabama: Journal of Mammalogy, v. 44, no. 2, p. 267–268.

Parmalee, P. W., and Oesch, R. D., 1972, Pleistocene and recent faunas from the Brynjulfson caves, Missouri: Illinois State Museum Reports of Investigations, v. 25, 52 p.

Patton, P. C., and Dibble, D. S., 1982, Archeologic and geomorphic evidence for the paleohydrologic record of the Pecos River in West Texas: American Journal of Science, v. 282, p. 97–121.

Patton, T. H., 1963, Fossil vertebrates from Miller's Cave, Llano County, Texas: Bulletin of the Texas Memorial Museum, v. 7, p. 1–41.

Phillips, A. M., 1984, Shasta ground sloth extinction; Fossil packrat midden evidence from the western Grand Canyon, *in* Martin, P. S., and Klein, R. G., eds., Quaternary extinctions; A prehistoric revolution: Tucson, University of Arizona Press, p. 148–158.

Purdue, J. R., 1980, Clinal variation of some mammals during the Holocene: Quaternary Research, v. 13, no. 2, p. 242–258.

Ray, C. E., Cooper, B. N., and Benninghoff, W. S., 1967, Fossil mammals and pollen in a late Pleistocene deposit at Saltville, Virginia: Journal of Paleontology, v. 41, no. 3, p. 608–622.

Ritchie, W. A., 1969, The archaeology of New York State: Garden City, Natural History Press, 357 p.

Rhodes, R. S., II., 1984, Paleoecology and regional paleoclimatic implications of the Farmdalian Craigmile and Woodfordian Waubonsie mammalian local faunas, southwestern Iowa: Illinois State Museum Reports of Investigations, v. 40, p. 1–51.

Roth, E. L., 1972, Late Pleistocene mammals from Klein Cave, Kerr County, Texas: Texas Journal of Science, v. 24, p. 75–84.

Sadek-Kooros, H., 1966, Jaguar Cave; An early man site in the Beaverhead Mountains of Idaho [Ph.D. thesis]: Cambridge, Harvard University, 438 p.

Saunders, J. J., 1977, Late Pleistocene vertebrates of the western Ozark Highland, Missouri: Illinois State Museum Reports of Investigations, v. 33, 118 p.

—— , 1987, Vertebrates of the San Pedro valley 11,000 B.P., *in* Haynes, C. V., ed., The Clovis hunters: Tucson, University of Arizona Press (in press).

Schmits, L. J., 1978, The Coffey site; Environmental and cultural adaptation at a prairie plains Archaic site: Mid-Continental Journal of Archaeology, v. 3, p. 69–185.

Schultz, G. E., 1967, Four superimposed late Pleistocene vertebrate faunas from southwest Kansas, *in* Martin, P. S., and Wright, H. E., Jr., ed., Pleistocene extinctions; The search for a cause: New Haven, Yale University Press, p. 321–336.

—— , 1978, Supplementary data on the Rex Rodgers site; Micromammals, *in* Hughes, J. T., and Willey, P. S., ed., Salvage archeology at Mackenzie Reservoir: Austin, Texas Historical Commission Archeological Survey Report 24, p. 114.

Semken, H. A., Jr., 1980, Holocene climatic reconstructions derived from the three micromammal cultural horizons of the Cherokee Sewer site, northwestern Iowa, *in* Anderson, D. C., and Semken, H. A., Jr., eds., The Cherokee excavations; Holocene ecology and human adaptations in northwestern Iowa: New York, Academic Press, p. 67–99.

—— , 1983, Holocene mammalian biogeography and climatic change in the eastern and central United States, *in* Wright, H. E., Jr., ed., Late Quaternary environments of the United States; Volume 2, The Holocene: Minneapolis, University of Minnesota Press, p. 182–207.

Slaughter, B. H., 1966, The vertebrates of the Domebo local fauna, Pleistocene of Oklahoma, *in* Leonhardy, F. C., ed., Domebo; A paleo-indian mammoth kill in the Prairie-Plains: Lawton, Oklahoma, Contributions of the Museum of the Great Plains, no. 1, p. 31–35.

—— , 1967, Animal ranges as a clue to late Pleistocene extinction, *in* Martin, P. S., and Wright, H. E., Jr., ed., Pleistocene extinctions; The search for a cause: New Haven, Yale University Press, p. 155–167.

—— , 1975, Ecological interpretation of the Brown Sand Wedge Local Fauna, *in* Wendorf, F., and Hester, J. J., eds., Late Pleistocene environments of the southern High Plains: Taos, Fort Burgwin Research Center Publication 9, p. 179–192.

Slaughter, B. H., and Hoover, R., 1963, Sulphur River Formation and the Pleistocene mammals of the Ben Franklin local fauna: Dallas, Southern Methodist University, Journal of the Graduate Research Center, v. 31, no. 3, p. 132–148.

Speer, R. D., 1978, Fossil *Bison* remains from the Rex Rodgers Site, *in* Hughes, J. T., and Willey, P. S., eds., Salvage archeology at Mackenzie Reservoir: Austin, Texas Historical Commission Archeological Survey Report 24, p. 68–94.

Stanford, D. J., 1983, Pre-Clovis occupation south of the ice sheets, *in* Shutler, R., Jr., ed., Early man in the New World: Beverly Hills, California, Sage Publications, p. 65–72.

Stuckenrath, R., Adovasio, J. M., Donahue, J., and Carlisle, R. C., 1982, The stratigraphy, cultural features, and chronology at Meadowcroft Rockshelter, Washington County, southwestern Pennsylvania, *in* Carlisle, R. C., and Adovasio, J. M., eds., Meadowcroft; Collected papers on the archaeology of Meadowcroft Rockshelter and the Cross Creek drainage: 47th Annual Meeting, Society of American Archaeology, p. 69–90.

Styles, B. W., 1986, Aquatic exploitation in the lower Illinois River valley; The role of paleoecological change, *in* Neusius, S. W., ed., Foraging, collecting, and harvesting; Archaic Period subsistence and settlement in the eastern Woodlands: Carbondale, Center for Archaeological Investigations Occassional Papers no. 6, p. 145–174.

Styles, B. W., Fowler, M. L., Ahler, S. R., King, F. B., Styles, T. R., 1981, Modoc Rock Shelter arachaeological project, Randolph County, Illinois, 1980–81: Completion Report to the Department of the Interior, Heritage Conservation and Recreation Service and Illinois Department of Conservation, 183 p.

Styles, B. W., Ahler, S. R., and Fowler, M. R., 1983, Modoc Rock Shelter revisited, *in* Phillips, J. L., and Brown, J. A., eds., Archaic hunters and gatherers in the American Midwest: New York, Academic Press, p. 261–297.

Thomas, D. H., 1982, The archaeology of Gatecliff Shelter and Monitor valley: New York, Anthropological Papers of the American Museum of Natural History, v. 59, pt. 1, 552 p.

Thompson, R. S., and Mead, J. I., 1982, Late Quaternary environments and biogeography in the Great Basin: Quaternary Research, v. 17, p. 39–55.

Thompson, R. S., Van Devender, T. R., Martin, P. S., Foppe, T., and Long, A., 1980, Shasta ground sloth (*Nothrotheriops shastense* Hoffstetter) at Shelter Cave, New Mexico; Environment, diet, and extinction: Quaternary Research, v. 14, p. 360–376.

Van Devender, T. R., and Mead, J. I., 1978, Early Holocene and late Pleistocene amphibans and reptiles in Sonoran Desert packrat middens: Copeia, v. 1978, p. 464–475.

Van Devender, T. R., and Spaulding, W. G., 1979, Development of vegetation and climate in the southwestern United States: Science, v. 204, p. 701–710.

Van Devender, T. R., Spaulding, W. G., and Phillips, A. M., III, 1979, Late Pleistocene plant communities in the Guadalupe Mountains, Culberson County, Texas, *in* Genoways, H. H., and Baker, R. J., eds., Biological investigations in the Guadalupe Mountains National Park, Texas: U.S. Department of Interior, National Park Service Proceedings and Transactions Series 4, p. 13–30.

Van Devender, T. R., Martin, P. S., Thompson, R. S., Cole, K., Jull, A.J.T., Long, A., Tollin, L. S., and Donahue, D. J., 1985, Fossil packrat middens and the Tandem Accelerator Mass Spectrometer: Nature, v. 317, no. 6038, p. 610–613.

Van Zant, K., 1979, Late glacial and postglacial pollen and plant macrofossils from Lake West Okoboji, northwestern Iowa: Quaternary Research, v. 12, p. 358–380.

Walker, D. N., 1975, A cultural and ecological analysis of the vertebrate fauna from the Medicine Lodge Creek site (48BH499) [M.S. thesis]: Laramie, University of Wyoming, 117 p.

—— , 1982, Early Holocene vertebrate fauna, *in* Frison, G. C., and Stanford, D., eds., The Agate Basin site; A record of paleoindian occupation on the northwestern High Plains: New York, Academic Press, p. 274–308.

—— , 1987, Late Pleistocene/Holocene environmental changes in Wyoming; The mammalian record, *in* Graham, R. W., Semken, H. A., Jr., and Graham, M. A., eds., Late Quaternary mammalian biogeography and environments of the Great Plains and prairies: Springfield, Illinois State Museum Scientific Papers (in press).

Watts, W. A., 1983, Vegetational history of the eastern United States 25,000 to 10,000 years ago, *in* Porter, S. C., ed., Late Quaternary environments of the United States; Volume 1, The late Pleistocene: Minneapolis, University of Minnesota Press, p. 294–310.

Weigel, R. D., Holman, J. A., Paloumpis, A. A., 1974, Part 5, Vertebrates from Russell Cave, *in* Griffin, J. W., ed., Investigations in Russell Cave: Washington, D.C., U.S. Department of the Interior, National Park Service, p. 81–85.

Whittaker, R. H., 1970, Communities and ecosystems: New York, The MacMillan Company, 158 p.

Williams-Dean, G. J., 1978, Ethnobotany and cultural ecology of prehistoric man in southwest Texas: College Station, Texas A & M University, Department of Anthropology, 286 p.

Wittry, W. L., 1959, The Raddatz rockshelter, Sk5, Wisconsin: The Wisconsin Archeologist, v. 40, p. 33–68.

Woodman, N., 1982, A subarctic fauna from the late Wisconsinan Elkader site, Clayton County, Iowa [M.S. thesis]: Iowa City, University of Iowa, 56 p.

Ziegler, A. C., 1963, Unmodified mammal and bird remains from Deer Creek Cave, Elko County, Nevada, *in* Shutler, M. E., and Shutler, R., Jr., eds., Deer Creek Cave, Nevada: Carson City, Nevada State Museum Anthropological Papers no. 11, p. 15–24.

MANUSCRIPT ACCEPTED BY THE SOCIETY APRIL 8, 1987

AKNOWLEDGMENTS

We wish to thank H. E. Wright, W. F. Ruddiman, H. A. Semken, Jr., D. K. Grayson, and W. W. Klippel for their constructive comments on an earlier draft of this manuscript. Bonnie W. Styles provided stimulating conversation about Holocene environmental and faunal changes in the Prairie Peninsula. I express my gratitude to Julianne Snider for drafting the illustrations. I am also grateful to Mary Ann Graham for her assistance in preparing the final draft of the manuscript.

Printed in U.S.A.

The Geology of North America
Vol. K-3, North America and adjacent oceans during the last deglaciation
The Geological Society of America, 1987

Chapter 18

Environmental change and developmental history of human adaptive patterns; The Paleoindian case

Robson Bonnichsen
Institute for Quaternary Studies, University of Maine at Orono, Orono, Maine 04469
Dennis Stanford
Smithsonian Institution, Washington, D.C., 20560
James L. Fastook
University of Maine at Orono, Orono, Maine 04469

INTRODUCTION

Could environmental change be the cause underlying the origin, internal restructuring, and extinction of human adaptive patterns? Recent syntheses in the Quaternary sciences present an opportunity to investigate the linkages and changes in the internal dynamics among the earth's climatic, glaciologic, geologic, oceanographic, biotic, and human adaptive systems (Denton and Hughes, 1981; Martin and Klein, 1984; Porter, 1983; Wright, 1983). Prehistorians have long been aware of the importance of environment as a potential catalyst for change in human adaptive systems and have traditionally emphasized local and regional relationships. With the increasing number of available syntheses documenting human adaptive responses to late Pleistocene and early Holocene environments (Bryan, 1986; Mead and Meltzer, 1986; Shutler, 1983), it is now possible to enlarge the survey and attempt a continental overview of how the end of the last major climate cycle (18 to 6 ka) was linked to terrestrial environments and human adaptive systems in North America. Because of the breadth of the topic, emphasis is placed on the period of greatest change between 12 and 10 ka.

THEORETICAL CONSIDERATIONS

Humans adapt to environment biologically and culturally. Biological adaptation operates slowly and entails microevolutionary changes. By using their cognitive facilities, however, humans can respond almost instantaneously to changing contexts. Environment should be understood as both limiting and creative; an interplay occurs between cultural and environmental variables (Vayda and Rapaport, 1968).

The effect of continental or global environmental change on human adaptive strategies cannot readily be investigated in the context of modern environments. Such studies are often very short term and focus almost exclusively on present-day systems. They do not provide documentation on how early peoples responded to long-range changes.

Fortunately, much can be learned about the relationship humans have held with the environment by reconstructing the contexts in which human adaptation has occurred. Of considerable importance to this inquiry is the Milankovitch hypothesis. It proposes that the earth is pulled out of precise equilibrium by the gravitational force of other planets, affecting the precession of the equinoxes, orbital eccentricity, and obliquity of the rotational axis. It is seen as the single most potent force causing changes in many other related environmental systems (Kutzbach, 1983; Ruddiman and Duplessy, 1985; Webb and others, 1985; Wright, 1984; Vrba, 1987). These cyclic changes are also seen as a driving force in cultural and biological evolution during early human history (*South African Journal of Science,* 1985) and of importance for understanding human origins, dispersion of humankind, and the extinction of old adaptive systems and development of new ones.

We develop here the theme that change in insolation driven by the Milankovitch cycle is a triggering mechanism affecting linked dependent subsystems at the end of the Pleistocene. Several propositions can be inferred from this hypothesis. (1) Extremes in the cyclic variation of the Earth's orbit can serve as a catalyst, restructuring climatic systems and leading to adjustments in natural and human adaptive systems at the local and regional levels. (2) Biological systems may respond to environmental extremes by community reorganization, radiation, extinction, and origination of new species(?). Humans may respond through their adaptive systems to environmental extremes by reorganizing the structure of their settlement, subsistence, and procurement systems, by creating or adopting innovations to enhance chances of survival, and/or by dispersion.

The above propositions have several predictive implications

Bonnichsen, R., Stanford, D., and Fastook, J. L., 1987, Environmental change and developmental history of human adaptive patterns; The Paleoindian case, *in* Ruddiman, W. F., and Wright, H. E., Jr., eds., North America and adjacent oceans during the last deglaciation: Boulder, Colorado, Geological Society of America, The geology of North America, v. K-3.

that can be tested against the environmental and archaeological record. (1) If humans formulate their adaptive strategies in light of local or regional environmental contexts, a variety of different adaptive patterns are likely to originate or go extinct at or near the same time on a continental or intercontinental scale. (2) If cyclic variation in climate forces changes in linked environmental and human adaptive systems, then concurrent changes can be anticipated in discrete environmental and archaeological records from different environmental regions at approximately the same time. (3) If environmental response to global climatic change is buffered by a lag effect created by the melting of continental ice sheets, then areas least affected by such environmental lag (e.g., Australia) should produce environmental and archaeological records reflecting reorganization earlier than buffered areas.

Testing of the above ideas concerning the global response of biological and cultural organisms requires examination of data sets from both environmentally buffered areas and nonbuffered areas. A simple methodology is used to investigate these ideas. Discrete human adaptive patterns dated by the radiocarbon method are summarized for North America. Only patterns with a significant number of radiocarbon dates are considered, and much relevant data from South America, Asia, and Australia has been omitted because of space limitations.

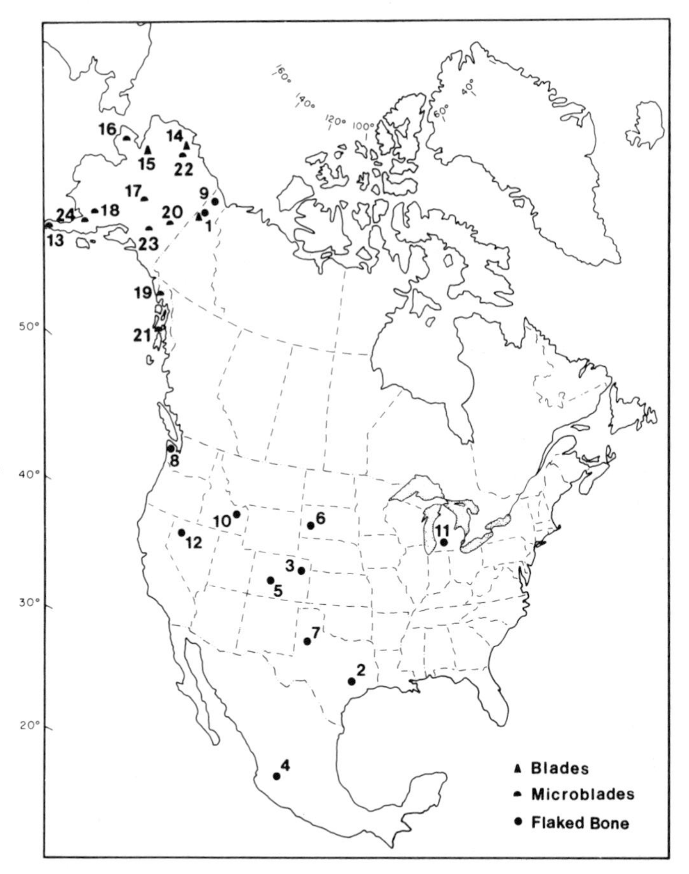

Figure 1. Localities yielding Rancholabrean bones that may have been flaked or modified in other ways by humans, as well as blades and Paleo-Arctic microblade and core assemblages: 1, Blue Fish Caves; 2, Duewall-Newberry; 3, Dutton and Selby; 4, Lago de Chapala; 5, Lamb spring; 6, Lange/Ferguson; 7, Lubbock Lake; 8, Manis; 9, Old Crow locality; 10, Owl Cave; 11, Pleasant Lake; 12, Rye Patch; 13, Anangula; 14, Gallager Flint Station; 15, Onion Portage, Akmak (blade) and Kobuck (microblade) levels; 16, Trail Creek; 17, Dry Creek, level 2; 18, Graveyard Point; 19, Ground Hog Bay; 20, Healy Lake; 21, Hidden Falls; 22, Putu; 23, Tangle Lakes; and 24, Ugashlik Narrows.

Terminology

The organizing unit most commonly employed to construct syntheses of early American prehistory is the concept of tradition. Archaeological components or specimens grouped within the same tradition are assumed to be historically related. Indeed this concept has heuristic value within a specific region with a variety of complex artifacts. But quite another problem arises with material remains dating 13 to 12 ka. Unrelated peoples may share fabrication techniques for simple stone and bone tools. For this reason, preference is given to the term *pattern*; it connotes redundancy of common features across regions and/or through time, avoiding unnecessary cultural-historical implications. At this juncture, archaeologists lack strong conceptual tools for distinguishing among several competing interpretive hypotheses to explain patterning, for example, migration, diffusion, and development in situ. For the present purpose, pattern is used as a convenient descriptive conceptual tool for organizing artifacts with shared shape and technology characteristics.

Early bone-flaking pattern

Over the last decade, several localities in the Beringia Refugium of Alaska and the Yukon Territory and in areas south of the continental ice sheet have yielded remains of extinct Pleistocene megamammals possibly modified by humans (Fig. 1). Of particular interest is the proposition that humans were present in the Old Crow Basin and south of the continental ice sheets prior to the end of the last glaciation. Part of their adaptive system involved shaping tools from the bones of large game animals. These early occupants undoubtedly had the knowledge to make stone tools, but recent discoveries suggest that the production of bone tools left a more visible archaeological record. Additional discoveries from south of the continental ice sheets, with context, suggest bone working may have been widespread across the continent during the late Pleistocene.

The 1966 discovery of a permineralized caribou tibia flesher along the lower Old Crow River, Yukon Territory, initiated a new era in northern archaeology (Irving, 1968). Radiocarbon assays on the apatite fraction of the tibia, along with two fragments of modified mammoth long bone, yielded dates ranging from 25 to 29 ka (Table 1). Research since then has shown that bone apatite is subject to contamination and is a poor material for dating (Hassan and Ortner, 1977). Collagen from the now-

TABLE 1. DATES FROM SITES WITH MODIFIED BONES

Site	Material*	^{14}C age (yr B.P.)	Lab No.	Source
Blue Fish Cave I	cl	15,500 ± 130	GSC-3053	Morlan and Cinq-Mars, 1982
	cl	12,900 ± 100	GSC-2881	"
Dutton and Selby	cl	16,630 ± 320	SI-5185	Stanford and Graham, 1985
	cl	13,600 ± 485	SI-5186	"
Dutton	cl	11,710 ± 150	SI-2877	"
Lamb Springs	cl	13,140 ± 1,000	M-1464	Rancier and others, 1982
	cl	11,735 ± 95	SI-4850	"
Old Crow Loc. 14N	ae	29,000 + 3,000 − 2,000	GX-1567	Irving and Harington, 1973
(flesher)	ae	27,000 + 3,000 − 2,000	GX-1640	"
	ae	25,750 + 1,800 − 1,500	GX-1568	"
(flesher)	cl	1350 ± 150	---	Mammoth Trumpet, 1985, no. 1, p. 1
MrVl-13	us	72,000 to 77,000	---	Mammoth Trumpet, 1984, no. 1, p. 1
Manis Mastodon	sw	12,000 ± 310	WSU 1866	Gustafson and others, 1979
	oc	11,850 ± 60	USGS-591	"
Pleasant Lake	wd	12,845 ± 165	Beta-1389	Fisher, 1984b
	wd	10,395 ± 100	Beta-1388	"
Rye Patch Reservoir	ae	29,790 ± 100	TX-2928	Firby and others, 1981
	ae	23,920 ± 730	TX-2929	"
	ae	23,000 ± 440	TX-3006	"

*Key to abbreviations:

ae = apatite	cp = carbonized plant	pt = peat
be = bone	he = humates	sc = soil carbonate
bo = bone organics	oc = organics	sw = seeds and wood
ce = carbonaceous earth	on = obsidian	ta = tephra
cg = collagen	pl = paleosol	wd = wood
cl = charcoal	pn = pine needles	us = uranium series
cm = clam		

famous flesher has been redated by the accelerator method as 1.35 ka (Nelson and others, 1986). Nonetheless, these initial Old Crow discoveries triggered a flurry of research covering more than a decade.

The combination of long-term research interests in the northern Yukon (Irving, 1968; Morlan, 1973), the initial promising radiometric dates (see Table 1), and the variety of altered bone found in existing paleontological collections (Bonnichsen, 1978, 1979; Morlan, 1980) stimulated the development of two major multidisciplinary research projects in 1975: the Yukon Refugium Project and the Yukon Research Programme. Unfortunately, not a single stratified site of Pleistocene age has been found in the Old Crow Basin.

Since the initial discovery, emphasis has been placed on understanding the stratigraphic context of the Old Crow Basin. Morlan (1980) suggests that the altered megamammal bones are from the upper part of stratigraphic unit 2, which lies below lake clays from proglacial Lake Kutchin. It is about 20 m thick and includes several weathered horizons, the most prominent being Disconformity A, which yielded fission-track dates of 60 ka. The team from the University of Toronto proposes that greater antiquity may be represented at Locality 12, where bones occur in 5 to 7 m of alluvial sediment well below Disconformity A (Irving and others, 1986, 1987). The bone bed is interpreted as a discard surface on the side of a pingo. Recent uranium-series dates on bone from this locality, however, tentatively suggest an age of 200 to 300 ka or perhaps earlier. Of the 8,000 bones and bone fragments recovered from excavations, some appear to have been modified by humans. These include mammoth bones broken while green, mammoth-bone cores, and flakes believed to be near their original site of deposition. More research is required to resolve the alternative stratigraphic interpretations.

The Rancholabrean remains from the Old Crow Basin were modified by flaking, whittling, polishing, slotting, scraping, and grinding to produce bone, antler, and ivory artifacts. Specimens proposed as candidates for artifacts include bone and ivory cores with flakes removed from the dorsal face, bone cores with flakes removed from the lateral edge, bone and ivory flakes, bone fragments with thinning flake scars and/or with shaping flakes removed from the distal end, a flesher, beveled antlers (wedges), expedient tools (Johnson, 1982), polished and ground bone artifacts, cut bone, and sawn bone. These occurrences led to the proposal that Pleistocene hunters and/or scavengers processed skeletal materials for marrow and tools.

Even though a variety of undeniable artifacts have been reported from Old Crow, caution must be used in accepting bone tools as evidence of Pleistocene human occupation. The recent advent of dating bone by accelerator mass spectrometry opens the possibility for dating very small pieces of bone (e.g., 1 gram) without destroying an entire specimen. Using this method Nelson and others (1986) report Holocene dates for bones, including the flesher that previously yielded a Pleistocene age. Mammoth-bone cores also dated by this method yielded Pleistocene dates (Table 1). Unfortunately, control tests have yet to demonstrate whether the amino acid method will produce consistent and reliable dates on bone.

The lack of stratigraphic context and definitive criteria used for identifying human workmanship continues to delay the widespread acceptance of modified bones from the Old Crow Basin as the product of human workmanship. Fortunately the case for early modified bones no longer rests exclusively on the evidence from Old Crow, for several new localities, both in the Beringia Refugium and south of the continental ice sheets, provide appropriate stratigraphic contexts for constructing more-compelling arguments.

At Blue Fish Caves, 58 km southwest of Old Crow, bone from the lower aeolian unit yielded ages of 12.9 and 15.5 ka (Table 1) (Cinq-Mars, 1979; Morlan and Cinq-Mars, 1982). The typical late Pleistocene fauna includes arctic fox (*Alopex* sp.), horse (*Equus* sp.), caribou (*Rangifer* sp.), elk (*Cervid* sp.), bison (*Bison* sp.), and mammoth (*Mammuthus* sp.). Archaeological remains from within the dated aeolian unit include stone tools made from materials exotic to the cave, bones that are split and polished and exhibit butchering marks, and a mammoth-bone core. The remains from Bluefish suggest that humans lived under harsh periglacial conditions during the last glacial maximum.

The most important sites producing modified bones of late Wisconsin age from south of the continental ice sheets include the Rye Patch, Lamb Spring, Dutton, and Selby sites. Rye Patch, located about 30 km south of Winnemucca, Nevada, may be the oldest of these sites. Lateral erosion by the Rye Patch Reservoir about 10 km north of Rye Patch dam exposed the top of a fossil spring complex buried beneath Lake Lahontan clays. The spring vents yielded Rancholabrean fauna (Green and Rusco, unpublished; Rusco and Davis, 1981; Rusco and others, 1979) and a stone tool. Three radiometric dates on the bone apatite from large mammals are 23, 23.92, and 29.79 ka (Table 1). Tephra from the throat of the spring vent is identified as Wono Tephra, dated elsewhere at 25 ka (Davis, 1981).

A taphonomic analysis of 374 modified bone fragments from the south spring vent at Rye Patch suggests a complex taphonomic history (Bonnichsen and others, 1987), including carnivore gnawing, root etching, bone degradation, split-line cracking, pitting, and rounding. Many specimens appear to have been broken while green; some exhibit impact marks suggesting the use of blows for extraction of marrow; others exhibit negative flake scars formed by percussion. The numerous possible bone cores are not unequivocal, but it is impossible to eliminate humans as a factor.

Excavation at Lamb Spring, south of Denver, Colorado, exposed the remains of over 25 mammoths associated with a spring deposit (Stanford and others, 1981; Rancier and others, 1982). Many of the long bones were fractured, although the majority of the axial skeletons were not damaged. Noteworthy is an intrusive 13.63 kg boulder, possibly used as a hammerstone for breaking limb bones. Several crudely flaked chert artifacts were also recovered with the bone.

Radiocarbon assays of bone collagen yielded dates of 11.735 and 13.14 ka (Table 1). Plant remains from below the mammoth bone were dated 12.75 ka. Although no projectile points were recovered, these dates suggest a Clovis to pre-Clovis age for Lamb Spring.

The Dutton and Selby sites are located in deflation basins in northeastern Colorado (Stanford, 1979; Graham, 1981; Stanford and Graham, 1985). Lacustrine sediments at the Dutton site are capped with gleysol, above which is a Clovis occupation layer. A Rancholabrean fauna was found on the surface of the associated Peorian loess and is buried in redeposited loess. Fragments of bones broken while green are both polished and flaked. No stone tools were found, however. The Peorian loess at Dutton yielded a radiocarbon dates of 13.6 ka, and the unit immediately below the Clovis horizon was dated to 11.71 ka (Table 1). The oldest dated bone from the Selby site, found in the Peorian loess, yielded a date of 16.63 ka.

A series of mastodon butchery sites of Clovis age has recently been reported for southern Michigan (Fisher, 1984a, 1984b; Shipman and others, 1984). At Pleasant Lake, wood from the tusk pulp cavity is dated at 10.395 ka, and wood from below the skeleton yielded an age of 12.845. Human butchery is suggested by articulated anatomical units surrounded by isolated bones and fragments that lack anatomical organization. Disarticulation of this sort suggests carcass separation into butchering units. The discovery of matching scar patterns on conarticular surfaces of disarticulated bone pairs, associated with bone fragments with a glossy polish, suggests that wedges were used for separating butchery units. Burned bone with cut marks and surfaces exhibiting glossy polish from the Pleasant Lake and Hudson sites, examined by scanning electron microscope, are additional types of modification attributed to humans. Impact fractures and secondary flaking were also observed on other elements.

The Manis Mastodon site on the Olympic Peninsula near Squim, Washington is a butchery locality associated with a possible living site (Gustafson and others, 1979; Gilbow, 1981). Initial excavations exposed the semiarticulated left side of an animal and 4,000 disarticulated, scattered bone fragments nearby. Many of the Manis bones were broken while fresh, and some exhibit cuts and scratches. The only stone tool reported is a flaked cobble spall. Recent work by Carl Gustafson (personal communication, 1987) has led to the recovery of additional butchered remains of mastodon and a significant number of bone tools. Radiocarbon dates place the site at about 12 ka (Table 1).

Owl Cave and Lange/Fergurson sites have recently yielded flaked mammoth bone and fluted points. Owl Cave is located on the Snake River Plains, southeastern Idaho. The lower levels produced a Rancholabrean fauna associated with parts of what have been classified as five Folsom points (Butler, 1971; Dort and Miller, 1977; Miller, 1983, 1987). These points could also be a Clovis variant. Heavy-walled, dense, cortical mammoth bones were shaped into tools by flaking. Fractured surfaces of mammoth bone broken while green were matched (Miller, 1983, 1987). Multiple overlapping flake scars are indicative of human flaking. Several specimens exhibit microflaking and polish, suggesting use as choppers and wedges. The initial bone dates of 12.85 and 12.2 ka (Table 5) were obtained without pretreatment for humic acid. An additional sample of mammoth bone with overlapping flake scars produced a collagen date of 10.92 ka. Green (unpublished) dated 27 samples of obsidian from the lower levels by the hydration method, corroborating the bone dates and implying that the Folsom points and mammoth bones were associated.

The Lange/Fergurson site in southwestern South Dakota (Fig. 1) has Clovis fluted points and artificially modified mammoth bone more than 3 m below the present surface (Hannus, 1983, 1987; Martin, 1983). Although small, the inventory of stone tools includes three Clovis points and a chalcedony flake. Apparently butchered adult and juvenile mammoths were buried in place without disturbance shortly after human occupation. Only the bones from the juvenile were modified by flaking; they include two bifacially flaked cleavers produced from scapulae, five bone cores with flakes removed from the ventral face (see Fig. 2 core with a matching flake), several bone flakes, long-bone diaphyseal fragments with cone scars (impact loading points), and ribs with impact scars.

Collectively, these data suggest the widespread human use of worked bone in a variety of different environmental contexts during the late Pleistocene. The association of flaked mammoth bone and fluted points suggests continuity in bone flaking from pre-Clovis to Clovis times. With the extinction of game animals with large bones, reliance on bone flaking diminished in importance.

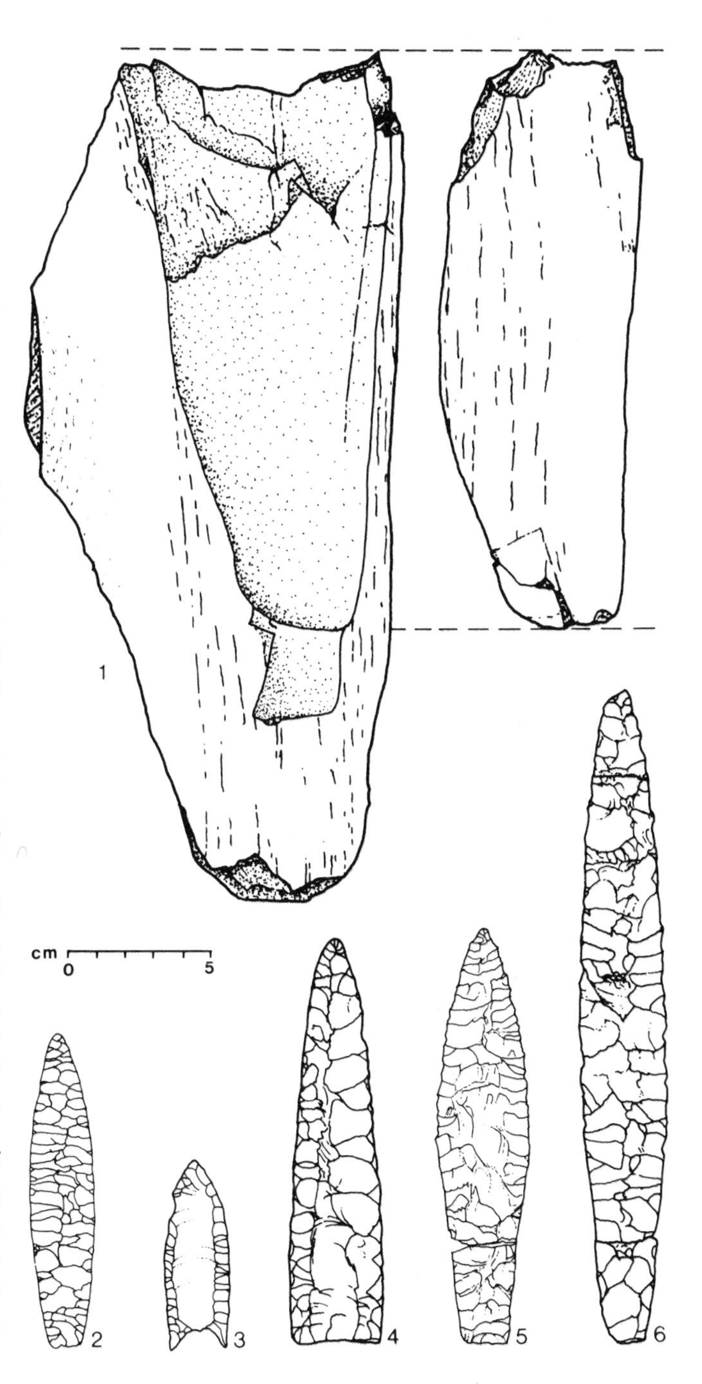

Figure 2. 1, Mammoth-bone core from Lange/Ferguson site, South Dakota (Hannus, 1983); 2, Agate Basin point from the Agate Basin site, Wyoming (Frison and Stanford, 1982); 3, Folsom point from Lindenmeier site, Colorado (Wilmsen and Roberts, 1978); 4, Clovis fluted point from the Anzick site, Montana; 5, Hell Gap point from Casper site, Wyoming (Frison, 1974); 6, Haskett point from Haskett site, Idaho (Butler, 1965).

Blade Pattern

The diverse physiography of Alaska presented a mosaic of habitats suitable for human use during the late Pleistocene (Fig. 1). While an overriding concern of Alaskan prehistory is the peopling of the Americas, the merging of well-established Asian and American patterns occurring there continues to confound attempts to develop a classification scheme acceptable to all concerned (Anderson, 1970; Dummond, 1980; West, 1981).

Macro-Blade. We examine the various manifestations in Alaskan prehistory in light of a series of patterns. An early macro-blade pattern is weakly represented in Alaska. It is known from Locality 1, level 2, of the Gallagher Flint Station (Dixon, 1975) and from Locality I, Anangula Island (Dummond, 1980). It may also be poorly expressed in the lower levels of Trail Creek (Larsen, 1968), in the Akmak level at Onion Portage (Anderson, 1968, 1970), and in the earliest level at Dry Creek (Powers and Hamilton, 1978). Favored techniques include the use of rotating blade cores for the removal of blades by the percussion method. A single date of 10.54 ka comes from the Gallagher Flint Station, and a series of dates from Anangula yields an average of 8.4 ka (Table 2).

Paleo-Arctic. Nelson (1935, 1937) first recognized parallels between American and Asian wedge-shaped microblade cores during the 1930s. Subsequent research demonstrates that microblade technologies are widely distributed in northern China (Gai Pei, 1985), Japan (Aikens and Higuchi, 1982), and Siberia (Michael, 1984; Yi and Clark, 1985). Detailed comparisons of artifacts and chronological relationships between America and Asia remain at an embryonic stage, however. The Paleo-Arctic pattern is one of the most important indicators of Asiatic-American ties. It appears at numerous localities, including the Blue Fish Caves, Healy Lake Village, Dry Creek II, Graveyard Point, Ground Hog Bay, Hidden Falls, Icy Straight, Kobuck, Putu, Tangle Lakes, and Trail Creek Caves (Fig. 1).

Characteristic artifacts of this pattern, recently summarized by Dixon (1985, Table 1), include: "Wedge-shaped microblade cores; blocky rotated blade and microblade cores; microblades and blades; elongate bifaces; scrapers; straight, concave, and convex base projective points; and spokeshaves." The oldest acceptable ^{14}C date in North America for this microlithic pattern is 10.69 ka from Dry Creek II (Table 2). The Paleo-Arctic pattern has a long continuum lasting until about 5 ka. It appears to be a response to megafaunal extinction and a technology primarily associated with the hunting of caribou.

FLUTED POINT PATTERN

Western Clovis

Fluted projectile points (Fig. 2) are widely distributed in North and South America (Fig. 3). Mammoth remains were first found with a Clovis triangular fluted point near Dent, Colorado (Figgins, 1931, 1933). Discoveries of fluted points and Rancholabrean fauna in the stratified Blackwater Draw site during the early 1930s (Hester, 1972) left no doubt about the early occurrence of Clovis. Many now regard the securely dated Clovis pattern as the first "clear-cut" evidence for human occupation in the Americas.

Clovis sites are identifiable on the basis of triangular blades with bifacial flaking, a concave base, one to several flutes on both faces of the base, and lateral and basal edge grinding. Associated artifacts include triangular bifaces with convex bases, triangular end scrapers, side scrapers, and bone foreshafts (Lahren and Bonnichsen, 1974) or bone points (Frison and Ziemens, 1980). The quality of Clovis craftsmanship is excellent. Great attention was given to the selection of colorful fine-grained siliceous materials for producing exquisite tools.

Through the efforts primarily of C. V. Haynes, western Clovis is the best-dated Paleo-Indian pattern in the Americas (Haynes and others, 1984). The 12 dated Clovis sites or occupation levels are distributed from the northern to the southern Plains and into the Southwest (Table 3). Clovis peoples are commonly depicted as mammoth-hunters. But in the Southern Plains, at Lubbock Lake and Lewisville, other game were utilized, including bison, horse, antelope, tapir, black bear, geocheolone, turtle, rodents, and reptiles (Johnson, 1974, 1977). Plant use is indicated by roasted hackberry seeds at the Lewisville site.

Three Clovis sites are known from caches of artifacts: the Anzick, Simon, and Drake (also known as the Piel site). The Anzick site (Lahren and Bonnichsen, 1974) occurs under a small collapsed rockshelter near Wilsall, Montana. Over 110 artifacts found in a burial cache covered with red ochre include fluted points, bifaces, bone foreshafts, utilized flakes, an end scraper, and the bones from two sub-adult humans dated at 10.6 ka (Table 3)—the oldest reliably dated human skeletal remains from the Americas (Taylor and others, 1985). The skilled artisans who made the flaked-stone burial offering used raw materials from several regional sources, including Phosphoria Formation chert from the Big Horn Mountains, porcellanite from the Powder River area, and chert from the flanks of the Rocky Mountains.

The other caches lack human skeletal remains. The Simon site (Butler, 1963; Butler and Fitzwater, 1965), located on an alluvial fan near Fairfield, Idaho, yielded handsome multicolored chert points (some with diagonal banding) and bifaces, including three green quartz crystal specimens. The Drake cache near Sterling, Colorado, included 13 carefully made fluted points from high-quality cherts from as far away as Texas.

Consistent production of high-quality tools entailed social, organizational, and behavioral commitments. Acquisition of chert for large bifacial tools must have required periodic visits to quarry areas. Craft specialization also must have developed as well as trade, for raw materials from the Simon and Drake sites are far removed from their source areas.

Clovis dates range between 12.65 ka for Lubbock Lake and 10.63 ka for Anzick (Table 3). The dated clam from Lubbock Lake is almost certainly too old, but the Anzick bone date errs on the young side. Most Clovis dates are between 11.8 and about 10.8 ka. Lewisville has a number of radiocarbon dates in excess of 35 ka, but samples collected during the original excavations were found to contain lignite.

With the extinction of large game at the end of the Pleistocene, the western Clovis pattern disappeared. Presumably the descendants of Clovis peoples developed alternate adaptive strategies in attempting to meet the challenge of new environmental conditions.

TABLE 2. BLADE RADIOCARBON DATES

Site	Material*	^{14}C age (yr B.P.)	Lab No.	Source
Onion Portage, Akmak Phase	cg	9857 ± 155	K-1583	Anderson, 1970, p. 2
Anangula	cl	8425 ± 715	SI-715	Dumond, 1980
	cl	8173 ± 87	P-1103	"
	cl	8129 ± 96	P-1104	"
	cl	7932 ± 497	P-1106	"
	cl	7895 ± 90	SI-1956	"
	cl	7796 ± 230	I-1046	"
	cl	7765 ± 95	SI-1955	"
	cl	7701 ± 93	P-1102	"
	cl	7660 ± 300	W-1180	"
	cl	7657 ± 95	Pn-1107	"
	cl	7287 ± 86	P-1108	"
Bluefish Cave	cg	15,500 ± 130	GSC-3053	Cinq-Mars, 1979
	cg	12,900 ± 100	GSC-2881	Morlan and Cinq-Mars, 1982
Dry Creek I	cl	11,120 ± 85	SI-2880	Thorson and Hamilton, 1977
	--	11,090 ± 170	GX-1341	Powers and Hamilton, 1978
Dry Creek II	cl	10,690 ± 250	SI-1561	"
	cl	9340 ± 95	SI-2329	"
Gallagher Flint Sta.	cl	10,540 ± 150	SI-974	Dixon, 1975
Graveyard Point	--	7895 ± 90	SI-1956	West, 1983
	--	7765 ± 95	SI-1955	"
Ground Hog Bay	cl	10,180 ± 800	WSU-412	Ackerman, 1973
Healy Lake Village		10,150 ± 250	SI-737	West, 1983
Hidden Falls	--	10,075 ± 75	SI-4359	Ackerman and others, 1979
	--	10,005 ± 75	SI-4352	West, 1983
	pt	9860 ± 75	SI-3776	"
	pt	9410 ± 70	SI-3778	"
Icy Strait	cl	9130 ± 130	I-6304	Ackerman, 1973
	cl	8230 ± 130	I-6395	"
Mount Hayes	--	9060 ± 925	UGA-941	West, 1983
Onion Portage, Kobuk Phase	--	8211 ± 84	P-985	Anderson, 1970
	--	8071 ± 84	P-984	"
Putu	cl	11,470 ± 500	SI-2382	Dumond, 1980
	sc	8450 ± 130	WSU-318	"
	sc	6090 ± 430	GAK-4939	"
Tangle Lakes	--	10,150 ± 280	UGA-572	West, 1983
	--	9060 ± 425	UGA-941	"
Trail Creek, Cave 2	--	15,750 ± 350	K-1210	Larsen, 1968
		13,070 ± 280	K-1327	"
		9070 ± 150	K-980	"
Ugashlik Narrows	cl	8995 ± 295	SI-2492	Dumond, 1980
XMH-297	--	8555 ± 380	GX-5998	West, 1983
	--	7190 ± 200	GX-6751	"

*Refer to Table 1 for key.

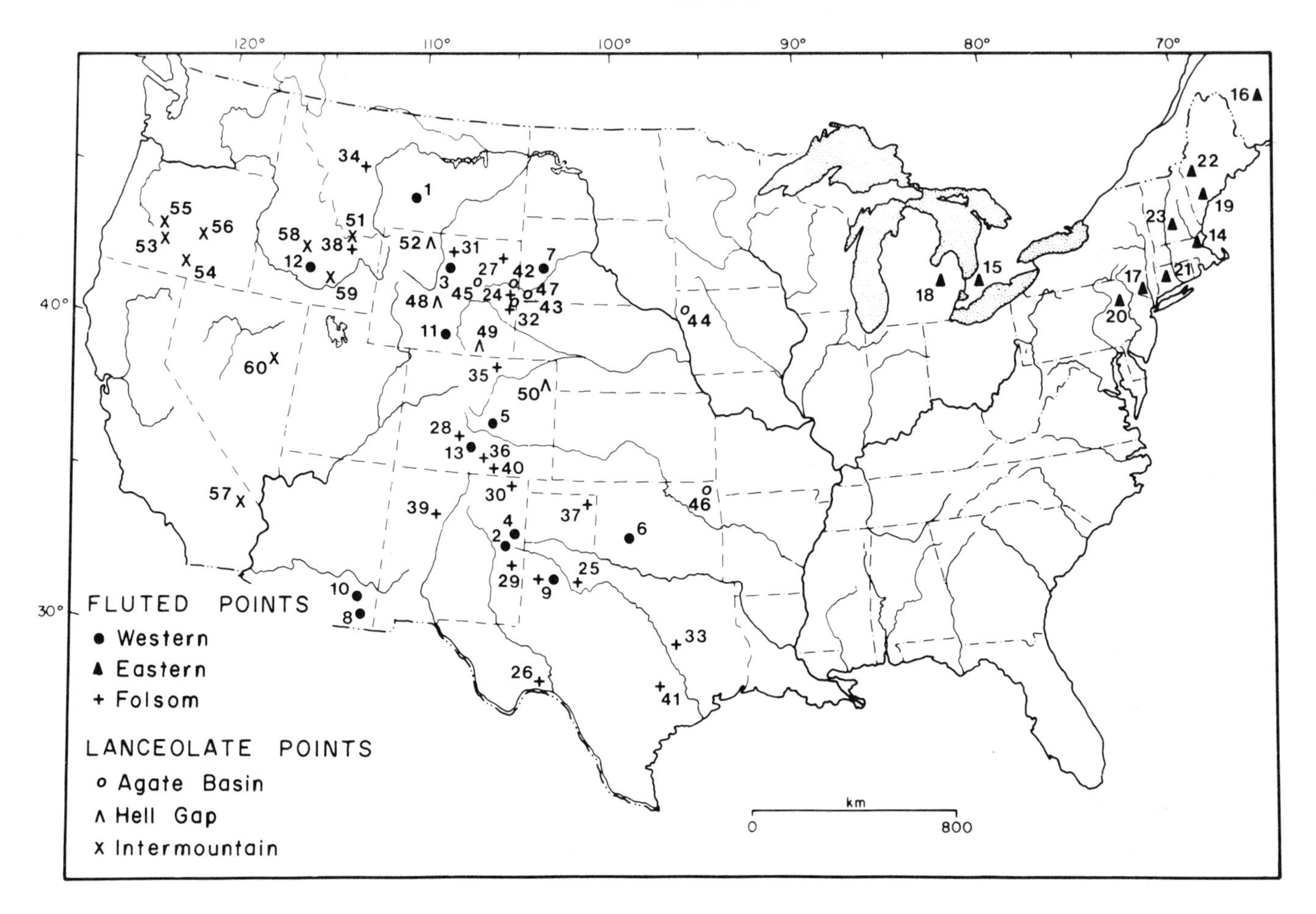

Figure 3. Localities of fluted and lanceolate points. 1, Anzick; 2, Blackwater Draw; 3, Colby; 4, Clovis; 5, Dent; 6, Domebo; 7, Lange/Ferguson; 8, Lehner; 9, Lubbock Lake; 10, Murray Springs; 11, Union Pacific Mammoth; 12, Simon; 13, Stoneham; 14, Bull Brook; 15, Crowfield; 16, Debert; 17, Duchess Cave Quarry; 18, Gainey; 19, Michaud; 20, Shawnee-Minisink; 21, Templeton; 22, Vail; 23, Whipple; 24, Agate Basin; 25, Adair-Steadman; 26, Bonfire; 27, Carter-Kerr-McGee; 28, Cattle Guard; 29, Elida; 30, Folsom; 31, Hanson; 32, Hell Gap; 33, Horn Shelter; 34, Indian Creek; 35, Lindenmeir; 36, Linger; 37, Lipscomb; 38, Owl Cave; 39, Rio Rancho; 40, Zapata; 41, 41BX52; 42, Agate Basin; 43, Brewster; 44, Cherokee Sewer; 45, Frazier; 46, Packard; 47, Ray Long; 48, Casper; 49, Seminoe Beach; 50, Jones-Miller; 51, Sister's Hill; 52, Bison and Veratic rockshelters; 53, Conley Caves; 54, Cougar Mountain Cave; 55, Dietz; 56, Fort Rock; 57, Lake Mohave; 58, Red Fish Rockshelter; 59, Haskett; 60, Smith Creek.

Eastern fluted points

Fluted points occur in every eastern state and adjacent Canadian province (*Archaeology of Eastern North America,* 1983, 1984, and 1985; Jackson, 1983; Meltzer, 1984; Chapdelaine, 1985). Large sites like Debert, Vail, and Bull Brook suggest communal caribou hunting (Bonnichsen and others, 1985). At smaller sites like Whipple, small groups may also have hunted caribou (Curran, 1984; Spiess and others, 1984). The Munsungun lithic source area implies that by Paleoindian times stone-tool materials were moved hundreds of miles. The undated feature at Crowfield, Ontario, which may be a cremation, yielded exquisitely made tools, implying the existence of a well-developed spiritual life.

Faunal remains are represented only by caribou (*Rangifer* sp.), beaver (*Castor canadensis*), and fish. Caribou is the most important and occurs at three sites: Whipple (Curran, 1984; Spiess and others, 1984), Duchess Cave Quarry No. 1 (Funk and others, 1970), and Holcombe (Cleland, 1965; Fitting and others, 1966). The lack of biological data from eastern fluted-point sites unfortunately hinders settlement and subsistence reconstructions.

The radiocarbon chronology of eastern fluted-point sites is ambiguous (Table 4). The oldest radiocarbon date on a caribou bone is from Duchess Cave Quarry No. 1. The bone may be in fortuitous association with stone tools. Carnivores often bring bones into cave sites. A more serious problem, however, is that dated charcoal may not be from human hearths but from forest fires, collecting in hearthlike depressions of tree throws and mix-

TABLE 3. WESTERN CLOVIS AND NORTHERN FLUTED POINT RADIOCARBON DATES

Site	Material*	^{14}C age (yr B.P.)	Lab No.	Source
Western Clovis				
Agate Basin,				
pre-Folsom level	cl	11,840 ± 130	I-10899	Haynes and others, 1984
	cl	11,700 ± 95	SI-3731	"
	cl	11,450 ± 110	SI-3734	"
(average of three)		11,650 ± 60		"
Anzick	cg	10,600 ± 300	---	Taylor and others, 1985
Blackwater Draw				
Location 1	cp	11,630 ± 350	A-491	Haynes and others, 1984
	cp	11,170 ± 110	A-481	"
	cp	11,040 ± 240	A-490	"
(average of three)		11,170 ± 100	---	"
Colby	cg	11,200 ± 200	RL-392	Frison, 1976; Frison and Todd, 1986
	ce	10,864	SMU-254	"
	cg	8719 ± 392	SMU-278	Frison and Todd, 1986
Dent Site	bo	11,200 ± 500	I-622	Haynes and others, 1984
Domebo Site	bo	11,220 ± 500	SI-172	"
	wd	11,045 ± 350	SM-695	"
(average of 2)		11,150 ± 400	---	"
Lange/Ferguson	oc	10,800 ± 530	I-13104	Hannus, 1987
	cg	10,670 ± 300	I-11710	"
Lehner	cl	11,470 ± 110	SMU-308	Haynes and others, 1984
	cl	11,170 ± 200	SMU-264	"
	cl	11,080 ± 230	SMU-196	"
	cl	11,080 ± 200	SMU-181	"
	cl	10,950 ± 110	SMU-194	"
	cl	10,950 ± 90	SMU-290	"
	cl	10,940 ± 100	A-378	"
	cl	10,860 ± 280	SMU-164	"
	cl	10,770 ± 140	SMU-168	"
	cl	10,710 ± 90	SMU-340	"
	cl	10,700 ± 150	SMU-297	"
	cl	10,620 ± 300	SMU-347	"
(average of 12)		10,930 ± 40	---	"
Lubock Lake	cm	12,650 ± 250	I-246	Holliday and others, 1983
Strat. 1	cm	12,150 ± 90	SMU-295	"
	wd	11,200 ± 100	SMU-548	"
	wd	11,100 ± 80	SMU-263	"
Murray Spring	cl	11,190 ± 180	SMU-18	Haynes and others, 1984
	cl	11,150 ± 450	A-80	"
	cl	11,080 ± 180	Tx-1413	"
	cl	10,930 ± 170	Tx-1462	"
	cl	10,890 ± 180	SMU-27	"
	cl	10,840 ± 70	SMU-41	"
	cl	10,840 ± 140	SMU-42	"
	cl	10,710 ± 160	Tx-1459	"
(average)		10,900 ± 50	---	"
U. P. Mammoth	oc	11,280 ±350	I-449	Haynes and others, 1984
Northern Fluted Points				
Charlie Lake	cg	10,460 ± 400	SFU-300	Fladmark and others 1984
	cg	10,300 ± 160	SFU-378	"
Sibbald Creek	cl	9570 ± 320	GX-8808	Gryba, 1983
	cl	7645 ± 260	GX-8810	"
	cl	5850 ± 190	GX-8809	"

*Refer to Table 1 for key.

TABLE 4. EASTERN FLUTED POINT RADIOCARBON DATES

Site	Material*	^{14}C age (yr B.P.)	Lab No.	Source
Bull Brook	cl	9300 ± 400	M-807	Byers, 1959
	cl	8940 ± 400	M-810	"
	cl	8720 ± 400	M-808	"
	cl	8560 ± 285	GX-6279	Grimes, 1979
	cl	7590 ± 255	GX-6278	"
	cl	6940 ± 800	M-809	Byers, 1959
	cl	5440 ± 160	GX-6277	Grimes, 1979
Debert				
(feature 17)	cl	11,011 ± 225	P-975	Stuckenrath, 1966
(feature 16)	cl	10,824 ± 119	P-974	"
(feature 11)	cl	10,758 ± 226	P-971	"
(feature 7)	cl	10,642 ± 134	P-739	"
(feature 15)	cl	10,637 ± 114	P-973	"
(feature 7)	cl	10,626 ± 244	P-967	"
(feature 7)	cl	10,557 ± 121	P-966	"
(feature 7)	cl	10,531 ± 126	P-741	"
(feature 11)	cl	10,503 ± 120	P-970	"
(feature 4)	cl	10,452 ± 128	P-743	"
(feature 11)	cl	10,452 ± 118	P-970A	"
(feature 12)	cl	10,496 ± 120	P-972	"
(feature 19)	cl	10,113 ± 275	P-977	"
(feature 3)	cl	7671 ± 92	P-740	"
(average of 13)		10,585 ± 47**	---	"
Duchess Cave				
Quarry No. 1	--	12,530 ± 370	I-4317	Funk and others, 1970
6LF-21/Templeton	cl	10,190 ± 300	W-3931	Moeller, 1980
Shawnee-Minisink	he	11,050 ± 1000	W-3391	McNett and others, 1979
	cl	10,750 ± 600	W-3134	"
	cl	10,590 ± 300	W-2994	"
	ce	9310 ± 1000	W-3388	"
(average of 4)		10,570 ± 250	---	"
Vail Site				
(feature 1)	cl	11,120 ± 180	Beta-183	Gramly, 1982
(feature 1)	cl	10,640 ± 400	AA-114	Haynes and others, 1984
(feature 2)	cl	10,550 ± 800	AA-115	"
(feature 2)	cl	10,500 ± 400	AA-117	"
(feature 1)	cl	10,300 ± 80	SI-4017	Gramly, 1982
(feature 1)	he	10,040 ± 400	AA-116	Haynes and others, 1984
Whipple Site	cl			
(feature 1)	cl	11,400 ± 360	AA-150c	Haynes and others, 1984
(feature 1)	cl	10,300 ± 500	AA-150A	"
(feature 1)	cl	9700 ± 700	AA-149b	"
(feature 1)	cl	9600 ± 500	AA-149a	"
(feature 1)	cl	9400 ± 500	AA-149a	"
(feature 1)	cl	8180 ± 360	GX-7496	Curran, 1984

*Refer to Table 1 for key.

**The average standard deviation of the 13 dates is 159, not 47.

ing with archaeological remains. At Bull Brook, for example, ^{14}C ages range from 9.57 to 5.44 ka (Table 4). At Whipple the two sets of dates, averaging 11.05 to 9.55 ka, suggest separate burning events. At Vail a concentration of charcoal (possibly in a tree throw) yielded dates of 11.12 and 10.3 ka, suggesting mixing of two populations of charcoal.

Debert, the best-dated eastern Paleoindian site, produced dates of 5.033 and 7.671 ka, as well as 13 dates ranging from 11.011 to 10.113 ka (Table 4) (MacDonald, 1964). Radiocarbon assays from this site traditionally have been averaged together (excluding the first two dates) as 10.585 ka. However, charcoal production at Debert may be coincident with the development of woodlands in southern Nova Scotia and may not be human. Until charcoal is found in clear-cut cultural features, all early dates from the Northeast must be regarded as suspect. Consequently the evidence is not clear whether eastern fluted points are as old as or post-date western Clovis.

Northern fluted points

Northern fluted points are restricted to Alberta, British Columbia, the Yukon Territory, and Alaska (Fig. 3). They have not been found at Siberian sites. Surface occurrences (Clark, 1984; Clark and Clark, 1983; Gryba, 1985) suggest that fluted points occur only beyond the 11 ka glacial limit of Denton and Hughes (1981).

The Sibbald Creek site in the southern Alterta foothills yielded one complete and parts of two other fluted points (Gryba, 1983). A date of 9.57 ka from the bottom occupation level was obtained from a charcoal scatter of uncertain origin (Table 3).

Charlie Lake Cave (Fladmark and others, unpublished) in rimrock overlooking the Peace River is just upstream from Fort St. John, British Columbia. Bison bone collagen dated at 10.46 and 10.36 ka (Table 3) was in association with an asymmetrical triangular point that exhibits multiple fluting scars on both faces. The faunal assemblage is dominated by snowshoe hare (*Lepus americanus*), ground squirrel (*Spermophilus columbianus*), and voles (*Microtus* sp.). Remains of fish and water birds suggest a lake and wetland. The earliest human occupation may have taken place prior to the full establishment of forest.

Fluted points from 26 localities in Alaska and the Yukon total between 50 and 55 specimens. Putu, on the north side of the Brooks Range, has a date of 11.47 ka (Table 2) on charcoal in association with fluted points and microcore and blade materials (Haynes, 1980). Soil samples, however, were later dated as 6.09 and 8.45 ka. Morlan and Cinq-Mars (1982), in attempting to reconcile these dates, suggest mixing of the relatively thin sequence from which the Putu complex was excavated.

Alaskan fluted points appear to be represented by several variants and may be associated with more than one archaeological complex of uncertain antiquity. Dates from Sibbald and Charlie Lake suggest that at least two of the fluted points in the north may date to the postextinction period.

FOLSOM

The 1924 discovery of fluted projectile points (Fig. 2) associated with an extinct form of bison (*Bison anqituus*) at Folsom, New Mexico, was the first acceptable evidence that humans were in the New World during the Pleistocene (Figgins, 1927, 1933). With the 1933 excavation of the stratified Blackwater Draw site, New Mexico, the Folsom artifacts were shown to be younger than Clovis (Antevs, 1935). Later the Folsom complex was dated to 10.78 ka (Table 5) at the Lindenmeier site (Haynes and Agogino, 1960).

The geographic distribution of Folsom sites is centered along the Rocky Mountains and the western Plains from Alberta and Saskatchewan (Forbis and Sperry, 1952; Wormington and Forbis, 1965) south to Bonfire Shelter (Dibble and Lorrain, 1967) and the Horn Shelter, southwest Texas (Forester, 1985; Redder, 1985). The eastern limit seems to be the Nebraska Sandhills and central South and North Dakota. Few Folsom points have been reported from Kansas. To the west, Folsom sites range to eastern Arizona, Utah, and Idaho.

The characteristic Folsom point is fluted on both sides, but unifacially fluted specimens are not uncommon. Unfluted points (i.e., Midland points) also occur at Folsom sites. In addition to projectile points, Folsom lithic assemblages are dominated by unifacial tools and utilized flakes, including knives, end and side scrapers, radial fracture burins, gravers, and a variety of perforators. Abraders and large chopper/anvil tools also occur. Bone tools include projectile points, fleshers, incised and notched bone disks, and eyed needles. Cherts in Folsom campsites come from as much as 350 km away from their original source.

Folsom campsites generally tend to be small. Larger sites like Lindenmeier and Adair-Steadman were probably reoccupied many times and perhaps inhabited for longer stretches of time. Circular habitation structures have been inferred from indirect evidence at the Hanson and Agate Basin sites. Circular depressions found at the Rio Rancho site were thought by Hibben to have been evidence for lodges (F. Hibben, personal communication, 1967). Two circular alignments of post holes in the Midland level at Hell Gap suggest two structures of 2 m and 4 m in diameter.

Faunal remains found at habitation sites include bison (*Bison* cf. *antiquus*), deer (*Odocoileus virginianus*), pronghorn (*Antilocapra americana*), two species of rabbit (*Lepus americanus* and *L. townsendii*), turtle (*Terrapena* cf. *ornata*), two species of wolf (*Canis*), prairie dog (*Cynomys ludovicianus*), peccary (*Platygonus compressus*), camel (*Camelops* sp.), mountain sheep (*Ovis canadensis*), marmot (*Marmota* sp.), and coot (*Fulica americana*).

Folsom bison-hunting techniques include cliff drives such as at Bonfire, arroyo knickpoint traps at Folsom and possibly Lipscomb, and especially simple ambush around springs and playa lakes such as at the Linger and Zapata sites. Compared to later Plano kill sites, Folsom kill sites have usually less than 10 animals. These small and under-utilized bison kills may reflect an opportunistic hunting strategy rather than the larger well-organized communal bison kills seen in the ensuing Paleoindian periods on the Plains.

The radiocarbon dates for Folsom sites are oldest in the north and youngest in the south (Fig. 3). The oldest dates are from Owl Cave in Idaho (Dort and Miller, 1977) and Indian Creek in Montana (Davis, 1984), and the youngest are from Bonfire Shelter, where three charcoal dates average 10.08 ka (Table 5). More dates are needed to confirm a north-to-south movement for the Folsom pattern.

LANCEOLATE POINT PATTERN

The lanceolate projectile point pattern has been termed the Plano Tradition (Sellards, 1952) and the Stemmed Point Tradition (Bryan, 1980). Lanceolate projectile points show differences in stem and base morphologies and include constricting stemmed (Agate Basin, 'Agate Basin–like' and intermountain lanceolate), shouldered points (Hell Gap, Alberta, and Cody), and those with concave bases (Plainview, Frederick, and Dalton, etc.). Figure 3 shows the distribution of sites with lanceolate points.

The lanceolate point pattern may have coexisted with Ice

TABLE 5. FOLSOM RADIOCARBON DATES

Site	Material*	^{14}C age (yr B.P.)	Lab No.	Source
Agate Basin, Folsom level	cl	10,780 ± 120	SI-3733	Haynes and others, 1984
	cl	10,655 ± 85	SI-3732	"
	cl	10,375 ± 700	I-472	"
(average of 3)		10,690 ± 70		
Bonfire Shelter	cl	10,230 ± 160	Tx-153	"
	cl	10,100 ± 300	Tx-658	"
	cl	9920 ± 150	Tx-657	"
(average of 3)		10,080 ± 100		
Carter/Kerr-McGee	--	10,400 ± 600	RL-917	Frison and Stanford, 1982
Blackwater Draw, Folsom	cp	10,490 ± 900	A-386	Haynes and others, 1984
	cp	10,490 ± 200	A-492	"
	cp	10,250 ± 320	A-380-379	"
	cp	10,170 ± 250	A-488	"
(average of 4)		10,380 ± 140		
Folsom	bo	10,260 ± 110	SMU-179	"
Hanson	cl	10,700 ± 67	RL-374	"
	cl	10,080 ± 330	RL-558	"
(average of 2)		10,200 ± 300		
Hell Gap, Folsom level	cl	10,840 ± 200	A-503	"
	cl	10,600 ± 500	A-504	"
	cl	10,200 ± 500	A-502	"
(average of three)		10,730 ± 170		
Indian Creek	--	11,125 ± 130	Beta-4951	Davis, 1984
	--	10,630 ± 280	Beta-13666	"
	--	10,980 ± 110	--	"
	--	10,160 ± 90	--	"
Lindenmeier	cl	10,780 ± 135	I-141	Haynes and others, 1984
	cl	11,200 ± 400	GX-1282	"
(average of 2)		10,820 ± 130		
Wasden (Owl Cave)	cg	12,850 ± 150	WSU-1281	Butler, 1971
	on	12,293 ± 435	ML-76437	Green, 1983
	cg	12,250 ± 200	WSU-1259	Butler, 1971
	on	12,078 ± 215	ML-D3	Green, unpublished
	on	12,008 ± 285	ML-76394	"
	on	11,983 ± 499	ML-76439	"
	on	11,733 ± 431	ML-76438	"
	on	11,657 ± 494	ML-76439b	"
	on	11,627 ± 267	ML-D2	"
	on	11,627 ± 235	ML-76465	"
	on	11,587 ± 422	ML-76439a	"
	on	11,242 ± 206	ML-76436	"
	on	11,206 ± 315	ML-D4	"
	cg	10,970 ± 150	WSU-1786	Plew and Pavesic, 1982
	cg	10,470 ± 100	WSU-2484	"
	cg	10,145 ± 170	WSU-2485	"
	cg	9735 ± 135	WSU-2483	"

*Refer to Table 1 for key.

Age mammals at Santa Isabel Istapan in the Valley of Mexico (Aveleyra, 1955, 1956; Aveleyra and Maldonado-Koerdell, 1952, 1953; Martinez del Rio, 1952), where points were found with the butchered remains of mammoth in a green clay of the upper Becerra Formation, which is thought to date between 11 and 16 ka. This find indicates contemporaneity of non-Clovis hunters and mammoths. Also, lanceolate points have been found with mastodon (*Mammut haplomastodon* sp.) and other megafaunal remains in a spring at the Tamia-tamia site in northern Venezuela (Gruhn and Bryan, 1984).

Goshen complex

More recently, Frison has tentatively reported the Milliron site, a single-component site located atop a heavily eroded butte in southeastern Montana, probably a small bison kill with nonfluted, concave-based points that are neither Clovis or Folsom (Dolzani, 1986). These points may fit into the Goshen complex initially identified at the Hell Gap site. The first of a series of radiocarbon samples yielded an age of 11.34 ka (Table 6). Work now in progress will further define this important discovery.

Nenana complex

R. W. Powers (personal communication, 1986) tentatively proposes a new taxonomic group of Alaskan sites in the Nenana Valley of the North Alaskan Range (see Table 2). Key locations are Walker Road (11.82 ka), Moose Creek Bluff (11.73 to 8.16 ka), Owl Ridge (11.34 ka), and Dry Creek (11.12 ka). Excavation of Nenana complex living floors (older than 11 ka) demonstrate the presence of a bifacial industry lacking microblades. This complex is characterized by unlined shallow hearths, red ochre, numerical dominance of bifacial tools, triangular bifaces with excurvate edges and convex bases, ovate bifaces, high-angle keeled scrapers, low-angle unkeeled scrapers, low-angle blades, and low-angle end-struck scrapers.

Emphasizing the central placement of the Nenana complex between Siberia and the United States, Powers regards it as pivotal for understanding the development of the fluted pattern (as defined here). Artifactual remains from the lowest level of Diuktai are seen as antecedent to the Nenana complex. The Nenana complex shares certain tool types with southern Clovis sites, and because it may slightly predate the Clovis complex of the south the Nenana complex is hypothesized as its progenitor. A brief examination of Table 3 reveals that the Nenana complex is contemporaneous with the earliest dated western Clovis sites.

Agate Basin

Agate Basin is best known from its type site in eastern Wyoming (Frison and Stanford, 1982). An Agate Basin occupation was found at Hell Gap (Irwin-Williams and others, 1973), Carter–Kerr McGee (Frison, 1978), Frazier (Cassells, 1983), on the Southern Plains at Blackwater Draw (Hester, 1972), and a small surface site in the Estancia Basin of New Mexico (Haynes, 1955). Although many Agate Basin–like points (Fig. 2) are found scattered throughout North America, specimens most closely matching those of the type specimens are rare and have a distribution limited to the Rocky Mountains and the western half of the Plains.

Agate Basin is recognized by long, slender, and finely pressure-flaked lanceolate projectile points. Ridge projections along the edges were removed with tiny pressure flakes. Edge grinding extends from the base to as much as half the length of the blade. Other Agate Basin tools resemble Folsom, such as gravers and end side scrapers made on flakes, as well as a variety of flake tools.

Few bone tools have been reported from Agate Basin sites, but an incised bone disk from Agate Basin itself resembles those from Lindenmeier (Frison and Stanford, 1982; Wilmsen and Roberts, 1978). A delicately made eyed needle of bone similar to those of Folsom was also found at the Hell Gap site (Irwin-Williams and others, 1973). The only evidence for Agate Basin house structures is suggested by three superimposed circles of post holes averaging slightly larger than 2 m in diameter at the Hell Gap site (Irwin-Williams and others, 1973).

Bison was the major mammalian species exploited by Agate Basin hunters. The type site contained a bison kill site and a processing station for over 100 animals killed during two different winter events. Concentrations of bones are thought to be the remains of butchered and frozen sections of bison carcasses. Tools scattered around the piles suggest a system of storage for processed frozen meat. Storage would imply a seasonally permanent camp (Frison and Stanford, 1982). Forty-three bison were also processed at the Frazier Agate Basin site in Colorado (Cassels, 1983).

Agate Basin hunters had large communal hunts, making it possible to kill many animals during a single hunt. The Folsom hunters, on the other hand, were organized in smaller groups, apparently to take advantage of opportune situations for killing a few animals at a time.

Agate Basin sites on the Plains date between 10.5 and 9.5 ka (Table 6). Other lanceolate projectile points termed "Agate Basin–like" date to later time periods. A radiocarbon date of 9.416 ka at the Packard site in northeastern Oklahoma suggests only a small interval between Agate Basin and similar forms on the Central Plains (Wyckoff, 1985). Despite similarities with Agate Basin projectile points, the Packard tool assemblage has many differences, especially the occurrence of notched points.

At the Cherokee Sewer site in western Iowa, "Agate Basin–like" points came from levels yielding dates from 10.05 to 8 ka (Table 6). Correlation with Campbell Beach occupation of Lake Agassiz, Manitoba (Pettipas, 1985), and of the barrenlands of the Northwest Territories (Gordon, 1976; Wright, 1976) suggest that an "Agate Basin–like" lanceolate projectile point pattern expanded to the east and northeast by 8 ka. Clearly the pattern was modified into a tool kit more adapted to the regions where it was utilized. An increase of woodworking tools and notched projectile points implies that the system was altered to accommodate activities in a forested environment, while even farther north the system was further modified for communal hunting of caribou.

Hell Gap

Hell Gap projectile points of southeastern Wyoming (Agogino, 1961) have broad blades with shoulders and constricted bases (Fig. 2). Usually they have ground lateral edges and flat to convex bases, but unground bases on rejuvenated specimens are also known. During resharpening the shoulders were often removed, causing the point to resemble Agate Basin projectiles. Hell Gap probably developed out of Agate Basin. Hell Gap sites date between 10.2 and 9.6 ka (Table 6).

As with Agate Basin, the distribution of Hell Gap sites is

TABLE 6. LANCEOLATE RADIOCARBON DATES

Site	Material*	^{14}C age (yr B.P.)	Lab No.	Source
		GOSHEN COMPLEX		
Milliron	--	11,340 ± 120	--	Dolzani, 1986
		NENANA COMPLEX		
Walker Road	--	11,820 ± 200	Beta-11254	Powers, personal communicatio
Moose Bluff	--	11,730 ± 250	GX-6281	"
	--	10,640 ± 280	I-11,277	"
	--	8940 ± 270	A-2144	"
	--	8160 ± 260	A-2168	"
Owl Ridge	--	11,340 ± 150	---	"
Dry Creek	--	11,120 ± 85	SI-2880	"
		AGATE BASIN		
Agate Basin	cl	10,430 ± 470	RL-557	Frison and Stanford, 1982
Brewster	cl	9990 ± 225	M-113	"
Boss Hill II	be	7875 ± 130	S-1251	Bryan, 1980
	cl	7750 ± 105	S-1371	"
Cherokee Sewer	--	10,050 ± 460	I-7783	Anderson and Semken, 1980
	--	8570 ± 200	UCLA-1877f	"
	--	8445 ± 250	UCR-490	"
	--	8000 ± 270	UCR-604	"
	--	6725 ± 115	I-10096	"
Packard	--	9416 ± 193	NI-478	Wyckoff, 1985
Ray Long	--	9380 ± 500	M-370	Bryan, 1980
		HELL GAP		
Agate Basin	cl	10,445 ± 110	SI-4430	Frison and Stanford, 1982
Casper	cl	9830 ± 350	RL-125	Frison, 1974
	cg	10,060 ± 170	RL-208	"
Hell Gap	cl	10,240 ± 300	A-500	Irwin, 1967
Jones-Miller	cl	10,020 ± 320	SI-1989	Stanford, 1984
Sister's Hill	--	9650 ± 250	I-221	Agogino and Galloway, 1965
	cl	9600 ±	A-372	Irwin, 1967
		INTERMOUNTAIN LANCEOLATE		
Connley Caves	--	11,200 ± 200	GAK-2141	Bedwell, 1973
	--	10,600 ± 190	GAK-2143	"
	--	10,100 ± 400	GAK-1742	"
	--	9800 ± 250	GAK-1743	"
Cal-S342	--	11,720 ± 145	---	Dolzani, 1987
Fort Rock Cave	cl	13,200 ± 720	GAK-1738	Carlson, 1983
	cl	10,200 ± 230	GAK-2147	"
Red Fish Rockshelter	cl	10,100 ± 300	WSU-1396	Bryan, 1980
Haskett Points	cl	9860 ± 300	WSU-1395	"
Cougar Mt. Cave #2	cl	11,950 ± 950	GAK-1751	Bedwell, 1973
Smith Creek	pn	12,600 ± 170	A-1565	Bryan, 1979, 1980
	wd	12,150 ± 120	BIRM-752	Thompson, 1985
	cl	11,680 ± 160	Tx-1421	Bryan, 1980
	cl	11,140 ± 120	Tx-1637	"
	cl	10,740 ± 130	BIRM-702	"
	wd	10,700 ± 180	BIRM-917	"
	cl	10,660 ± 220	GAK-5442	"
	cl	10,630 ± 190	GAK-5443	"
	cl	10,570 ± 160	GAK-5445	"
	cl	10,460 ± 260	GAK-5444b	"
	cl	10,330 ± 190	Tx-1638	"
	cl	9800 ± 190	GAK-5444	"
	cl	9280 ± 160	GAK-5446	"
		DALTON		
Rogers Shelter	cl	10,530 ± 650	ISGS-48	Coleman, 1972
		10,200 ± 330	M-2333	Ahler, 1976

*Refer to Table 1 for key.

confined to the Rocky Mountains and the western Plains of the United States and Canada (Wormington and Forbis, 1965). Although relatively uncommon, they are also found on the Southern Plains (Fig. 3). In Wyoming, Hell Gap habitation sites include the type site (Irwin-Williams and others, 1973) and the Agate Basin site (Frison and Stanford, 1982). A mixed Hell Gap–Agate Basin level was found at the Carter–Kerr McGee site (Frison, 1978), the Sister's Hill site (Agogino and Galloway, 1965), and the Seminoe Beach site (Miller, 1986). Kill sites include the Casper site of central Wyoming (Frison, 1974) and the Jones-Miller site in northeastern Colorado (Stanford, 1978, 1984).

At the Hell Gap site, Hell Gap points occur stratigraphically between the Folsom and Alberta levels. The site is situated in the Hartville uplift, a region with sources of high-quality quartzite and chert for flaked stone tools. The Hell Gap occupation appears to represent brief encampments for manufacture of lithic stone tools (Irwin-Williams and others, 1973). The Seminoe Beach site had a similar function.

The Sister's Hill site in north-central Wyoming appears to be a campsite (Agogino and Galloway, 1965). Faunal remains consist of ground squirrel (*Scuridae*), rabbit (*Lepus/Sylvilagus*), porcupine (*Erezthizon*), mule deer (*Odeocoleus*), and antelope (*Antilocapra americana*). Bison bone was also found (G. Frison, personal communication, 1985). Artifacts indicate projectile manufacture, hide working, and hunting.

The Hell Gap occupation of the Agate Basin site is found in three levels separated by thin soil lenses. Frison and Stanford (1982) believe that these levels represent several days rather than repeated occupation. Because much of the Agate Basin site was destroyed through erosion and vandalism, it is difficult to infer a site function. Bison processing appears to represent a major activity.

Hell Gap hunting strategies included large organized communal hunts as in the previous Agate Basin period, with more animals taken during a single event, reflecting improved hunting strategies, an increase in band sizes, or larger bison herds.

Hell Gap hunters used both constructed traps and natural topographic landforms. At Casper the animals were driven into a parabolic dune, while at Jones-Miller some sort of brush or snow impoundment was constructed. The Jones-Miller trap was used at least twice, once in the late fall and again in winter. About 150 animals were killed during each event. At Casper, approximately 100 animals were killed. The animals were totally butchered at the Jones-Miller site, while at Casper, major portions of the animals were removed for processing elsewhere.

Data from the Jones-Miller site suggest a winter impoundment probably constructed near an area where several Hell Gap bands shared a winter encampment. Excavation exposed a post mold and a number of associated artifacts, including a whistle, butchered bones of domestic dog, and numerous bison bones. Of the hundreds of bison butchered, fewer than a dozen skulls were found. Many tiny ear bones were recovered from the screenwash concentrate. The lack of skulls but the presence of ear bones suggest that the skulls were intentionally removed from the site for ceremonial purposes.

Hell Gap is followed on the Plains by the Alberta Complex (Wormington and Forbis, 1965). Nearly identical biface reduction strategies shared by these two complexes imply a common technological knowledge. Like Hell Gap hunters, Alberta hunters continued the trend of large organized communal bison hunts. One of the largest is the Hudson-Meng site, Nebraska, where Agenbroad (1978) estimates that approximately 600 animals were killed during several events over a short time.

Intermountain lanceolate

The term intermountain lanceolate refers to a series of very similar lanceolate points in western North America (Bryan, 1980; Carlson, 1983). These points have a leaf-shaped form with lateral edges tha expand to near the tip. Bases vary from slightly concave to convex. Slight shouldering occasionally occurs, and some broken points are shaped as burins. These lanceolate points were first found on strandlines of Pleistocene Lake Mohave (Bryan, 1980). Many more have been found on the surface along the margins of pluvial lakes throughout the Great Basin. They are also known from surface collections from the Snake River Plains of southern Idaho and the intermountain area of southwestern Montana (Fig. 3).

Points belonging to the Intermountain pattern are known by a variety of regional names, including Mohave (Campbell and Campbell, 1937), Birch Creek (Swanson and others, 1964), Haskett (Butler, 1965), Cougar Mountain (Bryan, 1980), and Moriah (Bryan, 1979). Stylistic variation noted among these points may in part be attributed to resharpening. At the Haskett site near American Falls, Idaho, one of the specimens is greater than 20 cm long (Fig. 2). Given the long, narrow form of this style, breakage would have been common, and resharpening of broken points would have led to variation in the shape of the blade and haft elements.

Only recently have these lanceolate points been placed in stratigraphic context. Fort Rock Cave produced a single resharpened convex specimen and a small concave-based specimen from an occupation layer resting on pluvial lake gravels along the margins of Lake Fort Rock (Bedwell, 1973). Charcoal associated with the occupation level yielded a date of 13.2 ka (Table 6), but the relationship between the charcoal and artifacts is not clear. At Connely caves, several Haskett points were located, but it is now impossible to determine which of several dates belong in association. More definitive is a cache of artifacts from Redfish Rockshelter in the Sawtooth Range of central Idaho. A hearth dating to 9.86 ka was associated with the cache level; the next lower level produced a date of 10.1 ka and the midsection of a Haskett point. While numerous Intermountain lanceolate points are known from the west, exactly what kind of animals these hunters preyed upon is still an open question. Butler (1965) reports only enamel fragments and calcined bone at the Haskett site. To the north in the Birch Creek valley, Swanson (1972) documents the occurrence of Birch Creek points at the deeply stratified Veratic and Bison rockshelters. Bison bone collagen from the lower levels of the Birch Creek phase dated 10.34 ka (Table 6). Intermountain lanceolates are not known to be in association with extinct fauna.

Excellent preservation at Smith Creek Cave, east-central Nevada, allowed Bryan (1979) to excavate a Mohave living surface with preserved organic materials and hearths. He argues that the "Stemmed Point Tradition" originated in the intermountain West prior to 12 ka (Table 6). The strongest evidence for this tradition is from Smith Creek, where a local variant of Mohave or Haskett points is associated in a compacted gray silt and ash overlying a unit of dung, silt, and rubble. A series of five hearths yielded dates ranging from 11.14 to 9.28 ka (Table 6). In addition to the charcoal recovered from the hearths, Bryan accepts a sample of scattered charcoal with a date of 11.68 ka. Thompson (1985) questioned Bryan's interpretation that the Mount Moriah occupation spans the period from 11.68 to 10.33 ka (Bryan rejects younger dates on samples that were not pretreated). Thompson argues for an occupation covering no more than 200 yr beginning around 10.65 ka. He points out that plant and animal macrofossils in association with Mount Moriah remains resemble Holocene rather than Pleistocene vegetation and that site occupants may have burned old wood, producing old dates. We agree that the 11.68 ka date based on scattered charcoal is not acceptable. The charcoal may be from a natural fire. Nonetheless, charcoal recovered from 12 hearths is an unusual archaeological record with excellent context. Thus the absence of Pleistocene fauna and the presence of mountain sheep remains suggest that the Mount Moriah occupation is a Holocene adaptive pattern no older than 11 to 11.14 ka.

The age of the Intermountain lanceolate point pattern is not as secure as would be desirable. McGee's (1889) discovery at Walker Lake, Nevada, of mammoth and ox (bison?) associated with a crudely flaked leaf-shaped point suggests that Intermountain points were used prior to extinction and may represent a cotradition with Clovis. Layton's (1972a, 1972b) obsidian-hydration study of Intermountain lanceolate points suggests that this pattern first appears in the West during the late Pleistocene and develops into a series of local traditions with the onset of the Holocene. At the Dietz site in Christmas Valley, Oregon, Fagan (1986) reports fluted points in the basal horizon with mammoth bone stratigraphically below and Intermountain lanceolate points above.

Dalton

The Dalton complex is widespread in midwestern and southeastern North America. Lanceolate points are characterized by a concave base and basal thinning; the lateral edges of point blades are often resharpened by serration technique, and many bifaces are altered into knives and drills. Because of their shape and technology, Dalton points are often seen as logical descendants from a southeastern fluted-point complex.

Although Dalton is widespread, remains from this complex are difficult to interpret. Goodyear (1982) notes that mixing is common, particularly at cave and rockshelter sites. Dalton points are frequently found in association with side-notched and corner-notched points, artifacts commonly attributed to Archaic peoples. Central to Goodyear's thesis is the Brand and Sloan sites, the Hawkins cache of Arkansas, and the Leopold site of southeast Missouri, which have produced unmixed Dalton assemblages. But the most important site is Rogers Shelter, Missouri. The terrace in front of the rockshelter yielded Dalton artifacts and hearths. Charcoal from the hearths dated at 10.53 and 10.2 ka (Table 6). These dates, in conjunction with dates from mixed sites, suggest an age for Dalton between 10.5 and 9.9 ka.

Dalton probably reflects hunting and gathering bands adapting to rapidly changing environmental circumstances underlying the transition from the Paleoindian to Archaic period. The Dalton transition appears to be weakly correlated with the loss of boreal species (jack pine and spruce) and the establishment of mesic forest dominated by oak, hickory, beech, and ironwood. Faunal exploitation is best known from Rogers Shelter. Dalton levels contained fish, aquatic and terrestrial turtle, rabbit, squirrel, raccoon, beaver/muskrat, deer, elk, bison, and turkey.

DISCUSSION

How was the last major climatic cycle linked to the environment and to human adaptive systems? The Milankovitch hypothesis proposes that variation in the earth's orbit is a significant factor that underlies cyclic variation in the earth's climates. Because climate, environment, and human adaptive systems are interdependent, climatic change can serve as a catalyst to force synchronous or delayed responses in linked biotic and human adaptive systems. Cyclic changes in the shape of Earth's orbit (eccentricity, obliquity of rotational axis, and precession of the equinoxes) are primary parameters responsible for cyclic climatic and seasonal variation (Kutzbach, this volume).

Vernekar (1968) calculated the amount of insolation arriving at the earth's upper atmosphere. We have transformed his tabular data into graphs for 65° and 45° north latitude to show the seasonality index, which is the average difference between summer and winter insolation divided by the standard deviation (Fig. 4). A positive value of the index corresponds to a high degree of seasonality.

The greatest values for seasonality since the Sangamon interglacial (about 80 ka) occurred about 11 ka and was greater at 65° than at 45°N. lat. Relative to the mean (solid line in center of graphs), summers were warmer and winters were colder than at present by as much as 5°C (Kutzbach, 1983).

The continental ice sheets buffered the climatic effects of radiation changes, causing a lag in ice-sheet retreat rates. Denton and Hughes' (1981) isochron retreat map for the Laurentide ice-sheet margin provides the data to reconstruct two transects covering the time interval of 22 to 8 ka. These were plotted, starting in North Dakota and New York and ending in Hudson Bay. Both

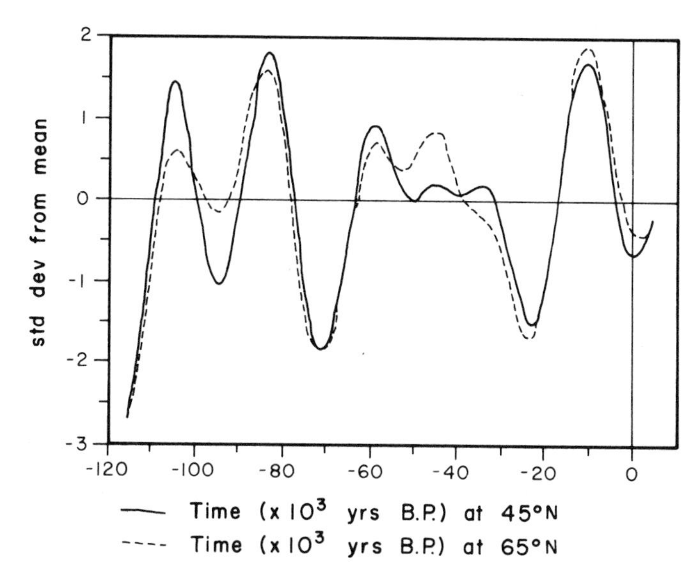

Figure 4. Seasonal variation of insolation during the last 120 ka at 65° and 45° N. Lat. (cf., Vernekar, 1968).

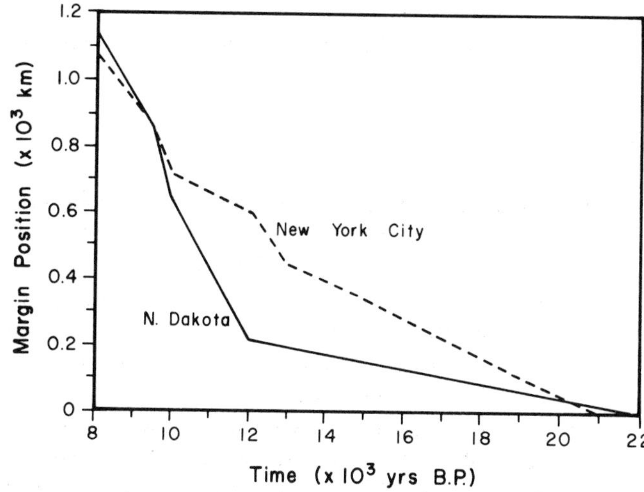

Figure 5. Ice-recession transects beginning in New York and North Dakota and terminating at Hudson Bay (cf., Denton and Hughes, 1981).

transects exhibit an increased recession rate at about 12 ka (Fig. 5). For the North Dakota transect the ice margin retreated approximately 450 km between 10 and 12 ka.

The jet stream was displaced to the south during the glacial maximum (Barry, 1983; Kutzbach, this volume). As the ice margin retreated and the ice height was reduced, the jet stream followed the ice margin northward. Ice-margin melting led to rising sea levels. These in turn led to ice streaming and rapid decrease in height and volume of continental glaciers (Denton and Hughes, 1983; Hughes, this volume).

Changing atmospheric circulation patterns could impact biotic environments of the terrestrial landscape in northern North America. Palynology offers insight into the effect of these changes on the vegetation record. Jacobson and others (this volume) measured the rates of vegetation change from pollen diagrams. By forming an east-west transect for sites located south of the Laurentide ice sheet and covering the period from 18 ka to the present, Jacobson and co-workers note that a substantial rate of change occurred from 12 to 10 ka. Rapid change in vegetation is likely related to the accelerating demise of the Laurentide ice sheet and to the climatic trends that caused it.

Most remarkable yet is what happened to the big-game animals. Martin (1984) notes that 33 genera of mostly large mammals met extinction near the end of the Pleistocene. In an analysis of radiocarbon-dated remains from 163 fossil sites (excluding Alaska), Mead and Meltzer (1985) conclude that the extinction process was complete by 10 and possible by 10.8 ka for well-dated genera. These include camel (*Camelops* sp.), horse (*Equus* sp.), mastodon (*Mammut* sp.), mammoth (*Mammuthus* sp.), ground sloth (*Nothrotheriops* sp.), and a large cat (*Panthera leo atrox*). Other taxa are not reliably dated in sufficient numbers for strong conclusions.

Several authors (see Martin and Klein, 1984; Graham and Mead, this volume) propose that climate change led to restructuring of Pleistocene vegetation communities. During the last glaciation and perhaps earlier, plant communities formed mosaic patterns, allowing many taxa of megafauna to coexist. At the end of the Pleistocene, these mosaic communities were replaced by more homogeneous communities. The individualistic response of each taxon led to a general reshuffling of plant and animal associations and the creation of new communities. Many of the large mammals were unable to adapt to the new vegetation associations. Some plants contain toxic chemicals used as antiherbivore defenses and these may have poisoned the animals subject to new dietary regimes (Guthrie, 1984).

A significant threshold occurs in the North American archaeological record in terminal Pleistocene times. It is a period when certain human adaptive systems died out, originated, and/or were markedly restructured to take advantage of new circumstances. The shaded vertical bar in Figure 6 highlights the extinction and origination period dating between 11.1 and 10.8 ka. The duration of the threshold period may have been even more abrupt, but resolution is limited by the accuracy of the radiocarbon dating method.

We propose that the threshold in human adaptive systems was triggered by changes in insolation, resulting in a chain of environmental changes. The most important include the demise of continental glaciers, new atmospheric circulation patterns, changing ocean levels, restructuring of biotic systems, and the extinction of 33 genera of megafauna. Human adaptive systems lie at the end of this chain of dependent relationships. Support of this model lies in the synchroneity of rapid rates of changes in linked systems.

CONCLUSIONS

A pronounced environmental threshold event between 11.2 and 10.8 ka is important for explaining the extinction and origination of human adaptive patterns and has a bearing on understanding the peopling of the Americas. The most important junctures in the correlated archaeological record include: (1) humanly flaked large mammal bones occurring in the Beringia Refugium and south of the continental ice sheet spanning the late Pleistocene to 10.8 ka, when megafaunal extinction was nearly complete; (2) western Clovis, appearing about 11.6 ka and disappearing about 10.8 ka; (3) Intermountain lanceolate points, still inadequately dated, but apparently beginning about 11.7 ka and continuing into early Holocene times; and (4) Paleo-Arctic microblades, macroblades, Dalton, Agate Basin, and Hell Gap, all apparently emerging immediately after megafauna extinction at about 10.6, although Folsom may be slightly earlier, about 10.9 ka. Eastern and northern fluted points are poorly dated and cannot be placed in an absolute chronological framework at this time.

An environmental-response model most adequately explains the synchronous origination and extinction of these human adaptive patterns. Simply stated, many different human groups responded to the same global environmental event by generating new strategies to ensure their welfare in the context of specific regional and local environments. Because the context of adaptation differed significantly from area to area, the details of specific responses varied.

Explanation of the patterns described above calls for positing several hypotheses concerning the peopling of the Americas. First, we propose that human populations were in North America, including the continental United States, prior to the end of the last glacial maximum. This is known from mid to late Pleistocene sites with humanly altered mammal bones; these include Blue Fish Cave, Old Crow, Rye Patch, Dutton, Selby, and several more. The Manis Mastodon site on the Olympic Peninsula and the complex of mastodon sites in southern Michigan, both in boreal environments, are dominated by bone implements that appear to be contemporaneous with but not part of the Clovis pattern. Elsewhere, at the Owl Cave and Lange/Ferguson sites, bone flaking is associated with the use of fluted points, implying that western Clovis is an addition to an earlier bone-flaking pattern. Exactly where the flaked bone pattern developed and when it spread across North America is poorly understood. Flaked megamammal bone occurs widely on the steppes of northern China and in the USSR (Soffer, 1985; Wu and Olson, 1985), suggesting possible ties with the North American archaeological record.

Secondly, we propose that the development of American projectile point patterns represents an intensification of big-game

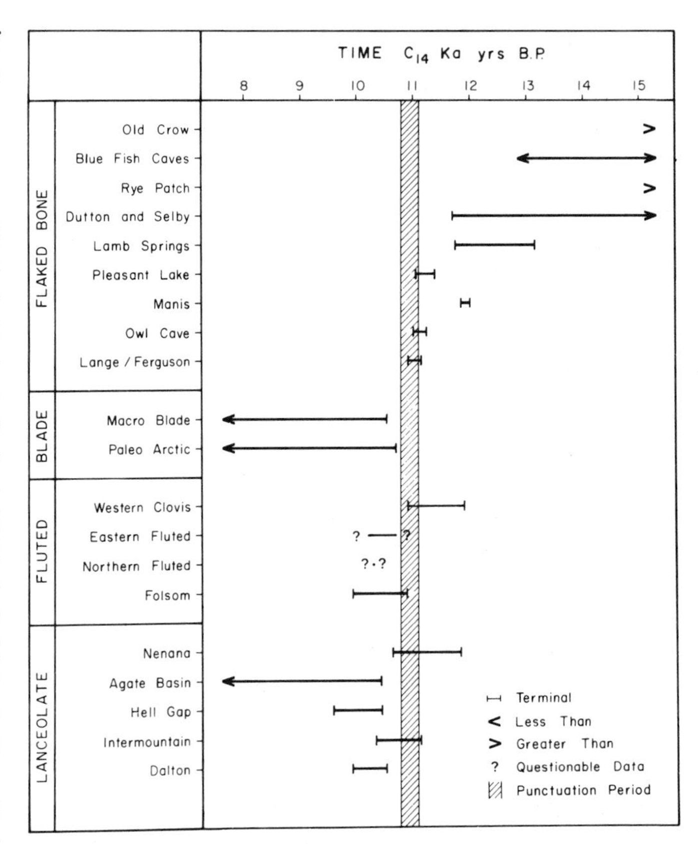

Figure 6. Temporal occurrences of North American human adaptative patterns that suggest a punctuation point between approximately 11.1 and 10.8 ka.

hunting and is a response to rapidly changing environments and declining megafaunal populations. A series of uniquely American projectile point patterns, such as Clovis, Nenana, Intermountain, Mill Iron, and El Jobo, all appear just prior to extinction. This preextinction phenomenon suggests that humans from geographically separate and discrete environments responded to changing conditions with similar adaptive strategies.

Communal hunting of big-game animals, bifacial technologies, lanceolate projectile points, and end scrapers all occur in Asia earlier than in North America. The late Pleistocene occurrence of these artifacts in North America and the cognitive repertoires that they represent suggest at least limited contact with Asia, if not an actual migration from there. Due to the earlier demise of the Eurasian ice sheet, extinctions are likely to have occurred slightly earlier in Asia than in North America, providing an impetus for migration. If human populations from Asia moved into North America during late Pleistocene times, common artifacts from the two continents are notably rare, making an assessment of common ties difficult.

Stimulus diffusion of the big-game hunting complex across indigenous American populations coupled with rapid population growth is a viable alternative to the often cited Clovis migration hypothesis. The stimulus diffusion hypothesis calls for widespread but small groups, such as nuclear or extended families, who subsisted by hunting and foraging for a diversity of taxa available in their local/regional environments. With the introduction and rapid spread of the big-game hunting complex, local groups adopted those parts of the pattern consistent with their own practices.

Thirdly, extinction of 70 percent of big-game genera, coupled with rapid environmental change, resulted in a cultural crisis at approximately 10.8 ka. Humans adjusted to the challenges of their local context by accepting ideas transmitted by diffusion and by generating new adaptive strategies. The postextinction period saw the nearly synchronous appearance of the Paleo-Arctic pattern in the western Arctic, Dalton in the southern and eastern United States, and Agate Basin, Hell Gap, and Folsom in the Great Plains. Many of the basic ingredients of these lithic patterns remained similar to preextinction patterns with a continued emphasis on the use of stone tools, quarry, and craft specialization for the creation of new projectile point styles. Hunting equipment was scaled down to take advantage of local fauna that survived extinction. Species that were preadapted to the new environmental conditions flourished shortly after megafaunal extinction and became the focus of new procurement systems. The economic focus shifted to bison in the Plains, mountain sheep in the intermountain West, and caribou in the far North and Northeast. In the South, Southwest, and parts of the Great Basin a broad spectrum of small game animals were hunted, and food gathering was given a new emphasis.

Our Paleoindian case study suggests a striking correlation between accelerated evolution of human adaptive systems and rapidly changing natural environments. Cyclic environmental change, driven at least in part by Milankovitch cycles, is seen as a catalyst. Environmental change occurs; linked adaptive systems must respond. Adjustments can take many forms, depending on the local circumstances. They may include dispersion by migration, acceptance of diffused ideas, technological innovation, and/or new or restructured settlement, subsistence, and lithic procurement systems. Thus, even if environment served as a forcing function, humans have always had some freedom of choice in how they respond to changing contexts.

REFERENCES CITED

Ackerman, R. E., 1973, Post Pleistocene cultural adaptations on the northern Northwest coast, *in* Raymond, S., and Schledermann, P., eds., International conference on the prehistory and paleoecology of western North America Arctic and Subarctic: Calgary, Calgary Archaeological Association, p. 1–20.

Ackerman, R. E., Hamilton, T. D., and Stuckenrath, T. D., 1979, Early cultural complexes on the northern Northwest Coast: Canadian Journal of Archaeology, no. 3, p. 195–210.

Agenbroad, L., 1978, The Hudson-Meng site; An Alberta bison kill in Nebraska High Plains: Washington, D.C., University Press of America, 230 p.

Agogino, G. A., 1961, A new point type from the Hell Gap valley, eastern Wyoming: American Antiquity, v. 26, p. 558–560.

Agogino, G. A., and Galloway, E., 1965, The Sister's Hill site; A Hell Gap site in north-central Wyoming: Plains Anthropologist, v. 10, p. 190–195.

Ahler, S. A., 1976, Material culture at Rodgers Shelter: a reflection of past activities, *in* Wood, R. W., and McMillan, R. B., Prehistoric Man and his environments; a case study in the Ozark highland: New York, Academic Press, p. 163–199.

Aikens, C. M., and Higuchi, T., 1982, Prehistory of Japan: New York, Academic Press, 354 p.

Anderson, D. C., and Semken, H. A., Jr., eds., 1980, The Cherokee excavations; Holocene ecology and human adaptations in northwestern Iowa: New York, Academic Press, 277 p.

Anderson, D. D., 1968, A stone age campsite at the gateway to America: Scientific American, v. 218, p. 24–33.

—— , 1970, Akmak: Acta Arctica, v. 16, 80 p.

Antevs, E., 1935, The occurrence of flints and extinct animals in pluvial deposits near Clovis, New Mexico; pt. II, Age of Clovis flake beds: Proceedings of the Philadelphia Academy of Natural Sciences, v. 87, p. 304–311.

Archaeology of Eastern North America, 1983, v. 11: Buffalo, Partner's Press, p. 1–80.

—— , 1984, v. 12: Buffalo's Partner's Press, 304 p.

—— , 1985, v. 13: Buffalo, Partner's Press, p. 1–92.

Aveleyra, A. de Anda, L., 1955, El Segundo mamut Fosil de Santa Isabel Iztapan, Mexico, y artefactos asociados: Instuto de Antropologia e Historia Publicaciones no. 1, p. 1–59.

—— , 1956, The second mammoth and associated artifacts at Santa Isabel, Iztapan, Mexico: American Antiquity, v. 22, p. 12–28.

Aveleyra, A. de Anda, L., and Maldonado-Koerdell, M., 1952, Asociacion de artefactos con Mamut en el Pleistocene superior de la Cuenca de Mexico: Estudios Antropologicos, v. 8, p. 3–29.

—— , 1953, Association of artifacts with mammoth in the Valley of Mexico: American Antiquity, v. 18, p. 332–340.

Barry, R. G., 1983, Late Pleistocene climatology, *in* Porter, S. C., ed., Late Quaternary environments of the United States, the late Pleistocene, v. 1: Minneapolis, University of Minnesota Press, p. 390–407.

Bedwell, S. F., 1973, Fort Rock basin: Eugene, University of Oregon Books, 189 p.

Bonnichsen, R., 1978, Critical arguments for Pleistocene artifacts from the Old Crow Basin, Yukon; A preliminary statement: Edmonton, Archaeological Researches International, p. 102–118.

—— , 1979, Pleistocene bone technology in the Beringian Refugium: National Museum of Manitoba Archaeological Survey of Canada, Mercury Series no. 89, p. 1–297.

Bonnichsen, R., Jacobson, G. L., Jr., Davis, R. B., and Borns, H. W., Jr., 1985, The environmental setting for human colonization of northern New England and adjacent Canada in late Pleistocene time, *in* Borns, H. W., Jr., LaSalle, P., Thompson, W. B., eds., Late Pleistocene history of northeastern New England: Geological Society of America Special Paper 197, p. 151–159.

Bonnichsen, R., Sorg, M. H., Rusco, M. K., Davis, J. O., and Tuohy, D., 1987, A possible association between evidence for humans and Rancholabrean fauna in northwestern Nevada; A preliminary analysis, *in* Bonnichsen, R., and Sorg, M. H., eds., Bone Modification: Orono, Center for the Study of Early Man (in press).

Bryan, A. L., 1979, Smith Creek Cave, *in* Tuohy, D. R., Rendal, D. L., eds., The archaeology of Smith Creek Canyon, eastern Nevada: Nevada State Museum Anthropological Papers no. 17, p. 162–253.

—— , 1980, The stemmed tradition; An early technological tradition in western North America, *in* Harten, L. B., Warren, C. N., and Tuohy, D. R., eds., Anthropological papers in memory of Earl H. Swanson, Jr.: Special Publication of the Idaho Museum of Natural History, p. 77–107.

—— , ed., 1986, New evidence for the Pleistocene peopling of the Americas: Orono, Maine, Center for the Study of Early Man, 386 p.

Butler, B. R., 1963, An early man site at Big Camas Prairie, south-central Idaho: Tebwia, v. 6, p. 22–33.

—— , 1965, A report on investigations on an early man site near Lake Channel, southern Idaho: Tebwia, v. 8, p. 1–132.

—— , 1971, The origin of the upper Snake country buffalo: Tebwia, v. 14, p. 1–20.

Butler, B. R., and Fitzwater, R. J., 1965, A further note on the Clovis site at Big Camas Prairie, south-central Idaho: Tebwia, v. 8, p. 38–40.

Byers, D. S., 1959, Radiocarbon dates for the Bull Brook site, Ipswich, Massachusetts: American Antiquity, v. 24, p. 427–429.

Carlson, R. L., 1983, The far West, *in* Shutler, R. S., Jr., ed., Early man in the New World: Beverly Hills, Sage Publications, p. 73–96.

Cassels, E. S., 1983, The archaeology of Colorado: Boulder, Johnson Publishing, 325 p.

Chapdelaine, C., ed., 1985, Des Éléphants, Des Caribous . . . et Des Hommes: Recherches Amerindiennes au Québec, v. 12, 184 p.

Cinq-Mars, J., 1979, Bluefish Cave I; A late Pleistocene eastern Beringian cave deposit in the northern Yukon: Canadian Journal of Archaeology, no. 3, p. 1–32.

Clark, D. W., 1984, Northern fluted points; Paleo-Eskimo, Paleo-Arctic, or Paleo-Indian: Canadian Journal of Anthropology, v. 4, p. 65–82.

Clark, D. W., and Clark, M. A., 1983, Paleo-Indians and fluted points: Subarctic Alternatives, v. 28, p. 283–292.

Cleland, C. E., 1965, Barren ground caribou (*Rangifer arcticus*): American Antiquity, v. 30, p. 350–351.

Coleman, D. D., 1972, Illinois State Geological Survey radiocarbon dates III: Radiocarbon, v. 14, p. 149–154.

Curran, M. L., 1984, The Whipple site and Paleoindian tool assemblage variation; A comparison of intrasite structuring: Archaeology of Eastern North America, v. 12, p. 5–50.

Davis, J. O., 1981, Geological studies at Rye Patch Reservoir, Phase IV, *in* Rusco, M. K., and Davis, J. O., eds., The Humbolt project, Rye Patch archeology phase IV; Final report: Carson City, Nevada State Museum, p. 15–31.

Davis, L. R., 1984, Late Pleistocene to mid-Holocene adaptations at Indian Creek, west-central Montana Rockies: Current Research, v. 1, p. 9–10.

Denton, G. H., and Hughes, T., eds., 1981, The last great ice sheet: New York, Wiley-Interscience, 484 p.

—— , 1983, Milankovitch theory of ice ages: Hypothesis of ice-linkage between regional insolation and global climate: Quaternary Research, v. 20, p. 125–144.

Dibble, D. S., and Lorrain, D., 1967, Bonfire Shelter; A stratified bison kill site, Val Verde County, Texas: Texas Memorial Museum, Miscellaneous Papers no. 1, 138 p.

Dixon, J. E., 1975, The Gallagher Flint Station; An early man site on the North Slope, Arctic Alaska, and its role in relation to the Bering Land Bridge: Arctic Anthropology, v. 12, p. 68–75.

—— , 1985, Cultural chronology of central interior Alaska: Arctic Anthropology, v. 22, p. 47–66.

Dolzani, M., 1986, The Mill Iron site; A point in Clovis time: Mammoth Trumpet, Orono, Center for the Study of Early Man, v. 2, no. 3, p. 1, 8.

—— ,1987, At home in the Pleistocene Mammoth Trumpet: Orono, Center for the Study of Early Man, v. 3, no. 2, p. 1, 3.

Dort, W., Jr., and Miller, S., 1977, Archaeological geology of the Birch Creek valley and the eastern Snake River Plain, Idaho: Twin Falls, Robco Printing, p. A-1, B1-B3, C1-C8, D1-D21, E1-E34, F1-F13, G1-G10, and H1-H4.

Dummond, D. E., 1980, The archaeology of Alaska and the peopling of America: Science, v. 209, p. 984–991.

Fagan, J., 1986, Western Clovis occupation in south-central Oregon; Archaeological research at the Dietz site 1983 to 1985: Current Research in the Pleistocene, v. 3, p. 3–5.

Figgins, J. D., 1927, The antiquity of man in America: Natural History, v. 27, p. 229–239.

—— , 1931, An additional discovery of the association a "Folsom" artifact and fossil mammal remains: Colorado Museum of Natural History Proceedings, v. 10, p. 23–24.

—— , 1933, A further contribution to the antiquity of man in America: Colorado Museum of Natural History Proceedings, v. 12, p. 4–8.

Firby, J. R., Mawby, J. E., Davis, J. O., 1981, Vertebrate paleontology and geology of paleontological site PPe23, *in* Rusco, M., and Davis, J. O., eds., The Humbolt project, Rye Patch archaeology phase IV; Final report: Carson City, Nevada State Museum Archaeological Services, p. 221–249.

Fisher, D. C., 1984a, Mastodon butchery by North American Paleo-Indians: Nature, v. 308, p. 271–272.

—— , 1984b, Taphonomic analysis of late Pleistocene mastodon occurrences; Evidence of butchery by North American Paleo-Indians: Paleobiology, v. 10, p. 338–357.

Fitting, J. E., DeVisscher, J., and Wahla, E. J., 1966, The Paleo-Indian occupation of the Holcombe Beach: Museum of Anthropology, University of Michigan, Anthropological Papers, no. 27, p. 1–59.

Fladmark, K., Alexander, D., and Driver, J., unpublished, Excavations at Charlie Lake Cave (HbRf 39), 1983: Burnaby, British Columbia, Department of Anthropology, manuscript, 150 p.

Forbis, R. E., and Sperry, J. D., 1952, An early man site in Montana: American Antiquity, v. 18, p. 127–137.

Forester, R. E., 1985, Horn Shelter number 2; The north end: Central Texas Archaeologist, no. 10, p. 25–36.

Frison, G. C., 1974, The Casper site: New York, Academic Press, 266 p.

—— , 1976, Cultural activity associated with prehistoric mammoth butchering and processing: Science, v. 194, p. 728–730.

—— , 1978, Prehistoric hunters of the High Plains: New York, Academic Press, 457 p.

Frison, G. C., and Stanford, D. J., 1982, The Agate Basin site: New York, Academic Press, 403 p.

Frison, G. C., and Todd, L. C., 1986, The Colby Mammoth site; taphonomy and archaeology of a Clovis kill in northern Wyoming: Albuquerque, University of New Mexico Press, 238 p.

Frison, G. C., and Zeimens, G. M., 1980, Bone projectile points; An addition to the Folsom complex: American Antiquity, v. 45, p. 231–237.

Funk, R., Fisher, D. W., and Reilley, E. M., Jr., 1970, Caribou and Paleo-Indians in New York State; A presumed association: American Journal of Sciences, v. 288 p. 245–256.

Gai Pei, 1985, Microlithic industries of China, *in* Wu, Rukang, and Olsen, J. W., eds., Paleoanthropology and paleolithic archaeology in the People's Republic of China: New York, Academic Press, p. 225.

Gilbow, D. W., 1981, Inference of human activity from faunal remains, Washington [M.A. thesis]: Pullman, Washington State University, 104 p.

Goodyear, A. C., 1982, The chronological position of the Dalton horizon in the southeastern United States: American Antiquity, v. 47, p. 382–395.

Gordon, B. H., 1976, Migod; 8,000 years of Barrenland prehistory: National Museum of Manitoba, Archaeological Survey of Canada Paper no. 56, 310 p.

Graham, R. W., 1981, Preliminary report on late Pleistocene vertebrates from the Selby and Dutton archaeological/paleontological sites, Yuma County, Colorado: Contributions to Geology, University of Wyoming, v. 20, p. 33–56.

Gramly, R. M., 1982, The Vail site; A Paleo-Indian encampment in Maine: Buffalo, Bulletin of the Buffalo Society of Natural Sciences, v. 30, 169 p.

Green, J. P., unpublished, Obsidian hydration chronology; The early Owl Cave remains: paper presented at the Eleventh Annual Conference of the Idaho Archaeological Society, Boise, 1983.

Green, R. G., and Rusco, M. K., unpublished, Report of activities at Paleontologic site PPe23: U.S. Department of the Interior Bureau of Reclamation Lahontan Basin Projects Office, Carson City, 1978, manuscript, 19 p. and two appendices.

Grimes, J. R., 1979, A new look at Bull Brook: Anthropology, v. 3, p. 109–130.

Gruhn, R., and Bryan, A. L., 1984, The record of Pleistocene megafaunal extinctions at Taima-taima, northern Venezuela, *in* Martin, P. S., Klein, R. G., eds., Quaternary extinctions; A prehistoric revolution: Tuscon, University of Arizona Press, p. 128–137.

Gryba, E. M., 1983, Sibbald Creek; 11,000 years of human use of the Alberta Foothills: Archaeological Survey of Alberta Occasional Paper no. 22, 219 p.

—— , 1985, Evidence of the fluted point tradition in Alberta, *in* Burley, D., ed., Contributions to Plains prehistory: Archaeological Survey of Alberta Occa-

sional Paper no. 26, p. 22–38.
Gustafson, C. E., Gilbow, D., and Daugherty, R. D., 1979, The Manis mastodon site; Early man on the Olympic Peninsula: Canadian Journal of Anthropology no. 3, p. 157–164.
Guthrie, R. D., 1984, Mosaics, allelochemics, and nutrients; An ecological theory of late Pleistocene megafaunal extinctions, *in* Martin, P. S., and Klein, R. G., eds., Quaternary extinctions; A prehistoric revolution: Tuscon, University of Arizona Press, p. 259–298.
Hannus, L. A., 1983, The Lange/Ferguson site; An event of Clovis mammoth butchery with the associated bone tool technology, Utah [Ph.D. thesis]: Salt Lake City, University of Utah, 222 p.
—— , 1987, Flaked mammoth bone from the Lange/Ferguson site; White River Badlands area, South Dakota, *in* Bonnichsen, R., and Sorg, M., eds., Bone modification: Orono, Maine, Center for the Study of Early Man (in press).
Hassan, A. A., and Ortner, D. J., 1977, Inclusions in bone material as a source of error in radiocarbon dating: Radiocarbon, v. 19, p. 131–135.
Haynes, C. V., 1955, Evidence of early man in Torrance County, New Mexico: Bulletin of the Texas Archaeological Society, v. 26, p. 44–165.
—— , 1980, The Clovis culture: Canadian Journal of Anthropology, v. 1, p. 115–121.
Haynes, C. V., and Agogino, G. A., 1960, Geological significance of a new radiocarbon date from the Lindenmeier site: Proceedings, Denver Museum of Natural History, no. 9, p. 1–23.
Haynes, C. V., Donahue, D. J., Jull, A.J.T., and T. H. Zabel, 1984, Application of accelerator dating to fluted point Paleoindian sites: Archaeology of Eastern North America, v. 12, p. 184–191.
Hester, J. J., 1972, Blackwater Locality No. 1; A stratified early man site in eastern New Mexico: Fort Burgwin Research Center Publication no. 8, 238 p.
Holliday, V. T., Johnson, E., Haas, H., and Stuckenrath, R., 1983, Radiocarbon ages from the Lubbock Lake site and ecological change on the southern High Plains: Plains Anthropologist, v. 28, p. 165–182.
Irving, W. N., 1968, Upper Pleistocene archaeology in Old Crow Flats, Yukon Territory: Arctic Circular, v. 17, p. 18–19.
Irving, W. N., and Harington, C. R., 1973, Upper Pleistocene radiocarbon-dated artifacts from the northern Yukon: Science, v. 179, p. 335–340.
Irving, W. N., Jopling, A. V., and Beebe, B., 1986, Indications of pre-Sangamon Humans near Old Crow, Yukon, Canada, *in* Bryan, A. L., ed., New evidence for the Pleistocene peopling of the Americas: Orono, Maine, Center for the Study of Early Man, p. 49–64.
Irving, W. N., Jopling, A. V., and Kritsch-Armstrong, I., 1987, Studies of paralithic technology and taphonomy, Old Crow Basin, Yukon Territory, *in* Bonnichsen, R., and Sorg, M., eds., Bone modification: Orono, Maine, Center for the Study of Early Man (in press).
Irwin, H.T.J., 1967, The Itama; late Pleistocene inhabitants of the Plains of the United States and Canada and the American Southwest [Ph.D. thesis]: Cambridge, Harvard University, 310 p.
Irwin-Williams, C., Irwin, H., Agogino, G., and Haynes, C. V., 1973, Hell Gap; Paleo-Indian occupation on the High Plains: Plains Anthropologist, v. 18, p. 40–53.
Jackson, L. J., 1983, Geochronology and settlement disposition in the early Paleo-Indian occupation of southern Ontario, Canada: Quaternary Research, v. 19, p. 388–399.
Johnson, E., 1974, Zooarchaeology of the Lubbock Lake site: The Museum Journal, v. 15, p. 107–122.
—— , 1977, Animal food resource of Paleoindians: The Museum Journal, v. 17, p. 65–77.
—— , 1982, Paleo-Indian bone expediency tools; Lubbock Lake and Bonfire Shelter: Canadian Journal of Anthropology, v. 2, p. 145–158.
Kutzbach, J. F., 1983, Modeling of Holocene climates, *in* Wright, H. E., Jr., ed., The late Quaternary environments of the United States, v. 2: Minneapolis, University of Minnesota Press, p. 252–268.
—— , 1984, The seasonal nature of climatic forcing and responses on Quaternary time scales, with emphasis on the period since the last glacial maximum: American Quaternary Association eighth biennial meeting program and abstracts, Boulder, p. 70–71.
Lahren, L., and Bonnichsen, R., 1974, Bone foreshafts from a Clovis burial in southwestern Montana: Science, v. 186, p. 147–150.
Larsen, H., 1968, Trail Creek; Final report on the excavation to two caves on Seward Peninsula, Alaska: Acta Arctica, fasc. XV, 79 p.
Layton, T., 1972a, A 12,000 year obsidian hydration record of occupation, abandonment, and lithic change from the northwestern Great Basin: Tebwia, v. 15, p. 22–28.
—— , 1972b, Lithic chronology in the Fort Rock valley, Oregon: Tebwia, v. 15, p. 1–21.
MacDonald, G. F., 1964, Debert; A Paleo-Indian site in central Nova Scotia: National Museum of Canada, Anthropology Papers no. 16, 207 p.
Mammoth Trumpet, 1984, Orono, Center for the Study of Early Man, v. 1, no. 1, p. 1 and 3.
—— , 1985, Orono, Center for the Study of Early Man, v. 2, no. 1, p. 1.
Martin, J. E., 1983, Archaeological and paleontological significance of the Lange/Ferguson Clovis kill site: Rapid City, South Dakota School of Mines and Technology, 119 p.
Martin, P. S., 1984, Prehistoric overkill; The global model, *in* Martin, P. S., and Klein, R. G., eds., Quaternary extinctions; A prehistoric revolution: Tucson, University of Arizona Press, p. 354–403.
Martin, P. S., and Klein, R. G., eds., 1984, Quaternary extinctions; A prehistoric revolution: Tucson, University of Arizona Press, 892 p.
Martinez del Rio, P., 1952, El mamut de Santa Isabel Iztapan: Cuadernos Americanos, v. 9, p. 149–170.
McGee, W. J., 1889, An obsidian implement from Pleistocene deposits in Nevada: American Anthropologist, v. 12, p. 301–312.
McNett, C. W., Jr., McMilliam, B. A., and Marshall, S. B., 1979, The Shawnee-Minisink site, *in* Newman, W. S., and Salwen, B., eds., Amerinds and their paleoenvironments in northeastern North America: Annals of the New York Academy of Sciences, v. 288, p. 282–296.
Mead, J. I., and Meltzer, D. J., eds., 1985, Environments and extinctions; Man in late glacial North America: Orono, Maine, Center for the Study of Early Man, 209 p.
Meltzer, D. J., 1984, Late Pleistocene human adaptation in eastern North America, [Ph.D. thesis]: Seattle, University of Washington, 440 p.
Michael, H. N., 1984, Absolute chronologies of late Pleistocene and early Holocene cultures of northeast Asia: Arctic Anthropology, v. 21, p. 1–68.
Miller, M. E., 1986, Preliminary investigations at Seminoe Beach site, Carbon County, Wyoming, *in* Walker, D. M., ed., The Wyoming archaeologist: Papers for George C. Frison, Wyoming State Archeologist, v. 29, p. 83–96.
Miller, S. J., 1983, Osteo-archaeology of the mammoth-bison assemblage at Owl Cave, the Wasden site, Idaho, *in* Le Moine, G. M., and MacEachern, A. S., eds., Carnivores, human scavengers, and predators; A question of bone technology, Proceedings of the 15th Annual Chacmool Conference: Calgary, Calgary Archaeological Association, p. 39–53.
—— , 1987, Characteristics of mammoth bone reduction at Owl Cave, Idaho, *in* Bonnichsen, R., and Sorg, M., eds., Bone modification: Orono, Maine, Center for the Study of Early Man (in press).
Moeller, R., 1980, 6LF21; A Paleo-Indian site in western Connecticut: American Indian Archaeological Institute Anthropology Paper no. 16, 160 p.
Morlan, R. E., 1973, The later prehistory of the middle Porcupine drainage, northern Yukon Territory: National Museum of Man, Archaeological Survey of Canada, Mercury Series no. 3, 583 p.
—— , 1980, Taphonomy and archaeology in the upper Pleistocene of the northern Yukon Territory; A glimpse of the peopling of the New World: National Museum of Man, Mercury Series no. 94, p. 1–398.
Morlan, R. E., and Cinq-Mars, J., 1982, Ancient Beringians; Human occupations in the late Pleistocene of Alaska and the Yukon Territory, *in* Hopkins, D. M., Mathews, J. V., Jr., Schweger, C. E., and Young, S. B., eds., Paleoecology of Beringia: New York, Academic Press, p. 353–382.
Nelson, D. E., Morlan, R. E., Vogel, J. S., Southon, J. R., and Harington, C. R., 1986, New dates on northern Yukon artifacts; Holocene not upper Pleisto-

cene: Science, v. 232, p. 749–751.
Nelson, N. C., 1935, Early migrations of man to North America: Natural History, v. 35, p. 356.
—— , 1937, Notes on cultural relations between Asia and America: American Antiquity, v. 2, p. 267–272.
Pettipas, L., 1985, Recent developments in Paleo-Indian archaeology in Manitoba, *in* Burley, D., ed., Contributions to Plains prehistory: Archaeological Survey of Alberta Occasional Paper no. 26, p. 39–63.
Plew, M. G., and Pavesic, M., 1982, A compendium of radiocarbon dates for southern Idaho archaeological sites: Journal of California and Great Basin Anthropology, v. 4, p. 113–122.
Porter, S., ed., 1983, Late Quaternary environments of the United States; The late Pleistocene, v. 1: Minneapolis, University of Minnesota Press, 407 p.
Powers, R. W., and Hamilton, T. D., 1978, A late Pleistocene human occupation in central Alaska, *in* Bryan, A. L., ed., Early man in America; From a Circum-Pacific perspective: Edmonton, Archaeological Researches International, p. 72–77.
Rancier, J., Haynes, G., and Stanford, D., 1982, 1982 investigations of Lamb Spring: Southwestern Lore, v. 48, p. 1–17.
Redder, A., 1985, Horn Shelter number 2; The south end: Central Texas Archaeologist, no. 10, p. 37–65.
Ruddiman, W. F., and Duplessy, J. C., 1985, Conference on the last deglaciation; Timing and mechanism: Quaternary Research, v. 23, p. 1–17.
Rusco, M. K., and Davis, J. O., 1981, The Humbolt project, Rye Patch archaeology, phase IV; Final report: Nevada State Museum Archaeological Service Reports, NSM-AS 14-6-2, 259 p. and 4 appendices.
Rusco, M. K., Davis, J. O., and Firby, R., Jr., 1979, The Humbolt project, Rye Patch archaeology, phase III; Final field report: Nevada State Museum Archaeological Service Reports, 147 p. and appendix.
Sellards, E. H., 1952, Early man in America; A study in prehistory: New York, Greenwood Press, 213 p.
Shipman, P., Fisher, D. C., and Rose, J. J., 1984, Mastodon butchery; Microscopic evidence of carcass processing bone tool use: Paleobiology, v. 1, p. 358–365.
Shutler, R., Jr., ed., 1983, Early man in the New World: Beverly Hills, Sage Publications, 223 p.
Soffer, O., 1985, The Upper Paleolithic of the Central Russian Plain: New York, Academic Press, 539 p.
South African Journal of Science, 1985, v. 9, 81 p.
Spiess, A. E., Curran, M. L., Grimes, J. R., 1984, Caribou (*Rangifer tarandus* L.) bones from New England Paleo-Indian sites: North American Archaeologist, v. 6, p. 145–159.
Stanford, D., 1978, The Jones-Miller sites; An example of Hell Gap bison procurement strategy, *in* Davis, L., and Wilson, M., eds., Bison procurement and utilization; A symposium: Plains Anthropologist Memoir 14, p. 90–97.
—— , 1979, The Selby and Dutton sites; Evidence for a possible pre-Clovis occupation of the High Plains, *in* Humphrey, R. L., and Stanford, D., eds., Pre-Llano cultures in the Americas; Paradoxes and possibilities: Washington, D.C., Anthropological Society of Washington, p. 101–125.
—— , 1984, The Jones-Miller site; A study of Hell Gap bison procurement and processing: National Geographic Research Projects, p. 615–635.
Stanford, D., and Graham, R. W., 1985, Archaeological investigations of the Selby and Dutton mammoth kill sites, Yuma, Colorado: National Geographic Research Reports, v. 20, p. 519–541.
Stanford, D., Wedel, W. R., and Scott, R., 1981, Archaeological investigation of the Lamb Spring site: Southwestern Lore, v. 47, no. 1, p. 14–27.
Stuckenrath, R., Jr., 1966, The Debert archaeological project, Nova Scotia; Radiocarbon dating: Quaternaria, v. 9, p. 75–80.
Swanson, E. H., Jr., 1972, Birch Creek: Pocatello, The Idaho State University Press, p. 1–237.
Swanson, E. H., Jr., Butler, B. R., and Bonnichsen, R., 1964, Birch Creek Papers No. 2, Natural and cultural stratigraphy in the Birch Creek valley of eastern Idaho: Occasional Papers of the Idaho State University Museum, no. 14, 120 p.
Taylor, R. E., Payen, L. A., Prior, C. A., Slota, P. J., Jr., Gillespie, R., Gowlett, J.A.J., Hedges, R.E.M., Jull, A.J.T., Zabel, T. H., Donahue, D. J., and Berger, R., 1985, Major revisions in the Pleistocene age assignments for North America human skeletons by C-14 accelerator mass spectrometry; None older than 11,000 years B.P.: American Antiquity, v. 50, p. 136–140.
Thompson, R. S., 1985, The age and environment of the Mount Moriah (Lake Mohave) occupation at Smith Creek Cave, Nevada: Orono, Maine, Center for the Study of Early Man, p. 111–120.
Thorson, R. M., and Hamilton, T. D., 1977, Geology of the Dry Creek sites; A stratified early man site in interior Alaska: Quaternary Research, v. 7, p. 149–176.
Vayda, A. P., and Rappaport, R., 1968, Ecology; Cultural and non-cultural, *in* Clifton, J. A., ed., Introduction to cultural anthropology: New York, Houghton Mifflin Company, p. 476–498.
Vernekar, A. D., 1968, Long-period global variation of incoming solar radiation, v. II, E; Final report to U.S. Department of commerce; Environmental Science Services Administration, v. III, contract no. E22-137-67(N): Hartford, Travelers Research Center Inc., 239 p.
Vrba, E. S., 1987, The environmental context of the evolution of early hominids and their culture, *in* Bonnichsen, R., and Sorg, M., 1987, Bone modification: Orono, Center for the Study of Early Man (in press).
Webb, T., III, Kutzbach, J., and Street-Perrott, F. A., 1985, 20,000 years of global climatic change; Paleoclimatic research plan, *in* Malone, T. F., and Roederer, J. G., eds., Global Change Symposium Series, No. 5: New York, ICSU Press, p. 182–219.
West, F. H., 1981, The archaeology of Beringia: New York, Columbia University Press, p. 268.
—— , 1983, The antiquity of man in America, *in* Porter, S. C., Late Quaternary environments of the United States; The late Pleistocene, v. 1: Minneapolis, University of Minnesota Press, p. 364–382.
Wilmsen, E. N., and Roberts, F.H.H., 1978, Lindemeier, 1934–1974, Concluding report on investigations: Smithsonian contributions to Anthropology no. 24, 187 p.
Wormington, H. M., and Forbis, R. G., 1965, An introduction to the archaeology of Alberta, Canada: Denver, Denver Museum of Natural History Proceedings no. 11, 248 p.
Wright, H. E., Jr., 1983, Introduction, *in* Late Quaternary environments of the United States; The Holocene, v. 1: Minneapolis, University of Minnesota Press, p. xi–xvii.
—— , 1984, Sensitivity and response time of natural systems to climatic change in the late Quaternary: Quaternary Science Reviews, v. 3, p. 91–131.
Wright, J. V., 1976, The Grant Lake site, Keewatin District, Northwest Territories: National Museum of Man, Mercury Series, Archaeological Survey of Canada Paper no. 47, 122 p.
Wu, R. K., and Olsen, J., eds., 1985, Palaeoanthropology and palaeolithic archaeology in the People's Republic of China: New York, Academic Press, 293 p.
Wyckoff, D. G., 1985, The Packard complex; Early Archaic, pre-Dalton occupation on the Prairie-woodlands border: Southeastern Archaeology, v. 4, p. 1–26.
Yi, S., and Clark, J., 1985, The "Dyuktai Culture" and New World origins: Current Anthropology, v. 26, p. 1–13.

MANUSCRIPT ACCEPTED BY THE SOCIETY FEBRUARY 2, 1987

ACKNOWLEDGMENTS

Support from Mr. William Bingham's Trust for Charity and the University of Maine allowed this research to be undertaken. James Payne and Karen Turnmire of the University of Maine (UM) assisted with the compilation of radiocarbon dates. Karen Hudgins typed the manuscript and Steve Bicknell drafted illustrations. We would like to thank Pegi Jodry for her assistance with the Folsom section. We gratefully acknowledge Ann Bonnichsen's domestic support, which allowed two of us (R.B. and D.S.) to collaborate on the manuscript without interruption. Collectively, we accept the responsibility for errors and omissions.

Printed in U.S.A.

The Geology of North America
Vol. K-3, North America and adjacent oceans during the last deglaciation
The Geological Society of America, 1987

Chapter 19

Model simulations of the climatic patterns during the deglaciation of North America

John E. Kutzbach
Center for Climatic Research, University of Wisconsin at Madison, Madison, Wisconsin 53706

INTRODUCTION

Atmospheric general circulation models (AGCMs) can be used to simulate climatic patterns—past, present, or future—provided that certain boundary and atmospheric conditions are specified. Examples of relevant boundary conditions are the solar radiation at the top of the atmosphere, composition of the atmosphere, height of the lower boundary (land, mountains, ice sheets), and characteristics of the lower boundary (albedo, roughness, sea-ice location, ocean temperature). Given this information, AGCMs simulate many (but not all) features of the present climate (Pitcher and others, 1983). These models are also used to estimate future conditions; for example, the sensitivity of the climate to increased concentration of carbon dioxide in the atmosphere (Washington and Meehl, 1984).

AGCMs have been used to simulate past climates in situations where geologic evidence or astronomical theory provides estimates of all or most of the appropriate past boundary conditions. Another application of AGCMs is to estimate the sensitivity of the simulated climate to likely changes of a single boundary condition (solar radiation, presence/absence of ice sheets, etc). Simulations of the climate of the last glacial maximum are reported by Alyea (1972), Williams and others (1974), Gates (1976a, b), Manabe and Hahn (1977), Hansen and others (1984), Rind and Peteet (1985, 1986), and Manabe and Broccoli (1985). Rind and others (1986) have studied the impact of cold North Atlantic sea-surface temperatures on the climate, with implications for understanding the Younger Dryas (11–10 ka). Climate simulations for 9 ka are found in Kutzbach (1981), Kutzbach and Otto-Bliesner (1982), Kutzbach and Guetter (1984), and Webb and others (1985). Kutzbach and Guetter (1986) report simulations for the period 18 ka to present at 3,000-year intervals; the experiments combine changes in solar radiation, atmospheric composition, and lower boundary conditions. Barry (1983) lists many of these climatic simulations of the last glacial maximum and provides an overview of the ice-age climatology of North America based upon observations.

This paper summarizes the results of Kutzbach and Guetter (1986) for the North American sector. It: (1) provides a context for interpreting the geologic records of the deglaciation of North America, (2) gives examples of the relative sensitivity of the climate to different boundary conditions, (3) gives examples of agreement or disagreement of results from different AGCMs, and (4) indicates some possible directions of future studies.

DESCRIPTION OF MODEL, PROCEDURES, BOUNDARY CONDITIONS

This section presents general information about the numerical model used for the experiments, the length and number of experiments, and the values used to prescribe boundary conditions.

Model

The simulation experiments were made with the Community Climate Model (CCM) of the National Center for Atmospheric Research (NCAR); see Pitcher and others (1983) and Ramanathan and others (1983) for a detailed description of the model. The model incorporates atmospheric dynamics, based upon the equations of fluid motion, radiative and convective processes, and condensation and evaporation. The surface-energy budget and surface temperature are computed over land, ice sheets, and sea ice. Orographic influences of mountains (and ice sheets) are included. Orbital parameters (eccentricity, tilt, date of perihelion), atmospheric CO_2 concentration, sea-surface temperature, sea-ice limit, snow cover, land albedo, and effective soil moisture are prescribed. For perpetual January and July simulations, orbital parameters are set to give the solar radiation appropriate for January 16 and July 16. The solar constant is 1370 W/m^2. The model has nine vertical levels (sigma coordinates) and uses a spectral representation to wavenumber 15 (rhomboidal) for the horizontal fields of wind, temperature, pressure, and moisture. When needed for certain calculations, the spectral rep-

Kutzbach, J. E., 1987, Model simulations of the climatic patterns during the deglaciation of North America, *in* Ruddiman, W. F., and Wright, H. E., Jr., eds., North America and adjacent oceans during the last deglaciation: Boulder, Colorado, Geological Society of America, The Geology of North America, v. K-3.

TABLE 1. LIST OF EXPERIMENTS AND WEIGHTING FACTORS*

ka	Perpetual January	Perpetual July	Annual Cycle	Ts	Ts/365.24
0	x	x	x	186.4	(0.510)
3	x	x		185.8	(0.509)
6	x	x		182.7	(0.500)
9	x	x	x	179.3	(0.491)
9 (no land ice)		x		179.3	(0.491)
12	x	x		178.1	(0.488)
15	x	x		180.1	(0.493)
18	x	x		183.9	(0.504)
18 (CO_2 = 200 ppmv)	x	x		183.9	(0.504)

*Notes:

1. The 0-, 9-, and 18-ka experiments are 450 days long for both July and January. The 3-, 6-, 12-, and 15-ka experiments are 450 days long for January and 150 days long for July.

2. The weights used to extimate weighted annual-averages are T_S (length of the summer half-year, vernal to autumnal equinox, in days) and T_W = 365.24 - T_S such that:

$$(\ )_{annual} = [T_S \times (\ )_{JULY} + T_W \times (\ )_{JAN}] / 365.24.$$

3. Annual cycle refers to full annual cycle experiments with a version of the CCM with interactive soil moisture.

resentation is converted to a grid of 4.4° latitude by 7.5° longitude.

Procedures

The experiments were started with all model variables set at values for a modern (control) simulation, but with the solar radiation and lower boundary conditions changed from modern to estimates of past values. In most experiments the model was run for 450 simulated days (see Table 1). In the first 60 days the model's circulation was adjusting to the changed boundary conditions, and these days were ignored. Three 90-day averages were then selected: days 61–150, 211–300, and 361–450. By omission of the two 60-day segments (151–210 and 301–360), the three 90-day averages are assumed to be independent (Blackmon and others, 1983). The three 90-day segments were then averaged and compared to a similarly constructed control (modern) simulation. The six independent 90-day averages, three from each experiment and three from the control, were used to estimate the model's inherent variability and assess the statistical significance of the simulated climatic changes (Chervin and Schneider, 1976).

While the 14 AGCM experiments were perpetual January and July simulations, two additional experiments were annual cycle simulations for 9 ka and 0 ka (Kutzbach and Otto-Bliesner, 1982; Kutzbach and Guetter, 1986). For these two annual-cycle simulations the annual-average precipitation, precipitation-minus-evaporation, and surface temperature, as based on the full annual cycle, were adequately approximated in most regions by the appropriately weighted January and July results from the perpetual experiments. This approximation was used to estimate certain annual-average conditions from the perpetual January and July results (Kutzbach and Guetter, 1986).

Boundary conditions (summarized in Table 2)

The boundary conditions for solar radiation, atmospheric composition, ice sheets, sea level, sea ice, sea-surface temperature, and land albedo, shown schematically in Figure 1, are summarized below for the North American sector; full details are in Kutzbach and Guetter (1986).

External conditions—solar radiation. Changes in solar radiation depend upon the earth's axial tilt, eccentricity of orbit, and longitude of perihelion (Berger, 1978). For July and January at 18 ka, the radiation differences between 18 ka and 0 ka were less than 1 percent of modern values at all latitudes (Fig. 1). Solar radiation was 3 to 4 percent more in March–May and 3 to 4 percent less in September–November when compared to present, because perihelion occurred near the vernal equinox at 18 ka. After 18 ka, the date of perihelion shifted toward the northern hemisphere summer solstice, and the axial tilt increased. At 12 ka, perihelion occurred in June, and at 9 ka it was in late July. The axial tilt was 24.2° at 9 ka, compared to 23.4° at 18 ka and present. At 9 ka, northern hemisphere solar radiation was 8 percent ($37 Wm^{-2}$) above the modern value in July and 8 percent ($-18 Wm^{-2}$) below the modern value in January; these changes increased the amplitude of the seasonal radiation cycle compared to present (Fig. 1). After 6 ka, the radiation regime gradually approached the present values (Fig. 1).

Atmospheric conditions—carbon dioxide concentration, aerosol. Evidence from ice cores (Oeschger and others, 1983; Lorius and others, 1984) indicates that atmospheric CO_2 concentration was about 200 ppmv at glacial maximum (around 18–15 ka); it increased to about 265–275 ppmv by about 9 ka, where it remained until recently (Neftel and others, 1982; Lorius and others, 1984; Stuiver and others, 1984; Fig. 1). The main

TABLE 2. SUMMARY OF BOUNDARY CONDITIONS

Orbital Parameters and Atmospheric Concentration of CO_2

ka	Eccentricity of orbit	Axial Tilt	Position of Perihelion*	CO_2 (ppmv)
0	.0167	23.44°	78°	330
3	.0178	23.82°	128.9°	330
6	.0187	24.11°	179.1°	330
9	.0193	24.24°	228.8°	330
12	.0196	24.15°	277.9°	330
15	.0196	23.87°	327.0°	330
18	.0195	23.45°	16.3°	330 (and 200)

Surface Boundary Conditions (see text for details)

ka	Ice Sheets	Sea Level	SST	Sea Ice	Land Albedo
0	control	control	control	control	control
3	"	"	"	"	"
6	"	"	"	"	"
9	Fig. 2	-10 m	"	"	"
12	Fig. 2	-40 m	see text	Fig. 2, 7	see text
15	Fig. 2	-100 m	Fig 3, 7	Fig. 2, 7	Fig. 2
18	Fig. 2	-100 m	Fig 3, 7	Fig. 2, 7	Fig. 2

*Note: Positon of perihelion (degrees of celestial longitude) measured clockwise from the vernal equinox.

series of experiments were made with the CO_2 concentration set at the control case value of 330 ppmv. However, the 18 ka experiment was also run with a CO_2 concentration of 200 ppmv. The high tropospheric loadings of land and marine aerosols that were probably characteristic of glacial times have not yet been included (Kolla and others, 1979; Thompson and Mosley-Thompson, 1981; Petit and others, 1981; Fig. 1).

Lower boundary conditions. CLIMAP project members' (1981) estimates of boundary conditions for ice sheet location and height, summer and winter sea-ice location, and summer and winter sea-surface temperature were used for the 18 ka experiments (Fig. 1, Fig. 2). Values for the NCAR CCM grid (4.4° latitude by 7.5° longitude) were obtained by averaging values from surrounding 2-degree squares of the CLIMAP grid. After 18 ka, these CLIMAP boundary conditions were adjusted toward modern boundary conditions (Fig. 1) as summarized below.

Ice sheets, sea level, land. At 18 ka the North American ice sheet covered 77 model grid points. A maximum elevation of about 3,300 m over Hudson Bay is assumed. The North American ice sheet forms a higher topographic barrier than the Rockies, because the spectral representation of the orography reduces the height of relatively narrow mountain ranges far more than that of the broad ice sheet (Fig. 2). Consistent with the large ice sheets on the continents, sea level was about 100 m lower, and certain ocean grid points in the control case were land grid points at 18 ka, for example between Alaska and eastern Siberia and in the Gulf of Mexico (Fig. 2).

For the experiments after 18 ka, the limits of the ice sheets were adjusted according to the analysis of Denton and Hughes (1981). At 15 ka the ice limits were very close to those at 18 ka, and they were kept identical in the model. At 12 ka the area covered by ice was reduced in western North America (Fig. 2).

Estimates of ice volume obtained from isotopic records in marine sediments (Mix and Ruddiman, 1985) suggest little change in ice volume between 18 and 14 ka and rapid melting between 14 and 12 ka. Thus at 15 ka, ice-sheet height was set equal to that of 18 ka, but at 12 ka it was reduced by one-half (to 1,650 m). The combination of reduced height and decreased area of ice approximated an ice volume for 12 ka of about 40 percent of glacial maximum. Sea level was adjusted to a value 40 m below present. At 9 ka, the North American ice sheet was reduced further in size (Fig. 2) and assigned a maximum height of 800 m. Sea level was 10 m lower than present. At 6 and 3 ka there was no excess land ice, and sea level was the same as at 0 ka. A small ice sheet still existed in Labrador at 6 ka, but it occupied less than one model grid rectangle and was not included in the model.

Sea ice. The CLIMAP-specified sea-ice boundary for 18 ka was shifted far south of its modern position in the North Atlantic in both January and July (Fig. 2). In the model experiments, sea-ice was kept at the 18 ka positions for 15 and 12 ka; thereafter it was placed at the modern position.

Sea-surface temperature. The charts of sea-surface temperature (SST) produced by CLIMAP Project Members

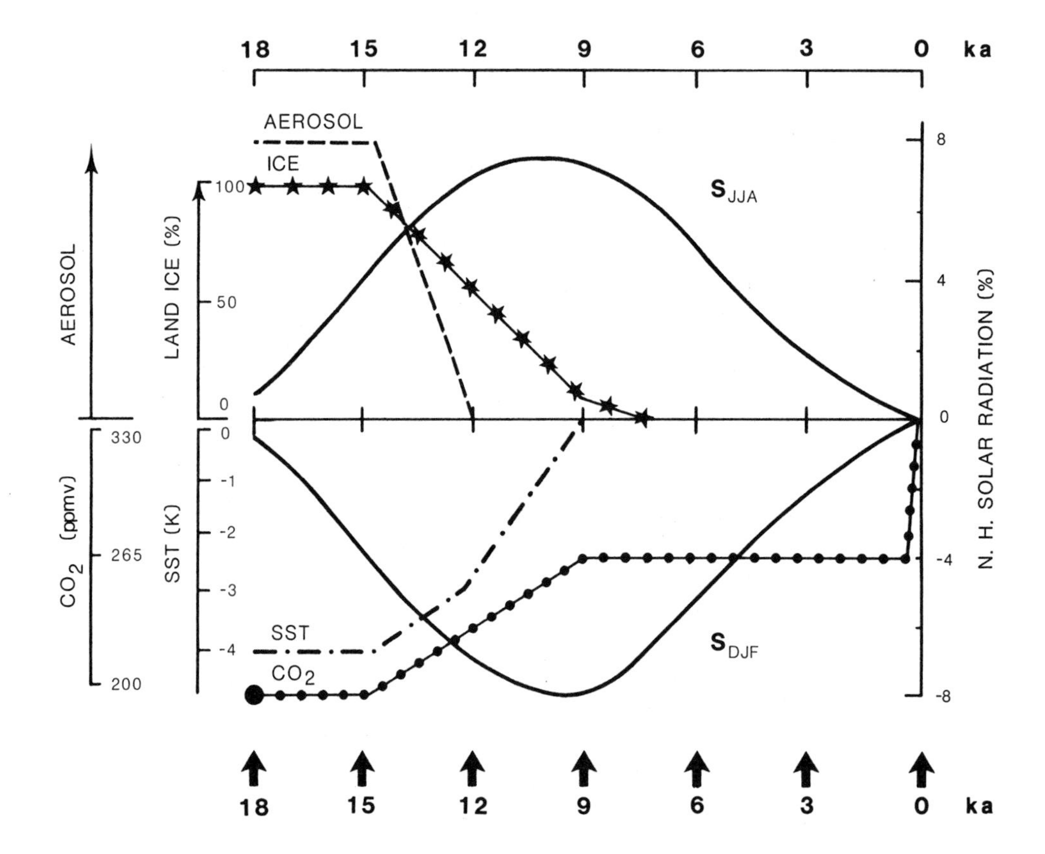

Figure 1. Schematic diagram of major changes since 18 ka in external forcing (Northern Hemisphere solar radiation in June–August (S_{JJA}) and December–February (S_{DJF}), as percent difference from present) and internal boundary conditions: land ice (ICE) as percent of 18-ka ice volume (CLIMAP Project Members, 1981; Denton and Hughes, 1981); global mean-annual sea-surface temperature (SST), including calculated surface temperature over sea ice, as departure from present, K (CLIMAP Project Members, 1981); excess glacial-age aerosol (AEROSOL), arbitrary scale (Petit and others, 1981; Thompson and Mosley-Thompson, 1981); and atmospheric CO_2 concentration (CO_2), in ppmv (Neftel and others, 1982; Lorius and others, 1984). The arrows correspond to the seven sets of simulation experiments with the CCM. One experiment for 18 ka included the lowered CO_2 concentration (220 ppmv, large solid circle); the main series of experiments used the same CO_2 concentration as the control case (330 ppmv) rather than the stepwise increase. Experiments incorporating the increased glacial-age aerosol loading are planned but not included here.

(1981) for 18 ka show substantially colder subpolar and middle-latitude waters immediately equatorward of the expanded sea-ice margin. The CLIMAP-specified SSTs were used in the model experiments at both 18 and 15 ka (see Figs. 3 and 7). At 12 ka the SST anomalies were reduced to one-half of their 18 ka values everywhere except in the North Atlantic, where SST's remained low (Ruddiman and McIntyre, 1981b); there at all grids from 46.4°northward the SSTs were set at 18 ka values. At 9 ka and thereafter, SSTs were set at 0 ka values; this boundary condition is consistent with evidence from ocean-sediment cores that SSTs were near modern values since about 9 ka, except perhaps near coastlines, where changes in upwelling occurred (W. Prell, personal communication, 1986).

Surface albedo. Surface albedo is changed in the model either by the addition of glacial ice on land, sea ice on ocean, or a change in bare-land albedo. Surfaces that were covered with glacial ice (at 18 ka and subsequently) were assigned an albedo of 0.80. The one exception was the North American ice sheet at 9 ka, which was assigned an albedo of 0.50 because it was assumed to be old ice, perhaps dirty and ponded with meltwater. Bare-land albedo for 18 ka was changed from that of the control case (0 ka) only if the CLIMAP-estimated albedo difference (18 ka minus modern) was greater than one-half of the albedo difference between the two surface types in the control simulation: nondesert albedo, 0.13; and desert albedo, 0.25 (or, in some cases, between a land-surface albedo and an ice or snow albedo, 0.80). Use of this rule kept the 18-ka land albedo the same as the 0-ka value at most locations in North America (see Fig. 2). At 12 ka, bare-land albedo was set halfway between the 18 ka and modern value. At 9 ka and thereafter, land albedos were the same as for 0 ka.

Snow cover is prescribed in the model so that snow-albedo temperature feedbacks are excluded (except insofar as they derive from the prescribed changes in glacial ice and sea ice). The land is snow-covered north of 68.9°N in July and 42.2°N in January for all experiments.

RESULTS

The simulation results for North America are presented in three levels of detail: a) selected maps for North America for 18 and 9 ka that illustrate the large-scale patterns of circulation and surface climate; b) a schematic summary of jet-stream locations and regional averages of temperature, precipitation, and precipitation-minus-evaporation for all seven time periods, 18 to 0 ka; and c) a discussion of model calculations of selected terms in the mass budget of the North American ice sheet. More detailed results, both for North America and (especially) for the globe, are in Kutzbach and Guetter (1986). Details for 18 ka for North America are in Kutzbach and Wright (1985).

Selected maps of climate for 18, 9, and 0 ka

In this section, charts for 18, 9, and 0 ka and for the 18-ka-minus-0-ka and 9-ka-minus-0-ka differences illustrate regional climatic features for July (the growing season) and January. In general the simulated changes in surface temperature are first discussed, and then the changes in sea-level pressure, winds aloft and at the surface, and precipitation.

18 ka—July. The prescribed changes of land ice, sea ice, and ocean-surface temperature produced major changes in the glacial-age climate of North America. Surface temperature was much lower over the North American ice sheet (by 20–30°C) and over the equatorward-extended North Atlantic sea ice (by 5–10°C) (Fig. 3). Immediately south of the ice sheet the temperature was about 10°C lower than modern, but in the southern U.S. and in Alaska the temperature was only a few degrees below present. Sea-level pressure was increased over the ice sheet and the North Atlantic sea ice and decreased over the North Pacific,

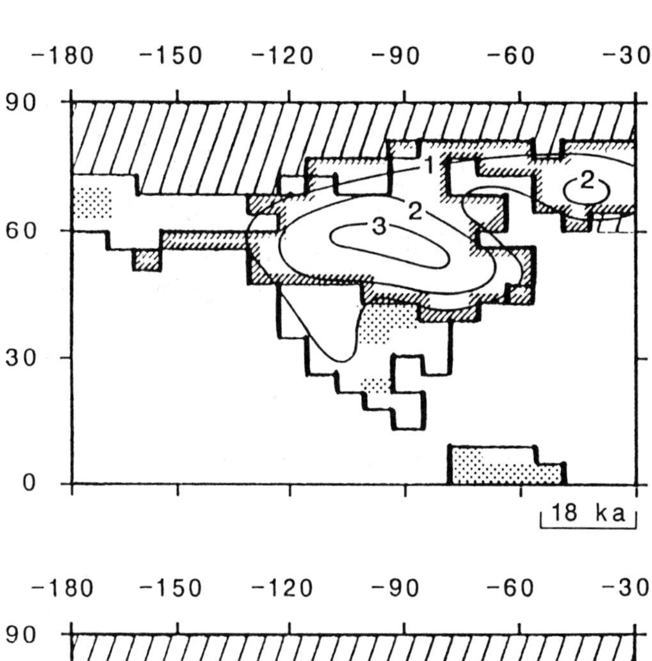

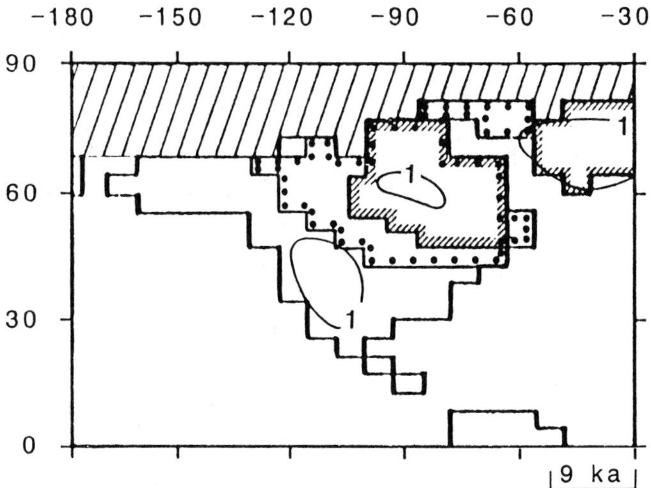

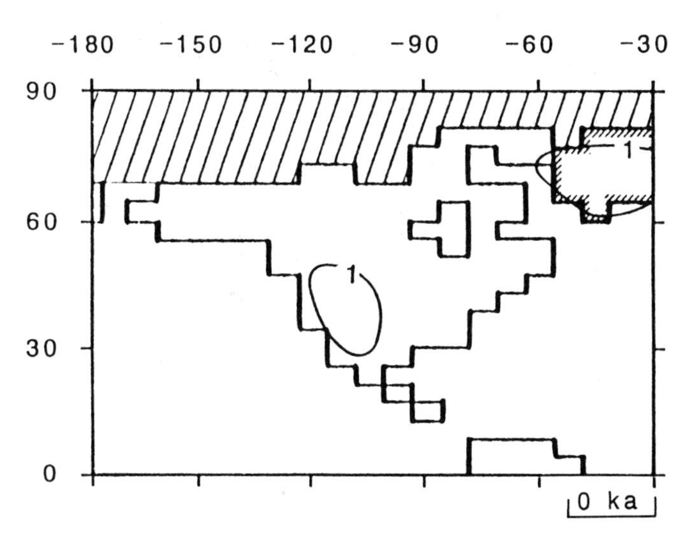

Figure 2. Lower boundary conditions for 18-ka (top), 9-ka (middle), and the 0-ka control (bottom): continental outlines, glacial ice (outlined with short hatch marks), July sea ice (diagonal stripes), topographic height (km), and bare-land albedo. Topographic height is shown for both mountains and ice sheets. Land albedo is either 0.13 (clear) or 0.25 (stippled). Albedo of glacial ice is 0.80, except 0.50 for the North American ice sheet at 9 ka. The 9-ka map also shows the ice-sheet boundaries for North America and Europe at 12 ka (dotted border). January sea-ice boundaries and January and July sea-surface temperatures are shown in Figures 5 and 10. Sea ice albedo is 0.70.

Surface Temperature

-180 -150 -120 -90 -60 -30
90
60
30
0
260
280
300
300
300
July 18 ka

-180 -150 -120 -90 -60 -30
90
60
30
0
280
300
300
July 9 ka

-180 -150 -120 -90 -60 -30
90
60
30
0
280
300
July 0 ka

Temperature Departure

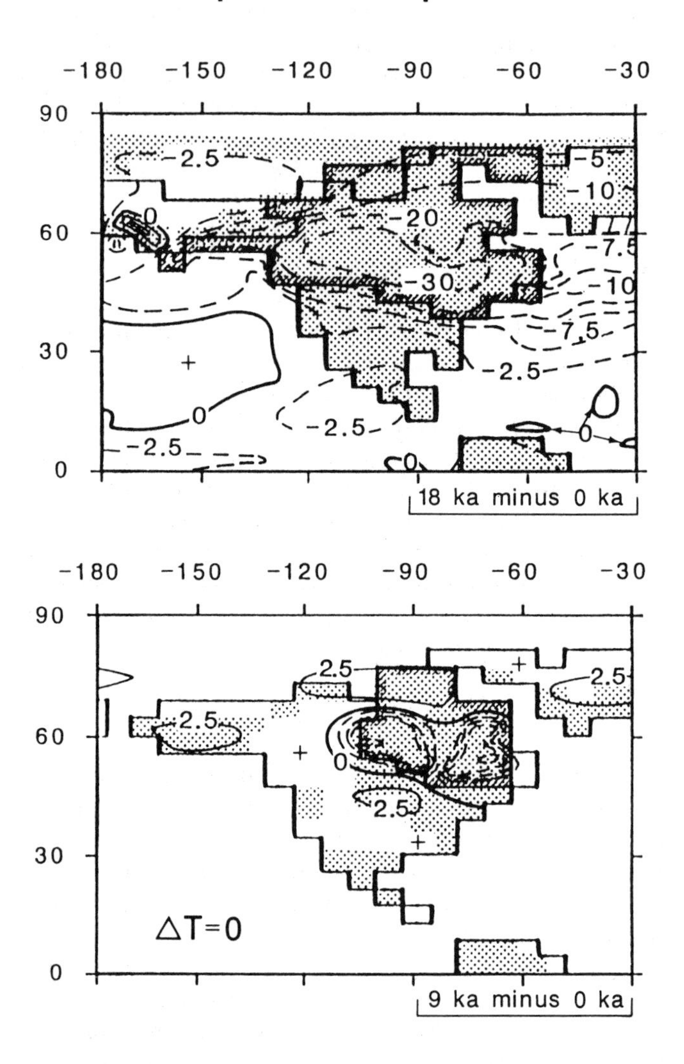

Figure 3. Surface temperatures (left, K) and temperature departures (right, experiment-minus-control, C) for July for 18-ka (top), 9-ka (middle), and the 0-ka control (bottom). For negative departures the contour lines are dashed. The shading on the departure maps indicates the departures are statistically significant (two-sided t test) at or above the 95 percent confidence level based on the model's inherent variability. See Figure 2 for explanation of glacial-ice and sea-ice boundaries. The departure maps (right) show the areas of increased glacial-ice and sea-ice extent at 18 and 9 ka. Ocean surface temperature is prescribed, and therefore no measure of statistical significance is computed for it.

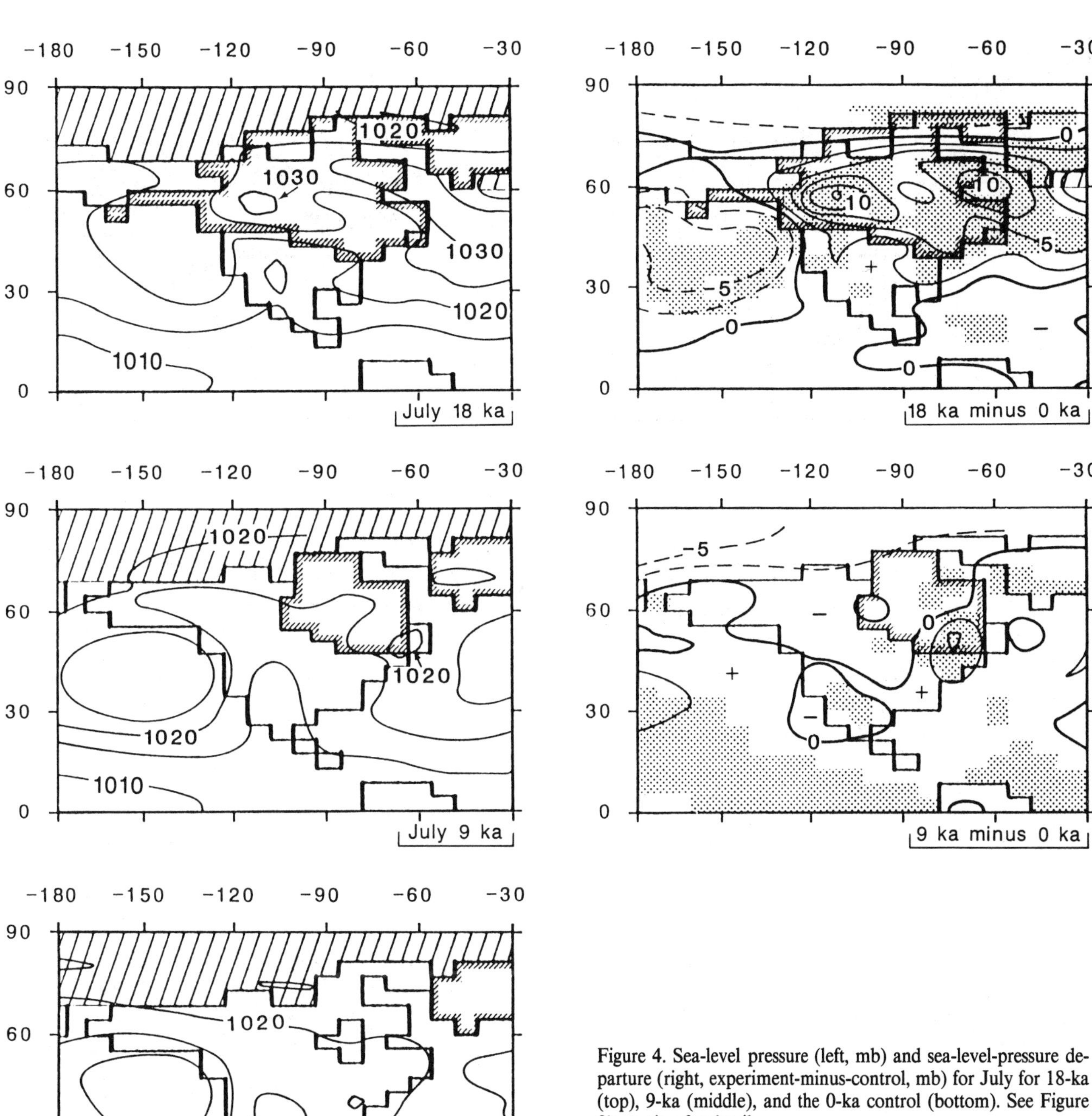

Figure 4. Sea-level pressure (left, mb) and sea-level-pressure departure (right, experiment-minus-control, mb) for July for 18-ka (top), 9-ka (middle), and the 0-ka control (bottom). See Figure 3's caption for details.

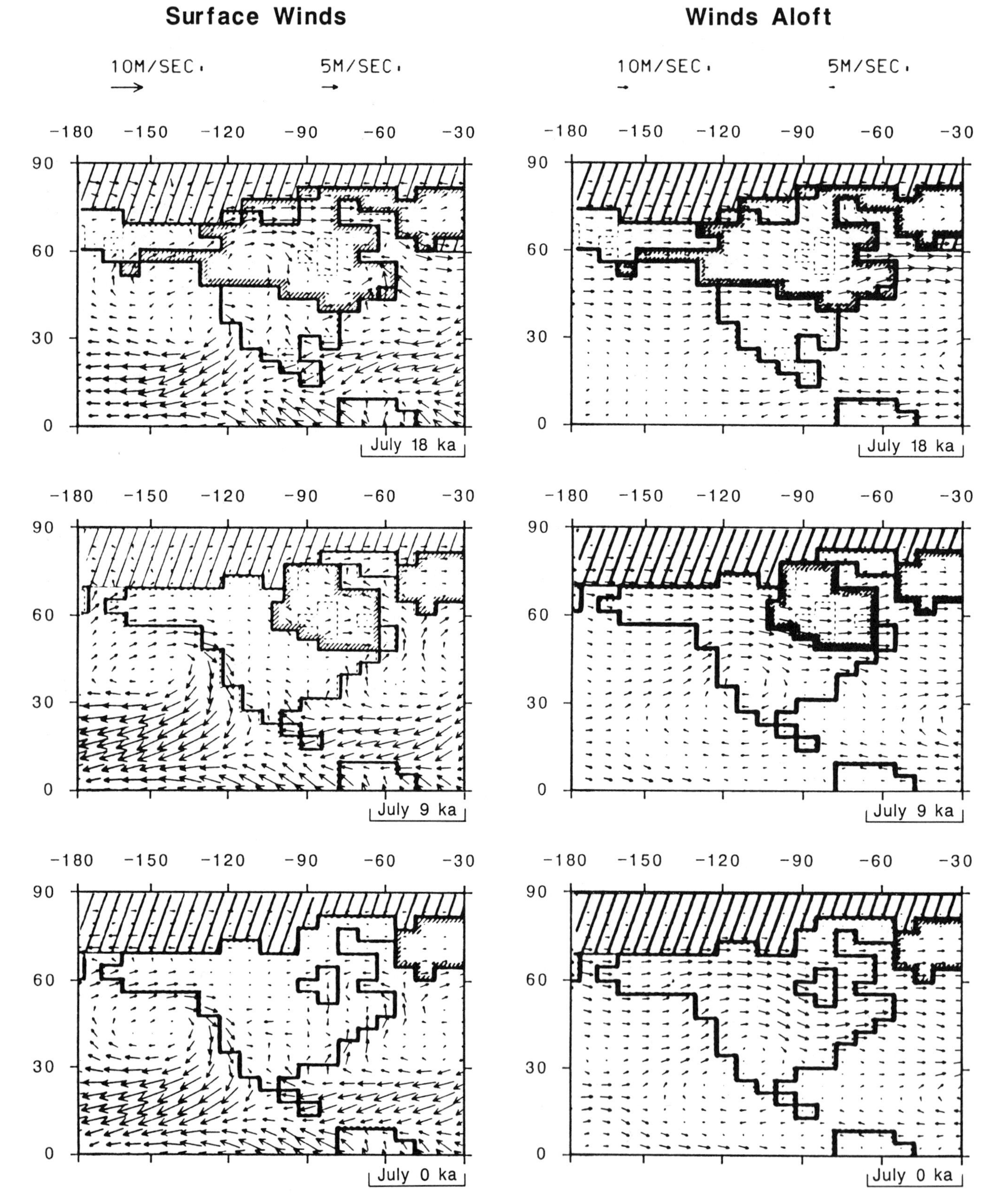

Figure 5. Surface winds (left) and winds at sigma levels 0.189 (about 190 mb or a height of about 12.2 km) (right) for July for 18-ka (top), 9-ka (middle), and the 0-ka control (bottom). Arrows denote direction and magnitude (according to key at top of figures).

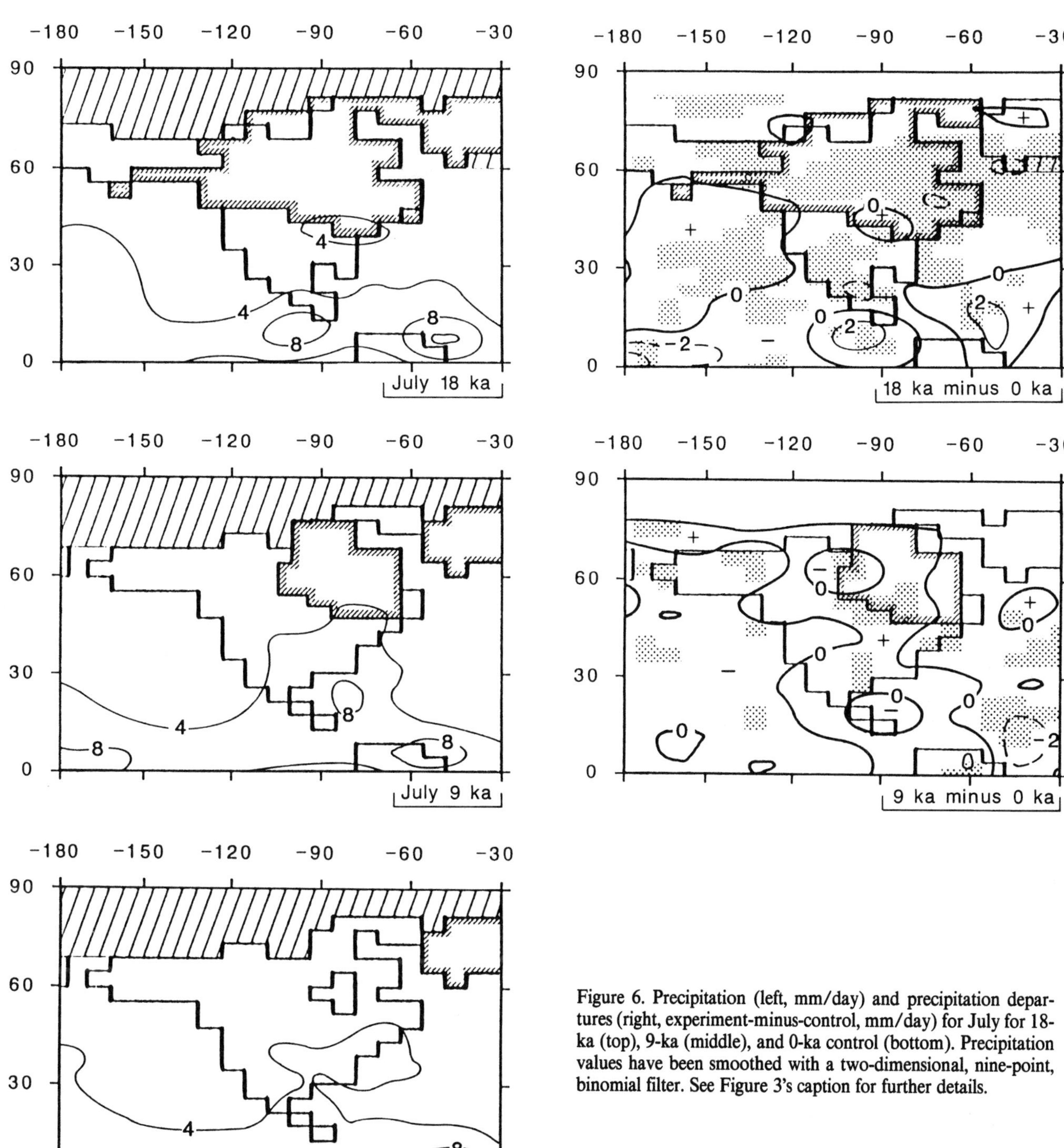

Figure 6. Precipitation (left, mm/day) and precipitation departures (right, experiment-minus-control, mm/day) for July for 18-ka (top), 9-ka (middle), and 0-ka control (bottom). Precipitation values have been smoothed with a two-dimensional, nine-point, binomial filter. See Figure 3's caption for further details.

where sea-surface temperature was warmer than at present (Fig. 4).

The westerly jet followed a split track around the North American ice sheet. The main branch was centered above the southern margin of the ice sheet and the North Atlantic sea ice (Fig. 5) and was about 5–10 m/s stronger than at 0 ka. Winds were weak over the ice sheet, and there was a secondary wind maximum along the northern flank. Storm tracks were shifted far south along the southern edge of the North American ice sheet and the North Atlantic sea-ice margin (see Kutzbach and Guetter, 1986). The surface winds show the anticyclonic circulation over the ice sheet. Precipitation (Fig. 6) was generally lower than at 0 ka across North America. The only exceptions were: (1) a small region along the jet core/storm track paralleling the southern edge of the ice sheet and (2) the west coast on the eastern fringe of the region of increased precipitation over the North Pacific.

9 ka—July. At 9 ka, the most important change in boundary conditions was the increased solar radiation associated with perihelion being in July and the axial tilt being increased. Surface temperature was increased everywhere in North America except over and near the residual ice sheet. The increase exceeded 2°C in places (Fig. 3). The large glacial anticyclone of 18 ka was now greatly reduced in size and centered along the southern edge of the ice (Fig. 4). Sea-level pressure was slightly lowered over the southwestern U.S., indicating a slightly strengthened summer monsoon. The surface winds had more southerly flow components along the Gulf Coast and in the south-central region (Fig. 5). The southern region experienced increased precipitation associated with the strengthened southerly flow (Fig. 6); whereas a small region along the southern edge of the ice sheet experienced increased storm-track precipitation.

18 ka—January. The prescribed major equatorward extension of sea ice in the North Atlantic to about 45°N was of particular importance for the simulated January climate. The replacement of open ocean by sea ice over this large area of the North Atlantic decreased the surface temperature by 30–40°C (it is 10–15°C at 0 ka and was -25 to -20°C at 18 ka; Fig. 7). The direct thermal influence of the ice sheet on surface temperature was small in January compared to July, because the region is snow covered and cold in the modern (control) simulation. The temperature was 5–10°C lower immediately south of the ice sheet, but only slightly lower in the southern U.S. and warmer than modern in Alaska (see below). An intense anticyclone covered the ice sheet and the frozen surface of the North Atlantic, but sea-level pressure was lower than now south of the ice sheet/sea-ice border and over the warmer North Pacific (Fig. 8).

The surface-temperature lowering increased the baroclinicity south of the ice sheet and along the sea-ice boundary and strengthened the North American jet and the equatorward-displaced North Atlantic jet (the maximum speed was 40 m/s at 18 ka, compared to 30 m/s at 0 ka; Fig. 9).

Another feature of major importance for the January circulation, similar to July, was the split in the westerly flow produced by the presence of the North American ice sheet. The southern branch of the jet developed over the eastern subtropical North Pacific and the southern U.S. The northern branch accounted for the advective warming of northwest North America at 18 ka and brought southerly flow across Alaska and westerly flow along the northern flank of the ice sheet (Fig. 9). The northern branch recurved southward between the eastern flank of the North American ice sheet and Greenland and rejoined the southern branch of westerlies over the North Atlantic. Because of the split flow, the minimum zonal wind at 18 ka over the ice sheet was at the latitude of the maximum zonal wind at 0 ka (Fig. 9). A similarly split flow pattern was simulated for 18 ka with the general circulation model of the Geophysical Fluid Dynamics Laboratory (Manabe and Broccoli, 1985).

The surface winds indicate an intense anticyclone over the North American ice sheet (Fig. 9). This anticyclone was also present in July of 18 ka (Fig. 5) but was less pronounced. Low-level outflow of cold air through the broad ice valley between North America and Greenland (wind speeds up to 15 m/s) and east of Greenland (10 m/s) could have helped to develop and maintain the extensive sea-ice cover in the North Atlantic (see Manabe and Broccoli, 1985). On the northern flank of the ice sheet, strong surface westerlies (5–10 m/s) occurred at 18 ka, whereas at 0 ka, winds are weak or easterly. South of the North American ice sheet, surface easterlies or weak westerlies occurred where winds today (0 ka) are strong westerly. This helped to produce more continental conditions at 18 ka, with surface temperature lowered by 5–10°C (Fig. 7).

The major eastern Pacific–North American wintertime storm track (not shown) was shifted about 20° of latitude south of its 0 ka position, roughly paralleling the southern branch of the jet (Fig. 9) and the band of lowered sea-level pressure (Fig. 8). Precipitation was increased in the central and eastern North Pacific (where CLIMAP-estimated SSTs were relatively warm at 18 ka), across the southwest and central U.S., and along the North Atlantic sea-ice border. This band of increased precipitation roughly paralleled the 18-ka storm track and the southern branch of the jet stream. Precipitation was decreased south of the ice sheet in the Pacific Northwest, where surface easterlies replaced the modern westerlies.

9 ka—January. The simulated January climate at 9 ka was the result of fairly subtle responses to the presence of the residual ice sheet and the decreased solar radiation. The changes were relatively small compared to those at 18 ka (and 9 ka July); in general they were not statistically significant. The decreased solar radiation caused lowered surface temperature over the southern U.S. (Fig. 7). From hydrostatic considerations, this led to increased surface pressure and the development of an anticyclone (Fig. 8). The increased westerly flow along the northern side of the anticyclone advected warmer air into the mid-continent (Fig. 7). A very similar but more pronounced response occurred in Europe (see Kutzbach and Guetter, 1986). The warmer conditions were also partly related to the anticyclonic flow associated with the residual ice sheet, i.e., warm advection on the west and north side of the glacial anticyclone; see the somewhat similar but

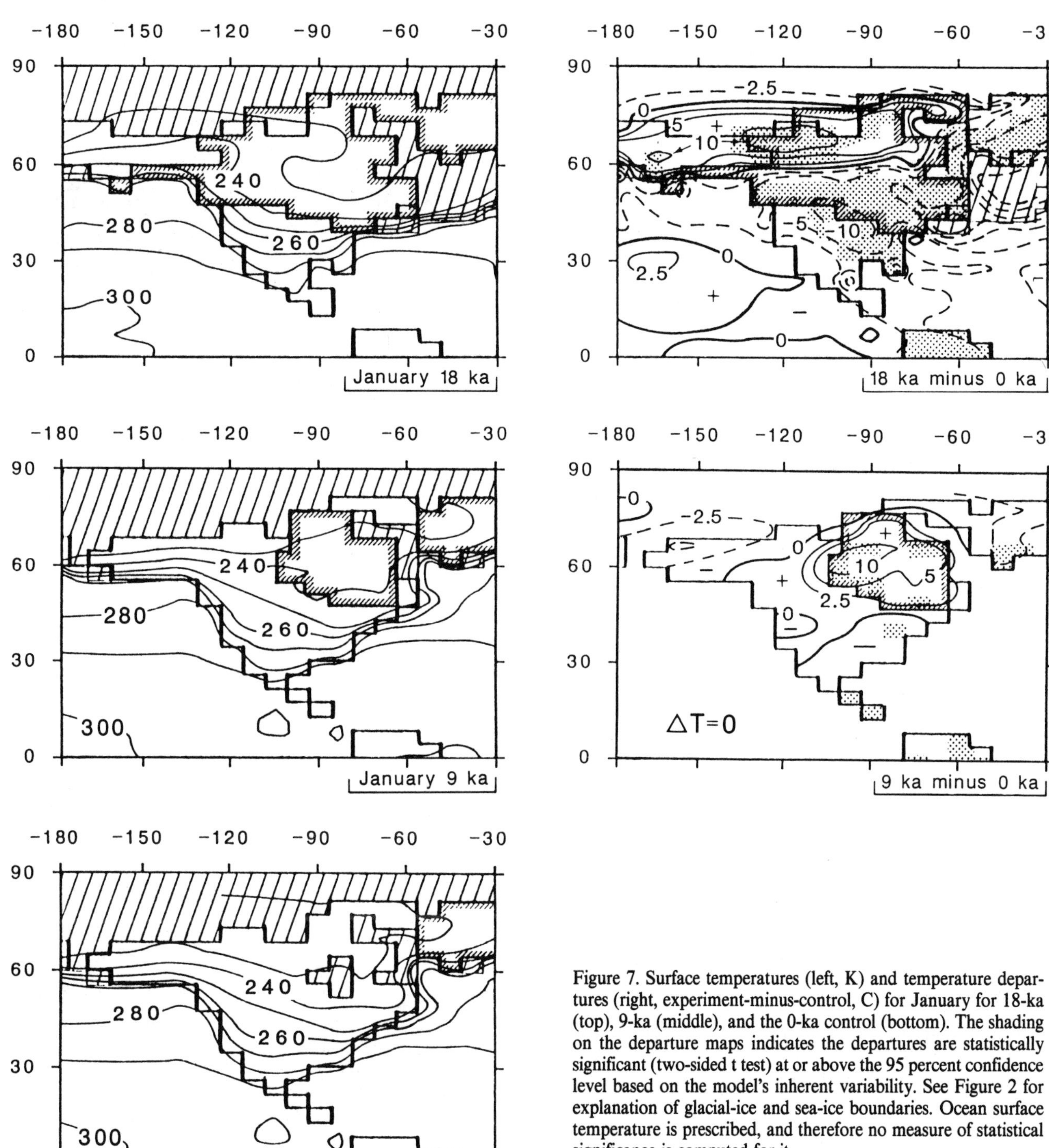

Figure 7. Surface temperatures (left, K) and temperature departures (right, experiment-minus-control, C) for January for 18-ka (top), 9-ka (middle), and the 0-ka control (bottom). The shading on the departure maps indicates the departures are statistically significant (two-sided t test) at or above the 95 percent confidence level based on the model's inherent variability. See Figure 2 for explanation of glacial-ice and sea-ice boundaries. Ocean surface temperature is prescribed, and therefore no measure of statistical significance is computed for it.

Sea-level Pressure

Sea-level Pressure Departure

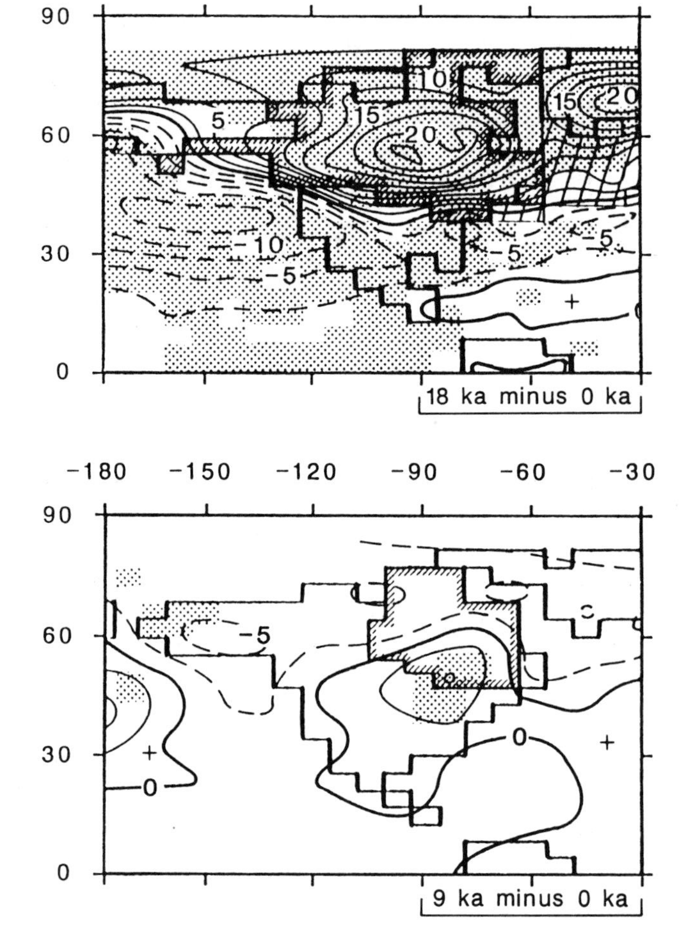

January 18 ka

January 9 ka

January 0 ka

Figure 8. Sea-level pressure (left, mb) and sea-level-pressure departure (right, experiment-minus-control, mb) for January for 18-ka (top), 9-ka (middle), and the 0-ka control (bottom). See Figure 3's caption for further details.

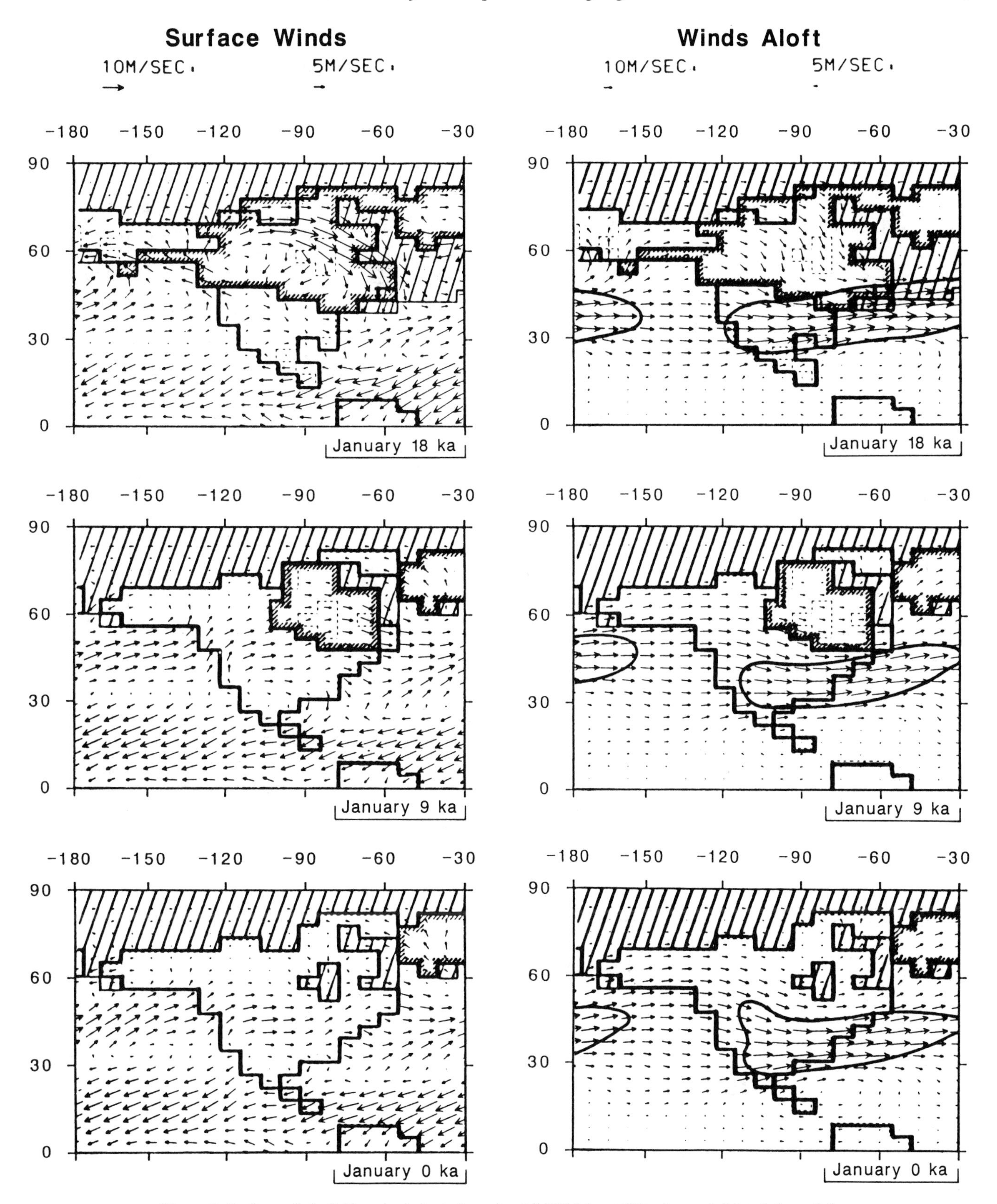

Figure 9. Surface winds (left) and winds at sigma level 0.500 (about 500 mb or a height of about 5.5 km) (right) for January for 18-ka (top), 9-ka (middle), and the 0-ka control (bottom). Arrows denote direction and magnitude (according to key at top of figures). For winds aloft, the 20 m/s isotach is shown.

Precipitation

−180 −150 −120 −90 −60 −30
90
60
30
0
4
4
4
8
4
January 18 ka

−180 −150 −120 −90 −60 −30
90
60
30
0
4
4
4
8
4
January 9 ka

−180 −150 −120 −90 −60 −30
90
60
30
0
4
4
4
8
8
January 0 ka

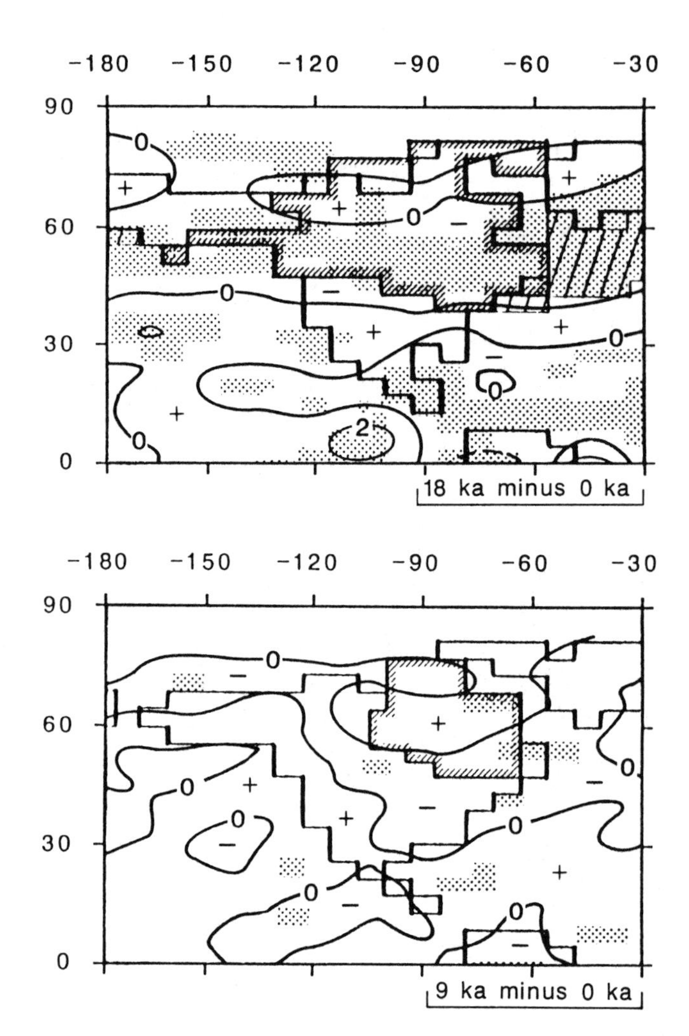

Figure 10. Precipitation (left, mm/day) and precipitation departures (right, experiment-minus-control, mm/day) for January for 18-ka (top), 9-ka (middle), and the 0-ka control (bottom). Precipitation values have been smoothed with a two-dimensional, nine-point, binomial filter. See Figure 3's caption for further details.

more intense pattern of 18 ka—January. In summary, the combination of the residual ice sheet and reduced insolation led to the (counterintuitive) result of slightly increased temperature in parts of northern North America. Changes in winds (Fig. 9) and precipitation (Fig. 10), compared to present, were small.

Summary of regional climates for 18, 15, 12, 9, 6, 3, and 0 ka

In this section, the results from the experiments between 18 and 9 ka (i.e., 15 and 12 ka) and between 9 and 0 ka (i.e., 6 and 3 ka) are included to give a more detailed summary of the temporal evolution of the climate. The behavior of the jet stream is first summarized, and then the regional time series of surface temperature, precipitation, and precipitation-minus-evaporation are described.

North American jet streams (Fig. 11). In July, the split flow around the North American ice sheet at 18 ka (Fig. 5) persisted at 15 ka with almost no change. By 12 ka, two important changes had occurred. First, the northern branch of the jet moved south, over and along the southern flank of the ice sheet, and merged with the southern branch over the northeastern U.S. This large change must have been related to the reduced size and height of the ice sheet. The second change was the reduced intensity of the North Atlantic extension of the jet. This was related to the (prescribed) retreat of the sea-ice border. The jet at 9 ka followed about the same track as at 12 ka, but with weakened intensity. At 6 ka and thereafter, only a single jet core was simulated over Alaska and northern Canada, and winds were weak compared to earlier times. In summary, the "modern" single jet core of July follows generally the same track as the northern branch of the split jet during July at the time of the glacial maximum. The southern branch of the split jet during glacial summers has no modern counterpart in July but resembles somewhat a modern January feature. The 12–9 ka patterns are transitional between glacial maximum and modern patterns (Fig. 11).

In January, the split flow around the North American ice sheet and the intense North American/North Atlantic jet core at 18 ka (Fig. 9) persisted at 15 ka with almost no change (Fig. 11). By 12 ka, the flow had adjusted to a single core of maximum winds that followed the west-coast-ridge, east-coast-trough pattern of today, i.e., the patterns of 12, 9, 6, 3, and 0 ka all have a single flow maximum rather than the split flow. At 12 ka, however, the jet maximum was almost as strong as at 18–15 ka (except in the southern U.S.), whereas by 9 ka the jet core strength was significantly weakened and similar to that of today. The decreased strength of the North Atlantic extension of the jet core (between 12 and 9 ka) is attributed primarily to the (prescribed) warming of the North Atlantic.

The major break in the January flow pattern between 15 ka (split jet flow and a minimum in the westerlies in western North America around 50°N) and 12 ka (combined jet flow and a maximum in the westerlies around 50°N) also had its counterpart in increased surface westerlies in the west and central areas of mid North America at 12 ka (not shown). The strong surface westerlies persisted at 9 ka and then weakened as jet-stream winds weakened.

Regional surface climates. The complicated changes in flow patterns in response to changes in solar radiation and lower boundary conditions led to a rich variety of regional climatic changes. These were summarized for North America for 18 ka and compared with the geologic record by Kutzbach and Wright (1985). Here similar regional summaries are presented, but in time-series fashion, to illustrate the times and magnitudes of major climatic changes, and regional differences. The eight areas are outlined in Figure 12. The climatic variables to be illustrated are July (growing season) temperature and annual precipitation and precipitation-minus-evaporation (Fig. 13).

In all areas, July temperature was lowest at 18 ka. Warming occurred at different rates in different areas, but by 9–6 ka the temperature was everywhere warmer than modern. The warming trend from 18 to 9 ka was due in part to the (prescribed) removal of glacial-age boundary conditions and in part to the increased summer solar radiation (Fig. 1). The maximum summer warmth, 9–6 ka, and the cooler summers after 6 ka, were caused by the radiation changes (Fig. 1).

Canada, east and west. Most of Canada was covered by glacial ice at 18 and 15 ka, and July surface temperature was below 0°C, or about 25°C lower than modern. Consistent with the earlier withdrawal of ice in the west, the west warmed earlier than the east. By 6 ka the temperature was 1–2°C warmer than now. Precipitation decreased at 18–12 ka due to the colder temperature and southward shift of storm tracks. See the discussion of aspects of the mass budget of the ice sheet below.

Alaska, central and north. This region (ice-free in the model) was only slightly colder than present in July of 18–15 ka and warmer than present at 9–6 ka. Both precipitation and P–E were consistently lower than today for the entire period from 18 ka.

Northwest. This region, immediately south of the ice sheet at 18–15 ka, had summer temperatures about 4°C colder than present. By 9–6 ka it was 2°C warmer than present. It was somewhat drier at 18–15 ka, when the storm track was shifted south and weak easterly winds prevailed. There was a brief return to wetter conditions at 12 ka, when the westerly flow resumed. P–E again became lower than present at 9–6 ka due primarily to increased evaporation (similar to the north-central region, discussed next).

North-Central. The north-central region, from the Rockies to the Appalachians and from immediately south of the ice sheet to 40°N, had summer temperatures of 16°C at 18–15 ka, about 7°C below present. The region was 2°C above present at 9–6 ka. Rainfall was less than present at 18–15 ka (colder, with the storm track shifted south of the region) and at 9–6 ka (less summer rain in the continental interior). Precipitation-minus-evaporation was slightly increased at 18–15 ka (evaporation decreased more than precipitation) and was more significantly decreased at 9–6 ka,

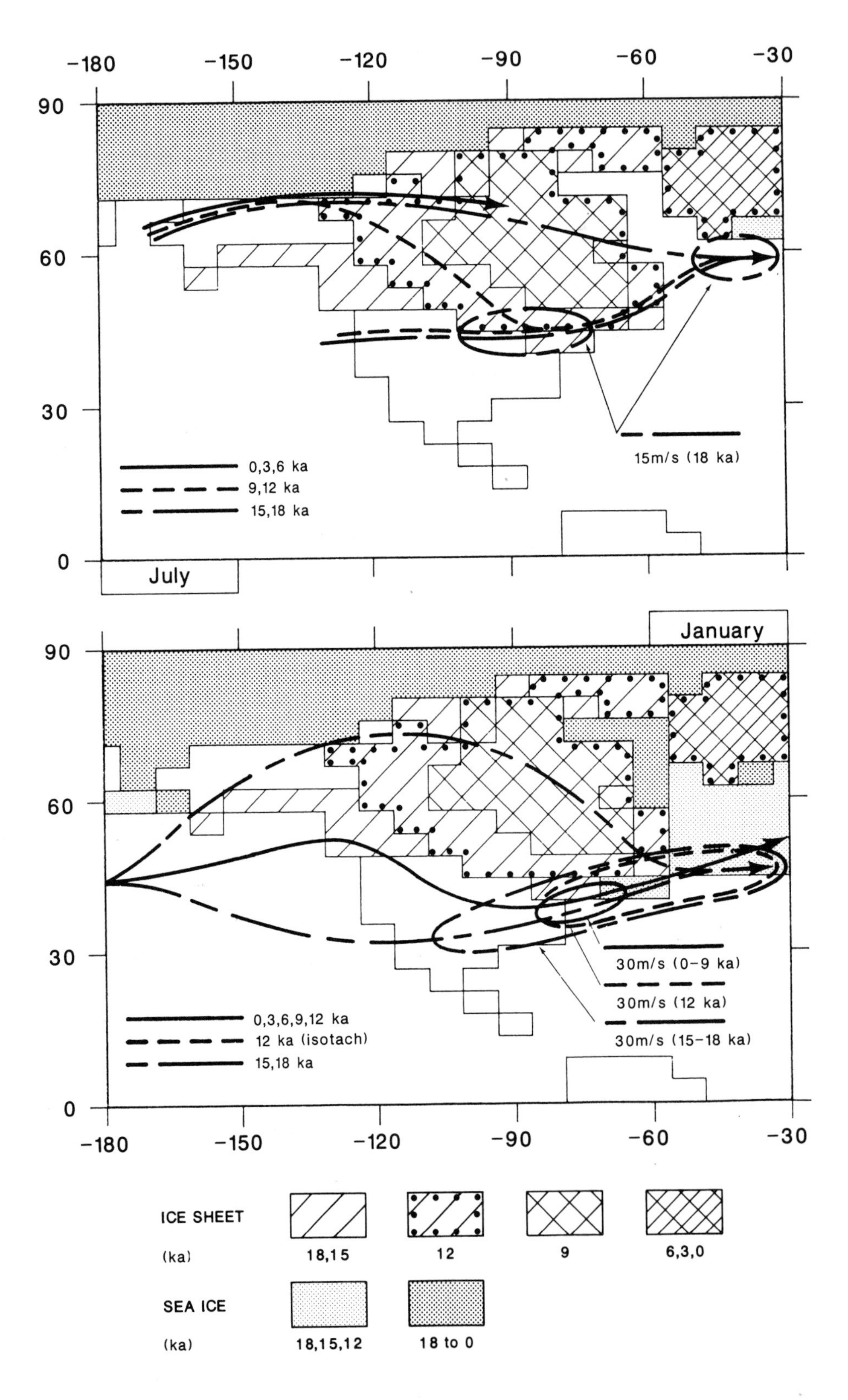

Figure 11. Schematic location of core of maximum winds and wind-speed maxima at sigma level 0.500 (about 500 mb or about 5.5 km) at 3,000-year intervals, 18 to 0 ka, July and January.

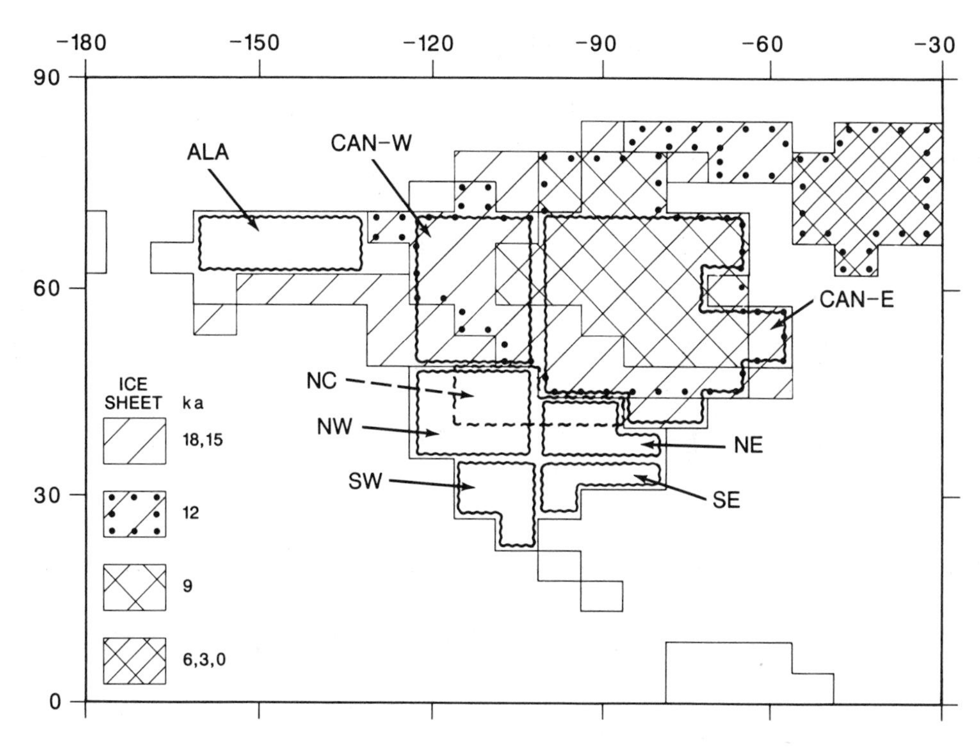

Figure 12. Location of eight areas for which area-average climatic time series are presented (see Fig. 13). Ice-sheet boundaries are indicated. Abbreviations are: CAN W (western Canada), CAN E (eastern Canada), ALA (Alaska), NW (northwest area—south of the 18-ka ice sheet), NC (north-central), NE (northeast), SW (southwest), SE (southeast).

when evaporation increased (associated in July with increased solar and net radiation at the surface and with higher temperature) and precipitation decreased.

Northeast. The region was 6°C colder than present at 18 ka and remained relatively cold even at 12 ka, reflecting the persistence of the ice sheet immediately to the north. It was 2°C warmer than present at 9–6 ka. Conditions were slightly wetter at 18–15 ka, associated with the increased storminess along the ice-sheet/sea-ice border. At 9 ka, precipitation was the same as present, but P–E was less than present, reflecting drier conditions associated with the increased summer warmth and increased evaporation (similar to the situation in the north-central region).

Southwest. The temperature was only 2°C lower than present at 18 ka. Precipitation and P–E were higher, reflecting both increased winter rains (associated with the southward shift of the storm track and jet) and reduced evaporation (associated with lower temperature). At 9–6 ka, it was somewhat wetter than present, associated with the intensified summer monsoon circulation.

Southeast. The southeast was 2°C colder than present at 18 ka but warmed to present levels by 15 ka. It was distinctly drier at 18–12 ka, associated primarily with reduced summer precipitation. The prescribed sea-surface temperature in the western North Atlantic was lower, and this probably contributed to the reduction of summer precipitation (see Imbrie and others, 1983). The region was also in an area of subsidence associated with the exit region of the summertime jet core (for details see Kutzbach and Wright, 1985). At 9–3 ka it was wetter than present, associated with the intensified summer monsoon.

Mass budget of the North American ice sheet

The deglaciation of North America has prompted questions and studies of the mass budget of the ice sheet, the rate of deglaciation, and the sources of energy for melting the ice. Andrews (1973) estimated that horizontal retreat rates of 200 m per year along the southern ice margin would require vertical ice wastage of between 10 and 50 m per year. The required energy for melting this ice would be on the order of 100 to 500 W/m^2. Andrews estimated that net radiation, downward flux of sensible heat (associated with a temperature inversion), and condensation of atmospheric moisture on the ice surface might possibly provide a total of about 75 W/m^2 (i.e., a considerable amount of energy but not even enough to fully explain the lowest wastage rate) and concluded that calving of ice into the sea or into glacial lakes might have been the most likely additional ablation process. Hare (1976) reviewed Andrew's estimates and again emphasized the need to study the energetics of deglaciation.

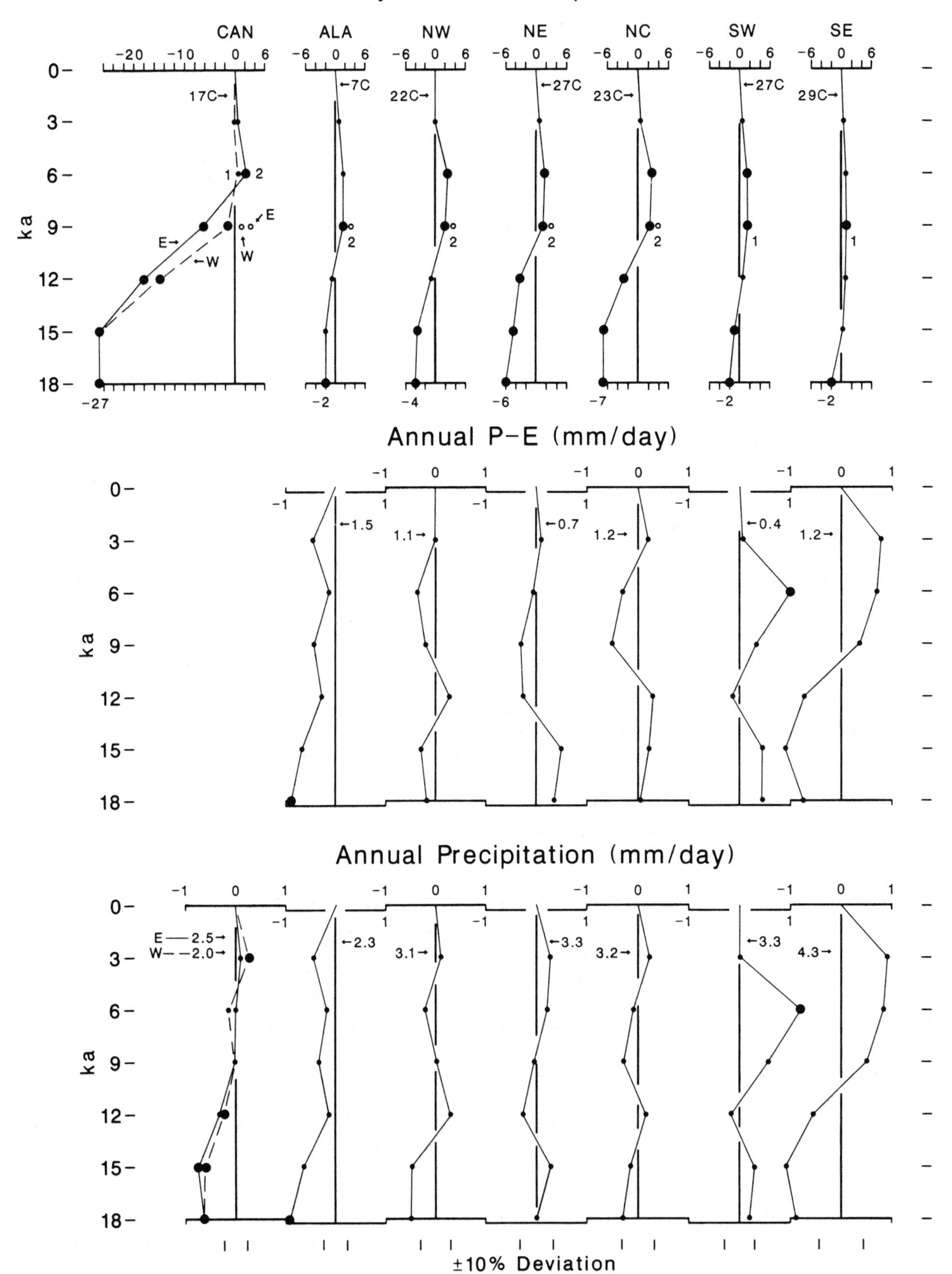
July Surface Temperature (C)
CAN
ALA
NW
NE
NC
SW
SE
ka
17C→
←7C
22C→
←27C
23C→
←27C
29C→
-27
-2
-4
-6
-7
-2
-2
Annual P−E (mm/day)
←1.5
1.1→
←0.7
1.2→
←0.4
1.2→
Annual Precipitation (mm/day)
E — 2.5→
W– –2.0→
←2.3
3.1→
←3.3
3.2→
←3.3
4.3→
±10% Deviation

General circulation models can provide estimates of at least some components of the ice sheet's mass and energy budget. Manabe and Broccoli (1985), in a simulation experiment for 18 ka, calculated an annual-average wastage rate of 0.2 cm per day (about 0.7 m per year) averaged for the entire ice sheet. This rate, if linearly extrapolated, would melt 2,000 m of ice in less than 3,000 years and was therefore viewed by them as an excessive rate for 18-ka conditions. They found wastage rates exceeding 1 cm per day (3.6 m per year) along some portions of the southern ice margin and identified downward sensible heat flux and condensation as the two major sources of heat for melting.

The version of the NCAR CCM used in our simulation experiments, unlike the model used by Manabe and Broccoli (1985), does not include an explicit calculation of snowmelt nor does it partition the calculated precipitation between snow or rain. It does, however, calculate the evaporation or condensation at the surface. Using this information and also estimating the snow/rain fraction based on the simulated air temperature, it is possible to roughly infer certain features of the ice sheet mass budget. Assuming that the January precipitation was snow and that the July precipitation was rain and therefore did not contribute to the ice mass budget (valid along the southern margin where most of the July precipitation occurred with the temperature above freezing), there is an approximate mass balance at 18, 15 and 12 ka (P–E approximately zero). At 9 ka when the prescribed ice sheet is relatively thin, wastage occurs (P–E negative), but at a rate that is less than that estimated explicitly by Manabe and Broccoli (1985) for 18 ka. Thus, the inferred area-average mass balances for the entire ice sheet do not fit the observed deglaciation sequence.

Perhaps of greater interest here than the overall mass budget, which the NCAR model did not calculate explicitly, is the magnitude of the downward sensible heat flux and condensational heating along the southern margin of the ice sheet in July, the likely time of maximum wastage. At 18 ka the southern ice margin is at 42N, the air temperature is several degrees above freezing over the ice surface (it is 10–15°C at the grid squares immediately south of the ice), the sensible heat flux to the ice surface is about 50 W/m^2 (i.e., a downward flux associated with the temperature inversion), and condensation is occurring with an energy gain to the ice surface of about 20 W/m^2. The net radiation is near zero. These model calculations imply that about 70 W/m^2 is available for melting ice during the peak summer season. (Additional thermal energy from warm rain falling on the ice would be less than 1 W/m^2.

Assuming, for discussion, that this summer melting rate might exist for three months per year, it would amount to a vertical wastage rate of about 1.5 m/year. These results, although based on an incomplete ice budget, are of the same sign and magnitude as those reported in the more complete analysis by Manabe and Broccoli (1985); the estimates also tend to confirm the order of magnitude estimates of Andrews (1973) and Hare (1976). Our model results for July for 15, 12, and 9 ka are similar to those for 18 ka except that the band of downward heat flux and condensation along the ice front moves north as the prescribed southern ice margins is moved north.

The overall climatic environment simulated by the model along the southern ice margin in eastern North America in July may be described as follows: the winds are weak southerly (upslope), there is a downward flux of heat because the air temperature is above freezing while the ice-surface temperature is at the freezing point, atmospheric moisture is condensing on the ice, and it is cloudy and rainy (associated with the juxtaposition of the jet stream and storm track). However, in spite of these simulated conditions being very favorable for ice wastage, our experiments do not completely answer the questions that have been raised about the detailed ice and energy budgets during deglaciation.

SENSITIVITY STUDIES AND SIMULATIONS WITH OTHER MODELS

Sensitivity studies with our model and the results of studies with other models provide additional insight on climatic mechanisms during deglaciation.

Sensitivity studies

There is considerable uncertainty associated with each of the atmospheric or lower boundary conditions that were prescribed in the experiments. Moreover, there is an intrinsic interest in knowing the effect of changing a single condition in contrast to the combined effect of changes in, for example, solar radiation, ice sheet, and sea-surface temperature. Several of our experiments have been aimed at estimating model sensitivity, but much more needs to be done. Broccoli and Manabe (1987) have reported a series of sensitivity experiments for the last glacial in which they determine the incremental effects of continental ice, reduced atmospheric CO_2 concentration, and changed land albedo. Related studies are reported by Rind and Peteet (1986). Rind and Peteet (1985) also considered the sensitivity of 18-ka simulations to changes of the CLIMAP-specified sea-surface temperatures.

9-ka ice sheet. The 9-ka climate with and without the residual ice sheet was simulated by Kutzbach and Otto-Bliesner (1982) and Kutzbach and Guetter (1984). The presence of the ice sheet significantly influenced the local climate but had little effect outside North America. Figure 13 illustrates the regional July

Figure 13. Area-average changes of surface temperature and estimated annual precipitation and precipitation-minus-evaporation for eight areas in North America for the period 18–0 ka in 3 ka intervals. See Figure 12 for location of the eight areas. The time series are departures from the control (0-ka) simulation. The control value is indicated at the top of each series. Large dots indicate that the departure is significant above the 95 percent level compared to the model's inherent variability. The open circles at 9 ka show the surface-temperature departure for the experiment without an ice sheet.

surface temperature at 9 ka with and without the ice sheet. In eastern Canada, for example, the simulated climate without the ice sheet was about 3°C warmer than present rather than 6°C colder than present.

Atmospheric CO_2. The 18-ka climate, as reported under Results but with the atmospheric CO_2 concentration lowered to 200 ppmv, was simulated by Kutzbach and Guetter (1986). Over North America, the main result was an additional decrease of temperature of about 1°C and a further small reduction in precipitation.

Soil moisture. The 9-ka climate was simulated with a version of the CCM that included interactive soil moisture and a full seasonal solar radiation cycle (see Kutzbach and Guetter, 1986). For North America, the primary difference was that the warming and drying of the continental interior was more extreme than was shown here.

Simulations with other models

All AGCMs have somewhat different parameterizations of physical processes, different horizontal and vertical resolutions, and, up until now, different paleoclimatic boundary conditions. These differences make it difficult to undertake detailed comparisons. Nevertheless, this will be an important task for future work.

Compared to the 18-ka experiment with the NCAR CCM, the 18-ka climate simulation of Manabe and Brocolli (1985) and Brocolli and Manabe (1987) produced a very similar split flow around the North American ice sheet, a similar drier condition south of the ice sheet in west–central U.S., a similar increase in precipitation in the southwest and northeast U.S., and a similar warmer region in Alaska in winter (see also Rind and Peteet, 1986).

Other model experiments for 9 ka have produced a similar summertime warming of 2–4°C for the northern midcontinents (Mitchell, 1977; Crowley and others, 1986). Rind and others (1986) have simulated possible climatic scenarios for the Younger Dryas (11–10 ka), using combinations of changes in solar-radiation and North Atlantic sea-ice and ocean temperature that might have been associated with the deglaciation of North America.

CONCLUSIONS

The sequence of 14 paleoclimate experiments (seven for July, seven for January) provides an overview of the large and diverse climatic changes that may have taken place in North America over the past 18,000 yr. From these experiments, we can draw several conclusions:

1) There were large changes in glacial-age circulation at both the surface and jet-stream levels and associated changes in temperature and moisture conditions. These climatic changes were primarily the result of prescribed glacial-age lower boundary conditions at 18–15 ka. At 12 ka, and to a much less extent at 9 ka, persistent effects of glacial-age boundary conditions influenced the climate. Beginning at 15 ka and culminating at 9–6 ka, the major changes in the seasonal solar-radiation cycle caused significant temperature changes over land and altered the land-ocean temperature and pressure contrasts and the seasonal monsoons.

2) Many of the simulated changes appear to be in agreement with the geologic record. Comparisons between the simulation for North America for 18 ka and the corresponding geologic record are in Kutzbach and Wright (1985). Spaulding and Graumlich (1986) compare model simulations and geologic observations for the Southwest, 18 to 0 ka. Ritchie and others (1983) report fossil–pollen and other evidence of an early Holocene thermal maximum in northwest Canada centered near 10–9 ka that is consistent with the model results. The simulated changes in precipitation-minus-evaporation are also in fair agreement with lake-level variations in North America during the past 18,000 yr (Street-Perrott, 1986; Harrison and Metcalfe, 1985; Winkler and others, 1986). Other comparisons are in this volume (e.g., Webb and others, Barnosky and others).

Although the comparisons between model and data show considerable agreement, there are also differences in magnitude, timing, and even the sense of the change. The process of reconciling such differences will involve further work on the models and on the geologic record. Present models have too coarse a grid resolution and involve too many simplifications to permit detailed comparison with local geologic records.

3) Perhaps more important than the specific detailed results are certain conceptual findings: a) The changes in orbital parameters produce a pattern of early- to mid-Holocene summertime warming and subsequent cooling that agrees with evidence. The orbital theory indicates, however, that this was primarily a summertime phenomenon and that mid-Holocene winters were cooler in some locations. During the period of maximum summer warmth, 9–6 ka, the Southwest and Southeast experienced increased P–E due to increased summer (monsoonal) precipitation. The increased precipitation did not in general extend to the northern U.S., and with increased evaporation there the P–E was decreased (especially in the north-central region).

b) The presence of the ice sheet and North Atlantic sea ice produced major shifts in jet stream patterns, storm tracks, and precipitation patterns. The simulations show that the results are rather sensitive to the prescribed lower boundary conditions, so that it is important to estimate ice-sheet height and area and sea-ice cover quite accurately (see below). While the significantly lower summer temperatures south of the ice sheet were to be expected, the temperature changes in the south and in Alaska were surprisingly small.

c) The simulated changes in tropospheric wind patterns and storm tracks have features that support the discussions of Ruddiman and McIntyre (1981a) concerning the role of extensions of North Atlantic sea-ice in shifting storm tracks (and precipitation) away from the North American ice sheet.

d) The stronger westerly flow at glacial maximum in the Arctic may have implications for the glacial-age circulation patterns of the Arctic Ocean.

4) Several of the results, along with new developments in climate modeling, point toward possible future studies: There is a need for additional sensitivity studies to understand how the timing of shifts in wind patterns is related in detail to the specified height, shape, and area of ice sheets and the percent sea-ice cover. There is a need to repeat similar experiments with models that incorporate more detailed calculations of surface-soil moisture and snow-cover budgets (see Kutzbach and Guetter, 1986) and interactive oceans (see Manabe and Broccoli, 1985). There is a need to use finer-mesh models in order to resolve greater spatial detail. (Work is beginning, jointly with NCAR, to use a model with a grid resolution of about 100 km.) There is a need for greater time resolution. It would be possible, using AGCMs, to reduce further the interval between "snapshots" (say, from 3,000 years to 1,000 years), but even this would not provide a true time-dependent simulation. Rather, coupled models with interactive ice sheets, oceans, land surface, and biosphere will ultimately be needed to study in greater detail the mechanisms whereby a more complete climate system responds to orbital forcing and shorter-term perturbations such as volcanic eruptions.

REFERENCES CITED

Alyea, F. N., 1972, Numerical simulation of an ice age paleoclimate: Fort Collins, Colorado State University, Atmospheric Science Paper no. 193, 120 p.

Andrews, J. T., 1973, The Wisconsin Laurentide ice sheet; Dispersal centers, problems of rates of retreat, and climatic implications: Arctic and Alpine Research, v. 5, p. 185–199.

Barry, R. G., 1983, Late Pleistocene climatology, *in* H. E. Wright, Jr., ed., Late Quaternary environments of the United States, v. 1, p. 390–407.

Berger, A. L., 1978, Long-term variations of caloric solar radiation resulting from the earth's orbital elements: Quaternary Research, v. 9, p. 139–167.

Blackmon, M. L., Giesler, J. E. and Pitcher, E. J., 1983, A general circulation model study of January climate anomaly patterns associated with interannual variation of equatorial Pacific sea surface temperatures: Journal of the Atmospheric Sciences, v. 40, p. 1410–1425.

Broccoli, A. J., and Manabe, S., 1987, The influence of continental ice, atmospheric CO_2, and land albedo on the climate of the last glacial maximum: Climate Dynamics, v. 1, p. 87–99.

Chervin, R. M., and Schneider, S. H., 1976, On determining the statistical significance of climate experiments with general circulation models: Journal of the Atmospheric Sciences, v. 33, p. 405–412.

CLIMAP Project Members, 1981, Seasonal reconstructions of the earth's surface at the last glacial maximum: Geological Society of America Map Chart Series, MC-36.

Crowley, T. J., Short, D. A., Mengel, J. G., and North, G. R., 1986, Role of seasonality in the evolution of climate during the last 100 million years: Science, v. 231, p. 579–584.

Denton, G. H., and Hughes, T. J., eds., 1981, The last great ice sheets: New York, John Wiley and Sons, 484 p., 28 folded maps.

Gates, W. L., 1976a, Modeling the ice-age climate: Science, v. 191, p. 1138–1144.

——, 1976b, The numerical simulation of ice-age climate with a global general circulation model: Journal of the Atmospheric Sciences, v. 33, p. 1844–1873.

Hansen, J., Lacis, A., Rind, D., Russell, G., Stone, P., Fung, I., Ruedy, R., and Lerner, J., 1984, Climate sensitivity; Analysis of feedback mechanisms, *in* Hansen, J. E. and Takahashi, T., eds., Climate processes and climate sensitivity: American Geophysical Union, Maurice Ewing Series no. 5, p. 130–163.

Hare, F. K., 1976, Late Pleistocene and Holocene climates; Some persistent problems: Quaternary Research, v. 6, p. 507–516.

Harrison, S. P., and Metcalfe, S. E., 1985, Spatial variations in lake levels since the last glacial maximum in the Americas north of the equator: Zeitschrift fur Gletscherkunde und Glazialgeologie, v. 21, p. 1–15.

Imbrie, J., McIntyre, A., and Moore, T. C., Jr., 1983, The ocean around North America at the last glacial maximum, *in* Wright, H. E., Jr., ed., Late Quaternary environments of the United States, *in* Porter, S. C., ed., v. 1, The Late Pleistocene: Minneapolis, University of Minnesota Press, p. 230–236.

Kolla, K., Biscaye, P. E., and Hanley, A. F., 1979, Distribution of quartz in late Quaternary Atlantic sediments in relation to climate: Quaternary Research, v. 11, p. 261–277.

Kutzbach, J. E., 1981, Monsoon climate of the early Holocene; Climatic experiment using the Earth's orbital parameters for 9,000 years ago: Science, v. 214, p. 59–61.

Kutzbach, J. E., and Guetter, P. J., 1984, The sensitivity of monsoon climates to orbital parameter changes for 9,000 years BP; Experiments with the NCAR general circulation model, *in* Berger, A., Imbrie, J., Hays, J., Kula, G., and Saltzman, B., eds., Milankovitch and climate, Part 2: Dordrecht, Netherlands, D. Reidel, p. 801–820.

——, 1986, The influence of changing orbital parameters and surface boundary conditions on climate simulations for the past 18,000 years: Journal of the Atmospheric Sciences, v. 43, p. 1726–1759.

Kutzbach, J. E., and Otto-Bliesner, B. L., 1982, The sensitivity of the African–Asian monsoonal climate to orbital parameter changes for 9,000 years B.P. in a low-resolution general circulation model: Journal of the Atmospheric Sciences, v. 39, p. 1177–1188.

Kutzbach, J. E. and Wright, H. E., Jr., 1985, Simulation of the climate of 18,000 yr BP; Results for the North American/North Atlantic/European Sector: Quaternary Science Reviews, v. 4, p. 147–187.

Lorius, C., Raynaud, D., Petit, J-R., Jousel, J., and Merlivat, L., 1984, Late-glacial maximum-Holocene atmospheric and ice-thickness changes from Antarctic ice-core studies: Annals of Glaciology, v. 5, p. 88–94.

Manabe, S., and Broccoli, A. J., 1985, The influence of continental ice sheets on the climate of an ice age: Journal of Geophysical Research, v. 90, p. 2167–2190.

Manabe, S., and Hahn, D. G., 1977, Simulation of the tropical climate of an ice age: Journal of Geophysical Research, v. 82, p. 3889–3911.

Mitchell, J.F.B., 1977, The effect on climate of changing the earth's orbital parameters; Two summer integrations with fixed sea surface temperatures: Bracknell, U.K., Meteorological Office 20 Technical Note II/100, 18 p. (Permission to cite this reference has been obtained from the Assistant Director of the Meteorological Office.)

Mix, A. C., and Ruddiman, W. F., 1985, Structure and timing of the last deglaciation; Oxygen isotope evidence: Quaternary Science Reviews, v. 4, p. 59–108.

Neftel, A., Oeschger, H., Schwander, J., Stauffer, B., and Zumbrunn, R., 1982, Ice core sample measurements give atmospheric CO_2 content during the past 40,000 yr: Nature, v. 295, p. 220–223.

Oeschger, H., Beer, J., Seigenthaler, V., Stauffer, B., Dansgaard, W., and Langway, C. C., 1983, Late-glacial climate history from ice cores, in Ghazi, A., ed., Palaeoclimatic models and research: Dordrecht, Netherlands, D. Reidel, p. 95–107.

Petit, J-R, Briat, M., and Royer, A., 1981, Ice age aerosol content from east Antarctic ice core samples and past wind strength: Nature, v. 293, p. 391–394.

Pitcher, E. J., Malone, R. C., Ramanathan, V., Blackmon, M. L., Puri, K., and Bourke, W., 1983, January and July simulations with a spectral general circulation model: Journal of the Atmospheric Sciences, v. 40, p. 580–604.

Ramanathan, V., Pitcher, E. J., Malone, R. C., Blackmon, M. L., 1983, The response of a spectral general circulation model to refinements in radiative processes: Journal of the Atmospheric Sciences, v. 40, p. 605–630.

Rind, D., and Peteet, D., 1985, Terrestrial conditions at the last glacial maximum and CLIMAP sea-surface temperature estimates; Are they consistent?: Quaternary Research, v. 24, p. 1–22.

—— , 1986, Comment on S. H. Schneider's Editorial "Can modeling of the ancient past verify prediction of future climates?": Climatic Change, v. 9, p. 357–360.

Rind, D., Peteet, D., Broecker, W., McIntyre, A., and Ruddiman, W., 1986, The impact of cold North Atlantic sea surface temperatures on climate; Implications for the Younger Dryas cooling (11–10K): Climate Dynamics, v. 1, p. 3–34.

Ritchie, J. C., Cwynar, L. C., and Spear, R. W., 1983, Evidence from north-west Canada for an early Holocene Milankovitch thermal maximum: Nature, v. 305, p. 126–128.

Ruddiman, W. F., and McIntyre, A., 1981a, Oceanic mechanisms for amplification of the 23,000-year ice-volume cycle: Science, v. 212, p. 617–627.

—— , 1981b, The North Atlantic Ocean during the last deglaciation: Palaeogeography, Palaeoclimatology, Palaeoecology, v. 35, p. 145–214.

Spaulding, W. G., and Graumlich, L. J., 1986, The last pluvial climatic episode in the deserts of southwestern North America: Nature, v. 320, p. 441–444.

Street-Perrot, F. A., 1986, The response of lake levels to climatic change; implications for the future, *in* Rosenzweig, C. and Dickinson, R., eds., Climate-Vegetation Interactions: Boulder, Colorado, Office for Interdisciplinary Earth Studies: University Corporation for Atmospheric Research, Report OIES-2, p. 77–80.

Stuiver, M., Burk, R. L., and Quay, P. D., 1984, $^{13}C/^{12}C$ ratios in tree rings and the transfer of biospheric carbon to the atmosphere: Journal of Geophysical Research, v. 89, p. 11731–11748.

Thompson, L. G., and Mosely-Thompson, E., 1981, Temporal variability of microparticle properties in polar ice sheets: Journal of Volcanology and Geothermal Research, v. 11, p. 11–27.

Washington, W. M., and Meehl, G. A., 1984, Seasonal cycle experiment on the climate sensitivity due to a double of CO_2 with an atmospheric general circulation model coupled to a simple mixed-layer ocean model: Journal of Geophysical Research, v. 89, p. 9475–9503.

Webb, T., III, Kutzbach, J., and Street-Perrott, F. A., 1985, 20,000 years of global climatic change; Paleoclimatic research plan, *in* Malone, T. F. and Roeder, J. G., eds., Global change: International Council of Scientific Unions Press, p. 182–218.

Williams, J., Barry, R. G., and Washington, W. M., 1974, Simulation of the atmospheric circulation using the NCAR global circulation model with ice age boundary conditions: Journal of Applied Meteorology, v. 13, p. 305–317.

Winkler, M. G., Swain, A. M., and Kutzbach, J. E., 1986, Middle Holocene dry period in the northern Midwestern United States; Lake levels and pollen stratigraphy: Quaternary Research, v. 25, p. 235–250.

Manuscript Accepted by the Society March 16, 1987

ACKNOWLEDGMENTS

Research grants to the University of Wisconsin–Madison from the National Science Foundation's Climatic Dynamics Program (grants ATM-8219070 and ATM-8412958) supported this work. The computations were made at the National Center for Atmospheric Research (NCAR), which is sponsored by the National Science Foundation, with a computing grant from the NCAR Computing Facility (no. 35381017). Warren Washington (NCAR) advised in the use of the NCAR Community Climate Model, Peter Guetter carried out the experiments and processed the results, Mary Sternitzky prepared the manuscript, and Bryan Richards and Pat Behling prepared the illustrations. The author was a summer visitor to the AAP Climate Section, NCAR, August 1984, 1985, and 1986.

Printed in U.S.A.

The Geology of North America
Vol. K-3, North America and adjacent oceans during the last deglaciation
The Geological Society of America, 1987

Chapter 20

Climatic change in eastern North America during the past 18,000 years; Comparisons of pollen data with model results

Thompson Webb III
Department of Geological Sciences, Brown University, Providence, Rhode Island 02912
Patrick J. Bartlein
Department of Geography, University of Oregon, Eugene, Oregon 97403
John E. Kutzbach
Center for Climatic Research, University of Wisconsin, Madison, Wisconsin 53706

INTRODUCTION

The Laurentide ice sheet expanded and retreated as part of the response of the climate system to the changes in solar radiation that are determined by the Earth's orbital variations (Berger and others, 1984). In turn, the area, height, and reflectivity of the ice sheet influenced the climate of the Northern Hemisphere. During the last deglaciation the changing size of the ice sheet and the changing latitudinal and seasonal distribution of solar radiation (Ruddiman and McIntyre, 1981) were a continuously varying set of boundary conditions for the climate of eastern North America. Fossil-pollen data from eastern North America record the past 18,000 years of vegetation changes that occurred in response to the consequent climatic changes (Jacobson and others, this volume). The purpose of our study is to examine how the changes in the boundary conditions during the past 18,000 years have governed the climatic changes of eastern North America that accompanied deglaciation and are recorded in the fossil-pollen data.

Kutzbach (this volume) and Kutzbach and Guetter (1986) used a general circulation model of the atmosphere—the Community Climate Model of the National Center for Atmospheric Research (NCAR CCM)—to simulate the response of climate to changing boundary conditions during the past 18,000 years. Models like the NCAR CCM illustrate the physically consistent response of individual climate variables to changes in the boundary conditions (Kutzbach, 1985). One focus of our paper is to use the model simulations to examine not only how atmospheric circulation and climate changed in eastern North America during deglaciation but also how these changes were controlled by the changing area, height, and reflectivity of the ice sheet and by the changes in the latitudinal and seasonal distribution of solar radiation. Thorough tests of such models are needed, however, before they can be routinely used to describe the nature and causes of past climatic changes (Webb and Wigley, 1985).

A second focus of this paper therefore is to compare selected model results for eastern North America with observations derived from pollen data. We use the pollen record from eastern North America to test the model results in two ways. First, we compare the summary maps of model output with isopoll maps and look for similarities in the timing and pattern of change in both sets of maps. Second, we use pollen/climate response surfaces (Bartlein and others, 1986) to transform model-simulated temperature and precipitation into estimates of pollen abundances for selected pollen types. We then compare isopoll maps of these simulated pollen values with maps of the observed values. Our study for eastern North America adds to the comparison between NCAR CCM results and paleoclimatic data for 18 ka by Kutzbach and Wright (1985) and complements other comparisons in Barnosky and others (this volume). Other comparisons of the NCAR CCM experiments with paleoclimatic data appear in Kutzbach and Street-Perrott (1985), Street-Perrott (1986), Prell and Kutzbach (1987), and COHMAP Members (in preparation).

Comparison of the model simulations with the fossil data can help to identify the proximate and ultimate causes of past climatic variations. Climate models provide a mechanism for translating hypotheses about the causes of a specific climatic change into a form testable by the paleoclimatic data. These data provide a standard to which climate-model experiments can be compared, that is, they indicate the paleoclimatic ground-truth (Schneider, 1986). Study of paleoclimatic data alone is insufficient to permit identification of causes, because the same climatic change in a particular region could owe its cause to a variety of

Webb, T., III, Bartlein, P. J., and Kutzbach, J. E., 1987, Climatic change in eastern North America during the past 18,000 years; Comparisons of pollen data with model results, *in* Ruddiman, W. F., and Wright, H. E., Jr., eds., North America and adjacent oceans during the last deglaciation: Boulder, Colorado, Geological Society of America, The Geology of North America, v. K-3.

different controls. The paleoclimatic data, in turn, may suggest some hypotheses that are testable by model experiments.

Before climate models can be routinely used for predicting the response to altered boundary conditions, they must be validated under a range of conditions. The paleoclimatic record provides one such set of alternative conditions. Our ultimate goal is to use the model to explain past climatic variations recorded by the data, and simultaneously to validate the ability of the model to simulate past climatic variations. By focusing on maps of pollen data from eastern North America and by transforming the model results into quantitative estimates of fossil-pollen abundance, our study provides a test of the model and highlights how the climatic patterns in North America were altered by the combination of changes (1) in the height, extent, and reflectivity of the Laurentide ice sheet (and concurrent changes in sea-ice cover and ocean surface temperature), and (2) in the latitudinal and seasonal distribution of solar radiation.

DATA AND METHODS

The two comparisons of model results with paleoclimatic data use data sets described by Jacobson and others (this volume) and Bartlein and others (1986) and model results described by Kutzbach (this volume) and Kutzbach and Guetter (1986). In the first comparison, a time sequence of isopoll maps for six major pollen types in eastern North America is compared to a time sequence of maps of surface winds simulated by the NCAR CCM. In the second comparison, response surfaces were used to convert model results into a form for direct comparison with observed pollen percentages for 18 ka to present.

The six pollen types used were sedge (Cyperaceae), spruce (*Picea*), the northern and southern pines (*Pinus*), oak (*Quercus*), and the prairie forbs. Forb pollen is the sum of sage (*Artemisia*), Compositae, and pigweed (Chenopodiaceae-Amaranthaceae) pollen. The northern pines were separated from southern pines by dividing the contemporary distribution of pine pollen at 40°N. Maps of these types represent many of the broad-scale patterns and changes in the vegetation of eastern North America during the past 18,000 years (Jacobson and others, this volume). Site density for the pollen data ranges from 3 sites/10^6 km^2 at 18 ka, to 12 sites/10^6 km^2 at 12 ka, to 25 sites/10^6 km^2 at 6 ka. A pollen sum of all tree, shrub, and herb pollen was used to calculate the pollen percentages that were the basis for the isopoll maps (Jacobson and others, this volume).

Bartlein and others (1986) and Bartlein and Webb (in preparation) describe the methods used to obtain the response surfaces, which estimate the percentages of the six pollen types (spruce, northern pines, southern pines, oak, sedge, and forbs) from values for mean July temperature, mean January temperature, and annual precipitation. These three climate variables represent general controls of plant distribution. Because response surfaces incorporating three predictor variables are difficult to display, we present two sets of surfaces for each pollen type. The first shows how the abundances of each pollen type are related to mean July temperature and annual precipitation, and the second shows the relationship to mean January and mean July temperature.

Several features of the CCM experiments that are especially important for the interpretation of the results are listed here and described in Kutzbach (this volume) and Kutzbach and Guetter (1986). The model calculates land surface temperature, precipitation, and evaporation as well as atmospheric variables such as wind, temperature, moisture, and clouds. Soil moisture, land albedo, snow cover, sea-ice cover, and ocean surface temperature are prescribed, that is, held at predetermined values. The model is run separately for January and July boundary conditions rather than for a full seasonal cycle. Papers documenting areas of agreement and disagreement between the climatology of the model and the observed modern climate are referenced in Kutzbach and Guetter (1986).

The conversion of the model results into estimated pollen percentages took several steps. Surface temperature and precipitation were obtained for all points on the 4.4° latitude by 7.5° longitude grid between 28.8° and 68.9°N and 60.0° and 105.0°W for the control case, which represents 0 ka, and for each of the model experiments from 18 to 6 ka. The density of the model grid is approximately 4 points/10^6 km^2. For most simulations, the model was run for the equivalent of 270 days, but for a few simulations it was run only for the equivalent of 90 days (Kutzbach and Guetter, 1986). For all January values, three 90-day means were averaged from available model runs to give one mean value at each grid point. The same was done for July values for the control case and for 9 and 18 ka, but single 90-day averages were used as July values for 3, 6, 12, and 15 ka. Annual values for precipitation (in mm) were then calculated from the weights in Table 1 of Kutzbach and Guetter (1986). This approximation of the annual average is an interim procedure until we complete experiments with a full seasonal cycle.

We interpolated model output from the 1 to 5 grid points adjacent to each fossil-pollen site using an inverse-distance–weighting method to obtain values of the climate variables there. For each site, the procedure thus produced a spatial average of the simulated climate variables. Only ice-free grid points on land were used. The number of grid points contributing to the average for each site ranged from 1 to 3 in coastal areas and along the ice sheet to 4 or 5 grid points in the interior. This interpolation procedure did not introduce any features into the resulting data not present in the original grid-point data, and features in the original data that are evident across several grid points are preserved in the interpolated data. The skill of general circulation models is generally proportional to the size of the area considered. Sampling theory, moreover, shows that the inherent variability of the simulated climate, which in turn affects the statistical significance of simulated climatic changes, decreases with the size of the area-average. Because of these sampling considerations, comparison of simulations and observations is often made in terms of broad zonal-averages or large area-averages (Kutzbach and Street-Perrott, 1985). In this paper, we are working near the probable lower limit of the potential resolving power of the NCAR CCM, and our study provides a test of new procedures for comparing simulations and observations.

Scatter diagrams display the relationships between the observed modern climate at the fossil-pollen sites and the modern or "control" simulations interpolated to those sites (Fig. 1). The model simulates a seasonal cycle of greater amplitude than is observed, with lower than observed January temperature in the model (~10°C), and higher than observed temperature in July (~5°C). The model also simulates more annual precipitation than is observed (~50 to 100 percent). These biases in the model are discussed further by Pitcher and others (1983).

Because the control experiment with the model has bias errors, the "sensitivity" results for the model, that is, the simulated climatic response to the changes in controls, also may contain bias errors (see, e.g., Palmer and Mansfield, 1986). Concern about bias errors in the response of the model necessitates the type of study we have completed. Our comparisons of the data with the model results are a first step toward assessing the correctness of the sensitivity of the model to changes in radiation and in glacial-age boundary conditions. To avoid incorporating the bias in the control experiment of the model into the paleoclimatic simulations, we simulated pollen percentages from values of climate variables that were obtained by applying the simulated anomalies to the observed modern values. We thus did not use the simulated paleoclimatic values directly. Simulated anomalies are the differences between the paleoclimatic experiment and control case (e.g., the 18 ka value minus the control value) or, for annual precipitation, percentages of the control-case values.

We mapped the observed and simulated fossil-pollen data to emphasize the agreement or disagreement between the two at a large spatial scale. The data were machine contoured by interpolating the observed and simulated fossil-pollen data onto an equal-area grid. The procedure resulted in some spatial smoothing of the patterns, and the pollen percentages were recoded to four levels: less than 1 percent, 1 to 5 percent, 5 to 20 percent, and greater than 20 percent. The resulting maps thus approximate the hand-contoured isopoll maps contained in Jacobson and others (this volume). The combined smoothing and generalization has the desirable effects of (a) reducing "palynological noise" (i.e., small-scale spatial and temporal variability unrelated to climatic controls); and (b) reducing the scale discrepancy between the coarse spatial scale of the model and the generally finer scale of the data.

We developed an experimental approach for quantitatively comparing individual pairs of observed and simulated fossil-pollen maps. This approach involved (1) summarizing the interpolated values from the contouring step above in an ordered contingency table, with categories corresponding to the map categories, and (2) computing two measures of association to describe the similarity of the two maps. The two measures used were the Goodman-Kruskal tau-(b) statistic (a nonparametric correlation coefficient applicable to contingency tables with ordered categories; Everitt, 1977, p. 62; Goodman and Kruskal,

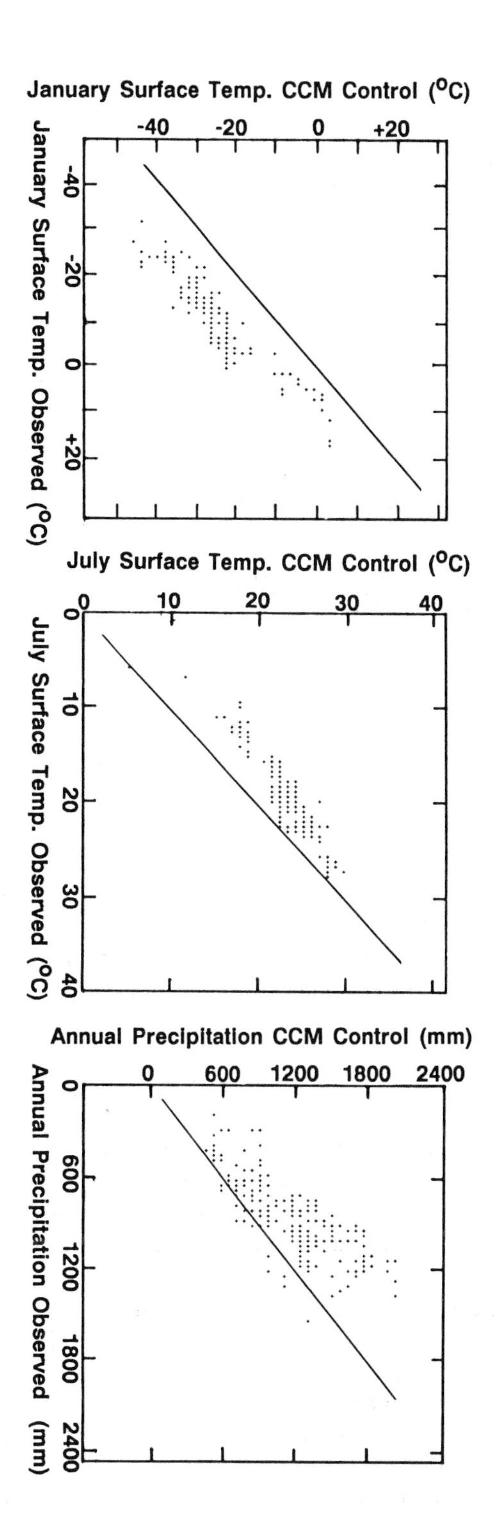

Figure 1. Scatter diagrams showing the relationship between observed modern values of the three climate variables at the fossil-pollen sites (Jacobson and others, this volume) and values simulated by the community climate model's (CCM) modern "control case" interpolated to those sites.

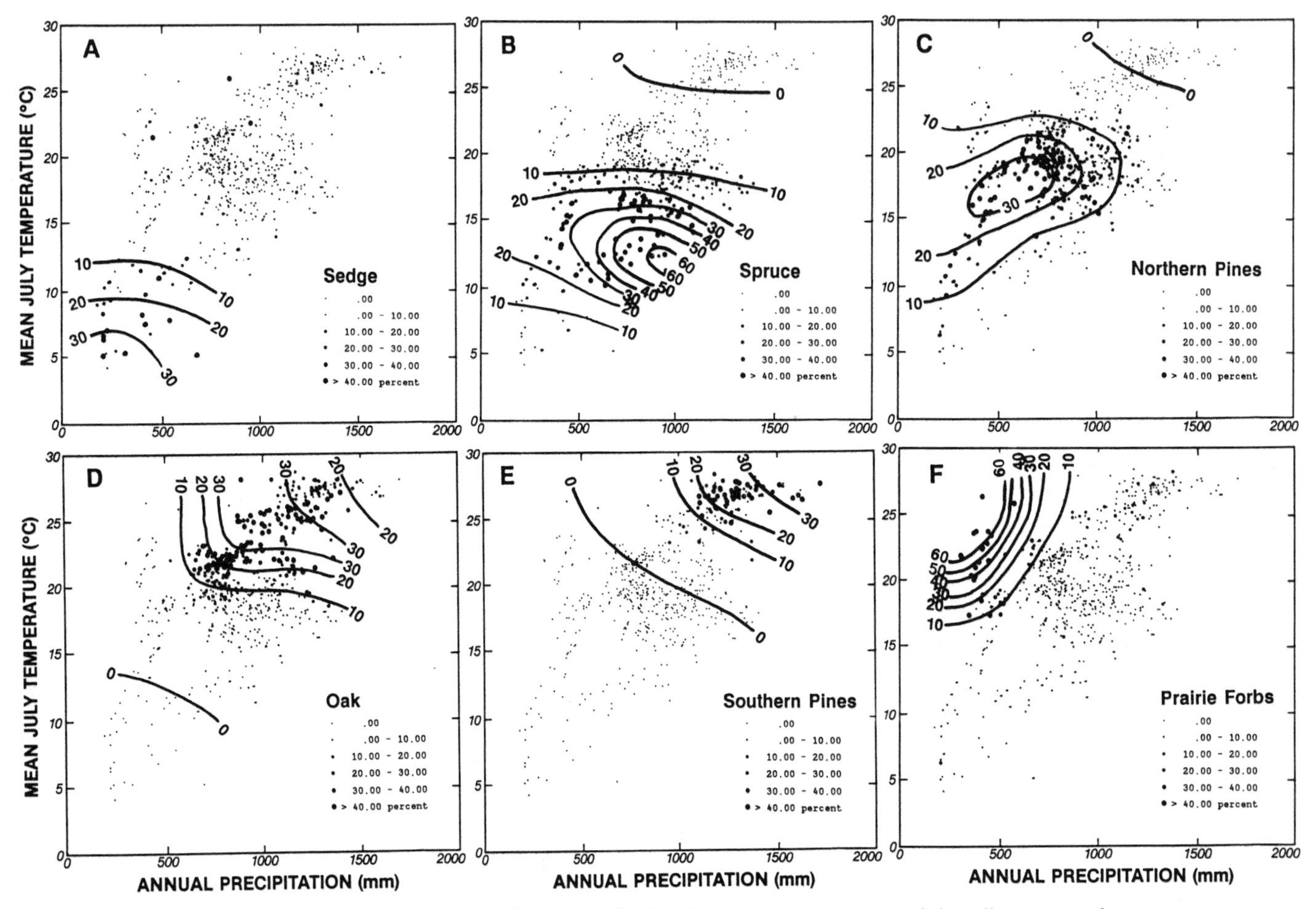

Figure 2. Response surfaces showing the relationship between the percentages of six pollen types and mean July temperature and annual precipitation.

1954), and the *P*roportion of the total area of the observed map *C*ategorized *C*orrectly by the simulated map (PCC). Tau-(b) measures similarity of pattern in two maps, but not necessarily similarity in magnitude, while PCC measures similarity in magnitude.

The significance of measures of association such as tau-(b) and PCC is difficult to assess analytically, because the map patterns are spatially highly autocorrelated (Cliff and Ord, 1981, Ch. 7). We therefore created an empirical reference distribution for each of these measures by computing it for all possible (unique) comparisons among the observed maps for all pollen types and times, but we excluded comparisons within each type at different times. This reference distribution gives an indication of how large tau-(b) and PCC could be when no association between maps is expected. For a specific comparison of an observed and simulated map (see Figs. 6 and 7), a large value of one of the measures, relative to the values in its reference distribution, indicates a stronger association between the simulated and observed map than that expected for the comparison of two observed maps for different pollen types. We selected the 90th, 95th, and 99th percentiles of the reference distributions to represent large relative values of the two measures. By showing how well the current pairs of simulated and observed maps agree (or disagree), the quantitative measures of association will allow us to gage how improvements to the data, model, or boundary conditions affect the agreement between the maps.

RESULTS

Response Surfaces

Response surfaces display the relationships between the individual pollen types and the climate variables (Figs. 2 and 3). Although general relationships can be inferred from isotherm, isohyet, and isopoll maps, the nature of the specific relationships are often more clearly displayed by response surfaces. In eastern North America the cold and relatively dry conditions that typify northern Canada contrast markedly with the warm and moist

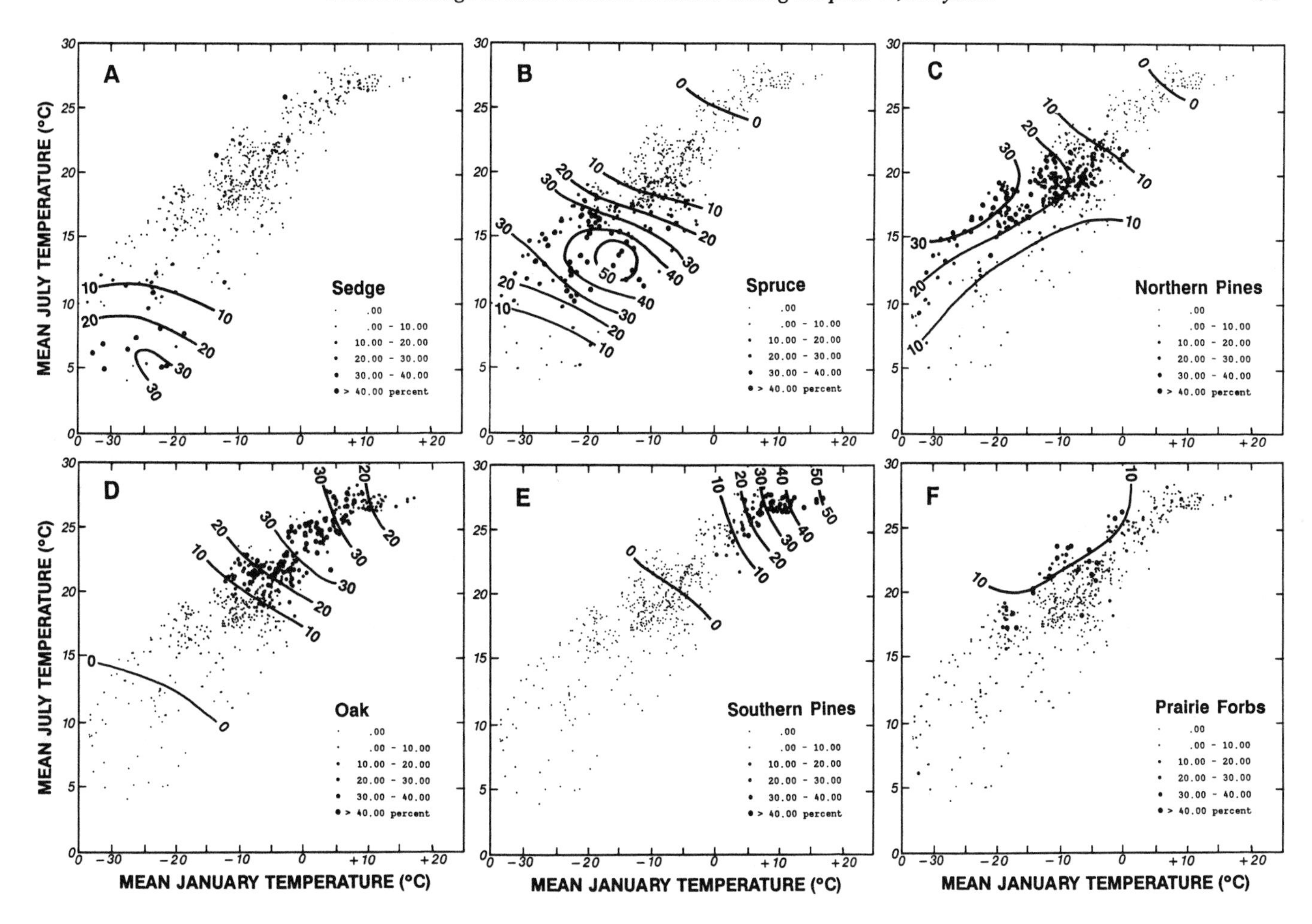

Figure 3. Response surfaces showing the relationship between the percentages of six pollen types and mean July and mean January temperature.

conditions of the southeastern U.S. Along a northwest-to-southeast transect, the vegetation and its associated pollen record reflect this climatic contrast as a latitudinal sequence of changes from a high abundance of sedge to spruce, northern pines, oak, and finally southern pines (see Fig. 6). Along a north-south transect from cold, dry conditions in north-central Canada to warmer but still relatively dry conditions in the Great Plains, the pollen assemblages change from a maximum of sedge to maximum abundances of spruce, northern pines, and then forbs. These variations are clearly evident in the plots of the response surfaces (Figs. 2 and 3). The highest values of forbs, for example, lie in the upper left corner of Figure 2F, the region of warm, dry conditions, whereas the highest values of spruce lie in the portion of the figure that represents cooler and wetter conditions.

The shapes of the response surfaces show how the pollen abundances vary in climate space, that is, along axes representing climate variables, and the modern geographic covariations between the pollen types and climate variables thus define the response surfaces. When we applied the response surfaces to the climate-model results to simulate temporal changes in the abundances of each pollen type, we made the implicit assumption that those temporal changes result from pollen-climate relationships comparable to the relationships represented by the geographic patterns of vegetation and climate today.

The shapes of the surfaces are useful for interpreting observed changes of pollen abundance in climatic terms. For example, the shape of the surface for spruce pollen (Fig. 2B) shows that the changes in spruce are relatively larger along the July temperature axis, and that spruce abundances first increase and then decrease as temperature increases. In contrast, the surface for prairie forbs (Fig. 2F) shows relatively greater changes along the annual precipitation axis. The surfaces for oak and southern pines, as functions of July and January temperatures, slope mainly parallel to the January temperature axis (when both temperatures are high). As the temperatures decrease, the abundance changes in oak become relatively more sensitive to changes in July temperature. These differences in the response of the different pollen types from place to place in the climate space (defined by the three climate variables) can serve as the basis for interpreting observed temporal changes of pollen abundances in climatic terms.

Climate simulations

Kutzbach (this volume) and Kutzbach and Guetter (1986) describe the prescribed changes in the boundary conditions of the NCAR CCM for the sequence fo model simulations representing the transition from full glacial conditions to present. Surface wind vectors (Fig. 4) and axes of midtropospheric wind maxima (Fig. 11 in Kutzbach, this volume) show the response of the model to these changes in controls by illustrating the changing pattern of atmospheric circulation. (Surface winds are simulated at the lowest atmospheric level in the model, generally about 100 m above the surface. The axes of midtropospheric wind maxima generally parallel the axes of the upper-tropospheric jet stream in the model, and we therefore refer to the midtropospheric wind features as reflecting "jet-stream" features.) Maps of the wind patterns show a significant change in atmospheric circulation between 15 and 12 ka, and an even larger change between 12 and 9 ka, induced both by the retreat and lowering of the Laurentide ice sheet (Denton and Hughes, 1981) and by the increased summer-season radiation and decreased winter radiation. The surface winds simulated for January at 18 ka blew anticyclonically around the ice sheet and were weak westerly to the south of the ice sheet in the central and eastern U.S. (Fig. 4). An anticyclonic circulation was also present in July but was somewhat weaker than in January. Aloft the westerlies split around the ice sheet; the southern wind maximum was located over the south-central U.S. in January and along the ice sheet border in July (Fig. 11 in Kutzbach, this volume).

For the north-central (NC) and northeast (NE) regions in the model at 18 ka, both January and July temperatures were 5° to 10°C below present just south of the ice sheet (Fig. 5). In the southeastern region (SE), July temperature was only 1° to 2°C below present. July precipitation was about the same as today or slightly lower except for a small zone of increased rainfall along the southern edge of the ice sheet paralleling the summer jet and storm track. (In the model results, we use maps of the variance of band-pass filtered surface pressure—variations with periods of two to six days—as a measure of synoptic-scale storminess. The mean storm track is taken to correspond to the band of maximum variance [Kutzbach and Guetter, 1986].) January precipitation (P) was generally lower than today except along the winter storm track over the south-central region. With the lower temperature, however, evaporation (E) was much lower than it is today, and annual precipitation minus evaporation (P-E) was near present values except in the SE, where it was distinctly lower.

At 12 ka the prescribed ice sheet covered a smaller area in western North America, its surface elevation was decreased to 50 percent of that at 18 ka, and July radiation was increased to about 8 percent above present (Kutzbach and Guetter, 1986). The January surface wind circulation changed dramatically to westerly flow across the ice sheet and along its southern edge (Fig. 4). The westerlies switched from split flow to a single wind maximum (Fig. 11 in Kutzbach, this volume), much like that of today (except winds were still very strong in the east). In July a weakened glacial anticyclone remained, but the flow in the west had become westerly rather than easterly as at 15 ka. The flow in the east remained much like it was at 15 ka. The July westerlies shifted north over western North America and followed a northwest-to-southeast track along the ice-sheet margin, from Alaska to the eastern U.S.

By 12 ka, July temperature in the NC and NE regions had increased considerably compared to 18 and 15 ka, but it was still 2° to 4°C lower than present (Fig. 5). In the SE region, the increased solar radiation produced conditions about 1°C warmer than present. The influence of the increased radiation was even more apparent in the south, west, and northwest (to the west of the ice sheet) where July temperatures were 1° to 2°C higher than earlier (Barnosky and others, this volume). This warming in ice-free areas helped maintain a strong northwest-to-southeast–trending baroclinic zone along the southwestern and southern flanks of the ice sheet. Precipitation remained generally less than present, except for a small area of increased precipitation in July along the mean baroclinic zone at the southern edge of the ice sheet. Annual P-E remained below present in the SE region, whereas P-E values were near present levels elsewhere.

At 9 ka, the residual ice sheet was further reduced in area (Fig. 4) and in height (to 800 m), and its albedo was lowered to reflect the presumed ponding and debris accumulation associated with melting. This, coupled with the increased July radiation (ca. 8 percent greater than today) produced major changes in surface winds, temperature, and precipitation (Figs. 4, 5). A very weak anticyclonic cell remained over the southern part of the ice sheet in July. The heating in the south and west also helped induce a weak heat low (summer monsoon) resulting in strengthened southerly components to the winds in the southeastern U.S. (east of the low). Aloft the January flow was now close to present, but in July a wind maximum, weaker than before, still persisted south of the ice sheet.

At 9 ka, July temperature was 1° to 2°C higher than present throughout all North America, except along coastal regions, over the ice sheet and along the ice-sheet border. Warming continued to be marked in the west and northwest (and Alaska) and continued to help maintain the northwest-to-southeast–oriented baroclinic zone along the southwestern edge of the ice sheet (as at 12 ka). In contrast, January temperature remained near or lower than present values throughout. July precipitation was increased (compared to 12 ka and today) in the region of more southerly flow. A small area of increased precipitation (as at 12 ka) persisted along the mean baroclinic zone at the southern border of the ice sheet. Annual P-E was lower than at present (and at 12 ka), however, because summertime temperature (and evaporation) also increased (except in the SE region, where conditions switched from dry to moist).

At 6 ka the prescribed residual ice sheet was gone. When

averaged for the Northern Hemisphere, July radiation, although less than at 9 ka, was still 5 percent greater than present. The increased radiation was no longer counteracted by the influence of the ice sheet in the east. The warmest anomalies were simulated in the continental interior, where temperatures were 2° to 3°C above present. This is probably an overestimate of the warming, because the model contains no Great Lakes, which have a moderating effect on summer temperatures in the midcontinent. With the ice sheet gone and the warmest anomalies in the continental interior, the weak continental heat low (monsoon low) shifted toward the midcontinent (from the southwest). In the surface wind field (Fig. 4), this produced a strengthened southwesterly and southerly flow over parts of the east and strengthened westerly and northwesterly flow to the west of 90°W. This flow pattern was consistent with increased precipitation in the east (similar to 9 ka, but greater than present) but decreased precipitation west of 90°W. Annual P-E was a little lower than present in the NC region and near present in the NE region. Only in the SE region was P-E significantly increased compared to present, because precipitation had increased more than evaporation.

After 6 ka, as the solar radiation geometry approached the present (control) configuration, July temperature differences from the control simulation of the model became small, and the isotherms shifted south as the summers cooled. The SE region dried as annual P-E decreased (Fig. 5), and the NC became slightly moister than it had been (in part due to the disappearance of the enhanced monsoon low).

The specific results summarized above illustrate the response of the simulated climate of eastern North America to changes in two large-scale controls: (1) the decrease in the height, area, and albedo of the Laurentide ice sheet (and concurrent changes in the North Atlantic sea-ice border and sea-surface temperature), and (2) the amplification of the seasonal cycle of solar radiation. Circulation patterns in eastern North America in both winter and summer were strongly influenced by the large ice sheet at 18 ka and 15 ka. At 12 ka and 9 ka, the smaller ice sheet influenced the circulation pattern mainly in summer. The response of the simulated climate to the increasing solar radiation in summer occurred earlier in regions distant from the ice, and the full effect of the summer radiation maximum that occurred around 10 ka was delayed in regions close to the ice.

In addition to these primary responses to the changing controls, several interrelated secondary responses are evident (Figs. 4, 5). These include (a) maintenance of a strong northeast-to-southwest temperature gradient between the ice-covered area and adjacent regions free to respond to increasing radiation (at 12 ka and 9 ka); (b) a shift in location of the largest positive anomalies of temperature in July toward the interior of the continent with the disappearance of the ice from 9 ka to 6 ka, and consequent shift in the surface wind patterns around the weakly developed heat low; (c) January temperature that remained lower than the model's control values until 6 ka, while July temperature increased to values higher than the model's control values by 9 ka; (d) generally negative precipitation anomalies when the "glacial" circulation patterns prevailed; (e) decreased precipitation in the interior at 6 ka associated with stronger northwesterly flow around the heat low; (f) higher than modern (control) precipitation at all times along the southern edge of the ice sheet, and in regions of strong southerly flow at 9 ka and 6 ka; and (g) lower than modern (control) P-E in the continental interior during the interval of maximum summer solar radiation (at 9 ka and 6 ka).

The sequence of simulated climates in eastern North America can be summarized as follows: From 18 ka to 12 ka, the influence of the ice sheet decreased and finally disappeared after 9 ka. The influence of the increasing summer solar radiation occurred earlier in regions distant from the ice, but the full response to the increased insolation was not registered until after the radiation maximum, and was delayed by the presence of the residual ice. The sequence of climatic changes in eastern North America includes (1) a "glacial" period, characterized by the simulations from 18 ka and 15 ka, when the influence of the ice sheet predominated; (2) a "transitional" period (12 ka and 9 ka), when the influence of the ice sheet decreased, that of the summer solar radiation increased, and atmospheric circulation patterns changed toward those at present; and (3) an "interglacial" period, illustrated by the simulations from 6 ka to present, when the influence of the greater summer solar radiation decreased.

Observed isopoll maps

The timing and direction of changes in the pollen record in eastern North America parallel the timing and nature of the model-simulated climatic changes (Fig. 6). At the continental scale the patterns in the pollen data change only gradually within a full-glacial configuration from 18 ka to 12 ka, then change considerably to a new orientation that was established by 9 ka, and finally change gradually until present. From 18 ka to 12 ka the isopoll maps illustrate that a broad north-south gradient existed from the spruce-sedge parkland across the central states to the pine and oak forests to the south (Fig. 6). By 9 ka that pattern was replaced by bands of tundra, spruce forest, mixed forest, and prairie that developed along a northeast-to-southwest gradient from northeastern Canada to the Midwest, and by bands of prairie, oak forest, and southern pine forests along a northwest-to-southeast gradient in the south. The pollen maps (Fig. 6) illustrate a sequence of changes similar in timing to those noted for the climate simulations, with "glacial," "transitional," and "interglacial" patterns prevailing at the same time as those in the maps for simulated climate (Fig. 4).

In addition to the broad-scale changes that parallel the primary sequence of simulated climatic responses to the changing controls, some of the details of the vegetational history seem attributable to the secondary responses. The development of the northeast-to-southwest gradient of the vegetation after 12 ka described above, for example, is consistent with the temperature gradient established then. Similarly, the development of the prairie as a coherent formation in the Midwest by 9 ka, and its

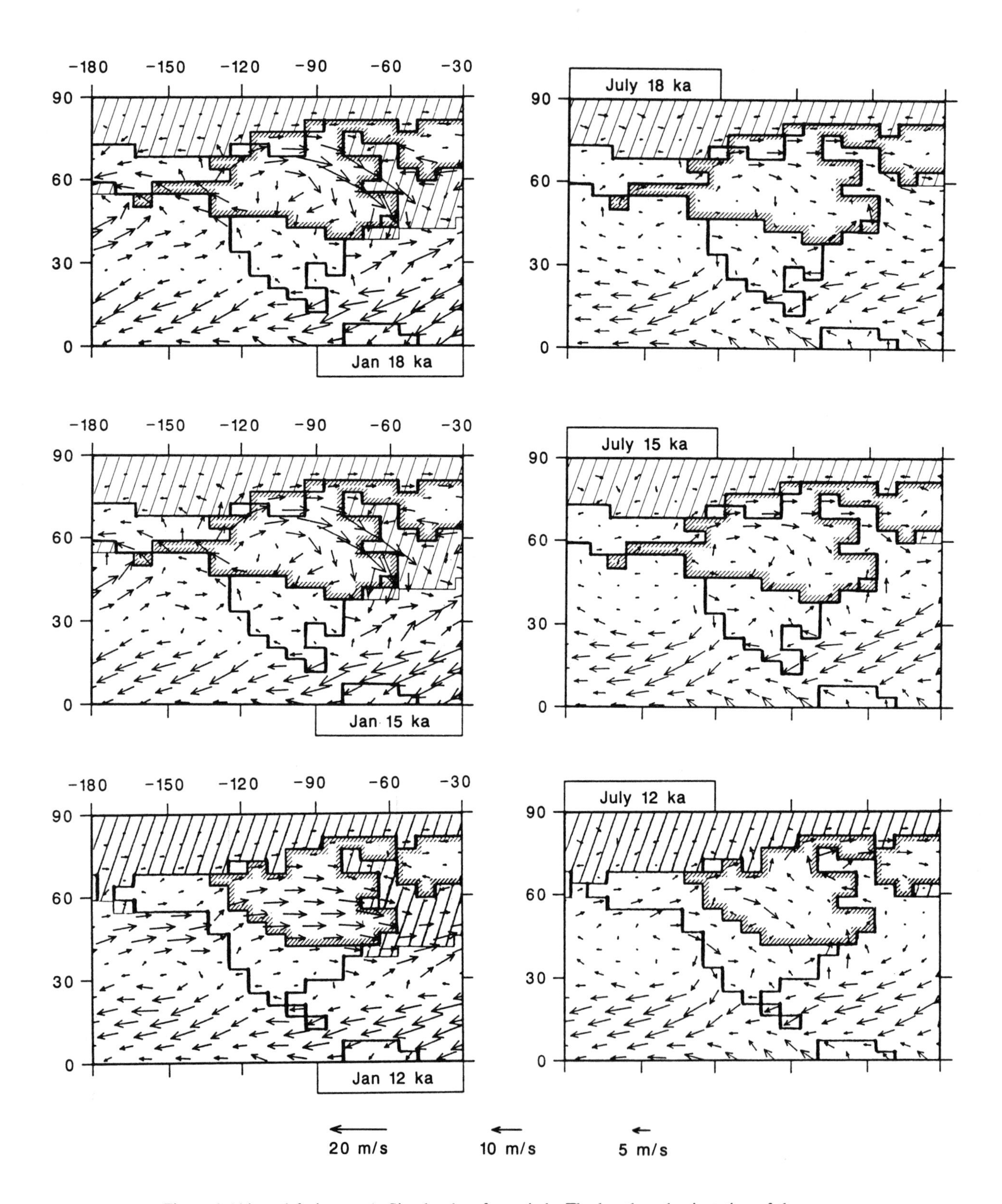

Figure 4 (this and facing page). Simulated surface winds. The length and orientation of the arrows indicate wind speed and direction.

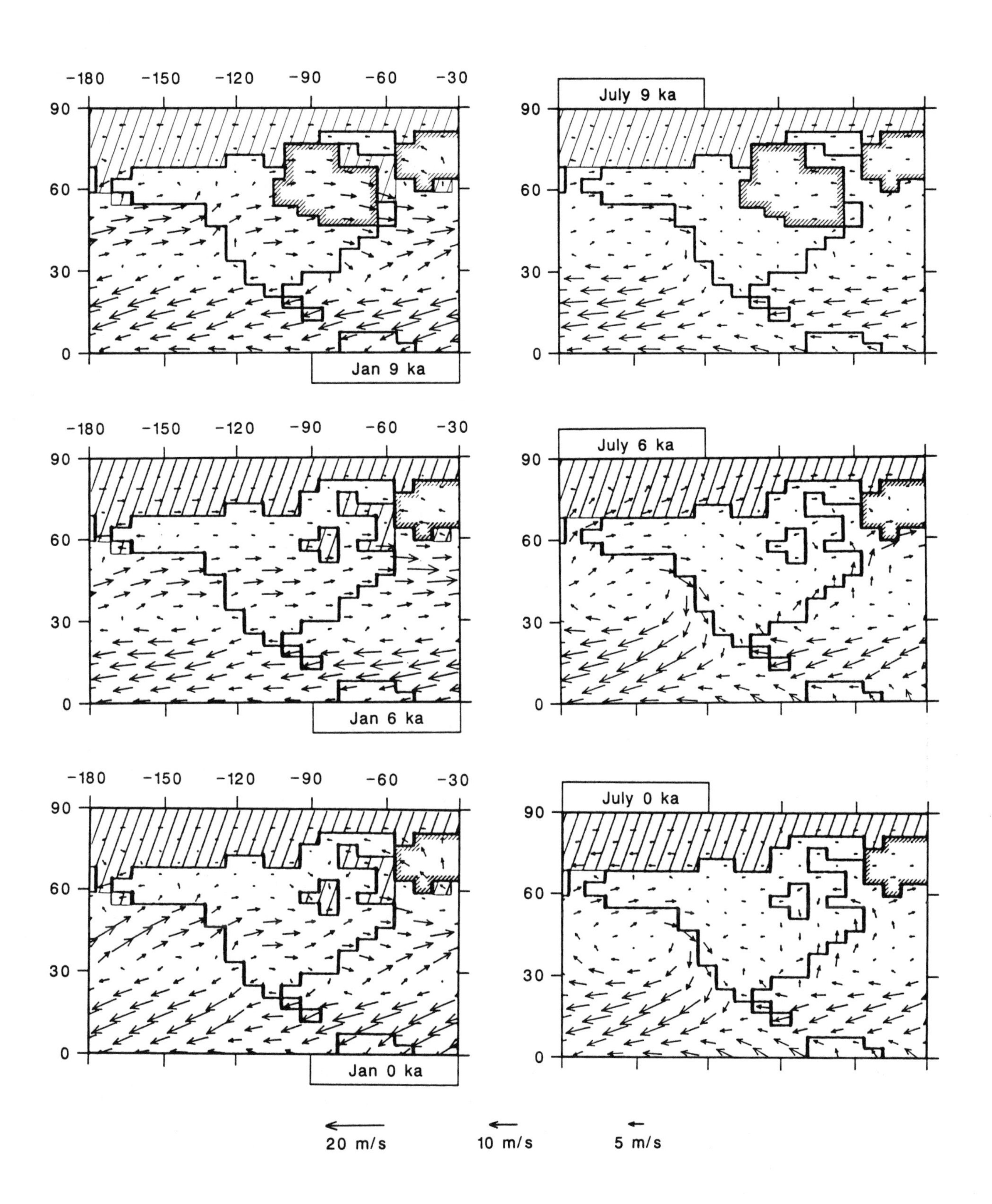

-180
-150
-120
-90
-60
-30
90
60
30
0
Jan 9 ka
July 9 ka
Jan 6 ka
July 6 ka
Jan 0 ka
July 0 ka
20 m/s
10 m/s
5 m/s

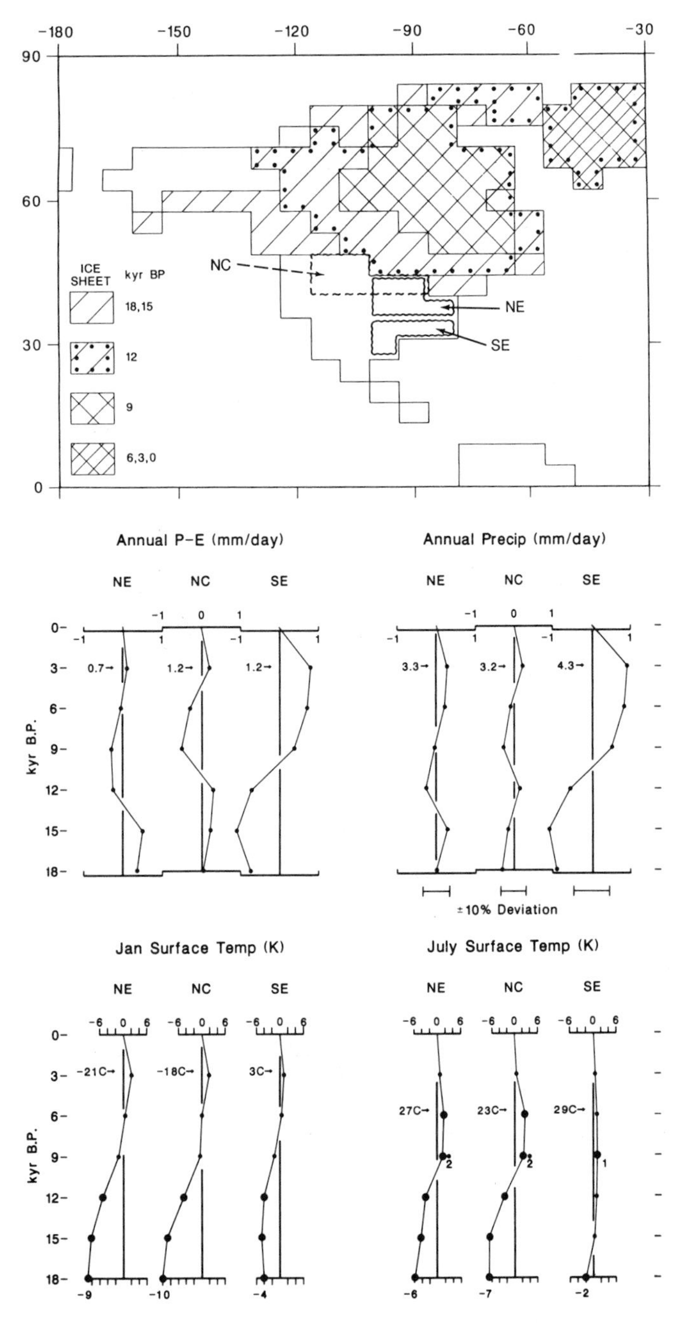

Figure 5. Area-average changes of January and July surface temperature, annual precipitation, and precipitation-minus-evaporation (P–E) for three regions in eastern North America. The time series are departures from the control (0 ka) simulation. For annual precipitation, plus or minus 10 percent deviations from the control values are shown. The control value is written beside each series. Large dots indicate that the departure is significant, above the 95 percent confidence level, compared to the model's inherent variability. The unconnected circle and temperature value at 9 ka show the results for the "no ice sheet" experiment of Kutzbach and Guetter (1984). Figure modified from Kutzbach (this volume).

subsequent eastward movement at 6 ka (Fig. 6F), is consistent with the decreases in precipitation and P-E simulated for these dates. The spruce-sedge parkland occupied the region south of the ice sheet (Figs. 6A and 6B) while the "glacial" circulation patterns prevailed and while that region was colder and drier than present. The progressive northward movement of the vegetation formations after 12 ka also matches the similar movement of the maximum temperature anomalies for July (Kutzbach, this volume).

The sequence of vegetation changes in individual regions is also consistent with the climatic changes simulated by the model. For example, on the response surfaces (Figs. 2 and 3) a trajectory from relatively cold and dry conditions toward warmer and moister ones (the changes that occurred from 18 ka to 6 ka in the eastern and southeastern U.S.) crosses in sequence the abundance maxima of sedge, spruce, northern pines, oak, and southern pines. A trajectory from cold and relatively moist conditions toward warmer and drier ones (such as that simulated for the NC region from 12 ka to 6 ka as shown on Figure 5) crosses in sequence the abundance maxima of spruce, northern pines, and prairie forbs. Finally, a change toward cooler conditions, as was simulated for the NC and NE regions in response to decreasing summer radiation after 6 ka, moves toward the abundance maximum of spruce.

Simulated isopoll maps

The qualitative comparisons above show a seemingly close tie between the vegetation changes and the response of the simulated climate to changes in the glacial-age boundary conditions and inputs of solar radiation. These comparisons can be supplemented by using the response surfaces and specific climate-model output to derive simulated isopoll maps. Comparison of simulated isopoll maps with maps of observed values provides a stronger test of the reasonableness of the simulated climate. For good agreement to occur, it is necessary to simulate correctly not only the direction of the change but also the magnitude of change in individual climatic variables.

The sequence of simulated isopoll maps for spruce (Fig. 7B) shows fair agreement with the observed maps at the continental scale (Fig. 6B; Table 1). On both sets of maps at 18 ka, spruce pollen was at its greatest abundance in the Midwest and decreased in abundance toward the east and southeast. From 18 ka to 12 ka the region of highest abundance for observed spruce pollen extended farther eastward, and by 9 ka the abundance of spruce was nearly everywhere decreased, while the region of greatest abundance moved northward in apparent response to increasing temperature. These variations are also evident in the simulated maps. At 6 ka the abundance of spruce increased but remained lower than at present. Finally the abundance of spruce increased again from 6 ka to present, and the region of greatest abundance advanced southward, possibly in response to reduced summer radiation. This agreement of the patterns of observed and simulated spruce is reflected by relatively large values of tau-(b) in Table 1. Inspection of the response surfaces for spruce (Figs. 2

TABLE 1. MEASURES OF ASSOCIATION BETWEEN OBSERVED AND SIMULATED ISOPOLL MAPS

Goodman-Kruskal tau-(b)							
Pollen Type/Age	18 ka	15 ka	12 ka	9 ka	6 ka	3 ka	0 ka
Sedge	-0.05	0.28	-0.29	0.48*	0.66**	0.22	0.55**
Spruce	0.59*	0.75***	0.81***	0.73***	0.81***	0.88***	0.91***
Northern Pines	0.04	-0.26	0.09	0.75***	0.70***	0.72***	0.87***
Oak	0.54**	0.77***	0.75***	0.89***	0.92***	0.91***	0.93***
Southern Pines	0.00	0.00	0.01	0.74***	0.78***	0.82***	0.94***
Prairie Forbs	0.14	-0.17	-0.08	0.56**	0.39	0.43	0.78***

Proportion of Observed Isopoll Map Correctly Categorized by Simulated Map (PCC)							
Pollen Type/Age	18 ka	15 ka	12 ka	9 ka	6 ka	3 ka	0 ka
Sedge	0.26	0.26	0.24	0.76**	0.76**	0.57*	0.57*
Spruce	0.34	0.29	0.53*	0.62**	0.62**	0.77**	0.85***
Northern Pines	0.25	0.17	0.21	0.45*	0.57*	0.61**	0.81***
Oak	0.13	0.17	0.57*	0.81***	0.85***	0.81***	0.90***
Southern Pines	0.61**	0.51*	0.70**	0.79**	0.83***	0.88***	0.95***
Prairie Forbs	0.30	0.41	0.27	0.57*	0.55*	0.49*	0.82***

*Values greater than the 90th percentile of the appropriate empirical reference distribution.
**Values greater than the 95th percentile of the appropriate empirical reference distribution.
***Values greater than the 99th percentile of the appropriate empirical reference distribution.
(See text for details.)

and 3) shows the strong influence of July temperature on the abundance of spruce; the fair agreement between the simulated and observed maps suggests that the model simulated a reasonable sequence of July temperature values for the region extending from the Midwest and northeastern U.S. northward.

Differences between the maps are evident when finer scale patterns are compared. At 18 ka the simulated values do not match the low-to-intermediate values of spruce pollen in the southern United States, and at 15 ka, higher values of spruce pollen are observed in the eastern United States than are simulated. The PCC values are small for these times (Table 1). The match at 9 ka is generally good, but at 6 ka the simulated climate values lead to higher estimates of spruce pollen in Canada than those observed. For the remaining pollen types we therefore focus our comparisons on the patterns involving several grid points and note whether the simulated pollen values reproduce the major patterns of change for each pollen type.

Comparisons of the simulated and observed isopoll maps for the northern pines (Figs. 6C and 7C) and oak (Figs. 6D and 7D) show less agreement than for spruce. After 12 ka in the case of oak, and 9 ka in the case of pine, the patterns and abundances on the observed and simulated maps are similar (Table 1), but before those times the discrepancies are large. At 18 ka and 15 ka, more oak and less pine than was observed were simulated in the southeast, although the pattern of oak on the simulated map was in general agreement with that on the observed. Inspection of the response surfaces for these types (Figs. 2C, 2D, 3C, and 3D) suggests such a result could occur if the simulated climate were either too warm or too moist. The simulation of more prairie-forb pollen in the southeast from 18 ka to 12 ka than was observed seems to preclude the latter possibility. Reductions of simulated temperatures in the southeast on the order of 2° to 4°C would bring the patterns for oak into greater agreement at these times.

The agreement between predicted and observed patterns of sedge and prairie forbs is relatively poor on the maps for 18 ka to 12 ka (Figs. 6A, 6F, 7A, and 7F; Table 1). The response surfaces for these types as well as that for the northern pines (Fig. 2) indicate their greater sensitivity to precipitation than is the case for spruce or oak, and the poor simulation of these types may be attributed to poor simulation of precipitation in the midlatitudes. At 9 ka and 6 ka the simulated and observed patterns for sedge show good agreement in the north, and for forbs the agreement in the Midwest is good. The simulated and observed patterns in the eastern U.S., however, do not agree (Table 1).

The sequence of patterns on the simulated isopoll maps for the southern pines is generally similar to those on the observed maps (Figs. 6E and 7E). From 18 ka to 12 ka, low values were simulated, and the southern pines are not present in significant abundances on the maps of observed values. Abundances increased in parallel on the two sets of maps following 12 ka. The replacement of oak in the southeast by southern pines is evidently related to the increase in January temperature simulated by the model (Fig. 5).

Collectively the patterns on the simulated and observed isopoll maps show substantial agreement at the continental scale (Table 1). Several factors could have reduced the extent of

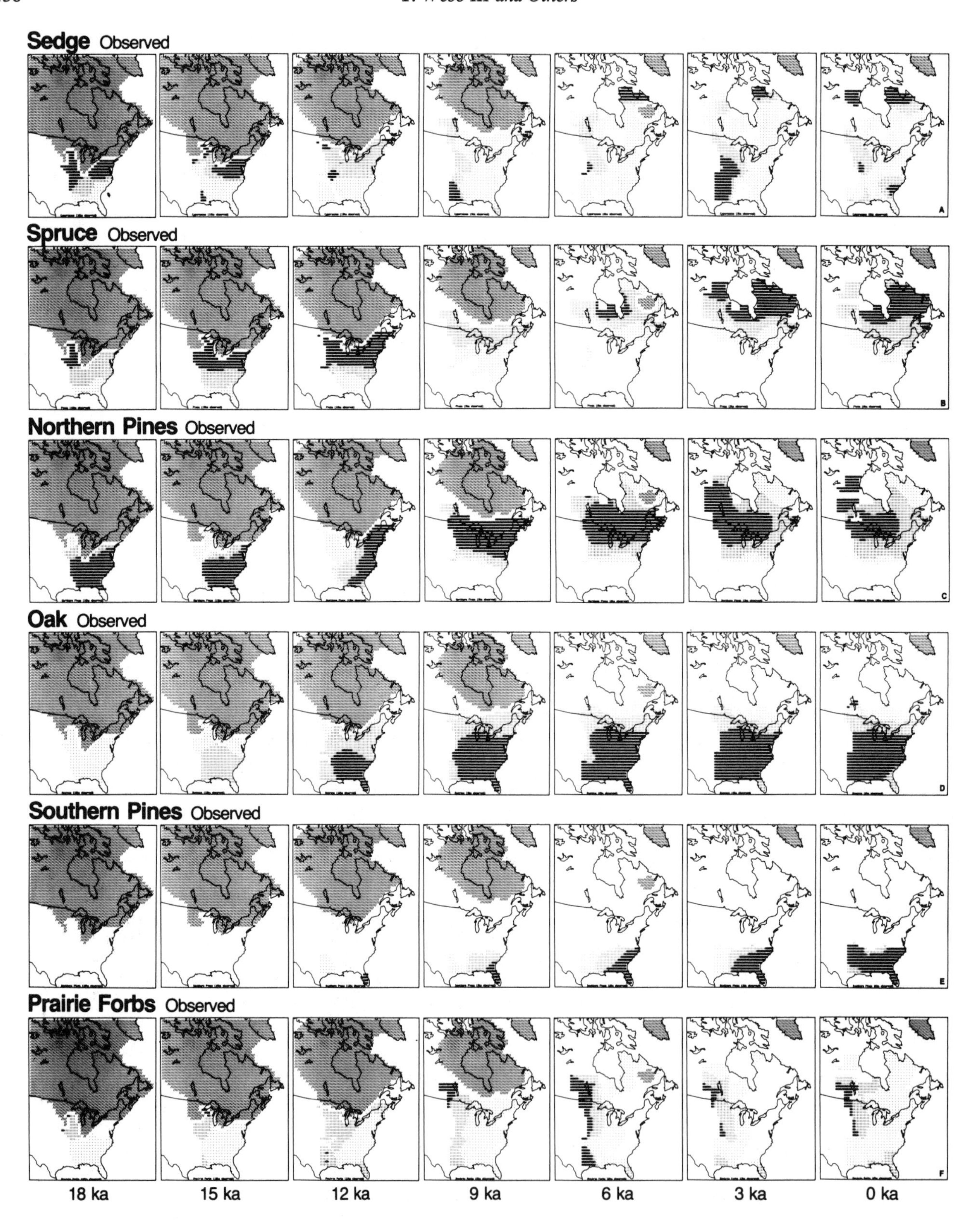

Figure 6. Isopoll maps of observed pollen percentages with the Laurentide ice sheet shown from 18 ka to 6 ka. Three levels of shading indicate pollen percentages greater than 1 percent (lightest), 5 percent and 20 percent (darkest). The isopoll maps were constructed by machine contouring the fossil-pollen data described in Jacobson and others (this volume).

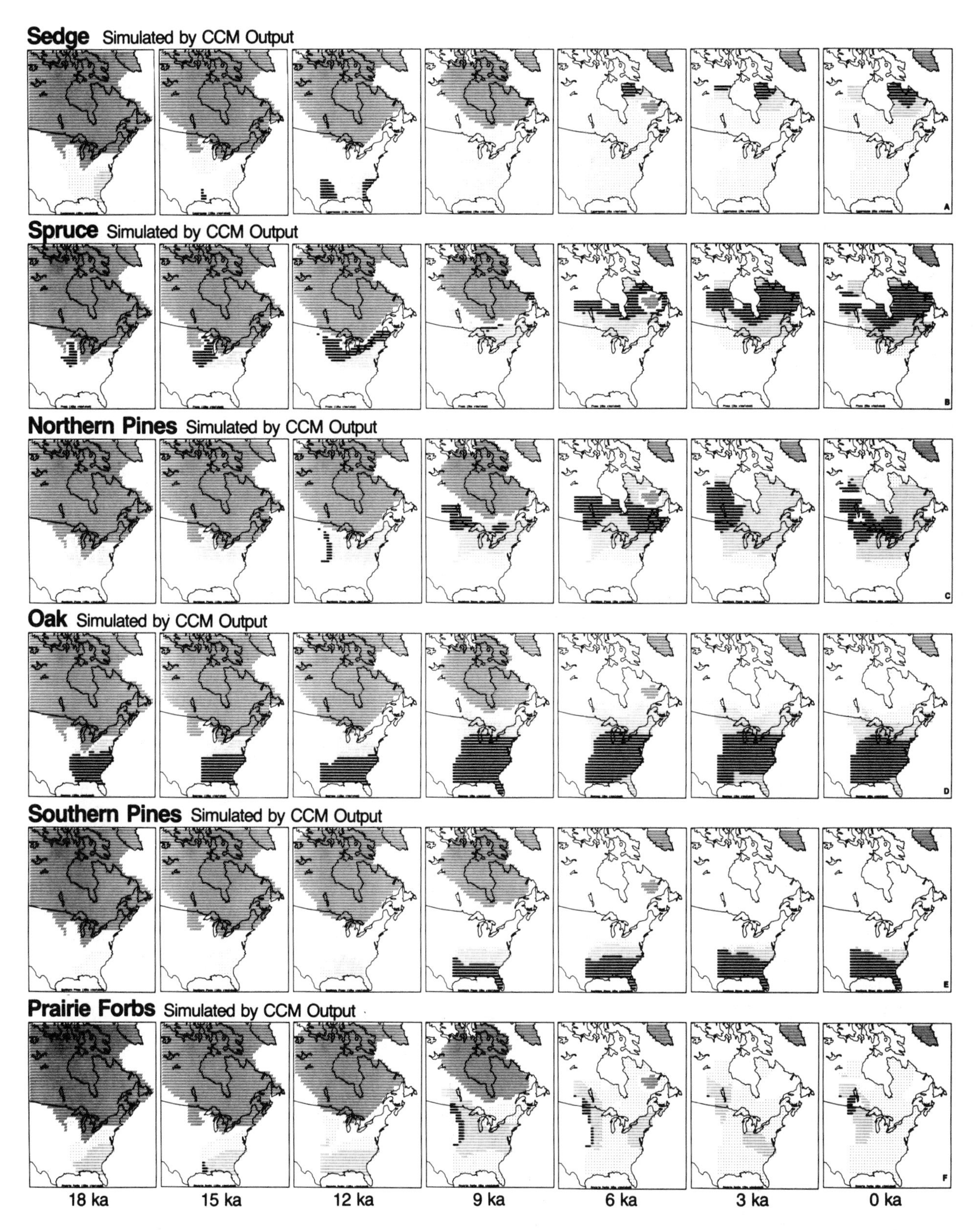

Figure 7. Isopoll maps of simulated pollen percentages with the Laurentide ice sheet shown from 18 ka to 6 ka. Three levels of shading indicate pollen percentages greater than 1 percent (lightest), 5 percent, and 20 percent (darkest). Response surfaces were used to derive estimates of fossil-pollen percentages from model-simulated values of mean July and January temperatures and annual precipitation. The resulting values were machine contoured.

agreement. These factors include (a) the lack of a dynamic equilibrium between the vegetation and climate (Webb, 1986), (b) inability of the response surfaces to portray adequately the true relationship between vegetation and climate, (c) inappropriate rescaling of the coarse spatial scale of the model's output to the scale of individual fossil-pollen sites, (d) the absence in the model of mesoscale climatic features, such as the Great Lakes and Appalachians, that have influenced the postglacial movement of plant taxa in eastern North America, and (e) other climate-model deficiencies that influence the accuracy of the simulation, such as the inability of the model to simulate the regional details of the present climate, inaccuracies of prescribed boundary conditions, errors or approximations in model parameterizations, and errors arising from the estimation of annual precipitation using only January and July values.

Despite the potential effects of the above factors, the location of the abundance maxima and associations among types are approximately correct for nearly all times and pollen types. After 9 ka the simulated and observed patterns appear alike at the continental scale, and both the location and relative abundances of the individual types are well-simulated (Table 1). From 18 ka to 12 ka, simulated temperature in the southeast and precipitation throughout may depart from the unobservable true values, but summer temperatures in a large part of the region were simulated accurately enough by the model that the simulated distribution of spruce during this interval matched the observed.

DISCUSSION

General approach

During the past 18,000 years, continent-wide vegetation patterns have responded to changing patterns of temperature and moisture and associated changes in atmospheric circulation. Each of these was part of the climatic response to changes in both seasonal radiation gradients and the surface boundary conditions. At the same time the circulation and radiation changes have both influenced the retreat of the Laurentide ice sheet and have been influenced by it. A full description of how the climate has influenced the vegetation is therefore not a straightforward story and will ultimately require several levels of explanation. The components of the climate system are interrelated, and the ultimate cause of a specific regional change may be difficult to trace. A complete explanation requires knowledge of what external factors (such as solar radiation) have caused climate to change and how certain "internal" components of the system, currently prescribed in most models (e.g., the ice-sheet size, height, and reflectivity, the sea-surface temperature, and the carbon-dioxide concentration), have influenced other climatic variables (e.g., circulation patterns, air temperature, etc.). A complete explanation of the history of climate and vegetation in eastern North America therefore requires the resolution of several questions: What were the past patterns of atmospheric circulation, and what caused them to change? How did they influence the patterns of temperature and precipitation? How also did the temperature and moisture patterns cause individual taxa, and thus the vegetation, to change? And finally, how did the vegetational changes affect surface roughness, albedo, and evapotranspiration and thus influence the climate?

Climate models provide an imperfect but improving tool to help in answering these questions. The NCAR CCM does not yet contain feedback loops for estimating how the vegetation influences the climate. The models are an elegant tool for translating well-formulated meteorological theory about atmospheric dynamics and radiation balance into spatial patterns of climatic variables, thus providing a set of testable hypotheses for comparison with data. For example, the organized series of model runs by Kutzbach and Guetter (1986) translate a changing sequence of lower boundary conditions and radiation conditions into a sequence of simulated changes in atmospheric circulation, temperature, and moisture balance.

The role of calibration functions (Howe and Webb, 1983; Bartlein and Webb, 1985) or response surfaces is to translate either the data or the model results into a form that permits direct comparison between the two. With response functions, for example, the model simulations for temperature and moisture were translated into changes in taxon location, abundance, and association and thus into estimated vegetational changes (Fig. 7).

Concordance between the observed and simulated data suggests that the simulated climate changes may resemble the past changes in climate, and differences between observed and simulated data suggest areas for improvement in the model, or in the interpretation of the data. The learning is iterative, and different model runs permit testing the consequences of separate factors that affect the climate, for example, the presence or absence of an ice sheet at 9 ka (Kutzbach and Guetter, 1984). The model results also suggest new ways to organize existing data and new data sets to collect and compile (Peterson and others, 1979). The initial results of the data/model comparisons are beneficial whether the comparisons reveal concordance or discordance. The analysis of the results teaches us much about the model, the climate system, and the data. It also teaches us about the methods used for relating the model results to the data and can suggest improvements for these methods.

Specific results

The results of our study show how two principal factors, namely the retreating ice sheet (and concurrent changes in prescribed sea-ice and ocean-surface temperature) and the changing seasonal cycle of solar radiation, affected the timing, magnitude, pattern, and speed with which atmospheric circulation changed in eastern North America. The high albedo of the ice sheet, as well as its great area and height, strongly influenced the climate of eastern North America. The slow retreat of the ice sheet acted to delay the period of peak summer warmth in much of this region until between 9 and 6 ka, whereas the direct effect of the radiational changes alone would place the maximum summer warming between 12 and 9 ka, as occurred in the Northwest (Ritchie and others, 1983; Heusser and others, 1985; Barnosky and others, this volume). This delay is apparent in both the model results and the data, and it contrasts with the immediate response to radiational heating and monsoon enhancement in the tropics of Africa and southern Asia (Kutzbach and Street-Perrott, 1985).

The maps of simulated surface winds and of pollen data give no clear indication that more warm, moist air in summer reached the southern border of the ice sheet at 15 ka and 12 ka than at 18 ka, as postulated by Delcourt and Delcourt (1984). A broad spruce parkland south of the ice sheet existed throughout this period. Increased values of oak pollen in the south at 12 ka indicate higher temperatures there than at 18 ka, but this region is far from the ice front (Fig. 4). Still, the advection of warm air across this region may have affected the region to the north. Percentages of ash and hornbeam pollen are higher south of the ice sheet at 12 than at 18 ka, but the climatic interpretation of this change is unclear (Jacobson and others, this volume). By 9 ka, both the simulated surface winds and the pollen maps suggest that northward circulation of maritime air masses in summer was more pronounced than it is today, and that these air masses would have enhanced melting along the southern edge of the ice sheet.

A band of high values of sedge and forb pollen is evident south of the ice sheet at 12 ka (Fig. 6a and 6f). This band may indicate the extent of anticyclonic outflow from the ice sheet. By 9 ka the band had disappeared, and its absence may signal the lowering of the ice sheet below a height necessary for a strong anticyclone and katabatic winds.

Past changes in the vegetation indicate the primacy of summer climatic conditions over winter conditions. Between 12 and 9 ka a northeast-to-southwest vegetational gradient from Baffin Island to Nebraska became established. During the same interval, the model shows a similar gradient in summer temperature becoming strengthened, with the summer maximum in upper-level winds becoming oriented northwest to southeast, which is consistent with the temperature gradient (Fig. 11 in Kutzbach, this volume). (This gradient of July temperature is the one observed on an isotherm map for July temperature today.) As summers cooled after 6 ka in response to decreased radiation, spruce increased in abundance and moved southward. The one major vegetation change indicative of a change in winter temperature is the increase in southern pines in the Southeast from 6 ka onward (Fig. 6). The response surface for pine pollen shows that the abundance gradient for southern pines parallels that for winter temperature and not for summer temperature (Fig. 3). This pattern suggests that the southern pines increased in abundance as winter temperature increased in the latter part of the Holocene. The northwest-to-southeast vegetational gradient from Nebraska to Florida is primarily a winter temperature gradient.

The model results and physical reasoning about the causes of past climatic variations lead to an expectation that spatial patterns will always be apparent in maps of temperature departures between today and any previous time (Williams and Wigley, 1983). The large positive departures in the center of the continent for the simulated July mean temperature at 9 and 6 ka illustrate one such pattern. As discussed previously by Kutzbach (1981) and Crowley and others (1986) in terms of energy budgets, the increased radiation in July should warm the land more than the oceans, and the center of the continents more than the coasts. In North America, however, this anomaly pattern did not develop before the mid-Holocene because of the presence of the Laurentide ice sheet. Between 18 and 6 ka the changing geometry of the ice sheet and intensity of radiational forcing led to a series of different temperature-anomaly patterns.

Hypsithermal conditions, which can be defined as occurring when local or regional July temperatures were higher than today, were therefore neither uniform nor synchronous across North America (Kutzbach, this volume), or even within the Midwest (Bartlein and others, 1984), and in some regions, for example, the northeast and north-central areas, they were probably associated in the early and mid-Holocene with lower January temperatures than today. The original chronostratigraphic definition of the Hypsithermal (Deevey and Flint, 1957), therefore, is not longer tenable (Watson and Wright, 1980). Any future use of this term in a climatostratigraphic sense should be accompanied by a clear definition of what types of climatic conditions are being described. Our revised definition of this term is given above, but further work is needed before our definition can be construed as formal. General understanding of how the climate system has behaved during the Quaternary has advanced during the past 30 years since Deevey and Flint (1957) coined the term hypsithermal. In its simple focus on temperature change, it may well have outlived its usefulness as an important description of late-Quaternary climate change.

Key results in our study are several, but the most important is the agreement of the sequence of model-simulated maps of spruce pollen with the observed maps. These maps illustrate the usefulness of response surfaces in comparing climate-model simulations to paleoclimatic data. The retreating Laurentide ice sheet (and concurrent changes in sea-ice and ocean-surface temperature) and the changes in solar radiation had profound effects on eastern North American climates. Our study has illustrated its effect on atmospheric circulation across the continent and on the expression of changes in radiation, and thus on the patterns of temperature and precipitation during the past 18,000 years. The qualitative and quantitative comparisons between model results and the fossil data have also shown differences between the simulated and observed patterns, and they suggest caution in using the current model results as a complete description of how climate changed in eastern North America during the last deglaciation.

REFERENCES CITED

Bartlein, P. J., and Webb, T., III, 1985, Mean July temperature estimates for 6,000 yr B.P. in eastern North America; Regression equations for estimates from fossil-pollen data, *in* Harrington, C. R., ed., Climatic Change in Canada, 5: Syllogeus, no. 55, p. 301–342.

Bartlein, P. J., Webb, T., III, and Fleri, E., 1984, Holocene climatic change in the northern Midwest; Pollen-derived estimates: Quaternary Research, v. 22, p. 361–374.

Bartlein, P. J., Prentice, I. C., and Webb, T., III, 1986, Climatic response surfaces from pollen data for some eastern North American taxa: Journal of Biogeography, v. 13, p. 35–57.

Berger, A., Imbrie, J., Hays, J., Kukla, G., and Saltzman, B., eds., 1984, Milankovitch and climate: Dordrecht, D. Reidel Company, 895 p.

Cliff, A. D., and Ord, J. K., 1981, Spatial processes; Models and applications: London, Pion Limited, 266 p.

Crowley, T. J., Short, D. A., Mengle, J. G., and North, G. R., 1986, Seasonality in the evolution of climate during the last 100 million years: Science, v. 231, p. 579–584.

Delcourt, P. A., and Delcourt, H. R., 1984, Late Quaternary paleoclimates and biotic responses in eastern North America and the western North Atlantic Ocean: Palaeogeography, Palaeoclimatology, and Palaeoecology, v. 48, p. 263–284.

Denton, G. H., and Hughes, T. J., eds., 1981, The last great ice sheets: New York, Wiley-Interscience, 484 p.

Deevey, E. S., and Flint, R. F., 1957, Postglacial Hypsithermal interval: Science, v. 125, p. 182–184.

Everitt, B. S., 1977, The analysis of contingency tables: London, Chapman and Hall, 128 p.

Goodman, L. A., and Kruskal, W. H., 1954, Measures of association for cross classifications: Journal of the American Statistical Association, v. 49, p. 732–764.

Heusser, C. J., Heusser, L. E., and Peteet, D. M., 1985, Late-Quaternary climatic change on the American North Pacific coast: Nature, v. 315, p. 485–487.

Howe, S., and Webb, T., III, 1983, Calibrating pollen data in climatic terms; Improving the methods: Quaternary Science Reviews, v. 2, p. 17–51.

Kutzbach, J. E., 1981, Monsoon climate of the early Holocene; Climate experiment with the earth's orbital parameters for 9,000 years ago: Science, v. 214, p. 59–61.

Kutzbach, J. E., 1985, Modeling of paleoclimates: Advances in Geophysics, v. 28A, p. 159–196.

Kutzbach, J. E., and Guetter, P. J., 1984, The sensitivity of monsoon climates to orbital parameter changes for 9,000 years B.P.; Experiments with the NCAR general circulation model, *in* Berger, A., Imbrie, J., Hays, J., Kukla, G., and Saltzman, B., ed., Milankovitch and climate; Understanding the response to astronomical forcing: Dordrecht, Holland, D. Reidel Company, p. 801–820.

——, 1986, The influence of changing orbital parameters and surface boundary conditions on climate simulations for the past 18,000 years: Journal of the Atmospheric Sciences, v. 43, p. 1726–1759.

Kutzbach, J. E., and Street-Perrott, F. A., 1985, Milankovitch forcing of fluctuations in the level of tropical lakes from 18 to 0 kyr B.P.: Nature, v. 317, p. 130–134.

Kutzbach, J. E., and Wright, H. E., Jr., 1986, Simulation of the climate of 18,000 yr B.P.; Results for the North American/North Atlantic/European sector: Quaternary Science Reviews, v. 4, p. 147–187.

Palmer, T. N., and Mansfield, D. A., 1986, A study of wintertime circulation anomalies during past El Nino events, using a high resolution general circulation model: Journal of the Royal Meteorological Society, v. 112, p. 613–638.

Peterson, G. M., Webb, T., III, Kutzbach, J. E., van der Hammen, T., Wijmstra, T., and Street, F. A., 1979, The continental record of environmental conditions at 18,000 yr B.P.; An initial evaluation: Quaternary Research, v. 12, p. 47–82.

Pitcher, E. J., Malone, R. C., Ramanathan, V., Blackmon, M. L., Puri, K., and Bourke, W., 1983, January and July simulations with a spectral general circulation model: Journal of the Atmospheric Sciences, v. 40, p. 580–604.

Prell, W. L., and Kutzbach, J. E., 1987, Monsoon variability over the past 150,000 years: Journal of Geophysical Research (in press).

Ritchie, J. C., Cwynar, L. C., and Spear, R. W., 1983, Evidence from northwest Canada for an early Holocene Milankovitch thermal maximum: Nature, v. 305, p. 126–128.

Ruddiman, W. F., and McIntyre, A., 1981, The mode and mechanism of the last deglaciation; Oceanic evidence: Quaternary Research, v. 16, p. 125–134.

Schneider, S. H., 1986, Response to Rind and Peteet: Climatic Change, v. 9, p. 361–362.

Street-Perrott, F. A., 1986, The response of lake levels to climatic change; Implications for the future, *in* Rosenzweig, C., and Dickenson, R., eds., Climate-vegetation interactions: Boulder, Colorado, University Corporation for Atmospheric Research, Office of Interdisciplinary Earth Science, Report OIES-2, p. 77–80.

Watson, R. A., and Wright, H. E., Jr., 1980, The end of the Pleistocene; A general critique of chronostratigraphic classification: Boreas, v. 9, p. 153–163.

Webb, T., III, 1986, Is the vegetation in equilibrium with climate? How to interpret late-Quaternary pollen data: Vegetation, v. 67, p. 75–91.

Webb, T., III, and Wigley, T.M.L., 1985, What past climates can indicate about a warmer world, *in* MacCracken, M. C., and Luther, F. M., eds., The potential climatic effects of increasing carbon dioxide: Washington, D.C., U.S. Department of Energy, Report DOE/ER-0237, p. 239–257.

Williams, L. D., and Wigley, T.M.L., 1983, A comparison of evidence for late Holocene summer temperature variations in the Northern Hemisphere: Quaternary Research, v. 20, p. 286–307.

MANUSCRIPT ACCEPTED BY THE SOCIETY APRIL 10, 1987

ACKNOWLEDGMENTS

Grants from the National Science Foundation Climate Dynamics Program for COHMAP (Cooperative Holocene Mapping Project) to Brown University (ATM-8406832) and the University of Wisconsin–Madison (ATM-8412958), and from the Department of Energy, Carbon Dioxide Research Program (DE-FG02-85ER60304) supported this study. We thank K. Anderson, P. Behling, P. Guetter, P. Klinkman, S. Klinkman, B. Molfino, and B. Richards for technical assistance and R. S. Bradley, J. T. Overpeck, W. F. Ruddiman, T.M.L. Wigley, and H. E. Wright, Jr. for critical readings of the manuscript.

Printed in U.S.A.

The Geology of North America
Vol. K-3, North America and adjacent oceans during the last deglaciation
The Geological Society of America, 1987

Chapter 21

Synthesis; The ocean ice/sheet record

William F. Ruddiman
Lamont-Doherty Geological Observatory of Columbia University, Palisades, New York 10964

INTRODUCTION

The timing of the last deglaciation has largely been inferred from two independent sources: (1) maps of the retreating ice margins, converted to ice volume by various hypothetical area/volume relationships, and (2) marine oxygen isotopic curves, thought to represent ice volume. This chapter first reevaluates the marine oxygen isotopic evidence by examining possible temperature overprints on the deglacial $\delta^{18}O$ curves. Three alternative models of the deglaciation are then evaluated in the light of all available paleoclimatic evidence from North America and adjacent oceans to arrive at a preferred model of the deglaciation. Finally, the importance of several ablation mechanisms in providing positive feedback to the primary forcing from orbital insolation variations is briefly reviewed.

TIMING OF THE DEGLACIATION: EVALUATION OF MODELS

The timing of ice-volume loss following the most recent North American ice-volume maximum is one major focus of this volume. This knowledge is needed to address the second major focus: the relative importance of ablation mechanisms and their times of greatest action.

Significant deglaciation did not begin until 14 ka and ended by 6 ka (all ages cited in this chapter will be in radiocarbon years). This conclusion is validated by maps of ice area (Andrews, Chapter 2), by marine $\delta^{18}O$ records (Mix, Chapter 6), and by numerous terrestrial and marine records in and around North America (Chapters 4, 5, 7, 13–17). Ice-sheet fluctuations for several thousand years prior to 14 ka were generally local in extent and apparently not indicative of net melting of the ice sheet. Strong meltwater runoff to the Gulf of Mexico also began near 14 ka (Broecker and others, 1987). Thus, the deglaciation is constrained within an interval of 8,000 years.

The structure of deglaciation within this 8,000-year interval is equivocal (Fig. 1). There is evidence supporting: (1) a "smooth deglaciation" model with fastest ice wastage centered on 11 ka; (2) a "two-step deglaciation" model with rapid ice wasting from 14 to 12 ka and 10 to 7 ka, and a mid-deglacial pause with little or no ice disintegration from 12 to 10 ka; and (3) a "Younger Dryas deglaciation" model with two rapid deglacial steps as in (2) above, interrupted by a mid-deglacial reversal with significant ice growth from 11 to 10 ka.

The critical data supporting the smooth deglaciation model are maps of Laurentide ice area based on ^{14}C-dated glacial deposits (Prest, 1969; Bryson and others, 1969). Although there are subtle suggestions of more rapid retreat at or near the time of the two steps mentioned above, these curves indicate a steady progressive retreat of North American ice, with significant oscillations in retreat rate only at local spatial scales (Andrews, Chapter 2; Teller, Chapter 2; Teller, Chapter 3). Some marine $\delta^{18}O$ curves also show a smooth progressive decrease toward Holocene values.

The step deglaciation model is also supported by some marine $\delta^{18}O$ records (Mix, Chapter 6). In addition, the distinctive patterns of change in sea-surface temperature of the North Atlantic Ocean and in Greenland ice-core $\delta^{18}O$ values also show abrupt steplike warmings at 10 ka and at approximately 13 ka (Ruddiman, Chapter 7; Paterson and Hammer, Chapter 5); these warmings might be associated with steplike decreases in Laurentide ice volume. These signals are in accord with European records of fossil pollen and insects and with records of lake geochemistry (see Rind and others, 1986), all of which show steplike changes during deglaciation. Regionally integrated rates of pollen change in eastern and central North America (Jacobson and others, Chapter 13) also show a rapid change centered on 10 ka, with less rapid changes centered on 13.7 and 12.3 ka.

Although the Younger Dryas deglaciation model is to some extent a "straw man," it is mentioned here to clarify all reasonable alternatives. The idea of mid-deglacial ice growth is not entirely new; small-scale reversals appeared in ice-volume curves published by Bloom (1971) and MacDonald (1971). The Younger Dryas model is also suggested by the strong signal of sea-surface temperature cooling between 11 and 10 ka in the North Atlantic Ocean (Ruddiman, Chapter 7) and in Greenland ice cores (Paterson and Hammer, Chapter 5). Because the surface temperature of the high-latitude North Atlantic Ocean is largely

Ruddiman, W. F., 1987, Synthesis; The ocean/ice sheet record, *in* Ruddiman, W. F., and Wright, H. E., Jr., eds., North America and adjacent oceans during the last deglaciation: Boulder, Colorado, Geological Society of America, The Geology of North America, v. K-3.

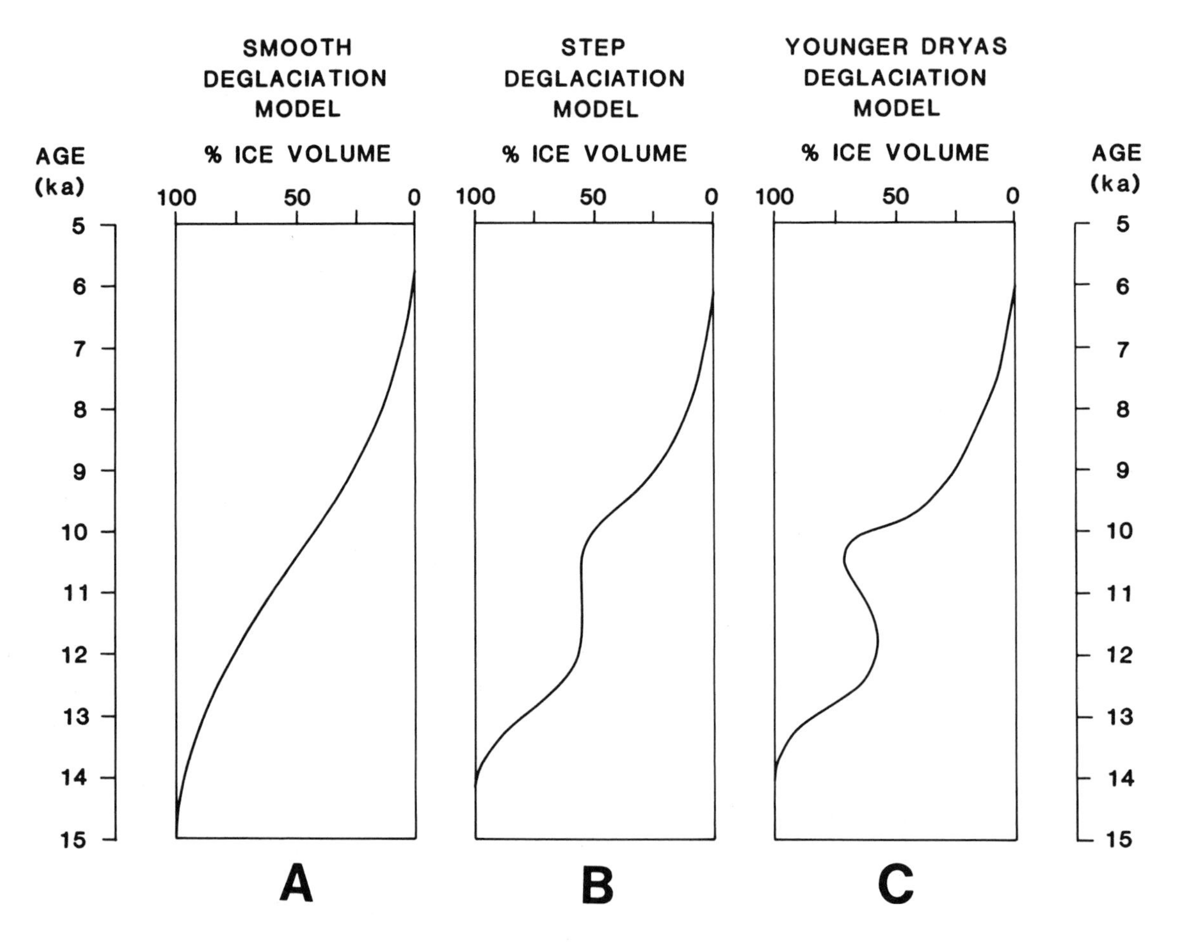

Figure 1. Conceptual models of the last deglaciation: A, Smooth deglaciation model; B, Step deglaciation model; C, Younger Dryas deglaciation model. Percentages show amount of Laurentide ice volume left versus time in ka.

controlled by the size of the Laurentide ice sheet over orbital time scales (Ruddiman and McIntyre, 1984), the Younger Dryas oceanic cooling could be interpreted as a response to a brief deglacial increase in Laurentide ice volume.

In summary, the basic form of ice-volume loss in North America is uncertain. Because some of the deglaciation models are largely based on marine $\delta^{18}O$ curves used as a proxy for global ice volume, this evidence is next examined in more detail.

The oxygen-isotopic record

A composite marine $\delta^{18}O$ record obtained by stacking numerous ^{14}C-dated records from the equatorial Atlantic (Mix and Ruddiman, 1985) shows muted steplike decreases in $\delta^{18}O$ values from 14 to 12 and 10 to 7 ka and a mid-deglacial pause from 12 to 10 ka. This curve is in close agreement with results from Berger and others (1985), but places the first step roughly 1,000 years later than Duplessy and others (1981). A new composite $\delta^{18}O$ curve, calculated after allowing small-scale (< 1 kyr) realignments of steps visible in individual curves, brings out more clearly the deglacial steps and mid-deglacial pause (Mix, Chapter 6; Fig. 2A). The ages of the steps are also close to those reconstructed from North Atlantic records (Bard and others, 1987).

Deglacial steps are evident in these marine $\delta^{18}O$ signals. But are they necessarily proof of steps in the decrease of global ice volume and of Laurentide ice volume? This raises two questions.

First, how much of the marine $\delta^{18}O$ signal is due to ice volume and how much to temperature? A total glacial/interglacial $\delta^{18}O$ amplitude of 1.7 ‰ is typical of many signals from benthic and planktonic foraminifera (Mix, Chapter 6). Chappell and Shackleton (1986) and Shackleton and Duplessy (1986) argued that about 0.4 ‰ of the 1.7 ‰ signal in Pacific Ocean benthic foraminifera represents a glacial cooling of deep water, leaving 1.3 ‰ as the ice-volume effect. Birchfield (1987) has arrived at a value of 1.3 to 1.4 ‰ for the ice-volume component using a different method.

$\delta^{18}O$ signals measured in extensively studied equatorial Atlantic cores also have a mean glacial-interglacial amplitude of 1.7 ‰ (Mix, Chapter 6). Because these records must contain the

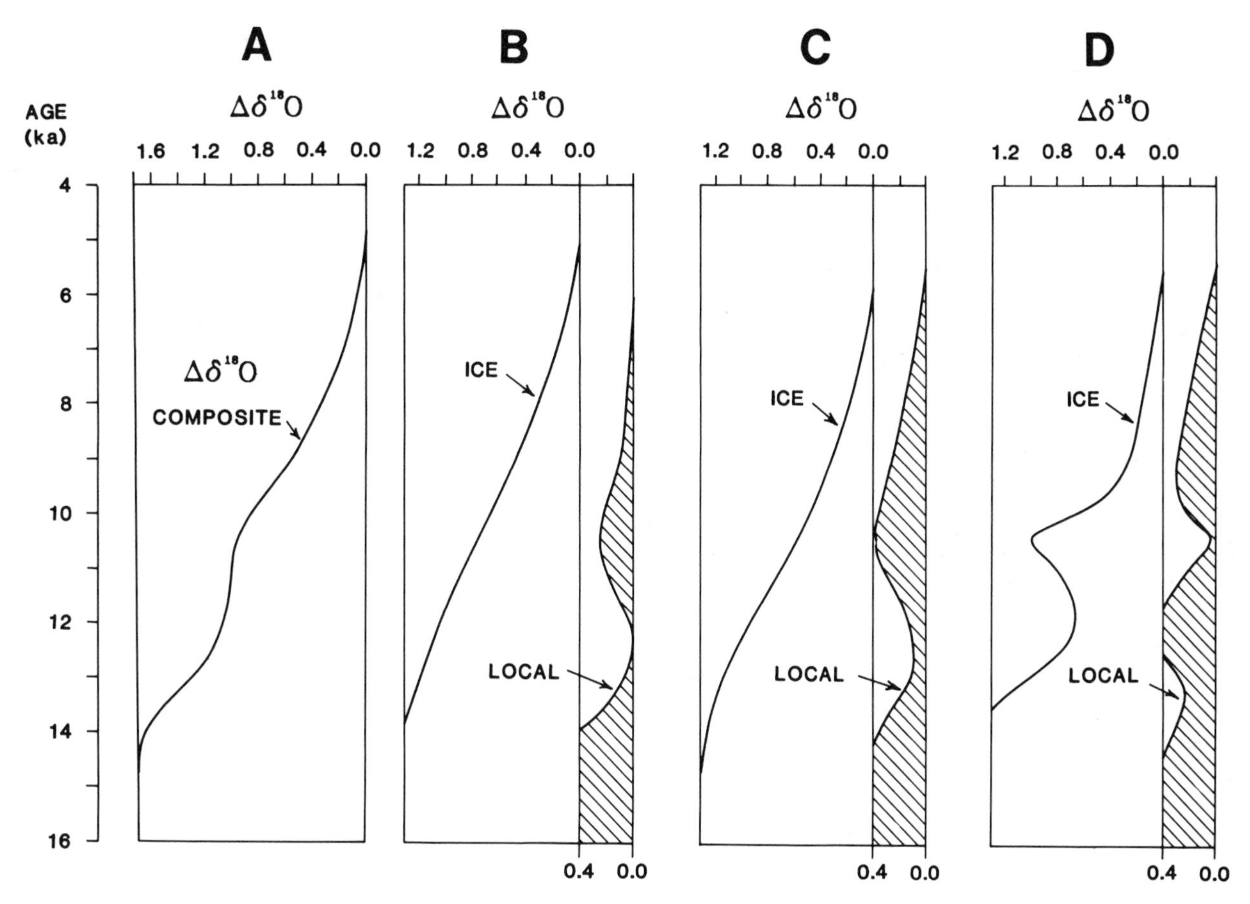

Figure 2. Possible allocations of ice-volume and temperature effects in $\delta^{18}O$ signals: A, composite of observed $\delta^{18}O$ signals with two steps (schematically after Mix, Chapter 6), B) and C), "Smooth" ice-volume signals produced by removing 0.4 ‰ temperature overprints shown by diagonal shading; D) "Younger Dryas" ice-volume signal produced by removing temperature overprint shown.

same ice-volume component as the Pacific Ocean benthic foraminifera (estimated at 1.3 ‰), they also presumably have the same temperature overprint of 0.4 ‰ during the last glacial maximum, relative to today.

Second, how much of the 1.3 ‰ signal in marine records that has been attributed to ice volume specifically represents North American ice? This can be evaluated from a compilation of individual ice-sheet sizes and their estimated $\delta^{18}O$ composition in Table 2 of Mix (Chapter 6); the "minimum" ice-volume reconstruction of Denton and Hughes (1981) best matches the 1.3 ‰ ice-volume signal in marine $\delta^{18}O$ records. For this reconstruction, 60 percent (0.8 ‰) of the 1.3 ‰ ice-volume signal represents North American ice. An additional effect of 0.2 to 0.35 ‰ may reflect changes in the Antarctic ice sheet, which is widely regarded as controlled by Northern Hemisphere ice volume via sea-level effects (Hollin, 1962). Laurentide ice should account for at least 0.15 to 0.2 ‰ of this Northern Hemisphere influence on Antarctic ice volume. Thus at least 75 percent (0.95 to 1.0 ‰) of the 1.3 ‰ ice-volume portion of the marine $\delta^{18}O$ signal in this compilation reflects North American ice volume, either directly or indirectly.

If the smaller estimate of Laurentide ice volume from Paterson (1972) noted in Mix (Chapter 6, Table 2) were used instead, Laurentide ice would still represent 50 percent (0.66 ‰) of the 1.3 ‰ ice-volume component of the marine $\delta^{18}O$ signal. Along with the Antarctic component, Laurentide ice would control more than 60 percent of the ice-volume portion of this $\delta^{18}O$ signal.

In addition, Laurentide ice probably has a direct influence on many other mid-latitude ice sheets of the Northern Hemisphere through its pervasive regional cooling effects (Boulton and others, 1985). This would tie still more of the ice-volume portion of the marine $\delta^{18}O$ signal directly to Laurentide ice, although changes in these other ice sheets might lag somewhat behind changes in Laurentide ice.

In summary, the 1.3 ‰ ice-volume portion of the marine $\delta^{18}O$ signal is at least 60 to 75 percent dominated by Laurentide ice. The following section analyzes possible ways in which temperature may have overprinted the "global" (dominantly Laurentide) ice-volume component of marine $\delta^{18}O$ signals.

Temperature overprints on marine $\delta^{18}O$ records

Although there is thought to be a positive temperature effect of 0.4 $^0/_{00}$ on marine $\delta^{18}O$ values at the glacial maximum, there is no a priori constraint as to when during deglaciation this effect should be removed from the marine $\delta^{18}O$ trends. Rather than try to make ad hoc assumptions about this problem, I return to the more specific question posed above: Does the existence of deglacial steps in marine $\delta^{18}O$ signals necessarily mean that there were steps in the ice-volume trend? More specifically, can the 0.4 $^0/_{00}$ local temperature overprint be used as a "degree of freedom" in such a way as to remove the steps from the marine $\delta^{18}O$ composite and produce a smooth deglacial $\delta^{18}O$ curve?

The answer is yes; by arbitrarily removing temperature effects as shown in Figure 2B and 2C, it is possible to produce a smooth sigmoidal deglacial ice-volume trend with a midpoint centered anywhere between 10 ka (Fig. 2B) and 11 ka (Fig. 2C). But, on the other, hand, the 0.4 $^0/_{00}$ temperature effect can also be added in such a way to enhance the steps in the "ice-volume" component and even to create a mid-deglacial reversal (Fig. 2D) equivalent to the Younger Dryas model (Fig. 1). These simple calculations suggest that, despite the numerous well-dated marine $\delta^{18}O$ time series of the deglaciation analyzed and published thus far, we cannot rule out the smooth deglaciation model, the step deglaciation model, or the Younger Dryas deglaciation model (Fig. 1).

There may, however, be a significant message in the shapes of the hypothetical temperature overprints required to produce the smooth deglaciation curves shown in Figures 2B and 2C. Both curves closely resemble curves of North Atlantic surface-ocean temperature (Ruddiman, Chapter 7) and Greenland ice-core $\delta^{18}O$ (Paterson and Hammer, Chapter 5): cold temperatures during glaciation, abrupt warming around 14 to 12.5 ka, cooling from 12.5 until 10.5 ka, and warming after 10.5 ka.

This similarlity may not be coincidental. It could suggest that the large-amplitude (5° to 10°C) Younger Dryas signal in the North Atlantic region was somehow transmitted to other regions and recorded in marine $\delta^{18}O$ curves with a reduced amplitude (about 1.5°C).

How plausible is such a signal transfer? The transfer to Atlantic benthic foraminiferal $\delta^{18}O$ records (Pastouret and others, 1978; Sarnthein and others, 1982) is probably the most plausible. Cd/Ca and $\delta^{13}C$ records from benthic foraminifera show that rates of formation of North Atlantic Deep Water are highly variable on glacial-interglacial time scales and have a major impact on chemical properties in the deep Atlantic (Boyle and Keigwin, 1985).

Cd/Ca and $\delta^{13}C$ records from a western subtropical North Atlantic core indicate slower rates of deep-water formation prior to 14 ka and during the Younger Dryas (Boyle and Keigwin, 1987). Because both of these geochemical signals closely resemble the deglacial sea-surface temperature trend in the subpolar North Atlantic, Boyle and Keigwin inferred that formation rates of North Atlantic Deep Water during deglaciation were tightly linked to North Atlantic surface temperatures. This confirms that at least some kinds of "Younger Dryas" signals were projected into the deep Atlantic from the subpolar North Atlantic.

In the absence of other information, the deglacial geochemical signals from Boyle and Keigwin (1987) can be proposed as one guide for removing the temperature overprint from $\delta^{18}O$ signals in the deep Atlantic. In this context, the close resemblance of these deglacial geochemical signals to the local temperature/$\delta^{18}O$ overprint independently calculated here (Figs. 2B, 2C) is a tantalizing hint that the deep-ocean temperature signal required to remove deglacial steps from these marine $\delta^{18}O$ curves is in fact realistic.

This is, however, only one possible reconstruction of the deglacial temperature overprint in deep Atlantic waters. Berger and Vincent (1986) inferred a very different scenario from that of Boyle and Keigwin, with formation of North Atlantic Deep Water slowing or shutting down between 13 and 11 ka. This would probably have very different implications for the form of the temperature overprint. In addition, complications resulting from changes in salinity and water mass (with Antarctic bottom water partially replacing North Atlantic Deep Water) would have to be addressed in a full analysis of this problem.

Benthic foraminiferal records from cores in the Indian and Pacific oceans generally show no consistent evidence of steps or a mid-deglacial reversal, although most cores are neither closely sampled nor dated. Evidence of steps occurs in one undated North Pacific core (Keigwin, 1987), but this may be an artifact of bioturbation and widely varying foraminiferal abundance. It is likely that any "Younger Dryas signal" emanating from the North Atlantic would be highly attenuated during transmittal to the Indian and Pacific oceans, given the slower mixing time of those deep oceans.

The existence both of steps and of a mid-deglacial reversal to heavier $\delta^{18}O$ values in planktonic foraminiferal records from the North Atlantic (Duplessy and others, 1981) could also be a local temperature overprint. The 10°C amplitude of deglacial temperature changes in North Atlantic surface water (Ruddiman and others, 1977) would equate to an $\delta^{18}O$ change of about 2.4 $^0/_{00}$, well in excess of the $\delta^{18}O$ variations observed during the Younger Dryas (Duplessy and others, 1981) and calculated from deconvolution techniques (Bard and others, 1987). On the other hand, potentially large salinity/meltwater changes during the Younger Dryas (discussed later) would oppose and might even cancel the temperature effects on the $\delta^{18}O$ signal. Given these uncertainties, as well as extensive complications associated with bioturbation and widely varying foraminiferal abundances, it is not possible to derive an accurate estimate of surface-ocean $\delta^{18}O$ values in the North Atlantic during the deglaciation, nor is it yet feasible to demonstrate that the steps and reversals in planktonic foraminiferal $\delta^{18}O$ records from the North Atlantic are artifacts of local factors.

Still more speculative is the question of transferring the kinds of local temperature signals shown in Figures 2B and 2C from the subpolar North Atlantic to planktonic foraminiferal $\delta^{18}O$ records measured in equatorial Atlantic and western equatorial Indian Ocean cores (Prell, 1984; Berger and others, 1985; Mix and Ruddiman, 1985). In the absence of coupled ocean-atmosphere models that incorporate ocean dynamics and upwelling, general-circulation modeling studies (Manabe and Broccoli, 1985; Broccoli and Manabe, 1987; Rind and others, 1986) are inconclusive on this point.

Attempts have been made to use sea-surface temperatures estimated by transfer functions to remove an array of local temperature signals from marine $\delta^{18}O$ curves in an extensive set of equatorial Atlantic cores, with the goal of deriving the common "ice volume" portion of the $\delta^{18}O$ signal (Mix, Chapter 6). Instead, the "corrected" $\delta^{18}O$ curves were more dissimilar than the "uncorrected" versions. Although this does not disprove the validity of the transfer-function estimates, it does indicate that this transfer function is not the proper basis for removing the local $\delta^{18}O$ effects.

As a result, there is presently no clear basis for removing the 0.4 ‰ temperature (or other local) overprint that appears to be embedded in equatorial Atlantic and Indian ocean planktonic foraminiferal $\delta^{18}O$ signals. And it is thus impossible to rule out our speculation that the Younger Dryas signal from the North Atlantic was transmitted to low latitudes and recorded in planktonic foraminiferal records as shown in Figures 2B and 2C.

Partial independent support for this kind of speculation is found in ice-core $\delta^{18}O$ trends from the Quelccaya Glacier in the Peruvian Andes, which show a strong correlation with measured and estimated surface temperature trends in Great Britain over the last 375 years (Thompson and others, 1986). The similarity between these two temperature-dominated signals suggests that there could be temperature linkage between the North Atlantic and low latitudes on time scales comparable to the Younger Dryas event.

In any case, the main conclusion here is that the deglacial steps visible in many marine $\delta^{18}O$ signals do not prove that steps must exist in the ice-volume component of that signal, and particularly in the Laurentide ice-volume portion. Based on isotopic evidence alone, it is not possible to disprove the smooth, step, or Younger Dryas deglaciation models shown in Figure 1. This puts the focus back on the regional North American evidence summarized in this volume in order to address the question of the timing of the deglaciation.

Evaluation of Younger Dryas deglaciation model

Despite the mid-deglacial cooling in and around the North Atlantic Ocean, the pervasive lack of North American evidence for a mid-deglacial reversal noted in this volume argues against the Younger Dryas model (Fig. 1). Ice-area trends show no regionally integrated ice readvance back toward glacial limits in North America (Andrews, Chapter 2); a pause in retreat on the southeastern Laurentide margin between 11 and 10 ka is balanced by very rapid retreat along the western margin. Although mid-deglacial increases in ice height conceivably could have occurred due to ice growth or to bedrock rebound during a pause in deglaciation, these changes should have been felt south of the ice sheet. General circulation model results show that the size of the Laurentide ice sheet is a key control on North American temperatures adjacent to and south of the ice sheet (Kutzbach, Chapter 19); any episode of significant ice growth is unlikely to have gone undetected south of the ice sheet, particularly in the climate-sensitive monitors near the ice margin.

The lack of a mid-deglacial reversal in most local-scale climatic signals from North America, such as individual pollen, plant macrofossil, or insect records (Chapters 11–16, 20), as well as in $\delta^{18}O$ trends measured in North American lake sediments (Fritz and others, 1987), argues against such a change. This does not appear to be a problem involving the response time of the vegetation; modeling studies indicate that cooling events with amplitudes of 2°C that last longer than 200 years leave a clear imprint on the fossil pollen record (Davis and Jacobson, 1985).

In all of North America, only the Canadian Maritimes and New England, as well as a small area in Ohio, suggest any degree of reversal back toward colder conditions (Jacobson and others, Chapter 13; Gaudreau and Webb, 1985). The oscillation recorded in the Canadian maritimes and New England can be entirely explained by a cooling event centered in the North Atlantic and involving no ice-sheet growth (Rind and others, 1986). The Younger Dryas deglaciation model is thus rejected.

Evaluation of the smooth deglaciation model

The smooth deglaciation model (Fig. 1) agrees with the evidence of a steady decrease of ice area through the deglaciation. And it agrees with the absence of a mid-deglacial cooling in most terrestrial climatic records from North America.

But this model also requires that the strongly coupled behavior of the Laurentide ice sheet and high-latitude North Atlantic surface ocean observed at orbital (20,000-year to 100,000-year) time scales (Ruddiman, Chapter 7) does not apply to shorter (< 2,000-year) time scales. In this case, not only the mid-deglacial cooling, but also the two steplike warmings in the North Atlantic, must be interpreted as local oceanic responses neither caused by, nor felt by, the North American Laurentide ice sheet immediately upwind.

This model disagrees, however, with the existence of one or more steps in the deglacial rate of pollen change in North America (Jacobson and others, Chapter 13; Webb and others, Chapter 20). These steps imply accelerated warming (and thus possibly faster deglaciation) at or just before 10 ka and possible 13.7 and 12.3 ka.

The smooth deglaciation model places the North Atlantic Younger Dryas event in the middle of the interval of most rapid deglaciation. This is in accord with hypotheses discussed subsequently that attribute the Younger Dryas event to influxes of meltwater and/or icebergs caused by rapid deglaciation. Finally, this model places the fastest ice-sheet wasting in the middle of the time of strongest insolation forcing (Fig. 1). This best matches the simplest predictions of the basic insolation theory of Milankovitch (1941).

In summary, the smooth deglaciation model agrees with much of the evidence reviewed in this volume and is the model preferred here.

Evaluation of the step deglaciation model

The step deglaciation model (Fig. 1) requires rapid ice-sheet wastage at 13 and 10 ka. Because Laurentide ice margins retreated relatively smoothly through these intervals or accelerated only slightly (Bloom, 1971; MacDonald, 1971), rapid decreases in Laurentide ice-sheet height would have to have occurred at 13 and 10 ka.

The second (10 ka) step in the step deglaciation model is supported by widespread evidence for rapid changes in North American pollen (Jacobson and others, Chapter 13; Webb and others, Chapter 20). This pollen response suggests a rapid terrestrial warming at a time equivalent or very close to the decrease in ice height required by the step deglaciation model. Modeling work suggests that the climatic (temperature) change responsible for this rapid pollen shift probably occurred no more than a few hundred years prior to the time of maximum change in fossil pollen (Davis and Jacobson, 1985). The smaller and less clearly documented pollen changes at 13.7 and 12.3 ka (Jacobsen an others, Chapter 13) may indicate smaller scale changes of North American climate within the time of the first step (14 to 12 ka).

The coincidence of the two steps in the step deglaciation model with the abrupt warmings of the North Atlantic also suggests that the size (height) of the Laurentide ice sheet could control surface-ocean temperatures even at short (1,000-year) time scales. There is, however, a problem in explaining the two-step warming of the North Atlantic by changes in Laurentide ice. If decreased ice height is suggested as the cause of the initial (13 ka) ocean warming, and if major Laurentide ice growth (or bedrock rebound) did not occur in North America during the Younger Dryas interval, how could there have been a second oceanic warming, part of which encompassed the same region that warmed at 13 ka and then cooled in the Younger Dryas (Ruddiman, Chapter 7)? This problem is addressed in a subsequent section.

In addition, the step deglaciation scenario places the Younger Dryas cooling within an interval of marked deceleration in the rate of deglaciation; this timing makes it more difficult to call on the products of deglaciation (icebergs, meltwater) to explain the Younger Dryas event in the Atlantic. Finally, the step deglaciation model places the slowest rates of deglaciation in the middle of the strongest insolation forcing, requiring strong negative feedbacks within the system to oppose the impetus toward deglaciation.

In summary, although some elements of the step deglaciation model are supported by evidence summarized in this volume, significant problems appear with this model that did not appear in the smooth deglaciation model.

UNIFIED MODEL OF NORTH AMERICAN DEGLACIATION

The bulk of the evidence from this volume (and elsewhere) suggests that the actual melting of North America ice was closer to the smooth than the step deglaciation model. For the unified deglaciation model preferred here (Fig. 3), melting is heavily concentrated between 13 and 9.5 ka and is fastest from 10.5 to 9.5 ka. This model rejects the rapid decrease in ice volume from 14 to 12 ka and the pause in melting from 12 to 10 ka required by the step deglaciation model, but retains a relatively rapid (steplike) rate of ice-sheet disintegration centered on 10 ka. In the following sections, the complete deglaciation sequence is summarized, and the ability of the unified model to explain this evidence is evaluated.

The glacial maximum

At the last glacial maximum, ice sheets in most regions appear to have been within 50 to 100 km of the margins of the "maximum model" of ice volume reconstructed by Denton and Hughes (1981); yet the previous discussion indicated that the oxygen isotopic compilations by Mix (Chapter 6), adjusted for a 0.4 ‰ temperature effect, are closer to the predictions of the Denton-Hughes "minimum model." If the substantial overestimation of Antarctic ice volume and smaller underestimation of several circum-Arctic marine ice sheets in the "minimum model" were updated (G. Denton, personal communication, 1987), this would reduce the 1.48 ‰ amplitude predicted for the marine $\delta^{18}O$ signal by the "minimum model" (Mix, Chapter 6) to a value near the estimated 1.3 ‰ ice-volume effect.

Taken together, these two observations suggest that some parts of the major glacial-maximum ice sheets must have been considerably thinner than shown in the Denton-Hughes "maximum model." With the dominance of North American ice, this must have included parts of the Laurentide ice sheet. One school of thought (Andrews, Chapter 2) has favored the existence of separate smaller ice-sheet domes at the glacial maximum, rather than a large central dome or ridge over Hudson Bay. Another school of thought (Hughes, Chapter 9) notes that postglacial isostatic rebound patterns and glacially induced gravity anomalies appear to require a central dome or northwest/southeast–trending ridge over Hudson Bay.

Several studies (summarized in Andrews, Chapter 2; Hughes, Chapter 9) suggest that an extensive ramp of thin ice existed along the northwestern, western, southwestern, and possibly southern margins of the ice sheet. Lowering the elevations used in the Denton-Hughes "maximum model" to the level of this proposed ramp would reconcile some of the problem of excess ice volume and yet retain a thick Laurentide ice sheet in the central

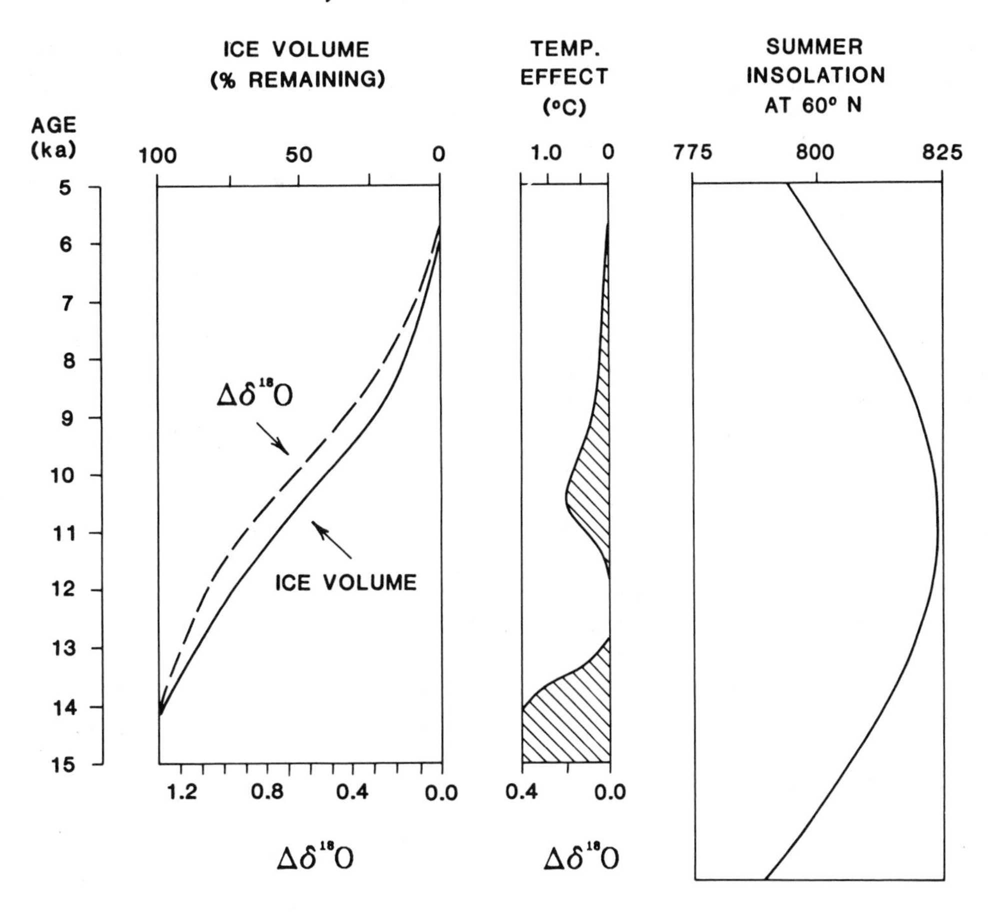

Figure 3. Preferred model of Laurentide ice-volume decrease. Melting mostly occurs between 13 and 9 ka and mainly follows smooth trend; melting is very rapid from 10.5 to 9.5 and then slows in the final stages of deglaciation. Schematic $\delta^{18}O$ ("ice-volume") curve shown lagging 0.5 to 1 ka behind true ice volume. Summer (calendric) insolation forcing shown to right.

and northeastern regions where geophysical evidence favors thicker ice. Andrews (1982) pointed out the possibility of this kind of "asymmetric" ice sheet.

Flow directions inferred from the composition of erratics also challenge the idea of a central dome or ridge trending northwest/southeast over Hudson Bay (Shilts, 1980). The "anomalous" flow of erratics eastward from the Dubawnt Beds in Keewatin may simply represent late-stage transport of debris down Hudson Strait from separate domes left after water invaded Hudson Bay (Hughes, Chapter 9).

The anomalous transport of Proterozoic sedimentary rocks and volcanics southwestward from the east side of Hudson Bay cannot be late-stage transport; these deposits appear to require a longer term transport path in that direction. This may be an indication that the southern axis of the ice divide veered eastward along the southeastern end of Hudson Bay. Alternatively, the Laurentide ice divide may have slowly migrated westward during long-term growth of the Laurentide ice sheet (Flint, 1971); this migration could sweep erratics southwestward throughout the ice-growth intervals (suggested by T. Hughes, personal communication, 1987). It might also account for the fact that the most extensive late-glacial ice margins become younger westward from the eastern United States and Great Lakes (21 to 17 ka) toward Iowa and the Prairies (15 to 14 ka).

Initial deglaciation (14 ka)

The evidence is reasonably clear that major deglaciation of North America began at 14 ka. Strong evidence comes from the initiation of rapid meltwater flux down the Mississippi; this onset, detected by negative anomalies in oxygen-isotopic records from planktonic foraminifera, has now been precisely dated by accelerator mass spectrometry to 14 ka (Broecker and others, 1987). This date is also consistent with the first rise in many marine $\delta^{18}O$ signals (Mix, Chapter 6), although it can be argued that the false temperature signal embedded in the marine $\delta^{18}O$ curves could disguise even earlier melting.

Temperate deciduous trees such as *Fraxinus* (ash) first appeared in the north-central Midwest near the southern ice-sheet margins at 14 ka (Jacobson and others, Chapter 13; Webb and others, Chapter 20), possibly due to amelioration in the climate at the ice margins. Integrated rates of change in pollen composition across east-central North America also began to increase around 14 ka (Jacobson and others, Chapter 13). Vertebrate faunas from the northern midcontinent document the withdrawal of arctic species at this time, although some boreal forest taxa remain in the late-glacial faunas (Graham and Mead, Chapter 17).

In addition, loess deposition ended in several areas of the midcontinent as early as 14 ka. This was probably due in part to the stabilizing effect of vegetation on outwash plains. Increased vegetation in turn requires greater levels of effective moisture, which implies the presence of warmer air masses.

Numerous oscillations in the southern margin of the Laurentide ice sheet occurred prior to 14 ka, and there is a continuing debate as to whether or not these were synchronous (Mickelson and others, 1983). Regardless of the answer, most parts of the margin were not much reduced at 14 ka from their glacial-maximum limits, and we interpret this as consistent with isotopic evidence indicating no appreciable *net* deglacial melting of the Laurentide ice sheet before 14 ka. The development of an ice-free embayment in Cabot Strait and up the St. Lawrence Seaway between 13.6 and 13.2 ka also indicates significant early melting on the southeastern ice margin just after 14 ka. Melting of both marine and terrestrial Cordilleran ice margins also began around 14 ka (Booth, Chapter 4).

Some climatic indicators on the North American continent do not mark 14 ka as the beginning of deglaciation, but the disagreements are tolerable. In the west, some changes in vegetation occurred before 14 ka, but these may reflect either modest adjustments in the position of the jet stream along the thin southwestern margin of the Laurentide ice sheet or the effects of gradually increasing summer insolation in regions more remote from the ice sheet (Barnosky and others, Chapter 14; Kutzbach, Chapter 19; Webb and others, Chapter 20). Lake-level and plant macrofossil evidence from the southwest (Benson and Thompson, Chapter 11; Forester, Chapter 12; Van Devender and others, Chapter 15) and pollen data from Alaska (Barnosky and others, Chapter 14) do not contradict the conclusion that the major changes associated with deglaciation occurred mainly after 14 ka.

Changes in insect assemblages occurred near the ice front in the midcontinent after 17 ka and before 15 to 14.5 ka (Morgan, Chapter 16). This may be a case in which an indicator with different sensitivity has recorded an early onset of the much stronger warming to follow. Many more well-dated insect records are needed to produce a regionally coherent view of the deglaciation.

In summary, none of the terrestrial data are in fundamental disagreement with the idea that significant net deglacial melting began at 14 ka and accelerated after that time. But why did deglaciation begin at 14 ka? There are several possible answers.

Summer insolation is widely regarded as the primary cause of long-term changes in the ice sheets (Milankovitch, 1941). The ice-volume response lags an average of 6,000 to 10,000 years behind the insolation forcing because of the inherently slow responses involved in the physics of ice sheets (Imbrie and Imbrie, 1980). The insolation minima responsible for the ice growth in North America leading to the last glacial maximum occurred at 24 ka, after which insolation began to rise toward the maximum at 11 ka (Fig. 3). The observed long-term lag of 6,000 to 10,000 years thus predicts that deglaciation should have begun around 18 to 14 ka.

But what specifically happened on the ice margin? The basic reason for the lag in deglaciation behind insolation forcing probably involves the self-sustaining nature of large ice sheets. Their high elevation keeps most of the ice in regions of low ablation and positive mass balance, while their size alters the large-scale atmospheric circulation in such a way as to chill the air and retard ablation on the lower ice margins. One explanation for the beginning of deglaciation around 14 ka is that a critical threshold of warmth on the ice margins was exceeded due to rising summer insolation. Modeling experiments show that increasing summer insolation progressively enhances the deglacial heating of land south of the ice margin (Kutzbach, Chapter 19). Perhaps at some point this warmed the lower ice margins past some threshold at which ablation accelerated and deglaciation began. This may have been aided by slow sinking of bedrock beneath thicker parts of the ice margins as part of the delayed isostatic response to the earlier interval of ice growth centered at or near 24 ka.

A second possibility is that an external factor triggered melting on the southern margins. One obvious candidate is an increase in atmospheric CO_2, although ice-core evidence indicates that the CO_2 rise was synchronous with ice melting, rather than preceding it.

Still another possibility is that even earlier melting in other regions (Britain and Scandinavia) triggered a sufficient rise in sea level to initiate calving on marine ice margins. This idea is supported by the rapid early deglaciation of Cabot Strait by calving.

The origin of the mid-deglacial Younger Dryas event

With the evidence available, the most difficult decision to make regarding the form of the deglaciation is whether or not it consisted of steps at 13 and 10 ka or a smooth deglaciation centered at 11 to 10 ka. Recent evidence in favor of the smooth deglaciation model comes from oxygen isotopic evidence from the Gulf of Mexico (Broecker and others, 1987). This evidence indicates that the influx of meltwater down the Mississippi River increased from 14 ka until at least 12 ka and possibly as late as 11 ka. An increasing meltwater influx past 12 ka is more consistent with the smooth deglaciation model than with the step model, because the latter predicts a slowing (or cessation) of meltwater influx by 12.5 or 12 ka. Unfortunately, a hiatus truncates the record and precludes a detailed interpretation of the later runoff signal. Of course, the Mississippi River is only one pathway for the products of the melting Laurentide ice sheet, and the meltwater signal in the Gulf of Mexico is thus not the entire story of the deglaciation.

As summarized above, there is other evidence both for and against the smooth and step deglaciation models. To help resolve this problem, it is necessary to return to the problem of the mid-deglacial Younger Dryas oscillation in the North Atlantic ocean, described as an unresolved "mystery" in a previous chapter (Ruddiman, Chapter 7).

If the Younger Dryas response in the North Atlantic Ocean was not driven by large-scale changes in the areal extent and height of the Laurentide ice sheet (Ruddiman, Chapter 7), what is its origin? Most explanations look to influxes of products of the deglacial melting: small icebergs calved from the marine margins of the Laurentide and Scandinavian ice sheets; large tabular icebergs from disintegrating ice shelves in the Arctic Ocean or around its margins (Mercer, 1969; Ruddiman and McIntyre, 1981); or fresh meltwater from the continents (Johnson and McClure, 1976; Rooth, 1982; Broecker and others, 1987).

The lack of a discernible influx of continental ice-rafted debris during the Younger Dryas (Ruddiman and McIntyre, 1981) appears to rule out an iceberg influx from the continents. Even though we do not know the entire history of deglacial calving into the Labrador Sea and regions proximal to Hudson Strait, the lack of ice-rafted quartz and feldspar in cores from the more distal central and eastern portions of the North Atlantic Ocean after 13 ka proves that no renewed influx of icebergs reached these areas to account for the ocean cooling. For this reason, this explanation of the Younger Dryas had previously been rejected (Ruddiman and McIntyre, 1981).

The marine ice shelves proposed to have existed in the Arctic Ocean or around its margins would have contained little or no sedimentary debris; as a result, it is not possible to rule them out as a factor despite the lack of ice-rafted debris in ocean cores during the Younger Dryas. In addition, Icelandic ash in North Atlantic sediments of Younger Dryas age trace a pathway suggesting delivery on ice that passed through the Denmark Straits from sources farther north (Ruddiman and McIntyre, 1981). This remains an interesting theory supported by some evidence but lacking firm proof.

The theory of freshwater influx from North America has important independent support: the meltwater drainage path diverted from the Gulf of Mexico to the Atlantic at 10.8 ka and then back to the Gulf of Mexico at 10 ka (Teller, Chapter 4). This diversion closely matches the radiocarbon age of the Younger Dryas event. But this theory also has a problem: it fails to explain why the rediversion of meltwater flow back to the Atlantic after 9.5 ka didn't produce a second ocean-cooling event later in the deglaciation.

I suggest here that the freshwater-diversion theory can explain the Younger Dryas oscillation, but only if combined with a second factor: the basic ice-sheet control of North Atlantic surface-ocean temperatures that is well demonstrated at orbital time scales (Ruddiman and McIntyre, 1984). This control of the North Atlantic by the size (area and height) of the North American ice sheet provides the basic explanation for the progressive surface-ocean warming and retreat of the polar front; the diversion of freshwater provides the impetus for the Younger Dryas reversal; together, both factors can explain the full North Atlantic sequence.

The first retreat of the polar front at and after 13 ka may have been a response to an early phase of Laurentide ice sheet shrinkage below some critical size (area/height) threshold. This ice wasting removed the downwind cooling effect from the remote eastern North Atlantic, but left the cooling of the more proximal western Atlantic and Labrador Sea intact (Ruddiman and McIntyre, 1973). The area of Laurentide ice was reduced by some 25 to 30 percent by 12 ka (Prest, 1969; Bryson and others, 1969); the accompanying reductions in Laurentide ice height (Bloom, 1971; MacDonald, 1971; Paterson, 1972) apparently were sufficient to allow the east-central part of the North Atlantic to warm.

This ice-sheet shrinkage does not appear to have been steplike. There was only a slight acceleration in the integrated rate of areal retreat of Laurentide ice around 13 ka (Bloom, 1971; MacDonald, 1971). Although it is possible that there were larger, more sudden changes in ice height at this time, the relatively small size of the two pulses of regional pollen change in east-central North America (Jacobson and others, Chapter 13; Webb and others, Chapter 20) does not support the idea of radically abrupt (steplike) changes in ice height. We suggest instead that the Laurentide ice sheet shrank through some sensitive threshold of height (and/or area) below which it had a substantially smaller downwind cooling impact on the North Atlantic.

Such a threshold may involve interactions with the jet stream, which was split by the glacial-maximum ice sheet in general circulation modeling tests (Kutzbach, Chapter 19). The Denton-Hughes "maximum model" of ice volume has broad areas in the central part of the ice sheet at a height of about 3,000 m (Denton and Hughes, 1981). The postulated threshold effect would probably appear at heights some 20 percent lower, around 2,400 m.

General circulation experiments published to date cannot be used to detect such a threshold effect. A 12-ka ice sheet with a 1,600-m central elevation showed significant reductions in ice-sheet impact on the jet stream relative to the 3,500-m central elevation of the glacial-maximum ice sheet (Kutzbach and Guetter, 1986), but this does not directly address the existence of a sensitive threshold near 2,200 to 2,400 m. In addition, because sea-surface temperatures were prescribed in the 12-ka model reconstruction, ice-sheet effects on the North Atlantic Ocean cannot be directly evaluated.

The subsequent cooling and polar-front readvance culminating around 10.5 ka in the North Atlantic can then be attributed to the diversion of North American runoff, causing a mid-deglacial influx of Laurentide meltwater to the North Atlantic. This influx was sufficient to flip the North Atlantic back toward a glacial mode for less than a millenium. It should also be noted that this meltwater influx was additive to the reduced (but still-present) chilling effect of the smaller Laurentide ice sheet on the atmosphere and the ocean. It is also possible that tabular icebergs calved around the periphery of the Arctic Ocean during the interval contributed to this cooling of the North Atlantic Ocean (Mercer, 1969).

At 10 ka, the rediversion of North American drainage back to the Gulf of Mexico removed the meltwater effect from the North Atlantic (Teller, Chapter 3), and North Atlantic temperatures were again determined largely by the size of the Laurentide ice sheet. By this time, however, the Laurentide ice sheet was some 50 to 60 percent smaller in size than at the glacial maximum (Prest, 1969; Bryson and others, 1969; Bloom, 1971; MacDonald, 1971), and thus was probably capable of imposing significant cooling effects only on the nearby Labrador Sea. This reduced cooling effect on the North Atlantic is supported by the diminished impact of small ice sheets on atmospheric circulation in model reconstructions of 12 and 9 ka (Kutzbach and Guetter, 1986).

The final rediversion of North American drainage back to the Atlantic after 9.5 ka had no obvious impact on the eastern and central portions of the North Atlantic. This raises a key point. Why would this diversion of flow at 9.5 ka not affect the North Atlantic in the same way as the prior diversion at 10.8 ka? The answer must lie in a combination of two factors: (1) a still smaller Laurentide ice sheet that had become even less capable of cooling the Atlantic directly through the atmosphere; and (2) a reduced rate of freshwater discharge from the Laurentide ice sheet after 9.5 ka compared to that from 11 to 10 ka.

One of these two factors directly requires, and the other strongly implies, a major decrease in size of the Laurentide ice sheet between 10.8 and 9.5 ka. Other evidence is consistent with this interpretation. Maximum rates of Laurentide ice sheet retreat occurred betweeen 10.5 and 9 ka (Bloom, 1971). Pollen records in east-central North America show the fastest rates of change within the entire deglaciation between 10.8 and 9 ka (Jacobson and others, Chapter 13; Webb and others, Chapter 20). And there are major warmings indicated in North Atlantic plankton (Ruddiman, Chapter 7), in Greenland ice-core $\delta^{18}O$ records (Paterson and Hammer, Chapter 5), and in virtually every paleoclimatic indicator from North America (Chapters 13–20).

In addition, the second deglacial step in many marine $\delta^{18}O$ records occurred at 10 ka (Mix, Chapter 6). Even though we earlier showed that this $\delta^{18}O$ step does not require an ice-volume step, this analysis did not rule out rapid deglaciation during this time. We suggest here that 10.8 to 9.5 ka was a time of particularly rapid ice-volume loss over North America (Fig. 3). If the ice-volume component of the $\delta^{18}O$ record lags some 500 to 1,000 years behind true ice volume (as discussed later), unusually rapid ice-volume loss between 10.8 and 9.5 ka would register in marine $\delta^{18}O$ records almost 1,000 years later, although it would be partly masked by the temperature overprint.

This explanation of the deglacial warming of the North Atlantic is also consistent with long-term orbital trends in the North Atlantic Ocean (Ruddiman, Chapter 7). Long-term control of ocean temperatures at the 100,000-yr and 41,000-yr periods by the size of the Laurentide ice sheet is reflected in the basic deglacial warming of the North Atlantic as ice sheets shrank; this control specifically appears as the 13-ka warming of the east-central North Atlantic and the subsequent late-deglacial warming of the western Atlantic and Labrador Sea (Ruddiman and McIntyre, 1973).

The Younger Dryas cooling appears to be associated with the 23,000-yr sea-surface temperature signal that is prominent between 40° and 50°N latitude; this signal has coldest temperatures occurring during mid-deglaciations and has thus been ascribed to influxes of deglacial meltwater (Ruddiman and McIntyre, 1984). In the surface ocean north of 50°N, the 23,000-yr signal is generally subordinate to the larger 41,000-yr and 100,000-yr responses, but the Younger Dryas event appears to be a brief interval during which the "cold-on-deglacial" signal characteristic of middle latitudes at a rhythm of 23,000 years overrode the "cold-glacial, warm-interglacial" signal characteristic of higher latitudes at rhythms of 41,000 and 100,000 years.

Mid-deglaciation (13 to 9.5 ka)

During the mid-deglacial interval, the pattern of areal retreat of the Laurentide ice sheet was asymmetrical (Andrews, Chapter 2). The entire western margin of the Laurentide ice sheet melted back very rapidly, whereas the northeastern margin of the ice sheet hardly retreated at all. The southeastern margin melted relatively slowly after the early marine incursion into the St. Lawrence Seaway, and it stabilized just north of the St. Lawrence valley from 11 to 10 ka.

What accounts for this asymmetric meltback of the Laurentide ice sheet? The basic answer appears to be ice thickness. The rapid meltback on the northwestern, western, and southwestern ice-sheet margins is consistent with the existence of a thin ramp of ice in these areas. This also appears to reconcile a problem raised by Andrews (1973), who suggested that the mid-deglacial retreat rates in those areas were too rapid to be explained even by postglacial atmospheric heat and insolation levels. Beget (1987) calculated that thin ice ramps on those margins would pose no such energy problem.

The slower retreat in the east was probably due to several factors. Because the eastern ice was thicker, its margins melted back more slowly. Ablation was also probably lower, particularly on the northeastern margin, due to cold temperatures caused by the perturbation of atmospheric circulation by the ice sheet itself (Manabe and Broccoli, 1985; Kutzbach and Guetter, 1986; Lindeman and Oerlemans, 1987).

The deglaciation model presented here suggests that the ice sheet was greatly reduced in size between 10.8 and 9.5 ka, including a substantial reduction in ice elevation. It is not entirely clear how this proposed reduction in height over the central portions of the ice was achieved. Melting and warming of the western and southern ice-sheet margins may have increased the flow rates of ice from the interior sufficiently to draw down the central ice dome (Peltier, Chapter 8).

Another possibility is ablation via calving through Hudson Strait, which could proceed without significant retreat of the ice margins and thus affect interior elevations of the ice sheet. Andrews (Chapter 2) summarizes evidence suggesting that ice shelves existed along the northeastern ice margin during the earlier parts of the deglaciation; these ice shelves may have suppressed rapid flow in the Hudson Strait ice stream until this critical interval later in the deglaciation (Hughes and others, 1977). A more accurate chronology of ice-shelf breakup and Hudson Strait outflow would clarify this aspect of the deglaciation.

Late deglaciation (9.5 to 6 ka)

The major question about the later stages of deglaciation is the height of the central ice sheet over Hudson Bay. Paterson and Hammer (Chapter 5) suggested that maximum rates of volumetric deglaciation occurred around 8 ka, using a conversion of ice area to ice volume in which high-elevation ice was assumed to have persisted over Hudson Bay until just before the marine invasion at 7.8 ka (Prest, 1969; Bryson and others, 1969). This late-deglaciation scenario was also suggested by Hughes and others (1977) and it finds some support in the possible existence of a termination "1c" at 8 to 7 ka in some marine oxygen isotopic records (Mix and Ruddiman, 1985).

The model proposed above, however, views the sudden clearance of ice from a large area of Hudson Bay at 8 to 7.8 ka as simply the end product of a much faster decrease in ice volume over the central portions of the Laurentide ice sheet some 2,000 years earler (10.8 to 9.5 ka). Somewhat similar views have been expressed by Andrews and Peltier (1976), who inferred a thinned ice shelf over northern Hudson Bay by 10 ka, but retained thicker ice over southern portions of Hudson Bay away from the ice stream draining through Hudson Strait. Earlier collapse is also at least broadly consistent with the basic view taken by Paterson (1972) that the ice sheet tended increasingly toward a stagnant condition in the later part of the deglaciation as it shrank below the height needed for active ice flow.

Support for this view also comes from the cessation of rapid change in pollen composition south of the ice sheet after 9 ka (Jacobson and others, Chapter 13). This suggests that the elevation of central areas of the ice sheet had decreased by 9.5 to 9 ka to the point that Laurentide ice was no longer a major influence on atmospheric flow over North America south of the ice sheet. In addition, Webb and others (Chapter 20) noted that the absence of herb pollen along the ice margin by 9 ka may indicate that the ice sheet was too low to generate katabatic winds by this time. Also, the apparent absence of a pulse of surface-ocean cooling in the Labrador Sea or North Atlantic Ocean after 9.5 ka argues against rapid influxes of meltwater due to late collapse of the Laurentide ice sheet. Final melting of separate ice masses in the Ungava, Keewatin, and Foxe basins from 9 to 6 ka represents the last phases of melting of increasingly stagnant, low-elevation ice.

The problem of isotopically inhomogeneous ice sheets

Next, I address a complication previously noted by Mix and Ruddiman (1984). Because early melting preferentially removed the less ^{18}O-depleted ice from the southern margins of the Laurentide ice sheet, actual loss of ice volume early in the deglaciation should be to some extent underrepresented in the marine δ^{18}O curves. Because of this effect, Mix and Ruddiman (1984) inferred possible lags of the δ^{18}O signal behind true ice volume by an average of 500 to 3,000 years.

Lags of 500 to 1,000 years are consistent with the deglaciation model proposed here. If true ice volume decreased as shown by the curve labeled "ice-volume" in Figure 3, then the portion of the δ^{18}O curve attributable to ice volume could rise somewhat later, as shown by the curve labeled "δ^{18}O". To reconcile this "ice-volume" version of the δ^{18}O curve with that actually observed (Fig. 2A) would require the temperature overprint shown in Figure 3. Based on previous discussions, this overprint is permissible, and in some regions even likely.

Mix and Ruddiman (1985) pointed out the lag between their observed deglacial δ^{18}O curve and various Laurentide and global deglacial ice-volume curves; the lag is irregular but averages about 1,000 years. The proposed deglaciation scheme shown in Figure 3 acknowledges this problem by placing the middle of the true ice-volume loss near 11 ka, but registering it in the isotopic record some 500 to 1,000 years later. This lag is consistent with early melting of the less isotopically depleted ice in the south (Mix and Ruddiman, 1984).

In summary, the model proposed here rejects significant deglaciation prior to 14 ka, as required both by recent models (Denton and Hughes, 1981; Duplessy and others, 1981; Ruddiman and McIntyre, 1981) and by earlier views of the deglaciation (Flint, 1971; Bloom, 1971). Conversely, the model suggested here also rejects the persistence of high-elevation (2 km) ice over Hudson Bay as late as 8 ka (Andrews, 1975). It also disagrees with the isotopically based two-step deglaciation model proposed by Duplessy and others (1981) and supported by evidence in Berger and others (1985) and Mix and Ruddiman (1985). It does, however, affirm that the most rapid rates of deglaciation occurred near the second step in these isotopically based models.

The proposed model is most in accord with the simple insolation forcing of Milankovitch (1941), with fastest melting occurring during the time of strongest forcing (Fig. 3). The model also attributes the mid-deglacial cooling in the North Atlantic and growth of ice in Europe during the Younger Dryas to changes in the direction of meltwater drainage during rapid ice disintegration in North America, perhaps aided by outflow of tabular icebergs from the Arctic Ocean.

DEGLACIAL ABLATION MECHANISMS

The second major thematic focus of this volume is an evaluation of the relative importance of the ablation mechanisms during deglaciation. Insolation forcing is the main impetus for deglaciation; the specific contributions of ablation mechanisms mentioned in the introduction are now reviewed.

Bedrock rebound

Delayed bedrock rebound may act as an ablation mechanism by keeping the terrestrial margins of a retreating ice sheet at low altitudes in a regime of warm air and strong ablation (Peltier, Chapter 8). This requires that the slow viscous component of mantle behavior be dominant relative to the fast elastic component. It also requires an ice sheet sufficiently thick and persistent to significantly depress the crust.

It is difficult to reconstruct the past elevation of ice margins. Along margins where the ice is in contact with the sea, raised beaches can be used to reconstruct bedrock rebound. However, a fairly wide range of ice-unloading histories will satisfy the (mostly Holocene) dates from raised beaches. In addition, uncertainties in lithospheric thickness, lower mantle viscosity, and other factors do not yet permit "unique" solutions of past ice loads (Peltier, Chapter 8).

On the critical southern margins of retreating ice sheets, proglacial lakes offer a crude measure of bedrock depression, because the lakes lie in troughs caused by as-yet unrealized rebound (Fig. 6 in Peltier, Chapter 8). T. Hughes (personal communication, 1986) suggested that these deep ice–marginal depressions should be filled with glacial meltwater and that the distribution of proglacial lakes through the deglaciation should thus indicate regions in which ablation due to unrealized bedrock rebound was occurring.

The areal extent of proglacial lakes in North America during the deglaciation shows a simple increasing trend from relatively low values early in the deglaciation (before 12 ka) to values an order of magnitude higher, culminating at 10 to 9 ka late in the deglaciation (Teller, Chapter 3). This could be taken to indicate that the Laurentide ice margins did not, for the most part, lie in a bedrock "hole" early in the deglaciation, and that there was little opportunity for this kind of ablation feedback to operate prior to 11 or 10 ka. Evidence already summarized for thin ice on the southwestern margin of the early-deglacial Laurentide ice sheet supports this idea. Thin ice is less likely to depress bedrock and cause extensive proglacial lakes.

In contrast, the extensive lakes present during the last half of the deglaciation suggest that delayed bedrock rebound did act as a major ablation agent later in the deglaciation, in full accord with this theory (Peltier, Chapter 8). This evidence places the strongest ablation action due to delayed bedrock rebound in the time of fastest deglaciation (Fig. 3) and is consistent with the idea that bedrock rebound was a major ablationary factor.

To some extent, the greater prevalence of lakes later in the deglaciation could partly reflect the fact that regional drainage patterns led southward away from the ice during the first part of deglaciation, thus tending to inhibit the formation of proglacial lakes, whereas the drainage led northward toward the ice sheet during the second part of the deglaciation, tending to dam the water against the ice. Other evidence cited previously, however, supports the existence of thin southwestern and western ice margins early in the deglaciation, supporting the idea that ablation caused by bedrock depression was minimal at this time. If rapid ablation occurred on these margins early in the deglaciation, it did so for different reasons (see section below on Terrestrial calving, surging, and slip).

Marine calving and surging

Calving of icebergs to the ocean is widely accepted as an important ablationary factor in modern ice sheets. Denton and Hughes (1981) noted that this mechanism has the potential to provide strong feedback to insolation forcing of Pleistocene ice sheets in the following way: rising sea level caused by insolation-induced melting may unpin ice shelves and increase the rate of calving of icebergs to the ocean; this additional calving accelerates the sea-level rise, and a feedback loop is engaged. Denton and Hughes (1981) hypothesized that this process may ultimately cause a partial collapse ("downdraw") of portions of the ice sheet remote from the ablationary region, because calving activates faster ice-stream flow that carries ice more quickly out of the ice-sheet interior (Hughes, Chapter 9).

Calving of icebergs to the Atlantic Ocean is widely accepted as an ablationary factor during the early parts of the last deglaciation (e.g., Chapters 2, 5, 7, and 9). There is less agreement among these chapters as to how strong a factor it was. This disagreement largely reflects the lack of evidence for or against iceberg calving on the marine margins of the Laurentide ice sheet.

Strong evidence for calving early in the deglaciation comes from the Cabot Strait, where a calving bay developed up into the St. Lawrence Seaway, separating increasingly stagnant New England ice from the main Laurentide ice sheet by 12 ka (Borns, 1973). This episode falls within the earliest rise of marine $\delta^{18}O$ values (Mix, Chapter 6), and it also occurs during the late stages of an interval of rapid deposition of ice-rafted sand in the subpolar North Atlantic (Ruddiman, Chapter 7).

Although this calving episode was clearly an important factor on the southeastern margin of the ice sheet, there is no direct evidence from the eastern margin that it had a large impact on elevations in the interior by 13 ka. As noted above, the initial retreat of the North Atlantic polar front from the east-central North Atlantic requires at least some shrinkage of the interior elevation of the Laurentide ice sheet, and calving to the ocean is one potential mechanism for removing ice from the interior of the ice sheet early in the deglaciation.

There is comparatively little direct evidence from the other major site of calving to the Atlantic—the Hudson Strait. In fact, there is very little retreat of the northeastern Laurentide ice margins at all until the final clearance of ice from most of Hudson Bay at 7.8 ka. Andrews and Peltier (1976) inferred a collapse of the ice sheet over northern Hudson Bay by 10 ka due to discharge of ice through Hudson Strait. The deglaciation model presented here, which calls on a more widespread drop in central ice elevation between 10.8 and 9.5 ka, is consistent with (but not proof of) increased calving from 10.8 to 9.5 ka. There is insufficient oceanic and coastal marine data as yet to constrain the rate of calving (and meltwater) outflow in the Hudson Strait region.

Terrestrial calving, surging, and slip

One form of ablation on the terrestrial ice-sheet margins proposed by Andrews (1973) involves calving of icebergs into proglacial lakes. This thins the ice-sheet margins and adds an additional heating component (water) to the atmospheric attack on the ice. Deglacial lake trends (Teller, Chapter 3) show a ten-fold increase in lake area late in the deglaciation, and even larger changes in lake area relative to the area of Laurentide ice remaining. This suggests that although calving of a very large ice sheet into small proglacial lakes was a minor ablationary factor early in the deglaciation, calving into very extensive lakes may have become more important later on.

Another source of significant ablation on the terrestrial margins of the Laurentide ice sheet may have involved slip and surging on soft substrates (subglacial sediments, lake beds, Paleozoic sediments, and rocks) and enhanced plastic flow of warmer parts of the ice sheet. Both of these processes would have had the effect of thinning the ice margins, and this seems necessary to account for the very thin southwestern, western, and northwestern margins of the Laurentide ice sheet at the beginning of deglaciation at 14 ka. The increasingly lobate form of the southern ice margins during deglaciation (Flint, 1971) may also reflect such processes. In addition, the evidence of numerous large-scale advances and retreats of the southern margin suggests vigorous slipping and/or surging (Clayton and others, 1985).

A thin ramp of ice around the southwestern and possibly southern margins of the ice sheet would be ideal for ablation: a very extensive and low-lying surface subject to the strong heating characteristic of the lower kilometer of the atmosphere.

Carbon dioxide

Evaluating the importance of CO_2 first requires knowing the shape of the glacial-to-interglacial rise in CO_2 concentrations from the 200 ppmv values typical of the glaciation to the 270 to 280 ppmv levels attained before the industrial revolution (Paterson and Hammer, Chapter 5). There are two ways to measure this trend, but they disagree.

Direct measurements of CO_2 in Greenland ice cores during the deglaciation show CO_2 trends very similar to North Atlantic sea-surface temperatures (Paterson and Hammer, Chapter 5). CO_2 values rise abruptly around 13 ka, decrease between 12 and 11 ka, and rise abruptly at 10 ka (radiocarbon years). This deglacial CO_2 increase is, to a first-order approximation, synchronous with the deglacial ice-volume change, and the last 10,000 years of record in the Dye 3 ice core from Greenland are directly dated by counting annual layering, although earlier layers are undated. However, Greenland ice-core CO_2 measurements are not yet accepted as representative of global CO_2 trends, because CO_2 values in the warmer Holocene layers may be invalidated by melting events.

The second method of estimating past CO_2 trends involves measuring the difference in $\delta^{13}C$ between planktonic foraminifera in surface waters depleted of nutrients and in benthonic foraminifera from deep waters. This difference is a measure of the sequestering of carbon in the deep ocean out of contact with the surface ocean and the atmosphere (Broecker, 1982), which may be a major factor in setting lower atmospheric CO_2 levels. This technique shows that storage of carbon in the deep ocean was higher during the last glaciation and apparently decreased rapidly several thousand years prior to the decrease in ice volume (Shackleton and others, 1983), suggesting that the deglacial increase in atmospheric CO_2 significantly preceded the decrease in ice volume. Recent studies suggest, however, that this $\delta^{13}C$ difference records only one of the components contributing to the deglacial change in CO_2 and accounts for only about half of the signal (Mix and Shackleton, 1986; Curry and Crowley, 1987). With the timing of other components unknown, the deglacial CO_2 signal cannot be reconstructed by this technique.

At this time, the Greenland ice-core records are the best indication of the deglacial CO_2 signal. These records indicate that the CO_2 rise was basically synchronous with, rather than leading, the ice-volume decrease.

What is the likely impact of the CO_2 increase on the ice sheets? General circulation modeling results from Broccoli and Manabe (1987) and from Kutzbach and Guetter (1986) indicate that a CO_2 increase of the magnitude of 70 to 100 ppm would cause a mean annual warming over the North American continent of about 1°C. Results from Broccoli and Manabe (1987) further indicate that the largest temperature effects (up to 2°C) would occur in winter, but that the summer warming would be substantially less than 1°C. Because summer is the primary ablation season, these modeling results suggest that CO_2 had a modest feedback effect in increasing ablation on the lower margins of the ice sheets.

Oceanic moisture

Both on long-term orbital time scales and specifically during the last deglaciation, the midlatitude North Atlantic ocean (40° to 55°N) has temperature response phased so as to deny moisture to adjacent ice sheets during insolation-driven deglaciation (Ruddiman, Chapter 7). Because coldest surface-ocean temperatures occur during mid-deglaciation, release of latent heat to the atmosphere is minimized, and denial of oceanic moisture could be a factor determining ice-sheet mass balance.

This role of oceanic moisture is most clearly applicable to

the Fennoscandian ice sheet lying downwind of the North Atlantic. Because it is more difficult for air masses from the Atlantic to turn westward against the prevailing westerly flow and penetrate far into the North American continent, Atlantic moisture probably affected mainly the easternmost margins of the Laurentide ice sheet. While Atlantic precipitation could still be critical for feeding a growing ice sheet at lower altitudes, denial of moisture probably had relatively little impact on the high-standing ice sheet during deglaciation. Because of the strong decrease of moisture with altitude, elevation is widely regarded as a more critical factor in denying moisture to large ice sheets.

In addition, removing the entire Laurentide ice sheet within 8,000 years, and about 70 percent of it within 4,000 years (Fig. 3), requires very strong ablation. Suppressing accumulation by denying moisture is a relatively weak means of influencing ice-mass balance compared to the vigorous action of ablation factors. As a result, moisture flux from the North Atlantic is regarded as having been a secondary ablation component during the last deglaciation.

An interesting problem for future modeling work is related to the position of the jet-stream maximum over eastern North America (Kutzbach, Chapter 19). This feature focuses precipitation on the southeastern margin of the ice sheet in both seasons, adding to the band of orographically induced precipitation on the southern ice margins. Changes in the jet stream, and in the location of this important precipitation maximum, occur with changes in height of the ice sheet and in extent of North Atlantic sea ice. General circulation models at this point do not reproduce modern precipitation over North America well enough to be used for a full evaluation of this problem.

The 100,000-year ice-volume cycle

Finally, there is the question raised in the introductory chapter about the origin of the dominant 100,000-year cycle during the late Pleistocene; one possible explanation of this cycle calls on nonlinear positive feedback from strong ablation mechanisms (Imbrie and Imbrie, 1980). Each of the ablation mechanisms suggested to be important during the last deglaciation may thus be part of the explanation of the 100,000-year cycle. The most important of these ablation factors appear to be delayed bedrock rebound, marine calving, CO_2, and possibly some form of slipping of the terrestrial ice margins.

To tie each ablationary mechanism more directly to the 100,000-year cycle, some kind of feedback link must be established in which the larger the size of the ice sheet, the stronger the action of the ablation mechanism, and the faster the resulting rate of deglaciation. Such a link is needed to explain the rapid deglaciations that have occurred every 100,000 years during the last 700,000 years, abruptly terminating long intervals of slower, oscillatory ice-sheet growth to large size.

In the case of the delayed bedrock rebound theory, the link is clearest. Larger ice sheets create deeper bedrock holes, which rebound more slowly (viscously) and thus accelerate ablation and deglaciation (Peltier, Chapter 8).

For the calving mechanism, larger ice sheets may have two effects (Denton and Hughes, 1981; Hughes, Chapter 9). First, large ice masses are capable of shedding larger volumes of icebergs and meltwater to the ocean, which makes sea level rise more rapidly, and rapidly destabilizes the marine margins of ice sheets, activating even faster calving. Second, larger ice sheets cause more depression of bedrock near marine margins and thus become more susceptible to effects of rising sea level.

In the case of terrestrial calving/surging/slip, the basic links have not yet been formulated. Do larger ice sheets inherently cause more calving into proglacial lakes or create more extensive ice ramps on the southern ice-sheet margins? Is the prevalence of basal slip on the ice-sheet margins somehow linked to the height of ice in the interior regions? How does ablation on the terrestrial margins provide the kind of positive feedback to the ice sheets that leads to still more ablation? Is the link through the marine margins, with ablation on terrestrial margins on leading to sea-level rise and calving?

CO_2 ablation raises similar problems, and additional ones as well. Although the direct temperature effects of CO_2 near the ice are small, even small changes may be important in initiating ablation. Rising CO_2 levels increase ablation on all the ice-sheet margins, which may enhance all of the feedbacks noted to date. But, to evaluate CO_2 as an ablationary factor will first require a better understanding of the entire CO_2 system, as well as a more reliable record of the last-deglacial increase in CO_2.

CONCLUSIONS

1. Marine oxygen-isotopic data cannot be used to choose between a smooth and step deglaciation model for the Laurentide ice sheet because of the strong temperature overprint. The trend of ice-volume loss could have been smooth (lacking in steps) if: (a) the temperature overprint on marine $\delta^{18}O$ records had the form of the deglacial North Atlantic sea-surface temperature curve, including the Younger Dryas cooling; and (b) the recording of the ice-volume decrease in marine $\delta^{18}O$ records lagged 0.5 to 1.0 ka behind true ice volume because of the inhomogeneous $\delta^{18}O$ composition of melting ice.

2. The evidence in this volume favors a smooth deglaciation model with the following key characteristics: initial melting beginning at 14 ka; increasing rates of melting throughout the middle glacial interval (13 to 9.5 ka), with no abrupt deglacial "step" at 13 ka and no pause in deglaciation from 12 to 10 ka; fastest rates of melting between 10.5 and 9.5 ka; and final melting of residual ice at lower elevations after 9.5 ka.

3. The progressive deglacial warming and polar-front retreat in the North Atlantic is explained by a diminishing ice-sheet cooling of the ocean; the Younger Dryas readvance is explained by diversion of Laurentide meltwater from the Mississippi to the North Atlantic, perhaps abetted by tabular icebergs from marine

ice shelves around the Arctic Ocean. Together, these two factors explain the full deglacial sequence in the North Atlantic, and the sequence in the North Atlantic in turn explains the mid-deglacial Younger Dryas cooling and ice-growth interval in Europe.

4. Insolation forcing explains both the timing and the (smooth) trend of ice-volume loss during deglaciation. The mid-point of deglaciation coincides with the strongest insolation forcing (the last summer insolation maximum at 11 to 10 ka).

5. Ablationary factors important during specific phases of the deglaciation include: slipping and surging of ice on the terrestrial margins during the slow first half of the deglaciation; marine calving during all but the final melting of residual ice; CO_2 warming during the entire sequence; and bedrock rebound, particularly during the rapid deglaciation after 11 ka. Additional work is needed to decide which are the key factors in the 100,000-year ice-volume cycle of the last 700,000 years.

REFERENCES CITED

Andrews, J. T., 1973, The Wisconsin Laurentide ice sheet; Dispersal centers, problems of rates and retreat, and climatic implications: Arctic and Alpine Research, v. 5, p. 185–199.

—— , 1975, Support for a stable late Wisconsin ice margin (14,000 to 9,000 B.P.); A test based on glacial rebound: Geology, v. 4, p. 617–620.

—— , 1982, On the reconstruction of Plesitocene ice sheets; A review: Quaternary Science Review, v. 1, p. 1–30.

Andrews, J. T., and Peltier, W. R., 1976, Collapse of the Hudson Bayice center and glacio-eustatic rebound: Geology, v. 5, p. 73–75.

Bard, E., Arnold, A., Duprat, J., Moyes, J., and Duplessy, J.-C., 1987, Reconstruction of the last deglaciation; Deconvolved records of $\delta^{18}O$ profiles, micropaleontological variations, and accelerator mass spectrometer ^{14}C dating: Climate Dynamics, v. 1, p. 101–112.

Beget, J., 1987, Low profile of the northwest Laurentide ice sheet: Arctic and Alpine Research, v. 19, p. 81–88.

Berger, W. H., and Vincent, E., 1986, Sporadic shutdown of North Atlantic deep water production during the glacial-Holocene transition?: Nature, v. 324, p. 53–55.

Berger, W. H., Killingley, J. S., Metzler, C. V., and Vincent, E., 1985, Two-step deglaciation; ^{14}C-dated high resolution $\delta^{18}O$ records from the tropical Atlantic Ocean: Quaternary Research, v. 23, p. 258–271.

Birchfield, G. E., 1987, Changes in deep ocean water $\delta^{18}O$ and temperature from the last glacial maximum to present: Paleoceanography (in press).

Bloom, A. L., 1971, Glacial-eustatic and isostatic controls of sea level since the last glaciation, *in* Turekian, K. K., ed., Late Cenozoic glacial ages: New Haven, Connecticut, Yale University Press, p. 355–379.

Borns, H. W., Jr., 1973, Late Wisconsin fluctuations of the Laurentide ice sheet in southern and eastern New England, *in* Black, R. F., Goldthwait, R. P., and Willman, H. B., eds., The Wisconsinan stage: Geological Society of America Memoir 166, p. 37–46.

Boulton, G. S., Smith, G. D., Jones, A. S., and Newsome, J., 1985, Glacial geology and glaciology of the last mid-latitude ice sheets: Journal of the Geological Society of London, v. 142, p. 447–474.

Boyle, E. A., and Keigwin, L. D., 1985, Comparison of Atlantic and Pacific paleochemical records for the last 215,000 years; Changes in deep ocean circulation and chemical inventories: Earth and Planetary Science Letters, v. 76, p. 135–150.

—— , 1987, North Atlantic thermohaline circulation during the last 20,000 years; Link to high-latitude surface temperature: Nature (in press).

Broccoli, A. J., and Manabe, S., 1987, The influence of continental ice, atmospheric CO_2, and land albedo on the climate of the last glacial maximum: Climate Dynamics, v. 1, p. 87–99.

Broecker, W. S., 1982, Ocean chemistry during glacial time: Geochimica et Cosmochimica Acta, v. 46, p. 1689–1750.

Broecker, W. S., and 6 others, 1987, The chronology of the last deglaciation; Implications to the cause of the Younger Dryas event: Paleoceanography (in press).

Bryson, R. A., Wendland, W. M., Ives, J. D., and Andrews, J. T., 1969, Radiocarbon isochrones on the disintegration of the Laurentide ice sheet: Arctic and Alpine Research, v. 1, p. 1–14.

Chappell, J., and Shackleton, N. J., 1986, Oxygen isotopes and sea level: Nature, v. 324, p. 137–138.

Chinzei, K., and 8 others, 1987, Postglacial environmental change of the Pacific Ocean off the coast of Japan: Marine Micropaleontology, v. 11, p. 273–292.

Clayton, L., Teller, J. T., and Attig, J. W., 1985, Surging of the southwestern part of the Laurentide ice sheet: Boreas, v. 14, p. 235–241.

Curray, W. B., and Crowley, T. J., 1987, $\delta^{13}C$ in equatorial Atlantic surface waters; Implications for ice ages pCO_22 levels: Paleoceanography (in press).

Davis, M. B., and Jacobsen, G. L., Jr., 1985, Sensitivity of cool-temperature forests and their fossil pollen record to rapid temperature change: Quaternary Research, v. 23, p. 327–340.

Denton, G. H., and Hughes, T. J., 1981, The last great ice sheets: New York, Wiley Interscience, 484 p.

Duplessy, J. C., Delibrias, G., Turon, J. L., Pujol, C., and Duprat, J., 1981, Deglacial warming of the northeastern Atlantic Ocean; Correlation with paleoclimatic evolution of the European continent: Palaeogeography, Palaeoclimatology, Palaeoecology, v. 35, p. 121–144.

Flint, R. F., 1971, Glacial and Quaternary geology: New York, John Wiley and Sons, 892 p.

Fritz, P., Morgan, A. V., Eicher, U., and McAndrews, J. H., 1987, Stable isotope, fossil Coleoptera, and pollen stratigraphy in last Quaternary sediments from Ontario and New York State: Palaeogeography, Palaeoclimatology, Palaeoecology, v. 58, p. 183–202.

Gaudreau, D. C., and Webb, T., III, 1985, Late Quaternary pollen stratigraphy and isochrone maps for the northeastern United States, *in* Bryant, V. M., Jr., and Holloway, R. G., eds., Pollen records of late-Quaternary North American sediments: American Association of Stratigraphic Palynologists, p. 247–280.

Hollin, J. T., 1962, On the glacial history of Antarctica: Journal of Glaciology, v. 4, p. 173–195.

Hughes, T., Denton, G. H., and Grosswald, M. G., 1977, Was there a late-Wurm Arctic ice sheet?: Nature, v. 266, p. 296–602.

Imbrie, J., and Imbrie, J. Z., 1980, Modeling the climatic response to orbital variations: Science, v. 207, p. 943–953.

Johnson, R. G., and McClure, B. T., 1976, A model for Northern Hemisphere continental ice sheet variation: Quaternary Research, v. 6, p. 325–353.

Keigwin, L., 1987, Late Quaternary paleoceanographic record of the northwesternmost Pacific: EOS Transactions of the American Geophysical Union, v. 68, p. 329.

Keigwin, L. D., Corliss, B. H., and Druffel, E.R.M., 1984, High resolution isotope study of the latest deglaciation based on Bermuda Rise cores: Quaternary Research, v. 22, p. 383–386.

Kutzbach, J. E., and Guetter, P. J., 1986, The influence of changing orbital parameters and surface boundary conditions on climate simulations for the past 18,000 years: Journal of Atmospheric Science, v. 43, p. 1726–1759.

Lindeman, M., and Oerlemans, J., 1987, Northern Hemisphere ice sheets and planetary waves; A strong feedback mechanism: Journal of Climatology, v. 7, p. 109–117.

MacDonald, B. C., 1971, Late Quaternary stratigraphy and deglaciation in eastern Canada, *in* Turekian, K. K., ed., Late Cenozoic glacial ages: New Haven, Connecticut, Yale University Press, p. 331–354.

Manabe, S., and Broccoli, A. J., 1985, A comparison of climate model sensitivity

with data from the last glacial maximum: Journal of Atmospheric Science, v. 42, p. 2643–2651.
Mathews, W. H., 1974, Surface profiles of the Laurentide ice sheet in its marginal areas: Journal of Glaciology, v. 13, p. 37–43.
Mercer, J. H., 1969, The Allerod oscillation; A European climatic anomaly?: Arctic and Alpine Reserach, v. 1, p. 227–234.
Mickelson, D. M., Clayton, L., Fullerton, D. S., and Borns, H. W., Jr., 1983, The late Wisconsin glacial record of the Laurentide ice sheet in the United States, *in* Porter, S. C., ed., Late Quaternary environments of the United States; Vol. 1, The late Pleistocene: Minneapolis, University of Minnesota Press, p. 3–37.
Milankovitch, M. M., 1941, Canon of insolation and the ice-age problem: Koniglich Serbische Akademie, Beograd. (English translation by the Israel program for scientific translation, published for the U.S. Department of Commerce and the National Science Foundation, Washington, D.C.)
Mix, A. C., and Ruddiman, W. F., 1984, Oxygen-isotope analyses and Pleistocene ice volumes: Quaternary Research, v. 21, p. 1–20.
——, 1985, Structure and timing of the last deglaciation; Oxygen isotope evidence: Quaternary Science Reviews, v. 4, p. 59–108.
Mix, A. C., and Shackleton, N. J., 1986, $\delta^{13}C$ analyses of foraminifera and atmospheric pCO_2 variations [abs.]: Woods Hole, Massachusetts, 2nd International Conference on Paleoceanography.
Pastouret, L., Chamley, H., Delibreias, G., Duplessy, J.-C., and Thiede, J., 1978, Late Quaternary climatic changes in western tropical Africa deduced from deep-sea sedimentation off the Niger delta: Oceanologica Acta, v. 1, p. 217–232.
Paterson, W.S.B., 1972, Laurentide ice sheet; Estimated volumes during late Wisconsin: Reviews in Geophysics and Space Physics, v. 10, p. 885–907.
Prell, W. L., 1984, Variation of monsoonal upwelling; A response to changing solar radiation, climate processes, and climate sensitivity: Geophysical Monographs, v. 29, p. 48–57.
Prest, V. K., 1969, Retreat of Wisconsin and recent ice in North America: Geological Survey of Canada Map 1257A.
Rind, D., Peteet, D., Broecker, W., McIntyre, A., and Ruddiman, W., 1986, The impact of cold North Atlantic sea-surface temperatures on climate; Implications for the Younger Dryas cooling (11–10K): Climate Dynamics, v. 1, p. 3–33.
Rooth, C., 1982, Hydrology and ocean circulation: Progress in Oceanography, v. 11, p. 131–149.
Ruddiman, W. F., and McIntyre, A., 1973, Time-transgressive deglacial retreat polar waters from the North Atlantic: Quaternary Research, v. 3, p. 117–130.
——, 1981, The North Atlantic during the last deglaciation: Palaeogeography, Palaeoclimatology, Palaeoecology, v. 35, p. 145–214.
——, 1984, Ice-age thermal response and climatic role of the surface Atlantic ocean, 40° to 63°N: Geological Society of America Bulletin, v. 95, p. 381–396.
Ruddiman, W. F., Sancetta, C. D., and McIntyre, A., 1977, Glacial/interglacial response rate of subpolar North Atlantic waters to climatic change; The record left in deep-sea sediments: Philosophical Transactions of the Royal Society of London, series b, v. 280, p. 119–142.
Sarnthein, M., Erlenkeuser, H., and Zahn, R., 1982, Termination I; The response of continental climate in the subtropics as recorded in deep-sea sediments: Bulletin de l'Institut Geologique d'Aquitaine, v. 31, p. 393–407.
Shackleton, N.J., and Duplessy, J.-C., 1986, Temperature changes in ocean deep waters during the late Pleistocene [abs.]: Woods Hole, Massachusetts, 2nd International Conference on Paleoceanography.
Shackleton, N. H., Hall, M. A., Line, J., and Shuxi, C., 1983, Carbon isotope data in core V19-30 confirm reduced carbon dioxide concentration in the ice age atmosphere: Nature, v. 306, p. 319–322.
Shilts, W. W., 1980, Flow patterns in the central North American ice sheet: Nature, v. 286, p. 213–218.
Thompson, L. G., Mosley-Thompson, E., Dansgaard, W., and Grootes, P. M., 1986, The Little Ice Age as recorded in the stratigraphy of the tropical Quelccaya ice cap: Science, v. 234, p 361–364.

MANUSCRIPT ACCEPTED BY THE SOCIETY JULY 27, 1987

ACKNOWLEDGMENTS

I thank: all DNAG chapter authors for timely reviews of this synthesis chapter under difficult time constraints; W. Broecker, A. McIntyre, and J. Overpeck for internal (LDGO) reviews; and Ann Esmay for help with the manuscript and figures. This effort was partly supported by Grants OCE-8521514 and OCE-8608328 from the Marine Geology and Geophysics Program of the National Science Foundation. This is Lamont-Doherty Geological Observatory Contribution no. 4195.

Printed in U.S.A.

The Geology of North America
Vol. K-3, North America and adjacent oceans during the last deglaciation
The Geological Society of America, 1987

Chapter 22

Synthesis; The land south of the ice sheets

H. E. Wright, Jr.
Department of Geology and Geophysics and Limnological Research Center, University of Minnesota, Minneapolis, Minnesota 55455

INTRODUCTION

The North American ice sheets reached their maximum extent on the south generally at about 20 to 21 ka, and a fluctuating retreat began soon thereafter. What was the environmental and biotic history of the land south of the ice sheets during the glacial maximum, and what changes occurred there and in the deglaciated area during the long period of ice retreat? Various aspects of these subjects are treated in the preceding chapters, and the synthesis that follows is an attempt to highlight some of the interrelations as well as to add certain points not otherwise considered.

THE PERIGLACIAL CLIMATE AT 18 KA

South of the glacial border in North America the climatic history during the last deglaciation was complicated by the fact that the climatic effects of the ice sheet are difficult to separate from the progressive changes in the distribution of solar radiation. Radiation relations during that time, as controlled by the Milankovitch orbital changes, were distinctly different from those of today, and the ice sheet was still large enough to affect the climate to the south. Thus modern analogues are hard to find for the climatic conditions of the deglaciation period. During the preceding glacial maximum, however, the radiation relations were not unlike those of today, and therefore they do not complicate the interpretations of the periglacial conditions. How far beyond the ice margin did these influences extend? If the relatively simple conditions during the glacial maximum are examined first, then the subsequent complications caused by radiation changes may be traced with greater clarity.

A periglacial climate is one that is influenced by proximity to an ice sheet. Although the term periglacial is commonly restricted to low-temperature conditions near an ice sheet, it is here used to refer to any climatic variables caused primarily by the existence of a glacier, near or far. The reason for adopting such a broad view of a periglacial environment in the context of the deglaciation history is to draw attention to the secondary climatic effects of the ice sheets in contrast to the direct climatic effects of changes in distribution of solar radiation during the same time.

Low mean annual temperatures are manifested particularly by permafrost features, and other climatic variables by the distribution of diagnostic plants and animals. Strong winds are recorded by sand dunes and loess deposits. But at the outset a distinction should be made between the climatic effects caused by local conditions next to the ice sheet and the subcontinental climatic changes brought about when the North American ice sheets caused a major shift in the general circulation of the atmosphere. Close to the ice sheet, katabatic or downslope winds are particularly strong where funneled down a valley. They dissipate where the slope flattens, as can be seen in the Victoria Dry Valleys of Antarctica, where such winds are strong enough to produce pebble ripple marks and then deposit sand dunes where they diminish at the lower end of the valley. Cold-air drainage down mountain valleys may be prevalent enough to cause reversed elevational zonation of vegetation on valley walls.

A second type of local periglacial wind is like the foehn winds of the Alps or the chinook winds of western America. These involve a much deeper air mass, which descends from a mountain mass and is adiabatically warmed. Today they occur occasionally in the winter when a low-pressure area exists in the lee of the mountains. In the case of the ice sheet, the air could descend perhaps 3,000 m and thus be warmed by 30°C. It would have a low relative humidity because of the higher moisture capacity of warm air.

Another relatively local factor in the character of periglacial winds results simply from the high albedo of the ice sheet, which cools the air and produces a high-pressure area over the ice sheet even in summer. In contrast, the periglacial region is subject to incursions of warm maritime air from the south. The contrast in temperature and thus pressure in these two regions results in strong winds at the surface.

On a broader scale, it can be shown by climate modeling (Kutzbach, Chapter 19) that the North American ice sheets caused a split in the westerly jet stream. One branch crossed Alaska and the Canadian arctic, largely replacing (or at least reducing) the easterly winds that prevail there today. The other branch crossed the United States at midlatitudes. It entered the

Wright, H. E., Jr., 1987, Synthesis; The land south of the ice sheets, *in* Ruddiman, W. F., and Wright, H. E., Jr., eds., North America and adjacent oceans during the last deglaciation: Boulder, Colorado, Geological Society of America, The Geology of North America, v. K-3.

continent in California, strongly affecting the Southwest, and it streamed across the continent to the North Atlantic. In contrast, the modern jet features a wave that is anchored in the Northern Rockies and in the winter it bends far south in the continental interior. The model results also show that anticyclonic circulation developed over the ice sheet, bringing easterly winds to the area just to the south. These altered circulation patterns were periglacial in the broad sense, for they were established by the presence of the ice sheet.

Permafrost

Today, continuous permafrost is found in areas with mean annual temperaturese below about -6° or -8°C. For the southern limit of discontinuous permafrost the limiting temperature is about -2°C—the discontinuities result primarily from such local variables as drainage conditions, for water movement is an effective mechanism of heat transfer. The most recent survey of such permafrost indicators as ice-wedge polygons and pingos (Péwé, 1983) indicates the presence of relic forms in New Jersey and from Illinois to Montana, generally within 200 km of the late Wisconsin glacial border. An especially favorable area for permafrost was the Driftless Area of southwestern Wisconsin, as well as adjacent parts of Illinois and Iowa in the broad reentrant between the Lake Michigan and Des Moines lobes. Ice-wedge features are also reported from the high Wyoming plains and in alpine portions of the Rocky Mountains.

The absence of organic matter in most permafrost features makes their development almost impossible to date directly by radiocarbon analysis, and thus to reveal the temperature trends during deglaciation. Most of them undoubtedly date from the Late Wisconsin maximum, for definite permafrost indicators on younger deposits are rare in the Great Lakes region. Ice-wedge casts, however, have been reported from southern Quebec on drift dated to 13 to 11 ka (Péwé, 1983) and from northern Wisconsin on drift dated to 13 ka (Attig and Clayton, 1986). Buried glacial ice is thermally equivalent to permafrost in the sense that it requires freezing temperatures to survive. It persisted in Late Wisconsin moraines in Minnesota at least until 11 ka, as indicated by radiocarbon dates on the basal sediments of kettle lakes.

Apart from its significance as a measure of mean annual temperature, the presence or absence of permafrost in the area immediately adjacent to the ice margin is important in glacier dynamics (Andrews, Chapter 2; Hughes, Chapter 9). If permafrost exists in front of an ice sheet it may occur beneath the toe of the ice as well, especially if the ice is thin. If the ice front is frozen to the bed, it may block the escape of subglacial water, allow the buildup of basal pore-water pressure, and facilitate surging. Also, it should favor compressive flow at the terminus, concentration of surficial rock debris, and the formation of stagnant ice. Actually, the subdued nature of the moraines dating to the glacial maximum in the Illinois area indicates none of these processes. Rather it is thought that the ice bed was thawed and that the glacier slid easily to its terminus (Mickelson and others, 1986). The incompatibility of this reconstruction with the evidence for permafrost in Illinois needs to be resolved.

Plants and animals

Is there fossil evidence for such frigid conditions in front of the ice sheet? Pollen and plant-macrofossil studies of sediments dating to 18 ka in the immediate periglacial area suggest nearly treeless tundralike vegetation, bordered to the south by open spruce forest in the Middle West or spruce/pine forest in the East (Jacobson and others, Chapter 13; Webb and others, Chapter 20). Certain beetles that are now restricted to arctic Alaska and the Yukon occurred in Iowa and Illinois until at least 16.5 ka (Morgan, Chapter 16). The distribution of permafrost features is highly correlated with the Late Wisconsin distribution of the collared lemming (*Dicrostonyx*), which has special adaptations to deal with severe arctic climates (King and Graham, 1986).

The integrated study of the Conklin Quarry site in eastern Iowa, involving analysis of fossil pollen, larger plant remains, small mammals, mollusks, and beetles, provides an excellent basis for postulating tundra conditions for the period 18.1 to 16.7 ka in this area, where the present July temperatures average about 23°C (Baker and others, 1986). Today the arctic treeline, which is formed by white spruce (*Picea glauca*), black spruce (*P. mariana*), and tamarack (*Larix laricina*), reflects the pattern primarily of summer temperatures and generally follows the 10° to 12°C July isotherm. The tundra to the north is broadly coincident with continuous permafrost today.

Thus the zone close to the full-glacial ice front from New England to Montana was probably characterized by both permafrost and tundra, and the narrowness of this zone suggests that the ice sheet itself had a primary influence on climatic conditions, especially in summer, when reduced temperatures caused by the prevalence of arctic air slowed the thawing of the ground and reduced the growing season for trees. Southwest of the ice sheet in the western Dakotas and in Montana and Wyoming, the treeless belt may have been broader, although the lack of paleoecological sites in this region makes it difficult to reconstruct the position of the full-glacial tree line. Still farther west in the Northern Rocky Mountains and the Columbia Basin, it appears that the upper (alpine) tree line was deeply depressed during full-glacial time, and that the lower (steppe/forest) treeline against the steppe was raised because of dry climatic conditions, so that the forest zone in the mountains was narrow and open, if present at all (Barnosky and others, Chapter 14). Here the dry conditions can be attributed to the easterly anticyclonic winds, which reduced the incidence of the dominant moisture-bearing westerly winds of today (Kutzbach, Chapter 19). The coast of Washington featured subalpine parkland rather than tundra when the Cordilleran ice sheet was approaching its maximum, apparently because of the ameliorating effects of the Pacific Ocean. Alaska, largely beyond the influence of any ice masses, was completely treeless—perhaps a mosaic of meadowlike tundra in the lowlands capable of supporting a fauna of large herbivores, combined with more xeric and barren slopes and ridgetops (Barnosky and others, Chapter 14).

The spruce forest of the midcontinent south of the treeless fringe was also narrow, compared to the present boreal forest of Canada, which has an average breadth of about 1,100 km. In full-glacial time the range of spruce apparently extended as far south as northern Louisiana and Texas. The forest was probably open over much of its expanse, as indicated by the relatively high proportion of herb pollen. Spruce also covered most of the Appalachian Highlands as far south as northern Georgia (except for the highest ridges, which were above treeline), as well as the Atlantic Coastal Plain south to South Carolina, but throughout the East, spruce was accompanied by jack pine (*Pinus banksiana*).

The absence of jack pine in the full-glacial spruce forest of much of the Middle West constitutes a real puzzle, for today the ranges of jack pine and white and black spruce essentially coincide, except that jack pine does not extend quite so far norther and is absent from the extreme eastern end (Labrador) and western end (Alaska) of the spruce ranges. Jack pine also does not extend quite so far south as spruce in the prairie border area of Saskatchewan, suggesting that spruce is altogether more tolerant of varied environmental conditions. Jack pine was common in the northern Middle West during the Farmdalian interstadial prior to the Late Wisconsin maximum but was absent from about 20 to 10 ka. Because radiation conditions at 20 ka were similar to those of today, some special periglacial component of the climate must have brought about its extirpation during the glacial maximum. Jack pine did not reoccupy the Middle West until about 10.5 ka. It is possible that natural fire, to which jack pine is adapted, was reduced during this period, perhaps because of the prevalence of arctic air in the summer throughout the Middle West, engendered by the high albedo of the ice sheet. Thunderstorms in this region today owe their incidence to moist maritime tropical air from the Caribbean; Pacific air loses its moisture in the western mountains, and arctic air has a low moisture content initially. Repeated invasions of cool, dry arctic air during summer could permit the southward expansion of spruce because of the decreased fire frequency.

In the Middle West south of the full-glacial treeless area, some of the pollen sites suggest a forest of almost pure spruce (*viz.* Muscotah and Arrington in northeastern Kansas, Boney Spring and other sites in the Ozark Highlands of Missouri, and Seminary School basin in southern Illinois), but these may have been wetlands covered with black spruce. Two sites, however (Pittsburg basin, close to Seminary School basin in southern Illinois, and Nonconnah near Memphis in southwestern Tennessee), show 10 to 20 percent oak and other temperate hardwoods—values much higher than can be found today in the boreal spruce forest (Webb and McAndrews, 1976). Farther south in northern Louisiana (Rayburn salt dome of Kolb and Fredlund, 1981), oak and hickory were dominant but spruce was still prominent. This mixture of boreal and temperate trees implies climatic conditions different from today's. Thus the pollen evidence for the vegetation during the glacial maximum indicates a narrow band of tundra or spruce-tundra bordered in the Middle West by spruce forest, with oak and other temperate hardwoods increasing southward.

The faunal record provides more specific evidence for full-glacial conditions with no modern counterparts (Graham and Mead, Chapter 17). Near the ice front in Iowa and Wisconsin were arctic shrews, lemmings, voles, and ground squirrels that live in tundra regions today. Yet in the same assemblage are species that do not occur today in tundra or boreal forest but have prairie and deciduous-forest affinities, for example, prairie vole, sagebrush vole, thirteenline ground squirrel, and eastern chipmunk, with increases in grassland species to the west. The cool summer temperatures required by the arctic species must have been the critical factor in maintaining permafrost as well as spruce trees. At the same time the more temperate species imply winter temperatures not much colder than today (Graham, 1986). Similarly, the fossil beetle assemblage at the southern ice margin includes mixtures of what are today subarctic (open-ground), boreal, and prairie species (Morgan, Chapter 16).

Another faunal anomaly comparable to the spruce/oak pine problem is the full-glacial distribution of the collared lemming along the ice front and the virtual absence of the brown lemming (*Lemnus*), compared to their overlapping distributions today (R. W. Graham, 1986, personal communication).

How do the model results compare with this reconstruction of a narrow frigid belt bordering the ice sheet? Simulated temperatures over the ice sheet are of course extremely low in both winter and summer, because of the high albedo and the high elevations (Kutzbach, Chapter 19). South of the ice sheet the simulated temperatures are about 10° lower than today in summer, but the grid spacing is too large to show the narrowness of the frigid fringe inferred from the permafrost and paleoecological data. The temperature gradient to the south is steep enough so that on the Gulf Coast and Florida the temperatures in both July and January were depressed only about 2°C, as controlled by the sea-surface temperatures in the Gulf. The pollen-based reconstructions in the Southeast, however, suggest a greater temperature depression (Webb and others, Chapter 20).

Winds

Another manifestation of periglacial conditions in the broad sense is strong winds capable of transporting sand and silt. Such winds are caused by the steep gradient in atmospheric pressure between the very cold ice-sheet area and the less cold land area to the south. The model results illustrate the relations. The area concerned is much broader than that affected by permafrost or strong frost processes.

Under certain conditions the geologic record of past winds can differentiate between prevailing seasonal winds and the strong winds necessary to produce sand dunes and loess deposits. Prevailing winds may be responsible for importing moisture to an area and thus for strongly affecting the vegetation. For example, today in the Pacific Northwest the moist Pacific air mass brings moisture to produce a heavy forest cover. But at 18 ka the great reduction in forest cover in the Puget Lowland implies diminution if not cessation of westerly winds in this area. The model results provide an explanation for these relations, for at 18 ka an anticyclonic circulation pattern developed over the ice sheet, bringing easterly air flow to this segment (Barnosky and others, Chapter 14).

On the other hand, the sand dunes and linear deflation features throughout the northern Great Plains indicate strong northwesterly winds (Kutzbach and Wright, 1985). Although the linear erosional features cannot be dated, the great dune fields downwind in western Nebraska and the loess still farther to the southeast are believed to date from the Late Wisconsin (Wright and others, 1984). Climatic conditions are thought to have been very dry, for dunes of this magnitude are currently forming only in semidesert regions with annual precipitation 15 to 30 percent of modern values in the Nebraska Sandhills (Wells, 1983). Northwesterly strong winds now prevail during the winter half year in the northern Great Plains, although the southern Great Plains show more southwesterly winds. Summer winds are weaker and variable in direction. During the glacial maximum, with the strong contrasts in atmospheric temperature and pressure between the ice sheet and the periglacial area, the strong northwesterly winds may have lasted throughout the year, causing sand and silt to be deflated wherever the ground was not too firmly frozen. Even in the permafrost regions of the arctic today, the areas close to major streams are not frozen, and it is in these areas where burrowing rodents such as ground squirrels live (R. W. Graham, 1986, personal communication).

THE PERIGLACIAL CLIMATE DURING DEGLACIATION

In many respects the environmental history of the periglacial area is more difficult to interpret for the phases of deglaciation than it is for the glacial maximum, at least as far as causal mechanisms are concerned. Even though more data are available from paleoecological sites, the summer radiation in midlatitudes steadily increased, culminating at about 10 ka, so that the changes in landscape and biota could reflect climatic conditions resulting from the combined influence of radiation changes and ice-sheet morphology.

Moraines

The southern margin of the Laurentide ice sheet began its retreat soon after it reached its maximum, but the retreat was interrupted by literally dozens of pauses and even by readvances. The most detailed history is that documented for the Lake Michigan lobe. About 30 mapped moraines in Illinois cover the interval 20 to 14 ka, although some are barely visible. Recent lithostratigraphic studies in this area now assemble these moraines into five groups, defined largely by the particle-size distribution and clay mineralogy of the tills (Hansel and others, 1985). The lithologic differences in some cases can be attributed to readvance of the ice front rather than simply to a pause during retreat, for proglacial sediments of a particular texture (especially fine-grained lake beds) were apparently incorporated in the ice during readvance. Such a readvance implies a more significant climatic perturbation than by a simple pause during retreat—unless surging was involved (see below). A recent synthesis of glacial fluctuations in the area from Illinois to Ohio (Mickelson and others, 1983) assumes synchronous advances of the various adjacent ice lobes at about 21, 20, 19, 18.1, 17.2, 16.7, and 16.1 ka—before the Erie interstade (16.1–15.5 ka)—and several others thereafter. In some cases the advance was preceded by an inferred retreat of as much as 600 km. The dating evidence for this synchrony may be weak, but the several distinct fluctuations of the ice margin imply changes in contemporaneous ablation or in prior accumulation, and thus in climate.

Distant retreat and readvance are particularly well documented as the ice lobes withdrew well into the Great Lakes basins. For example, readvance of ice across lake beds in the Lake Erie basin after the Erie interstade produced the clayey till of the Powell and Defiance moraines to the southwest (Mickelson and others, 1983; Teller, Chapter 3). Several readvances of the Lake Michigan and Superior lobes over lake beds resulted in clay-rich tills. In one case the retreat of the Lake Michigan lobe allowed partial drainage of the proglacial lake eastward into the Lake Huron basin; the famous Two Creeks forest bed developed on the drained lake floor and was later overridden when the ice readvanced. This particularly conspicuous sequence was initially correlated with the Alleröd/Younger Dryas fluctuation of northwestern Europe, but radiocarbon dates now show that it was almost 1,000 years too early. This was apparently only one of several significant ice-margin fluctuations of the Lake Michigan lobe between 13 and 11 ka (Acomb and others, 1982).

Farther west the Late Wisconsin deglacial sequence is complicated by the well-documented nonsynchronous behavior of adjacent ice lobes. For example, the Superior lobe in Minnesota reached its Late Wisconsin maximum perhaps as early as 20 ka. It was fed by ice from the Labradorean sector southeast of Hudson Bay, in the same manner as ice lobes in the other Great Lakes basins to the east. The Superior lobe subsequently retreated into the Lake Superior basin. Meanwhile the Des Moines lobe advanced from the Keewatin sector along the Red River/Minnesota River lowland, then downslope to the northeast as the Grantsburg sublobe across terrain bared by the Superior lobe, which at that time extended barely out of the Lake Superior basin. This advance may have occurred as early as 16 ka, but then as the Des Moines lobe thickened it spilled southward across a low divide on the Minnesota/Iowa border and reached central Iowa at 14 ka. This late dominance of the Des Moines lobe over the Superior lobe suggests an increasing importance of western sectors of the Laurentide ice sheet as the Late Wisconsin glaciation progressed—an interpretation long ago posed by Leverett (1932) and recently supported indirectly by the Canadian reconstructions discussed by Andrews (Chpater 2). The shift is probably not a result of differential response times for the two ice lobes, as inferred for the great asynchrony in ice advance between the Cordilleran ice sheet and the nearby but much smaller alpine glaciers of the Cascade Range (Booth, Chapter 4).

As the several ice lobes withdrew into the Great Lakes basins after about 14 ka (plus the Des Moines lobe into the Red River lowland), phases in the retreat may have been influenced by the presence of large proglacial lakes (Teller, Chapter 3). The shallower lake basins themselves may owe their origin in part to delays in isostatic uplift after crustal depression (Peltier, Chapter 8). In deeper basins, such as that of Lake Superior, glacial erosion must also have been a factor. In these cases, the great water depth could facilitate rapid ice retreat by calving.

In contrast to the evidence for multiple fluctuations during the general retreat of the southern margin, the northern margin provides little record. The maximum extent there on land was not reached until as late as 10 to 8 ka; either the earlier history was obliterated, or the ice margin was stable (Andrews, Chapter 2). Despite the increasing sophistication of models that incorporate insolation, climate, ice sheets, oceans, and isostasy, the instabilities of the southern margin do not emerge as output. Either some unexplained climatic perturbations at roughly 1,000-yr intervals may account for the succession, or one must postulate surging or other quasiperiodic rapid movements not directly related to climatic change. Similarly, in the case of the Puget lobe of the Cordilleran ice sheet, the late-glacial Sumas readvance has no clear climatic cause (Booth, Chapter 4).

In the case of surging, the dynamics may be facilitated when the substratum consists of fine-grained sediments of proglacial lakes. The instabilities may come from the buildup of basal pore-water pressure according to a mecahnism documented by Clarke and others (1984) for the surging of a Yukon glacier. Thus the later ice advances in the Great Lakes area have been attributed to surging on the evidence of favorable substrate conditions, thin ice, low slope, and rapid wastage (Clayton and others, 1985; Teller, Chapter 3). In the Lake Michigan/Huron/Erie basins, the evidence for reworked lake sediments in some of the tills has already been mentioned. Here the case for periodic rapid movement after increase of pore-water pressure may apply. The Des Moines lobe, however, crossed two divides and advanced downslope into central Iowa, and along this course there is no evidence for proglacial lakes. Surging also seems unlikely for the earlier phases of the Rainy and Superior lobes of eastern Minnesota, which have permeable stony deposits derived from the crystalline rocks of the Precambrian Shield. Later in the deglaciation sequence, however, a proglacial lake in the Lake Superior basin may have provided the appropriate setting for one or more surges of the Superior lobe and adjacent lobes in northern Wisconsin. Still later, as the ice retreated into northwestern Ontario and northern Manitoba between 10 and 8 ka, it formed several distinct moraines on the Canadian Shield. The ice margin then was generally formed by Glacial Lake Agassiz, and kame moraines and other glaciofluvial features were deposited in association with relatively thin and rapidly wasting ice. The ice was intermittently active enough to form recessional moraines, however, with only local evidence for surging (Dredge, 1983).

Winds

Strong wind action in the Middle West during deglaciation is best recorded in the loess stratigraphy. Decreasing thickness and grain size for loess deposits east from the Mississippi and Missouri river valleys as well as the Illinois, Wabash, and smaller valleys indicate that glacial outwash was the principal source of the loess (and of sand dunes as well in some areas). Stratigraphy of loess deposits in the Iowa/Illinois area reflects the activity of the outwash streams that provided the source materials during deglaciation, and it also reflects the existence of winds strong enough to transport the silt. Dates on organic remains in the Peoria Loess range from 22 to 14 ka. The termination of major loess deposition as early as 14 ka in the deglacial sequence—best documented by the absence of loess on moraines and outwash terraces of that age or younger—must mean either that winds were substantially reduced in intensity by that time or that outwash plains were more widely stabilized by vegetation, or both, because outwash was still being deposited after 14 ka along the Mississippi River as well as along tributaries leading from the Des Moines lobe and contemporaneous and younger ice lobes not terminating in lakes. Reduction in wind strength may be a response to some major change in the atmospheric circulation related to the ice-sheet morphology. Model results do not help here, at least thus far—a finer-mesh grid will be necessary to show why winds apparently decreased in the Middle West at 14 ka, and sensitivity tests with different ice-sheet heights might prove instructive.

On the other hand, some significant wind action lasted longer. Dune formation in the Nebraska Sandhills apparently prevailed until 12 to 9 ka (Wright and others, 1984). The absence of annual ice layers in the Greenland ice core before 10,750 years ago (as dated by ice-layer counts) is attributed to deposition of calcareous dust until that time, for the layers are preserved only under acidic conditions (Paterson and Hammer, Chapter 5). The attribution of the dust source to calcareous marine deposits on emergent continental shelves seems unnecessary, however, for virtually all loess from midcontinental periglacial America is calcareous and is an adequate source for the glacial dust. The very fine particle size makes possible even more distant sources, for example, interior Asia. Also, emergent coastal areas would presumably be revegetated before they could serve as a supplementary source of dust.

Mississippi River

The glacial Mississippi River system drained almost the entire southern margin of the Laurentide ice sheet. It received numerous major nonglacial tributaries from the Great Plains on the west and the Appalachian Highlands on the east, and it built a huge delta in the Gulf of Mexico. The headwaters were controlled by great volumes of glacial meltwater and glacial sediment. The middle course was affected by periglacial wind action, for it provided a major source for the Midwestern loess deposits. Its nonglacial tributaries drained areas with vegetation and soil conditions that influenced runoff and sediment supply in different ways. The downstream end was affected by changes in base level (sea level) and by crustal subsidence beneath the delta. Stream aggradation by outwash sediments might be expected in the upstream area, decreasing downstream and changing to degradation near the mouth in response to lowered sea level. The postglacial trends should be the opposite, with upstream dissection after the end of glacial outwash, and downstream aggradation related to sea-level rise. Thus the glacial-age longitudinal profile should cross the postglacial profile. How does the Mississippi River history fit this model?

During the early phases of deglaciation, great floods of outwash aggraded previously existing valleys, causing backwater flooding up the major and minor nonglacial tributaries (Schumm and Brakenridge, Chapter 10). The outwash valley train in the upper Mississippi River starts in Minnesota as a terrace 45 m above the present flood plain and dates to about 14 ka. It gradually descends over a river distance of more than 1,000 km to a height of only 5 m, and in the Lower Mississippi Valley below the entrance of the Ohio River it exists as sandy braided-stream surfaces only slightly above the modern flood plain, which in contrast is characterized by a meandering channel with banks of silt and clay (Saucier, 1981). But even there the aggradation was sufficient to cause backwater flooding up tributaries like the Ouachita River (Saucier and Fleetwood, 1978). Toward the delta the terrace remnants are completely buried by the fine-grained Holocene sediments of the modern flood plain, deposition of which was facilitated by the rise of base level.

Dissection of this Mississippi River terrace in its middle reaches may have started as early as the time of the Kankakee floods in Illinois, which resulted from drainage of proglacial lakes formed during the retreat of the Lake Michigan lobe in Illinois about 14 ka. But the deepest dissection of the entire main stream probably began about 11.7 ka, when the Glacial River Warren started to drain Glacial Lake Agassiz. The glacial Great Lakes added a further increment of outflow not burdened with sediment. Lake Agassiz rapidly increased in size, and a whole string of ice-marginal lakes in Saskatchewan, Manitoba, and North Dakota drained progressively and catastrophically into Lake Agassiz (Teller, Chapter 3). In the upper reaches the depth of dissection by the River Warren at this time reached 85 m in southeastern Minnesota, producing a gradient too flat for transport of all the sediment subsequently supplied by tributaries after the termination of the River Warren outflow about 9.5 ka. Accordingly the entire valley has aggraded as an adjustment to the reduced water volume. Deposition of alluvial fans by tributaries has created a series of river lakes in the upper reaches in Minnesota. The amount of post–River Warren fill in southeastern Minnesota has exceeded 50 m.

Thus the glacial/interglacial model for river profiles was not fully followed by the Mississippi River. Glacial aggradation in the upper reaches certainly occurred, with decreasing gradient downstream, but the effect of sea-level depression probably did not extend far upstream beyond the head of the present delta near Baton Rouge (Saucier, 1981). The theoretical postglacial dissecton in the upper reaches, as predicted by the model, was preempted by the over-deep erosion in the late-glacial by the River Warren, which presumably affected the tributaries as well by the local base-level control. Since the end of River Warren the system has slowly recovered, but whether the rate of alluviation has slowed is uncertain. In any case the postglacial has been marked by aggradation rather than dissection.

The Mississippi River had an additional role during deglaciation, for it carried isotopically light glacial meltwater from most of the southern margin of the Laurentide ice sheet to the Gulf of Mexico. Distinct peaks in the curve for sediments dated between 14 and 11.7 ka can be attributed to this influx (Broecker and others, 1987). With further retreat of the ice to the north, most of the meltwater was shifted from the Mississippi River to the St. Lawrence River (Teller, Chapter 3).

Vegetation

If ice-lobe fluctuations resulted from climatic changes during deglaciation, to what extent are they reflected in the vegetation history? The rates of vegetational change, as manifested in a representative group of pollen diagrams for eastern North America, accelerated at intervals centering at about 13.7, 12.3, and particularly 10 ka (Jacobson and others, Chapter 13), whether at the center or near the edges of major vegetation formations. Although these times may compare favorably with certain steps or accelerations in the ice-volume reductions as recorded by oxygen-isotope profiles of ocean sediments (Mix, Chapter 6), they fail to coincide with major ice-lobe fluctuations in the Great Lakes region, for example, the Erie interstade (ca. 16.1–15.5 ka) and the Two Creeks interstade (ca. 11.8 ka). Similarly, in the Pacific Northwest the phases of ice retreat cannot be closely correlated with phases in the vegetational history (Booth, Chapter 4).

The major trends in the deglacial vegetational patterns for central and eastern North America are graphically shown in the sequential maps of individual pollen types (Webb and others, Chapter 20) or of groups of types (Jacobson and others, Chapter 13). Although the coverage of sites early in the deglaciation is insufficient to be sure of the details, it is clear that dramatic shifts occurred between 12 and 9 ka as the Laurentide ice sheet was rapidly retreating.

The pollen record of vegetation change is particularly sensitive to climatic change near ecotones between major vegetational formations, or at least near the range limits of important indicator pollen types (Wright, 1984). Thus the southern limit of the spruce forest might have been quite sharp, as it is today, and when the climate warmed this ecotone transgressed northward. The pollen profiles for spruce at various sites in the Middle West were quite complacent until the ecotone passed the site, transgressing from 16.5 ka in central Missouri northward to 10 ka in central Minnesota. The passage of the ecotone was marked by the reduction in spruce pollen and the increase in pollen of temperate hardwoods, or in the north by pine. The vegetation within the spruce forest did change at the same time, especially in the latter part of the sequence as the herbs were reduced and the forest canopy closed (Webb, 1987).

The relations of pine and spruce during the deglacial period are as perplexing as they were during the full-glacial. Jackpine was a major component of the spruce forest in the Appalachian Highlands and Atlantic Coastal Plain at 18 ka, increasing in proportion to spruce southward to Georgia and South Carolina. During deglaciation it diminished in the highlands (Webb and others, Chapter 20); starting about 15 ka (Anderson Pond in Tennessee), jack pine was apparently replaced gradually by oak, while spruce maintained its importance until about 10 ka (or later in Virginia). The vegetation of the southern Appalachians thus became an oak/spruce forest, as did that of the lower Middle West during the full-glacial. In the Ozark Highlands, where pine was not a factor during most of the deglaciation, the spruce/oak assemblage developed as early as 16.5 ka (King, 1973).

For most locations the late-glacial vegetational sequence suggests a unidirectional climatic change. The only pollen diagrams that show a distinct reversal in the inferred climatic trend are those from western Ohio and adjacent areas, where spruce and other boreal conifers returned briefly about 10.8 to 10 ka, after gradual replacement by ash and other hardwoods starting about 13 ka (Shane, 1987). This fluctuation is contemporaneous with the Younger Dryas interval in Europe (Watts, 1980) and the equivalent oscillation in North Atlantic surface temperatures (Ruddiman, Chapter 7). Apparently the spruce/hardwoods ecotone was located in the Ohio area during this part of the deglaciation and moved back and forth in response to a significant climatic change. The northeastern ecotone of the spruce forest—the forest/tundra border in Nova Scotia and adjacent areas—showed contemporaneous fluctuations (Mott and others, 1986), leading to the previous conception that the Younger Dryas event west of the Atlantic was confined to the Canadian Maritime Provinces (Rind and others, 1986). The Ohio results suggest that this restriction does not hold. Identification of a comparable fluctuation in areas to the east (e.g., New England) is hampered by the dominance of pine in the pollen sequence.

The ice margin in the Great Lakes area did not advance during the Younger Dryas event; rather, between 10.8 and 10 ka, it retreated to north of Lake Superior. This retreat allowed the huge Lake Agassiz system to drain eastward to the Great Lakes and thereby to the North Atlantic, before the ice briefly readvanced to northern Michigan at 10 to 9.5 ka to close the outlet (Teller, Chapter 3). It has already been suggested that the diversion of the Great Lakes drainage from the Mississippi River to the St. Lawrence added a freshwater cap to the North Atlantic, thereby permitting more extensive formation of sea ice (Johnson and McClure, 1976). The effects of this cap on the production of North Atlantic deep water are discussed by Broecker and others (1987). The extra discharge of Lake Agassiz more than doubled the volume of meltwater released to the North Atlantic at this time (Teller, Chapter 3). The chronology of the event is closely controlled at 10.8 to 10 ka. It is here speculated that this huge discharge of fresh cold water to the North Atlantic south of Newfoundland might divert the Gulf Stream southward enough so that the influx of tabular icebergs from the breakup of arctic ice shelves and sea-based ice fronts could be more effective in cooling the North Atlantic surface waters and could permit the formation of sea ice. Such an influx of shelf ice, first proposed by Mercer (1969) to explain the Younger Dryas oscillation in Europe, thus might be more effective with this large supplement of Laurentide proglacial lake waters, which in addition provides the chronology to support the correlation. If this modified hypothesis is correct, then the local *retreat* of the Laurentide ice sheet in northwestern Ontario, caused perhaps by some glaciological instability, was the ultimate cause for the *advance* of the polar front in the North Atlantic and, downwind, for the well-documented Younger Dryas *readvance* of the Scandinavian and Scottish ice and the distinct vegetational change in western Europe.

Regardless of the possible causes for the Younger Dryas fluctuation—in Europe or the North Atlantic or even in North America—such an event cannot be accommodated by the Milankovitch model, which makes no provision for such brief events. Appeal must therefore be made to instabilities in the ice sheet mentioned above, or to feedback mechanisms among ice sheet, ocean, and atmosphere (Ruddiman and Wright, Chapter 1).

The Southwest

The American Southwest constitutes an area of special interest because of the wealth of data documenting climatic conditions at 18 ka. This includes plant macrofossils from packrat middens (Van Devender and others, Chapter 15), changes in lake levels (Benson and Thompson, Chapter 11), and the biostratigraphy of semidesert lake sediments (Forester, Chapter 12). With respect to the problem of differentiating the primary effects of radiational changes from the secondary effects of the ice sheet, one might postulate that the Southwest is far enough removed from the ice sheet—and to a certain extent upwind from it—that the immediate periglacial effects would be subordinate, although the general circulation of the atmosphere might be affected. What evidence supports this hypothesis?

High lake levels in the Southwest during the glacial maximum have long been known, but a long-standing debate has existed as to whether they were maintained by increased precipitation or by decreased temperature and evaporation. Benson and Thompson (Chapter 11) conclude that increased winter precipitation is the key to high lake levels, according to analysis of energy budgets and comparison with recent weather patterns. Because winter moisture comes primarily from the Pacific, they postulate that the southward displacement of the jet stream in winters during the glacial maximum was the ultimate cause for the high lake levels. More frequent Pacific storms might be anticipated in summer as well, for the same reason (Forester, Chapter 12). Increased summer precipitation is not so effective in maintaining high lake levels, however, because of high evaporation, although cloudiness may shade the lake sufficiently to lessen the evaporative effects.

The paleoecological record provided by fossil packrat middens in much of the Southwest offers little evidence for gradual climatic changes during the deglacial period (Van Devender and others, Chapter 15), although on the Colorado Plateaus, where the conifer flora is larger, distinct changes are evident as early as 15 ka (J. Betancourt, 1987, personal communication). In the more southerly desert regions, plant distributions changed little until 11 ka, when the previously cooler summers and mild and moist winters shifted gradually to the modern climatic regime, which is characterized by monsoonal summer rains, especially at lower elevations and more easterly locations (Van Devender and others, Chapter 15). Maximum continentality was not attained until the mid-Holocene. Mild winters during the glacial maximum are attributed in Chapter 15 to blockage of arctic air masses by the ice sheets, and increased precipitation is attributed to the southward displacement of the westerly jet. Cooler summers are indicated by the downward expansion of subalpine trees like bristlecone pine, and milder winters are suggested by the upward expansion of desert scrub. The result was ecological mixtures not seen today—a biogeographic situation reminiscent of that proposed for the Middle West, for a similar climatic reason, that is, cooler summers and milder winters. Thus even though the Southwest was far from the North American ice sheets, the periglacial influence was still the main climatic factor during most of the deglacial period, in the sense that the ice sheet caused a major shift in the westerly jet stream.

CONCLUSIONS

One conclusion stands out in all the chapters dealing with the fossil flora and fauna, namely that the Late Wisconsin fossil assemblages have at best only imperfect analogues today. The message is clearest in the case of fossil mammals (Graham and Mead, Chapter 17), which show assemblages of species that now have widely separate ranges, for example, tundra forms mixed with temperate woodland and prairie forms. These "disharmonious" faunas are attributed to environmental conditions not matched today, either because of changes in plant-food resources, which reflected the climatic conditions, or directly because of changes in climatic variables critical to reproduction and survival. Various components of seasonality may have been particularly important, such as seasonal extremes in temperature, length of the growing season, and seasonal distribution of precipitation. All these factors may have been affected by the patterns of solar radiation as well as by changing circulation controlled by ice-sheet morphology.

The fossil mammal assemblages suggest a greater climatic equability than today, that is, cooler summers combined with winters that lacked the extremely low temperatures often experienced in the midcontinent today. Cooler summers are to be expected as a consequence of the proximity of the ice sheet, with the more frequent penetration of arctic air even to the southern United States. The model results (Kutzbach, Chapter 19; Webb and others, Chapter 20) provide support for this interpretation. For winters, one might expect the mountain of ice to deliver even colder air masses to the Middle West than does the more distant Arctic today. But a case can be made that the mountain of ice actually blocked the arctic air from sweeping across the plains, and that the air that did descend from the heights of the ice sheet was warmed adiabatically and arrived on the plains dry and windy (Bryson and Wendland, 1967), like the chinook winds that occasionally descend from the Rocky Mountains to the High Plains today. The arctic air instead was swept eastward by the northern branch of the split jet stream and then curved southward around Greenland into the North Atlantic, and into Siberia as well (Kutzbach, Chapter 19).

The elimination of winter blizzards with extreme cold might permit the survival of plants and animals sensitive to these conditions. For example, the full-glacial and late-glacial spruce forest of the lower Middle West apparently contained a significant component of north-temperate hardwoods like ash and oak, which are only sparingly represented at the very southern edge of the present boreal spruce forest. White and black spruce are limited on the south by high summer temperatures, whereas the hardwoods mentioned literally freeze and die under conditions of very low winter temperatures. The overlap in ranges for these taxa implies a greater climatic equability than today. Similarly, the combination of small arctic, temperate, and prairie mammals in the Iowa area can be understood only if summers were cooler than today and winters less severe. For 18 ka this condition cannot be attributed to seasonal solar-radiation relations, which were about the same as today. Instead it must be attributed to periglacial effects. In the Southwest a comparable mixture of now-separated taxa characterized the full-glacial and late-glacial.

As the ice sheet diminished in size, the periglacial effects in the southern part of the continent might be expected to have decreased. At the same time the radiational changes increased until about 11 to 10 ka, with enhanced seasonality resulting from increased summer heating and winter cooling. Model experiments show such a development in Eurasia, causing increased monsoonal rainfall in India and North Africa (Kutzbach and Otto-Bliesner, 1982). Although North America is a smaller landmass and still had an ice sheet over much of Canada at 10 ka to counteract the summer warming, the early decline of boreal conifers in the Southeast at 13 ka and expansion of mesic hardwoods like oak, hickory, and beech may reflect warmer summers there, as well as an increase in monsoonal rainfall.

The combination of these two trends—reduction in the periglacial factor and increase in the radiation factor—thus resulted in a rapid increase in seasonality, culminating about 10 ka. This climatic shift is believed by many to be the major cause for the spectacular mammalian extinction at this time, for many large herbivores could not adapt to the rapidly changing vegetational patterns (Graham and Mead, Chapter 17). Human hunters may have been an additional factor when the Laurentide ice sheet withdrew enough in Alberta to open a corridor for their immigration from Beringia (Bonnicksen and others, Chapter 18; Martin and Klein, 1984).

In Alaska and adjacent Yukon, "upwind" of the ice sheets, the early expansion of spruce over tundra at about 11 ka coincided with the maximum of summer radiation, which has the greatest effects at high latitudes (Ritchie and others, 1983; Barnosky and others, Chapter 14). Increased summer warming and aridity may also explain the relatively abrupt transformation from spruce forest to prairie in the northeastern Great Plains at 12 to 10 ka (Webb and others, Chapter 20). In this case it must be assumed that the area is too far from the Gulf moisture source for enhanced monsoonal rains, or that the greater evaporation brought by higher summer temperatures exceeded rainfall. This relation decreases eastward, closer to the retreating ice sheet, for in Minnesota the prairie did not develop until 8 to 7 ka (Jacobson and others, Chapter 13).

The increased radiation in summer was matched by a decrease in winter. These differences began to diminish after 10 ka; by 6 ka the trend was reversed, and the increased winter temperatures in the Southeast led to expansion of southern pines (Webb and others, Chapter 20).

REFERENCES CITED

Acomb, L. J., Mickelson, D. M., and Evenson, E. B., 1982, Till stratigraphy and late-glacial events in the Lake Michigan lobe of eastern Wisconsin: Geological Society of America Bulletin, v. 93, p. 389–296.

Attig, J. W., and Clayton, L., 1986, History of late Wisconsin permafrost in northern Wisconsin: American Quaternary Association, 9th Biennial Meeting, Program and Abstracts, p. 115.

Baker, R. G., Rhodes, R. S., II, Schwert, D. P., Ashworth, A. C., Frest, T. J., Hallberg, G. R., and Janssens, J. A., 1986, A full-glacial biota from southeastern Iowa, U.S.A.: Journal of Quaternary Science, v. 1, p. 91–107.

Broecker, W. S., Andree, J., Wolfli, W., Oeschger, H., Bonani, G., Peteet, D., and Kennett, J., 1987, The chronology of the last deglaciation; Implications to the cause of the Younger Dryas event: Paleoceanography (in press).

Bryson, R. A., and Wendland, W. M., 1967, Tentative climatic patterns for some late-glacial and postglacial episodes in central North America, *in* Mayer-Oakes, W. J., ed., Life, land, and water: Winnipeg, University of Manitoba Press, p. 271–298.

Clarke, G.K.C., Collins, S. G., and Thompson, D. E., 1984, Flow, thermal structure, and subglacial conditions of a surge-type glacier: Canadian Journal of Earth Sciences, v. 21, p. 232–240.

Clayton, L., Teller, J. T., and Attig, J. W., 1985, Surging of the southwestern part of the Laurentide ice sheet: Boreas, v. 14, p. 235–241.

Dredge, L. A., 1983, Character and development of northern Lake Agassiz and its relation to Keewatin and Hudsonian ice regimes: Geological Association of Canada Memoir 26, p. 117–132.

Graham, R. W., 1986, Plant-animal interactions and Pleistocene extinctions, *in* Elliott, D. K., ed., Dynamics of extinctions: New York, John Wiley and Sons, p. 131–154.

Hansel, A. K., Johnson, W. H., Socha, B. J., and Follmer, L. R., 1985, Introduction; Depositional environments and correlation problems of the Wedron Formation (Wisconsinan) in northeastern Illinois: Illinois State Geological Survey Guidebook 16, p. 1–11.

Johnson, R. G., and McClure, B. T., 1976, A model for Northern Hemisphere continental ice sheet variation: Quaternary Research, v. 6, p. 325–353.

King, J. E., 1973, Late Pleistocene palynology and biogeography of the western Missouri Ozarks: Ecological Monographs, v. 43, p. 539–565.

King, J. E., and Graham, R. W., 1986, Vertebrates and vegetation along the southern margin of the Laurentide ice sheet: American Quaternary Association, 9th Biennial Meeting, Program and Abstracts, p. 43–45.

Kolb, C. R., and Fredlund, G. G., 1981, Palynological studies, Vacherie and Rayburn's domes, North Louisiana Salt Dome Basin: Baton Route, Louisiana State University, Institute of Environmental Studies Topical Report E530-02200-T-2.

Kutzbach, J. E., and Otto-Bliesner, B., 1982, The sensitivity of the African-Asian monsoonal climate to orbital parameter changes for 9,000 years B.P. in a low-resolution general circulation model: Journal of Atmospheric Sciences, v. 39, p. 1177–1188.

Kutzbach, J. H., and Wright, H. E., Jr., 1985, Simulation of the climate of 18,000 years BP; Results for the North American/North Atlantic/European sector and comparison with the geologic record of North America: Quaternary Science Reviews, v. 4, p. 147–187.

Leverett, F. B., 1932, Quaternary geology of Minnesota and parts of adjacent states: U.S. Geological Survey Professional Paper 161, 149 p.

Martin, P. S., and Klein, R. G., eds., 1984, Quaternary extinctions; A prehistoric revolution: Tucson, University of Arizona Press. (Review by H. E. Wright, Jr., 1986, Faunal extinctions at the end of the Pleistocene: Reviews in Anthropology, v. 13, p. 223–235.)

Mercer, J. H., 1969, The Alleröd oscillation; A European climatic anomaly?: Arctic and Alpine Research, v. 1, p. 227–234.

Mickelson, D. M., Clayton, L., Fullerton, D. S., and Borns, H. W., Jr., 1983, The late Wisconsin glacial record of the Laurentide ice sheet in the United States, *in* Porter, S. C., ed., Late-Quaternary environments of the United States; Vol. 1, The late Pleistocene: Minneapolis, University of Minnesota Press, p. 3–37.

Mickelson, D. W., Clayton, L., and Muller, E. W., 1986, Contrasts in glacial landforms, deposits, glacier-bed conditions, and glacier dynamics along the southern edge of the Laurentide ice sheet: American Quaternary Association, 9th Biennial Meeting, Program and Abstracts, p. 28–30.

Mott, R. J., Grant, D. R., Stea, R., and Ochietti, S., 1986, Late-glacial climatic oscillation in Atlantic Canada equivalent to the Alleröd/Younger Dryas event: Nature, v. 123, p. 247–250.

Pélé, T. L., 1983, The periglacial environment in North America during Wisconsin time, *in* Porter, S. C., ed., Late-Quaternary environments of the United States; Vol. 1, The late Pleistocene: Minneapolis, University of Minnesota Press, p. 157–189.

Rind, D., Peteet, D., Broecker, W. G., McIntyre, A., and Ruddiman, W., 1986, The impact of cold North Atlantic sea surface temperatures in climate; Implications for the Younger Dryas cooling (11–10k): Climate Dynamics, v. 1, p. 3–33.

Ritchie, J. C., Cwynar, L. C., and Spear, R. W., 1983, Evidence from northwest Canada for an early Holocene Milankovitch thermal maximum: Nature, v. 305, p. 126–128.

Saucier, R. T., 1981, Current thinking on riverine processes and geologic history as related to human settlement in the Southeast: Geoscience and Man, v. 22, p. 7–18.

Saucier, R. T., and Fleetwood, A. R., 1978, Origin and chronologic significance of late Quaternary terraces, Ouachita River, Arkansas and Louisiana: Geological Society of America Bulletin, v. 81, p. 889–890.

Shane, L.C.K., 1987, Late-glacial and climatic vegetational history of the Allegheny Plateau and till plains of Ohio and Indiana, U.S.A.: Boreas, v. 16,

p. 1–20.
Watts, W. A., 1980, Regional variations in the response of vegetation to Lateglacial climatic events in Europe, *in* Lowe, J. J., Gray, J. M., and Robinson, J. E., eds., The Lateglacial of north-west Europe: New York, Pergamon Press, 205 p.
Webb, T., III, 1987, The appearance and disappearance of major vegetational assemblages; Long-term vegetational dynamics in eastern North America: Vegetatio, v. 69, p. 177–187.
Webb, T., III, and McAndrews, J. H., 1976, Corresponding patterns of contemporary pollen and vegetation in central North America: Geological Society of America Memoir 145, p. 267–299.
Wells, G. L., 1983, Late-glacial circulation over central North America revealed by aeolian features, *in* Street-Perrott, A., Beran, M., and Ratcliffe, R., eds., Variations in the global water budget: Boston, Reidel, p. 317–330.
Wright, H. E., Jr., 1984, Sensitivity and response time of natural systems to climatic change in the late Quaternary: Quaternary Science Reviews, v. 3, p. 91–132.
Wright, H. E., Jr., Almendinger, J. C., and Grüger, J., 1984, Pollen diagram from the Nebraska Sandhills, and the age of the dunes: Quaternary Research, v. 24, p. 115–120.

MANUSCRIPT ACCEPTED BY THE SOCIETY JULY 29, 1987
CONTRIBUTION NO. 351 OF THE LIMNOLOGICAL RESEARCH CENTER, UNIVERSITY OF MINNESOTA

ACKNOWLEDGMENTS

Reviews of a draft of this chapter were made by C. W. Barnosky, P. J. Bartlein, J. L. Betancourt, R. W. Graham, G. L. Jacobson, A. V. Morgan, W. F. Ruddiman, R. S. Thompson, T. R. VanDevender, and T. Webb III. Support from NSF Grant ATM 84-12959 is acknowledged.

Printed in U.S.A.

Index

[Italic page numbers indicate major references]

Typeset by WESType Publishing Services, Inc., Boulder, Colorado
Printed in U.S.A. by Malloy Lithographing, Inc., Ann Arbor, Michigan